INVERTEBRATES IN FRESHWATER
WETLANDS OF NORTH AMERICA

INVERTEBRATES IN FRESHWATER WETLANDS OF NORTH AMERICA

Ecology and Management

edited by

Darold P. Batzer, Russell B. Rader, and Scott A. Wissinger

JOHN WILEY & SONS, INC.

New York · Chichester · Weinheim · Brisbane · Singapore · Toronto

Library of Congress Cataloging-in-Publication Data:
Invertebrates in freshwater wetlands of North America : ecology and
    management / edited by Darold P. Batzer, Russell B. Rader, Scott A.
    Wissinger.
        p.   cm.
    ISBN 0-471-29258-3 (cloth : alk. paper)
    1. Freshwater invertebrates—Ecology—North America.   2. Wetland ecology—North
America.   3. Wetlands—North America.   4. Wildlife conservation—North America.
I. Batzer, Darold P.   II. Rader,
Russell Ben.   III. Wissinger, Scott A.
QL365.4.A1I58   1999
592.1768′097—dc21   98-39322

Printed in the United States of America.

10 9 8 7 6 5 4 3 2 1

CONTENTS

                                                                          v

PREFACE

Scientists who study the ecology and management of wetland invertebrates suffer from an identity crisis. Because wetlands are unique, being neither truly aquatic nor terrestrial habitats (or alternatively being both), it is often hard for scientists working in other habitats to relate to wetland studies. On the other hand, the narrow field of wetland ecology has largely been developed by botanists, hydrologists, and vertebrate ecologists, who have difficulty relating to invertebrate studies. Thus, to the majority of invertebrate ecologists, we work in an unfamiliar habitat, and to the majority of wetland ecologists, we work on unfamiliar organisms.

Even among wetland invertebrate ecologists communication is poor. Most contributors to this volume (including its editors) knew only a handful of the other contributors before this project began. This is probably because we are all to some extent products of other fields—we were trained as aquatic ecologists, limnologists, entomologists, waterfowl biologists, or in other specialties, but seldom as strictly wetland invertebrate ecologists. As a group, we were astounded (although pleased) that there were enough of us to put together such an extensive volume.

Despite feelings of scientific isolation, we have watched our field gain in stature. Invertebrate ecologists from other habitats are taking notice of wetland invertebrates because of increasing public awareness of the importance of endangered wetland habitats. Suddenly our habitat of choice is popular. Among those scientists who have a longstanding interest in wetlands, the ecological importance of invertebrates has become apparent. Much of the recent focus on wetland invertebrates stems from the finding that they are crucial components in the diets of many vertebrates, especially ducks.

We have organized this text with two major goals in mind. First, we wanted to inform those in other fields about the ecology and management of wetland invertebrates. Even recent texts on wetland ecology still virtually ignore the invertebrate fauna, and hence a central aspect of wetland ecology has not been adequately presented. Second, we wanted to provide focus for our own growing field. Our sincere hope is that the field of wetland invertebrate ecology will progress so rapidly that a new edition of this text will be needed in the not too distant future.

DAROLD P. BATZER

Department of Entomology
University of Georgia

RUSSELL B. RADER

United States Forest Service
Rocky Mountain Experiment Station

SCOTT A. WISSINGER

Departments of Biology and Environmental Science
Allegheny College

CONTRIBUTORS

CHESTER ANDERSON, Department of Entomology, Cornell University, Ithaca, NY 14853

BRIAN J. ARMITAGE, Ohio Biological Survey, 1315 Kinnear Road, Columbus, OH 43212

DAROLD P. BATZER, Department of Entomology, University of Georgia, Athens, GA 30602

ARTHUR C. BENKE, Aquatic Biology Program, Department of Biological Sciences, University of Alabama, Tuscaloosa, AL 35487

AMY BERGSTEDT, Institute of Ecology, University of Georgia, Athens, GA 30602

DAVID C. A. BLADES, Royal British Columbia Museum, 675 Belleville St., Victoria, BC V8V 1X4

ANDY BOHANAK, Rocky Mountain Biological Laboratory, Crested Butte, CO 81224

PAMELA SILVER BOTTS, School of Science, Pennsylvania State University—Erie, Erie, PA 16563

WENDY BROWN, Rocky Mountain Biological Laboratory, Crested Butte, CO 81224

THOMAS BURTON, Department of Zoology, Michigan State University, East Lansing, MI 48824

DEBORAH CORTI, Center for Aquatic Ecology, Illinois Natural History Survey, 607 E. Peabody, Champaign, IL 61820

THOMAS L. CRISMAN, Center for Wetlands, University of Florida, Gainesville, FL 32611

KENNETH CUMMINS, Ecosystem Restoration Department, South Florida Water Management District, West Palm Beach, FL 33416

ADRIENNE DeBIASE, Drawer E, Savannah River Ecology Laboratory, Aiken, SC 29802

FERENC A. DE SZALAY, Department of Biological Sciences, 256 Cunningham Hall, Kent State University, Kent, OH 44242

ROBERT J. DIAZ, School of Marine Science, Virginia Institute of Marine, College of William and Mary, Gloucester Point, VA 23062

WALTER G. DUFFY, California Cooperative Fisheries Research Unit, Humboldt State University, Arcata, CA 95521

SALLY ENTREKIN, Jones Ecological Research Center, Newton, GA 31770

NED H. EULISS, JR., U.S. Geological Survey, Northern Prairie Wildlife Research Center, Jamestown, ND 58401

DAVID L. EVANS, Water and Air Research, Inc., 6821 S.W. Archer Rd., Gainesville, FL 32608

G. WINFIELD FAIRCHILD, Biology Department, West Chester University, West Chester, PA 19383

ANN M. FAULDS, Biology Department, West Chester University, West Chester, PA 19383

ALBERT T. FINNAMORE, Provincial Museum of Alberta, 12845—102nd Ave., Edmonton, AB T5N 0M6

ERNEST B. FISH, Department of Range, Fisheries, and Wildlife Management, Texas Tech University, Lubbock, TX 79409

MARY C. FREEMAN, USGS Patuxent Wildlife Research Center, University of Georgia, Athens, GA 30602

DEBBIE FOLKERTS, Department of Botany and Microbiology, Auburn University, Auburn, AL 36849

LEIGH H. FREDRICKSON, Gaylord Laboratory, University of Missouri, Puxico, MO 63960

LESLIE J. GALLAGHER, Department of Biology, Allegheny College, Meadville, PA 16335

GARY R. GASTON, Biology Department, University of Mississippi, University, MS 38677

JOSEPH P. GATHMAN, Department of Zoology, Michigan State University, East Lansing, MI 48824

K. ELIZABETH GIBBS, Department of Biological Sciences, University of Maine, Orono, ME 04469

DONNA GIBERSON, Department of Biology, University of Prince Edward Island, Charlottetown, PEI C1A 4P3

STEPHEN W. GOLLADAY, Jones Ecological Research Center, Newton, GA 31770

DIANNE L. HALL, Department of Entomology, University of Missouri, Columbia, MO 65211

BRENDA J. HANN, Department of Zoology, University of Manitoba, Winnipeg, MB, Canada, R3T 2N2

M. LORETTA HARDWICK, Watershed Ecosystems Program, Trent University, Petersborough, ON K9J 7B8

E. ANDREW HART, Department of Zoology, University of Wyoming, Laramie, Wyoming 82071

ANNE HERSHEY, Department of Biology, University of North Carolina, Greensboro, NC 27402

MICHAEL HIGGINS, Department of Entomology, Michigan State University, East Lansing, MI 48824

ALEXANDER HURYN, Department of Biological Sciences, University of Maine, Orono ME 04469

STEVEN KOHLER, Center for Aquatic Ecology, Illinois Natural History Survey, 607 E. Peabody, Champaign, IL 61820

DOUGLAS LEEPER, Southwest Florida Water Management District, 2379 Broad St., Brooksville, Florida 34609

ANDREA J. LESLIE, Center for Wetlands, University of Florida, Gainesville FL 32611

ANN R. LIMA, Natural Resources Research Institute, University of Minnesota-Duluth, Duluth, MN 55811

JAMES R. LOVVORN, Department of Zoology, University of Wyoming, Laramie, WY 82071

PATRICK A. MAGEE, Biology Program, Sciences Department, Western State College, Gunnison, CO 81231

STEPHEN A. MARSHALL, Department of Environmental Biology, University of Guelph, Guelph, ON, N1G 2W1

MORGAN McCLURE, Carolina Ecological Services, 2411 Savannah Highway, Charleston, SC 29407

RICHARD W. MERRITT, Department of Entomology and Department of Fisheries and Wildlife, Michigan State University, East Lansing, MI 48824

TONY R. MOLLHAGEN, Water Resources Center, Department of Civil Engineering, Texas Tech University, Lubbock, TX 79409

DARYL L. MOORHEAD, Department of Biological Sciences, Texas Tech University, Lubbock, TX 79409

HENRY R. MURKIN, Institute for Wetland and Waterfowl Research, c/o Ducks Unlimited Canada, Stonewall, MB R0C 2Z0

DAVID M. MUSHET, U.S. Geological Survey, Northern Prairie Wildlife Research Center, Jamestown, ND 58401

MARILYN J. MYERS, Department of Environmental Science, Policy and Management, University of California, Berkeley, CA 94720

GERALD J. NIEMI, Natural Resources Research Institute, University of Minnesota-Duluth, Duluth, MN 55811

RACHEL O'MALLEY, Department of Environmental Studies, San Jose State University, San Jose, CA 95192

BARBARA PECKARSKY, Department of Entomology, Cornell University, Ithaca, NY 14853

KAREN G. PORTER, Institute of Ecology, University of Georgia, Athens, GA 30602

JOSEPH P. PRENGER, Center for Wetlands, University of Florida, Gainesville FL 32611

RUSSELL B. RADER, United States Forest Service, Rocky Mountain Experiment Station, Laramie WY 82070

RONALD R. REGAL, Department of Mathematics and Statistics, University of Minnesota—Duluth, Duluth, MN 55812

FREDRIC A. REID, Ducks Unlimited, Western Regional Office, Sacramento, CA 95827

VINCENT H. RESH, Department of Environmental Science, Policy and Management, University of California, Berkeley, CA 94720

TERRY D. RICHARDSON, Department of Biology, University of North Alabama, Florence, AL 35632

LISETTE C.M. ROSS, Institute for Wetland and Waterfowl Research, c/o Ducks Unlimited Canada, Stonewall, MB R0C 2Z0

LYNNETTE L. SAUNDERS, Biology Department, West Chester University, West Chester, PA 19383

DANIEL W. SCHNEIDER, Department of Urban and Regional Planning, Illinois Natural History Survey, University of Illinois at Urbana-Champaign, Champaign, IL 61820

LYLE SHANNON, Department of Biology, University of Minnesota—Duluth, Duluth, MN 55812

REBECCA SHARITZ, Drawer E, Savannah River Ecology Laboratory, Aiken, SC 29802

KATHERINE A. SION, Department of Biology, Canisius College, Buffalo, NY 14208

ROBERT W. SITES, Enns Entomology Museum, Department of Entomology, University of Missouri, Columbia, MO 65211

MICHAEL C. SLAMECKA, Department of Entomology, University of Illinois at Urbana-Champaign, Urbana, IL 61801

LEONARD A. SMOCK, Department of Biology, Virginia Commonwealth University, Richmond, VA 23284

WILLIAM STREEVER, Center for Wetlands, University of Florida, Gainesville FL 32611

BARBARA TAYLOR, Drawer E, Savannah River Ecology Laboratory, Aiken, SC 29802

BRAD W. TAYLOR, Jones Ecological Research Center, Newton, GA 31770

BRIGITTE VANDENEEDEN, Pymatuning Laboratory of Ecology, University of Pittsburgh, Linesville, PA 16424

G. MILTON WARD, Aquatic Biology Program, Department of Biological Sciences, University of Alabama, Tuscaloosa, AL 35487

HOWARD H. WHITEMAN, Rocky Mountain Biological Laboratory, Crested Butte, CO 81224

D. DUDLEY WILLIAMS, Division of Life Sciences, University of Toronto at Scarborough, Scarborough, ON, M1C 1A4

NANCY E. WILLIAMS, Division of Physical Sciences, University of Toronto at Scarborough, Scarborough, ON, M1C 1A4

MICHAEL R. WILLIG, Department of Biological Sciences, Texas Tech University, Lubbock, TX 79409

SCOTT A. WISSINGER, Departments of Biology and Environmental Science, Allegheny College, Meadville, PA 16335 and Rocky Mountain Biological Laboratory, Crested Butte, CO 81224

CRAIG WOLF, Department of Biological Sciences, Texas Tech University, Lubbock, TX 79409

WILFRED M. WOLLHEIM, The Ecosystems Center, Marine Biological Laboratory, Woods Hole, MA 02543

DALE A. WRUBLESKI, Institute for Wetland and Waterfowl Research, c/o Ducks Unlimited Canada, Stonewall, MB R0C 2Z0

DAVID J. YOZZO, Barry A.Vittor and Associates, Inc., 668 Aaron Ct., Willow Park Office Complex, Kingston, NY 12401

1 An Introduction to Freshwater Wetlands in North America and Their Invertebrates

REBECCA R. SHARITZ and DAROLD P. BATZER

INTRODUCTION

North America has an abundance and diversity of wetlands. The ecological values of these systems, their complexity, the ecosystem services they provide, and their connection with the surrounding landscape are still not well understood. The wetlands that now occur in North America may represent only a fraction of those existing 200 or more years ago. Although estimates of the extent of wetlands in the United States vary widely, most studies indicate a rapid rate of loss, at least prior to the mid-1970s. The most accurate estimate of present wetland area in the conterminous United States is 42 million hectares (Dahl and Johnson 1991), of which 80 percent are inland wetlands and 20 percent are coastal (Frayer et al. 1983). In Alaska there are about 69 million hectares (Dahl 1990) and in Canada about 127 million hectares of wetland area (Zoltai 1988). Historically, policies of the United States government were intended to encourage or subsidize the conversion of wetlands to filled or drained lands that could be used for agriculture or other purposes. Only recently have the unique properties and multiple values of wetlands become better appreciated, especially ecosystem services such as improving water quality, reducing effects of floodwaters, and providing habitat for many groups of animals and plants (Daily 1997). Wetland protection is now a common objective of laws, regulations, and management plans.

Invertebrates are one of the most diverse and abundant groups of organisms in inland and coastal freshwater wetlands. They are also crucially important to the overall functioning of wetland ecosystems because of their central position in wetland food webs. They feed upon wetland plants and plant detritus, as well as on algae, and in turn are consumed by wetland birds and

Invertebrates in Freshwater Wetlands of North America: Ecology and Management, Edited by Darold P. Batzer, Russell B. Rader, and Scott A. Wissinger
ISBN 0-471-29258-3   © 1999 John Wiley & Sons, Inc.

fish (Murkin and Batt 1987, Murkin 1989, Batzer and Wissinger 1996). Many researchers interested in understanding the ecological functions of wetlands or conserving the biological diversity and ecosystem health of endangered wetland habitats are now focusing on the invertebrate fauna. The chapters of this book highlight and synthesize many of these research efforts on invertebrates in freshwater wetlands of North America. Here we review some general aspects of wetland ecology by defining wetlands in a landscape context, describing some of the different kinds of wetlands, and then briefly discussing their hydrology, chemistry, plant communities, and invertebrate communities.

DEFINING WETLANDS

There is no single, indisputable, ecologically sound definition for wetlands because of the diversity of wetland types, hydrologic conditions, landscape positions, and geomorphologies. In general, wetlands are distinguished by the presence of water above or near the surface of the soil, or slightly below. The depth and timing of inundation vary considerably among types of wetlands and among seasons and years within a single wetland type. Some wetlands are continually flooded, whereas others are inundated or saturated with water only briefly. Since the hydrologic conditions are so variable, the boundaries of wetlands often cannot be distinguished by the presence of water at any one time. However, the presence of wetland soils and/or wetland plants provides clues to the extended presence of water at some time during the growing season.

Wetlands often occur as intermediate communities along a soil moisture gradient between aquatic and terrestrial habitats. Although wetlands have frequently been described as transitional areas or ecotones between dry land and open water, this view may lead to the incorrect perception that wetlands are temporary and evolve naturally into either upland or deep water ecosystems (Tiner 1993). The majority of wetlands are relatively distinct and long-lived features on the landscape, defined by a characteristic combination of water, substrate (including physicochemical features), and biota (Fig. 1.1). For example, wetlands usually have unique soil conditions that differ from adjacent uplands because of the seasonal or continuous presence of water. Wetland soils often are more aerobic than are permanently flooded sediments in deep water ecosystems, especially when inundation is seasonal. Wetlands plants, animals, and microbes range from those that can live in either wet or dry conditions (facultative) to those adapted only to a wet environment (obligate).

Although a precise wetlands definition is difficult, they are usually described as areas where the water table is at, near, or above the land surface long enough each year to promote the formation of hydric soils and the growth of wetland plants (hydrophytes) (Cowardin et al. 1976). In the United States, jurisdictional definitions of wetlands usually include (1) vegetation dominated by hydrophytes; (2) undrained hydric (wet) soils; and (3) inundation or saturation with water at some time during the growing season of

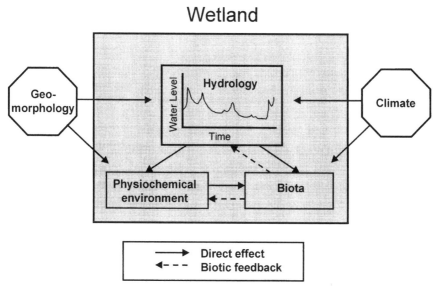

Fig. 1.1 The relationshps among hydrology, biogeochemical processes, and biota in wetlands. Plants provide feedback to hydrology through evapotranspiration and increase in flow resistance and to biogeochemical processes by affecting soil properties (organic content, dissolved oxygen) and elevation (accumulating organic matter, trapping sediment). Animals such as beavers and muskrats can also affect hydrology and soils. (Modified from and reprinted with permission from *Wetlands Characteristics and Boundaries.* Copyright 1995 by the National Academy of Sciences. Courtesy of the National Academy Press, Washington, D.C.)

each year (Cowardin et al. 1979), although the application of these criteria differs among federal agencies and among states. Freshwater wetlands are all those in which salinity from ocean-derived salts is less than 0.5 parts per thousand. The upland boundary of a wetland is technically considered to be the boundary between hydrophytic and terrestrial vegetation cover, between hydric and nonhydric soils, or between land that is flooded or saturated at some time during the growing season and land that is not. The jurisdictional boundary between wetland and deepwater habitat lies at a depth of 2 meters below low water level; however, if trees or other emergent vegetation grow beyond this depth, their deepwater edge is considered to be the wetland boundary (Cowardin et al. 1979). Although these boundaries between wetlands and adjacent upland or open water habitats may be easy to distinguish conceptually, they are often difficult to delineate, particularly because of the fluctuating hydrologic conditions in many wetlands. Ecologically, the demarcation is even more difficult since wetlands often lie along a continuum of hydrologic (and other physical and biochemical) conditions.

Considerable emphasis has been placed on the difference between wetlands and terrestrial environments (see NRC 1995), but very little attention has focused on differences between wetlands and deep water lentic habitats, or between wetlands and lotic systems. This emphasis results in part from federal regulations that apply to conversion of wetlands to dry conditions but not to aquatic conditions. From the perspective of invertebrates, however, these differences between wetlands and aquatic habitats are extremely important. Two major features, water depth and rate of flow, influence environmental factors that control wetland invertebrate habitat. Shallow wetlands can have fluctuating water levels and even dry periodically, whereas lakes are considered permanent and usually do not dry. Wetlands usually have rooted emergent and submergent vegetation that forms the base of primary production, whereas lakes often are dominated by phytoplankton production. Wetlands also may have higher rates of primary production and slower rates of decomposition (because of refractory lignin and humic compounds in macrophytes). Another important distinction between wetlands and open lake habitats results from stratification of temperate lakes during the summer (into an epilimnion, a metalimnion, and a hypolimnion), which limits the availability of nutrients from the bottom sediments into the photic zone. Ions and nutrients will typically decrease in the lake epilimnion as the summer progresses and biological processes remove them. Wetlands may be similar to the epilimnions of lakes, except that stratification seldom forms a barrier to the movement of nutrients from the sediments. Furthermore, wave action can temporarily mix the whole water column in wetlands. In littoral zone wetlands along lake shorelines, oxygen-rich water from the lake mixes with the water in the shallow wetland and dilutes ionic concentrations, limiting seasonal increases in ions that occur in more stagnant wetlands. A final distinction is that shallow wetlands can freeze to the bottom in cold winter climates, whereas lakes do not.

Wetlands associated with flowing water systems often differ functionally from adjacent streams and rivers because of differences in flow rates and degree of mixing of elements through the water column. In swift-flowing streams there are no depth gradients of ionic concentrations, and even in large rivers only gradual or temporary gradients occur through the water column. In riverine wetlands that become isolated from the main stream channel, stagnant conditions can result in steep chemical gradients over small spatial scales (centimeters of depth) and short temporal scales (diel fluctuations). Ionic concentrations also vary seasonally in wetlands as drying concentrates ions and reflooding again dilutes them. The ecological consequences of these abiotic differences between wetlands and open water systems in affording suitable habitat for aquatic invertebrates can be substantial.

FRESHWATER WETLAND TYPES

Freshwater wetlands occur across a range of geomorphic and hydrologic conditions. The national wetland classification developed by the U.S. Fish and Wildlife Service for purposes of inventorying and describing wetlands and deepwater habitats identifies three classes of freshwater wetlands: palustrine, riverine, and lacustrine (Cowardin et al. 1979). Palustrine wetlands are dominated by trees, shrubs, persistent emergent herbaceous plants, and emergent mosses or lichens. They are nontidal or occur in tidal areas where salinity is below 0.5 parts per thousand. According to Cowardin et al. (1979), wetlands without vegetation can also be classified as palustrine if they are small (less than 8 hectares), are less than 2 meters deep, and meet several other criteria. Palustrine wetlands are usually bounded by uplands or by other types of wetlands, such as riverine or lacustrine. Riverine wetlands are contained within channels and usually have flowing water. They are often bounded by palustrine wetlands established on the river floodplain or by levees or uplands. Lacustrine wetlands occur in topographic depressions or dammed river channels, and water is not usually flowing. Typically there are extensive areas of deep water in these systems, and they are often bounded by marshes or wetland shrubs and trees (palustrine wetlands).

The majority of the wetlands discussed in this book are palustrine. They include vegetated wetlands traditionally called by such names as *marsh, swamp, bog, fen, prairie pothole, playa, limesink,* and *Carolina bay,* as well as small, shallow permanent or intermittent ponds. Many managed or constructed wetlands are also palustrine, including marsh impoundments managed for waterfowl, green-tree reservoirs and moist-soil impoundments, and marshes constructed for wastewater treatment or to mitigate losses of other wetlands. The national wetland classification system also denotes wetlands that have been excavated, impounded, diked, partly drained, or farmed or that occur on artificial substrates (Cowardin et al. 1979).

Marshes

Freshwater marshes are ubiquitous throughout North America and cover an estimated 7 million hectares in the United States (Shaw and Fredine 1956). They range in size from small marshes of less than a hectare to the immense saw grass stands of the Everglades, and they differ in geological origins and in their driving hydrological forces. The largest single freshwater marsh system in North America is the Everglades of Florida (see Chapter 2). These saw grass marshes lie over a flat limestone formation with a low slope so that water from Lake Okeechobee flows slowly southward as a broad sheet during the wet season. The Everglades marshes originally occupied about 10,000 km^2 (Kushlan 1991), but about half of the original area has been lost to agriculture and urban development (Gunderson and Loftus 1993).

Tidal freshwater marshes develop in flat areas inland of estuaries and coastal salt marshes where there is adequate rainfall or river flow to maintain fresh conditions but where there is still a significant tide range (see Chapters 36 and 37). They are most abundant along the middle and south Atlantic and Gulf coasts. Extensive shoreline marshes around large lakes often formed originally in the deltas of rivers flowing into the lakes and in protected shallow areas (see Chapters 23, 38, 39, 40). Many of the remaining marshes of the

Great Lakes are now managed and protected from water level changes by artificial dikes (Herendorf 1987). River floodplains and deltas throughout North America also contain vast intermittently flooded marshes and marshy lakes (see Chapters 3, 16, 28, 29, 34).

In the central plain of the United States and Canada, small marshes can be very abundant. On the northern prairie, marshes occur in the millions of small isolated depressions formed by glacial action and include the well-known prairie potholes (see Chapters 21, 22, 25). The southern high plain of Texas and New Mexico supports numerous small depressional marshes called *playas,* probably formed by an assortment of integrated geological and biological processes (see Chapters 26, 27). Other isolated small wetlands (of various origins) that support freshwater marshes include limesink wetlands (see Chapter 9), flatwoods marshes (see Chapter 4), and Carolina bays (see Chapter 8) of the Southeast, montane depressions (see Chapters 30, 31), desert rock pools (see Chapter 32), and assorted spring-formed wetlands (see Chapters 20, 33). Rice fields and other agricultural wetlands are essentially marshes, albeit highly manipulated ones (see Chapter 35). Because marshes are such valuable habitats for wetland birds and mammals, many wildlife management areas contain extensive marsh complexes that are managed to benefit those animals (see Chapters 21, 23, 28, 34).

Peatlands

The vast northern peatlands are distributed in cold-temperate climates of high humidity, where precipitation exceeds evapotranspiration and the soil is saturated. In North America they occur primarily in tundra and boreal forest regions of Canada and Alaska, but they dip into northern Minnesota, Maine, and Michigan and occur in high-elevation mountain ranges (see Chapters 16, 17, 18, 31). Ombrotrophic bogs (raised bogs) are isolated from mineral-bearing groundwater and have lower nutrients and minerals and lower pH than do minerotrophic peatlands (fens), which receive surface or groundwater inputs. Since peat can accumulate wherever drainage is impeded and anoxic conditions predominate, scattered bogs also are found farther south in the north-central United States, in the Appalachian and western mountain ranges, and in flat, poorly drained areas (called *pocosins*) and depressions on the southeastern Coastal Plain (see Chapters 5, 8). Peat can also be found in portions of large southern wetlands such as the Everglades (Chapter 2) and the Okefenokee Swamp (Chapter 6).

Wetland Forests

Forested freshwater wetlands are common across North America, but especially in the southeastern United States. Large cypress swamps (e.g., parts of the Okefenokee Swamp, see Chapter 6) occur in areas where there is extensive rainfall and restricted overland flow or drainage. Forested wetlands dominated by bottomland hardwood trees occur along the floodplains of streams and rivers. They range from the low-lying, extensive floodplains of the large rivers of the Southeast (see Chapters 7, 28, 36, 37), which tend to have strong seasonal hydrologic pulses, to the riparian systems of the arid West, which often occur on low-order streams (see Chapter 33) and are characterized by extreme and variable fluvial conditions. In central and northern Florida and southern Georgia there are numerous small depressional wetlands that support trees (e.g., cypress domes, see Chapter 5; Carolina bays, see Chapter 8; and limesinks, see Chapter 9). Ephemeral woodland ponds and pools occur in northern forests when rainfall or snowmelt fill shallow depressions (see Chapters 12, 13, 14) and in high-elevation montane areas (see Chapter 31). Wetlands created by beavers often begin as flooded forests, but if they hold water for long periods they develop into marshes and open ponds (see Chapters 10, 15).

Restored and Man-made Wetlands

Over a period of 200 years, from the 1780s to the 1980s, about 53 percent of the original wetlands (freshwater and coastal) in the lower 48 states were lost (Dahl 1990). The most significant historical loss has resulted from agricultural practices that involved wetland drainage (87 percent, Tiner 1984). Physical and chemical alterations associated with urban and industrial development have also resulted in substantial declines in freshwater wetland area and habitat quality. Recently, efforts have intensified to restore wetlands destroyed by past human development. Techniques for restoring wetlands include (1) reestablishing or managing wetland hydrology; (2) eliminating or controlling chemicals or other contaminants affecting wetlands; and (3) reestablishing and managing native biota.

Small marshes are frequently given high priority for restoration because of their importance to waterfowl and other birds and their ease of manipulation (see Chapters 4, 19). However, except for efforts to restore natural drainages in the prairie pothole region, most drained depressional wetlands in other parts of the country have received little restoration attention. Many of the larger, altered wetland complexes, such as the Florida Everglades, pocosins of the southeastern Coastal Plain, wide river floodplains such as in the lower Mississippi drainage, and lands adjacent to the Great Lakes, retain wetland soils and viable seeds even though the hydrology has been substantially altered. To a major degree, success in restoring wetland habitat in these areas is dependent upon restoring or managing their hydrology. Sometimes, in efforts to mitigate wetland losses, new wetland habitats are created from former uplands (see Chapter 19). Wetlands may also be created for wastewater treatment, remediation of acid mine drainage, and retention of urban runoff. Many of these new wetlands take the form of small depressions.

Research on wetland restoration and creation has focused chiefly on techniques of species establishment and development of species composition and wetland community structure. The functional values of wetlands, although widely recognized, are seldom evaluated. Mitigation efforts cannot yet claim to have duplicated lost wetland functional values or ecosystem services (Daily 1997), and it has not been shown that restored wetlands maintain regional biodiversity (NRC 1992).

HYDROLOGY

Hydrology is the single most important factor determining the establishment and maintenance of specific types of wetlands and wetland processes. Hydrology controls the abiotic and biotic characteristics of wetlands, although it in turn can be influenced by the biota (Fig. 1.1). Hydrologic conditions can directly modify or change chemical and physical properties such as nutrient availability, degree of substrate anoxia, sediment properties, and pH. Many wetlands are relatively "open" systems that have water entering and leaving them; others are more "closed" and have limited hydrologic association with the surrounding area. Water inputs are the major source of nutrients to many wetlands, and water outflows often remove biotic and abiotic materials from wetlands. When hydrologic conditions change even slightly, major biotic changes in species composition and richness and in productivity can occur (see Chapters 2, 9, 13, 22, 23, 29, 31).

Hydroperiod

The hydroperiod is the seasonal pattern of the water level in a wetland and is like a hydrologic signature of each wetland type (Mitsch and Gosselink 1993). It defines the rise and fall of a wetland's surface and subsurface water. The hydroperiod is an integration of all inflows and outflows of water, but it is also influenced by physical features of the terrain and by the wetland vegetation (Fig. 1.1). Two terms, *flood duration* and *flood frequency,* are commonly used to describe wetland hydroperiods. Flood duration is the amount of time that a wetland has standing water, and flood frequency indicates the average number of times that a wetland is flooded in a given period. The wetlands described in this book range from regularly flooded tidal freshwater marshes; to temporarily or seasonally flooded pools and ponds; to semipermanently flooded wetlands; to more permanently saturated or flooded peatlands, swamps, and sloughs. Another important aspect of hydroperiod, especially for wetland biota, is the timing of flooding or drying. For example, vernal pools and snowmelt ponds (see Chapters 12, 13) that hold water in early spring are substantially different habitats from autumnal woodland pools (see Chapter 14, discussion in Chapter 31). For most wetlands, of course, the hydroperiod is not the same each year but varies with interannual variations in climatic conditions. This is especially the case in wetlands that are greatly influenced by precipitation, such as prairie potholes or Carolina bays.

Water Budgets

Hydroperiod is influenced by inflows and outflows of water and by the capacity of the wetland to store water. Water inputs include precipitation, surface inflows, and groundwater inflows. Water losses from wetlands are in the form of evapotranspiration, surface runoff, or groundwater outflows. Wetlands occur most extensively in regions where precipitation inputs (including rainfall and snowfall) exceed evapotranspiration and surface runoff. Precipitation generally has well-defined yearly patterns, although variations among years can be great. In parts of North America (e.g., the Northeast), precipitation is relatively uniform throughout the year; whereas in other areas it varies seasonally (e.g., west coastal areas with wet winters and dry summers). Surface inflows to wetlands can be channelized streamflow from a drainage basin (e.g., riverine swamps) or nonchannelized overland flow following rainfall or a spring thaw (e.g., prairie potholes, snowmelt ponds). Wetlands in the floodplains of unregulated rivers or streams (often called *riparian wetlands*) receive floods that vary in number, intensity, and duration from year to year.

Groundwater can heavily influence some wetlands, whereas in others it may have very little effect (Carter and Novitzki 1988; see Chapter 21). Groundwater inflow (= discharge) occurs when the water level of a wetland is at or below the water table of the surrounding upland. This often occurs in limesink wetlands (see Chapter 9), those associated with springs or groundwater seeps (see Chapters 20, 23), and those in the floodplains of rivers (see Chapter 7). When the water level in the wetland is higher than the surrounding water table, water can flow out of the wetland and enter (= recharge) the groundwater. A wetland can alternate between having inflows and outflows of groundwater, as is often the case in riparian wetlands. Wetlands that appear to be little influenced by groundwater include depressional wetlands on soils with poor permeability, such as Carolina bays, playas, and many prairie potholes. These wetlands often have fluctuating hydroperiods related to precipitation patterns.

Evapotranspiration is the water that vaporizes from water or soil in a wetland (evaporation), together with the moisture that passes through vascular plants to the atmosphere (transpiration). In general, evapotranspiration is enhanced by meteorological conditions such as higher temperatures, increased solar radiation, decreased humidity, and increased wind speeds. Evapotranspiration is difficult to measure, and the actual effects of wetland vegetation on evapotranspiration are not well understood. The presence of vegetation retards evaporation from the water surface, but the transpiration of water through the plants can equal or exceed the difference (Kadlec et al. 1988).

Hydrology and Ecosystem Function

Hydrology influences many features of wetlands, including plant species composition, primary productivity, decomposition, mineralization, export of organic material, nutrient cycling, and nutrient availability (Fig. 1.1; see reviews in Chapters 12, 23). Waterlogged soils and the subsequent changes in oxygen content and other chemical conditions significantly limit the number and types of rooted plants that can survive. In general, plant species richness declines along the gradient from moist to flooded conditions in wetlands. Studies have also shown that wetlands in stagnant (nonflowing) or continuously deep water have low productivities compared with wetlands that have flowing water, although this relationship has been demonstrated more often in forested wetlands than in marshes (Mitsch and Ewel 1979, Megonigal et al. 1997).

The effects of hydrology on decomposition are even less clear, and it cannot be assumed that increased frequency or duration of flooding will necessarily increase or decrease decomposition rates (Brinson et al. 1981). The importance of hydrology for import and export of organic material is obvious, however. Riverine and lacustrine wetlands often contribute and receive large amounts of organic detritus to or from streams and lakes, whereas hydrologically isolated wetlands such as northern peatlands have much lower organic import and export. Nutrients are carried into wetlands by precipitation, surface and groundwater inflows, and tides (freshwater tidal wetlands); nutrient outflows are controlled primarily by the outflow of water. The flooding of wetland soils, by altering both the pH and the redox potential, also influences the availability of nutrients.

CHEMICAL PROCESSES

Biogeochemical processes in wetlands are complex because these ecosystems often fluctuate dramatically between aerobic and anaerobic conditions, in both time and space. In addition, some wetlands, such as riverine swamps and tidal marshes, have a significant exchange of minerals with their surrounding areas. Other wetlands that are more isolated hydrologically, such as ombrotrophic bogs, cypress domes, and Carolina bays, have little exchange of nutrients across their boundaries.

Oxygen

The depletion of oxygen in water and soils that is associated with flooding is probably the most important chemical feature of the wetland environment. Although the water in many shallow wetlands and those with flowing water generally remains well oxygenated, isolated and deeper wetlands may become stagnant and oxygen-depleted. When soils are flooded, water fills the pore spaces and limits the rate at which oxygen can diffuse, usually creating anaerobic, or reduced, conditions. This condition of oxygen depletion can occur within several hours to a few days after inundation (Gambrell and Patrick 1978). The lack of oxygen strongly affects the availability of plant nutrients and toxic materials in the soil. For example, availability of major ions such as phosphorus, potassium, and magnesium, and several trace nutrients such as iron, manganese, and sulfur, is affected by the hydrologic conditions. A small layer of oxidized soil, sometimes only a few millimeters thick, usually remains at the soil-water interface, however, and this thin layer is often very important in the chemical transformations that occur in wetlands. Ferric and manganic ions, as well as nitrate, sulfate and other oxidized ions, are found in this microlayer, whereas the deeper anaerobic soils are dominated by reduced forms such as ferrous and manganous salts, ammonia, and sulfides (Mohanty and Dash 1982).

Nutrients

The water chemistry in freshwater wetlands ranges from nutrient-rich (eutrophic) at one extreme to nutrient-poor (oligotrophic) at the other, and differences are related chiefly to the magnitude of nutrient and other chemical inputs. Although many types of wetlands have more than adequate supplies of nutrients and consequently are highly productive (e.g., tidal marshes and riparian forests), there are other wetlands that have low supplies of nutrients (e.g., ombrotrophic bogs, cypress domes, subalpine pools). Thus, concentrations of dissolved ions can range from relatively high in wetlands that receive substantial inputs from surface water runoff and groundwater inflows to low in wetlands that are fed primarily by rainwater. The availability of nitrogen and phosphorus is significantly altered by anoxic conditions, but these nutrients do not necessarily limit plant production. Ammonium nitrogen is high in most wetland soils, and soluble inorganic phosphorus is often abundant. Intrasystem cycling of nutrients is often limited by the degree to which processes such as primary productivity and decomposition are controlled by the wetland environment (Lockaby and Walbridge 1997). Thus, some wetlands have extremely rapid nutrient cycling (e.g., high-nutrient freshwater marshes), whereas in others nutrient cycling is extremely slow (e.g., low-nutrient ombrotrophic bogs). Similarly, pH ranges from relatively neutral or slightly alkaline in wetlands with mineral soils to acidic in rainwater-fed wetlands with organic soils.

Nutrient cycling in wetlands differs from that in both aquatic and terrestrial ecosystems. For example, more nutrients are tied up in sediments and peat in wetlands than in most terrestrial systems. Wetlands often are more open to nutrient fluxes (inputs and outputs) than are upland systems, and are coupled to adjacent ecosystems through chemical exchanges that occur with water inflows and outflows. Many wetlands act as sinks for particular inorganic nutrients, and many wetlands are sources of organic material to downstream or adjacent ecosystems. In most deep aquatic systems the retention of nutrients in sediments is probably longer than in wetlands. Most wetland plants obtain their nutrients from the sediments, whereas phytoplankton in the autotrophic zone of lakes depend on nutrients dissolved in the water column. Wetland plants are often described as "nutrient pumps" that bring nutrients from the anaerobic sediments to the aboveground strata; whereas phytoplankton in lakes take up nutrients in the aerobic zone and deposit them in anaerobic sediments through death and settling.

Chemical Transformations

Various chemical and biological transformations take place in a predictable sequence as soils become oxygen-depleted or reduced. The redox potential, or oxidation-reduction potential, is used to quantify the degree of electrochemical reduction in wetland soils. One of the first reactions that occurs as wetland soils become anaerobic is the reduction of nitrate (NO_3^-), first to nitrite (NO_2^-), and ultimately to gaseous nitrous oxide (N_2O) or molecular nitrogen (N_2), which are lost from the wetland. As the redox potential continues to decrease, manganese is transformed from manganic (Mn^{+4}) to manganous (Mn^{++}) compounds, iron is transformed from ferric (Fe^{+++}) to ferrous (Fe^{++}) form and sulfate ($SO_4^=$) is reduced to sulfide ($S^=$). Many of these transformations are mediated by microorganisms. Iron and manganese in their reduced forms can reach toxic concentrations in wetland soils, as can sulfides. In the most reduced conditions, organic matter itself becomes reduced, producing methane gas (CH_4). Although phosphorus is not directly altered by changes in redox potential, it is indirectly affected by its association with elements that are. Under aerobic conditions, phosphorus becomes relatively unavailable to plants and microconsumers through precipitation of insoluble phosphates with ferric iron, calcium, and aluminum and through adsorption onto clay particles, organic peat, and ferric and aluminum hydroxides and oxides. When soils become anaerobic, phosphorus is released into solution through the hydrolysis of ferric and aluminum phosphates and the release of phosphorus adsorbed onto clays (Faulkner and Richardson 1989, Ponnamperuma 1972). A detailed description of these transformations is given in Mitsch and Gosselink (1993).

PLANT COMMUNITIES

The freshwater wetland environment is in many ways physiologically harsh. Major stresses are anoxia and the wide fluctuations in water conditions characteristic of an environment that is neither aquatic nor terrestrial. Strictly aquatic organisms might not be adapted to deal with the periodic drying that occurs in many wetlands, whereas terrestrial organisms are stressed by long periods of flooding. The most severe stress to plants, however, is caused by the absence of oxygen in flooded wetland soils, which prevents them from respiring through normal metabolic pathways. In the absence of oxygen, the supply of nutrients available to plants is also modified, and concentrations of certain elements can reach toxic levels, as described previously. Vascular plants have evolved both structural and physiological adaptations to deal with these stresses (see review by McKevlin et al. 1997). For example, one important adaptation is the development of pore space in the cortical tissues, which allows oxygen to diffuse from the aerial parts of the plants to the roots to supply root respiratory demands. The magnitude of oxygen diffusion is large enough in many wetland plants to allow oxygen to diffuse out of the roots and oxidize the adjacent anoxic soil, producing an oxidized rhizosphere. This can be an important mechanism for moderating the toxic effects of soluble reduced ions such as manganese in anoxic soil (Armstrong 1975). Some wetland plants also have complex metabolic adaptations enabling them to minimize the accumulation in their roots of toxic compounds produced during anaerobic respiration (MacMannon and Crawford 1971, Mendelssohn et al. 1981, 1982) or to shift to an alcoholic fermentation pathway to provide a source of energy to the roots. Some wetland plants utilize the C_4 biochemical pathway of photosynthesis, characteristic of plants in water-stressed environments.

Many features influence the types of plant communities that occur in freshwater wetlands, including the hydrologic conditions, geomorphic features, climate, nutrient richness, and other abiotic and biotic factors. Most wetland plant communities can be classified as one of the following structural types: aquatic beds, emergent marsh, scrub-shrub, and forest (Cowardin et al. 1979; see Fig. 38.2 in Chapter 38). Tiner (1993) makes the point that wetlands, while not appropriately considered to be ecotones between aquatic and terrestrial ecosystems, may have ecotonal or transitional plant communities at their edges. For example, at the landward edges of many wetlands, obligate hydrophytes (such as aquatic bed species) are absent and the community is dominated by facultative species that occur in both wetlands and nonwetlands. Similarly, the waterward edges of permanently and semipermanently flooded wetlands may have both submersed aquatic and wetland species and be considered ecotonal between the wetland and the aquatic community (Tiner 1993). Finally, it should be remembered that plant composition in wetlands can change depending on water conditions that may vary from year to year or over greater periods of time.

Aquatic Beds

Beds of aquatic macrophytes occur in wetlands that usually have relatively permanent water or repeated flooding, and the plants grow principally on or below the surface of the water. Rooted vascular plants occur in palustrine, riverine, and lacustrine wetlands at all depths within the photic zone. They often are most abundant in sheltered areas where there is little water movement, but some species are also found in flowing water in streams and rivers. These macrophytes include submersed species such as pondweed (*Potamogeton* spp.), water milfoil (*Myriophyllum* spp.), naiads (*Najas* spp.), ditch grasses (*Ruppia* spp.), and waterweed (*Elodea* spp.). Common rooted vascular plants that have floating leaves on the surface are water lilies (*Nymphea odor-*

ata or *N. tuberosa*), spatterdock (*Nuphar advena*), water shield (*Brasenia schreberi*), and lotus (*Nelumbo lutea*). Aquatic plants such as the duckweeds (*Lemna* spp., *Spirodela* spp.), water lettuce (*Pistia stratiotes*), and water hyacinth (*Eichhornia crassipes*) float on the water surface and may be moved about by wind or water currents. Other submersed plants such as bladderworts (*Utricularia* spp.) and coontails (*Ceratophyllum* spp.) float in beds beneath the water surface. These aquatic beds are found in many of the wetlands described in this book, especially in depressional wetlands (see Chapters 8, 25) and natural or managed impoundments with relatively stable water levels (see Chapters 6, 10, 27, 34). They are also common along the sheltered shorelines of larger lakes (see Chapters 38, 39, 40).

Emergent Marshes

Emergent wetlands have erect, rooted, herbaceous hydrophytes present for most of the growing season in most years. Marshes of this type are found associated with many palustrine, riverine, and lacustrine wetlands. The dominant species vary from place to place, but many genera are common to marshes throughout the temperate zone. Some of these wetlands are considered nonpersistent, dominated by plants which die back below the water surface or to the substrate at the end of the growing season. Typical nonpersistent emergent plants are pickerelweed (*Pontederia cordata*), arrow arum (*Peltandra virginica*), and arrowheads (*Sagittaria* spp.). Persistent emergent wetlands contain a vast array of grasslike plants, such as cattails (*Typha* spp.), bulrushes (*Scirpus* spp.), saw grass (*Cladium jamaicense*), sedges (*Carex* spp.), bur reed (*Sparganium eurycarpum*), and spike rush (*Eleocharis* spp.) and true grasses such as reed (*Phragmites australis*), wild rice (*Zizania aquatica*), and panic grasses (such as *Panicum hemitomum*). There are also many broad-leaved persistent emergents such as dock (*Rumex* spp.), purple loosestrife (*Lythrium salicaria*), smartweeds (*Polygonum* spp.), and water willow (*Decocon verticillatus*). Some persistent emergent plants also die back each year, but they decompose slowly and therefore add considerable structure and detritus to the wetland (see Chapter 24).

Many freshwater emergent marshes have zones of vegetation that occur across water depth gradients (including ecotones between the marsh and adjacent upland and aquatic systems). For example, a typical prairie pothole, limesink, or Carolina bay might have beds of floating-leaved and submerged aquatics in the deeper central part of the basin, followed shoreward by emergent species distributed according to their flooding tolerances (see Chapter 23). Cattails often grow in water up to a meter deep, whereas sedges, rushes, and arrowheads might occupy the shallowly flooded edge of the basin. Emergent marshes are found in many of the wetlands described in this book, including the Everglades (see Chapter 2), various kinds of depressional wetlands (see Chapters 4, 8, 9, 21, 22, 26, 30, 31, 32), beaver impoundments (see Chapters 10, 15), lake shorelines (see Chapters 23, 38, 39, 40), river floodplains (see Chapters 3, 16, 28, 29), and the bayous and tidal freshwater marshes (see Chapters 36, 37).

The tidal freshwater wetlands deserve a special note. The periodic and predictable tidal inundation is a major hydrologic feature of these wetlands, causing a repeated shift from anaerobic to aerobic conditions. Zonation is usually very apparent in these marshes, and elevational differences correspond with different plant associations (Simpson et al. 1983, Odum et al. 1984; see Chapter 36). Submerged vascular plants grow in the streams and permanent ponds (see Chapter 37). The low marsh has broad-leaved plant species typical of the nonpersistent emergent marsh type. The higher marsh has diverse populations of annual and perennial grasses, sedges, and herbaceous plants and is typically persistent.

Scrub-shrub

Shrub wetlands are dominated by woody vegetation less than 6 meters tall (Cowardin et al. 1979). The species include true shrubs, saplings of trees, and trees or shrubs that are small or stunted because of environmental conditions. These wetlands may be a successional stage leading to forested wetlands, or they may be relatively stable communities. They are widespread across North America and are characterized by a wide array of plant species. Typical deciduous shrubs are alders (*Alnus* spp.), willows (*Salix* spp.), and buttonbush (*Cephalanthus occidentalis*) (see Chapters 28, 33). Common evergreen shrubs of northern bogs and peatlands are heather (*Calluna yelgaris*), leatherleaf (*Chamaedaphne calyculata*), cranberry and blueberry (*Vaccinium* spp.), and Labrador tea (*Ledum palustre*) (see Chapter 17). In southeastern pocosins, zenobia (*Zenobia pulverulenta*) is essentially endemic and is abundant along with fetterbush (*Lyonia lucida*), titi (*Cyrilla racemiflora*), inkberry (*Ilex glabra*), and doghobble (*Leuchthoe axillaris*). In northern bogs spruce (*Picea* spp.) and tamarack (*Larix* spp.) often occur with the shrubs, and in pocosins pond pine (*Pinus serotina*) and bay trees (e.g., swamp redbay [*Persea palustris*], loblolly bay [*Gordonia lasianthus*], and sweetbay [*Magnolia virginiana*]) can overtop the shrubs on richer sites.

Forests

Forested wetlands have woody vegetation that is 6 meters tall or taller (Cowardin et al. 1979). They are most common in the eastern United States and in sections of the West where moisture is relatively abundant, such as along rivers and in the mountains. These forests typically have an overstory of trees, an understory of saplings or shrubs, and an herbaceous layer. In the Southeast, river floodplains often are dominated in seasonally flooded areas by bottomland hardwood forests containing broad-leaved deciduous species such as red maple (*Acer rubrum*), American elm (*Ulmus americana*), ashes (*Fraxinus pennsylvanica* and *F. nigra*), sweetgum (*Liquidambar styraciflua*), and oaks, including laurel (*Quercus laurifolia*), water (*Q. nigra*), cherrybark (*Q. pagoda*), swamp chestnut (*Q. michauxii*), and overcup (*Q. lyrata*) (see Chapters 7, 28). In more permanently flooded areas, water tupelo (*Nyssa aquatica*) occurs, along with bald cypress (*Taxodium distichum*), a deciduous conifer (see Chapter 7). Evergreen forests of broad-leaved species such as red bay (*Persea borbonia*), loblolly bay and sweet bay occur on organic wetland soils in the Southeast, and black spruce (*Picea mariana*) or white cedar (*Thuja occidentalis*) forests grow on organic soils in the northern peatlands. Red maple swamps occur in poorly drained areas of the Northeast, and depressions dominated by pond cypress (*Taxodium distichum* var. *nutans*) are a characteristic feature of the southeastern Coastal Plain (see Chapters 5, 9). Many temporary wetlands that result from snowmelt or autumn rains also occur in forests (see Chapters 12, 13, 14).

INVERTEBRATE COMMUNITIES

Wetlands typically support diverse invertebrate communities consisting of aquatic, semiaquatic, and terrestrial species. Some invertebrates are obligate wetland residents, including certain aquatic species that require periods of desiccation to complete development (certain mosquitoes or fairy shrimp) and specialized terrestrial herbivores that feed only on certain wetland plants (some beetles, moths, and butterflies). However, many invertebrates that evolved primarily in lake, rivers, or streams or in upland habitats will opportunistically and successfully exploit wetland environments.

Aquatic Invertebrates

Many aquatic insects, crustaceans, annelids, and mollusks live in the ponded surface waters of freshwater wetlands. However, the diversity of this fauna is typically lower than in lakes, rivers, and streams (but see Chapter 36), possibly because of the stressful environments created in wetlands. Just as plants can, aquatic invertebrates can be stressed by cycles of wetting and drying and the periodic depletion of dissolved oxygen.

How aquatic invertebrates deal with habitat drying is a common theme of wetland invertebrate research. Wiggins et al. (1980) examined some of the evolutionary and ecological strategies that different wetland invertebrate taxa use to live in temporary wetland pools and concluded that their success was ultimately dictated by how the organisms dealt with the drought phase (see Table 12.1 and review in Chapter 12). However, Williams (1996) challenged the notion that drought is a significant barrier to the success of insects in temporary waters. He described how virtually all of the aquatic insect orders contain some species able to live in habitats that are only temporally flooded, and many are well represented. He suggested that exploiting temporary waters may not be particularly difficult for invertebrates to accomplish and that the concept of faunas being constrained by temporary water conditions may exist more as a human perception than as a biological reality. In some cases invertebrates (amphipods and isopod crustaceans) without any apparent adaptation for resisting or avoiding drought can flourish in temporary water wetlands (see Chapters 9, 12, 14). Wissinger (1997) suggests that some of the most successful aquatic invertebrates in temporary water wetlands cannot tolerate drought but instead employ fairly predictable migrations between temporary and permanent waters (see Chapters 12, 15, 26); he refers to them as "cyclic colonizers." Past histories of biotic interaction may be integrated into the apparent responses of invertebrates to varying hydroperiods (see Chapters 13, 31). An inability to tolerate predation restricts some taxa to temporary water habitats where many predators, especially fish or salamanders, cannot survive (Wellborn et al. 1996, Chapters 13, 31; but see Chapter 29).

Wetland invertebrates often are adapted to tolerate the low-oxygen conditions that frequently develop in stagnant flooded wetlands (see above discussion of wetland oxygen dynamics). Many of the aquatic insect groups that abound in lakes, rivers, or streams (e.g., all stoneflies, most families of mayflies and caddisflies, and many families of beetles and flies) are absent or poorly represented in wetlands, even permanently flooded wetlands, presumably because of an inability to cope with low oxygen supplies. Among all aquatic invertebrates, many of the most exotic modes of oxygen extraction can be found among the wetland species. Some benthic midges and oligochaete worms possess hemoglobin in their hemolymph, which helps them tolerate low oxygen conditions, at least over short periods (see Chapters 24, 37). Some insect larvae (*Coquillettidia* mosquitoes, *Notophila* brine flies, and *Donacia* beetles) attach to plant roots with modified respiratory structures to extract oxygen. Rat-tailed maggots (family Syrphidae) lie along benthic substrates and extend a long telescoping respiratory siphon to the water's surface to breathe. However, the majority of air-breathing aquatic invertebrates in wetlands (most beetles, hemipterans, and fly larvae) periodically venture to the water's surface to replenish their oxygen supplies. These surface-breathing insects are much more conspicuous in wetlands than in lakes, rivers, or streams. Migrating up and down through the water column is probably a less risky endeavor for invertebrates in wetlands than in other aquatic habitats because of shallow depths, minimal water current, low densities of predatory fish, and abundant plant cover. Some wetland invertebrates (clams, dragonfly naiads, tube-building midge larvae) cope with stagnant water conditions by pumping water over their respiratory structures. However, many aquatic invertebrates (microcrustaceans, most midge larvae, damselfly naiads) surprisingly flourish in wetlands despite possessing fairly rudimentary respiratory structures and often simply absorb oxygen across their cuticles.

Many aquatic invertebrates in wetlands can tolerate stressful levels in water quality parameters other than just oxygen. In California a midge species (*Chironomus stigmaterus*) that is most commonly reported from freshwater marshes also flourishes in moderately saline marshes and even in sewage lagoons (Batzer et al. 1997, see Chapter 34). Even wetlands with extremely high water conductivities (saline or alkaline ponds) can support an abundance of aquatic invertebrates (see Chapters 21, 25, 26, 27, 34).

Aquatic invertebrates in wetlands rely heavily on plants for food and shelter. Although few aquatic invertebrates can consume living macrophyte tissue, many will readily consume the detritus (or at least the associated microorganisms) created when plants die (see Chapters 7, 12). Algae is another important food for some aquatic invertebrates (see Chapters 19, 21, 24, 34, 40), and it often grows on surfaces of submersed vascular wetland plants. The importance of plants as structural habitat for aquatic invertebrates, apart from being a food source, should not be overlooked. In stressful freshwater tidal habitats invertebrate diversity is typically low except where macrophyte beds are extensive (see Chapter 37). In nontidal marshes the highest diversity and abundance of aquatic invertebrates is often associated with emergent or submersed plants or complexes of plants and open water (see Chapters 23, 24, 25, 27, 34, 40). In the case of pitcher plants, habitat for some specialized aquatic fly larvae is created when tiny pools of water collect inside the leaves (see Chapters 11, 18).

Semiaquatic and Terrestrial Invertebrates

While most wetland invertebrate research has focused on the aquatic component of the fauna, a diversity of wetland invertebrates do not live in the water itself, but rather around wetland perimeters, on wetland plants, or on the water's surface. In fact, the terrestrial invertebrate fauna dwarfs the aquatic fauna in terms of diversity (see Chapter 17). A marsh study by Bergey et al. (1993) found that aphids and collembolans were by far the most abundant invertebrates in beds of the pond weed *Potamogeton,* despite the fact that they occupied only the water's surface and those few plant sprigs that were not submersed.

The terrestrial invertebrate fauna of wetlands can be very important ecologically, as it is in upland habitats (see Chapters 11, 35). Feeding of herbivorous beetles and moths on the aerial stems and leaves or the stem and root pith can strongly impact wetland plants and potentially regulate plant community structure. In fact, virtually all of the important invertebrate consumers of vascular plants in wetlands are typical terrestrial insects (aphids and leaf hoppers, moth and butterfly larvae, beetle larvae and adults) (Batzer and Wissinger 1996). On the other hand, many wetland plants rely on flying insects as pollinators (mostly bees or butterflies) (see Chapter 11).

Many wetland insects with aquatic immatures have terrestrial adult stages (midges, mayflies, odonates), and upon emergence they are readily consumed by birds, spiders, and other terrestrial predators (Orians 1980, Sjoberg and Danell 1982). In the case of dragonflies and mosquitoes, the terrestrial adults can be important aerial predators in their own right. In a unique turnabout, several wetland plants (pitcher plants, sundews) prey upon the terrestrial invertebrate fauna of wetlands (see Chapter 11). Clearly, invertebrates of the terrestrial arena are important from the perspectives of both biodiversity and ecosystem functioning in wetlands and merit considerably more research attention.

CONCLUSION

The following chapters will introduce most of the major categories of freshwater wetland in North America and provide the most current knowledge on the ecology and management of their invertebrate faunas. Because this information is presented in a habitat-specific manner, it is hoped that the concepts will be more accessible to readers who may know much about certain wetland habitats but little about invertebrates.

ACKNOWLEDGMENTS

Preparation of this chapter was partly supported by Financial Assistance Award Number DE-FC09-96SR18546 between the U.S. Department of Energy and the University of Georgia.

LITERATURE CITED

Armstrong, W. 1975. Waterlogged soils. Pages 181–218 *in* J. R. Etherington (ed.), Environment and Plant Ecology. John Wiley & Sons, London.

Batzer, D. P., F. de Szalay, and V. H. Resh. 1997. Opportunistic response of a benthic midge (Diptera:Chironomidae) to management of California seasonal wetlands. Environmental Entomology 26:215–222.

Batzer, D. P., and S. A. Wissinger. 1996. Ecology of insect communities in nontidal wetlands. Annual Review of Entomology 41:75–100.

Bergey, E. A., S. F. Balling, J. N. Collins, G. A. Lamberti, and V. H. Resh. 1993. Bionomics of invertebrates within an extensive *Potamogeton pectinatus* bed of a California marsh. Hydrobiologia 234:15–24.

Brinson, M. M., A. E. Lugo, and S. Brown. 1981. Primary productivity, decomposition and consumer activity in freshwater wetlands. Annual Review of Ecology and Systematics 12:123–161.

Brown, S. L. 1981. A comparison of the structure, primary productivity, and transpiration of cypress ecosystems in Florida. Ecological Monographs 51:403–427.

Carter, V., and R. P. Novitzki. 1988. Some comments on the relation between ground water and wetlands. Pages 68–86 *in* D. D. Hook et al. (eds.), The Ecology and

Management of Wetlands. Vol. 1: Ecology of Wetlands. Timber Press, Portland, OR.

Cowardin, L. M., V. Carter, F. C. Golet, and E. T. LaRoe. 1976. Interim Classification of Wetlands and Aquatic Habitats of the United States. U.S. Fish and Wildlife Service.

Cowardin, L. M., V. Carter, F. C. Golet, and E. T. LaRoe. 1979. Classification of Wetlands and Deepwater Habitats of the United States. U.S. Fish and Wildlife Service, Pub. FWS/OBS-79/31, Washington, DC.

Dahl, T. E. 1990. Wetlands Losses in the United States, 1780s to 1980s. U.S. Department of the Interior, Fish and Wildlife Service, Washington, DC.

Dahl, T. E., and C. E. Johnson. 1991. Wetlands Status and Trends in the Conterminous United States, Mid-1970s to Mid-1980s. U.S. Department of the Interior, Fish and Wildlife Service, Washington, DC.

Daily, G. C. (ed.). 1997. Nature's Services: Societal Dependence on Natural Ecosystems. Island Press, Washington, DC.

Faulkner, S. P., and C. J. Richardson. 1989. Physical and chemical characteristics of freshwater wetland soils. Pages 41–72 in D. A. Hammer (ed.), Constructed Wetlands for Wastewater Treatment. Lewis, Chelsea, MI.

Frayer, W. E., T. J. Monahan, D. C. Bowden, and F. A. Graybill. 1983. Status and Trends of Wetlands and Deepwater Habitats in the Conterminous United States, 1950s to 1970s. Department of Forest and Wood Sciences, Colorado State University, Fort Collins, CO.

Gambrell, R. P., and W. H. Patrick, Jr. 1978. Chemical and microbiological properties of anaerobic soils and sediments. Pages 375–423 in D. D. Hook and R. M. M. Crawford (eds.), Plant Life in Anaerobic Environments. Ann Arbor Science Publishers, Ann Arbor, MI.

Gunderson, L. H., and W. F. Loftus. 1993. The Everglades. Pages 199–255 in W. H. Martin, S. C. Boyce, and A. C. Echternacht (eds.), Biodiversity of the Southeastern United States: Lowland Terrestrial Systems. John Wiley & Sons, New York.

Heilman, P. E. 1968. Relationship of availability of phosphorus and cations to forest succession and bog formation in interior Alaska. Ecology 49:331–336.

Herendorf, C. E. 1987. The Ecology of the Coastal Marshes of Western Lake Erie: A Community Profile. U.S. Fish and Wildlife Service, Biological Report 85, Washington, DC.

Kadlec, R. H., R. B. Williams, and R. D. Scheffe. 1988. Wetland evapotranspiration in temperate and arid climates. Pages 146–160 in D. D. Hook et al. (eds.), The Ecology and Management of Wetlands. Timber Press, Portland, OR.

Kushlan, J. A. 1991. The Everglades. Pages 121–142 in R. J. Livingston (ed.), The Rivers of Florida. Springer-Verlag, New York.

Lockaby, B. G., and M. R. Walbridge. 1997. Biogeochemistry. Pages 149–172 in M. G. Messina and W. H. Conner (eds.), Southern Forested Wetlands: Ecology and Management. Lewis Pub., Boca Raton, LA.

MacMannon, M., and R. M. M. Crawford. 1971. A metabolic theory of flooding tolerance: the significance of enzyme distribution and behavior. New Phytologist 10:299–306.

McKevlin, M. R., D. D. Hook, and A. A. Rozelle. 1997. Adaptations of plants to flooding and soil waterlogging. Pages 173–203 in M. G. Messina and W. H. Conner (eds.), Southern Forested Wetlands. Ecology and Management: Lewis, Boca Raton, FL.

Megonigal, J. P., W. H. Conner, S. Kroeger, and R. R. Sharitz. 1997. Aboveground production in southeastern floodplain forests: a test of the subsidy-stress hypothesis. Ecology 78:370–384.

Mendelssohn, I. A., K. L. McKee, and W. H. Patrick, Jr. 1981. Oxygen deficiency in *Spartina alterniflora* roots: metabolic adaptation to anoxia. Science 214:439–441.

Mendelssohn, I. A., K. L. McKee, and M. L. Postek. 1982. Sublethal stresses controling *Spartina alterniflora* productivity. Pages 223–242 *in* B. Gopal, R. E. Turner, R. G. Wetzel, and D. F. Whigham (eds.), Wetlands Ecology and Management. Natural Institute of Ecology and International Scientific Publications, Jaipur, India.

Mitsch, W. J., and K. C. Ewel. 1979. Comparative biomass and growth of cypress in Florida wetlands. American Midland Naturalist 101:417–426.

Mitsch, W. J., and J. G. Gosselink. 1993. Wetlands (2nd ed.). Van Nostrand Reinhold, New York.

Mitsch, W. J., and B. C. Reeder. 1992. Nutrient and hydrologic budgets of a Great Lakes coastal freshwater wetland during a drought year. Wetlands Ecology and Management 1:211–223.

Mohanty, S. K., and R. N. Dash. 1982. The chemistry of waterlogged soils. Pages 389–396 *in* B. Gopal, R. E. Turner, R. G. Wetzel, and D. F. Whigham (eds.), Wetlands Ecology and Management. Natural Institute of Ecology and International Scientific Publications, Jaipur, India.

Murkin, H. R. 1989. The basis for food chains in prairie wetlands. Pages 316–338 *in* A. G. van der Valk (ed.), Northern Prairie Wetlands. Iowa State University Press, Ames, IA.

Murkin, H. R., and B. D. J. Batt. 1987. The interactions of vertebrates and invertebrates in peatlands and marshes. Pages 15–30 *in* D. M. Rosenberg and H. V. Danks (eds.), Aquatic Insects of Peatlands and Marshes of Canada. Memoirs of the Entomological Society of Canada 140:1–174.

NRC (National Research Council). 1992. Restoration of Aquatic Ecosystems: Science, Technology and Public Policy. National Academy Press, Washington, DC.

NRC (National Research Council). 1995. Wetlands Characteristics and Boundaries. National Academy Press, Washington, DC.

Odum, W. E., T. J. Smith III, J. K. Hoover, and C. C. McIvor. 1984. The Ecology of Tidal Freshwater Marshes of the United States East Coast: A Community Profile. U.S. Fish and Wildlife Service, FWS/OBS-87/17, Washington, DC.

Orians, G. H. 1980. Some Adaptations of Marsh-nesting Blackbirds. Princeton University Press, Princeton, NJ.

Ponnamperuma, F. N. 1972. The chemistry of submerged soils. Advances in Agronomy 24:29–96.

Shaw, S. P., and C. G. Fredine. 1956. Wetlands of the United States, Their Extent, and Their Value for Waterfowl and Other Wildlife. U.S. Fish and Wildlife Service, Circular 39, Washington, DC.

Simpson, R. L., R. E. Good, M. A. Leck, and D. F. Whigham. 1983. The ecology of freshwater tidal wetlands. BioScience 33:255–259.

Sjoberg, K., and K. Danell. 1982. Feeding activity of ducks in relation to diel emergence of chironomids. Canadian Journal of Zoology 60:1383–1387.

Tiner, R. W., Jr. 1984. Wetlands of the United States: Current Status and Recent Trends. National Wetlands Inventory. U.S. Department of the Interior, Fish and Wildlife Service, Washington, DC.

Tiner, R. W. 1993. Wetlands are ecotones: Reality or myth? Pages 1–15 in B. Gopal, A. Hillbricht-Ilkowska, and R. G. Wetzel (eds.), Wetlands and Ecotones: Studies on Land-Water Interactions. National Institute of Ecology, New Delhi, India.

Wellborn, G. A., D. K. Skelly, and E. E. Werner. 1996. Mechanisms creating community structure across a freshwater habitat gradient. Annual Review of Ecology and Systematics 27:337–363.

Wiggins, G. B., R. J. Mackay, and I. M. Smith. 1980. Evolutionary and ecological strategies of animals in annual temporary pools. Archiv für Hydrobiologie, Supplement 58:97–206.

Williams, D. D. 1996. Environmental constraints in temporary fresh waters and their consequences for the insect fauna. Journal of the North American Benthological Society 15:634–650.

Wissinger, S. A. 1997. Cyclic colonization in predictably ephemeral habitat: a template for biological control in annual crop systems. Biological Control 10:4–15.

Zoltai, S. C. 1988. Wetland environments and classification. Pages 1–26 in National Wetlands Working Group (ed.), Wetlands of Canada, Ecological Land Classification Series. No. 24. Environment Canada, Ottawa, Ontario, and Polyscience, Montreal, Quebec.

PART 1
Marshes and Swamps of the Southeast

2 The Florida Everglades

Natural Variability, Invertebrate Diversity, and Foodweb Stability

RUSSELL B. RADER

*P*rior to discussing the ecological factors (hydrology, macrophyte density, and fish predation) affecting the distribution, diversity, and abundance of invertebrates, this chapter begins with a description of the origin, hydrology, and mosaic of habitat types that defines the Everglades ecosystem. Although the potential impact of eutrophication and invasion of introduced species on invertebrates can have important localized effects, hydrology is the single most important factor in maintaining foodweb stability. The length of inundation, water depth, and rate of water subsidence can each favor some predator guilds more than others. The full range of natural hydrological variation is important in maintaining all members of the Everglades community. Long periods (years) of inundation, long dry periods, and even long periods of a constant cycle of inundation and drying without intermittent variation (prolonged dry or wet seasons) could create the risk of the eventual collapse of one or more invertebrate or vertebrate guilds. Attempting to engineer optimal conditions for all species at once is not possible, whereas a holistic view that allows for natural environmental dynamics and subsequent interannual variability in species success and ecosystem processes can accomplish long-term sustainability.

GEOMORPHOLOGY AND HYDROLOGY

The freshwater portion of the Florida Everglades is part of a large watershed that begins about 20 km south of Orlando at the headwaters of the Kissimmee

Invertebrates in Freshwater Wetlands of North America: Ecology and Management, Edited by Darold P. Batzer, Russell B. Rader, and Scott A. Wissinger
ISBN 0-471-29258-3   © 1999 John Wiley & Sons, Inc.

River and ends in the southwestern corner of the Florida peninsula. The overall drainage basin (28,200 km²) stretches about 450 km north to south and 100 km east to west and includes the Kissimmee River, Lake Okeechobee, and the Everglades (SFWMD 1992, Light and Dineen 1994). The Everglades basin itself, one of the largest freshwater wetlands in the world, is approximately 11,000 km² and is bounded to the west by slightly elevated sand flats and Big Cypress Swamp and to the east by the Atlantic Coastal Ridge (Light and Dineen 1994). Gradient is one of the most important factors accounting for the existence of the Everglades ecosystem. There is an average slope of only 3.0 cm km⁻¹ from the southern rim of Lake Okeechobee to the Florida Bay (Kushlan 1990, SFWMD 1992). Steeper gradients might have resulted in the cutting of a distinct outflow channel and the Kissimmee River being both the primary inflow and outflow from Lake Okeechobee. As it is, water in the Everglades is unchannelized and flows as large, broad sheets in a general north to south direction (Gleason 1974). Ground elevations range from approximately 6.0 m above sea level in the northern Everglades to 10 or 30 cm above sea level in the southern Everglades (SFWMD 1992).

Formation of the Everglades as we know it today began about 5000 years ago (Gleason 1974). The Pleistocene epoch was punctuated by four major ice ages that caused the oceans to recede and rise, covering and uncovering the Florida peninsula. Between periods of peak glaciation, south Florida was a large reef in a shallow sea. The last time Florida was submerged was prior to the Wisconsin glaciation about 60,000 years ago. Sea level was approximately 80 m higher than the present mark. At the peak of the Wisconsin glaciation (about 18,000 years ago), sea level was about 130 m below the present mark (Wanless et al. 1994). As glaciers from the Wisconsin ice age began to melt, the sea level rose until reaching its present position between 5,500 and 3,200 years ago (Gleason et al. 1984, Wanless et al. 1994). Only small sea level fluctuations have occurred since then. Underlying soils in the Everglades is a porous layer of sand and calcium carbonate limestone produced by bryozoans and corals during periods of oceanic inundation. Unlike many wetlands, the Everglades has no underlying impermeable bedrock layer (Fernald and Patton 1984). Fresh groundwater reserves are held in porous limestone aquifers by the surrounding saltwater. As more water is extracted from the Everglades ecosystem, saltwater intrusion into freshwater aquifers threatens to reduce water supplies to south Florida residents (U.S. Army Corps of Engineers 1992).

Most soils in the Everglades are 90 percent organic, being composed of dead and decaying vegetation (Gleason 1974). For much of the Everglades, the dominant vegetation is sawgrass (*Cladium jamaicense*), which produces a highly fibrous recalcitrant stalk (Davis 1991). Soil saturation and inundation cause anaerobic conditions that inhibit oxidation/decomposition, allowing for the accumulation of peat. Historic (prior to 1970) peat accretion rates in the northern Everglades averaged 2.2 mm yr⁻¹ (Craft and Richardson 1993). Peat soils were deepest (5.0 m) in the northern Everglades just south of Lake

Okeechobee and declined to an average depth of 0.75 m in the south in Everglades National Park (McCollum et al. 1978). Thick peat soils drained for agriculture (northern Everglades) have subsided (2.5 cm yr^{-1}) by as much as 1.5 m in the past 60 years (Snyder and Davidson 1994). In some parts of the southern Everglades, dry conditions have exposed the underlying bedrock, which is covered by a thin (<30 cm) layer of peat filling small depressions between jagged projections of calcareous limestone. In addition to peat-based habitats, short-hydroperiod marshes in the southern Everglades have developed a marl-type soil (calcitic mud) formed by the deposition of calcium carbonate precipitating mats (2–10 cm thick) of filamentous blue-green algae (Gleason 1972, Gleason and Stone 1994).

The Everglades hydroperiod is driven by rainfall and maintained by sheet-flow from adjacent upstream wetlands. Prior to human construction, water levels and flows peaked during the wet season (May–October) and receded during the dry months of winter (November–April). Afternoon thundershowers can occur somewhere in the Everglades almost every day during the rainy season, but especially in late May–June and September–October. On average, south Florida receives about 135 cm of rain annually, 75 percent of which falls in the wet season (MacVicar and Lin 1984). Nearly 90 percent of the water supply to the Everglades is derived from rainfall within the basin, whereas 10 percent comes from rainfall-induced runoff from adjacent regions (SFWMD 1992). Groundwater additions to surface flows are almost nonexistent and contribute less than 1 percent of the total inflow (Finkl 1995). Approximately 66 percent of water losses are attributed to evapotranspiration, while surface flow and groundwater seepage to the sea amount to another 22 percent (SFWMD 1992). Water losses are recycled within the Everglades, since evapotranspiration from one area may form the clouds that rain on another. Current velocities vary depending on the season (dry versus rainy) and density of emergent and submergent vegetation. Average water velocities measured biweekly in the southern Everglades (Shark River Slough) during the dry period and into the peak of the rainy season ranged from 0.1 cm s^{-1} in areas with high emergent macrophyte densities to 1.0 cm s^{-1} in open-water habitats (Rosendahl and Rose 1982). However, Hunt (1961) recorded maximum water velocities of 3.0 cm s^{-1} in Taylor Slough, another open-water habitat east of Shark River Slough.

Today water flow and hydroperiod in the Everglades are regulated by a series of dikes and canals. Efforts to regulate flow began with the destruction and flooding of newly developed agricultural communities by hurricanes in 1926 and 1928. In response, the Army Corp of Engineers constructed a dike around the southern rim of Lake Okeechobee. Since that time approximately 50 percent of the Everglades, including the Everglades Agricultural Area (3,059 km^2), has been drained and dried, while the remainder has been divided into three hydrologically isolated sections called Water Conservation Areas (3,554 km^2), plus Everglades National Park (4,363 km^2; Holling et al. 1994). (The freshwater portion of Everglades National Park is smaller than this estimate.) Surface water flow into and through each section is controlled by some of the largest pumping stations in the world. Over the past 10 years, approximately four times more water has been pumped from the northern Everglades east to the Atlantic Ocean than has been allowed to proceed downstream to Everglades National Park (Light and Dineen 1994).

HABITAT

Authors of early ecological research in the Everglades described four basic habitat types, as determined by the dominance and density of emergent vegetation: (1) sawgrass plains; (2) wet prairies; (3) tree islands/hummocks; and (4) sloughs (Davis 1943, Loveless 1959). Sawgrass plains are composed of tall (2–3 m), dense, monospecific stands of *Cladium jamaicense*. Sloughs are the wettest habitats with the longest hydroperiod and include associations of floating (e.g., *Nymphaea*), submerged (*Utricularia, Chara*), and short emergent (*Eleocharis, Panicum*) macrophytes. The deepest sloughs (2.0 m) are like ponds and, in addition to the above macrophytes, often contain abundant growths of *Potamogeton*. Wet prairies, like sawgrass plains, are dominated by *C. jamaicense*, but the sawgrass is much shorter (<1.5 m) and more sparse, resulting in more open-water. In contrast to sawgrass plains, submergent macrophytes and 14 other species of emergent vegetation (e.g., *Rynchospora, Panicum, Eleocharis, Sagittaria, Pontederia*) occur in wet prairie habitat (Gunderson 1994). Tree islands are patches of taller broad-leaved hydrophytic hardwoods (willows, various bayhead species, cypress, and pond apple) surrounded by shorter vegetation or sloughs (Gunderson 1994). Gunderson (1990) examined the interaction between hydroperiod and habitat by comparing long-term hydrological records among each habitat type. He found that hydroperiod and habitat type were only loosely correlated because of high year-to-year variation and difficulties of integrating the effects of droughts and extended floods on plant species composition. Nonetheless, his analysis did confirm a general trend often observed by other investigators (e.g., Goodrick 1974, Wood and Tanner 1990, Jordan 1996): sloughs have the longest hydroperiod, followed by wet prairies, then sawgrass plains.

Based on the relative abundance of these four habitats at the landscape scale, Davis et al. (1994) identified four physiographic types presently extant within the Everglades basin: sawgrass plains, slough/tree island/sawgrass mosaic, sawgrass-dominated mosaic, and marl-forming marshes. Sawgrass plains are the same as described above. The slough/tree island/sawgrass mosaic is a combination of three of the four basic habitats and is characterized by a matrix of open-water sloughs surrounding sparse strands of sawgrass and an occasional tree island. The sawgrass-dominated mosaic is primarily composed of wet prairie habitat where sloughs and hummocks are islands surrounded by a matrix of intermediately dense *C. jamaicense*. Marl-based marshes are short-hydroperiod open-water habitats of the southern Everglades where the soils (calcitic mud) are formed by the deposition of calcareous periphyton (Gleason 1972). Dominant macrophytes in marl-based marshes are similar to some of the vegetation commonly found in shallow sloughs (e.g., *Utricularia, Nymphaea, Eleocharis*). Of these four physiographic types, the slough/tree island/sawgrass mosaic (44 percent), and marl based marshes (31 percent) are the most abundant, followed by the sawgrass-dominated mosaic (15 percent) and sawgrass plains (10 percent; Davis et al. 1994). Open-water habitats (sloughs, wet prairies, and marl marshes) that lack dense overstories of emergent vegetation allow for the growth of short emergent (e.g., *Eleocharis* spp.) and submergent (*Utricularia* spp.) macrophytes and algae, which, like overstory emergents, provide cover but also are important sources of food and oxygen for invertebrates.

Several water chemistry parameters vary across both spatial and temporal scales in the Everglades (e.g., Waller and Earle 1975). I will discuss those that are of direct importance to invertebrates. At large spatial scales (km^2), oxygen concentrations are highest in open-water habitats where light penetrates to the soil-water interface, allowing for the growth of algae (Kolipinski and Higer 1969, Van Meter-Kasanof 1973, Belanger et al. 1989, Rader and Richardson 1992). Algae and some free-floating submergent macrophytes (e.g., *Utricularia* spp.) are the primary sources of oxygen in the Everglades, where diel fluctuations (1.0 mg l^{-1} at night to 30 mg l^{-1} during day) vary as a function of photosynthesis versus respiration (Gleason and Spackman 1974, Belanger et al. 1989, Rader and Richardson 1992). Supersaturation and the formation of oxygen bubbles in periphyton commonly occur during periods of peak photosynthesis. In contrast, rooted macrophytes conserve oxygen, "pumping" it to the roots, so limiting the amount released to the surrounding water. Shading by tall, dense sawgrass inhibits the growth of algae, resulting in consistently low oxygen concentrations (0.0–4.0 mg/L) that vary only slightly (Belanger et al. 1989) or not at all (Rader and Richardson 1992) over a 24-hour period. Atmospheric diffusion is the primary source of oxygen (approximate average of 1.94 g m^{-2} day^{-1}) in shaded habitats with a dense overstory of emergent vegetation (Belanger et al. 1989). Low oxygen concentrations in areas of dense emergent vegetation also occur in wet prairie marshes along the St. Johns River in central Florida (e.g., Jordan et al. 1996a). The processes (photosynthesis versus respiration) determining oxygen availability and resulting in the relationship among dense emergent vegetation, light penetration, algal growth, and oxygen availability should prevail in most wetland environments (Wetzel 1996). For example, Rose and Crumpton (1996) found that dense emergent vegetation (*Typha glauca, Typha angustifolia,* and *Scirpus fluviatilis*) reduced light availability to less than 2 percent of unshaded areas, causing a significant reduction of oxygen in prairie pothole marshes.

Oxygen concentrations can also vary on a seasonal basis with fluctuations in hydroperiod. Several studies reported reduced diel fluctuations and overall lower O$_2$ concentrations when winter drying or drought conditions reduced

Everglades marshes to small, shallow, turbid pools (5–6 m diameter, <25 cm depth) with high concentrations of fish and invertebrates (e.g., Kolipinski and Higer 1969, Waller and Earle 1975, Kushlan 1979). Diffusion is the primary means of chemical dispersal in poorly mixed, shallow, low-flow environments (Carlton and Wetzel 1987). Slow current velocities produce a minimal amount of mixing and create steep chemical gradients that can vary orders of magnitude over short distances (e.g., Wetzel 1996). On a small spatial scale (m^2), oxygen concentrations in the Everglades follow a general vertical gradient being highest at the air-water interface (atmospheric diffusion) and declining to anaerobic conditions near the bottom because of microbial respiration (Rader and Richardson 1992). (This assumes that epipelic algae are not abundant.) Superimposed on this general vertical gradient is horizontal variation attributed to algal mats and clusters of metaphyton or loosely aggregated algae (often filamentous) that are not attached to substrate surfaces (epipelon or periphyton) nor truly suspended in the water column (Wetzel 1996). Rader and Richardson (1992) found that oxygen concentrations measured 35 cm apart at midday varied from 30 mg l^{-1} (200 percent saturation) in an algal clump to 2–3 mg l^{-1} at the sediment-water interface. Efforts to characterize oxygen availability (e.g., between-site comparisons) must account for and include this temporal and spatial variability.

Because the Everglades is underlain by calcareous limestone, surface water pH is rarely acidic and is much more constant across different habitat types than oxygen. Calcium and bicarbonate are two of the most abundant ions in the Everglades ecosystem. However, on smaller spatial scales (m^2) and over a diel cycle, pH, like oxygen, will often fluctuate in response to rates of photosynthesis (CO_2 depletion) and respiration (CO_2 production). pH within algal clumps can vary between 6.9 (night) and 10.5 (day) over a 24-hour period (Kolipinski and Higer 1969, Rader and Richardson 1992). Although there are no microspatial nutrient data for the Everglades, phosphate and nitrate can also show large diurnal fluctuations in concentration within algal aggregations (Thybo-Christesen et al. 1993).

ECOLOGY OF INVERTEBRATES

The Invertebrate Community

Information on invertebrates in the Everglades is lacking in the scientific literature. None of the books or book chapters that summarize the flora and fauna of the Everglades includes a section on invertebrates (e.g., Gleason 1974, Myers and Ewel 1990, Davis and Ogden 1994). In fact, only seven studies (most are government reports) provide information on the community composition of macroinvertebrates in the Everglades (Table 2.1). The number of taxa identified from each study varies because of differences in the taxonomic level of identification, sampling techniques, and habitat types sampled.

TABLE 2.1. Summary of Invertebrate Studies in the Everglades[a]

Study	Habitat	Subhabitat	Sample Device	Mesh Size	Sample Size	Richness
Kolipinski and Higer 1969	Southern wet prairie	Swimming in the water column	Pull-up trap	3 mm	40	7
Van Meter-Kasanof 1973	Southern slough	Periphyton mat	Plant stem segments	—	62	13
Waller 1976	Northern and southern sloughs	Peat soils	Eckman dredge	?	10	?
Terczak 1980	Northern slough, wet prairie, and cattails	Macrophytes, periphyton, and soils	Tub and PVC cores	3 mm	96–160?	84
Loftus et al. 1986	Southern sloughs and wet prairie	Macrophytes, periphyton mat, and soils	Throw and funnel traps + PVC cores	5 mm, and 75 μm	480	20?
Davis 1994	Northern sawgrass and cattails	Decaying litter	Litter bags	—	54	11?
Rader 1994	Northern sloughs	Macrophytes, periphyton, and soils	Sweep net	2 mm	320	148

[a] "Habitat" is the general location (north or south) and type of habitat sampled (see text for habitat descriptions). The subhabitat component, sampling device, and mesh size are also indicated. "Sample Size" refers to the total number of samples (all sampling devices combined) used to determine taxonomic richness. "Richness" is the total number of taxa identified. Dashes indicate categories that did not apply to a specific study, and a "?" represents categories with unspecified or insufficient information.

Only three studies have attempted to describe all macroinvertebrates to the lowest feasible taxonomic level (Van Meter-Kasanof 1973, Terczak 1980, Rader 1994). Rader's (1994) work found most invertebrates from earlier studies, plus an additional 103 taxa. Terczak (1980), however, identified 26 invertebrates (2 clams, 8 beetles, 4 midges, 8 dragonflies, and 4 caddisflies) not collected in Rader's study. When these 26 taxa are added to Rader's list, a total of 174 taxa currently comprise the known invertebrate community in the Everglades. This number undoubtedly underestimates actual macroinvertebrate richness because the large mesh size (≈2 mm) used by both Rader (1994) and Terczak (1980) failed to collect smaller species of chironomids, ostracods, oligochaetes, and nematodes. Also, numerous Diptera, especially chironomids (Epler 1994), remain undescribed. Furthermore, these two studies were restricted to the same general location (Water Conservation Areas of the northern Everglades). Future investigations, especially in the southern Everglades, will undoubtedly reveal a long list of macroinvertebrate species that inhabit the Everglades. The actual number of taxa may eventually exceed 200 to 250 species. Currently, Diptera (49 taxa), Coleoptera (48 taxa), Gastropoda (17 taxa), Odonata (14 taxa), and Oligochaeta (11 taxa) are the most diverse groups (Rader 1994). Most invertebrate research (Kushlan 1975, Kushlan and Kushlan 1979, Kushlan and Kushlan 1980, Jordan 1996) has focused on the ecology of larger invertebrates (crayfish, shrimp, and apple snails) often cited for their importance to higher trophic levels (e.g., bass, wading birds, and the snail kite). Despite the importance of these taxa, they should not be the sole focus of invertebrate investigations. Other taxa are also important in the Everglades foodweb (e.g., dragonflies, predaceous beetles, mayflies) and undoubtedly play an important role in ecosystem function (e.g., decomposition).

Macroinvertebrates of the Everglades include mostly indigenous species from the southeastern United States with a few colonists from Central and South America (Rader 1994). However, Young (1954) suggested that many of the beetles were synonymous with species found in the Caribbean/West Indies. Some invertebrates of the Everglades might have a broader regional range, extending to the Caribbean and Central America, than we currently realize. Rader (1994) found only three endemic species, the most abundant snail (*Planorbella duryi*), *Planorbella scalaris,* and the mayfly, *Callibaetis floridanus*. Animal endemics are rare in the Everglades probably because it is only about 5000 years old and because endemism is uncommon in temporary environments (Pielou 1979). Further details on the biogeography and species composition of Everglades invertebrates have been summarized elsewhere (Rader 1994, Rader and Richardson 1994) and will not be included in this chapter.

Although information on macroinvertebrates is sparse, it appears rather substantial compared to the number of investigations describing zooplankton. The best taxonomic information on zooplankton in the Everglades involves the Copepoda (Reid 1992). Reid (1992) identified 13 copepod species from

Everglades National Park, including two which were previously undescribed (*Thermocyclops parvus* and *Eucyclops conrowae*), a neotropical species previously unknown from the United States (*Eucyclops bondi*), and a new southern range extension for *Mesocyclops americanus*. She suggested that the zooplankton assemblage from the southern Everglades may be characterized by a high diversity of copepods, with a substantial representation of tropical/Caribbean species (Reid 1992). In addition to Reid's work, I am aware of only two other studies that provide information on zooplankton, and they were also restricted to the southern Everglades. Van Meter-Kasanof (1973) collected 17 samples (plankton net tows) from the water column in a shallow, open-water habitat and identified two Cladocera (*Daphnia* sp., *Bosmina* sp.) and a single cyclopoid copepod (*Cyclops* sp.); abundance estimates were not included (Van Meter-Kasanof 1973). In the second study, Loftus et al. (1990) used a "multiple-funnel trap" designed to capture benthic organisms that show diel movements from periphyton mats and peat sediments during the day into the water column at night and found that Copepoda nauplii (2,032 m^{-2}), cyclopoid copepods (1,106 m^{-2}), and chydorid (2,747 m^{-2}) and macrothricid (1,385 m^{-2}) Cladocera were the most abundant organisms collected. Calanoid and harpaticoid copepods and two additional cladoceran families (Sididae and Daphniidae) were also present but much less abundant (Loftus et al. 1990). Fish gut analyses also indicate the presence of a diverse assemblage of zooplankton comprising an important part of the diet of some taxa (e.g., Kolipinski and Higer 1969). These studies raise several questions concerning the diversity and abundance of zooplankton and how and where to sample them in the Everglades. Although conclusions are not appropriate with so little information, it does appear that an abundant microcrustacean community may be associated with periphyton in the Everglades.

Factors Affecting Invertebrate Abundance and Diversity

As in other natural communities, numerous factors that operate across various spatial and temporal scales can have both direct and indirect effects on community structure and function in the Everglades. DeAngelis (1994) discussed some of the primary driving forces that maintain the Everglades ecosystem. I will focus on those forces important to invertebrates at fast (1–10 year) and intermediate (10–100 years) temporal scales and across small (10–10,000 m^2), intermediate (1–100 km^2), and large (100–10,000 km^2) spatial scales. The effects of gradually changing forces (e.g., sea level rise) that operate over hundreds to thousands of years and explain the formation of the Everglades and its primary physical and chemical features (geology, topography, and climate) have been briefly discussed in preceding sections.

The annual pattern of inundation and drying (hydroperiod) is probably the most important event affecting the fauna and flora across large and small scales in the Everglades. Jordan (1996) found that variation in the relative abundance of prawns and crayfish resulted from factors that affected the entire landscape, especially hydrology. For example, prawn (*Palaemonetes paludosus*) densities were positively correlated with extended hydroperiods and deep water (Jordan 1996). Loftus et al. (1990) found that total macroinvertebrate density in nine slough/open-water marshes of the southern Everglades was three to five times greater in long compared to short hydroperiod habitats. Over the course of seven years, the average depth of short hydroperiod sloughs dropped below 10 cm five times, whereas water depth over the same time period in long hydroperiod sloughs remained above 20 cm (Loftus et al. 1990). Similarly, Kushlan (1975) found that apple snail populations (*Pomacea paludosa*) in a southern, marl-based marsh were two to four times greater during two years of high water and continuous inundation compared to two previous and two subsequent years of annual drying. Snails buried themselves in the moist, decaying layers of periphyton during periods of exposure. Large adult snails were better than smaller adults at surviving dry conditions. Perhaps more important than hydroperiod's effects on adult snails is its impact on eggs and hatchlings. Adult apple snails climb above the water (nighttime) on emergent vegetation (broad leaves, > 6 mm in diameter, are used more often than narrow) to lay clutches of approximately 20 pearl-sized, calcareous-shelled eggs (Turner 1996). Incubation ranges from two to three weeks. During this time, egg submersion will result in increased embryonic mortality (Turner 1994). After hatching, the young snails drop from the emergent vegetation to the water below. Hatchlings can withstand stranding and desiccation for only 24 to 96 hours, depending on the amount of remaining moisture (Turner 1996). Both water level increases (egg mortality) and decreases (hatchling mortality) can have an important detrimental impact on apple snail populations.

During dry periods, invertebrates inhabiting ephemeral marshes can be exposed, be stranded in small shallow pools, enter quiescent or diapause stages, or be forced to migrate to deep water refugia (canals, air boat trails, and ponds). Although wetland invertebrates are adapted to periods of exposure, it is undoubtedly stressful and can produce high mortality rates. As invertebrates concentrate in small pools, high community respiration can combine with turbidity to produce rapid rates of oxygen consumption with low rates of primary production, causing an increase in oxygen stress (e.g., Kushlan 1979). Also, concentrated invertebrates are subject to intense predation pressure from wading birds (e.g., Hoffman et al. 1994). Clearly, long hydroperiods that provide sufficient time for both small and large invertebrates to complete their life cycles are advantageous for most aquatic invertebrates.

Although drying may have a detrimental impact on most invertebrate populations, not all taxa will achieve highest densities or select habitats with the longest hydroperiod and deepest water. For example, crayfish (*Procambarus alleni*) in the northern Everglades preferred drier sawgrass and wet prairie habitats over deeper sloughs (Jordan 1996). The abundance of large predatory fish (e.g., largemouth bass and gar) tends to increase with increasing hydroperiod (Weichman 1987, Loftus and Eklund 1994), which may cause a decrease in some invertebrates, depending upon the complex interaction between habitat structural complexity/refuge availability and predation efficiency (Jordan 1996). Although information on the influence of fish predation in wetlands is sparse (Batzer and Wissinger 1996), recent studies suggest that fish can reduce both the diversity and abundance of invertebrates (e.g., Mallory et al. 1994, Hanson and Riggs 1995). Jordan et al. (1996a, b) found that *P. alleni* were most abundant in wet prairies with a short hydroperiod and annual drying because the increased habitat complexity provided by emergent vegetation reduced predation on small crayfish by largemouth bass. Ovigerous (egg-laying) females were only collected from emergent vegetation and were apparently absent from deep sloughs where bass were abundant (Jordan 1996b). Larger crayfish appeared to utilize slough habitats, but almost exclusively at night. Kushlan and Kushlan (1979, 1980) found that both crayfish (*P. alleni*) and shrimp (*Palaemonetes paludosus*) were less abundant during years of continuous inundation than in years with an annual period of drying and exposure. They also suggested that increased fish predation accounted for the decline of decapods in continuously inundated marshes.

Davis (1943) and Reark (1961) were the first to recognize that open-water sloughs contain the highest plant and animal diversity in the Everglades. Dense sawgrass plains are not only a monoculture of one emergent plant (e.g., Jordan 1996) but also contain a reduced diversity and abundance of invertebrates. Rader (1994) found that invertebrate density in the northern Everglades was almost five times greater in open-water slough/wet prairie habitats than in sawgrass. (This study did not distinguish sloughs from wet prairies. Goodrick (1974) found extensive overlap in macrophyte species composition between these two habitats.) Compared to dense sawgrass plains, open-water habitats (sloughs and wet prairie) offer more food and oxygen because of an abundance of algae and submersed macrophytes (Reark 1961, Wood and Tanner 1990). Generally, submersed macrophytes are more labile (e.g., *Nymphaea odorata*) than recalcitrant sawgrass leaves and release more oxygen to the surrounding water (e.g., *Utricularia*). Such generalities concerning the preferred habitat of invertebrates in the Everglades can be confused by imprecise habitat definitions and by the proximity and interaction between adjacent habitat types. For example, Turner and Trexler (1997) found a high abundance of invertebrates in sawgrass habitat bordering open-water sloughs and wet prairie marshes. They concluded that, contrary to previous research from the interior portion of dense sawgrass and cattail habitats (e.g., Loftus and Kushlan 1987, Rader 1994), "emergent vegetation (cattails and sawgrass) can contain higher densities of fish and invertebrates than spikerush sloughs." Emergent vegetation surrounding deeper, open-water habitats may provide refuge from fish predation without the associated harsh abiotic conditions (low algae and oxygen) characteristic of the interior portion of dense sawgrass or cattail stands (Rader and Richardson 1992, Rader 1994). High invertebrate diversity and abundance are often associated with the edge habitat between open water and emergent vegetation (e.g., Murkin and Ross, this volume). As shown by more extensive investigations (Reark 1961, Kushlan 1974, Loftus and Kushlan 1987, Rader and Richardson 1992, Davis 1994, Rader 1994), interior portions of sawgrass or cattail stands (monocultures) do not contain greater invertebrate and fish diversity and abundance than open-water habitats (e.g., sloughs).

Figure 2.1 is a diagram showing the general influence of hydroperiod, percentage of open-water habitat, and intensity of fish predation on the abundance and diversity of invertebrates in the Everglades. The diversity and abundance of invertebrates is negatively correlated with the intensity of fish predation (Kushlan and Kushlan 1980, Loftus et al. 1990) and positively correlated with hydroperiod (Loftus et al. 1990, Kolipinski and Higer 1969). The percentage of open-water habitat or macrophyte density (emergent and submergent) probably has a nonlinear, quadratic relationship with invertebrate

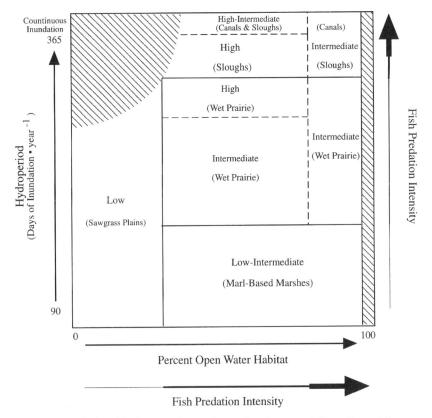

Fig. 2.1. The relationship between invertebrate abundance and diversity and three controlling factors (hydroperiod, percent open water habitat and fish predation intensity) in the Everglades. Hatched areas represent the unlikely occurrence of dense sawgrass with long hydroperiods (upper left) and near 100 percent open water (right margin). See text for a complete explanation.

diversity and abundance. At one extreme, shading caused by dense emergent or floating vegetation can diminish algal food resources and oxygen (Morris and Barker 1977, Rose and Crumpton 1996, Euliss et al., this volume), whereas sparse vegetation (reduced structural complexity) at the other extreme should intensify the effects of fish predation. Both should result in lower invertebrate abundance and diversity. Intermediate levels of macrophyte density should support the highest invertebrate abundance and diversity by providing ample food, oxygen, and refuge from fish predation. In lake and marine environments, dense macrophytes are often not associated with harsh abiotic conditions (e.g., low oxygen), and large invertebrates (e.g., decapods) may be most abundant in the most densely vegetated habitat available (Rozas and Odum 1987, Savino and Stein 1989, Sogard and Able 1991). The precise length of inundation, intensity of fish predation, or macrophyte percent cover causing a shift among high, intermediate, and low diversity and abundance is not known and probably fluctuates depending on the influence of other factors. In general, however, the highest diversity and abundance (Fig. 2.1) should occur in habitats with a long hydroperiod (upper section) and intermediate densities of macrophytes (middle section), whereas the lowest diversity and abundance should occur in dense emergent vegetation with a short hydroperiod (lower left corner). Plus, invertebrate abundance and diversity will decline as long hydroperiods extend into years of continuous inundation, allowing for an increase in fish predation. These areas of varying invertebrate diversity and abundance roughly correspond to dense sawgrass plains (low), marl-based marshes (low to intermediate), wet prairies (high and intermediate), sloughs (high and intermediate), and canal habitats similar to lake littoral environments (high to intermediate). Hatched areas represent unlikely conditions; the occurrence of long hydroperiods combined with dense sawgrass stands (upper left corner) and near 100 percent open-water habitat (right margin). Some submergent or emergent vegetation will occur in virtually all wet prairies, marl-based marshes, and sloughs.

In addition to hydroperiod, percent open-water, and fish predation, disturbances should also influence invertebrate abundance and diversity. DeAngelis (1994) recognizes four types of natural disturbances in the Everglades: fires, freezes, hurricanes and drought. Gunderson and Snyder (1994) calculate a return interval of 3 years for smaller burns (4000–8000 ha) and 10 to 15 years for large burns (>10,000 ha). Fire plays an important role in vegetation dynamics and creating and maintaining the mosaic of habitat types that characterize the Everglades landscape (sawgrass plains, wet prairies, sloughs). Although data on the impact of fire on invertebrates in the Everglades are not available, the impact of fire on plant communities should indirectly influence invertebrates. For example, fires may provide a subsidy for invertebrates by releasing nutrients and removing the effects of shading (low algal growth and oxygen) in areas with a dense emergent canopy (de Szalay and Resh 1997). Freezes (<0°C) are rare events of short duration (an average of approximately three days per year; Duever et al. 1994) that probably have little direct effect on invertebrates. But like fires, freezes can influence vegetation and habitat structure. Hurricanes are also rare events and are often associated with increased water levels and flooding. Except for localized areas where high winds can uproot vegetation, hurricanes are also probably more of a subsidy (increased flooding) than a stress to invertebrates. Droughts probably have the greatest impact on invertebrate populations. From 1940 to 1963, prior to current levels of water regulation, there were 12 droughts sufficiently severe to dry most wet prairies of the southern Everglades (Kolipinski and Higer 1969). Seven of these droughts were so severe so as to cause drying of many of the deeper sloughs and alligator holes. Davis et al. (1994) calculated that a severe drought could dry as much as 90 percent of the Everglades basin from Lake Okeechobee to the southern tip of Florida. In 1965 a severe drought lasted from April through July and dried most of the sloughs and alligator holes in the southern Everglades. As water levels began to rise in August, Kolipinski and Higer (1969) found that small fish and invertebrate densities were vastly lower than in previous wet seasons.

Role of Invertebrates in the Everglades Ecosystem

Invertebrates are an important link between primary producers and higher trophic levels. The Everglades foodweb is extremely complex and characterized by (1) abundance and diversity of predators; (2) numerous omnivorous taxa, and (3) ontogenetic shifts in food selection and prey preferences. The challenge is to organize hundreds of taxa in a way that will provide some insight into the flow of energy and factors (hydroperiod) that influence seasonal shifts in predator-prey interactions.

Foodweb Structure. Figure 2.2 is a depiction of the Everglades foodweb where taxa are organized into guilds based on body size and feeding attributes. Classification of organisms into ecological aggregates that are functionally similar and transcend taxonomic boundaries (functional groups or guilds) can condense a large amount of species-specific information into a few categories that have generalized attributes (e.g., Cummins 1973, Hawkins and MacMahon 1989). I used a variety of sources to assign functional feeding attributes, feeding habits, and prey preferences (e.g., Hunt 1952, Kolipinski and Higer 1969, Fogarty 1974, Kushlan 1978, Rader 1994, Merritt et al. 1996).

Perhaps the most unusual characteristic of the Everglades community is the presence of a large variety of both invertebrate and vertebrate predators. The abundance of top predators (wading birds, alligators, turtles, otters) a trophic level above fishes, plus the variety of different fish guilds, make the Everglades, and other wetlands as well, unique among freshwater ecosystems. Fishes are the top predators in many freshwater foodwebs (Lamberti 1996). Furthermore, the types of predators at the top of the Everglades foodweb (small fish, large fish, or fish-eating birds and reptiles) may vary with habitat

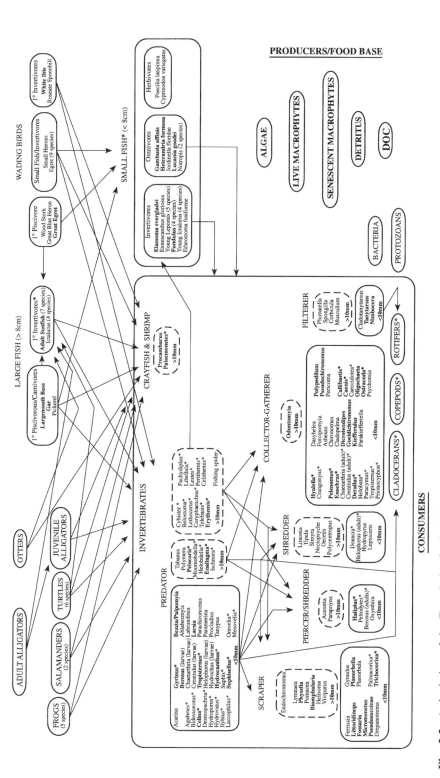

Fig. 2.2. A depiction of the Everglades foodweb by grouping species into guilds based on size and prey preferences/functional-feeding classification. The most abundant and mobile taxa within each guild are identified with bold lettering and an asterisk, respectively. Arrows point from predators to prey guilds that contain taxa (asterisk) that are likely to concentrate in small, shallow bodies of water or in deeper refugia (canals, air boat trails, ponds, and alligator holes) during the dry season. Unless otherwise specified, taxa names represent both immature and adult stages.

type and hydroperiod. For example, ephemeral marshes are dominated by large invertebrate predators that prey on small fish that, in turn, prey on smaller invertebrates, whereas deeper, more permanent habitats (canals, ponds, air boat trails) support both large fish (piscivores and invertivores) and predators that prey on large fish (alligators, turtles, etc.).

Foodweb Interactions. Figure 2.2 identifies most of the aquatic predators in the Everglades and highlights the influence of hydroperiod on prey availability. Organisms that are likely to concentrate in small, shallow bodies of water or in deeper refugia (canals, air boat trails, ponds, and alligator holes) during the dry season are noted with an asterisk. All predator guilds show multiple arrows pointing to prey guilds that contain some members that will move and concentrate. The concentration of prey in shallow habitats during the winter months preceding spring nesting activities has long been recognized as one of the most important factors determining the breeding success of wading birds in the Everglades (Kushlan 1976, Frederick and Spalding 1994). However, drying and drawdown can also benefit other predators (large fish, frogs, salamanders, turtles, adult alligators) by concentrating fishes and mobile invertebrates (e.g., shrimp) in deeper, more permanent refuge habitat.

Although members of the Everglades community that are most likely to migrate in and out of ephemeral marshes have been poorly investigated, it seems clear that many macroinvertebrates are not sufficiently mobile to track receding water levels. Invertebrates that are capable of swimming (e.g., shrimp, amphipods, odonates, leeches, mayflies), flight (e.g., adult beetles, hemipterans), or crawling over dry land are the most likely to track receding water levels and the least likely to become stranded. Clearly, the poorest swimmers and nonfliers (e.g., zooplankton, ostracods, and isopods) will only be able to concentrate over short distances, whereas good swimmers and fliers may search for refugia (and concentrate) over several kilometers. For example, Jordan (1996) found that shrimp density in a northern Everglades slough increased from an average of 50 m^{-2} prior to drawdown to 425 m^{-2} following water recession. In contrast, crayfish numbers did not fluctuate with the same water level fluctuations (Jordan 1996). Some prey probably do not migrate into and out of ephemeral habitats because they create their own refugia and remain hidden, quiescent, or in a state of diapause during the dry season. *Procambarus alleni,* like other burrowing crayfish, remain in their burrows as water levels recede (Pennak 1989, Loftus et al. 1992). Several taxa in addition to crayfish may seek refuge in such burrows. Although there are no studies from the Everglades, Creaser (1931) found more than 6000 ostracods, copepods, and amphipods (mobile taxa that can migrate over short distances) in a single crayfish burrow in a dried slough in the Ozark mountains.

In addition to the ability to track receding water levels, another prerequisite to the food concentration concept is the ability of taxa to expand from refuge areas and exploit newly inundated habitats as water levels increase. Densities of prey that remain in and around deep water refugia do not change during a normal dry period, because their habitat remains fairly constant in size. Their numbers diminish only slightly and then increase as they expand into and retreat from ephemeral marshes. Thus, while gar and bass (drought-sensitive taxa) are very important prey for alligators, they probably do not venture far from deep-water habitats (e.g., canals; Jordan 1996) and therefore will show strong tendencies to concentrate only during severe drought or drawdown when their deep-water habitat is diminished. Therefore, there is no arrow in Fig. 2.2 connecting adult alligators with such prey. Large piscivorous fish (and alligators; Mazzotti and Brandt 1994) remained year-round in deep refuge habitats even when water regulation caused continuous marsh inundation (Loftus and Eklund 1994). Frogs, salamanders, and turtles, like adult alligators and large piscivorous fish, may remain in and around deep habitats. In contrast, the small fish assemblage and large invertivorous sunfish and ictalurids increase in shallow marshes with continuous inundation and as water levels rise (Loftus and Eklund 1994). Therefore, their numbers in refuge habitat should increase as water levels recede. Similarly, juvenile alligators will migrate into shallow marshes in order to avoid predation by adult alligators, thus concentrating as marshes recede (Mazzotti and Brandt 1994). Many taxa (e.g., small fish) may seek to avoid large predators by moving into more shallow, complex (dense emergent and submergent macrophytes) areas provided by newly inundated marshes. Some of the most important prey taxa continuously reproduce during the summer (shrimp and small fish) and, by exploiting newly available resources in freshly inundated marshes, can greatly increase the numbers that concentrate in deep water refugia and smaller pools when water levels again decline.

Perhaps the most important conclusion from this type of analysis is the need to restore the full range of hydrological variation in order to maintain community structure and function. For example, abnormally dry seasons favor food acquisition by wading birds and other large predators (e.g., bass, gar, adult alligators) that benefit from the concentration of prey in small pools and permanent deep-water habitats. Conversely, long hydroperiods and deeper water depths may limit wading bird feeding efficiency but favor invertebrates and small fishes and large invertivorous sunfish that have time to exploit ephemeral marshes. Hydroperiod may also influence the impact of different fish guilds on the invertebrate community. For example, small fishes (e.g., *Gambusia holbrooki* and *Heterandria formosa*) that readily colonize short hydroperiod marshes will primarily feed on smaller invertebrates (midges and mosquitoes), whereas large invertivores (e.g., sunfish) that can feed on almost all sizes of invertebrates (shrimp, odonates, etc.) likely require longer periods (two or more years) of continuous inundation in order to exploit ephemeral marsh resources. In addition to the length of inundation and water depth, the rate of water subsidence during the dry season can also favor some predator guilds more than others. A rapid recession will strand more taxa in shallow pools, making them more available to wading birds, while a slow recession should give mobile taxa time to find refuge in deeper habitats, making them more available to such predators as large piscivorous fish, alligators, and frogs. In short, the full range of natural hydrological variation (hydroperiod, depth, rate of recession) is undoubtedly important in maintaining all members of the Everglades community. Long periods (years) of inundation, long dry periods, and even long periods of a constant cycle of inundation and drying without intermittent variation (prolonged dry or wet seasons) could create the risk of the eventual collapse of one or more invertebrate or vertebrate guilds. Attempting to engineer optimal conditions for all species at once is not possible, whereas a holistic approach that allows for natural environmental dynamics and subsequent interannual variability in species success and ecosystem processes can accomplish long-term sustainability.

Research Needs

Further information on the role of invertebrates in the Everglades ecosystem is limited by a scarcity of ecological data. For example, very little is known about how invertebrates may influence rates of decomposition and algal or macrophyte production. Both shrimp and small fish (*Gambusia holbrooki* and *Heterandria formosa*) will bite and tear small pieces of floating senescent *Nymphaea* during the winter (Rader, personal observations). *Nymphaea*, along with other labile plants (*Utricularia, Chara*, periphyton), may constitute an important wintertime food resource. In addition, the importance of detritus and senescing macrophytes versus algal resources and live macrophytes in supporting secondary production in the Everglades is virtually unknown (Browder et al. 1994). The relative importance of algae versus macrophytes and living versus dead primary production will require a seasonal analysis using current techniques (e.g., stable isotopes). Little is also known about the ability of grazers to affect algal production, biomass, and physiognomy. Can grazers (invertebrates and fish) regulate algal standing crop in the Everglades? Many grazers may be unable to utilize the filamentous Cyanophyta prominent in Everglades open-water habitats (Rader and Richardson 1992, Browder et al. 1994). However, these plants plus macrophytes are substrate for a wide variety of other algae (diatoms, etc.). Despite its obvious importance, another completely neglected area is secondary production. How does invertebrate production vary with habitat type, hydroperiod, and disturbances? Currently we only vaguely understand the impact of various types of predator guilds on invertebrate community structure and function. How hydroperiod and habitat complexity affects the prevalence of fish and invertebrate predators must strongly influence the diversity and abundance of invertebrate populations (Jordan 1996). For example, invertebrates collected from Hillsboro Canal in the northern Everglades showed habitat-specific differences in abundance (Rader 1994). Hillsboro Canal connects agricultural sources of nutrient inputs with interior marshes and contains a large number and variety of different fish predators. Shrimp and similar-sized fish (*G. holbrooki* and *H. formosa*) were most abundant in dense, shallow grasses (*Panicum* spp.), whereas amphipods, leeches, beetles, and odonates were most abundant on floating vegetation (water lettuce and water hyacinth). Depending on the size of prevalent fish predators and various structural characteristics of both emergent and submergent vegetation, different habitats may provide different invertebrates with different levels of refuge from fish predation. Also, can fish predation limit aquatic invertebrate production available to terrestrial predators (birds, bats, etc.)? How, where, and when are invertebrates primarily controlled by bottom-up resource availability (primary production), disturbances, and/or top-down regulation by predators? Answers to these questions should play an important role in forming the foundation for future management decisions.

CONSERVATION AND MANAGEMENT

The Everglades is beset by a variety of environmental problems. One of the most important is habitat loss due to draining and development. To date, roughly 50 percent of the historic Everglades, mostly higher-elevation ephemeral marshes, has been lost to development (Davis et al. 1994). Kushlan et al. (1975) suggest that the loss of this type of habitat may in part account for a decline in wood storks because these ephemeral marshes allowed many invertebrate populations to flourish during the summer and concentrate into shallow accessible pools prior to spring nesting activities. Invertebrates are also important in the conservation efforts associated with other threatened and endangered species. For example, the apple snail (*Pomacea paludosa*) is the primary food resource of the endangered snail kite (*Rostrhamus sociabilis*). In fact, all aquatic threatened and endangered species in the Everglades (e.g., alligators, roseate spoonbill, Cape Sable seaside sparrow) rely directly or indirectly on the invertebrate foodbase. As previously discussed, restoration of the natural hydroperiod is probably the single most important means of maintaining and sustaining the extant Everglades ecosystem (e.g., Davis and Ogden 1994). However, other problems (eutrophication and the invasion of exotic species) also pose a threat to constituents of the Everglades community, including humans. Elevated levels of mercury in fish (e.g., largemouth bass; Hand and Friedman 1990) and the deaths of three panthers, believed to have been caused by mercury poisoning (Roelke et al. 1991), have prompted state officials to issue a Health Advisory urging the limitation of human fish consumption in the Everglades (Fleming et al. 1995). Although invertebrates play a critical role in the biomagnification process, I am unaware of any invertebrate-related work with mercury. Based on their importance in the Everglades foodweb, it might be useful to know which species of invertebrates harbor the greatest concentrations of this heavy metal and why.

Eutrophication

Upstream from the Water Conservation Areas (WCAs) in the northern Everglades lie 280,000 ha of agricultural land used primarily for growing sugar cane (Coale et al. 1994). Because agricultural runoff adds approximately 1814 metric tonnes of nitrogen (N) and 60 tonnes of phosphorus (P) each year (SFWMD 1992), N and P in WCA-2A have increased 12.4 and 10.0 times, respectively, over the past 35 years (Craft and Richardson 1993). Phosphorus is the primary limiting nutrient in the Everglades (e.g., Stewart and Ornes 1975, Vymazal et al. 1994). Alteration or disruption of the Everglades food-web by excess additions of P is a major concern. Rader and Richardson (1994) sampled eight sites (three enriched, two intermediate and 3 unen-riched) along a nutrient enrichment gradient in WCA-2A and found that enriched open-water habitats had a greater diversity and abundance of invertebrates than unenriched sloughs. This result was unexpected because of the known harmful effects of nutrient enrichment in other freshwater environments and because other studies indicate that the shorter hydroperiod of the enriched habitats compared to the continuously inundated, unenriched sloughs should have resulted in fewer, not more, invertebrates. These results, however, were substantiated by more samples collected in the highly enriched Hillsboro Canal, which runs along the northern border of WCA-2A. The canal data revealed extraordinarily high densities of amphipods, shrimp, leeches, beetles, and odonates (Rader 1994). Contrary to results for some lakes and streams, these data suggest that current levels of nutrient enrichment in the Everglades may not have a direct harmful impact on invertebrates.

What are the mechanisms associated with P enrichment that could affect invertebrates, and how do they relate to the unexpected findings of higher diversity and density in enriched open-water habitats (Rader and Richardson 1994)? Elevated concentrations of P could harmfully affect invertebrates in at least three ways: reduced food availability, reduced oxygen, and alteration of the macrophyte community, resulting in a reduction or elimination of open-water habitat (sloughs and wet prairies). Several studies report that increased P concentrations will increase primary production and algal biomass in the Everglades (e.g., Hall and Rice 1990, Vymazal et al. 1994). Increased algal growth will result in increased food availability for grazers and detritivores. Therefore, P enrichment is not likely to have a harmful effect on invertebrates in the Everglades by reducing food resources. Commensurate with increases in algal production are increases in algal oxygen production and in microbial respiration and oxygen consumption (Drake et al. 1996). The overall effects of P enrichment on oxygen availability will depend on other factors that might limit photosynthesis versus respiration. For example, shading (emergent vegetation, turbidity) might tip the balance (photosynthesis versus respiration) in favor of respiratory consumption, causing a decline in oxygen and therefore a potential decline in invertebrates. Conversely, if microbial respiration is limited, say by the quantity or quality of carbon, then there should be a surplus of oxygen when primary production is stimulated by increased nutrient availability. Wetzel (1996) described how photosynthesis can rapidly overwhelm respiration, causing the oxygenation of interstitial waters and bubble formation in metaphyton. Both can be nighttime reservoirs of oxygen when daytime photosynthesis stops. All of the enriched open-water habitats sampled by Rader and Richardson (1994) contained abundant growths of epiphytic periphyton and clusters of metaphyton. Although enriched, these sites probably produced ample oxygen reserves, which, when combined with enhanced food resources, may account for the increased abundance and diversity of invertebrates in these sites compared to unenriched sloughs. Additionally, most wetland invertebrates, unlike lotic taxa, are adapted to low fluctuating levels of oxygen (see Rader 1994). For example, shrimp, crayfish, and small fishes can utilize oxygen-rich water from the air-water interface as anaerobic conditions increase (Kushlan 1979, Jordan et al. 1996a). It should be noted, however, that not all enriched open-water habitats contained an abundance of algae. Some enriched open-water areas adjacent to air boat trails and along the margin of Hillsboro Canal do not sustain abundant, visible growths of algae (McCormick, personal communication). We might expect reduced invertebrate abundance and diversity in these enriched low-oxygen habitats because invertebrates that are less adapted to anaerobiosis are unable to survive.

Phosphorus enrichment may also cause a shift in emergent macrophyte vegetation, resulting in the overgrowth and shading of important open-water habitats. The nutrient-enriched portion of WCA-2A is overgrown by a dense stand of cattails. Recent research indicates that cattail is better adapted than sawgrass to exploit high P concentrations, especially if combined with long periods of continuous inundation (Urban et al. 1993, Kludze and DeLaune 1996, Newman et al. 1996). Cattails are probably the most inhospitable habitat for some invertebrates (especially benthic taxa), even poorer than dense sawgrass because a thick anaerobic layer of partially decomposed cattail flocculence eliminates the distinct sediment-water interface. Although crayfish can burrow into sawgrass peat and pump oxygenated water, if available, from a distinct sediment-water interface, the same is probably not possible in continuously inundated and enriched cattail habitat. Many benthic invertebrates (e.g., isopods, amphipods, snails) may be adversely affected by this layer of anaerobic flocculence. Therefore, P enrichment (especially if coupled with continuous inundation) can have an adverse impact on invertebrates in the Everglades if open-water habitats are replaced by dense cattail stands. It is interesting, however, that small pelagic fish (primarily *H. formosa* and *G. holbrooki*) of the northern Everglades reached their highest densities (four to five times greater than unenriched sites) in enriched cattail habitat (Jordan 1996). Overall, however, plants (algae and macrophytes) may be better indicators of nutrient enrichment effects than invertebrates or fish because invertebrates and fish are only indirectly affected, depending on the factors that influence photosynthesis versus respiration and the availability of suitable habitat. Also, plants are better at providing an early warning to potential changes caused by excess nutrients. Raschke (1993) found that changes in diatom community structure suggested the first signs of P enrichment in Everglades National Park, located 70 to 80 km downstream from agricultural inputs.

Exotic Species

Florida has been invaded by a variety of exotic species (SFWMD 1992). Of the 217 taxa of introduced plants identified from Everglades National Park, approximately 70 percent have centers of distribution in the tropics (Doren and Whiteaker 1990). Fortunately, only a few species are sufficiently successful to displace native vegetation, alter native habitats, and threaten the integrity of the Everglades community. With respect to invertebrates, the primary threat of invading vegetation is the overgrowth and destruction of open-water areas. The tree, *Melaleuca quinquenervia,* is particularly well adapted to the Everglades climate and is capable of reaching high densities in ephemeral marshes. Although its effect on invertebrates has not been investigated, its ability to shade open-water areas and increase rates of evapotranspiration might push sawgrass, wet prairie, and marl-based marsh habitats toward drier, more terrestrial conditions, which must have an overall detrimental impact on aquatic invertebrates. The exotic water hyacinth (*Eichhornia crassipes*) and water lettuce and duck weed are nuisance floating aquatics capable of reaching extremely high densities in canals and air boat trails. Although they are presently poorly established in interior marsh areas, their ability to shade open-water habitats, causing a decline in water column and benthic photosynthesis with subsequent impacts on oxygen, makes them a potential threat to invertebrates (e.g., Morris and Barker 1977, Pokorny and Rejmankova 1983). In addition to plants, there are 17 introduced and established fish (mostly cichlids), all of which originate in the tropics (Metzger and Shafland 1984). Two taxa, the Mayan cichlid and the walking catfish, are particularly important because of their ability to colonize a variety of habitats and withstand fluctuating water levels (Loftus 1987, 1988, Courtenay and Miley 1975). The Mayan cichlid is a large (10–20 cm in length) omnivorous fish that preys heavily on snails and submersed macrophytes (e.g., *Chara;* Loftus 1987). The walking catfish (20–25 cm in length) is an air-breathing voracious predator (invertebrates and small fish) that can thrive in shallow pools as water levels recede (Courtnay and Miley 1975). Although a potential prey item for some birds, it is more likely to compete with them by consuming prey in shallow pools during the dry season. The eventual impact of both species on invertebrates is unknown but potentially important because both taxa exploit ephemeral marshes where there are few large predators (alligators, gar, etc.) that might control their populations. In addition, there are several exotic invertebrates, including three of the most abundant snails (*Biomphalaria havanensis, Marissa cornuaurietus, Pomacea bridgesi*). Their impact on native invertebrate populations is not known but is potentially important given their extensive distributions and high abundances.

Invertebrates as Indicators of Everglades Health and Recovery

Are invertebrates useful indicators of ecosystem health and integrity in the Everglades? Despite their response to nutrient enrichment, the answer is yes.

First, individual taxa might yet prove useful in indicating habitat degradation resulting from eutrophication. Second, invertebrates occupy an intermediate position in the foodweb between primary producers and higher trophic levels. Depending on the type of disturbance, primary producers and invertebrates may rapidly respond as the ecosystem begins to function improperly. Some invertebrates will respond (increase or decrease) to both bottom-up (e.g., enrichment, exotic plants) and top-down (e.g., heavy metal biomagnification, exotic fish) disturbances and alterations that impact both (flow regulation). Merritt et al. (1996 and this volume) describe the relationship between ecosystem function in the Kissimmee River-Floodplain ecosystem and the relative abundance of different invertebrate guilds as defined by feeding, habit, and life history traits. They suggest that such ecosystem attributes as gross primary production versus community respiration (P/R), changes in top-down regulation, and habitat stability can be predicted using the invertebrate community. This approach (classifying invertebrates based on functional traits that reflect changes in ecosystem attributes) should also prove useful in the Everglades. Clearly, alterations to the Everglades will be reflected through the invertebrates as they respond to changes in ecosystem function.

ACKNOWLEDGMENTS

The author appreciates the input and suggestions provided by Darold Batzer and Paul McCormick and also thanks the lunch-time discussion group at Florida International University, especially Joel Trexler and Bill Loftus for their review of an early version of this chapter.

LITERATURE CITED

Batzer, D. P., and S. A. Wissinger. 1996. Ecology of insect communities in nontidal wetlands. Annual Review of Entomology 41:75–100.

Belanger, T. V., D. J. Scheidt, and J. R. Platko II. 1989. Dissolved oxygen budgets in the Everglades Water Conservation Area 2A. Florida Institute of Technology, Report to the South Florida Water Management District, West Palm Beach, FL.

Browder, J. A., P. J. Gleason, and D. R. Swift. 1994. Periphyton in the Everglades: spatial variation, environmental correlates, and ecological implications. Pages 379–418 in S. M. Davis and J. C. Ogden (eds.), Everglades: The Ecosystem and Its Restoration. St. Lucie Press, Delray Beach, FL.

Carlton, R. G., and R. G. Wetzel. 1987. Distributions and fates of oxygen in periphyton communities. Canadian Journal of Botany 65:1031–1037.

Coale, F. J., F. T. Izuno, and A. B. Bottcher. 1994. Sugarcane production impact on nitrogen and phosphorus in drainage water from an Everglades histosol. Journal of Environmental Quality 23:116–120.

Courtenay, W. R., Jr., and W. W. Miley II. 1975. Range expansion and environmental impress of the introduced walking catfish in the United States. Environmental Conservation 2:145–148.

Craft, C. B., and C. J. Richardson. 1993. Peat accretion and phosphorus accumulation along a eutrophication gradient in the northern Everglades. Biogeochemistry 22: 133–156.

Creaser, E. P. 1931. Some cohabitants of burrowing crayfish. Ecology 12:243–244.

Cummins, K. W. 1973. Trophic relations of aquatic insects. Annual Review of Entomology 18:183–206.

Davis, J. H., Jr. 1943. The natural features of southern Florida. Florida Geological Survey Bulletin No. 25.

Davis, S. M. 1991. Growth, decomposition, and nutrient retention of *Cladium jamaicense* Crantz and *Typha domingensis* Pers. in the Florida Everglades. Aquatic Botany 40:203–224.

———. 1994. Phosphorus inputs and vegetation sensitivity in the Everglades. Pages 357–378 *in* S. M. Davis and J. C. Ogden (eds.), Everglades: The Ecosystem and Its Restoration. St. Lucie Press, Delray Beach, FL.

Davis, S. M. and J. C. Ogden. 1994. Toward ecosystem restoration. Pages 769–796 *in* S. M. Davis and J. C. Ogden (eds.), Everglades: The Ecosystem and Its Restoration. St. Lucie Press, Delray Beach, FL.

Davis, S. M., L. H. Gunderson, W. A. Park, J. R. Richardson, and J. E. Mattson. 1994. Landscape dimension, composition, and function in a changing Everglades ecosystem. Pages 419–444 *in* S. M. Davis and J. C. Ogden (eds.), Everglades: The Ecosystem and Its Restoration. St. Lucie Press, Delray Beach, FL.

DeAngelis, D. L. 1994. Synthesis: spatial and temporal characteristics of the environment. Pages 307–322 *in* S. M. Davis and J. C. Ogden (eds.), Everglades: The Ecosystem and Its Restoration. St. Lucie Press, Delray Beach, FL.

de Szalay, F. A., and V. H. Resh. 1997. Responses of wetland invertebrates and plants important in waterfowl diets to burning and mowing of emergent vegetation. Wetlands 17:149–156.

Doren, R. F., and L. D. Whiteaker. 1990. Effects of fire on different size individuals of *Schinus terebinthifolius*. Natural Areas Journal 10:107–113.

Drake, H. L., N. G. Aumen, C. Kuhner, C. Wagner, A. Griefshammer, and M. Schmittroth. 1996. Anaerobic microflora of Everglades sediments: effects of nutrients on population profiles and activities. Applied and Environmental Microbiology 62: 486–493.

Duever, M. J., J. F. Meeder, L. C. Meeder, and J. M. McCollom. 1994. The climate of south Florida and its role in shaping the Everglades ecosystem. Pages 225–248 *in* S. M. Davis and J. C. Ogden (eds.), Everglades: The Ecosystem and Its Restoration. St. Lucie Press, Delray Beach, FL.

Epler, J. H. 1992. Identification manual for the larval Chironomidae (Diptera) of Florida. Biology Section, Florida Department of Environmental Regulation, Tallahassee, FL.

Fernald, E. A., and D. J. Patton. 1984. Water resources: Atlas of Florida. Florida State University, Tallahassee, FL.

Finkl, C. W. 1995. Water resource management in the Florida Everglades: Are lessons from experience a prognosis for conservation in the future? Journal of Soil and Water Conservation 63:592–600.

Fleming, L. E., S. Watkins, R. Kaderman, B. Levin, D. R. Ayyar, M. Bizzio, D. Stephens, and J. A. Bean. 1995. Mercury exposure in humans through food consumption from the Everglades of Florida. Water, Air and Soil Pollution 80:41–48.

Fogarty, M. J. 1974. The ecology of the Everglades alligator. Pages 208–221 *in* P. J. Gleason (ed.), Environments of South Florida, Present and Past. Miami Geological Society, Miami, FL.

Frederick, P. C., and M. G. Spalding. 1994. Factors affecting reproductive success of wading birds (Ciconiiformes) in the Everglades ecosystem. Pages 659–692 *in* S. M. Davis and J. C. Ogden (eds.), Everglades: The Ecosystem and Its Restoration. St. Lucie Press, Delray Beach, FL.

Gleason, P. J. 1972. The origin, sedimentation, and stratigraphy of a calcite mud located in the southern freshwater Everglades. Ph.D. Dissertation, Pennsylvania State University, University Park, PA.

———. 1974. Environments of South Florida: Present and Past. Miami Geological Society, Miami, FL.

Gleason, P. J., and W. Spackman, Jr. 1974. Calcareous periphyton and water chemistry in the Everglades. Pages 146–181 *in* P. J. Gleason (ed.), Environments of South Florida, Present and Past. Miami Geological Society, Miami, FL.

Gleason, P. J., and P. Stone. 1994. Age, origin, and landscape evolution of the Everglades peatland. Pages 149–198 *in* S. M. Davis and J. C. Ogden (eds.), Everglades: The Ecosystem and Its Restoration. St. Lucie Press, Delray Beach, FL.

Gleason, P. J., A. D. Cohen, P. Stone, W. G. Smith, H. K. Brooks, R. Goodrick, and W. Spackman. 1984. The environmental significance of holocene sediments from the Everglades and saline tidal plains. Pages 297–351 *in* P. J. Gleason (ed.), Environments of South Florida, Present and Past II. Miami Geological Society, Coral Gables, FL.

Goodrick, R. L. 1974. The wet prairies of the northern Everglades. Pages 185–189 *in* P. J. Gleason (ed.), Environments of South Florida, Present and Past. Miami Geological Society, Miami, FL.

Gunderson, L. H. 1990. Historical hydropatterns in vegetation communities of Everglades National Park. Pages 1099–1112 *in* R. R. Sharitz and J.W. Gibbons (eds.), Freshwater Wetlands and Wildlife. Savannah River Ecology Laboratory, Aiken, SC.

———. 1994. Vegetation of the Everglades: Determinants of community composition. Pages 323–340 *in* S. M. Davis and J. C. Ogden (eds.), Everglades: The Ecosystem and Its Restoration. St. Lucie Press, Delray Beach, FL.

Gunderson, L. H., and J. R. Snyder. 1994. Fire patterns in the southern Everglades. Pages 291–306 *in* S. M. Davis and J. C. Ogden (eds.), Everglades: The Ecosystem and Its Restoration. St. Lucie Press, Delray Beach, FL.

Hall, G. B., and R. G. Rice. 1990. Response of the Everglades marsh to increased nitrogen and phosphorus loading. Part III: Periphyton community dynamics. Everglades National Park Publication, Homestead, FL.

Hand, J., and M. Friedmann. 1990. Mercury in largemouth bass and water quality. Report submitted to Florida Department of Environmental Protection, Tallahassee, FL.

Hanson, M. A., and M. R. Riggs. 1995. Potential effects of fish predation on wetland invertebrates: A comparison of wetlands with and without fathead minnows. Wetlands 15:167–175.

Hawkins, C. P., and J. A. MacMahon. 1989. Guilds: the multiple meanings of a concept. Annual Review of Entomology 34:423–451.

Hoffman, W., G. T. Bancroft, and R. J. Sawicki. 1994. Foraging habitat of wading birds in the Water Conservation Areas of the Everglades. Pages 585–614 *in* S. M.

Davis and J. C. Ogden (eds.), Everglades: The Ecosystem and Its Restoration. St. Lucie Press, Delray Beach, FL.

Holling, C. S., L. H. Gunderson, and C. J. Walters. 1994. The structure and dynamics of the Everglades system: Guidelines for ecosystem restoration. Pages 741–756 *in* S. M. Davis and J. C. Ogden (eds.), Everglades: The Ecosystem and Its Restoration. St. Lucie Press, Delray Beach, FL.

Hunt, B. P. 1952. Food relationships between Florida spotted gar and other organisms in the Tamiami Canal, Dade County, Florida. Transactions of the American Fisheries Society 82:13–33.

———. 1961. A preliminary survey of the physico-chemical characteristic of Taylor slough with estimates of primary productivity. Everglades National Park, Homestead, FL.

Jordan, F. 1996. Spatial ecology of decapods and fishes in a northern Everglades wetland mosaic. Ph.D. Dissertation, University of Florida, Gainesville, FL.

Jordan, F., K. J. Babbitt, C. C. McIvor, and S. J. Miller. 1996a. Spatial ecology of the crayfish *Procambarus alleni* in a Florida wetland mosaic. Wetlands 16:134–142.

Jordan, F., C. J. DeLeon, and A. C. McCreary. 1996b. Predation, habitat complexity, and distribution of the crayfish *Procambarus alleni* within a wetland habitat mosaic. Wetlands 16:452–457.

Kolipinski, M. C., and A. L. Higer. 1969. Some aspects of the effects of the quantity and quality of water on biological communities in the Everglades National Park. U.S. Geological Survey Report 69007, Tallahassee, FL.

Kludze, H. K., and R. D. DeLaune. 1996. Soil redox intensity: effects on oxygen exchange and growth of cattail and sawgrass. Soil Science Society American Journal 60:616–621.

Kushlan, J. A. 1974. Quantitative sampling of fish populations in shallow, freshwater environments. Transactions of the American Fisheries Society 103:348–352.

———. 1975. Population changes of the apple snail (*Pomacea paludosa*) in the southern Everglades. Nautilus 89:21–23.

———. 1976. Wading bird predation in a seasonally fluctuating pond. Auk 93:86–94.

———. 1978. Feeding ecology of wading birds. Pages 248–297 *in* A. Sprunt, J. C. Ogden, and S. Winckler (eds.), Wading Birds, National Audubon Society Research Report 7.

———. 1979. Temperature and oxygen in an Everglades alligator pond. Hydrobiologia 67:267–271.

———. 1990. Freshwater marshes. Pages 324–363 *in* R. L. Myers and J. J. Ewel (eds.), Ecosystems of Florida. University of Central Florida, Orlando, FL.

Kushlan, J. A., and M. S. Kushlan. 1979. Observations on crayfish in the Everglades, Florida. Crustaceana supplement 5:115–120.

———. 1980. Population fluctuations of the prawn, *Palaemonetes paludosus,* in the Everglades. American Midland Naturalist 103:401–403.

Kushlan, J. A., J. C. Ogden, and J. L. Tilmant. 1975. Relation of water level and fish availability to wood stork reproduction in the southern Everglades, Florida. Open File Report 75–434, U.S. Geological Survey.

Lamberti, G. A. 1996. The role of periphyton in benthic food webs. Pages 533–572 *in* R. J. Stevenson and M. L. Bothwell, and R. L. Lowe (eds.), Algal Ecology. Academic Press, San Diego, CA.

Light, S. S., and J. W. Dineen. 1994. Water control in the Everglades: A historical perspective. Pages 47–84 *in* S. M. Davis and J. C. Ogden (eds.), Everglades: The Ecosystem and Its Restoration. St. Lucie Press, Delray Beach, FL.

Loftus, W. F. 1987. Possible establishment of the Mayan cichlid, *Cichlasoma urophthalmus* (Günther) (Pisces:Cichlidae), in Everglades National Park, Florida. Florida Scientist 50:1–6.

———. 1988. Distribution and ecology of exotic fishes in Everglades National Park. Pages 24–34 *in* L. K. Thomas (ed.), Proceedings of the 1986 conference on science in the national parks, July 1986, Fort Collins, CO.

Loftus, W. F., J. D. Chapman, and R. Conrow. 1990. Hydroperiod effects on Everglades marsh food webs, with relation to marsh restoration efforts. *In* Proceedings of the Fourth Triennial Conference on Science in the National Parks. Fort Collins, CO.

Loftus, W. F., and A. Eklund. 1994. Long-term dynamics of an Everglades small-fish assemblage. Pages 461–484 *in* S. M. Davis and J. C. Ogden (eds.), Everglades: The Ecosystem and Its Restoration. St. Lucie Press, Delray Beach, FL.

Loftus, W. F., R. A. Johnson, and G. H. Anderson. 1992. Ecological impacts of the reduction of groundwater levels in short-hydroperiod marshes of the Everglades. Pages 199–208 *in* J. A. Stanford and J. J. Simon (eds.), Proceedings of the First International Conference of Groundwater Ecology. American Water Resources Association, Bethesda, MD.

Loftus, W. F., and J. A. Kushland. 1987. Freshwater fishes of southern Florida. Bulletin of the Florida State Museum, Biological Sciences 31:147–344.

Loveless, C. M. 1959. A study of the vegetation of the Florida Everglades. Ecology 40:1–9.

MacVicar, T. K., and S. S. T. Lin. 1984. Historical rainfall activity in central and southern Florida: Average return period estimates and selected extremes. Pages 477–509 *in* P. J. Gleason (ed.), Environments of south Florida: Present and Past. Miami Geological Society, Coral Gables, FL.

Mallory, M. L., P. J. Blancher, P. J. Weatherhead, and D. K. McNicol. 1994. Presence or absence of fish as a cue to macroinvertebrate abundance in boreal wetlands. Hydrobiologia 279/280:345–351.

Mazzotti, F. J., and L. A. Brandt. 1994. Ecology of the American alligator in a seasonally fluctuating environment. Pages 485–506 *in* S. M. Davis and J. C. Ogden (eds.), Everglades: The Ecosystem and Its Restoration. St. Lucie Press, Delray Beach, FL.

McCollum, S. H., O. E. Cruz, L. T. Stem, W. H. Wittstruck, R. D. Ford, and F. C. Watts. 1978. Soil survey of Palm Beach County Area, Florida. U.S. Soil Conservation Service, Washington, D.C.

Merritt, R. W., J. R. Wallace, M. J. Higgins, M. K. Alexander, M. B. Berg, W. T. Morgan, K. W. Cummins, and B. Vandeneeden. 1996. Procedures for the functional analysis of invertebrate communities of the Kissimmee River-Floodplain ecosystem. Florida Scientist 59:216–274.

Metzger, R. J., and P. L Shafland. 1984. Possible establishment of *Geophagus surinamensis* (Cichlidae) in Florida. Florida Scientist 47:201–203.

Meyers, R. L., and J. J. Ewel. 1990. Ecosystems of Florida. University Presses of Florida, Gainesville, FL.

Morris, P. F., and W. G. Barker. 1977. Oxygen transport rates through mats of *Lemna minor* and *Wolffia* sp. and oxygen tension within and below the mat. Canadian Journal of Botany 55:1926–1932.

Newman, S., J. B. Grace, and J. W. Koebel. 1996. Effects of nutrients and hydroperiod on *Typha, Cladium,* and *Eleocharis:* Implications for Everglades restoration. Ecological Applications 6:774–783.

Pennak, R. W. 1989. Freshwater invertebrates of the United States (3rd ed). John Wiley & Sons, New York.

Pielou, E. C. 1979. Biogeography. John Wiley & Sons, New York.

Polorny, J., and E. Rejmankova. 1983. Oxygen regime in a fish pond with duckweeds (Lemnaceae) and *Ceratophyllum.* Aquatic Botany 17:125–137.

Rader, R. B. 1994. Macroinvertebrates of the northern Everglades: Species composition and trophic structure. Florida Scientist 57:22–33.

Rader, R. B., and C. J. Richardson. 1992. The effects of nutrient enrichment on algae and macroinvertebrates in the Everglades: A review. Wetlands 12:121–135.

———. 1994. Response of macroinvertebrates and small fish to nutrient enrichment in the northern Everglades. Wetlands 14:134–146.

Raschke, R. L. 1993. Diatom (Bacillariophyta) community response to phosphorus in the Everglades National Park, USA. Phycologia 32:48–58.

Reark, J. B. 1961. Ecological investigations in the Everglades. Pages 1–9 *in* Second Annual Report, Everglades National Park, Homestead, FL.

Reid, J. W. 1992. Copepoda (Crustacea) from fresh waters of the Florida Everglades, USA, with a description of *Eucyclops conrowae* n. sp. Transactions of the American Microscopical Society 111:229–254.

Roelke, M. E., D. P. Schutz, C. F. Facemire, S. F. Sundlof, and H. E. Royals. 1991. Mercury contamination in Florida Panthers. Report submitted to the Florida Panther Interagency Committee prepared by the Florida Panther Technical Subcommittee, West Palm Beach, FL.

Rose, C., and W. G. Crumpton. 1996. Effects of emergent macrophytes on dissolved oxygen dynamics in a prairie pothole wetland. Wetlands 16:495–502.

Rosendahl, P. C., and P. W. Rose. 1982. Freshwater flow rates and distribution within the Everglades marsh. Pages 385–401 *in* R. D. Cross and D. L. Williams (eds.), Proceedings of the National Symposium on Freshwater Inflow to Estuaries. Coastal Ecosystems Project, U.S. Fish and Wildlife Service, Washington, D.C.

Rozas, L. P., and W. E. Odum. 1987. Fish and macrocrustacean use of submerged plant beds in tidal freshwater marsh creeks. Marine Ecology and Progress Series 38:101–108.

Savino, J. F., and R. A. Stein. 1989. Behavior of fish predators and their prey: Habitat choice between open water and dense vegetation. Environmental Biology of Fishes 24:287–293.

SFWMD (South Florida Water Management District). 1992. Surface water improvement and management plan for the Everglades, supporting information document. Technical Report. SFWMD, West Palm Beach FL.

Snyder, G. H., and J. M. Davidson. 1994. Everglades agriculture: past, present, and future. Pages 85–116 *in* S. M. Davis and J. C. Ogden (eds.), Everglades: The Ecosystem and Its Restoration. St. Lucie Press, Delray Beach, FL.

Sogard, S. M., and K. W. Able. 1991. A comparison of eelgrass, sea lettuce macroalgae, and marsh creeks as habitats for epibenthic fishes and decapods. Estuarine, Coastal, and Shelf Science 33:501–519.

Stewart, K. K., and W. H. Ornes. 1975. The autecology of sawgrass in the Florida Everglades. Ecology 56:162–171.

Terczak, E. F. 1980. Aquatic macrofauna of the Water Conservation Areas. Technical Report, South Florida Water Management District, West Palm Beach, FL.

Thybo-Christesen, M., M. B. Rasmussen, and T. H. Blackburn. 1993. Nutrient fluxes and growth of *Cladophora sericea* in a shallow Danish bay. Marine Ecology: Progressive Series 100:273–281.

Turner, A. M., and J. C. Trexler. 1997. Sampling aquatic invertebrates from marshes: Evaluating the options. Journal of the North American Benthological Society 16: 694–709.

Turner, R. L. 1994. The effects of hydrology on the population dynamics of the Florida applesnail (*Pomacea paludosa*). Special Publication SJ94-SPE, St. Johns River Water Management District, Palatka, FL.

———. 1996. Use of stems of emergent plants for oviposition by the Florida applesnail, *Pomacea paludosa*, and implications for marsh management. Florida Scientist 59:34–49.

Urban, N. H., S. M. Davis, and N. G. Aumen. 1993. Fluctuations in sawgrass and cattail densities in Everglades Water Conservation Area 2A under varying nutrient, hydrologic and fire regimes. Aquatic Botany 46:203–223.

U.S. Army Corps of Engineers. 1992. Water control plan for Water Conservation Areas—Everglades National Park, and ENP—South Dade Conveyeance System Central and Southern Florida Project, October 1992.

Van Meter-Kasanof, N. 1973. Ecology of the micro-algae of the Florida Everglades, Part I: Environmental and some aspects of freshwater periphyton, 1959, 1963. Nova Hedwigia 24:619–664.

Vymazal, J., C. B. Craft, and C. J. Richardson. 1994. Periphyton response to nitrogen and phosphorus additions in Florida Everglades. Algological Studies 73:75–97.

Waller, B. G. 1976. Analysis of selected benthic communities in the Florida Everglades with reference to their physical and chemical environments. U.S. Geological Survey Water Resources Investigation 76–75, Tallahassee, FL.

Waller, B. G., and J. E. Earle. 1975. Chemical and biological quality of water in part of the Everglades, southeastern Florida. Water Resources Investigations, U.S. Geological Survey, Tallahassee, FL.

Wanless, H. R., R. W. Parkinson, and L. P. Tedesco. 1994. Sea level control on stability of Everglades wetlands. Pages 199–223 *in* S. M. Davis and J. C. Ogden (eds.), Everglades: The Ecosystem and Its Restoration. St. Lucie Press, Delray Beach, FL.

Wetzel, R. G. 1996. Benthic algae and nutrient cycling in lentic freshwater ecosystems. Pages 641–667 *in* R. J. Stevenson, M. L. Bothwell, and R. L. Lowe (eds.), Algal Ecology. Academic Press, San Diego, CA.

Wiechman, J. D. 1987. Abundance, condition, and size structure of fishes in relation to water level and water quality in a Florida Everglades impoundment. Masters Thesis, University of Florida, Gainesville, FL.

Wood, J. M., and G. W. Tanner. 1990. Graminoid community composition and structure within four Everglades management areas. Wetlands 10:127–149.

Young, F. N. 1954. The water beetles of Florida. University of Florida Press, Gainesville, FL.

3 The Kissimmee River-Riparian Marsh Ecosystem, Florida

Seasonal Differences in Invertebrate Functional Feeding Group Relationships

RICHARD W. MERRITT, MICHAEL J. HIGGINS,
KENNETH W. CUMMINS, and BRIGITTE VANDENEEDEN

*H*istorically, the Kissimmee River was a complex braided channel with an extensive floodplain wetland that was channelized and converted to a series of five impoundments by the U.S. Army Corps of Engineers from 1962 through 1971. Restoration of the Kissimmee River system has begun and includes a South Florida Water Management District program to evaluate the success of the restoration and provide input for adaptive management during the recovery of the system. Invertebrate community composition can serve as a useful system attribute for evaluating the success of the Kissimmee River ecosystem restoration project. The research reported examines and compares seasonal differences in invertebrate functional relationships of the Kissimmee River-floodplain ecosystem and relates these to differences in general ecosystem attributes as predicted by invertebrate community characteristics. Spring samples taken of invertebrate functional groups indicated that the remnant run of the Kissimmee River-floodplain ecosystem is heterotrophic, similar to what was found in the fall. Two important invertebrates, the grass shrimp (Palaemonetes paludosus) and the side-swimmer (Hyallela azteca), which probably can be characterized as keystone species, are indicative of the differences between an autotrophically based and a heterotrophically based system. Palaemonetes, a facultative shredder, feeds primarily on CPOM derived from parts of vascular hydrophytes, while Hyallela, a facultative scraper, feeds primarily on periphyton attached to submerged stems and leaves of vascu-

Invertebrates in Freshwater Wetlands of North America: Ecology and Management, Edited by Darold P. Batzer, Russell B. Rader, and Scott A. Wissinger
ISBN 0-471-29258-3 © 1999 John Wiley & Sons, Inc.

lar hydrophytes. Invertebrate community succession is in the early stages dominated by short life-cycle, rapidly moving species, suggesting that newly established habitats will be colonized quickly as the restoration proceeds. In addition, drift-feeding juvenile fishes may have abundant food supplies along the expanded riparian marsh, but food could be limiting for benthic-feeding fish and wading birds.

INTRODUCTION

The Kissimmee River originates in the Kissimmee Lakes area in south-central Florida and flows south into Lake Okeechobee (Fig. 3.1). The waters that feed the Kissimmee River form the headwaters of the greater Kissimmee-Okeechobee-Everglades ecosystem. Historically, the river was a 166-km-long braided channel system with an extensive floodplain wetland (National Archives 1856, Koebel 1995, Toth 1996). These data suggest that the Kissimmee basin underwent a seasonal wet-dry cycle, typical of subtropical regions, with only peripheral areas of the floodplain undergoing consistent annual seasonal drying (Koebel 1995). Periods of high flow and floodplain inundation generally occurred in December–May after the onset of winter rains, with low flows and floodplain drying occurring during late summer and autumn (Koebel 1995).

In 1962, in response to catastrophic flooding, the U.S. Army Corps of Engineers initiated a project that channelized the river. It consisted of a series of flow control structures that impounded the river, forming a series of five pools. Thus, the broad 166-km-long meandering river was transformed into a 90-km-long, 9-m-deep, and 100-m-wide canal. The project, completed in 1971, resulted in the loss of approximately 14,000 ha of floodplain habitat (Koebel 1995). Coincident with the loss of meandering river and floodplain wetland habitat, many species of plants and animals were restricted in their distribution or eliminated (Toth et al. 1995). By inference, associated invertebrate (Vannote 1971, Harris et al. 1995) and vertebrate communities (Trexler 1995, Weller 1995) were significantly altered as well, and migratory waterfowl populations declined by as much as 90 percent (Perrin et al. 1982).

Since channelization, research has focused on the best way to reverse the damage and restore the river/floodplain ecosystem while continuing to provide flood protection and navigation (Dahm et al. 1995). In 1984 the South Florida Water Management District (SFWMD) initiated a demonstration project to evaluate the effects of increased flow and floodplain inundation on the channelized river. This was accomplished by a series of three weirs that directed additional flow through three remnant channel systems along the impounded section of river known as Pool B, and by stage manipulations in the

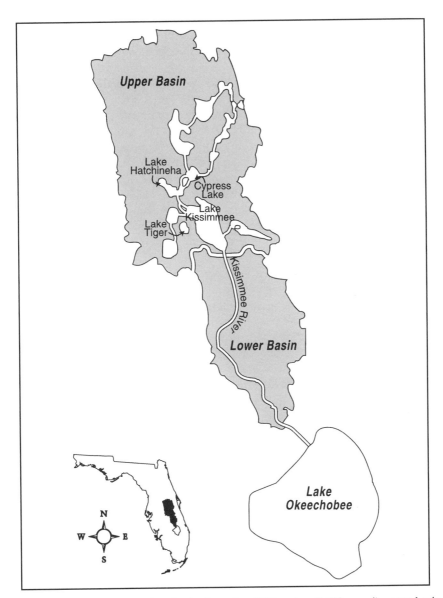

Fig. 3.1. The Kissimmee River basin from the 4229 sq km (1600 sq mi) upper basin headwaters in the Kissimmee Lakes through the 1200 sq km (750 sq mi) lower basin and into Lake Okeechobee. The upper basin includes Lake Kissimmee and many smaller lakes. The lower basin includes the tributary watersheds of the river (excluding Lake Istapoga) from Lake Kissimmee to Lake Okeechobee. Location of the basin in south-central Florida is indicated in the inset. Figure modified from South Florida Ecosystem Restoration Plan, Ecosystem Restoration Department, South Florida Water Management District, West Palm Beach, FL.

pool that caused inundation of the floodplain (Fig. 3.2). In 1990 the State of Florida endorsed the SFWMD plan to backfill approximately 35 km of channel (C-38 Canal) and restore 11,000 ha of floodplain wetland over a 15-year period.

The overall management goal for the Kissimmee River Restoration Project is improvement and protection of the "health" of the Kissimmee River ecosystem. "Health" is defined as a desirable and sustainable ecosystem structure and function, or ecosystem integrity (Toth 1993, Merritt et al. 1996). Thus, effectiveness of the restoration will be viewed in terms of the recovery of ecological integrity. A key component of the project is a research plan to evaluate the effects of the restoration on structure and function of the river-floodplain ecosystem (Toth 1993, SFWMD 1995, Dahm et al. 1995).

When the Kissimmee River was channelized and placed under stage-fluctuation management, invertebrate life in the river-floodplain ecosystem was altered from a predominantly riverine to a more lacustrine fauna. Rheophilic invertebrate taxa typical of many large lotic systems (e.g., hydropsychid caddisflies and heptageniid mayflies) were largely replaced by species common to lentic systems (e.g., chaoborid and chironomid midges, hemipterans, many odonates, hydrophilid beetles and other Coleoptera). The heterogeneous plant communities that were present before channelization supported a diverse invertebrate fauna, including caddisflies, dragonflies, damselflies, water bugs, water beetles, isopods, amphipods, freshwater shrimp, midges, sphaeriid clams, and unionid mussels (Vannote 1971). Therefore, with the conversion from a meandering, braided floodplain-river system to a canal came a shift in the dominant invertebrate taxa and the trophic dynamics of the system (Toth 1990, 1991, 1993). Stabilized water levels and reduced flow eliminated prechannelization river-floodplain interactions (Koebel 1995, Harris et al. 1995). The reintroduction of flow through remnant river channels of Pool B has resulted in colonization and/or expansion by invertebrate taxa characteristic of river or lotic communities rather than lentic ecosystems. Thus, the prediction is that a relinkage of the Kissimmee River with the floodplain following restoration will result in numerous changes to such ecologically important factors as streamflow, substrate composition, food quality and quantity, and water quality, all of which will influence invertebrate communities (Harris et al. 1995). The restoration will convert the future floodplain area from its present condition dominated by pasture plants (approximately 38 percent cover), with about 10 percent riparian marsh (defined as the littoral zone plus broadleaf marsh) species, to a river-floodplain ecosystem having about 70 percent riparian marsh and little or no pasture land. The riparian marsh littoral zone is dominated by *Nuphar* and *Polygonum* and the broadleaf marsh portion by *Pontederia* and *Sagittaria* in the Pool B remnant channels (Fig. 3.3).

In an earlier paper, Merritt et al. (1996) described procedures for the functional analysis of invertebrate communities in the Kissimmee River-floodplain ecosystem by (1) identifying the regional species pool of aquatic invertebrates

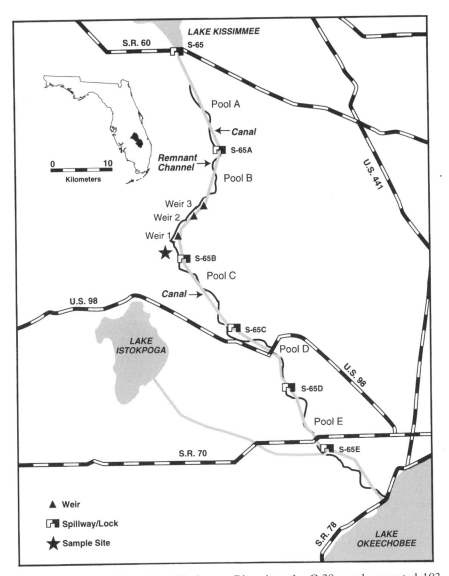

Fig. 3.2. Channelization of the Kissimmee River into the C-38 canal converted 103 mi (166 km) of meandering river into a series of five impoundments (Pools A–E) behind structures S-65A–S-65E. Flow out of Lake Kissimmee is regulated by structure S-65. Restoration of the river-floodplain system will involve four phases of backfilling 22 mi (35 km) in Pools B–D, beginning in 1998. The invertebrate samples were taken in the partially restored remnant channels created by Weirs 1–3. The star indicates the sampling area of remnant channel. Major highways in the region and the location of the basin in south-central Florida (inset) also are shown. Figure modified from South Florida Ecosystem Restoration Plan, Ecosystem Restoration Department, South Florida Water Management District, West Palm Beach, FL.

Fig. 3.3. Riparian marsh plant communities and other habitats sampled for macroinvertebrates from the reach of the lower remnant meander channel of Pool B of the Kissimmee River. Species pictured are as follows: (*a*) *Nuphar luteum;* (*b*) *Polygonum densiflorum;* (*c*) *Pontederia cordata;* (*d*) *Sagittaria lancifolia;* (*e*) snag (dead live oak, *Quercus virginiana*); (*f*) sediments.

that potentially occur in the Kissimmee River basin; and (2) showing that invertebrate community composition could serve as a useful surrogate for system attributes important for evaluating the success of the restoration project. This initial study on a partially restored remnant channel of the river (Fig. 3.2) indicated that the riparian marsh floodplain habitats could be readily distinguished from each other on the basis of invertebrate functional relationships. Further, these relationships were a good predictor of general ecosystem attributes, such as the autotrophy-heterotrophy index (P/R) and the availability of invertebrate food for drift-feeding fish and benthic-feeding wading birds and fish.

The objective of this study was to examine and compare seasonal differences in invertebrate functional relationships of the Kissimmee River-floodplain ecosystem and relate these to differences in ecosystem attributes as predicted by invertebrate community characteristics. In particular, functional organization of invertebrate communities has been related to differences in the autotrophy-heterotrophy index measured directly in the fall dry (September) and the spring wet season (May) (Brock and Cummins, unpublished data). These data are examined for their potential use in evaluating the success of the Kissimmee River-floodplain restoration (e.g., Dahm et al. 1995).

THE USE OF INVERTEBRATES IN RIVER-FLOODPLAIN ANALYSIS

Desirable ecosystem conditions are most often judged against historical standards as chronicled by the condition of particular groups of organisms, such as invertebrates. Invertebrates are an integral part of all aquatic ecosystems, serving as a link between primary producers and the plant-derived detritus and higher trophic levels. Many riverine-wetland vertebrates, including fishes, amphibians, wading birds, and waterfowl, forage predominantly or exclusively on aquatic invertebrates. Because of these fundamental trophic linkages, the richness and abundance of the invertebrate fauna in a given aquatic ecosystem have a long history of being used as indicators of the overall health and status of aquatic environments (Rosenberg and Resh 1993) and therefore should be useful in predicting ecosystem attributes. In addition, because many invertebrates are sessile or limited in mobility, the presence or absence of certain taxa can be used to evaluate both short- and long-term environmental conditions.

Invertebrate Functional Feeding Groups

The biological assessment of running water ecosystems has advanced considerably in the past decade and now includes methods for both fish and macroinvertebrates (Karr et al. 1986, Plafkin et al. 1989, Rosenberg and Resh 1993, Kerans and Karr 1994, Davis and Simon 1995). Recently, Barbour et al. (1996) have developed an approach to biological assessment that would document effects of pollution on benthic macroinvertebrate assemblages in Florida streams using biological metrics and aggregated indices for biocriteria and assessment. Among these metrics, they chose trophic relationships as surrogates of complex processes such as trophic interactions, production, and food source availability (Barbour et al. 1996).

Thus, recognition of stream macroinvertebrate functional feeding groups has proved to be a useful tool for assessing the ecological state of running water communities (Cummins 1974, Cummins and Wilzbach 1985, Merritt and Cummins 1996a, b, Barbour et al. 1996), including the Kissimmee River

(Merritt et al. 1996). The value of such an approach is that these functional attributes can be related more directly to system ecology and ecosystem integrity than can taxonomy per se. Analysis of functional feeding groups, for example, is particularly useful because it focuses on linkages between food resources (e.g., periphyton, vascular hydrophytes, coarse plant litter [CPOM], fine particulate organic matter [FPOM], and prey) and the special morphological and behavioral adaptations of invertebrates for the acquisition of these resources (Merritt and Cummins 1996a, b).

In addition to feeding, other functional attributes of invertebrate taxa that bear directly on restoration success include (1) habit, defined here as mode of attachment, locomotion, and/or concealment (Merritt and Cummins 1996a); and (2) generation time (voltinism). Information on feeding modes and habits that allow for the correct placement of many genera of southern Florida's aquatic invertebrates into functional feeding groups is available and summarized in taxonomically organized ecological tables in Merritt and Cummins (1996a) and for the Kissimmee River system in Merritt et al. (1996).

Use of Functional Feeding Group Ratios to Evaluate Ecosystem Attributes

Invertebrate faunal composition and density can be estimated by using dimensionless ratios of components from semi-quantitative collections within each of the river and riparian marsh habitat types (Cummins 1992, Cummins and Wilzbach 1985, Harris et al. 1995, Merritt et al. 1996). The ratios of invertebrate functional feeding and habit groups, as well as voltinism characteristics, are proposed as analogs of different ecosystem attributes (Table 3.1). An example of an ecosystem-level prediction based on invertebrate functional group analysis would be the use of the ratio scrapers plus live vascular hydrophyte shredders as a proportion of CPOM (detrital vascular plant tissue) shredders plus total collectors (Table 3.1). This ratio can serve as an analog of P/R, that is, the ratio of gross primary production to community respiration, or more generally, the ratio of autotrophy to heterotrophy. Because P/R provides an evaluation of the balance between autotrophy and heterotrophy, it is an excellent indicator of overall ecosystem organization and function and is useful for predicting whether there has been a change in the basic taxa of functional groups due to some perturbation. For example, turbidity, shading, and/or certain types of nutrient limitation would restrict autotrophy, whereas the quantity and quality (i.e., degradability) of vascular plant detritus and nitrogen and phosphorous levels are known to regulate heterotrophy (e.g., Cummins 1992, Hauer and Lamberti 1996).

An example of the relationship between functional habit groups and ecosystem function would be the availability of invertebrates to primarily sight-feeding wading birds and benthic fish in the Kissimmee River-floodplain ecosystem (Table 3.1). The ratio of more exposed invertebrates, such as sprawlers (clingers and climbers), to those less exposed or vulnerable inver-

TABLE 3.1. Relationships between Invertebrate Functional Groups (Feeding, Habit, and Voltinism) and Ecosystem Attributes for Which They Can Serve as Analogs

Ecosystem Parameters Measured Directly	Methods	Functional Group Ratios	Expected Ratios[a]
Gross primary production *as a proportion of* community respiration P/R	P/R measurements per unit area on a daily basis	Shredders (live vasc. plants) + scrapers *as a proportion of* shredders (CPOM detritivores) + total collectors	Autotrophic system > 0.75
Coarse particulate organic matter *as a proportion of* fine particulate organic matter CPOM/FPOM	CPOM/FPOM measurements per unit area of storage in and on the sediments (benthic) on a seasonal basis (wet and dry seasons)	Total shredders *as a proportion of* total collectors	Normal shredder riparian system by season Dry: fall–winter > 0.50 Wet: spring–summer > 0.25
Suspended particulate organic matter *as a proportion of* deposited (benthic) particulate organic matter SPOM/BPOM	SPOM measured per unit volume and BPOM measured per unit area	Filtering collectors *as a proportion of* gathering collectors	Enriched in suspended particulate organic matter > 0.50
Top-down control TOP-DOWN	High ratio of slow turnover predators indicates high proportion of fast turnover predator taxa	Predators *as a proportion of* total of all other functional feeding groups	Normal top-down predator control < 0.15

TABLE 3.1. (Continued)

Ecosystem Parameters Measured Directly	Methods	Functional Group Ratios	Expected Ratios[a]
Habitat (substrate) stability HABITAT STABILITY	Measures available surfaces for stable attachment (sediment coarser than moved by maximum transport velocity, large woody debris, rooted vascular hydrophytes)	Scrapers + filtering collectors *as a proportion of* total shredders + gathering collectors or clingers + climbers *as a proportion of* burrowers + sprawlers + swimmers	Stable substrates not limiting in rivers > 0.50 Stable substrates not limiting in rivers, littoral zones, wetlands > 0.60
Predictable food supply for water column-feeding fish DRIFT FOOD	Measures the predictable dawn and dusk availability of food for drift-feeding fish	Behavioral drifters *as a proportion of* accidental drifters	Good food supply for water column-feeding fish > 0.50
Most available food supply for wading birds and benthic-feeding fish BENTHIC FOOD	Measures the vulnerability of food to wading birds and benthic feeding fish	Sprawlers *as a proportion of* clingers + climbers + burrowers + swimmers	Good food supply for wading birds and benthic-feeding fish > 0.60
Short life cycle vs. Long life cycle LIFE CYCLE	Measures the relative dominance of more rapidly colonizing species	Generations per year > 1 *as a proportion of* generations per year ≤ 1	Pioneer, early successional (less stable) community > 0.75

Source: modified after R. W. Merritt et al., Procedures for the functional analysis of invertebrate communities of the Kissimmee River-floodplain ecosystem, Florida Scientist 59:216–274, 1996.

[a] Expected ratios are based on biomass, but usually agree with numerical ratios. The proposed ratios are based on values calculated from the literature for lotic habitats and those directly measured by the authors in other studies. Ecosystem metabolism is a special case in which P/R was measured directly in the Kissimmee River system. Further studies are underway to directly test these ratios.

tebrates (burrowers and swimmers) would indicate that either a better or worse food supply was available for wading birds and benthic feeding fish, such as darters, some cyprinids, and catfish feeding on the bottom, and centrarchids feeding on vascular plant surfaces. Invertebrates in the water column would be available to drift-feeding fish, primarily juveniles of most fish species.

Therefore, based on the linkages that exist among trophic levels, the ratios of various functional groups should reflect the nature of the organic food resources available. This information provides us with a picture of whether or not a river and the riparian marsh portion of its floodplain are functioning normally within a general conceptual model of ecological structure and function (Cummins and Klug 1979, Merritt and Cummins 1996b, Merritt et al. 1996).

METHODS AND MATERIALS

Sampling Design

Freshwater macroinvertebrates were sampled from the 1000-meter reach of the lower remnant meander channel of Pool B of the Kissimmee River (Fig. 3.2) (Merritt et al. 1996). In the 1980s flow was augmented through construction of a series of weirs to partially restore this and two other remnant runs along Pool B (Toth 1996). Sampling was conducted on May 29–30, 1996, by boat or wading, depending on water depth, in the lower remnant run. We used a D-frame aquatic net with an 800-μm mesh size. Five standard time (45-second) collections were made in each of six habitats (for a total of 30 samples). As previously described by Merritt et al. (1996), samples were taken from four riparian marsh (i.e., littoral zone plus broadleaf marsh) plant communities (*Nuphar luteum, Polygonum densiflorum* in the littoral portion and *Pontederia cordata, Sagittaria lancifolia* in the broadleaf marsh) and in two channel habitat types (snags and bottom sediments) (Fig. 3.3). The four aquatic plant habitats of the riparian marsh that were sampled occurred along gradients of decreasing depth from *Nuphar* and *Polygonum* to *Pontederia* and *Sagittaria*. Snag and sediment (a mix of sand and organic muck) samples were collected along the channel margin. The woody debris snag that was sampled consisted of a large dead live oak (*Quercus virginiana*) with a portion of the bole and branches in the channel and the root wad embedded in the bank. Large woody debris snags are rare along the partially restored remnant channels associated with Pool B. For the present analysis, the samples taken in Spring 1996 (wet season) were compared with those taken in September 1994 (dry season) from the same habitats (Merritt et al. 1996).

After each sample was collected, the net was washed into an enamel tray, where most of the coarse plant material was carefully brushed, rinsed, and discarded. The remaining organic detritus and sediment were washed onto a 500-μm sieve, labeled, and preserved in sample bottles in 95 percent ethanol, then diluted to approximately 75 percent by the sample wash water. Samples were sorted in the laboratory under a dissecting microscope. Taxonomic determinations were made using Merritt and Cummins (1996a), Thorp and Covich (1991), Pennak (1989), and other taxonomic references given in Merritt et al. (1996). Verification of the identifications was accomplished by comparing determinations made at Michigan State University (Merritt and Higgins), the South Florida Water Management District (Cummins), and the University of Pittsburgh Pymatuning Laboratory of Ecology (Vandeneeden).

Invertebrate taxa that were collected in the Kissimmee River samples were enumerated and assigned to the following functional feeding groups using Tables 1 and 5 from Merritt et al. (1996):

1. *shredders*, which feed on CPOM (either live aquatic macrophyte tissue or coarse plant litter);
2. *scrapers*, which harvest periphyton and associated particulate material from substrate surfaces;
3. *filtering* and *gathering collectors* of FPOM;
4. *plant piercers*, which imbibe cell fluids from macroalgae and vascular hydrophytes; and
5. *predators*, which capture live prey.

General functional group (feeding and habit) and voltinism assignments were given in Merritt et al. (1996, Table 5) and were based on Merritt and Cummins 1996a, b, and other references cited in Merritt et al. (1996). The ratios of invertebrate functional feeding groups that are proposed as analogs for ecosystem attributes are summarized in Table 3.1.

RESULTS AND DISCUSSION

The diversity and abundance of Kissimmee River-floodplain invertebrates collected from semi-quantitative samples in the littoral, floodplain, and channel sediment habitats in May 1996 are shown in Table 3.2. Compared to the taxa found in fall samples (Table 3; Merritt et al. 1996), diversity was lower in spring (Table 3.2). This was most likely due to the early spring emergence of some species and the presence of eggs and small instars of other species which would have been missed due to the mesh size of the sampling nets. In most habitats two abundant invertebrates were present, the side-swimmer amphipod, *Hyalella azteca,* and the grass shrimp, *Palaemonetes paludosus. Hyalella* populations in the spring were similar to those in the fall, while *Palaemonetes* abundance was significantly greater during the spring (Table 2 and Table 3 in Merritt et al. 1996). *Hyalella* is classified as a facultative scraper, feeding on periphyton attached to the stems and submerged leaves of vascular hydrophytes, submerged woody debris, and, in some stable lo-

TABLE 3.2. Numbers of Animals in Semi-quantitative samples from the Littoral Zone (*Nupar, Polygonum*) and Broadleaf Marsh (*Pontederia, Sagittaria*) Portions of the Riparian Marsh and Channel Sediment and Snag Habitats of the Kissimmee River in May 1996[a]

Taxa	Nuphar Bed		Polygonum Bed		Pontederia Stand		Sagittaria Stand		Snags		Sediment	
	No.	Tot.	No.	Tot.	No.	Tot.	No.	Tot.	No.	Tot.	No.	Tot.
Annelida												
Oligochaeta	—	—	0.4	2	—	—	—	—	—	—	13.6	68
Hirudinea	—	—	—	—	—	—	—	—	0.4	2	0.4	2
Helobdella stagnalis	0.2	1	5.4	27	0.2	1	0.6	3	2.4	12	0.2	1
Placobdella	0.2	1	1.0	5	—	—	0.2	1	—	—	0.2	1
Mollusca												
Gastropoda												
Ancylidae												
Undetermined	—	—	0.2	1	—	—	—	—	—	—	—	—
Planorbidae	—	—	0.2	1	—	—	—	—	—	—	0.2	1
Pomacea	0.2	1	0.6	3	0.2	1	0.6	3	—	—	4.0	20
Limnaeidae	—	—	0.6	3	—	—	—	—	—	—	0.2	1
Helisomidae												
Helisoma	—	—	0.2	1	—	—	—	—	—	—	—	—
Physidae												
Physella	0.2	1	2.4	12	0.6	3	0.2	1	0.2	1	—	—
Bivalvia												
Corbiculidae												
Corbicula fluminea	0.4	2	—	—	4.0	20	1.8	9	—	—	9.8	49
Crustacea												
Maxillopoda												
Argulus	—	—	—	—	—	—	—	—	—	—	0.2	1

TABLE 3.2. (Continued)

Taxa	Nuphar Bed		Polygonum Bed		Pontederia Stand		Sagittaria Stand		Snags		Sediment	
	No.	Tot.	No.	Tot.	No.	Tot.	No.	Tot.	No.	Tot.	No.	Tot.
Amphipoda												
Hyalella	10.6	53	49.4	247	6.0	30	26.0	130	16.4	82	4.8	24
Isopoda												
Asellus	—	—	1.2	6	—	—	—	—	—	—	—	—
Decapoda												
Cambaridae	0.6	3	2.6	13	2.2	11	2.2	11	0.2	7	0.6	3
Palaemonidae												
Palaemonetes	5.2	26	20.2	101	4.2	21	16.6	83	0.8	4	7.0	35
Acari	—	—	0.2	1	—	—	—	—	—	—	—	—
Insecta												
Odonata												
Anisoptera												
Corduliidae												
Didymops	—	—	—	—	—	—	—	—	—	—	0.6	3
Libellulidae												
Undetermined	0.2	1	1.4	7	—	—	0.2	1	—	—	—	—
Pachydiplax	—	—	0.4	2	—	—	—	—	0.2	1	—	1
Gomphidae	—	—	—	—	—	—	—	—	—	—	0.2	1
Zygoptera												
Coenagrionidae												
Undetermined	1.2	6	1.2	6	0.6	3	0.4	2	0.6	3	0.2	1
Nehalennia	—	—	0.6	3	—	—	—	1	—	—	—	—
Enallagma	1.8	9	0.4	2	0.4	2	0.6	3	1.0	5	1.0	5
Ischnura	—	—	5.8	29	—	—	0.2	1	—	—	—	—

Taxon												
Ephemeroptera												
Baetidae												
Undetermined	—	—	—	—	0.2	1	0.2	1	—	—	—	—
Baetis	0.2	1	0.2	1	—	—	—	—	—	—	—	—
Cleon	0.2	1	0.2	1	—	—	—	—	—	—	—	—
Centroptilum	—	—	—	—	—	—	0.4	2	—	—	—	—
Caenidae												
Caenis	0.4	2	3.0	15	1.0	5	1.4	7	1.0	5	0.6	3
Trichoptera												
Polycentropodidae												
Cyrnellus	—	—	—	—	1.8	9	0.4	2	8.8	44	3.0	15
Cernotina	—	—	0.4	2	—	—	—	—	—	—	—	—
Leptoceridae												
Undetermined	—	—	—	—	—	—	0.4	2	—	—	—	—
Oecetis	—	—	0.4	2	—	—	0.4	2	0.2	1	0.2	1
Hydroptilidae												
Undetermined	—	—	—	—	0.2	1	—	—	—	—	—	—
Hemiptera												
Belostomatidae												
Belostoma	—	—	0.4	2	—	—	—	—	—	—	—	—
Naucoridae												
Pelocoris	—	—	0.2	1	—	—	—	—	—	—	—	—
Lepidoptera												
Pyralidae												
Parapoynx	—	—	0.4	2	—	—	—	—	—	—	—	—
Noctuidae												
Bellura	0.2	1	—	—	—	—	—	—	—	—	—	—
Coleoptera												
Terrestrial—Adults	0.2	1	0.2	1	—	—	—	—	—	—	—	—
Dytiscidae												
Celina	—	—	0.2	1	—	—	—	—	—	—	—	—

TABLE 3.2. (Continued)

Taxa	Nuphar Bed		Polygonum Bed		Pontederia Stand		Sagittaria Stand		Snags		Sediment	
	No.	Tot.	No.	Tot.	No.	Tot.	No.	Tot.	No.	Tot.	No.	Tot.
Ilybius	—	—	0.4	2	—	—	—	—	—	—	—	—
Haliplidae												
Peltodytes	—	—	—	—	0.2	1	—	—	—	—	—	—
Hydrophilidae												
Undetermined	—	—	0.2	1	—	—	—	—	—	—	—	—
Tropisternus	—	—	0.4	2	—	—	—	—	—	—	—	—
Noteridae	—	—	0.2	—	—	—	—	—	—	—	—	—
Diptera												
Chaoboridae												
Chaoborus	0.4	2	—	—	—	—	—	—	—	—	—	—
Ceratopogonidae												
Undetermined	—	—	—	—	—	—	—	—	0.2	1	—	—
Probezzia	0.4	2	—	—	—	—	—	—	—	—	2.8	14
Chironomidae												
Pupae	—	—	—	—	0.2	1	0.4	2	0.4	2	1.0	5
Tanypodinae	3.4	17	0.2	1	—	—	—	—	1.4	7	0.8	4
Orthocladiinae	—	—	—	—	—	—	—	—	0.2	1	3.2	16
Chironominae												
Chironomini	8.4	42	3.8	19	5.4	27	4.4	22	13.4	67	8.4	42
Tanytarsini	1.0	5	—	—	0.4	2	—	—	2.4	12	10.6	53
TERRESTRIAL												
Arachnida	—	—	1.0	5	0.2	1	—	—	—	—	—	—
Hemiptera												
Homoptera (nymphs & adults)	—	—	0.2	1	—	—	—	—	—	—	—	—
Cicadellidae	0.2	1	—	—	—	—	—	—	—	—	—	—

[a]Collections in all habitats were conducted using standard-time (45-sec) sweep collections with D-frame aquatic nets (mesh size 0.8 mm). Numbers are mean per sample (no.) and total number for all samples (tot.). Number of samples: five for each habitat.

cations, the sediments. *Palaemonetes* is classified as a facultative shredder that feeds primarily on coarse dead vascular plant tissue (CPOM), which is derived largely from dead leaves, stems, and reproductive parts that continually slough off of the vascular hydrophytes of the littoral (*Nuphar* and *Polygonum*) and floodplain (*Sagittaria* and *Pontederia*) riparian marsh. This feeding and habitat functional group division between these two invertebrates is reflected in the higher densities and biomass values for each of the above categories (shredders and scrapers) in Table 3.3. The combined total mass collected in all samples of the grass shrimp and amphipod show reversed patterns when the *Nuphar* and *Polygonum* samples are compared (Table 3.3). Among the insects, several genera of damselflies belonging to the family Coenagrionidae, which are generally climbers on stems of vascular hydrophytes, were fairly abundant in the littoral zone of the riparian marsh, while collector-gatherer mayflies of the family Caenidae were present in sediments and on plant surfaces (Table 3.3). Behavioral (i.e., active) drifters dominated all plant habitats except *Nuphar* and the sediments. The littoral zone habitats (*Nuphar* and *Polygonum*) of the riparian marsh supported a lower biomass of long life-cycle species than did the broadleaf marsh portion or snag, and sediment habitats. Except for the periphyton-rich stems of *Nuphar,* where clingers and climbers (largely scrapers) constituted most of the biomass, habitat functional group biomass was dominated by burrowers, especially odonates.

Invertebrate analyses indicated that the spring (May) macroinvertebrate community was predominantly heterotrophic (Table 3.4), similar to that in the fall (September). CPOM/FPOM and SPOM/BPOM (suspended particulate organic matter/deposited particulate organic matter) ratios indicated the importance of shredder activity and a moderate transported particulate load in the spring (Table 3.4). Filter-feeding bivalves also were more abundant in the shallow water sediments during the spring than during the fall (Table 3.2), suggesting an ample supply of suspended particulate matter in transport. This assessment of the system as being heterotrophic is in agreement with direct measures of community metabolism (P/R) made by Brock and Cummins (unpublished data).

Based on functional feeding groups, available surfaces for stable attachment (Habitat Stability) on rooted vascular hydrophytes and coarse sediments did not appear limiting in either spring or fall (Table 3.4). Predators (Top-down), which normally constitute about 15 percent or less of the total fauna (Table 3.1), were within this range in most habitats. The exception to this would be in *Nuphar* beds where an abundance of larval damselflies occurred, likely feeding extensively on *Hyallela*.

The Drift Food category indicated that there were more invertebrates in transport during the spring than during the fall (Table 3.4), leading to the prediction that food supply for juvenile fish in the water column would be better in the spring. In contrast to the Drift Food ratio, which was greater in the spring, the Benthic Food ratio for wading birds and fish was lower for

TABLE 3.3. Density and Dry Biomass^a of Kissimmee River Littoral and Floodplain Invertebrates^b

	Nuphar Beds		Polygonum Beds		Pontederia Stands		Sagittaria Stands		Snags		Sediment	
	No.	Mass.	No.	Mass.	No.	Mass.	No.	Mass.	No.	Mass.	No.	Mass.
Functional Feeding Groups												
Shredders												
Live vasc. plants	0	0	2	10.45	0	0	0	0	0	0	0	0
CPOM Detritus (shrimp)	31	236.3	120	752.9	33	578.5	94	1289.4	5	77.5	38	285.7
Collectors												
Gathering FPOM	46	9.4	40	30.9	31	5.0	35	6.4	88	17.7	115	38.7
Filtering FPOM	5	.35	2	6.3	11	12.4	2	.51	56	18.1	68	34.4
Scrapers (side-swimmer)	57	25.7	271	87.2	174	69.0	143	60.3	83	18.6	135	696.5
Piercers—Plant	0	0	3	1.7	0	0	0	0	0	0	0	0
Predators	37	43.6	57	141.6	5	5.5	10	7.4	20	34.1	31	65.6
Nonfeeding	0	0	0	0	2	.32	2	.28	2	.28	5	2.01
Habitat Functional Groups												
Clingers	5	164.3	26	19.4	34	18.5	15	12.1	47	22.5	128	724.8
Climbers	69	58.7	290	132	156	69.2	137	55.6	90	43.9	30	17.6
Sprawlers	22	8.8	28	40.7	5	.63	12	3.3	15	4.7	23	10.0
Burrowers	76	81.5	141	757	59	581.3	117	1291.3	100	94.9	206	368.5
Swimmers	2	1.55	10	82	2	1.04	5	2.1	2	.28	5	2.01
Skaters	0	0	2	1.3	1	32.7	0	0	0	0	0	0
Drift Fish Food												
Behavioral drift	131	106.3	388	833.0	210	645.3	250	1341.9	182	113.4	157	310.5
Accidental drift	48	216.7	146	206.6	48	58.6	40	23.5	84	55.1	237	812.6
Voltinism												
One year or less	174	315.1	494	989.3	200	112.0	175	92.5	239	58.7	188	141.1
More than one year	5	9.4	40	50.5	58	588.5	115	1272.8	27	109.9	206	978.9

^{a}Total number and mass in mg collected in all samples from each habitat type.
^{b}Totaled by feeding, habit, and drift functional groups and voltinism designations for each of six habitat types in Spring 1996.

TABLE 3.4. Comparison between Fall and Spring Sampling Periods for Calculated Functional Group (feeding and habit) and Voltinism Ratios for Each Ecosystem Attribute[a] for Each of the Six Habitat Types[b]

Ecosystem Attributes	Nuphar Beds		Polygonum Beds		Pontederia Stands		Sagittaria Stands		Snags		Sediment	
	No.	Mass.	No.	Mass.	No.	Mass.	No.	Mass.	No.	Mass.	No.	Mass.
P/R												
Fall (Sept.)	0.42	1.16	0.39	0.33	1.50	0.67	0.74	0.69	1.63	0.21	0.12	0.02
Spring (May)	0.51	0.17	2.00	0.55	2.84	0.54	1.15	0.18	0.46	0.47	0.14	0.08
CPOM/FPOM												
Fall (Sept.)	0.12	10.17	0.05	0.49	0.07	0.33	0.11	0.69	0.04	3.31	< 0.01	< 0.01
Spring (May)	0.58	22.48	3.25	20.24	0.73	29.96	3.55	102.70	0.12	11.66	0.28	0.94
SPOM/BPOM												
Fall (Sept.)	0.22	0.09	0.02	<0.01	0.37	0.89	0.51	0.39	0.39	2.35	0.03	0.03
Spring (May)	0.12	0.05	0.11	1.19	2.02	6.16	0.40	1.89	0.95	1.36	1.88	15.72
Habitat Stability[c]												
Fall (Sept.)	0.67	0.08	0.39	0.15	2.28	1.43	1.49	0.97	2.59	0.45	0.15	0.05
	0.50	0.18	0.37	0.21	1.68	0.54	0.98	2.01	2.42	0.41	0.39	0.11
Spring (May)	0.64	0.17	2.05	0.56	4.23	0.58	1.53	0.20	1.45	1.54	2.01	13.88
	0.03	0.67	0.06	0.01	0.18	0.02	0.05	0.01	0.24	0.29	0.65	8.89
Top-down												
Fall (Sept.)	0.19	0.17	0.18	0.94	0.09	1.07	0.09	0.52	0.05	0.23	0.43	0.13
Spring (May)	0.29	0.40	0.12	0.51	0.01	0.02	0.04	0.02	0.16	0.54	0.09	0.06
Life Cycle												
Fall (Sept.)	2.63	2.07	4.18	6.17	2.51	7.44	3.06	1.53	19.29	0.86	5.53	2.66
Spring (May)	30.25	40.40	22.90	244.52	4.96	0.63	2.85	.022	14.82	0.06	1.27	0.07
Drift Food												
Fall (Sept.)	0.99	0.46	0.87	0.37	0.89	0.35	1.90	1.18	3.51	2.45	0.79	1.11
Spring (May)	3.30	2.30	3.17	10.01	9.54	35.42	7.23	87.70	2.07	1.75	0.77	0.57
Benthic Food												
Fall (Sept.)	1.61	0.24	0.84	3.87	1.33	4.69	0.84	2.80	2.84	0.57	0.32	0.16
Spring (May)	0.16	0.09	0.06	0.04	0.02	<0.01	0.04	0.01	4.49	0.02	0.44	0.01

[a] See R. W. Merritt et al., Procedures for the functional analysis of invertebrate communities of the Kissimmee River-floodplain ecosystem, Florida Scientist 59: 216–274, 1996, Table 3.6.

[b] Expected ratios and ranges are given in Table 3.1.

[c] First pair of ratios based on functional feeding groups, second pair based on habit groups (see Table 3.1).

TABLE 3.5. Comparison of Numbers and Mass (mg) of the Facultative Shredder *Palaemonetes* **to That of the Facultative Scraper** *Hyallela* **by Habitat Type**

	Nuphar	Polygonum	Pontederia	Sagittaria	Snags	Sediment
Palaemonetes						
Total no. all samples	26	101	21	83	4	35
Mean no./ sample	5.2	20.2	4.2	16.6	0.8	7.0
Percent no. in samples	14.5	18.9	8.1	28.6	1.5	8.9
Total mass all samples	48.57	709.82	377.44	1258.17	67.72	230.83
Mean mass/ sample	9.71	141.86	75.49	251.63	13.54	46.17
Percent mass in sample	15.0	68.2	53.6	92.1	40.2	20.6
Hyallela						
Total no. all samples	53	247	30	130	82	24
Mean no./ sample	10.6	49.4	6.0	26.0	16.4	4.8
Percent no. in sample	29.6	46.3	11.6	44.8	30.8	6.1
Total mass all samples	202.28	74.09	63.04	48.76	18.56	8.60
Mean mass/ sample	4.06	14.82	12.61	9.75	3.71	1.72
Percent mass in sample	7.2	7.1	9.0	3.6	11.0	0.1

the same time period (Table 3.4). This was due primarily to greater numbers of climbers and burrowers that would not be as available to benthic sight-feeding species (Table 3.3). Similar to the case in fall, the Life Cycle ratio indicates that rapid colonizers dominated the system, suggesting again, as in the fall, that recolonization of new habitats created by the restoration will be rapid.

CONCLUSIONS

The spring (May) samples of invertebrate functional groups taken in the lower remnant run of Pool B of the Kissimmee River-floodplain ecosystem clearly indicate that the system is heterotrophic. These results are similar to those reported previously for the fall season (Merritt et al. 1996) and are confirmed by the direct measure of P/R (daily gross primary production/daily total community respiration) made by Brock and Cummins (unpublished data).

Two important invertebrates probably can be characterized as keystone species (i.e., they are important food items in almost all fish species [Bond 1994]): the grass shrimp, *Palaemonetes paludosus,* and the side-swimmer, *Hyallela azteca.* They are indicative of the difference between an autotrophically based and a heterotrophically based system (Fig. 3.4). *Palaemonetes,* a facultative shredder, feeds primarily on CPOM derived from different parts of the vascular hydrophytes, especially the decaying leaves of smartweed (*Polygonum densiflorum*). CPOM is continually converted to FPOM by shredder feeding and microbial breakdown. These two fractions constitute the very large detrital pool that is the primary base of the Kissimmee River heterotrophic food webs. *Palaemonetes* is considered a "facultative shredder" because it can survive in culture on FPOM, and some fine material is usually found in the gut with CPOM fragments. However, it readily consumes CPOM and grows well in culture.

In contrast, *Hyallela* is a "facultative scraper," feeding on periphyton attached to stems and submerged leaves of vascular hydrophytes, especially the periphyton-rich stems of *Nuphar,* and other substrates. *Hyallela* is considered a facultative scraper because, unlike species of *Gammarus* amphipods, it does not possess mouthparts specialized for shredding nor does it skeletonize

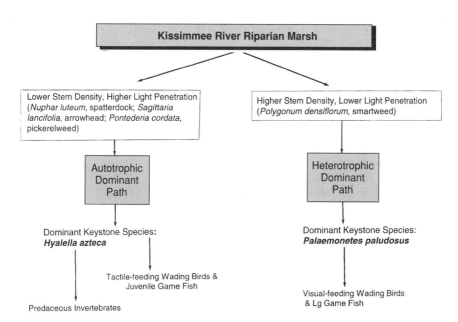

Fig. 3.4. Proposed conceptual model of dominant autotrophic and heterotrophic habitats and associated keystone invertebrate species and suggested links to higher trophic levels.

CPOM in culture. In addition, all field-collected specimens contain significant amounts of epiphytic algae in their guts. However, like *Palaemonetes,* it can survive in culture on FPOM. Further laboratory experiments are needed to confirm the degree of facultative feeding in these two species by comparing temperature-normalized growth rates on different food resources (Cummins and Klug 1979).

Palaemonetes reaches a much larger size (the specimens collected ranged from 2 to 22 mm in length) than does *Hyallela* (2 to 7 mm). Although *Hyallela* was often the more abundant of the two, *Palaemonetes* accounted for over twice as much biomass (Table 3.3). Because of its larger size and habit as a sprawler in accumulations of CPOM, *Paleamonetes* should be relatively more available to benthic-feeding fish species and wading birds. On the other hand, *Hyallela* has been shown to drift, that is, periodically migrate in the water column, and would be expected to be readily available to drift-feeding fish.

Since riparian marsh (i.e., littoral plants plus broadleaf marsh) is predicted to be the major river-floodplain habitat that greatly expands as a consequence of the Kissimmee River Restoration Project (Toth et al. 1995), the autotrophic versus heterotrophic nature of this expanded habitat, and therefore the concomitant abundance of *Palaemonetes* versus *Hyallela,* will be a critical issue because of the consequences for the foodwebs that develop associated with each habitat type. The samples analyzed to date (Table 3.4) and the data collected by Brock and Cummins (unpublished data) show that the dominant riparian marsh is capable of supporting dominant autotrophic foodwebs. If increased flows resulting from new flow-regulation schedules as part of the restoration (Toth 1996) reduce the organic sediment load on community respiration, communities of P/R > 1 may be more prevalent and support greater coverage of periphyton-based foodwebs. The relative dominance of *Nuphar,* with its low stem density (16 stems/m^2) and dense periphyton cover, as compared to *Polygonum,* with its greater stem density (39 stems/m^2), will also be important in shaping overall P/R and the linked invertebrate associations.

In general, other conclusions based on samples taken in the fall (September) (Merritt et al. 1996) have been confirmed for the spring (May) samples. For example, invertebrate community succession is in its early stages, dominated by short life-cycle, rapidly moving species, suggesting that newly established habitats will be colonized quickly as the restoration proceeds. In addition, drift-feeding juveniles of most fish species and smaller species, such as mosquito fish and killifish, may have abundant food supplies in the expanded broadleaf marsh, but food could be limiting for benthic-feeding fish and wading birds. As shown previously (Merritt et al. 1996), the type of invertebrate analysis reported here can provide a valuable tool for tracking important ecosystem attributes of altered habitats as the largest river restoration ever attempted proceeds.

ACKNOWLEDGMENTS

The field, laboratory, and synthesis work was supported under Contract C-6631-0202 from the South Florida Water Management District. The authors would like to thank Scott Merritt for field work and Kelly J. Wessell and Becky J. Blasius for reading earlier drafts of the MS. The authors would also like to acknowledge the Riverwoods Field Laboratory, Lorida, Florida, for allowing the use of their facilities while conducting this study.

LITERATURE CITED

Barbour, M. T., J. Gerritsen, G. E. Griffith, R. Frydenborg, E. McCarron, J. S. White, and M. L. Bastian. 1996. A framework for biological criteria for Florida streams using benthic macroinvertebrates. Journal of the North American Benthological Society 15:185–211.

Bond, W. J. 1994. Keystone species. Pages 237–253 in E. D. Schulze and H. A. Mooney (eds.), Biodiversity and Ecosystem Function. Springer-Verlag, New York.

Cummins, K. W. 1974. Stream ecosystem structure and function. BioScience 24:631–641.

———. 1992. Catchment characteristics and river ecosystems: a functional view. Pages 247–262 in P. J. Boon, P. Calow, and G. F. Petts (eds.), River Conservation and Management. John Wiley & Sons, New York.

Cummins, K. W., and M. J. Klug. 1979. Feeding ecology of stream invertebrates. Annual Review of Ecology and Systematics 10:147–172.

Cummins, K. W., and M. A. Wilzbach. 1985. Field procedures for analysis of functional feeding groups of stream macroinvertebrates. Pymatuning Laboratory of Ecology, University of Pittsburgh, PA.

Dahm, C. N., K. W. Cummins, H. M. Valett, and R. L. Coleman. 1995. An ecosystem view of the restoration of the Kissimmee River. Restoration Ecology 3:225–238.

Davis, W. S., and T. P. Simon (eds.). 1995. Bioligical Assessment and Criteria: Tools for Water Resource Planning and Decision Making. Lewis, Boca Raton, FL.

Harris, S. C., T. H. Martin, and K. W. Cummins. 1995. A model for aquatic invertebrate response to Kissimmee River restoration. Restoration Ecology 3:181–194.

Hauer, F. R., and G. A. Lamberti (eds.). 1996. Methods in stream ecology. Academic Press, San Diego, CA.

Karr, J. R., K. D. Fausch, P. L. Angermeier, P. R. Yant, and I. J. Schlosser. 1986. Assessing Biological Integrity in Running Waters: A Method and Its Rationale. Special Publications 5. Illinois Natural History Survey, Urbana, IL.

Kerans, B. L., and J. R. Karr. 1994. A benthic index of biotic integrity (B-IBI) for rivers of the Tennessee Valley. Ecological Applications 4:768–785.

Koebel, J. W., Jr. 1995. An historical perspective on the Kissimmee River restoration project. Restoration Ecology 3:160–180.

Merritt, R. W., and K. W. Cummins (eds.). 1996a. An Introduction to the Aquatic Insects of North America (3rd ed.). Kendall/Hunt, Dubuque, IA.

————. 1996b. Trophic relations of macroinvertebrates. Pages 453–474 *in* F. R. Hauer and G. A. Lamberti (eds.), Methods in Stream Ecology. Academic Press, San Diego, CA.

Merritt, R. W., J. R. Wallace, M. J. Higgins, M. K. Alexander, M. B. Berg, W. T. Morgan, K. W. Cummins, and B. Vandeneeden. 1996. Procedures for the functional analysis of invertebrate communities of the Kissimmee River-floodplain ecosystem. Florida Scientist 59:216–274.

National Archives. 1856. Military Map of the Peninsula of Florida South of Tampa Bay, compiled by Lieut. J. C. Ives, Topographical Engineers, by order of Jefferson Davis, Secretary of War. Scale 1:400,000. Record Group 77, Drawer 128, Sheet 26.

Pennak, R. W. 1989. Fresh-water Invertebrates of the United States (3rd ed.). John Wiley & Sons, New York.

Perrin, L. S., M. J. Allen, L. A. Rowse, F. Montalbano, III, K. J. Foote, and M. W. Olinde. 1982. A Report on Fish and Wildlife Studies in the Kissimmee River Basin and Recommendations for Restoration. Florida Game and Fresh Water Fish Commission, Tallahassee, FL.

Plafkin, J. L., M. T. Barbour, K. D. Porter, S. K. Gross, and R. M. Hughes. 1989. Rapid Bioassessment Protocols for Use in Streams and Rivers: Benthic Macroinvertebrates and Fish. EPA/440/4-89-001. Office of Water, U.S. EPA, Washington, DC.

Rosenberg, D. M., and V. H. Resh (eds.). 1993. Freshwater Biomonitoring and Benthic Macroinvertebrates. Chapman & Hall, New York.

South Florida Water Management District. 1995. South Florida Ecosystem Restoration Plan. South Florida Water Management District, Ecosystem Restoration Department, West Palm Beach, FL.

Thorp, J. H., and A. P. Covich. 1991. Ecology and Classification of North American Freshwater Invertebrates. Academic Press, San Diego, CA.

Toth, L. A. 1990. Impacts of channelization on the Kissimmee River ecosystem. Pages 47–56 *in* M. K. Loftin, L. A. Toth, and J. T. B. Obeysekera (eds.), Proceedings of the Kissimmee River Restoration Symposium, South Florida Water Management District, October 1988, Orlando, FL.

————. 1991. Environmental Responses to the Kissimmee River Demonstration Project. South Florida Water Management District, Technical Publication No. 91-02.

————. 1993. The ecological basis of the Kissimmee River restoration plan. Florida Scientist 56:25–51.

————. 1996. Restoring the hydrogeomorphology of the channelized Kissimmee River. Pages 369–383 *in* A. Brookes and F. D. Shields, Jr. (eds.), River Channel Restoration: Guiding Principles for Sustainable Projects. John Wiley & Sons, New York.

Toth, L. A., D. A. Arrington, M. A. Brady, and D. A. Muszic. 1995. Conceptual evaluation of factors potentially affecting restoration of habitat structure within the channelized Kissimmee River restoration. Restoration Ecology 3:160–180.

Trexler, J. C. 1995. Restoration of the Kissimmee River: A conceptual model of past and present fish communities and its consequences for evaluating restoration success. Restoration Ecology 3:195–210.

Vannote, R. J. 1971. Field Evaluation No. 15: Kissimmee River (Central and Southern Florida Project). United States Army Corps of Engineers, Jacksonville, FL.

Weller, M. W. 1995. Use of two waterbird guilds as evaluation tools for the Kissimmee River restoration. Restoration Ecology 3:211–224.

4 Natural Flatwoods Marshes and Created Freshwater Marshes of Florida

Factors Influencing Aquatic Invertebrate Distribution and Comparisons between Natural and Created Marsh Communities

DAVID L. EVANS, WILLIAM J. STREEVER, and
THOMAS L. CRISMAN

*F*latwoods marshes occur as small, isolated depressions in an otherwise flat landscape in peninsular Florida. Although many flatwoods marshes are seasonally flooded, the majority of marshes created to replace them are continually flooded. This chapter discusses influencing factors of macroinvertebrate communities in flatwoods marshes and their created counterparts, including hydroperiod, soil and water pH, type and quantity of macrophyte cover, and the amount of silt and coarse organic material in soils. Macroinvertebrate densities peaked with reflooding, perhaps due to ample algal food resources resulting from a pulse of inorganic nutrients. Invertebrate predators were found to be significantly more abundant in seasonally flooded natural marshes than in continually flooded created wetlands. Natural flatwoods marshes tended to be acidic, and gastropod populations were depauperate where pH was less than 5.5, perhaps due to an inability to construct shells. Invertebrate abundance and diversity were lowest in created marshes with a dense cover of floating plants, where dissolved oxygen tended to be limited. Among created wetlands, highest organism densities were attained in marshes with the highest plant litter accumulation, suggesting that vascular plant production

Invertebrates in Freshwater Wetlands of North America: Ecology and Management, Edited by
Darold P. Batzer, Russell B. Rader, and Scott A. Wissinger
ISBN 0-471-29258-3   © 1999 John Wiley & Sons, Inc.

may be an important factor influencing invertebrate productivity via the detrital foodweb. Although significant differences between created and natural marshes were found for pH, conductivity, and sediment quality, dipteran communities occurring in pickerelweed areas of natural and created marshes were not significantly different, perhaps indicating that dipterans have a wide tolerance for environmental conditions. In spite of the apparent adaptability of dipterans, this study indicates that created wetland design criteria, including bank slope, water elevation, and soil types, may have negative consequences for some aspects of biological community structure and function, particularly in relation to macrophyte communities and non-dipteran invertebrates.

INTRODUCTION

Florida's freshwater marshes provide numerous ecological and anthropogenic functions and benefits, but relatively few publications have considered the wetland invertebrates of these systems. For example, Davis and Ogden's (1994) collection of papers on the Everglades and its restoration gave little consideration to invertebrates, despite its broad coverage of other issues. Most wetland invertebrate studies in Florida have been associated with wastewater impact sites (Brightman 1976, Haack 1984, Pezeshki 1987) and wetland creation by the phosphate mining industry (Evans and Sullivan 1988, Evans 1989, Erwin 1988, Erwin 1990, Streever and Bloom 1993, Streever and Crisman 1993, Streever and Portier 1994, Streever et al. 1995, Streever et al. 1996).

This chapter synthesizes current knowledge of aquatic macroinvertebrate communities in central Florida's flatwoods marshes and created marshes. First, the chapter briefly describes the environmental setting: marsh classification and geographical distribution, climate and hydrology, and associated geology and soils. Second, it describes macroinvertebrate community species composition and factors that influence macroinvertebrate communities. Third, because wetland mitigation requirements have led to creation of a large number of wetlands, the chapter compares macroinvertebrate populations of created and natural marshes. The chapter closes with a discussion of information gaps and suggestions for future studies.

ENVIRONMENTAL SETTING

This brief overview provides a context for the discussion of macroinvertebrates in flatwoods marshes and created marshes of peninsular Florida. Other authors, notably Kushlan (1990), have provided more complete descriptions.

Marsh Classification and Geographic Distribution

Florida's freshwater marshes can be categorized into five major groups, based on elevation, rainfall, evaporation, and geology (Fig. 4.1). Kushlan (1990) named them from highest to lowest elevation: highland marshes, flatwoods marshes, the Kissimmee marsh complex, the St. Johns marshes, and the Everglades. The largest freshwater marshes in Florida are found south of Lake Okeechobee in the Everglades and north of Lake Okeechobee in association with the Kissimmee and St. Johns Rivers. Smaller isolated marshes are prev-

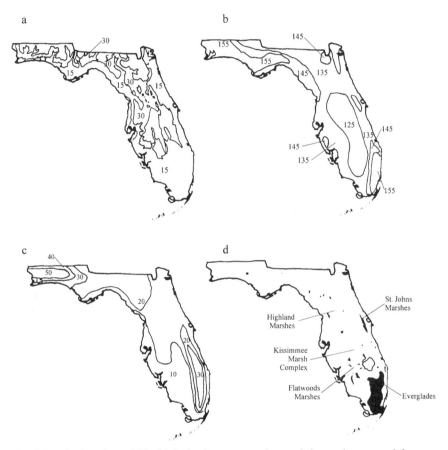

Fig. 4.1. The location of Florida's freshwater marshes and the environmental features that affect them: (*a*) topography in 15-m increments; (*b*) annual rainfall (cm); (*c*) difference between annual rainfall and potential evaporation (cm); (*d*) present distribution of marshes. (Redrawn from E. A. Fernald and D. J. Patton [eds.], Water Resources Atlas of Florida, Inst. Sci. Public Aff., Florida State University, Tallahassee, FL, 1984; modified from J. A. Kushlan, Freshwater marshes, pages 324–363 R. L. Myers and J. J. Ewel eds., Ecosystems of Florida, University of Central Florida Press, Orlando, FL. © 1990. Reprinted with the permission of the University Press of Florida.)

alent in peninsular Florida, but are relatively scarce in the northern panhandle. Highland marshes are distributed along the central highland ridge. Flatwoods marshes occur throughout Florida's pine flatwoods community, but they are concentrated at intermediate elevations between the central highlands ridge and the east and west coasts. Most of the results discussed in this chapter are based on investigations of natural flatwoods marshes and freshwater marshes designed and constructed by Florida's phosphate mining industry.

Climate and Hydrology

Florida's climate is influenced by the drying effects of the Bermuda high pressure cell and the humidifying nature of air masses moving landward off the Caribbean Sea and the Gulf of Mexico. Fall and winter are relatively dry when the Bermuda high prevents the development of thunderstorms from convective cloud cover. In the spring and summer the Bermuda high weakens, creating favorable conditions for late afternoon thunderstorms. Seasonal oscillations in precipitation have a major influence on Florida's freshwater marshes, particularly small isolated marshes, which often undergo drydown during the fall and winter (Kushlan 1990). In addition to human-induced fire management regimes, the high frequency of lightning strikes in Florida gives rise to fires in freshwater marshes and other habitats (Doren and Rochefort 1984).

Geology and Soils

A permeable sand or limestone layer underlies most of Florida. Marshes occur where spodic and clay horizons render soils impermeable, in isolated areas where the water table is at or above the ground surface, or along river fringes. Many of the isolated marshes of central Florida sit in characteristically saucer-shaped basins that appear to have formed following the dissolution of underlying limestone.

Soil types with peat, marl, and sand are associated with marshes. The formation of these soils is influenced by hydroperiod, water table depth during the dry season, and flood elevation. Peat generally occurs in areas that are flooded more than nine months of the year and in which water does not recede far below the surface during the dry season, a situation that results in anoxic conditions and subsequent slow decomposition rates. The ability of peat to retain water is an important characteristic of deep-water marshes because it affords a moist environment for hydrophytic plants, even during the dry season. Marl is primarily formed by encrusting algae where hydroperiod is short enough to permit oxidation of organic material (Noble 1989). Sandy substrates occur where flooding duration is very short and soils dry completely during periods of low rainfall. Under such conditions peat and marl cannot form because of rapid oxidation and frequent fires (Kushlan 1990).

MACROINVERTEBRATE FAUNA AND FACTORS INFLUENCING MACROINVERTEBRATE COMMUNITIES IN FLATWOODS MARSHES

Most of the evidence for relationships described in this section is based on an extensive study of 8 natural flatwoods marshes and 11 created marshes in central Florida (Evans 1996). Two sites were visited quarterly within each wetland to collect five core samples (diameter = 5 cm) and two sweep net samples per site. Each sweep net sample was composed of five sweeps with a U.S. Standard No. 30 mesh dip net.

As in most freshwater systems, chironomids and oligochaetes are numerically dominant in Florida marshes. Worm assemblages are dominated by the families Tubificidae, Naididae, and Lumbriculidae. Naidids are most diverse in marshes with relatively high chlorophyll *a* content (Evans 1996). Diverse aquatic beetle communities occur in high densities, suggesting the presence of ample prey. Dragonflies and damselflies are also an important component of the invertebrate fauna, the most abundant genera being *Pachydiplax, Ischnura,* and *Enallagma.* Caddisflies, including *Oecetis, Orthotrichia,* and *Oxyethira,* usually occur in small numbers, although the latter two genera can be locally abundant. The most important mayfly genera are *Caenis* and *Callibaetis,* partially because they are adapted to low-dissolved-oxygen conditions (Berner 1950, Berner and Pescador 1988). The amphipods *Hyalella azteca* and *Crangonyx floridana* and the decapods *Palaemonetes* and *Procambarus* are abundant scavengers in most Florida marshes.

Many of the environmental factors that influence macroinvertebrate communities are linked. For example, decomposition and siltation rates, nutrient levels, pH, dissolved oxygen, and chlorophyll *a* are all affected by rainfall (Kiefer 1991). Similarly, inorganic nutrients in the water column influence algal production and chlorophyll *a,* while dissolved oxygen and pH undergo daily fluctuations primarily driven by photosynthetic activity and respiration.

Hydrology

During the drydown phase in late summer and early winter, fish populations in isolated marshes increase in density due, in part, to reduction in wetland size. This property of isolated, ephemeral wetlands renders them excellent feeding grounds for wood storks, which depend on concentrated food resources during the breeding season (Kushlan 1986). Reflooding of marshes after drydown can be followed by an increase in macroinvertebrate densities (Evans 1996). Algal blooms often develop shortly after inundation of soils because inorganic nutrients are released (Ponnamperuma 1972, Evans and Sullivan 1984, Hossner and Baker 1988) and can provide an abundant food supply for many macroinvertebrate species. This perhaps explains observed pulses in macroinvertebrate production following reflooding. In contrast, increases in macroinvertebrate densities during periods of prolonged flooding have also been observed. Murkin and Kadlec (1986) attributed these increases to increased availability of coarse plant litter, which serves as both habitat and a source of macroinvertebrate food. Water level fluctuation in Florida freshwater wetlands may be the most important factor influencing seasonal abundance of fish and macroinvertebrates.

Some invertebrate taxa may be reliable indicators of marsh hydrological regimes. For example, *Limnophyes* sp. and *Pseudosmittia* sp. seem to occur in semiterrestrial habitats (Cranston et al. 1983) and may indicate that a marsh is ephemerally flooded or has recently reflooded. The beetle *Desmopachria* also is commonly found in temporarily flooded marshes. Other taxa commonly occurring in small isolated marshes that are subject to frequent dry-down include *Psectrocladius* (*Monopsectrocladius*) sp. and *Corynoneura* sp. Many of these taxa may have special adaptations for living in temporarily flooded habitats, including the ability to migrate, short development time, and resistance to drought. Some species lay eggs in moist soil rather than in water, and development takes place when conditions are favorable (Wiggins et al. 1980).

Soils

pH. Soil pH can be an important characteristic controlling biological communities. Ambient hydrogen ion concentration is controlled primarily by soils, organic decomposition, and algal photosynthesis and can have direct effects on macroinvertebrate populations (Ponnamperuma 1972, Hudson et al. 1990). The characteristically low pH of flatwoods soils has been attributed to the downward translocation of materials, which lowers cation exchange capacity and increases hydrogen ion concentration at the soil surface (Abrahamson and Hartnett 1990). Hydrogen ion concentration of natural flatwoods marshes can be significantly higher than in created marshes (Evans 1996, Streever et al. 1996).

For unknown reasons, *Labrundinia* spp. tend to be acidophilic (Fittkau and Roback 1983, Hudson et al. 1990). Evans (1996) recorded significantly higher population densities of *Labrundinia neopilosella* where pH was lowest in Florida flatwoods marshes.

The distribution of the amphipod *Hyalella azteca* also may be influenced by pH. Evans (1996) recorded highest densities of *Hyalella* in Florida marshes with relatively high pH. Few to none were collected in more acidic marshes, where *H. azteca* was replaced by *Crangonyx floridanus*. Grapentine and Rosenberg (1992) reported that *Hyalella* populations declined in Canadian lakes with pH less than 5.8 and that no organisms were observed in lakes with pH less than 5.4. In Florida, Evans (1996) collected *Hyalella* in waters with pH as low as 4.6. Reasons why *Hyalella* appears to be adapted to lower pH in Florida marshes are unknown. However, calcium availability can influence the distribution of some crustaceans (Grapentine and Rosenberg 1992) and may explain in part the differences in temperate and subtropical pH tolerance ranges for some taxa. Limestone and soils rich in calcium are characteristic of some Florida freshwater marshes, but many flatwoods soils are relatively low in calcium (Abrahamson and Hartnett 1990).

Because calcium carbonate, a key compound required for shell formation (Pennak 1989), is characteristically unavailable in acidic waters, shell formation in mollusks may not be possible in some freshwater marshes in Florida. Snail populations are typically depauperate in acidic marshes of central Florida. Highest snail populations have been recorded where pH was above 6.0, and no snails were observed where pH was less than 4.7 (Evans 1996).

Apparent invertebrate community responses to pH and calcium availability may be confounded by other variables, including competitive and predator-prey interactions. Also, because field measurement of pH is typically done by sampling water at one or a few points in time, and because pH can fluctuate dramatically during the day, apparent relationships between pH and invertebrate community composition should not be overinterpreted. Intensive laboratory and field experimentation, such as that conducted by Grapentine and Rosenberg (1992), is required to gain a more complete understanding of causative relationships influencing distribution of *Hyalella* and molluscan species in Florida wetlands.

Silt. DeMarch (1976) observed that silt was important in influencing the distribution of aquatic insects. Chutter (1969) recorded lower abundance and species richness in streams having large amounts of silt and sand. Other investigators have made similar observations (Nuttall and Beilby 1973). Silt can affect feeding success and rate of growth by either making food less manageable or diluting organic content (Wotton 1994). Reice (1974, 1977) reported slower processing of leaf litter and fewer kinds and abundances of leaf-processing insects on silt than on other substrates.

In flatwoods marshes of Florida, Evans (1996) found little correlation between invertebrate densities and percent silt content in sediments, perhaps because siltation in most Florida marshes is not severe enough to cause a detectable reduction in organism density. However, taxa richness and Shannon-Weiner species diversity index values of core samples in the wet season tended to be negatively correlated with percent silt content. If siltation is responsible for a decline in invertebrate species richness and species diversity, the effect is expected to be greatest during the wet season, as the results suggest.

Some filter-feeding organisms may be particularly sensitive to siltation. Even moderate increases in percent silt content may render sediments a less favorable habitat for these organisms. For example, the abundance of *Chironomus* (*Lobochironomus*) sp., a filter-feeder, tended to be negatively correlated with silt content in Florida (Evans 1996). Another chironomid that feeds by collecting organic materials, *Kiefferulus dux/pungens,* exhibited a similar relationship to silt. These relationships may be due to interference of silt with feeding success. Undoubtedly some organisms prefer to live in silty substrates, but positive correlations with silt in Florida marshes have not been identified.

Organic Content. Organic materials in sediments, especially fine particulate organic matter (FPOM) (<1 mm) and the associated microflora, are an important source of food for macroinvertebrates, also providing case-building materials and physical habitat. The increase in growth rate associated with improved food quality is most often related to microbial populations and the nitrogen content of detritus (Ward and Cummins 1979, Anderson and Cummins 1979). In some systems detritivores (Gallepp 1977, Colbo and Porter 1979), herbivores (Collins 1980), and predators (Macan 1977, Fox and Murdoch 1978, Lawton et al. 1980) have been observed to grow faster with increasing food quantity, suggesting that food is a limiting resource in these systems. Many invertebrate species that occur in Florida freshwater marshes are detritivores. Thus, seasonal peaks in invertebrate production may in part be driven by a response to fall and winter pulses of decomposing plant material, although in Florida this effect is dampened by mild winters.

Evans (1996) found no significant relationships between sediment organic content (FPOM) and organism abundance, species richness, or diversity in created marshes of central Florida. In Florida freshwater marshes plant production exceeds decomposition and organic peat accumulates over time, and thus food may not be a limiting resource for organisms that feed on sediment. Similar situations have been observed in lotic systems when ample organic matter is available (Minshall and Minshall 1977, Peckarsky 1980, Williams 1980).

Availability of coarse material also can influence organism distribution. Organisms that feed by shredding coarse organic matter can be significantly more abundant in Florida marshes where the food resource is prevalent (Evans 1996). This observation, coupled with the apparent absence of correlation between organism abundance and FPOM, suggests that availability of plant material food resources may depend on particle size distribution. Perhaps plant production can be limiting for some trophic guilds, such as shredders, but not for others, such as detritivores grazing on FPOM.

Chlorophyll **a.** Phytoplankton can provide an important food resource for organisms that feed by filtering particles from the water column. Consequently filterers are often abundant where phytoplankton populations are most plentiful. However, available data do not demonstrate a high correlation between filter-feeding organisms and chlorophyll *a* concentration in Florida marshes (Kiefer 1991, Evans 1996). In some cases absence of a positive response to increasing chlorophyll *a* concentration may represent a preponderance of toxic, unmanageable, or unpalatable algae (Porter 1977). Alternatively, the absence of a relationship between filter feeders and chlorophyll *a* concentrations could be related to the presence of suspended inorganic particulate matter (silt) that can interfere with filtering mechanisms (Wotton 1994) or to difficulties associated with obtaining representative chlorophyll *a* samples from marshes.

With the exception of investigations in the Everglades (Swift 1981, Swift and Nicholas 1987, Vymazal et al. 1994, McCormick et al. 1996, McCormick and Odell 1996), few direct measurements of periphyton communities in Florida marshes are available. In a study of created and natural marshes in central Florida, filamentous algal mats, or periphyton, primarily consisted of blue-green algae. Using chlorophyll *a* as an indicator of periphyton growth, Evans (1996) and Kiefer (1991) showed that the abundance of scraping organisms, consisting primarily of snails, tended to decline with decreasing periphyton growth. Also, low macroinvertebrate densities tend to be associated with blue-green algal mats in at least some Florida marshes (Evans 1996). Lower macroinvertebrate production among blue-green algal mats may be attributed to a number of factors; blue-green algae may inhibit growth or cause mortality in invertebrates (Porter 1973).

Influence of Macrophytes and Filamentous Algal Mats

The influence of macrophytes on invertebrate community structure is extremely complex and can depend, in part, on the plant structure (e.g., physical space for attachment, availability as food resource), invertebrate habitat preference (e.g., infauna versus epifauna), and the species of vertebrate predators present. Macrophytes can affect food availability, vertebrate predation, physical environment (e.g., site attachment, protection of sediments, shading), and chemical environment (e.g., dissolved oxygen availability).

Macrophytes as a Food Resource. In a study of Florida created marshes, Kiefer (1991) found evidence for the occurrence of a "trophic surge" in above-ground macrophyte biomass approximately three years after construction, presumably brought on by high nutrient availability in newly flooded soils. The increase in living plant biomass was followed by marked accumulations of plant litter and detritus. Evans (1996) observed peaks in benthic macroinvertebrate densities in association with plant litter accumulation in the same wetlands. These observations support the hypothesis that plant production is perhaps one of the most important factors influencing invertebrate productivity via the detrital food web (Murkin and Wrubleski 1988). Epiphytic plant growth and "aufwuchs" attached to plant surfaces also can provide an important food resource for grazing invertebrates (Cattaneo 1983), and senescing vegetation can provide an important detrital food resource (Smock and Stoneburner 1980).

Influence on Vertebrate Predation. Nest-building centrarchid sunfish create open patches among littoral vegetation when they construct nests. Thorp (1988) reported that macroinvertebrate densities can be significantly lower in centrarchid nests than in adjacent vegetated areas. Beckett et al. (1992) found significantly higher invertebrate densities in sediments underlying macrophytes than in sediments in unvegetated patches. Temperature, dissolved oxygen, and organic content of sediments were eliminated as causative factors. Both Thorp (1988) and Beckett et al. (1992) attributed lower macroinvertebrate densities in open patches to sediment disturbance and enhanced predation rates.

Fish in vegetated areas have been observed to consume benthic invertebrates at half the rate observed in open areas, suggesting that vegetation serves to reduce predation pressure by interfering with the search and capture of prey organisms (Hershey 1985). However, in a created Florida marsh, abundances of common dipteran genera were higher in unvegetated areas than in vegetated areas (Streever et al. 1995). It is possible that vegetation has a positive effect on some vertebrate predators (e.g., mosquitofish), thereby causing an increase in predation on small invertebrates in vegetated areas.

Influence on Dissolved Oxygen. Dissolved oxygen availability is influenced by temperature, atmospheric exchange rate, and metabolic consumption and production. Photosynthetic and respiratory activities of algae have a marked influence on dissolved oxygen concentration in the water column. Macrophytes also can stimulate microbial respiration by providing a carbon source, substrate for attachment, and transfer of oxygen to flooded soils near root channels.

In wetlands where dissolved oxygen can be limiting, particularly in sediments, rooted emergent plants offer oxygenated refugia for aquatic macroinvertebrates in the thin microlayer associated with plant root surfaces. Macrophytes with extensive root systems can oxidize sediments (Carpenter and Lodge 1986, Reddy et al. 1989), perhaps rendering them more habitable for benthic fauna. Taxa requiring relatively high oxygen availability, such as *Corynoneura,* commonly occur in marshes in spite of low oxygen concentrations in the water column. These species may find sufficient oxygen in the oxygen-rich microlayer produced by periphytic algae attached to surfaces of vascular plants.

Aquatic plants, particularly floating-leaved plants, can cause hypoxia by reducing light availability and interfering with gas exchange at the water surface (Morris and Barker 1977). Rose and Crumpton (1996) attributed low oxygen concentrations in dense macrophyte stands to physical effects influenced by aquatic plants, such as limitations on gas exchange at the water-air interface and reduced oxygen production via photosynthesis because of shading.

In Florida dense mats of floating vegetation can eliminate light required for photosynthesis and interfere with mixing and oxygenation at the air-water interface, causing hypoxia in underlying waters. Oxygen availability tends to be low in wetlands with dense floating vegetation. For example, one created marsh located in central Florida with complete cover by floating plants had a mean dissolved oxygen concentration during daylight hours of 2.1 mg/l at the surface and 0.7 mg/l at the bottom (Kiefer 1991). Lack of sufficient oxygen can reduce macroinvertebrate community diversity (Hynes 1960, Wiederholm 1984).

In some created marshes of central Florida, dominance by floating plants is probably a result of relatively deep standing water (>0.5 m) and ample inorganic nutrients in the water column. Duckweeds, water fern (*Azolla*), and water hyacinth are typically abundant. Floating mats of emergent vegetation, such as cattail or pickerelweed, have been observed in some deep created marshes. Such conditions can be probably be avoided by designing and constructing relatively shallow wetlands with gradual slopes. In some cases planting and use of herbicides may be necessary to establish desirable plant communities.

Summary of Evidence from Florida Flatwoods Marshes

Numerous studies have demonstrated higher benthic invertebrate abundance in vegetated habitats than in nonvegetated habitats (Minckley 1963, Lindegaard et al. 1975, Gregg 1981). These positive relationships with macrophytes have been attributed to greater surface area for attachment, more abundant detrital food resources, and enhanced availability of refugia, allowing successful avoidance of predation in vegetated areas.

Although investigations conducted by Evans (1996) were not specifically designed to address the influence of plant communities on aquatic invertebrates, some results may be of interest. Evans (1996) observed that, in contrast to previous studies indicating a positive relationship between macrophytes and invertebrate abundance, organism abundance in sweep samples collected in winter was negatively correlated with total percent plant cover for Florida marshes. During the wet season macroinvertebrate density and diversity in Florida flatwoods marshes were also negatively correlated with percent cover of filamentous algae. Although the negative relationships with filamentous algae appear to be significant, data points were not evenly distributed across the regressions, causing some doubt to be cast on their validity. Since the wetlands with the highest total plant cover also exhibited the highest algal cover, the relationship between plant cover and invertebrate density is confounded by the relationship with filamentous algae.

Although invertebrate densities may not increase with macrophyte cover, macroinvertebrate community diversity may (Evans 1996). Habitat diversity generally increases with increasing vegetation cover. In Florida flatwoods marshes strongest correlations were observed between macroinvertebrate diversity and filamentous blue-green algae (negative), total percent plant cover (positive), and percent cover of floating plants (negative). The observed relationships between plant cover and macroinvertebrate diversity support the tenets of niche theory, which predicts an increase in species richness with increasing habitat diversity (Hutchinson 1959). Negative relationships with filamentous algae and floating plant cover may be a result of a combination of mechanisms, including shading, reduced dissolved oxygen, release of toxins, and reduced lability. Although it is not possible to draw strong conclusions regarding the influence of macrophytes on invertebrates of Florida marshes, there is limited indirect evidence that dominant cover of floating plants and/or filamentous algae can cause a reduction in invertebrate density and diversity. The mechanisms involved may include shading, oxygen reduction, release of toxins, and reduction of food resources.

Factors Influencing Invertebrate Predator Populations

Beetles, odonates, and hemipterans comprise the majority of invertebrate predators in Florida marshes. Beetles and hemipterans are known to undergo migratory flights, enabling them to capitalize quickly on available aquatic food resources. In contrast, dragonflies and damselfies must deposit eggs and early life stages must develop before dragonflies and damselflies can make use of aquatic food resources. In essence, beetles and hemipterans have the qualities of successful opportunists in spite of their relatively long life cycles.

Invertebrate predators are often observed in high densities in seasonal wetlands (Neckles et al. 1990, Batzer et al. 1993). Invertebrate predators that feed by engulfing prey were significantly more abundant in seasonally flooded Florida flatwoods marshes than in continually flooded created wetlands (Evans 1996). Abundance of invertebrate predators may, in part, be affected by (1) the probability of being eaten by vertebrate predators and (2) the available food supply. The absence of major fish predators in small, temporarily flooded marshes makes it unlikely that fish predation has a large influence on invertebrate predator populations. Centrarchid sunfish, which may feed selectively on larger macroinvertebrates (invertebrate predators), are limited or absent in ephemeral marshes. Invertebrate prey may be more available in seasonally flooded wetlands due to pulses in production during reflooding. Invertebrate predators may have an advantage in seasonally flooded marshes due to limited predation pressure posed by vertebrate predators and to an abundance of small invertebrate prey as a food resource.

COMPARING CREATED AND NATURAL INVERTEBRATE COMMUNITIES

Driven by regulations that stem from no-net-loss wetland conservation policies, wetland creation has become a common practice in Florida. Typically, wetland creation entails excavation or construction of earthworks to support wetland hydrology, followed by planting of selected wetland macrophytes (Streever and Crisman 1996). Some creation projects import substrate, or "muck," from a donor marsh, which may introduce invertebrates to a newly created system, but in general invertebrates immigrate into created wetlands on their own.

Currently, there is no central authority overseeing wetland creation projects in Florida and no central record of creation sites in Florida. Instead, records are scattered among various federal, state, and local agencies, and because individual sites may be identified by more than one name, collating records from all agencies would be difficult. Nevertheless, it is clear that created wetlands are fast becoming a major habitat type in Florida. Despite the large number of created wetlands in Florida and other states, debate continues regarding the ability of created wetlands to replace the structure and function of lost natural wetlands (e.g., National Research Council 1992, Kentula et al. 1993, Zedler 1996).

Comparison of various attributes of natural "reference wetlands" and created wetlands provides one means of assessing the similarity of natural and created wetlands. Plant communities are probably the most commonly addressed attribute (e.g., Broome et al. 1988, Kentula et al. 1993, Galatowitsch and van der Valk 1994, Streever and Crisman 1996), but a need for comparison of other attributes has long been recognized (Ewel 1987, Evans 1989, Erwin 1990, Zedler and Langis 1991, Streever and Crisman 1993). Monitoring programs required by government permits have resulted in collection of invertebrate data from created wetlands, but these data are usually archived in consultancy reports that are not readily available to the public. Because these reports are not subject to peer review, the information they contain should be viewed with caution. However, published literature on wetland creation also contains comparisons of invertebrate communities from natural and created wetlands, including both estuarine and freshwater systems (Minello and Zimmerman 1992, Streever and Crisman 1993, Sacco et al. 1994, Streever et al. 1996).

Comparison of natural and created wetlands requires careful consideration of sampling issues. All too often, comparisons only look at a single created wetland and a single natural wetland. This approach, which lacks true replication (Hurlbert 1984), does not consider variability among created and natural wetlands. In studies that sample multiple created and natural wetlands, other potential problems include collection of too few samples to fairly assess the invertebrate community of a site and collection of samples from different habitats at different sites. Because different macrophyte communities in both natural and created wetlands harbor different invertebrate assemblages (Krecker 1939, Calow 1973, Pip and Stewart 1976, Murkin and Kadlec 1986, Wrubleski and Rosenberg 1990, Streever et al. 1995), samples should be collected from the same type of vegetation from all sites, or sample sites should be stratified by vegetation to insure that different communities are fairly represented. Numerous methods have been developed to assist with sampling design in benthic invertebrate studies, and at least one method has been developed specifically for comparison of natural and created wetland invertebrates (Streever and Portier 1994). This method suggests sampling designs based on variances from pilot data, target levels for type I and type II error rates (that is, the target probabilities of committing type I and type II statistical errors), sampling budget, and cost of sample collection, but it does not incorporate recommendations for stratification across vegetation communities.

Several attempts have been made to compare created and natural freshwater marsh invertebrate communities in Florida. Florida Department of Environmental Protection (1994) researchers sampled invertebrates from a number of created wetlands and stated that "most of the created marshes we studied were, both structurally and functionally, at least moderately similar to the reference wetland." However, because only a single natural wetland was sampled, a comprehensive comparison was not possible in the Florida Department of Environmental Protection study. An attempt to compile information from consultant reports to the Florida phosphate mining industry initially showed promise, but differences between sampling methods, taxonomic resolution, and sampling sites rendered interpretation difficult (T. L. Crisman, personal communication).

One study that was designed specifically to allow comparison of created and natural central Florida freshwater marshes relied on collection of dipteran samples from 10 created and 10 natural wetlands, with five sites visited in each wetland and seven core samples (core diameter = 5 cm) collected at each station (Streever et al. 1996). Many of these wetlands were the same wetlands investigated by Evans (1996). All samples were collected from areas with pickerelweed (*Pontederia cordata*) in July and August 1993. This study considered only dipterans, the assumption being that dipterans would rapidly colonize new habitats (Nursall 1952, Driver 1977, Danell and Sjoberg 1982) and that therefore any differences which might occur would reflect important functional differences between natural and created wetlands instead of incomplete colonization of created sites.

Fifty-seven dipteran taxa were represented in samples, but only 19 taxa occurred with sufficient abundance to justify statistical comparison (Table 4.1). There were no statistically significant differences between mean abundances of the dipteran taxa in created and natural wetlands based on the results of nested ANOVA tests, and further analysis using ordination methods also failed to show convincing differences in dipteran community structure for created and natural marshes. However, significant differences between natural and created wetlands were found for pH, conductivity, and sediment quality, indicating that similar dipteran communities can occur in marshes with different environmental conditions. Because of the variable water levels and other environmental conditions found within individual marshes, it might be expected that marsh invertebrates would have a wide tolerance to environmental conditions, which might in turn explain why similar communities can occur under demonstrably different conditions.

It should be emphasized that results from Streever et al. (1996) do not indicate total functional similarity between created and natural wetlands. Samples were collected only from patches of pickerelweed, and there may be differences between the macrophyte communities of created and natural wetlands that could lead to differences in invertebrate communities. Invertebrate

TABLE 4.1. Dipteran Taxonomic Composition in 10 Created and 10 Natural Freshwater Herbaceous Wetlands in Central Florida[a]

Taxon	Mean Individuals m^{-2}		Number of Sites Having at Least One Occurrence	
	Created Wetlands	Natural Wetlands	Created Wetlands	Natural Wetlands
Chironomidae				
Larsia*	2,136.1	3,749.9	10	8
Monopelopia*	877.4	873.1	9	9
Polypedilum*	774.1	1,628.3	10	10
Chironomus*	343.4	139.7	9	7
Glyptotendipes*	240.1	10.2	7	4
Tanytarsus*	164.4	90.2	9	7
Labrundinia*	106.2	52.4	6	4
Natarsia*	96.0	141.2	8	7
Goeldichironomus*	88.8	8.7	7	4
Lauterborniella	74.2	0	1	0
Ablabesmyia*	58.2	80.0	6	8
Dicrotendipes*	45.1	33.5	2	4
Tanytarsini A Roback	40.7	14.6	5	4
Kiefferulus*	34.9	133.9	6	6
Tanypus	32.0	5.8	5	3
Pseudochironomus	30.6	4.4	6	1
Unknown chironomid 4	29.1	0	1	0
Unknown chironomid 2	23.3	0	1	0
Zavreliella	17.5	30.6	6	6
Paratanytarsus	17.5	1.5	1	1
Fittkauimyia	11.6	23.3	2	5
Parachironomus	10.2	1.5	3	1
Paratendipes	8.7	29.1	2	4
Clinotanypus	7.3	0	2	0
Nimbocera	4.4	0	3	0
Pseudosmittia	2.9	0	1	0
Paramerina	2.9	4.4	2	2
Cladotanytarsus	1.5	0	1	0
Djalmabatista	1.5	0	1	0
Guttipelopia	0	23.3	0	1
Procladius	0	5.8	0	1
Unknown chironomid 1	0	4.4	0	2
Unknown chironomid 3	0	2.9	0	2
Cladopelma	0	1.5	0	1
Ceratopogonidae				
Ceratopogoninae*	1,655.9	2,278.7	10	10
Dasyhelea*	922.6	211.0	10	8
Forcipomyia	46.6	109.1	8	9
Atrichopogon	1.5	0	1	0

TABLE 4.1. (Continued)

| Taxon | Mean Individuals m^{-2} | | Number of Sites Having at Least One Occurrence | |
	Created Wetlands	Natural Wetlands	Created Wetlands	Natural Wetlands
Tipulidae				
Tipulid 2	74.2	33.5	5	3
Tipulid 5	23.3	20.4	7	5
Tipulid 3	18.9	7.3	4	4
Tipulid 4	11.6	27.7	3	7
Tipulid 1	5.8	5.8	3	2
Tipulid 6	1.5	5.8	1	2
Culicidae				
Uranotaenia*	97.5	82.9	9	6
Mansonia*	80.0	101.9	6	7
Culex	65.5	27.7	6	5
Anopheles	0	1.5	0	1
Chaoboridae				
Corethrella*	21.8	42.2	3	6
Chaoborus	1.5	4.4	1	2
Stratiomyidae*	53.8	27.7	7	7
Ephydridae	34.9	7.3	6	2
Tabanidae*	14.6	52.4	5	9
Muscidae	2.9	4.4	2	2
Empididae	1.5	0	1	0
Psychodidae	1.5	0	1	0
Syrphidae	1.5	4.4	1	2

[a] All samples were collected from stands of pickerelweed (*Pontederia cordata*). Means are from 35 core samples collected in each wetland. For the 19 taxa that occurred frequently enough to justify statistical comparison, there were no significant differences for mean individuals m^{-2} between created and natural marshes.

Asterisks mark taxa that occurred frequently enough to justify statistical comparison.

Source: adapted from W. J. Streever, K. M. Portier, and T. L. Crisman, A comparison of dipterans from ten created and ten natural wetlands, Wetlands 16:416–428, 1996.

samples collected from another plant community or during another time of year might show differences between created and natural wetlands. Similarly, only dipterans were addressed, and close examination of another invertebrate group might show differences, as suggested by Evans (1996). However, these results do show that certain attributes of created and natural marsh function and structure may be similar, while others may differ. Data sets from other created and natural wetlands have shown a similar pattern. For example, clear differences in soil organic content were found between created and natural salt marshes of North Carolina, but only limited differences were found for infaunal communities from the same sites (Sacco et al. 1994).

In many of the same wetlands studied by Streever, Evans (1996) observed significantly higher nektonic invertebrate abundance in natural flatwoods marshes than in created marshes, particularly as reflooding occurred following the dry season (Table 4.2). Although taxa richness of nektonic invertebrates was also significantly higher in natural marshes, no significant difference in Shannon-Weiner diversity values of natural and created marshes was found.

Based on ancillary data (Erwin 1988, Evans and Sullivan 1988, Kiefer 1991), created wetlands were classified according to age to allow comparisons between newly established sites (<3 years old) and older sites (>3 years old). Abundance of nektonic and benthic invertebrates was not significantly different between wetland age classes (Tables 4.2 and 4.3). However, taxa richness and Shannon-Weiner species diversity of benthic invertebrates were significantly higher at older sites. This finding, in combination with the observation that shredder abundance increases with the amount of available coarse organic material, suggests that soil development may play an important part in the functional development of created marshes, as proposed by Kiefer (1991). These results indicate that we should continue efforts to design and manage created wetlands in ways that maximize establishment of diverse, persistent vegetation communities and minimize siltation.

FUTURE RESEARCH NEEDS

Journal articles and book chapters contain a plethora of information on the invertebrates of Florida aquatic ecosystems, and more unpublished data exist in scattered locations. However, little of this information directly addresses the invertebrate communities of Florida's freshwater marshes. Compilation of data from numerous sources, followed by interpretation, including application of meta-analytical methods, might yield interesting insights. However, the usefulness of existing data may be limited, as was found when attempts were made to assess macroinvertebrate data from consultancy reports generated for Florida phosphate mining companies.

Even in the absence of problems that typically arise when attempts are made to analyze existing data sets, it would be incorrect to believe that there is no need for additional collection of field samples from Florida's freshwater marshes, which, with cypress domes, may represent one of Florida's least studied invertebrate habitats. The use of macroinvertebrates in water quality assessment is commonplace in some systems, but further work on the reliability of freshwater marsh invertebrates as biological indicators in Florida is needed. Future studies should consider biomass as well as abundance in order to provide insight on thermodynamics.

In addition to a need for more collection of field samples, there is a need for manipulative experiments capable of establishing firm cause and effect links. As a start, researchers could assess the influence of various environmental variables on invertebrate communities. Efforts to manage marshes for

TABLE 4.2. Comparison of Benthic Macroinvertebrates in Eight Natural Flatwoods Marshes, Four Constructed Marshes <3 Years Old, and Seven Constructed Marshes >3 Years Old Located in Central Florida[a]

	Wetland Type	Max	Min	Mean
Density (No./m^2)	Natural	17472	6639	9721
	All constructed	16510	5737	9006
	<3 years old	10500	5738	6982
	>3 years old	16510	7675	10741
Number of taxa	Natural	52	30	40
	All constructed	48	20	31
	<3 years old	30	20	23
	>3 years old	48	29	38
Diversity	Natural	4.50	3.94	4.23
	All constructed	4.50	2.49	3.43
	<3 years old	4.00	3.30	3.10
	>3 years old	4.50	2.49	3.77

[a] Means are from 40 core samples per wetland.

Source: adapted from D. L. Evans, Aquatic macroinvertebrate communities of constructed and natural marshes in central Florida, Ph.D. Dissertation, University of Florida, Gainesville, FL, 1996.

TABLE 4.3. Comparison of Nektonic Macroinvertebrates in Eight Natural Flatwoods Marshes, Four Constructed Marshes <3 Years Old, and Seven Constructed Marshes >3 Years Old Located in Central Florida[a]

Parameter	Wetland Type	Max	Min	Mean
Abundance	Natural	966	111	457
	All constructed	586	81	194
	<3 years old	586	81	219
	>3 years old	274	95	176
Number of taxa	Natural	55	26	43
	All constructed	39	20	28
	<3 years old	38	20	22
	>3 years old	39	24	33
Diversity	Natural	4.40	3.76	4.10
	All constructed	4.14	2.67	3.62
	<3 years old	4.01	2.67	3.28
	>3 years old	4.14	3.07	3.81

[a] Means are from 16 sweep net samples per wetland.

Source: adapted from D. L. Evans, Aquatic macroinvertebrate communities of constructed and natural marshes in central Florida, Ph.D. Dissertation, University of Florida, Gainesville, FL, 1996.

the improvement of water quality and enhancing wildlife habitat point to the need for studies that focus on the effects of hydroperiod, vascular plant and algal communities, and various water quality parameters. Little is known of the role that periphyton and bacteria play in controlling benthic invertebrate community structure in Florida marshes. Experiments are also needed to investigate the influence of invertebrates on their environment, and in particular on nutrient cycling and soil conditions.

ACKNOWLEDGMENTS

Support for this work came from a National Science Foundation Graduate Research Fellowship awarded to W. J. Streever and from a grant awarded by the Florida Institute of Phosphate Research.

LITERATURE CITED

Abrahamson, W. G. and D. C. Hartnett. 1990. Pine flatwoods and dry prairies. Pages 103–149 *in* R. L. Myers and J. J. Ewel (eds.), Ecosystems of Florida. University of Central Florida Press, Orlando, FL.

Anderson, N. H., and K. W. Cummins. 1979. Influences of diet on the life histories of aquatic insects. Journal of the Fisheries Research Board of Canada 36:335–342.

Batzer, D. P., M. McGee, V. H. Resh, and R. R. Smith. 1993. Characteristics of invertebrates consumed by mallards and prey response to wetland flooding schedules. Wetlands 13:41–49.

Beckett, D. C., T. P. Aartila, and A. C. Miller. 1992. Contrasts in density of benthic invertebrates between macrophyte beds and open littoral patches in Eau Galle Lake, Wisconsin. American Midland Naturalist 127:77–90.

Berner, L. 1950. The Mayflies of Florida. University of Florida Presses, Gainesville, FL.

Berner, L., and M. L. Pescador. 1988. The Mayflies of Florida. University of Florida Presses, Gainesville, FL.

Brightman, R. S. 1976. Benthic macroinvertebrate response to secondarily treated sewage effluent in north central Florida cypress domes. Masters Thesis, University of Florida, Gainesville, FL.

Broome, S. W., E. D. Seneca, and W. W. Woodhouse. 1988. Tidal salt marsh restoration. Aquatic Botany 32:1–22.

Calow, P. 1973. Gastropod associations within Malham Tarn, Yorkshire. Freshwater Biology 3:521–34.

Carpenter, S. R., and D. M. Lodge. 1986. Effects of submersed macrophytes on ecosystem processes. Aquatic Botany 26:341–370.

Cattaneo, A. 1983. Grazing on epiphytes. Limnology and Oceanography 28:124–32.

Chutter, F. M. 1969. The effects of silt and sand on the invertebrate fauna of streams and rivers. Hydrobiologia 34:57–76.

Colbo, M. H., and G. N. Porter. 1979. Effects of the food supply on the life history of Simuliidae (Diptera). Canadian Journal of Zoology 57:301–306.

Collins, N. C. 1980. Developmental responses to food limitation as indicators of environmental conditions for *Ephydra cinerea* Jones (Diptera). Ecology 61:650–661.

Cranston, P. S., D. R. Oliver, and O. A. Saether. 1983. The larvae of Orthocladiinae (Diptera: Chironomidae) of the Holarctic region—keys and diagnoses. Lund, Sweden. Ent. Scand. Suppl. 19:149–291.

Danell, K., and K. Sjoberg. 1982. Successional patterns of plants, invertebrates and ducks in a man-made lake. Journal of Applied Ecology 19:395–409.

Davis, S. M., and J. C. Ogden (Eds.). 1994. Everglades: The Ecosystem and its Restoration. St. Lucie Press, Delray Beach, FL.

De March, B. G. E. 1976. Spatial and temporal patterns in macrobenthic stream diversity. Journal of Fisheries Research Board of Canada 33:1261–1270.

Doren, R. F., and R. M. Rochefort. 1984. Summary of fires in Everglades National Park and Big Cypress National Reserve, 1981. Everglades National Park South Florida Research Center Report No. SFRC-84/01.

Driver, E. A. 1977. Chironomid communities in small prairie ponds: some characteristics and controls. Freshwater Biology 7:121–133.

Erwin, K. L. 1988. Agrico Fort Green Reclamation Project, Sixth Annual Report. Agrico Mining Company, Mulberry, FL.

———. 1990. Freshwater marsh creation and restoration in the southeast. Pages 223–248 *in* J. A. Kusler and M. E. Kentula (eds.), 1990. Wetland Creation and Restoration: The Status of the Science. Island Press, Washington, DC.

Evans, D. L. 1989. A comparison of the benthic macroinvertebrate fauna of a newly created freshwater marsh with natural emergent marshes in South Central Florida. Pages 71–86 *in* Proceedings of the 16th Annual Conference on Wetlands Restoration and Creation. Tampa, FL.

———. 1996. Aquatic macroinvertebrate communities of constructed and natural marshes in central Florida. Ph.D. Dissertation, University of Florida, Gainesville, FL.

Evans, D. L., and J. H. Sullivan. 1984. Annual Report for Horse Creek Wetland Creation. August 1983–August 1984. Water and Air Research, Inc., Gainesville, FL.

———. 1988. Horse Creek Wetland Creation Fifth and Final Annual Report September 1987–June 1988. Water and Air Research, Inc., Gainesville, FL.

Ewel, J. J. 1987. Restoration is the ultimate test of ecological theory. *In* W. R. Jordan III, M. E. Gilpin, and J. D. Aber (eds.), Restoration Ecology: A Synthetic Approach to Ecological Research. Cambridge University Press, Cambridge, UK.

Fernald, E. A., and D. J. Patton (eds.). 1984. Water Resources Atlas of Florida. Institute of Science and Public Affairs, Florida State University, Tallahassee, FL.

Fittkau, E. J., and S. S. Roback. 1983. The larvae of Tanypodinae (Diptera: Chironomidae) of the Holarctic region—Keys and diagnoses. Lund, Sweden. Entomologica Scandanavica Supplement 19:33–110.

Florida Department of Environmental Protection. 1994. The Biological Success of Created Marshes in Central Florida. Florida Department of Environmental Protection, Tallahassee, FL.

Fox, L. R., and W. W. Murdoch. 1978. Effects of feeding history on short-term and long-term functional responses in *Notonecta hoffmanni*. Journal of Animal Ecology 47:945–959.

Galatowitsch, S. M., and A. G. van der Valk. 1994. Restoring Prairie Wetlands. Iowa University Press, Ames, IA.

Gallepp, G. W. 1977. Responses of caddisfly larvae (*Brachycentrus* spp.) to temperature, food availability and current velocity. American Midland Naturalist 98:59–84.

Grapentine, L. C., and D. M. Rosenberg. 1992. Responses of the freshwater amphipod *Hyalella azteca* to environmental acidification. Canadian Journal of Fisheries and Aquatic Sciences 49:52–64.

Gregg, W. W. 1981. Aquatic macrophytes as a factor affecting the microdistribution of stream invertebrates. Master's Thesis, Idaho State University, Pocatello, ID.

Haack, S. K. 1984. Aquatic macroinvertebrate community structure in a forested wetland: Interrelationships with environmental parameters. Master's Thesis, University of Florida, Gainesville, FL.

Hershey, A. E. 1985. Effects of predatory sculpin on the chironomid communities in an arctic lake. Ecology 66:1131–1138.

Hossner, L. R., and W. H. Baker. 1988. Phosphorus transformations in flooded soils. *In* D. D. Hook, W. H. McKee, H. K. Smith, J. Gregory, V. G. Burrell, M. R. Devoe, R. E. Sojka, S. Gilbert, R. Banks, L. H. Stolzy, C. Brooks, T. D. Matthews, and T. H. Shear (eds.), The Ecology and Management of Wetlands. Volume 2. Timber Press, Portland, OR.

Hudson, P. L., D. R. Lenat, B. A. Caldwell, and D. Smith. 1990. Chironomidiae of the Southeastern United States: A Checklist of Species and Notes on Biology, Distribution, and Habitat. United States Department of the Interior, Fish and Wildlife Service, Washington, DC.

Hurlbert, S. H. 1984. Pseudoreplication and the design of ecological field experiments. Ecological Monographs 54:187–211.

Hutchinson, G. E. 1959. Homage to Santa Rosalia; or why are there so many kinds of animals? The American Naturalist 93:145–159.

Hynes, H. B. N. 1960. The Biology of Polluted Waters. Liverpool University Press, Liverpool, UK.

Kentula, M. E., R. P. Brooks, S. E. Gwin, C. C. Holland, A. D. Sherman, and J. C. Sifneos. 1993. An Approach to Improving Decision Making in Wetland Restoration and Creation. CRC Press, Boca Raton, FL.

Kiefer, J. H. 1991. Chemical Functions and Water Quality in Marshes Reclaimed on Phosphate Mined Lands in Central Florida. Master's Thesis, University of Florida, Gainesville, FL.

Krecker, F. H. 1939. A comparative study of the animal population of certain submerged aquatic plants. Ecology 20:553–562.

Kushlan, J. A. 1986. Responses of wading birds to seasonally fluctuating water levels: strategies and their limits. Colonial Waterbirds 9:155–162.

———. 1990. Freshwater marshes. Pages 324–363 *in* R. L Myers and J. J. Ewel (eds.), Ecosystems of Florida. University of Central Florida Press, Orlando, FL.

Lawton, J. H., B. A. Thompson, and D. J. Thompson. 1980. The effects of prey density on survival and growth of damselfly larvae. Ecological Entomology 5:39–51.

Lindegaard, C., J. Thorup, and M. Bahn. 1975. The invertebrate fauna of a moss carpet in the Danish spring Ravnkilde and its seasonal, vertical, and horizontal distribution. Archiv für Hydrobiologie 75:109–139.

Macan, T. T. 1977. The influence of predation on the composition of fresh-water animal communities. Biological Reviews of the Cambridge Philosophical Society 52:45–70.

McCormick, P. V., P. S. Rawlik, K. Lurding, E. P. Smith, and F. H. Sklar. 1996. Periphyton-water quality relationships along a nutrient gradient in the northern Florida Everglades. Journal of the North American Benthological Society 15:433–449.

McCormick, P. V. and M. B. O'Dell. 1996. Quantifying periphyton responses to phosphorus in the Florida Everglades: A synoptic experimental approach. Journal of the North American Benthological Society 15(4):450–468.

Minckley, W. L. 1963. The ecology of a spring system: Doe Run, Meade County, Kentucky. Wildlife Monographs 11:1–124.

Minello, T. J., and R. J. Zimmerman. 1992. Utilization of natural and transplanted Texas salt marshes by fish and decapod crustaceans. Marine Ecology Progress Series 90:273–285.

Minshall, G. W., and J. N. Minshall. 1977. Microdistribution of benthic invertebrates in a Rocky Mountain (U.S.A.) stream. Hydrobiologia 55:231–249.

Morris, P. F., and W. G. Barker. 1977. Oxygen transport rates through mats of *Lemna minor* and *Wolfia* sp. and oxygen tension within and below the mat. Canadian Journal of Botany 55:1926–1932.

Murkin, H. R., and J. A. Kadlec. 1986. Responses by benthic macroinvertebrates to prolonged flooding of marsh habitat. Canadian Journal of Zoology 64:65–72.

Murkin, H. R., and D. A. Wrubleski. 1988. Aquatic invertebrates of freshwater wetlands: function and ecology. Pages 239–249 *in* D. D. Hook, W. H. McKee, H. K. Smith, J. Gregory, V. G. Burrell, M. R. Devoe, R. E. Sojka, S. Gilbert, R. Banks, L. H. Stolzy, C. Brooks, T. D. Matthews, and T. H Shear (eds.), The Ecology and Management of Wetlands. Volume 1. Timber Press, Portland, OR.

National Research Council. 1992. Restoration of Aquatic Ecosystems. National Academy Press, Washington, DC.

Neckles, H. A., H. R. Murkin, and J. A. Cooper. 1990. Influences of seasonal flooding on macroinvertebrate abundance in wetland habitiats. Freshwater Biology 23:311–322.

Noble, C. V. 1989. Marl soils in south Florida. Soil Survey Horizons 30(1).

Nursall, J. R. 1952. The early development of a bottom fauna in a new power reservoir in the rocky mountains of Alberta. Canadian Journal of Zoology 30:387–409.

Nuttall, P. M., and G. H. Beilby. 1973. The effect of china-clay wastes on stream invertebrates. Environmental Pollution 5:77–86.

Peckarsky, B. L. 1980. Influence of detritus upon colonization of stream invertebrates. Canadian Journal of Fisheries and Aquatic Sciences 37:957–963.

Pennak, R. W. 1989. Freshwater Invertebrates of the United States: Protozoa to Mollusca (3rd ed.). John Wiley & Sons, New York.

Pezeshki, C. 1987. Response of benthic macroinvertebrates of a shrub swamp to discharge of treated wastewater. Master's Thesis, University of Florida, Gainesville, FL.

Pip, E., and J. M. Stewart. 1976. The dynamics of two aquatic plant-snail associations. Canadian Journal of Zoology 54:1192–1205.

Ponnamperuma, F. N. 1972. The chemistry of submerged soils. Advances in Agronomy 24:29–96.

Porter, K. G. 1973. Selective grazing and differential digestion of algae by zooplankton. Nature 244:179–180.

———. 1977. The plant-animal interface in freshwater ecosystems. American Scientist 65:159–170.

Reddy, K. R., W. H. Patrick, and C. W. Lindau. 1989. Nitrification-denitrification at the plant root-sediment interface in wetlands. Limnology and Oceanography 34:1004–1013.

Reice, S. R. 1974. Environmental patchiness and the breakdown of leaf litter in a woodland stream. Ecology 55:1271–1282.

———. 1977. The role of animal association and current velocity in sediment-specific leaf litter decomposition. Oikos 29:357–365

Rose, C., and W. G. Crumpton. 1996. Effects of emergent macrophytes on dissolved oxygen dynamics in prairie pothole wetland. Wetlands 16:495–502.

Sacco, J. N., E. D. Seneca, and T. R. Wentworth. 1994. Infaunal community development of artificially established salt marshes in North Carolina. Estuaries 17:489–500.

Smock, L. A., and D. L. Stoneburner. 1980. The response of macroinvertebrates to aquatic macrophyte decomposition. Oikos 35:397–403.

Streever, W. J., and S. A. Bloom. 1993. The self-similarity curve: A new method of determining the sampling effort required to characterize commnities. Journal of Freshwater Ecology 8:401–403.

Streever, W. J., and T. L. Crisman. 1993. A preliminary comparison of meiobenthic cladoceran assemblages in natural and created wetlands in central Florida. Wetlands 13:229–336.

———. 1996. Constructing freshwater wetlands to replace impacted natural wetlands: A subtropical perspective. Pages 295–303 in F. Schiemer and K. T. Boland (eds.), Perspectives in Tropical Limnology. SPB Academic Publishing, Amsterdam, The Netherlands.

Streever, W. J., D. L. Evans, C. Keenan, and T. L. Crisman. 1995. Chironomidae (Diptera) and vegetation in a created wetland and implications for sampling. Wetlands 14:285–289.

Streever, W. J., and K. M. Portier. 1994. A computer program to assist with sampling design in the comparison of natural and constructed wetlands. Wetlands 14:199–205.

Streever, W. J., K. M. Portier, and T. L. Crisman. 1996. A comparison of dipterans from ten created and ten natural wetlands. Wetlands 16:416–428.

Swift, D. R. 1981. Preliminary Investigation of Periphyton and Water Quality Relationships in the Everglades Water Conservation Areas, Technical Publication 81-5, South Florida Water Management District, West Palm Beach, FL.

Swift, D. R., and R. B. Nicholas. 1987. Periphyton and Water Quality Relationships in the Everglades Water Conservation Areas, 1978–1982. Technical Publication 87-2, South Florida Water, West Palm Beach, FL.

Thorp, J. H. 1988. Patches and the responses of lake benthos to sunfish nest-building. Oecologia 76:168–174.

Vymazal, J., C. B. Craft, and C. J. Richardson. 1994. Periphyton response to nitrogen and phosphorus additions in the Florida Everglades. Algological Studies 73:75–97.

Ward, G. M., and K. W. Cummins. 1979. Effects of food quality on growth of a stream detritivore, *Paratendipes albimanus* (Meigen) (Diptera: Chironomidae). Ecology 60: 57–64.

Wiederholm, T. 1984. Responses of aquatic insects to environmental pollution. Pages 508–557 *in* V. H. Resh and D. M. Rosenberg (eds.), The Ecology of Aquatic Insects. Praeger Scientific, New York.

Wiggins, G. B., R. J. Mackay, and I. M. Smith. 1980. Evolutionary and ecological strategies of animals in annual temporary pools. Archiv für Hydrobiologie Supplement 58:97–206.

Williams, D. D. 1980. Some relationships between stream benthos and substrate heterogeneity. Limnology and Oceanography 25:166–172.

Wotton, R. S. 1994. The biology of particles in aquatic systems (2nd ed.). CRC Press, Boca Raton, FL.

Wrubleski, D. A., and D. M. Rosenberg. 1990. The Chironomidae (Diptera) of Bone Pile Pond, Delta Marsh, Manitoba, Canada. Wetlands 10:243–275.

Zedler, J. B. (ed.). 1996. Wetland Mitigation (special issue). Ecological Applications 6:33–101.

Zedler, J. B., and R. Langis. 1991. Comparison of created and natural salt marshes of San Diego Bay. Restoration and Management Notes 9:21–25.

5 Cypress Domes in North Florida

Invertebrate Ecology and Response to Human Disturbance

ANDREA J. LESLIE, JOSEPH P. PRENGER, and
THOMAS L. CRISMAN

P*ondcypress swamps (cypress domes, cypress ponds and cypress heads) are important features of the southeastern coastal plain of Florida and southern Georgia. Those in north Florida and Georgia are characterized by a canopy of pondcypress* (Taxodium distichum var. nutans), *deep organic matter, and unpredictable hydrological cycles, while some of those found in south Florida have shallower organic layers and more regular hydroperiods. These wetlands are often exploited as a source of cypress mulch during clearcutting of the pine plantations in which they are frequently found. Aquatic invertebrate communities of these swamps are highly variable throughout Florida. In north Florida, those examined were dominated by Amphipoda and Diptera, with the amphipod* Crangonyx *contributing 40 percent and midges* (Chironomidae) *34 percent of overall density. Authority: The Audubon Society Field Guide to North American Trees and Leslie, et al 1997. Qualitative data from south Florida swamps suggest that midges are less dominant than in north Florida. Pondcypress swamps can host a relatively large number of aquatic invertebrate taxa. A survey of three small swamps in north Florida noted 104 taxa over two and a half years. Dipterans and beetles were taxonomically rich, accounting for 36 and 32 percent of total taxa, respectively. Communities of pondcypress swamps can be highly variable; although a study of three swamps within a 1-km² area found consistently lower taxon richness in one, total density showed no predictable cycle even after two and a half years of monitoring. Total densities in swamps having no standing water for over one month were similar to those reported for wet sampling periods. Clear-*

Invertebrates in Freshwater Wetlands of North America: Ecology and Management, Edited by
Darold P. Batzer, Russell B. Rader, and Scott A. Wissinger
ISBN 0-471-29258-3   © 1999 John Wiley & Sons, Inc.

cutting of pondcypress swamps resulted in increased densities and obvious taxonomic shifts in macroinvertebrate communities, including a pronounced increase in select midge and odonate taxa.

CYPRESS DOME HABITATS

Forested wetlands are important features of the southeastern coastal plain and include diverse systems such as bottomland hardwoods, deepwater cypress and tupelo swamps, and pocosins. Although swamps have been reduced in area and number by drainage and filling, they still comprise 10 percent of Florida's land area and are widely distributed throughout the state (Wharton et al. 1982). Pondcypress swamps (also known as cypress domes, cypress ponds, and cypress heads) are small (<10 ha) basin wetlands found throughout the pine flatwoods of Florida and southern Georgia. These wetlands are characterized by a canopy of pondcypress (*Taxodium distichum* var. *nutans*), but often have considerable numbers of black gum (*Nyssa sylvatica* var. *biflora*), red maple (*Acer rubrum*), wax myrtle (*Myrica cerifera*), and slash pine (*Pinus elliotii*) (Monk and Brown 1965). Other hardwoods typical of cypress domes include swamp red bay (*Persea palustris*) and sweet bay (*Magnolia virginiana*) (Brown 1981). A striking characteristic of these wetlands is the dome-like shape resulting from shorter, young cypress trees at the periphery and larger, older trees near the center (Monk and Brown 1965). The understory of pondcypress swamps is quite variable, some being dominated by different emergents (e.g., *Carex* spp., *Woodwardia virginica, Lacnanthes*) and others having a preponderance of shrubby vegetation (e.g., *Lyonia lucida, Myrica cerifera*). *Sphagnum* mosses are prevalent in some domes, while floating macrophytes such as *Lemna, Nuphar and Nymphaea* spp. are present in the deeper swamps. In general there is low algal productivity due to shading.

Standing water is present in pondcypress swamps 6–12 months per year (Ewel and Smith 1992), but water levels fluctuate unpredictably (Fig. 5.1). These areas have low relief and poor drainage, are characterized by organic soils up to a meter in depth (Ewel and Wickenheiser 1988), and are underlain by sand, clay, and/or hardpan (Spangler 1984). Rainfall is the primary hydrologic input (Ewel and Smith 1992), although groundwater flow may be significant (Heimburg 1984, Crownover et al. 1995). Although evapotranspiration is the major hydrologic output (Ewel and Smith 1992), groundwater flow does occur (Crownover et al. 1995), and some overland flow may take place at high water. The water in north Florida systems has a low buffering capacity, with low alkalinity and concentrations of dissolved ions. A study of one north Florida pondcypress swamp noted surface water with a mean conductivity of 52 μmho/cm, a range of pH from 3.5 to 5.4 and low levels

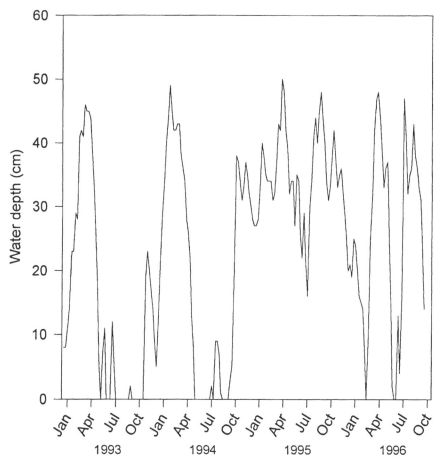

Fig. 5.1. Water depth, continuously recorded in a 0.6 ha cypress pond near Gainesville, FL (H. Riekirk and L. Korhnak, unpublished data).

of major cations (Dierberg and Brezonik 1984). In contrast, the water in some south Florida pondcypress swamps has neutral pH and high concentrations of calcium ions due to contact with the underlying limestone (Smock 1995). Low pH is partly due to humic acids produced within the swamps, which confer a distinctive color to the water. The high level of dissolved organic carbon (16–25 mg/l) contributes to the high biological oxygen demand (BOD) and low dissolved oxygen (DO mean 2.03 mg/l; Dierberg and Brezonik 1984). Nutrient concentrations are also low due to the relative hydrologic isolation of the swamps, with average phosphorus levels of 50–240 μg/l (Brown 1981, Mitsch 1984, Dierberg and Brezonik 1984).

    Even though these systems are very common, there has been little research on the aquatic invertebrate communities of pondcypress swamps, with the exception of Brightman's (1984) study of the invertebrate response to wastewater application. Most of these swamps were logged for their valuable cypress logs in the past, and are now often exploited as a source of cypress mulch during harvest of surrounding pine plantations. Since these systems undergo natural disturbance such as unpredictable drawdown, the role and ramifications of human-induced disturbances on ecosystem structure are important to understand.

INVERTEBRATE ABUNDANCE AND RICHNESS

Aquatic invertebrate communities of pondcypress swamps are highly variable throughout Florida; however, annual densities are similar to those of other southern forested wetlands, ranging from 5,690 individuals/m^2 in a Louisiana cypress-tupelo swamp to 442 individuals/m^2 in a Florida cypress slough (Sklar 1985, Haack 1984). In north Florida, where swamps are typically characterized by deep organic matter and unpredictable cycles of drying and rewetting, total annual densities of benthic invertebrates range from 1,127 individuals/m^2 in a deep swamp (Brightman 1984) to 5,320 individuals/m^2 in a shallow swamp (Prenger et al., manuscript in preparation). Although there has been no quantitative sampling of aquatic invertebrates in south Florida swamps (shallow organic matter), qualitative sampling suggests that annual densities may be lower (Smock 1995, D. Ceilley, personal communication).

In north Florida swamps Diptera and Amphipoda dominate total density. During two years of bimonthly benthic sampling in three small pondcypress swamps, amphipods of the genus *Crangonyx* accounted for 40 percent of total density (Prenger et al., manuscript in preparation). Midges (Chironomidae) and biting midges (Ceratopogonidae) were the most abundant dipterans, accounting for 34 percent and 8 percent of overall density, respectively. *Chironomus* and *Polypedilum* were the dominant midges in these swamps, with tanypods *Monopelopia* and *Larsia* abundant in some months. Brightman (1984) reported that midges were dominant in a deep pondcypress dome, with *Chironomus, Strictochironomus,* and *Tanypus* accounting for 77 percent of the total annual density. Midges (often *Chironomus* or *Polypedilum*), amphipods or isopods, and oligochaetes often account for a majority of individuals sampled in other southern swamps (e.g., Haack 1984, Sklar 1985, Camp et al. 1992, Duffy and LaBar 1994).

Qualitative data from south Florida swamps suggest that midges are less dominant than in more northerly locations. Dip net samples have contained beetles, aquatic bugs, amphipods (*Hyallela azteca*), oligochaetes, and dragonflies, but few midges (Smock 1995, D. Ceilley, personal communication). *Callibaetis,* a mayfly, can occur in great densities after flooding in the rainy season (D. Ceilley, personal communication). The dip net method used to sample these swamps may have been biased against sediment-associated invertebrates, however, because scraping the thin organic layer into the net is difficult.

Pondcypress swamps can host a relatively large number of aquatic invertebrate taxa. A benthic invertebrate survey of three small swamps in north Florida found 104 taxa over two and a half years (Prenger et al., manuscript in preparation). Dipterans and beetles were taxonomically rich, accounting for 36 and 32 percent of total taxa, respectively (Fig. 5.2a). As in other wetlands (e.g., Duffy and LaBar 1994, Rader et al. 1994), a relatively high number of midge taxa (16 genera) were sampled. Predaceous diving beetles (Dytiscidae) and water scavenger beetles (Hydrophilidae), often taxonomically rich in temporary water bodies (e.g., Nilsson 1986, Williams 1987), accounted for the majority of beetle taxa. Not surprisingly, there were few or no taxa of caddisflies (Trichoptera), mayflies (Ephemeroptera), or stoneflies (Plecoptera); those that were collected (*Caenis, Oecetis, Oxyethira*) are typically found in lentic waters and thus tolerant of low oxygen concentrations (Wiggins 1977, Merritt and Cummins 1984). The taxonomic richness of aquatic bugs (Hemiptera) and dragonflies and damselflies (Odonata) may have been underestimated because a small benthic corer was used for invertebrate collection. This sampling device did not efficiently sample the water column and likely allowed readily mobile invertebrates such as bugs and odonates to escape. There are no detailed year-round data available from pondcypress swamps of south Florida, so annual taxonomic richness cannot be estimated. Isolated sampling events, however, suggest that these systems may host fewer taxa than those of north Florida. Dip net samples of five swamps on one collection date yielded only 28 taxa (Smock 1995).

INVERTEBRATE ECOLOGY

Small pondcypress swamps in north Florida are characterized by unpredictable changes in invertebrate abundance. In two and a half years of monitoring of benthic invertebrate communities in north Florida pondcypress swamps, no predictable cycle of total density was noted (Fig. 5.2b; Prenger et al., manuscript in preparation). Peaks in total density often corresponded with irregular density peaks of predominant taxa (e.g., *Crangonyx, Chironomus*). Perhaps the lack of predictability in pondcypress swamps of north Florida is driven in part by variable patterns of drying and rewetting observed in these swamps (Fig. 5.1).

Response to Drawdown

Although many wetlands undergo dry periods, most aquatic invertebrate studies have sampled only during wet periods (e.g., White 1985, Duffy and LaBar 1994). In our research in north Florida pondcypress swamps, benthos were sampled in both wet and dry periods (Leslie et al. 1997). Total density of benthic invertebrates in swamps having no standing water for over one month was surprisingly similar to that reported for wet sampling periods. Ninety

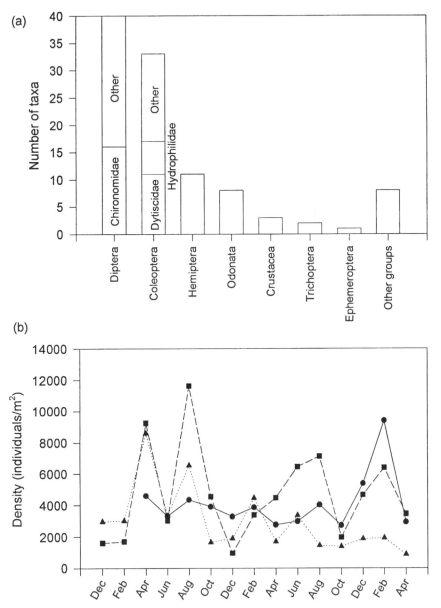

Fig. 5.2. Benthic macroinvertebrates of three cypress ponds in north Florida. (*a*) Taxon richness of major invertebrate groups over 2.5 years. (*b*) Total density in individual ponds over time.

percent of the individuals sampled belonged to taxa sampled during previous wet months; these taxa were able to survive drawdown *in situ,* either as larvae or by burrowing into moist sediments (McLachlan and Cantrell 1980, Wiggins et al. 1980, Williams 1987). This strategy would be beneficial to invertebrates that inhabit systems with unpredictable cycles of drying and rewetting.

A sizable proportion (33 percent) of the aquatic invertebrate taxa sampled in dry swamps were not sampled in wet periods. Terrestrial species were not considered in this study; however, many semiaquatic taxa were found, including the beetles *Hydraena, Stenus,* and Carabidae and larvae of the Dolichopodidae (Diptera). These taxa exclusive to the dry swamps contributed significantly to total taxon richness for the one-and-a-half-year period of the study. Sampling wetland invertebrate communities only during wet periods may lead to substantial underestimation of taxon richness.

Hydrological patterns can be significant factors in shaping aquatic invertebrate communities (Wiggins et al. 1980, Williams 1987). Pondcypress swamps with unpredictable regimes of drying and rewetting can host only invertebrate taxa with adaptations to drawdown such as burrowing or aestivating larvae, or those that can leave the system as adults before drawdown. As noted previously, aquatic invertebrate communities of these systems show no predictable yearly pattern of density. On the other hand, pondcypress swamps with a permanent central pool can host taxa that have no adaptations to withstand drawdown. Many pondcypress swamps of south Florida have more predictable cycles of drying and rewetting than those of north Florida and are therefore likely to be characterized by more predictable cycles of aquatic invertebrate abundance (D. Ceilley, personal communication).

Habitat Variation

Aquatic invertebrate communities of pondcypress swamps within a small geographic area can be highly variable. In a 17-month study of three swamps within a 1-km^2 study area in north Florida, one swamp was consistently lower in taxon richness (Leslie et al. 1997), and in particular, tanypod midges and biting midges were rarely sampled in this swamp. The most taxonomically rich invertebrate community in this study occurred in the wetland with the most diverse emergent vegetation. This is consistent with other studies which demonstrate that vegetation can influence both density and taxonomic composition of aquatic invertebrate communities by providing both a food source and substrate (e.g., Voigts 1976, Beckett and Aartila 1992).

Benthic invertebrate communities also display a high degree of intraswamp variability. Estimates of the amount of variance in benthic core samples contributed by different sources (sampling month, swamp, and core) were calculated for three swamps (Table 5.1; Leslie et al. 1997). Core-to-core variability was the largest source of variance for densities of all dominant invertebrates. High core-to-core variance has been noted for midges in Florida marshes as well (Streever and Portier 1994). Benthic invertebrate populations

TABLE 5.1. Variance among Transformed Macroinvertebrate Counts and Densities from Pondcypress Swamps of North Florida[a]

Taxon	Variance Source (core counts)				
	Pond	Time	Time × Pond	Core	Total
Total taxa	0.00	2.61	2.43	5.62	10.66
Cecidomyidae	0.00	0.01	0.13	0.50	0.64
Ceratopogonid larva 1	0.29	0.27	0.39	0.74	1.69
Chironomus	0.02	0.35	0.53	1.60	2.50
Crangonyx	0.72	3.90	2.18	5.91	12.71
Culicoides	0.01	0.02	0.04	0.38	0.45
Monopelopia	0.16	0.14	0.10	0.61	1.01
Polypedilum	0.49	0.46	0.21	2.69	3.85

[a]The $x^{1/2} + (x + 1)^{1/2}$ transformation, used for data sets with zero counts (Freeman and Tukey 1959), was applied. Variance estimates were determined from macroinvertebrate counts of 420 core samples (20 cores taken in three pondcypress swamps in each of seven months).

are often characterized by clumped distributions (Elliot 1971), which lead to high core-to-core variability. These studies underscore the importance of taking a large number of replicate samples within a wetland in order to obtain a good description of the benthic community.

Biotic Interactions

Vertebrate predators in pondcypress swamps are likely to be important in determining the structure of the aquatic invertebrate community. Invertebrate predators may also shape benthos community structure. Dragonflies (Aeshnidae, Libellulidae), damselflies (Coenagrionidae, Lestidae), predaceous beetles (Dytiscidae, Hydrophilidae, Noteridae), and aquatic bugs (Noteridae) are relatively common in both south and north Florida swamps (Leslie 1996, D. Ceilley, personal communication). Benthic samples of north Florida swamps often have high numbers of tanypod midges and biting midges as well (Brightman 1984, Leslie 1996).

Generalists dominate functional feeding guilds in pondcypress swamps of north Florida (Brightman 1984, Leslie 1996). This group can exploit a variety of food resources, including detritus, plants, epiphytic algae, and sometimes other animals. In addition, generalists may be able to withstand disturbance (e.g., drawdown) when food resources change (Pennak 1978, Cummins and Merritt 1984). Due to the shade provided by tree canopy and the high color in the water column, there is little algal production (Leslie 1996). Algal production is discussed in more detail below with regard to impacts of clearcutting.

Pondcypress swamps, like many wetlands, are detritus-based ecosystems, and most aquatic invertebrates in these systems likely feed on autochthonous detritus. Although abscised pondcypress needles are low in nitrogen and phosphorous (Dierberg et al. 1986), macrophytes and leaves of hardwoods such as *Nyssa* may provide detritus of better nutritional value (e.g., Day 1982). Bacteria and fungi condition detritus (Brinson 1977), making it more palatable for invertebrates, and serve as a high-quality food themselves (Lamberti and Moore 1984). The quality of detritus may be increased by drying and rewetting characteristic of pondcypress swamps. Detrital protein levels increase with fungal and bacterial colonization during drawdown (Barlocher et al. 1978) and could facilitate high invertebrate production upon reflooding.

HUMAN DISTURBANCE

Societal demands for resources are increasing the frequency and severity of disturbances to wetlands. Numerous human activities impact pondcypress swamps, including landscape scale activities such as the harvesting of timber, mining, and withdrawal of groundwater. Smaller, more isolated disturbances include application of wastewater to swamps and accidental events such as oil spills and leakage of brine from oil-production facilities. Very little work has been done on the effects of these activities on the benthic communities of swamps; however, several studies have shown considerable changes following disturbance (Brightman 1984, Pezeshki 1987, Leslie 1996).

Logging

Pondcypress swamps are often found in pine flatwoods and have become common features of pine plantations. These wetland areas are thus heavily influenced by logging and forestry maintenance practices. Swamps in these areas are impacted directly by logging for mulch and timber and indirectly by the removal of pine forests on the surrounding uplands. Logging activity in the wetlands not only removes and destroys the overstory, but also temporarily removes much of the understory and herbaceous vegetation. In addition, the heavy equipment used in modern logging greatly mixes and alters the soil stratification, and large amounts of coarse organic material are laid down on the soil surface.

The impacts of deforestation on sprouting and growth of pondcypress trees have been documented (Ewel et al. 1989). With the exception of bird populations (Workman 1996), changes in the fauna of these systems have not been previously evaluated. Many of the changes that we have observed in benthic invertebrate communities of logged pondcypress swamps appear to be responses to environmental changes. These changes include a transient increase in open water surface; increased light availability; and an increase in both coarse organic material (bark, etc.) and fine inorganic sediments (e.g., ruts caused by heavy machinery) in surface sediments. Alterations in the benthic macroinvertebrate community are likely to be secondary and tertiary to these changes.

Removal of the canopy results in increased ambient light, producing an expansion of periphyton and phytoplankton populations. Large blooms of filamentous green algae were noted in clearcut domes, predominantly of the genera *Microspora* and *Oedogonium* (Leslie 1996). Because of the small scale of this study, only data from a single sampling period had sufficient statistical power to compare clearcut and control cypress domes. This comparison showed a significant difference ($p < 0.05$). The increase in algae would provide high-quality food for algivores. Concomitantly, changes in soil organic content may result in alterations in the ratio of detritivores utilizing coarse and fine particulate matter (Leslie 1996). *Chironomus* and *Kiefferulus* prefer silty sediments (McLachlan 1969, Lindegaard-Petersen 1971, McLachlan and McLachlan 1975) like those found in ruts caused by logging equipment; both were present in high densities in the clearcut domes (Fig. 5.3, Prenger et al., manuscript in preparation). Similar changes in food quality and habitat in streams have been correlated with logging impacts (Newbold et al. 1980, Hawkins et al. 1982, Gurtz and Wallace 1984, Wallace and Gurtz 1986).

Clearcutting of the pondcypress swamps resulted in increased densities and obvious taxonomic shifts in the macroinvertebrate communities, including a pronounced increase for selected midge and odonate taxa (Fig. 5.3, Prenger et al., manuscript in preparation). Loss of canopy and temporary expansion of open water provide increased opportunity for colonization and utilization by groups such as odonates while reducing substrate and protection for others (Leslie 1996). Odonate assemblages were represented by taxa typical of uncanopied waters, including *Pachydiplax longipennis* and *Lestes disjunctus* (Leslie 1996). Chironomid communities were characterized by high densities and taxonomic richness, and many taxa were found exclusively in logged swamps. *Polypedilum* spp. dominated the chironomid fauna prior to harvest but declined in numbers afterwards. This decline was probably due to competition with *Chironomus, Procladius,* and *Kiefferulus,* which became dominant after cutting (Fig. 5.3, Prenger et al., manuscript in preparation). High densities of some taxa may be indicative of increased food quality and abundance, and increased substrate heterogeneity (e.g., elevated silt content in ruts caused by heavy machinery and coarse organic material from fallen trees) may be responsible for the higher taxonomic richness observed in harvested wetlands.

Other impacts resulting from logging may also affect benthic communities. The extended hydroperiod and increased average depth observed for clearcut swamps (approximately one growing season in duration, data not shown) would in turn alter oxygen levels in sediments and reduce benthic communities dependent on shallow water or dry periods. Forestry management practices such as fertilizer and herbicide application may affect invertebrates directly through toxicity or indirectly through increased productivity, changes in food quality, or oxygen stress.

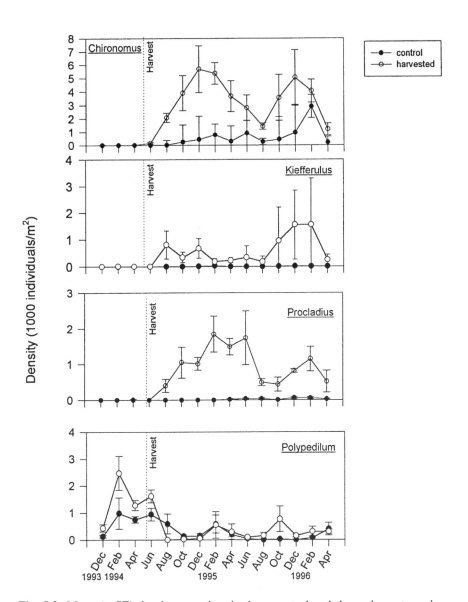

Fig. 5.3. Mean ($\pm$ SE) density over time in three control and three clearcut pondcy-press swamps in north Florida (Prenger et al., manuscript in preparation). All means were calculated for three ponds except December 1993, when only two control and two harvested ponds were sampled, and February 1994, when two control and three harvested ponds were sampled.

Withdrawal of Groundwater

In many areas of Florida severe depletion of groundwater is becoming an issue of critical importance. For example, well fields serving large urban areas result in a depressed water table and decreased hydroperiod in nearby wetlands (Rochow 1985, Corral and Thompson 1988), eventually eliminating invertebrates dependent on longer hydroperiods. An invertebrate community more closely resembling a terrestrial fauna may result from such human-induced events, as observed in natural drawdown (Leslie et al. 1997).

Wastewater Application

A number of studies in the 1970s and 1980s examined the possible use of pondcypress swamps for tertiary treatment of wastewater (Ewel and Odum 1984). The only published study on the effects of wastewater application on benthic invertebrate communities of these systems demonstrated that this addition resulted in decreased diversity and biomass (Brightman 1984). These changes were attributed to high biological oxygen demand and low dissolved oxygen.

Other Disturbances

While many other human activities affect pondcypress swamps, few have been studied. For example, unpublished work by T. Crisman examined the effects of oil and brine spillage from an oil-production facility in the Big Cypress Swamp of Florida. This study concluded that these systems are naturally oxygen-stressed and that traditional macroinvertebrate indices of water pollution are not applicable.

RESEARCH NEEDS

The only long-term quantitative studies of aquatic invertebrates in pondcypress systems have been in north Florida swamps. Research on other pondcypress-dominated systems would provide a better understanding of the role of biotic and abiotic factors in the establishment of community structure. For example, south Florida swamps characterized by shallow organic matter, deep central pools, neutral pH, and more regular hydroperiod should be quantitatively sampled to elucidate the role of geomorphology and hydrology. In order for us to understand the total aquatic invertebrate community of swamps and its role in the foodweb, quantitative measurements are needed of surface- and water column-dwelling invertebrates, fish, and amphibians in addition to those of benthos. The role of vertebrate predators such as fish in shaping pondcypress swamp invertebrate communities should also be examined.

The role of drawdown on benthic invertebrate communities should be further explored. As swamps dry, do migration to wet areas and increased competition for food and habitat occur? How severe does a drawdown have to be in order to wipe out the aquatic invertebrate community? Climate zones in Florida range from tropical in the Florida Keys to warm-temperate in the panhandle region. Comparisons of systems throughout this range are important, particularly with regard to temperature gradients and variations in precipitation.

LITERATURE CITED

Barlocher, F., R. J. MacKay, and G. B. Wiggins. 1978. Detritus processing in a temporary vernal pool in southern Ontario. Archiv für Hydrobiologie 81:269–295.

Beckett, D. C., and T. P. Aartila. 1992. Contrasts in density of benthic invertebrates between macrophyte beds and open littoral patches in Eau Galle Lake, Wisconsin. American Midland Naturalist 127:77–90.

Brightman, R. S. 1984. Benthic macroinvertebrate response to secondarily treated wastewater in north-central Florida cypress domes. Pages 186–196 *in* K. C. Ewel and H. T. Odum (eds.), Cypress Swamps. University Presses of Florida, Gainesville, FL.

Brinson, M. M. 1977. Decomposition and nutrient exchange of litter in an alluvial swamp forest. Ecology 58:601–609.

Brown, S. 1981. A comparison of the structure, primary productivity, and transpiration of cypress ecosystems in Florida. Ecological Monographs 51:403–427.

Camp, Dresser, and McKee, Inc. 1992. Experimental Wetland Exemption System Monitoring and Research Program. Third Annual Summary Report. Public Utilities Division, Orange County, FL.

Corral, M. A., and T. H. Thompson. 1988. Hydrology of the Citrus Park Quadrangle, Hillsborough County, Florida. Tallahassee, FL, Water-Resources Investigations Report 87-4166.

Crownover, S. H., N. B. Comerford, D. G. Neary, and J. Montgomery. 1995. Horizontal groundwater flow patterns through a cypress swamp-pine flatwoods landscape. Soil Science Society of America Journal 59:1199–1206.

Cummins, K. W., and R. W. Merritt. 1984. Ecology and distribution of aquatic insects. Pages 59–65 *in* R. W. Merritt and K. W. Cummins (eds.), An Introduction to the Aquatic Insects of North America. Kendall/Hunt, Dubuque, IA.

Day, F. P., Jr. 1982. Litter decomposition in the seasonally flooded Great Dismal Swamp. Ecology 63:670–678.

Diehl, S. 1992. Fish predation and benthic community structure: The role of omnivory and habitat complexity. Ecology 73:1646–1661.

Dierberg, F. E., and P. L. Brezonik. 1984. Water chemistry of a Florida cypress dome. *In* K. C. Ewel and H. T. Odum (eds.), Cypress Swamps. University Presses of Florida, Gainesville, FL.

Dierberg, F. E., P. A. Straub, and C. D. Hendry. 1986. Leaf-to-twig transfer conserves nitrogen and phosphorous in nutrient poor and enriched cypress swamps. Forest Science 32:900–913.

Duffy, W. G., and D. J. LaBar. 1994. Aquatic invertebrate production in southeastern USA wetlands during winter and spring. Wetlands 14:88–97.

Elliot, J. M. 1977. Some Methods for the Statistical Analysis of Samples of Benthic Invertebrates (2nd ed.). Freshwater Biological Association, Ambleside (The Librarian, The Ferry House, Ambleside, Cumbria, LA22 0LP).

Ewel, K. C. and H. T. Odum. 1984. Cypress Swamps. University Presses of Florida, Gainesville, FL.

Ewel, K. C., and L. P. Wickenheiser. 1988. Effect of swamp size on growth rates of cypress (*Taxodium distichum*) trees. American Midland Naturalist 120:362–370.

Ewel, K. C., H. T. Davis, and J. E. Smith. 1989. Recovery of Florida cypress swamps from clearcutting. Southern Journal of Applied Forestry 13:123–126.

Ewel, K. C., and J. E. Smith. 1992. Evapotranspiration from Florida pondcypress swamps. Water Resources Bull. 28:299–304.

Freeman, M. F., and J. W. Tukey. 1959. Transformations related to the angular and the square root. Annals Mathematical Statistics 21:607–611.

Gurtz, M. E., and J. B. Wallace. 1984. Substrate-mediated response of stream invertebrates to disturbance. Ecology 65:1556–1569.

Haack, S. K. 1984. Aquatic macroinvertebrate community structure in a forested wetland: Interrelationships with environmental parameters. Master's Thesis, University of Florida, Gainesville, FL.

Harris, L. D., and C. R. Vickers. 1984. Some faunal community characteristics of cypress ponds and the changes induced by perturbances. Pages 171–185 *in* K. C. Ewel and H. T. Odum (eds.), Cypress Swamps. University Presses of Florida, Gainesville, FL.

Hawkins, C. P., M. L. Murphy, and N. H. Anderson. 1982. Effects of canopy, substrate composition, and gradient on the structure of macroinvertebrate communities in Cascade Range streams of Oregon. Ecology 63:1840–1856.

Heimburg, K. 1984. Hydrology of north-central Florida cypress domes. *In* K. C. Ewel and H. T. Odum (eds.), Cypress Swamps. University Presses of Florida, Gainesville, FL.

Heuner, J. D., and J. A. Kadlec. 1992. Macroinvertebrate response to marsh management strategies in Utah. Wetlands 12:72–78.

Lamberti, G. A., and J. W. Moore. 1984. Aquatic insects as primary consumers. Pages 164–195 *in* V. H. Resh and D. M. Rosenberg (eds.), The Ecology of Aquatic Insects. Praeger, New York.

Leslie, A. J. 1996. Structure of benthic macroinvertebrate communities in natural and clearcut cypress ponds of north Florida. Master's Thesis, University of Florida, Gainesville, FL.

Leslie, A. J., T. L. Crisman, J. P. Prenger, and K. C. Ewel. 1997. Benthic macroinvertebrates of small Florida pondcypress domes and the influence of dry periods. Wetlands 17:447–455.

McLachlan, A. J., and M. A. Cantrell. 1980. Survival strategies in tropical rain pools. Oecologia 47:344–351.

Merritt, R. W., and K. W. Cummins. 1984. An Introduction to the Aquatic Insects of North America. Kendall/Hunt, Dubuque, IA.

Mitsch, W. J. 1984. Seasonal patterns of a cypress dome in Florida. Pages 25–33 *in* K. C. Ewel and H. T. Odum (eds.), Cypress Swamps. University Presses of Florida, Gainesville, FL.

Monk, C. D., and T. W. Brown. 1965. Ecological consideration of cypress heads in northcentral Florida. American Midland Naturalist 74:126–140.

Newbold, J. D., D. C. Erman, and K. B. Roby. 1980. Effects of logging on macroinvertebrates in streams with and without buffer strips. Canadian Journal of Fisheries and Aquatic Sciences 37:1076–1085.

Nilsson, A. N. 1986. Community structure in the Dytiscidae (Coleoptera) of a northern Swedish seasonal pond. Annales Zoologici Fennici 23:39–47.

Pezeshki, C. 1987. Response of benthic macroinvertebrates of a shrub swamp to discharge of treated wastewater. Master's Thesis, University of Florida, Gainesville, FL.

Prenger, J. P., A. J. Leslie, T. L. Crisman, and K. C. Ewel. Impacts of clearcutting on benthic macroinvertebrate communities of small Florida Pondcypress Swamps. (Manuscript in preparation.)

Rader, R. B., and C. J. Richardson. 1994. Response of macroinvertebrates and small fish to nutrient enrichment in the northern Everglades. Wetlands 14:134–146.

Rochow, T. F. 1985. Hydrologic and vegetational changes resulting from underground pumping at the Cypress Creek well field, Pasco County, Florida. Technical Report 1985-5, Southwest Florida Water Management District, Brooksville, FL.

Sklar, F. H. 1985. Seasonality and community structure of the backswamp invertebrates in a Louisiana cypress-tupelo wetland. Wetlands 5:69–86.

Smock, L. A. 1995. Characterization of macroinvertebrate communities in isolated wetlands of south Florida. Virginia Commonwealth University, Richmond, VA.

Spangler, D. P. 1984. Geologic variability among six cypress domes in north-central Florida. Pages 60–66 in K. C. Ewel and H. T. Odum (eds.), Cypress Swamps. University Presses of Florida, Gainesville, FL.

Streever, W. J., and K. M. Portier. 1994. A computer program to assist with sampling design in the comparison of natural and constructed wetlands. Wetlands 14:199–205.

Voigts, D. K. 1976. Aquatic invertebrate abundance in relation to changing marsh vegetation. American Midland Naturalist 113:56–68.

Wallace, J. B., and M. E. Gurtz. 1986. Response of *Baetis* mayflies (Ephemeroptera) to catchment logging. American Midland Naturalist 115:25–41.

Wharton, C. H., W. M. Kitchens, E. C. Pendleton, and T. W. Sipe. 1982. The Ecology of Bottomland Hardwood Swamps of the Southeast: A community profile. U.S. Fish and Wildlife Service, Biological Services Program FWS/OBS-81/37.

Wiggins, G. B. 1977. Larvae of the North American Caddisfly Genera (Trichoptera). University of Toronto Press, Toronto.

Wiggins, G. B., R. J. MacKay, and I. M. Smith. 1980. Evolutionary and ecological strategies of animals in annual temporary pools. Archiv für Hydrobiologie Supplement 58:97–206.

Williams, D. D. 1987. The Ecology of Temporary Waters. Croon Helm, London.

Workman, T. 1996. Bird communities in managed pondcypress swamps. Master's Thesis, University of Florida, Gainesville, FL.

6 The Okefenokee Swamp

Invertebrate Communities and Foodwebs

KAREN G. PORTER, AMY BERGSTEDT, and
MARY C. FREEMAN

*T*he Okefenokee Swamp is one of the largest wetlands in North America. It is the home of the largest population of the American alligator, a breeding site for the Sandhill crane, and a major stopover on the Atlantic flyway for songbirds and waterfowl. Eleven species listed as endangered or threatened reside in Okefenokee habitats. Despite the obvious importance of the Okefenokee, remarkably little is known about the structure and ecological function of its animal communities, especially the invertebrates. Most research on Okefenokee invertebrates has been as a part of process- and ecosystem-oriented studies, and only a handful of studies address the invertebrate fauna in any detail. Much of the information is available only from difficult-to-access dissertations, theses, reports, or unpublished manuscripts. This chapter collates and synthesizes information on Okefenokee invertebrates and discusses invertebrate roles in swamp foodwebs.

Distinct aquatic invertebrate communities are associated with open water, littoral/macrophyte marsh, and sedge prairie habitats. Algae and non-lignocellulose plant material are probably at the base of the foodweb for aquatic invertebrates. Plant detritus composed primarily of lignocellulose does not appear to be a major contributor, either directly or indirectly, through bacterial production, to the carbon budget of the aquatic foodwebs in the swamp. Breakdown of "refractory DOM" by UV light may enhance bacterial production sufficiently to make it significant as well. Particle-associated bacteria are collected more efficiently than free-living bacteria by filter-feeding microcrustaceans. Metazoans, such as crustacean microzooplanton, amphipods, isopods, and decapods, are considered to be the

Invertebrates in Freshwater Wetlands of North America: Ecology and Management, Edited by
Darold P. Batzer, Russell B. Rader, and Scott A. Wissinger
ISBN 0-471-29258-3   © 1999 John Wiley & Sons, Inc.

major consumers of algae, bacteria, and detritus, although the role of cili-
ate and flagellate protozoans remains to be investigated. Predatory aquatic
insects, small fishes, and amphibians top the aquatic food chain and can be
fed upon by spiders, birds, raccoons, and other nonaquatic vertebrates.

HABITAT CHARACTERISTICS

Geomorphology

The Okefenokee Swamp watershed (Fig. 6.1) covers over 3781 km^2 of the
lower Atlantic Coastal Plain in south Georgia and north Florida. It is a mosaic
of habitats, 1754 km^2 (46 percent) of which is swamp, making it one of the
largest freshwater wetlands in North America. The Okefenokee is bounded
on the east by Trail Ridge (Fig. 6.1), a Pleistocene sand dune, which rises to
47 m at sea level from the Coastal Plain. Water surfaces rise from 37 m
adjacent to Trail Ridge to 47 m on the northwest. Standing-water lakes, up
to 3 m in depth, occur throughout the swamp and are connected by maintained
canals. Although the hydrology of the eastern region of the swamp is not
well known, it is likely that Trail Ridge plays a critical role by obstructing
the flow of groundwaters east to the Atlantic Ocean. Due to the differential
elevation of the Ridge, the formation makes a contribution to the surface water
input to the eastern regions of the Okefenokee, where the greatest extent of
wetlands exists. The northwest side of the watershed is primarily forested
upland and rises to 56 m.

The gross hydrology of the swamp is known. There are at least five major
basins within the swamp (Loftin 1997). Water enters the swamp through precipitation (61 percent) and surface drainage (39 percent) (Blood 1980, Patten
and Matis 1984) and leaves through evapotranspiration (80 percent) and riv-
erflow and streamflow (20 percent) (Rykiel 1984). Major drainage from the
swamp occurs by sheetflow to the south (Patten and Matis 1984), with approximately 85 percent flowing southwest through the Suwannee River, 11
percent to the southeast through the St. Mary's River, and 4 percent through
Cypress Creek, which joins the Suwannee. The influences of groundwater
movements are largely unknown, and understanding local and regional linkages will be essential to the restoration and management of the Okefenokee.

Aerial photographs (Fig. 6.1) reveal complex arc-shaped sand and vegetation structures resembling oxbows and banks, presumably formed by ancient
drainage patterns when the surface of the swamp lay behind coastal dunes
(ca 10,000 B.P.). They may be similar in structure to wind-formed basins of
Carolina bays (see Taylor et al., this volume). The swamp's geological origins
began 200,000 years B.P., but existence as a peat-accumulating wetland began
only 6,500 years ago (Cohen et al. 1984).

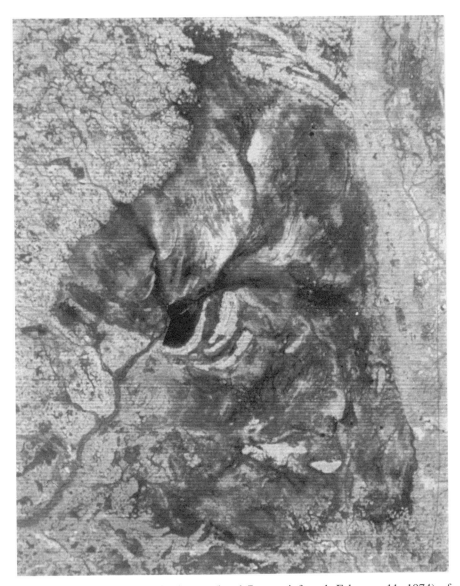

Fig. 6.1. LANDSTAT satellite image (band 7, near infrared, February 11, 1974) of the Okefenokee Swamp. Distinct features of the Okefenokee include Trail Ridge on the eastern border, the Suwannee River draining the southwest side (channel first evident below the Suwanee River Sill), and pine uplands on the northwest border that constitute the major watershed.

The Okefenokee is a mosaic of aquatic habitats including open-water lakes and channels, beds of water weeds, sedge prairies, and shrub and wooded tree islands. These communities are in a dynamic state of succession, or "disclimax," maintained by disturbance. Major forms of disturbance are associated with variation in hydroperiod: short-term seasonal cycles of drying and inundation and longer-term wet and dry cycles. Fires are an important part of long-term cycles and maintain the character of the swamp by periodically releasing nutrients and altering the topography by lowering the surface of the peat. Decomposition within the peat also results in gas-driven flotation and upwelling of "peat batteries" that become new substrate for plant succession.

Biogeochemistry

The Okefenokee is a southern blackwater swamp (Mitsch and Gosselink 1993), defined by its acidic, tea-colored water with an average pH of 3.5 to 3.9 (Blood 1980, Auble 1982). Swamp waters have 46–58 mg C/l of dissolved organic matter (DOM), with 63–78% of the dissolved carbon occurring as humic and fulvic acids originating from decomposing plant matter (Bano et al. 1997). Due to the immobilization of nutrients in the peat and to refractory DOM, the Okefenokee Swamp is poor in available nutrients (Flebbe 1980, Benner et al. 1985, 1986). As a result, plant and animal life in the swamp relies on abiotic processes to release nutrients, such as cycles of drying and inundation (Blood 1980, Duever and Riopelle 1984), fire (Rykiel 1984, Patten and Matis 1984, Institute of Natural Resources 1987), and UV-light-mediated decomposition of DOM (Bushaw et al. 1996). The nutrients released by these processes stimulate microbial and plant production. Decomposition of peat can also release toxins such as mercury that bioaccumulate in the food chain (Arnold-Hill et al. ms.). Nutrient regeneration in the swamp is rapid, as evidenced by the fact that the Okefenokee has high microbial biomass and production (Murray and Hodson 1984) and high fish production (Freeman and Freeman 1985).

BIOTA

Plant Communities

The Okefenokee Swamp has developed into a dynamic mosaic of modern habitats (Patten and Matis 1984). Today 46 percent of the 3,781 km² Okefenokee watershed is wetland and 54 percent is pine upland (Fig. 6.1). Hamilton (1982) indicates that the wetland habitats are more botanically diverse than the surrounding uplands. The following vegetative habitats occur: shrub swamps (34 percent of total wetland area), black gum forests (<6 percent, mainly in the northwest), bay forests (<6 percent), mixed cypress forests (23 percent), *Carex* sedge prairies (21 percent), tree islands (about 70 of them making up 12 percent of the area), and open-water lakes with aquatic macrophytes. The majority of the peat (80 percent) is formed from water lilies (*Nymphae odorata*) and cypress (*Taxodium*).

The successional climax community would be southern mixed hardwoods or pine, but it is rarely realized due to continuous natural and anthropogenic disturbance. The model for plant community succession is from open marsh to cypress, or from shrub swamp to broad-leafed evergreen or mixed hardwood forests (Hamilton 1982, Glasser 1986). Plant succession in the Okefenokee is routinely set back by such factors as historically frequent fires, the upwelling of peat batteries due to outgassing of methane from peat decomposition (King et al. 1981), and the influence of the fluctuating water table (Greening and Gerritsen 1987). In the early 1900s canals were dug and the swamp was logged of its cypress communities, further altering evapotranspiration, water flow, and community structure. These recurring disturbance regimes led to a heterogeneous and ever-changing "disclimax" ecosystem with a mosaic of habitats.

Invertebrate Communities

Open-water areas in the swamp support microscopic bacterial, algal, and zooplankton communities that live in association with the sediments, peat, and aquatic plants (Moran et al. 1988, Freeman et al. 1984, Schoenberg 1986). These particle-associated bacteria are more easily ingested by microzooplankton than are free-living bacteria (Schoenberg and Maccubbin 1985). Periphyton and phytoplankton production is highest in temporarily inundated areas (Schoenberg and Oliver 1988), due presumably to release of nutrients from oxidized peat. The overall diversity of some of the smaller, fragile algal and invertebrate species in the Okefenokee may be low relative to the typical biota for the southeastern United States (Cohen et al. 1984), perhaps because of harsh swamp conditions such as oligotrophy, low oxygen concentration, low pH, and flooded organic soil.

During periods of high water, algal-based foodwebs consisting mostly of aquatic microinvertebrates spread into the prairies and provide food for higher trophic levels, which include predatory insects, fishes, birds, turtles, and alligators (Laerm et al. 1980). Moran et al. (1988) determined through ingestion rate measurements and modeling that use of decomposing plant material by aquatic foodwebs was insignificant. The majority of foodweb carbon originates directly from the consumption of microbial production, algae, and living plants rather than detritus.

There is little available data describing abundances or species richness of Okefenokee invertebrates. Research conducted in the early 1980s (Freeman et al. 1984 and unpublished data) provides some quantitative data for selected macroinvertebrates in Little Cooter Prairie (LCP), an open-water site within the Okefenokee. A metal 1-m^2 drop trap was used to sample because that device provided a more accurate description of large invertebrate and fish communities (Fig. 6.2) and was more efficient than either seining or drop nets (Freeman et al. 1984). Animals and detritus were removed from drop traps by sweeping with a 16-mm mesh dip-net. Drop-trap samples were collected at LCP every six to eight weeks from November 1982 to September 1985. These data at least partially describe the seasonal and annual variation in abundances of large nektonic invertebrates in the Okefenokee.

Macroinvertebrate taxa collected from LCP (Table 6.1, Fig. 6.2) included nine orders of crustaceans and insects (although this does not represent a complete taxa list for Okefenokee wetland prairies). While *Pelocoris, Acentropus, Laccornis* (and other dytiscid beetles), *Hydrocanthus, Caecidotea,* and *Procambarus fallax* (Fig. 6.4) were most abundant in summer, overall invertebrate catches tabulated from July 1983 to August 1984 displayed no pronounced seasonal trends (Fig. 6.3). Collections were numerically dominated by amphipods, *Crangonyx* spp. (Fig. 6.3), and anisopteran odonates were also well represented (Fig. 6.5).

Analysis of seasonal and interannual variation for the most common odonate taxa revealed some differing patterns. Three libellulid species, *Erythemis simplicicollis, Pachydiplax longipennis,* and *Libellula axilena,* were almost continuously abundant from November 1982 through September 1985 (Fig.

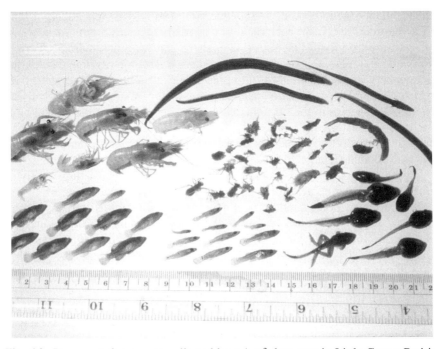

Fig. 6.2. Representative groups collected by a 1-m² drop trap in Little Cooter Prairie, Okefenokee Swamp.

**TABLE 6.1. Large Nektonic Invertebrates Collected with a 1-m²
Drop Trap from Little Cooter Prairie, Okefenokee**[a]

Crustacea	*Nannothemis*
AMPHIPODA	*Pachydiplax longipennis*
Crangonyx	*Tramea*
	Coenagrionidae
DECAPODA	*Ishnura*
Procambarus fallax	*Nehalennia gracilis*
Procambarus pygmaeus	Lestidae
	Lestes
ISOPODA	
Caecidotea	HEMIPTERA
	Belostomatidae
Insecta	*Lethocerus*
	Corixidae
ODONATA	Naucoridae
Aeshnidae	*Pelocoris*
Anax junius	
Coryphaeschna ingens	LEPIDOPTERA
Corduliidae	Pyralidae
Epicordulia	*Acentropus*
Tetragoneuria semiaquea	
Libellulidae	COLEOPTERA
Celithemis amanda	Dytiscidae
Celithemis elisa	*Cybister*
Celithemis fasciata	*Laccornis*
Celithemis ornata	Hydrophilidae
Celithemis verna	*Hydrobius*
Erythemis simplicicollis	Noteridae
Erythrodiplax minuscula	*Hydrocanthus*
Ladona deplanata	Curculionidae
Libellula spp.	
Libellula auripennis	DIPTERA
Libellula axilena	Tabanidae
Libellula semifasciata	*Tabanus*
Libellua vibrans	Chironomidae

[a]Only taxa collected on more than one occasion are listed.
Source: unpublished data from H. Greening and odonate data from M. Freeman,
see also Fig. 6.2.

6.5). Occasional declines in nymph abundances for these species were probably associated with emergence events, most often in spring. Nymphal populations of some other species, including *Erythrodiplax minuscula* and
Ishnura spp., fluctuated interannually, with greatest abundances during 1982
(Fig. 6.5). These patterns may indicate strong influences of hydrology or
community interactions on populations, and deserve further investigation. Be-

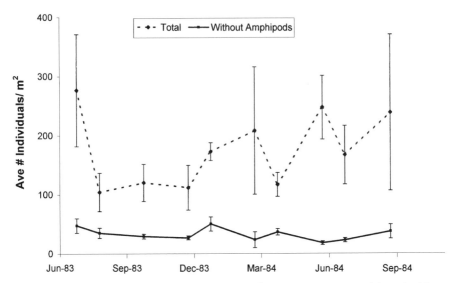

Fig. 6.3. Total numbers of invertebrates per 1-m² drop trap sample (with and without amphipods) in Little Cooter Prairie, Okefenokee Swamp.

cause they are abundant and speciose predators and appear sensitive to hydrologic variation, the odonates might be useful indicators of altered hydrology on community dynamics in the Okefenokee.

A lack of seasonality in invertebrate population change is in line with observations from other southern wetlands (e.g., cypress domes; see Leslie et

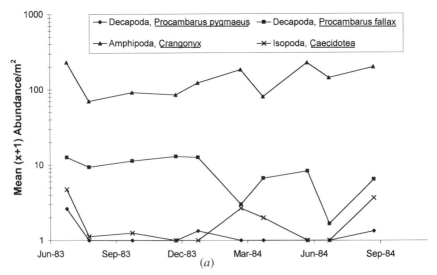

Fig. 6.4. (*a*), (*b*). Numbers of common crustaceans and nonodonate insects per 1-m² drop trap sample in Little Cooter Prairie, Okefenokee Swamp.

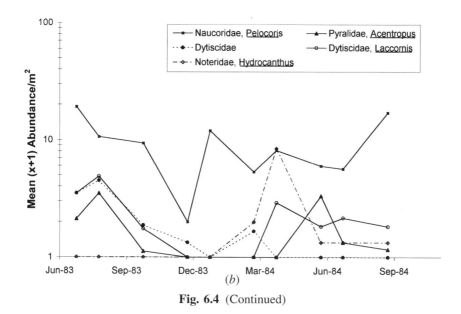

Fig. 6.4 (Continued)

al., this volume). Pronounced seasonal changes in the numbers of invertebrates may not occur because of the mild climate of the area. Those peaks and declines in average population numbers that do exist (Figs. 6.3–5) may be linked to fluctuations in hydroperiod.

One unique aquatic subcommunity is the *Utricularia*-periphyton association (Bosserman 1979). This self-contained system includes carnivorous plants commonly known as "bladderworts" and their associated microorganisms: algae, rotifers, crustaceans, insects, annelids, protists, platyhelminthes, and gastrotrichs. Tight nutrient recycling within this microecosystem allows it to become abundant in the nutrient-poor open waters of the Okefenokee.

Terrestrial insects play an important role in the cycling of nutrients in the swamp (Auble 1982, Gist and Risley 1982). Insects account for a projected 8 percent of total standing-crop consumption of *Carex* sedge in the prairies. From 12 to 40 percent of the forest canopy is consumed, primarily by noctuid Lepidoptera larvae. Feeding by insects converts vegetation to frass that can be readily recycled through various Okefenokee foodwebs (Auble 1982, Gist and Risley 1982, Hamilton 1982).

Vertebrates

Like the plants, vertebrates exhibit a higher diversity in wetland than upland habitats of the Okefenokee, presumably because of the greater habitat heterogeneity (Laerm and Freeman 1986). In a 1978–1980 survey, 420 different vertebrate species were recorded, although none were endemic. Thirty-six fishes, 37 amphibians, 66 reptiles, 233 birds, and 48 mammals were found. Eleven of these species are threatened or critically listed species, including

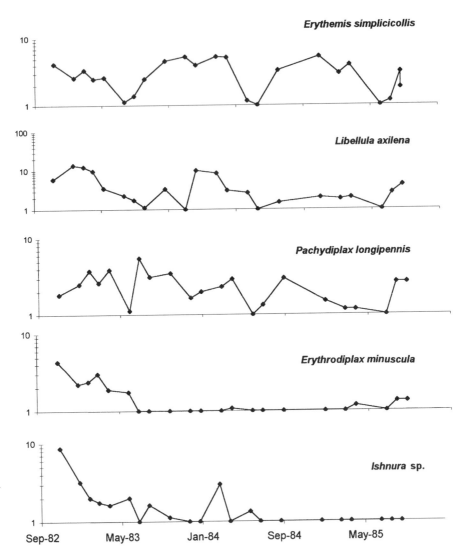

Fig. 6.5. Numbers of common odonate species per 1-m² drop trap sample in Little Cooter Prairie, Okefenokee Swamp.

the red wolf and the Florida panther (Laerm et al. 1984). Once found within this landscape, the Carolina parakeet and ivory-billed woodpecker are now extinct.

There is an especially high diversity of reptiles in the Okefenokee. The American alligator may be particularly important ecologically. Due to the softness of peat coupled with their feeding behavior, alligators create "gator holes" that eventually become sanctuaries for many aquatic animals, especially during dry periods. The significant impact of alligators on the landscape and structure of the Florida Everglades is well known (Mitsch and Gosselink 1993; see Rader, this volume), and their presence is correlated with physical modifications of habitat in the Okefenokee Swamp as well (Laerm et al. 1980).

Common fish species observed within the boundaries of the Okefenokee include sunfish and largemouth bass, topminnows, Florida gar, bowfin, pickerels, and the American eel (Laerm and Freeman 1986). Minnows have a low tolerance for high acidity and consequently are not observed within the swamp's open-water habitats. Fluctuations in temperature and depth of shallow waters are especially harsh for the fish populations. In the natural cycle of hydroperiod, conversion of peat from anaerobic to aerobic conditions produces a rapid release of nutrients. The reinundation of prairies and increased nutrient levels stimulate production of algae and zooplankton, the basis of the fish's foodweb. The body size of many swamp fishes (such as the pygmy sunfish) is small, and life cycles include stages dependent on isolated shallow-water environments. Smaller fish can reproduce rapidly to exploit newly flooded habitat (Freeman and Freeman 1985). However, sustained higher water levels allow larger predator species, such as turtles and alligators, to establish widespread populations in the swamp. This has a direct adverse effect on the usually dominant smaller prey species. Drought periods lower water levels and subject fish to the predatory effects of odonates, hemipterans, and coleopterans in the drying prairies and alligators and turtles in the subsiding lakes and canals.

While relatively few upland bird species occur in the Okefenokee (Meyers 1982), aquatic bird diversity is high. Sandhill cranes, herons, and storks are drawn to the Okefenokee National Wildlife Refuge. Aquatic birds also have a unique impact on nutrient cycling in the swamp, similar to that of insects but in some cases greater in magnitude. Nutrient-rich bird waste is concentrated at the site of bird colonies, or "rookeries." The concentrations of nutrients in the water surrounding an active rookery alter water chemistry and the distribution and production of plant life. The continued influence of guano even in abandoned rookeries is seen when peat-trapped nutrients are flushed out during high-water periods. Many of the bird species in the swamp are limited by the availability of suitable nesting habitat, which is at least as important as food resources to bird survival.

Human Influences and Management Issues

The Seminoles named the Okefenokee Swamp "land of trembling earth." They hunted game and sought refuge within its boundaries. Settlement in the mid-nineteenth century by "swampers" drove out the Native Americans. Since the early 1900s logging, canals, and the building of the Suwannee River Sill in 1960–1962 have significantly altered the face of the swamp (Trowell

1984, Cohen et al. 1984, Patten and Matis 1984, Malcolm et al. 1994). Today 80 percent of the 1754-km^2 swamp is protected as the Okefenokee National Wildlife Refuge, and the U.S. Fish and Wildlife Service manages water levels.

Although protected, the Okefenokee is far from pristine, having been logged since the late 1800s. Today water levels remain influenced by maintained channels and the Suwannee River Sill, which has stabilized water levels in the southwestern region of the swamp (Yin 1992). The importance of variability in hydroperiod to the dynamic nature of the swamp is evident near the sill, where reduced vegetational diversity and production occurs in stabilized areas. The Fish and Wildlife Service is currently conducting research to determine whether to repair or remove the sill.

Another possible influence on the swamp hydrology is the proposed mining of adjacent areas for titanium. In 1994 E.I. DuPont de Nemours & Co. began acquiring 36,000 acres along Trail Ridge, an ancient sand dune formation on the eastern boundary of the swamp, for the purpose of extending titanium mining operations currently underway in northeast Florida. The hydrology of ground and surface waters on and in close proximity to Trail Ridge remains poorly known. The impact of the mining operation might extend beyond Trail Ridge into the Okefenokee if groundwater, surface drainage, and hydrological cycles became altered. If hydrology changed, one would expect changes on the biota of the Okefenokee (Porter 1997; see discussion of the interrelationships of invertebrates and hydrology in Wissinger, this volume).

CONCLUSIONS

The community structure, nutrient cycling, and production dynamics of the Okefenokee Swamp are maintained in a dynamic disequilibrium. Water depth and cycles of flood and drought (hydroperiod) determine rates of nutrient cycling, productivity, and population growth. Anything that changes the hydrology can therefore have a major influence on the aquatic communities of the Okefenokee Swamp. Such examples of human influence as damming and canals have altered water levels and maximum and minimum fluctuations. Logging has changed the plant composition and evapotranspiration rates. Global climate change might additionally influence the Okefenokee if predicted temperature elevation and increased rainfall develop. More immediate problems can arise from altered water table and flow regimes. These may occur if the sill is altered, allowing increased drainage to the south, or if proposed titanium mining on Trail Ridge allows significant amounts of water to flow toward the east. Careful assessment of small-scale drainage patterns should be made and a water budget constructed that includes recent climate, vegetation, and evapotranspiration patterns. This can then be used to predict effects on the mosaic of aquatic communities in the Okefenokee.

ACKNOWLEDGMENTS

The authors thank Jeroen Gerritsen, Holly Greening, Lee Carrubba, Bud Freeman, Kevin Hiers, Jim Porter, and Bernard Patten for their help and comments.

LITERATURE CITED

Arnold-Hill, B., C. H. Jagoe, and P. V. Winger. Ms. Biomagnification of Hg in the Okefenokee Swamp.

Auble, G. T. 1982. Biogeochemistry of Okefenokee Swamp: Litterfall, litter decomposition, and surface water dissolved cation concentrations. Ph.D. Thesis, University of Georgia, Athens, GA.

Bano, N., M. A. Moran, and R. E. Hodson. 1997. Bacterial utilization of dissolved humic substances from a freshwater swamp. Aquatic Microbial Ecology 12:233–238.

Benner, R., M. A. Moran, and R. E. Hodson. 1985. Effects of pH and plant source on lignocellulose biodegradation rates in two wetland ecosystems, the Okefenokee Swamp and a Georgia salt marsh. Limnology and Oceanography 30:489–499.

———. 1986. Biogeochemical cycling of lignocellulose carbon in marine and freshwater ecosystems: Relative contribution of procaryotes and eucaryotes. Limnology and Oceanography 31:89–100.

Blood, E. R. 1980. Surface water hydrology and biogeochemistry of the Okefenokee Swamp watershed. Ph.D. Thesis, University of Georgia, Athens, GA.

Bosserman, R. W. 1979. The hierarchial integrity of *Utricularia*-periphyton microecosystems. Ph.D. Thesis, University of Georgia, Athens, GA.

Bushaw, K. L., R. G. Zepp, M. A. Tarr, D. Schulz-Jander, R. A. Bourbonniere, R. E. Hodson, W. L. Miller, D. A. Bronk, and M. A. Moran. 1996. Photochemical release of biologically available nitrogen from aquatic dissolved organic matter. Nature 381:404–407.

Cohen, A. D., M. J. Andrejko, W. Spackman, and D. Corvinus. 1984. Peat deposits of the Okefenokee Swamp. Pages 493-553 *in* A. D. Cohen, D. J. Casagrande, M. J. Andrejko, and G. R. Best, (eds.), The Okefenokee Swamp: Its Natural History, Geology, and Geochemistry. Wetland Surveys, Los Alamos, NM.

Duever, M. J., and L. A. Riopelle. 1984. Tree-ring analysis in the Okefenokee Swamp. Pages 180–188 *in* A. D. Cohen, D. J. Casagrande, M. J. Andrejko, and G. R. Best, (eds.), The Okefenokee Swamp: Its Natural History, Geology, and Geochemistry. Wetland Surveys, Los Alamos, NM.

Flebbe, P. A., 1982. Biogeochemistry of carbon, nitrogen and phosphorus in the aquatic subsystems of the Okefenokee Swamp sites. Ph.D. Thesis. University of Georgia, Athens, GA.

Freeman, B. J., H. S. Greening, and J. D. Oliver, 1984. Comparison of three methods for sampling fishes and macroinvertebrates in a vegetated freshwater wetland. Journal of Freshwater Ecology 2:603–609.

Freeman, B. J., and M. C. Freeman, 1985. Production of fishes in a subtropical black-water ecosystem: The Okefenokee Swamp. Limnology and Oceanography 30:686–692.

Gist, C. S., and R. A. Risley, 1982. Role of insects in the nutrient dynamics of the Okefenokee Swamp. Technical Report. Oak Ridge Associated Universities, Oak Ridge, TN.

Glasser, J. E. 1986. Pattern, diversity and succession of vegetation in chase prairie, Okefenokee Swamp: A hierarchial study. Ph.D. Thesis, University of Georgia, Athens, GA.

Greening, H. S., and J. Gerritsen, 1987. Changes in macrophyte community structure following drought in the Okefenokee Swamp, Georgia, USA. Aquatic Botany 28: 113–128.

Hamilton, D. B. 1982. Plant succession and the influence of disturbance in the Oke-fenokee Swamp, Georgia. Ph.D. Thesis, University of Georgia, Athens, GA.

Institute of Natural Resources. 1987. Modern History of Fires in the Okefenokee for the Role of Fires in Maintaining the Natural Conditions of the Okefenokee Swamp. Progress Report No. 2. For the Okefenokee Wildlife Refuge United States Fish and Wildlife Service, Institute of Natural Resources, University of Georgia, Athens, GA.

King, G. M., T. Berman, and W. J. Wiebe, 1981. Methane formation in the acidic peats of Okefenokee Swamp, Georgia. American Midland Naturalist 105:386–389.

Laerm, J., and B. J. Freeman, 1986. Fishes of the Okefenokee Swamp. University of Georgia Press, Athens, GA.

Laerm, J., B. J. Freeman, L. J. Vitt, J. M. Meyers, and L. Logan. 1980. Vertebrates of the Okefenokee Swamp. Brimleyana 47:47–73.

Loftin, C. 1997. Okefenokee swamp hydrology. *In* Kathyryn J. Hatcher (ed.), Pro-ceedings of the 1997 Georgia Water Resources Conference. Institute of Ecology, The University of Georgia, Athens, GA.

Malcolm, R. L., D. M. McKnight, and R. C. Averett, 1994. History and description of the Okefenokee Swamp-Origins of the Suwannee River. Pages 1–12 *in* R. C. Averett, J. A. Leenheer, D. M. McKnight, and K. A. Thorn (eds.), Humic Sub-stances in the Suwannee River, Georgia: Interactions, Properties, and Proposed Structures. United States Government Printing Office, Washington, DC.

Meyers, J. M. 1982. Community structure and habitat associations of breeding birds in the Okefenokee Swamp. Ph.D. Thesis, University of Georgia, Athens, GA.

Mitsch, W. J., and J. G. Gosselink, 1993. Wetlands (2nd ed.). Van Nostrand Reinhold, New York.

Moran, M. A., T. Legovic, R. Benner, and R. E. Hodson. 1988. Carbon flow from lignocellulose: A simulation analysis of a detritus-based ecosystem. Ecology 69: 1525–1536.

Moran, M. A., A. E. Maccubbin, R. Benner and R. E. Hodson. 1987. Dynamics of microbial biomass and activity in five habitats of the Okefenokee Swamp ecosys-tem. Microbial Ecology 14:203–217.

Murray, R. E., and R. E. Hodson. 1984. Microbial biomass and utilization of dissolved organic matter in the Okefenokee Swamp ecosystem. Applied Environmental Mi-crobiology 47:685–692.

Patten, B. C., and J. H. Matis. 1984. The macrohydrology of Okefenokee Swamp. Pages 246–263 *in* A. D. Cohen, D. J. Casagrande, M. J. Andrejko, and G. R. Best, (eds.), The Okefenokee Swamp: Its Natural History, Geology, and Geochemistry. Wetland Surveys, Los Alamos, NM.

Porter, K. G. 1997. The Okefenokee Swamp: The waters run deep. Panorama 27:1–4.

Rykiel, E. J., Jr. 1984. General hydrology and mineral budget for Okefenokee Swamp: Ecological significance. Pages 212–228 *in* A. D. Cohen, D. J. Casagrande, M. J. Andrejko, and G. R. Best, (eds.), The Okefenokee Swamp: Its Natural History, Geology, and Geochemistry. Wetland Surveys, Los Alamos, NM.

Schoenberg, S. A. 1986. Field and laboratory investigations of the interactions between zooplankton and microbial resources in the Okefenokee Swamp. Ph.D. Thesis, University of Georgia, Athens, GA.

Schoenberg, S. A., and A. E. Maccubbin. 1985. Relative feeding rates on free and particle-bound bacteria by freshwater macrozooplankton. Limnology and Oceanography 30:1084–1090.

Schoenberg, S. A., and J. D. Oliver. 1988. Temporal dynamics and spatial variation of algae in relation to hydrology and sediment characteristics in the Okefenokee Swamp, Georgia. Hydrobiologia 162:123–133.

Trowell, C. T. 1984. Indians in the Okefenokee Swamp. Pages 38–57 *in* A. D. Cohen, D. J. Casagrande, M. J. Andrejko, and G. R. Best, (eds.), The Okefenokee Swamp: Its Natural History, Geology, and Geochemistry. Wetland Surveys, Los Alamos, NM.

Yin, Z. 1992. The impact of the Suwannee River Sill on the surface hydrology of Okefenokee Swamp, USA. Journal of Hydrology 136:193–217.

7 Riverine Floodplain Forests of the Southeastern United States

Invertebrates in an Aquatic-Terrestrial Ecotone

LEONARD A. SMOCK

*F*loodplain forests occurring along streams and rivers are the predominant freshwater wetland in the southeastern United States. They are characterized by a seasonally inundated hydrologic regime that results in the floodplains being an ecotone along a spatial and temporal continuum from continually flooded to terrestrial habitat. These floodplains support a wide variety of macroinvertebrate species whose abundance, production, and life history and behavioral traits are closely tied to the seasonal cycle of flooding and drying. This chapter first provides an overview of the geomorphology, hydrology, and plant community structure of the floodplains. It then follows with discussions of macroinvertebrate community composition and abundance, spatial and temporal distribution patterns, adaptations for life on floodplains, and trophic relationships. The continuing focus throughout the chapter is on how the physical environment, especially hydrology, shapes community structure, function, and interactions with other organisms and how both the aquatic and terrestrial communities of the floodplains relate to population-, community-, and ecosystem-level processes.

RIVERINE FLOODPLAIN FORESTS

Riverine floodplain forests bordering low-gradient streams and rivers are the predominant freshwater wetlands in the southeastern United States. They

Invertebrates in Freshwater Wetlands of North America: Ecology and Management, Edited by
Darold P. Batzer, Russell B. Rader, and Scott A. Wissinger
ISBN 0-471-29258-3 © 1999 John Wiley & Sons, Inc.

range from tens of meters to kilometers in width and harbor extensive stands of highly productive bottomland hardwood forests with distinct aquatic and semiaquatic faunal assemblages. The defining characteristic and driving force of these systems is their hydrologic regime. Many of the floodplains are inundated for extended periods, often including permanent swamp habitat. Others range from seasonal inundation, typically from late autumn through spring, to areas that are inundated only briefly during periods of overbank flooding. The floodplain forests considered in this chapter thus exist as a broad ecotone along a spatial and temporal continuum from continually flooded areas to terrestrial systems (Figs. 7.1 and 7.2).

A considerable body of information exists concerning the general ecology of riverine floodplain forests; excellent reviews can be found in Wharton et al. (1981, 1982), Haack et al. (1989), Brinson (1990), and Sharitz and Mitsch (1993). Early studies of spatial heterogeneity of plant species distribution were followed by many investigations of the factors influencing the distribution patterns of vegetation, plant adaptations, and primary production. Also, numerous studies have been published on nutrient cycling and energy flow in southeastern floodplains, focusing in particular on exchanges between terrestrial and aquatic systems (e.g., Yarbro 1983, Walbridge and Lockaby 1994). Among the fauna, much attention has focused on the species composition and ecology of fish and terrestrial animals. Studies that have investigated aquatic macroinvertebrate communities inhabiting floodplains suggest that aquatic life can be diverse and productive, function as an important contributor to nutrient

Fig. 7.1. The inundated floodplain at Colliers Creek, Virginia.

Fig. 7.2. The floodplain and channel at Colliers Creek, Virginia, during the dry period.

and organic matter cycling, and be a potentially important link in the trophic structure of these systems.

This chapter describes the structure and function of macroinvertebrate communities of southeastern United States bottomland hardwood floodplains and the interrelationship between those communities and environmental factors. The chapter begins with a general description of the geomorphology, hydrology, and plant community structure of the floodplains since an understanding of those factors is crucial for developing insights into the factors controlling macroinvertebrate communities. The main body of the chapter discusses macroinvertebrate community composition and abundance, spatial and temporal distribution patterns, adaptations for life on floodplains, and trophic relationships. Throughout the chapter the focus is on how the physical environment, especially hydrology, shapes macroinvertebrate community structure, function, and interactions with other organisms.

River channels and their floodplains do not exist in isolation from each other, but rather are highly integrated components of river systems. Floodplains add an important lateral dimension to many streams and rivers (Ward 1989a). An understanding of the ecology of both channels and floodplains cannot develop unless both are considered as interconnected systems. While this chapter discusses the impact of this connectivity on the structure and function of the macroinvertebrate communities of the floodplains, it does not delve into the ecology of the river channels. Comprehensive reviews of the ecology of the channels of lowland streams and rivers in the southeastern United States, including discussions of their macroinvertebrate communities, can be found in Garman and Nielsen (1992), Felley (1992), Livingston (1992), and Smock and Gilinsky (1992). In addition, Merritt et al. (this volume) discuss macroinvertebrate communities in the Kissimmee River's nonforested marsh floodplain in Florida.

GEOMORPHOLOGY, HYDROLOGY, AND PLANT COMMUNITY STRUCTURE

Riverine floodplains occur along lowland streams and rivers flowing throughout the southeastern United States, primarily on the Coastal Plain physiographic province, but also extending farther inland along some Piedmont waterways. They are most prevalent as broad, flat floodplains along meandering, large rivers, but also occur along headwater streams, their size being positively correlated with watershed size (Bedinger 1981). The general extent of the historical distribution of these wetlands in the southeastern United States is shown in Fig. 7.3. The actual extent of this habitat, however, is less than that shown in the figure because of continual loss to clearing for timbering and agriculture, various water regulation projects, and other incursions converting these wetlands to other habitats and uses.

Floodplains are a spatial mosaic of erosional and depositional areas extending laterally from the channel to terrestrial habitat. Their geomorphic

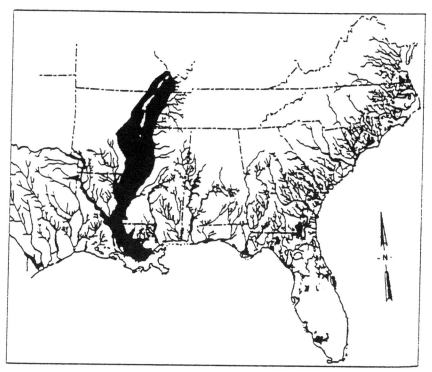

Fig. 7.3. Potential extent of bottomland hardwood forests of the southeastern United States. (After W. J. Mitsch and J. G. Gosselink, Wetlands, Van Nostrand Reinhold, New York © 1993. Reprinted by permission of John Wiley & Sons, Inc.)

structure is determined primarily by the energy of flowing water from the channel and the sediment load of that water (Décamps 1996). There usually is little topographic relief, but even small differences in elevation have important hydrological and ecological implications. Side-channels typically run through the floodplains, some draining directly from higher-elevation areas, others resulting from scouring during high flow, and still others being active or remnant parts of the main channel. These side-channels often form a highly braided and anastomosing network of waterways transporting water, nutrients, and organisms through the floodplains. Isolated depressions of various sizes also occur, ranging from root wad holes from fallen trees to ponds and sloughs on larger floodplains. Along with the natural grading of the floodplain from the main channel to higher elevation, these different areas provide the geomorphic framework for an often highly spatially and temporally heterogeneous hydrologic regime.

Floodplain hydroperiod is characterized by the timing, amplitude, duration, frequency, and predictability of the "flood pulse" each year (Junk et al. 1989), all of which are important factors affecting macroinvertebrate communities. The timing of inundation is variable, but usually occurs from late autumn through spring or early summer. Flooding typically begins with the autumnal decrease of evapotranspiration and the recharge of groundwater; it ceases when evapotranspiration lowers the water table below the floodplain surface. Other floodplains, especially those occurring along smaller streams, are flooded irregularly only during storms. The duration of inundation thus varies from long, continuous periods to short, storm-driven periods; the frequency, duration, and predictability of annual flooding each are greater with increasing stream order and hence river discharge (Bedinger 1981).

Along with temporal inundation patterns, a spatial mosaic of inundation exists. Whereas some floodplains are almost entirely inundated, resulting in a completely interconnected aquatic habitat, others have water in only the lowest areas. In particular, areas such as side-channels and pools often remain wet when much of the remaining floodplain is dry, forming persistent aquatic habitats surrounded by terrestrial habitat. Those aquatic habitats are hydrologically linked only during periodic floods and often provide the only refugia for aquatic invertebrates on the floodplain during the summer.

A continuum of lotic to lentic conditions also provides a variety of habitats for aquatic macroinvertebrates. Water velocity generally decreases away from the main channel as frictional drag from vegetation and the soil slows water movement. Erosive power of the water on floodplains is considerably less than in the main channel, resulting in low shear stress at the soil surface. Floodplains fringing smaller streams and rivers tend to be more lotic in their flow regime, whereas larger rivers often have extensive areas that are highly lentic and frequently isolated from other aquatic habitats, especially during the drier months.

Besides controlling the hydrologic regime, flooding also impacts other characteristics of floodplains that affect invertebrates. Soils, which typically are dense silts and clays with a high organic matter content, quickly become anoxic once waterlogged. Anoxia can extend upward throughout much of the water column in areas with low flushing and reaeration. Nutrients sequestered in aerated soils are more readily dissolved during flooding, and flood waters from channels provide a well-known nutrient subsidy to floodplains, together increasing nutrient availability to primary producers and decomposers (see Ward 1989b).

Primary producers provide structure and energy that have an important influence on the composition and function of macroinvertebrate communities. Tree species richness is high, and the forests are highly productive. It has been generally accepted that seasonally inundated floodplains have higher rates of primary production than do terrestrial or permanently flooded areas, although Megonigal et al. (1997) have challenged that hypothesis. The vegetation shows a zonation pattern determined mainly by hydroperiod and the resulting anoxia (Larsen et al. 1981, Wharton et al. 1982). The wettest areas, with near-continuous flooding, are least diverse and often dominated by bald cypress (*Taxodium distichum*) and water tupelo (*Nyssa aquatica*), species highly adapted for existence in flooded areas. Diversity of vegetation increases and species composition grades to that of a mixed hardwood forest as conditions become drier (Sharitz and Mitsch 1993).

The high productivity of the forests provides an important source of detritus, primarily as leaves and wood, to the floodplain floor. Input and storage of this material follow the normal seasonal pattern for temperate forests, with storage peaking in late autumn and early winter (Smock 1990). Processing of detritus by leaching, microbes, and invertebrates occurs under both terrestrial and aquatic conditions. The duration of inundation and dissolved oxygen concentrations are the primary factors determining processing rates (Brinson et al. 1981). Rates are slowest under permanently inundated, anoxic conditions and fastest under continually wet, aerobic conditions. Low pH, which often occurs on floodplains because of the leaching of organic acids, also slows processing rates.

Exchange of detritus between floodplains and stream channels occurs in both directions; the floodplains in particular provide an important energy subsidy to the channels. Retention of detritus within the floodplain, however, is high because of the low transport power of water flow and the trapping effect of the vegetation (Jones and Smock 1991). As a result, much of the particulate organic matter enters decomposer and consumer foodwebs on the floodplain.

The abundance of aquatic primary producers on floodplains is highly variable both spatially and temporally. Algal growth generally is limited by low light levels beneath the tree canopy, but algae can become abundant during the late winter and spring as light levels increase before leaf-out. Filamentous green algae can carpet shallow inundated areas at that time, but the algae are highly susceptible to being flushed from the system or being buried by silt deposited during storms. Local areas of high algal growth occur during the summer where gaps in the canopy due to wind-throw and other factors allow light to penetrate to inundated areas.

Macrophytes are common locally in open areas with sufficient sunlight and permanent inundation. They can be especially abundant in ponds and sloughs on larger floodplains, where full sunlight, low flushing, and long periods of inundation support high productivity of both macrophytes and algae. Duckweed (*Lemna*) can form extensive mats on the surface of the water, particularly in deep-water cypress and tupelo swamps. Together, algae and macrophytes provide additional habitat and food resources for invertebrates.

MACROINVERTEBRATE SPECIES COMPOSITION AND DISTRIBUTION

The forested floodplains of the southeastern United States support a wide variety of macroinvertebrate species. Studies show high species richness, diversity, density, and production, depending on environmental characteristics of the system (e.g., Parsons and Wharton 1978, Wharton et al. 1981, 1982, Sklar 1985, Gladden and Smock 1990, Duffy and LaBar 1994). Species composition varies geographically over the southeast, but locally is dependent primarily on the duration of flooding, the presence of lotic and lentic habitats, and the dominant vegetation within the floodplain (Parsons and Wharton 1978, Wharton et al. 1981, 1982, Batema et al. 1985, Haack et al. 1989). Floodplains often have many species in common with their river channels, depending primarily on the extent of their hydrologic connectivity and particularly on the extent to which the two areas are directly connected by water flow paths.

Species richness in southeastern floodplains is related to habitat heterogeneity and the spatial scale and perspective used to define system boundaries. Species richness in specific habitats on floodplains tends to be lower than in comparable habitats in neighboring streams and ponds, limited by the intolerance of many species to the cycle of flooding and drying. Also, high organic matter loading and low flushing in swamps often induce anaerobic conditions that limit species richness. If a broader spatial and temporal perspective is considered, however, floodplains often have a greater species richness than other aquatic environments. The mosaic of lotic and lentic habitats on floodplains harbors a wide variety of species. When these habitats are considered together, and both the aquatic and terrestrial species that occur annually are included, floodplains can be seen as species-rich systems. Also, present theory holds that river systems should have the highest species diversity in mid-order sections (River Continuum Concept, Vannote et al. 1980). However, if river systems are defined to include their floodplains, the lower sections of rivers, with broad floodplains harboring a variety of species, may have diversity comparable to or greater than that in upstream river sections constrained to the channel.

Wharton et al. (1981, 1982) identified six generalized habitat zones, based on inundation pattern and the resulting environmental characteristics, ranging from permanent bodies of water to nearly permanent terrestrial areas with a low probability and duration of flooding. They used these zones to define distinct assemblages of macroinvertebrates on floodplains, but in reality there is a continuum of inundation that, along with high dispersal rates and reproductive potential of many species, makes attempts at characterizing the macroinvertebrate fauna according to inundation zones difficult.

The most ubiquitous species on southeastern floodplains are those capable of withstanding both wet and dry years. Noninsect species typically are predominant, especially Copepoda, Cladocera, Ostracoda, Oligochaeta, Isopoda, Amphipoda, and sphaeriid clams. Among the insects, Chironomidae are abundant throughout floodplains; other insect taxa can be locally and seasonally common. In general, the number of taxa and the abundance of insects increase as conditions become more lentic.

Crustaceans often comprise a significant portion of the numbers, biomass, and secondary production on floodplains. Copepods and cladocerans, the latter primarily in more lentic areas, are ubiquitous and often abundant. Some of the more frequently encountered copepod genera include the cyclopoid *Acanthocyclops,* the harpacticoids *Attheyella* and *Bryocamptus,* and the calanoids *Diaptomus* and *Osphranticum.* Common cladocerans include species in the genera *Alona, Bosmina, Ceriodaphnia, Chydorus, Daphnia, Eurycercus,* and *Leydigia.* Bryan et al. (1975, 1976) present an overview of the zooplankton occurring in southeastern swamps, but in general few studies have specifically addressed their population ecology. The Ostracoda occurring on floodplains can be abundant but are poorly studied; species of *Candona, Cypridopsis, Herpetocypris,* and *Potamocypris* have been reported.

Isopods and amphipods are among the most conspicuous macroinvertebrate taxa inhabiting floodplains. Species of the isopod genera *Caecidotea* and *Lirceus* inhabit some floodplains together, whereas in others only one genus occurs to the exclusion of the other. Similarly, although a variety of species of *Caecidotea* occur in floodplains throughout the southeast, including *C. communis, C. laticaudatus, C. obtusus,* and *C. racovitzai,* the cooccurrence of two or more species has been rarely reported. The mechanisms segregating genera and species have not been studied.

Amphipods typically are represented by various species of *Gammarus* and *Crangonyx,* especially the *C. obliquus* complex and *C. pseudogracilis. Gammarus tigrinus,* a highly salt-tolerant species, is frequently found in areas fringing the mouths of streams and rivers near estuaries. *Hyalella azteca* can be abundant in open ponds and sloughs on floodplains. Several undescribed species of amphipods with highly reduced eyes and pale pigmentation were reported from groundwater within the soil of a North Carolina floodplain (Sniffen 1981). Because of a near-complete lack of sampling of groundwater habitats, the extent of occurrence in southeastern floodplains of subterranean species and those with adaptations for a hypogean life is unknown.

Many species of crayfish, primarily of the genera *Procambarus* and *Cambarus,* are important components of swamp macrofauna. Wharton et al. (1981) provide a list of some of the more commonly occurring species in the southeast. Their abundance fluctuates with a variety of factors, including multiyear patterns in the extent of flooding and drought. Floodplains along two streams in Virginia supported almost no crayfish during a four-year period when unusually dry summers and autumns led to long periods with dry soil and groundwater levels far below the surface (L. A. Smock, unpublished data). Recolonization of the floodplains by crayfish was slow following the return of wetter, more normal hydrologic conditions, taking about four years to achieve a stable density.

Gastropods are occasionally important in areas where macrophyte and periphyton growth is heavy, but usually they constitute only a minor portion of the biota on floodplains. However, bivalves, particularly species of fingernail clams (Sphaeriidae), typically are abundant and make a significant contribution to macroinvertebrate biomass. Species of *Eupera, Musculium, Pisidium,* and *Sphaerium* all occur throughout various floodplain areas, increasing in numbers as inundation period increases. Unionid mussels are rare, but species of *Anodonta* and *Ligumia,* as well as the Asiatic clam, *Corbicula fluminea,* have been reported from permanent waters.

As in most aquatic environments, dipterans are the most species-rich and abundant insects occurring on floodplains. Many species of aquatic and semi-aquatic midge flies (Chironomidae) and crane flies (Tipulidae) occur throughout floodplains, showing distinct seasonal and spatial patterns of abundance. The abundance of other insect taxa varies considerably and often unpredictably, depending especially on the duration and timing of inundation, water velocity, types of substratum present, and dissolved oxygen concentrations. Most mayflies present are typical of lentic conditions, including species of *Caenis, Leptophlebia,* and *Siphlonurus,* although stream-dwelling species of the genera *Paraleptophlebia* and *Eurylophella* also can be found. Diverse assemblages of Odonata, Coleoptera and Hemiptera occur on floodplains, primarily in permanent bodies of water, but also in seasonally inundated areas. Plecoptera, Megaloptera, and Neuroptera are poorly represented on floodplains.

BEHAVIORAL AND LIFE HISTORY ADAPTATIONS OF MACROINVERTEBRATES

It is clear that the constraining factor for the existence of macroinvertebrates in floodplain swamps is the degree of their tolerance to flooding and drying. This is especially critical on those floodplains with low predictability of the timing and duration of flooding. Few of the aquatic species commonly occurring on floodplains are constrained to life in temporary habitats, but rather nearly all of the aquatic species are found in both permanent lotic or lentic bodies of water. Similarly, most terrestrial species that inhabit floodplains during dry periods also occur in adjoining fully terrestrial environments. In contrast to those species confined to permanent water, the invertebrate inhabitants of floodplains thus primarily are opportunistic residents of a temporary environment. Exploitation of the floodplain requires the ability to tolerate and respond quickly to changing hydrology, both when inundation occurs and with the onset of drying.

Much has been written about life in temporary habitats, especially focusing on the various morphological, physiological, behavioral, and life history adaptations that allow organisms to exist in aquatic habitats that seasonally dry (see Wiggins et al. 1980, Williams 1987, 1996). Little attention, however, has been given specifically to species dwelling on floodplains. As with other temporary habitats, these species can be grouped into several broad categories in terms of their strategies to cope with and take advantage of the alternating cycle of flooding and drying.

Resistance to Desiccation

The majority of aquatic species that are permanent residents on floodplains have a desiccation-resistant stage, typically in the egg or immature stage but for a few species as adults. Aquatic species that are temporary occupants of floodplains have at least one life history stage capable of terrestrial life or are sufficiently mobile to move between floodplains and permanent water. Terrestrial species are faced with the opposite problem: they require adaptations to survive flooding for extended periods or the capability to move to surrounding dry areas as flooding occurs. How the general strategies of resistance to and avoidance of desiccation or flooding are met varies considerably among species.

Many aquatic species permanently dwelling on floodplains are capable of surviving dry periods by using mechanisms that vary extensively in form and complexity. Beetles and mollusks take advantage of their thick exoskeletons or shells, respectively, to reduce water loss. Adult beetles can bury themselves into moist mud and aestivate for months until flooding reoccurs, although they are better known for their flights to permanent waters (Macan 1939, Jeffries 1994). Sphaeriid clams bury into the soil, following groundwater, or close their valves and aestivate (Way et al. 1980). Gastropods also retreat into their shells, then further inhibit water loss by means of mucosal secretions around their bodies. Many dipterans, especially chironomids, have the ability to aestivate as larvae (Williams and Hynes 1977, Pritchard 1983, Wiggins et al. 1980); isopods and amphipods can aestivate as adults for long periods in moist areas.

Copepods, cladocerans, ostracods, and chironomids are well known for producing resting stages in lentic habitats, typically either as eggs or juvenile instars. Presumably this occurs in floodplains, although it has not been examined. Hairston and Munns (1984) suggest that in an optimal model dormant stages should be produced one generation before complete drying occurs. The warm water temperatures occurring in late spring may make life histories sufficiently short that crustaceans can shift to producing resting eggs in a matter of days, probably in response to decreasing water levels or associated environmental cues (B. E. Taylor, personal communication). A bet-hedging strategy also is likely (Hairston et al. 1985) whereby resting eggs are produced by part of the population as the dry period approaches, perhaps cued by temperature, with the remainder of the population continuing to grow and reproduce as long as the floodplain is inundated.

Not all aquatic species on floodplains necessarily stop activity during non-flooded periods. Evidence from several studies suggests that a variety of species are capable of maintaining activity and growth within moist soil and litter. Isopods on a Missouri floodplain grew an average of 8–16 percent of their maximum potential growth from April to October, when the floodplain was not inundated but the surface litter remained moist (White 1985). Sphaeriid clams also reproduced throughout the year on that floodplain. Gladden and Smock (1990), working in Virginia floodplains, noted an increase in the mean sizes of populations of several chironomids (*Polypedilum, Paratendipes*), ceratopogonids, tabanids, isopods (*Caecidotea*), and amphipods (*Crangonyx*) from the time of drying to first inundation, a period of about three months.

The extent of this growth, and its impact on floodplain production and trophic dynamics, are likely to vary considerably with the extent of drying and the ability of organisms to actively seek moist areas. During the summers of several drought years, the top 5–10 cm of soil in the Virginia floodplain studied by Gladden and Smock (1990) completely dried; no active organisms were found in the soil until the onset of autumn rains, when aestivating forms quickly became active. In other years the surface litter remained moist throughout the summer and autumn; individuals of many aquatic taxa were active in the litter. Studies of aquatic macroinvertebrates in seasonally inundated floodplains thus should not be limited to inundated periods, but should incorporate year-long sampling.

Initial inundation of floodplains in the autumn causes the rapid appearance of many species. Nematodes, copepods, ostracods, and chironomids all can become active within days after flooding; amphipods and especially isopods typically appear within a week (Sniffen 1981, White 1985, Gladden and Smock 1990). Copepods in Carolina bay ponds show a continuum of the time needed after flooding before individuals become active, ranging from one day to months (Taylor and Mahoney 1990, Wyngaard et al. 1991). Laboratory studies involving flooding of soil collected during October from a noninundated Virginia floodplain resulted in the appearance of a wide diversity of primarily meiofaunal taxa (Table 7.1). Several species of nematodes appeared within four hours of flooding. Euglenoid protozoans, gastrotrichs, rotifers, and copepods were present within 4 days; cladocerans, ostracods, chironomids, ephydrids, and isopods all were active within 10 days. Since no external recolonization source existed in this study, all of these taxa must have originated from resting stages.

Aestivation of the adults or larvae of many species typically can be broken simply by flooding, but inundation alone may not be a sufficient cue to break dormancy of other resting stages (Danks 1987). For example, the copepod *Diaptomus stagnalis* requires exposure to cold temperatures before flooding stimulates activity of resting eggs (Taylor et al. 1990). These multiple cues keep the majority of the population from responding to short, irregular periods of flooding during the summer, when the likelihood of mortality from drying is high. Although a similar model seems plausible for species in forested floodplains, this has not yet been tested.

Summer floods usually do not inundate an entire floodplain. Also, considerable variation exists in the length of inundation required to break the dormancy of individuals within a population. Perhaps short-duration summer or early autumn flooding breaks dormancy of only a fraction of a population,

TABLE 7.1. Time of First Appearance of Invertebrate Taxa Following Flooding of Floodplain Soils Collected During the Dry Period (October) from the Colliers Creek, Virginia, Floodplain[a]

Taxon	Time to First Appearance (days)
Nematoda *Miconchus* sp.	0.1
Nematoda *Labronema* sp.	0.2
Protozoa *Chilomonas* sp.	0.4
Nematoda *Prismatolaimus* sp.	2
Rotifera *Monostyla* sp.	2
Gastrotricha *Chaetonotus* sp.	3
Copepoda *Attheyella illinoisensis*	4
Copepoda *Bryocamptus hiatus*	6
Cladocera *Leydigia leydigi*	6
Ostracoda *Candona sharpei*	6
Nematoda *Dorylaimus* sp.	7
Chironomidae *Polypedilum* sp.	7
Chironomidae *Tanytarsus* sp.	7
Ephydridae	9
Chironomidae *Paratendipes* sp.	10
Isopoda *Caecidotea racovitzae*	10

[a]Results indicate the first appearance of a taxon in any of 10 replicate soil cores flooded with Colliers Creek water in the lab and held under ambient water and light conditions. Samples were examined for organisms hourly during the first 12 hours and daily thereafter for 14 days.

thus providing a spatial mosaic of patches where some individuals exploit the temporarily flooded environment and other individuals persist in resting stages, ensuring that the population will respond to future flooding.

Immigration and Emigration

Immigration to and emigration from floodplains provide another mechanism of survival during unfavorable conditions and for recolonization when favorable conditions return. As water levels drop on floodplains during spring, some of the more mobile species move to permanently flooded channels or ponds. Species capable of flight, such as hemipterans and coleopterans, have an obvious survival advantage, as do those species with life histories timed such that adult emergence occurs prior to drying. A variety of mobile species, particularly isopods, amphipods, and crayfish, can move deep into the soil during the summer, following receding groundwater in crayfish burrows and rotted root holes (Creaser 1931, Sniffen 1981). Survival in groundwater, however, may be limited during the summer by low dissolved oxygen concentrations.

A number of species recolonize newly inundated floodplains through aerial dispersal of adults that emerge from nearby permanent bodies of water. Adult dipterans, and in particular chironomids, often commence ovipositing almost immediately after initial inundation. The overall extent and effectiveness of recolonization through aerial dispersal, however, are likely to depend on the timing of both flooding and the life history of a species. In warmer areas to the south, multivoltinism of many species in permanent bodies of water provides a fairly continuous source of dispersing adults. In more northern floodplains, fewer species of aquatic insects have aerial adults present during much of the inundated period, which coincides with colder weather, thereby limiting aerial recolonization.

Because of the unpredictability of the timing of drying, species ovipositing in floodplains in the spring may not be able to complete development from the egg to adult emergence before drying occurs. Mortality of these species will be high unless the eggs or immature stages can aestivate through the summer. Adults of stream-dwelling Odonata (Gomphidae, Calopterygidae) have been observed ovipositing in Virginia floodplains as long as water existed in side channels, but no nymphs of these taxa were found on the floodplain (L. A. Smock, personal observation). Ovipositing by three lotic mayfly species throughout the floodplain also was observed late in the spring. No survival of the obligate rheophilic mayfly *Stenonema modestum* (Heptageniidae) was observed. A few early instars of the more lentic mayflies *Eurylophella doris* (Ephemerellidae) and *Leptophlebia* sp. (Leptophlebiidae), however, were found shortly after autumnal flooding, suggesting survival over the summer. Both species produce eggs that diapause through the summer in adjoining permanent streams. Their eggs may be sufficiently desiccation-resistant to survive drying on the floodplain.

For most nonpermanent species the primary mechanism of recolonization of floodplains is through drifting or crawling into the floodplain from adjoining stream channels. Smock (1994) noted 51 taxa that drifted into a Virginia floodplain from its first-order stream channel over one year. Taxa included all of the common lotic species inhabiting the stream, including species of nearly all orders of aquatic insects and a number of noninsects. There was no indication that these organisms were actively dispersing into the floodplain; rather, they were passively transported there, drifting in water spreading from the stream channel.

Drifting organisms displayed typical drift behavior, peak densities occurring just after sunset and being approximately an order of magnitude higher than during the day. Highest input of organisms into the floodplain occurred from February through April, drift density over the year being positively correlated with discharge. The extent to which individuals survived once they entered the floodplain is unknown. Survivorship of these primarily rheophilic species probably is largely determined by the extent and maintenance of flow on the floodplain.

Smock's (1994) study showed that far fewer species of invertebrates drifted out of the floodplain than into it, but in terms of numbers, more individuals drifted to the channel than entered the floodplain. Most of the species drifting from the floodplain were small, primarily copepods, ostracods, and chironomids. Large numbers of these species rose into the water column during the early evening and passively drifted out of the floodplain. Lower water velocity on the floodplain than in the channel probably inhibits drifting of larger individuals, causing a primarily one-way connectivity between the channel and floodplain for many of the larger species. Further study of the connectivity of floodplains and their channels is needed to enable better understanding of population and community dynamics, especially at critical hydrologic and life history periods, as well as the implications of that connectivity for ecosystem-level processes (Naiman and Décamps 1990).

Whereas drifting seems to be passive dispersal into floodplains, some species move into floodplains through directed crawling, primarily keyed to habitat shifts associated with life history events that are synchronized with floodplain drying. Smock (1994) showed that final instars of the caddisflies *Ironoquia parvula* (Limnephilidae) and *Phryganea* sp. (Phryganeidae), as well as the fishfly *Chauliodes rastricornis* (Corydalidae), crawled from channels onto floodplains in Virginia during March and April. *Ironoquia* aestivated in the floodplain and fringing areas until pupation and emergence the following autumn, whereas *Phryganea* and *Chauliodes* pupated in crevices in saturated logs and emerged that summer.

Large numbers of several species of mayflies also move onto floodplains prior to emergence. Late instars of *Leptophlebia* sp. and *Paraleptophlebia* sp. (Leptophlebiidae) appear on floodplains from February through April and emerge before summer drawdown (Parsons and Wharton 1978, Smock 1994). Females then oviposit primarily in streams. Movements of mayflies from streams to fringing wetlands have been hypothesized as occurring to provide refuge from unfavorable environmental conditions, reduce predation pressure, or allow completion of life cycles (Hayden and Clifford 1974, Gibbs and Mingo 1986, Söderström and Nilsson 1987). No systematic study of the influence of these factors, however, has been conducted.

The above discussion indicates that many species of aquatic macroinvertebrates have mechanisms to survive the seasonal drying that occurs in most areas of floodplains, including desiccation-resistant stages, the ability to emigrate to permanent water, and life history stages appropriately synchronized with seasonal patterns of inundation. Even with these various survival mechanisms, however, mortality of aquatic organisms as a result of drying likely is heavy. As drawdown occurs, organisms aggregate in ever-shrinking pools, where densities can become exceedingly high, predation and competition pressure is intense, and low dissolved oxygen concentrations can become lethal (e.g., Sniffen 1981, White 1985).

Mass mortality of *Leptophlebia* sp. occurred in several isolated floodplain pools, where densities of late instars of this mayfly reached 3730 m^2, when the pools dried before the mayflies could emerge (L. A. Smock, unpublished data). Individuals emerged successfully from nearby areas of the floodplain that remained inundated for two to three additional weeks. White (1985) estimated an 80 percent mortality of isopods over a summer dry period, even though conditions were sufficiently favorable to allow surviving individuals to grow. Thus, although many species have adaptations that allow them to take advantage of the floodplain habitat, they clearly live in jeopardy in this unpredictable environment.

Terrestrial Macroinvertebrates

The above discussion omitted an often ignored but important component of the macroinvertebrate inhabitants of seasonally inundated floodplains. Terrestrial species can be just as diverse and functionally active during dry periods as are aquatic species during flooded periods (Brinson 1993). Flooding can have an effect on their distribution, population dynamics, and life histories (e.g., Goff 1952, Gasdorf and Goodnight 1963, Beck 1972, Uetz et al. 1979). For example, some terrestrial arthropods have defined reproductive periods and are univoltine on floodplains, but polyvoltine in nearby terrestrial areas (Adis and Mahnert 1985, Adis and Sturm 1987).

Physiological adaptations and the wetability of body surfaces are primary considerations in allowing survival of terrestrial species during flooding (e.g., Crisp and Lloyd 1954, Gifford 1968, Wallwork 1976). Survival can vary among life history stages and among closely related taxa. Carabid beetles and some millipedes, for example, survive flooding in either immature or pupal stages, but adults are intolerant of flooding (Bell 1968, Adis 1986). Some species of lumbricid and enchytraeid oligochaetes tolerate inundation, whereas other species are intolerant (Merritt and Lawson 1992). Overall, the majority of terrestrial species cannot tolerate long-term inundation.

Rather than surviving during inundation, most terrestrial species are sufficiently vagile to escape flooding or have terrestrial stages during wet periods. Species such as crickets, staphylinids, spiders, and carabids move quickly to high ground with flooding and then are among the first taxa to recolonize with receding waters (Beck 1972, Irmler 1979). Mass migrations of earthworms to fringing dry areas occur with flooding of Amazonian floodplains (Adis and Righi 1989). Some species of macroinvertebrates move up tree trunks with flooding and dwell above flood waters (Adis et al. 1988). Although many species are capable of reaching dry areas during flooding, the proportion of individuals that do so is unknown. Most likely many of the less mobile species experience high rates of mortality, especially from less predictable summer storms.

MACROINVERTEBRATE ABUNDANCE AND PRODUCTION

Few quantitative studies have been published on the abundance and production of macroinvertebrates inhabiting floodplains. This is surprising given the extent and importance of floodplains throughout the southeastern United States and the presumed central position of invertebrates in the trophic structure of these systems. Wharton et al. (1982), in their extensive review of bottomland hardwood forests of the southeast, provide only a brief account of the more common invertebrate taxa, indirectly alluding to the near-complete lack of quantitative data on abundance and function. Since that review, a few studies have quantified densities, production, and distribution, but large gaps still remain in our understanding of both those parameters and the general structure and function of macroinvertebrate communities.

General statements concerning macroinvertebrate species abundance and levels of production in southeastern floodplains are difficult to make because of the paucity of studies and the variety of methods used for sampling and for calculation of production. Species of copepods, isopods, amphipods, and sphaeriid clams are among the most abundant and productive species (Sniffen 1981, White 1985, Gladden and Smock 1990, Smock et al. 1992, Duffy and LaBar 1994). Although the production of individual species of chironomids does not reach the levels of the above taxa, chironomid production as a group likely predominates on floodplains. Current knowledge on the production of terrestrial invertebrates on floodplains is inadequate, especially at the community level.

It is clear that aquatic invertebrate abundance, biomass, and production are closely linked to the duration and stability of inundation, all increasing as the period of flooding increases. Highest values occur in permanently flooded swamps, although only if the water remains oxic; numbers and production can drop to very low in anoxic areas. The two Virginia floodplains studied by Gladden and Smock (1990) had different inundation patterns. The Colliers Creek floodplain was inundated continuously for 255 days. The low floodplain at nearby Buzzards Branch was inundated 248 days over one year, though not continuously since short periods of drying occurred between inundation. High floodplain areas, contiguous with the low floodplain and just 20–50 cm higher in elevation, were inundated for a total of only 32 nonconsecutive days when storms caused overbank flooding. Density, biomass and production were four to five times greater in the low floodplain than in the high floodplain at Buzzards Branch (Table 7.2), showing the importance of the duration of flooding. Also, biomass was more than double, and production 50 percent greater, at Colliers Creek than at the Buzzards Branch low floodplain, even though the latter was inundated only seven fewer days over the year. The short dry periods that occurred at Buzzards Branch eliminated many species that required continuous flooding and also resulted in many species spending much added time in dormant rather than actively growing stages.

Sniffen's (1981) study in North Carolina showed a substantial difference in invertebrate production between two consecutive years, 9.3 g/m^2 and 19.6 g/m^2, which he attributed to the longer period of inundation in the second year. While illustrating the impact of inundation period, both production values likely are high because inaccurate assumptions were used in calculating production of some taxa. The lowest production measured in a southeastern United States forested wetland was in a Mississippi greentree reservoir, a shallow man-made impoundment in a forested wetland flooded to provide habitat for waterfowl (Duffy and LaBar 1994). Production at that site, inundated for less than four months from late November until early March, was only 1.1 g/m^2.

An interesting question concerns the trade-off of secondary production during aquatic and terrestrial periods. It is unknown whether the rate of production of aquatic or terrestrial invertebrates on floodplains is greater. Differences in their rates of production would impact their potential for providing food resources to higher consumers and their effect on mineral cycling. If rates of secondary production are similar between dry and wet periods, then the duration of flooding may be immaterial to ecosystem-level processes. Large differences in rates between dry and wet periods, however, could result in large changes in system energetics dependent on the duration of inundation.

Little is known about the growth rates of invertebrates on floodplains. Anderson and Benke (1994) measured somatic biomass growth rates as high as 61 percent per day for the cladoceran *Ceriodaphnia dubia,* collected from an Alabama floodplain and fed swamp seston. They concluded that the quantity and quality of the seston was capable of sustaining growth rates similar to those reported for other *Ceriodaphnia* species in other aquatic habitats. Annual macroinvertebrate community production to biomass (P/B) ratios, which are an indirect measure of development time, were inversely related to the length of floodplain inundation (Table 7.2). Greater P/B ratios corresponded to the need for species living in briefly inundated habitats to have rapid growth rates and short development times to complete their life cycles.

Most invertebrates on floodplains occur among the litter on the soil surface, although wood and macrophytes can be important substrata under some con-

TABLE 7.2. Invertebrate Density, Biomass, Production, and Annual Production-to-Biomass (P/B) Ratios for Virginia Floodplains with Differing Duration and Stability of Inundation

Floodplain	Days of Inundation	Stability of Inundation	Density (no./m^2)	Biomass (g/m^2)	Production (g m^{-2} y^{-1})	P/B
Buzzards Branch (high floodplain)	32	periodic	7,852	0.87	1.08	12.4
Buzzards Branch (low floodplain)	248	noncontinuous	31,320	0.40	4.30	10.7
Colliers Creek	255	continuous	24,646	0.96	6.12	6.4

Source: J. E. Gladden and L. A. Smock, Macroinvertebrate distribution and production on the floodplains of two lowland headwater streams, Freshwater Biology 24:533–545, 1990.

ditions (Sklar 1985, Thorp et al. 1985). Gladden and Smock (1990) found that 99 percent of the individuals and 96 percent of macroinvertebrate production occurred on the soil surface; the remainder occurred on submerged logs and tree trunks. This is in contrast to the streams and rivers associated with these floodplains, where snags in the water column are critical sites for invertebrate production (Benke et al. 1984, Smock et al. 1985). Whereas snags provide the only stable substratum in channels, where the sand and silt sediment is easily scoured, floodplain soils have a low scour potential, and hence the presence of other stable structure is not a critical requirement for invertebrates.

Macrophytes can be an important local habitat for invertebrates, especially in low-flow, deep-water sites (Krull and Hubert 1973, Ziser 1978). They provide a complex habitat that likely serves as a refuge from predators and also may allow organisms to move up into the oxygenated water column, away from anoxic conditions that frequently occur at the flooded soil in deep-water areas. Floating-leaved duckweeds often harbor very high densities of invertebrates, although the biomass associated with these plants is low (Ziser 1978, Sklar 1985). Levels of production of the invertebrates associated with duckweed and other macrophytes on floodplains are unknown, but probably are fairly high, given that the most abundant invertebrate taxa among these plants are small species with fast growth rates.

Temporal patterns of invertebrate abundance are affected by a variety of factors. The timing of inundation and drying and the occurrence of anoxia are the primary abiotic factors affecting abundance; densities change rapidly as organisms break or enter dormant stages and otherwise respond to the opportunities and stresses brought on by their changing environment. During periods of stable environmental conditions, densities change primarily in response to life history events (Gladden and Smock 1990). The effects of predation and competition on floodplains have not been investigated; predation probably has an important effect, given the numbers of invertebrate and vertebrate predators occurring on floodplains.

TROPHIC ECOLOGY

Macroinvertebrates are generally viewed as having a central role in the trophic structure of forested floodplains during inundated periods, but their function in the trophic dynamics of floodplains has been largely unstudied except for some work examining their use as food by fish and ducks. Species of detritivores predominate during both wet and dry periods, responding to the large autumnal input of leaves to the floodplain floor. Gladden and Smock (1990) found that 75–79 percent of the aquatic invertebrate production on the two floodplains they studied was by detritivores and the remainder was by predators. About 85–90 percent of the detritivore production was by species of the collector functional feeding group, species that feed on fine particles of detritus. Most of that production was by collector-gatherers, consumers of deposited detritus, and only 8–17 percent of the production was by collector-filterers, primarily sphaeriid clams. In addition, terrestrial detritivores such as mites, springtails, and earthworms predominate on floodplain floors during dry periods (Grey 1973).

The taxonomic diversity, abundance, and production of aquatic shredding invertebrates, species that feed on large particles of detritus, is surprisingly low (Cuffney and Wallace 1987, McArthur et al. 1994). Shredders are an important component of the aquatic fauna of streams that have high inputs of detritus; however, Gladden and Smock (1990) found shredders to be contributing only 7–13 percent of the primary consumer production on two floodplains. The predominant aquatic shredders typically are isopods and amphipods, but those species are facultative shredders that also consume large amounts of fine particulate detritus, especially as early instars. Omnivorous crayfish, which can shred large quantities of leaf litter, may have an important impact on detritus processing where they are abundant.

Spatially heterogeneous rates of leaf breakdown are expected across floodplains, dependent on the inundation regime (Merritt et al. 1984, Cuffney and Wallace 1987, Merritt and Lawson 1992), but the actual impact of invertebrates on leaf processing on floodplains of the southeast has been poorly studied. The rate of processing by invertebrates depends greatly on the nutritional quality and palatability of leaves, which vary among tree species and also are affected by inundation history. Since inundation varies both spatially and temporally across floodplains, the floodplain landscape may consist of adjoining patches of detritus with different nutritional quality, resulting in patches with different rates of consumption by invertebrates and hence with a potential effect on their population ecology.

Algae and macrophytes may be locally important as food for invertebrates. High numbers of macroinvertebrates, in particular isopods, amphipods, and the mayfly *Leptophlebia,* feed on patches of algae that can be abundant during periods of stable flow in the late winter and spring (L. A. Smock, unpublished data). Having a source of high-quality food such as algae just prior to metamorphosis to the reproductive adult stage may be a reason for the movement of *Leptophlebia* nymphs from the channel, where algae is uncommon, to the floodplain. Most algae and macrophytes, however, likely enter the foodweb primarily upon their senescence.

The position of floodplains as ecotones between aquatic and terrestrial habitats facilitates a strong linkage of floodplain foodwebs with both river channel and terrestrial food webs. Large amounts of detritus that are partially processed on floodplains are exported to channels (Cuffney 1988, Smock 1990, Wainwright et al. 1992), making the floodplains the functional headwaters of the channels (Smock et al. 1985, Smock and Roeding 1986). It has been suggested that detritus as well as large numbers of bacteria flushed from floodplains are partially responsible for supporting highly productive proto-zoan and macroinvertebrate communities in the channels (Edwards 1987, Wallace et al. 1987, Carlough and Meyer 1989).

Invertebrates are an important food for vertebrate predators on floodplains, although the extent of energy flow from invertebrates to higher consumers has generally not been quantified. Many species of fish move onto floodplains to spawn and forage (e.g., Ross and Baker 1983, Killgore and Baker 1996), feeding on both aquatic and terrestrial organisms. Many smaller species of invertebrates, such as the abundant copepods and chironomids, likely are important in the diets of larval fish. Fish move quickly from the channel to exploit invertebrates trapped during initial flooding; later they also feed on the many invertebrates that fall into the water over the year. Overall, much of the fish production in river-floodplain systems, especially of fish in larger rivers, is supported by consumption of floodplain invertebrates (Welcomme 1979). Largely unexplored is the importance of aquatic invertebrates to the many amphibians and birds that are abundant throughout floodplains. Much of the diet of wood ducks may be aquatic invertebrates, which can provide an important source of protein required for migration (Drobney and Fredrickson 1979, Heitmeyer 1985).

Although many aquatic invertebrates make use of the terrestrial environment during some portion of their life history, there is almost no information on the transfer of energy from aquatic to terrestrial foodwebs. Insectivorous birds are important predators of the adult invertebrates that emerged from immature aquatic stages in some wetland habitats (Murkin and Batt 1987, Blancher and McNicol 1991). In forested floodplains large numbers of adult insects that emerge from the water probably are consumed by a wide variety of birds, bats, spiders, and other terrestrial predators. The extent and significance of this linkage deserves detailed study because of the potential effects on energy flow at the population, community, and ecosystem levels.

CONCLUSION

The recurring theme throughout this chapter has been that the macroinvertebrate community inhabiting southeastern forested floodplains is dependent on the hydrologic regime, and more specifically the flood pulse, of the floodplains. The seasonal cycle of flooding and drying in large part determines the species composition, abundance, production, and life history and behavioral traits of the macroinvertebrates inhabiting the floodplains. These factors in turn have important ramifications not only at the population level, but also for community- and ecosystem-level processes. The floristic structure of the floodplains, which consists of productive hardwood forests that provide large inputs of detritus and hence affect the base of the foodweb, also is tied to system hydrology. The key to maintaining the ecological integrity of the macroinvertebrate community and the processes upon which it has an effect (e.g.,

Cuffney, T. F., and J. B. Wallace. 1987. Leaf litter processing in Coastal Plain streams and floodplains of southeastern Georgia, U.S.A. Archiv für Hydrobiologie Supplement 76:1–24.

Danks, H. V. 1987. Insect dormancy: an ecological perspective. Biological Survey of Canada Monograph Series No. 1. Biological Survey of Canada, Ottawa.

Décamps, H. 1996. The renewal of floodplain forests along rivers: A landscape perspective. Internationale Vereinigung für Theoretische und Andgewandte Limnologie, Verhandlungen 26:35–59.

Drobney, R. D., and L. H. Fredrickson. 1979. Food selection by wood ducks in relation to breeding status. Journal of Wildlife Management 43:109–120.

Duffy, W. G., and D. J. LaBar. 1994. Aquatic invertebrate production in southeastern USA wetlands during winter and spring. Wetlands 14:88–97.

Edwards, R. T. 1987. Sestonic bacteria as a food source for filtering bacteria in two southeastern blackwater rivers. Limnology and Oceanography 32:221–234.

Felley, J. D. 1992. Medium-low-gradient streams of the Gulf coastal plain. Pages 233–270 in C. T. Hackney, M. Adams, and W. Martin (eds.), Biodiversity of the southeastern United States: Aquatic Communities. John Wiley & Sons, New York.

Garman, G. C., and L. A. Nielsen. 1992. Medium-sized rivers of the Atlantic coastal plain. Pages 315–350 in C. T. Hackney, M. Adams, and W. Martin (eds.), Biodiversity of the Southeastern United States: Aquatic Communities. John Wiley & Sons, New York.

Gasdorf, E. C., and C. J. Goodnight. 1963. Studies on the ecology of soil arachnids. Ecology 44:261–268.

Gibbs, K. E., and T. M. Mingo. 1986. The life history, nymphal growth rates, and feeding habits of *Siphlonisca aerodroma* Needham (Ephemeroptera: Siphlonuridae) in Maine. Canadian Journal of Zoology 64:427–440.

Gifford, D. R. 1968. The significance of soil water to soil invertebrates. Pages 175–179 in R. M. Wadsworth (ed.), The Measurement of Environmental Factors in Terrestrial Ecology. British Ecological Society Symposium, vol. 8. Blackwell Scientific Publishing, Oxford, UK.

Gladden, J. E., and L. A. Smock. 1990. Macroinvertebrate distribution and production on the floodplains of two lowland headwater streams. Freshwater Biology 24:533–545.

Goff, C. C. 1952. Flood plain animal communities. American Midland Naturalist 47:478–486.

Grey, W. F. 1973. An analysis of forest invertebrate populations of the Santee-Cooper Swamp, a floodplain habitat. Master's Thesis, University of South Carolina, Columbia, SC.

Haack, S. K., G. R. Best, and T. L. Crisman. 1989. Aquatic macroinvertebrate communities in a forest wetland: interrelationships with environmental gradients. Pages 437–454 in R. R. Sharitz and J. W. Gibbons (eds.), Freshwater Wetlands and Wildlife. CONF-8603101, DOE Symposium Series No. 61. USDOE Office of Scientific and Technical Information, Oak Ridge, TN.

Hairston, N. G., Jr., and W. R. Munns, Jr. 1984. The timing of copepod diapause as an evolutionarily stable strategy. American Naturalist 123:733–751.

Hairston, N. G., Jr., E. J. Olds, and W. R. Munns. 1985. Bet-hedging and environmentally cued strategies of diaptomid copepods. Internationale Vereinigung für Theoretische und Angewandte Limnologie, Verhandlungen 22:3170–3177.

Hayden, W., and H. F. Clifford. 1974. Seasonal movements of the mayfly *Leptophlebia cupida* (Say) in a brown-water stream of Alberta, Canada. American Midland Naturalist 191:90–102.

Heitmeyer, M. E. 1985. Wintering strategies of female mallards related to dynamics of lowland hardwood wetlands in the upper Mississippi Delta. Ph.D. Dissertation, University of Missouri, Columbia, MO.

Irmler, U. 1979. Considerations on structure and function of the "Central-Amazonian Inundation Forest Ecosystem" with particular emphasis on selected soil animals. Oecologia 43:1–18.

Jeffries, M. J. 1994. Invertebrate communities and turnover in wetland ponds affected by drought. Freshwater Biology 32:603–612.

Jones, J. B., Jr., and L. A. Smock. 1991. Transport and retention of particulate organic matter in two low-gradient headwater streams. Journal of the North American Benthological Society 10:115–126.

Junk, W. J., P. B. Bayley, and R. E. Sparks. 1989. The flood pulse concept in river-floodplain systems. Canadian Special Publication of Fisheries and Aquatic Sciences 106:110–127.

Killgore, K. J., and J. A. Baker. 1996. Patterns of larval fish abundance in a bottomland hardwood wetland. Wetlands 16:288–295.

Krull, J. N., and W. A. Hubert. 1973. Seasonal abundance and diversity of benthos in a southern Illinois swamp. Chicago Academy of Sciences Natural History Miscellanea No. 190:1–4.

Larsen, J. S., M. S. Bedinger, C. F. Bryan, S. Brown, R. T. Huffman, E. L. Miller, D. G. Rhodes, and B. A. Touchet. 1981. Transition from wetlands to uplands in southeastern bottomland hardwood forests. Pages 225–273 *in* J. R. Clark and J. Benforado (eds.), Wetlands of Bottomland Hardwood Forests. Elsevier, Amsterdam, The Netherlands.

Livingston, R. J. 1992. Medium-sized rivers of the Gulf coastal plain. Pages 351–385 *in* C. T. Hackney, M. Adams, and W. Martin (eds.), Biodiversity of the southeastern United States: aquatic communities. John Wiley & Sons, New York.

Macan, T. T. 1939. Notes on the migration of some aquatic insects. Journal of the Society of British Entomologists 2:1–6.

MacArthur, J V., J. M. Aho, R. B. Rader, and G. L. Mills. 1994. Interspecific leaf interactions during decomposition in aquatic and floodplain ecosystems. Journal of the North American Benthological Society 13:57–67.

Megonigal, L. P., W. H. Conner, S. Kroeger, and R. Sharitz. 1997. Aboveground production in southeastern floodplain forests: A test of the subsidy-stress hypothesis. Ecology 78:370–384.

Merritt, R. W., W. Wuerthele, and D. L. Lawson. 1984. The effect of leaf conditioning on the timing of litter processing on a Michigan woodland floodplain. Canadian Journal of Zoology 62:179–182.

Merritt, R. W., and D. L. Lawson. 1992. The role of leaf litter macroinvertebrates in stream-floodplain dynamics. Hydrobiologia 248:65–77.

Mitsch, W. J., and J. G. Gosselink. 1993. Wetlands. Van Nostrand Reinhold, New York.

Murkin, H. R., and B. D. J. Batt. 1987. The interactions of vertebrates and invertebrates in peatlands and marshes. Pages 15–30 *in* D. M. Rosenberg and H. V. Danks (eds.),

Aquatic Insects of Peatlands and Marshes of Canada. Memoirs of the Entomological Society of Canada 140.

Naiman, R. J., and H. Décamps. 1990. The Ecology and Management of Aquatic-Terrestrial Ecotones. Man and the Biosphere Series, vol. 4. Unesco, Paris, and Parthenon, UK.

Parsons, K., and C. H. Wharton. 1978. Macroinvertebrates of pools on a piedmont river floodplain. Georgia Journal of Science 36:25–33.

Ross, S. T., and J. A. Baker. 1983. The response of fishes to periodic spring floods in a southeastern stream. American Midland Naturalist 109:1–14.

Sharitz, R. R., and W. J. Mitsch. 1993. Southern floodplain forests. Pages 311–372 *in* W. H. Martin, S. G. Boyce, and A. C. Echternacht (eds.), Biodiversity of the Southeastern United States: Lowland Terrestrial Communities. John Wiley & Sons, New York.

Sklar, F. H. 1985. Seasonality and community structure of the backswamp invertebrates in a Louisiana cypress-tupelo wetland. Wetlands 5:69–86.

Smock, L. A. 1990. Spatial and temporal variation in organic matter storage in low gradient, headwater streams. Archiv für Hydrobiologie 118:169–184.

———. 1994. Movements of invertebrates between stream channels and forested floodplains. Journal of the North American Benthological Society 13:524–531.

Smock, L. A., and E. Gilinsky. 1992. Coastal plain blackwater streams. Pages 271–314 *in* C. T. Hackney, M. Adams, and W. Martin (eds.), Biodiversity of the Southeastern United States: Aquatic Communities. John Wiley & Sons, New York.

Smock, L. A., E. Gilinsky, and D. L. Stoneburner. 1985. Macroinvertebrate production in a southeastern United States blackwater stream. Ecology 66:1491–1503.

Smock, L. A., and C. E. Roeding. 1986. The trophic basis of production of the macroinvertebrate community of a southeastern U.S.A. blackwater stream. Holarctic Ecology 9:165–174.

Smock, L. A., J. E. Gladden, J. L. Riekenberg, L. C. Smith, and C. R. Black. 1992. Lotic macroinvertebrate production in three dimensions: Channel surface, hyporheic, and floodplain environments. Ecology 73:876–886.

Sniffen, R. P. 1981. Benthic invertebrate production during seasonal inundation of a floodplain swamp. Ph.D. Dissertation, University of North Carolina, Chapel Hill, NC.

Söderström, O., and A. N. Nilsson. 1987. Do nymphs of *Parameletus chelifer* and *P. minor* (Ephemeroptera) reduce mortality from predation by occupying temporary habitats? Oecologia 74:39–46.

Taylor, B. E., and D. L. Mahoney. 1990. Zooplankton in Rainbow Bay, a Carolina bay pond: Population dynamics in a temporary habitat. Freshwater Biology 24:597–612.

Taylor, B. E., G. A. Wyngaard, and D. L. Mahoney. 1990. Hatching of *Diaptomus stagnalis* eggs from a temporary pond after a prolonged dry period. Archiv für Hydrobiologie 117:271–278.

Thorp, J. H., E. M. McEwan, M. F. Flynn, and F. R. Hauer. 1985. Invertebrate colonization of submerged wood in a cypress-tupelo swamp and blackwater stream. American Midland Naturalist 113:56–68.

Uetz, G. W., K. L. van der Laan, G. F. Summers, P. A K. Gibson, and L. L. Getz. 1979. The effects of flooding on floodplain arthropod distribution, abundance and community structure. American Midland Naturalist 101:286–299.

Vannote, R. L., G. W. Minshall, K. E. Cummins, J. R. Sedell, and C. E. Cushing. 1980. The river continuum concept. Canadian Journal of Fisheries and Aquatic Sciences 37:130–137.

Wainwright, S. C., C. A. Couch, and J. L. Meyer. 1992. Fluxes of bacteria and organic matter into a blackwater river from river sediments and floodplain soils. Freshwater Biology 28:37–48.

Walbridge, M. R., and B. G. Lockaby. 1994. Effects of forest management on biogeochemical functions in southern forested wetlands. Wetlands 14:10–17.

Wallace, J. B., A. C. Benke, A. H. Lingle, and K. Parsons. 1987. Trophic pathways in subtropical blackwater streams: contribution to production of macroinvertebrate primary consumers. Archiv für Hydrobiologie Supplement 74:423–451.

Wallwork, J. A. 1976. The Distribution and Diversity of Soil Fauna. Academic Press, London, UK.

Ward, J. V. 1989a. The four-dimensional nature of lotic ecosystems. Journal of the North American Benthological Society 8:2–7.

———. 1989b. Riverine-wetland interactions. Pages 385-400 in R. R. Sharitz and J. W. Gibbons (eds.), Freshwater Wetlands and Wildlife. CONF-8603101, DOE Symposium Series No. 61. USDOE Office of Scientific and Technical Information, Oak Ridge, TN.

Way, C. M., D. J. Hornbach, and A. J. Burky. 1980. Comparative life history tactics of the sphaeriid clam, *Musculium partumeium* (Say), from a permanent and a temporary pond. American Midland Naturalist 104:319–327.

Wellcomme, R. L. 1979. Fisheries Ecology of Floodplain Rivers. Longman, London, UK.

Wharton, C. H., V. W. Lambour, J. Newsom, P. V. Winger, L. L. Gaddy, and R. Mancke. 1981. The fauna of bottomland hardwoods in southeastern United States. Pages 87–160 in J. R. Clark and J. Benforado (eds.), Wetlands of Bottomland Hardwood Forests. Elsevier, Amsterdam, The Netherlands.

Wharton, C. H., W. M. Kitchens, E. C. Pendleton, and T. W. Sipe. 1982. The Ecology of Bottomland Hardwood Swamps of the Southeast: A Community Profile. U.S. Fish and Wildlife Service, Biological Services Program, Washington, DC. FWS/OBS-81/37.

White, D. C. 1985. Lowland hardwood wetland invertebrate community and production in Missouri. Archiv für Hydrobiologie 103:509–533.

Wiggins, G. B., R. J. Mackay, and I. M. Smith. 1980. Evolutionary strategies of animals in annual temporary pools. Archiv für Hydrobiologie Supplement 58:97–206.

Williams, D. D. 1987. The Ecology of Temporary Waters. Timber Press, Portland, OR.

———. 1996. Environmental constraints in temporary fresh waters and their consequences for the insect fauna. Journal of the North American Benthological Society 15:634–650.

Williams, D. D., and H. B. N. Hynes. 1977. The ecology of temporary streams II. General remarks on temporary streams. Internationale Revue der Gesamten Hydrobiologie 62:53–61.

Wyngaard, G. A., B. E. Taylor, and D. L. Mahoney. 1991. Emergence and dynamics of cyclopoid copepods in an unpredictable environment. Freshwater Biology 25: 219–232.

Yarbro, L. A. 1983. Influence of hydrologic variations on phosphorus cycling and retention in a swamp stream ecosystem. Pages 223-246 *in* T. D. Fontaine and S. M. Bartell (eds.), Dynamics of Lotic Ecosystems. Ann Arbor Science, Ann Arbor, MI.

Ziser, S. W. 1978. Seasonal variations in water chemistry and diversity of the phytophilic macroinvertebrates of three swamp communities in southeastern Louisiana. Southwestern Naturalist 23:545–562.

8 Carolina Bays

Ecology of Aquatic Invertebrates and Perspectives on Conservation

BARBARA E. TAYLOR, DOUGLAS A. LEEPER,
MORGAN A. McCLURE, and ADRIENNE E. DeBIASE

*C*arolina bays are geomorphically distinctive basins of the Atlantic Coastal Plain of North America. Most contain wetland ponds, and a few contain shallow lakes. They are abundant, and they constitute an important type of natural lentic habitat in the region. Fluctuating water level is a primary factor influencing composition and dynamics of the invertebrates. The occurrence of fish is also an important factor: occasional drying combined with absence of surface inlets or outlets eliminates fish from many bays. Many bays and much of their surrounding landscapes have been heavily altered by human activity. Because invertebrate assemblages are diverse among as well as within bays, maintenance of the diversity of invertebrates (and other animals) probably depends on protecting groups of these habitats, as well as the other aquatic habitats that can serve as seasonally alternate habitats for transient members of the bay assemblages.

INTRODUCTION

Carolina bays are shallow, isolated, oval basins that occur in the Atlantic Coastal Plain of North America, mainly in North and South Carolina. The basins generally contain palustrine wetland habitats, which we will refer to as *wetland ponds*. A few of the larger basins hold shallow permanent lakes.

Invertebrates in Freshwater Wetlands of North America: Ecology and Management, Edited by
Darold P. Batzer, Russell B. Rader, and Scott A. Wissinger
ISBN 0-471-29258-3 © 1999 John Wiley & Sons, Inc.

Aquatic invertebrates are abundant and diverse in Carolina bays. As habitat for aquatic invertebrates, Carolina bays have several notable attributes. The first is their hydrology. Water levels fluctuate widely in most Carolina bays, and many dry out seasonally, or at least occasionally. Their aquatic inhabitants must therefore have some capacity to resist desiccation or to disperse and recolonize as the wetlands dry and refill. The second is their isolation. Bays typically lack surface inlets and outlets, restricting the exchange of aquatic animals among bays or between bays and other aquatic habitats. Fish are absent from most bays that dry regularly. A third is the chemistry of their waters, which are typically acidic, soft, and moderately to heavily colored.

This chapter provides an overview of the composition, natural history, and ecology of invertebrates in Carolina bays and also discusses conservation issues and research needs.

DISTRIBUTION AND DESCRIPTION OF CAROLINA BAYS

Carolina bays occur in areas of sandy surficial sediments on Atlantic Coastal Plain from New Jersey to northern Florida (Fig. 8.1). Typically, the long axis of the oval basin has a northwest-southeast orientation (Johnson 1942, Prouty 1952), and an elevated sand rim may be present (Fig. 8.2). The largest bay, Lake Waccamaw in North Carolina, has a length of 8 km and an area of 3600 ha. Most bays are much smaller. On the Upper Coastal Plain on the Savannah River Site (SRS) in South Carolina, the median size of Carolina bays is 1 ha, with a range of 0.1–50 ha (Schalles et al. 1989). The basins are shallow. The seasonal maximum water depth in bays on the SRS is typically <1 m (Mahoney et al. 1990). For bay lakes in North Carolina, Frey (1949) reported maximum water depths of 2.2–3.6 m. Estimates of the number of bays are as high as 500,000, but the number is more probably 10,000–20,000 (Richardson and Gibbons 1993).

The distinctive shape and orientation of Carolina bays have been attributed to meteor impacts, solution depressions, and a variety of other causes (e.g., Johnson 1942, Savage 1982, Ross 1987). The most generally accepted explanation entails modification of shallow ponds through the action of waves generated by westerly winds (Thom 1970, Kaczorowski 1977, Grant et al. 1997); elongation of the basin occurs perpendicular to the direction of the prevailing wind. Basal dates from organic sediments in the basins range from 10,000 to more than 20,000 years B.P. The wetland habitats of the bays are thus probably at least as old as most North American lakes, although paleoenvironmental and archaeological records suggest that these habitats have been dynamic, with changes driven by climatic and geologic process as well as human activity (e.g., Frey 1951a, Watts 1980, Bliley and Burney 1988, Brooks et al. 1997, Gaiser 1997).

The substrate of the basin may be either peat or clay. Peat-based bays are common in the Lower Coastal Plain of North Carolina and the adjacent coun-

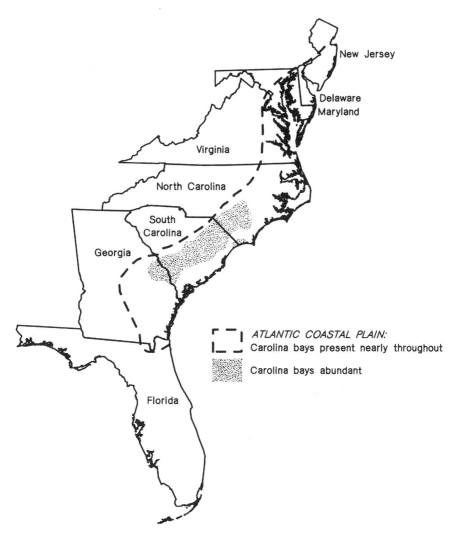

Fig. 8.1. Distribution of Carolina bays. Boundaries of the Atlantic Coastal Plain are based on Isphording and Fitzpatrick (1992). Regions of abundant bays are based on Johnson (1942) and Prouty (1952).

ties of South Carolina. These bays have thick deposits (1–2 m or more) of peat, and water is perched on an aquitard layer of humate-impregnated sand (Thom 1970). Clay-based or hard-bottomed bays are common in the Upper Coastal Plain of North Carolina (Nifong 1982) and throughout most of the Coastal Plain of South Carolina (Bennett and Nelson 1991). In these bays the upper layer of organically enriched sediment is shallow (often <20–30 cm), peat is usually absent, and a clay layer forms the aquitard. Clay-based bays may have hydrologic histories of more frequent or prolonged drying, which

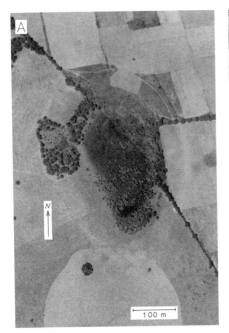

Fig. 8.2. Aerial photographs of two Carolina bays. The photographs illustrate characteristic oval shape, orientation, and sand rims, as well as recent human modifications. (*a*) Flamingo Bay, Aiken County, South Carolina, in 1943 (Cartographic and Architectural Branch of the National Archives, Washington, DC). The eastern side of the bay was plowed; the western side was probably pasture. The dark lines are trees along fence lines. Flamingo Bay is now protected as a DOE Research Set-aside Area on the Savannah River Site. (*b*) Woods Bay, Sumter and Clarendon Counties, South Carolina, in 1994 (U.S.G.S. Aerial Photography Field Office, Salt Lake City, Utah). The small impoundment on the eastern edge was constructed as a mill pond in the latter part of the nineteenth century, and cypress trees were harvested from the interior of the bay in the 1910s and 1920s (S. Wolfe, Woods Bay State Park, personal communication). Woods Bay is now protected as a state park.

would promote oxidation of organic material by biological processes or fire and thus retard accumulation of peat.

The Coastal Plain has a mild, moist temperate climate with rainfall distributed, on average, fairly evenly throughout the year. In the Coastal Plain of South Carolina the average monthly temperature ranges from 8°C in January to 27°C in July. The average annual rainfall is 123 cm, the wettest months occurring in summer and the driest months occurring in fall (30-year averages for divisions 4, 6, and 7, National Oceanic and Atmospheric Administration 1992). Snowfall is unusual, and ponds rarely ice over.

Water regimes of Carolina bays range generally from seasonally flooded to permanently flooded. Seasonally flooded bays typically fill in winter and dry in late spring. A few bays are known to be spring-fed, and groundwater may contribute substantially to the hydrologic budgets of others (Lide et al. 1995). Generally, however, fluctuations in water level are highly correlated with precipitation and evapotranspiration, and year-to-year variation can be large. At Rainbow Bay, a seasonally flooded wetland pond on the SRS in South Carolina, the median annual hydroperiod (main filling only) was 160 days, with a range of 3–391 days over 16 years (Semlitsch et al. 1996). Among nearly a hundred bays and other wetland ponds on the SRS for which some hydrologic record exists, most dry in most years, and only a few have never been observed to dry (Savannah River Ecology Laboratory, unpublished data).

The waters of Carolina bays are typically acidic, soft, and moderately to heavily colored. From a survey of 49 bays in North and South Carolina, Newman and Schalles (1990) reported a median pH of 4.6 (range 3.4–6.7); from a survey of 75 bays and other wetland ponds in South Carolina, Gaiser (1997) reported a mean pH of 4.8 (range 3.9–5.8). Dissolved organic carbon is moderate. Newman and Schalles (1990) reported a mean of 17.11 mg DOC L^{-1}. Levels of calcium are generally low. Newman and Schalles (1990) reported a median of 1.69 mg L^{-1} (range 0.16–11.75 mg L^{-1}); Gaiser (18 bays and other wetland ponds, 1997) reported a mean of 1.02 mg L^{-1} (range 0.15–3.09 mg L^{-1}). Other solutes are also generally low (see Newman and Schalles 1990, Pickens and Jagoe 1996). Most of these data come from clay-based, rather than peat-based, bays.

Less information is available about nutrient chemistry, but the concentrations of major nutrients appear to fall into ranges that would indicate meso-eutrophic or eutrophic conditions in lakes (see Wetzel 1983). In a winter and spring survey of 19 Carolina bays and other wetland ponds in South Carolina, DeBiase and Taylor (unpublished data) found median total Kjeldahl nitrogen concentrations of 0.83 mg L^{-1} (range 0.22–5.01 mg L^{-1}) and median total phosphorus concentrations of 0.037 mg L^{-1} (range 0.008–0.243 mg L^{-1}). Again, most of these data come from clay-based bays.

Wetland habitats of Carolina bays range from forests to wetland meadows to open water (Sharitz and Gresham 1997). Bennett and Nelson (1991) described plant communities of bays in South Carolina that are relatively undisturbed by human activity (according to their estimate, <20 percent of the bays are greater than 0.8 ha in area). The commonest types are: pond cypress pond, which has a closed canopy of pond cypress (*Taxodium ascendens*); nonalluvial swamp, which is dominated by hardwoods, such as swamp tupelo (*Nyssa sylvatica biflora*), red maple (*Acer rubrum*), and sweetgum (*Liquidambar styraciflua*), and may be codominated by pond cypress; and pocosin, which is dominated by a dense growth of shrubs such as fetter-bush (*Lyonia lucida*), titi (*Cyrilla racemiflora*), inkberry (*Ilex glabra*), blueberries (*Vaccinium* spp.), and loblolly bay (*Gordonia lasianthus*), vines such as greenbrier or "bamboo" (*Smilax laurifolia*), and stunted trees such as pond pine (*Pinus serotina*). Pond cypress savanna, which has an open canopy of pond cypress, is also common. Pond cypress pond, pond cypress savanna, and nonalluvial swamp vegetation are associated with clay-based bays; pocosin vegetation is associated with peat-based bays. Depression meadow, the only herbaceous palustrine community described by Bennett and Nelson, is uncommon among relatively undisturbed bays. It is dominated by grasses (*Panicum* spp. and *Leersia hexandra*) and sedges (*Carex* spp.). Open-water lakes are rare among bays in South Carolina. They support extensive floating and emergent aquatic vegetation, including water lily (*Nymphaea odorata*), water shield (*Brasenia schreberi*), and heartleaf (*Nymphoides* spp.), as well as many grasses and sedges.

The vegetation of most bays has been disturbed by human activity (Bennett and Nelson 1991, Kirkman et al. 1996). Row crops, pastures, and pine plantations represent the extremes. The short-term legacy of logging or clearing for agriculture is often the development of a herbaceous or shrub community (see Kirkman et al. 1996). Common species of these communities on the SRS include grasses, sedges, and the shrub buttonbush (*Cephalanthus occidentalis*) (De Steven, University of Wisconsin at Milwaukee, unpublished data).

Pond-breeding amphibians are the most abundant and productive of the vertebrates in Carolina bays (Richardson and Gibbons 1993). Common species in bays and other wetland ponds on the SRS (Gibbons and Semlitsch 1991) include the mole salamander (*Ambystoma talpoideum*), marbled salamander (*Ambystoma opacum*), dwarf salamander (*Eurycea quadridigitata*), red-spotted newt (*Notophthalmus viridescens*), southern cricket frog (*Acris gryllus*), southern toad (*Bufo terrestris*), eastern narrow-mouthed toad (*Gastrophryne carolinensis*), green treefrog (*Hyla cinerea*), spring peeper (*Pseudacris crucifer*), ornate chorus frog (*Pseudacris ornata*), southern leopard frog (*Rana sphenocephala*), and spadefoot toad (*Scaphiopus holbrooki*). In temporary ponds, including Carolina bays, of the sandhills region of North Carolina, the broken-striped newt (*Notophthalmus viridescens dorsalis*) and tiger salamander (*Ambystoma tigrinum*) are common (Morin 1983).

Hydrologic fluctuations limit the development of fish populations in many bays, but species such as lake chubsucker (*Erimyzon sucetta*), dollar sunfish (*Lepomis marginatus*), and mud sunfish (*Acantharchus pomotis*) occasionally colonize these habitats (Snodgrass et al. 1996). During times of high water, fish may gain access through ditches or overflows to bays that are otherwise isolated. Larger and more diverse populations of fish appear in continuously inundated habitats. The bay lakes of North Carolina support common southeastern pond species such as lake chubsucker, yellow bullhead (*Ameiurus natalis*), warmouth (*Lepomis gulosus*) and other centrarchids, and yellow perch (*Perca flavescens*); Lake Waccamaw also harbors several endemic fishes (Frey 1951b, Richardson and Gibbons 1993).

Reptiles, birds, and mammals also use Carolina bays (Clark et al. 1985, Schalles et al. 1989, Richardson and Gibbons 1993, Sharitz and Gresham 1997). Turtles such as the eastern mud turtle (*Kinosternon subrubrum*) and chicken turtle (*Deirochelys reticularia*) are common, as are various snakes and lizards (Gibbons and Semlitsch 1991). Alligators (*Alligator mississippiensis*) reside in some bays. More than a hundred species of nesting and summering birds were found at bays in North Carolina (Lee 1987); similar numbers of species were observed at two small bays on the SRS (SREL 1980). The larger nesting birds include wood duck (*Aix sponsa*), great blue heron (*Ardea herodias*), little blue heron (*Egretta cerulea*), great egret (*Casmerodius albus*), and anhinga (*Anhinga anhinga*) (Lee 1987; C. Eldridge, SREL, personal communication). Mammals such as oppossum (*Didelphis marsupialis*), raccoon (*Procyon lotor*), and white-tailed deer (*Odocoileus virginianus*) visit bays. Beaver (*Castor canadensis*) activity has been observed in a few bays on the SRS. Small mammals such as mice, shrews, moles, and bats are common.

INVERTEBRATE BIOTA

We provide an overview of composition and diversity, feeding, and life cycles, including adaptations to fluctuating water levels. Aquatic invertebrates in temporary habitats can be classified either as residents, adapted to persist during the dry season, or transients, adapted to recolonize during the wet season (see Wiggins et al. 1980, Batzer and Wissinger 1996). We classify life cycles, exclusive of the resting stage, as very short (a few days to a week), short (a week to a month), or long (months). Unless noted otherwise, information is derived from Barnes (1980), Brigham et al. (1982), Pennak (1989), Thorp and Covich (1991), Williams and Feltmate (1992), Merritt and Cummins (1996), and Batzer and Wissinger (1996).

Most of the research on invertebrates of Carolina bays has been conducted on the Savannah River Site (SRS), a large federal facility in Aiken and Barnwell Counties, South Carolina. The bays that have been studied are mainly clay-based, and there is little information on the invertebrates of peat-based bays.

The data discussed here are not all from wetlands whose basins have the characteristic shape and orientation of Carolina bays. In the field and in customary usage the distinction between Carolina bays and other types of isolated, depression wetlands is often unclear (Lide 1997). Of the best-studied sites on the SRS, Thunder Bay is unambiguously a bay, while Rainbow Bay is arguably not a bay. While we try to qualify the sources of data, our intuition is that the geomorphic distinction is not important ecologically.

The bay lakes of Bladen and Columbus Counties in North Carolina are shallow open-water lakes with extensive wetland margins. Because published information on their wetlands is sparse, we do not include them in the review below. Early accounts describe crustaceans of White Lake (Coker 1938) and benthic and planktonic invertebrates of Lake Waccamaw (Frey 1948, 1949). Lake Waccamaw, where the acidity of the water has been buffered by a natural outcrop of limestone, is the only bay or bay lake in which mollusks are abundant. The aquatic assemblage includes several endemic species, and an endemic land snail is found in the environs. Porter (1985) provides a detailed review of the older literature and collections. The evolution and ecology of these mollusks are topics of active research (e.g., Fuller 1977, Davis et al. 1981, Kat 1983, Johnson 1984, Cahoon et al. 1992, Stiven and Alderman 1992, Cahoon and Owen 1996).

Insects

A great variety of aquatic and semiaquatic insects live in Carolina bays (Table 8.1). A single bay may support more than 100 species (McClure 1994, Leeper and Taylor 1998, in press), and well over 300 species have been collected. Among the orders commonly found in aquatic habitats, only the Plecoptera (stoneflies), which live mainly in lotic habitats, are absent. Aquatic insects of Carolina bays range in length from about 0.3 mm for the earliest instars of odonates and 1 mm for the smallest dipteran larvae to >4 cm for the elongate hemipteran *Ranatra* (Leeper and Taylor 1998, in press).

For most aquatic insects, the complex life cycle with the winged adult stage facilitates dispersal into terrestrial and other aquatic habitats. In bays where the water is impermanent, many species persist as resting stages or as terrestrial adults when the bay dries; many others colonize seasonally or opportunistically. Batzer and Wissinger's (1996) review suggests that such patterns of "cyclic colonization" are generally important for the maintenance of insect populations in wetlands.

Dipterans. Larvae of dipterans, mainly chironomids, dominate the insect assemblages of Carolina bays. At Rainbow Bay on the SRS, where the most detailed studies have been conducted, 79 taxa, including 65 taxa of chironomids, were collected (SREL 1980, Leeper and Taylor 1998). Dipterans constituted 97 percent of the insects collected from benthic substrates and the water column (Leeper and Taylor, in press), and chironomids accounted for 93 percent of insect emergence (Leeper and Taylor 1998). Chironomids were the dominant insects at five other bays in the region, accounting for 63 percent of the insects collected (McClure 1994). Most dipterans, including the Chironominae, the most diverse and abundant subfamily of chironomids in these wetlands, are collector-gatherers as larvae. Larvae of *Chaoborus* and chironomids of the subfamily Tanypodinae consume small invertebrates, including other midge larvae. Life cycles of dipterans are typically short. Some species have desiccation-resistant eggs or larvae; others probably colonize from nearby aquatic habitats.

Coleopterans. Aquatic coleopterans, particularly the Dytiscidae and Hydrophilidae, are moderately diverse in Carolina bays. Fifty-three species of coleopterans were collected from Ashleigh Bay in Barnwell County, South Carolina (McClure 1994). At Rainbow Bay on the SRS, 23 genera were

TABLE 8.1. Aquatic and Semiaquatic Insects of Carolina Bays and Other Depression Wetlands in South Carolina

Order	Number of Species	Common Taxa
Ephemeroptera mayflies	5	Baetidae: *Callibaetis* Caenidae: *Caenis*
Odonata dragonflies and damselflies	37	Aeschnidae: *Anax* Libellulidae: *Erythemis, Erythrodiplax, Libellula, Pachydiplax, Sympetrum, Tramea* Coenagrionidae: *Anomalagrion, Argia, Enallagma, Ischnura* Lestidae: *Lestes*
Hemiptera true bugs	30	Corixidae: *Hesperocorixa, Sigara* Notonectidae: *Buenoa, Notonecta* Naucoridae: *Pelocoris*
Megaloptera dobsonflies	1	Corydalidae: *Chauliodes rastricornis*
Coleoptera beetles	86	Dytiscidae: *Coptotomus, Hydroporus, Thermonectus* Hydrophilidae: *Berosus, Enochrus, Tropisternus* Noteridae: *Hydrocanthus* Haliplidae: *Peltodytes*
Trichoptera caddisflies	35[a]	Leptoceridae: *Oecetis, Triaenodes*
Lepidoptera moths and butterflies	2	Pyralidae: *Synclita*
Diptera true flies	165	Chironomidae: *Ablabesmyia, Chironomus, Dicrotendipes, Kiefferulus, Polypedilum, Procladius, Psectrocladius* Culicidae: *Aedes, Culex* Chaoboridae: *Chaoborus* Ceratopogonidae: *Forcipomyia*

[a] May include lotic species

Source: Lists are based on surveys of Rainbow Bay and Sun Bay on the SRS (SREL 1980), longer-term studies at Thunder Bay on the SRS (Schalles 1979, Schalles and Shure 1989) and Rainbow Bay (Leeper and Taylor 1998, in press), and a survey of five bays in South Carolina, including two on the SRS (McClure 1994).

collected (Leeper and Taylor 1998). Generally, both larvae and adults are aquatic. Larval and adult dytiscids and larval hydrophilids are predacious; adult hydrophilids are largely herbivorous. Life cycles are generally long. Beetles produce desiccation-resistant eggs, some beetles pupate in terrestrial habitats, and the winged adults can disperse widely.

Odonates. Odonates are typically diverse and abundant in southeastern wetlands. However, their occurrence in Carolina bays may be limited by seasonal drying of aquatic habitats. In a study of five bays in South Carolina, McClure (1994) found 3–7 odonate species at three bays that dried during summer and 18–20 species at two bays that held water throughout the year. The aquatic odonate nymphs are predators, consuming micro- and macroinvertebrates, as well as larval amphibians and small fishes. Life cycles are long, and most species probably complete only one generation annually. Some nymphs can survive desiccation, but populations in ponds that have dried and refilled are probably more commonly reestablished by adults from other habitats. Adult odonates are strong fliers that range widely and oviposit freely. On the SRS the flight seasons for most anisopteran odonates are between April and September (Cross 1955; see also Kondratieff and Pyott 1987). Many species lay eggs in aquatic habitat; a few lay diapausing eggs on moist soil or in wetland plants (endophytic oviposition).

Hemipterans. The notonectids are the most abundant and conspicuous of the bugs in Carolina bays. Most of the aquatic bugs, including notonectids, are aggressive predators of invertebrates and small vertebrates. Corixids feed generally on a mix of microorganisms, detritus, and microinvertebrates. The habitat and habits of juveniles and adults are similar. Life cycles are typically long. Most aquatic bugs are winged as adults, and dispersal is probably good.

Trichopterans. Trichopterans are generally not abundant in Carolina bays. Some taxa may be limited by seasonal drying or lack of suitable materials for case construction (McClure 1994). The common trichopterans of bays are case-makers and are presumably collector-gatherers. Life cycles are typically long.

Other Orders. Other aquatic insect groups are less diverse or common in Carolina bays. The number of species of ephemeropterans in bays is low, although their abundances may be high (McClure 1994, Leeper, personal observation). The single species of *Chauliodes* is the only megalopteran recorded from Carolina bays. Aquatic and semiaquatic lepidopterans, including the pyralids *Synclita* and *Vogtia,* and several noctuids have been collected from bays in South Carolina (SREL 1980, McClure 1994, Ford, SREL, unpublished data). A few species of plecopterans (SREL 1980), which probably represent capture of transient adults from lotic habitats, have also been reported from bays. Aquatic collembolids are common, but only *Smithnurius* (McArthur, SREL, unpublished data) has been identified.

Crustaceans

Crustaceans in Carolina bays include members of the subclasses Branchiopoda, Copepoda, Ostracoda, and Malacostraca (Table 8.2). At least 90 species of cladocerans, other branchiopods, and calanoid copepods have been collected from Carolina bays in South Carolina. The assemblages rank among the richest in the world for temporary ponds (see Mahoney et al. 1990).

Most of the crustaceans of Carolina bays are small. The largest species of copepods, cladocerans, and ostracods attain only 3–4 mm in length; most range from <0.1 mm (for early naupliar stages of copepods) or 0.2 mm (for early stages of small cladocerans and ostracods) to 2 mm in length. Amphipods and isopods may reach 5–10 mm in length. Conchostracans and anostracans may reach 1.3 and 2.5 cm, respectively, in length. The crayfishes, which attain lengths of 5 cm or more, are the largest of the aquatic invertebrates found in bays.

Except for some decapods, most of the freshwater crustaceans have limited mobility in terrestrial habitats. Thus, most are permanent residents of bays. Many can produce resting eggs; others enter dormancy as juveniles or adults.

Branchiopods. Three of the major groups of branchiopods, the Cladocera, the Conchostraca, and the Anostraca, are found in Carolina bays; the Notostraca (tadpole shrimps) are absent. Cladocerans occur in nearly all lentic freshwater habitats, while anostracans and conchostracans, as well as notostracans, occur mainly in temporary ponds.

Cladocerans are present in virtually all Carolina bays. All of the families, except for Leptodoridae and Holopedidae (Mahoney et al. 1990, DeBiase and Taylor, unpublished data), are represented, although members of the Polyphemidae and Moinidae are rare. The greatest diversity occurs among the chydorids. In a typical bay on the SRS, one might find eight species of chydorids, five daphnids, two macrothricids, a sidid, and a bosminid (Mahoney et al. 1990). The only *Daphnia* species, *D. laevis,* collected in South Carolina surveys appears to be restricted to Carolina bays and other wetland ponds.

Most cladocerans are filter-feeders or scrapers, consuming algae and other fine particulate material or periphyton. An exception, *Polyphemus,* preys on small invertebrates. Some, including many daphnids and sidids, are planktonic or free-swimming in habit; others, including most chydorids and macrothricids, are associated more closely with benthic and littoral substrates. Life cycles of cladocerans are short, less than one week under warm temperatures, perhaps one to two weeks at cool temperatures, and they produce resting eggs to survive the dry season.

Anostracans and conchostracans are common and sometimes abundant, but not diverse, in Carolina bays. They are filter-feeders, consuming algae and

TABLE 8.2. Crustaceans of Carolina Bays and Other Depression Wetlands in South Carolina

Subclass and Order	Number of Species	Common Taxa
Subclass Branchiopoda		
Order Anostraca fairy shrimp	2	Streptocephalidae: *Streptocephalus seali* Chirocephalidae: *Eubranchipus holmani*
Order Conchostraca clam shrimps	2	Lynceidae: *Lynceus gracilicornis* Limnadiidae: *Limnadia lenticularis*
Order Cladocera water fleas	45[a]	Sididae: *Diaphanosoma, Pseudosida bidentata* Daphnidae: *Ceriodaphnia, Daphnia laevis, Scapholeberis armata armata, Simocephalus* Bosminidae: *Neobosmina tubicen* Macrothricidae: *Ilyocryptus, Macrothrix* Chydoridae: *Alona, Alonella, Chydorus, Ephemeroporus, Pseudochydorus* Polyphemidae: *Polyphemus pediculus*
Subclass Ostracoda seed shrimps	unknown	unknown
Subclass Copepoda copepods		
Order Calanoida	11	Diaptomidae: *Aglaodiaptomus, Leptodiaptomus moorei, Onychodiaptomus sanguineus* Centropagidae: *Osphranticum labronectum*
Order Cyclopoida	12	Cyclopidae: *Acanthocyclops robustus, Diacyclops, Macrocyclops fuscus, Tropocyclops*
Order Harpacticoida	unknown	unknown

TABLE 8.2. (Continued)

Subclass and Order	Number of Species	Common Taxa
Subclass Malacostraca Order Isopoda aquatic sow bugs	unknown	Asellidae: *Caecidotea*
Order Amphipoda scuds	unknown	Crangonyctidae: *Crangonyx* Gammaridae: *Gammarus* Talitridae: *Hyalella*
Order Decapoda shrimps, crayfishes	unknown	Palaemonidae: *Palaemonetes* Cambaridae: *Procambarus*

[a]Includes distinct but unidentified or undescribed species.

Source: Lists for Branchiopoda and Calanoida are based on surveys of 23 bays and depression wetlands on the SRS (Mahoney et al. 1990), 88 bays and depression wetlands on the SRS (DeBiase and Taylor 1993, and unpublished data), and three bays elsewhere in the Coastal Plain (DeBiase and Taylor, unpublished data). The list for the Cyclopoida is based on long-term studies at Rainbow Bay on the SRS (Wyngaard et al. 1991, Medland 1997), as are the lists for the Amphipoda and Isopoda (Leeper and Taylor, in press). Decapoda on the SRS were surveyed by Hobbs et al. (1978), but results were not reported specifically for Carolina bays.

other fine particulate material. Their life cycles are probably short, perhaps one to three weeks, depending on temperature. Both appear to be univoltine in bays, producing only resting eggs (Mahoney et al. 1990). Within a pond they may appear only sporadically, apparently persisting as resting eggs over multiple years (Taylor, unpublished data).

Copepods. All three orders of free-living freshwater copepods, the Calanoida, the Cyclopoida, and the Harpacticoida, are represented in Carolina bays. The calanoids are the best studied.

Calanoid copepods are present in many Carolina bays. As many as six species have been recorded from a single bay on the SRS, but the number is more commonly one or two (Mahoney et al. 1990, DeBiase and Taylor, unpublished data). About half of the Diaptomidae belong to the genus *Aglaodiaptomus,* including a locally common, newly described species, *Aglaodiaptomus atomicus* (DeBiase and Taylor 1997). Many of the bay calanoids, including all of the *Aglaodiaptomus,* are brightly pigmented with blue or red carotenoid pigments.

The diaptomid calanoids are free-swimming and planktonic in habitat; they feed on algae, other fine particulate material, and small invertebrates; the centropagid *Osphranticum labronectum* is epibenthic. The life cycle from hatching to egg production for *Aglaodiaptomus stagnalis,* the largest species, is completed in five to six weeks (Taylor and Mahoney 1990); smaller species probably require two to three weeks, depending on temperature. Calanoid copepods produce resting eggs to survive desiccation and other adversities.

Aglaodiaptomus stagnalis is strictly univoltine, producing only resting eggs, while *A. atomicus* and some other species are multivoltine.

Cyclopoid copepods are virtually ubiquitous in lentic habitats. Carolina bays are no exception (Mahoney et al. 1990), but their cyclopoids have received little attention. With the exception of work by Coker (1938) in North Carolina, detailed studies have been made only at Rainbow Bay on the SRS. Cyclopoid copepods are epibenthic or planktonic, feeding generally on algae, bacteria, detritus, and microinvertebrates. Most probably shift ontogenetically from herbivory in the early juvenile stages to omnivory in the late juvenile and adult stages. The life cycle of the active stages is short: one to several weeks, depending on temperature. Cyclopoid copepods survive adverse seasons in dormancy at a late juvenile stage (copepodid instar IV, the antepenultimate instar). Maturation and egg production may occur within a few days of emergence (Wyngaard et al. 1991, Medland 1997). Most species of cyclopoids at Rainbow Bay appeared to complete several generations before producing dormant stages (Medland 1997).

Harpacticoid copepods are also common and abundant in Carolina bays (Mahoney et al. 1990, Taylor and Mahoney 1990, Leeper and Taylor, in press), but have received even less attention than cyclopoids. They live in benthic microhabitats and feed generally on detrital material. Life cycles are probably short. Harpacticoid copepods also survive adverse seasons in dormancy at a late juvenile stage.

Ostracods. Ostracods are found in most Carolina bays (Mahoney et al. 1990). They may be rich in species (see Ebert and Balko 1987 and King et al. 1996 for accounts of their diversity in California vernal pools), but their composition has not been studied. They feed generally on fine particulate material, and life cycles are probably short. They can produce desiccation-resistant resting eggs.

Malacostracans. Amphipods and isopods are common but not ubiquitous in Carolina bays; decapods are less frequently encountered. Mahoney et al. (1990) found amphipods and isopods in about one third of 23 bays and other wetland ponds surveyed on the SRS. In their survey of decapods on the SRS, Hobbs et al. (1978) collected half a dozen species from lentic habitats, but did not specify which were found in bays. Members of all three orders are common in acidic depression wetlands on the Lower Coastal Plain of South Carolina (DeBiase and Taylor, unpublished data). Amphipods, isopods, and decapods are all benthic animals, variously feeding on periphyton or plants or scavenging. Their life cycles are typically long. They generally lack special adaptations, such as resting eggs, for surviving desiccation, but have some capacities to persist in moist substrates.

Annelids

Aquatic annelids, including oligochaetes and leeches, are common and abundant in Carolina bays (Mahoney et al. 1990). At least 14 species of oligochaetes, including an enchytraeid, the naidids *Dero, Nais, Pristina,* and *Stephensoniana,* the tubificids *Limnodrilus, Tasserkidrilus,* and *Tubifex,* and the lumbriculid *Eclipidrilus,* have been reported from bays on the SRS (McArthur, SREL, unpublished data, Leeper and Taylor, in press). Many of the aquatic oligochaetes in bays are small (<1–2 cm in length), and they are mainly benthic deposit feeders. Leeches are scavengers, ectoparasites, and predators. Life cycles are probably short. Aquatic oligochaetes and leeches can aestivate in mucus-lined cysts.

Nematodes

Nematodes are also common and abundant in Carolina bays (Mahoney et al. 1990, Leeper and Taylor, in press). Although the assemblages are probably quite diverse, only one genus, *Dorylaimus,* has been identified (McArthur, SREL, personal communication). These nematodes are small (1–2 cm in length). Nematode nutrition is diverse; they include predators, scavengers, and deposit feeders. Life cycles are probably very short. Eggs, larvae, and adults can survive desiccation.

Rotifers

Rotifers, like cyclopoid copepods, are virtually ubiquitous in lentic habitats. At Rainbow Bay on the SRS (Taylor and Mahoney 1990), more than a dozen taxa were present, of which *Polyarthra* sp. and *Conochilus unicornis,* a colonial form, were the most common. Rotifers are typically 0.1–0.2 mm in length. Most of the rotifers feed as collector-gatherers on algae and bacteria; others prey on very small invertebrates. Their life cycles are very short. Rotifers produce desiccation-resistant eggs.

Mollusks

Mollusks generally do not thrive in waters of low pH or low calcium concentration, which are typical conditions in Carolina bays. Only a few small gastropods, including the limpet *Ferrissia,* have been reported from bays other than Lake Waccamaw (Schalles 1979, Schalles and Shure 1989, Mahoney et al. 1990, McArthur, SREL, unpublished data). These mollusks feed as scrapers on epiphytic material. Life cycles are probably short, and the animals probably aestivate to survive dry seasons.

Other Aquatic Invertebrates

Other aquatic invertebrates, including poriferans, bryozoans, hydrozoans, tardigrades, turbellarians, and water mites, have also been reported from Carolina bays (Mahoney et al. 1990, Leeper and Taylor, in press, McArthur, SREL, unpublished data). Sponge spicules are frequently found in the surficial sediments (Stager and Cahoon 1987, Gaiser 1997), even in bays that dry completely. These occurrences contradict Williams's (1987) observation that sponges do not occur in temporary ponds.

Terrestrial Invertebrates

Assemblages of terrestrial invertebrates occurring at Carolina bays have received less attention than aquatic assemblages. At Sun Bay on the SRS a spring survey yielded specimens from 75 families of terrestrial insects, 10 families of spiders, 1 superfamily of harvestmen, 5 families of millipedes, and 1 family of centipedes (SREL 1980). Twenty-three species of ants were collected from the environs of three other bays on the SRS (Van Pelt and Gentry 1985). Semlitsch (1986) reported on the life history of the mole cricket, *Neocurtilla hexadactyla,* which is common in mesic and hydric habitats of bays of the SRS. Haddad (SREL, unpublished data) surveyed the SRS for butterflies and identified at least two hesperids, *Ancyloxypha numitor* and *Panoquina ocola,* which were associated with bays and swampy areas. Draney (SREL, personal communication) found that wolf spiders (Lycosidae) dominated the assemblage of ground spiders in the basin of Rainbow Bay on the SRS and that orb-weavers (Araneidae) were also common.

ECOLOGY

Assemblages

Most phyla of freshwater invertebrates are well represented in Carolina bays, and the invertebrate assemblages are rich in species. The only well-studied groups are insects and crustaceans. A single bay might support a more than 100 species of aquatic insects (see **Insects** above) and more than 30 species of crustaceans (see **Crustaceans** above; this tally omits speciose but unstudied groups such as ostracods), as well as unknown numbers of species of oligochaetes, rotifers, nematodes, and other invertebrates. We note that because hydroperiod and other environmental conditions are highly variable among years in bays, species lists based on a single year's study will almost certainly be incomplete, even if the hydroperiod during the study year seems "typical."

Microcrustacean assemblages are diverse among bays, as well as within bays (Fig. 8.3). Nearby ponds tend to be more similar than distant ones, but this effect is weak for ponds separated by more than 1 km. The pattern might reflect greater exchange of immigrants among nearby ponds, according to a stepping stone model, or greater similarity of habitat among nearby ponds. Corresponding data are unavailable for other invertebrates. Particularly for insects with cyclic colonization patterns, it seems plausible that assemblages would be influenced by proximity to other aquatic habitats, as well as to other bays.

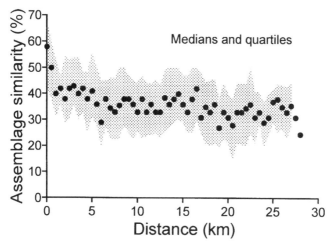

Fig. 8.3. Similarity versus distance for microcrustacean assemblages of Carolina bays and wetland ponds on the Savannah River Site in South Carolina. The Jaccard index of percentage similarity was computed for 88 ponds sampled in February, April, May, August, and November (DeBiase and Taylor, unpublished data). The comparisons are based on a list of 34 cladocerans (except chydorids and macrothricids), other branchiopods, and calanoid copepods that were consistently identified to species level. Medians (filled circles) are shown for 0.5-km distance classes; quartiles (shaded band) are shown if N ≥ 10 observations. Among all pairs, the average similarity was 36 percent.

Genetic data indicate high levels of differentiation among populations of calanoid copepods and of the cladoceran *Daphnia laevis* (Boileau and Taylor 1991). The extent to which these patterns reflect persistent founder effects, ecological divergence, or low exchange of migrants is unresolved. Undoubtedly, wide-ranging vertebrates, including wood ducks, herons, egrets, deer, and humans, function as agents of passive dispersal for invertebrates among bays and other aquatic habitat. The frequency and magnitude of such transfers have not been measured.

Isolation among bays has not resulted in extensive speciation among the resident invertebrates. The only recognized endemics among the bay invertebrates are some of the mollusks of Lake Waccamaw, where the isolation is enhanced by the locally atypical water chemistry of the lake. Some of the calanoid copepods, such as *Aglaodiaptomus atomicus,* may also be endemic to the region, if not to Carolina bays, but too little is presently known about their distributions to make that judgement. The microcrustacean groups that yielded the greatest numbers of rare or undescribed species in California vernal pools (King et al. 1996) remain unstudied (ostracods and harpacticoid copepods) or poorly resolved (chydorid cladocerans and cyclopoid copepods) in Carolina bays.

Among temporary ponds, species richness often increases with duration of the hydroperiod (e.g., vernal and autumnal temporary pools in Ontario, Canada, Wiggins et al. 1980; vernal pools in California, Ebert and Balko 1987; temporary ponds in Wisconsin, Schneider and Frost 1996). A substantial amount of variation in invertebrate assemblages of Carolina bays seems to be associated with hydroperiod. Among 88 bays and other wetland ponds on the SRS, the median number of microcrustacean species (cladocerans except chydorids and macrothricids, other branchiopods, and calanoid copepods) ranged from 6 in ponds that hold water for only a few months annually to 12 in ponds that seldom dry (DeBiase and Taylor, unpublished data). Because all of these microcrustaceans have life cycles short enough to be completed in any of the ponds, the number of species is not determined simply by duration of the hydroperiod. Some of the microcrustaceans are distinctly seasonal in occurrence (see **Population Dynamics and Seasonal Succession** below). The longer hydroperiods, spanning more seasons, can support a greater variety of phenologies and thus more species.

Occurrences of invertebrates with longer life cycles may be affected directly by duration, as well as timing, of the hydroperiod. McClure (1994) observed that insect assemblages of three bays with intermittent hydrologies were dominated by multivoltine species with rapid larval development (mainly dipterans), while two bays with more persistent water were dominated by univoltine species with slow development (odonates, hemipterans, and coleopterans). A similar pattern occurs with pond-breeding amphibians in Carolina bays (Pechmann et al. 1989). The number of species of metamorphosing larvae increases with length of the hydroperiod, both among ponds within years and within ponds among years.

The size of the pond has a modest influence on the number of species. Species-area relations have been examined only for microcrustaceans. Among 23 Carolina bays and other wetland ponds on the SRS, Mahoney et al. (1990) found that the number of cladoceran taxa, but not of calanoid copepod taxa, was positively correlated with area of the wetland. The relationship is approximately log-linear. Because the effect is due mainly to chydorids, we speculate that it is associated with diversification of the littoral habitats favored by members of this family.

There are striking differences between the assemblages of Carolina bays and nearby manmade permanent ponds and reservoirs on the SRS, particularly among the free-swimming or "planktonic" microcrustaceans (DeBiase and Taylor, unpublished data). Relatively large or brightly pigmented microcrustaceans, such as *Daphnia laevis, Aglaodiaptomus* spp., anostracans, and conchostracans, are conspicuous in the bays but absent from the impoundments. Assemblages of these impoundments consist of species that are small or transparent, fitting a pattern that has been repeatedly associated with intense predation by fish or other vertebrates. Few or none of the bays on the SRS hold water permanently. Although there are other differences, such as water chemistry and morphometry, between bays and impoundments, we hypothesize that differences in assemblages are due mainly to the exclusionary effect of bay hydrology and isolation on fish populations (see Wellborn et al. 1996). Because the relatively large or brightly pigmented microcrustaceans do coexist with salamanders in many bays, we must also hypothesize that predation on microcrustaceans by salamanders in the bays is less intense than predation by fish in the impoundments.

Among the smaller epibenthic and littoral taxa, there are no obvious differences between assemblages of Carolina bays and impoundments. In a survey of 88 bays and other wetland ponds and 8 impoundments on the SRS, about half the cladoceran species found in Carolina bays were also found in permanent ponds and reservoirs (DeBiase and Taylor, unpublished data); chydorids and macrothricids were not identified in this study.

The bright carotenoid pigmentation of some calanoid copepods and other microcrustaceans in the Carolina bays may confer protection against ultraviolet radiation (Hairston 1976). However, because ultraviolet radiation should attenuate very rapidly in the dark-stained waters typical of Carolina bays (see Wetzel 1983), the photoprotective benefit is probably less important in bays than in other types of shallow ponds. Its other benefits are as yet unknown.

Population Dynamics and Seasonal Succession

In the annual hydrologic cycle of Carolina bays, winter is the time when water levels generally increase. Among bays that dry, refilling often occurs in winter, and even among bays that do not dry entirely, substantial areas may become reinundated. For most invertebrates in Carolina bays, the active aquatic phase of the life cycle is shorter than the hydroperiod, and many species are distinctly seasonal or even ephemeral in occurrence. The initial dominants are invertebrates that emerge from resting stages in the basin. For example, within hours after the sediments are inundated, calanoid, cyclopoid, and harpacticoid copepods appear; at Rainbow Bay they constituted the largest part of the zooplankton for about a month after the pond filled (Taylor et al. 1989). Most of the calanoid species are active only during the cooler months, but a few, such as *Aglaodiaptomus atomicus,* remain active through summer if the pond holds water (Mahoney et al. 1990, DeBiase and Taylor, unpublished data). Although cladocerans also hatch from resting eggs soon after the pond fills, their abundances are initially low (Taylor and Mahoney 1990). Abundances and diversity of cladocerans increase seasonally. Among bays and other wetland ponds on the SRS, Mahoney et al. (1990) found that the median number of species per pond increased from 7 in February to 13.5 in June; most of the increase was among chydorids and macrothricids, which are mainly epibenthic or littoral. Abundances in most other groups, including oligochaetes, dipterans, and other insects, at Rainbow Bay also increased seasonally, reaching maxima in mid- to late spring (Leeper and Taylor, in press).

For resident aquatic invertebrates that produce resting stages, breaking and reentering dormancy are critical features of population dynamics. These processes have been examined through detailed field and laboratory studies for some of the microcrustaceans at Rainbow Bay on the SRS; both ecological and environmental conditions seem, variously among species, to control them. The cyclopoid copepods are divided nearly equally between species that appear whenever the pond fills and those that seem to require more specific conditions, such as seasonal cues (Medland 1997). Resting eggs of the calanoid copepod *Aglaodiaptomus stagnalis* hatch only if the pond fills between late fall and early spring; if the pond fills earlier and retains water through the late fall and early spring, the eggs remain dormant (Taylor et al. 1990, and unpublished data). Some of the larger microcrustaceans, including *A. stagnalis* and the fairy and clam shrimps, produce only resting eggs. For many other species the return to dormancy seems to depend on ecological or environmental conditions. In the cladoceran *Daphnia laevis,* production of males and resting eggs coincided with declining fecundities and food resources, which occurred at different times in successive years (Taylor and Mahoney 1990, and unpublished data).

The resting stage itself is not exempt from demographic processes. The mortality, as well as the passive dispersal, that can occur during the resting stages may have important effects on dynamics of a population. After several years of drought at Rainbow Bay, numbers of emerging cyclopoid copepodids were reduced by about an order of magnitude, and numbers of hatching *Aglaodiaptomus stagnalis* nauplii were reduced by about two orders of magnitude (Taylor et al. 1990, Wyngaard et al. 1991). Populations of the cyclopoids, which include multivoltine species, recovered to predrought abundances within two months, but *A. stagnalis,* which is univoltine, did not recover until a subsequent year.

The populations of transient aquatic invertebrates in bays are subject to regulation in their alternate habitats, as well as in the Carolina bays. Phenology of dispersal, as well as success in alternate habitats, is obviously important to their success in bays. Bays that are dry during the flight seasons for odonates, for example, may be ignored by species that oviposit only in aquatic habitats.

Trophic Structure and Production

In Carolina bays many of the large aquatic invertebrates, including odonates, hemipterans, and some coleopterans, are predators, feeding generally on other invertebrates but occasionally on larval amphibians or small fishes. Insects and larval salamanders are probably the main aquatic predators of invertebrates in most bays, with fish assuming importance only in those that are permanently flooded.

Larval salamanders in Rainbow Bay on the SRS feed mainly on microcrustaceans and chironomid larvae (Taylor et al. 1988). Predation by salamander larvae has been shown experimentally to depress populations of microcrustaceans (Ginger's Bay on the SRS, Scott 1990; artificial ponds in

North Carolina, Morin et al. 1983b). Pond-breeding salamanders can be active as predators in bays during much of the year. Adults, depending on species, return to breed from autumn through late spring, and the larvae, depending again on species, are present from winter through late summer (Gibbons and Semlitsch 1991).

Most of the small aquatic invertebrates in Carolina bays are collector-gatherers, feeding on smaller particles of food, which may be algae, bacteria, detritus, protozoans, or smaller invertebrates. Shredders are underrepresented (McClure 1994, Leeper and Taylor, in press). With the exception of some beetles, the common aquatic invertebrates of Carolina bays do not consume the living macrophytes. Thus, the main support for their production is probably derived from periphyton, phytoplankton, and detritus, including the remains of both aquatic and terrestrial plants.

Invertebrate biomass has been estimated at only two Carolina bays, both on the SRS, and the data offer contrasting pictures of the relative importance of the two general classes of invertebrate consumers (Fig. 8.4). In Thunder Bay the invertebrate biomass is dominated by macroinvertebrate predators, mainly odonate nymphs. Schalles and Shure (1989) hypothesized that turnover among microinvertebrate prey must be extraordinarily high to support the biomass of predators. In Rainbow Bay the invertebrate biomass is dominated by collector-gatherers, mainly oligochaete worms (the bulk of which are tubificids and naidids), chironomid larvae, cladocerans, and isopods (Leeper and Taylor, in press). Odonates were notably sparse at Rainbow Bay, perhaps because the pond had dried during preceding summers (see *Odonates* above).

At Rainbow Bay demographic analyses indicated that populations of microcrustacean collector-gatherers, such as the cladoceran *Daphnia laevis,* were usually limited by food rather than by predation (Taylor and Mahoney 1990, and unpublished data), and observations of fecundities at other ponds suggest that the phenomenon is widespread. At both Rainbow and Flamingo Bays rapid declines in fecundity coincided with increases in cladoceran populations and great decreases in biovolume of phytoplankton (Taylor and Mahoney, unpublished data). Whether this apparent importance of algal resources to microcrustaceans applies to bays generally or extends to other invertebrate consumers is an open question. Batzer and Wissinger (1996) comment that "the importance of algae to wetland food webs has probably been underestimated."

There are only a few estimates of invertebrate production from bays. Annual production of the planktonic microinvertebrates at Rainbow Bay in 1984 was 6.2 g dry mass m^{-2}, a value that ranks as moderate in comparison with other shallow lakes and ponds (Taylor et al. 1989). In the same year, production of larval salamanders was 1.5 g dry mass m^{-2}, the bulk of which was probably sustained by dipteran larvae (Taylor et al. 1988). In a subsequent study at Rainbow Bay, annual production by oligochaetes, crustaceans, and rotifers was 37 and 15 g dry mass m^{-2} during two consecutive years (1992–

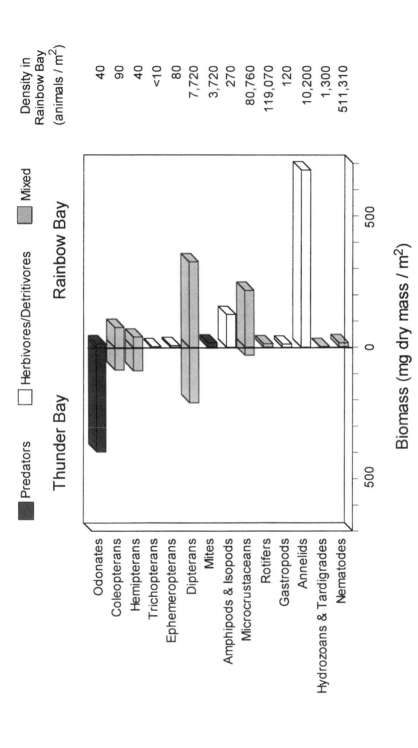

Fig. 8.4. Composition of the invertebrate assemblages at two Carolina bays on the Savannah River Site in South Carolina. Biomass density is shown for both bays. Population density is given only for Rainbow Bay; corresponding data for Thunder Bay were unavailable. Thunder Bay held water continuously during the one-year study by Schalles and Schure (1989); Rainbow Bay held water for 250 and 175 days annually during the two-year study by Leeper and Taylor (in press). In both studies quantitative sampling included sediments, the water column, and aquatic macrophytes. Micro- and macroinvertebrates were sampled at Rainbow Bay; only macroinvertebrates were sampled at Thunder Bay.

93, Leeper and Taylor, in press). The difference between these two years was due primarily to differences in production by benthic oligochaetes and chironomids. For Thunder Bay, Schalles and Shure (1989) argued that 15 g dry mass m^{-2} in annual production of invertebrates would be required to support the observed biomass of odonate and amphibian predators.

The magnitudes of trophic connections with terrestrial consumers remain unexplored. Birds such as wood ducks, herons, and egrets feed on aquatic invertebrates, as well as on the amphibian and piscine predators of aquatic invertebrates (e.g., Landers et al. 1977). Insectivorous birds and bats feed on adult insects emerging from the bays; these insects are also consumed by predatory invertebrates such as adult odonates and spiders.

CONSERVATION

Although Bennett and Nelson (1991) write that "bays have been historically regarded as very uninviting to humans," Carolina bays have a long record of human activity. Archaeological artifacts indicate that bays were used by the earliest people known to have lived in this region (Eberhard et al. 1994). Their activities included habitation on the rims during some prehistoric periods. With European settlement of the Coastal Plain during the eighteenth and nineteenth centuries, human usage and modification of the bays intensified. Many bays were ditched and drained for agriculture or silviculture (Ash et al. 1983, Gilliam and Skaggs 1981). Timber (pond cypress or pine) was harvested from some bays, and a few larger bays were used as mill ponds (see Fig. 8.2). Three additional threats to bays emerged in the 1980s: peat mining, tertiary wastewater treatment, and urbanization (see Richardson and Gibbons 1993, Kadlec and Knight 1995). Today only about 40 percent of the bays in South Carolina remain relatively intact (Bennett and Nelson 1991).

Four major federal programs regulate or protect Carolina bays. The Wetlands Advance Identification Program (ADID), jointly implemented by the United States Army Corps of Engineers and the Environmental Protection Agency, delineates and evaluates wetlands, designating those that are unsuitable for filling or other modification and recommending those that are suitable for restoration or mitigation. The Carolina bay ADID identifies approximately 250 bays (24,700 ha) in South Carolina. The National Environmental Policy Act applies to federal lands, including military reservations and national forests, which include more than 230 bays (5000 ha). The Wetland Reserve Program provides financial incentives to private landowners to protect and restore wetlands. Because so many bays were previously converted to agriculture, South Carolina gives high priority to bays in this program. Under the Emergency Wetlands Resources Act, the United States Fish and Wildlife Service evaluated wetlands to determine which were eligible for federal or federally assisted state acquisition. The Southeast Regional plan listed 28 eligible Carolina bays or bay complexes.

State and private programs also protect Carolina bays. Coordinated efforts between The Nature Conservancy and state natural resource agencies have purchased bays in Georgia (1 bay, 405 ha), South Carolina (32 bays, 3870 ha), and North Carolina (8 bays, 246 ha). These sites have been chosen mainly for their populations of rare or endangered plants. State parks protect all or part of five bay lakes in North Carolina and a large bay and a bay complex in South Carolina. Other bays are located in state forests and wildlife refuges. A few bays in North and South Carolina have been restored and protected under wetlands mitigation banking programs.

Conservation efforts often focus on a few representative examples of a type of habitat. The great diversity of invertebrate assemblages among Carolina bays suggests that this strategy would not protect all or even most of the species within an area, and preservation of insect diversity may depend on the maintenance of other types of aquatic habitats. Circumstances that afford protection of groups of bays are thus particularly fortunate. An outstanding example is the Lewis Ocean Bay preserve in Horry County, South Carolina, which is administered by the South Carolina Heritage Trust. Federal facilities, including Fort Gordon in Georgia and the Savannah River Site in South Carolina, also protect complexes of bays and other wetland and aquatic habitats.

Many Carolina bays on the Savannah River Site, now a National Environmental Research Park, had been severely altered by agricultural activity before 1951, when the land was acquired by the federal government. After more than four decades of passive, benign management, these bays support microcrustacean assemblages that rank among the richest in the world, as well as thriving populations of other invertebrates and amphibians. Whether the communities have returned to the predisturbance states of the late eighteenth or early nineteenth century is unknown, probably unknowable, and in most cases unlikely. In the long view, paleoenvironmental data suggest that the modern era of anthropogenic disturbance represents yet another change in a history that stretches back for 10 or more millennia. In the short view, ecological results reinforce the opinion that disturbed, as well as undisturbed, wetland habitats are worthy of protection.

Few invertebrates from Carolina bays appear on conservation lists. Some of the endemic mollusks of Lake Waccamaw in North Carolina are considered species of special concern, threatened, or endangered (Fuller 1977, Porter 1985). Lake Waccamaw also harbors three endemic fishes, one of which appears on the federal list of threatened species (Rohde et al. 1994). Lake Waccamaw seems to be unique among bays in its endemic invertebrate fauna. Two calanoid copepods found in South Carolina rated as "vulnerable" on 1996 IUCN Red List of Threatened Animals (IUCN 1996). *Hesperodiaptomus augustaensis* occurs in a few Carolina bays and other wetland ponds on the SRS, and *Aglaodiaptomus marshianus* occurs in wetland ponds on the Lower Coastal Plain of South Carolina (DeBiase and Taylor, unpublished data). Their recommender, Janet Reid of the Smithsonian Institution, comments that these and other narrowly distributed species in temporary ponds may become endangered because "ephemeral habitats are under particular pressure for human alteration worldwide" (Reid 1997).

RESEARCH NEEDS

Much of the present knowledge of the ecology of Carolina bays, particularly of the animals, is based on studies on the Savannah River Site. Additional basic research is needed to characterize the communities of bays, particularly elsewhere on the Atlantic Coastal Plain. Critical questions for further research have been summarized by Richardson and Gibbons (1993) and Sharitz and Gresham (1997). The potential effects of fire, hydrologic alteration, and other management techniques on the invertebrates are poorly known. The use of bay habitats by terrestrial invertebrates during both wet and dry seasons, as well as the ecology of aquatic invertebrates during dry seasons, deserves further attention. Paramount is the need to evaluate the roles, current and potential, of these wetlands in the functioning of the larger ecosystem of the Coastal Plain.

ACKNOWLEDGMENTS

Preparation of this chapter was supported in part by the United States Department of Energy under contract DE-FC09-96SR18546 with the Research Foundation of the University of Georgia. We thank M. J. Brooks, S. A. Wissinger, R. R. Sharitz, and J. W. Gibbons for comments on the manuscript. We also thank the researchers and natural resource managers who assisted us in the search for information.

LITERATURE CITED

Ash, A. N., C. B. McDonald, E. S. Kane, and C. A. Pories. 1983. Natural and Modified Pocosins: Literature Synthesis and Management Options. FWS/OBS-83/04. U.S. Fish and Wildlife Service, Division of Biological Services, Washington, DC.

Barnes, R. D. 1980. Invertebrate Zoology (4th ed.). Saunders College/Holt, Rinehart & Winston, Philadelphia, PA.

Batzer, D. P., and S. A. Wissinger. 1996. Ecology of insect communities in nontidal wetlands. Annual Review of Entomology 41:75–100.

Bennett, S. H., and J. B. Nelson. 1991. Distribution and Status of Carolina Bays in South Carolina. Nongame and Heritage Trust Publications, No. 1. South Carolina Wildlife and Marine Resources Department, Columbia, SC.

Bliley, D. J., and D. A. Burney. 1988. Late Pleistocene climatic factors in the genesis of a Carolina bay. Southeastern Geology 29:83–101.

Boileau, M. G., and B. E. Taylor. 1994. Chance events, habitat age, and the genetic structure of pond populations. Archiv für Hydrobiologie 132:191–202.

Brigham, A. R., W. U. Brigham, and A. Gnilka, (eds.). 1982. Aquatic Insects and Oligochaetes of North and South Carolina. Midwest Aquatic Enterprises, Mahomet, IL.

Brooks, M. J., B. E. Taylor, and J. A. Grant. 1996. Carolina bay geoarchaeology and Holocene landscape evolution on the Upper Coastal Plain of South Carolina. Geoarchaeology 11:481–504.

Cahoon, L. B., W. D. Grater, and D. L. Covington. 1992. Phenotypic and genotypic differences between two adjacent populations of the Asiatic clam, *Corbicula fluminea*. Journal of the Elisha Mitchell Scientific Society 108:57–63.

Cahoon, L. B., and D. A. Owen. 1996. Can suspension feeding by bivalves regulate phytoplankton biomass in Lake Waccamaw, North Carolina? Hydrobiologia 325: 193–200.

Clark, M. K., D. S. Lee, and J. B. Funderburg, Jr. 1985. The mammal fauna of Carolina bays, pocosins, and associated communities in North Carolina: An Overview. Brimleyana 11:1–38.

Coker, R. E. 1938. Notes on the peculiar crustacean fauna of White Lake, North Carolina. Archiv für Hydrobiologie 34:130–133.

Cross, W. H. 1955. Anisopteran Odonata of the Savannah River Plant, South Carolina. Journal of the Elisha Mitchell Scientific Society 71:9–17.

Davis, G. M., W. H. Heard, S. L. H. Fuller, and C. Hesterman. 1981. Molecular genetics and speciation in *Elliptio* and its relationships to other taxa of North American Unionidae (Bivalvia). Biological Journal of the Linnean Society 13:131–150.

DeBiase, A. E., and B. E. Taylor. 1993. New occurrences of *Eurytemora affinis* and *Epischura fluviatilis*, freshwater calanoid copepods of the family Temoridae, in South Carolina. American Midland Naturalist 130:386–392.

DeBiase, A. E., and B. E. Taylor. 1997. *Aglaodiaptomus atomicus*, a new species (Crustacea: Copepoda: Calanoida: Diaptomidae) from freshwater wetland ponds in South Carolina, U.S.A., and a redescription of *A. saskatchewanensis* (Wilson 1958). Proceedings of the Biological Society of Washington 110:569–580.

Eberhard, K., K. E. Sassaman, and M. J. Brooks. 1994. Crosby Bay (38AK682): Paleoindian and Early Archaic occupations at a Carolina bay in Aiken County, South Carolina. South Carolina Antiquities 26:33–46.

Ebert, T. A., and M. L. Balko. 1987. Temporary pools as islands in space and in time: The biota of vernal pools in San Diego, Southern California, USA. Archiv für Hydrobiologie 110:101–123.

Frey, D. G. 1948. A biological survey of Lake Waccamaw. Wildlife in North Carolina (July), 4–6, 23. Raleigh, NC.

———. 1949. Morphometry and hydrography of some natural lakes of the North Carolina Coastal Plain: The bay lake as a morphometric type. Journal of the Elisha Mitchell Scientific Society 65:1–37.

———. 1951a. Pollen succession in the sediments of Singletary Lake, North Carolina. Ecology 32:518–533.

———. 1951b. The fishes of North Carolina's bay lakes and their intraspecific variation. Journal of the Elisha Mitchell Scientific Society 67:1–44.

Fuller, S. L. H. 1977. Freshwater and terrestrial mollusks. Pages 143–194 *in* J. E. Cooper, S. S. Robinson, and J. B. Funderburg (eds.), Endangered and Threatened

Plants and Animals of North Carolina. North Carolina State Museum of Natural History, Raleigh, NC.

Gaiser, E. E. 1997. Paleolimnological reconstruction of Holocene environments in wetland ponds of the Upper Atlantic Coastal Plain using siliceous microfossils. Ph.D. Dissertation, University of Georgia, Athens, GA.

Gibbons, J. W., and R. D. Semlitsch. 1991. Guide to the Reptiles and Amphibians of the Savannah River Site. University of Georgia Press, Athens, GA.

Gilliam, J. W., and R. W. Skaggs. 1981. Drainage and agricultural development: Effects on drainage waters. Pages 109–124 *in* C. J. Richardson (ed.), Pocosin Wetlands: An Integrated Analysis of Coastal Plain Freshwater Bogs in North Carolina. Hutchinson Ross, Stroudsburg, PA.

Grant, J. A., M. J. Brooks, and B. E. Taylor. 1997. New constraints on the evolution of Carolina Bays from ground penetrating radar. Geomorphology 22:325–345.

Hairston, N. G., Jr. 1976. Photoprotection by carotenoid pigments in the copepod *Diaptomus nevadensis*. Proceedings of the National Academy of Sciences USA 73: 971–974.

Hobbs, H. H., III, J. H. Thorp, and G. E. Anderson. 1978. The Freshwater Decapod Crustaceans (Palaemonidae, Cambaridae) of the Savannah River Plant, South Carolina. Savannah River Ecology Laboratory, Aiken, SC.

International Union for Conservation of Nature and Natural Resources. 1996. J. Baillie and B. Groombridge (eds.), IUCN Red List of Threatened Animals. IUCN Publications Service Unit, Cambridge, UK.

Isphording, W. C., and J. F. Fitzpatrick, Jr. 1992. Geologic and evolutionary history of drainage systems in the southeastern United States. Pages 19–56 *in* C. T. Hackney, S. M. Adams, and W. H. Martin (eds.), Biodiversity of the Southeastern United States / Aquatic Communities. John Wiley & Sons, New York.

Johnson, D. 1942. The Origin of the Carolina Bays. Columbia University Press, New York.

Johnson, R. I. 1984. A new mussel, *Lampsilis* (*Lampsilis*) *fullerkati* (Bivalvia: Unionidae) from Lake Waccamaw, Columbus County, North Carolina, with a list of the other unionid species of the Waccamaw River system. Occasional Papers on Mollusks 4:305–319.

Kaczorowski, R. T. 1977. The Carolina bays and their relationship to modern oriented lakes. Ph.D. Dissertation, University of South Carolina, Columbia, SC.

Kadlec, R. H., and R. L. Knight. 1995. Treatment Wetlands: Theory and Implementation. Lewis, Boca Raton, FL.

Kat, P. W. 1983. Morphological divergence, genetics, and speciation among *Lampsilis* (Bivalvia: Unionidae). Journal of Molluscan Studies 49:133–145.

King, J. L., M. A. Simovich, and R. C. Brusca. 1996. Species richness, endemism and ecology of crustacean assemblages in northern California vernal pools. Hydrobiologia 328:85–116.

Kirkman, L. K., R. F. Lide, G. Wein, and R. R. Sharitz. 1996. Vegetation changes and land-use legacies of depression wetlands of the western Coastal Plain of South Carolina: 1951–1992. Wetlands 16:564–576.

Kondratieff, B. C., and C. J. Pyott. 1987. The Anisoptera of the Savannah River Plant, South Carolina, United States: Thirty years later. Odonatologica 16:9–23.

Landers, J. L., T. T. Fendley, and A. S. Johnson. 1977. Feeding ecology of wood ducks in South Carolina. Journal of Wildlife Management 41:118–127.

Lee, D. S. 1987. Breeding birds of Carolina bays: Succession-related density and diversification on ecological islands. The Chat 51:85–102.

Leeper, D. A., and B. E. Taylor. 1998. Insect emergence from a South Carolina (USA) temporary wetland pond, with emphasis on the Chironomidae (Diptera). Journal of the North American Benthological Society 17:54–72.

———. Abundance, biomass and production of aquatic invertebrates in Rainbow Bay, a temporary wetland in South Carolina, U.S.A. Archiv für Hydrobiologie. (In press.)

Lide, R. F. 1997. When is a depression wetland a Carolina bay? Southeastern Geographer 37:90–98.

Lide, R. F., V. G. Meentemeyer, J. E. Pinder, III, and L. M. Beatty. 1995. Hydrology of a Carolina bay located on the Upper Coastal Plain of western South Carolina. Wetlands 15:47–57.

Mahoney, D. L., M. A. Mort, and B. E. Taylor. 1990. Species richness of calanoid copepods, cladocerans and other branchiopods in Carolina bay temporary ponds. American Midland Naturalist 123:244–258.

McClure, M. A. 1994. Aquatic insects of five clay-based Carolina bays in South Carolina. Master's Thesis, Clemson University, Clemson, SC.

Medland, V. L. 1997. Impact of environmental variability on the success of dormancy strategies of cyclopoid copepods. Ph.D. Dissertation, University of Georgia, Athens, GA.

Merritt, R. W., and K. W. Cummins, (ed.). 1996. An Introduction to the Aquatic Insects of North America (3rd ed.). Kendall/Hunt, Dubuque, IA.

Morin, P. J. 1983. Competitive and predatory interactions in natural and experimental populations of *Notophthalmus viridescens dorsalis* and *Ambystoma tigrinum*. Copeia 1983:628–639.

Morin, P. J., H. M. Wilbur, and R. N. Harris. 1983. Salamander predation and the structure of experimental communities: Responses of *Notophthalmus viridescens* and microcrustacea. Ecology 64:1430–1436.

National Oceanic and Atmospheric Administration. 1992. Climatological Data Annual Summary, South Carolina, Volume 95, Number 13. National Climatic Data Center, Asheville, NC.

Newman, M. C., and J. F. Schalles. 1990. The water chemistry of Carolina bays: A regional survey. Archiv für Hydrobiologie 118:147–168.

Nifong, T. D. 1982. The "clay subsoil" Carolina bays of North Carolina. Report submitted to the North Carolina Nature Conservancy.

Pechmann, J. H. K., D. E. Scott, J. W. Gibbons, and R. D. Semlitsch. 1989. Influence of wetland hydroperiod on diversity and abundance of metamorphosing juvenile amphibians. Wetlands Ecology and Management 1:3–11.

Pennak, R. W. 1989. Fresh-Water Invertebrates of the United States: Protozoa to Mollusca (3rd ed.). John Wiley & Sons, New York.

Pickens, R. M., and C. H. Jagoe. 1996. Relationships between precipitation and surface water chemistry in three Carolina bays. Archiv für Hydrobiologie 137:187–209.

Porter, H. J. 1985. Molluscan census and ecological interrelationships. North Carolina Wildlife Resources Commission, Raleigh, NC.

Prouty, W. F. 1952. Carolina bays and their origin. Bulletin of the Geological Society of America 63:167–224.

Reid, J. 1997. Copepod species in the 1996 IUCN Red List of Threatened Animals. Monoculus 33:23–24.

Richardson, C. J., and J. W. Gibbons. 1993. Pocosins, Carolina bays, and mountain bogs. Pages 257–310 *in* W. H. Martin, S. G. Boyce, and A. C. Echternacht (eds.), Biodiversity of the Southeastern United States/Lowland Terrestrial Communities. John Wiley & Sons, New York.

Rohde, F. C., R. G. Arndt, D. G. Lindquist, and J. F. Parnell. 1994. Freshwater Fishes of the Carolinas, Virginia, Maryland, and Delaware. University of North Carolina Press, Chapel Hill, NC.

Ross, T. E. 1987. A comprehensive bibliography of the Carolina Bays literature. Journal of the Elisha Mitchell Scientific Society 103:28–42.

Savage, H., Jr. 1982. The Mysterious Carolina Bays. University of South Carolina Press, Columbia, SC.

Savannah River Ecology Laboratory. 1980. A Biological Inventory of the Proposed Site of the Defense Waste Processing Facility on the Savannah River Plant in Aiken, South Carolina. SREL-7. Savannah River Ecology Laboratory, Aiken, SC.

Schalles, J. F. 1979. Comparative limnology and ecosystem analysis of Carolina Bay ponds on the Upper Coastal Plain of South Carolina. Ph.D. Dissertation, Emory University, Atlanta, GA.

Schalles, J. F., R. R. Sharitz, J. W. Gibbons, G. J. Leversee, and J. N. Knox. 1989. Carolina Bays of the Savannah River Plant, Aiken, South Carolina. SRO-NERP-18. Savannah River Ecology Laboratory, Aiken, SC.

Schalles, J. F., and D. J. Shure. 1989. Hydrology, community structure, and productivity patterns of a dystrophic Carolina bay wetland. Ecological Monographs 59:365–385.

Schneider, D. W., and T. M. Frost. 1996. Habitat duration and community structure in temporary ponds. Journal of the North American Benthological Society 15:64–86.

Scott, D. E. 1990. Effects of larval density in *Ambystoma opacum:* An experiment in large-scale field enclosures. Ecology 71:296–306.

Semlitsch, R. D. 1986. Life history of the northern mole cricket, *Neocurtilla hexadactyla* (Orthoptera: Gryllotalpidae), utilizing Carolina-bay habitats. Annals of the Entomological Society of America 79:256–261.

Semlitsch, R. D., D. E. Scott, J. H. K. Pechmann, and J. W. Gibbons. 1996. Structure and dynamics of an amphibian community: Evidence from a 16-year study of a natural pond. Pages 217–248 *in* M. L. Cody and J. A. Smallwood (eds.), Long-Term Studies of Vertebrate Communities. Academic Press, San Diego, CA.

Sharitz, R. R., and C. A. Gresham. 1997. Pocosins and Carolina bays. Pages 343–377 *in* M. G. Messina and W. G. Conner (eds.), Southern Forested Wetlands: Ecology and Management. Lewis/CRC Press, Boca Raton, FL.

Snodgrass, J. W., A. L. Bryan, Jr., R. F. Lide, and G. M. Smith. 1996. Factors affecting the occurrence and structure of fish assemblages in isolated wetlands of the upper coastal plain, U.S.A. Canadian Journal of Fisheries and Aquatic Sciences 53:443–454.

Stager, J. C., and L. B. Cahoon. 1987. The age and trophic history of Lake Waccamaw, North Carolina. Journal of the Elisha Mitchell Scientific Society 103:1–13.

Stiven, A. E., and J. Alderman. 1992. Genetic similarities among certain freshwater mussel populations of the *Lampsilis* genus in North Carolina. Malacologia 34:355–369.

Taylor, B. E., R. A. Estes, J. H. K. Pechmann, and R. D. Semlitsch. 1988. Trophic relations in a temporary pond: Larval salamanders and their microinvertebrate prey. Canadian Journal of Zoology 66:2191–2198.

Taylor, B. E., and D. L. Mahoney. 1990. Zooplankton in Rainbow Bay, a Carolina bay pond: Population dynamics in a temporary habitat. Freshwater Biology 24:597–612.

Taylor, B. E., D. L. Mahoney and R. A. Estes. 1989. Zooplankton production in a Carolina bay. Pages 425–435 *in* R. R. Sharitz and J. W. Gibbons (eds.), Freshwater Wetlands and Wildlife. CONF-8603101. DOE Symposium Series No. 61. Office of Scientific and Technical Information, United States Department of Energy, Oak Ridge, TN.

Taylor, B. E., G. A. Wyngaard, and D. L. Mahoney. 1990. Hatching of *Diaptomus stagnalis* eggs from a temporary pond after a prolonged dry period. Archiv für Hydrobiologie 117:217–278.

Thom, B. G. 1970. Carolina bays in Horry and Marion counties, South Carolina. Geological Society of America Bulletin 81:783–814.

Thorp, J. H., and A. P. Covich (eds.). 1991. Ecology and Classification of North American Freshwater Invertebrates. Academic Press, San Diego, CA.

Van Pelt, A., and J. B. Gentry. 1985. The ants (Hymenoptera: Formicidae) of the Savannah River Plant, South Carolina. SRO-NERP-14. Savannah River Ecology Laboratory, Aiken, SC.

Watts, W. A. 1980. Late-Quaternary vegetation history at White Pond on the Inner Coastal Plain of South Carolina. Quaternary Research 13:187–199.

Wellborn, G. A., D. K. Skelly, and E. E. Werner. 1996. Mechanisms creating community structure across a freshwater habitat gradient. Annual Review of Ecology and Systematics 27:337–363.

Wetzel, R. G. 1983. Limnology (2nd ed.). Saunders College, Philadelphia, PA.

Wiggins, G. B., R. J. Mackay, and I. M. Smith. 1980. Evolutionary and ecological strategies of animals in annual temporary pools. Archiv für Hydrobiologie Supplement 58:97–206.

Williams, D. D. 1987. The Ecology of Temporary Waters. Timber Press, Portland, OR.

Williams, D. D., and B. W. Feltmate. 1992. Aquatic Insects. CAB International, Wallingford, UK.

Wyngaard, G. A., B. E. Taylor, and D. L. Mahoney. 1991. Emergence and dynamics of cyclopoid copepods in an unpredictable environment. Freshwater Biology 25:219–232.

9 Forested Limesink Wetlands of Southwest Georgia

Invertebrate Habitat and Hydrologic Variation

STEPHEN W. GOLLADAY, SALLY ENTREKIN, and
BRAD W. TAYLOR

*L*imesink wetlands are a common aquatic habitat in southwest
Georgia. These wetlands are nonalluvial, occupying shallow depressions formed from dissolution of limestone bedrock and collapse of surface sands. Limesink wetlands are seasonally
inundated with a typical hydroperiod extending from late February to early
July. Little is known about factors influencing invertebrate assemblages in
limesink wetlands. We had an opportunity to examine the influence of hydrologic variation on invertebrates forested limesink wetlands over a three-
year period. Quantitative samples of invertebrates were taken monthly on
benthic and wood surfaces from March 1994 through July 1996. This included a period of unusually heavy precipitation in summer and autumn of
1994, when the wetlands would normally have been dry. Immediately following inundation, benthic samples were dominated by amphipods (Crangonyx sp.), isopods (Caecidotea sp.), cladocerans, and copepods. Maximum
total densities (1000–4000 individuals per m²) were observed within three
months of inundation. During summer and autumn densities declined (<500
individuals per m²) and the benthos was dominated by larval chironomids.
Wood surfaces were dominated by chironomids with greatest densities
(1000–3000 individuals per m²) observed in summer and autumn. Although
not quantified, freshwater sponge became very abundant on wood surfaces
during autumn. During the spring of 1995 and 1996 invertebrate densities
in benthic samples remained low. Chironomids remained very abundant on
wood. Our results suggest that extended inundation is a disturbance to

Invertebrates in Freshwater Wetlands of North America: Ecology and Management, Edited by
Darold P. Batzer, Russell B. Rader, and Scott A. Wissinger
ISBN 0-471-29258-3 © 1999 John Wiley & Sons, Inc.

some wetland invertebrates. It may cause short-term reductions in popula-
tions by reducing summer refugia or altering environmental cues necessary
for the completion of life cycles.

LIMESINK WETLANDS

In the southeastern United States, seasonally inundated wetlands are a common feature of the landscape (Tansey and Cost 1990). These wetlands are
often shallow basins isolated from streams or other permanent water bodies.
The seasonal wetlands of the southeast have been recognized as sites of moderate productivity and diversity and an important wildlife habitat (e.g., Sklar
1985, Taylor et al. 1988, 1989, Dodd 1992, 1995). They are also a habitat
threatened by regional development (Tansey and Cost 1990). To date there
have been few ecological studies on invertebrate assemblages in these systems.

Limesink wetlands are a type of seasonal wetland on the Gulf Coastal Plain
of southwest Georgia. They are shallow surface depressions ranging from
small sinks with steep sides to shallow, flat wetlands of several hectares
(Wharton 1978). Depths range from <1 m to 8 m (Torak et al. 1991). Within
the region the broad range of limesink wetland size and widely fluctuating
water levels provide diverse habitats. Vegetation ranges from open herbaceous
meadows to pond cypress savannas to dense pond cypress-tupelo forests (Sutter and Kral 1994). Numerous rare plants are associated with limesink wetlands, and they are also an important breeding habitat for rare or threatened
amphibians (Lynch et al. 1986, Sutter and Kral 1994, Dodd 1992).

Little is known about factors influencing invertebrate assemblages in limesink wetlands or many other southeastern seasonal wetlands. Although the
length of the hydroperiod has been acknowledged as an important influence
(e.g., Sutter and Kral 1994), there have been few multiyear studies of hydrologic variation and its effect on invertebrates. In addition, the role of wood
debris as a habitat has been virtually ignored, even though many southeastern
wetlands are heavily forested (Mitsch and Gosselink 1993). Wood debris has
been recognized as a site of high invertebrate activity, diversity, and productivity in southeastern Coastal Plain streams (Benke et al. 1985) and in forested
wetlands on their floodplains (Thorp et al. 1985). To date, most studies of
southeastern wetland invertebrates have focused on planktonic or benthic assemblages.

This chapter reviews habitat use and response to extended inundation in
forested wetlands of southwest Georgia over a three-year period (1994–1996).
During the summer and autumn of 1994 a series of tropical depressions resulted in unusually heavy precipitation in southwest Georgia. Many limesink

wetlands remained continually inundated from March 1994 through July 1995, substantially extending the normal hydroperiod. We examined habitat use and responses of aquatic invertebrates during these unusual hydrologic conditions.

Geological Setting

Limesink wetlands occur on the Dougherty Plain, an area of about 11,400 km² located in southwest Georgia. The boundaries of the Dougherty Plain are the Chattahoochee River and Fall-line Hills to the west and north and Pelham Escarpment to the south and east (Beck and Arden 1983). It is a region of low topographic relief, ranging in elevation from approximately 91 m above sea level in the northwest to 15 m above sea level in the south (Hayes et al. 1983).

The geology of the Dougherty Plain is formed from three major geologic units. The surface layer is residuum, that is, porous sand and clay ranging in depth from a 1–40 m with an average of 15 m (Hayes et al. 1983). Beneath the residuum lies the Ocala limestone, which ranges in thickness from 1–100 m and is highly fractured and riddled with irregular pores, giving it a sponge-like character (Beck and Arden 1983, Hayes et al. 1983). Beneath the Ocala Limestone lies the Lisbon Formation, an unfractured and dense limestone layer (Hayes et al. 1983). The Ocala limestone is the primary rock layer, containing the Floridan aquifer, underlying southern South Carolina, parts of Georgia, Alabama, and Florida (Hicks et al. 1981, Hayes et al. 1983).

Limesink wetlands occupy shallow basins formed from the dissolution of limestone bedrock, followed by the collapse of the overlying residuum (Hendricks and Goodwin, 1952). It has been hypothesized that depressions become sealed with clay as they age, allowing them to hold water for varying lengths of time (Hendricks and Goodwin 1952). Stream drainage density is relatively low on the Dougherty Plain, and most limesink wetlands are isolated from streams. Regional water movement is primarily subsurface. Rainfall rapidly enters near-surface groundwater where it discharges into streams and wetlands or moves more slowly into the underlying aquifer.

Average Climatic Conditions

Southwest Georgia has a subtropical climate, with hot, humid summers and mild winters. Average daily temperatures range from 10°C in winter to 27°C in summer (National Climate Data Center, Asheville, NC). Annual precipitation averages 130 cm (range 64–196, 1895–1996, National Climate Data Center, Asheville, NC). Overall, rainfall is evenly distributed throughout the year. However, minor peaks in average rainfall occur during March and July (~12–16 cm), and the lowest amount of rainfall occurs in October (6 cm). Winter precipitation (January through March) is usually associated with frontal weather systems and tends to be long in duration and moderate in intensity

(Hayes et al. 1983). Summer precipitation (June through August) is usually associated with thunder showers which are generally short in duration and relatively intense (Hayes et al. 1983). Deviations from this pattern can occur during years when tropical storms and hurricanes result in heavy summer rainfalls.

Seasonal moisture surpluses and deficits, that is, the balance between rainfall and evapotranspiration, control the availability of water to limesink wetlands. High summer temperatures result in high rates of evapotranspiration. Under average conditions, moisture deficits occur from April through September as average daily temperatures increase above 15°C (National Climate Data Center, Asheville, NC). During the rest of the year moisture surpluses result from milder temperatures. Inundation of limesink wetlands generally occurs during late winter, and they generally dry completely during summer (Hayes et al. 1983).

Recent Climatic Conditions

The hydroperiod of limesink wetlands can vary greatly from year to year, depending on patterns of rainfall. For example, during 1993 precipitation was near average (121 cm, National Climate Data Center, Asheville, NC). Late-winter rains caused inundation of wetlands in late February (Fig. 9.1). Low rainfall in combination with increasing temperatures resulted in late spring moisture deficits. In 1993 the hydroperiod for limesink wetlands ended in late June/early July.

1994 also began as an "average year," with the hydroperiod beginning in February in response to average amounts of rainfall and seasonal moisture surpluses. Wetlands began to dry in late spring. However, during early July Tropical Storm Alberto spread heavy rains (13–71 cm) over the region (Hippe et al. 1994). In addition, a series of named (four total, Neumann et al. 1993) and unnamed tropical cyclones occurred through October 1994 creating large moisture surpluses (Fig. 9.1). The hydroperiod of wetlands was extended through the end of the year. With the total estimated at 192 cm, 1994 had the greatest rainfall in 100 years of records in southwest Georgia (National Climate Data Center, Asheville, NC).

Conditions in 1995 were closer to average (Fig. 9.1), with 119 cm of rainfall. However, because of water surpluses the previous year, wetlands remained inundated through July, when normal drying occurred. The unusual conditions of 1994 resulted in the extension of the average hydroperiod (typically about 6 months) to almost 16 months. During 1996 limesink wetlands filled in February and dried in late June.

MACROINVERTEBRATE SAMPLING

We have been sampling aquatic macroinvertebrates since March 1994 (Fig. 9.1). Our study sites are three forested limesink wetlands located on the Ichau-

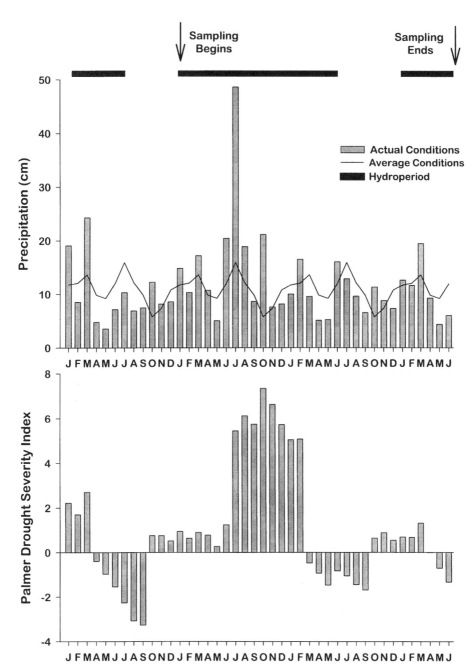

Fig. 9.1. Precipitation and drought severity for southwest Georgia since 1993. The length of the hydroperiod of limesink wetlands is indicated by horizontal bars. Data are from the National Climate Data Center, Drought Series Data, Asheville, NC.

way Ecological Reserve, near Newton, Georgia. The reserve is a 10,500-ha remnant tract of longleaf pine/wire-grass savanna. The wetlands (Collins Pond, King Pond, and Predest Pond) range in size from 7.5–15 ha and show little evidence of recent human disturbance, having mature forest stands (Lynch et al. 1986). Their vegetation is dominated by an open to dense canopy of pond cypress (*Taxodium ascendens*) over a subcanopy of swamp blackgum (*Nyssa biflora*) (Lynch et al. 1986, Watt, J. W. Jones Ecological Research Center, unpublished data). Based on canopy vegetation, they can be classified as swamp tupelo forests, 1 of 18 nonalluvial wetland communities recognized across the Coastal Plain (Allard 1990, cited in Sutter and Kral 1994). Swamp tupelo wetlands are generally nutrient-poor, with low alkalinity, and like other limesink wetlands they have a hydroperiod driven by rainfall (Mitsch and Gosselink 1993, Sutter and Kral 1994). The wetlands we sampled are mostly shallow (<1 m), although a few small deeper depressions (~2 m) are present (Golladay, personal observation). The primary habitats in the wetlands are the benthos (i.e., leaf litter and fine organic detritus with little inorganic matter) and wood debris. Scattered patches of emergent aquatic plants (primarily *Carex* sp.) occur in the wetlands, but these were not sampled.

At each wetland three permanent sampling transects were established at equal intervals around the perimeter. At roughly monthly intervals during wetland inundation, we collected one sample from the benthic surface and from wood debris at each transect (i.e., three samples from each habitat per pond on each date). All collections were made within 5–10 m of the edge of the water at depths of 30–50 cm. Samples were collected randomly, and as the ponds filled and dried we moved along the transect to consistently sample at a 30- to 50-cm depth. Benthic samples were collected using a 25.4-cm diameter core and hand-operated bilge pump. Coarse organic matter (primarily whole leaves and leaf fragments) was removed from the core by hand and placed in a container. Water and fine material (primarily organic matter) were pumped from the core through a 500-μm mesh net. The contents of the net were placed in a container, preserved in ethanol, and taken to the laboratory for processing and counting. The samples included organisms present in the water column, on the sediment surface, and within the sediment.

On each sampling date a single piece of wood debris was sampled at each transect. Submerged wood was placed in a bucket and taken to shore. On shore, invertebrates were washed from wood surfaces into a bucket using a soft bristle brush and wash bottle. Wood was carefully inspected following washing, and any remaining invertebrates were picked from the surface. The contents of the bucket were rinsed onto a 500-μm mesh sieve, preserved in ethanol, and taken to the laboratory for processing and counting. The length and diameter of wood were recorded and used to estimate area sampled.

In the laboratory, invertebrates were sorted by hand and identified using Merritt and Cummins (1984) and Pennak (1989). Most invertebrates were identified to family or genus. Based on information in taxonomic references, food habits of each taxon were determined and assigned to broad feeding categories (i.e., predator, detritivore, scavenger). Larval chironomids were excluded from the classification because this group contains both predators and detritivores. Predators were assumed to feed on other macroinvertebrates. While some taxa classified as detritivores derive substantial nutrition from bacteria, protozoans, or other microinvertebrates, we assumed most were feeding on the detrital aggregate. Scavengers were assumed to be opportunistic feeders on either plant or animal matter.

Invertebrate densities were expressed as number of individuals per m² of wood or benthic surface sampled. To improve homogeneity of variances, means for each pond/substrate/date combination were calculated using a natural log + 1 transformation; back transformed means were used in subsequent analyses (Sokal and Rohlf 1995). Average density of dominant taxonomic groups was calculated for February–June 1994, 1995, 1996 for each habitat. These averages represent population means during the usual period of inundation prior to, immediately following, and one year following the unusual hydrologic conditions of 1994. Densities of major taxonomic groups were compared using a two-way ANOVA (using taxonomic group and year as class-level factors). This analysis allowed us to statistically examine the responses of dominant invertebrate groups to extended inundation. However, because we sampled only one extended hydroperiod, differences in invertebrate density cannot be statistically attributed to extended inundation.

MACROINVERTEBRATE FAUNA

Forested limesink wetlands support an abundant invertebrate fauna. In all, 57 taxa from 17 invertebrate groups were collected on wood and benthic surfaces in our study (Table 9.1). Most taxa were present in about the same frequency on benthic and wood surfaces. However, several exceptions existed. Larval chaoborids and crayfish (Cambaridae) were found in greater frequency in benthic than wood samples. An aquatic neuropteran (*Sisyra* sp.) was found with greater frequency on wood than the benthos. *Sisyra* is a predator of freshwater sponges (Merritt and Cummins 1984). Although not quantified, freshwater sponge was found almost exclusively on wood debris or on the submerged trunks of trees. Larval *Scirtes* (Coleoptera) and *Chauloides* (Megaloptera) were also found exclusively on wood.

In benthic samples the greatest number of taxa were usually observed in late winter or early spring, particularly those following a dry period (1994 and 1996) (Fig. 9.2). Taxa numbers were reduced during the period of extended inundation (summer and autumn 1994). Numbers of taxa were more variable on wood, with peaks in taxa abundance observed during spring (1994, 1995, 1996) and during extended inundation (autumn 1994) (Fig. 9.2).

Of the 57 taxa collected, 6 dominated the community. Amphipods, oligochaetes, cladocerans, copepods, chironomids, and isopods were present on 60–100 percent of days sampled and accounted for >90 percent of

TABLE 9.1. Taxa List of Invertebrates Found in Association with Benthic and Wood Habitats in Forested Wetlands of Southwest Georgia[a]

Group	Family	Genus	Classification	Benthos	Wood
Amphipoda	Crangonyctidae	*Crangonyx*	detritivore	80.8	88.5
Oligochaeta			detritivore	76.9	88.5
Cladocera			detritivore	57.7	80.8
Coleoptera	Curculionidae		shredder	0.0	7.7
	Dytiscidae	*Agabus*	predator	11.5	11.5
		Dytiscus	predator	15.4	11.5
		Hydroporus	predator	3.8	0.0
		Celina	predator	3.8	0.0
		Acilius	predator	0.0	7.7
	Gyrinidae	*Gyrinis*	predator	3.8	15.4
		Dineutus	predator	0.0	19.2
	Dryopidae		predator	3.8	0.0
	Elmidae		collector	3.8	0.0
	Haliplidae	*Peltodytes*	herbivore	3.8	7.7
	Scirtidae	*Scirtes*	detritivore	0.0	26.9
	Noteridae	*Hydrocanthus*	predator	3.8	0.0
	Hydrophilidae		predator	7.7	7.7
		Berosus	predator	0.0	7.7
		Hydrochara	predator	0.0	11.5
		Tropisternus	predator	3.8	0.0
	Staphylinidae		predator	3.8	0.0
	Unknown	Adults	unknown	7.7	11.5
		Pupae	unknown	0.0	3.8
		Larve	unknown	15.4	11.5
		Unknown	unknown	19.2	23.1
			collector	7.7	23.1
Collembola	Smintheridae		collector	7.7	11.5
Copepoda			detritivore	88.5	92.3
Decapoda	Cambaridae		scavenger	11.5	0.0
		Faxonella	scavenger	38.5	11.5
Diptera	Ceratopogonide	*Bezzia*	predator	61.5	69.2
	Chaoboridae		predator	65.4	30.8
	Chironomidae		unknown	92.3	92.3
	Culicidae		detritivore	3.8	19.2
	Dolichopodidae		predator	19.2	11.5
	Stratiomyidae		detritivore	0.0	3.8
	Unknown	Larvae	unknown	15.4	30.8
		Pupae	unknown	0.0	7.7
Ephemeroptera	Baetidae	*Callibaetis*	collectors	3.8	7.7
Hemiptera	Belostomatidae		predator	0.0	7.7
	Corixidae		predator	7.7	3.8
		Corisella	predator	3.8	7.7
		Hesperocorixa	predator	30.8	23.1
	Gerridae	*Gerris*	predator	3.8	3.8
	Notonectidae	*Notonecta*	predator	7.7	3.8
		Buenoa	predator	23.1	11.5

TABLE 9.1. (Continued)

Group	Family	Genus	Classification	Benthos	Wood
Isopoda	Ascellidae	*Caecidotea*	detritivore	73.1	84.6
Megaloptera	Corydalidae	*Chauloides*	predator	0.0	19.2
Mollusca	Ancylidae		scraper	7.7	7.7
		Laevapex	scraper	0.0	7.7
Neuroptera	Sisyridae	*Sisyra*	predator	15.4	57.7
Odonata	Coenagrionidae	*Enallagma*	predator	3.8	11.5
	Lestidae	*Lestes*	predator	3.8	0.0
	Libellulidae		predator	26.9	46.2
Ostracoda			scavenger	15.4	11.5
Unknown		Larvae	unknown	11.5	11.5
		Adults	unknown	15.8	57.7

[a]Classification is by the dominant food habits of the group. Numbers are frequencies with which invertebrates were collected during monthly sampling from March 1994 through July 1996. To be counted, a taxon had to be present on 1 of 9 samples taken from the ponds on a particular date. Frequency was calculated as: (months present/months sampled)*100.

invertebrate density on both substrates. These taxa were predominantly detritivorous (Table 9.1), suggesting that the foodweb of forested limesink wetlands is detritus-based.

Within a few weeks of inundation in 1994, densities of invertebrates in benthic samples were 1000–2000 individuals per m² in all ponds (Fig. 9.3). Early benthic communities were dominated by amphipods (*Crangonyx* sp.) and isopods (*Caecidotea* sp.) (Fig. 9.3). Most individuals collected were mature. After several months of inundation, cladocerans and copepods became abundant and densities of amphipods and isopods declined. During 1994 maximum densities of invertebrates in benthic samples were observed within four months of inundation. Invertebrate densities were low during the usually wet summer and autumn of 1994, and the benthic community was dominated by larval chironomids. Although small increases in invertebrate densities were observed during the spring of 1995, amphipod, isopod, copepod, and cladoceran densities were substantially lower than during spring of 1994. A rapid response of benthic invertebrates was also observed following wetland inundation in 1996. However, densities of invertebrates were lower than those observed prior to extended inundation.

Invertebrate densities on wood debris increased more slowly than on the benthos following wetland inundation (Fig. 9.4). Wood surface invertebrate densities peaked at 2000–3000 individuals per m² during summer and autumn of 1994. Larval chironomids were almost always the most abundant invertebrate collected from wood surfaces (Fig. 9.4). Amphipods, isopods, copepods, and cladocerans, while present on wood surfaces, were never as abundant as in benthic samples. Invertebrate densities on wood declined during the winter of 1995 and increased during the spring of 1995. Invertebrate densities on

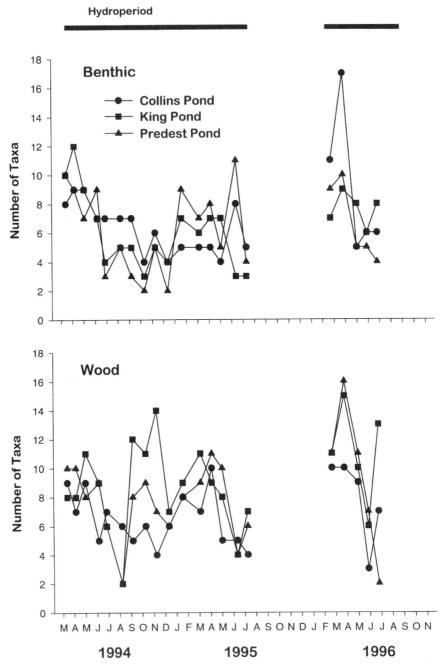

Fig. 9.2. Numbers of taxa in benthic and wood samples taken from forested limesink wetlands in southwest Georgia.

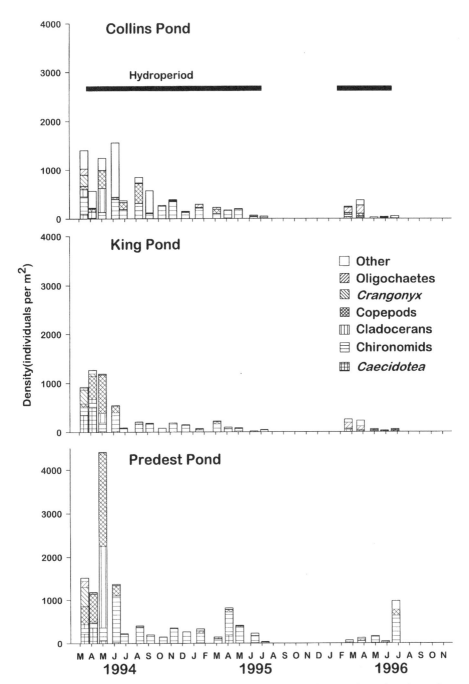

Fig. 9.3. Densities of dominant invertebrate taxa in benthic samples taken from forested limesink wetlands in southwest Georgia. Densities were calculated using a natural log *n* + 1 transformation, then back transformed.

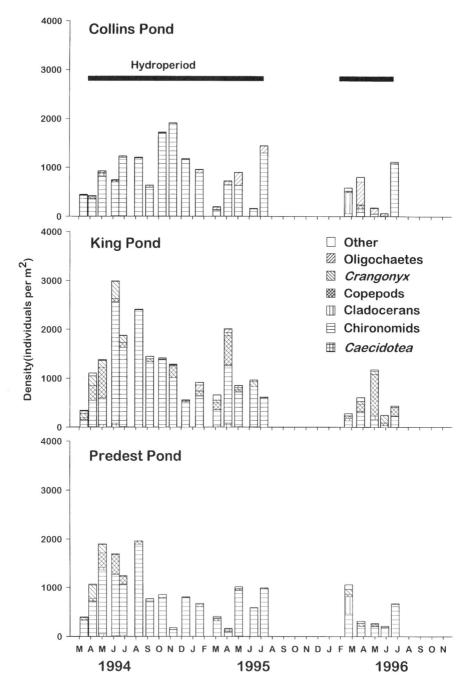

Fig. 9.4. Densities of dominant invertebrate taxa in wood samples taken from forested limesink wetlands in southwest Georgia. Densities were calculated using a natural log $n + 1$ transformation, then back transformed.

wood during the spring of 1995 and 1996 were comparable to those observed prior to extended inundation.

On benthic surfaces densities of amphipods, cladocerans, copepods, isopods, and taxa in the "other" category were significantly reduced during February–June 1995 following extended inundation (Table 9.2). Densities of larval chironomids were not significantly affected. Only "other" taxa showed evidence of recovery one year following extended inundation (February–June 1996). Responses of invertebrates on wood surfaces were more variable. Densities of chironomids, copepods, and oligochaetes were not significantly

TABLE 9.2. Average Densities of Dominant Taxa in Limesink Wetlands[a]

Substrate	Taxa	Pre-extended Inundation	Post-extended Inundation	
		Feb–Jun 1994	Feb–Jun 1995	Feb–Jun 1996
Benthic	*Caecidotea*	113(47) a	21(16) b	9(3) b
	Chironomidae	250(57) a	159(66) a	63(50) b
	Cladocera	205(102) a	1(1) b	3(1) b
	Copepoda	369(163) a	19(4) b	20(7) b
	Crangonyx	71(14) a	1(1) b	0(0) b
	Oligochaeta	23(8) b	7(3) b	35(14) a
	Other	158(134) a	12(5) b	55(13) a
Wood	*Caecidotea*	23(8) a	14(12) a	3(1) b
	Chironomidae	894(106) a	540(109) a	258(48) a
	Cladocera	10(3) b	3(2) c	69(20) a
	Copepoda	142(65) a	71(63) a	96(78) a
	Crangonyx	99(48) a	26(20) b	30(15) b
	Oligochaeta	7(3) a	48(23) a	44(40) a
	Other	10(4) b	16(7) a	67(15) a

[a]Values are seasonal means (s.e.) from three wetlands. Row means with different letters are significantly ($p < 0.05$) different.

affected by extended inundation, whereas densities of cladocerans and amphipods on wood surfaces were reduced following extended inundation (Table 9.2). "Other" taxa increased on wood surfaces following extended inundation. Densities of copepods appeared to recover during February–June 1996, being significantly greater than either of the previous years.

MACROINVERTEBRATE ASSEMBLAGES

To date, much of what is known about invertebrate assemblages in nonalluvial forested wetlands of the southeast originates from surveys. While differences in sampling methods and intensity make direct comparisons difficult, general patterns emerge. Dominant taxa, either in density or biomass, are similar to those we collected and include crustaceans (amphipods, isopods, copepods, and cladocerans), dipterans (particularly larval Chironomidae), and oligochaetes (e.g., Moore 1970, Ziser 1978, and Sklar 1985). A majority of dominant taxa are detritivores, demonstrating the importance of litter inputs from wetland vegetation in supporting aquatic foodwebs. A number of aquatic insect groups may also be occasionally abundant, including coleopterans (larvae and adults), odonates, and hemipterans. Many of the additional taxa are also predators, feeding on other aquatic invertebrates or larval amphibians (Moore 1970, Ziser 1978, Sklar 1985).

The unique physical conditions of nonalluvial forested wetlands probably strongly influence the invertebrate assemblages in those habitats. Defining physical conditions include a variable and unpredictable hydroperiod, high environmental temperatures, and low dissolved oxygen, particularly during the latter portions of the hydroperiod (Moore 1970, Ziser 1978, Sklar 1985). Our results suggest that invertebrates of forested limesink wetlands have the ability to respond rapidly to the presence of surface water and have adaptations which permit them to avoid adverse physical conditions. The ability to respond rapidly to the presence of surface water is characteristic of many wetland invertebrates. Often, extremely high population densities are observed within a few weeks of inundation (Wiggins et al. 1980, Williams 1985, Watson et al. 1995). Our results are consistent with this observation, particularly for benthic samples. Maximum invertebrate densities were observed within one to three months of inundation. Invertebrate densities on wood surfaces responded more slowly to inundation. However, there have been few other studies of wood surfaces in wetlands for comparison. In our study, the primary benthic substrate was leaf litter and fine organic detritus. Slower colonization of wood compared to leaf surfaces has been reported in streams and has been attributed to more gradual microbial conditioning (e.g., Hax and Golladay 1993). Once conditioning occurred, wood surface densities exceeded those on leaves (Hax and Golladay 1993). We observed a similar pattern, where invertebrate densities on wood exceeded those on the benthos after an initial conditioning period.

Several factors have been proposed as contributing to the rapid response of wetland invertebrates to seasonal inundation. In forested wetlands the rapid response may be caused by the presence of a high-quality detrital food resource (e.g., Sklar 1985, Lake et al. 1989). Late-winter standing crops of organic matter (leaves and wood) ranging from 2,000–5,400 g DW m^{-2} have been reported in seasonally inundated wetlands of the Atlantic Coastal Plain (Day 1979). Wiggins (1980) suggests that alternate drying and wetting may enhance detrital quality, although this has never been examined. We did not quantify litter inputs during this study, but leaf litter and other detritus were very abundant throughout the study (Golladay, personal observation). The early abundance of detritivores provides additional evidence that detrital availability may be driving this system.

Sklar (1985) suggests that water temperature and dissolved oxygen regulate assemblage abundance and composition in the forested wetlands of the Gulf Coastal Plain. As temperatures increase during spring, dissolved oxygen concentrations decline. Sklar (1985) suggests that high temperature and low oxygen stimulate behavioral or physiologic adaptations (discussed below) for avoiding stresses associated with summer. Rapid response to inundation would maximize opportunities for population growth prior to adverse summer conditions. In limesink wetlands decreasing benthic population densities corresponded with increasing temperature (personal observation). Water temperature and dissolved oxygen were not measured regularly during the study. However, during summer, water temperatures of 28–32°C and dissolved oxygen concentrations ranging from <1 mg/L at the sediment/water interface to 2 mg/L at the water surface have been measured (Golladay, personal observation). To date, the role of environmental stress in regulating wetland invertebrate assemblages has not been systematically investigated in southeastern wetlands. Also, little is known about how each invertebrate population responds to stressful summer conditions. Potential responses include diapausing life cycle stages, migration to more favorable microhabitats (e.g., moist but not saturated sediments, algal mats), and migration to more favorable aquatic habitats (e.g., well-aerated streams). Clearly, experimental research is needed on biological and physical factors influencing community organization in these systems. Our ongoing research is addressing some of these issues.

There has been considerable interest in adaptations which permit invertebrates to survive seasonal drying in seasonally inundated wetlands (e.g., Wiggins et al. 1980, Williams 1985, Batzer and Wissinger 1996). A general consensus has emerged that seasonal wetlands support a unique assemblages compared to perennial aquatic habitats (Williams et al. 1980, Batzer and Wissinger 1996). Organisms lacking a desiccation-resistant life history stage or the ability to migrate tend to be eliminated by drying (Williams 1985). Less is known about the effects of extended inundation on seasonal wetland invertebrate assemblages, although recently it has been suggested that unusually long hydroperiods may represent a greater disturbance than seasonal drying (Neckles et al. 1990). Inundation extending into the summer appears to adversely affect the invertebrate fauna of forested limesink wetlands. Densities of several dominant taxa, including amphipods, isopods, cladocerans, and copepods, were substantially reduced following a summer and autumn of continuous inundation. These groups appear to have different adaptations for persisting in seasonal wetlands, so the effect of extended inundation may be different for each.

Amphipods and isopods lack a terrestrial life history stage and are not known to produce desiccation-resistant eggs or diapausing developmental stages (Wiggins et al. 1980). Because they appear to lack adaptations for surviving drying, it has been assumed that they are not commonly found in seasonal wetlands (Williams 1985). However, they are common in southeastern wetlands (e.g., Sklar 1985) and probably survive summer conditions by aestivating in moist, unsaturated sediments (Wiggins et al. 1980, Sklar 1985). In forested limesink wetlands drying typically occurs in early summer, leaving sediments moist but not saturated. When sediments remain saturated during summer, anoxic conditions probably greatly reduce their suitability as a summer refuge for amphipods and isopods. We attribute low population densities following the unusual hydrologic conditions of 1994 to mortality associated with the reduction in summer refuges within the wetlands.

Cladocerans and copepods survive seasonal drying by producing drought-resistant eggs (both groups) or by having drought-resistant developmental stages (copepods) (Wiggins et al. 1980). In these groups environmental cues such as seasonal temperature changes and sediment drying may be essential for development (Wiggins et al. 1980). Egg hatching or emergence is dependent not only on the presence of water but also on receiving the right sequence of seasonal cues (Wiggins et al. 1980). If the cues do not occur in the appropriate sequence, development is arrested. Eggs of some taxa within the groups are very long-lived and can persist in the sediments for many years until the correct sequence of environmental stimuli occurs (Wiggins et al. 1980, Hairston et al. 1995). In forested limesink wetlands extended inundation undoubtedly altered the typical seasonal progression of temperature and moisture conditions in sediments. We attribute low population densities following extended inundation to delayed development of desiccation-resistant life history stages.

Extended inundation did not appear to adversely affect all of the dominant taxonomic groups in our study. Larval chironomid densities were highest in benthic and wood samples during late spring and summer. Larval chironomids are known to be tolerant of environmental extremes. They can survive in temporary aquatic habitats by aestivating in cocoons or tubes during dry periods (Wiggins et al. 1980). They are also capable of aerial dispersal, and that ability, along with relatively short generation times, enables them to colonize temporary aquatic habitats from adjacent perennial wetlands (Williams and Hynes 1976). Some larval chironomids are also capable of tolerating low oxygen conditions (Wiggins et al. 1980, Butler and Anderson 1990). In this study the ability to tolerate low oxygen conditions undoubtedly contributed to persistence of larvae chironomids during extended inundation. It was also interesting that chironomid densities were highest on wood debris during summer, presumably a period of greatest environmental stress. Much of the wood we sampled was in the water column above the sediment surface, and its proximity to the water's surface may provide higher dissolved oxygen concentrations than in the sediments. Thus, wood debris may be an important refuge for some elements of the invertebrate community in seasonal wetlands during unusual hydrologic conditions.

CONCLUSIONS AND INFORMATION NEEDS

Our results, although preliminary, clearly indicate the importance of hydrology in regulating invertebrate populations and have a variety of implications for wetland management. Alterations of hydrology due to human activity will impact invertebrate assemblages. Extending hydroperiods (e.g., conversion to ponds) will probably adversely impact dominant crustaceans if anoxia occurs during summer or if environmental cues necessary for the completion of life cycles are altered. If fish are introduced, many invertebrate populations will probably be reduced and the assemblage altered (Wiggins et al. 1980). Shortening hydroperiods (e.g., aquifer withdrawals or drainage) may result in inadequate time to complete life cycles. There are also other potential adverse impacts due to human activity. Clearing or altering wetland vegetation may decrease litter inputs or affect detrital quality, adversely impacting wetland foodwebs. Finally, disturbance to the surrounding uplands without protective buffers may result in sedimentation or other non-point source pollution, degrading wetland water quality.

Forested limesink wetlands of southwest Georgia support an abundant and dynamic invertebrate fauna. Yet little is known about the factors which regulate invertebrate populations or other ecological characteristics in these and similar southeastern wetlands. Greater information is needed because these wetlands are threatened by regional development (Bennett and Nelson 1990, Tansey and Cost 1990). Historically, nonalluvial wetlands smaller than 4 ha have been developed with minimal review. On the Dougherty Plain this has resulted in estimated losses of up to 14 percent of limesink wetlands (Blood et al. 1997). The extent of subtle degradation due to urban development, upland land use (agriculture and forestry), or aquifer withdrawals has not been documented. Future research needs include a greater understanding of the role of hydrologic variation influencing limesink wetland structure and function, documentation of acceptable levels of human activity in and around wetlands, and the development of strategies for identifying and restoring wetlands that are degraded.

ACKNOWLEDGMENTS

We thank Frankie Hudson for her assistance with coordinating the processing and sorting of invertebrate samples. Dorthy Garrett and Cookie Hudson also assisted with sample processing. Matt Lauck and Kevin Watt assisted with the field work. Funding for this research was provided by the J. W. Jones Ecological Research Center and the R. W. Woodruff Foundation.

LITERATURE CITED

Allard, D. J. 1990. Southeastern United States Ecological Community Classification. Interim Report, Version 1.2. The Nature Conservancy, Southeast Regional Office, Chapel Hill, NC.

Batzer, D. P., and S. A. Wissinger. 1996. Ecology of insect communities in nontidal wetlands. Annual Review of Entomology 41:75–100.

Beck, B. F., and D. D. Arden. 1983. Hydrology and Geomorphology of the Dougherty Plain. Southwest Georgia. Southeastern Geological Society, Tallahassee, FL.

Benke, A. C., R. L. Henry, III, D. M. Gillespie, and R. J. Hunter. 1985. Importance of snag habitat for animal production in southeastern stream. Fisheries 10:8–13.

Bennett, S. H., and J. B. Nelson. 1990. Distribution and Status of Carolina Bays in South Carolina. South Carolina Wildlife and Marine Resources Department, Columbia, SC.

Blood, E. R., J. S. Phillips, D. Calhoun, and D. Edwards. 1997. The role of the Floridan aquifer in depressional wetlands hydrodynamics and hydroperiod. Pages 273–279 in K. J. Hatcher (ed.), Proceedings of the 1997 Georgia Water Resources Conference. University of Georgia, Athens, GA.

Butler, M. G., and D. H. Anderson. 1990. Cohort structure, biomass, and production of a merovoltine *Chironomus* population in a Wisconsin bog lake. Journal of the North American Benthological Society 9:180–192.

Day, F. P., Jr. 1979. Litter accumulation in four plant communities in the Dismal Swamp, Virginia. American Midland Naturalist 102:281–289.

Dodd, C. K. 1992. Biological diversity of a temporary pond herpetofauna in north Florida sandhills. Biodiversity and Conservation 1:125–142.

————. 1995. Reptiles and amphibians in the endangered longleaf pine ecosystem. Pages 129–131 in E. T. LaRoe, G. S. Farris, C. E. Puckett, P. D. Doran, and M. J. Mac (eds.), Our Living Resources: A Report to the Nation on the Distribution, Abundance, and Health of U.S. Plants, Animals, and Ecosystems. U.S. Department of Interior—National Biological Service, Washington, DC.

Hairston, N. G. Jr., R. A. Van Brunt, C. M. Kearns, and D. Engstrom. 1995. Age and survivorship of diapausing eggs in a sediment egg bank. Ecology 76:1706–1711.

Hax, C. L., and S. W. Golladay. 1993. Macroinvertebrate colonization and biofilm development on leaves and wood in a boreal river. Freshwater Biology 29:79–87.

Hayes, L. R., M. L. Maslia, and W. C. Meeks. 1983. Hydrology and Model Evaluation of the Principal Artesian Aquifer, Dougherty Plain, Southwest Georgia. United States Geological Survey, Atlanta, GA.

Hendricks, E. L., and M. H. Goodwin. 1952. Water-level Fluctuations in Limestone Sinks in Southwestern Georgia. Geological Survey Water Supply Paper 1110-E. United States Printing Office, Washington, DC.

Hicks, D. W., R. E. Krause, and J. S. Clarke. 1981. Geohydrology of the Albany Area, Georgia. U.S. Geological Survey, Atlanta, GA.

Hippe, D. J., D. J. Wangsness, E. A. Frick, and J. W. Garrett 1994. Water quality of the Appalachicola-Chattahoochee-Flint and Ocmulgee river basins related to flooding from tropical storm Alberto; Pesticides in urban and agricultural watersheds; and nitrate and pesticides in ground water, Georgia, Alabama, and Florida. U.S.G.S. Water Resources Investigations Report 94-4183. Atlanta, GA.

Lake, P. S., A. E. Bayly, and D. W. Morton. 1989. The phenology of a temporary pond in western Victoria, Australia, with special reference to invertebrate succession. Archiv für Hydrobiologie 115:171–202.

Lynch, J. M., A. K. Gholson Jr., and W. W. Baker. 1986. Natural Features Inventory of Ichauway Plantation, Georgia: Volume I. The Nature Conservancy Southeast Regional Office, Chapel Hill, NC.

Merritt, R. W., and K. W. Cummins. 1984. An Introduction to the Aquatic Insects of North America. Kendall/Hunt, Dubuque, IA.

Mitsch, W. J., and J. G. Gosselink. 1993. Wetlands. Van Nostrand Reinhold, New York.

Moore, W. G. 1970. Limnological studies of temporary ponds in southeastern Louisiana. The Southwestern Naturalist 15:83–110.

Neckles, H. A., H. R. Murkin, and J. A. Cooper. 1990. Influences of seasonal flooding on macroinvertebrate abundance in wetland habitats. Freshwater Biology 23:311–322.

Neumann, C. J., B. R. Jarvinen, C. J. McAdie, and J. D. Elms. 1993. Tropical Cyclones of the North Atlantic Ocean, 1871–1992. National Climate Data Center Historical Climatology Series 6-2, Asheville, NC.

Pennak, R. W. 1989. Fresh-water Invertebrates of the United States: Protozoa to Mollusca. John Wiley & Sons, New York.

Sklar, F. H. 1985. Seasonality and community structure of the backswamp invertebrates in a Louisiana cypress-tupelo wetland. Wetlands 5:69–86.

Sokal, R. R., and F. J. Rohlf. 1995. Biometry. W. H. Freeman, New York.

Sutter, R. D., and R. Kral. 1994. The ecology, status, and conservation of two non-alluvial wetland communities in the South Atlantic and Eastern Gulf Coastal Plain, USA. Biological Conservation 68:235–243.

Tansey, J. B., and N. D. Cost. 1990. Estimating the forested-wetland resource in the southeastern United States with forest survey data. Forest Ecology and Management, 33/34:193–213.

Taylor, B. E., R. A. Estes, J. H. K. Pechmann, and R. D. Semlitsch. 1988. Trophic relations in a temporary pond: Larval salamanders and their microinvertebrate prey. Canadian Journal of Zoology 66:2191–2198.

Taylor, B. E., D. L. Mahoney, and R. A. Estes. 1989. Zooplankton production in a carolina bay. Pages 425–435 in R. R. Sharitz and J. W. Gibbons (eds.), Freshwater Wetlands and Wildlife. USDOE Office of Scientific and Technical Information, Oak Ridge, TN.

Thorp, J. H., E. M. McEwan, M. F. Flynn, and F. R. Hauer 1985. Invertebrate colonization of submerged wood in a cypress-tupelo swamp and blackwater stream. The American Midland Naturalist 113:56–68.

Torak, L. J., G. S. Davis, G. A. Strain, and J. G. Herndon. 1991. Geohydrology and Evaluation of Water-resource Potential of the Upper Floridan Aquifer in the Albany

Area, Southwest Georgia. U.S. Geological Survey, Open-File Report 91-52, Atlanta, GA.

Watson, G. F., M. Davies, and M. J. Tyler. 1995. Observations on temporary waters in northwestern Australia. Hydrobiologia 299:53–73.

Wharton, C. H. 1978. The Natural Environments of Georgia. Georgia Department of Natural Resources, Atlanta, GA.

Wiggins, G. B., R. J. Mackay, and I. M. Smith. 1980. Evolutionary and ecological strategies of animals in annual temporary pools. Archiv für Hydrobiologie Supplement 58:97–206.

Williams, D. D., and H. B. N. Hynes. 1976. The recolonization mechanisms of stream benthos. Oikos 27:265–272.

Williams, W. D. 1985. Biotic adaptations in temporary lentic waters, with special reference to those in semi-arid and arid regions. Hydrobiologia 125:85–110.

Ziser, S. W. 1978. Seasonal variations in water chemistry and diversity of the phytophilic macroinvertebrates of three swamp communities in southeastern Louisiana. The Southwestern Naturalist 23:545–562.

10 Beaver-Impounded Wetlands of the Southeastern Coastal Plain

Habitat-Specific Composition and Dynamics of Invertebrates

ARTHUR C. BENKE, G. MILTON WARD, and
TERRY D. RICHARDSON

*B*eaver dams constructed on low-gradient streams of the southeastern United States create small wetlands that have a heterogeneous mixture of habitats, ranging from pondlike to littoral to semiaquatic. Quantitative studies of invertebrates were conducted in such a wetland located in the Talladega National Forest, Alabama. Invertebrates were separately sampled from several habitats in each of three major zones: a small, unvegetated, open-water zone nearest the beaver dam, a moderately shallow area dominated by the white water lily Nymphaea odorata (Nymphaea zone), and a very shallow semiaquatic area dominated by the emergent rush Juncus effusus (Juncus zone). The open-water zone contained a relatively simple benthic assemblage of dipteran larvae, oligochaetes, and benthic microcrustaceans and a relatively simple zooplankton assemblage. The greatest richness and density of invertebrates from the Nymphaea zone occurred in the benthos, in contrast to that found on surfaces of submerged wood and Nymphaea leaves. As in the open-water benthos, chironomids dominated the Nymphaea benthos, but typical pond-dwelling invertebrates such as mayflies, dragonflies, damselflies, beetles, hemipterans, and benthic crustaceans were common as well. A high richness of microcrustaceans was found both in the water column and associated with the Nymphaea leaves. In the Juncus habitat tussocks and rivulets had distinctly different invertebrate assemblages, with many

Invertebrates in Freshwater Wetlands of North America: Ecology and Management, Edited by Darold P. Batzer, Russell B. Rader, and Scott A. Wissinger
ISBN 0-471-29258-3 © 1999 John Wiley & Sons, Inc.

semiaquatic and terrestrial taxa dominating in the tussocks. Invertebrate densities (especially chironomid midges) were high (>15,000/m²) in all three major zones. Density of chironomids tended to be highest in the spring and declined throughout the summer in both open-water and Nymphaea *zones, whereas abundance was highest in the summer in* Juncus *tussocks. Insect emergence occurred throughout most of the year, but was highest in the spring in* Nymphaea *and highest in the summer in* Juncus. *Taxon richness, as seen from benthic sampling and from insect emergence, increased from the open-water zone to the* Nymphaea *zone to the* Juncus *zone. The* Nymphaea *plants, by increasing the heterogeneity of the zone, were probably responsible for the addition of several taxa. However, the high richness of the* Juncus *zone was largely attributed to the addition of several semiaquatic and terrestrial insect families.*

INTRODUCTION

When one thinks of wetland ecosystems of the southeastern United States, extremely large systems typically come to mind: the Everglades, the Okefenokee Swamp, and the Atchafalaya Swamp, just to name a few (e.g., Mitsch and Gosselink 1993). However, small wetland systems of diverse origin are widespread throughout the southeast: Carolina bays, cypress domes, sag ponds, pitcher plant marshes, oxbow lakes, and small temporary pools of forested river swamps (e.g., Crisman 1992, Mitsch and Gosselink 1993). Small wetlands also are created by the activities of beavers, particularly on small streams where there is a low gradient, such as in the Coastal Plain. Some streams may have a series of small beaver dams resulting in a cascade of small wetland pools and extensive emergent and submergent wetland vegetation. Although such beaver-impounded wetlands are increasing in the southeast (Hackney and Adams 1992), the invertebrates of such wetlands are not well studied, and quantitative analyses are particularly scarce.

Research on wetland invertebrates has increased in the last decade (Batzer and Wissinger 1996), but has been conducted primarily at north-temperate or boreal latitudes. While much work on wetland invertebrates in the southeastern USA is nonquantitative, some quantitative studies on benthic and planktonic invertebrates have been conducted in Carolina bays (e.g., Schalles and Shure 1989, Taylor and Mahoney 1990). Furthermore, studies of invertebrate population or community production have been conducted in a few other southeastern wetlands (Martien and Benke 1977, White 1985, Gladden and Smock 1990, Duffy and LaBar 1994). In one of the few invertebrate studies of a southeastern beaver-impounded wetland, Duffy and LaBar (1994) estimated secondary production. However, their study was somewhat limited because they only used a core sampler and did not appear to sample semiaquatic habitats.

A major reason why quantitative invertebrate studies in wetlands have lagged behind those in streams, ponds, and lakes is likely because of their heterogeneous habitat structure. Beaver-impounded wetlands often include shallow zones with dense stands of emergent and submergent macrophytes and extensive submerged woody debris. These provide habitats for aquatic, semiaquatic, and terrestrial invertebrates beyond the normally sampled benthic sediments. Standard sampling equipment such as Ekman grabs and plankton nets is very difficult, if not impossible, to operate under such conditions. Thus, adequate sampling of many wetland systems requires the use of several non-standard sampling devices necessary to deal with various habitat types (see Sharitz and Batzer, this volume). There is often a gradient of habitat complexity going from low heterogeneity in the deeper water to higher heterogeneity with submerged and floating macrophytes to the highest heterogeneity where land and water meet. At this interface, we expect to find terrestrial invertebrates, as well as semiaquatic and aquatic taxa. Attempts to understand both the spatial and temporal dynamics of wetlands communities (de Szalay and Resh 1996) must consider such habitat heterogeneity.

The objectives of this chapter are to:

1. describe some of the physical, chemical and biological features of a beaver-impounded wetland of the southeastern USA;
2. qualitatively describe the distribution of invertebrate populations among major habitats of the wetland;
3. present some quantitative data on the major invertebrate taxa, particularly focusing on benthic densities and emergence of dipterans;
4. provide an example of production dynamics for one of the common species (*Hyalella azteca*); and
5. describe the influence of a leaf beetle on one of the major wetland plants.

This chapter is intended to provide an initial overview of the invertebrate community and is part of a larger ecosystem study (e.g., Mann and Wetzel 1995, Johnson and Ward 1997, Ward 1998).

STUDY SITE

General Description

The study was conducted in a beaver-impounded wetland within the Oak-mulgee District of the Talladega National Forest. This Talladega wetland ecosystem (TWE) was formed approximately 50 years ago as a series of small beaver impoundments in the valley of a second-order Coastal Plain stream (Payne Creek). These impoundments and the surrounding wetlands occupy 10.3 ha of a 384-ha catchment (Fig. 10.1). Geologically, the TWE lies in the Gordo Formation, which consists of massive beds of cross-bedded sands, gravelly sands and beds of moderate-red and pale-red purple clay of upper Carboniferous age. In the area of the TWE the Gordo Formation contains chert and quartz pebbles (Szabo et al. 1988).

The upland forest vegetation in the basin consisted primarily of loblolly/ long leaf pine, plus a diverse mixture of deciduous vegetation, including red and white oak, sweet gum, and yellow poplar. Vegetation within the wetland itself was a spatially complex mixture of semiaquatic and aquatic species, primarily consisting of tag alder (*Alnus serrulata*), willow (*Salix* spp.), and the soft rush (*Juncus effusus*). Much of this vegetation was flooded in winter, but in drier months the wetland was criss-crossed by an anastomosing network of small channels.

Several small impoundments occurred within the TWE, the largest of which was a 1.36-ha pond where the research herein described was conducted. Within this pond, vegetation was dominated by *Juncus effusus* and floating-leafed white water lily (*Nymphaea odorata*) (Fig. 10.1). *Nymphaea* occurred

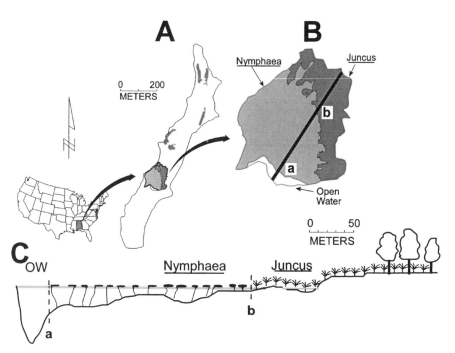

Fig. 10.1. (*a*) Talladega Wetland Ecosystem. (*b*) The major ponded area of the TWE. (*c*) Schematic diagram of cross section of the ponded area showing the three major zones.

at depths <1 m that were inundated year-round, and covered 72 percent of the pond. *Juncus* inhabited slightly higher elevations at the northern border of the pond, covering approximately 25 percent of the pond area, was inundated in winter, but often became dry during summer. At somewhat higher elevations vegetation was dominated by *Alnus/Juncus*, willow (*Salix* spp.), red maple (*Acer rubrum*), sweet gum (*Liquidambar styraciflua*), and tulip poplar (*Liriodendron tulipifera*). In a small area of the pond where water depth exceeded 1 m (3 percent), no rooted vegetation existed. Mean depth of the pond was 0.2 m.

Precipitation and Hydrology

The TWE, fed by several perennial and intermittent tributaries, received water almost entirely as runoff from recent precipitation. Surface waters appeared to be separated from the regional water table in this area by a thick bed of iron-rich clay. As a consequence, the TWE exhibited marked hydrologic seasonality. Surface flow into the TWE varied from a low of 8.7 L/s in July and August to over 65 L/s in March. Much of the annual precipitation occurred from January through March, when 100–200 mm/mo was typical. Summer tended to be driest, but precipitation as high as 150 mm/mo was not unusual (Fig. 10.2*a*). Mean annual precipitation for WY (water year) 1993 and WY 1994 was 1318 mm.

The TWE pond stage and pond volume varied with runoff and evapotranspiration (E-T) demand (Ward 1998). During winter, when E-T demand was low and precipitation was high, pond surface elevation was relatively constant at approximately 90 m, as determined by the elevation of the beaver dam. Stage varied much more in summer, with reduced runoff and increased E-T. During hot and dry periods daily shrinkage of the pond was evident. Surface area expanded with each significant rain event, only to be reversed after runoff had again been reduced to summer base flow. Such expansion and shrinkage of the inundated portions of the pond sediments was documented on a daily basis (Ward 1998). Water depths in the *Nymphaea* zone varied spatially from a maximum of approximately 1 m to <5 cm near the edge of the *Juncus* zone. During summer, as water levels fell and E-T remained high, the sediments in the *Juncus* zone slowly dried, leaving behind a more terrestrial-like habitat.

Temperature

Air and surface water temperatures in this shallow wetland, measured continuously at a meteorological station established within the pond, tracked each other quite closely (Fig. 10.2*b*). Mean annual water temperature for WY 1993 and WY 1994 was 17.5°C, and mean annual air temperature was 15.8°C. Maximum air, water, and sediment temperatures occurred in June and July and minima occurred in January. Sediment temperatures were more buffered

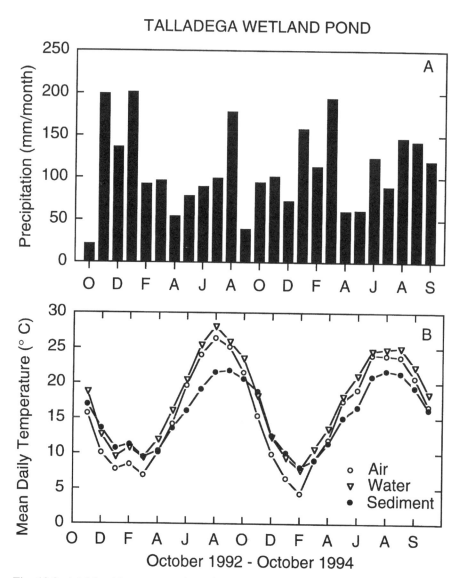

Fig. 10.2. (*a*) Monthly patterns of precipitation and (*b*) mean monthly temperature at the Talladega Wetland Ecosystem for WY 1993 and WY 1994 (WY = water year from October 1 to September 30).

from temperature fluctuations than the water or air. Mean sediment temperature measured in the *Juncus* zone was 15.2°C (8–24°C), approximating mean annual air temperature. Maximum mean daily sediment temperatures during summer were lower than air or water temperatures and reached their peak slightly later than did either air or water temperatures.

Water Chemistry

Because of the deeply weathered nature of the sandy terrain of the upper Coastal Plain, surface water in the TWE was relatively ion-poor. Alkalinity was low through the year (≈ 0.1 meq/L), with only slight increases in summer (summary of all water chemistry data from Roden and Wetzel 1996, Johnson and Ward 1997, Stanley and Ward 1997). Surface water was somewhat acid, with a pH of approximately 6.1. Orthophosphorus concentrations were <10 μg P/L through winter and increased to 30–50 μg P/L in summer and fall. Dissolved inorganic nitrogen (DIN) was relatively low throughout most of the year (<50 μg N/L). However, during August concentrations of NH_4-N increased within the soil water of the *Juncus* sediments, and in isolated pools within the *Juncus* zone DIN concentrations in surface water exceeded 150 μg N/L. Oxygen saturation in *Nymphaea* zone surface water averaged approximately 50 percent annually. Saturations >90 percent occurred throughout winter, but were depressed during summer to <20 percent.

BIOLOGICAL COMMUNITIES

Vegetation Zones

Our investigation of invertebrates in the TWE focused primarily on the major wetland pond, which contained three distinct zones related to water depth and vegetation. The deepest zone (>1 m) was the unvegetated open-water zone (0.11 ha) found closest to the main beaver dam (Fig. 10.1). The benthic habitat of the open-water zone consisted of a shallow detrital layer only about 1 cm deep, covering a sandy substratum.

Away from the dam, water depth rapidly declined to <1 m (Fig. 10.1c), at which point the *Nymphaea* zone began and covered the largest area of the pond (0.72 ha). Although *Nymphaea* rhizomes were found throughout this zone, their leaves did not appear until March, increasing in density until they covered much of the pond surface by midsummer (Pickard and Benke 1996). The benthic habitat of the *Nymphaea* zone contained a thicker layer of detritus (4–8 cm) than that found in the open-water zone. Various other aquatic plants could be found in this zone, but *Nymphaea* predominated. Because beaver-impounded wetlands such as the TWE were once riparian forests, submerged wood and stumps from fallen trees represented a substantial benthic habitat.

The *Juncus* zone occurred where the water was very shallow (<5 cm) during winter inundation or could disappear altogether during the summer. The *Juncus* zone was made up primarily of two distinct habitat types: standing water and semiaquatic *Juncus* tussocks. *Juncus* roots tightly bound the wetland sediments, forming raised tussocks of grass and soil. These tussocks varied throughout the year from being dry in the top several centimeters to saturated during wet periods. Between the tussocks were found small pools connected by narrow rivulets which in some cases also connected to the

Nymphaea zone. Although very shallow, the rivulets persisted throughout most of the year except during extensive dry periods in the summer. The benthic habitat in these rivulets contained an extensive organic layer 7–13 cm deep.

Sampling Strategy in a Multihabitat System

As with most wetlands, the heterogeneous nature of the TWE presented difficulties for obtaining quantitative samples of invertebrates, particularly as related to the macrophytes in the *Nymphaea* and *Juncus* zones. Even within a zone, several habitats needed to be sampled. Therefore, several types of sampling gear were used that were specific for both vegetation zone and habitats within a zone.

The open-water zone contained no macrophytes and its relatively uniform bottom was the easiest to sample. Preliminary sampling suggested a relatively simple community composed of chironomids, oligochaetes, and microcrustaceans, with few large macroinvertebrates such as dragonfly larvae. Because the benthic habitat appeared to have relatively small-sized invertebrates, we used an Ogeechee River corer (Wildlife Supply Company) with a relatively small diameter (5 cm) rather than the larger Ekman grab used in other benthic habitats (see below). Zooplankton were sampled from the entire water column using a 6-cm-diameter PVC tube with a valve that could be closed near the bottom. The zooplankton from the sampler were then concentrated on a 106-μm sieve. Replicated monthly samples of both zooplankton and benthic invertebrates were taken from randomly selected 10 × 10-m quadrats arrayed as a grid across the zone.

The *Nymphaea* zone required sampling three different "benthic" habitats (see Pickard and Benke 1996 for details) as well as the zooplankton. A pole-mounted Ekman grab (15 × 15 cm) was used to sample the sediment habitat. The larger and less abundant macroinvertebrates present here would not be sampled well with the smaller corer. We also collected core samples, as in the open-water zone, for meiobenthos, such as microcrustaceans. The wood habitat was sampled by placing precut pieces of natural wood (collected from the pond) on the pond bottom and retrieving them monthly. Invertebrates from leaves of the water lilies were sampled with a hinged "suitcase" sampler (Downing 1986) in which one or more leaves near the water surface were enclosed. For both wood and leaf samples, habitat surface areas of each substratum were measured so that numbers could be presented per surface area of leaf or wood habitat (Pickard and Benke 1996). Zooplankton in the *Nymphaea* zone was sampled with the same water column sampler that was used in the open-water zone. Replicated monthly samples from all habitats were randomly selected from 10 × 10-m or 5 × 5-m quadrats in the *Nymphaea* zone.

Rivulets or pools in the *Juncus* zone were sampled with a pole-mounted Ekman grab. However, the dense root mass of the *Juncus* tussocks precluded using the Ekman grab in this habitat. Therefore, a macro-corer (15-cm inside diameter and 40-cm height) was constructed from heavy-gage metal pipe. Handles were welded to the pipe near the top, and the bottom edge was sharpened to cut through the root mass. A cylinder of *Juncus* roots and soil up to 30 cm in length was collected. The root masses were sufficiently tight to form a seal and prevent the sample contents from falling out of the sampler as it was raised. Replicated samples of rivulets and tussocks were taken from within randomly selected 5 × 5-m quadrats located in the *Juncus* zone.

Emerging insects were quantitatively sampled using 0.25-m^2 conical floating emergence traps in the open-water and *Nymphaea* zones (Stagliano et al. 1998). The same conical trap was used in the *Juncus* zone, except that it was mounted on top of a plastic cylinder (a cut-out garbage can) that was sealed into the sediment. Each *Juncus* trap partially enclosed some emergent vegetation (tussock) and partially covered water standing in a rivulet. Emerging insects were trapped continuously from fixed stations within each zone for an entire year. Insects were removed from the traps at least weekly.

Open-Water Zone Invertebrates

The zooplankton of the open-water zone was characterized by a relatively simple assemblage, dominated by the water fleas (cladocerans) *Diaphanosoma brachyurum, Ceriodaphnia dubia,* and *Bosmina longirostris* (Table 10.1). Two copepod taxa were common and found in benthic as well as plankton samples. The harpaticoid copepod *Canthocamptus robertcokeri* was abundant and exclusively benthic. The rest of the benthic invertebrate assemblage was also relatively simple, primarily consisting of midge larvae (Chironomidae, Table 10.2) and earthworms (Lumbricidae: *Lumbricina,* not presented in tables). The chironomids were dominated by *Cladopelma* sp., *Einfeldia* sp., *Tanytarsus* sp., and *Procladius* sp. The phantom midge *Chaoborus punctipennis* was commonly collected in the benthic samples during the day, but is well known to be a predator of zooplankton at night (Hilsenhoff 1991). Biting midge larvae (Ceratopogonidae), as well as *Procladius* sp., were probably the most significant predaceous invertebrates of the open-water benthic assemblage, based on their relative abundance (Table 10.2).

Nymphaea Zone Invertebrates

The *Nymphaea* zone contained greater habitat heterogeneity and a much higher richness of taxa than the open water. Chironomids were still the most diverse and abundant group of invertebrates. Most major taxa of chironomids were found to varying degrees in benthic, *Nymphaea* leaf, and wood habitats (Table 10.2). However, *Chironomus* spp., *Cladopelma* sp., *Dicrotendipes* spp., *Polypedilum* spp., *Tanytarsus* spp., *Ablabesmyia* spp., and *Procladius* sp. were clearly the dominant taxa in the benthic habitat, which contained by far the greatest densities of invertebrates (see below). As in the open water, cerato-

TABLE 10.1. Habitat-Specific Relative Abundance of Microcrustaceans among the Major Zones of the Pond of the Talladega Wetland Ecosystem[a]

		Nymphaea		Open-Water		Juncus	
Family	Genus/Species	Be	Pl/Le	Be	Pl	Ri	Tu
Cladocera Sididae	Diaphanosoma brachyurum	***	***		***		
Chydoridae	Eurycercus vernalis	**				*	
	Camptocercus macrurus		*				
	Euryalona sp.	*					
	Leydigia leydigia						
	Kurzia lattisimi	**				*	**
	Alona spp.		*				
	Alonella spp.		*				
	Pleuroxus striatus		**				
	Chydorus spp.		*				
Daphniidae	Daphnia ambigua	**			**	*	*
	Simocephalus serratulus		***				
	Ceriodaphnia dubia		**		***		
	Scapholeberis mucronata		***		***	*	
Bosminidae	Bosmina longirostris					**	
Macrothricidae	Ilyocryptus spinifer	**					
Copepoda Cyclopoida	Ectocyclops phaleratus		*			***	***
	Paracyclops fimbriatus poppe		*				
	Eucyclops speratus	**					
	Eucyclops agilis	**					
	Tropocyclops prasinus	***	***	***	***		
	Macrocyclops albidus	***	**	**	**		
	Mesocyclops sp.	*	*				
Harpacticoida	Canthocamptus robertcokeri	***		***		**	***
	Attheyella sp.	**		**			
Ostracoda Cypridoidea	Cypridopsis sp.	*				***	**

[a] Be = benthic, Pl = planktonic, Le = leaf habitat, Ri = rivulets, Tu = tussocks. *** = abundant, ** = common, * = rare.

Source: *Nymphaea* and open-water data provided by A. Maria Miller.

TABLE 10.2. Habitat-Specific Relative Abundance of Dipterans from the Major Zones of the Main Pond of the Talladega Wetland Ecosystem[a]

Genus/Species	Open-Water		Nymphaea Zone				Juncus Zone		
	Be	Em	Be	Le	Wo	Em	Ri	Tu	Em
Chironomidae									
Chironominae									
Chironomini									
Chironomus spp.	**	*	***	*	*	**	**	*	*
Cladopelma sp.	***	****	***	*	*	***	**	*	
Dicrotendipes spp.	*	**	**	*	**	***	*	*	*
Einfeldia sp.	***	**	*	*		*	*	*	
Endochironomus sp.		*		*	*			*	
Glyptotendipes sp.							*		
Goeldichironomus sp.		*				*			
Harnischia sp.		*				*			
Hyporhygma sp.		*	*			*			
Kiefferulus sp.		*				*			
Lauterborniella sp.	*	*							
Micropsectra sp.	*								**
Microtendipes sp.	*	*	**	**	**	**	*	*	*
Parachironomus sp.		*	**	*	*	**		**	*
Paratendipes sp.			*				**		*
Phaenopsectra sp.			*	*	*		*		*
Polypedilum spp.	**	**	***	**	****	***	***	**	***
Pseudochironomus sp.					*	*			
Stenochironomus sp.					*				
Tribelos sp.							*		
Zavreliella sp.	**		*	*	*	*	*		
Tanytarsini									
Cladotanytarsus sp.			*		*	*	*		
Paratanytarsus sp.			*	*	*	***	*	**	*
Tanytarsus spp.	****	****	***	*	*	***	***	**	***

227

TABLE 10.2. (Continued)

Genus/Species	Open-Water		Nymphaea Zone				Juncus Zone		
	Be	Em	Be	Le	Wo	Em	Ri	Tu	Em
Orthocladiinae									
Corynoneura sp.		**		*	*	**	*		***
Cricotopus sp.		*				*			*
Nanocladius sp.			*	*	*		**	***	**
Psectrocladius sp.			*	*	*			*	*
Pseudorthocladius sp.							*	***	**
Pseudosmittia sp.				*	*				
Rheocricotopus sp.				*					
Thienemanniella sp.			*	*			*		
Unniella multivirga			*	*					
Zalutschia sp.			*	*	*				*
Tanypodinae									
Ablabesmyia spp.	**	**	***	***	***	***	**		*
Clinotanypus sp.			*						
Fittkauimyia sp.	*				*				
Guttipelopia sp.		**	*	*	*		**		*
Larsia sp.			***			**	**	**	***
Monopelopia sp.		*				*			
Psectrotanypus sp.		**	***			**			
Procladius sp.	***	**	***	*	*	**	***	**	*
Tanypus spp.		*	*	*			**	**	**
Anthomyiidae									
Bibionidae									*
Cecidomyiidae (9 genera)	***	*	***	*	*	**	*	*	***
Ceratopogonidae (11 genera)	**	***	**	*	*	**	***	***	***
Chaoboridae *Chaoborus punctipennis*	**	***	**						

	Be	Le	Wo	Ri	Tu	Em
Chloropidae (2 genera)						**
Culicidae *Aedes, Anopheles*	*	**				*
Diastidae						*
Dixidae *Dixella* sp.	*					*
Dolichopodidae (5 genera)	*	*				**
Drosophilidae						**
Empididae (7 genera)		*				**
Muscidae						*
Mycetophilidae *Mycetophila* sp.		*				**
Opomyzidae						*
Otitidae *Pseudotephritis* sp.						*
Phoridae						*
Psychodidae *Psychoda* sp.	*	*				*
Ptychopteridae *Bittacomorpha* sp.						*
Scatopsidae						
Sciaridae (5 genera)	*	***				***
Sepsidae *Sepsis* sp.						*
Sphearoceridae						**
Stratiomyidae	*	**				
Syrphidae (4 genera)	*	**		***		*
Tabanidae (5 genera)	*	**		*		*
Tipulidae (11 genera)	*	**				**

a Be = benthic, Le = leaf, Wo = wood, Ri = rivulet, Tu = tussock, Em = emergence. *** = abundant, ** = common, * = rare.

pogonids were commonly found. However, unlike the open-water benthos, the *Nymphaea* zone contained taxa from several other invertebrate groups (Table 10.3). Naidad species were the most common oligochaetes, but other families occurred there as well (relative abundance not yet determined). Larvae of two mayfly (Ephemeroptera) taxa, *Caenis diminuta* and *Callibaetis* sp., were commonly found, with the former being the most abundant. Beetle (Coleoptera) adults and larvae were commonly collected, with dytiscids predominating. Larvae of the chrysomelid beetle *Donacia* spp. were abundant, being closely associated with *Nymphaea* rhizomes. The most notable predators of the *Nymphaea* benthos were the larvae of Odonata, particularly coenagrionid damselflies and several species of libellulid dragonflies. Several species of caddisfly larvae (Hydroptilidae and Phryganeidae) were observed, but were not abundant compared to most other insects. The amphipod *Hyalella azteca* was by far the most abundant noninsect macroinvertebrate. Crayfishes (*Procambarus* spp.) were common in the *Nymphaea* zone, but were rarely collected by our Ekman grab. The wood habitat contained many of the same taxa found in the *Nymphaea* benthic habitat, but some of the larger taxa (e.g., several dragonfly species) were not collected, probably because they escaped during retrieval of the wood sample.

Many of the invertebrates found in the *Nymphaea* benthos were also found on the *Nymphaea* leaves themselves, although they were usually less abundant. This was particularly true of most chironomid taxa (see below). Coenagrionid damselflies were common on leaves, but most of the dragonflies were early instars or were confined to the benthic habitat. The mayflies and *Hyalella* were among those taxa that were relatively common on the leaves. Insects that were specifically associated with leaves were the adults of *Donacia* spp., which grazed directly on the upper leaf surfaces, and pyralid (Lepidoptera) moth larvae, which constructed protective cases out of the leaf material. Various mobile hemipteran taxa such as corixids and notonectids, although not abundant, were collected more in leaf samples than from the sediments.

A high species richness of microcrustaceans was found in both benthic and leaf/planktonic habitats of the *Nymphaea* zone (Table 10.1). Several species, particularly the chydorid water flea *Eurycercus vernalis* and the harpacticoid copepod *Canthocamptus robertcokeri,* were exclusively found in the benthos, whereas others, such as *Diaphanosoma* sp., *Ceriodaphnia dubia,* and *Bosmina longirostris,* were mostly found in the water column or associated with the leaves. A few taxa, such as the copepods *Tropocyclops prasins* and *Macrocyclops albidus,* could be found in both habitats.

Although this chapter focuses on the invertebrates, several vertebrate species inhabited the TWE pond, besides beavers. Most of the species were primarily collected or observed in the *Nymphaea* zone. The most abundant fishes were the warmouth (*Lepomis gulosus*) and the bluegill (*Lepomis macrochirus).* Other fishes present were the creek chubsucker (*Erimyzon oblongus*), pirate perch (*Aphredoderus sayanus*), chain pickerel (*Esox niger*),

TABLE 10.3. Habitat-Specific Relative Abundance of Nondipteran Invertebrates from *Nymphaea* and *Juncus* Zones of the Talladega Wetland Ecosystem, Based on Benthic Sampling or Emergence[a]

Order	Family	Genus/Species	*Nymphaea*			*Juncus*	
			Be	Le	Wo	Ri	Tu
Collembola							
Coleoptera	Carabidae		***			**	***
	Chrysomelidae	*Donacia cincticornis*	***	**		*	**
		Donacia rufescens	***	**			*
		Pyrrhalta nymphaeae		*			
	Curculionidae	*Bagous* sp.		*		*	**
	Dermestidae						**
	Dytiscidae	*Celina* sp.	**	*		*	**
		Deronectes sp.	**				
		Hydroporus sp.	*				
	Gyrinidae	*Dineutus* sp.	*		*		*
	Hydrophilidae	*Berosus* sp.	*	*	*	*	**
	Haliplidae		*			*	
	Noteridae	*Suphisellus* sp.	*			*	**
	Staphylinidae					*	
Ephemeroptera	Baetidae	*Callibaetis* sp.	**	***	*	*	*
	Caenidae	*Caenis diminuta*	***	***	***		
	Ephemerellidae	*Eurylophella* sp.			*		
Hemiptera	Corixidae	*Trichocorixa* sp.	**	**	*	*	*
	Belostomatidae			*			
	Notonectidae	*Notonecta* sp.		*		*	
Homoptera	Aphididae						**
	Cicadellidae						**

TABLE 10.3. (Continued)

Order	Family	Genus/Species	Be	Le	Wo	Ri	Tu
			Nymphaea			*Juncus*	
			Be	Le	Wo	Ri	Tu
Hymenoptera	Braconidae						**
Lepidoptera	Coleophoridae						*
	Cosmopterigidae						***
	Nepticulidae						**
	Noctuidae						*
	Pterophoridae						*
	Pyralidae	*Acentria* sp.	*				
		Parapoynx sp.	**	*		*	**
Odonata	Coenagrionidae	*Enallagma dubium*	**	*			
		Enallagma sp.	**	*	**		
		Ischnura kellicotti	**	**			
		Ischnura spp.	**	**	**		
	Lestidae	*Lestes inaequalis*	*				
	Aeschnidae	*Corphaeschna ingens*				*	*
	Corduliidae	*Epitheca cynosura*	**			*	
	Libellulidae	*Celithemis fasciata*	**	**			
		Celithemis sp.	*	*			
		Erythemis simplicicollis	*				
		Erythrodiplax minuscula	**				
		Ladona deplanata	*				
		Libellula auripennis	**				
		Libellula cyanea	**				
		Libellula incesta	***				
		Pachydiplax longipennis		*	**		
		Perithemis tenera	*				
		Sympetrum vicinum					

232

Order	Family	Species	Be	Le	Wo	Ri	Tu
Orthoptera	Gryllidae	*Gryllus* sp.					**
Trichoptera	Phryganeidae	*Ptilostomis* sp.	**		**	*	**
	Hydroptilidae	*Hydroptila* spp.		*			**
		Orthotrichia spp.			*		
		Oxyethira spp.	*	*	*	*	**
	Limnephilidae	*Limnephilus* sp.			*	*	
Amphipoda	Gammaridae	*Gammarus* sp.	*		*		**
	Hyalellidae	*Hyalella azteca*	***	***	***	***	**
Isopoda	Asellidae		**				**
Decapoda	Cambaridae	*Procambarus* spp.	**	**	**		**
Araneae	Lycosidae				*		
Pseudoscorpiones							
Hydracarina			*	*	*		***
Gastropoda	Ancylidae			**			***
Oligochaeta	Aelosomatidae			**			***
	Lumbricidae			**			***
	Lumbriculidae			**			***
	Naididae	*Dero nivea*	***				
		Dero digitata	***				
		Nais spp.	***				
		Pristina leidyi	**				
		Uncinais uncinata	**				
	Tubificidae		***				***

[a]Be = benthic, Le = *Nymphaea* leaves, Wo = wood, Ri = rivulets, Tu = tussocks. * = rare, ** = common, *** = abundant. Where no asterisk is given for a taxon, it either was found only as an adult or in a less common habitat.

spotted bass (*Micropterus punctulatus*), blackspotted topminnow (*Fundulus olivaceus*), yellow bullhead (*Ameiurus natalis*), weed shiner (*Notropis texanus*), and brook lamprey (*Lampetra appendix*). Amphibians included the lesser siren (*Siren intermedia*), red-spotted newt (*Notophthalmus viridescens*), and unidentified frogs. Also present were muskrats and several snake species.

Juncus Zone Invertebrates

The semiaquatic and aquatic nature of the *Juncus* zone tussocks and rivulets was reflected in the invertebrate taxa. While more invertebrate taxa were found in this zone (>100 taxa total), chironomids were the most common group, just as they were in the *Nymphaea* and open-water zones. Five families of oligochaetes were found in both the tussocks and rivulets (Table 10.3). The aquatic Naididae and Tubificidae were the dominant earthworms in the rivulets, while the semiaquatic Lumbriculidae were dominant in the tussocks. Many of the dominant chironomid taxa found elsewhere, such as *Polypedilum* sp., *Tanytarsus* sp., and *Procladius* sp., were also dominant in the *Juncus,* but primarily in rivulets (Table 10.2). Semiaquatic/terrestrial orthoclad midges, such as *Nanocladius* sp. and *Pseudorthocladius* sp., dominated in the tussocks. Among the Tanypodinae, *Procladius* was abundant in rivulets, as it was in *Nymphaea* and open-water zones, while *Larsia* and *Tanypus* were the most common tanypods in the tussocks. Ceratopogonids were abundant in the *Juncus,* as they were in other zones.

While many taxa were common to both the *Juncus* and *Nymphaea* zones, several were unique to the *Juncus.* Fungus gnats (Sciaridae) were limited to the *Juncus* zone and were abundant only in the tussocks. The semiaquatic/terrestrial springtails (Collembola) were commonly collected in the rivulets and tussocks and were second in abundance to the chironomids (Table 10.3). Several beetle taxa were commonly sampled in the *Juncus,* but the weevils (Curculionidae) were the dominant group in the tussocks. Mites (Hydracarina) were abundant in the *Juncus* zone, particularly in the tussocks. The most common predators in the *Juncus* zone were predaceous diving beetles (Dytiscidae), ground beetles (Carabidae), rove beetles (Staphylinidae), and wolf spiders (Lycosidae).

Microcrustaceans were common in the *Juncus* zone, but the copepods were not separated into species as they were for the other two zones. Nonetheless, based upon the Cladoceran taxa richness, it did not appear that there were as many microcrustacean taxa in the *Juncus* as in the *Nymphaea* zone. The chydorid water flea *Alona* sp. was commonly sampled in the tussocks, and Macrothricidae were common in the rivulets (Table 10.1). The dominant microcrustacean taxa in the *Juncus* were the cyclopoid and harpacticoid copepods and seed shrimps (Ostracoda). Cyclopoids were abundant in both the rivulets and the tussocks. However, ostracods were most abundant in the rivulets, and harpacticoids were abundant in the tussocks.

INVERTEBRATE DYNAMICS

Temporal and Spatial Patterns of Chironomid Abundance

Our description of benthic invertebrate abundance focuses on the chironomids in each of the three major habitats. However, it should be kept in mind that other invertebrate groups may also be relatively abundant in each of the habitats; that is, microcrustaceans in all zones, large macroinvertebrates in the *Nymphaea* zone, and many semiaquatic/terrestrial taxa in the *Juncus* zone.

Chironomids in the open-water zone showed considerable annual variation, with peaks in winter and midsummer of greater than $13,000/m^2$ (Fig. 10.3a). However, during most of the summer chironomid densities were $<5,000/m^2$. The tribe Chironomini was clearly the dominant group throughout the year, with densities usually $>2,000/m^2$. The Tanypodinae were consistently found at densities of 1,000 to $2,000/m^2$, and the tribe Tanytarsini had densities up to about $3,000/m^2$, but only during winter and early spring. Orthoclads were rarely found in the open-water zone.

Densities of chironomids in the *Nymphaea* zone were relatively high and evenly distributed among the Chironomini, Tanytarsini, and Tanypodinae (Fig. 10.3b). Peak densities in *Nymphaea* benthos approached $20,000/m^2$ during the colder months. Tanypodinae densities showed little seasonal variability, ranging from only 2,000 to $5,000/m^2$. In contrast, both Chironomini and Tanytarsini each had their highest densities during the colder months and declined to less than $2,000/m^2$ during the summer. Chironomids were not nearly as abundant on the wood and leaf habitats of the *Nymphaea* zone (Figs. 10.3c and 10.3d) as they were in the benthic habitat (Fig. 10.3b). Densities rarely reached $>4,000/m^2$ of leaf or wood surface. Chironomini and Tanypodinae predominated in both habitats, with relatively few Tanytarsini. Only on the *Nymphaea* leaves did orthoclads begin to appear in substantive numbers, and only in the colder months.

In the *Juncus* zone striking differences occurred among the chironomids between the rivulet and tussock habitats (Figs. 10.3e and 10.3f). In the shallow water of the rivulets, chironomid densities were usually relatively low ($<5,000/m^2$) and consisted primarily of Chironomini and Tanypodinae, with no strong seasonal patterns (Fig. 10.3e). Both Tanytarsini and Orthocladiinae were present throughout the year in lower densities, except in November, when there was a rise in Tanytarsini of $>7,000/m^2$.

The *Juncus* tussocks had a very different pattern of chironomid composition and abundance than any of the other habitats (Fig. 10.3f). Here the Orthocladiinae were the most abundant group, often exceeding $20,000/m^2$. Densities of Chironomini and Tanypodinae, although relatively less abundant, had densities comparable to those found in both open-water and *Nymphaea* benthic habitats. The temporal pattern of the orthoclads was also distinctly different than that of the other groups. Although orthoclads are often described as a cold-water group, these semiaquatic taxa were clearly most abundant during the warmest months.

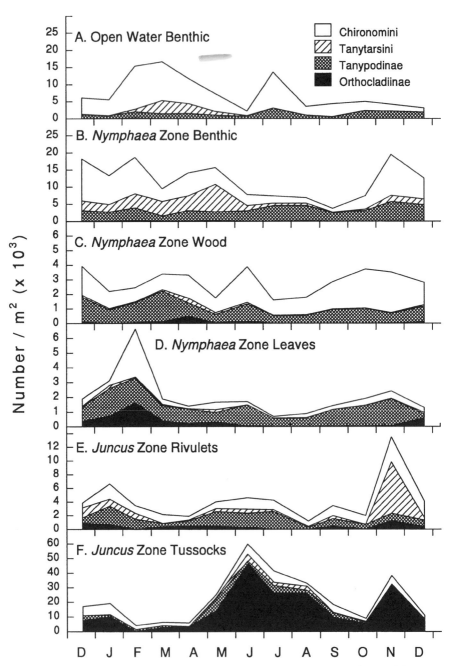

Fig. 10.3. Temporal patterns of larval abundance for the major chironomid groups from the three habitats of the *Nymphaea* zone (wood, *Nymphaea* leaves, and benthic), from benthos of the open-water zone, and from the two habitats of the *Juncus* zone (rivulets and tussocks). Each line is cumulative. Note that the scales of the y-axis differ.

In summary, total chironomid densities were highest in the *Juncus* tussocks, represented primarily by the semiaquatic Orthocladiinae (Table 10.4). Orthoclads occurred in much lower numbers in other habitats. Chironomids also were relatively high in the *Nymphaea* habitat, where Chironomini predominated. The total for the *Nymphaea* zone would include the addition of insects from the wood and leaf habitats to the benthos, but these additions would represent much lower numbers than seen for their habitat-specific densities. The lowest densities in the open-water zone were probably due to lower habitat heterogeneity than found in the other two habitats.

Temporal and Spatial Patterns of Insect Emergence

Emergence of insects occurred in every month of the year in each of the three major zones (Fig. 10.4). For taxon-specific temporal patterns and biomass estimates of the open-water and *Nymphaea* zones, see Stagliano et al. (1998). Chironomid emergence in the open-water zone was very low until March, when Chironomini and Tanytarsini began appearing with total densities >20 m^{-2} d^{-1} (Fig. 10.4a). By April, Tanypodinae also began comprising a significant component of the open-water emergence. By June, Tanytarsini emergence declined to relatively low numbers. Small numbers of orthoclads emerged throughout most of the year, in higher numbers than would have been expected by their larval abundance (Tables 10.4 and 10.5). Total chironomid emergence was usually <30 m^{-2} d^{-1} throughout the year, with only 3466/m^2 for the year.

Chironomid emergence in the *Nymphaea* zone clearly was dominated by the Chironomini and Tanytarsini (Fig. 10.4b, Table 10.5). Emergence began to exceed 20 m^{-2} d^{-1} in February, and from mid-March through May it exceeded 100 m^{-2} d^{-1}. The remainder of the year, emergence in the *Nymphaea* zone did not climb above 50 m^{-2} d^{-1}, but even during the summer it

TABLE 10.4. Mean Densities of Chironomid Groups from Each Major Habitat in the TWE, Averaged across 13 Sample Dates for a Year[a]

Habitat	Chironomini	Tanytarsini	Tanypodinae	Orthocladiinae	Total
Open water	5,664	764	1,428	18	7,875
Nymphaea zone					
Benthos	5,900	2,505	3,441	39	11,886
Wood	1,546	92	1,076	90	2,804
Leaves	558	62	1,109	339	2,067
Juncus zone					
Rivulets	1,507	1,014	1,323	387	4,240
Tussocks	4,264	851	2,432	14,620	22,167

[a] All densities are individuals/m^2 of habitat surface area. Densities are not adjusted for amount of habitat surface area in the case of *Nymphaea* zone wood and leaves.

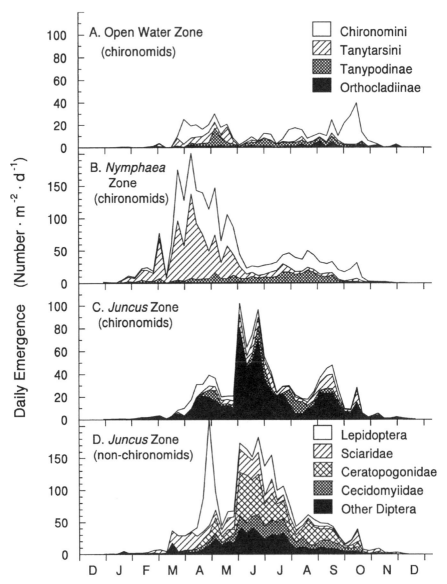

Fig. 10.4. Temporal patterns of mean daily emergence for each of the three major zones of the Talladega Wetland Ecosystem. Emergence from (*a*) the *Nymphaea* and (*b*) open-water zones is represented only by chironomids. Emergence from the *Juncus* zone is represented by (*c*) chironomids and (*d*) nonchironomids. Note that the scales of the *y*-axis differ.

TABLE 10.5. Annual Insect Emergence (± 1 SE) in Each of the Three Major Zones of the Major Pond of the Talladega Wetland Ecosystem[a]

	Open-Water	*Nymphaea*	*Juncus*
Diptera			
Chironomidae	3466 $\pm$ 902	15561 $\pm$ 1441	7596 $\pm$ 1021
Chironomini	1804 $\pm$ 616	6836 $\pm$ 505	964 $\pm$ 292
Tanytarsini	584 $\pm$ 276	6542 $\pm$ 1302	1106 $\pm$ 260
Orthocladiinae	358 $\pm$ 54	249 $\pm$ 39	4661 $\pm$ 647
Tanypodinae	720 $\pm$ 64	1934 $\pm$ 246	865 $\pm$ 133
Chaoboridae	750 $\pm$ 302	263 $\pm$ 65	0
Cecidomyiidae	0	0	2270 $\pm$ 227
Ceratopogonidae	86 $\pm$ 14	112 $\pm$ 55	4722 $\pm$ 631
Sciaridae	0	0	3384 $\pm$ 542
Other Diptera	0	0	3652 $\pm$ 362
Lepidoptera	0	0	3577 $\pm$ 404
Trichoptera	84 $\pm$ 20	243 $\pm$ 82	146 $\pm$ 39
Total Emergence	4386 $\pm$ 594	16179 $\pm$ 1367	25347 $\pm$ 2010

[a] All values are in the units of number of individuals emerging m^{-2} yr^{-1}. Insects were trapped continuously in replicated traps at fixed stations ($n = 2$ for open water, $n = 6$ for *Nymphaea*, and $n = 5$ for *Juncus*). For details of emergence in open water and *Nymphaea*, see Stagliano et al. (1998).

was usually several times higher than emergence from the open water. Tanypodinae tended to become a more consistent component of emergence during the summer months. Emergence of orthoclads was quite low in the *Nymphaea* zone. The general pattern of emergence decline in June (Fig. 10.4*b*) was consistent with a decline in benthic densities (Fig. 10.3*b*).

Emergence of chironomids in the *Juncus* zone was distinctly different than in the other zones, being dominated by the Orthocladiinae and showing a summer, rather than a spring, peak in abundance (Table 10.5, Fig. 10.4*c*). The relative abundance and temporal pattern of the orthoclads was quite consistent with their larval abundance pattern observed in the tussocks (Fig. 10.3*f*). Although annual emergence of *Juncus* chironomids was only about half that found in the *Nymphaea* zone, emergence of nonchironomids in the *Juncus* was substantial (Fig. 10.4*d*, Table 10.5). Ceratopogonidae was the most abundant nonchironomid family, with emergence >4700/m^2 for the year, much higher than found in the other zones. However, several other families of dipterans emerged from the *Juncus* zone in high numbers, particularly the sciarid fungus gnats (Sciaridae) and gall gnats (Cecidomyiidae). Lepidopterans, particularly the cosmopterygid moths, were also abundant, with an annual emergence of 3577/m^2. The great majority of the insects emerging from the *Juncus* were either semiaquatic insects living in the tussocks or terrestrial insects that developed on the emergent plants. Most of the semiaquatic taxa were found in both benthic tussock samples and emergence samples (Table 10.2). Emergence of most nonchironomid groups peaked in early summer, when emergence was >100 individuals m^{-2} d^{-1} (Fig. 10.4*d*).

Temporal and Spatial Patterns of Production for *Hyalella azteca* (Amphipoda)

The ultimate objective of the invertebrate analyses in the TWE pond is to describe the temporal and spatial patterns of production for all major taxa and for the major feeding groups as a whole. As an example of such analyses, we summarize results for the scud *Hyalella azteca* (Pickard and Benke 1996), a ubiquitous amphipod found in many freshwater habitats throughout North America. In the TWE *Hyalella* was found primarily in the *Nymphaea* zone and only rarely in open-water benthos or *Juncus* zone (Table 10.3). Production was calculated using the instantaneous growth method (see Benke 1993, 1996), in which growth rates are determined under controlled temperature conditions with natural foods (Pickard and Benke 1996). Growth rates were predicted from an equation incorporating both temperature and individual mass. Mean daily production on each field sampling date was estimated as the product of mean size-specific biomass and growth rates. Production was estimated in each of the three *Nymphaea* habitats and corrected for habitat area.

Hyalella biomass had its maximum values during the winter months at the TWE, but both density and production peaked during May (Fig. 10.5). Daily production tended to decline throughout the summer, much as did total chironomid benthic densities and emergence in this zone. After correction for habitat abundance, it was clear that the wood and *Nymphaea* habitats contributed relatively small amounts to total *Hyalella* production. *Hyalella* had a mean annual density of $901/m^2$ (three habitats combined), a mean annual biomass of 85 mg dry mass/m^2, and an annual production of 675 mg m^{-2} y^{-1} (somatic only). Independent estimates of egg production showed that eggs contributed 27 percent (252 mg m^{-2} y^{-1}) to total production. Production of *Hyalella* was not especially high in comparison to other studies (Pickard and Benke 1996), and it did not appear to be the dominant primary consumer in the wetland, with densities much lower than chironomids.

Herbivory on the White Water Lily, *Nymphaea odorata*

Various insects were observed feeding upon the leaves of *Nymphaea odorata* during the warmer months. Adult leaf beetles (*Donacia cincticornis* and *D. rufescens;* Chrysomelidae) were the major herbivores, but other beetles (*Pyrrhalta nymphaeae* and *Bagous* sp.), moth larvae (*Parapoynx* sp.), and orthopterans (*Gryllus* sp.) also were observed. Adults of both *Donacia* species resided on and fed upon the upper surfaces of the *Nymphaea* leaves from April through October.

Damage by *Donacia* feeding on *Nymphaea* leaves, including the effects on leaf longevity and production, was determined experimentally in the field using an exclosure design (Carter 1995). Nine permanent quadrats were established. One set of three quadrats was completely open and allowed un-

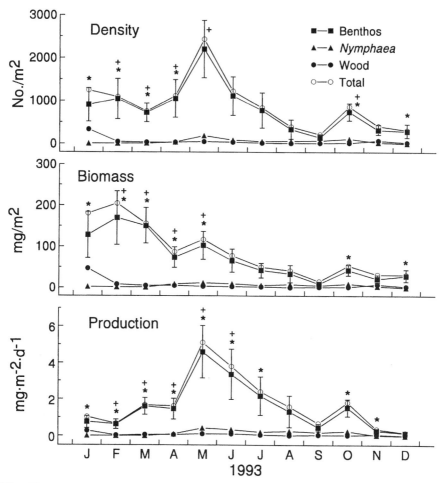

Fig. 10.5. Temporal patterns of density, biomass, and daily production of *Hyalella azteca* for each habitat of the *Nymphaea* zone (benthic, *Nymphaea* leaves, and wood) and total per m² pond. Standard errors (±1 SE) are shown only for benthic habitat. For each date, * indicates that the benthic value is significantly higher than the *Nymphaea* value and + indicates that the benthic value is significantly higher than the wood value. (From D. P. Pickard and A. C. Benke, Production dynamics of *Hyallela azteca* (Amphipoda) among different habitats in a small wetland in the southeastern USA, Journal of the North American Benthological Society 15:537–558, 1996. Reprinted by permission of the Journal of the North American Benthological Society.)

impeded access to *Nymphaea* leaves, while a second set excluded *Donacia* adults. The third set was a shade control, which established the same shading of the *Nymphaea* leaves as did the exclosure device, but permitted access by beetles to the leaves.

Although *Donacia* damage to the leaves was clearly evident, densities of beetles at TWE (1–$2/m^2$) had no statistically significant effect on leaf longevity, biomass, or production (Carter 1995). The primary damage by *Donacia* to *Nymphaea* leaves was small circular patches cut through the upper epidermis of the leaf, and sometimes through the entire leaf. Often female beetles passed their ovipositors through these holes in order to attach egg masses to the under sides of the leaves. Over the course of the growing season approximately 1.7 percent of leaf surface area contained beetle damage, with the greatest damage (4.2 percent) observed in October.

CONCLUSIONS

Dam-building by beavers in Coastal Plain streams of the southeastern United States creates wetlands with considerable habitat heterogeneity. However, there are relatively few quantitative studies on the ecological role of benthic and planktonic invertebrates in these or similar wetlands in the southeastern United States. Our study at the TWE focused on habitat-specific patterns of invertebrate abundance and taxonomic richness. Abundance was particularly high among several dipteran taxa in all three major zones. Taxonomic richness of insects, as assessed through both benthic and emergence sampling, appeared directly related to habitat heterogeneity within each zone. Richness was lowest in the open-water zone, which was lacking in vegetation. Richness was intermediate in the *Nymphaea* zone, where *Nymphaea* (leaves, stems, rhizomes) and submerged wood provided considerable physical heterogeneity. Richness was highest in the semiaquatic *Juncus* zone, at the upstream margin of the wetland where conditions are seasonally wet and dry (Ward 1998), and included a semiaquatic habitat of the tussock soils and the terrestrial dimension of emergent leaves. On the other hand, microcrustacean richness was highest in the *Nymphaea* zone, suggesting that both planktonic and benthic species may play an important role in trophic pathways. Similarly, the benthic crustacean *Hyalella azteca* was common only in the *Nymphaea* zone, where it was moderately productive. High abundance of larval chironomids and their emergence throughout the year suggested that growth of invertebrates in general (including crustaceans) may occur almost continuously in this warmwater wetland where temperatures rarely fall below 10°C. Chironomid temporal patterns were distinctly different between zones, with abundance and emergence highest during early spring in the *Nymphaea* zone, but highest in summer in the *Juncus* zone.

Given the importance of invertebrates as organic matter processors and as a food source for fish and waterfowl (e.g., Batzer and Wissinger 1996), it is essential that we better understand the factors affecting their diversity and abundance in the many small wetlands scattered across the southeastern United States. Adequate understanding of the invertebrate community and its role in wetland ecology requires quantitative data, consideration of habitat heterogeneity, assessment of temporal dynamics, and examination of patterns of taxa richness and trophic interactions. Such considerations are much better understood through the simultaneous quantification of larval populations and emergence than through either type of information by itself. Unfortunately, considering all these factors at the same time, as we have attempted to do at the TWE, has rarely been attempted in wetland or freshwater systems of any kind. Nonetheless, conservation and wise management of natural wetlands and evaluation of created wetlands depend on as thorough an understanding of the invertebrate community as possible.

ACKNOWLEDGMENTS

Special thanks to those individuals who provided data for this paper: Maria Miller provided the data on microcrustaceans from the open-water and *Nymphaea* zones; David Stagliano provided summaries of emergence data from all zones; Jeff Converse provided data for the benthic macroinvertebrates from the open-water and *Nymphaea* zones; Chris Bevis provided information on fish composition; Michael Wade prepared the map of the field site using the facilities of the GIS laboratory in the Department of Biological Sciences. Steve Harris identified or verified the identifications of several Trichoptera taxa. We also thank several persons who helped with field collections and laboratory processing: David Stagliano, Jeff Converse, Dan Pickard, Maria Miller, Chris Bevis, David Lancaster, Ashley Robinson, Jeff Selby, and Joseph Lee. This research was supported by grants from the National Science Foundation EPSCoR program (OSR 91-8761) and the University of North Alabama. This paper is contribution 246 of the Aquatic Biology Program, University of Alabama.

LITERATURE CITED

Benke, A. C. 1993. Concepts and patterns of invertebrate production in running waters. Verhandlungen der Internationalen Vereinigung für theoretische und angewandte Limnologie 25:15–38.

———. 1996. Secondary production of macroinvertebrates. Pages 557–578 *in* F. R. Hauer and G. A. Lamberti (eds.), Methods in Stream Ecology. Academic Press, New York.

Batzer, D. P., and S. A. Wissinger. 1996. Ecology of insect communities in nontidal wetlands. Annual Review of Entomology 41:75–100.

Carter, S. M. 1995. Herbivory by *Donacia rufescens* Lacordaire and *Donacia cincticornis* Newman on the white water-lily, *Nymphaea odorata* Aiton. Master's Thesis, University of Alabama, Tuscaloosa, AL.

Crisman, T. L. 1992. Natural lakes of the southeastern United States: Origin, structure, and function. Pages 475–538 *in* C. T. Hackney, S. M. Adams, and W. H. Martin (eds.), Biodiversity of the Southeastern United States: Aquatic Communities. John Wiley & Sons, New York.

de Szalay, F. A., and V. H. Resh. 1996. Spatial and temporal variability of trophic relationships among aquatic macroinvertebrates in a seasonal marsh. Wetlands 16: 458–466.

Downing, J. A. 1986. A regression technique for the estimation of epiphytic invertebrate populations. Freshwater Biology 16:161–173.

Duffy, W. G., and D. J. LaBar. 1994. Aquatic invertebrate production in southeastern USA wetlands during winter and spring. Wetlands 14:88–97.

Gladden, J. E., and L. A. Smock. 1990. Macroinvertebrate distribution and production on the floodplain of two lowland headwater streams. Freshwater Biology 24:533–545.

Hackney, C. T., and S. M. Adams. 1992. Aquatic communities of the southeastern United States: Past, present, and future. Pages 747–760 *in* C. T. Hackney, S. M. Adams, and W. H. Martin (eds.), Biodiversity of the Southeastern United States: Aquatic Communities. John Wiley & Sons, New York.

Hilsenhoff, W. L. 1991. Diversity and classification of insects and Collembola. Pages 593–663 *in* J. H. Thorp and A. P. Covich (eds.), Ecology and Classification of North American Freshwater Invertebrates. Academic Press, New York.

Johnson, M. D., and A. K. Ward. 1997. Influence of phagotrophic protistan bacterivory in determining the fate of dissolved organic matter (DOM) in a wetland microbial food web. Microbial Ecology 33:149–162.

Mann, C. J., and R. G. Wetzel. 1995. Dissolved organic carbon and its utilization in a riverine wetland ecosystem. Biogeochemistry 31:99–120.

Martien, R. F., and A. C. Benke. 1977. Distribution and production of two crustaceans in a wetland pond. American Midland Naturalist 98:162–175.

Mitsch, W. J., and J. G. Gosselink. 1993. Wetlands (2nd ed.). Van Nostrand Reinhold, New York.

Pickard, D. P., and A. C. Benke. 1996. Production dynamics of *Hyalella azteca* (Amphipoda) among different habitats in a small wetland in the southeastern USA. Journal of the North American Benthological Society 15:537–550.

Roden, E. E., and R. G. Wetzel. 1996. Organic carbon oxidation and suppression of methane production by microbial Fe (III) oxide reduction in vegetated and unvegetated freshwater wetland sediments. Limnology and Oceanography 41:1733–1748.

Schalles, J. F., and D. J. Shure. 1989. Hydrology, community structure, and productivity patterns of a dystrophic Carolina bay wetland. Ecological Monographs 59: 365–385.

Stagliano, D. M., A. C. Benke, and D. H. Anderson. 1998. Emergence of aquatic insects from two habitats in a small wetland of the southeastern USA: Temporal patterns of numbers and biomass. Journal of the North American Benthological Society 17:37–53.

Stanley, E. H., and A. K. Ward. 1997. Inorganic nitrogen regimes in an Alabama Wetland. Journal of the North American Benthological Society 16:820–832.

Szabo, W. M., W. E. Osborne, and C. W. Copeland. 1988. Geologic Map of Alabama. Special Map 220. Geological Survey of Alabama, Tuscaloosa, AL.

Taylor, B. E., and D. L. Mahoney. 1990. Zooplankton in Rainbow Bay, a Carolina bay pond: Population dynamics in a temporary habitat. Freshwater Biology 24:597–612.

Ward, G. M. 1998. A preliminary analysis of hydrodynamic characteristics of a small lotic wetland ecosystem. Verhandlungen der Internationalen Vereinigung für theoretische und angewandte Limnologie 26:1373–1376.

White, D. C. 1985. Lowland hardwood wetland invertebrate community and production in Missouri. Archiv für Hydrobiologie 103:509–533.

11 Pitcher Plant Wetlands of the Southeastern United States

Arthropod Associates

DEBBIE FOLKERTS

*S*eventeen or more species of obligate arthropod associates and numerous facultative associates exist in addition to pollinators, prey, and casual associates of pitcher plants. All of these contribute to a complex community of organisms not found elsewhere in the world. Interaction types, including pollinators, prey, prey consumers, capture interrupters, herbivores, and parasites and predators of primary associates, occur. This probably represents a greater complexity of ecological interactions than found elsewhere. Several pitcher plant insects are known to impact one another in a variety of ways. Most interactions are commensalistic (e.g., Wyeomyia and Metriocnemus, Isodontia and Exyra), while others are probably exploitative (e.g., Bradysia and sarcophagids). The entire community is at risk as habitat destruction for pitcher plants occurs in the southeastern United States.*

INTRODUCTION

Pitcher plant wetlands of the southeastern United States harbor a high diversity of carnivorous plant types. Presumably, carnivorous plants are successful in these habitats because of the advantage of obtaining materials from prey in sites where many nutrients are in short supply or unobtainable from the soil (Givnish 1989). Types of carnivorous plants found in these habitats include six species of sundews (*Drosera* spp., Droseraceae), six species of butterworts (*Pinguicula* spp., Lentibulariaceae), three species of bladderworts

Invertebrates in Freshwater Wetlands of North America: Ecology and Management, Edited by Darold P. Batzer, Russell B. Rader, and Scott A. Wissinger
ISBN 0-471-29258-3   © 1999 John Wiley & Sons, Inc.

(*Utricularia* spp., Lentibulariaceae), Venus flytrap (*Dionaea muscipula*, Droseraceae), and eight species of pitcher plants (*Sarracenia* spp., Sarraceniaceae). Several additional species of *Utricularia* occur in aquatic habitats in the region. Southeastern North America supports a greater diversity of carnivorous plant types than anywhere else in the world. Australia has more species but fewer families and genera.

Although a few insects, such as the spittlebug, *Lepyronia angulifera* (Hanna and Moore 1966), the grasshoppers *Gymnoscirtetes pusillus* and *G. morsei* (Dakin and Hays 1970, G. Folkerts, personal communication), and a number of crayfishes, are known to be endemic or nearly restricted to pitcher plant habitats, little is known of the general arthropod fauna of these habitats. However, a number of arthropods interact directly with pitcher plants in a variety of ways. These are the major focus of this chapter.

The arthropod food resource attracted and accumulated by carnivorous plants provides an unusual resource for other arthropods. Relatively few arthropods are associated, other than as prey, with adhesive- and snap-trap carnivorous plants (Antor and Garcia 1995). However, pitcher-trap carnivorous plants, including both Old World and New World species, support a variety of symbiotic arthropods (Juniper et al. 1989). Several species have evolved to feed solely on pitcher plant prey. Others are herbivores which feed on pitcher plant tissues. The container-like nature of the pitchered leaves may serve as a refuge from parasites, predators, and physical stresses, and is used by several arthropods as well an occasional vertebrate. The fact that pitcher plants are attractive to insects has provided an opportunity for predaceous types to intercept the prey drawn to the plants. Numerous novel plant-animal interactions have evolved, including predator-prey relationships, prey theft, prey consumption after capture, prey mass habitation, and a number of forms of herbivory, often involving herbivores with remarkable entrapment-escaping adaptations. The community of pitcher plants and their associated arthropods may be more complex than many other plant/arthropod communities. Clearly there are a number of interaction types not found elsewhere. A few general works (Fish and Beaver 1978, Rymal and Folkerts 1982, Juniper et al. 1989) have aggregated and added to information on many of the component species in pitcher plant communities of the southeastern United States, but much remains to be discovered.

PITCHER PLANTS AND THEIR HABITATS

The plant communities are distinctive and are most recognizable by the conspicuous pitcher plants of the genus *Sarracenia* (Sarraceniaceae). As a result, they are often referred to as "pitcher plant bogs," even though a number of other endemic plant species may sometimes be more abundant. Eight species are traditionally recognized in the genus (McDaniel 1971), although subspecific taxa have been used and other taxonomic viewpoints exist. One of these species, *S. purpurea,* has an extensive range which extends northward into Labrador and west to British Columbia. In the northern portion of its range this species inhabits peatlands ecologically different from its typical habitats in the Southeast. Because it is the only pitcher plant species present in northerly areas, and because only in this species do pitchers normally hold long-lasting pools of water, many of the pitcher plant-associated arthropods of southeastern pitcher plant wetlands are absent. Considerable research has been dedicated to the study of arthropods associated with *S. purpurea* in northern fens and bogs. Only those works necessary to understanding of the southeastern communities are mentioned here. A more detailed discussion of northern pitcher plant wetlands and associated arthropods can be found in the chapter by Giberson and Hardwick (this volume).

Although a number of different habitat types or variants of types may support pitcher plants, most have several features in common. They are characterized by saturated soil, low pH, low nutrient availability, and frequent fire. The source of water varies from precipitation to groundwater sources. Soils are characteristically nutrient-poor, acidic sands or sandy loams (Plummer 1963). Eleven different subtypes of pitcher plant habitat have been characterized (G. Folkerts 1991). These include stream and river terraces, depressed drainageways, sinkhole pond edges, Carolina bays, sphagnous mat bogs, seepage bogs, springhead seepages, muck bogs, savannas, swales, and anthropogenic habitats. The floristics and ecology of these areas have been studied by various authors (Wells and Shunk 1928, Eleutarius and Jones 1969, Gaddy 1982, G. Folkerts 1982, Walker and Peet 1983, MacRoberts and MacRoberts 1988, 1990, G. Folkerts 1991). Species composition varies but may include more than 100 plant species at a particular site. A total of 260 or more vascular plant species are characteristic of pitcher plant habitats in the southeast (G. Folkerts 1991). As many as five pitcher plant species may occur naturally at a single site. Under conditions of frequent fire the communities are self-perpetuating. These habitats were once very common in the southeastern United States but have been drastically reduced in acreage in recent years. The reasons for decline include absence of fire, forestry practices, road construction, agriculture, pond construction, housing development, and plant poaching. It has been estimated (G. Folkerts 1982) that only 3 percent of the pre-Columbian area of pitcher plant habitat along the Gulf Coast remains in relatively natural condition.

Pitcher plants of the genus *Sarracenia* possess tubular leaves which function as passive pitfall traps in most species. The plants are perennial and produce rosettes of upright or decumbent tubular leaves from underground or surface rhizomes (G. Folkerts 1989). Rhizome branching occurs in all species, with a rosette of leaves usually forming at each apical meristem. What may appear to be a group of adjacent independent plants usually consists of the ramets of a single genet, still attached or recently severed by dieback of rhizome portions beneath the soil. Leaves are produced in the spring and also as a more or less distinct second crop of pitchers in the summer. All species also produce nonpitchered leaves (phyllodia) to varying extents in response to changes in light intensity.

The trap leaves are highly modified tubular structures. The morphological upper surface of the leaf forms the inner wall of the tube. An ala, or wing, representing the fused leaf margins, appears on the outer surface for the length of the tube. A lobe at the top of the pitcher forms the hood. Downward-pointing hairs on the hood serve to direct prey movement toward the orifice. A rolled lip surrounding the outer edge of the orifice is covered with down-ward-pointing cellular projections and waxy secretions (Adams and Smith 1977). This slippery covering continues along the inside of the tube for a variable distance until it is replaced by a smooth glandular area of variable extent and eventually by a region of long, stiff, downward-pointing hairs. This trap structure has been described as having four functional zones (Lloyd 1942):

1. attractive (hood);
2. conductive (slippery lip and inner tube);
3. glandular; and
4. detentive (long, stiff hairs) (Fig. 11.1).

It is likely that *Sarracenia* pitchers produce chemicals involved in olfactory attraction of prey. Miles et al. (1975) identified compounds that may function in this way in one species. Some workers have speculated that the odor of decomposing prey may be an attractant for certain kinds of prey insects (Jones 1904, Lloyd 1942). It has also been thought that the odor of the nectar produced by certain species may be attractive to a certain spectrum of prey. However, nectar production probably functions primarily to retain insects on or in the vicinity of pitchers long enough to increase the likelihood of capture (Juniper et al. 1989). *Sarracenia* pitchers possess nectaries on the outside of the pitchers as well as on the hood and just inside the orifice. Visual attractants include ultraviolet light-absorbing patterns (Joel et al. 1985) and red or purple pigmentation (Juniper et al. 1989). A few species possess unpigmented window-like areoles between the pigmented areas.

An insect-paralyzing agent, coniine, may be produced by several *Sarracenia* species and has been identified from *S. flava* (Mody et al. 1976). The digestive pool in *S. purpurea* has low surface tension, probably due to organic acids that act as wetting agents (Hepburn et al. 1927, Juniper et al. 1989). Digestive enzymes are produced by two types of glands (Joel 1986, Juniper et al. 1989). Many authors have verified the release of proteases and other enzymes by *Sarracenia* pitchers (Mellichamp 1875, Zipperer 1885, 1887, Robinson 1908, Hepburn et al. 1920, 1927). Once open, however, the pitchers develop a bacterial flora which probably also contributes to enzymatic activity (Plummer and Jackson 1963, Albrecht 1974, Lindquist 1975, Juniper et al. 1989). Dinitrogen-fixing bacteria have been isolated from pitchers of *S. pur-*

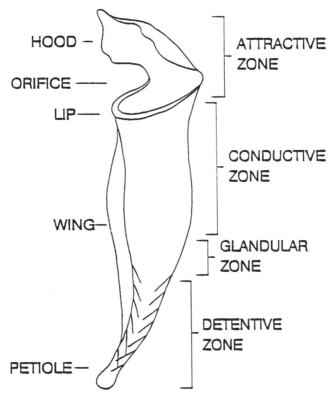

Fig. 11.1. Generalized *Sarracenia* pitcher showing functional zones as described by Lloyd (1942). (From D. R. Folkerts, Interactions of pitcher plants (Sarracenia: Sarraceniaceae) with their anthropod prey in the southeastern United States, Ph.D. Dissertation, University of Georgia, Athens, GA.)

purea (Prankevicius and Cameron 1989). Absorption of prey-digestion products has been demonstrated by the disappearance of nitrogenous compounds in pitchers (Hepburn et al. 1927) and through the use of radioactive tracers (Plummer and Kethley 1964, Williams 1966, Christensen 1976).

The flower structure in *Sarracenia* is unique. Although self-pollination produces viable seeds (Mandossian 1965), the flower morphology helps to insure that cross pollination will take place. Flower structure and related biology will be discussed in more detail in a later discussion of pollinators.

The general morphology of *Sarracenia* species has been described in detail by McDaniel (1971). Eight species (*S. alata, S. flava, S. leucophylla, S. minor, S. oreophila, S. psittacina, S. purpurea, S. rubra*) occur in the southeastern United States. A brief overview of their morphologies can be found in Table 11.1. Hybrids are produced between most pairs of *Sarracenia* species (Bell 1952, Bell and Case 1956) and usually possess characteristics that are intermediate to the parental types.

TABLE 11.1. Characteristics and Ranges of *Sarracenia* Species Occurring in the Southeastern United States

Scientific Name	Common Name	Range	Pitcher Color and Morphology	Flower Color
S. alata	Winged pitcher plant	Gulf Coastal Plain from southwest Alabama to east Louisiana; west Louisiana and east Texas	Tall-form (upright pitchers with conductive zone approximately half length, minimal glandular zone, and extensive detentive zone), green with purple reticulate above or deep red above	Yellow
S. flava	Yellow pitcher plant	Coastal Plain from southwest Alabama to northeast Florida and north to southeast Virginia	Largest tall-form, green with red spot at base of hood	Yellow
S. leucophylla	White-topped pitcher plant	Gulf Coastal Plain from Florida panhandle to east Mississippi	Tall-form, red/white reticulate above, flowery appearance	Red
S. minor	Hooded pitcher plant	South-central Florida north to southeast North Carolina and west to Apalachicola River valley	Tall-form, hood curved over orifice, prominent wing, green/purple reticulate above with translucent areoles	Yellow

S. oreophila	Green pitcher plant	Scattered populations in northeast Alabama, northwest Georgia and southwest North Carolina	Tall-form, green or with red above	Yellow
S. psittacina	Parrot pitcher plant	South Mississippi across Florida panhandle to southeast Georgia	Decumbent, globose hood with small orifice, large wing, minimal conductive zone, long detentive zone, green with purple reticulate and areoles	Red
S. purpurea	Northern pitcher plant	West Mississippi to Florida panhandle and north along Coastal Plain to Virginia, west to Iowa and north to Manitoba and Labrador	Semi-decumbent, urceolate, erect hood, prominent wing, short and wide pitchers, short conductive zone, short detentive zone, extensive glandular zone, green with red/purple above	Pink to red
S. rubra	Red pitcher plant	Coastal Plain from east Mississippi to Florida panhandle and south-central Georgia, N to North Carolina; scattered populations in central Alabama and mountains of North and South Carolina	Smallest tall-form, green or with red or purple above.	Red with green/yellow under petals

The physical environment of pitcher interiors has not been well studied. There is some evidence that pH is usually low (Juniper et al. 1989). Temperature is more consistent and humidity higher than ambient temperature and humidity.

The only North American pitcher plant with a long-lasting phytotelm (plant-held water) is *S. purpurea*. This distinction has consequences in the plant's mode of capturing prey and process of digestion and in the types of pitcher inhabitants it harbors. Physical characteristics of the phytotelm may vary considerably from the external environment (see Giberson and Hardwick, this volume). Other *Sarracenia* species secrete small amounts of fluids, and pitchers normally hold moist masses of decomposing prey. Temporary or partial filling of pitchers with water occurs occasionally in most species as a result of heavy rains or flooding. Two species, *S. minor* and *S. psittacina*, almost never fill with water because the pitcher hoods completely prevent rainwater from entering pitcher orifices.

PITCHER PLANT/ARTHROPOD INTERACTIONS

A diverse array of interaction types exists among the many arthropods that are adapted to life in or on pitcher plants. These include both terrestrial and aquatic types as well as facultative and obligate types. Although a number of different classification schemes are possible, they are grouped here into the following categories according to their potential influences upon the plants: pollinators; prey; prey consumers; capture interrupters; herbivores; and parasites and predators of other associates.

Within each of these categories further subdivisions can be made according to whether associates are facultative or obligate, are inhabitants of pitchers or able to avoid entrapment, are exploitative or beneficial, and are host-specific or associate with several members of the genus. A few species or groups can be placed in more than one category. Most of these species are illustrated in Fig. 11.2. For a complete listing of commonly encountered species, see Table 11.2. The term *inquilines* has often been used to refer to insects that live in pitchers. However, I feel that the term *pitcher inhabitants* is more appropriate (Folkerts and Folkerts 1996).

Pollinators

The unusual morphology of *Sarracenia* flowers requires that insects of sufficient size and manipulative abilities be present for adequate pollen transfer. These unique flowers have been described by several authors (Macfarlane 1908, Russell 1918, Uphoff 1936, McDaniel 1971). Flowers are supported singly on long, nodding scapes. The style is modified, the distal half being expanded into an umbrella-shaped structure which catches the pollen after anther dehiscence. Stigmas are located on each of five cleft lobes of the

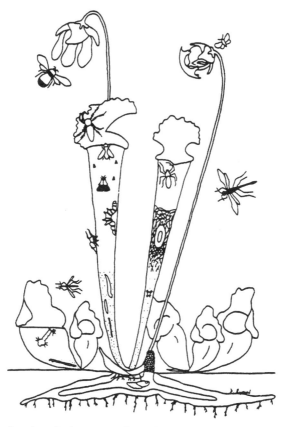

Fig. 11.2. This drawing depicts a number of facultative and obligate arthropods associated with pitcher plants as pollinators, herbivores, capture interrupters, and prey consumers. Representing the pollinators is the bee near the flower in the upper left. Queen bumblebees of the genus *Bombus* are the primary pollinators of the large-flowered species of pitcher plants. The small-flowered species are pollinated by native bees, primarily megachilids, which may have declined since the introduction of the honeybee. In the upper right is a fruit with a caterpillar inside and a nearby moth. Frugivory occurs occasionally by caterpillars of the pitcher plant moth (*Exyra* spp.) but primarily by one or two species of *Endothenia* (Tortricidae). Three different spiders can be seen in the drawing, representing the several spiders that are frequent visitors of pitcher plants. *Peucetia viridans* (hood of pitcher on left) often captures prey from ambush sites at the pitcher hood. *Strotarchus piscatoria* (inside pitcher on right) often attacks another pitcher inhabitant (*Exyra semicrocea*) from inside pitchers. Salticid spiders of the genus *Phidippus* (on the side of the pitcher on the left) capture prey on and around pitchers and often use pitchers as retreats. Within the pitcher on the right can be seen the nest of *Isodontia mexicana*, with an adult wasp flying nearby. This grass-carrier wasp facultatively utilizes pitchers as nest cavities and provisions offspring with orthopteran prey. At the orifice of the left pitcher is a large fly representing several species of Sarcophagidae that are obligate inhabitants. Adult females larviposit from the pitcher orifice and larvae consume pitcher plant prey until emerging to pupate

(*Continued on next page*)

umbrella. Elongate, rounded petals hang between the pointed lobes, creating a floral chamber with the expanded style as the floor, petals as walls, and the ovary with nectar glands, anthers, and other flower parts at the roof. Entry requires manipulation of the petals and style lobes in such a way that the insect comes into contact with a stigmatic lobe (Jones 1908, Schnell 1976). While gathering nectar inside the chamber, the insect brings its dorsum into contact with pollen held by the umbrella. Exit from the chamber requires simply pushing a petal outward and does not usually result in further contact with a stigma. This floral structure encourages cross pollination but limits pollen transfer to large insects with manipulative powers, such as bees.

No insects are known to be entirely dependent on pitcher plants as a source of pollen or nectar. However, the conservation status of the plants is significantly affected by interactions with pollinators. Reproduction in small remnant populations of the green pitcher plant has been found to be pollinator-limited (unpublished data). However, most interactions with pollinators have not been intensively investigated. Numerous insects visit the flowers, which produce large amounts of nectar and pollen, but most are probably nectar thieves. Schnell (1983) concluded that queen *Bombus* were the only effective pollinators of *S. flava*. *Bombus pennsylvanicus* is a common pollinator in populations of *S. oreophila* and other large flowered species (*S. alata, S. flava,*

Fig. 11.2. (Continued) in the soil. Just inside the pitcher orifice are three small flies representing three species, two of which are obligate inhabitants of the tall-form pitcher plants. Adults of all three cling to the interior pitcher walls, and the larvae consume pitcher plant prey. *Dohrniphora cornuta* (Phoridae) is a facultative associate of *S. flava*. *Bradysia macfarlanei* (Sciaridae) occurs as yellow maggots within the prey mass of pitchers. The adults are only occasionally seen. A third, undescribed fly species of the family Chloropidae occurs abundantly in pitchers of *S. leucophylla* and other tall-form species. Maggots of these flies are depicted within the prey mass of the tall pitcher on the left. Just below the flies, in the pitcher on the left, can be seen a moth and caterpillar of *Exyra semicrocea* (Noctuidae). This species occurs in all *Sarracenia,* while two other species of the genus are obligate associates of single *Sarracenia* species. Caterpillars feed on leaf tissue in such a manner as to provide for themselves a protective feeding chamber within the leaves. Adults spend practically their entire lives within pitchers and are among the few species which can walk on interior leaf surfaces. Within the prey mass of the tall pitcher on the right is a mite depicting a group of species that is probably more diverse than is known. One phytoseiid mite preys upon several species of detritus-feeding histiostomatid mites. Within the water held in a pitcher of *Sarracenia purpurea* (bottom left) can be seen larvae of two aquatic insect species that are obligate inhabitants, *Wyeomyia smithii* (Culicidae) and *Metriocnemus knabi* (Chironomidae). An adult mosquito hovers above. A pupa of the pitcher plant rhizome borer *Papaipema appassionata* is depicted within the rhizome. A frass tube can also be seen arising from the base of the plant. This species is probably the rarest of all the pitcher plant associates, and although it has been reported from *S. flava,* it probably occurs only in *S. purpurea.*

TABLE 11.2. Commonly Encountered Pitcher Plant Associates and the *Sarracenia* Species in Which they Occur[a]

Pollinators
    Bombidae (several species)—large-flowered *Sarracenia* spp.
    Megachilidae (several species)—small-flowered *Sarracenia* spp.

Prey
    High diversity of arthropod types, including insects, spiders, millipedes, etc., with some degree of specialization by *Sarracenia* species.

Prey Consumers
    Wyeomyia smithii (Culicidae)—*S. purpurea*
    Metriocnemus knabi (Chironomidae)—*S. purpurea*
    Fletcherimyia fletcheri (Saracophagidae)—*S. purpurea*
    F. celerata (Sarcophagidae)—unknown host specificity
    F. jonesi—unknown host specificity
    F. rileyi—unknown host specificity
    Sarcophaga sarraceniae—unknown host specificity
    Bradysia macfarlanei (Sciaridae)—possibly all *Sarracenia*
    Aphanotrigonum nov. sp. (Chloropidae)—primarily in *S. leucophylla* and other tall-form species.
    Sarraceniopus gibsoni (Acari: Histiostomatidae)—*S. purpurea*
    S. hughesi—all *Sarracenia* except *S. purpurea*
    Dohrniphora cornuta (Phoridae)—*S. flava*

Capture Interrupters
    Isodontia mexicana (Sphecidae)—tall-form *Sarracenia* spp.
    Phidippus spp. (Araneae: Salticidae)
    Peucetia viridans (Araneae: Oxyopidae)
    Other spiders and occasional Mantidae, etc.

Herbivores
Fruit
    Endothenia daeckeana (Tortricidae)—*S. purpurea*
    E. hebesana—all *Sarracenia* spp.?
Pitchers
    Exyra semicrocea (Noctuidae)—all *Sarracenia* spp.
    E. ridingsii—*S. flava* only
    E. fax—*S. purpurea* only
    Choristoneura parallela (Tortricidae)—tall-form *Sarracenia* species
    Macrosiphum jeanae (Aphidae)—*S. purpurea*
    ?*Aspidiotus* sp. (Diaspididae)—*S. psittacina, S. purpurea*
Rhizomes
    Papaipema appassionata (Noctuidae)—*S. purpurea* and possibly *S. flava*.

Parasites and Predators of Pitcher Plant Associates
    Strotarchus piscatoria (Araneae: Clubionidae)—common predator of *Exyra*, etc.
    Macroseius biscutatus (Acari: Phytoseiidae)—predator in all *Sarracenia* except *S. purpurea*
    ?*Senotainia trilineata* (Sarcophagidae)—parasite of *I. mexicana*
    ?*Eurytoma bicolor* (Chalcidae)—parasite of *I. mexicana*
    ?unidentified Tachinidae and Braconidae

Obligate species are indicated with *. Those of unknown host specificity are indicated by ?.

S. leucophylla, S. purpurea). Small megachilid bees are the major pollinators of the small flowered species (*S. minor, S. psittacina, S. rubra*) (G. Folkerts, personal communication). It is possible that populations of these bees are dependent on pitcher plants for most of their resources during periods in April and May. Although honeybees frequently attempt to visit pitcher plant flowers, they are probably not efficient pollinators and often fail to gain entrance.

Prey

It has long been known that carnivorous plants are capable of digesting their prey (Darwin 1878). A number of studies have elucidated the physiological mechanisms involved in trapping, digestion, and absorption of prey (see Luttge 1983 for a review). Foliar absorption of macronutrients and micronutrients has been demonstrated by radioactive tracer studies in *Sarracenia* (Plummer and Kethley 1964, Christensen 1976).

Although it is generally assumed that the insect resource is beneficial, if not essential, to carnivorous plants, it has been difficult to provide evidence that insect prey enhances growth and reproduction. Several authors have concluded that insects are not required by carnivorous plants (Beccarri 1904, Christensen 1976, Small et al. 1977). Others have determined that insect prey is clearly advantageous to the plants and enhances growth (Darwin 1878, Plummer 1966). In spite of the mixed and confusing results of experiments in which plants are fed or deprived of prey, it is difficult to imagine that such elaborate mechanisms for trapping prey would have evolved in carnivorous plants if the prey resource were not important in fitness. Some of the ambiguity may result from the fact that experiments have usually been done under greenhouse conditions and for short time periods. Because the plants are capable of storing large amounts of nutrients in their rhizomes, long-term field experiments may be necessary to provide a definitive answer. Hermann and Platt (personal communication, S. Hermann, Tall Timbers Research Station, Tallahassee, FL) gathered data on starved versus fed *S. flava* plants in the field. Greater growth in the fed plants was observed in the first year, but greater reproductive output was not seen until the third year.

Pitcher plants may not always attract sufficient quantities of prey to maximize fitness. D. Folkerts (1992) compared prey capture in plots of varying pitcher plant density and determined that the prey resource was limited and that the plants competed for prey. Others have reported that number and biomass of prey captured by *S. purpurea* were dependent on pitcher age and size (Wolfe 1981, Cresswell 1993). In some populations, where lack of fire has resulted in overgrowth by fire-intolerant species, capture of arthropod prey may be seriously reduced.

No pitcher plant or other carnivorous plant depends on a single species or group of arthropods as prey; however, most probably specialize to some degree (D. Folkerts 1992, Gibson 1983, Porch 1989). Partial lists of prey types have been compiled for several pitcher plant species (e.g., Goodnight 1940,

Wray and Brimley 1943, Judd 1959, Swales 1969, 1972, Beaver 1979, Cresswell 1991). Rare or unusual insects have often been noted as prey (Ferge and Kuehn 1976, Larochelle 1976, 1977, 1978, Chew 1978, Valtonen 1980). Gibson (1983) speculated that the ratio of flying to crawling insects trapped increases with trap height in at least some carnivorous plant genera. He also found that trap size and insect body size influenced insect escape from *Sarracenia leucophylla* and two other carnivorous plants (Gibson 1991). Fish (1976) found that *Sarracenia minor* trapped disproportionately large numbers of ants in a northern Florida bog. I found (D. Folkerts 1992), by comparing prey captured by *S. flava* to flight-intercept trap and pitfall trap samples, that this pitcher plant is selective in capturing prey. A lower diversity and species richness of prey were captured by the plants than by the artificial traps. The plants captured primarily flying insects, with nectarophilous types of Coleoptera, Diptera, and Hymenoptera being most common. Prey capture declined sharply after June, but a second crop of pitchers was produced, which corresponded with a second peak of insect availability in the fall.

A comparison of prey taken from different pitcher plant species occurring at the same site allowed characterization of the prey trapped by six *Sarracenia* species (D. Folkerts 1992). Differences in the types and numbers of prey trapped seem to be correlated with morphological differences in leaf structure. *Sarracenia flava* and *S. leucophylla*, which have the largest leaves, trapped a greater number, diversity, and taxon richness of prey than did other species. The tall form and large orifice of the pitchers of these species allow them to capture a broader spectrum of prey than do the smaller-leaved species. Nectarophilous forms are common prey of both species. Both produce large quantities of nectar on pitchers. *Sarracenia leucophylla*, however, captures a larger component of dipterans and lepidopterans. The white portion of the hood may be attractive to night-flying moths.

The smallest-leaved species, *S. rubra,* captures small flies and wasps more abundantly than do other species. The small diameter of its pitchers effectively restricts the mobility of small flying insects, allowing this species to trap a spectrum of the potential prey generally unavailable to larger-leaved species.

Sarracenia minor captures a larger proportion of ants than do other species. Small dipterans are typically the only other major component of its prey. This species possesses two unique pitcher characteristics which are probably responsible for the trapping of these two types of prey. First, the minute, antrorse hairs on the wing probably serve to direct cursorial insects such as ants toward the pitcher orifice. Second, *S. minor* pitchers possess translucent areoles on the back of the hood. These probably serve to confuse small flying insects, causing them to be retained because they fly toward the light entering the areoles rather than toward the orifice. Additionally, the curvature of the hood over the orifice inhibits access by large flying insects.

Sarracenia psittacina pitchers seem to trap fewer prey than other species. However, the unique trapping characteristics of this species are designed for trapping ground-dwelling cursorial arthropods and allow it to capture types of prey largely unavailable to the other species. The decumbent position of the pitchers results in the orifice being positioned near the substrate surface. The large wing probably serves to block insect movement over pitchers, directing movement toward the orifice. The small orifice, globose areolate hood, and extremely long retrorse hairs of the detentive zone retain insects that have entered the pitchers. Ants, ground-dwelling beetles, spiders, and diplopods are the major forms captured.

Sarracenia purpurea also possesses a unique trapping mechanism. Undamaged pitchers typically contain large quantities of water. Entrapment results mainly from immobilization in the pool of fluid, although the downward-pointing hairs on the hood often prevent escape. In southeastern North America this species traps more spiders than do other species.

Prey Consumers

A number of arthropods, mainly dipterans, have evolved to be able to utilize as a resource the accumulating prey remains within pitcher plant leaves. It has often been assumed that these species, by consuming prey, are simply exploitative and that they have a negative influence, even if not detectable, on the host plants. However, Bradshaw and Creelman (1984) showed that midge larvae and mosquito larvae inhabiting pitchers of *S. purpurea* have a beneficial effect in accelerating the breakdown of prey. It is possible that other species of prey consumers have similarly beneficial effects. In any case, most prey-consuming species seem to be highly adapted to pitcher plants and can occur in extremely high numbers without apparent harmful effects. Those that are obligate associates are dependent upon the survival of the plants and would risk extinction were they to lower plant fitness to a significant degree.

Bradysia macfarlanei, a sciarid fly, is an obligate associate, occurring in all *Sarracenia*. Although first noted by Macfarlane in 1909 and described in 1920 (Jones 1920), the species has been poorly studied and seldom reported. This delicate, long-legged, dusky-winged fly is sometimes seen clinging to the inner walls of pitchers. Larvae are encountered more often. They are yellowish orange, have prominent head capsules, and occur in the prey mass. Pupation occurs within a frothy tangle of silk attached to interior pitcher walls just above or within the prey mass. Several generations occur in a season. This species is most common in *S. leucophylla* and *S. flava*. Larvae are rarely encountered in pitchers of *S. psittacina*. Wray and Brimley (1943) reported larvae from *S. purpurea* in North Carolina, but it almost never occurs in this species. A negative interaction between this species and pitcher plant sarcophagids is apparent. Although both forms may occur abundantly at the same site, they are rarely found together in a single pitcher (unpublished data).

A chloropid fly, *Aphanotrigonum* sp., currently being described (Folkerts and Folkerts, unpublished) is an obligate inhabitant of the pitchers of several *Sarracenia* species. It is most common in the pitchers of *S. leucophylla*, although it also occurs commonly in other species with tall-form pitchers. It is the sister species of *Aphanotrigonum macfarlanei,* which inhabits pitchers of the cobra pitcher plant (*Darlingtonia californica,* Sarraceniaceae) in western North America. Its range extends from southeastern Louisiana to central Georgia. Larvae are small, whitish maggots which inhabit and feed on the prey mass. Pupation occurs in the prey mass. Adults are often abundant on the interior pitcher walls below the orifice. Several generations occur during a season. Overwintering takes place in the prey mass in the pupal stage. Without close inspection, adults of *Dohrniphora cornuta* (a facultative associate) can be confused with this fly. The two species are not known to co-occur.

A phorid fly, *Dohrniphora cornuta,* is a frequent inhabitant of the pitchers of *S. flava* at sites along the Atlantic Coastal Plain. It is a facultative associate that occurs in a wide variety of habitats throughout the New World. Jones (1918), a naturalist who described much of the biology of pitcher plant insects, first reported this species as *D. venusta.* The flattened, brownish-white larvae feed in the prey mass.

Because the pitchers hold water, *S. purpurea* provides habitat for aquatic insects. A variety of aquatic insects are occasionally found alive in pitchers (Mather 1980, De La Rosa and Nastase 1987, Hamilton et al. 1996, Turner et al. 1996), although only three species are obligate inhabitants of *S. purpurea* phytotelmata. All three of these occur in the northern portion of the range of *S. purpurea* in addition to occurring in southeastern pitcher plant habitats. These species are included here because they are important components of the community of pitcher plant arthropods in the Southeast. They are covered in greater detail by Giberson and Hardwick (this volume).

The pitcher plant mosquito, *Wyeomyia smithii,* is unquestionably the best-known arthropod associated with carnivorous plants. It has been widely used to elucidate ideas about selection, population phenomena, life history strategies, and physiological ecology (Bradshaw 1976, Goins 1977, Lounibos et al. 1982, Miller et al. 1994, and see Giberson and Hardwick, this volume). Adults are often observed flying in the vicinity of pitchers of *S. purpurea,* the only species in which the mosquito occurs. Females oviposit on pitcher walls just above the water line. Larvae feed on protozoa and other microorganisms in the pitcher fluid (Addicott 1974, Cochran-Stafira 1994). Although they do not feed directly on pitcher plant prey, this species is included in this category because it is part of a food chain supported by pitcher plant prey.

Larvae of the pitcher plant midge, *Metriocnemus knabi* (Chironomidae), inhabit the lower portions of *S. purpurea* pitchers and feed directly on the prey carcasses. The small, white, thread-like larvae are difficult to observe unless the contents of the pitcher are agitated and poured into a container. Paterson and Cameron (1982) hypothesized that the larvae sometimes migrate to new leaves, although they did not propose a mechanism for this dispersal. This species exhibits a complex life cycle, with three overlapping generations every two years that vary from 3 to 12 months in development time, depending on season and leaf condition (Paterson and Cameron 1982, and see Giberson and Hardwick, this volume). Pupation takes place in a gelatinous mass attached to pitcher walls above the water level.

Feeding activity by this species and the pitcher plant mosquito have been shown to be beneficial to the plants because of acceleration in the breakdown of prey (Bradshaw and Creelman 1984). It has also been shown that feeding by *M. knabi* facilitates pitcher plant mosquitoes (Heard 1994). This species has been involved in a number of other ecological studies covering topics such as oxygen relationships, photoperiodism, niche separation, and other community dynamics (Parris and Jenner 1959, Buffington 1970, Cameron et al. 1977, Heard 1994, Harvey and Miller 1996).

Five described fly species and apparently a number of undescribed forms (Robert Naczi, Northern Kentucky University, personal communication) of the family Sarcophagidae are associated with *Sarracenia* (Riley 1874, Aldrich 1916, Jones 1935, San Jean 1957, Judd 1959, Fish and Hall 1978, Farkas and Brust 1986). The conspicuous, whitish maggots of these species inhabit the prey mass. Pupation occurs in soil or moss near the plant (Forsyth and Robertson 1975). Several overlapping generations probably occur for most species in the southeastern United States. Adults are often conspicuous in the habitat, frequently resting on the external portions of pitchers. *Fletcherimyia fletcheri,* an obligate inhabitant of *S. purpurea,* appears to be common throughout much of the range of the plant. The abundance of the other three species of *Fletcherimyia* (*F. celerata, F. jonesi,* and *F. rileyi*) and of *Sarcophaga sarraceniae* is unknown. Adults and larvae of these species are nearly impossible to differentiate in the field. The degree to which others are specifically associated with certain *Sarracenia* species also needs investigation.

Although sarcophagid maggots may co-occur within pitchers along with other prey-consumer species, this is a relatively rare occurrence. An adult female may larviposit several individuals into a single pitcher, but usually only one will reach maturity. They seem to be cannibalistic on other sarcophagid maggots (Forsyth and Robertson 1975) and may have an additional negative influence on other fly species (unpublished data). Hepburn and Jones (1919) found that larvae have antiproteases. Sarcophagid maggots may have an advantage over other pitcher inhabitants in deep water because they are able to float to the surface and exchange gases. Fish (1976) reported their occurrence in 64 percent of functional pitchers of *S. minor* examined in Florida.

Mites of two distantly related families have evolved to inhabit the pitchers of *Sarracenia* species. Of these, only the histiostomatids are prey consumers. Phytoseiids, discussed later, are predators on other mites. A number of species of mites of the genus *Sarraceniopus* (Histiostomatidae) inhabit pitchers of *Sarracenia*. Fashing and O'Connor (1984) discussed the systematics of this genus. Some species are apparently undescribed (Naczi and Fredrickson 1985). All are scavengers in the prey mass. *Sarraceniopus gibsoni* inhabits only *S. purpurea* leaves, where it occurs on the pitcher walls beneath the surface of the water (Hughes and Jackson 1958). *Sarraceniopus hughesi* was described from *S. flava* and *S. minor* (Hunter and Hunter 1964) and seems to occur in all species except *S. purpurea*. In the southeastern United States, mites occur in essentially all pitchers, within the mass of prey as well as on pitcher walls above the prey mass. Despite their almost ubiquitous presence, the method of dispersal of pitcher plant mites is not well known (Mather and Catts 1980). They have been found attached to the legs and body of specimens of three other pitcher plant associates: *Exyra semicrocea, Dohrniphora cornuta,* and the undescribed *Aphanotrigonum* species (G. Folkerts, personal communication).

Capture Interrupters

It is not uncommon to find predaceous insects (e.g., mantids) and spiders exploiting the prey-attracting characteristics of pitcher plants. Several species commonly await prey on the hood or near the orifice of pitchers, capturing insects which otherwise might have become plant prey. In addition, some wasp species build nests in pitchers, preventing further trapping. This group of species is considered here as capture interrupters. The nature of the benefits gained from association with the pitcher plant varies. However, all can be considered as exploiters. Cresswell (1993) referred to instances in which spiders built webs over the orifice of *S. purpurea* pitchers as resource parasitism. Although capture interrupters are not restricted to life with pitcher plants, some are very typical of pitcher plant habitats and appear to be highly adapted in their association with the plants.

The sphecid wasp, *Isodontia mexicana,* commonly nests in the leaves of tall-form *Sarracenia* species. This is one of a group of wasps commonly called "grass-carriers" because they use dried grass to construct their nests. They are also often referred to as "trap-nesting" wasps because much of what we know of them comes from studies in which they nest in artificial cavities supplied by investigators. Relatively few studies have been conducted in the natural habitat of pitcher plant bogs (Carithers 1998).

Wasps often chew near the orifice of unopened leaves and cause them to wilt and bend. This apparently produces a more suitable nest cavity opening, as nests in such modified leaves are more successful (Carithers 1998). They also often use pitchers that have been previously fed on by caterpillars of the pitcher plant moth. Carithers (1998) found that nests in these moth-altered pitchers were less often parasitized than other nests. Nests consist of wadded coils and tufts of dry grass with one to nine cells containing prey insects on which the larvae feed. This species preys primarily on tettigoniid orthopterans, especially *Conocephalus* nymphs, and occasionally on gryllids (Rymal and Folkerts 1982, Carithers 1998).

Nest construction by this wasp prevents further insect capture by that pitcher. Furthermore, female wasps may clean a pitcher of previously trapped prey before preparing a nest. However, population levels are probably rarely higher than the 2.5 percent of functional leaves reported by Fish (1976), and thus this insect appears to present no threat to the health of pitcher plant populations. What is considered to be the same species nests in a variety of plant cavities throughout the continent (Engelhardt 1928, Lin 1962, Medler 1965). However, these wasps appear to be behaviorally adapted to nesting in pitcher plant habitats, where nesting sites may be more available than in other habitats. It is possible that this species is negatively influenced by the destruction of pitcher plant habitats.

Other wasp nests are less frequently encountered in pitcher plant leaves. Bernon (1969) reported an active nest of a vespid paper wasp in a pitcher of *S. purpurea* in Massachusetts. In pitcher plant bogs of the southeastern United States, pitchers of *S. alata* and *S. rubra* have been found filled with dirt and apparently provisioned with paralyzed salticid and clubionid spiders by unidentified sphecid wasps (Rymal and Folkerts 1982).

Several species of spider are frequently encountered on and around pitcher plants. One of the most common is the green lynx spider, *Peucetia viridans* (Oxyopidae). These large, green spiders wait cryptically on pitcher hoods and seize attracted insects. Although this species has not been investigated in pitcher plants, Louda (1982) discovered that it consumed enough flower-visiting insects to decrease the amount of pollination occurring in flowers they exploited. A similarly negative influence on amounts of pitcher plant prey is possible, since this species is known to hunt in ambush on the same pitcher for considerable periods of time. Green lynx spiders also use pitchers as refuges while they guard their egg sacs and newly hatched young.

Other spiders commonly seen in and on pitchers include a number of jumping spiders of the genus *Phidippus*. These spiders lurk about on pitcher exteriors and leap considerable distances to seize prey. They often also build silken retreats within pitchers and probably exclude further trapping by the pitchers they inhabit. Occasionally, funnel web spiders (*Agelenopsis* sp.: Agelenidae) have been found with the tubular retreat portion of their webs entering pitcher orifices. Several species of crab spiders (Thomisidae) are commonly found in and near pitcher openings. Cresswell (1991, 1993) reported webs of linyphiid spiders covering the orifices of *S. purpurea* pitchers. Although spiders make up a small component of the prey trapped by pitcher plants, most are capable of escaping entrapment by using a dragline of silk to escape from pitchers.

Herbivores

Several specialized herbivores have evolved to consume the tissue of pitcher plants. Some of these attack the leaf tissue and are able to inhabit pitchers without being trapped or consumed by the plants. Others attack plant parts not directly involved in the carnivorous function of the plant and are typical herbivores without unique anti-entrapment adaptations. A few herbivorous arthropods associated with pitcher plants also feed on other plant species. Those that are obligate associates are significantly affected by the conservation status of the plants.

Choristoneura parallela (Lepidoptera: Tortricidae), the spotted fireworm, is a facultative associate which occasionally feeds on pitcher tissue (Jones 1908). The larva, described and illustrated by MacKay (1962), feeds primarily on immature leaves, often constructing a tangle of silk within the pitcher tube. External signs of its feeding are visible as irregularly browned and chewed leaf tips. Although it feeds on a number of pitcher plant species, mainly *S. alata, S. flava, S. leucophylla,* and *S. minor,* it is not abundant at any sites. This lepidopteran feeds on many plant species (Freeman 1958) and is not likely to be affected by the conservation status of pitcher plants. Because it is only occasionally encountered in pitcher plants, neither can it be considered to be a threat to the survival of any *Sarracenia* population.

Tortricid caterpillars of the genus *Endothenia* feed in fruits of all *Sarracenia* species (Jones 1908, Forbes 1923, Hilton 1982, G. Folkerts 1992). As yet there is no clear taxonomic picture of the types involved. Two names, *Endothenia hebesana* and *E. daeckeana,* have been used for the forms found on pitcher plants. Miller (1983) thought that these two names were synonyms. However, life history differences among populations suggest the possibility of several species. The existence of host-specific types is possible. These moths often damage up to 90 percent of fruits and were at one time considered to present a problem in populations of the endangered green pitcher plant. However, they rarely damage all carpels and do not present a threat to the reproductive potential of the plants (Hilton 1982, G. Folkerts 1992).

Endothenia daeckeana may be restricted to *S. purpurea.* The larvae bore into fruits and consume the seeds within. Pupation occurs within the flower stalk. Adults emerge in the spring after the pupae push through an exit hole made by the larvae (Hilton 1982). Life history traits of the moths which attack other *Sarracenia* species differ considerably. Pupation occurs within fruits, and larvae never bore into flower stalks. Larvae consume flower parts as well as seeds and may exhibit more than one generation per year. Their presence can be detected by observation of frass and plant material fastened by silk to wilted flower parts. The species that occurs in pitcher plant habitats of the southeastern United States is probably *E. hebesana,* a species which attacks several different plant hosts and was first reported from pitcher plants by Forbes (1923).

The pitcher plant rhizome borer, *Papaipema appassionata* (Lepidoptera: Noctuidae), is probably the rarest of all pitcher plant insects, especially in the southerly portion of its range. It lives as a larva in the rhizome pith. Frass, which is often extruded through a hole in the rhizome surface, forms a vertical tube that is usually the only sign of its presence. At times a larva may extrude frass through a hole in the basal portion of the flower stalk or through a hole in the interior of a dry pitcher, and it is then very difficult to detect. Jones (1908) reported that the boring activities killed *S. flava* through destruction of the apical meristem. However, Bird (1903), Forbes (1954), and Brower and Brower (1970) reported this species solely from *S. purpurea,* the only species in which it has been recently found in the southeast (personal communication, G. Folkerts). When boring in *S. purpurea,* the larva seldom kills the plant, confining its activities to the pith and usually making an external frass tube without damaging the apical meristem. Apparently there is only one generation per year, even in the most southerly areas. They emerge in the late summer or fall and are sometimes attracted to light (Brower and Brower 1970).

This species has become very rare, or perhaps was always rare, in the southeastern United States. Two factors seem to be involved in its decline. First, the red imported fire ant (*Solenopsis invicta*) and the black imported fire ant (*Solenopsis richteri*), introduced from South America early in the century, prey on the larvae of this species. A second exotic predator possibly impacting rhizome borer populations is the nine-banded armadillo (*Dasypus novemcinctus*). Armadillos also rip open pitchers to obtain prey or pitcher inhabitants. The pitcher plant rhizome borer is currently known from only one site in the southeastern United States. Outside that area it seems to be more common, but should be considered to be seriously threatened by habitat destruction and invasion of exotic species.

The pitcher plant aphid, *Macrosiphum jeanae,* is known from populations of *S. purpurea* in Canada (Robinson 1972). Although it has not been reported on pitcher plants in the southeastern United States, it is difficult to observe and may have been missed. An undescribed species of diaspidid scale insect of the genus *Aspidiotus* (Michael L. Williams, Dept. of Entomology, Auburn University) occurs on pitchers of *S. psittacina* and *S. purpurea*. It is encountered only rarely.

Of all pitcher plant insects, moths of the genus *Exyra* (Noctuidae) have made the most remarkable adaptations to life within pitcher plant leaves. The larvae of the three species of this genus are herbivores on leaf and occasionally flower tissue of *Sarracenia* species. *Exyra semicrocea* feeds on all *Sarracenia* species, although it is most abundant in *S. leucophylla* and does not occur in *S. purpurea* north of the range of *S. flava*. The other two species are both host-specific, *E. ridingsii* occurring only on *S. flava* and *E. fax* on *S. purpurea* (Jones 1935). A number of authors have described the biology of these moths (Riley 1874, Jones 1893, 1904, 1907, 1921, Hubbard 1896, Judd 1957, Brower and Brower 1970, Rymal 1980, Rymal and Folkerts 1982, Folkerts and Folkerts 1996). Although life histories of all three species are similar, they differ enough to be distinguished by signs of their feeding and other activities even in the absence of specimens (Folkerts and Folkerts 1996).

Moths usually emerge in early April along with the appearance of new pitcher plant leaves. Adults are found resting singly or in small groups on interior pitcher walls. As a result of behavioral and morphological adaptations, they are among the few insects capable of walking on the leaf surface of the conductive zone. This ability is due in part to the fact that they maintain a head-up position and walk backward and forward rather than turning around within pitchers. They also have unusually positioned tarsal claws that allow them to walk among the downward cellular projections (Rymal 1980). Adults spend nearly their entire lives within pitchers, leaving only at dusk to locate mates or new pitchers for oviposition. Copulation occurs inside of pitchers, with male and female at nearly right angles to one another so that neither is positioned head-down (Rymal 1980). One to several eggs are laid just inside the orifice.

Larvae feed by eating the internal layers of pitcher tissue, leaving the outer epidermis intact as a protective enclosure. Tops of pitchers are closed by a protective sheet of silk or by wilting of the upper portion. This wilting results from larval feeding in a channel that encircles the pitcher. Larvae are almost always found singly in pitchers and may migrate from pitcher to pitcher several times during development. Rarely are leaves entirely consumed. Observations of *E. semicrocea* in *S. leucophylla* in Alabama (Rymal 1980) indicate that a single larva feeds on and inhabits an average of four pitchers before pupation, and that four or more overlapping generations occur. Larvae of instars 3 through 5 overwinter without a true diapause (Rymal 1980). Small amounts of feeding and development may occur during warm spells in the winter. Larvae of *E. semicrocea* have laterally projecting lappets on thoracic and abdominal segments that may help prevent entrapment in the narrowed bottom portion of pitchers (Jones 1921). *Exyra ridingsii* larvae also possess lappets, but they do not occur on *E. fax*.

In all three species larvae usually migrate to undamaged pitchers before pupation. A drainage hole is usually cut near the bottom of pitchers. Larvae of *E. fax*, inhabiting the water-filled pitchers of *S. purpurea*, cut drainage holes in nearly every pitcher inhabited. Cocoons are constructed of silk or silk and frass.

Visually, the damage to pitcher plants by populations of *Exyra* can be alarming. Occasionally populations reach high enough levels that every pitcher at a site may show some signs of *Exyra* herbivory. However, rarely are all leaves of a single plant attacked, and rarely are entire leaves consumed. It appears that populations of *Sarracenia* can withstand heavy infestations of these herbivores. It is possible that the harmful effects of herbivory are partially alleviated by the effect of accumulating frass, from which the plants may possibly absorb nutrients.

On the other hand, being obligate associates, *Exyra* species depend on the survival of pitcher plant populations for their own survival. *Exyra semicrocea* is very common at many sites in the southeastern United States. *Exyra ridingsii* is abundant at a number of sites, but seldom as abundant as *E. semicrocea*. *Exyra fax*, associated only with *S. purpurea*, is declining rapidly in the southern portion of its range. Fire ant and armadillo predation seem to be involved in its decline. Its host plant has also decreased in numbers in the southern portion of its range in recent years. Currently, no known sites exist for this moth species in the southeastern United States, even though relatively undisturbed populations of the host plant remain. In the northern portion of its range it is occasionally encountered but is less common than its congeners and most likely is suffering decline also.

Parasites and Predators

The existence of parasites and predators on pitcher plant arthropods completes the complex web of interactions found in pitcher plants. The influence of this group on the plants is difficult to decipher and depends on the influence of the primary associate each preys upon. As pointed out earlier, the influence of the primary associates is not always clear-cut. Prey consumers may superficially seem to be detrimental, yet some have been shown to be beneficial. Most species that exploit pitcher plants seem to be tolerated by pitcher plants with little or no apparent harm to populations of the plants. *Exyra* and *Endothenia* moths, for instance, cause obvious damage, but relatively stable populations are consistently found. It may be that a rich diversity of parasites and predators is in part responsible for this stability.

As important as they may be, little is known of the parasites of pitcher plant insects. Unidentified tachinid flies and braconid wasps have been reported as parasites of *Exyra semicrocea* (Rymal 1980). *Senotainia trilineata,* other sarcophagids (Carithers 1998), and a hymenopterous parasite, *Eurytoma bicolor* (Rau 1935), have been reared from nests of *Isodontia mexicana* wasps.

Predation on pitcher plant arthropods is perhaps easier to observe. The large predaceous mite, *Macroseius biscutatus* (Phytoseiidae), occurs in abundance in pitchers of tall-form pitcher plants in the southeastern United States. This species appears to be a predator on histiostomatid mites and nematodes (Muma and Denmark 1967). The green lynx spider is commonly seen preying on *Exyra* caterpillars and *Isodontia* wasps. A common predator of *Exyra semicrocea*, observed in a life history study of the species in Alabama, was the clubionid spider *Strotarchus piscatoria* (Rymal 1980). Fire ants, although exotic, are prevalent predators on *Exyra, Isodontia, Papaipema,* and other species and may present a threat to the survival of some populations.

CONSERVATION

Pitcher plants and associated arthropods continue to be threatened by habitat destruction and invasion by exotic species. Obligate pitcher plant associates are a group of arthropods that almost consistently exist in stable populations with pitcher plants in spite of apparent negative interactions. The community of pitcher plants and associates is obviously highly coevolved. However, the entire community is likely to disappear within the near future. Two pitcher plant species are federally protected as endangered species, and others are candidates for protection or appear on state lists. Although none of the arthropods discussed here are currently protected, several of them have declined rapidly in recent years. All of the obligate species are at risk of extinction as habitat destruction for pitcher plants occurs rapidly in the southeastern United States.

ACKNOWLEDGMENTS

Parts of the information presented here are the result of studies funded by the U.S. Fish and Wildlife Service. Much of this information was gathered and compiled with the assistance of numerous students, friends, and associates. The author is gratefully indebted to Reagan Booth, Paige Carithers, Karen Carte, Peggy Corbitt, D. A. Crossley, Jr., Amy Daigle Carroll, George Folkerts, Patrick Murphy, Cary Norquist, and Allison Teem.

LITERATURE CITED

Adams, R. M., and G. W. Smith, 1977. An S. E. M. survey of the five carnivorous pitcher plant genera. American Journal of Botany 64:265–272.

Addicott, J. F. 1974. Predation and prey community structure: An experimental study of the effect of mosquito larvae on the protozoan communities of pitcher plants. Ecology 55:475–492.

Albrecht, R. M. 1974. Microbial activity in *Sarracenia purpurea*. Carnivorous Plant Newsletter 3:32.

Aldrich, J. M. 1916. Sarcophaga and Allies in North America. Entomological Society of America, Lafayette, IN.

Antor, R. J., and M. B. Garcia. 1995. A new mite-plant association: Mites living amidst the adhesive traps of a carnivorous plant. Oecologia 101:51–54.

Beaver, R. A. 1979. Fauna and foodwebs of pitcher plants in West Malaysia. The Malayan Nature Journal 33:1–10.

Beccarri, O. 1904. Wanderings in the great forests of Borneo. Constable, London, UK.

Bell, C. R. 1952. Natural hybrids in the genus *Sarracenia*. Journal of the Elisha Mitchell Scientific Society 68:55–80.

Bell, C. R., and F. W. Case. 1956. Natural hybrids in the genus *Sarracenia*. II. Current notes on distribution. Journal of the Elisha Mitchell Scientific Society 72:142–152.

Bernon, G. L. 1969. Paper wasp nest in pitcher plant, *Sarracenia purpurea* L. Entomological News 80:148.

Bird, H. 1903. New histories in *Papaipema (Hydroecia)*. Canadian Entomologist 35: 91–94.

Bradshaw, W. E. 1976. Geography of photo-periodic response in diapausing mosquitoes: Control of dormancy in *Wyeomyia smithii*. Nature 262:384–385.

Bradshaw, W. E., and R. A. Creelman. 1984. Mutualism between the carnivorous purple pitcher plant and its inhabitants. American Midland Naturalist 112:294–304.

Brower, J. H., and A. E. Brower. 1970. Notes on the biology and distribution of moths associated with the pitcher plant in Maine. Proceedings of the Entomological Society of Ontario 101:79–83.

Buffington, J. D. 1970. Ecological considerations of the cohabitation of pitcher plants by *Wyeomyia smithii* and *Metriocnemus knabi*. Mosquito News 30:89–90.

Cameron, C. J., G. L. Donald, and C. G. Paterson. 1977. Oxygen-fauna relationships in the pitcher plant *Sarracenia purpurea* L. with reference to the chironomid *Metriocnemus knabi* Coq. Canadian Journal of Zoology 55:2018–2023.

Carithers, T. P. 1998. The nesting biology of the pitcher plant wasp, *Isodontia mexicana* (Sphecidae). Master's Thesis, Auburn University, Auburn, AL.

Case, F. W., and R. B. Case 1974. *Sarracenia alabamensis*: A newly recognized species from central Alabama. Rhodora 76:650–652.

———. 1976. The *Sarracenia rubra* complex. Rhodora 78:270–325.

Chew, F. S. 1978. *Pieris rapi oleracea* (Pieridae) caught by insectivorous plant. Journal of the Lepidopterist Society 32:129.

Christensen, N. L. 1976. The role of carnivory in *Sarracenia flava* L. with regard to specific nutrient deficiencies. Journal of the Elisha Mitchell Scientific Society 92: 144–147.

Cochran-Stafira, D. L. 1994. Food web dynamics and microbial community structure in *Sarracenia purpurea* pitchers. Ph.D. Dissertation, Northern Illinois University, De Kolb, IL.

Cresswell, J. E. 1991. Capture rates and composition of insect prey of the pitcher plant *Sarracenia purpurea*. American Midland Naturalist 125:1–9.

———. 1993. The morphological correlates of prey capture and resource parasitism in pitchers of the carnivorous plant *Sarracenia purpurea*. American Midland Naturalist 129:35–41.

Dakin, M. E., and K. L. Hays. 1970. A synopsis of Orthoptera (sensu lato) of Alabama. Bulletin 404. Alabama Agricultural Experiment Station, Auburn, AL.

Darwin, F. 1878. Experiments on the nutrition of *Drosera rotundifolia*. Journal of the Linnean Society of Botany 17:17–32.

De La Rosa, C., and A. J. Nastase. 1987. Larvae of *Metriocnemus* cf. *fuscipes, Limnophyes* sp., Pentaneurini (Diptera: Chironomidae) and *Culicoides* (Diptera: Ceratopogonidae) from pitcher plants, *Sarracenia purpurea*. Journal of the Kansas Entomological Society. 60:339–341.

Eleutarius, L. N., and S. B. Jones, Jr. 1969. A floristic and ecological study of pitcher plant bogs in south Mississippi. Rhodora 71:29–34.

Farkas, M. J., and R. A. Brust. 1986. Pitcher plant sarcophagids from Manitoba and Ontario. Canadian Entomologist 118:1307–1308.

Fashing, N. J., and B. M. O'Connor. 1984. *Sarraceniopus*—a new genus for histiostomatid mites inhabiting the pitchers of the Sarraceniaceae (Astigmata: Histiostomatidae). International Journal of Acarology 10:217–277.

Ferge, L. A., and R. M. Kuehn. 1976. *Boloria frigga,* new record Nymphalidae in Wisconsin, USA. Journal of the Lepidopterist Society 30:233–234.

Fish, D. 1976. Insect-plant relationships of the insectivorous pitcher plant *Sarracenia minor*. Florida Entomologist 59:199–203.

Fish, D., and R. Beaver. 1978. A bibliography of the aquatic fauna inhabiting bromeliads (Bromeliaceae) and pitcher plants (Nepenthaceae and Sarraceniaceae). Proceedings of the Florida Anti-mosquito Association, 49th Meeting: 11–19.

Fish, D., and D. W. Hall. 1978. Succession and stratification of aquatic insects inhabiting the leaves of the insectivorous pitcher plant, *Sarracenia purpurea*. American Midland Naturalist 99:172–183.

Folkerts, D. R. 1992. Interactions of pitcher plants (Sarracenia: Sarraceniaceae) with their arthropod prey in the southeastern United States. Ph.D. Dissertation, University of Georgia, Athens, GA.

Folkerts, D. R., and G. W. Folkerts. 1996. Aids for field identification of pitcher plant moths of the genus *Exyra* (Lepidoptera: Noctuidae). Entomological News 107:128–136.

Folkerts, G. W. 1982. The Gulf Coast pitcher plant bogs. American Scientist 70:261–267.

———. 1989. Facultative rhizome dimorphism in *Sarracenia psittacina* Michx. (Sarraceniaceae): An adaptation to deepening substrate. Phytomorphology 39:285–289.

———. 1991. A preliminary classification of pitcher plant habitats in the southeastern United States. Journal of the Alabama Academy of Science 62:199–225.

———. 1992. Identification and Measurement of Damage Caused by Flower and Seed Predators Associated with Sarracenia Oreophila, and Recommended Management Control Measures Deemed Appropriate. Final report submitted to the Endangered Species Office, U.S. Fish and Wildlife Service, Jackson, MS.

Forbes, W. T. M. 1923. The Lepidoptera of New York and Neighboring States. Part I. Cornell University Agricultural Experiment Station Memoir 68.

———. 1954. The Lepidoptera of New York and Neighboring States. Part III. Noctuidae. Cornell University Agricultural Experiment Station Memoir 329.

Forsyth, A. B., and R. J. Robertson. 1975. K-reproductive strategy and larval behavior of the pitcher plant sarcophagid fly, *Blaesoxipha fletcheri*. Canadian Journal of Zoology 53:174–179.

Gaddy, L. L. 1982. The floristics of three South Carolina pine savannas. Castanea 47:393–402.

Gibson, T. C. 1983. Competition, disturbance, and the carnivorous plant community in the southeastern United States. Ph.D. Dissertation. University of Utah, Salt Lake City, UT.

———. 1991. Differential escape of insects from carnivorous plant traps. American Midland Naturalist 125:55–62.

Givnish, T. J. 1989. Ecology and evolution of carnivorous plants. Pages 243–290 *in* W. G. Abramson (ed.), Plant-Animal Interactions. McGraw-Hill, Toronto.

Goins, A. E. 1977. Observations on the life history and ecology of the southern pitcher-plant mosquito, *Wyeomyia haynei* Dodge. Master's Thesis, Auburn University, Auburn, AL.

Goodnight, C. J. 1940. Insects taken by the southern pitcher plant. Transactions of the Illinois Academy of Science 33:213.

Hamilton, R., IV, M. Whitaker, T. C. Farmer, A. A. Benn, and R. M. Duffield. 1996. A report of *Chauliodes* (Megaloptera: Corydalidae) in the purple pitcher plant, *Sarracenia purpurea* L. (Sarraceniaceae). Journal of the Kansas Entomological Society 69:257–259.

Hanna, M., and T. E. Moore. 1966. The spittlebugs of Michigan. Papers of the Michigan Academy of Science, Arts and Letters 51:39–73.

Heard, S. B. 1994. Pitcher-plant midges and mosquitoes: A processing chain commensalism. Ecology 75:1647–1660.

Hepburn, J. S., and F. M. Jones. 1919. Occurrence of antiproteases in the larvae of the Sarcophaga associates of *Sarracenia flava*. Contributions of the Botanical Laboratory of the University of Pennsylvania 4:460–463.

Hepburn, J. S., F. M. Jones, and E. Q. St. John. 1927. Biochemical studies of the North American Sarraceniaceae. Transactions of the Wagner Free Institute of Science of Philadelphia 11.

Hepburn, J. S., E. Q. St. John, and F. M. Jones. 1920. The absorption of nutrients and allied phenomena in the pitchers of the Sarraceniaceae. Journal of the Franklin Institute 189:147–184.

Hilton, D. F. J. 1982. The biology of *Endothenia daeckeana* (Lepidoptera: Olethreutidae), an inhabitant of the ovaries of the northern pitcher plant, *Sarracenia purpurea* (Sarraceniaceae). Canadian Entomologist 114:269–274.

Hubbard, H. G. 1896. Some insects which brave the dangers of the pitcher plant. Proceedings of the Entomological Society of Washington 3:314–318.

Hunter, P. E., and C. A. Hunter. 1964. A new *Anoetus* mite from pitcher plants (Acarina: Anoetidae). Proceedings of the Entomological Society of Washington 66:39–46.

Joel, D. M. 1986. Glandular structures in carnivorous plants: Their role in mutual exploitation of insects. Pages 219–234 *in* B. E. Juniper and T. R. E. Southwood (eds.), Insects and the Plant Surface. Edward Arnold, London.

Joel, D. M., B. E. Juniper, and A. Dafni. 1985. Ultraviolet patterns in the traps of carnivorous plants. New Phytologist 101:585–593.

Jones, F. M. 1893. Two weeks in Richmond County, North Carolina. Entomological News 4:189–191.

Jones, F. M. 1904. Pitcher plant insects I. Entomological News 15:14–17.

———. 1907. Pitcher plant insects II. Entomological News 18:413–420.

———. 1908. Pitcher plant insects III. Entomological News 19:150–156.

———. 1918. *Dohrniphora venusta* Coquillet (Diptera) in *Sarracenia flava*. Entomological News 29:299–302.

———. 1920. Another pitcher-plant insect (Diptera: Sciarinae). Entomological News 31:91–94.

———. 1921. Pitcher plants and their moths. Natural History 21:297–316.

———. 1935. Pitcher plants and their insect associates. Pages 25–34 *in* Walcott, Illustrations of North American Pitcher Plants, The Smithsonion Institution, Washington, DC.

Judd, W. W. 1957. Studies of the Byron bog in southwestern Ontario. I. Description of the bog. Canadian Entomologist 89:235–238.

———. 1959. Studies of the Byron bog in southwestern Ontario. X. Inquilines and victims of the pitcher plant, *Sarracenia purpurea* L. Canadian Entomologist 91:171–180.

Juniper, B. E., R. J. Robins, and D. M. Joel. 1989. The Carnivorous Plants. Academic Press, San Diego, CA.

Larochelle, A. 1976. Sarracenie pourpre, predateur de carabidae. Fabreries 2:66.

———. 1977. The insectivorous plants of the genus *Drosera* as predators of Odonata (dragonflies). Cordulia 3:37–48.

———. 1978. Les plantes insectivores, predateurs de coleopteres carabidae (les cicindelini compris). Cordulia 4:65.

Lin, C. S. 1962. Biology and nesting habits of the hunting wasp *Isodontia harrisi*. Texas Journal of Science 14:429–430.

Lindquist, J. 1975. Bacterial and ecological observations on the northern pitcher plant, *Sarracenia purpurea* L. Master's Thesis, University of Wisconsin, Madison, WI.

Lloyd, F. E. 1942. The carnivorous plants. Chronica Botanica, Waltham, MA.

Louda, S. M. 1982. Inflorescence spiders: A cost/benefit analysis for the host plant, *Haplopappus venetus* Blake (Asteraceae). Oecologia 55:185–191.

Lounibos, L. P., C. Van Dover, and G. F. O'Meara. 1982. Fecundity, autogeny and larval environment of the pitcher plant mosquito, *Wyeomyia smithii*. Oecologia 55: 160–164.

Luttge, U. 1983. Ecophysiology of carnivorous plants. Pages 489–517 *in* O. L. Lange P. S. Nobel, C. B. Osmond, and H. Zeigler (eds.), Encyclopedia of Plant Physiology. Vol. 12C. Springer-Verlag, Berlin.

MacFarlane, J. M. 1908. Sarraceniaceae. *In* A. Engler (ed.), Das Pflanzenreich. 4: pt.110.

MacKay, M. R. 1962. Larvae of the North American Tortricinae (Lepidoptera: Tortricidae). Canadian Entomologist Supplement 28.

MacRoberts, B. R., and M. H. MacRoberts. 1988. Floristic composition of two west Louisiana pitcher plant bogs. Phytologia 65:184–190.

———. 1990. Vascular flora of two west Louisiana pitcher plant bogs. Phytologia 68: 271–275.

Mandossian, A. J. 1965. Some aspects of the life history of *Sarracenia purpurea*. Ph.D. Dissertation, Michigan State University, East Lansing, MI.

Mather, T. N. 1980. Larvae of alderfly (Megaloptera: Sialidae) from pitcher plant. Entomological News 92:32.

Mather, T. N., and E. P. Catts. 1980. Seasonal dispersal of pitcher plant mites (Acarina: Anoetidae). Journal of the New York Entomological Society 88:60–61.

McDaniel, S. 1971. The genus *Sarracenia* (Sarraceniaceae). Bulletin of the Tall Timbers Research Station 9:1–36.

Medler, J. T. 1965. Biology of *Isodontia* (*Murrayella*) *mexicana* in trap-nests in Wisconsin (Hymenoptera: Sphecidae). Annals of the Entomological Society of America 58:137–142.

Mellichamp, J. H. 1875. Notes on *Sarracenia variolaris*. Proceedings of the American Association for the Advancement of Science, 23rd meeting. 1875B:113–133.

Miles, D. H., U. Kokpol, and N. V. Mody. 1975. Volatiles in *Sarracenia flava*. Phytochemistry 14: 845-846.

Miller, T., D. Cassill, C. Johnson, C. Kindell, J. Leips, D. McInnes, T. Bevis, D. Mehlman, and B. Richard. 1994. Intraspecific and interspecific competition of *Wyeomyia smithii* (Coq.) (Culicidae) in pitcher plant communities. American Midland Naturalist 131:136–145.

Miller, W. E. 1983. Neararctic *Endothenia* species: A new synonymy, misidentification, and a revised status (Lepidoptera: Tortricidae). The Great Lakes Entomologist 16: 5–12.

Mody, N. V., R. Henson, P. A. Hedin, V. Kokpol, and D. H. Miles. Isolation of the insect paralysing agent coniine from *Sarracenia flava*. Experimentia 32:829–830.

Muma, M. H., and H. A. Denmark. 1967. Biological studies on *Macroseius biscutatus* (Acarina: Phytoseiidae). Florida Entomologist 50:249–255.

Naczi, R., and R. W. Fredrickson. 1985. Pitcher mites of the red pitcher plant (*Sarracenia rubra*). Proceedings of the Pennsylvania Academy of Science 59:82.

Parris, O. H., and C. E. Jenner. 1959. Photoperiodic control of diapause in the pitcher-plant midge, *Metriocnemus knabi*. Pages 601–624 *in* R. B. Withrow (ed.), Photoperiodism and Related Phenomena in Plants and Animals. American Association for the Advancement of Science Publication 55, Washington, DC.

Paterson, C. G., and C. J. Cameron. 1982. Seasonal dynamics and ecological strategies. of the pitcher plant chironomid, *Metriocnemus knabi* Coq. (Diptera: Chironomidae), in southeast New Brunswick. Canadian Journal of Zoology 60:3075–3083.

Plummer, G. L. 1963. Soils of the pitcher-plant habitats in the Georgia Coastal plain. Ecology 44:727–734.

————. 1966. Foliar absorption in carnivorous plants. I. Carolina Tips 29:25–26.

Plummer, G. L., and T. H. Jackson. 1963. Bacterial activities within the sarcophagus of the insectivorous plant *Sarracenia flava*. American Midland Naturalist 69:462–469.

Plummer, G. L., and J. B. Kethley. 1964. Foliar absorption of amino acids, peptides, and other nutrients by the pitcher plant, *Sarracenia flava*. Botanical Gazette 125:245–260.

Prankevicius, A. B., and D. M. Cameron. 1989. Free-living dinitrogen-fixing bacteria in the leaf of the northern pitcher plant (*Sarracenia purpurea* L.). Naturaliste Canadien 116:245–249.

Rau, P. 1935. The grass-carrying wasp, *Chlorion* (*Isodontia*) *harrisi* Fernald. Bulletin of the Brooklyn Entomological Society 30(2):65–68.

Riley, C. V. 1874. On the insects more particularly associated with *Sarracenia variolaris* (spotted trumpet-leaf). Proceedings of the American Association for the Advancement of Science 1874:18–25.

Robinson, W. J. 1908. A study of the digestive power of *Sarracenia purpurea*. Torreya 8:181–194.

Robinson, A. G. 1972. A new species of aphid (Homoptera: Aphididae) from a pitcher plant. Canadian Entomologist 104:955–957.

Rymal, D. E. 1980. Observations on the life history of the pitcher plant moth, *Exyra semicrocea* (Guenee) (Lepidoptera: Noctuidae) Master's Thesis, Auburn University, Alabama.

Rymal, D. E., and G. W. Folkerts. 1982. Insects associated with pitcher plants (*Sarracenia:* Sarraceniaceae) and their relationship to pitcher plant conservation: a review. Journal of the Alabama Academy of Science 53:131–151.

San Jean, J. 1957. Taxonomic studies of *Sarcophaga* larvae of New York, with notes on the adults. Memoirs of the Cornell University Agricultural Experiment Station 349.

Schnell, D. E. 1976. Carnivorous Plants of the United States and Canada. John F. Blair, Winston-Salem, NC.

————. 1977. Infraspecific variation in *Sarracenia rubra* Walt.: Some observations. Castanea 43:1–20.

————. 1983. Notes on the pollination of *Sarracenia flava* L. (Sarraceniaceae) in the piedmont province of North Carolina. Rhodora 85:405–420.

Small, J. G. C., A. Onraet, D. S. Grierson, and G. Reynolds. 1977. Studies on insect-free growth, development and nitrate-assimilating enzymes of *Drosera aliciae* Hamet. The New Phytologist 79:127–133.

Swales, D. E. 1969. *Sarracenia purpurea* as host and carnivore at Lac Carre, Terrebonne Co., Quebec. Part I. Naturaliste Canadiene 96:759–763.

———. 1972. *Sarracenia purpurea* L. as host and carnivore at Lac Carre, Quebec. Part II. Naturaliste Canadiene 99:41–47.

Turner, T. S., et al. 1996. An unusual occurrence in West Virginia of stoneflies (Plecoptera) in the pitcher-plant, *Sarracenia purpurea* L. (Sarraceniaceae). Proceedings of the Entomological Society of Washington 98:119–121.

Uphoff, J. C. T. 1936. Sarraceniaceae. Pages 1–24 *in* A. Engler and H. Harms (eds.), Die Natürlichen Pflanzenfamilien (2nd ed.).

Valtonen, P. 1980. Natural causes of death in dragonflies. Luonnon Tutkija 84:88.

Walker, J., and R. K. Peet. 1983. Composition and species diversity of pine-wiregrass savannas of the Green Swamp, North Carolina. Vegetatio 55:163–179.

Wells, B. W., and I. V. Shunk. 1928. A southern upland grass-sedge bog: An ecological study. Ecological Monographs 1:465–520.

Williams, R. M. 1966. Utilization of animal protein by the pitcher plant, *Sarracenia purpurea*. Michigan Botanist 5:14–17.

Wolfe, L. M. 1981. Feeding behavior of a plant: Differential prey capture in old and new leaves of the pitcher plant (*Sarracenia purpurea*). American Midland Naturalist 106:352–359.

Wray, D. L., and C. S. Brimley. 1943. The insect inquilines and victims of pitcher plants in North America. Annals of the Entomological Society of America 36:128–137.

Zipperer, P. 1887. Beiträge zur Kenntnis der Sarraceniaceen (Contributions to the Knowledge of the Sarraceniaceae). Botanisches Centralblatt 29:358–359.

PART 2
Woodland Ponds, Peatlands, and Marshes of the North and Northeast

12 Temporary Woodland Ponds in Michigan

Invertebrate Seasonal Patterns and Trophic Relationships

MICHAEL J. HIGGINS and RICHARD W. MERRITT

*T*emporary woodland ponds are relatively small, shallow wetlands that retain water for a few weeks to several months out of the year. Most of the energy flow within these habitats stems from microbial degradation of leaf litter deposited by surrounding trees and shrubs. The composition of the invertebrate community found within any particular pond is related to its size and the duration of flooding. The invertebrates that inhabit these ponds show varying degrees of adaptation to ephemeral habitats, but nearly all are characterized by rapid larval growth. Medium- to large-sized ponds exhibit a predictable seasonal succession of species, a pattern that has evolved in response to both physical constraints and biotic interactions. The early-season inhabitants are particularly well adapted to the ephemeral habitat and cold temperatures characteristic of early spring in Michigan. These animals feed primarily on the abundant microbial community present on the leaf litter and within the water column, and avoid heavy predation pressure by beginning development before the appearance of most of the predators. Most of the ponds' inhabitants that are not specifically adapted to ephemeral habitats are predators. These are generally insects that overwinter in permanent water and recolonize temporary ponds each spring. By consuming a high-quality food source such as animal protein, these migrants are able to develop rapidly and thus ensure completion of the larval phase before the ponds dry.

Invertebrates in Freshwater Wetlands of North America: Ecology and Management, Edited by Darold P. Batzer, Russell B. Rader, and Scott A. Wissinger
ISBN 0-471-29258-3 © 1999 John Wiley & Sons, Inc.

TEMPORARY PONDS

Prior to Euro-American settlement, the wooded landscape of southern Michigan was dotted by innumerable ephemeral ponds formed millennia ago in vast glacial outwash plains. Although only a small percentage of these temporary woodland ponds remains today, they represent a fairly common yet remarkably understudied aquatic habitat. Woodland ponds can range in size from a few square meters to over a hectare, although most are probably less than 0.4 ha. Maximum water depth is generally less than 1.5 m, and the average depth is usually under 1 m. These small woodland ponds are generally unsuitable for waterfowl production compared to larger, more open habitats, a factor which may explain the relative inattention these wetlands have received in the scientific literature. Ponds may begin to flood in late autumn or winter, reaching maximum size in the early spring as a result of snowmelt and spring rains. A distinction has been made between vernal pools (Fig. 12.1a), which flood only in the spring, and autumnal ponds (Fig. 12.1b), which flood in the autumn and remain wet until the following summer (Wiggins et al. 1980). It should be pointed out that the flooding which occurs in autumn often only covers the deepest parts of these ponds, and much of the pond area remains dry until the following spring. While the flooded area of an autumnal pond may provide important overwintering habitat for some aquatic invertebrates that lack specific adaptations for drying and freezing (see Batzer and Sion, this volume), much of the well-adapted temporary pond fauna remains unaffected by this flood event. Thus, while the presence or absence of water during the autumn and winter will influence faunal composition somewhat (Kenk 1949, Wiggins et al. 1980, Batzer and Sion, this volume), we believe that the size and duration of a particular pond during the vernal phase has an even more pronounced influence on community composition.

The duration of flooding is directly related to area and depth. All of these water bodies are closed depressions and dry from evaporation and groundwater outflow. Most are dry by the middle of summer, and some may undergo a second, somewhat accelerated cycle of flooding and drying in the mid- to late summer as a result of heavy precipitation from thunderstorms. Flooding during this aestival (summer) phase is generally smaller in areal extent than during the vernal phase. Small vernal pools flood only in the spring.

Temporary woodland ponds occur in forested landscapes and are thus bordered on all sides by trees (e.g., red maple, silver maple, elm, cottonwood) and shrubs (e.g., dogwoods, alder, spicebush). Trees frequently occur within the flooded portions of the ponds as well as the borders (Fig 12.1b). Because of the intense shading from trees and shrubs along the margins of these ponds, often very little emergent vegetation is present, or the emergent plants may occur only in relatively small areas that receive sufficient sunlight. In addition, submerged aquatic vegetation (including submerged macrophytes and mats of filamentous algae) generally does not occur in these ponds. After the trees

Fig. 12.1. Examples of temporary woodland ponds. (*a*) Small vernal pool (approximately 4 m in diameter). (*b*) Autumnal-vernal pond during the vernal phase. Note the abundant leaf litter in each pond.

leaf out in the spring, the ponds themselves may receive little direct sunlight, a factor which may limit algal growth compared to more open types of wetlands (e.g., Kenk 1949). The surrounding woodland vegetation is also important for the input of substantial amounts of leaf litter into the dry basins in the fall. Barlocher et al. (1978) recorded an average of 132.8 g/m² of leaf litter (ash-free dry mass) falling into Ontario pond basins in the autumn.

In southern Michigan water temperatures range from 3°C in the early spring when ice is still present, to 27°C in the summer. Because of the shallow nature of these bodies of water, daily water temperature fluctuations of 5°C are not uncommon in the spring, particularly in the small ponds (Fig. 12.2). In addition, a thin layer of ice frequently covers the surface at night during the early spring. Early in the season pH generally ranges from 7–7.5 and gradually becomes more alkaline (7.5–8) as the ponds shrink in size during the late spring and summer. Due to the large surface-to-volume ratio, ice-free ponds do not often become anoxic, but anaerobic conditions exist in the underlying sediments. Dissolved oxygen and pH also undergo diel fluctuations as a result of increase algal respiration at night (Williams 1987).

SEASONAL PATTERNS—VERNAL PHASE

In a landmark paper on temporary pond ecology, Wiggins et al. (1980) divided temporary pond breeding inhabitants into four groups (Table 12.1), based on their adaptations (or lack thereof) to the dry phase of these habitats and also on their oviposition/colonization habits. Animals that are particularly well adapted to life in these temporary environments (Groups 1–3) are basically year-round residents, spending the dry phase in some drought resistant stage (often the egg stage). Organisms which lack drought resistance (Group 4) must move to permanent water before the ponds dry and then recolonize the temporary ponds the following spring. This latter category is characterized by species with excellent colonizing abilities and rapid larval development.

The composition of the invertebrate community present in any given pond is related to its size and duration of flooding (Schneider and Frost 1996; see also Schneider, this volume). Small ponds of only a few weeks' duration contain relatively few species and are dominated by ostracods and mosquito larvae of the genus *Aedes*. In addition, gastropods, triclad turbellarians (planarians), cladocerans, and occasional predatory beetle larvae (family Dytiscidae) may be present. Amphipods (*Crangonyx*) have been surprisingly abundant in several small vernal pools we have sampled. Amphipods are not thought to be particularly well adapted to the dry and frozen conditions that characterize vernal ponds during the autumn and winter (Wiggins et al. 1980, Batzer and Sion, this volume), and their presence in these short-duration (8–10 weeks) vernal pools remains an enigma. Isopods, which are also poorly adapted to temporary ponds, were occasionally recovered from a few vernal pools. Their presence may be explained, however, by the cooccurrence of

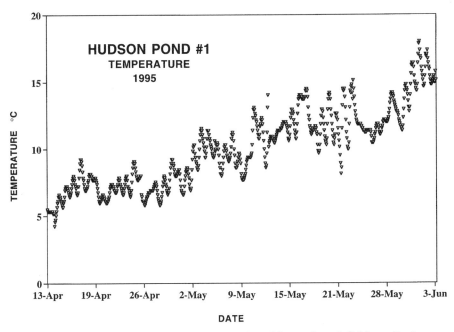

Fig. 12.2. Temperature data for a small vernal pool in southern Michigan. Each symbol represents a reading taken at approximately 1.5-hour intervals.

crayfish in these particular ponds. Crayfish burrows may provide a refuge in temporary habitats for poorly adapted organisms such as isopods (Wiggins et al. 1980).

The relatively few number of species present in these small pools suggests relatively simple trophic relationships (Schneider and Frost 1996). Medium-to large-sized ponds, persisting for four to six months in the spring and summer, generally contain a large diversity of invertebrate organisms and are characterized by a fairly predictable succession of species. While there may be some invertebrates active during the winter in autumnal ponds—generally small crustaceans such as copepods, ostracods, and cladocerans—the primary season of invertebrate activity begins when ice melts along the margins of the ponds, usually in late February or early March in southern Michigan. If the ponds are sufficiently flooded, *Aedes* mosquito eggs that were deposited in the dry basins by females the previous season begin to hatch. Analyses of soil samples from temporary ponds indicate that the vast majority of mosquito eggs are laid near the margins of maximum flooding extent (James 1966, Iversen 1971). Thus, if insufficient precipitation occurs, resulting in lower than normal water levels, little hatching will take place.

In addition to *Aedes* mosquito larvae, other early-season inhabitants of medium to large-sized temporary woodland ponds include fairy shrimp (order

TABLE 12.1. Summary of Life History Groups for Invertebrates Inhabiting Temporary Woodland Ponds

Group[a]	General Characteristics	Examples
1. Overwintering residents	Permanent residents with drought-resistant stage Passive dispersal only Generally noninsect invertebrates	Cladocerans, copepodos, ostracods, fairy shrimp, planarians, gastropods, fingernail clams
2. Overwintering spring recruits	Nearly permanent residents with drought-resistant stage (usually the egg stage) Adults oviposit in water before pond dries	Some dytiscid beetles (e.g., *Agabus*), *Polycentropis crassicornis* caddisfly, soldier flies (Stratiomyidae)
3. Overwintering summer recruits	Nearly permanent residents with drought-resistant stage (usually egg stage) Adults oviposit in dry basin	*Aedes* and *Psorophora* mosquitoes, caddisflies (*Limnephilus*), damselflies (*Lestes*), dragonflies (*Sympetrum*)
4. Nonwintering spring migrants	No drought-resistant stage, must recolonize pond each spring Excellent dispersal capabilities Overwinter in permanent water	All hemipterans (e.g., Corixidae, Gerridae, Notonectidae, Veliidae), some dytiscid beetles (e.g., *Acilius, Colymbetes, Dytiscus*), green darner dragonfly (*Anax junius*)

[a]From G. B. Wiggins, R. J. Mackay, and I. Smith, Evolutionary and ecological strategies of animals in annual temporary pools, Archiv für Hydrobiologie Supplement 58:97–206, 1980.

Anostraca), small crustaceans (copepods, cladocerans, ostracods), midge larvae (Chironomidae), phantom midge larvae (genus *Mochlonyx*), caddisfly larvae (primarily *Limnephilus*), and gastropods (e.g., *Physella*). Except for *Mochlonyx*, which preys upon small crustaceans and perhaps some first-instar mosquito larvae, there are few predators during this early part of the season. Conditions at this time are harsh, with cold water temperatures (<10°C) and frequent ice formation on the surface, factors which do not appear to harm the organisms listed above. Mosquito larvae have been shown to survive 10 days of ice cover on a woodland pond in Ontario (Westwood et al. 1983), and Walker (1995) demonstrated that larvae continue to feed, albeit more slowly, at temperatures down to 0°C. Most of the predators in temporary woodland ponds do not appear until water temperatures exceed 10°C on a regular basis. One of the few predators that appears to have solved physiological problems associated with cold and ice is the dytiscid beetle larva, *Agabus erichsoni,* the eggs of which hatch almost simultaneously with those of mosquitoes. Cold early spring temperatures and short-term ice cover do not seem to have a negative impact on this species, whose primary prey appears to be mosquito larvae (James 1961, Higgins, unpublished data).

As temperatures warm in mid- to late April, more species—chiefly predators—make their appearance within the ponds. Dragonflies and damselflies (*Sympetrum* and *Lestes*), present as eggs laid within the basin the previous summer, hatch out as very small individuals and consume small crustaceans during their early instars. Eggs of a predatory caddisfly, *Polycentropis crassicornis* (Polycentropidae), hatch in late April, and the larvae initially feed on small crustaceans that become trapped in the caddisflies' silken retreats constructed within the leaf litter (Higgins, unpublished data). Adult water striders (Gerridae) and backswimmers (Notonectidae) that overwintered in permanent water recolonize these temporary environments and begin breeding at this time, as do several species of beetles (e.g., Dytiscidae: *Acilius, Colymbetes, Dytiscus, Rhantus;* Hydrophilidae: *Hydrochara, Hydrochus, Tropisternus;* Gyrinidae: *Gyrinus*). Eggs of chorus frogs (*Pseudacris triseriata*) and salamanders (*Ambystoma*) that were laid in the ponds in late March or early April hatch by mid to late April. *Ambystoma* larvae are the only important vertebrate predators in temporary woodland ponds.

Adult mosquitoes begin to emerge from the ponds in early to mid-May of most years. By this same time, fairy shrimp have completed development, deposited eggs, and died. Limnephilid caddisflies, which have been feeding continuously on leaf detritus for two months, reach their final instar by the middle of the month and begin pupating in late May. With the emergence of mosquitoes, the pupation of caddisflies, and the disappearance of fairy shrimp, the invertebrate fauna becomes dominated by predatory species in late May. In addition to the species listed above, other migrants arrive, including giant water bugs (Belostomatidae), broad-shouldered water striders (Veliidae), water boatmen (Corixidae), water scorpions (Nepidae), and some additional beetle species. In addition, migratory green darner dragonflies (*Anax junius*) oviposit in large temporary ponds in April and May, and the voracious predatory larvae can become quite abundant in some ponds by mid-May. Phantom midge larvae (*Chaoborus*) also become abundant at this time. Many of these migrants are opportunistic and are not particularly adapted for ephemeral habitats, except that they are all characterized by rapid larval development.

There are few changes in the invertebrate faunal composition in June. With the advent of warm temperatures and less precipitation, the surface area and volume of these ponds begin to shrink, with concomitant increases in organism densities and nutrient concentrations. As the abundant predatory species increase their body sizes, shifts in their preferred prey may drastically alter relative abundances within the faunal assemblages. The general paucity of nonpredatory macroinvertebrates at this time of the year means that predators are feeding on predators, and foodwebs may become very complex.

Adult *Sympetrum* dragonflies and *Lestes* damselflies begin emerging from the ponds by July 1, and *Anax* dragonflies emerge in mid- to late July from the larger ponds. By mid-July most of the medium-sized ponds have dried and the larger ponds have shrunk to only a small fraction of their maximum size. By this time almost all insects have completed larval development and emerged as adults, and cladocerans have produced abundant epiphia, or drought-resistant eggs. The active invertebrate fauna at this time is characterized by adult insects (primarily bugs and beetles) capable of flying to permanent water, as well as other invertebrates that can burrow into the moist soil and/or form a drought-resistant stage (e.g., gastropods, planarians, and ostracods).

Even the largest of the temporary woodland ponds usually lose all surface water by early August. Undoubtedly there are some insect larvae that do not complete development by this time and perish. In drought years even insects that are well adapted to ephemeral habitats may become stranded. In most years, however, insects that perish from desiccation are either typical temporary pond migrants (i.e., Group 4 of Wiggins et al. 1980) that failed to complete development, or they represent oviposition mistakes by insects more typical of permanent water. An example of this latter category is the presence of early-instar dragonfly larvae of the genera *Libellula* and *Aeshna* in some of the larger temporary ponds during the summer. These insects are typical residents of permanent ponds, and most species require at least one year for larval development. In years of high precipitation in which some of the usually temporary ponds do not dry, these insects may survive and complete development the following year. The usual consequence of such oviposition mistakes, however, is complete larval mortality (Higgins, unpublished data).

SEASONAL PATTERNS—AESTIVAL PHASE

In most years heavy precipitation during the mid- to late summer can cause dry (or nearly dry) basins to flood again, triggering another cycle of invertebrate activity. The surface area of flooding during this aestival phase is generally less than half that of the much more extensive vernal phase. A few invertebrates appear to be specifically adapted to this later period of flooding. The floodwater mosquitoes *Aedes vexans* and *Aedes trivittatus,* as well as mosquitoes in the genus *Psorophora,* are particularly well adapted to summer rain pools. Although some eggs may hatch in the spring along with other species of *Aedes,* most *A. vexans* and *A. trivittatus* eggs, and all those of *Psorophora,* hatch following reflooding in the summer (Carpenter and La-Casse 1955). Unlike spring species of *Aedes* that oviposit primarily near the margins of the vernal extent of flooding, *A. vexans* also oviposits extensively in the interior portions of pond basins (Enfield and Pritchard 1977), a strategy that ensures hatching during summer flood events. Development is extremely rapid, with first-instars appearing within a few hours of flooding and adults emerging in less than a week. Densities of *A. vexans* larvae can reach several hundred per liter in these habitats (Dixon and Brust 1972), and the large number of biting adult females that emerge make this species a serious pest of humans during the summer (Carpenter and LaCasse 1955, Wood et al. 1979). Another species of mosquito, *Psorophora ciliata,* which may have coevolved with *A. vexans,* is predatory in larval instars II–IV, feeding primarily on *A. vexans* larvae (Breeland et al. 1961).

Other inhabitants of these aestival pools are either permanent residents (e.g., small crustaceans, planarians, gastropods) or opportunistic migrants (e.g., *Anopheles* mosquitoes, several species of beetles and bugs). This latter group includes adult insects of species typical of more permanent water, some of which may oviposit and attempt to complete an additional generation in these summer rain pools. While some of these migrants appear within one or two days of flooding (e.g., *Anopheles* mosquitoes), predatory beetle larvae (e.g., *Acilius*), as well as most other predators, do not appear until several days after inundation. This lag time between inundation and the appearance of predators allows the rapidly developing mosquito larvae to feed and grow relatively unmolested. Drought-resistant eggs deposited by insects and other arthropods that are well adapted to temporary ponds do not hatch at this time because they require a cold period followed by a warm-up in order to break their diapause (Horsefall and Fowler 1961, Wiggins et al. 1980). In addition, most of these eggs are deposited near the margins of the vernal extent of flooding and are not inundated by summer flood events.

The aestival phase is usually very brief, with surface water persisting for only a month or less. Animals that are specifically adapted to aestival pools, such as the mosquito species listed above, must be capable of extremely rapid development for this life history strategy to be successful. This strategy can be viewed as an evolutionary trade-off between risks and benefits. Although there is the risk of desiccation before larval development is completed, the larvae occupy a warm, nutrient-rich, and relatively predator-free environment in which development can occur rapidly.

TROPHIC RELATIONSHIPS

Temporary woodland ponds are detritus-based, heterotrophic habitats, with energy flow stemming predominantly from the leaf litter that falls into the basins. Emergent, submergent, and floating vascular plants are not common and thus contribute little to the overall energy budget. Although primary production in the form of algal photosynthesis takes place, the intense shading by the surrounding woods in these ponds reduces its input compared with more open bodies of water, particularly later in the spring (Moore 1970). Algal production that occurs in the early spring prior to tree leaf-out, however, may provide a significant food source for filter-feeding organisms (e.g., cladocerans).

Leaf litter that falls into the dry basins in the autumn is initially colonized by terrestrial microbes (principally fungi) that begin the process of decomposition. Barlocher et al. (1978) examined protein and fungal biomass levels in experimental leaf packs placed in vernal pools in Ontario. Higher protein levels (corresponding to higher levels of fungal biomass) were observed in leaf packs that were exposed to terrestrial microbes and aerobic decomposition compared with leaf packs that were submerged in water for the same period of time. All protein levels declined rapidly, however, following submergence in the spring. The authors concluded that the protein-rich detritus of temporary ponds supports the required rapid development of animals that inhabit these ephemeral environments (Barlocher et al. 1978).

The rapid decline in leaf litter protein levels following submergence observed by these researchers suggests, however, that there is more than just high-protein detritus supporting temporary pond fauna through larval development. The relative paucity of shredding detritivores in these habitats (Fig. 12.3), compared with many lotic situations, also suggests that other trophic pathways may be more important than direct feeding on leaf litter. Indeed, the early spring fauna is characterized by a diverse filter-feeding guild comprised of cladocerans, ostracods, fairy shrimp, and—at least part-time—mosquito larvae. Larvae of *Aedes* mosquitoes, in addition to filtering microorganisms and detritus from the water column, are also known to graze biofilm from the surfaces of leaf litter (Merritt et al. 1992). These larvae can apparently grow as well filtering pond water alone as they can when provided with detritus on which to graze. In a field experiment conducted in 1996, we examined the growth of 30 first-instar mosquito larvae (*Aedes stimulans*) in each of 15 microcosms provisioned with either conditioned leaves, nonconditioned leaves, or no leaves. No difference in time to adult emergence or adult weight was observed among the three treatments (Table 12.2; see also Walker and Merritt [1988] for similar results with treehole mosquitoes). These results suggest an abundance of planktonic food sources of importance to the trophic hierarchy at least equal to that of the enriched leaf detritus. But are the two related?

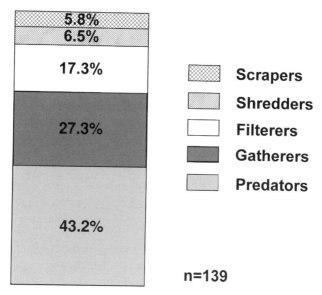

Fig. 12.3. Functional feeding group composition (by genera) of invertebrates in temporary woodland ponds.

It has been demonstrated that the degree of microbial colonization of leaf litter plays an important role in subsequent growth rates of invertebrate detritivores in both terrestrial and aquatic habitats (e.g., Barlocher et al. 1978, Suberkropp et al. 1983, Lawson et al. 1984, Merritt et al. 1984, Arsuffi and Suberkropp 1989, Walker et al. 1997). The dry phase of temporary ponds appears to be the most important period of microbial colonization of leaf litter, due to the prevalence of terrestrial fungi as the major decomposers (Barlocher et al. 1978). This microbial colonization process begins as soon as water levels begin to decline in the late spring or summer. As a pond loses surface water and shrinks in size, the leached-out and highly recalcitrant leaf detritus from the previous year becomes exposed and the moist surfaces are rapidly colonized by a host of fungi, bacteria, and protozoa. The celluloses and hemicelluloses contained in the leaves are principally exploited by terrestrial fungi, which depolymerize these complex compounds via cellulolytic enzymes into simpler carbohydrates (Ljungdahl and Eriksson 1985). Terrestrial fungi, principally white-rot fungi, are also important in breaking down lignin, which binds to the celluloses and hemicelluloses in leaves (Ljungdahl and Eriksson 1985). Although some lignin degradation occurs during the aquatic phase (Chamier 1985), the process is slow compared to that which occurs during the terrestrial phase, and many of the celluloses and hemicelluloses remain bound and unavailable for enzymatic degradation (Webster and Benfield 1986). The leaf surfaces are also colonized during this terrestrial

TABLE 12.2. Growth of First-Instar *Aedes stimulans* Mosquito Larvae in Field Microcosms Supplied with no Leaves or with 3 g (Initial Dry Mass) of Conditioned or Unconditioned Leaves, Wild Ginger Pond, Lansing, Michigan, 1996

Microcosm Treatment	n	Days to Emergence				Adult Mass (mg)			
		Males		Females		Males		Females	
		mean	p	mean	p	mean	p	mean	p
Conditioned leaves	5	47.43	0.913	49.22	0.876	0.583	0.999	0.641	0.978
Nonconditioned leaves	5	46.93		48.89		0.587		0.636	
No leaves	5	47.41		49.39		0.586		0.662	

phase by bacteria and their protozoan predators (Bamforth 1977), many of which secrete extracellular compounds (Nalewajko 1977). A similar colonization occurs on leaves that fall into the dry basins in the autumn. In this way the complex structure of the leaves is slowly converted into microbial biomass.

It is suggested here that when the ponds reflood in the spring to their maximum size (and when they partially reflood in the summer to form aestival ponds), the heavily enriched surfaces of the leaf detritus provide a nutrient broth in the form of dissolved and fine particulate organic matter stemming from the inundated fungal biomass, byproducts of lignin degradation (Kirk 1984), partially degraded celluloses, and the biomass and extracellular compounds of bacteria and protozoa. These dissolved substances, principally dissolved organic carbons and nitrogens (DOC and DON), are used by planktonic and attached heterotrophic bacteria, and a microbial bloom ensues. At this same time, algal growth may also be stimulated from DON and the sunlight available before trees leaf out in the spring. The resulting growth in the microbial community in turn supports the vast filter-feeding guild characteristic of the early spring fauna of temporary woodland ponds. In addition, following inundation, aquatic hyphomycetes readily colonize and degrade the dead terrestrial fungal cells (which contain little or no lignin), resulting in a substantially higher biomass of hyphomycetes and associated microorganisms than would be possible with the recalcitrant leaf detritus alone (Chamier 1985). This helps support invertebrate scrapers and gathering collectors as well as shredding detritivores.

Trophic relationships within temporary woodland ponds are closely tied to the seasonal succession and adaptational strategies (Table 12.1) of the associated fauna. Animals with drought-resistant stages may begin/resume development soon after the ponds reflood, triggered by physiochemical cues within the water (e.g., Horsefall 1956, Horsefall and Fowler 1961). The organisms that initially appear early in the spring are generally those that take advantage of the microbially enriched water and detritus, i.e., filter-feeders, gathering collectors (e.g., certain chironomid midge larvae), and detritivorous shredders (primarily limnephilid caddisflies). These species can apparently tolerate the low temperatures and frequent ice cover characteristic of this time of the year. Even though growth is slower in this cold environment (Atkinson 1994), development occurs in a relatively nutrient-rich and predator-free environment. This strategy of beginning development early in the spring may have initially evolved as an adaptation for survival in small, ephemeral habitats. A reinforcing factor may have been predator avoidance in time. By hatching early, organisms such as mosquito larvae and fairy shrimp—animals with few defensive mechanisms—feed and grow relatively unmolested by predators. The majority of predators do not appear until weeks later. By the time most of the predatory larvae of the odonates and beetles make their appearance, these potential prey species are generally too large for the small, early-instar predators to catch effectively. Insects that develop in cold temperatures reach larger body size than those reared at higher temperatures (Brust 1967, Atkinson 1994), and the larvae of spring species of *Aedes* are some of the largest in this genus (Wood et al. 1979). Thus, the early spring inhabitants of temporary ponds may escape predation in time by beginning development early and achieving larger body sizes than their potential predators. As previously stated, only some dytiscid beetles in the genus *Agabus* that overwinter in the egg stage have apparently evolved to take advantage of the abundant prey available early in the spring. By beginning development early in the spring, larvae of these beetles are able to exploit even small ponds of only a few weeks' duration, habitats unavailable for other beetle species that appear later in the season.

Although there are relatively few predators early in the spring, the number of predatory species increases substantially as the water temperature rises. Predator diversity reaches its peak in mid-May, with predators often becoming the dominant functional group within the ponds at this time. Most predators in temporary woodland ponds can be classified as either opportunistic migrants (Group 4 of Wiggins et al. 1980) or cyclic colonizers (Batzer and Wissinger 1996, Wissinger 1997), i.e., they generally lack specific adaptations to survive drought and overwinter in permanent-water habitats. Figure 12.4 illustrates that over half the predatory genera found in these ponds belong to this life history category. More striking, perhaps, is the very high proportion of predators within the Group 4 category itself (Fig. 12.5). The dominance of predators in this category is not terribly surprising given that animals in this group must recolonize ponds every year, breed, and complete larval de-

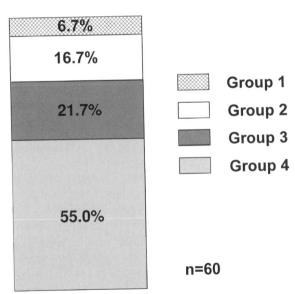

Fig. 12.4. Predator composition by life history strategy in temporary woodland ponds.

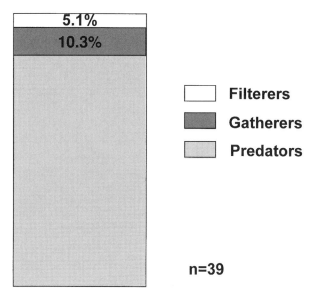

Fig. 12.5. Functional feeding group composition for Wiggins et al. (1980) Group 4 life history strategy.

velopment before the ponds dry. The consumption of animal protein allows the larvae of these relatively late arrivals to develop very quickly, the assimilation efficiency of predators being highest among trophic groups (Wotton 1994). Relatively rapid development is a characteristic of most animals breeding in ephemeral habitats, particularly those organisms that lack adaptations for coping with the dry phase.

Conversely, this probably explains the total lack of shredding detritivores exhibiting a Group 4 strategy (Fig. 12.5). Although the leaf litter of temporary ponds may be enriched by the microbial processes described above, it is still a relatively poor food source, and detritivorous animals must consume large quantities to maintain growth. One of the few shredding insects, the caddisfly *Limnephilus indivisus,* begins development early in the spring—sometimes hatching the previous fall in autumnal ponds (Wiggins 1973)—a strategy which helps to ensure that the 8–10 week development will be completed before the pond dries. For a migrant species that arrives at a pond later in the spring, a feeding strategy based on consuming leaf detritus would be risky; growth would be slow and there would often be insufficient time to complete development. Interestingly, limnephilid caddisfly larvae in temporary ponds are known to be occasionally cannibalistic (Wissinger et al. 1996) and have been reported to be infrequent predators of mosquito larvae (Downe and West 1954, Baldwin et al. 1955). Cargill et al. (1985) demonstrated the importance of lipids in the diet of final-instar shredding caddisflies. The accumulation of triglyceride reserves during the last larval instar is apparently essential for completion of the larval stage and subsequent adult reproduction (Cargill et al. 1985). Although lipids are available in aquatic hyphomycete fungi (Cargill et al. 1985), incidences of cannibalism and predation in otherwise detritivorous insects may be a means of quickly acquiring lipids as well as protein. This diet supplementation of protein and essential lipids may provide a boost in growth rates and ensure survival through the larval period (Wissinger et al. 1996).

CONCLUSIONS

In summary, temporary woodland ponds are small, microbially driven wetlands that may be flooded from a few weeks to several months out of the year. What may initially appear a harsh and uncertain habitat actually exhibits a reasonably predictable cycle of flooding and drying, as well as a fairly predictable seasonal succession of species. Invertebrates that are well adapted to this ephemeral habitat have evolved within the constraints of the physical environment and in response to community interactions (Schneider and Frost 1996, Williams 1996). Both abiotic and biotic factors appear to have shaped the observed succession of species, with early-season, cold-adapted inhabitants simultaneously taking advantage of the microbially enriched environment and minimizing predation pressure by beginning development before most of the predators appear. Many of the insects that arrive later in the spring are migrants with no special adaptations to the dry phase of these ponds. The vast majority of these species are predatory in at least the larval stage, employing a trophic strategy that permits rapid growth in a shrinking environment.

ACKNOWLEDGMENTS

This study was funded by a United States Department of Agriculture National Needs Water Science Fellowship through Michigan State University (USDA Grant 93-38420-8798). Additional support was provided by National Institutes of Health Grant AI21884, as well as the College of Natural Science and the Department of Entomology at Michigan State University. The authors wish to thank K. W. Cummins, E. D. Walker, and M. G. Kaufman for their helpful comments on an earlier draft.

LITERATURE CITED

Arsuffi, T. L., and K. Suberkropp. 1989. Selective feeding by shredders on leaf-colonizing fungi: Comparison of macroinvertebrate taxa. Oecologia 79:30–37.

Atkinson, D. 1994. Temperature and organism size—a biological law for ectotherms? Advances in Ecological Research 25:1–57.

Baldwin, W. F., H. G. James, and H. E. Welch. 1955. A study of mosquito larvae and pupae with a radio-active tracer. Canadian Entomologist 87:350–356.

Bamforth, S. S. 1977. Litters and soils as freshwater ecosystems. Pages 243–256 *in* J. Cairns, Jr. (ed.), Aquatic Microbial Communities. Garland, New York.

Barlocher, F., R. J. Mackay, and G. B. Wiggins. 1978. Detritus processing in a temporary vernal pool in southern Ontario. Archiv für Hydrobiologie 81:269–295.

Batzer, D. P., and S. A. Wissinger. 1996. Ecology of insect communities in nontidal wetlands. Annual Review of Entomology 41:75–100.

Breeland, S. G., W. E. Snow, and E. Pickard. 1961. Mosquitoes of the Tennessee valley. Journal of the Tennessee Academy of Science 36:249–319.

Brust, R. A. 1967. Weight and development time of different stadia of mosquitoes reared at various constant temperatures. Canadian Entomologist 99:986–993.

Cargill, A. S., II et al. 1985. The role of lipids as feeding stimulants for shredding aquatic insects. Freshwater Biology 15:455–464.

Carpenter, S. J., and W. J. LaCasse. 1955. Mosquitoes of North America. University of California Press, Berkeley, CA.

Chamier, A.-C. 1985. Cell-wall-degrading enzymes of aquatic hyphomycetes: a review. Botanical Journal of the Linnean Society 91:67–81.

Dixon, R. D., and R. A. Brust. 1972. Mosquitoes of Manitoba. III. Ecology of larvae in the Winnipeg area. Canadian Entomologist 104:961–968.

Downe, A. E. R., and A. S. West. 1954. Progress in the use of the precipitin test in entomological studies. Canadian Entomologist 86:181–184.

Enfield, M. A., and G. Pritchard. 1977. Estimates of population size and survival of immature stages of four species of *Aedes* (Diptera: Culicidae) in a temporary pond. Canadian Entomologist 109:1425–1434.

Horsefall, W. R. 1956. Eggs of floodwater mosquitoes (Diptera: Culicidae). III. Conditioning and hatching of *Aedes vexans*. Annals of the Entomological Society of America 49:66–71.

Horsefall, W. R., and H. W. Fowler, Jr. 1961. Eggs of floodwater mosquitoes VIII. Effect of serial temperatures on conditioning of eggs of *Aedes stimulans* Walker (Diptera: Culicidae). Annals of the Entomological Society of America 54:664–666.

Iversen, T. M. 1971. The ecology of a mosquito population (*Aedes communis*) in a temporary pool in a Danish beech wood. Archiv für Hydrobiologie 69:309–332.

James, H. G. 1961. Some predators of *Aedes stimulans* (Walk.) and *Aedes trichurus* (Dyar) (Diptera: Culicidae) in woodland pools. Canadian Journal of Zoology 39: 533–540.

———. 1966. Location of univoltine *Aedes* eggs in woodland pool areas and experimental exposure to predators. Mosquito News 26:59–63.

Kenk, R. 1949. The Animal Life of Temporary and Permanent Ponds in Southern Michigan. University of Michigan, Museum of Zoology, Miscellaneuous Publication 71.

Kirk, T. K. 1984. Degradation of lignin. Pages 399–437 *in* D. T. Gibson (ed.), Microbial Degradation of Organic Compounds. Marcel Dekker, New York.

Lawson, D. L., M. J. Klug, and R. W. Merritt. 1984. The influence of the physical, chemical, and microbiological characteristics of decomposing leaves on the growth of the detritivore *Tipula abdominalis* (Diptera: Tipulidae). Canadian Journal of Zoology 62:2339–2343.

Ljungdahl, L. G., and K.-E. Eriksson. 1985. Ecology of microbial cellulose degradation. Pages 237–299 *in* K. C. Marshall (ed.), Advances in Microbial Ecology, vol. 8. Plenum, New York.

Merritt, R. W., R. H. Dadd, and E. D. Walker. 1992. Feeding behavior, natural food, and nutritional relationships of larval mosquitoes. Annual Review of Entomology 37:349–376.

Merritt, R. W., W. Wuerthele, and D. L. Lawson. 1984. The effect of leaf conditioning on the timing of litter processing on a Michigan woodland floodplain. Canadian Journal of Zoology 62:179–182.

Moore, W. G. 1970. Limnological studies of temporary ponds in south-eastern Louisiana. Southwestern Naturalist 15:83-110.

Nalewajko, C. 1977. Extracellular release in freshwater algae and bacteria: Extracellular products of algae as a source of carbon for heterotrophs. Pages 589–624 *in* J. Cairns, Jr. (ed.), Aquatic Microbial Communities. Garland, New York.

Schneider, D. W., and T. M. Frost. 1996. Habitat duration and community structure in temporary ponds. Journal of the North American Benthological Society 15:64–86.

Suberkropp, K., T. L. Arsuffi, and J. P. Anderson. 1983. Comparison of degradative ability, enzymatic activity, and palatability of aquatic hyphomycetes grown on leaf litter. Applied and Environmental Microbiology 46:237–244.

Walker, E. D. 1995. Effect of low temperature on feeding rate of *Aedes stimulans* larvae and efficacy of *Bacillus thuringiensis* var. *israelensis* (H-14). Journal of the American Mosquito Control Association 11:107–110.

Walker, E. D., M. G. Kaufman, M. P. Ayers, M. H. Riedel, and R. W. Merritt. 1997. Effects of variation in quality of leaf detritus on growth of the eastern tree-hole mosquito, *Aedes triseriatus* (Diptera: Culicidae). Canadian Journal of Zoology 75: 706–718.

Walker, E. D., and R. W. Merritt. 1988. The significance of leaf detritus to mosquito (Diptera: Culicidae) productivity from treeholes. Environmental Entomology 17: 199–206.

Webster, J. R., and E. F. Benfield. 1986. Vascular plant breakdown in freshwater ecosystems. Annual Review of Ecology and Systematics 17:567–594.

Westwood, A. R., G. A. Surgeoner, and B. V. Helson. 1983. Survival of spring *Aedes* spp. mosquito (Diptera: Culicidae) larvae in ice-covered pools. Canadian Entomologist 115:195–197.

Wiggins, G. B. 1973. A contribution to the biology of caddisflies (Trichoptera) in temporary ponds. Life Science Contributions of the Royal Ontario Museum 88:1–28.

Wiggins, G. B., R. J. Mackay, and I. Smith. 1980. Evolutionary and ecological strategies of animals in annual temporary pools. Archiv für Hydrobiologie, Supplement 58:97–206.

Williams, D. D. 1987. The Ecology of Temporary Waters. Croom Helm, Timber Press, Portland, OR.

———. 1996. Environmental constraints in temporary fresh water and their consequences for the insect fauna. Journal of the North American Benthological Society 15:634–650.

Wissinger, S. A. 1997. Cyclic colonization in predictably ephemeral habitats: A template for biological control in annual crop systems. Biological Control 10:4–15.

Wissinger, S. A., et al. 1996. Intraguild predation and cannibalism among larvae of detritivorous caddisflies in subalpine wetlands. Ecology 77:2421–2430.

Wood, D. M., P. T. Dang, and R. A. Ellis. 1979. The Mosquitoes of Canada (Diptera: Culicidae). The Insects and Arachnids of Canada, Part 6. Publication 1686, Biosystematics Research Institute, Research Branch, Agriculture Canada, Ottawa, Ontario, Canada.

Wotton, R. S. 1994. Particulate and dissolved organic matter as food. Pages 235–288 *in* R. S. Wotton (ed.), The Biology of Particles in Aquatic Systems. Lewis, Boca Raton, FL.

13 Snowmelt Ponds in Wisconsin

Influence of Hydroperiod on Invertebrate Community Structure

DANIEL W. SCHNEIDER

*H*ydroperiod structures the invertebrate communities of snowmelt ponds in Wisconsin. Ponds fill with snowmelt in April and dry over the course of the summer, from less than one week to more than three months after filling. Some ponds dry only in particularly dry years. Short-duration ponds are structured by the adaptations of the invertebrates to short duration and drying. Long-duration ponds are increasingly structured by biotic interactions—competition and predation—among the invertebrates. Variability in hydroperiod also causes changes in the invertebrate communities of these ponds. A drought caused changes in the invertebrate fauna of these ponds. Some taxa, such as snails, caddisflies, and dragonflies, were reduced in abundance, while others, such as mosquitoes and fairy shrimp, increased. The most hydrologically variable ponds changed the most biologically. Hydroperiod duration and variability interact with life histories and biotic interactions to produce complex spatial and temporal patterns of abundance for some species of wetland invertebrates, such as fairy shrimp. Conservation of these habitats thus requires the maintenance of all aspects of the hydroperiod, including its variability.

SNOWMELT PONDS

Temporary ponds in northern Wisconsin fill with snowmelt in early to mid-April. Depending on the size of the pond and its position in relation to the groundwater table, the ponds last from a few days to several months before

Invertebrates in Freshwater Wetlands of North America: Ecology and Management, Edited by Darold P. Batzer, Russell B. Rader, and Scott A. Wissinger
ISBN 0-471-29258-3 © 1999 John Wiley & Sons, Inc.

drying. Some ponds are perennial and dry only in unusually dry years. The ponds form in the many closed depressions in the glacial topography. Because these depressions are of varying size and elevation in relation to the groundwater, a region will contain a wide variety of ponds with different hydroperiods. The species pool for all of the ponds in a region will be similar, as will the soils and other edaphic factors. This situation allows for an analysis of the effect of hydroperiod on wetland communities.

Hydroperiod has been one of the organizing principles in wetland ecology. While the role of hydroperiod in determining distribution of plant species has been well studied (Mitsch and Gosselink 1993), its impact on the distribution of aquatic invertebrates has received less attention. Wetland biologists have often focused on the autecological aspects of how hydroperiod controls wetland communities (e.g., van der Valk 1981). Physiological and life-history adaptations are important aspects of the ecology of wetland organisms. However, community-level effects of hydroperiod may also be important. In this chapter I consider the effects of hydroperiod on the invertebrate communities of temporary pond wetlands in northern Wisconsin.

To evaluate the effects of hydroperiod on wetland communities, I apply some of the insights gained from the study of ecological disturbances. Disturbances can be defined as any "relatively discrete event in time that disrupts ecosystem, community, or population structure and changes resources, substrate availability, or the physical environment" (White and Pickett 1985). The dry phase of a wetland hydroperiod may thus be thought of as an ecological disturbance. The effect of a disturbance on a community depends on the characteristic disturbance regime, which includes the frequency, magnitude, and predictability of the disturbance event. Wetland disturbance regimes can be characterized by hydroperiod or duration, the amount of time a wetland holds water. Duration, or time between drying, is the inverse of the disturbance frequency: the longer the duration, the lower the disturbance frequency. Predictability is related to the variation around the mean disturbance frequency: the higher the variability, the less predictable the hydroperiod.

Studies of disturbance, primarily in marine systems, have suggested that disturbance frequency affects community composition by mediating the relative importance of the processes that structure communities (Lubchenco 1986, Menge et al. 1986). Hydroperiod may affect wetland communities in a similar manner (Smith 1983, Wilbur 1987, Schneider and Frost 1996). These studies, both on marine and wetland systems, all suggest that in frequently disturbed habitats physical stress and the adaptation of organisms to the physical environment exert a dominant influence over community composition (Lubchenco 1986, Wiggins et al 1980). As the time between disturbances increases, biotic interactions among the species increase in importance. In intermediate-duration habitats physical constraints are relaxed and organisms achieve higher densities, leading to density-dependent effects on growth rate, emergence, or reproduction (Wilbur 1987). As habitat duration increases further, predators increase in abundance (Smith 1983, Schneider and Frost 1996, Wellborn et al. 1996) and can control species distributions or reduce populations of potential competitors to levels below those at which competition can occur.

The life histories of temporary pond taxa play an important role in the effect of duration on community structure. Life history adaptations to surviving the dry phase of the pond involve desiccation-resistant stages, including eggs, immature stages or adults, timing of oviposition, and adaptations for rapid growth and development (Wiggins et al. 1980). Duration acts as a template (Southwood 1977), selecting from among the potential species pool for taxa with life histories appropriate to particular disturbance regimes.

Because ponds do not last the same amount of time every year, the disturbance regime may also be characterized by the variability around a mean pond duration. This variability may affect wetlands in different ways from mean duration or hydroperiod. While the effects of variance around habitat duration have been explored by evolutionary biologists examining the selective pressures on life histories of temporary pond organisms (Wilbur and Collins 1973, Ritland and Jain 1984, Hairston et al. 1985, Newman 1989, Dodd 1993, Wissinger 1997), the community-level effects have not been examined. Understanding the role of natural variability will be important for understanding how wetland communities respond to anthropogenic alteration of their hydroperiod.

One component of hydroperiod variability is the occurrence of unusually severe drought. The magnitude of community change following drought may be related to the disturbance regime of the particular pond or wetland. One possibility might be that community change would be related to the disturbance frequency of a habitat. Rarely disturbed communities, with taxa not adapted to a disturbance, might show greater change following a disturbance than communities that were regularly disturbed. Alternatively, Denslow (1980) has suggested that the change in a community following a disturbance will depend on how closely the present disturbance regime corresponds with the historical average, not on a community's mean disturbance frequency or intensity. In other words, habitats with a more variable disturbance regime should be biologically more variable as well, and the magnitude of change should be more closely correlated with variance in the disturbance regime than the mean.

The relation between the disturbance regime and community change is likely mediated by the life history characteristics of resident taxa. In temporary pond habitats, taxa that leave resting stages in the pond basin may be more likely to be represented in the community following a disturbance. As pond duration increases, the proportion of taxa that leave desiccation-resistant resting stages within the pond basin should decrease (Wiggins et al 1980, Schneider and Frost 1996). Because short-duration ponds should be composed of a higher proportion of desiccation-resistant species, they might be expected to change less drastically following an unusually dry period.

In this chapter I present data on the invertebrate communities of seven snowmelt ponds in northern Wisconsin during two years: in 1985 before a regional drought and in 1989 following the drought. I evaluate the importance of life histories and biotic interactions in the response of the communities to hydroperiod and to variability in the hydroperiod.

STUDY SITE AND RESEARCH METHODS

The ponds discussed here all lie in a terminal moraine of the Wisconsin glaciation. There are many closed depressions within this region, but not all fill with water and support wetland communities. The ponds are located in a maple/birch-dominated hardwood forest. Under the Cowardin classification system they are considered palustrine, unconsolidated bottom wetlands. Their hydrological modifier varies with duration and ranges from temporarily flooded to intermittently exposed (Table 13.1). The area receives 76 cm of precipitation each year, including 127–152 cm of snow. Pond duration appeared to be chiefly dependent on the initial volume of snowmelt and less sensitive to spring and summer rains (personal observation). The previous drought as severe as the drought under study (defined as a winter with less than 95 mm of snow) was in 1968, although the winter of 1981 was also very dry. The mean annual temperature is less than 5°C (North Temperate Lakes, Long Term Ecological Research program, unpublished data).

Ponds were sampled in order to develop taxa lists for the entire season. The biota was sampled in 1985, before the drought, and in 1989, following the drought. Ponds were sampled weekly, from snowmelt in April through August, and approximately biweekly from September to ice-on. To determine the presence or absence of taxa, the ponds were sampled with a D-frame insect net (mesh size = 1 mm). Sweeps were taken through the water column, aquatic vegetation, and along the bottom of the ponds for approximately 45 minutes or until repeated sampling no longer yielded new taxa. Taxa were sorted in the field, and the samples were pooled. Organisms that could not be identified with certainty were preserved in 70 percent ethanol for later identification. Zooplankton, fairy shrimp (*Eubranchipus*), and floodwater mosquitoes (*Aedes*) were sampled quantitatively with a 3-l dip bucket. Typically, 24 l were collected and filtered through a 53-μm nylon mesh. Animals were preserved in sugar formalin for identification and enumeration. Population sizes of *Eubranchipus* and *Aedes* were calculated from density and pond volume. All taxa, with the exception of midges (Chironomidae), leeches, fingernail clams (Sphaeridae), and tadpoles (*Rana* and *Hyla*), were identified to genus. Cyclopoid copepods included both *Acanthocyclops vernalis* and *Diacycops navus* and were grouped. Data from intensive sampling in 1985 and 1989 were supplemented by observations of taxa in ponds in 1984 and 1986–1988, although systematic sampling was not performed.

I classified the life histories of temporary pond taxa into four groups on the basis of timing of dispersal, timing of oviposition, and the presence of desiccation-resistant resting stages, according to the scheme of Wiggins et al. (1980):

TABLE 13.1. Physical/Chemical Features of Temporary Ponds, Measured on Day of Maximum Depth, 1985

Pond	Cowardin Classification	Mean Duration (d)	CV of Duration (%)	Maximum Depth (m)	Volume (m³)	Area (m²)	Temperature (°C)	Conductivity (µS/cm)	Dissolved Oxygen (mg/L)	pH
1	P UB A	3.4	140	0.292	4	33	4.8	17	3.2[a]	5.2
2	P UB A	8.0	99	0.416	12	90	4.0	19	4.1[a]	5.1
3	P UB C	29.1	64	0.639	54	233	12.5	28	5.1	5.4
4	P UB C	67.0	50	0.563	41	157	12.5	37	3.6	5.6
5	P UB C	75.0	42	1.07	144	302	12.1	28	6.2	5.8
6	P UB F	190.2	77	1.65	474	598	12.0	29	4.1	5.9
7	P UB G	330.2	16	1.89	1675	1803	16.5	29	6.9	6.1

[a] Measured five days after maximum depth.

Group 1: overwintering residents that spend their entire life in the ponds, and leave resting stages resistant to desiccation in the pond basin;

Group 2: overwintering spring recruits that deposit eggs in the ponds before they dry in the spring;

Group 3: overwintering summer recruits that deposit eggs in the dried pond basin;

Group 4: nonwintering spring migrants that do not possess drought-resistant resting stages and must overwinter either on land or in permanent waters.

I also used data on the duration of the aquatic stage of the taxa compiled by Schneider and Frost (1996).

To compare community similarity, I used information measures of diversity (Pielou 1969). These metrics measure the increase in information created by combining the taxa lists of two communities. If these two communities have identical taxa composition, there is no increase in information and the metric is 0. It is thus a dissimilarity index, and is calculated by

$$J = \frac{2u \ln(2)}{2s \ln(2)}$$

where u is the number of taxa which are not held in common by the two communities and s is the total number of taxa in both communities. This metric ranges from 0 for complete similarity to 1 for complete dissimilarity.

I evaluated between-pond and between-year similarity among communities. To analyze the between-pond similarity in each of the two years of biological sampling, I calculated dissimilarity indexes of all pair-wise combinations of the taxa compositions of Ponds 1–7. I then used these 21 comparisons to create a tree diagram which clustered the ponds according to their dissimilarity using the single-linkage method. I used the algorithms of SYSTAT (Wilkinson 1989) to construct cluster tree diagrams, in which trees were uniquely ordered by the ecological distance between ponds, such that a pond on the edge of a cluster was placed adjacent to the pond outside the cluster to which it was most similar (Gruvaeus and Wainer 1972).

HYDROPERIOD AND INVERTEBRATES

Physical and Chemical Characteristics

The hydroperiod of the study ponds, defined as the mean number of days per year a pond is wet, ranges from 3 to 330 days (Table 13.1). Water chemistry changes rapidly following initial filling, and then stabilizes until drying. The pH of the ponds is initially moderated by reactions with the substrate in the ponds. As more material from the pond basin is dissolved, conductivity increases, along with pH. In ponds that had been dry in 1984 (Ponds 1–6), conductivity in 1985 increased by approximately 20 μS over the first two weeks while pH increased from values below 5.5 to approximately 6. Conductivity did not increase as the ponds dried out, indicating that loss of water is dominated by seepage into the soil and groundwater rather than by evaporation.

During 1987–1989 a drought affected the upper Midwest, with impacts on many wetland habitats throughout the region (Bataille and Baldassarre 1994, Winter and Rosenberry 1995). During this period pond duration was markedly shorter than in 1984–1986, and many ponds did not even form in some years (Schneider 1997). Short-duration ponds had greater year-to-year variability in their duration than did long-duration ponds (Table 13.2; $r^2 = 0.754$, 6 d.f., $p = 0.011$). However, Pond 6 was more variable than expected, and identified as an outlier (Cook's D statistic = 0.839 (Rawlings 1988)).

Hydroperiod and Invertebrate Community Structure

In 1985 the ponds supported a diversity of primarily invertebrate taxa (Fig. 13.1). At the generic level the most diverse group of taxa in the ponds was beetles, followed by dragonflies, true bugs, flies and rotifers. As pond duration increased, taxa number also increased, ranging from 4 in short-duration ponds to 65 in long-duration ponds. Taxa were not added randomly to ponds; rather, they exhibited a continuity in their distribution. Once a pond lasted long enough to support a particular taxon, that taxon tended to occur in all longer-duration ponds as well (Schneider and Frost 1996).

Cluster analysis of community dissimilarity showed that in each year pond duration was important in structuring temporary pond communities along a

TABLE 13.2. Community Dissimilarity between 1985 and 1989 for All Taxa and the Fraction of the Community Composed of Taxa with (Groups 1–3) or without (Group 4) Desiccation-Resistant Life History Stages[a]

Pond	All Taxa	Dissimilarity Groups 1–3	Group 4
2	0.60	0.43	1.00
3	0.68	0.53	0.83
4	0.31	0.10	0.60
5	0.36	0.29	0.47
6	0.47	0.36	0.56
7	0.22	0.22	0.22

[a]Paired t-test (t = 3.312, 5 d.f., p = 0.0212) shows that dissimilarity was higher for Group 4 taxa. Pond 1 is not included because it was dry in 1989.

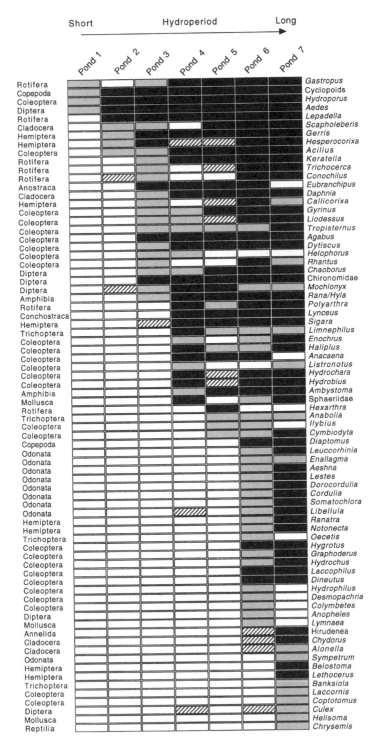

Fig. 13.1. Distribution of taxa across the habitat duration gradient in 1985, before the drought, and 1989, after the drought. Taxa are arranged in order of the shortest-duration pond in which they appeared in 1985. Solid bars represent presence in both 1985 and

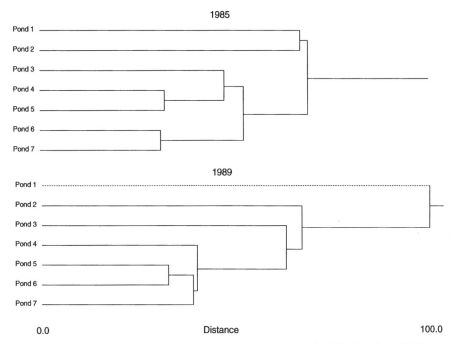

Fig. 13.2. Cluster analysis of pond dissimilarity in 1985 and 1989. Pond 1, which was dry in 1989, shows complete dissimilarity and is indicated with a dashed line. Ponds are uniquely ordered such that ponds on the edge of a cluster are placed adjacent to the pond they most resemble (Gruvaeus and Wainer 1972).

gradient of habitat duration (Fig. 13.2). In both 1985 and 1989 the ponds were ordered exactly according to duration, indicating that ponds were most similar to those ponds adjacent to them along the gradient of habitat duration. Three clusters of ponds were suggested in 1985. Ponds 6 and 7 were most similar to each other and formed a cluster of long-duration ponds. Ponds 4 and 5 formed a cluster of intermediate-duration ponds, and Ponds 1 and 2 a cluster of short-duration ponds. Pond 3 clustered with ponds 4 and 5 as an intermediate-duration pond.

In 1989 the ponds were still ordered exactly by duration, but the clusters were less clearly defined. The most notable difference was that Pond 6 clustered with a shorter-duration pond following the drought. Pond 6, a long-

Fig. 13.1. (Continued) 1989, hatched bars in only 1989, and stippled bars in only 1985. Open bars denote absence in both years. (Data from D. W. Schneider and T. M. Frost, Habitat duration and community structure, Journal of the North American Benthological Society 15:64–86. © 1996. Used by permission of the Journal of the North American Benthological Society; and Schneider [1997].)

duration pond in 1985, clustered with Pond 5, an intermediate-duration pond. This cluster was joined by Ponds 7 and 4. This cluster was most similar to Pond 3 and then to Pond 2.

Following the drought the number of taxa in each pond was substantially lower (Fig. 13.1). The drought affected the taxa in two ways. Some taxa were greatly reduced in all ponds studied. The snails *Lymnaea* and *Helisoma,* for instance, were common in ponds 6 and 7, respectively, prior to the drought, but were not observed in these ponds in 1988 (personal observation) or 1989. Similarly, caddisflies, particularly *Limnephilus*, were a common component of ponds 4–7 in 1984 (personal observation), 1985, and 1986 (personal observation). In 1989, however, I collected only one individual, from Pond 4.

In contrast, other, previously rare taxa became common in certain ponds following the drought. Fairy shrimp were a rare component of Pond 6, never appearing in quantitative samples from that pond. However, they were abundant in ponds of shorter duration. In 1988 (personal observation) and 1989 Pond 6, which had an unusually short duration in 1987, had large populations of *Eubranchipus*. In 1989 *Eubranchipus* reached a total population of >56,000 individuals in Pond 6, comparable to populations in short-duration ponds prior to the drought. *Aedes* mosquitoes showed a similar response. They were never sampled quantitatively from Pond 6 in 1985. However, in 1989 they had an estimated population size of >49,000.

Short-duration ponds were the most variable, both in their hydroperiod and their communities, while the longest-duration pond showed the least variability in these characteristics. But community change was not uniformly related to duration. Pond 6 had unusually high hydrological variability, and its community changed to a greater degree than would be expected based on its mean duration. Overall, the most hydrologically variable ponds changed the most biologically. The similarity of the taxa composition of ponds before and after the drought was significantly related to the coefficient of variation in pond duration (Table 13.2; $r^2 = 0.662$, $p = 0.049$), but not to the log of the mean duration ($r^2 = 0.551$, $p = 0.091$). Including both the coefficient of variation and mean duration in a multiple regression does not significantly increase the predictive power of the coefficient of variation alone, explaining only 70.4 percent of the variance in the similarity of taxa composition.

Hydroperiod and Life History Characteristics

The continuity of taxa distribution across the habitat duration gradient suggests that duration acts as a template or filter for taxa with appropriate life history characteristics. Data on the life histories of the temporary pond taxa support this. There was a significant correlation between the duration at which a taxon first occurred in the ponds and the length of the aquatic phase of its life cycle (Schneider and Frost 1996). In addition, the proportion of taxa with adaptations to drying (diapause, desiccation-resistant eggs, etc.) was higher in shorter-duration ponds (Schneider and Frost 1996).

The change in communities following the drought was also related to the taxa's life history characteristics. Taxa tended to disappear from ponds at the short-duration edge of their range following the drought (Fig. 13.1). Of 49 taxa that occurred in both 1985 and 1989 but whose distributions differed between years, 31 (63 percent) were lost only from the short-duration edge of their range. However, with the exception of Pond 6, the mean duration of the aquatic phase of the life cycle of taxa before and after the drought did not change. The taxa in Pond 6 following the drought had a significantly shorter aquatic phase than the taxa present before the drought. In 1985 the mean duration of the aquatic phase of all of the taxa in Pond 6 was 147.6 days; in 1989 it was 26 days (t test, log transformed data, t = 2.074, 43 d.f, p = 0.041). This decline in the mean length of the life cycle mostly represents the loss of long-lived predators such as dragonfly nymphs. The proportion of species with a desiccation-resistant stage was significantly correlated in 1989 with pond duration (Spearman rank correlation, Z = –2.074, p = 0.038). All taxa in Pond 2 following the drought possessed desiccation-resistant stages. As expected, taxa with desiccation-resistant stages tended to be less affected by the drought. The fraction of the community composed of desiccation-resistant taxa (Groups 1–3) was more similar between years than the fraction composed of taxa without adaptations for surviving desiccation (Group 4) (Paired t-test, t = 3.312, 5 d.f., p = 0.021) (Table 13.2).

Hydroperiod and Biotic Interactions

As pond duration increased, interactions among the taxa—competition and predation—became more important in determining the abundance of the taxa.

Competition. Competition between *Daphnia* and rotifers depended on habitat duration. When *Daphnia* were at high density, they were able to competitively exclude rotifers. *Daphnia* abundance peaked in intermediate duration ponds. In contrast, rotifer diversity and abundance were at their lowest in these ponds. This competitive exclusion was likely the result of interference competition between *Daphnia* and rotifers (Schneider and Frost 1996). As *Daphnia* feed, they entrain rotifers in their feeding currents and mechanically damage them (Gilbert 1988, Schneider 1990, MacIsaac 1991). The only abundant rotifers in intermediate duration ponds were *Hexarthra* and *Polyarthra,* taxa that have powerful swimming appendages and are capable of avoiding interference by *Daphnia.* Other potential competitive interactions that have been demonstrated in other systems include inter- and intraspecific competition between species of tadpoles (Wilbur 1987) and between mosquito larvae (Pritchard and Scholefield 1983), but I have not experimentally verified these interactions in the northern Wisconsin ponds.

Predation. Rates of predation (the percentage of a pond cleared of a prey taxa per day by the entire predator community), measured by summing the clearance rates of individual predator taxa multiplied by their abundance, increased with increasing pond duration. This increase was the result of increases in three components of predation: individual predation rates, predator diversity, and predator abundance. Some individual taxa had higher predation rates in long-duration ponds than in short-duration ponds (Schneider and Frost 1996). But most important in terms of the overall predation rate was the increase in overall diversity of predators as pond duration increased. The diversity increased both taxonomically and in terms of feeding guilds. In short duration ponds the abundant predators were predacious diving beetles and dipteran larvae such as *Mochlonyx*. As pond duration increased, the diversity of predacious diving beetles increased, and they were joined by predators such as backswimmers (*Notonecta*), giant water bugs (Belostomatidae), phantom midges (Chaoboridae), dragonfly larvae, and salamander larvae (*Ambystoma*). In 1985 the number of predator taxa increased from 1 in the shortest-duration pond to 34 in the longest-duration pond (Schneider 1997). Over the same change in duration, total taxa number increased from 4 to 65, so predators made up a greater portion of all taxa in the longest-duration ponds. Abundance also increased, from 3.3 m^{-3} in Pond 3 to 76.7 m^{-3} in Pond 7.

While predation rates increased with increasing duration, only in long-duration ponds were predation rates high enough to control abundances of prey populations. In long-duration ponds the predator community could clear 16–31 percent of the pond of clam shrimp (*Lynceus*) per day, 7–12% of *Daphnia,* 7–11 percent of mosquitoes, and 24–28 percent of fairy shrimp. These values were all ≤1 percent in the short-duration ponds.

Preference of predators changed across the habitat-duration gradient. Predators in long-duration ponds preferred prey that typically inhabited shorter duration ponds. When fairy shrimp and mosquitoes were available in the prey pool, they were the preferred prey of back-swimmers, dragonfly nymphs, and predacious diving beetles. When these prey were not available, these predators preferred *Daphnia* and clam shrimp (Conchostraca). The decline in *Daphnia* in long-duration ponds is due to this switch in preference under different prey environments. In the absence of mosquitoes and fairy shrimp in the long-duration ponds, *Daphnia* is the preferred prey.

Following the drought many predators were lost from the ponds, leading to a change in foodweb structure. The number of links in the foodwebs of the ponds was reduced, as were the links per species, suggesting a decline in the importance of predation (Schneider 1997).

The Interaction of Hydroperiod Variability, Life Histories, and Predation

The complex linkages among hydroperiod, hydroperiod variability, life history adaptations, and biotic interactions is illustrated by the changing populations of fairy shrimp in Pond 6, one of the most hydrologically variable ponds. *Eubranchipus* showed enormous changes in its density between years in Pond

6, increasing from near 0 to 53,000 individuals between 1988 and 1989. Because of this rapid population response, it is unlikely that *Eubranchipus* colonized Pond 6 following the drought. Rather, it was probably present as eggs in the pond sediment.

The variance in duration in Pond 6 appears to have created a spatiotemporal refuge from predation for *Eubranchipus*. *Eubranchipus* is very susceptible to predation and is preyed upon preferentially by predators from long-duration ponds (Schneider and Frost 1996). These predators represent a significant source of mortality to the fairy shrimp, with community predation rates on the order of twice the maximal rate of population increase (Schneider and Frost 1996). Fairy shrimp possess eggs which can withstand drying and can remain viable for over 10 years (Donald 1983). In fact, Eubranchipus eggs require the increased oxygen levels associated with dried sediments to break diapause and hatch (Hartland-Rowe 1972). These conditions are met both in ponds which dry out entirely every year, like Ponds 1–5, and in ponds which do not dry every year but whose area fluctuates, exposing sediments, like Ponds 6 and 7. *Eubranchipus* was common in the shorter-duration ponds, yet typically rare in the long-duration ponds, although the habitat requirements for egg hatching were present. During wet years, predators were common in the long-duration ponds (Schneider and Frost 1996), and over repeated cycles of change in pond area without drying, predators would be able to remove any fairy shrimp that hatched from eggs in the exposed sediments. Because of predation, no fairy shrimp would reproduce and there would be no new source of eggs to the sediments on the periphery of the pond. Over these repeated cycles, then, the only eggs which would remain unhatched would be in the center of the pond which had not dried. In unusually dry years, when the entire pond dries, the eggs in the center of the pond would be exposed to the conditions necessary for hatching. Because long-lived predators such as backswimmers and salamander and dragonfly larvae would have been killed when the pond dried, Eubranchipus would hatch into a fundamentally different community and would be able to reproduce, leaving another generation of eggs. Variation around mean pond duration thus creates a spatiotemporal refuge from predation for this species.

ECOLOGICAL IMPLICATIONS

Pond duration structures temporary pond communities in a hierarchical fashion. The cluster analysis of between-pond similarity suggested that while the exact taxa composition of pond communities might vary between years, at any one time duration imposed an ordering on pond communities. Ponds of similar mean duration were more similar to each other than to ponds of different mean duration. The way in which duration structures pond communities is illustrated in Fig. 13.3. Duration acts as a template (Southwood 1977) for taxa with appropriate life histories (Fig. 13.3*a*). Taxa that can complete their

a) Short-duration pond

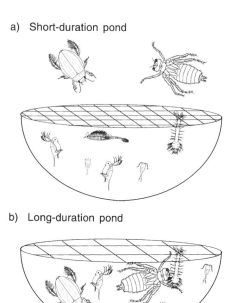

b) Long-duration pond

Fig. 13.3. A model of community organization for temporary pond invertebrates. The grid over the pond symbolizes pond duration, which acts like a filter or template for community organization. Taxa with life histories compatible with pond duration are capable of passing through the filter and inhabiting the pond. (*a*) Short-duration pond: Only species with rapid growth and development or desiccation-resistant resting stages are capable of inhabiting the pond. Taxa include mosquitoes, copepods, fairy shrimp, *Daphnia,* and rotifers. Predators such as predaceous diving beetles and dragonfly larvae have life cycles too long to complete development and are precluded from the pond. As a result, predation is a minor process in determining community structure. High populations of *Daphnia* may reduce the diversity of rotifers. (*b*) Long-duration pond: The pond lasts long enough to permit colonization by large predators. Predation on susceptible taxa such as mosquitoes and fairy shrimp severely reduces population size or precludes them from ponds. Predation on water fleas reduces their populations, in turn allowing rotifers to increase in diversity and abundance. (Invertebrate drawings modified from R. W. Pennak, Fresh-water Invertebrates of the United States [2nd ed.], John Wiley & Sons. © 1978. Reprinted by permission of John Wiley & Sons, Inc.)

aquatic phase in a pond of a particular duration will be present in that pond and all ponds of longer duration as well. Thus, taxa, at least at the generic level, are added in a cumulative manner as pond duration increases. However, as pond duration increases, and with it species diversity, the ecological guilds of the component taxa may also change as well (Fig. 13.3*b*).

As predators become more diverse and abundant, predation exerts a more profound influence on the community. Predation rates are higher and can exert control on prey communities in longer-duration ponds. Wellborn at al. (1996)

suggest that a critical transition occurs in ponds along a habitat permanence gradient. They suggest that once ponds become permanent, invertebrate predation imposes a different structure on communities. Analysis of predation along the habitat duration gradient in temporary ponds in Wisconsin suggests aspects of both continuous gradients and critical transitions in the importance of predation. As ponds last longer, there is a continuous increase in the diversity and abundance of invertebrate and vertebrate (but not fish) predators, and a consequent increase in community predation rates. However, only in the longest-duration ponds (those that stay wet for more than one year, at least on occasion) does predation affect the distribution and abundance of the prey taxa. This transition occurs because many of the particularly effective predators, such as back-swimmers and dragonfly nymphs, must overwinter in permanent waters. There are keystone effects of predation on the invertebrate community as well. As *Daphnia* are controlled by predation in long-duration ponds, their effect on rotifer species diversity and abundance is removed (Fig. 13.3*b*). With this suppression lifted in long-duration ponds, rotifers reach high diversity and abundance.

Finally, superimposed on the effect of mean habitat duration on community structure is the effect of variability in habitat duration. Year-to-year changes in taxa composition appear related to hydroperiod variability. Many taxa were lost from the ponds following the drought, while others increased in population. Because of the effect of variability on the abundance of predators, these taxa, such as mosquitoes and fairy shrimp, appear capable of surviving in ponds that have high variability in duration. Communities in ponds that have a highly variable hydroperiod have characteristics of more temporary waters following a dry period and of more permanent waters following wet years. Variability may be important for the protection of other wetland species as well. Donald (1983) observed similar patterns of large year-to-year variation in the abundance and species composition of Anostraca in a temporary pool in Alberta, and related it to variation in annual weather and pond volume. Kirkman and Sharitz (1994) showed that rare plant species were also dependent on periodic drought and disturbance for regeneration. Disturbance removed competitively dominant plants and allowed establishment from the seed bank of rare species.

Short-duration ponds were in general hydrologically more variable than long-duration ponds. However, there was also a site-specific component to variation as well. Winter and Rosenberry (1995) noted local variation in hydroperiod in a series of prairie potholes. The interaction of a wetland and the water table along an elevational gradient did not correspond precisely to the wetland's vertical position; rather, local variation in geologic materials played an important role in hydrology of the wetland. Regardless of its source, hydrological variability appears to be important in determining the patterns of community change following a drought.

Overall, the most hydrologically variable ponds changed the most biologically. Jeffries (1994) also showed that following a drought ponds had a fauna more typical of temporary habitats and shorter-duration ponds had a greater turnover of species than permanent or semipermanent ponds. In contrast, Barratsegretain and Amoros (1996) found that plant communities in more frequently disturbed (flooded) floodplain channels were more similar between years than less frequently flooded sites. However, neither study evaluated hydrologic variability at a single site among years. This study suggests that turnover may be more closely related to hydrologic variability than mean hydroperiod.

This study was based on community composition data from only two years of systematic sampling, making it difficult to attribute changes in community composition following the drought to the effects of hydroperiod variability, and not background year-to-year variation, perhaps in colonization or extinction. However, the response of some taxa seems clearly related to the drought. Systematic sampling in two years, combined with observations of the more conspicuous taxa, suggested that taxa such as snails, caddisflies, and dragonflies, which were consistently present in ponds prior to the drought, were reduced. Conversely, taxa consistently rare in some ponds, such as mosquitoes and fairy shrimp, became abundant. Long-term data on community composition and its year-to-year variability would help place the changes that occurred in these wetlands in context.

As predicted, the pattern of change in the taxon composition of the ponds appeared related to life history characteristics of the taxa. In particular, characteristics associated with surviving the dry phase of the pond appeared to be more important than characteristics associated with absolute pond duration. The overall similarity of communities before and after the drought was not related to a change in the mean length of the aquatic phase of taxa in the ponds, even though pond duration had varied considerably over the period. The length of the mean life cycle changed in only one pond, Pond 6.

Ponds that had a higher proportion of species with adaptations for surviving the dry phase of a pond were predicted to show a smaller change in their species composition following the drought. These ponds tended to be the shorter-duration ponds because they supported a higher proportion of Group 1–3 taxa (Schneider and Frost 1996). Indeed, the relation between duration and the proportion of taxa with desiccation-resistant stages was strengthened following the drought, and as predicted, the composition of the fauna with desiccation-resistant resting stages did change less than the fauna with no such adaptations (Table 13.2). However, in short-duration ponds the taxa without desiccation-resistant stages changed greatly after the drought (Table 13.2), and as a result short-duration ponds were more dissimilar between years than long-duration ponds. When the entire community was considered, similarity between years was most correlated with hydroperiod variability, not mean habitat duration.

It is perhaps not surprising that the changes in species composition in a community should be more related to characteristics of changes in the environment than to any static measure. Forests subject to frequent fires are examples of communities which have incorporated disturbance into their normal functioning (Allen and Starr 1982). Many of the plant species are fire-adapted, dependent on periodic fires for germination (Heinselman 1981). Rather, change in the habitat duration is experienced by these communities as environmental change. Changes in the historic frequency of burning created by forest management may be causing changes in species diversity of these forest communities (Kilgore and Taylor 1979, Denslow 1985). Likewise, in temporary pond communities change in community composition appears to be related to variation around the disturbance regime because the community has incorporated the disturbance of drying in the life histories of the taxa. Drought (Jeffries 1994) and unusually wet periods (Neckles et al. 1990) both lead to changes in wetland communities that are adapted to a particular mean hydroperiod. Thus, the proper template variable at which to look for the causes of change in communities subject to disturbance may be variability in disturbance characteristics, rather than the mean.

Different habitats, however, may have different site-specific processes that govern variability across a disturbance frequency gradient. Kratz et al. (1991) have shown that along a landscape gradient such as elevation, the variance in components of wetland and aquatic ecosystems are related to patterns of hydrologic variance along the landscape gradient. The particular relation between the landscape gradient and environmental variance is site- and parameter-specific. Locations high in the watershed of an estuarine wetland site had high variance in parameters dominated by freshwater inflow, but low variance in parameters controlled by tidal and oceanic processes.

IMPLICATIONS FOR MANAGEMENT

The disturbance regime of temporary ponds is responsible for the maintenance of diversity in these systems. Thus, in order to protect the diversity, one must maintain the disturbance regime or hydroperiod. Further, it is important to maintain all aspects of the disturbance regime—both mean disturbance frequency and the variability in disturbance frequency.

Although this study was based on only two years of data on community composition, it suggests that an analysis of the year-to-year variability in the physical environment can yield insight into the processes which structure wetlands and provide guidance for their management. Ecologists have recognized the conservation value of short hydroperiods and that temporary ponds support many rare taxa not found in more permanent habitats (Collinson et al. 1995, Kirkman and Sharitz 1994). However, in order to maintain the diversity of wetland communities, it may be important to maintain variance in disturbance frequencies as well (DeHorter and Tamisier 1996). Management of wetlands with an annual drawdown and flood, for instance, but no year-to-year variation in the timing and duration of the drawdown, may diminish the diversity of the habitat. Environmental variance itself, as distinct from the mean condition, is an important characteristic of a system (Kratz et al. 1987) and is in need of greater research.

LITERATURE CITED

Allen, T. F. H., and T. B. Starr. 1982. Hierarchy: Perspectives for ecological complexity. University of Chicago, Chicago, IL.

Barratsegretain, M. H., and C. Amoros. 1996. Recovery of riverine vegetation after experimental disturbance: A field test of the patch dynamics concept. Hydrobiologia 321:53–68.

Bataille, K. J., and G. A. Baldassarre. 1993. Distribution and abundance of aquatic macroinvertebrates following drought in three prairie pothole wetlands. Wetlands 13:260–269.

Collinson, N. H., J. Biggs, A. Corfield, M. J. Hodson, D. Walker, M. Whitfield, and P. J. Williams. 1995. Temporary and permanent ponds: An assessment of the effects of drying out on the conservation value of aquatic macroinvertebrate communities. Biological Conservation 74:125–133.

Dehorter, O., and A. Tamisier. 1996. Wetland habitat characteristics for waterfowl wintering in Camargue, southern France: Implications for conservation. Revue d'ecologie: la terre et la vie. 51:161–172.

Denslow, J. S. 1980. Patterns of plant species diversity during succession under different disturbance regimes. Oecologia 46:18–21.

———. 1985. Disturbance mediated coexistence of species. Pages 307–323 *in* S. T. A. Pickett and P. S. White (eds.), The Ecology of Natural Disturbance and Patch Dynamics. Academic Press, New York.

Dodd, C. K., Jr. 1993. Cost of living in an unpredictable environment: The ecology of striped newts *Notophthalmus perstriatus* during a prolonged drought. Copeia 1993:605–614.

Donald, D. B. 1983. Erratic occurrence of anostracans in a temporary pond: Colonization and extinction or adaptation to variation in annual weather. Canadian Journal of Zoology 61:1492–1498.

Gilbert, J. J. 1988. Susceptibilities of ten rotifer species to interference from *Daphnia pulex*. Ecology 69:1826–1838.

Gruvaeus, G., and H. Wainer. 1972. Two additions to hierarchical cluster analysis. British Journal of Mathematical and Statistical Psychology 25:200–206.

Hairston, N. G., Jr., E. J. Olds, and W. R. Munns, Jr. 1985. Bet-hedging and environmentally cued diapause strategies of diaptomid copepods. Internationales Vereinigung für theoretische und ongewandte Limnologie, Verhandlungen 22:3170–3177.

Hartland-Rowe, F. 1972. The limnology of temporary waters and the ecology of Euphyllopoda. Pages 15–31 *in* R. B. Clark and R. J. Wooton (eds.), Essays in Hydrobiology. University of Exeter, Exeter, UK.

Heinselman, M. L. 1981. Fire and succession in the conifer forests of northern North America. Pages 374–405 *in* D. C. West, H. H. Shugart, and D. B. Botkin (eds.), Forest Succession: Concepts and Application. Springer-Verlag, New York.

Jeffries, M. 1994. Invertebrate communities and turnover in wetland ponds affected by drought. Freshwater Biology 32:603–612.

Kilgore, B. M., and D. Taylor. 1979. Fire history of a sequoia-mixed conifer forest. Ecology 60:129–142.

Kirkman, L. K., and R. R. Sharitz. 1994. Vegetation disturbance and maintenance of diversity in intermittently flooded Carolina bays in South Carolina. Ecological Applications 4:177–188.

Kratz, T. K., B. J. Benson, E. Blood, G. L. Cunningham, and R. A. Dahlgren. 1991. The influence of landscape position on temporal variability in four North American ecosystems. American Naturalist 138:355–378.

Kratz, T. K., T. M. Frost, and J. J. Magnuson. 1987. Inferences from spatial and temporal variability in ecosystems: Long-term zooplankton data from lakes. American Naturalist 129:830–846.

Lubchenco, J. 1986. Relative importance of competition and predation: Early colonization by seaweeds in New England. Pages 537–555 *in* J. Diamond and T. J. Case (eds.), Community Ecology. Harper & Row, New York.

MacIsaac, H. J., and J. J. Gilbert. 1991. Discrimination between exploitative and interference competition between cladocera and *Keratella cochlearis*. Ecology 72: 924–937.

Menge, B. A., J. Lubchenco, S. D. Gaines, and L. R. Ashkenas. 1986. A test of the Menge-Sutherland model of community organization in a tropical rocky intertidal food web. Oecologia 71:75–89.

Mitsch, W. J., and J. G. Gosselink. 1993. Wetlands (2nd ed.). Van Nostrand Reinhold, New York.

Neckles, H. A., H. R. Murkin, and J. A. Cooper. 1990. Influences of seasonal flooding on macroinvertebrate abundance in wetland habitats. Freshwater Biology 23:311–322.

Newman, R. A. 1989. Developmental plasticity of *Scaphiopus couchii* tadpoles in an unpredictable environment. Ecology 70:1775–1787.

Pennak, R. W. 1978. Fresh-water Invertebrates of the United States (2nd ed.). John Wiley & Sons, New York.

Pielou, E. C. 1969. An Introduction to Mathematical Ecology. Wiley-Interscience, New York.

Pritchard, G., and P. J. Scholefield. 1983. Survival of *Aedes* larvae in constant area ponds in southern Alberta (Diptera: Culicidae). Canadian Entomologist 115:183–188.

Rawlings, J. O. 1988. Applied Regression Analysis: A Research Tool. Wadsworth & Brooks, Pacific Grove, CA.

Ritland, K., and S. Jain. 1984. The comparative life histories of two annual *Limnanthes* species in a temporally variable environment. American Naturalist 124:656–679.

Schneider, D. W. 1990. Direct assessment of the independent effects of exploitative and interference competition between *Daphnia* and rotifers. Limnology and Oceanography 35:916–922.

———. 1997. Predation and food web structure along a habitat duration gradient. Oecologia 110:567–575.

Schneider, D. W., and T. M. Frost. 1996. Habitat duration and community structure in temporary ponds. Journal of the North American Benthological Society 15:64–86.

Smith, D. C. 1983. Factors controlling tadpole populations of the chorus frog (*Pseudacris triseriata*) on Isle Royale, Michigan. Ecology 64:501–510.

Smith, L. M., and J. A. Kadlec. 1985. The effects of disturbance on marsh seed banks. Canadian Journal of Botany 63:2133–2137.

Southwood, T. R. E. 1977. Habitat: The templet for ecological strategies? Journal of Animal Ecology 46:337–365.

van der Valk, A. G. 1981. Succession in wetlands: A Gleasonian approach. Ecology 62:688–696.

Wellborn, G. A., D. K. Skelly, and E. E. Werner. 1996. Mechanisms creating community structure across a freshwater habitat gradient. Annual Review of Ecology and Systematics 27:337–363.

White, P. S., and S. T. A. Pickett. 1985. Natural disturbance and patch dynamics: An introduction. Pages 3–13 *in* S. T. A. Pickett and P. S. White (eds.), The Ecology of Natural Disturbance and Patch Dynamics. Academic Press, New York.

Wiggins, G. B., R. J. Mackay, and I. M. Smith. 1980. Evolutionary and ecological strategies of animals in annual temporary pools. Archiv für Hydrobiologie, Supplement 58:97–206.

Wilbur, H. M. 1987. Regulation of structure in complex systems: experimental temporary pond communities. Ecology 68:1437–1452.

Wilbur, H. M., and J. P. Collins. 1973. Ecological aspects of amphibian metamorphosis. Science 182:1305–1314.

Wilkinson, L. 1989. SYSTAT: The system for statistics. SYSTAT, Evanston, IL. USA.

Winter, T. C., and D. O. Rosenberry. 1995. The interaction of ground water with prairie pothole wetlands in the Cottonwood Lake area, east-central North Dakota, 1979–1990. Wetlands 15:193–211.

Wissinger, S. A. 1997. Cyclic colonization in predictably ephemeral habitats: A template for biological control in annual crop systems. Biological Control 10:4–15.

14 Autumnal Woodland Pools of Western New York

Temporary Habitats That Support Permanent Water Invertebrates

DAROLD P. BATZER and KATHERINE A. SION

*S*pecialized adaptation to survive or avoid drought is generally considered a requirement for invertebrate success in temporary wetlands. However, in some New York autumnal pools we collected numerous invertebrates that lacked specific adaptations for life in temporary water. Unexpectedly, amphipods and isopods that typically reside in large permanent water bodies such as rivers and lakes occurred in large numbers in autumnal pools that dried for several months. These habitats were surrounded by old-growth beech-maple forest. The intense shading and abundant leaf litter prevented these pools from drying excessively in summer, and probably permitted invertebrates to survive that otherwise would perish from desiccation. In forested areas outside of the old-growth stands, autumnal pools supported fewer amphipods and isopods. This study indicates that (1) invertebrates not specifically adapted for life in ephemeral waters can still succeed in some temporary water wetland settings; (2) surrounding forest condition can affect invertebrate ecology in temporary pool habitats; and (3) anthropogenically induced changes in forest condition may have important consequences for associated wetlands.

AUTUMNAL POOL HABITATS

Wetland ecologists use the terms *temporary* and *vernal* somewhat interchangeably when describing small wetlands that periodically dry. Although

Invertebrates in Freshwater Wetlands of North America: Ecology and Management, Edited by Darold P. Batzer, Russell B. Rader, and Scott A. Wissinger
ISBN 0-471-29258-3 © 1999 John Wiley & Sons, Inc.

vernal pools are widespread and well studied (see Wiggins et al. 1980 and chapters by Schneider, Higgins and Merritt, and Wissinger et al. in this volume), they make up only a subset of all temporary pool habitats. Many of the temporary wetlands in western New York are correctly classified as autumnal rather than vernal pools. Whereas vernal pools first fill from snowmelt and/or seasonal rains in spring (hence the name), autumnal pools partially fill from autumn rains and remain flooded over the winter (Fig. 14.1, Wiggins et al. 1980). Spring snowmelt and rain further fill autumnal pools and allow them to retain water into the summer. When water losses from summer evaporation and groundwater recharge exceed inputs from rain, water levels recede and autumnal pools typically will dry completely for much of the summer and early autumn. In New York's late autumn the onset of heavier rains and cooler temperatures begins the flooding cycle anew. Thus, the hydrological cycles in autumnal and vernal pools differ significantly (Fig. 14.1), and this difference undoubtedly affects the ecologies of resident invertebrates (see Wiggins et al. 1980).

We investigated invertebrates in a set of autumnal pools located in Dr. Victor Reinstein Woods Nature Preserve in suburban Buffalo, New York (managed by the New York Department of Environmental Conservation). This

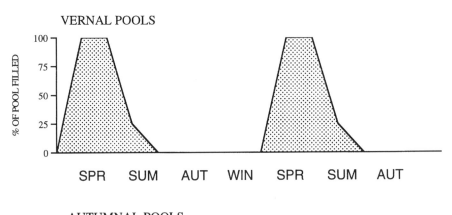

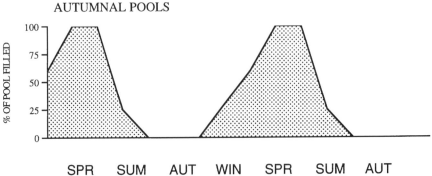

Fig. 14.1. Schematic of the hydroperiods in autumnal verses vernal pools.

preserve was established to protect remnant stands of old-growth beech-maple forest (Kershner 1993), and many autumnal pools exist within portions of this old-growth forest. We selected the four largest pools for intensive study. In terms of hydrology and morphology, each of the four habitats was unique. The smallest was only 25 m² when full and had a maximum depth of 20 cm. The next largest pool was 40 m² and 50 cm deep when full. A third pool was 60 m² with a 30-cm maximum depth. None of these three sites had any obvious connection to permanent water sites in the area. The largest of the four pools, Hemlock Swale, covered approximately 500 m² when fully flooded. Most of this site was fairly shallow (<30 cm) except for a small deep-water section (80 cm deep) that existed along one edge (this deeper hole possibly resulted from a tree uprooting, or might have been anthropogenically created). Unlike the other three habitats, Hemlock Swale had a small seepage outlet that emptied into a small permanent water pond and from there into a nearby permanent marsh. The deciduous trees surrounding each pool contributed an abundance of leaf litter into each basin, which was probably the major source of plant energy for these habitats. The old-growth beech-maple canopies that shaded the pools were probably responsible for the virtual absence of herbaceous plant cover or algal growth in the pools.

In 1995 all pools had dried by early June, except for the small deeper section of Hemlock Swale, which did not dry completely until mid-August. Monitoring throughout the summer indicated that two of the habitats partially refilled for a short period after a mid-June rain, and one of those pools also held water briefly after an early August rain. A particularly heavy rain in early October partially refilled one site and filled the residual pool in Hemlock Swale. These two habitats were filled further by subsequent autumn rains, which also partially refilled the remaining two sites. All four pools continued to fill into the winter months.

Thus, the duration of the dry phase varied greatly among sites. Most of Hemlock Swale was dry from early June through early November, except for the small deeper portion, which was dry only from mid-August through early October. The overall wettest site was completely dry only from mid-June through early August and again from mid-August through early October. A third site was dry from mid-June through early November. The driest habitat was free of surface water from early June until mid-November. During September, when pools were at their driest, digging indicated that free water was not present in the upper 50 cm of the underlying pool substrata. However, even then the leaf litter and surface mud of all four sites remained visibly damp.

INVERTEBRATE COMMUNITIES

We sampled aquatic invertebrate communities in each of the four sites using a D-frame net (1-mm mesh). At representative transects in each site, 1-m horizontal sweeps along the bottom substrate and through the water column and suspended leaf litter were collected at three randomly selected locations. Sampling was conducted on April 15 and May 17, 1995, when all pools held water. In the summer period qualitative samples were collected in areas that had previously dried but temporarily held water after rains. In autumn, after sites had refilled, we again collected sweep net samples in October in the one site that held water and in November when all four sites had at least partially refilled. These samples consisted of two representative 0.5-m sweeps per site; we reduced our sample volume to prevent excessive harvest of invertebrates in these small, partially filled habitats.

Although we collected numerous macroinvertebrate species (40+ taxa) in the four woodland pools, only five species were widespread and abundant (i.e., occurred in at least three of the four sites and comprised >3 percent of the total numbers of invertebrates collected) (Table 14.1). *Crangonyx pseudogracilis* amphipods (Amphipoda: Crangonyctidae), *Pisidium casertanum* fingernail clams (Bivalvia: Pisidiidae), and *Aedes stimulans* mosquito larvae (Diptera: Culicidae) occurred in every site. *Sphaerium occidentale* fingernail clams and *Caecidotea (= Asellus) racovitzai* isopods (Isopoda: Asellidae) were collected in three sites each. The remainder of the invertebrate fauna in the pools consisted primarily of assorted species of chironomid midge larvae, oligochaete worms, snails, microcrustaceans, and water beetles.

The invertebrate communities in our study habitats were in many ways atypical compared to those in temporary wetlands described in overviews by

TABLE 14.1. Invertebrate Numbers per 1-m Sweep Sample from Four Autumnal Pools in Reinstein Woods Preserve, Cheektowaga, New York[a]

Invertebrate Taxa	April 15 Mean (1 SE)	May 17 Mean (1 SE)	October 21 or November 22 Mean (1 SE)
Mollusks			
Bivalvia—Pisidiidae			
Sphaerium occidentale	10 (9)	32 (18)	16 (11)
Pisidium casertanum	15 (8)	25 (20)	4 (2)
Crustaceans			
Amphipoda—Crangonyctidae			
Crangonyx pseudogracilis	31 (17)	36 (7)	1 (0)
Isopoda—Asellidae			
Caecidotea racovitzai	3 (2)	5 (3)	1 (1)
Insects			
Diptera—Culicidae			
Aedes stimulans	23 (9)	35 (21)	0

[a]Included are those taxa that were found in at least three of the four pools and comprised >3 percent of the total invertebrates collected. Collections consisted of three 1-m sweeps per site in spring (April 15 and May 17, 1995) and two 0.5-m sweeps per site in autumn (within two weeks after sites refilled, October 22 in one site and November 17 in the other three sites).

Wiggins et al. (1980), Williams (1987, 1996), and Schneider and Frost (1996). Only two of the five most common taxa, *Aedes* mosquito larvae and *Sphaerium occidentale* fingernail clams, are temporary pool specialists (Wiggins et al. 1980, McKee and Mackie 1981, Schneider and Frost 1996). *Pisidium casertanum* fingernail clams are ecological generalists that use both permanent and temporary water habitat (Bailey and Mackie 1986). Our finding that *C. pseudogracilis* amphipods and *C. racovitzai* isopods were abundant in temporary pools was an unusual occurrence (at least for those habitats that are physically isolated from permanent water bodies). Neither of these crustaceans is thought to be adapted for temporary waters because amphipods and isopods do not possess desiccation-resistant stages and are poorly adapted to disperse into newly created habitat (Wiggins et al. 1980). Both of these crustaceans more typically occur in large permanent water bodies rather than small temporary pools. *Crangonyx pseudogracilis* commonly resides in rivers and lakes (Bousfield 1958, Holsinger 1976). *Caecidotea racovitzai* typically occurs in assorted permanent water habitats, including rivers, ponds, lakes, and swamps (Williams 1976). In western New York it is the dominant isopod of the eastern Great Lakes, where it occurs down to 42 m (Williams 1976). One other noteworthy though relatively rare taxon in the autumnal pools was *Chauliodes* sp. (Megaloptera: Corydalidae). Lentic dobsonfly larvae commonly occur in permanent water wetlands (Evans and Neunzig 1984), but are not considered to be residents of temporary waters (Williams 1996). Because megalopteran larvae require from two to five years to mature (Evans and Neunzig 1984), the large mature individuals that we encountered probably had endured several wet-dry cycles.

PERMANENT WATER INVERTEBRATES IN TEMPORARY POOLS

The presence of typically permanent water species in the small autumnal pools of Reinstein Woods was a curiosity, and serves as the primary theme for the remainder of this chapter.

How Did Amphipods and Isopods Access Autumnal Pools?

Wiggins et al. (1980) suggest that the presence of amphipods or isopods in temporary pools may indicate a recent connection to nearby permanent water habitats. However, we observed no obvious connections between three of the four autumnal pools and any permanent water source, even during high-water periods of autumn, winter, and spring. As mentioned previously, the fourth site, Hemlock Swale, was attached to permanent water via a seepage area. To investigate possible similarities between autumnal pool communities and those in nearby permanent waters, we surveyed the invertebrate fauna of permanent sites using sweep net sampling. These permanent habitats included a 1-ha cattail marsh and five open ponds ranging in size from 20 m^2 to 3 ha (including the Hemlock Swale outlet pool). These habitats would be the most likely sources of permanent water organisms for the four temporary pools. However, only two of the five species that dominated autumnal pools were collected in these permanent water habitats. We found numerous *P. casertanum* fingernail clams and small numbers of *C. racovitzai* isopods. Although *C. pseudogracilis* amphipods were not collected in these permanent water habitats, another amphipod, *Hyalella azteca,* was very abundant. If the amphipods had recently arrived in the autumnal pools of Reinstein Woods via flood waters from permanent water habitats, we would have expected to find *H. azteca* amphipods in those sites rather than *C. pseudogracilis*. Clearly the *C. pseudogracilis* amphipods and probably the *C. racovitzai* isopods that abounded in Reinstein Woods autumnal pools did not simply arrive in overland flood waters from permanent habitats.

An alternative mechanism for amphipods and isopods to access autumnal pools is via phoretic transfer on the feathers or feet of water birds (Wiggins et al. 1980, Swanson 1984). However, this seemed an improbable way for large numbers of crustaceans to enter the Reinstein Woods autumnal pools. During two years of observation we observed aquatic birds in an autumnal pool on only one occasion. Yet we discovered amphipods and/or isopods not only in the 4 primary study pools, but in 21 other autumnal pools in the immediate area (see below). Most of these sites lay deep within the forest, and some of these were only a few square meters in area. For aquatic birds to find and successfully transfer amphipods or isopods into all of these sites seemed a remote possibility. Additionally, for reasons discussed above, the most likely amphipod to be transferred by birds was *H. azteca,* rather than the *C. pseudogracilis* that actually occurred.

A third potential way that amphipods and/or isopods might have gained entry to the pools was via ground water (Covich and Thorp 1991). For example, *Crangonyx* and the closely related genus *Stygonectes* contain many species that are specially adapted for existence in groundwater or groundwater-related habitats (Holsinger 1976, 1978, Covich and Thorp 1991). However, subterranean taxa are typically eyeless, and *C. pseudogracilis* has well-developed eyes. Furthermore, published accounts of amphipod distributions report *C. pseudogracilis* only from surficial habitats like lakes and rivers (Bousfield 1958, Holsinger 1976).

Although flooding from permanent waters, phoretic transfer by birds, and/or groundwater transport might have been responsible for the initial entry of amphipods and isopods into Reinstein Woods autumnal pools, we believe these crustaceans are now successfully persisting and reproducing in these habitats without the need for yearly reintroduction.

How Can Amphipods and Isopods Persist in Autumnal Pools?

Several factors might allow populations of amphipods and isopods to persevere indefinitely in the autumnal pools of Reinstein Woods, even if they lack a desiccation-resistant stage. Perhaps these typically permanent water crustaceans were surviving in residual pockets of water that still persisted in the autumnal pools during the summer. This mechanism clearly was important in Hemlock Swale. After the majority of that site dried, we consistently encountered large numbers of *C. pseudogracilis* and *C. racovitzai* in the small residual pool that remained flooded over most of the summer. However, in the remaining three sites we found no residual pockets of surface water during the summer dry season, except after a couple of particularly heavy rains. Kenk (1949) and Wiggins et al. (1980) found that tunnels of burrowing crayfish could provide other invertebrates with a connection to the water table beneath dry sites. However, neither crayfish nor animal burrows of any kind were discovered in the Reinstein Woods autumnal pools. By September the underlying aquifer beneath dry pools was >50 cm below habitat surfaces. Amphipods and isopods would probably have had difficulty burrowing themselves down through the compact sand substrates to this water table (but see abstract by Holsinger and Dickson [1978]).

Alternatively, we had preliminary indications that amphipods and isopods were spending the dry period near basin surfaces. Soon after most significant rains we revisited sites to monitor water levels. On several of these occasions we collected amphipods or isopods swimming in the pools of water that temporarily developed from these rains. Their rapid appearance might indicate that they were persisting in close proximity to the basin surface. We believed that rather than using hypogean habitat, both *C. pseudogracilis* and *C. racovitzai* were simply persisting in damp organic detritus in dry basins. Holsinger and Dickson (1978) report that a cave-dwelling amphipod (*Crangonyx antennatus*) burrows into surficial mud substrates to survive temporary drying.

To test this hypothesis we waited until the driest portion of the late summer and then collected samples of leaf litter and underlying substrate (upper 5 cm) from 30-cm-by-30-cm quadrats located in the deepest portions of each of two dried pools (N = 3 per site). These samples were returned to the laboratory where we flooded them with deionized water. Over a one-week period we collected any aquatic invertebrates that became active in this water (Wiggins et al. 1980). We also collected qualitative samples of substrate and leaf detritus from beneath fallen logs in both sites and processed those samples as above. We retrieved, among other invertebrates, six *C. pseudogracilis,* eight *C. racovitzai,* and one megalopteran larvae from detritus collected from dry basins—most from a basin that had been completely dry for almost three months. We found these individuals primarily in the samples collected from beneath fallen logs. Kenk (1949) also isolated *C. gracilis* amphipods from a dry substrate sample collected in a Michigan autumnal pool. However, because conventional wisdom indicated that amphipods cannot survive drying, he still suspected that their overall success was probably linked to some sort of a hypogean existence. Smock (this volume) isolated *C. racovitzai* isopods, but not amphipods, from dry substrates taken from a southeastern floodplain.

In further laboratory tests we established that both *C. pseudogracilis* and *C. racovitzai* could survive for extended periods without free water. We maintained field-collected amphipods and isopods in covered petri dishes containing moist, crushed leaf detritus (10 gm of oven-dried detritus + 8 ml of water) for three weeks at room temperature. While the leaf detritus remained damp throughout the three-week period, free water was never present. Ninety percent of these amphipods and isopods survived these trials. Because amphipods and isopods began actively swimming or crawling within minutes after detritus was reflooded, the crustaceans apparently had not entered a quiescent physiological state to survive drying. They simply persisted by burrrowing into the moist detritus. As detritivores, they might even be able to feed and develop in the absence of free water, although we have not tested that hypothesis. Gladden and Smock (1990) found that isopods and amphipods grew during a floodplain's dry phases (although it was not clearly established that they were persisting in the absence of water).

Under natural field conditions the damp organic substrates in autumnal pools surrounded by old-growth forest may be particularly conducive to amphipod and isopod survival. However, sampling in the four study pools during autumn suggested that mortality from summer drying was still extensive. We were able to collect *C. pseudogracilis* and *C. racovitzai* during autumn in most of the same sites where they occurred the previous spring. However, amphipod numbers in autumn were only 2 percent of their predrawdown levels, and isopod numbers had declined to only 25 percent of their previous densities (Table 14.1). *Sphaerium occidentale* fingernail clams, which are well adapted to survive habitat drying, showed a negligible decline from spring to autumn. In contrast, the generalist clam *P. casertanum* suffered higher mortality, with autumn numbers being only 14 percent of spring levels. *Crangonyx pseudogracilis, C. racovitzai,* and *P. casertanum* apparently lack a well-developed mechanism of desiccation resistance. However, populations may still be able to persevere because the few individuals that do survive probably proliferate rapidly once habitats reflood.

INFLUENCE OF SURROUNDING FOREST CONDITIONS

The old-forest conditions surrounding the Reinstein Woods pools might enable typically permanent water invertebrates to persist in temporary water settings. Three other references report numerous amphipods and isopods from hydrologically isolated temporary pools, and in all cases the habitats were associated with mature forest. One study was conducted in south Georgia pools containing old-growth cypress trees (Golladay et al., this volume), and the other two in Michigan pools surrounded by mature hardwood forest (Kenk 1949, Higgins and Merritt, this volume). To further investigate a possible link between old-growth forest and the presence of amphipods and isopods in autumnal pools of Reinstein Woods, we used a descriptive survey. We identified 18 autumnal pools in three separate blocks of old-growth forest of Reinstein Woods Preserve. We identified another 32 pools in various portions of Reinstein Woods that had been logged during the 1800s (Kershner 1993).

The forest regrowth in these latter areas was dominated by black cherry, white ash, and sugar maple in various stages of succession.

We sampled populations of *C. pseudogracilis* and *C. racovitzai* in these 50 pools and compared percent occurrence and densities of crustaceans between old growth and previously logged areas. For this survey we employed a stratified random sampling regime where perimeter areas of pools were dipped using 150-ml measuring cups and the individual 20 to 100 ml collections of water were pooled until a 1-l sample was accumulated. Preliminary sampling had already indicated that this technique effectively collected both amphipods and isopods in pools because they tended to congregate on leaf litter in shallow perimeter areas. The 1-l water samples were transferred to white enameled pans, where amphipods and isopods were identified and their numbers tabulated. Of 18 sites in the old growth area, 15 supported *C. pseudogracilis* and/or *C. racovitzai*. Of 32 sites in the previously logged area, only 9 supported either crustacean. No other amphipod or isopod species was discovered in any site. Contingency testing indicated that the difference in percent occurrence between old-growth and historically logged habitats was highly significant ($\chi^2 = 14.1$, df $= 1$, P $= 0.0002$). Densities of amphipod and isopod crustaceans also were higher in old-growth pools than in non-old-growth pools (Mann-Whitney U-test, Z $= 2.91$, df $= 48$, P $= 0.0036$). Apparently, surrounding forest condition influences the suitability of autumnal pool habitats for *C. pseudogracilis* and *C. racovitzai*. However, amphipods and isopods were still on occasion very abundant in habitats outside of the old-growth forest. Here conditions for these crustaceans might have remained suitable even after the surrounding forest was cut, or possibly appropriate conditions had redeveloped over the years.

We believe that the environmental conditions in old-growth forest probably benefitted crustaceans by ensuring that sites remained damp even during the driest periods. Amphipods have been previously reported to survive temporary habitat drying in caves (Holsinger and Dickson 1978); the conditions in caves and under dense tree canopies will both reduce drying intensity. In Reinstein Woods dense tree canopies provided shade and an abundance of insulating leaf litter (the ability of *C. pseudogracilis* and *C. racovitzai* to live in moist detritus was shown above). The summer of 1995 was one of the driest on record, yet the waterless basins of Reinstein Woods autumnal pools remained moist throughout that period, at least in the old-growth forest areas. In other hydrologically similar autumnal pools that were outside of the old growth area, we observed visible ground cracks developing at the peak of the 1995 drought.

VARYING ECOLOGICAL STRATEGIES OF TEMPORARY POOL FAUNAS

In their classic discussion of the strategies that aquatic animals use to live in temporary pools, Wiggins et al. (1980) argue that successful residents must have structural, behavioral, and/or physiological adaptations for tolerating or avoiding drought. Research in temporary water wetlands has typically focused on the unique adaptations that invertebrates use in these often extreme environments (Wiggins et al. 1980, McLeod et al. 1981, Williams 1987, Batzer and Wissinger 1996, Schneider and Frost 1996, Wissinger 1997). However, experimental work indicates that certain responses by invertebrates to temporary pool conditions are not intrinsic adaptations but simply a response to local environmental conditions. Intuitively, one would expect that reproductive strategies of temporary pool animals would be shaped by the short period of inundation. Bailey and Mackie (1986) tested this hypothesis by comparing reproductive strategies of a fingernail clam in temporary versus permanent water. However, their experiments indicated that habitat-specific reproductive responses probably were more related to different food levels than to varying habitat hydrologies. Similarly, experiments by Berte and Pritchard (1986) suggested that rapid growth of temporary pool caddisfly larvae may simply reflect the habitat's inherent warm water temperatures rather than adaptation by caddisflies to avoid desiccation. In terms of caddisfly growth rates, they found that a permanent water species actually outperformed a temporary water caddisfly under the same environmental conditions.

While Wiggins et al. (1980) primarily focus on the various strategies of drought resistance by temporary pool invertebrates, Wissinger (1997) indicates that many successful temporary pool insects lack drought-resistant stages. Some of these species, called "cyclic colonizers" by Wissinger, flourish by migrating between temporary and permanent water sites. Soon after temporary pools flood, reproductively mature adults enter habitats to lay eggs. Eggs hatch, progeny develop rapidly, and many emerge before the sites dry. Those individuals then move to permanent water refugia. Many researchers have discovered that some of the most successful taxa of temporary wetlands also flourish in nearby permanent water habitats; temporary habitats included seasonally flooded marshes (Neckles et al. 1990, Duffy and LaBar 1994, Batzer et al. 1997), seasonal prairie potholes (Driver 1977, Battaille and Baldassarre 1993), temporary bog pools (Larson and House 1990), ephemeral tropical pools (Muthukrishnan and Palavesam 1992, Batzer 1995), and stream floodplains (Gladden and Smock 1990). Certain insects are probably behaviorally adapted to seek out temporary wetlands to breed, or, alternatively, might simply deposit their eggs in an assortment of aquatic or wetland habitats, both temporary and permanent, as a bet-hedging strategy.

In Reinstein Woods autumnal pools, however, *C. pseudogracilis* amphipods and *C. racovitzai* isopods were succeeding despite their lack of specific adaptations for life in temporary waters (except possibly a propensity to burrow). Their vulnerability to prolonged drought was evidenced by high summer mortality rates. A strategy of cyclic colonization to avoid drought could not be used, because they lacked aerial dispersal. We suggest that at an unknown time and via an unknown mechanism these invertebrates were introduced into the autumnal pools from the permanent water habitats where they had evolved. However, conditions in the pools apparently never become harsh enough for them to be completely eliminated. Even after the unusually dry summer of 1995, we collected *C. pseudogracilis* or *C. racovitzai* in 24 autumnal pools in Reinstein Woods the following spring. The fact that these pools were autumnal rather than vernal was probably crucial to survival of permanent water taxa. Organisms were not forced to concurrently survive drying and freezing winter temperatures, because habitats reflooded in autumn (see Wiggins et al. 1980).

Another curiosity about the Reinstein Woods pools was the absence of some invertebrates that typically occur in temporary pools, such as fairy shrimp (Anostraca). The fairy shrimp that typically reside in eastern woodland ponds require habitat drying for egg hatch and thus exist only in temporary water wetlands (Wiggins et al. 1980). Possibly the autumnal pools of Reinstein Woods do not dry sufficiently to allow fairy shrimp to hatch. The dry period in Reinstein Woods autumnal pools may actually be too wet for certain taxa.

The fact that *C. pseudogracilis* was found only in autumnal pools and not in permanent water habitats of Reinstein Woods might suggest that conditions in the temporary pools were actually more conducive to their success than those in local permanent water sites. Other cases of apparent "specialization" by amphipods or isopods for temporary waters have been reported from Michigan autumnal pools (Kenk 1949) and Missouri temporary streams (Hubricht and Mackin 1940). Many amphipods are very susceptible to fish predation, and perhaps living in habitats that occasionally dry is less problematic than trying to coexist with fish. Wissinger et al. (this volume) describe how a caddisfly that is able to survive in both temporary and perennial water wetlands is found almost exclusively in the temporary water settings because it is poorly adapted to coexist with vertebrate predators.

CONCLUSIONS AND SPECULATIONS

1. *Environmental conditions vary widely among different kinds of temporary wetlands, and this variation undoubtedly affects invertebrate ecological functioning within individual habitats.* Our results indicate that differences between autumnal and vernal habitats may make it inappropriate to adapt many of the ecological paradigms developed in well-studied vernal pools to virtually unstudied autumnal pools (see also Wiggins et al. 1980). For example, while dealing with the dry phase may be of paramount importance to invertebrate success in vernal pools (Wiggins et al. 1980), the dry phase was a much more benign environmental disturbance in autumnal pools.

2. *Habitat-specific adaptations were not required for invertebrate success in autumnal pools.* Some of the most successful taxa had evolved else-

where. However, once introduced to autumnal pools, these taxa were not only persisting but flourishing. Serendipity strongly shaped the invertebrate community structure in Reinstein Woods autumnal pools.

3. *Forest condition influences invertebrates in woodland pools, and forest alteration may have long-term consequences for associated wetlands.* Even after many decades of forest regrowth, we were still able to detect differences in the invertebrate communities between wetlands in formerly deforested versus old-growth areas.

4. *The unique nature of the invertebrate fauna in autumnal pools of Reinstein Woods is probably related to their continuous existence in old-growth forest.* While currently these pools and their faunas might be anomalies, historically that was not the case. Before eastern North America's old-growth forests were eliminated, the faunas now occurring in Reinstein Woods autumnal pools were probably the norm rather than the exception. Perhaps some of our ideas about the ecological functioning of temporary pools are biased by past human interventions.

ACKNOWLEDGMENTS

We especially appreciated the cooperation of Jeffrey Liddle of Dr. Victor Reinstein Woods Preserve. Support from the Biology Department at Canisius College for D. Batzer and a Hughes Foundation award to K. Sion were used for this study. The help of Canisius College students Michelle Newman and Christopher Pusateri was invaluable. We thank Wayne Gall and Bohdan Bilyj for providing taxonomic expertise.

LITERATURE CITED

Bailey, R. C., and G. L. Mackie. 1986. Reproduction of a fingernail clam in contrasting habitats: Life-history tactics? Canadian Journal of Zoology 64:1701–1704.

Battaille, K. J., and G. A. Baldassarre. 1993. Distribution and abundance of aquatic macroinvertebrates following drought in three prairie pothole wetlands. Wetlands 13:260–269.

Batzer, D. P. 1995. Aquatic macroinvertebrate response to short-term habitat loss in experimental pools in Thailand. Pan-Pacific Entomologist 71:61–63.

Batzer, D. P., F. de Szalay, and V. H. Resh. 1997. Opportunistic response of a benthic midge (Diptera: Chironomidae) to management of California seasonal wetlands. Environmental Entomology 26:215–222.

Batzer, D. P., and S. A. Wissinger. 1996. Ecology of insect communities in nontidal wetlands. Annual Review of Entomology 41:75–100.

Berte, S. B., and G. Pritchard. 1986. The life histories of *Limnephilus externus* Hagen, *Anabolia bimaculata* (Walker) and *Nemotaulius hostilis* (Hagen) (Trichoptera: Limnephilidae) in a pond in southern Alberta. Canadian Journal of Zoology 64:2348–2356.

Bousfield, E. L. 1958. Fresh-water amphipod crustaceans of glaciated North America. Canadian Field-Naturalist 72:55–113.

Covich, A. P., and J. H. Thorp. 1991. Crustacea: Introduction and Peracarida. Pages 665–689 *in* J. H. Thorp and A. P. Covich (eds.), Ecology and Classification of North American Freshwater Invertebrates. Academic Press, New York.

Driver, E. A. 1977. Chironomid communities in small prairie ponds: Some characteristics and controls. Freshwater Biology 7:121–133.

Duffy, W. G., and D. J. LaBar. 1994. Aquatic invertebrate production in Southeastern USA wetlands during winter and spring. Wetlands 14:88–97.

Evans, E. D., and H. H. Neunzig. 1984. Megaloptera and Aquatic Neuroptera. Pages 261–270 *in* R. W. Merritt and K. W. Cummins (eds.), An Introduction to the Aquatic Insects of North America (2nd ed.). Kendall/Hunt, Dubuque, IA.

Gladden, J. E., and L. A. Smock. 1990. Macroinvertebrate distribution and production on the floodplains of two lowland headwater streams. Freshwater Biology 24:533–545.

Holsinger, J. R. 1976. The Freshwater Amphipod Crustaceans (Gammaridae) of North America. U.S. Environmental Protection Agency, Water Pollution Control Research Series 18050 ELDO4/72 (2nd ed.).

———. 1978. Systematics of the subterranean amphipod genus *Stygobromus* (Crangonyctidae). Part 2: Species of the eastern United States. Smithsonian Contributions, Zoology 266:1–144.

Holsinger, J. R., and G. W. Dickson. 1978. Burrowing activity in the troglobitic crustacean *Crangonyx antennatus* Packard (Crangonyctidae). National Speleological Society Bulletin 40:85(abstract).

Hubricht, L., and J. G. Mackin. 1940. Descriptions of nine new species of fresh-water amphipod crustaceans with notes and new localities of other species. American Midland Naturalist 23:187–218.

Kenk, R. 1949. The animal life of temporary and permanent ponds in southern Michigan. Miscellaneous Publication of the Museum of Zoology, University of Michigan 71:1–66.

Kershner, B. S. 1993. Buffalo's Backyard Wilderness. Western New York Heritage Institute, Buffalo.

Larson, D. J., and N. L. House. 1990. Insect communities of Newfoundland bog pools with emphasis on the Odonata. Canadian Entomologist 122:469–501.

McKee, P. M., and G. L. Mackie 1981. Life history adaptations of the fingernail clams *Sphaerium occidentale* and *Musculium securis* to ephemeral habitats. Canadian Journal of Zoology 59:2219–2229.

McLeod, M. J., D. J. Hornbach, S. I. Guttman, C. M. Way, and A. J. Burky. 1981. Environmental heterogeneity, genetic polymorphism, and reproductive strategies. American Naturalist 118:129–134.

Muthukrishnan, J., and A. Palavesam. 1992. Secondary production and energy flow through *Kiefferulus barbitarsis* (Diptera: Chironomidae) in tropical ponds. Archiv für Hydrobiologie 125:207–226.

Neckles, H. A., H. R. Murkin, and J. A. Cooper. 1990. Influences of seasonal flooding on macroinvertebrate abundance in wetland habitats. Freshwater Biology 23:311–322.

Schneider, D. W., and T. M. Frost. 1996. Habitat duration and community structure in temporary pools. Journal of the North American Benthological Society 15:64–86.

Swanson, G. A. 1984. Dissemination of amphipods by waterfowl. Journal of Wildlife Management 48:988–991.

Wiggins, G. B., R. J. Mackay, and I. M. Smith. 1980. Evolutionary and ecological strategies of animals in annual temporary pools. Archiv für Hydrobiologie Supplement 58:97–207.

Williams, D. D. 1987. The Ecology of Temporary Waters. Croom Helm, Timber Press, Portland, OR .

Williams, D. D. 1996. Environmental constraints in temporary fresh waters and their consequences for the insect fauna. Journal of the North American Benthological Society 15:634–650.

Williams, W. D. 1976. Freshwater Isopods (Asellidae) of North America. U.S. Environmental Protection Agency, Water Pollution Control Research Series 18050 ELDO5/72 (2nd ed.).

Wissinger, S. A. 1997. Cyclic colonization and predictable disturbance: a template for biological control in ephemeral crop habitats. Biological Control 10:4–15.

15 Beaver Pond Wetlands in Northwestern Pennsylvania

Modes of Colonization and Succession after Drought

SCOTT A. WISSINGER and LESLIE J. GALLAGHER

W*etlands associated with beaver dams have become increasingly abundant in North America as beaver populations have recovered during the past century. The ponds, marshes, wet meadows, and swamps associated with active and abandoned dams are important components of the wetland resources in northwest Pennsylvania and integral to the management and conservation of waterfowl and nongame species by state and federal agencies. In this chapter we first describe the invertebrate fauna from two permanent and two autumnal wetlands associated with active and abandoned impoundments, respectively. Although several groups of taxa exhibit hydroperiod-specific patterns of distribution and abundance, the community composition of the permanent and autumnal wetlands is quite similar. This is in part due to the presence of seasonally inundated marshes and shrub thickets in addition to the main permanent pond at active dam sites. A second goal of this chapter is to describe the results of an experimental and comparative study of recolonization of three of these wetlands after drought. During the drought we rehydrated sediments from two autumnal and one previously permanent basin to assess which taxa were present in one or more drought-resistant stages (egg, larvae, pupae, adult) in the substrate. These laboratory data, combined with field surveys after the basins refilled, suggest that a greater proportion of the predrought fauna in the two autumnal wetlands (71 and 63 percent) recolonized from drought-resistant stages than in the permanent wetland (38 percent). These initial recolonists were mainly flies (Diptera), water bugs (Hemiptera), and beetles (Coleoptera), many of which occur in*

Invertebrates in Freshwater Wetlands of North America: Ecology and Management, Edited by Darold P. Batzer, Russell B. Rader, and Scott A. Wissinger.
ISBN 0-471-29258-3 © 1999 John Wiley & Sons, Inc.

habitat margins when the basins are filled. The second most important mode of recolonization (22, 25, and 27 percent for the two autumnal ponds and the permanent pond, respectively) was via the immigration of winged adults of species that seasonally cycle between permanent and temporary habitats such as gerrid, notonectid, and corixid water bugs and dytiscid diving beetles. Finally, a third group of permanent-pond taxa, especially dragonflies, continued to aerially colonize the permanent wetland during the following summer and fall. One year after the drought, about 90 percent of the predrought fauna had returned to the two autumnal wetlands, as compared to 77 percent in the previously permanent basin. Our results suggest that desiccation resistance and local migration between a mosaic of permanent and temporary habitats are responsible for the resiliency of these faunas after disturbance by drought and that an understanding of patterns of invertebrate diversity and species distributions in beaver-pond wetland complexes will require a metapopulation and perhaps even metacommunity approach.

INTRODUCTION

As beaver populations have recovered from near-extinction at the turn of the last century, the abundance of wetlands associated with their dams and foraging activities has steadily increased throughout North America (Naiman et al. 1988, Naiman et al. 1986). In northwestern Pennsylvania beaver impoundments are now ubiquitous in low-gradient reaches of headwater and middle-order streams. We distinguish between two general types of habitats created by beaver dams in this region. On 2nd to 4th order streams, beaver impoundments create upstream pools of relatively slow-moving water with a lotic vertebrate and invertebrate fauna that is characteristic of other pool habitats in these low-gradient streams (e.g., Fig. 15.1). The lodges are typically in or on the stream banks, and the extent to which lentic habitats are formed laterally into the floodplain is variable depending on the geomorphologic context and size of the dam. Although impoundments on these streams are often damaged during winter and spring floods, they are usually repaired at the same locations year after year. There has been considerable research that describes how beavers in this geomorphologic context can dramatically change the composition and dynamics of stream invertebrate and fish communities and alter ecosystem-level processes, including hydrology, channel geomorphology, sedimentation, nutrient budgets, and detritus retention and processing (Sprules 1940, Naiman and Melillo 1984, Naiman et al. 1984, Naiman et al. 1986, McDowell and Naiman 1987, Naiman et al.1988, Ford and Naiman 1988, Johnston and Naiman 1990, Smith et al. 1991, Schlosser 1995).

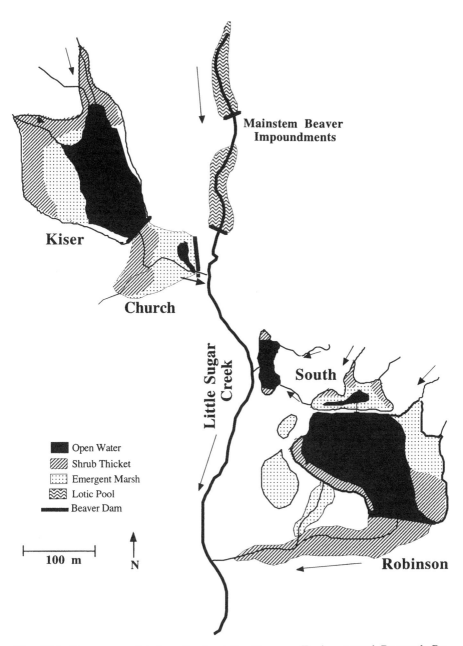

Fig. 15.1. Beaver-associated wetlands at the Bousson Environmental Research Reserve. Scale is approximate and is intended to provide general orientation to the habitats described in this chapter.

A second geomorphologic context for beaver activities is in low-gradient reaches of *headwater* (<2nd order) streams. Poorly drained glacial soils, gently rolling topography, and abundant year-round precipitation create the ideal conditions for beaver dam development in even the smallest of tributaries. Where human activities do not exclude beavers, low-gradient reaches of many headwater streams in the region contain active or abandoned beaver dams and associated wetland habitats. Beaver impoundments on these headwater streams create ponds and associated swamps and marshes that are distinctly lentic and that persist as wooded wetlands, marshes, and wet meadows long after the impoundments are abandoned (see Naiman et al. 1988, Cirmo and Driscoll 1993). It is well established that the rich food resources (invertebrates and plants) and cover for nesting and foraging in these wetland complexes are important for game and nongame wildlife (e.g., Beard 1953, Deifenbach and Owen 1989, Ermer 1984, Reinecke and Owen 1980, McDowell and Naiman 1987, Rottella and Ratti 1992*a*, 1992*b*, Grover and Baldassarre 1995, McCall et al. 1996). It is therefore not surprising that many of the largest concentrations of beaver-associated wetlands in the region have been purchased and are now managed by state and federal agencies for wildlife conservation and waterfowl propagation (Erie National Wildlife Refuge, numerous Pennsylvania State Game Lands).

Literature on the population and community ecology of invertebrates in North American beaver wetlands is surprisingly sparse (Sprules 1940, Hodkinson 1975, France 1997). In this chapter we first describe the invertebrate communities in permanent and semipermanent wetlands associated with two active and two abandoned headwater beaver dams, respectively. One goal for collecting these baseline data was to establish species-specific patterns of distribution and abundance across wetlands with differing hydroperiods. Hydroperiod is among the most important determinants of species composition in lentic freshwater habitats, and shifts in species composition along a gradient from permanent to temporary basins have been observed for many invertebrate and vertebrate taxa (Batzer and Wissinger 1996, Schneider and Frost 1996, Wellborn et al. 1996). Experimental studies with amphibians have shown that in some cases these shifts are related to species-specific differences in drought tolerance, whereas in others they are causally related to a shift from invertebrate to vertebrate predation along a gradient from temporary to permanent habitats (Wellborn et al. 1996, Skelly 1997, Wilbur 1997). Only a few studies have experimentally disentangled the relative importance of abiotic and biotic mechanisms that might explain shifts in invertebrate community composition along this gradient (McPeek 1990*a*, 1990*b*, 1996, Wellborn 1994, Wissinger et al. 1999). One of the goals of our work has been to document correlations between hydroperiod and invertebrate community composition as a basis for designing future experimental studies. Understanding the role that fish play in changing invertebrate community composition should be of great interest to wildlife managers, given that fish are often viewed as competitors for the dominant taxa in the diets of waterfowl (Mallory et al. 1994, Batzer 1998).

A second goal of this chapter is to describe the recolonization of three of these wetlands following drought conditions in fall 1995. We were particularly interested in documenting the relative importance of desiccation resistance versus aerial recolonization in semipermanent versus permanent basins. We hypothesized that the previously permanent wetland that dried in 1995 would be least resilient to drought and most dependent on aerial immigrants for recolonization. Because of the variety of different types of water bodies in close association at our study site (see Jeffries 1994), we also hypothesized that all of the habitats should be quickly colonized by species whose life histories involve seasonal cycles of migration between permanent and temporary habitats (group 4 colonists of Wiggins et al. 1980, "cyclic colonizers" of Batzer and Wissinger 1996 and Wissinger 1997).

HABITAT DESCRIPTION

The wetlands at our study site are all associated with headwater tributaries of Little Sugar Creek and located in the Bousson Environmental Research Reserve (BERR) of Allegheny College in Crawford County, Pennsylvania (Fig. 15.1). Kiser Pond and Church Marsh occur on the same unnamed headwater stream that flows east through the reserve, and South Marsh and Robinson Pond on separate headwater streams that flow south and west to Little Sugar Creek (Fig. 15.1). Prior to 1995 Kiser and Robinson Ponds had been permanent for at least the previous 15 years. During the study (1994–1997) both had active beaver lodges and the associated wetlands included open shallow ponds as well as extensive areas of back-flooded marshes, shrub thickets, and woodlands (Table 15.1). The other two wetlands, Church Marsh and South Marsh, did not have active lodges during the study. Since 1993 Church Marsh and South Marsh have dried partially or completely in late summer but then refilled in early fall, and are therefore best described as semipermanent and autumnal (after Wiggins et al. 1980).

Since 1990 we have been monitoring the water chemistry of these habitats to assess the impacts of acid precipitation on the amphibian fauna at BERR. By virtue of its downwind proximity to the industrial midwest, this region has among the most acidic rainfall in North America (e.g., Barrie and Hales 1984, Shannon and Sisterson 1992). During the period from 1990–1997, precipitation pH was typically 3.5–4.5. Despite this continuous acidic input, the water in these wetlands was almost always circumneutral (6.8–7.4) as a result of the extremely high buffering capacity associated with sedimentary bedrock and derived soils (Yaworski et al. 1979). The water in all of the wetlands had high conductivity (100–150 μS cm^{-1} TDS) and alkalinity (100–150 mg/l). Thus, even under extreme runoff conditions during snowmelt, pulses of episodic acidification (5.0–5.5) last only a few days before a return to circumneutral conditions (Wissinger, unpublished data). All of the habitats are relatively nutrient-rich, with high levels of nitrogen (total nitrogen = 2–4 mg/l) and phosphorus (total reactive phosphorous = 0.1–0.4 mg/l).

TABLE 15.1. Physical Habitat Characteristics of Four Wetlands Associated with Beaver Activity on the Bousson Environmental Reserve

Wetland	Area (ha)	Max. Depth (m)	Hydroperiod	Vegetation Zones	Vertebrate Predators
Kiser	6.55	1.5	Permanent	40% open pond (*Potamogeton, Nuphar*) 25% shrub thicket (*Salix, Cornus, Alnus*) 35% emergent marsh (*Typha, Carex, Juncus, Scirpus*)	bass, bluegill, redear sunfish
Robinson	1.05	2.0	Permanent	50% open pond (*Ceratophyllum, Nuphar*) 20% shrubs (*Salix, Cornus, Vaccinium*) 30% emergent marsh (*Typha, Carex*)	green sunfish, bullhead
Church	3.25	0.5	Autumnal	10% open pond (*Potamogeton, Ceratophyllum*) 50% emergent marsh (*Carex, Juncus, Typha, Scirpus*) 40% shrub thicket (*Spirea, Alnus, Cornus*)	stickleback, mudminnow newts, marble salamanders
South	1.62	0.6	Autumnal	30% open pond (*Utricularia, Polytrichium*) 50% emergent vegetation (*Carex, Sparganium, Scirpus*) 20% shrubs (*Salix, Alnus, Vaccinium*)	spotted salamanders, stickleback

As is often true in lentic habitats, the abundance and type of vertebrates at our study site covary with hydroperiod. Fish are the top predators in all of the permanent ponds (Table 15.1). Largemouth bass (*Micropterus salmoides*), sunfish (*Lepomis macrochirus, Lepomis gibbosus, Lepomis cyanellus*), and bullhead (*Ictalurus natalis*) were present in varying numbers, depending on the severity of winterkill in these shallow ponds. Since 1992 there have been only stickleback (*Culaea inconstans*) and mudminnows (*Umbra limi*) in Church Marsh, and only stickleback in South Marsh. However, both marshes are the breeding sites of several predatory amphibians. This is especially true in South Marsh, where larvae of spotted and marble salamanders (*Ambystoma maculatum, A. opacum*) and adults and larvae of red-spotted newts (*Notopthalmus viridescens*) are abundant. Salamanders are less abundant in Church Marsh, where a few spotted and marble salamanders have begun to breed in spring and fall, respectively, since the dam was abandoned in 1992. These habitats are also the breeding sites of several amphibians with herbivorous/detritivorous larval stages. Bullfrogs (*Rana catesbeiana*) and toads (*Bufo americana*) were the dominant larval anurans in the main basins of Kiser and Robinson Ponds, with spring peepers (*Hyla crucifer*), chorus frogs (*Pseudacris triseriata*), pickerel frogs (*Rana palustris*), and leopard frogs (*Rana pipiens*) in the surrounding marshes and wet meadows. The most common larval anurans in Church Marsh were spring peepers, chorus frogs, and green frogs (*Rana clamitans*), whereas those in South Marsh were spring peepers, chorus frogs, gray tree frogs (*Hyla versicolor*), and wood frogs (*Rana sylvatica*).

The plant community in Kiser Pond is extremely heterogeneous, with an open-water zone that becomes grown over by midsummer with the floating pads of spatterdock (*Nuphar variegata*) (Fig. 15.1, Table 15.1). This open-water zone is surrounded (except at the dam) by emergent marsh, shrub thickets, and flooded woodlands. From wet meadow to open water, the zones of vegetation are:

1. wet meadow dominated by rushes (*Juncus effusus*), grasses (*Phalaris arundinacea*), sedges (*Carex hystericina, C. lurida*) and other herbaceous species;
2. shallow (<0.5 m) marsh dominated by woolgrass (*Scirpus cyperinus*) and bur reed (*Sparganium eurycarpum*); and
3. deep (0.5–1.0 m) marsh dominated by cattails (*Typha latifolia*).

Shrub thickets in the backwaters are dominated by alder (*Alnus incaca*), dogwoods (*Cornus ammomum, C. stolonifera*), meadowsweet (*Spirea alba*), and willows (*Salix discolor*), which grade into seasonally inundated woodlands of mainly red maple (*Acer rubrum*) and elm (*Ulmus americana*). The open-water zone of Church Marsh is much smaller, and most of the former pond basin is now dominated by shallow marsh species, especially sedges (*Carex comosa, C. hystericina*) and bulrushes (*Scirpus cyperinus, S. validus*). A variety of wet meadow plants have invaded the central basin of this former beaver pond during the past five years, including marsh marigold (*Caltha palustris*), joe-pye weed (*Eupoatorium maculatum*), goldenrod (*Solidago graminifolia*), vervain (Verbena hastata), bur marigold (*Bidens cernua*), and willow herb (*Epilobium leptophyllum*). Shrub thickets of meadowsweet (*Spirea alba*), willow (*Salix discolor*), and sweet gale (*Myrica gale*) are also rapidly invading the former pond basin. The upland forest adjacent to Kiser Pond and Church Marsh is second-growth deciduous forest, mainly sugar maple (*Acer saccharum*), black cherry (*Prunus serotina*), beech (*Fagus grandifolia*), and white pine (*Pinus alba*). Aspen (*Populus tremuloides*) occur in disturbed areas, which, along with alder, are the main forage and construction material of the beavers.

South Marsh and Robinson Pond are more completely surrounded by moist forest (elm, red maple, hemlock) and contain more standing snags and fallen logs than Church Marsh and Kiser Pond. The vegetation in South Marsh and Robinson Pond differs in a manner similar to that described for Church Marsh and Kiser Pond in that the active beaver pond (Robinson Pond) has all of the same shallow-water habitats as the abandoned marsh (South Marsh), plus an extensive open-water zone that becomes choked by spatterdock during summer. Shrub thickets include highbush blueberry (*Vaccinium corymbosum*), in addition to alder, willows, and dogwoods.

INVERTEBRATE COMMUNITIES

Quantitative invertebrate samples using a drop box (0.25 m^2, after Wissinger 1988, 1989) in open-basin and shoreline habitats were accompanied by qualitative surveys using D-net sweep samples between 1994 and 1997. Peak densities typically occurred in early fall in the permanent habitats (1994 September data [n = 6] for Kiser Pond, 2288 $\pm$ 604/m^2, Robinson, 2060 $\pm$ 854/m^2) and in early summer in the autumnal basins (1994 July data [n = 6] for South Marsh, 1194 $\pm$ 488/m^2, Church Marsh, 597.6 $\pm$ 293/m^2). The numerically dominant taxa in the open-water zone were chironomid and ceratopogonid midges, dytiscid and hydrophilid beetles, clams, snails, oligochaete worms, and amphipods. Shoreline samples were similar but more diverse, with increased numbers and species of water bugs (gerrids, notonectids, corixids, belostomatids), beetles, caddisfly larvae, and odonate nymphs.

Across all habitats, the dominant taxonomic groups were dipteran flies (48 species), dragonflies and damselflies (32 species), beetles (31 species), water bugs (16 species), mollusks (11 species), caddisflies (9 species), and a variety of benthic (ostracods, amphipods, isopods, crayfish) and planktonic (cladocerans, copepods) crustaceans (Table 15.2). Several groups that were not especially diverse, such as mayflies, amphipods, clams, and oligochaetes, were nonetheless extremely abundant and at times among the dominant contributors to total biomass. The taxonomic composition of these wetlands is similar to

TABLE 15.2. Distributions of Invertebrates in Wetlands Associated with Beaver Activity at the Bousson Environmental Research Reserve[a]

	Autumnal		Permanent	
	Church	South	Kiser	Robinson
Ephemeroptera				
Caenidae				
Caenis hillaris	J96C	J96C	J96C	C
Baetidae				
Callibaetis		J96C	J96C	C
Odonata				
Aeshnidae				
Anax junius	J96C	J96C	J96	
Aeshna tuberculifera		N95C	N95	
Gomphidae				
Gomphus cornutus			N96C	C
Arigomphus lucifer		J96C		
Corduliidae				
Epitheca cynosura			C	C
Libellulidae				
Celithemis elisa			R	R
Erythemis simplicicollis		N96C	N96C	C
Leucorrhinia intacta			C	C
Leucorrhinia glacialis				C
Libellula lydia	N95C	95C	N95	R
Libellula julia	J96R	J96R		
Libellula pulchella	J96C	*C	N96R	C
Libellula luctuosa			N96C	C
Libellula incesta			R	R
Pachydiplax longipennis			C	C
Pantala flavescens	R	R		
Perithemis tenera			C	R
Tramea lacerata	J96R	J96	J96	
Sympetrumvicinum	J96C	J96C	J96C	C
Coenagrionidae				
Enallagma aspersum	J96C	R	J96	
Enallagma basidens	J96C	J96C		
Enallagma boreale	J96C	J96C	J96	
Enallagma carnuculatum			C	C
Enallagma civile	N95C	J96C	N95R	R
Enallagma signatum		R	J96C	C
Ischnura posita			C	C
Ischnura verticalis	J96C	J96C	J96C	C
Lestidae				
Lestes congener			R	C
Lestes disjunctus	J96C	J96C		
Lestes dryas			J96C	C
Lestes eurinus	J96C	J96C		
Lestes unguiculatus			J96C	C

TABLE 15.2. (Continued)

	Autumnal		Permanent	
	Church	South	Kiser	Robinson
Hemiptera				
Belostomatidae				
Belostoma	J96R	J96C	J96C	C
Corixidae				
Callicorixa			C	C
Hesperocorixa	N95C	J96C	N95C	C
Sigara	N95C	N96	J96	
Gelastocoridae				
Gelastocoris	*		*	C
Gerridae				
Gerris	J96C	J96C	J96C	C
Hydrometridae				
Hydrometra	J96C	J96C	J96C	C
Hebridae				
Hebris curtis	*	*	*	
Mesoveliidae				
Mesovelia	J96C		J96C	C
Naucoridae				
Pelocoris	*	*	*	
Nepidae				
Ranatra	C		C	C
Notonectidae				
Notonecta	N95C	J96C	J96C	C
Pleidae				
Neoplea	*C	*C		
Paraplea			C	C
Saldidae				
Micracanthia	*C		*	C
Veliidae				
Microvelia	*C	*	*C	C
Coleoptera				
Carabidae	*	*	*	
Chrysomelidae				
Donacia	*C	*C	*C	C
Curculionidae	*		*C	C
Dytiscidae				
Acilius	95C	95C	N96	
Agabetes	J96C	J96C		
Agabus	*C	*C	*C	C
Bidessonotus			R	C
Coptotomus			R	R
Hydrocanthus		96C		
Hydroporus	N95*C	95*C	N95*C	C
Hydrovatus			J96C	C

TABLE 15.2. (Continued)

	Autumnal		Permanent	
	Church	South	Kiser	Robinson
Hygrotus laccophilinus	N95*C	*C	*	R
Ilybius				C
Laccophilus fasciatus	N95*C	N95*C	N96	
Liodessus				R
Neoporus	N95*C	95*C		
Potomonectes			N96C	C
Rhantus			J96C	C
Gyrinidae				
Gyrinus	*C	J96C	J96C	C
Haliplidae				
Haliplus	N95*C	*	J96C	C
Peltodytes	N95*C	N95*C	N95*C	C
Hydraenidae				
Hydraena	*	*	*	C
Hydrochidae				
Hydrochus	*C	*	J96C	C
Hydrophilidae				
Berosus	J96C	J96C	N96	
Enochrus	*R	*C	*	C
Hydrobius	J96		R	C
Laccobius				
Tropisternus	N95*C	*	N95C	C
Noteridae				
Pronoterus	*	*	*	C
Scirtidae				
Cyphon	*		*	C
Elodes	*		*	C
Staphylinidae				
Stenus	*	*	*	C
Diptera				
Ceratopogonidae				
Bezzia	J96C	J96C	J96C	C
Ceratopogon	N95*C	N95*C	N95*C	C
Culicoides		C		
Chaoboridae				
Chaoborus	*C		J96C	C
Chironomidae				
Chironomini				
Chironomus riparius	*R	*C	J96	C
Cladopelma	J96C	J96C		
Cryptochironomus		R		
Dicrotendipes	N95*C	*C	R	C
Endochironomus			J96R	C
Microtendipes	N95*C	*C	J96	
Polypedilum	*C	R	*	C
Paratendipes			J96	C

TABLE 15.2. (Continued)

	Autumnal		Permanent	
	Church	South	Kiser	Robinson
Orthocladinae				
Cricotopus	*C	J96C	J96	
Psectrocladius sp. 1	N95*C	J96C	R	C
Psectrocladius sp. 2		R		
Rheocricotopus	*C	J96C	R	C
Tanypodinae				
Ablabesmyia sp.1	J96C	J96C	J96C	
Albabesmyia sp. 2				C
Larsia			R	R
Paramerina				C
Procladius	*C	*C	*C	C
Psectrotanypus			J96C	C
Tanytarsini				
Cladotanytarsus	*C	*C	N96C	C
Microspecta		R	J96C	C
Paratanytarsus	*C	*C		
Culicidae				
Aedes		*C	*	C
Anopheles	J96C	*C	C	C
Culex	*C	*C	*C	C
Dixidae				
Dixa	J96	*C	J96C	C
Dolichopodidae				
Argyra	*C	*C	N96C	C
Gymopternus	*	*	*	C
Empididae				
Euhybus	*		*	C
Ephydridae	*	*	*	C
Psychodidae				
Threticus	*	*	*	C
Ptychopteridae				
Bittacomorpha	*	*	*	C
Ptychoptera			J96C	C
Scyiomyzidae				
Sepedon	*		*	C
Stratiomyidae				
Odontomyia	*	*	N96C	C
Stratiomys	*C	*C	J96C	C
Syrphidae				
Chrysogaster			*	
Tabanidae				
Chrysops #1	*C	*C	*C	C
Chrysops #2		R	R	C
Tabanus			C	C

TABLE 15.2. (Continued)

	Autumnal		Permanent	
	Church	South	Kiser	Robinson
Tipulidae				
Dicranoptycha		*	*	C
Neoclatura	*C	*	*	C
Tipula # 1	*C	*C	*C	C
Tipula # 2		R		
Trichoptera				
Leptoceridae				
Oecetis			J96C	C
Limnephilidae				
Anabolia consocia			N96C	C
Nemotaulis hostilis		N96C	N96C	C
Platycentropus radiatus	N95C	N95C	N95C	C
Pycnopsyche subfasciata	C		C	C
Phryganeidae				
Arypnia vestita			N96C	R
Phryganea sayi			J96C	R
Ptilostomis ocellifera	N95C	N95C	N95C	C
Fabria inornata	C			
Lepidoptera				
Nymphulinae				
Nymphula	J96C		R	C
Paraponynx			J96R	C
Hydracarina				
Arrenuridae				
Arrenurus		N96C	N96	C
Limesiidae				
Limnesia	N96C	N96C	N96C	C
Amphipoda				
Crangonyx			R	C
Hyallela	J96C	J96C	N95C	C
Isopoda				
Caecidotea forbesii	*C	*C	*C	C
Ostracoda				
Cypridodopsis	96C	*C		C
Cladocera				
Chydoridae				
Alona			*C	C
Pleuroxus	*C	R	*C	C
Chydorus sphaericus	*C	*R	*C	C
Daphniidae				
Daphnia pulex	*C	*C	*C	C
Ceriodaphnia			J96C	C
Scapholebris		R	J96C	C
Simocephalus vetulis		J96C	*C	C

TABLE 15.2. (Continued)

	Autumnal		Permanent	
	Church	South	Kiser	Robinson
Bosminidae				
Bosmina			J96C	C
Copepoda				
Cyclopoida				
Mesocyclops edax	J96C	*C	*C	C
Acanthocyclops vernalis	*C	C	*J96C	C
Decapoda				
Cambarus	J96R		J96C	C
Gastropoda				
Lymnaeidae				
Pseudosuccinea columella	J96C	N96C	J96C	
Fossaria	C		C	C
Physidae				
Physella integra			C	
Physella heterostropha	J96C		C	C
Planorbidae				
Gyraulus circumstriatus	*C	*C	N95*C	C
Gyraulus parvus	N96		C	C
Helisoma anceps	C	R	N95*C	C
Helisoma trivolvis	J96C	J96C	C	C
Bivalvia				
Sphaeriidae				
Musculium	J96C	J96C	N95C	C
Pisidium			J96C	
Sphaerium	*N95C	*N95C		
Oligochaeta				
Lumbriculidae	*C	*C	*C	C
Naididae	*C	*C	*C	C
Hirudinea				
Erpobdellidae				
Erpodella punctata		C	R	C
Mooreobdella fervida	J96C	R	R	C
Glossophonidae				
Helobdella stagnalis	*C	J96C		
Placobdella ornata			C	C

[a]Species present prior to the drought in summer 1995 are indicated by R = Rare and C = Common. * indicates taxa that emerged from rehydrated substrate. The first appearance after the drought in field samples is indicated by N95 (November 95), J96 (June–July 1996), and N96 (November 1996). Robinson Pond did not completely dry in 1995.

that reported from other shallow permanent and autumnal marshes in north-eastern North America (e.g., Paterson and Fernando 1969, Wiggins et al. 1980, Williams et al. 1996, Bogo 1997, Brown et al. 1997), and distinctly more lentic than the invertebrate communities described from beaver pond impoundments on larger streams (e.g., McDowell and Naiman 1987, Smith et al. 1991).

Hydroperiod-specific patterns of distribution and abundance were observed in most of the species-rich groups. We will discuss only those taxa (odonates and caddisflies) for which we have species-level taxonomic resolution and are therefore confident that distributional patterns are not confounded by species-specific differences in distribution within a genus (Table 15.1). Among the odonates, the distributions of different species of *Enallagma* and *Ischnura* damselflies are probably related to differences in vulnerability to different types of predators. McPeek (1990a) has experimentally shown that many of the permanent-pond *Enallagma* species at our study site exhibit behaviors that are more effective than those of temporary-pond species in avoiding fish predators and less effective in avoiding invertebrate predators (McPeek 1990b, McPeek et al. 1996). The nymphs of large vegetation-dwelling dragonflies (e.g., *Tramea, Anax, Aeshna, Pantala*) are one such group of invertebrate predators that are probably restricted to nonpermanent habitats by fish predation. Whether hydroperiod-specific patterns of distribution and abundance in *Lestes* damselflies and *Libellula* dragonflies (Table 15.2) are related to similar trade-offs is not known. However, at least for *Lestes*, these patterns are probably not simply the result of differences in tolerance to drying, given that most *Lestes* have rapid larval development and desiccation-resistant eggs (Lutz 1968, Sawchyn and Church 1973, Sawchyn and Gillott 1974). Vulnerability to desiccation is probably a sufficient explanation for the restricted distributions of several dragonflies that occur only or mainly in permanent habitats (*Celithemis elisa, Leucorrhinia intacta, Epitheca cynosura, Epitheca princeps, Pachydiplax longipennis, Perithemis tenera, Libellula luctuosa*). Larval development in these species requires 11 months to 2 years (*Epitheca*) at this latitude, and neither eggs nor larvae are drought-tolerant (Wissinger 1988, 1989 and references therein).

Caddisflies are a second group for which we currently have sufficient taxonomic precision to identify species-specific patterns of abundances. Although rarely encountered during summer, the larvae of limnephilid and phryganeid caddisflies are among the most conspicuous large benthic invertebrates in these wetlands from late fall to early spring. Life history data for these caddisflies suggest that adults emerge in spring and early summer, but we do not yet know whether they oviposit dessication-resistant eggs then, or enter an ovarian diapause until fall (Swaney 1998). Regardless, the phenology of larval development could be timed to:

1. the flush of allochthonous and autochthonous detritus that is available in autumn (all species are shredders of and build cases from coarse particulate organic matter);

2. predator avoidance (e.g., *Ambystoma maculatum* larvae are present during summer but metamorphose in August; similarly, predatory fish become relatively inactive in late autumn); and/or

3. surviving the dry phase of autumnal hydroperiods.

Because these caddisflies are abundant only in winter, when northern wetlands are least likely to be sampled, their abundance and importance as shredders of coarse detritus might be underestimated in studies that conclude that detritus shredders are rare in wetlands (see review by Batzer and Wissinger 1996). Regardless, all of these caddisflies can apparently complete their life cycles in both permanent and autumnal basins, and explanations for hydroperiod-specific patterns of distribution are worthy of additional study (Swaney 1998).

Taxa for which distributional patterns are not as clearly related to habitat permanence, include ground beetles (Carabidae), burrowing water beetles (Noteridae), moss beetles (Hydraenidae), rove beetles (Staphylinidae), marsh beetles (Scirtidae = Helodidae), marsh bugs (Hebridae), crane flies (Tipulidae), dixid midges (Dixidae), horse and deer flies (Tabanidae), long-legged flies (Dolichopodidae), dance flies (Empididae), shore flies (Ephyridae), marsh flies (Sciomyzidae), and a variety of marsh-dwelling spiders. Many of these species were not collected in aquatic samples before we conducted the experimental rehydration study (see below), and subsequent field collecting has revealed that the immature and adult stages of these taxa are most abundant in the marsh vegetation and soils of Church Marsh and in backwater marshes of Kiser Pond. The occurrence of these taxa in marshes of a "permanent" pond suggests that establishing correlations between species distributions and hydroperiod in wetlands associated with beaver activity will be confounded by the diversity of habitats within basins. The main basin of Kiser Pond is permanent, but this wetland also includes autumnally drying marshes and shrub thickets and backwater moist-soil habitats that are only inundated during spring and after extraordinary runoff events. There is an obvious need to quantify the habitat-specific nature of invertebrate distributions in these spatially complex habitats (as in Benke et al., this volume). The immature stages of many genera within these groups live in saturated soils or in the vegetation and are typically underestimated with standard aquatic sampling techniques (see summaries and reference lists in Hilsenhoff 1991, Williams and Feltmate 1992, Turner and Trexler 1997). In addition to sampling with standard aquatic equipment, a host of alternative techniques such as collecting soil samples with modified Berlese funnels, sweeping with aerial nets, rearing from plant material, funnel and activity traps, and employing traps with lights or bait will be necessary to estimate accurately the abundance of all of the taxa in the backwater and seasonally wet habitats in wetland complexes associated with beaver impoundments (see Williams et al. 1996, Turner and Trexler 1997, Benke et al., this volume).

COLONIZATION AFTER DROUGHT

Substrate Rehydration Experiment

Methods. In summer 1995 the region experienced short-term drought conditions—less than 17 cm of precipitation fell between June 1 and October 1 as compared to the long-term average for this time period of about 45 cm (NOAA Regional Data Summary). The driest period was between August 7 and September 7, when less than 1 cm of rain fell in the region. During this period South Swamp, Church Marsh, and Kiser Pond dried completely, and the main basin in Robinson Pond was reduced to about 20 percent of its total area. Although several minor precipitation events in mid-September temporarily rewetted the dried ponds, the first significant areas of standing water did not appear before late October, when all of the basins began to refill.

Between September 21 and October 9 we removed rectangular (60 × 40 × 18-cm depth) divots of substrate from the dried basins of the three wetlands that dried completely and placed them in plastic storage boxes (61 × 41 × 22-cm depth). Three replicates were taken from each of the following three microhabitats:

1. "lowest point in the basin" and inferred last area to dry;
2. substrate under woody debris, including woody debris; and
3. open meadow habitats that were covered by emergent and invading upland plants.

None of the 27 pits filled with water, suggesting that all of the substrate samples were above the water table at the time of removal.

The 27 boxes with the divots were placed in a walk-in controlled environmental chamber with temperature, humidity, and photoperiod control. On October 9 all of the samples were rehydrated with seasoned, filtered tapwater until divot surfaces were covered by 5 cm of water. Each experimental unit was gently bubbled with a single airstone and the water maintained at the 5 cm depth. The storage box lids were modified with (1) screening to allow air circulation and light penetration; (2) the insertion of an airline and airstone to provide aeration, and c) the attachment of an emergence trap for the collection of adult insects as they emerged (Fig. 15.2). We began the experiment at 12.5°C and 12:12 L:D photoperiod cycle to mimic the fall conditions, and then reduced temperature by 2°C and one hour of daylight per month until late December, when we began to increase temperature and photoperiod by the same increments through May, when we terminated the experiment. Aquatic stages were sampled from the water column with small dip nets and adults via the emergence traps on the container lids on a biweekly basis until the end of February and monthly thereafter until May.

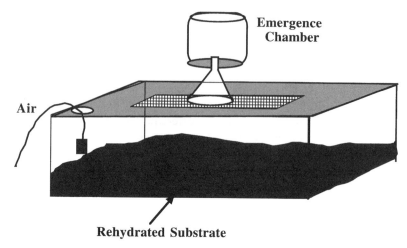

Fig. 15.2. Emergence chamber used in the substrate rehydration experiments. See text for details.

Invertebrate Responses to Rehydration

Invertebrate taxa began to appear immediately after rehydration, and taxa that emerged during the experiment are denoted with an asterisk on the species list (Table 15.2). The sequence of taxa that appeared during the experiment can be divided into the following four groups based on their order of appearance after rehydration.

1. *Terrestrial arthropods.* Immediately upon inundation, adult terrestrial arthropods emerged into the lid traps, including several groups of spiders (Erigonidae—dwarf spiders, Lycosidae—wolf spiders, Pisauridae—fishing spiders, and Salticidae—jumping spiders), homopterans (Aphididae, Eriosomatidae, and Delphacidae), orthopterans (Gryllidae, Acrididae), hymenopterans (Ichneumonidae), and parasitic dipterans (Cecidomyidae). Although the homopterans and spiders can be present on the above-water stems of emergent vegetation during inundation, the other insects are probably invaders from adjacent uplands.

2. *Flightless invertebrates and wetland insects.* During the first month after inundation several flightless groups appeared on the surface of the substrate, including adult snails, clams, amphipods, isopods, and oligochaetes. The rapid appearance of fully mature individuals suggests that adults of these taxa were present in the dried sediments and that they are species with life histories similar to Group 1 taxa in Wiggins et al. (1980). Concomitantly, a host of adult bugs and flies characteristic of shallow wetland habitats were collected from the water column or in the emergence cones. These taxa emerged from samples from all three

wetlands but were clearly most abundant and diverse in substrates taken from Church Marsh. They included shore bugs (Saldidae), velvet water bugs (Hebridae), deer flies (Tabanidae), crane flies (Tipulidae), biting midges (Ceratopogonidae), mosquitoes (Culicidae), and moth flies (Psychodidae). The shore bugs and velvet water bugs could have originated from either aestivating adults or nymphs, whereas the others species were likely to have emerged from aestivating eggs, larvae, or pupae. During the first month we also found several final instars of the dragonfly *Libellula pulchella* in the substrates from South Marsh. Many of the invertebrates that appeared during the first month after rehydration were taxa that are restricted to, or occur mainly in, wetland habitats and can therefore be considered "wetland invertebrates" (see synthesis chapter by Wissinger, this volume).

3. *Desiccation-resistant aquatic insects.* During the second to fourth months (December to February) emergence was dominated by chironomid midges, biting midges, and several other dipteran taxa (Chaoboridae, Dolichopodidae, Stratiomyidae). The chironomids were by far the most abundant and diverse group of taxa and, based on the presence of larvae and pupae in the water column, had apparently hatched from eggs. During this third phase of invertebrate emergence after inundation we also found dytiscid beetle larvae, immature leeches, and an abundance of small snails, all of which also appeared to have hatched from eggs. The eggs of these species were probably deposited before and/or after the basins dried in late summer (groups 2 and 3 of Wiggins et al. 1980).

4. *Crustacean zooplankton with photoperiod-controlled diapause.* In early spring cladocerans and copepods appeared in chambers from both the permanent and autumnal wetlands. The ability of these taxa to resist desiccation as diapausing eggs is well documented, and their early spring emergence in permanent habitats is thought to be a predator-avoidance strategy cued by photoperiod (see Hairston 1987, Hairston et al. 1990 and references therein).

We used two-way ANOVA (wetland × microhabitat) to compare the total number of individuals and number of species that emerged from the rehydrated substrates. On average, across all microhabitats a significantly greater number of invertebrates emerged from substrates taken from Church Marsh than from Kiser Pond and South Marsh (wetland main effect $F_{2,18} = 49.5$, $p < 0.001$) (Fig. 15.3). Interpreting the significant main effect for microhabitat ($F_{2,18} = 192$, $p << 0.001$) is complicated by a wetland × microhabitat interaction ($F_{4,18} = 15.2$, $p < 0.001$). In Church Marsh, the total number of invertebrates emerging from the last areas to dry was significantly greater than in the wet meadow vegetation, which was significantly greater than under woody debris (Scheffe's *a posteriori* test $p < 0.05$; Fig. 15.3). In the other

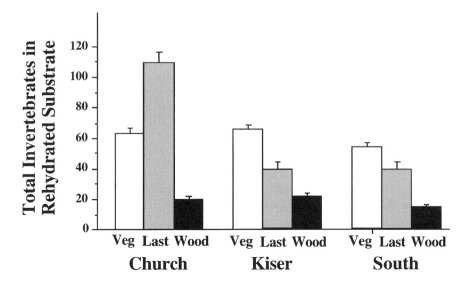

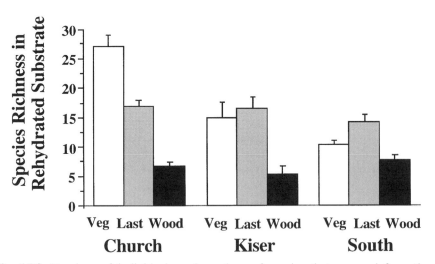

Fig. 15.3. Numbers of individuals and numbers of species that emerged from the rehydration chambers with substrates from vegetated meadow (Veg), last pool to dry (Last), and covered by woody debris (Wood). Bars are means (n = 3) + 1 S.E. See text for ANOVA statistics.

two wetlands the greatest number of invertebrates emerged from substrates taken from wet meadow, followed by last areas to dry and woody debris (Fig. 15.3). The low numbers of organisms that emerged from sediments taken from under woody debris in all three basins was surprising in light of several previous studies indicating that such debris is important for the survival of a variety of taxa during the dry phase of temporary wetlands (Wiggins et al. 1980, Maher 1984, Lake et al. 1989, Jeffries 1994).

Patterns of species richness from the rehydrated samples roughly paralleled those for abundance and again exhibited wetland-specific patterns, as indicated by a significant two-way interaction between wetland and microhabitat ($F_{4,18}$ = 6.13, p = 0.002). Overall, a significantly greater number of taxa were recovered from substrates from Church Marsh than from the other two wetlands ($F_{2,18}$ = 13.2, p < 0.001, Scheffe's *a posteriori* test p < 0.05). Church Marsh also exhibited the greatest differences among microhabitats, with the numbers of species recovered from vegetated meadow samples greater than in the other two microhabitats (overall microhabitat main effect $F_{2,18}$ = 53.9, p < 0.001, Scheffe's *a posteriori* contrast for Church Marsh p < 0.05). To a large extent, the high diversity of organisms that emerged from the vegetated meadow substrates in Church Marsh can be attributed to the presence of the numerous bugs, beetles, and dipteran flies that emerged soon after inundation. One unanticipated outcome of this experiment was that it alerted us to the presence of numerous species that were not encountered with the standard aquatic sampling techniques for benthic and planktonic invertebrates (see discussion above).

Relative Importance of Different Modes of Wetland Colonization

To complement the data from the rehydration experiment, we collected six replicate drop box samples and conducted qualitative surveys in each wetland in November 1995, July 1996, and November 1996 in order to monitor post-drought changes in community composition and invertebrate abundance. By combining the data from the rehydration experiment with field observations, we were able to estimate the relative importance of desiccation resistance versus aerial colonization for each of the three wetlands that dried in 1995 (Table 15.3).

Desiccation Resistance. Included in the calculation for the percentage of species that recolonized from one or more tolerant stages in the dried bases were:

1. All species that hatched in the rehydration emergence chambers, excluding adults of terrestrial taxa that crawled into the emergence chambers within a few days after inundation (spiders, aphids, ants, etc.). This included species that apparently aestivated as adults, immatures, and/or eggs; see results of rehydration experiment.

TABLE 15.3. Postdrought Recolonization of Two Autumnal Wetlands (Church and South) and One Permanent Wetland (Kiser) Associated with Beaver Impoundments That Dried in August–October 1995[a]

	Predrought Taxa	Desiccation Tolerance	Aerial Recolonization		Total Returns One Year after Drought	New Taxa
			Fall 1995–Spring 1996	Summer–Fall		
Church Marsh	109	71%	22%	3%	91%	3
South Marsh	110	63%	25%	2%	90%	2
Kiser Pond	135	38%	27%	12%	77%	16

[a]See text for explanation.

2. Flightless species that appeared in the June 1996 field samples (cladocerans, copepods, crayfish, snails, clams, leeches).

3. Species that appeared in November 1995 or June 1996 field samples that probably hatched from eggs deposited before or during the drought (e.g., the caddisflies *Platycentropus radiatus* and *Ptilostomis ocellifera*) (Swaney 1998). We also included several species of dragonflies in this group for two reasons. First, for *Aeshna tuberculifera, Libellula lydia, Libellula julia,* and *Enallagma civile,* all appeared as early-instar larvae in early November. The flight seasons for these species ended at least a month before the basins refilled, and thus we inferred they hatched from desiccation-resistant eggs. Second, early-instars of *Lestes* spp. and *Sympetrum vicinum* that first appeared in spring 1996 samples were also probably from desiccation-resistant eggs that were deposited during the previous summer and fall, respectively (see discussions and references in Wiggins et al. 1980, Wissinger 1988, 1989).

Taken together, these data suggest that about 71 percent and 63 percent of all taxa that were present before the drought had recolonized Church and South Marshes from one or more drought-tolerant stages, as compared to only 38 percent for Kiser Pond. This is consistent with the general notion that a greater fraction of species in temporary than in permanent habitats should possess one or mechanisms for surviving predictable or unpredictable drying (Wiggins et al. 1980, Williams 1987, Jeffries 1994). A caveat is that we do not have sufficient replication in this study to conclude that degree of permanence is the variable that is responsible for differences in recolonization among these wetlands. Clearly, a manipulative study with several permanent and several nonpermanent habitats would provide a more rigorous test of this notion for wetlands in general than our incompletely replicated "natural experiment" (see Wilbur 1997).

Although desiccation resistance was less important in Kiser Pond than in the two autumnal habitats, we were nonetheless surprised that a third of all taxa in this permanent habitat recolonized from drought-resistant stages. In part this result is related to presence of species that diapause as eggs in the sediments, regardless of habitat hydroperiod. For example, cladoceran and copepod zooplankton rely on diapausing egg stages to seasonally avoid fish predators (Hairston 1987, Hairston et al. 1990). A second explanation is that many of the taxa from Kiser Pond that appeared in the rehydration chamber were the same water bugs, beetles, and flies that were also abundant in the autumnal pond substrate samples. We have since learned that these taxa live mainly in the shallow marshes along the edges and backwaters of Kiser Pond, and we speculate that one or more stages of these species migrated with the waterline as it receded into the main basin.

Aerial Colonization. Aerial dispersal resulted in a second wave of recolonization, which accounted for 25, 27, and 39 percent of the first-year recovery of the predrought fauna in Church Marsh, South Marsh, and Kiser Pond, respectively. We can distinguish three groups of aerial colonizers. The first group of species included water bugs (Belostomatidae, Gerridae, Notonectidae, Veliidae) and beetles (Dytiscidae, Hydrophilidae, Gyrinidae) that colonized in fall and early spring as adults. Many species within these groups are "cyclic colonizers" that overwinter in permanent habitats and migrate to temporary habitats in spring, where they may complete one or more generations, before returning to permanent habitats when temporary basins dry (Group 4 of Wiggins et al. 1980; also see reviews by Batzer and Wissinger 1996, Wissinger 1997). The appearance of winged adults of these taxa in Kiser Pond in fall 1995, soon after reinundation, suggests that these species also move among permanent basins during fall, when they are searching for overwintering habitats (see Wissinger 1997).

A second group of early colonizers were opportunistic species that had previously only been encountered in seasonal habitats. Chironomids are often among the earliest colonizers of newly inundated habitats (see review by Batzer and Wissinger 1996), and *Chironomus riparius,* a species abundant only in the autumnal wetlands before the drought, was among several chironomids to colonize Kiser Pond in spring 1996. By July *C. riparius* larvae were extremely abundant (1000–1500 m²) in decaying grasses and other meadow vegetation that had been reinundated along the edges of Kiser Pond. Because of the population explosion of this and other chironomid species, total densities in June 1996 were considerably higher than those before the drought in June 1995 (see above) for all three wetlands (mean densities ± 1 S.E., n = 6, Kiser Pond = 3200 ± 840/m², South Marsh = 2570 ± 740/ m², Church Marsh = 1714 ± 260/m²). The higher invertebrate densities after as compared to before drying were probably due to both increased detrital resources and decreased predation by fishes (Batzer 1998 and references therein).

In late spring 1996 Kiser Pond was also colonized by two species of dragonflies (*Anax junius* and *Tramea lacerata*) that were not present before the drought. At this latitude, *Anax* adults can be either local or long-distance migrants, whereas *Tramea* were undoubtedly long distance migrants from subtropical latitudes (see Wissinger 1989 and references therein). These 2 species and *Pantala* spp. are opportunistic colonizers that are typically only abundant in shallow autumnal or winterkill habitats, and it will be interesting to document their continued presence in Kiser Pond when fish recolonize.

A third group of aerial colonizers consisted of the adults of mainly permanent-pond species that did not recolonize Kiser Pond until summer and fall 1996, including several caddisflies, dytiscid beetles, and odonates (Table 15.1). These species had all been present before the drought and constituted the last wave of immigrants to Kiser Pond. During this same time period several species not encountered before the drought were found in Kiser Pond, including several odonates (*Aeshna tuberculifera, Enallagma civile, Enallagma boreale*), beetles (*Acilius, Laccophilus, Berosus*), a water boatman (*Sigara*), several chironomid flies (*Microtendipes, Paratendipes*), and a water mite (*Arrenurus*).

In summary, 90, 91, and 77 percent of the predrought taxa had recolonized the two autumnal and permanent wetlands by fall 1996, respectively (Table 15.3). Drought-resistant stages were most important in Church Marsh and South Marsh, whereas rapid aerial dispersal by cyclic colonizers and delayed dispersal by permanent pond species were as important as desiccation resistance for the recolonization of Kiser Pond. Most species that had not returned one year after drought were either rare before the drought or were permanent pond odonates or several groups of taxa that rely on passive dispersal (Table 15.2). Overall, a total of 16 new taxa appeared in Kiser Pond during the first year after drought, as compared to only two and three in the autumnal wetlands.

The results of this study are consistent with several previous studies that suggest colonization of wetlands is rapid during the first six months after drought, and is the result of emergence from desiccation-resistant stages combined with aerial colonization, mainly by cyclic colonizers (see review by Batzer and Wissinger 1996). The timing of drought is undoubtedly an important determinant of how quickly and completely the invertebrate fauna of dried wetlands recover. In this study drought occurred at the same time of the year that autumnal wetlands in this area typically dry. Thus, life history and other adaptations should have and were effective at providing colonists when these basins refilled in autumn (also see Bataille and Baldassarre 1993).

Invertebrate Communities in Wetland Complexes Associated with Beaver Activity: Metapopulations and Metacommunities?

Wetlands associated with beaver activities (either active or abandoned impoundments) are extremely heterogeneous environments both within and among the sites. Within basin habitat, heterogeneity will be greatest in relatively flat terrain, where slight changes in the height of beaver dams can result in the inundation of large areas of seasonally back-flooded habitats. A single impoundment can create permanent ponds, semipermanent marshes and/or swamps, and vernal wet meadows, shrub thickets, and/or wet woodlands. Thus, these wetland complexes are an amalgam of several of the different wetland types and hydroperiods used in conventional classification systems (Cowardin et al. 1979, Tiner 1995). This spatial diversity of habitats and hydroperiods should support greater diversities of plants and invertebrates than would be found in structurally simpler wetland basins, and this diversity might in part explain why they are particularly attractive to game and non-game wildlife (Naiman et al. 1988, Johnston and Naiman 1990, Grover and Baldassare 1995, Brown et al. 1996, McCall et al. 1996, Benke et al., this volume). Future studies that document the composition and diversity of invertebrate communities in the natural wetlands created by beaver impoundments are of obvious importance for restoration strategies and the design of constructed wetland habitats.

Habitat heterogeneity among impoundment sites should also contribute to the diversity and stability of invertebrate communities associated with beaverdam wetlands. Beaver activities within drainages often result in a mosaic of active and abandoned impoundments with different wetland types and hydroperiods, depending on their geomorphologic setting and successional stage after abandonment (Naiman et al. 1988). Our results suggest that there is considerable dispersal among these habitats and that recolonization after drought may be enhanced by an abundance of local source populations of migrants (also see Jeffries 1994). Some of these immigrants are opportunistic species that rely on high rates of dispersal and rapid proliferation in disturbed habitats, whereas others move cyclically on a seasonal basis between permanent and temporary basins.

Metapopulation theory predicts that such populations should be more stable in terms of their resistance to and resilience after disturbance than those that are isolated from other populations (see reviews by Gilpin and Hanski 1991, Hanski and Gilpin 1997). We hypothesize that the overall resiliency of a particular basin embedded in a beaver-pond wetland complex should be greater than the resiliency of isolated habitats because the former should have higher rates of dispersal and colonization. The loss of permanent habitats from such a complex should reduce the pool of cyclic colonizers that seasonally invade temporary habitats, thus reducing diversity in those temporary habitats. Conversely, the loss of temporary habitats should reduce diversity in permanent habitats if such habitats serve as seasonal refugia for species that would otherwise be driven extinct by predators (e.g., Soderstrom and Nilsson 1987). The loss of temporary habitats should also reduce the rate and sequence of recolonization after drought or other disturbances because they serve as a local source of disturbance-adapted, opportunistic species (as in this study; also see examples in Batzer and Wissinger 1996). For cyclic colonizers permanent and temporary habitats should act alternatively in different seasons as sources and sinks (see Pulliam 1988, Pulliam and Danielson 1988), depending on whether a species is more vulnerable to drought or to vertebrate predation. The presence of both habitats may be critical for the long-term viability of populations, as suggested for metapopulations of water beetles (Nilsson 1984).

Understanding the degree to which the invertebrate community within a particular wetland is dependent on dispersal connections with nearby wetlands will undoubtedly require a metapopulation and even metacommunity approach. Current metapopulation and source-sink models that predict the effects of dispersal on population and community stability (see reviews by Pulliam 1988, Pulliam and Danielson 1988, Hanski and Gilpin 1997) assume unidirectional rather than cyclic migrations among habitats. Future models that test the importance of migration in wetland complexes should account for the cyclic as well as linear movements of invertebrates between the different types of habitats (Wissinger 1997). Federal and state regulations and policies that protect jurisdictional wetlands should take into account not only the values of individual basins, but also the metacommunity context and degree to which other wetlands would be affected by the loss of a basin to which they are connected by both active and passive dispersal.

ACKNOWLEDGMENTS

Darold Batzer, Stephanie Swaney, Leslie Hunter, Jessica Kundman, Jason Mickey, Allison Roy, and Carrie Lyle provided constructive comments that improved an earlier version of the manuscript. Thanks to Scott Ingmire for lending his expertise with wetland plants, to Stephanie Swaney and Susan Bolden for unpublished data and taxonomic verification for caddisflies and mollusks, respectively, Jason Paulovich for water chemistry data, and especially Wendy Brown for reviewing our taxonomy. SAW gratefully acknowledges support from the Acid Rain Research Fund of the Pennsylvania Electric Company and from the National Science Foundation (DEB-9407856).

LITERATURE CITED

Barrie, L. A., and J. M. Hales. 1984. The spatial distributions of precipitation acidity and major ion wet deposition in North America during 1980. Tellus 36B333–335.

Bataille, K. J., and G. A. Baldassarre. 1993. Distribution and abundance of aquatic macro-invertebrates following drought in three prairie potholes. Wetlands 13:260–269.

Batzer, D. P. 1998. Trophic interactions among detritus, benthic midges, and predatory fish in a freshwater marsh. Ecology 79:1688–1698.

Batzer, D. P., and S. A. Wissinger. 1996. Ecology of insect communities in nontidal wetlands. Annual Review of Entomology 41:75–100.

Beard, E. 1953. The importance of beaver in waterfowl management at Seney National Wildlife Refuge. Journal of Wildlife Management 17:398–436.

Bogo, J. L. 1997. Invertebrate communities in natural versus constructed wetlands in northwestern Pennsylvania. Bachelor's Thesis, Allegheny College, Meadville, PA.

Brown, D. J., W. A. Hubert, and S. H. Anderson. 1996. Beaver ponds create wetland habitat for birds in mountains of southeastern Wyoming. Wetlands 16:127–133.

Brown, S. C., K. Smith, and D. P. Batzer. 1997. Macroinvertebrate responses to wetland restoration in northern New York. Environmental Entomology 26:1016–1024.

Cirmo, C. P., and C. T. Driscoll. 1993. Beaver pond biogeochemistry: Acid neutralizing capacity generation in a headwater wetland. Wetlands 13:277–292.

Cowardin, L. M., V. Carter, G. C. Gloet, and E. T. LaRoe. 1979. Classification of wetlands and deepwater habitats of the United States: U.S. Fish and Wildlife Service Report FWS/OBS-79-31.

Dieffenbach, D. R., and R. B. Owen. 1989. A model of habitat use by breeding American black ducks. Journal of Wildlife Management 53:383–389.

Ermer, E. M. 1984. Analysis of benefits and management costs associated with beaver in western New York. New York Fish and Game Journal 31:119–132.

France, R. L. 1997. The importance of beaver lodges in structuring littoral communities in boreal headwater lakes. Canadian Journal of Zoology 75:1009–1013.

Ford, T. E., and R. J. Naiman. 1988. Alteration of carbon cycling by beaver: Methane and evasion rates from boreal forest streams and rivers. Canadian Journal of Zoology 66:529–533.

Gilpin, M., and I. Hanski (eds.). 1991. Metapopulation dynamics: empirical and theoretical investigations. Academic Press, New York.

Grover, A. M., and G. A. Baldassarre. 1995. Bird species richness within beaver ponds in south-central New York. Wetlands 15:108–118.

Hairston, N. G., Jr. 1987. Diapause as a predator avoidance adaptation. *In* W. C. Kerfoot and A. Sih (eds.), Predation: Direct and Indirect Impacts on Aquatic Communities. University Press of New England, Hanover, NH.

Hairston, N. G., Jr., T. A. Dillon, and B. T. DeStasio. 1990. A field test for the cues of diapause in a freshwater copepod. Ecology 71:2218–2223.

Hanski, I. A., and M. E. Gilpin. 1997. Metapopulation biology: Ecology, genetics, and evolution. Academic Press, New York.

Hilsenhoff, W. L. 1991. Diversity and classification of insects and Collembola. Pages 593–663 *in* J.H. Thorp and A. P. Covich (eds.), Ecology and Classification of North American Freshwater Invertebrates. Academic Press, San Diego, CA.

Hodkinson, I. D. 1975. A community analysis of the benthic insect fauna of an abandoned beaver pond. Journal of Animal Ecology 44:533–552.

Jeffries, M. 1994. Invertebrate communities and turnover in wetland ponds affected by drought. Freshwater Biology 32:603–612.

Johnston, C. A., and R. J. Naiman. 1990. Aquatic patch creation in relation to beaver population trends. Ecology 71:1617–1621.

Lake, P. S., I. A. E. Bayly, and D. W. Morton. 1989. The phenology of a temporary pond in western Victoria, Australia, with special reference to invertebrate succession. Archiv für Hydrobiologie 115:171–202.

Lutz, P. E. 1968. Life history studies on *Lestes eurinus* (Odonata). Ecology 49:576–570.

Maher, M. 1984. Benthic studies of waterfowl breeding habitat in southwestern New South Wales. I. The fauna. Australian Journal of Marine and Freshwater Research 35:85–96.

Mallory, M. L., P. J. Blancer, P. J. Weatherhead, and D. K. McNicol. 1994. Presence or absence of fish as a cue to macroinvertebrate abundance in boreal wetlands. Hydrobiologia 279:345–351.

McCall, T. C., T. P. Hodgman, D. R. Diefenbach, and R. B. Owen. 1996. Beaver populations and their relation to wetland habitat and breeding waterfowl in Maine. Wetlands 16:163–172.

McDowell, D. M., and R. J. Naiman. 1987. Structure and function of a benthic invertebrate stream community as influenced by beaver (*Castor canadensis*). Oecologia 68:481–489.

McPeek, M. A. 1990a. Determination of species composition in the *Enallagma* damselfly assemblages of permanent lakes. Ecology 71:83–98.

———. 1990b. Behavioral differences between *Enallagma* species (Odonata) influencing differential vulnerability to predators. Ecology 71:1714–1726.

———. 1996. Tradeoffs, food web structure, and the coexistence of habitat specialists and generalists. American Naturalist 148:S124–S138.

Naiman, R. J., C. A. Johnston, and J. C. Kelley. 1988. Alteration of North American streams by beaver. Bioscience 38:753–762.

Naiman, R. J., D. M. McDowell, and B. S. Farr. 1984. The influence of beaver (*Castor canadensis*) on the production dynamics of aquatic insects. Verhandlungen Internationale Vereinigung Limnologie 22:1801–1210.

Naiman, R. J., and J. M. Mellilo 1984. Nitrogen budget of a subarctic stream altered by beaver (*Castor canadensis*). Oecologia 62:150–155.

Naiman, R. J., J. M. Mellilo, and J. E. Hobbie. 1986. Ecosystem alteration of boreal forest streams by beaver (*Castor canadensis*). Ecology 67:1254–1269.

Nilsson, A. N. 1984. Species richness and succession of aquatic beetles in some kettle-hole ponds in northern Sweden. Holarctic Ecology 7:149–156.

Paterson, C. G., and C. H. Fernando. 1969. Macroinvertebrate colonization of the marginal zone of a small impoundment in eastern Canada. Canadian Journal of Zoology 47:1229–1238.

Pulliam, H. R. 1988. Sources, sinks, and population regulation. American Naturalist 132:652–661.

Pulliam, H. R., and B. J. Danielson. 1988. Sources, sinks, and habitat selection: A landscape perspective on population dynamics. American Naturalist 137:S50–S66.

Reinecke, J. J., and R. B. Owen. 1980. Food use and nutrition of black ducks in Maine. Journal of Wildlife Management 44:549–558.

Rottella, J. J., and J. T. Ratti. 1992a. Mallard brood survival and wetland habitat conditions in southwestern Manitoba. Journal of Wildlife Management 56:499–507.

———. 1992b. Mallard brood movements and wetland selection in southwestern Manitoba. Journal of Wildlife Management 56:508–515.

Sawchyn, W. W., and N. S. Church. 1973. The effects of temperature and photoperiod on diapause development in the eggs of four species of *Lestes* (Odonata: Zygoptera). Canadian Journal of Zoology 51:1257–1266.

Sawchyn, W. W., and C. Gillott. 1974. The life histories of three species of *Lestes* (Odonata: Zygoptera) in Saskatchewan. Canadian Entomologist 106:1283–1293.

Schlosser, I. J. 1995. Dispersal, boundary processes, and trophic-level interactions in streams adjacent to beaver ponds. Ecology 76:908–925.

Schneider, D. W., and T. M. Frost. 1996. Habitat duration and community structure in temporary ponds. Journal of the North American Benthological Society 15:64–86.

Shannon, J. D., and D. L. Sisterson. 1992. Estimation of S and NO_x deposition budgets for the United States and Canada. Water, Air, and Soil Pollution 63:211–215.

Skelly, D. M. 1997. Tadpole communities. American Scientist 85:36–45.

Smith, M. E., C. T. Driscoll, B. J. Wyskowski, C. M. Brooks, and C. C. Costentini. 1991. Modification of stream ecosystem structure and function by beaver in the Adirondack Mountains, New York. Canadian Journal of Zoology 69:55–61.

Soderstrom, O., and A. N. Nilsson. 1987. Do nymphs of *Parameletus chelifer* and *P. minor* (Ephemeroptera) reduce mortality from predation by occupying temporary habitats? Oecologia 74:39–46.

Sprules, W. M. 1940. The effect of a beaver dam on the insect fauna of a trout stream. Transactions of American Fisheries Society 70:2236–2248.

Swaney, S. L. 1998. Biotic and abiotic factors that affect the distributions of lentic caddisflies (Trichoptera: Limnephilidae, Phyrganeidae). Bachelor's Thesis, Allegheny College, Meadville, PA.

Tiner, R. W. 1995. Wetland definitions and classifications in the United States. U.S. Geological Survey Water-Supply Paper 2425, Washington, DC.

Turner, A. M., and J. C. Trexler. 1997. Sampling aquatic invertebrates from marshes: Evaluating the options. Journal of the North American Benthological Society 16: 694–709.

Wellborn, G. A. 1994. Size-biased predation and the evolution of prey life histories: A comparative study of freshwater amphipod populations. Ecology 75:2104–21017.

Wellborn, G. A., D. K. Skelly, and E. E. Werner. 1996. Mechanisms creating community structure across a freshwater habitat gradient. Annual Review of Ecology and Systematics 27:337–363.

Wiggins, G. B., R. J. Mackay, and I. M. Smith. 1980. Evolutionary and ecological strategies of animals in annual temporary pools. Archiv für Hydrobiologie Supplement 58:97–206.

Wilbur, H. M. 1997. Experimental ecology of food webs: Complex systems in temporary ponds. Ecology 78:2279–2302.

Williams, D. D. 1987. The ecology of temporary waters. Timber Press, Portland, OR.

Williams, D. D., and B. W. Feltmate. 1992. Aquatic insects. CAB International, Oxford, UK.

Williams, R. N., M. S. Ellis, and D. S. Fickle 1996. Insects in the Killbuck Marsh Wildlife Area, Ohio. 1994 Survey. Ohio Journal of Science 96:34–40.

Wissinger, S. A. 1988. Spatial distribution, life history, and estimates of larval survivorship in a 14-species assemblage of larval dragonflies (Odonata: Anisoptera). Freshwater Biology 20:329–340.

———. 1989. Life history and size variation in larval dragonfly populations. Journal of the North American Benthological Society 7:13–28.

———. 1997. Cyclic colonization and predictable disturbance: a template for biological control in ephemeral crop systems. Biological Control 10:1–15.

Wissinger, S. A., and J. McGrady. 1993. Intraguild predation and competition between larval dragonflies: Direct and indirect effects on shared prey. Ecology 74:207–218.

Wissinger, S. A., G. B. Sparks, G. L. Rouse, W. S. Brown, and H. Steltzer. 1996. Intraguild predation and cannibalism among larvae of detritivorous caddisflies in subalpine wetlands. Ecology 77:2421–2430.

Wissinger, S. A., H. H. Whiteman, G. B. Sparks, G. L. Rouse, and W. S. Brown. 1999. Tradeoffs between competitive superiority and vulnerability to predation in caddisflies along a permanence gradient in subalpine wetlands. Ecology (in press).

Yaworski, M., D. Recotor, J. Eckenrode, E. Jensen, and R. Grubb, 1979. Soil Survey of Crawford County, Pennsylvania. 1979. USDA, National Resources Conservation Service.

16 Riparian Sedge Meadows in Maine

A Macroinvertebrate Community Structured by River-Floodplain Interaction

ALEXANDER D. HURYN and K. ELIZABETH GIBBS

S *mall rivers with extensive tracts of riparian sedge meadows as floodplains are a conspicuous part of the Maine landscape. During spring increasing river discharge because of melting snow inundates these floodplains. Although the period of flooding is short (April–May), a rich macroinvertebrate community develops. Members of the community come from two general sources: floodplain and river. Floodplain fauna complete their entire life cycle on the floodplain, whereas river-floodplain fauna have life cycles with both river and floodplain phases. Conspicuous floodplain fauna includes mosquitoes (Aedes), caddisflies (Anabolia and Limnephilus), and fingernail clams (Pisidium). River-floodplain fauna includes the mayflies Leptophlebia, Siphlonisca, and Siphlonurus. These mayflies are able to use the stream as a refuge during the dry period of the summer and the freezing temperatures of winter, and the floodplain during a short but critical period of rapid growth and development during spring. Approximately 75 percent of macroinvertebrate biomass on the floodplain during inundation is composed of the mayflies Leptophlebia, Siphlonisca, and Siphlonurus. Compared to this assemblage, contributions by other fauna to community biomass are minor. Both proximity and physical interconnection of the river and floodplain are required for development of the characteristic floodplain community structure during the short period of inundation.*

Invertebrates in Freshwater Wetlands of North America: Ecology and Management, Edited by
Darold P. Batzer, Russell B. Rader, and Scott A. Wissinger
ISBN 0-471-29258-3 © 1999 John Wiley & Sons, Inc.

FLOODPLAIN SEDGE MEADOWS

Once generally ignored by aquatic ecologists, the importance of floodplain wetlands to the ecology of rivers now seems almost self-evident. Much of what has been discovered about links between these systems, however, is based upon studies conducted in the tropics (Junk et al. 1989, Hamilton et al. 1992, Power et al. 1995, Goulding et al. 1996), where the physical structure of large rivers and their floodplains tends to be less altered by human activities compared to systems of similar scale in temperate regions (Ward 1989, Bayley 1995, Sparks 1995). These studies have provided a conceptual base for examining relationships between rivers and their floodplains in the form of energy and material flow, habitat formation and maintenance, and system biodiversity (primarily fishes, e.g., Goulding et al. 1996). It is now clear that this "lateral dimension" to river ecosystems must be seriously considered in designing studies of river ecosystem structure and function (Junk et al. 1989, Ward 1989).

Probably because of logistical difficulties in working in large rivers and the paucity of intact river-floodplain complexes outside of the tropics, quantitative data describing the ecology of these systems are generally lacking (Ward 1989, Power et al. 1995). There are few good examples of relatively unaltered large river-floodplain systems remaining for study in temperate North America (Ward 1989, Sparks 1995). There are, however, many examples of small systems that remain relatively unaltered (e.g., floodplains that receive floodwater from rivers with discharges $<<50$ m^{-3} s^{-1}). Small river-floodplain complexes should be of particular interest to stream ecologists because they may provide small-scale analogues of larger systems that are both more abundant and more tractable for study. Almost all small river-floodplain complexes that have received comprehensive study, however, are forested systems of the southeastern United States (see review by Gladden and Smock 1990). In this chapter we introduce the macroinvertebrate community of a floodplain meadow in the extreme northeastern United States and attempt to show how the strong physical link between river and floodplain results in its characteristic structure.

Alluvial sedge meadows that border small low-gradient and meandering streams are a conspicuous part of the landscape of eastern and northern Maine ($\sim$24,000 ha in Maine, Gibbs et al. 1991). The invertebrates of these sedge meadows have received continuous attention since 1980, when the unusual carnivorous mayfly *Siphlonisca aerodromia* (Siphlonuridae) was discovered at Tomah Stream in eastern Maine (Gibbs 1980). Interest in the conservation biology of *S. aerodromia* has resulted in numerous studies of the invertebrate communities of sedge meadows throughout Maine (e.g., Gibbs and Mingo 1986, Gibbs and Siebenmann 1996). This species is now known from a number of locations in Maine, but appears to be most abundant in the river-floodplain complex at Tomah Stream.

THE RIVER-FLOODPLAIN COMPLEX AT TOMAH STREAM

Tomah Stream is a 4th-order tributary of the St. Croix River in Washington County, Maine. Along a 10-km reach of the stream south of Route 6 are extensive alluvial sedge meadows that function as floodplains. These floodplains gently slope toward the active river channel, or toward oxbow swales that drain into the active channel. The width of the floodplain varies depending on the geomorphology of the channel. Along freely meandering reaches, the floodplain may be >1 km wide, although widths of ~100 m are more common. Along reaches that are constrained by topography, the floodplain is often <30 m wide, or may be entirely absent.

Sediments of the Tomah Stream catchment are glacial till, aqueo-glacial outwash, and materials of marine origin (International Joint Commission Report 1957). The extensive floodplains of lower Tomah Stream are formed from aqueo-glacial outwash, e.g., sediments deposited by streams associated with melting glaciers at the end of the Pleistocene. These sediments were graded as they were transported away from the front of the retreating ice to form the alluvial deposits that dominate the present-day landscape of the lower Tomah Stream. These basement sediments are now deeply buried by peat (International Joint Commission Report 1957).

The study site used for much of the information provided in this chapter is 1.6 km south of Route 6 (45°26′42″N, and 67°34′50″W), just north of the confluence of Tomah Stream and Beaver Creek (= "Tomah Stream study site"). Here the stream has a straight-walled channel ~6–8 m wide. Depths during low flow condition during midsummer range from ~0.3 to 2+ m. Substrata are sand-and-gravel. The floodplain at this site is on the west bank of the stream and has an area of 2.6 ha.

Hydrology and Thermal Regime

During much of the year (June–March) Tomah Stream is confined to its channel. During this period discharge may be as low as 0.3 m^{-3} s (measured on July 21, 1997). Between December and March the stream is usually covered by ice and snow; ice cover over the upper floodplain may be continuous with the soil. During late March or early April an increase in discharge, caused by the melting snow pack, causes the river to inundate its floodplain. The inundated area gradually decreases from April through May (although it may fluctuate dramatically depending on weather conditions). The stream usually returns to its channel by early June, and the floodplain dries. In some years, however, heavy rains may cause the floodplain to be briefly inundated at other times of the year. Maximum water depth on the floodplain during peak inundation is <1.5 m, and usually well below 1.0 m.

The long-term pattern of flooding at the Tomah Stream study site was predicted from a regression equation relating extent of inundation to discharge of the St. Croix River. Based on a 34-year record of discharge obtained from the USGS, the probability of inundation of the upper floodplain (e.g., within 20 horizontal m of the high-water mark observed in 1997) on any given day of the year was estimated (Fig. 16.1). The probability of inundation was 70–80 percent from March to May, but was usually below 20 percent during the remainder of the year, and well below 10 percent during summer. Fauna that require the continuous presence of surface water will generally be able to occupy the floodplain for a maximum period of ~2.5 months (late March–early June).

Floodplain water temperature was measured at hourly intervals at the Tomah Stream study site during 1997. The mean daily temperature from April 28 to June 9 was 10.0 ± 0.6°C (±SE). The minimum mean daily temperature was 3.0°C, measured on April 29; the maximum was 19.6°C, measured on June 9. The increase in mean daily temperature over the period of inundation was approximately linear. Daily fluctuation of temperature was low, ranging to only 3.4°C. Stream temperatures measured from June 4–9 were usually equal to or slightly warmer (<1°C) than floodplain temperatures.

Water Chemistry

Floodplain and stream water at the Tomah Stream study site are similar in chemistry, both being acidic, moderately to poorly buffered, and oligotrophic.

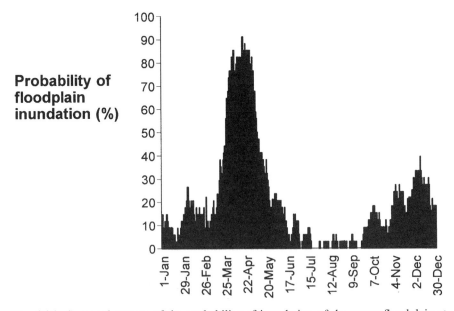

Fig. 16.1. Seasonal pattern of the probability of inundation of the upper floodplain at Tomah Stream.

From May 5 to June 10, 1997, pH was 6.3 ± 0.2 (mean + SE, four dates) on the floodplain and 6.2 ± 0.1 in the stream. Conductivity was 33 ± 8 μmho/cm on the floodplain and 25 ± 3 in the stream. Total alkalinity was 6.3 ± 0.2 mg $CaCO_3$/L on the floodplain and 6.5 ± 1.5 in the stream. Concentrations of soluble reactive phosphorous and of nitrate and ammonium in both floodplain and stream water were usually below 0.05 mg/L. Total phosphorus was slightly higher in floodplain (0.040 ± 0.013 mg/L) compared to stream water (0.023 ± 0.003).

Vegetation

At least 68 species of plants occur on the Tomah Stream floodplain: 34 forbs, 12 sedges and rushes, 5 grasses, 10 woody plants, and 7 ferns and allies. Sedges and rushes almost always comprise >50 percent of overstory vegetation, with woody plants comprising <10 percent. The remaining plants are primarily forbs, ferns, and grasses (Gibbs 1991). The understory vegetation consists of *Sphagnum* moss and detritus.

At the Tomah Stream study site tussock sedge (*Carex stricta*) is the most important (e.g., percent cover) plant species. The sedge *Carex vesicaria* is also abundant. Other major plant species are wool grass (*Scirpus cyperinus*), soft rush (*Juncus effusus*), and horsetails (*Equisetum*) (Gibbs 1989). The narrow leaves and stems of the tussock sedge form dense, broad tussocks that persist as dead leaves among new annual growth. During inundation the sedge tussocks tend to protrude from the water, with prostrate leaf detritus forming a dense meshwork between tussocks. During 1997 new growth from the sedge tussocks was not apparent until late May and was not a visually dominant feature until early June, when surface water on the floodplain was reduced to isolated pools.

FLOODPLAIN MACROINVERTEBRATES

Following initial inundation in March or April, until the floodplain dries in June, aquatic macroinvertebrates are abundant among clumps of living and dead sedge, with mayflies most obvious. Species of mayflies documented from the Tomah Stream floodplain include the siphlonurids *S. aerodromia* and *Siphlonurus mirus, S. alternatus,* and *S. quebecensis,* the metretopodid *Siphloplecton basale,* the arthropleid *Arthroplea bipunctata,* the ephemerellids *Eurylophella temporalis, Ephemerella subvaria,* and *E. septentrionalis,* the baetiscid *Baetisca laurentina,* and the leptophlebiids *Leptophlebia cupida, L. nebulosa,* and *L. johnsoni* (Gibbs and Mingo 1986, Burian and Gibbs 1991, Gibbs and Siebenmann 1996). Larval chironomids, Trichoptera (Limnephilidae: *Limnephilus*), Coleoptera (Dytiscidae: *Hydroporus,* Helophoidae: *Helophorus*), Hemiptera (Corixidae: *Sigara*), and the amphipod *Hyalella azteca* have also been variously reported (Gibbs and Mingo 1986, Gibbs and Sie-benmann 1996).

Community Structure

During 1997 ice over the floodplain had melted sufficiently to allow sampling at the Tomah Stream study site by April 25. Macroinvertebrates were sampled at weekly intervals thereafter with a cylinder sampler (area = 0.13 m², n = 5 samples per date). The mesh size of the net used to clear the cylinder was 500 μm and defines the lower size limit for macroinvertebrates in this study. Information about community structure reported here is from samples collected on April 27 ($\sim$ two weeks after ice melt), May 13 ($\sim$ four weeks after ice melt), and June 3 (shortly before complete drying of surface water). On June 3 surface water occurred as isolated pools on the upper floodplain and an oxbow swale in the lower floodplain. Data describing community structure are reported as biomass.

On April 27 total macroinvertebrate biomass was 745 mg dry mass (= DM) m^{-2}. *Leptophlebia* spp. contributed 85 percent of the total biomass (Table 16.1). The mayflies *Siphlonurus* spp., and *S. aerodromia* and mosquito larvae *Aedes* sp. were also major contributors (range = 2–6 percent of total biomass; Table 16.1). On May 13 total macroinvertebrate biomass was 4082 mg DM m^{-2}. *Leptophlebia* spp. remained the major contributor (61 percent) (Table 16.1). The mayflies *Siphlonurus* spp., and *S. aerodromia,* mosquito larvae *Aedes* sp., caddisfly larvae *Limnephilus* cf. *indivisus,* and pea clams *Pisidium* were also important contributors (range = 4–6 percent of total biomass; Table 16.1). On June 3 total macroinvertebrate biomass was 11,828 mg DM m^{-2}. Biomass was dominated by the mayfly *Siphlonurus* spp, which contributed 45 percent (Table 16.1). The mayflies *Leptophlebia* spp., *S. aerodromia,* and *Eurylophella* and limnephilid caddisflies (*Anabolia, Limnephilus* cf. *indivisus*) were also major contributors (range = 9–17 percent of total biomass; Table 16.1). Large numbers of *Leptophlebia* emerged during the week prior to sampling, which may explain the relatively lower contribution of this taxon to total biomass on this date.

Macroinvertebrate biomass at the Tomah Stream study site showed a dramatic 16-fold increase from April to June, which translates into substantial daily instantaneous growth rate of 7.5 percent. The meaning of these statistics in terms of production dynamics of floodplain fauna is difficult to interpret, however, because the area of surface water fluctuated greatly during the study. During late April the floodplain was completely inundated and the abundance of individuals on a m^{-2} basis was relatively diluted. On May 13, however, the wetted perimeter of the inundated floodplain had moved $\sim$16 m toward the stream channel (measured along a permanent transect placed perpendicular to the stream channel). By June 3 the wetted perimeter had moved $\sim$68 m toward the stream channel compared to its position on April 27. This resulted in a concentrating effect toward the end of the season as larvae

TABLE 16.1. Biomass of Common Macroinvertebrates on the Tomah Stream Floodplain (Maine) during Inundation in the Spring of 1997[a]

	April 27, 1997		May 13, 1997		June 3, 1997	
	mg DM m^{-2}	Percent	mg DM m^{-2}	Percent	mg DM m^{-2}	Percent
River-floodplain fauna						
Leptophlebia (E)	629.0 (108.9)	84.5	2475.2 (988.5)	60.6	2017.3 (587.8)	17.1
Siphlonisca (E)	23.7 (15.7)	3.2	243.8 (105.8)	6.0	1418.5 (386.6)	12.0
Siphlonurus (E)	44.8 (29.4)	6.0	261.2 (92.3)	6.4	5377.1 (3104.9)	45.5
Floodplain fauna						
Aedes (D)	14.7 (6.5)	2.0	186.7 (138.5)	4.6	— (—)	
Anabolia (T)	— (—)		19.0 (11.0)	0.5	216.3 (64.5)	1.8
Donacia (larvae) (C)	— (—)		3.1 (3.1)	0.1	1.7 (1.7)	0.0
Limnephilidae (T)	— (—)		10.3 (7.2)	0.3	180.1 (89.2)	1.5
Limnephilus (T)	2.5 (0.8)	0.3	174.5 (75.2)	4.3	1173.9 (196.4)	9.9
Phalacrocera (D)	— (—)		1.0 (1.0)	0.0	— (—)	
Pisidium (B)	1.3 (0.8)	0.2	222.5 (64.8)	5.5	28.6 (13.2)	0.2
Podonominae (D	0.2 (0.2)	0.0	38.4 (22.1)	0.9	— (—)	
Source habitat uncertain						
Acarina	0.2 (0.0)	0.0	7.5 (2.7)	0.2	0.5 (0.2)	0.0
Agabus (larvae) (C)	— (—)		— (—)		6.5 (6.5)	0.1
Arthroplea (E)	2.0 (0.8)	0.3	71.2 (69.7)	1.7	141.7 (19.1)	1.2

TABLE 16.1. (Continued)

	April 27, 1997		May 13, 1997		June 3, 1997	
	mg DM m^{-2}	Percent	mg DM m^{-2}	Percent	mg DM m^{-2}	Percent
Atrichopogon (D)	— (—)		4.6 (4.6)	0.1	— (—)	
Banksiola (T)	2.4 (1.8)	0.3	3.4 (2.1)	0.1	62.8 (49.6)	0.5
Ceratopogonidae (D)	— (—)		10.1 (4.7)	0.2	1.1 (1.1)	0.0
Chironominae (D)	0.1 (0.1)	0.0	2.2 (1.4)	0.1	60.8 (16.0)	0.5
Eurylophella (E)	1.9 (0.8)	0.3	58.2 (38.8)	1.4	1058.6 (110.2)	9.0
Fossaria (P)	7.5 (3.6)	1.0	96.0 (29.9)	2.4	— (—)	
Hirudinea	— (—)		— (—)		25.7 (15.7)	0.2
Hydrocanthus (adult) (C)	— (—)		6.3 (3.9)	0.2	— (—)	
Laccophilus (adult) (C)	— (—)		— (—)		86.2 (86.2)	0.7
Lymnaea (P)	— (—)		— (—)		47.3 (28.9)	0.4
Nematoda	— (—)		2.5 (1.3)	0.1	0.4 (0.2)	0.0
Oligochaeta	7.1 (3.8)	1.0	26.8 (10.8)	0.7	0.7 (0.7)	0.0
Orthocladiinae (D)	6.9 (2.4)	0.9	101.4 (30.1)	2.5	6.4 (4.0)	0.1
Physa (P)	— (—)		— (—)		25.9 (15.9)	0.2
Polycentropus (T)	— (—)		— (—)		60.2 (51.3)	0.5
Psychodidae (D)	— (—)		2.3 (2.3)	0.1	— (—)	
Ptilostomis (T)	— (—)		33.7 (21.6)	0.8	— (—)	
Sigara (adult) (H)	— (—)		2.1 (2.1)	0.1	10.9 (6.9)	0.1
Sympetrum (O)	0.6 (0.2)	0.1	3.9 (3.3)	0.1	14.8 (14.8)	0.1
Tanypodinae (D)	0.2 (0.2)	0.0	1.7 (0.6)	0.0	67.7 (22.4)	0.6
Total biomass	744.7 (127.3)		4,082.3 (1,110.7)		11,827.5 (3,428.2)	

[a] Only macroinvertebrates that contributed 0.1% or more to total biomass on at least 1 date are included. Taxa are arranged by probable source habitat. Values in parentheses are standard errors (n = 5). DM = dry mass. B = Bivalvia, C = Coleoptera, D = Diptera, E = Ephemeroptera, H = Hemiptera, P = Pulmonata, O = Odonata, T = Trichoptera.

crowded into the diminishing surface water. The concentration of mayfly larvae in floodplain pools during early June was striking, being sufficient to produce turbulence at the water's surface when the mayflies were disturbed. Rapid fluctuations in area of surface water in floodplain wetlands are an important and dynamic factor whose effects must be interpreted simultaneously with individual biomass and rates of growth and mortality when production of floodplain invertebrate communities is studied.

Regardless of problems in interpreting population dynamics from these data, it is clear that the community biomass is dominated by the mayflies *Leptophlebia, Siphlonurus,* and *Siphlonisca,* which together comprised 73–94 pecent of total macroinvertebrate biomass over the entire period of inundation.

Trophic Structure

The trophic structure of the macroinvertebrate community at the Tomah Stream study site is simple, consisting of collector-gathers and shredders, as primary consumers, and their predators. The majority of primary consumers are collector-gatherers (Merritt and Cummins 1996) and so feed on biofilms and fine particles of organic matter that accumulate among submerged sedge leaves and on the benthos. This includes two of the dominant mayflies, *Leptophlebia* and *Siphlonurus,* as well as other relatively important contributors to biomass such as *Eurylophella, Aedes,* and *Pisidium. Aedes* can filter small particles from the water, but usually feeds as a collector-gatherer by brushing particles from substrata (Clements 1992). Like *Aedes, Pisidium* is also a filter-feeder, but as infauna it probably feeds on organic detritus or interstitial bacteria from sediments rather than the water column and is probably best considered a collector-gatherer as well (McMahon 1991). Collector-gatherers contributed 74 percent (June 3) to 96 percent (April 27) to community biomass. The major shredders on the floodplain are the limnephilid caddisflies *Limnephilus* and *Anabolia,* which consume vascular plant detritus, presumably decaying sedge leaves. Shredders contributed <1 percent (April 27) to 13 percent (June 3) to community biomass.

The major macroinvertebrate predator was the mayfly *S. aerodromia,* which apparently functions as a collector-gatherer early in larval development but becomes predacious as development proceeds (Gibbs and Mingo 1986). The most common prey are larvae of *Siphlonurus,* but *Leptophlebia, Eurylophella,* and midge larvae are also consumed (Gibbs and Mingo 1986). *Siphlonurus* is apparently omnivorous (Edmunds et al. 1976), but the relative importance of animal prey compared to biofilm and organic particles is unknown, and it is considered a collector-gatherer for present purposes. Adult and larval dytiscids, larvae of phryganeid caddisflies, *Polycentropus,* Ceratopogonidae, and Tanypodinae, and leeches and mites are also predators. Their biomass was relatively small compared with that of *S. aerodromia* (Table 16.1), however. Predators collectively contributed 4 percent (April 27) to 15 percent (June 3) to community biomass. *Siphlonisca aerodromia* alone contributed 3.2–12.0 percent to total biomass on these dates.

Vertebrates that feed on aquatic macroinvertebrates are conspicuous during floodplain inundation at the Tomah Stream study site. Their quantitative effect on the floodplain macroinvertebrate fauna, however, is unknown. The common shiner (*Notropis cornutus*), the three-spine stickleback (*Gasterosteus aculeatus*), the chain pickerel (*Esox niger*), and the common white sucker (*Catostomus commersoni*) have all been reported from the Tomah Stream study site during April and May. The direct examination of gut contents from these species and from brook trout (*Salvelinus fontinalis*) captured from the stream shows that they feed heavily on macroinvertebrates from the floodplain, especially mayflies (Gibbs and Mingo 1986).

Terrestrial vertebrates also feed on invertebrates from the floodplain. Common snipe (*Capella gallinago*), for example, were almost continuously present at the Tomah Stream study site during inundation in 1997. These birds are predators of wetland invertebrates and have been specifically reported to feed on mayfly larvae (Terres 1980). The black duck (*Anas rubripes*) is also a predator of mayfly larvae. As much as 50 percent of the diet of adult female black ducks during egg laying in Maine consists of mayflies (Reinecke 1977). Although present elsewhere along Tomah Stream during April and May 1997, black ducks were not observed at the study site.

Energy Base

Sources of carbon used for macroinvertebrate production at the Tomah Stream study site were investigated by stable isotope analysis (Hershey and Peterson 1996). This method assumes that ratios of stable isotopes of carbon ($^{12}C:^{13}C$) composing the tissue of consumers reflects their food sources. This assumption is generally valid, and changes in ratios of stable carbon isotopes through successive trophic levels appear to be minor ($\sim 1\%o$, Hershey and Peterson 1996). Providing there is sufficient discrimination among potential food sources—terrestrial versus aquatic primary production, for example—isotope ratios of consumers will indicate which food source is actually being assimilated. Stable isotope ratios are reported relative to a standard, and the units of measure (δ-^{13}C) represent departures ($\%o$) from the standard. Negative δ-values indicate a lower proportion of ^{13}C compared to the standard (^{13}C "depleted"); positive δ-values indicate relative "enrichment" (Hershey and Peterson 1996). δ-^{13}C values were measured for potential food sources (sedge and grass detritus, sediment [particle size <100 μm], and periphyton) and consumers from the Tomah Stream study site during April and May 1997 (Fig. 16.2). Samples were analyzed at the Sawyer Environmental Laboratory at the University of Maine, Orono.

Consumers at the Tomah Stream study site were substantially depleted in ^{13}C, compared with obvious food sources—sedge and grass detritus, sediment, and periphyton (Fig. 16.2a). *Leptophlebia* larvae collected from above the ice layer before floodplain inundation and within two weeks following ice melt were substantially less depleted in ^{13}C compared to larvae collected

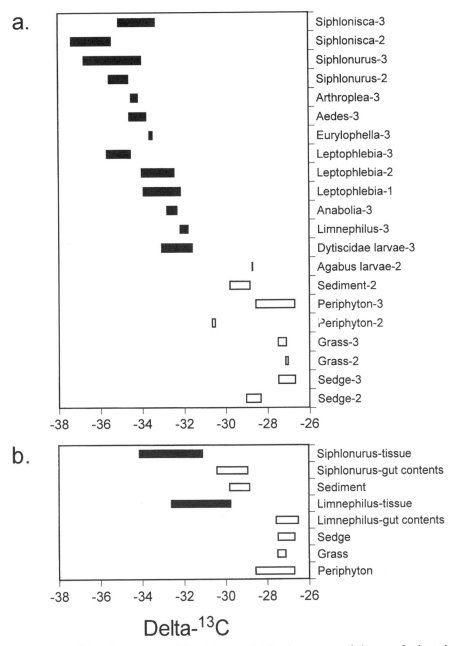

Fig. 16.2. δ-^{13}C values measured for (*a*) potential food sources and tissues of selected consumers and (*b*) potential food sources, gut contents, and tissues of *Limnephilus* and *Siphlonurus* collected at the Tomah Stream study site. Numbers associated with individual categories refer to date sampled during 1997 (1 = April 5, 2 = April 25, 3 = May 27). Specimens used for analysis of gut contents were collected May 27. Bars represent the maximum and minimum values measured. Sample size ranged from 1 to 7. Individual samples consisted of 3–10 specimens.

during the later phase of floodplain inundation (Fig. 16.2*a*). As the season progressed, larvae were apparently using a carbon source on the floodplain that was substantially more depleted than obvious food sources such as sedge or grass detritus, which are the most abundant source of organic matter on the floodplain, or periphyton, as shown for wetland foodwebs both in tropical (Hamilton et al. 1992) and temperate (Keough et al. 1996) regions. This result is almost identical to the results of Bunn and Boon (1993), who studied the foodweb in Australian oxbow lakes (billabongs).

For determining whether invertebrates were ingesting abundant sources of organic matter, but perhaps selectively assimilating material that is depleted in ^{13}C, both tissue and gut contents were analyzed from *Siphlonurus* and *Limnephilus*. The gut content of *Siphlonurus* had δ-^{13}C values that matched those of sediments, a likely food source for a collector-gatherer (Fig. 16.2b). The δ-^{13}C value for the gut content of *Limnephilus,* a shredder, also matched that of the appropriate food source—vascular plant material and attached periphyton (Fig. 16.2*b*). The tissues of these consumers, however, were substantially more depleted than the gut contents (Fig. 16.2*b*). These results, although preliminary, indicate that a small fraction of the ingested material is assimilated and that some of this material is highly depleted in ^{13}C. This material would have to make up a small proportion of the ingested material, otherwise the overall stable isotope ratio would be more strongly skewed. Providing that larval gut contents are turned over rapidly, assimilation of only a small proportion of the ingested food could result in tissue-stable carbon ratios that varied from food ratios.

Although this is speculative, a likely candidate for highly ^{13}C-depleted material in the "assimilated fraction" are bacteria that scavenge biogenic methane released from wetland sediments (methanotrophic bacteria) (Bunn and Boon 1993). Since biogenic methane is both abundant in wetlands and highly depleted in ^{13}C (δ-^{13}C < −52), Bunn and Boon (1993) suggested that methanotrophic bacteria should be depleted in ^{13}C to the degree that assimilation of relatively small quantities of their carbon by consumers could significantly influence isotope ratios of their tissues. Calculations based on a simple two-source mixing model (Araujo-Lima et al. 1986), indicate that δ-^{13}C values measured for *Limnephilus* (~ −31) would result if larvae obtained 20 percent of their tissue carbon from methanotrophic bacteria (assuming δ-^{13}C values = −52) and 80 percent from vascular plant tissue (δ-^{13}C ~ −27). A similar ratio was obtained for *Siphlonurus* feeding on sediment.

Sources of Community Members

The floodplain community at Tomah Stream floodplain is derived from two general sources: the floodplain and the stream. Macroinvertebrates that are major contributors to community biomass and permanent residents of the floodplain ("floodplain fauna") include the mosquito *Aedes,* the caddisflies *Limnephilus* cf. *indivisus* and *Anabolia,* and the pea clam *Pisidium* (Table

16.1). Macroinvertebrates that colonize the floodplain from the stream ("river-floodplain fauna") include the mayflies *Leptophlebia* spp., *S. aerodromia,* and *Siphlonurus* spp. (Table 16.1).

Life Cycles of Floodplain Fauna

Floodplain fauna life cycles contrast with those of river-floodplain fauna because they must survive the dry period during summer, which may be punctuated by brief periods of inundation and a long period of ice and snow during winter, before rapidly completing growth, development, and reproduction during April and May. The general life histories of macroinvertebrates in intermittent aquatic habitats have been well documented (Wiggins 1973, Wiggins et al. 1980, Ward 1992, Williams 1987), but detailed information about the life cycles of many taxa found at the study site is lacking. Information about the life cycles of *Aedes, Limnephilus,* and *Pisidium,* however, is abundant and indicates that these taxa are probably permanent residents of the floodplain.

Aedes, *Limnephilus,* and *Anabolia,* as well as *Pisidium,* contain species that are able to pass through summer and winter as eggs (*Aedes*), as aestivating juveniles or adults (*Pisidium*), or as terrestrial adults in reproductive diapause (*Limnephilus* and possibly *Anabolia*) (Wiggins 1973, Clarke 1981, Ward 1992). In the latter case, adults survive the summer dry period and deposit egg masses in areas that are likely to be flooded in the spring. Winter is passed in the egg stage (Wiggins 1973). *Pisidium* is abundant in damp *Sphagnum* moss and sedge detritus on the Tomah Stream floodplain throughout the dry period. Although not major contributors to total biomass, *Lasiodiamesa* (Podonominae) and *Phalacrocera* (Tipulidae) occur in pockets of *Sphagnum* moss (Alexander and Byers 1981, Brundin 1983) and so form a special group within the floodplain fauna.

Life Cycles of River-Floodplain Fauna

Although remarkable in their complexity, life histories that span both river and floodplain habitats are not unusual for mayflies on the Tomah Stream floodplain. The life history of *S. aerodromia* has been intensively studied (Gibbs and Mingo 1986, Gibbs and Siebenmann 1996) and provides a general example of the life cycle of river-floodplain fauna (Fig. 16.3). Larvae of *S. aerodromia* first appear beneath the ice of the stream channel during November. Larvae remain in the stream, growing slowly at water temperatures ~0°C until snowmelt during March or April. Following snowmelt, larvae migrate onto the inundated floodplain (Fig. 16.3), where they become closely associated with patches of tussock sedge (Gibbs 1991). Most larval growth and development occurs during this period, and larvae enter their final instar in late May and early June (Fig. 16.3). The emergence period in early June is short (9–10 days) and synchronous and occurs only when water temperature is >11°C (Gibbs and Siebenmann 1996). Oviposition by females is completed

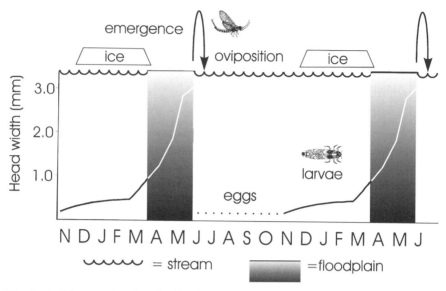

Fig. 16.3. Diagram showing details of the life cycle of *Siphlonisca aerodromia* at the Tomah Stream study site (for more information see Gibbs and Mingo 1986).

by mid-June. Standing water has usually disappeared from the floodplain at this time, and females return to the stream channel, where they deposit eggs on the water's surface. Eggs hatch in the stream the following November and December (Fig. 16.3).

The life cycles of other common mayflies at the study site, *Siphlonurus mirus* and *Leptophlebia cupida,* have been documented from river-floodplain complexes similar to Tomah Stream (Neave 1930, Hayden and Clifford 1974, Voshell 1982). The life cycles of these mayflies generally follow the pattern described for *S. aerodromia*. Adult females of both *S. mirus* and *L. cupida* oviposit in river channels, following emergence from floodplain marshes and ponds in early summer. Larvae appear in the river channel during autumn and migrate onto floodplains to complete growth and development during the spring. Upon emergence and mating, females return to the river channels to oviposit. *Siphlonurus* is also known to survive dry periods as resistant eggs (Wiggins et al. 1980, Voshell 1982). It is not known if any of the species occurring at Tomah Stream are able to use this strategy. Larvae collected shortly after inundation were relatively large (7–9 mm in length), and it is assumed that these migrated from the stream. Early instars became increasingly abundant as the season progressed, however, and it is possible that these were derived from eggs deposited on the floodplain the previous year.

Synchronized migrations of *L. cupida,* from river to floodplain following snowmelt, have been well documented (Neave 1930, Hayden and Clifford 1974). During early April in 1997 *Leptophlebia* nymphs were observed in water over ice that covered the floodplain at Tomah Stream. At this time ice on the floodplain was ~0.5-m thick and was continuous with the floodplain soil. The larvae had apparently escaped from the stream channel through holes in the ice and migrated over the ice toward the river margins as described by Hayden and Clifford (1974) for a population in Alberta. Similar migratory behavior was described for the mayfly *Paramaletus chelifer* in a Norway river-floodplain system (Olsson and Soderstrom 1978, Soderstrom and Nilsson 1987), and we assume that *S. aerodromia* and *Siphlonurus* spp. show migratory behavior at Tomah Stream.

MACROINVERTEBRATE COMMUNITIES AND RIVER-FLOODPLAIN INTERACTION

The biomass of the macroinvertebrate community on the Tomah Stream floodplain is largely composed of mayflies that are able to use the stream as a refuge during the dry period of the summer and the freezing temperatures of winter, and the floodplain during a short but critical period of rapid growth and development during spring. Compared to the contributions by this river-floodplain fauna, those by floodplain fauna to community biomass seem remarkably small. Because nothing is presently known about possible competitive interactions among members of these two assemblages, however, speculation about this factor seems unwarranted. The proximity and interconnection of the river and floodplain, however, are clearly required for the development of the characteristic floodplain community structure during the short period of inundation.

Given that river-floodplain life cycles of mayflies are relatively common, it is not surprising that there has been long interest in factors contributing to the evolution of this strategy. Most authors agree that the major risk for organisms that colonize floodplains during inundation is death because of desiccation before completing development (Neave 1930, Soderstrom and Nilsson 1987, Gibbs and Siebenemann 1996). Compared to proposed risks, explanations of the advantages of this strategy are more diverse and generally fall into the following categories (modified from Soderstrom and Nilsson 1987):

1. Avoidance of fish and invertebrate predators (Neave 1930, Olson and Soderstrom 1978, Soderstrom and Nilsson 1987, Gibbs and Siebenemann 1996);

2. Avoidance of high water velocity and ice in channel during spring floods (Neave 1930, Hayden and Clifford 1974, Olson and Soderstrom 1978);

3. Food supplies on floodplain abundant compared to river (Neave 1930, Gibbs and Siebenmann 1996);

4. Water temperatures on the floodplain more optimal for growth (Olson and Soderstrom 1978, Gibbs and Siebenmann 1996, Siebenmann and Gibbs 1996); and

    5. Emergence sites in river inadequate (Soderstrom and Nilsson 1987, Gibbs and Siebenmann 1996).

Soderstrom and Nilsson (1987) tested the first of these explanations by using a combination of field observations and laboratory experiments. These authors showed that larvae of *Parameletus* that migrated from river channels to floodplain ponds in Norway were subject to lower risk of fish predation than were larvae that remained in the river channel. Although this study indicated lower risk from fish predation, Soderstrom and Nilsson (1987) conceded that the risk of predation by other invertebrates (Dytiscidae) and wetland birds might have actually been higher on the floodplain. Tomah Stream provides a similar situation because *S. aerodromia,* one of the major mayfly taxa that colonize the floodplain, is a predator of the other mayflies. The remaining explanations remain speculative, presumably because of logistical difficulties in devising experiments required for rigorous testing, and also because the different categories of explanation are not mutually exclusive. Perhaps they are also overly complex.

    In the review of those factors suggested as being important in the evolution of the river-floodplain life cycle, it was apparent that there was perhaps an overemphasis on proposing adaptive explanations for movements from river-to-floodplain rather than floodplain-to-river. There is at least one critical evolutionary advantage for movements from floodplain to river by adults. If the eggs of a given species are not desiccation resistant or cannot withstand freezing, oviposition and eventual hatching of larvae in the stream channel may simply provide the only option for completing the life cycle for organisms otherwise adapted to wet meadow habitats.

    Other research concerning macroinvertebrates of river-floodplain systems has concentrated on their potential role in translocating biomass between the stream and floodplain (Smock 1994). In a study of movements of macroinvertebrates in forested river-floodplain system in Virginia, Smock (1994) shows that movements across the channel-floodplain boundary were minimal and that these low numbers suggest that movements of invertebrates between the channels and floodplains would have little effect on the trophic dynamics of these systems. Although probably true when only the aquatic life history stages are considered over relatively short time intervals (<1 year), this conclusion overlooks other properties of the river-floodplain community.

    At Tomah Stream the migrations of mayfly larvae onto the floodplain are rapid and extensive. Although most of the growth and production probably occurs on the floodplain (e.g., Fig. 16.3), the individuals that cross the river-floodplain boundary to colonize the floodplain eventually dominate community biomass and almost certainly do play a major role in the trophic dynamics of the floodplain community. Consideration of the exchange of biomass alone overlooks this important qualitative factor. The fate of the biomass produced on the floodplain involves aerial movements to the stream channel that are also overlooked when only larval movements are considered (e.g., Smock 1994). During the emergence of *S. aerodromia* at Tomah Stream, Gibbs and

Siebenemann (1996) observed intense fish predation on ovipositing female imagos in the stream channel and striking numbers of dead imagos of *Leptophlebia* spp. originating from the floodplain accumulated on the surface of Tomah Stream following oviposition in late May and June of 1997. The quantitative importance of aerial translocations of biomass from floodplain to river and their effect on trophic dynamics of the entire complex are unknown, but should not be ruled insignificant without further study.

MANAGEMENT AND CONSERVATION ISSUES

Prior to European colonization, streams with extensive riparian sedge-meadows were abundant in Maine and elsewhere in the Northeast. Where records exist, these systems are reported to have supported a rich and diverse fauna of wildlife (Widoff 1988). The damming of rivers for mills, water storage, and transporting timber has been a central feature in the history and development of Maine since the earliest days of European settlement, and thousands of dams have been built (Hasbrouck 1984). These dams have created numerous lakes on reaches of rivers formerly bordered by meadows and drastically altered the normal seasonal flow patterns of rivers (Widoff 1988).

Because of a complex life cycle that spans both river and floodplain habitats, *S. aerodromia* and other members of the sedge-meadow community are vulnerable to activities which alter the seasonal discharge patterns of rivers. Increased flow following snowmelt is essential to produce the inundated floodplain habitat in April and May. Threats to habitat by the building of dams are ongoing and are exemplified by a recent proposal to construct a dam on Tomah Stream (Anonymous 1990). Such a dam would threaten the most abundant and predictable population of *S. aerodromia* known. The disappearance of *S. aerodromia* from the Sacandaga River in New York following construction of the Sacandaga Reservoir confirms that dam construction is a serious threat to the species and to the entire river-floodplain community (Gibbs and Siebenmann 1993, McCafferty and Edmunds 1997).

Seasonally inundated floodplains support a wide variety of plants and animals, and their value as habitats is little understood (Ward 1992). The discovery of *S. aerodromia* has provided a focus for efforts to conserve these habitats in Maine. *Siphlonisca aerodromia* has been officially recognized as a threatened species in Maine (M. McCollough, Maine Department of Inland Fisheries and Wildlife, personal communication), and as such it may act as an umbrella that will provide protection for other members of the riparian sedge meadow community.

ACKNOWLEDGMENTS

The authors thank J. Beaudin, B. Drummond, J. Dupres, D. Introne, M. McCullough, A. Pakulski, and D. Tucker for assistance in field, laboratory, and office. Financial support came from the Maine Department of Inland Fisheries and Wildlife, the Maine Agriculture and Forestry Experiment Station, and the United States Fish and Wildlife Service Office of Endangered Species (Section 6).

LITERATURE CITED

Alexander, C. P., and G. W. Byers. 1981. Tipulidae. Pages 153–190 *in* J. F. McAlpine (ed.), Manual of Nearctic Diptera. Vol. 1. Research Branch Agriculture Canada, Monograph No. 27.

Anonymous. 1990. Tomah flowage lease enables 30-year management. News, Maine Department of Inland Fisheries and Wildlife, Public Information Division. February.

Araujo-Lima, C. A. R. M, B. R. Forsberg, R. Victoria, and L. Martinelli. 1986. Energy sources for detritivorous fishes in the Amazon. Science 234:1256–1258.

Bayley, P. B. 1995. Understanding large river-floodplain ecosystems. BioScience 45: 153–158.

Brundin, L. 1983. The larvae of Podonominae (Diptera: Chironomidae) of the Holarctic region—keys and diagnoses. Entomologica Scandinavica Supplement 19.

Bunn, S. E., and P. I. Boon. 1993. What sources of organic carbon drive food webs in billabongs? A study based on stable isotope analysis. Oecologia 96:85–94.

Burian, S. K., and K. E. Gibbs. 1988. A redescription of *Siphlonisca aerodromia* Needham (Ephemeroptera: Siphlonuridae). Aquatic Insects 10:237–248.

———. 1991. Mayflies of Maine: An annotated faunal list. Maine Agricultural Experiment Station Technical Bulletin 142.

Clarke, A. H. 1981. The freshwater molluscs of Canada. National Museum of Natural Sciences, National Museums of Canada, Ottawa.

Clements, A. N. 1992. The Biology of Mosquitoes. Vol. I. Chapman & Hall, London.

Edmunds, G. F., Jr., S. L. Jensen, and L. Berner. 1976. The mayflies of North and Central America. University of Minnesota Press, Minneapolis, MN.

Gibbs, J. P., W. G. Shriver, and S. M. Melvin. 1991. Spring and summer records of the yellow rail in Maine. Journal of Field Ornithology 62:509–512.

Gibbs, K. E. 1980. The occurrence and biology of *Siphlonisca aerodromia* Needham (Ephemeroptera: Siphlonuridae) in Maine, U.S.A. Pages 167–168 *in* J. F. Flannagan and K. E. Marshall (eds.), Advances in Ephemeroptera Biology. Plenum, New York.

———. 1989. Distribution and Habitat Requirements of the Mayfly *Siphlonisca aerodromia* in Maine. Unpublished report to the Nature Conservancy.

———. 1991. 1990 Studies on Siphlonisca aerodromia. Unpublished report to the Maine Department of Inland Fisheries and Wildlife, Endangered and Nongame Wildlife Program, Bangor, ME.

Gibbs, K. E., and T. M. Mingo. 1986. The life history, nymphal growth rates, and feeding habits of *Siphlonisca aerodromia* Needham (Ephemeroptera: Siphlonuridae) in Maine. Canadian Journal of Zoology 64:427–430.

Gibbs, K. E., and M. Siebenmann. 1996. Life history attributes of the rare mayfly *Siphlonisca aerodromia* Needham (Ephemeroptera: Siphlonuridae). Journal of the North American Benthological Society 15:95–105.

Gladden, J. E., and L. A. Smock. 1990. Macroinvertebrate distribution and production on the floodplains of two lowland headwater streams. Freshwater Biology 24:533–545.

Goulding, M., N. J. H. Smith, and D. J. Mahar. 1996. Floods of fortune: Ecology and economy along the Amazon. Columbia University Press, New York.

Hamilton, S. K., W. M. Lewis, Jr., and S. J. Sippel. 1992. Energy sources for aquatic animals in the Orinoco River floodplain: Evidence from stable isotopes. Oecologia 89:324–330.

Hasbrouck, S. 1984. Maine rivers and streams. Resource Highlights, August, The Land and Water Resources Center, University of Maine, Orono, ME.

Hayden, W., and H. F. Clifford. 1974. Seasonal movements of the mayfly *Leptophlebia cupida* (Say) in a brown-water stream of Alberta, Canada. American Midland Naturalist 91:90–102.

Hershey, A. E., and B. J. Peterson. 1996. Stream food webs. Pages 511–532 *in* F. R. Hauer and G. A. Lamberti (eds.), Methods in Stream Ecology. Academic Press, San Diego, CA.

International Joint Commission Report. 1957. Water Resources of the St. Croix River Basin, Maine and New Brunswick. Report on preliminary investigation to the International Joint Commission under the reference of 10 June 1955 by the International St. Croix River Engineering Board, June.

Junk, W. J., P. B. Bayley, and R. E. Sparks. 1989. The flood pulse concept in river-floodplain systems Pages 110–127 *in* D. P. Dodge (ed.), Proceedings of the International Large River Symposium. Canadian Special Publications in Fisheries and Aquatic Sciences 106.

Keough, J. R., M. E. Sierszen, and C. A. Hagley. 1996. Analysis of a Lake Superior coastal food web with stable isotope techniques. Ecology 41:136–146.

McCafferty, W. P., and G. F. Edmunds, Jr. 1997. Critical commentary on the genus *Siphlonisca* (Ephemeroptera: Siphlonuridae). Entomological News 108:141–147.

McMahon, R. 1991. Mollusca: Bivalvia. Pages 315–399 *in* J. H. Thorp and A. P. Covich (eds.), Ecology and Classification of North American Freshwater Invertebrates. Academic Press, San Diego, CA.

Merritt, R. W., and K. W. Cummins. 1996. An Introduction to the Aquatic Insects of North America. Kendall/Hunt, Dubuque, IA.

Neave, F. 1930. Migratory habits of the mayfly, *Blasturus cupidus* Say. Ecology 11:568–576.

Olsson, T., and O. Soderstrom. 1978. Springtime migration and growth of *Paramaletus chelifer* (Ephemeroptera) in a temporary stream in northern Sweden. Oikos 31:284–289.

Power, M. E., G. Parker, W. E. Dietrich, and A. Sun. 1995. How does floodplain width affect floodplain river ecology? A preliminary exploration using simulations. Geomorphology 13:301–317.

Reinecke, K. J. 1977. The importance of freshwater invertebrates and female energy reserves for black ducks breeding in Maine. Ph.D. Dissertation, University of Maine, Orono, ME.

Siebenmann, M., and K. E. Gibbs. 1996. 1995 and 1996 Studies on *Siphlonisca aerodromia*. Unpublished report to the Endangered and Nongame Wildlife Fund, Endangered and Threatened Species Group, Maine Department of Inland Fisheries and Wildlife, Bangor, ME.

Smock, L. A. 1994. Movements of invertebrates between stream channels and forested floodplains. Journal of the North American Benthological Society 13:524–531.

Soderstrom, O., and A. N. Nilsson. 1987. Do nymphs of *Parameletus chelifer and P. minor* (Ephemeroptera) reduce mortality from predation by occupying temporary habitats? Oecologia 74:39–46.

Sparks, R. E. 1995. Need for ecosystem management of large rivers and their floodplains. BioScience 45:168–182.

Terres, J. K. 1980. Audubon Society Encyclopedia of North American Birds. Alfred A. Knopf, New York.

Voshell, J. R. 1982. Life history and ecology of *Siphlonurus mirus* Eaton (Ephemeroptera: Siphlonuridae) in an intermittent pond. Freshwater Invertebrate Biology 1: 17–26.

Ward, J. V. 1989. The four-dimensional nature of lotic ecosystems. Journal of the North American Benthological Society 8:2–8.

———. 1992. Aquatic Insect Ecology. John Wiley & Sons, New York.

Widoff, L. 1988. Maine Wetlands Conservation Plan. Maine State Planning Office, Augusta, ME.

Wiggins, G. B. 1973. A contribution to the biology of caddisflies (Trichoptera) in temporary pools. Life Sciences Contribution 88, Royal Ontario Museum, Toronto, Canada.

Wiggins, G. B., R. J. Mackay, and I. M. Smith. 1980. Evolutionary and ecological strategies of animals in temporary pools. Archiv für Hydrobiologie Supplement 58: 97–206.

Williams, D. D. 1987. The Ecology of Temporary Waters. Timber Press, Portland, OR.

17 Canadian Peatlands

Diversity and Habitat Specialization of the Arthropod Fauna

STEPHEN A. MARSHALL, ALBERT T. FINNAMORE, and DAVID C. A. BLADES

*T*his chapter reviews Canadian peatland habitats with a discussion of the main groups of arthropods characteristic of peatlands. Canada's 170 million ha of peatland support a diverse and specialized fauna dominated by beetles, flies, wasps, and moths. Although no single peatland has been completely inventoried and the partial surveys available are not directly comparable, it is clear that individual peatlands support thousands of arthropod species, of which a significant proportion are peatland specialists. This chapter give an overview of many of the known peatland specialists.

CANADIAN PEATLAND HABITATS

Wetlands are generally divided into two groups, based on their nutrient regime and rate of accumulation of organic sediments (see Fig. 17.1). One group, the eutrophic (nutrient-rich) wetlands, contains two subgroups: marshes and swamps. Organic sediment decomposition equals or exceeds plant production, and aquatic life abounds in these areas. Wetlands in the other main group, peatlands, are characterized by nutrient-poor (oligotrophic) conditions and a decomposition rate much lower than the rate of plant production. Peatlands also tend to have more stable water levels than either swamps or marshes (Vitt 1994).

Invertebrates in Freshwater Wetlands of North America: Ecology and Management, Edited by Darold P. Batzer, Russell B. Rader, and Scott A. Wissinger
ISBN 0-471-29258-3   © 1999 John Wiley & Sons, Inc.

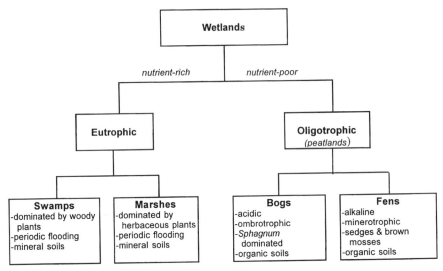

Fig. 17.1. Canadian Peatland Habitats

Peatlands form in poorly drained areas and are fed by nutrient-poor water sources. The water source ultimately affects the development of peatlands and their floras. The two main types of peatlands are bogs and fens. Fens form in areas fed by Ca^{++} and Mg^{++} ion-rich groundwater and precipitation (minerotrophic), while bogs derive moisture mostly from precipitation (ombrotrophic). A typical continental bog forms when *Sphagnum* mosses and associated species grow inward from the shore and create a spongy floating mat over the water surface. Continuous deposition of plant remains eventually fills the lake and raises the bog surface above the water level, isolating the living layer from the water table. Fens generally form in flat areas where mineral-rich springs produce shallow, slow-moving water bodies often containing marl deposits. High alkalinity and ion-rich waters encourage only specialized plant species, typically sedges, brown mosses, and orchids. Fens can develop into bogs as the accumulating organic matter isolates the surface layer from the groundwater and *Sphagnum* growth is favored. Peatland types exist along a gradient from the highly acidic (pH 4 or less) *Sphagnum*-dominated *bog* to less acidic *poor fen* to *transitional fen* (near neutral pH) to *rich fen* (alkaline; to pH 8 or more) (Vitt 1994). Within this gradient many distinct kinds of bogs and fens are recognized (Harris et al. 1996). Keys to peatland types generally use a combination of physical characteristics (hydrology, topography, pH, mineral content, and organic substrate composition) and biological indicator species to classify peatlands along the gradient of bog to rich fen.

PEATLAND ARTHROPOD FAUNAS

Canada's 170 million ha of peatlands support a diverse arthropod fauna including several hundred aquatic species and several thousand terrestrial and semiaquatic species. About 10 percent of the arthropods known from peatlands are restricted to, or strongly associated with, peatlands. The magnitude of this fauna, combined with exceptionally high habitat specificity, renders peatland arthropods of particular importance and interest. Despite this importance, many arthropod taxa abundant in peatlands remain poorly known. The study of peatland faunas is strongly hindered by a "taxonomic impediment" imposed by the inadequacy of the taxonomic literature for many groups of arthropods and exacerbated by the scarcity of specialists able to identify key peatland taxa. Most of the recent attempts to document the arthropods of Canadian peatlands have been associated with Biological Survey of Canada projects, one on the aquatic insects of peatlands and marshes (Rosenberg and Danks 1987) and another on the terrestrial insects of peatlands (Finnamore and Marshall 1994).

Characterizing arthropods as associated with one main type of peatland, or even as fen-versus bog-associated, is often difficult because collection data on specimens not taken as part of specific peatland studies are vague and unreliable, with the term *bog* often used for any wetland. This problem is exacerbated by the fact that Canadian peatlands often occur as complexes of bogs, fens, and eutrophic wetlands, called "mires" (National Wetlands Working Group, 1988). It is assumed in the following discussion that *bog* refers to a *Sphagnum*-dominated habitat (bog or poor fen), while *fen* refers to transitional and rich fens with a flora dominated by sedges and brown mosses. Perhaps in the future it will be possible to discuss insect faunas in terms of the 23 distinct kinds of bogs, fens, and other wetlands recognized and keyed by Harris et al. (1996).

The following sections overview the main groups of arthropods that form characteristic parts of the Canadian peatland fauna and for which data are available. Frequent reference is made to extensive surveys by Blades and Marshall (1994) in four southern Ontario peatlands and by Finnamore (1994) in an Alberta fen. Although numbers of taxa are often given for both studies to show diversity and distribution, these studies are not directly comparable because of differing sampling methodologies and different treatment of resulting samples due to limitations of resources and expertise. For example, a high number of Hymenoptera was found in the Alberta fen (Wagner Fen) as a result of the use of a Malaise trap in addition to pantraps and spot collections. This added a large number of ichneumonid species. The southern Ontario collections relied entirely on pantrap collections and showed a bias toward taxa living on or near the ground surface.

Arachnida

Arachnids, especially spiders (Araneae) and mites (Acari), are among the most numerous and diverse of Canada's peatland arthropods. Spiders are invariably the most conspicuous peatland predators, and have been extensively sampled in bogs and fens of Quebec, Ontario, and Manitoba. As is true for most arthropod groups, little information is available on spiders of peatlands in British Columbia or the Maritime provinces. Aitchison-Benell (1994) lists 105 species from Manitoba, Dondale and Redner (1994) list 198 species from Ontario peatlands, Koponen (1994) lists 169 species from Quebec, and Finnamore (1994) records 52 spider species from a fen in Alberta. Although these studies differ in technique and extent, some general patterns are evident. The family Linyphiidae (including the Erigonidae) dominates Canadian peatlands in numbers of species, generally making up around half the species collected. Lycosidae and Gnaphosidae are also diverse, and sometimes exceed Linyphiidae in numbers of individuals taken in pan or pit traps. Some families, such as the Salticidae, are significant in southern peatlands but become scarce in more northerly sites. Just over half the species have transcontinental ranges, and 20–30 percent of Canada's peatland spider species are holarctic, the proportion increasing in more northerly sites. Boreal and subarctic peatlands differ markedly in overall spider fauna from southern bogs, but seven characteristically northern species are known to have disjunct populations in southern bogs (Koponen 1994). Dondale and Redner (1994) suggest that 10.1 percent (19 species) of the spiders known from Ontario peatlands are peatland-restricted, about half of which are known only from bogs. No fen-restricted species were identified. Koponen found 80 percent of the above peatland-associated spiders in Quebec and added another 11 species considered bog-associated.

Although peatland-associated species have been collected in many Canadian peatlands, they are not necessarily the most abundant peatland spiders. For example, pan traps in southern Ontario peatlands took more than 700 specimens of the widespread lycosid *Pirata insularis* Emerton but only 9 specimens of the peatland species *Pirata canadensis* (Dondale and Redner 1994), and the same traps took 275 specimens of the widespread lycosid *Pardosa moesta* Banks but only 42 specimens of the bog species *Pardosa hyperborea* (Thorell) (Blades and Marshall 1994). All of the specimens of *P. hyperborea* and 200 of the specimens of *P. moesta* were taken in the same bog.

Acari (mites) are abundant in both terrestrial and aquatic environments within peatlands, but little is known about the ecology or taxonomy of most groups of mites. Mites in the genus *Sarraceniopus* live among the debris in the bottoms of pitcher plants. Although there are several species in this genus, mostly undescribed, the common species in Canadian peatlands is *S. gibsoni* (Nesbitt). Smith (1987) reviews the water mites of peatlands and marshes in Canada, including more than 100 species, and Behan-Pelletier and Bissett

(1994) overview the dominant group of peatland mites, the Oribatida or beetle mites, recording 71 species from Canadian peatlands. Only a few of these species are currently recognized as characteristic peatland species. *Hypochthonius rufulus* C. S. Koch is a common bacteria-eating species considered to be characteristic of peatlands in Europe and is one of many holarctic mites found in European and Canadian peatlands (Behan-Pelletier and Bissett 1994). Behan-Pelletier and Bissett (1994) record nine species of *Limnozetes* from Canadian peatlands, describing this genus of limnozetid mite as the most species-rich genus of oribatid mite in *Sphagnum* in eastern Canada. *Limnozetes* are the dominant microarthropods in waterlogged peat and wet *Sphagnum*, and *Limnozetes* species composition apparently reflects the type and condition of peatland, as well as microhabitat within the peatland. Although the *Limnozetes* of northeastern North America are relatively wellknown (Behan-Pelletier 1989), taxonomy of most other mite groups is poorly known, and much remains to be done with peatland mite faunas in Canada.

Arachnids other than spiders and mites appear to play only minor roles in Canadian peatlands. Koponen (1994) records eight species of Opiliones and one pseudoscorpion from peatlands, and Blades and Marshall (1994) record seven species of Opiliones from peatlands in Ontario. Only three species of Opiliones are recorded in both of the above studies, two of which were also found by Aitchison-Bennell (1994) in Manitoba peatlands. One of these, *Odiellus pictus* (Wood), was one of the two Opiliones found in an Alberta fen (Finnamore 1994). Most of these species are widespread, and none is thought to be characteristic of bogs.

Myriapoda

Centipedes and millipedes are not conspicuous or particularly abundant members of the peatland fauna, although Blades and Marshall (1994) record 10 species of millipede and three species of centipede from pan trap samples in southern Ontario bogs. Of these, the widespread parajulid *Allajulus latestriatus* (Curtis) was most abundant (141 specimens), and the peatland-associated parajulid *Aniulus paludicolens* Causey was also common (55 specimens) in fen samples and in samples from a bog damaged by draining and mining.

Hexapoda

Collembola. Springtails are conspicuously abundant in peatlands. They occur by the hundreds or thousands in most pan trap samples from Canadian peatlands and are known to occur at densities of around 33,000 individuals per m^2 in a blanket bog in the U.K. (Heal and Perkins 1978). Thum (1986) found Collembola to be the group most frequently captured by *Drosera rotundifolia* in German bogs (although flying insects were more frequently caught by *Drosera intermedia*), and Norgaard (1951) found springtails to be the primary prey of lycosid spiders in a Danish bog. Despite the tremendous abundance of Collembola in peatlands and their importance in peatland ecology, little is known about the taxonomic composition of springtail communities in Canadian peatlands. Judd (1972) records only one species of springtail, the widespread *Tomocerus flavescens* (Tullberg), from the *Sphagnum* mat in a southwestern Ontario bog.

Odonata. Except for the butterflies, the dragonflies and damselflies are the best-known peatland insects. Many peatland species were recognized in Dragonflies of Canada and Alaska (Walker 1953, 1958, Walker and Corbet 1975), and these species have been further documented in treatments of the Canadian odonate fauna at the provincial (e.g., Cannings and Stuart 1977, Robert 1963) and regional (e.g., Holder 1996) levels. Odonata are currently of great interest to a large and growing group of naturalists, so our detailed knowledge of distribution and biology is in a period of rapid growth. Despite this, many deficiencies remain in our knowledge of peatland Odonata. Little has been published about the phylogeny or historical zoogeography of the peatland taxa, and in many cases larval biologies remain incompletely known. There is a much higher proportion of peatland-restricted species (around 20 percent) among the peatland-inhabiting Odonata than in other aquatic insect groups, and the order includes peatland-restricted genera with affinities to southern faunas, presumably reflecting pre-Pleistocene relationships. This phenomenon stands in marked contrast to most other peatland arthropods, which exhibit species-level habitat restriction and northern affinities correlated with Pleistocene zoogeography. There is no doubt that special insights into the ecology and history of our peatlands will follow the current high level of interest in this group.

Hilton (1987) reviews the Odonata of peatlands and marshes in Canada, pointing out the existence of 1 damselfly (*Nehalennia gracilis* Morse) and 13 dragonflies (*Gomphaeschna furcillata* [Say], *Aeshna septentrionalis* Burmeister, *A. sitchensis* Hagen, *A. subarctica* Walker, *Williamsonia fletcheri* Williamson, *Somatochlora brevicincta* Robert, *S. franklini* [Selys], *S. incurvata* Walker, *S. sahlbergi* Trybom, *S. septentrionalis* [Hagen], *S. whitehousei* Walker, *Nannothemis bella* [Uhler], and *Leucorrhinia patricia* [Walker]) restricted to bogs. Only one odonate, the damselfly *Coenagrion interrogatum* (Hagen), is listed as occupying "mossy bogs or fens." Cannings and Cannings (1994) include 2 additional species, *Aeshna tuberculifera* Walker, and *Somatochlora sahlbergi* Trybom, in their list of peatland obligates in northern cordilleran peatlands. They record 40 species from northern cordilleran peatlands, of which 8 are considered peatland obligates and another 4 are almost always found in peatlands. Most (21 species) of the northern cordilleran species are in the genera *Aeshna* and *Somatochlora,* and all have northern affinities (unlike some eastern peatland dragonflies).

Hilton (1987) treats almost all the peatland-associated odonates as bog-associated, but most literature from which his records were taken is based on adult collection records rather than larval collections or observation of oviposition. Furthermore, descriptions of peatlands on specimen labels or in the literature tend to be inadequate to recognize a particular type of peatland reliably. Cannings and Cannings (1994) suggest that, from the odonate perspective, there are no clear differences between bogs and fens, and dragonflies seem to respond to the habitat's form and structure rather than to its acidity or nutrient levels. Similar generalizations seem to apply to the characteristic peatland dragonflies of eastern Canada. For example, two of the putative bog-restricted genera, *Nanothemis* and *Williamsonia*, are common in fens in southern Ontario but do not occur in any of the bogs with which we have experience. Clearer understanding of where peatland Odonata occur, and why they occur there, awaits detailed observations on oviposition and larval biology, investigation into critical population sizes and dispersal abilities, and careful study of the peatlands known to support species of interest.

Ephemeroptera, Plecoptera, Trichoptera. Excepting the Odonata, the aquatic orders include relatively few characteristic peatland species. No mayfly or stonefly is restricted to peatlands, although a few widespread species of Ephemeroptera are found in bogs and fens and some Plecoptera are common in fens with flowing water. Blades and Marshall (1994) report five species of mayfly from southern Ontario peatlands, three of which (*Leptophlebia cupida* Say, *Caenis amica* Hagen, and *Arthroplea bipunctata* McDunnough), along with the additional species *Callibaetis simulans* McDunnough, are reported from Canadian peatlands by Flannagan and Macdonald (1987). All these species are common in other habitats.

Flannagan and Macdonald (1987) list eight species of Trichoptera recorded from Canadian peatlands, of which most, if not all, occur in other habitats. Blades and Marshall (1994) record 15 species of caddisfly from pan traps in southern Ontario peatlands, of which only 3 (*Banksia crotchi* Banks, *Limnephilus ornatus* Banks, and *Limnephilus submonilifer* Walker) are listed by Flanagan and Macdonald (1987) as known from peatlands. None of these species is characteristic of peatlands. Larson and House (1990) find *Banksiola dossuaria* (Say), *Polycentropus smithae* Denning, and *Beothukus complicata* Wiggins to be common in Newfoundland bog pools, with the latter species characteristic of this habitat and known only from bog pools in Newfoundland and Ontario (Wiggins and Larson 1989). Brower and Brower (1970) record larvae of *Frenesia difficilis* Walker as living in pitcher plants in Maine. This northeastern species is normally found in spring seepage areas, feeding on leaves and decaying wood (Wiggins 1977).

Orthopteroids. Although introduced Mantodea and Dermaptera have been recorded from Canadian peatlands, only the Orthoptera s.s. (including Grylloptera) include characteristic peatland species. Blades and Marshall (1994) record 12 species of Orthoptera from southern Ontario peatlands, of which only the two crickets *Neonemobius palustris* (Blatchley) and *Allonemobius fasciatus* (DeGeer) are considered characteristic peatland species. The former species, which feeds and oviposits on *Sphagnum,* is restricted to *Sphagnum-*dominated bogs and poor fens in eastern Canada in the northeastern United States; the latter is also known from transitional fens and occurs across southern Canada and the northern United States (Vickery and Kevan 1985). Several other orthopterans regularly occur in peatlands and other habitats. *Melanoplus borealis borealis* (Fieber) occurs throughout Canada, including arctic and southern areas, where it is associated with cool, damp habitats such as peatlands, and *M. fasciatus* Walker occurs across Canada in association with heath plants such as *Vaccinium* (Vickery and Kevan 1985). Other orthopteran species we have found to be common in peatlands, although primarily associated with other habitats, include the acridids *Booneacris glacialis canadensis* (Walker) and *Melanoplus dawsoni* (Scudder) and the gryllid *Eunemobius carolinus* (Scudder).

Hemipteroids. The aquatic and semiaquatic Hemiptera of peatlands and marshes in Canada are reviewed by Scudder (1987), who records 32 species from fens and 33 from bogs, none of which are restricted to peatlands. One peatland-restricted saldid, *Saldoida turbaria* Schuh, has been described from *Sphagnum* bogs in Michigan (Schuh 1967), but is not yet recorded from Canada. None of the 43 species of terrestrial Hemiptera recorded from southern Ontario peatlands (Blades and Marshall 1994) are considered peatland specialists. The Psocoptera, although diverse and abundant in peatlands, similarly lack any particular peatland elements and are also more of a reflection of the surrounding habitats than the peatlands themselves. Although one aphid (*Macrosiphum jeanae* Robinson) is known only from pitcher plants in a Manitoba bog (Robinson 1972), and there are probably other Sternorrhyncha restricted to peatland flora, most of the few hemipteroids known to be associated with peatlands are found in the Auchenorrhyncha, especially the Cicadellidae and Cercopidae.

Leafhoppers are abundant and species-rich in peatlands, with 87 species of Cicadellidae recorded from southern Ontario peatlands (Blades and Marshall 1994), 69 from an Alberta fen (Finnamore 1994), and 68 from two Ohio fens (Cwickla 1987). The leafhopper genera *Draeculacephala, Cicadula, Forcipata, Macrosteles,* and *Paraphlesius* include single peatland species or pairs of peatland species, and the genus *Limotettix* includes a large group (about 14 species) characteristic of peatlands (Hamilton 1994). Hamilton (1994) analyzed the evolution of the peatland species of *Limotettix* in a phylogenetic revision of the world species of the genus, providing evidence that the bog-inhabiting species in the genus are descended from swale- and fen-inhabiting ancestors and that inhabitants of drier habitats are descended from bog species that fed on ericaceous plants. This scenario is in marked contrast to the usual idea that peatland specialists are recently evolved, terminal lineages, and suggests that bogs have been an important component of North American biogeography since well before the Pleistocene.

Cercopidae, or spittlebugs, are also abundant in peatlands. The most abundant of the nine species found in Ontario peatlands by Blades and Marshall (1994) was the widespread and common introduced meadow spittlebug, *Philaenus spumarius* (L.). The second-most common species was a peatland-associated cercopid, the heath spittlebug, *Clastoptera saintcyri* (Provancher), which feeds on a variety of heath plants (Hamilton 1982). Only three species of spittlebug were taken in Finnamore's (1994) study of an Alberta fen, none of which were peatland-associated. The rare spittlebug *Paraphilaenus parallelus* (Stearns) was recently found breeding in a rich fen on Ontario's Bruce Peninsula. Still undescribed as a nymph, this species seems to occur only in a few peatlands in Ontario and Wisconsin (Hamilton, 1982).

Lepidoptera. Many butterflies and moths are associated with Canadian peatlands. Bog coppers (*Lycaena epixanthe* Boisduval and LeConte), for example, are found only in eastern North American bogs with their cranberry host plants (*Vaccinium macrocarpon* and *V. oxycoccos*). Populations are often quite small, and bog coppers may differ visibly from bog to bog because of this species' poor dispersal abilities and the isolation of some bogs (Opler and Krizek 1984). The bog fritillary (*Boloria eunomia* [Barnes and McDunnough]) and the Jutta Arctic (*Oeneis jutta* [Hubner]) are holarctic species restricted to acid bogs. These species are primarily northern, belong to otherwise mostly northern genera, and occur in the southern part of the country only in isolated bogs. Other characteristic peatland butterflies include the bog elfin (*Incisalia lanoraieensis* Sheppard), which feeds on *Picea mariana,* and the pink-edged sulphur (*Colias interior* Scudder), which feeds on *Vaccinium myrtilloides.*

The best known of the peatland-associated moths are those associated with pitcher plants, especially *Sarracenia purpurea*. Species in the noctuid genus *Exyra* live in the pitchers of species of *Sarracenia,* with *Exyra fax* (Grote) occurring in the common Canadian pitcher plant, *S. purpurea,* throughout the host plant range, from northeastern British Columbia to Labrador and south into the northeastern United States (Folkerts and Folkerts 1996). The pitcher plant borer moth (*Papaipema appassionata* [Harvey]), on the other hand, is found only in peatlands of eastern North America, where it is considered uncommon and has very local distributions (Brower and Brower 1970). One tortricid, *Endothenia hebesena* (Walker), can be found feeding on pitcher plant seed heads wherever pitcher plants occur.

Although Lepidoptera, especially microlepidoptera, have not been adequately surveyed in any peatland, indirect evidence suggests that thousands of species may occur in individual peatlands. Finnamore's (1994) figures on hymenopterous parasitoids in one Alberta fen suggest that more of the Hymenoptera taken in that study feed on Lepidoptera than any other group. He extrapolates from his parasitoid figures to an estimate of about 6000 species of terrestrial arthropods, 31.3 percent of which are Lepidoptera, 29.3 percent Diptera, 22.6 percent Hymenoptera, and 5.6 percent Coleoptera.

Coleoptera. Although extrapolation from Finnamore's (1994) parasitoid data suggests that beetles represent only 5.6 percent of the fauna in an Alberta fen, most of the species so far documented from Canadian peatlands are either Coleoptera, Diptera, or Hymenoptera, and hundreds of species in each of these orders have been recorded from Canadian peatlands. Marshall and Blades (1994) record 459 beetle species from pan traps in southern Ontario peatlands, and Finnamore (1994) records 338 terrestrial species from a fen in Alberta.

Of 485 aquatic beetle species known from peatlands and marshes in Canada, 107 are recorded from peatlands (Larson 1987). Some aquatic beetles, such as *Ilybius discedens* Sharp (Dytiscidae) and *Gyrinus minutus* Fabricius (Gyrinidae), are recognized by Larson (1987) as characteristic inhabitants of bog pools, but most aquatic beetles found in bogs are found in other habitats. Distributions of aquatic beetles are influenced more by hydrology, vegetation structure, sediments, and temperature than by pH, flora, or nutrient regime (Larson 1987).

Bog-specific terrestrial beetles occur in several families, but the Carabidae, Pselaphidae, and Silphidae have probably received the most study. Benninger and Peck (1992) draw attention to the dependence of the silphid beetle *Nicrophorus vespilloides* Herbst on peatlands, stating that this species will bury carrion only on soft, *Sphagnum* substrates. Runtz and Peck (1994) found only one silphid, the boreal forest species *Nicrophorus defodiens* Mannerheim, in their study of Coleoptera in a bog in Algonquin Park, Ontario. Of the 14 species of Carabidae taken in the latter study, only one, *Platynus mannerheimi* Dejean, was considered a peatland associate. Runtz and Peck conclude that, in general, the beetle fauna of the peatland under study reflects the nature of the surrounding habitats rather than the peatland itself. Other studies have recorded larger numbers of carabid species from peatlands, and the ecology of peatland ground beetles has been carefully studied by several authors (although mostly outside North America). Of the 66 species of Carabidae taken by Blades and Marshall (1994) in southern Ontario peatlands, 5 (*Agonum mutatum* Gemminger and Harold, *A. darlingtoni* Lindroth, *Platynus mannerheimi, Carabus maender* Fisher von Waldheim, and *Pterostichus patruelis* Dejean) are considered peatland specialists. The latter 3 species were among the 62 species of carabid taken in an Alberta fen (Finnamore 1994). An additional peatland-associated northeastern North American species of *Platynus, P. indecentus* Liebherr and Will, has recently been described based on specimens previously confused with *P. decentus* (Say) (Liebherr and Will 1996). Bergdahl (1997) records 46 species of carabid from bogs in the Puget Sound region of Washington and British Columbia, of which the most common is the flightless *Agonum belleri* (Hatch), a diurnally active peatland specialist restricted to the Pacific Northwest. Bergdahl (1997) suggests that *A. belleri* is a "parasite" of sundew plants because it regularly steals insects trapped by the sticky leaves of these insectivorous plants.

Most of the literature dealing with ecology of peatland Carabidae is European (i.e., Butterfield and Coulson 1983, Dufrene 1987, Holmes et al. 1993), although Frambs (1994) compares bogs in Sweden and northeastern North America and concludes that open and wooded peatlands have different carabid faunas and that a habitat mosaic made up of *Sphagnum* hummocks and hollows is essential to the development of a diverse carabid fauna. Holmes et al. (1993) suggest that the major factors influencing distribution of Carabidae in Welsh peatlands are nutrient status and saturation of the substrate, altitude, and grazing of stock animals. Although the Carabidae are the best-studied group of beetles in peatlands (as in most habitats), the Staphylinidae are the largest and most common group of beetles in peatlands. Rove beetles, however, are difficult to identify, and their degree of association with the peatland habitat is poorly documented.

The family Pselaphidae is relatively small, but includes more peatland-restricted species than any other terrestrial family, and seven peatland-restricted species (*Pselaphus bellax* Casey, *Reichenbachia borealis* Casey, *R. scabra* (Brendel), *R. spatulifer* Casey, *Rybaxis mystica* Casey, *R. transversa* Fall, and *R. varicornis* (Brendel)) were found in southern Ontario peatlands (Blades and Marshall 1994). Reichle (1966) considers the distributions of these and other Pselaphidae which occur along the post glacial fringe and concludes that over 20 pselaphid species are restricted to *Sphagnum* bogs along the southern limit of the Wisconsonian glaciation, representing elements of a relict fauna left in boreal-like pockets with the retreat of the glaciers.

Hymenoptera. The Hymenoptera, including sawflies, ants, wasps, and bees, is one of the four most diverse orders of insects found in terrestrial ecosystems (the others are Coleoptera, Lepidoptera, and Diptera). The order includes at least 300,000 species worldwide, with about 16,000 of those occurring in Canada (Goulet and Huber 1993). Canadian (Blades and Marshall 1994, Finnamore 1994) and British studies (Benson 1932, Hancock 1932, Kerrich 1932, Nevinson 1932) document a rich diversity of Hymenoptera in peatlands.

Although aculeate wasps and bees (Chrysidoidea, Vespoidea, and Apoidea) are the most conspicuous Hymenoptera in peatlands and have a prominent role as pollinators of many peatland plants (Reader 1975), they exhibit low species richness and few may be peatland-restricted. Finnamore (1994) reports 82 species of aculeate Hymenoptera (excluding bees) from a rich fen in Alberta. This represents about 17 percent of the aculeate fauna (excluding bees) known from all ecosystems in the province. The low species richness of aculeates in peatlands is probably related to the relative scarcity of suitable nesting sites. Blades and Marshall (1994) list only three bog-restricted aculeate species, all ants: *Hypoclinea pustulatus* Mayr, *Lasius minutus* Emery, and *Myrmica lobifrons* Pergande.

The bulk of Hymenoptera species collected in peatlands have been parasitoids. Finnamore (1994) lists 1410 species of Hymenoptera from a rich fen in Alberta, 85 percent of which are parasitoids dominated by the superfamily Ichneumonoidea, with 913 species, including 248 species of Braconidae and 665 species of Ichneumonidae. The overwhelming number of parasitic species in that study are parasitoids emerging from host larvae (71 percent), while about 17 percent species are parasitoids emerging from host eggs and about 10 percent are parasitoids emerging from host pupae. The principal host groups attacked in peatlands are Lepidoptera (29 percent), Diptera (27 percent), and other Hymenoptera (20 percent). The hymenopteran hosts include the Symphyta (sawflies), particularly those that have an exposed caterpillar-like larva.

Blades and Marshall (1994) listed 584 species of Hymenoptera from peatlands in Ontario, of which only 49 species were also listed in the Alberta study. The low number of species common to both studies is partly attributable to the difference in sampling protocols, the large number of specimens identified only to morphospecies, and the large number of yet unidentified ichneumonid specimens from Ontario bogs.

One of the most interesting implications of the diversity of Hymenoptera parasitoids from Canadian peatlands is the degree to which that fauna may reflect the overall diversity of insects and spiders that serve as hosts. Finnamore (1994) suggests that the hymenopteran diversity in one rich fen in Alberta implies a total insect and spider species richness of about 6000 species, mostly Diptera, Hymenoptera, and Lepidoptera. Unfortunately, the "taxonomic impediment" prevents the greater use of Hymenoptera in comparative studies. In the Alberta study 70 percent of hymenopteran species were recognized as morphospecies. Generic revisions and the development of keys to our fauna are necessary before the Hymenoptera can be used to full advantage in ecosystem studies.

Diptera. The order Diptera is arguably the largest order of living things in Canada, and numbers of species and individual flies in Canadian peatlands can be impressive. Wrubleski (1987) estimates the density of larval Chironomidae in a fen in northern Alberta to be up to 12,400 individuals per square meter, and Blades and Marshall (1994) record 522 species of Diptera from southern Ontario peatlands. The latter number is only a fraction of the total Diptera fauna, as only pan traps were used, several significant families (including the Chironomidae and other Nematocera) were excluded from the study, and several groups were identifiable only to the generic level. Rosenberg et al. (1988) collected 84 species of Chironomidae, 37 of which are considered to be true peatland fauna, using emergence traps on three poor fens in northwestern Ontario. Most of the peatland chironomids taken by Rosenberg et al. were either undescribed or new to Canada or North America, reflecting a taxonomic impediment in this group. One peatland chironomid, *Metriocnemus knabi* (Coquillett), is a well-known inhabitant of pitcher plants. Other well-known dipteran inhabitants of pitcher plants include the sarcophagids *Sarraceniomyia sarraceniae* (Riley) and *Fletcherimyia fletcheri* (Aldrich) and the culicid *Wyeomyia smithi* (Coquillett) (see Giberson and Hardwick, this volume).

Some 50 species of biting flies are known from Canadian peatlands (Lewis 1987), with 45 known from bogs and 22 from fens. Although this is not a large number of species, some occur at high densities and several species are bog-restricted. This is especially true for the Tabanidae. Of the 43 species recorded from peatlands and marshes, 32 are found in bogs, 11 in fens, and only 16 in marshes.

Several other groups of Diptera are diverse in peatlands, particularly the Mycetophilidae, Empidoidea, Syrphidae, Sciomyzidae, Sarcophagidae, Tachinidae, and Sphaeroceridae. The syrphids *Eristalis compactus* Walker, *Sericomyia transversa* (Osburn), and *Platycheirus rosarum* (Fabricius) are peatland-associated species, and the monotypic sciomyzid genus *Pherbectia* Steyskal appears to be restricted to northeastern North American fens. The larvae of *P. limenitis* Steyskal presumably feed on mollusks but have not yet been associated with specific hosts. Peatland species of Sphaeroceridae were examined by Marshall (1994), who recorded 66 species from Canada, 15 of which are characteristic of peatlands. Sphaerocerids are abundant acalyptrate flies that live as microbial grazers in the moist peat and are effectively sampled by the pan traps used in most major peatland arthropod studies. Since most of Canada's genera of Sphaeroceridae have been revised in the last 15 years, it was possible not only to identify specimens in this family, but also to retrieve recent information on distribution and phylogenetic affinity. Most of the species found in peatlands have been described as new in the past 15 years, and most of the described species have only recently been recognized as part of the Canadian fauna; this is thus another example of a group previously inaccessible because of a "taxonomic impediment." Although any attempts to assess distributional patterns of peatland insects are subject to what Marshall and Blades (1988) term the "faunistic impediment" due to the lack of samples from several parts of the country, some clear distributional patterns emerged from this study. The isolated peatlands of southern Canada serve as refugia for species that survived the Pleistocene glaciations south of the ice sheets. Many of these species now have disjunct distributions, with more continuous populations occurring in the north. Many peatland-associated species are Holarctic, including the very abundant characteristic peatland species *Pullimosina dahli* (Duda). Widespread, synanthropic species of Sphaeroceridae, especially *Spelobia clunipes* (Meigen), are usually more numerous in peatland samples than characteristic peatland species.

PATTERNS OF RELATIONSHIP AND DISTRIBUTION AMONG CANADIAN PEATLAND ARTHROPODS

Despite the taxonomic impediment created by the lack of an adequate systematic infrastructure for the study of many peatland taxa, and the faunistic impediment created by the dearth of adequately inventoried peatlands, some general patterns have emerged from recent work on arthropods of Canadian peatlands. First, there is an enormous fauna. About 3600 species have been recorded from Canadian peatlands so far, but some estimates suggest that individual peatlands actually have arthropod faunas of around 6000 species. About 10 percent of these species are restricted to, or strongly associated with, the peatland habitat (Marshall and Finnamore 1994). Individual peatlands thus support hundreds of highly habitat-specific arthropod species, even though these peatland-restricted species are usually not the most abundant species present. The occurrence of habitat-specific species is very important in considering human impact on peatlands through drainage, mining, climate change, and forestry or agricultural activities in the surrounding watershed. Work on most peatland arthropod groups has shown that peatland faunas are highly influenced by nearby habitats and habitat structure, chemical factors, and physical factors such as vegetation architecture and topography.

Except for peatlands in parts of the Yukon and possible refugia on the east coast, Canadian peatlands are relatively recent in origin. For this reason, and because of the significant disjunctions between isolated southern peatlands and more continuous peatland habitats in the north, history plays a significant role in explaining the occurrence and distribution of peatland-restricted arthropods. Many species show strong boreal and Holarctic affinities, and southern peatlands serve as isolated refugia for some species. Most peatland arthropods are associated with peatlands at the species or species-group level, but others, such as the dragonfly genera *Williamsonia* and *Nannothemis,* show generic level restriction and probably reflect older historical patterns. Some studies, such as Hamilton's (1994) analysis of peatland leafhoppers, show that peatland habitats have played a significant role in the development of the Nearctic fauna since well before the Pleistocene.

FUTURE NEEDS IN PEATLAND ARTHROPOD STUDIES

Perhaps the most general conclusion to be drawn from an overview of peatland arthropods is that much remains to be done. The lack of baseline inventory data is a glaring problem, highlighted by the degree to which this review pivots on the data from only two inventories. Ecological studies of peatland arthropods are scarce, which is not surprising given the paucity of even the most basic descriptions of peatland arthropod faunas. Much of the basic descriptive work remains incomplete pending further systematic study of common peatland taxa.

LITERATURE CITED

Aitchison-Benell, C. W. 1994. Bog arachnids (Araneae, Opiliones) from Manitoba taiga. Pages 21–31 *in* A. T. Finnamore and S. A. Marshall (eds.), Terrestrial Arthropods of Peatlands, with Particular Reference to Canada. Memoirs of the Entomological Society of Canada 169.

Behan-Pelletier, V. M. 1989. *Limnozetes* (Acari: Oribatida: Limnozetidae) of northeastern North America. Canadian Entomologist 121:453–506.

Behan-Pelletier, V. M., and B. Bissett. 1994. Oribatida of Canadian peatlands. Pages 73–88 *in* A. T. Finnamore and S. A. Marshall (eds.), Terrestrial Arthropods of Peatlands, with Particular Reference to Canada. Memoirs of the Entomological Society of Canada 169.

Beninger, C. W., and S. B. Peck. 1992. Temporal and spatial patterns of resource use among *Nicrophorus* in a Sphagnum bog and adjacent forest near Ottawa, Canada. Canadian Entomologist 124:79–86.

Benson, R. B. 1932. A preliminary account of the sawflies of Wicken Fen. Pages 313–323 *in* J. S. Gardiner (ed.), The Natural History of Wicken Fen. Bowes & Bowes, Cambridge, UK.

Bergdahl, J. 1997. Carabid beetles (Coleoptera: Carabidae) of Puget Sound lowland Sphagnum bogs. Pages 20–23 *in* G. Hayslip (ed.), Proceedings of the Eighth Northwestern Biological Assessment Workgroup. U.S. Environmental Protection Agency.

Blades, D. C. A., and S. A. Marshall. 1994. Terrestrial arthropods of Canadian peatlands: Synopsis of pan trap collections at four southern Ontario peatlands. Pages 221–284 *in* A. T. Finnamore and S. A. Marshall (eds.), Terrestrial Arthropods of Peatlands, with Particular Reference to Canada. Memoirs of the Entomological Society of Canada 169.

Brower, J. H., and A. E. Brower. 1970. Notes on the biology and distribution of moths associated with the pitcher plant in Maine. Proceedings of the Entomological Society of Ontario 101:79–83.

Butterfield, J., and J. C. Coulson. 1983. The carabid communities on peat and upland grasslands in northern England. Holarctic Ecology 6:163–174.

Cannings, R. A., and K. M. Stuart. 1977. The Dragonflies of British Columbia. Handbook of the British Columbia Provincial Museum 35.

Cannings, S. G., and R. A. Cannings. 1994. The Odonata of the northern cordilleran peatlands of North America. Pages 89–110 *in* A. T. Finnamore and S. A. Marshall (eds.), Terrestrial Arthropods of Peatlands, with Particular Reference to Canada. Memoirs of the Entomological Society of Canada 169.

Cwikla, P. S. 1987. Annotated list of leafhoppers from two Ohio fens with a description of a new *Chlorotettix*. Ohio Journal of Science 87:134–137.

Dondale, C. D., and J. H. Redner. 1994. Spiders (Araneae) of six small peatlands in southern Ontario or southwestern Quebec. Pages 33–40 *in* A. T. Finnamore and S. A. Marshall (eds.), Terrestrial Arthropods of Peatlands, with Particular Reference to Canada. Memoirs of the Entomological Society of Canada 169.

Dufrene, M. 1987. Distribution of carabid beetles in a Belgian peat bog: Preliminary results. Acta Phytopath. Entom. Hung. 22:349–345.

Finnamore, A. T. 1994. Hymenoptera of the Wagner Natural Area, a boreal spring fen in central Alberta. Pages 181–220 *in* A. T. Finnamore and S. A. Marshall (eds.), Terrestrial Arthropods of Peatlands, with Particular Reference to Canada. Memoirs of the Entomological Society of Canada 169.

Finnamore, A. T., and S. A. Marshall (eds.), 1994. Terrestrial Arthropods of Peatlands, with Particular Reference to Canada. Memoirs of the Entomological Society of Canada 169.

Flannagan, J. F., and S. R. Macdonald. 1987. Ephemeroptera and Trichoptera of peatlands and marshes in Canada. Pages 47–56 *in* D. M. Rosenberg and H. V. Danks (eds.), Aquatic Insects of Peatlands and Marshes in Canada. Memoirs of the Entomological Society of Canada 140.

Folkerts, D. R., and G. W. Folkerts. 1996. Aids for field identification of pitcher plant moths of the genus *Exyra* (Lepidoptera: Noctuidae). Entomological News 107:128–136.

Främbs, H. 1994. The importance of habitat structure and food supply for carabid beetles (Coleoptera, Carabidae) in peat bogs. Pages 73–88 *in* A. T. Finnamore and S. A. Marshall (eds.), Terrestrial Arthropods of Peatlands, with Particular Reference to Canada. Memoirs of the Entomological Society of Canada 169.

Goulet, H., and J. T. Huber (eds.). 1993. Hymenoptera of the World: An Identification Guide to Families. Research Branch Agriculture Canada publication 1894/E, Ottawa, ON.

Hancock, G. L. R. 1932. A preliminary account of the Ichneumonidae. Pages 122–139 *in* J. S. Gardiner (ed.), The Natural History of Wicken Fen. Bowes & Bowes, Cambridge, UK.

Hamilton, K. G. A. 1982. The Spittlebugs of Canada. Insects and Arachnids of Canada, Part 10. Agriculture Canada Publication 1740.

Hamilton, A. 1994. Evolution of *Limotettix* Sahlberg (Homoptera: Cicadellidae) in peatlands, with descriptions of new taxa. Pages 111–133 *in* A. T. Finnamore and S.A. Marshall (eds.), Terrestrial Arthropods of Peatlands, with Particular Reference to Canada. Memoirs of the Entomological Society of Canada 169.

Harris, A. G., S. C. McMurray, P. W. C. Uhlig, J. K. Jeglum, R. F. Foster, and G. D. Racey. 1996. Field Guide to the Wetland Ecosystem Classification for Northwestern Ontario. Northwest Science and Technology Field Guide FG-01.

Heal, O. W., and Perkins D. F. (eds.). 1978. Production Ecology of British Moors and Montane Grasslands. Springer-Verlag, Berlin.

Hilton, D. F. J. 1987. Odonata of peatlands and marshes in Canada. Pages 57–63 *in* D. M. Rosenberg and H. V. Danks (eds.), Aquatic Insects of Peatlands and Marshes in Canada. Memoirs of the Entomological Society of Canada 140.

Holder, M. 1996. The Dragonflies and Damselflies of Algonquin Provincial Park. Algonquin Park Technical Bulletin 11.

Holmes, P. R., D. C. Boyce, and D. K. Reed. 1993. The ground beetle (Coleoptera: Carabidae) fauna of Welsh peatland biotopes: Factors influencing the distribution of ground beetles and conservation implications. Biological Conservation 63:153–161.

Judd, W. W. 1972. Studies of the Byron Bog in Southwestern Ontario. XLVII. Insects of the Sphagnum mat. Entomological News 83:145–151.

Kerrich, G. J. 1932. Additions to the ichneumonoid fauna of Wicken Fen. Pages 560–566 *in* J. S. Gardiner (ed.), The Natural History of Wicken Fen. Bowes & Bowes, Cambridgem, UK.

Koponen, S. 1994. Ground-living spiders, opilionids, and pseudoscorpions of peatlands in Quebec. Pages 41–60 *in* A. T. Finnamore and S. A. Marshall (eds.), Terrestrial Arthropods of Peatlands, with Particular Reference to Canada. Memoirs of the Entomological Society of Canada 169.

Larson, D. J. 1987. Aquatic Coleoptera of peatlands and marshes in Canada. Pages 99–132 *in* D. M. Rosenberg and H. V. Danks (eds.), Aquatic Insects of Peatlands and Marshes in Canada. Memoirs of the Entomological Society of Canada 140.

Larson, D. J., and N. L. House. 1990. Insect communities of Newfoundland bog pools with emphasis on the Odonata. Canadian Entomologist 122:469–501.

Lewis, D. 1987. Biting flies (Diptera) of peatlands and marshes in Canada. Pages 133–140 *in* D. M. Rosenberg and H. V. Danks (eds.), Aquatic Insects of Peatlands and Marshes in Canada. Memoirs of the Entomological Society of Canada 140.

Liebherr, J. K., and K. W. Will. 1996. New North American *Platynus* Bonelli (Coleoptera, Carabidae), a key to species north of Mexico, and notes on species from the southwestern United States. The Coleopterists Bulletin 50:301–320.

Marshall, S. A. 1994. Peatland Sphaeroceridae (Diptera) of Canada. Pages 173–179 *in* A. T. Finnamore and S. A. Marshall (eds.), Terrestrial Arthropods of Peatlands, with Particular Reference to Canada. Memoirs of the Entomological Society of Canada 169.

Marshall, S. A. and D. C. A. Blades. 1989. The Biological Survey of Canada and Arthropods of Peatlands. Pages 360–365 *in* Bardecki and Patterson (eds.), Wetlands: Inertia or Momentum? Toronto, ON.

National Wetlands Working Group. 1987. The Canadian Wetland Classification System (Provisional Edition). Environment Canada. Canadian Wildlife Service. Lands Conservation Branch. Ecological Land Classification Series No. 21.

Nevinson, E. B. 1932. Hymenoptera Aculeata. Pages 253–254 *in* J. S. Gardiner (ed.), The Natural History of Wicken Fen. Bowes & Bowes, Cambridge. UK.

Norgaard, E. 1951. The ecology of two lycosid spiders (*Pirata piraticus* and *Lycosa pullata*) from a Danish *Sphagnum* bog. Oikos 3:1–21.

Opler, P. A., and G. O. Krizek, 1984. Butterflies East of the Great Plains. Johns Hopkins University Press, Baltimore.

Reader, R. 1975. Competitive relationships of some bog ericads for major insect pollinators. Canadian Journal of Botany 53:1300–1305.

Reichle, D. E. 1966. Some pselaphid beetles with boreal affinities and their distribution along the postglacial fringe. Systematic Zoology 15:330–344.

Robert, A. 1963. Les libellules du Quebec. Service de la faune, Bulletin 1.

Rosenberg, D. M., and H. V. Danks (eds.), 1987. Aquatic Insects of Peatlands and Marshes in Canada. Memoirs of the Entomological Society of Canada 140.

Rosenberg, D. M., A. P. Wiens, and B. Bilyj. 1987. Chironomidae (Diptera) of peatlands in northwestern Ontario, Canada. Holarctic Ecology 11:19–31.

Runtz, M. P. W., and S. B. Peck 1994. The beetle fauna of a mature spruce-sphagnum bog, Algonquin Park, Ontario: Ecological implications of the species composition. Pages 161–171 *in* A. T. Finnamore and S. A. Marshall (eds.), Terrestrial Arthropods of Peatlands, with Particular Reference to Canada. Memoirs of the Entomological Society of Canada 169.

Schuh, T. 1967. The shore bugs (Hemiptera, Saldidae) of the Great Lakes region. Contributions of the American Entomological Institute 2:1–35.

Scudder, C. G. E. 1987. Aquatic and semiaquatic Hemiptera of peatlands and marshes in Canada. Pages 65–98 *in* D. M. Rosenberg and H. V. Danks (eds.), Aquatic Insects of Peatlands and Marshes in Canada. Memoirs of the Entomological Society of Canada 140.

Thum, M. 1986. Segregation of habitat and prey in two sympatric carnivorous plant species, *Drosera rotundifolia* and *Drosera intermedia*. Oecologia 70:601–605.

Vickery, V. R., and D. K. M. Kevan. 1985. The Grasshoppers, Crickets, and Related Insects of Canada and Adjacent Regions. The Insects and Arachnids of Canada 14.

Vitt, D. H. 1994. An overview of factors that influence the development of Canadian peatlands. Pages 7–20 *in* A. T. Finnamore and S. A. Marshall (eds.), Terrestrial Arthropods of Peatlands, with Particular Reference to Canada. Memoirs of the Entomological Society of Canada 169.

Walker, E. M. 1953. The Odonata of Canada and Alaska. Vol. 1, Part 1: General, Part II. University of Toronto Press, Toronto.

———. 1958. The Odonata of Canada and Alaska. Vol. 2, Part III. University of Toronto Press, Toronto, ON.

Walker, E. M., and P. S. Corbet. 1975. The Odonata of Canada and Alaska. Vol. 3, Part III. University of Toronto Press, Toronto.

Wiggins, G. B. 1977. Larvae of the North American Caddisfly Genera. University of Toronto Press, Toronto, ON.

Wiggins, G. B., and D. J. Larson. 1989. Systematics and biology of a new Nearctic genus in the caddisfly family Phryganeidae (Trichoptera). Canadian Journal of Zoology 67:1550–1556.

Wrubleski, D. 1987. Chironomidae (Diptera) of peatlands and marshes in Canada. Pages 141–162 *in* D. M. Rosenberg and H. V. Danks (eds.), Aquatic Insects of Peatlands and Marshes in Canada. Memoirs of the Entomological Society of Canada 140.

18 Pitcher Plants (*Sarracenia purpurea*) in Eastern Canadian Peatlands

Ecology and Conservation of the Invertebrate Inquilines

DONNA GIBERSON and M. LORETTA HARDWICK

*I*n this chapter we review the relationship between the northern pitcher plant (Sarracenia purpurea) and its inquilines in northeastern North America. Three species of Diptera (flies) dominate the insect community of the northern pitcher plant. All three are detritivores and feed on the arthropod prey attracted to the pitcher plant. They coexist by partitioning the habitat and food resource spatially. Wyeomyia smithii, the pitcher plant mosquito, lives in the water column of the pitcher and feeds by filtering microorganisms (bacteria and protozoa) from the water. Metriocnemus knabi, *the pitcher plant midge, feeds on the dead organisms that have accumulated at the bottom of the pitcher.* Blaesoxipha fletcheri, *the pitcher plant sarcophagid, feeds at the surface on floating prey items that have drowned in the pitcher fluid. The action of these inquilines apparently speeds up the release of nutrients (mainly nitrogen and carbon dioxide) to the plant, and in turn the plant removes potentially toxic metabolic wastes (ammonia, CO_2) from the water and infuses oxygen. Pitcher plant habitats, mainly Sphagnum bogs, are at risk in some areas of the region, particularly in the more populated zones in the south, to urbanization, agriculture, aforestation, and peat harvest. However, this represents only a small part of the total available peatland habitat in Canada.*

Invertebrates in Freshwater Wetlands of North America: Ecology and Management, Edited by Darold P. Batzer, Russell B. Rader, and Scott A. Wissinger
ISBN 0-471-29258-3 © 1999 John Wiley & Sons, Inc.

PITCHER PLANTS

Pitcher plants have held a fascination for biologists for centuries because of their carnivorous habits (Juniper et al. 1989). The pitcher-shaped leaves of New World pitcher plants (Sarraceniaceae) collect water and serve as passive insect traps. The insects are attracted to the leaf by UV reflectance and nectar, then slip into the water and eventually drown, providing a nutrient supplement to the plant (Givnish 1989, Chapin and Pastor 1995).

Even more interesting than the predatory habits of the plant is the fact that the plant is also home to an entire community of inquilines (i.e., symbionts which live inside another organism without causing harm to it; Lincoln et al. 1982), including insects, mites, rotifers, protozoa, and bacteria. The water-filled pitcher is referred to as a *phytotelm,* a small aquatic ecosystem formed from a part of a terrestrial plant (Juniper et al. 1989). The ecosystem, although small, supports the full range of foodweb interactions of any other type of aquatic habitat, based upon the breakdown of the arthropod prey attracted to the pitcher by the plant. The small size of the habitat and its isolation from other pitcher habitats provide a relatively simplified system for ecological study or manipulation (Paterson 1971, Addicott 1974). There have been a wealth of studies investigating interactions between the plant and the inquilines (e.g., Bradshaw 1983, Bradshaw and Creelman 1984), interactions among and between inquilines (e.g., Buffington 1970, Addicott 1974, Fish and Hall 1978, Heard 1994a), and the life history and ecology of individual inquilines (e.g. Smith 1902, Paterson 1971, Smith and Brust 1971, Forsyth and Robertson 1975, Farkas and Brust 1986a). The purpose of this chapter is to review the information on the relationship between *Sarracenia purpurea* L., the northern or purple pitcher plant, and its inquilines, particularly with respect to eastern Canada, and to place this relationship into the context of peatland conservation.

THE NORTHERN PITCHER PLANT (*Sarracenia purpurea* L.)

Geographic Range

The northern pitcher plant (*Sarracenia purpurea*) is the most widespread of the eight New World species of *Sarracenia* occurring from Brazil to northern Canada (Pietropaolo and Pietropaolo 1974, Juniper et al. 1989). It is found from the Gulf Coast of Florida to Labrador in eastern North America, and from Newfoundland to British Columbia and the Northwest Territories in the northern boreal zone in Canada (Juniper et al. 1989). The species has now also become naturalized in Ireland and Sweden (Juniper et al. 1989). *Sarracenia purpurea* is generally found in moderately to intensely acidic soils (Juniper et al. 1989), although it may live in neutral or alkaline soils in some areas (especially in the south; Pietropaolo and Pietropaolo 1974). The plant usually grows in *Sphagnum* and sedge wetlands and is generally intolerant of competition or shade (Juniper et al. 1989).

Pitcher plants (*S. purpurea*) are widespread in eastern Canada wherever there is suitable habitat (Wells 1996). They are commonly associated with *Sphagnum* spp. in acid bogs, bog meadows, and boggy lake margins in Nova Scotia (Comeau 1971, Roland and Smith 1969), New Brunswick (Hinds 1986), Prince Edward Island (Erskine 1960), Newfoundland (Pollett and Meades 1970), and Quebec (Marie-Victorin 1964). Frère Marie-Victorin called them "the main ornamentation from our bogs" in Quebec (Marie-Victorin 1964) and pointed out that the local Amerindians called them "grass-toad" because the plant, like the toad, ate insects. The plant is well known for its unique showy leaf and long-stalked flower and has been recognized as the floral emblem of the province of Newfoundland and Labrador.

Life History of *S. Purpurea*

The purple pitcher plant is a herbaceous perennial which produces pitcher-shaped leaves in a rosette (Pietropaolo and Pietropaolo 1974, Chapin and Pastor 1995). The plant begins producing new pitchers in early spring and on average produces a new leaf every 20 days or so (Fish and Hall 1978). In southern Ontario new leaves appear in early May and open by mid-June (Judd 1959). New leaves do not appear until late May and early June in Manitoba (Farkas and Brust 1986a), New Brunswick (Paterson and Cameron 1982), and Prince Edward Island (Hardwick and Giberson 1996), and these generally do not open until early July (Fig. 18.1). Leaves begin collecting water and attracting insects as soon as they are open, and several new leaves may be produced over a summer (Fish and Hall 1978). Leaves overwinter as intact fluid-filled pitchers. In northern populations (e.g., those throughout Canada) the fluid freezes solid for four or more months inside the pitcher and thaws again in spring (Paterson 1971, Farkas and Brust 1986). Overwintered leaves remain intact for several weeks in the second year, but begin to show signs of rot by mid- to late summer. The rotted leaves lose their integrity and no longer hold water or digest prey (Paterson 1971, Farkas and Brust 1986a).

Insects are attracted to the leaves by visual and chemical cues (UV light guides and bright purple streaks at the top of the pitcher, and nectar produced in nectaries around the rim of the pitchers; Pietropaolo and Pietropaolo 1974, Juniper et al. 1989). Insects feeding on the nectar may be enticed into the pitcher by accumulation of nectar below the rim, and lose their footing and fall into the fluid. They are prevented from escaping by downward-pointing hairs on the pitcher walls and eventually drown (Pietropaolo and Pietropaolo 1974). Although several species of *Sarracenia* produce digestive enzymes in specialized gland cells (Juniper et al. 1989, Givnish 1989), plant-produced digestive enzymes have not been conclusively shown in *S. purpurea* (Bradshaw and Creelman 1984). Digestion occurs mainly as a result of proteolytic activity from bacteria and autolytic enzymes from the victims themselves

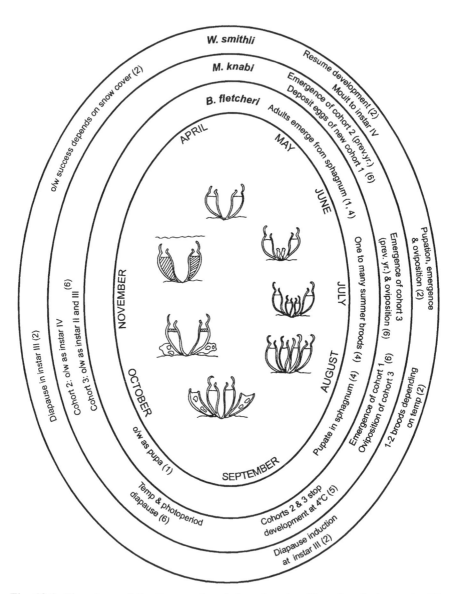

Fig. 18.1. Phenology of the three major pitcher plant inquilines in relation to the life history of the plant. (1) Farkas and Brust 1986b; (2) Farkas and Brust 1986a; (3) Evans and Brust 1972; (4) Forsyth and Robertson 1975; (5) Paterson 1971; (6) Paterson and Cameron 1982.

(Bradshaw 1983). Prey capture is slow at first, but then peaks between 10 and 20 days (after the pitcher fills with water) and then slowly declines. This results in only one to two leaves actively attracting insects at any time (Fish and Hall 1978).

As early as 1875, Charles Darwin noted that carnivorous plants are associated with nutrient-poor conditions (Juniper et al. 1989). However, some researchers have questioned the extent to which carnivorous plants depend on insect prey for nutrient supplementation (e.g., Christensen 1976, Stewart and Nilsen 1971). Chapin and Pastor (1995) showed experimentally that *S. purpurea* was nitrogen- and phosphorus-limited, despite the fact that it was growing in soil nutrient conditions that should not have been limiting. They suggested that soil microbes may compete for nitrogen and other nutrients (especially early in the season), resulting in carnivory being most important to the plant in spring and early summer. Another important competitor may be the *Sphagnum* with which the pitcher plants are usually associated; van Breeman (1995) indicated that *Sphagnum* and other mosses efficiently intercept nutrients from atmosphere, leachates, and litter, reducing the supply of nutrients for other plants. Bradshaw and Creelman (1984) suggested that other benefits of carnivory may be metallic ions and nutrients such as sulphur and phosphorus, rather than just nitrogen alone.

THE PITCHER PLANT AS A HABITAT FOR AQUATIC INVERTEBRATES

Limnological Characteristics

Pitchers from pitcher plants form small (generally about 20–30-ml volume), temporary aquatic habitats, characterized by fairly extreme conditions, especially relating to pH and temperature. Each "habitat" (leaf) goes through a recognizable cycle, in which it opens, begins to attract prey and inquilines, then persists for a period of weeks or months with little or no additional food input (Fish and Hall 1978). However, because leaves open throughout the summer, the timing of the cycle may vary among plants or leaves.

Most of the research on the organisms living inside pitchers has focused on ecological interactions between and among inquilines, and relatively little attention has been given to the characteristics of the pitcher habitat. Several researchers have investigated oxygen conditions in the leaf (e.g., Istock et al. 1975, Cameron et al. 1977) or recorded pH over time (Fish and Hall 1978), but we could find no studies that conducted a limnological type of investigation of the pitcher habitat, collecting data on a wide variety of physical factors at the same time. However, some characteristics seem to be common to the pitcher habitat.

Despite the wide geographical range over which northern pitcher plants are found, conditions inside the pitcher leaf apparently do not vary as much as equivalent habitats outside the pitcher (Juniper et al. 1989). For example, temperature inside the leaf is less variable than in open bog pools (Juniper et al. 1989), and oxygen levels are usually fairly high, about 5–10 ppm (Istock et al. 1975) or at least 77 percent of saturation (Cameron et al. 1977). However, large diurnal and seasonal fluctuations in all habitat parameters can occur at individual sites, and these result in a highly variable habitat for prospective inquilines.

Midsummer temperatures can vary within the leaf from a nighttime low of about 10°C to >35°C under full sun, even at northern latitudes (Istock et al. 1975, Bradshaw 1980, Paterson and Cameron 1982, Farkas and Brust 1986a). In a single bog, microhabitat features such as shade and sheltering from wind can affect temperature, so that considerable differences in accumulated temperatures (degree days) occur between pitchers usually in shade and those in direct sun (Kingsolver 1979). Temperature patterns should be measured directly in the field because leaf temperatures do not always reflect ambient conditions. In one study in an open bog in New Brunswick (Paterson and Cameron 1982), fluid temperatures during summer and early autumn approximated air temperatures, but in northern Manitoba pitcher temperatures in open bogs in Manitoba were usually warmer than ambient temperatures (Evans and Brust 1972). In late autumn the water in the pitcher begins to freeze (Paterson 1971), and in most locations the pitcher fluid is frozen solid by late November (Fig. 18.1). Any organisms still remaining in the pitchers must be able to tolerate freezing into the solid core. Snow cover at this time plays a crucial role in maintaining the pitcher temperature at just a few degrees below freezing, even in the most northerly locations (Paterson 1971, Smith and Brust 1971, Evans and Brust 1972, Farkas and Brust 1986).

Experimental studies have shown that oxygen conditions are maintained at higher levels than can be accounted for by simple diffusion from the atmosphere, particularly considering the high potential biological oxygen demand (BOD) in pitchers with decaying prey (Cameron et al. 1977, Bradshaw and Creelman 1984). Occasionally, pitchers with high prey volumes do go anoxic (Bradshaw and Creelman 1984), but this is relatively rare in nature (Cameron et al. 1977). Cameron et al. (1977) measured oxygen levels in pitchers in New Brunswick and found that the DO (dissolved oxygen) was usually near saturation, and even their lowest recorded value of 77 percent saturation occurred just above the "wad" of detritus at the bottom of the pitcher. Similarly, Istock et al. (1975) found DO to range between 5–10 ppm at temperatures that fluctuated from 10 to 35°C in a bog in New York State. Cameron et al. (1977) believed that the primary pathway for oxygen to enter the leaf was diffusion through the leaf tissue, since oxygen levels were maintained even in the dark. However, Bradshaw and Creelman (1984) showed that, in light, plants take up CO_2 produced in the fluid by the breakdown of prey and infuse O_2 into the water.

Water in the pitcher is generally very acid (about pH 3.8), although pH may range widely from highly acidic to nearly neutral (Juniper et al. 1989).

Fish and Hall (1978) monitored leaf pH over a period of time and reported that pH increased from 5.8 to 6.3 by day 8 (after filling with water from a sprinkler system). After that the pH declined steadily to pH 3.5 by day 35, after which it was maintained at that level, possibly by physiological control by the plant or by the pattern of prey decomposition.

Nutrient levels in the leaf fluid also depend on time and the amount of prey captured. Levels are low at first, but the process of breaking down the prey increases ammonia and other dissolved nutrients. Action of bacteria and autolytic enzymes provides a steady supply of soluble nitrogen to the plant, even when the inquilines are absent. However, the presence of living insects in the pitchers enhances the rate of nitrogen production and prevents nitrogen from being sequestered in bacteria and protozoa (Bradshaw and Creelman 1984). Despite high ammonia production when inquilines are present, the pitcher habitat does not become toxic, since the plant actively takes up ammonia from the water, especially when temperatures are high and light is bright (Bradshaw and Creelman 1984).

In summary, the pitcher plant provides an aquatic habitat that is often extremely acidic, with large temperature fluctuations both diurnally and seasonally. However, oxygen levels are generally sufficient to support a variety of inquilines that have become specialized to this container habitat, and the plant itself maintains a clean water environment by actively removing potentially toxic waste products from the water.

Inquilines in Pitcher Plants

A number of organisms live facultatively or obligately in the pitcher fluid of *Sarracenia purpurea*. These range from microscopic organisms (e.g., bacteria, protozoa, and rotifers) to several types of arthropods, although only a few of these are common (Table 18.1). Protozoa identified in a study in Michigan by Addicott (1974) showed a diverse group of nearly 40 taxa, although no information is available on how widespread these are in pitchers in other locations. The arthropods are somewhat less diverse; fewer than 20 taxa have been associated with the pitchers to date (Table 18.1).

Most of the research on pitcher plant inquilines in North America has focused on three flies which have an obligate relationship with *Sarracenia purpurea* (e.g., Judd 1959, Forsyth and Robertson 1975, Fish and Hall 1978, Heard 1994a, Nastase et al. 1995). These are *Wyeomyia smithii*, the pitcher plant mosquito (Diptera: Culicidae), *Metriocnemus knabi*, the pitcher plant midge (Diptera: Chironomidae), and *Blaesoxipha fletcheri*, the pitcher plant sarcophagid (Diptera: Sarcophagidae) (Juniper et al. 1989). These three insects partition the pitcher spatially (Fig. 18.2), and all are detritivores. The mosquitoes filter-feed on fine particulates and microorganisms throughout the water column and return to the surface to obtain oxygen (Bradshaw 1983). The midges feed at the bottom of the pitcher, boring into prey items that have fallen from the surface (Fish and Hall 1978, Bradshaw 1983). However, they

TABLE 18.1. Organisms Reported to Inhabit Pitchers of *Sarracenia purpurea*

Protozoa[b]	Rotifers	Arthropoda
Flagellates	unknown loricates[a] and illoricates[a]	Acarina, *Histiostoma* sp.[b,c]★
Chrysomonadida	*Habrotrocha rosa*[k]	*Anoetus gibsoni*[c]
Monas vulgaris		*Anoetus hughesi*[c]
Monas sp.		Crustacea, Copepoda[d]★
Cryptomonadida		Insecta, Plecoptera, *Leuctra maria*[e]
Cryptomonas sp.		Odonata, Libellulidae nymph[f]
Chilomonas paramecium		Megaloptera, *Sialis joppa*[g]
Euglenida		Lepidoptera, *Exyra rolandiana*[c] (consume the leaves)
Euglena sp		*E. semicrocea*[c] (consume the leaves)
Astasia klebsi		Diptera, Culicidae, *Wyeomyia smithii*[b,c]★
Scytomonas pusilla		Chironomidae *Metriocnemus knabi*[b,c]★
Anisonema emarginatum		*Metriocnemus* cf *fuscipes*[h]
Notosolenus sp.		*Lymnophyes* sp.[h]
Volvocida		unknown Pentaneurini[h]
Chlamydomonas sp.		Ceratopogonidae, *Culicoides* sp.[h]
Chlorogonium sp.		Sarcophagidae *Bleaesoxipha fletcheri*[c,i,j]★
Carteria sp.		*Sarcophaga sarraceniae*[j]
Polytomella agillis		Phoridae, *Dohrniphora cornuta*[c]
Kinetoplastida		unknown phorid[h]
Bodo sp.		Sciaridae, *Bradysis macfarlanei*[c]
Cercomonas sp.		unknown sciarid[i]
Trichomonadida		Hymenoptera, *Polistes fascatus pallipes* (paper wasp, nests
Urophagus rostrattus		in dry pitchers)[c]
Ciliates		
Gymnostomatida		
Urotricha ovata		
Urotricha agilis		
Platyophyra spumacola		
Prorodon sp.		

Rhopalophrya sp.
Spathidium sp.
Litonotus sp.
Chilodonella sp.
Dysteria sp.
Trichostomatida
Colpoda inflata
Colpoda steini
Colpoda sp.
Bresslaua sp.
Leptopharynx sphagnetorum
unknown microthoracid
Hymenostomatida
Colpidium campylum
Pseudoglaucomo muscorum
Paramecium bursaria
Cyrtolophosis elongata
Cyclidium elongatum
Peritrichida, Vorticella sp.
Oligotrichida, Halteria sp.
Hypotrichia, unknown spp.

★Refers to taxa that are frequently reported.
[a] Addicott 1974 (Mich.).
[b] Judd 1959 (Ont.).
[c] Juniper et al. 1989 (several locations).
[d] Hardwick and Giberson 1995 (P.E.I.).
[e] Turner et al. 1996 (W. Virg.)

[f] Wray and Brimley 1943 (N. Carolina).
[g] Mather 1981 (New Jersey).
[h] de la Rosa and Nastase 1987 (Penn.).
[i] Forsyth and Roberston 1975 (Ont.).
[j] Farkas and Brust 1986b (Man.).
[k] Bateman 1987 (Nfld.).

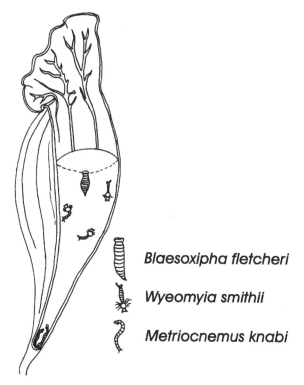

Blaesoxipha fletcheri

Wyeomyia smithii

Metriocnemus knabi

Fig. 18.2. The spatial distribution of the three major inquilines in the pitcher of the northern pitcher plant.

respire cutaneously and are dependent upon adequate oxygen supplies in the water at the bottom of the pitcher (Cameron et al. 1977). The sarcophagid feeds on newly drowned insect prey at the surface of the pitcher by piercing the prey cuticle and macerating the tissue. Like the mosquitoes, they obtain their oxygen at the surface (Forsyth and Robertson 1975).

Wyeomyia smithii (Coquillett)

Wyeomyia smithii is the best studied of the pitcher plant inquilines. It has been reported from the Florida Gulf Coast to Labrador and westward to northern Manitoba (Juniper et al. 1989). The life cycle is considered to be univoltine in the north (Farkas and Brust 1986a), although bivoltinism and multivoltinism have been reported farther south (Kingsolver 1979, Bradshaw and Holzapfel 1983). *Wyeomyia smithii* is autogenous (i.e., it does not require a blood meal for maturation of the eggs) for the first ovarian cycle throughout its range. In the north females produce only one clutch of eggs and are not known to bite, but those in southern latitudes will take a blood meal to mature further batches of eggs (Bradshaw 1986). *Wyeomyia smithii* appears to be restricted to the northern pitcher plant in the north, but will oviposit in other species of pitcher plant in the south (Juniper et al. 1989).

The phenology of *W. smithii* is shown in relation to the plant phenology in Fig. 18.1. Females preferentially oviposit into new pitchers (Fish and Hall 1978, Mogi and Mokri 1980, Bradshaw 1986), but will lay in older pitchers as well (Farkas and Brust 1986a). Istock et al. (1975) have shown that the mosquitoes distinguish natural pitchers from artificial ones by using a chemical cue, and believe that this cue may also help them distinguish between new and old pitchers. Over much of the north the timing of the first opening of new pitchers corresponds to the period when the adults are on the wing and looking for oviposition sites (Judd 1959, Farkas and Brust 1986a, Heard 1994a). A preference for new leaves at this time ensures that eggs are laid in pitchers that have the highest probability of (1) attracting prey to provide food for developing larvae (Fish and Hall 1978, Bradshaw 1986) and (2) successfully overwintering, since previously overwintered pitchers begin to rot within a few weeks of oviposition (Farkas and Brust 1986a).

Adults are highly susceptible to desiccation. Wallis and Frempong-Boadu (1967) found that adults did not survive long enough to oviposit at humidities of less than 70 percent, and observed that females did not travel far from the pitcher they emerged from. Many pitchers can be found in a small area, however, and Bradshaw (1983) has observed females visiting, and even rejecting, several pitchers before ovipositing. Eggs are laid singly, either on the surface of the water or on the sides of the pitcher (Smith 1902, Wallis and Frempong-Boadu 1967, Fish and Hall 1978), but are susceptible to desiccation (Fish and Hall 1978). Oviposition rates are very low, with many females depositing only a single egg in a pitcher (Heard 1994c).

Eggs are deposited about a week after emergence and hatch in four to six days at 23°C (Wallis and Frempong-Boadu 1967). Larvae require about three weeks to develop at 23°C and under optimum food conditions, but will slow or stop development when food is limiting (Wallis and Frempong-Boadu 1967). *Wyeomyia smithii* is resource-limited over most of its range, with larvae having higher growth rates and more synchronous development when given a nutritional supplement than when grown under natural food conditions (Farkas and Brust 1985, Heard 1994a). Larval development is also affected by temperature. Kingsolver (1979) notes that pitchers growing in open sun accumulated more degree days than those growing in the shade in the same bog, and that differences in the accumulated degree days determined whether there would be one or two generations of mosquitoes in a year in Michigan.

Wyeomyia smithii overwinter as third-instar larvae in the northern part of their range (Smith and Brust 1971, Evans and Brust 1972) and in the fourth instar in the south (Bradshaw and Lounibos 1972). Diapause is induced by photoperiod, but the critical photoperiod varies with latitude, ranging from 13.8–14.6 hours per day in Massachusetts (Bradshaw 1980) to 14.5–15 hours in southern Manitoba (Smith and Brust 1971) and about an hour later in northern Manitoba (Farkas and Brust 1986a). The latitudinal differences in diapause induction result in larvae entering the third-instar diapause at about the same calendar date in northern and southern Manitoba (Farkas and Brust 1986a). Interestingly, *W. smithii* require the same critical photoperiod for terminating diapause as they do for diapause induction (Smith and Brust 1971). Photoperiod is a well-known cue for inducing diapause but is less common in ending diapause (Beck 1968).

The diapausing larvae overwinter encased in ice in the pitcher leaf, but are not tolerant of very cold temperatures. Most diapausing larvae cannot survive any subzero temperatures for more than a few months (Paterson 1971, Evans and Brust 1972), yet they are found in climates (e.g. northern Manitoba, Labrador) where air temperatures may dip to –30°C and colder for extended periods. Evans and Brust (1972) found that only about 50 percent of larvae could supercool to –5°C, although a few larvae could supercool to ≤ –10°C. Paterson (1971) found that all larvae died in less than a month at –16.5°C. Survivorship in the northern parts of the range appears to be highly dependent on snow cover, which lays an insulating blanket over the pitcher, preventing the pitcher temperature from falling below about –3 to –5°C. Paterson (1971) reported nearly zero mortality in a population in New Brunswick where snow covered the bog early in the winter and winter temperatures persisted for only four months. Mortality was considerably higher (30–91 percent) in Manitoba where winter temperatures persisted for up to six months. Interestingly, survivorship was higher in northern locations than in the south in Manitoba (Farkas and Brust 1986a), probably because of earlier and more extensive snow cover in the northern site, which insulated the pitchers from cold temperatures in early winter. Heard (1994b) found that wind exposure was the most important factor affecting distribution of pitcher plant mosquitoes in Newfoundland, and he also attributed this pattern to snow cover, since sites with high wind exposure experienced high winter mortality from lack of snow.

In the spring diapause is terminated by a resumption of long days (Smith and Brust 1971) and warm enough temperatures for development to be resumed. Mosquitoes moult to the fourth instar in early spring, and pupation occurs (in northern populations) around the time that new leaves are just opening on the plant (Fig. 18.1).

Metriocnemus knabi Coquillet

Metriocnemus knabi, the pitcher plant midge, has not been as well studied as *W. smithii,* the pitcher plant mosquito. Their geographical ranges overlap, but the midge appears to be more widespread and abundant in the north than the mosquito. The midge lives in the mixed detritus that accumulates in the bottom of the pitcher (Fish and Hall 1978) and scavenges for particulate material or bores into drowned insects (Bradshaw 1983). It obtains its oxygen directly from the water and so is restricted to pitchers with sufficient oxygen for them to respire. Cameron et al. (1977) found that it was relatively intolerant of low oxygen, but observed that oxygen conditions in the leaves was generally high enough that strategies for dealing with low oxygen would rarely be needed.

Paterson and Cameron (1982) give the seasonal phenology of the midge for a location in New Brunswick. This is summarized, along with the phenology of the plant and other inquilines, in Fig. 18.1. In New Brunswick overwintered fourth-instar larvae began developing into pupae and adults by early May. Adults from this generation oviposited into overwintered leaves, and first-instar larvae were noted about a month after adult emergence. A second generation, which had overwintered as second- and third-instar larvae, matured by late June and July, giving a second emergence peak at that time, although low numbers of adults were seen throughout the rest of the summer. These adults oviposited in both old and new leaves, and early-instar larvae from this generation began to appear in late July. Larvae from the spring generation grew through the summer and began pupating by August, giving another emergence peak in late summer. Pupation occurred in a gelatinous mass on the sides of the pitcher, above the water line. Pupae were still observed as late as September, but development to the pupal stage was arrested by a critical photoperiod (Paris and Jenner 1959). There were apparently two cohorts present in New Brunswick (Paterson 1971), each producing three generations every two years; i.e., they emerged in May and August one year and July the next year. Some overlap of these cohorts was considered likely.

Development of the midges, like that of the mosquitoes, likely depends strongly on temperature, since in Newfoundland, which is farther north than New Brunswick, Heard (1994a) found that the midges were univoltine (one generation per year) and showed a similar life cycle to that of the mosquito, presumably overwintering as late-stage larvae. Paterson (1971) recorded high tolerance of *M. knabi* larvae to cold; his field populations showed very little overwintering mortality, and in the lab 75 percent of larvae tested were still alive following freezing to 16.5°C for 116 days. In contrast, mosquito larvae tested at the same time showed 100 percent mortality at that temperature after only 26 days. Both the mosquitoes and midges stopped feeding and voided their guts completely before their habitats froze, and Paterson (1971) felt that this strategy aided supercooling, since gut contents may serve as a site for ice-crystal formation in supercooled insects. The critical temperature for cessation of feeding in *M. knabi* was 4°C.

Since *M. knabi* show little or no preference for ovipositing in new pitchers, they may oviposit in previously overwintered pitchers that cannot overwinter a second time in an intact state. Paterson and Cameron (1982) note that midge larvae can migrate (by crawling) from older, rotten pitchers into adjacent new ones before entering diapause to overwinter. In contrast, mosquito larvae oviposited into old pitchers are not capable of migrating to new habitats and will not survive the winter (Farkas and Brust 1986a).

Abundance of pitcher plant midges appears to be related to pitcher volume (Paterson and Cameron 1982, Nastase et al. 1995), although not in all cases (Hardwick and Giberson 1996). This pattern is probably related to intraspecific competition, since larval mass (Heard 1994a) and pupation success (Bradshaw 1983) decline with increasing larval density in pitchers. Competition is probably food-related, since midges were found to be nutrient-limited in a study in Newfoundland (Heard 1994a).

Blaesoxipha (*Fletcherimyia*) *fletcheri* (Aldrich)

Blaesoxipha fletcheri larvae are apparently restricted to the leaves of the northern pitcher plant, and as with the other inquilines of this species, its range is potentially as large as the range of the plant. Its reported range, however, is from Georgia to Newfoundland and westward to Manitoba (Forsyth and Robertson 1975). The larvae are maggot-like and are the largest of the pitcher plant inquilines, with third-instar larvae reaching 15–20 mm in length. They breathe at the water surface through posterior spiracles. The spiracle area is enlarged, with a cup-like structure that can be spread to take advantage of the surface tension layer and keep them suspended (Fish and Hall 1978, Juniper et al. 1989). Larvae will drown if submerged for any period of time. They feed voraciously at the surface on newly drowned insect prey still floating on the water surface (Forsyth and Robertson 1975).

Adult females are viviparous (i.e., bear live young), and "larviposit" directly into the water while sitting on the rim of the pitcher (Juniper et al. 1989). In Ontario *B. fletcheri* begins development in early June in synchrony with the initial openings of new pitchers (Fig. 18.1) (Forsyth and Robertson 1975). Adults prefer to larviposit in new pitchers, and larvae are only found in older pitchers at peak densities (Forsyth and Robertson 1975, Farkas and Brust 1986b). Their strategy to colonize new pitchers may relate to their pattern of feeding, since pitchers attract most of their prey very soon after they open (Fish and Hall 1978) and the larvae require fresh, floating food to survive. The larvae leave the pitchers to pupate in the surrounding *Sphagnum* (Farkas and Brust 1986b). In Ontario (Forsyth and Robertson 1975) development from first instar to pupation occurred in less than two weeks, so there is potential for several generations to occur during the summer, depending on temperature and food availability. They apparently overwinter as pupae in the sphagnum moss (Fig. 18.1).

Females produce a clutch of 8–14 larvae, but larviposit single larvae into pitchers, and rarely is more than one larva found in a pitcher. This pattern relates to the aggressive behaviour of the larva; when a *B. fletcheri* larva encounters another, it coils around it and/or pierces it with its mouthparts to try to submerge and drown it (Forsyth and Robertson 1975). The larger larvae are most successful in these encounters, either killing the smaller ones or driving them from the pitcher. Larvae are able to migrate limited distances to move to an unoccupied pitcher. Density estimates of this species tend to be somewhat misleading (since densities are rarely more than one per leaf), and it may be more appropriate to refer to frequency of occurrence in pitchers, rather than density (Hardwick and Giberson 1996).

Interactions between Pitcher Plant Inquilines

The three major pitcher plant inquilines mentioned above are all detritivores and depend on the prey captured by the plant. Some early researchers (e.g., Buffington 1970, Paterson 1971) speculated on how communities are maintained despite the potential for competitive interactions. Both Buffington (1970) and Paterson (1971) suggested that the various inquilines fed on different stages of the detritus breakdown process and therefore partition the resource in order to avoid competitive exclusion. Fish and Hall (1978) further noted that the inquilines also partitioned the resource temporally, timing their life cycles to take advantage of the stage of decomposition of the prey. However, these interactions were not quantitatively tested until relatively recently. Bradshaw (1983) confirmed experimentally that the three inquilines tended to be "self-limited" (i.e., showing intraspecific competition) rather than being limited by each other (interspecific competition).

In fact, there is evidence that the presence of the midges actually enhances the mosquito populations. Bradshaw (1983), Nastase et al. (1995), and Heard (1994a) found that mosquito abundance was positively correlated to midge abundance. Heard (1994a) showed that this was an example of a processing-chain commensalism, where consumers specialize on food at different stages of processing. The feeding action of the midges speeds the breakdown of the prey and accelerates production of the bacteria and protozoa that make up the major food of the mosquito, so that the presence of midges enhances production of the mosquitoes. Mosquito abundance had no effect on midge abundance, however, indicating that the chain is unidirectional (Heard 1994a).

Another question concerns whether competition occurs between the inquilines and the plant itself, for example, in a sort of resource parasitism. Although the plant can survive without prey, growth rates are higher when prey are present (Chapin and Pastor 1995). It is clear that the insect inquilines that live in the plant intercept some of these resources, and potential energy is lost to the plant when the insects respire and emerge out of the pitcher. However, the insect larvae may benefit the plant by speeding up the decomposition of the resource and/or converting the nutrients into a form more easily used by the plant (Paterson and Cameron 1972). Bradshaw and Creelman (1984) confirmed experimentally that the relationship between the plant and the inquilines was mutualistic. The inquilines inject nitrogen (as ammonia) into the leaf fluid, which is taken up by the plant. This in turn prevents the build-up of toxic waste products in the water, which would have a harmful effect on the inquilines. Further, the leaves take up CO_2 (in light) and infuse oxygen into the water. Although bacterial and enzyme breakdown of prey occurs even in the absence of the inquilines, the inquilines speed up the process and also prevent the dissolved nutrients from becoming sequestered in microbial populations (by feeding on the bacteria and protozoa). Since pitcher plants in open bogs can experience very high temperatures and light conditions, even in the far north (e.g., Evans and Brust 1972), plant growth may become carbon-limited (Bradshaw and Creelman 1984). These same physical conditions may increase O_2 demand (and subsequent CO_2 release through respiration) by the inquilines in the water, so Bradshaw and Creelman (1984) suggest that the mutualistic interaction relating to O_2 and CO_2 may be as important as the ammonia/nitrogen one.

There is still no evidence, however, that the interaction between pitcher plants and its inquilines results in higher growth rates or production by the plant. The three main inquilines in northern pitcher plants require the plant to survive, but the degree of importance of the inquilines to the plant remains to be quantified.

CONSERVATION ISSUES RELATING TO PITCHER PLANT INQUILINES

Many pitcher plant inquilines are obligate inhabitants of pitcher plants and so are obviously dependent on the presence of the pitcher plant to survive. Pitcher plants in turn are dependent on the survival of the bog habitat. Peatlands may be disturbed by agriculture, urban development, aforestation, and peat harvest (Keys 1992).

Wetlands in Canada

European settlers in North America long considered wetlands wasted land, but that viewpoint is gradually being replaced by an understanding of the value of wetlands as wildlife and biodiversity refuges and carbon sinks and in maintenaning water quality (Keys 1992). More than 70 percent of the wetland areas in southern Canada have been altered for agriculture, urban expansion, and forestry, and the percentages are much higher in some heavily populated areas of the country (Keys 1992). However, peatlands, the habitats where pitcher plants flourish, are far more common in the sparsely populated north, where there has been relatively little disturbance. This is in sharp contrast to northern Europe, where few peatlands have not been altered. Peat in those areas is harvested for fuel and for horticultural purposes, and drained peatlands may also be used to support forestry or agriculture (Keys 1992).

Status of Peatlands in Atlantic Canada

Peatlands cover 111 million ha (12 percent) of Canada's territory, and occur mostly in the Boreal Forest Biome (Keys 1992). Total peatland area and its percentage of the total land area are given for each of the eastern Canadian provinces in Table 18.2. Less than 0.02 percent of the total peatland area in Canada is currently being harvested, and most of the harvest operations have been centered in Quebec, Alberta, and New Brunswick (Keys 1992, Price 1996). Peat harvest operations are important employers in these regions, and

TABLE 18.2. Area Classified as Peatlands in the Five Eastern Provinces of Canada

Province	Peatland Area (ha)	Percent of Total Area
New Brunswick	120,000	2
Newfoundland and Labrador	6,429,000	17
Nova Scotia	158,000	3
Ontario	22,555,000	25
Prince Edward Island	8,000	1
Quebec	11,713,000	9

the value of the industry (>$180 million in sales per year; Price 1996) has prompted a steady growth in harvesting operations over the past decade (Keys 1992). Although the area involved is small, peatlands tend not to be managed as sustainable resources, since they are simply abandoned once they are exhausted and are left to regenerate naturally. However, fewer than 10 percent of the abandoned bogs in Quebec show *Sphagnum* regeneration even after up to 20 years (Price 1996). Interest in conservation and restoration strategies has been increasing in recent years in Canada, resulting in studies on the effects of peat extraction on bog vegetation (Jonsson-Ninniss and Middleton 1991), physical characteristics of cutover bogs (Price 1996), and natural revegetation of harvested peatlands (Lavoie and Rochfort 1996). Machine-harvested peatlands (as are usually found in Canada; Keys 1992) do not readily return to their preharvest conditions after cutting, because the harvesters destroy the capillarity of the organic layer, making it difficult for *Sphagnum* to reestablish (Price 1996). During peat extraction the water table drops and the compression of the drained peat alters the pore structure, so that even when these areas are reflooded the water relations are changed (Schouwenaars 1993, Price 1996).

The changes in the hydrology of cutover bogs result in a change in vegetation as well, favoring well-rooted plants rather than *Sphagnum* moss, which relies on capillary action to obtain water (Jonsson-Ninniss and Middleton 1991, Price 1996). We had difficulty finding information on regeneration of pitcher plants (*Sarracenia*), mainly because most of the work that has been done on revegetation of harvested bogs comes from Europe, where *Sarracenia* is not native. However, no pitcher plants were found in revegetated bogs in Quebec (Lavoie and Rochfort 1996) or Ontario (Jonsson-Ninniss and Middleton 1991) after 20 years postharvest, although they were apparently present, in the Ontario bog at least, prior to harvest.

Pitcher plants (Sarraceniaceae) possess numerous small, winged seeds (Zomlefer 1994) and so may be dispersed on wind currents, but presumably they require a nearby source of colonists to recolonize revegetated areas. Peat harvesting generally removes all the surface vegetation and destroys any seeds that may be in the peat (Lavoie and Rochfort 1996), so recolonization of cutover bogs by *Sarracenia* may be very slow. Lavoie and Rochfort (1996) suggest that restoration attempts should include addition of fertilizer and deliberate dispersal of seeds and spores from bog plants. In addition, pitcher plants are easily transplanted from existing bogs and will reestablish in bogs as long as hydrological conditions are suitable (Hardwick and Giberson 1996).

Reestablishment of the Inquiline Community Following Disturbance

Bogs are generally isolated and localized habitats, a tendency which results in discrete, isolated populations of inquilines (Forsyth and Robertson 1975). What this means for dispersal of inquilines into bogs which have been disturbed and subsequently restored is difficult to say, as we could not find published studies of bog revegetation attempts that considered the invertebrate communities associated with the bog plants in North America. Recently, in Prince Edward Island (P.E.I.), we had the opportunity to evaluate how transplanting of pitcher plants *prior* to peat harvest (rather than during postharvest recovery) affected subsequent performance of the inquilines (Hardwick and Giberson 1996).

Due to impending drainage and harvest of an important bog near Miscouche, P.E.I., a local environmental group organized a "rescue" of "rare and/ or interesting" bog plants from the bog to a number of other bogs in the region (Keys 1992). The pitcher plants appeared healthy and vigorous two years following the transplant, but the inquilines did not fare as well (Hardwick and Giberson 1996). Mosquitoes were extremely rare or absent from all transplanted pitchers, although they were present in at least some pitchers in all other bogs investigated in the province. Midge numbers were also reduced, though not to the same degree as mosquito numbers, and only in the bogs where there was not a local source of colonizing adults. Only the sarcophagid appeared to be unaffected by transplanting. We concluded that mortality of inquilines was very high during the transplant process since most of the transplanted pitchers lost their fluid during transport, and that the subsequent differences in their abundance were related to recolonization abilities. Adult pitcher plant mosquitoes are apparently very poor dispersers (Wallis and Frempong-Boadu 1967), and Heard (1994b) has also noted that they do not readily return to sites where they have been displaced due to poor overwintering success. The midges appear to be able to disperse short distances to colonize empty pitchers (within the bog; Hardwick and Giberson 1996), but may be slow at colonizing isolated habitats. Sarcophagids are strong flyers, however, and should recolonize quickly after the pitcher plants themselves reestablish following disturbance.

CONCLUSIONS

Very little of Canada's total peatland habitat is at risk at the present time, largely because of the general distribution of this habitat away from major population centers. However, because most of the harvest activities are concentrated in the southern regions of the country, where suitable habitat is relatively rare, these activities have the potential to threaten both the pitcher plants and their associated inquilines, at least on a local scale. Transplanting has proven to be a successful method of preserving pitcher plants, given suitable habitat, but may not be adequate for preserving pitcher plant invertebrates. More experimental work is needed to determine the success of various restoration strategies following peat harvest.

ACKNOWLEDGMENTS

Funding for the study on the Pitcher Plant transplant was provided by a University of Prince Edward Island Senate Research Grant, and the Island Nature Trust of Prince Edward Island was very helpful in providing information on the transplant. Sylvie Desormeaux translated the information on Quebec pitcher plants from the French in the book by Frère Marie-Victorin. Pat Crawford helped guide the figures through CorelDraw.

LITERATURE CITED

Addicott, J. F. 1974. Predation and prey community structure: An experimental study of the effect of mosquito larvae on the protozoan communities of pitcher plants. Ecology 55:475–492.

Bateman, L. E. 1987. A bdelloid rotifer living as an inquiline in leaves of the pitcher plant, *Sarracenia purpurea*. Hydrobiologia 147:129–133.

Beck, S. D. 1968. Insect Photoperiodism. Academic Press, New York.

Bradshaw, W. E. 1980. Thermoperiodism and the thermal environment of the pitcher-plant mosquito, *Wyeomyia smithii*. Oecologia 46:13–17.

———. 1983. Interaction between the mosquito *Wyeomyia smithii*, the midge *Metriocnumus knabi* and their carnivorous host *Sarracenia purpurea*. *In* J. H. Frank and L. P. Lounibos (eds.), Phytotelmata: Terrestrial Plants as Hosts for Aquatic Insect Communities. Plexus, New Jersey.

———. 1986. Variable iteroparity as a life-history tactic in the pitcher-plant mosquito *Wyeomyia smithii*. Evolution 40:471–478.

Bradshaw, W. E., and R. A. Creelman. 1984. Mutualism between the carnivorous purple pitcher plant and its inhabitants. American Midland Naturalist 112:294–304.

Bradshaw, W. E., and C. M. Holzapfel. 1989. Life-historical consequences of density-dependent selection in the pitcher-plant mosquito, *Wyeomyia smithii*. American Naturalist 133:869–887.

Bradshaw, W. E., and L. P. Lounibos. 1972. Photoperiod control of development in the pitcher-plant mosquito, *Wyeomyia smithii*. Canadian Journal of Zoology 50: 713–719.

Buffington, J. D. 1970. Ecological considerations of the cohabitation of pitcher plants by *Wyomyia smithii* and *Metriocnemus knabi*. Mosquito News 30:89–90.

Cameron, C. J., G. L. Donald, and C. G. Paterson. 1977. Oxygen-fauna relationships in the pitcher plant *Sarracenia purpurea* L. with reference to the chironomid *Metriocnemus knabi* Coq. Canadian Journal of Zoology 55:2018–2023.

Chapin, C. T., and J. Pastor. 1995. Nutrient limitations in the northern pitcher plant, *Sarracenia purpurea.* Canadian Journal of Botany 73:728–734.

Christensen, N. L. 1976. The role of carnivory in *Sarracenia flava* L. with regard to specific nutrient deficiencies. Journal of the Elisha Mitchell Scientific Society 92: 144–147.

Comeau, P. L. 1971. A study of five raised bogs on the Cape Breton Plateau. Master's Thesis, Acadia University, Wolfville, NS.

de la Rosa, C., and A. J. Nastase. 1987. Larvae of *Metriocnemus* cf. *Fuscipes, Limnophyes* sp., Pentaneurini (Diptera: Chironomidae) and *Culicoides* (Diptera: Ceratopononidae) from pitcher plants, *Sarracenia purpurea.* Journal of the Kansas Entomological Society 60:339–341.

Erskine, D. S., P. M. Catling, and R. B. MacLaren 1985. The Plants of Prince Edward Island, with New Records, Nomenclatural Changes, and Corrections and Deletions. Publication 1798, Agriculture Canada Research Branch.

Evans, K. W., and R. A. Brust. 1972. Induction and termination of diapause in *Wyeomyia smithii* (Diptera: Culicidae), and larval survival studies at low and subzero temperatures. Canadian Entomologist 104:1937–1950.

Farkas, M. J., and R. A. Brust. 1985. The effect of a larval diet supplement on development in the mosquito *Wyeomyia smithii* (Coq.) under field conditions. Canadian Journal of Zoology 63:2110–2113.

———. 1986a. Phenology of the mosquito *Wyeomyia smithii* (Coq.) in Manitoba and Ontario. Canadian Journal of Zoology 64:285–290.

———. 1986b. Pitcher-plant sarcophagids from Manitoba and Ontario. Canadian Entomologist 118:1307–1308.

Fish, D., and D. W. Hall. 1978. Succession and stratification of aquatic insects inhabiting the leaves of the insectivorous pitcher plant, *Sarracenia purpurea.* American Midland Naturalist 99:172–183.

Forsyth, A. B., and R. J. Robertson. 1975. Reproductive strategy and larval behavior of the pitcher plant sarcophagid fly, *Blaesoxipha fletcheri.* Canadian Journal of Zoology 53:174–179.

Givnish, T. J. 1989. Ecology and evolution of carnivorous plants. Pages 243–290 *in* W. G. Anderson (ed.), Plant-Animal Interactions. McGraw-Hill, Toronto, ON.

Hardwick, M. L., and D. J. Giberson. 1996. Aquatic insect populations in transplanted and natural populations of the purple pitcher plant, *Sarracenia purpurea,* on Prince Edward Island, Canada. Canadian Journal of Zoology 74:1956–1963.

Heard, S. B. 1994a. Pitcher-plant midges and mosquitoes: A processing chain commensalism. Ecology 75:1647–1660.

———. 1994b. Wind exposure and distribution of pitcher plant mosquito (Diptera: Culicidae). Environmental Entomology 23:1250–1253.

———. 1994c. Imperfect oviposition decisions by the pitcher plant mosquito (*Wyeomyia smithii*). Evolutionary Ecology 8:493–502.

Hinds, H. R. 1886. Flora of New Brunswick. Primrose Press, Fredericton, NB.

Istock, C. A., S. S. Wasserman, and H. Zimmer. 1975. Ecology and evolution of the pitcher-plant mosquito: 1. Population dynamics and laboratory responses to food and population density. Evolution. 29:296–312.

Jonsson-Ninniss, S., and J. Middleton. 1991. Effect of peat extraction on the vegetation in Wainfleet Bog, Ontario. Canadian Field-Naturalist 105:505–511.

Judd, W. W. 1959. Studies of the Byron Bog in Southwestern Ontario X. Inquilines and victims of the pitcher plant, *Sarracenia purpurea.* Canadian Entomologist 91: 171–180.

Juniper, B. E., R. J. Robins, and D. M. Joel. 1989. The Carnivorous Plants. Academic Press, London, UK.

Keys, D. 1992. Canadian Peat Harvesting and the Environment. Wetlands Issues Paper No. 1992–3, North American Wetlands Conservation Council (Canada).

Kingsolver, J. G. 1979. Thermal and hydric aspects of environmental heterogeneity in the pitcher plant mosquito. Ecological Monographs 49:357–376.

Lavoie, C. and L. Rochfort. 1996. The natural revegetation of a harvested peatland in southern Quebec: A spatial and dendroecological analysis. Ecoscience 3:101–111.

Lincoln, R. J., G. A. Boxshall, and P. F. Clark. 1982. A Dictionary of Ecology, Evolution and Systematics. Cambridge University Press, Cambridge, UK.

Lounibos, L. P., and W. E. Bradshaw. 1975. A second diapause in *Wyeomyia smithii*: Seasonal incidence and maintenance by photoperiod. Canadian Journal of Zoology 53:215–221.

Marie-Victorin, Frère. 1964. Flore Laurentienne (2nd ed.). University of Montreal Press, Montreal, QUE.

Mather, T. N. 1981. Larvae of alderfly (Megaloptera: Sialidae) from pitcher plant. Entomology News 92:32.

Mogi, M., and J. Mokry. 1980. Distribution of *Wyeomyia smithii* (Diptera, Culicidae) eggs in pitcher plants in Newfoundland, Canada. Tropical Medicine 22:1–12.

Nastase, A. J., C. de la Rosa, and S. J. Newell. 1995. Abundance of pitcher plant mosquitoes, *Wyeomyia smithii* (Coq.) (Diptera: Culicidae) and midges, *Metriocnemus knabi* Coq. (Diptera: Chironomidae), in relation to pitcher characteristics of *Sarracenia purpurea* L. American Midland Naturalist 133:44–51.

Paris, O. H., Jr., and C. E. Jenner. 1959. Photoperiodic control of diapause in the pitcher plant midge, *Metriocnemus knabi.* Pages 601–624 *in* Photoperiodism and Related Phenomena in Plants and Animals. Publication No. 55 of the American Association for the Advancement of Science, Washington, DC.

Paterson, C. G. 1971. Overwintering ecology of aquatic fauna associated with the pitcher plant *Sarracenia purpurea* L. Canadian Journal of Zoology 49:1455–1459.

Paterson, C. G., and C. J. Cameron. 1982. Seasonal dynamics and ecological strategies of the pitcher plant chironomid, *Metriocnemus knabi* Coq. (Diptera: Chironomidae) in southeast New Brunswick. Canadian Journal of Zoology 60:3075–3083.

Pietropaolo, J., and P. Pietropaolo. 1974. The World of Carnivorous Plants. R. J. Stoneridge, Shortsville, NY.

Pollett, F. C., and W. J. Meades. 1970. A Preliminary Checklist of the Peatland Flora in Newfoundland. Forest Research Lab Information Report N-X-54, Canadian Forest Service, November, St. John's, NFLD.

Price, J. S. 1996. Hydrology and microclimate of a partly restored cutover bog, Quebec. Hydrological Processes 10:1263–1272.

Roland, A. E., and E. C. Smith. 1969. The Flora of Nova Scotia. Nova Scotia Museum, Halifax, NS.

Schouwenaars, J. M. 1993. Hydrological differences between bogs and bog-relicts and consequences for bog restoration. Hydrobiologia 265:217–224.

Smith, J. B. 1902. Life history of *Aedes smithii* Coq. Journal of the New York Entomological Society 10:10–15.

Smith, S. M., and R. A. Brust. 1971. Photoperiodic control of the maintenance and termination of larval diapause in *Wyeomyia smithii* (Coq.) (Diptera: Culicidae) with notes on oogenesis in the adult female. Canadian Journal of Zoology 49:1065–1073.

Stewart, C. N., and Nilsen, E. T. 1991. *Drosera rotundifolia* growth and nutrition in a natural population with special reference to the significance of insectivory. Canadian Journal of Botany 70:1409–1416.

Turner, T. S., J. L. Pittman, M. E. Poston, R. L. Petersen, M. MacKenzie, C. H. Nelson, and R. M. Duffield. 1996. An unusual occurrence in West Virginia of stoneflies (Plecoptera) in the pitcher-plant, *Sarracenia purpurea* L. (Sarraceniaceae). Proceedings of the Entomological Society of Washington 98:119–121.

van Breeman, N. 1995. How *Sphagnum* bogs down other plants. Trends in Ecology and Evolution 10:270–275.

Wallis, R. C., and J. Frempong-Boadu. 1967. Colonization of *Wyeomyia smithii* (Coquillett) from Connecticut. Mosquito News 27:9–11.

Wells, E. D. 1996. Classification of peatland vegetation in Atlantic Canada. Journal of Vegetation Science 7:847–878.

Wray, D. L., and C. S. Brimley. 1943. The insect inquilines and victims of pitcher plants in North Carolina. Annals of the Entomological Society of America 36:128–137.

Zomlefer, W. B. 1994. Guide to Flowering Plant Families. University of North Carolina Press, Chapel Hill, NC.

19 Constructed Marshes in Southeast Pennsylvania

Invertebrate Foodweb Structure

G. WINFIELD FAIRCHILD, ANN M. FAULDS, and
LYNNETTE L. SAUNDERS

We compared the taxonomic and trophic structure of aquatic invertebrate communities in 11 recently created freshwater marshes to communities at 7 nearby reference sites. Our purpose was to determine which features of the created sites most influence the invertebrate community and to evaluate the rates at which key elements of the community are acquired during early succession following site construction. The created and reference sites were all semipermanent, shallow ponds (<1.5 m mean depth) in the Northern Piedmont ecoregion of southeast Pennsylvania. We sampled the invertebrates using both a 0.2-m^2 quadrat and 82-cm^2 core tube at 10 locations in each wetland. Invertebrates were identified and weighed, then allocated to 1 of 57 families and to 1 of 7 trophic groups. We also obtained concurrent measurements of fish presence, vegetation abundance, physical habitat features, and water chemistry. Based on canonical correspondence analysis, the presence of fish had the greatest effect of any habitat variable on the invertebrate community. Mean invertebrate biomass was ca four times higher in wetlands with few or no fish (n = 6) than at sites with a well-established fish community (n = 12) (p < 0.001). Macroinvertebrate predators were the most strongly reduced of the 7 trophic groups in wetlands with fish. The created wetlands were rapidly colonized by most trophic groups, but macrocrustacean scavengers (amphipods and isopods) and consumers of aquatic macrophytes (primarily chrysomelid and curculionid beetles) were notably absent at several created sites. Using family-level taxonomic richness as one criterion of successful site construction, we found that "failures" (with fewer than the 95 percent confidence interval of 18.8–28.9 families based on the reference

Invertebrates in Freshwater Wetlands of North America: Ecology and Management, Edited by
Darold P. Batzer, Russell B. Rader, and Scott A. Wissinger
ISBN 0-471-29258-3   © 1999 John Wiley & Sons, Inc.

sites) also supported few macrophytes. Establishment of aquatic vegetation may thus be an important precondition for assembly of the invertebrate community.

CREATED WETLANDS

Under Section 404 of the Clean Water Act, unavoidable destruction of existing wetlands must be accompanied by replacement in kind through wetland restoration or creation (Kusler and Kentula 1990). Replacement in kind implies that the functions of the natural site are maintained in its newly constructed replacement. One such function is the "life support" of an aquatic invertebrate community (Hammer 1992). Many species of aquatic insects and a wide range of other invertebrates require wetland habitat for part or all of their life cycles and are critical elements in foodwebs supporting waterfowl (Peterson et al. 1989, Batzer et al. 1993) and fish (Hanson and Butler 1994, Mallory et al. 1994, Hanson and Riggs 1995).

Establishment of invertebrates at newly created sites can be expected to depend on the dispersal capabilities of potential colonists, wetland size and physicochemical features, the duration of standing water, and site position within the regional landscape (see review by Batzer and Wissinger 1996). All of these individual effects, and their potential interactions, are poorly understood at present. Whether a given created site will be "successful" in supporting an invertebrate community typical of a natural wetland in its region thus cannot yet be predicted with accuracy.

A number of studies have recently begun to compare the aquatic invertebrate communities of newly created wetlands to reference sites from the same region. Most of these studies have been taxonomically based considerations of organismal densities, with the degree of taxonomic resolution typically inversely related to the range of taxa under consideration (e.g., Street and Titmus 1979, Barnes 1983, Streever and Crisman 1993a, Streever et al. 1996, Scatolini and Zedler 1996). Yet functional success of a created site implies, at least in part, the acquisition of an invertebrate foodweb with all major trophic elements and linkages present (Pimm 1982). Studies by Murkin et al. (1983), Gladden and Smock (1990), and de Szalay and Resh (1996) have been notable in examining trophic structure as an alternative to taxonomic composition of wetland invertebrate communities, and Jones et al. (1996) have recently applied such an approach to evaluating the macroinvertebrate assemblages of wetlands created in surface mine depressions.

This study evaluates the functional success of created marshes in sustaining invertebrate foodwebs. The data set differs from most previous work not only in considering the trophic as well as taxonomic structure of the invertebrate assemblage, but also in presenting abundances in terms of biomass rather than densities. We address two questions:

1. Which habitat features of these marshes most influence invertebrate taxonomic composition and foodweb structure?

2. How rapidly are the taxonomic and trophic components of the invertebrate community assembled by colonization of newly created marshes, and do the invertebrate communities of the created marshes resemble those of reference sites in the region?

RESEARCH METHODS

Study Sites

Eighteen freshwater marsh ponds were chosen for study within the Northern Piedmont ecoregion (Omernik 1987) of southeast Pennsylvania (Fig. 19.1).

Fig. 19.1. Locations of 18 wetlands in southeast Pennsylvania. Names and site descriptions corresponding to the two-letter codes are given in Table 19.1.

Eleven of these were newly created, including six sites created as highway construction mitigation near large cities (CO, BE, BR, EX, GR, RA), two projects undertaken by the U.S. Fish and Wildlife Service for Chester County Parks and Recreation (EM, H1-3), and three sites providing mitigation associated with development by private corporations (FI, OA, OV). Herbaceous marsh or upland vegetation surrounded the ponded areas at most sites, but several sites were partially bordered by deciduous forest.

The created sites varied in age from one to eight years at the time of sampling. The seven reference sites were chosen to span the same geographic region as the constructed sites, and included four older constructed ponds ranging from 35–170 years in age and three oxbow ponds of unknown age adjacent to Brandywine Creek in Chester County. Sites were selected based on preliminary visits as fulfilling the criteria of having (a) permanent water during years of normal precipitation and (b) an average depth of <1.5 m.

With one exception, sites were sampled just once during early summer (typically June or July) of one year during the three-year study. Sampling at Hibernia Park (H1-3), however, was repeated in each of the three summers in order to investigate colonization and early succession of the invertebrate community at a newly constructed site.

Each wetland was mapped to determine the total ponded area, and areas dominated by vegetation versus open water if substantial open water was present. A portion of a pond was considered to be "open water" if no vegetation was seen at or near the pond surface, although smaller plants were later found at the bottom during sampling in some instances. Mapping was initially performed using a Sonin Combo Pro distance measurer, a Brunton compass affixed to a tripod, and standard surveying procedures. We later walked the perimeter of each site with a Trimble Pro XL backpack GPS unit (accuracy ±1 m) to confirm our estimates of pond area and to add the site location to a regional GIS database.

Invertebrate Sampling

Sampling was intended to characterize the invertebrate community of the wetlands as a whole, with representation of all major microhabitats. We chose 10 locations for sampling, based on the maps of each wetland. In well-vegetated, shallow wetlands with little depth variation and little open water (n = 11), the locations were positioned in a systematic arrangement, subject to the constraints of irregular pond shape, such that all locations were approximately equally distant from neighboring locations. We used stratified sampling in wetlands (n = 7) where the ponded area could be easily partitioned into vegetated versus open-water areas, and established seven locations in the vegetated portion, usually at equal distances along the band of fringing vegetation. Three locations were positioned in as systematic an arrangement as possible in the more homogeneous open-water stratum. In these stratified wetlands we estimated overall macroinvertebrate abundance and biomass by weighting the samples from the two strata according to the relative surface areas of the vegetated and open-water habitats.

At each of the 10 sampling locations a square acrylic quadrat, with walls 70 cm high and enclosing an area of 0.20 m^2, was pushed into the sediments. All entrained vegetation was removed with hedge clippers, any associated invertebrates were removed, and the plants were wet-weighed as fresh material using a spring scale. Additional invertebrates were removed from the water column and surficial sediments in the quadrat with successive (n = 3) complete sweeps of the enclosed space with a D-frame dipnet (300-μm mesh). We estimated percent capture of selected taxa in the three sweeps using a maximum-likelihood estimator for removal-depletion sampling (Van Deventer and Platts 1983): coenagrionid damselflies (n = 2 estimates), libellulid dragonflies (n = 4), caenid mayflies (n = 1), and adult noterid beetles (n = 1). The samples were preserved in 70 percent ethanol prior to identification and weighing (below).

We collected two smaller-bodied taxa (the dipteran families Chironomidae and Ceratopogonidae) from the surficial sediments at each of the 10 sampling locations using a PVC core (82 cm^2) pushed 2.0 cm into the substratum (vol = 163 mL). The sediments were mixed thoroughly, then partitioned for estimation of microinvertebrate densities (133 mL) and sediment analyses (30 mL; see below). The sediments were refrigerated for a maximum of 48 hours, and invertebrates were then separated from the sediments by the addition of a 1.12 specific gravity sugar solution containing Rose Bengal. The procedure was repeated three times. The organisms were then either counted immediately or placed in 4 percent formalin before counting.

Invertebrates were typically identified to genus using a Wild MZ-8 stereoscope (100× maximum magnification). Several macroinvertebrate groups (sphaeriid clams, water mites, *Chaoborus,* and oligochaetes) were ineffectively sampled or enumerated and are not included in the analysis. Biomasses of the larger taxa were determined as alcohol wet weights using a Cahn 31 electrobalance or American Scientific S/P 180 analytical balance, depending on organism size. Our choice of wet weights in estimating biomass allowed retention of the specimens for later taxonomic verification. To check for precision of the wet weight estimates, we randomly chose individuals of five taxa and reweighed each three times, obtaining an average coefficient of variation of 1.6 percent (range 0.5–4.9 percent).

Chironomids were slide-mounted with a mixture of CMCP-9 and CMCP-9AF (Beckett and Lewis 1982) to obtain generic identifications. Body volume was assumed to be that of a curved cylinder (Smit et al. 1993), with mean body diameter and axial length determined using image analysis (NIH Image 6.10). Wet weights were then estimated from the volume determinations assuming a tissue density of 1 mg/mm^3.

Trophic Group Designations

The families were initially classified as predators, macrophyte consumers, or algivores-detritivores based upon general feeding mode or predominant diet using Merritt and Cummins (1996) and Brigham et al. (1982) for the aquatic insects and Pennak (1989) or a variety of primary references for other taxa. The predators were further subdivided into two groups. Members of the chironomid subfamily Tanypodinae and ceratopogonid larvae were collected from the sediment cores and considered here to be small-bodied "Microinvertebrate predators," although utilization of non-animal food is known to occur particularly in the earlier instars of many species (e.g., Roback 1969). "Macroinvertebrate predators," obtained from the quadrat sampling, included a wide range of taxa known to capture prey from the water column, plant surfaces, sediments, or air-water interface.

We found that we could not consistently classify the wide range of taxa consuming small particles of sedimentary or plant-associated algae and detritus as algivores or detritivores. We did, however, partition the algivores-detritivores (AD) into three trophic groups of approximately equal biomass. "AD (non-tanypodine) chironomids," belonging almost entirely to the subfamily Chironominae, were collected in the sediment cores. "Snails" were tabulated separately as a second AD group. Finally, "other AD" included a taxonomically diverse assemblage of other algae-detritus feeders obtained in the quadrat samples. Corixid bugs (most of them immature and not identifiable to genus) were assigned to this latter category.

A sixth group, "macrophyte consumers," was narrowly defined as including invertebrates which feed predominantly on live vascular plant tissue. Leaf/stem mining species were not captured using our sampling methods, and most individuals were members of the beetle families Curculionidae or Chrysomelidae, leptocerid caddisflies, or lepidopteran larvae. Shredders known to process macrophyte detritus were allocated to the group "other AD." Finally, macrocrustacean "scavengers" (amphipods and isopods) were considered as a separate trophic group, as their opportunistic consumption of a wide range of living and dead material prevented their assignment to any of the above categories. Seven trophic groups or guilds are thus evaluated in our analyses.

Some genera within families are known to differ in terms of their predominant food source, and generic rather than family biomasses were thus allocated to the appropriate guilds in those cases. Likewise, where ontogenetic shifts in trophic group were known, such as in many hydrophilid beetle genera, biomasses of the life history stages were tabulated separately and added to the appropriate guilds.

Habitat Measurements

We estimated water depth, sediment organic content, and sediment algal biomass at each of the 10 sampling locations in each wetland. Sediment organic content was obtained from the core samples by weighing a 25-mL aliquot dried at 105°C for 24 hours in a tared crucible, ashing the sample at 500°C for three hours, then reweighing to obtain percent ash-free dry mass. Sediment photopigments were extracted from a second, 5-mL aliquot with 15 mL hot 95 percent EtOH, then measured with a Bausch and Lomb Spectronic 501 spectrophotometer following recommendations by Nusch (1980). Chlorophyll-a and phaeopigment densities were measured separately, but were later combined in estimating sediment algal biomass to avoid bias due to possible degradation of the chlorophyll-a during handling (Hansson 1988). We then computed mean estimates for the three habitat variables for each wetland, applying different weighting coefficients for vegetated versus open-water sites in the stratified wetlands, as we had for invertebrate abundances.

Water chemistry estimates were means based on four water samples collected at each wetland. We determined pH and temperature with an Orion Model SA250 pH meter, and used a YSI Model 51B oxygen meter held at middepth to measure dissolved oxygen. We measured conductivity and turbidity with a YSI 33 SCT meter and Hach Model 16800 Portalab Turbidimeter, respectively.

We assessed fish communities in each wetland in two ways. We deployed six standard double-aperture minnow traps overnight and identified and measured all captured individuals for total length. We also frequently encountered fish while site-mapping, or in the quadrats while collecting macroinvertebrates. We classified wetlands as having either (1) a substantial fish community or (2) few or no fish (two sites, in which we captured no fish but observed one or a few individuals after much searching, were allocated to the latter category).

Analysis

We compared the 11 newly created and 7 reference wetlands in terms of major habitat features (ponded area, conductivity, depth, dissolved oxygen, sediment organic matter, sediment chlorophyll-a, turbidity, temperature, and mean vegetation mass) using a one-way multivariate analysis of variance (MANOVA).

To evaluate the relationships of habitat features to the taxonomic composition of the invertebrate communities at the 18 wetlands (Question 1), we used Canonical Correspondence Analysis (CCA), available as version 3.10 of CANOCO (ter Braak and Verdonschot 1995). We used a stepwise forward procedure of habitat variable selection to determine the percentage of variation in the invertebrate community associated with particular habitat variables. We determined the significance of a variable by treating all other variables as covariates, then performing a Monte Carlo procedure with 99 permutations. The family-level biomass estimates were $\log(X + 1)$ transformed prior to analysis. We excluded taxa found at only one site or at region-wide biomass densities <1 mg/m^2 to reduce the importance of rare taxa in the ordination. Although we included just the third year of sampling at H1-3 in the calculations to maintain statistical independence of the sites, we included the first- and second-year data in the ordination plot as "passive" samples to show successional trends at the site.

Based on the CCA results, we singled out the presence or absence of fish and ponded area for further analysis. We compared the $\log(X + 1)$-transformed biomass and mean wet weight per individual for each trophic group in wetlands with fish (n = 12) versus few or no fish (n = 6) using MANOVA. We determined the effects of area on biomass of the trophic groups using multivariate analysis of covariance (MANCOVA), with fish included as an ANOVA factor and area considered as the covariate.

We examined the acquisition of taxonomic richness over successional time (Question 2) by relating the total number of invertebrate families sampled to site age for the 11 created wetlands. We further related the total number of families sampled at the created sites to a 95 percent confidence interval for family richness at the seven reference sites. We evaluated the degree of resemblance of mean biomass of the seven trophic groups in the created versus reference sites using MANOVA on $\log(X + 1)$-transformed data.

RESULTS

The sites ranged from 0.03–2.49 ha, and from 7–149 cm in mean depth (Table 19.1), with no significant differences between newly created and reference sites (p > 0.05). Specific conductance varied from 81–602 μS, likewise showing broad overlap between the created and reference sites. Mean percent organic matter of the surficial (2-cm depth) sediments was slightly, but not significantly, higher among reference sites (9.9 percent) than among newly created sites (6.7 percent), while sediment chlorophyll-a (including phaeophytin) was slightly higher in the newly created sites (p > 0.05). Water temperature was particularly low at Fricks Lock (FL), a very small, shallow spring-fed pool. In effect, the created sites were very similar in most habitat features to reference sites in the same region.

Based on maximum-likelihood estimates of capture efficiency, coenagrionid damselflies were captured with ca 70–75 percent efficiency per sweep (nearly 100 percent capture in three sweeps). Caenid mayflies and noterid beetles were likewise captured with 96 percent and 95 percent success using three sweeps, respectively. In marked contrast, capture of libellulid dragonfly larvae was highly variable, with estimates ranging from nearly 100 percent capture to values sufficiently low to preclude calculation of estimated densities. Substrate roughness (e.g., presence of aquatic plants and rocks) may thus have influenced our capture of this primarily sediment-associated family. We also noted evasion of the quadrats by gyrinid beetles prior to sampling, likewise causing probable underestimation of their biomass.

The taxonomic and trophic components of the invertebrate community are compiled in Table 19.2. Aquatic insects represented 50 of the 57 families

TABLE 19.1. Habitat Features of 18 Wetlands in Southeast Pennsylvania[a]

Wetland	Type	Age (years)	Area (ha)	Depth (cm)	Organic Matter (percent)	Conductance (μS)	Temperature (°C)	Vegetation Weight (kg/m²)	Fish (P/A)
Bergdoll (BE)	C	7	0.32	26	8.4	304	26.0	0.0	P
Blue Route (BR)	C	3	0.40	40	4.3	269	33.5	0.8	P
Brandywine Conserv. N (BN)	N	?	1.44	42	8.2	175	21.1	0.4	P
Brandywine Conserv. S (BS)	N	?	0.68	26	7.9	251	27.4	1.4	P
Conestoga (CO)	C	3	0.24	28	8.5	463	25.9	0.4	P
Embreeville (EM)	C	2	2.49	33	14.0	279	35.9	2.6	A
Exton (EX)	C	8	0.48	30	5.9	158	21.6	0.9	P
Fireschool (FI)	C	4	0.53	149	5.2	126	26.8	0.1	P
Fricks Lock (FL)	O	170	0.02	11	4.6	172	14.6	0.9	A
Grammes Lane (GR)	C	7	0.45	45	4.5	375	28.2	0.1	P
Great Marsh (GM)	O	90	0.62	19	22.6	212	21.7	0.5	P
Hanks (HA)	N	?	3.59	21	11.0	374	17.4	0.7	P
Hibernia Park (H1-3)	C	1-3	1.02	46-60	3.4-8.9	81-110	22.8-27.0	0.7-3.5	A
Oaklands (OA)	C	5	0.04	26	4.6	234	23.1	1.0	A*
Oxford Valley (OV)	C	2	0.03	7	8.6	195	25.1	0.5	A
Raiders Lane (RA)	C	7	0.28	20	6.3	254	22.0	0.6	P
Safe Harbor (SH)	O	70	0.40	18	5.1	602	24.8	1.3	A*
Smedley (SM)	O	35	0.94	60	9.5	106	16.4	1.0	P

[a]Wetlands were classified as C = created, O = older, or N = natural. Fish at two sites (marked *) were classified as Absent, although one or a few individuals were observed. Ponded area, depth, percent organic matter, conductivity, temperature, and vegetation weight are mean estimates based on 4–10 replicate samples. The range of annual means is provided for Hibernia Park.

TABLE 19.2. Taxonomic Composition and Trophic Structure of the Macroinvertebrate Communities at 18 Wetlands[a]

Taxon (Trophic Guild)	Number Genera	Frequency	Biomass
Leeches			
Glossiphoniidae (MPR)	4	13	167 ± 73
Crustacea			
Asellidae (SCA)	2	6	183 ± 123
Crangonyctidae (SCA)	1	2	49 ± 48
Gammaridae (SCA)	1	3	10 ± 8
Talitridae (SCA)	1	12	79 ± 53
Snails			
Lymnaeidae (SNA)	2	3	31 ± 24
Physidae (SNA)	1	16	1821 ± 493
Planorbidae (SNA)	4	12	460 ± 156
Ephemeroptera			
Baetidae (OAD)	2	15	32 ± 10
Caenidae (OAD)	1	14	145 ± 51
Ephemeridae (OAD)	1	1	7 ± 7
Odonata			
Aeshnidae (MPR)	2	11	432 ± 183
Coenagrionidae (MPR)	2	15	143 ± 44
Corduliidae (MPR)	2	6	18 ± 7
Gomphidae (MPR)	2	9	90 ± 38
Lestidae (MPR)	2	7	17 ± 7
Libellulidae (MPR)	10	15	1638 ± 500
Macromiidae (MPR)	1	1	7 ± 7
Hemiptera			
Belostomatidae (MPR)	1	11	110 ± 52
Corixidae (OAD)	?	15	236 ± 103
Gerridae (MPR)	3	11	10 ± 4
Hebridae (MPR)	1	1	<1
Hydrometridae (MPR)	1	2	<1
Mesoveliidae (MPR)	1	9	2 ± 1
Naucoridae (MPR)	1	2	31 ± 31
Nepidae (MPR)	1	2	1 ± 1
Notonectidae (MPR)	2	15	86 ± 27
Pleidae (MPR)	1	6	14 ± 9
Saldidae (MPR)	1	2	< 1
Veliidae (MPR)	2	9	< 1
Megaloptera			
Sialidae (MPR)	1	2	< 1
Trichoptera			
Hydroptilidae (OAD)	1	1	< 1
Leptoceridae (MPR/OAD/MCO)[b]	3	11	22 ± 9
Polycentropodidae (MPR)	1	1	<1

TABLE 19.2. (Continued)

Taxon (Trophic Guild)	Number Genera	Frequency	Biomass
Coleoptera			
Carabidae (MPR)	1	1	< 1
Chrysomelidae (MCO)	4	3	20 ± 14
Curculionidae (MCO)	3	3	< 1
Dytiscidae (MPR)	20	16	112 ± 52
Elmidae (OAD)	2	4	< 1
Gyrinidae (MPR)	1	4	5 ± 3
Haliplidae (OAD)	2	18	30 ± 10
Helophoridae (MCO)	1	2	1 ± 1
Hydrochidae (MCO)	1	2	< 1
Hydrophilidae (OAD/MPR)[c]	3	15	120 ± 66
Noteridae (MPR)	2	6	14 ± 11
Scirtidae (OAD)	1	1	< 1
Lepidoptera			
Noctuidae (MCO)	1	1	< 1
Diptera			
Ceratopogonidae (MIP)	1	17	206 ± 87
Chironomidae (ADC/MIP)[d]	34	18	1452 ± 397
Culicidae (OAD)	5	3	< 1
Dixidae (OAD)	1	1	< 1
Ephydridae (OAD)	1	3	< 1
Muscidae (MPR)	1	1	< 1
Sciomyzidae (MPR)	1	2	3 ± 2
Stratiomyidae (OAD)	1	3	2 ± 2
Syrphidae (OAD)	1	2	3 ± 2
Tabanidae (MPR)	2	10	70 ± 26
Tipulidae (OAD)	1	1	4 ± 4
Trophic Groups			
"AD chironomids" (ADC)	23	18	1240 ± 316
"Snails" (SN)	7	16	2235 ± 473
"Other AD" (OAD)	≥24	18	911 ± 246
"Microinvertebrate predators" (MIP)	12	18	617 ± 184
"Macroinvertebrate predators" (MPR)	≥74	18	3554 ± 742
"Macrophyte consumers" (MCO)	9	11	25 ± 12
"Scavengers" (SCA)	5	4	289 ± 159
TOTAL	≥149	18	8634 ± 1232

[a]The last seven lines of the table define the abbreviations used to describe the trophic guilds. Number of identified genera are given for most families, while ? indicates that some specimens were not identified beyond the family level. Frequencies indicate the number of wetlands where a taxon was found. Biomass is summarized as the mean and standard error of wet weights in mg per m^2, evaluated over all 18 sites.

[b]Includes *Oecetis* (MPR), *Mystacides* (OAD), *Triaenodes* (MCO).

[c]Larvae of all genera except *Berosus* considered predators (Brigham et al. 1982).

[d]Subfamily Tanypodinae considered predators, all other taxa allocated to AD chironomids.

collected. Chironomidae constituted much of the biomass at most sites. Physid snails and libellulid dragonflies were likewise present at almost all sites and contributed substantially to total biomass. "Macroinvertebrate predators," "snails," and "AD chironomids" comprised the greatest biomasses of the seven trophic groups. At least 152 genera were identified (individuals in some families were not all identified to genus), of which ≥74 genera were considered "macroinvertebrate predators."

In the CCA ordination (Fig. 19.2a) the environmental variables collectively accounted for 61 percent of total variation in taxon abundances among sites, with the first two canonical axes portraying 50 percent of the species-environment relation. The presence of fish was the most important of the habitat variables, accounting for 9 percent of total taxon variation ($p = 0.01$). The centroid of wetlands without fish is shown in the upper right portion of the ordination plots, while the centroid of sites with fish is shown in the lower left quadrant. Invertebrate families commonly found in wetlands without fish included many of the macrocrustacea (especially Asellidae and Crangonyctidae), syrphid flies, and noterid beetles (Fig. 19.2b). Wetlands with fish often had an abundance of gomphid dragonflies among many other taxa.

Of the continuous habitat variables shown as vectors in Fig. 19.2, wetland area was most important, accounting for 9 percent ($p = 0.04$) of total variation. Vegetation weight, which was slightly positively correlated with wetland area ($r = 0.29$, $p > 0.05$) as indicated by the acute angle formed by the two vectors, accounted for 7 percent of total variation ($p > 0.05$). Effects of depth, sediment chlorophyll-a, and dissolved oxygen (shown as shorter vectors in the figure) were less discernible, and three variables (temperature, turbidity, conductivity: not shown) had even less apparent influence on the invertebrate community. One site (FL) was identified as differing substantially from other wetlands in terms of both habitat variables and species composition. Succession during the first three years at Hibernia Park is indicated by a trend of the points H1, H2, and H3 upward and to the left in the ordination diagram.

The fish community at 9 of the 12 sites with fish was dominated by one to three species of the sunfish and bass family Centrarchidae. *Lepomis macrochirus* (bluegill), found at 9 sites, and *L. gibbosus* (pumpkinseed), found at 7 sites, were especially common. *Micropterus salmoides* (largemouth bass) and *Enneacanthus gloriosus* (bluespotted sunfish) were captured at 3 and 2 sites, respectively. Two sites were dominated by the minnow family Cyprinidae. In one of these we sampled large numbers of *Pimephales promelas* (fathead minnow) and *Carassius auratus* (goldfish), both exotic species commonly raised for sale to the baitfish industry. Finally, Fireschool (FI) was populated only by *Ameiurus nebulosus* (brown bullhead).

The presence of fish dramatically reduced overall biomass (Table 19.3). Mean biomass of macroinvertebrate predators in wetlands with few or no fish was more than five times the mean estimate for sites with fish. The taxonomically diverse guild of algivores-detritivores, "other AD" (excluding snails and chironomids), and macrocrustacean "scavengers" were likewise signifi-

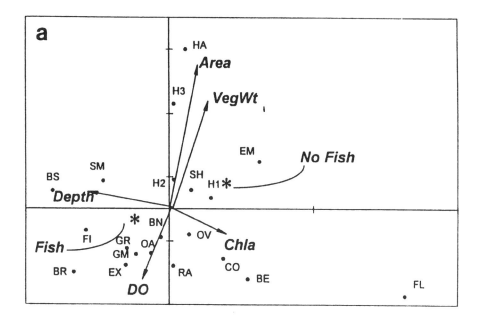

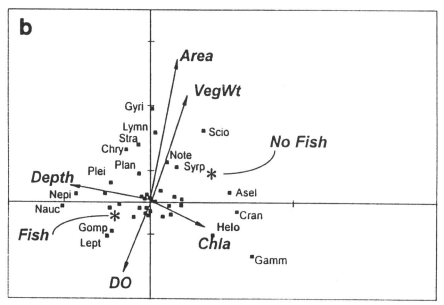

Fig. 19.2. CCA ordination of environmental variables: (*a*) sites; (b) families. Wetlands with fish versus without fish are shown as group centroids, indicated with asterisks. Vectors for the five continuous variables shown have been multiplied by three, and much shorter vectors for temperature, turbidity, and conductivity were deleted to improve the clarity of the crowded origin in the figure. Wetlands corresponding to the two-letter codes (*a*) are identified in Table 19.1. The three years of data for Hibernia Park are indicated as H1, H2, and H3. Four-letter codes for selected taxa on the fringes of (*b*) correspond to the first four letters of the family name in Table 19.2.

TABLE 19.3. Means ± SE of Wet Weight Biomass (g/m²) and Wet Weight per Individual (mg) in Wetlands with Abundant Fish (n = 12) versus with few or no fish (n = 6), and F-test for Effects of Ponded area on Wet Weight Biomass, for Trophic Groups[a]

Trophic Group	Biomass		Individual Wet Weight		
	Fish	No/Few Fish	Fish	No/Few Fish	Ponded Area (F)
AD Chironomids	0.72 ± 0.23	1.69 ± 0.79	0.26 ± 0.06	0.62 ± 0.26*	2.18
Snails	2.07 ± 0.53	2.73 ± 1.19	43.71 ± 12.31	28.45 ± 14.71	0.70
Other AD	0.41 ± 0.14	1.77 ± 0.56**	0.77 ± 0.08	1.11 ± 0.17	6.96*
Microinvertebrate predators	0.75 ± 0.27	0.35 ± 0.09	0.37 ± 0.04	0.29 ± 0.01	1.14
Macroinvertebrate predators	1.27 ± 0.39	6.59 ± 1.02***	29.77 ± 6.40	50.42 ± 11.74	1.64
Macrophyte consumers	0.03 ± 0.02	0.02 ± 0.01	5.96 ± 1.25	7.88 ± 3.30	13.20**
Scavengers	0.04 ± 0.02	0.87 ± 0.46*	1.79 ± 0.26	4.82 ± 0.60*	0.00
Total biomass	5.31 ± 1.04	14.03 ± 1.63***	2.05 ± 0.67	5.28 ± 2.09	0.93

[a]Significant effects are indicated as * = $p < 0.05$, ** = $p < 0.01$, *** = $p < 0.001$.

cantly more abundant in wetlands with few or no fish. "Microinvertebrate predator" biomass, by contrast, was slightly higher in wetlands with fish. Invertebrates considered to be specialists on live macrophyte tissue were invariably a very minor component of total invertebrate biomass and were not significantly affected by the presence of fish. Mean wet weights per individual for "scavengers" and "AD chironomids" were significantly greater at sites with few or no fish. Effects of ponded area on invertebrate biomass were less pronounced, and varied among trophic groups. Biomasses of "macrophyte consumers" and "other AD" were both significantly greater at larger sites.

Trophic group succession during the first three years after construction at H1-3 (Fig. 19.3*a*) indicated early dominance by deposit-feeding chironomids. *Chironomus* larvae carpeted the sediments, particularly at the deeper sampling stations during the first summer, but declined to very low levels by the third year. The biomass of "other AD" peaked during the second summer. "Macroinvertebrate predators" formed the largest trophic group in the third year of the study. By contrast, "macrophyte consumers" (not shown) were found for the first time in low densities (0.03 g/m^2) in the third summer, while "scavengers" were not collected during any of the three years.

Mean biomass of most trophic groups did not differ significantly between the 11 recently created versus 7 reference sites (Fig. 19.3b). The decline in deposit-feeding chironomids at H1-3 is clearly not a general or long-lived pattern, as biomass of this guild was slightly higher in the older reference sites (p > 0.05). Macrophyte specialists, though rare, were approximately an order of magnitude more abundant (p = 0.04) in the reference sites (mean = 0.06 g/m^2) than in the created sites (mean = 0.007 g/m^2).

Rapid colonization of the created wetlands is further indicated by the rapid rise in the total number of taxonomic families with site age (Fig. 19.3*c*). The number of families sampled at H1-3 rose from 14 in the first year to 27 by the third year, well within a 95 percent confidence interval of 18.8–28.9 families based on the 7 reference sites. The total number of invertebrate families recorded in wetlands with few or no fish (mean 24.0) was slightly but not significantly higher than in wetlands with fish (mean 21.8).

DISCUSSION

Caveats Regarding Sampling Protocols

Quantitative analysis of wetland invertebrate communities has been relatively recent, and there is little agreement on sampling protocols. A distinguishing feature of our effort is the estimation of mean biomass per unit area for each wetland as a whole. Our results are influenced by this approach in several important ways. First, we chose sampling locations spaced throughout the ponded area, including the deeper, open-water areas present at some sites.

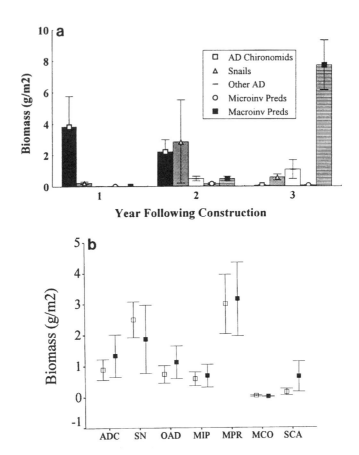

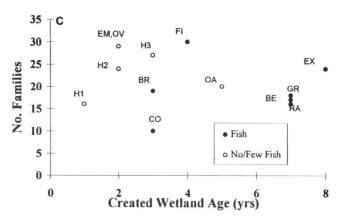

Most other studies have focused instead on a smaller, often vegetated segment of the wetland (e.g., Growns et al. 1992, Barnes 1983, Jeffries 1989). Consideration of the entire wetland influenced our estimates of invertebrate biomass in some instances. FI, for example, was in comparison to most sites more like a steep-sided pond with a narrow fringe of mixed submergent and emergent vegetation on the margins, and was sampled using a stratified approach. Although the seven samples from this vegetated fringe yielded high invertebrate abundances and taxonomic richness (30 families, the highest of any of the 18 wetlands), our estimate of overall mean biomass was relatively low, owing to the fewer invertebrates encountered in the three deep-water samples from the large central portion of the pond.

Secondly, expressing invertebrate abundance in terms of wet weight biomass, rather than simple densities, provided a different view, particularly of the trophic structure of the community. For example, biomass of larger-bodied macroinvertebrate predators typically exceeded that of the tanypodine chironomids, whereas the tanypodines were found at much higher densities. Our comparisons with other studies of trophic structure are thus made with caution.

Thirdly, we have shown that the use of multiple sweeps within a confined space did not completely eliminate the frequently encountered problem (e.g., Euliss et al. 1992, Anderson and Smith 1996, Brinkman and Duffy 1996, Turner and Trexler 1997) of differential capture efficiency of the various taxa (e.g., plant-associated or weakly swimming species were probably sampled more effectively than sediment sprawlers or burrowers). Likewise, the use of shallow (2-cm) cores probably led to underestimation of total sediment chironomid biomass, as some species can occur at depths >2.5 cm (Panis et al. 1996). The same protocols, however, were followed at all sites.

Effects of Habitat Variables

Of those habitat variables measured, the presence of fish most strongly affected invertebrate biomass and community structure. As discussed by Streever and Crisman (1993b), the fish assemblages of created marshes may be unstable, constrained by the source of colonists, and subject to periodic

---

Fig. 19.3. (*a*) Change in mean biomass of five of the seven trophic guilds during early succession at Hibernia Park. Scavengers were not found at the site, while macrophyte consumers were found only during the third summer at 0.03 g/m². (*b*) Mean ± SE of trophic guild biomass estimates (three-letter codes defined in Table 19.2) in created wetlands (n = 11, open squares) versus reference wetlands (n = 7, closed squares). (*c*) Successional change in the number of invertebrate families in created wetlands. Sites containing fish versus few or no fish are indicated with closed and open squares, respectively.

removal by the drying out of some sites. In a wide range of studies involving sites differing greatly in fish species composition, however, fish predation has had a pronounced impact on the invertebrate community. Hanson and Riggs (1995), for example, have shown that invertebrate biomass and taxon richness were both reduced in prairie marshes with fish communities dominated by fathead minnows compared to wetlands of comparable size and water chemistry without fish. Similarly, Weir (1972) found very dramatic reductions in species richness and abundance of Hemiptera and Coleoptera in pools with versus without fish. One indirect effect of fish predation in freshwater marshes is the reduction in macroinvertebrate prey otherwise available to predators such as waterfowl (Hanson and Butler 1994) and salamanders (Holomuzki et al. 1994). Mallory et al. (1994) have suggested that the reduction of invertebrate densities by fish is so consistent that waterfowl might use fish presence as a cue for choosing among alternative foraging and nesting sites. Our data also support the suggestion (Strayer 1991, Rasmussen 1993) that fish predation, by eliminating larger-bodied taxa, may reduce mean macroinvertebrate body size. As discussed by Strayer (1991), predation-imposed changes in the invertebrate size-spectrum can be expected to have major effects on secondary production and foodweb efficiencies.

The effects of fish on invertebrate trophic structure in wetlands remain poorly understood. Fish are known to reduce the abundances of larger and more active macroinvertebrate predators (McPeek 1990, Blois-Heulin et al. 1990, Batzer and Wissinger 1996), which may otherwise control the abundances of smaller members of the invertebrate food web. Our data indicate the possibility of a positive indirect effect of fish removing larger macroinvertebrate predators on the abundances of sediment-associated microinvertebrate predators, although Gilinsky (1984) has shown that high-intensity predation by bluegills depressed tanypodine abundances in experimental pond enclosures. What does seem clear, however, is that assessment of fish presence must be a prime concern in evaluating the success of invertebrate community establishment at a created site.

The importance of ponded area, identified in the CCA as the only other significant environmental variable, is harder to interpret, in part because of the positive correlation between ponded area and vegetation weight. Pond size may indeed influence both colonization and persistence of some invertebrates, but the nature of the influence cannot be inferred without further study. Effects of other habitat variables, especially vegetation and depth, which were less important in the CCA and are not discussed here, are more likely to influence the invertebrate community at the microsite level than when comparing wetlands as a whole. Further analyses comparing invertebrate abundances in relation to microhabitat variation at the 10 sampling locations within each wetland are anticipated as a means of better evaluating these habitat factors.

Succession

Although natural lakes and ponds are infrequent in southeast Pennsylvania, the landscape is dotted with large numbers of farm or recreational ponds and runoff basins, providing many source areas for potential colonists in newly created marsh sites. Barnes (1983), in her comparison of ball-clay ponds of differing age, found very rapid accumulation of macroinvertebrate taxa during the first two years after creation, as in our study, but also showed that taxonomic richness continued to increase more slowly over a 15-year period. A part of this continued increase must be reckoned the gradual addition of more poorly dispersed taxa.

Our observation at H1-3 of rapid establishment by *Chironomus* is consistent with previous work by Street and Titmus (1979), who also documented early dominance by *Chironomus* in their two-year study of colonization of a British gravel pit. Similarly, Layton and Voshell (1991) found substantial colonization by a wide range of taxa, with early dominance by chironomids, during the first year following construction of 12 experimental ponds in Virginia. The macrocrustacean amphipods and isopods, not found during the first three years after construction of H1-3 and absent in most of the younger sites in our study, are presumably limited in their dispersal by lack of an aerial adult stage (Barnes 1983, Batzer and Wissinger 1996). Physid snails, also dependent on passive dispersal, were by contrast characteristically present even at early successional sites. The number of invertebrate families sampled, used here as an indication of acquired taxonomic structure over time, appears to reach reference levels in approximately three years. Interestingly, many of the older created sites in the region (BE, GR, RA) have slightly lower family-level richness than more recent sites. This may largely reflect recent improvements in engineering design features adopted at the later sites. BE and GR, in particular, were relatively deep, turbid water bodies with little macrophyte cover and low sediment organic content. The cattail-dominated vegetation at CO had likewise inexplicably almost completely died and decomposed prior to our sampling, and the presence of little live macrophyte cover may again be a principal reason for the low number of invertebrate families observed at that site. By contrast, portions of the preexisting herbaceous terrestrial vegetation were intentionally retained in the construction of the more recent sites H1-3 and EM, contributing macrophyte-like physical structure and high sediment organic content presumably usable by early invertebrate colonists.

Less attention has been directed toward successional acquisition of the major trophic components of the community. Detritivores were early colonists in a seasonal saline marsh (de Szalay and Resh 1996), followed by predators and algivores. In contrast, Barnes (1983) found a shift from early dominance by predators and algivores toward dominance by detritivores, and the successional pattern may likely depend as much on dispersal properties of the organisms as on their trophic group. We have shown, however, that species specializing on macrophyte tissues arrive much later in succession. Lower colonization rates in these species are likely related to their lower densities in source areas and their particularly narrow feeding requirements (many taxa are associated with a very limited range of macrophyte species; Brigham et al. 1982). The three-yr successional pattern at H1-3 indicating more even representation of the seven trophic groups with time, however, hides much year-to-year change in the relative abundances of species comprising the guilds. For example, predatory beetles dominated the macroinvertebrate predator guild during the second summer, while odonate larvae were most abundant during the third summer. Jeffries' (1989, 1994) elegant studies of invertebrate colonization and extinction in wetlands ranging in hydroperiod have shown considerable species turnover even in ponds with permanent water. High rates of species replacement within trophic groups should thus be expected.

Evaluating the Functional Success of Created Sites

Given the observed rates of colonization at the newly created marsh sites, it may be reasonable to compare the communities of a created site with reference communities in the region as early as three years after site construction, although some elements of the community clearly require additional time to become established. Taxonomic richness, measured at the family level, proved an effective means of indicating potential problems in engineering design at the created sites. Identification to the family level has been shown to be a time-efficient and repeatable approach in stream biomonitoring (e.g., Hilsenhoff 1988, Plafkin et al. 1989), and the development of similar methods for evaluating wetland success seems reasonable.

Whereas the measurement of taxonomic richness may provide a direct measure of biodiversity preserved at a site, it provides at best an indirect measurement of foodweb structure and secondary productivity upon which fish and waterfowl may depend. Our analysis of trophic groups at the created sites indicated few consistent differences from wetlands used as reference conditions in the region. The regional persistence of invertebrates requiring particular species of macrophytes to complete their life cycles, however, is likely to be endangered by the mitigation process, as natural rates of dispersal appear to be low. Transplantation from the site to be destroyed to the newly created site of both macrophytes and their associated invertebrate herbivores, supplementing the use of plantings from nurseries, might be an effective tool in promoting the long-term persistence of these invertebrate specialists.

Whereas species-level turnover rates at individual wetlands are known to be high, the relative abundances of major trophic guilds may be a much more stable property of the community, as partly evidenced by the relatively low standard errors associated with the mean estimates for most groups. Macroinvertebrate predators contributed the largest portion of total invertebrate biomass and ca 50 percent of all genera in both the created and reference sites, and similarly strong representation by predators compared to other guilds has been mentioned by both de Szalay and Resh (1996) based on organismal densities and Jones et al. (1996) based on numbers of predator taxa. Because the predator guild is less arbitrarily defined than other groups, its structure may be a particularly effective tool for comparing research efforts and wetland sites.

ACKNOWLEDGMENTS

Undergraduates Felicia Chirico, Rose D'Amico, Mike Donaldson, Scott Fine, Laurie Fortis, Sue Hyder, Andy Raus, and Stephanie Steinmacher assisted with both field and laboratory work. We greatly appreciate the taxonomic assistance provided by Michael May and Clark Shiffer (Odonata), Jim Matta and David Wooldridge (Coleoptera), Bill Kintzer (Chironomidae), Jane O'Donnell and Eric Larsen (Hemiptera), Steve Good (Gastropoda), and Bob Waltz (Ephemeroptera). Mapping of the wetlands using GPS was conducted by Alan Clemson. The project was supported by grant R822457-01-1 from the the Office of Research and Development, U.S. Environmental Protection Agency.

LITERATURE CITED

Anderson, J. T., and L. M. Smith. 1996. A comparison of methods for sampling epiphytic and nektonic aquatic invertebrates in playa wetlands. Journal of Freshwater Ecology 11:219–224.

Barnes, L. E. 1983. The colonization of ball-clay ponds by macroinvertebrates and macrophytes. Freshwater Biology 13:561–578.

Batzer, D. P., and S. A. Wissinger. 1996. Ecology of insect communities in nontidal wetlands. Annual Review of Entomology 41:75–100.

Batzer, D. P., M. McGee, V. H. Resh, and R. R. Smith. 1993. Characteristics of invertebrates consumed by mallards and prey response to wetland flooding schedules. Wetlands 13:41–49.

Beckett, D. C., and P. A. Lewis. 1982. An efficient procedure for slide mounting of larval chironomids. Transactions of the American Microscopy Society 101:96–99.

Blois-Heulin, C., P. H. Crowley, M. Arrington, and D. M. Johnson. 1990. Direct and indirect effects of predators on the dominant invertebrates of two freshwater littoral communities. Oecologia 84:295–306.

Brigham, A. R., W. U. Brigham, and A. Gnilka. 1982. Aquatic Insects and Oligochaetes of North and South Carolina. Midwest Aquatic Enterprises, Mahomet, IL.

Brinkman, M. A., and W. G. Duffy. 1996. Evaluation of four wetland aquatic invertebrate samplers and four sample sorting methods. Journal of Freshwater Ecology 11:193–200.

de Szalay, F. A., and V. H. Resh. 1996. Spatial and temporal variability of trophic relationships among aquatic macroinvertebrates in a seasonal marsh. Wetlands 16:458–466.

Euliss, N. H., Jr., G. A. Swanson, and J. Mackay. 1992. Multiple tube sampler for benthic and pelagic invertebrates in shallow wetlands. Journal of Wildlife Management 56:186–191.

Gilinsky, E. 1984. The role of fish predation and spatial heterogeneity in determining benthic community structure. Ecology 65:455–468.

Gladden, J. E., and L. A. Smock. 1990. Macroinvertebrate distribution and production on the floodplains of two lowland headwater streams. Freshwater Biology 24:533–545.

Growns, J. E., J. A. Davis, F. Cheal, L. G. Schmidt, R. S. Rosich, and S. J. Bradley. 1992. Multivariate pattern analysis of wetland invertebrate communities and environmental variables in Western Australia. Australian Journal of Ecology 17:275–288.

Hammer, D. A. 1992. Creating freshwater wetlands. Lewis, Ann Arbor, MI.

Hanson, M. A., and M. G. Butler. 1994. Responses to food web manipulation in a shallow waterfowl lake. Hydrobiologia 279/280:457–466.

Hanson, M. A., and M. R. Riggs. 1995. Potential effects of fish predation on wetland invertebrates: A comparison of wetlands with and without fathead minnows. Wetlands 15:167–175.

Hansson, L.-A. 1988. Chlorophyll a determination of periphyton on sediments, identification of problems and recommendation of method. Freshwater Biology 20:347–352.

Hilsenhoff, W. L. 1988. Rapid field assessment of organic pollution with a family-level biotic index. Journal of the North American Benthological Society. 7:65–68.

Holomuzki, J. R., J. P. Collins, and P. E. Brunkow. 1994. Trophic control of fishless ponds by tiger salamander larvae. Oikos 71:55–64.

Jeffries, M. 1989. Measuring Talling's "element of chance in pond populations." Freshwater Biology 21:383–393.

———. 1994. Invertebrate communities and turnover in wetland ponds affected by drought. Freshwater Biology 32:603–612.

Jones, D. H., R. B. Atkinson, and J. Cairns, Jr. 1996. Macroinvertebrate assemblages of surface mine wetlands of southwest Virginia, USA. Journal of Environmental Science 8:1–14.

Kusler, J. A., and M. E. Kentula (eds.). 1990. Wetland Creation and Restoration: The Status of the Science. Island Press, Washington, DC.

Layton, R. J., and J. R. Voshell, Jr. 1991. Colonization of new experimental ponds by benthic macroinvertebrates. Environmental Entomology 20:110–117.

Mallory, M. L., P. J. Blancher, P. J. Weatherhead, and D. K. McNicol. 1994. Presence or absence of fish as a cue to macroinvertebrate abundance in boreal wetlands. Hydrobiologia 279/280:345–351.

McPeek, M. A. 1990. Behavioral differences between Enallagma species (Odonata) influencing differential vulnerability to predators. Ecology 71:1714–1726.

Merritt, R. W., and K. W. Cummins. 1996. Aquatic Insects of North America (3rd ed.). Kendall/Hunt, Dubuque, IA.

Murkin, H. R., P. G. Abbott, and J. A. Kadlec. 1983. A comparison of activity traps and sweep nets for sampling nektonic invertebrates in wetlands. Freshwater Invertebrate Biology 2:99–106.

Nusch, E. A. 1980. Comparison of different methods for chlorophyll and phaeopigment determination. Archiv für Hydrobiologie Supplement 14:14–36.

Omernik, J. M. 1987. Ecoregions of the conterminous United States. Annals of the Association of American Geographers 77:118–125.

Panis, L. I., B. Goddeeris, and R. Verheyen. 1996. On the relationship between vertical microdistribution and adaptations to oxygen stress in littoral Chironomidae (Diptera). Hydrobiologia 318:61–67.

Pennak, R. W. 1989. Fresh-water Invertebrates of the United States (3rd ed.). John Wiley & Sons, New York.

Peterson, L. P., H. R. Murkin, and D. A. Wrubleski. 1989. Waterfowl predation on benthic macroinvertebrates during fall drawdown of a northern prairie marsh. Pages 681–689 in R. R. Sharitz, and J. W. Gibbons (eds.), Freshwater Wetlands and Wildlife. USDOE Office of Scientific and Technical Information, Oak Ridge, TN.

Pimm, S. L. 1982. Food Webs. Chapman & Hall, New York.

Plafkin, J. L., M. T. Barbour, K. D. Porter, S. K. Gross, and R. M. Hughes. 1989. Rapid Bioassessment Protocols for Use in Streams and Rivers: Benthic Macroinvertebrates and Fish. U.S. Environmental Protection Agency. EPA/440/4-89/001.

Rasmussen, J. B. 1993. Patterns in the size structure of littoral zone macroinvertebrate communities. Canadian Journal of Fisheries and Aquatic Science 50:2192–2207.

Roback, S. S. 1969. Notes on the food of Tanypodinae larvae. Entomology News 80: 13–18.

Scatolini, S. R., and J. B. Zedler, 1996. Epibenthic invertebrates of natural and constructed marshes of San Diego Bay. Wetlands 16:24–37.

Smit, H., E. D. Van Heel, and S. Wiersma. 1993. Biovolume as a tool in biomass determination of Oligochaeta and Chironomidae. Freshwater Biology 29:37–46.

Strayer, D. L. 1991. Perspectives on the size structure of lacustrine benthos, its causes, and its consequences. Journal of the North American Benthological Society 10: 210–221.

Street, M., and G. Titmus. 1979. The colonisation of experimental ponds by chironomidae (Diptera). Aquatic Insects 4:233–244.

Streever, W. J., and T. L. Crisman. 1993a. A preliminary comparison of meiobenthic cladoceran assemblages in natural and constructed wetlands in central Florida. Wetlands 13:229–236.

———.1993b. A comparison of fish populations from natural and constructed freshwater marshes in central Florida. Journal of Freshwater Ecology 8:149–153.

Streever, W. J., K. M. Portier, and T. L. Crisman. 1996. A comparison of dipterans from ten created and ten natural wetlands. Wetlands 16:416–428.

ter Braak, C. J. F, and P. F. M. Verdonschot. 1995. Canonical correspondence analysis and related multivariate methods in aquatic ecology. Aquatic Science 57:255–289.

Turner, A. M., and J. C. Trexler. 1997. Sampling aquatic invertebrates from marshes: Evaluating the options. Journal of the North American Benthological Society 16: 694–709.

Van Deventer, J. S., and W. S. Platts. 1983. Sampling and estimating fish populations from streams. Transactions of the North American Wildlife and Natural Resources Conference 48:349–354.

Weir, J. S. 1972. Diversity and abundance of aquatic insects reduced by introduction of the fish *Clarius gaviepinus* to pools in Central Africa. Biological Conservation 4:169–174.

20 Canadian Springs

Postglacial Development of the Invertebrate Fauna

D. DUDLEY WILLIAMS and NANCY E. WILLIAMS

*C*anada has many different spring types supporting animal communities that are, in general, poorly known. In 1982 the Biological Survey of Canada identified springs as worthy of special study, and a project was launched to identify researchers who were working on these habitats in isolation and develop a forum through which they could share their common interest. A summary of the subsequent project findings is presented here.

Three insect groups (Trichoptera, Coleoptera, and Chironomidae) and the hydrachnid mites have been studied in some detail. Species richness within individual springs varies according to taxon (e.g., an average of 3.9 caddisfly species but up to 40 species of mite/spring), as does specificity to spring type (rheocrene, helocrene, limnocrene). Microhabitat diversity within a spring strongly influences its fauna, as do water temperature, substrate composition, and several other environmental factors. With the exception of the mites, the incidence of true crenophiles is low. Canadian springs commonly support limnephilid caddisflies, dytiscid beetles, orthocladine chironomids, and five families of mites. Community composition has quite a strong east/west dichotomy, and has been influenced markedly by the recent glacial history of the country in combination with differential dispersal abilities among species; springs in nonglaciated regions of the United States, for example, have more diverse faunas. In southern Ontario permanent and temporary springs share the same taxa but dominance differs. The present-day spring fauna can be expected to change in the near geological future as the Canadian climate responds to global perturbations. Only a

Invertebrates in Freshwater Wetlands of North America: Ecology and Management, Edited by
Darold P. Batzer, Russell B. Rader, and Scott A. Wissinger
ISBN 0-471-29258-3   © 1999 John Wiley & Sons, Inc.

few degrees rise in mean annual air temperature will likely have serious consequences for cold stenothermic species.

INTRODUCTION

Springs have always held a magical place in human culture. Throughout the Middle Ages and as far back as the Romans, and probably beyond, they were revered as the sacred dwelling places of spirits. Special status was given to those springs that were the sources of important rivers, and it is interesting to speculate on the origins of this practice. Did, for example, the secluded and picturesque nature of many springs conjure up images of tranquility that befitted deities, or was there a more practical reason for not wanting the sources of drinking water to be affected by human presence and activity? Indeed, the Romans believed that water from springs such as the Camenae and the Juturna had the capability of actually restoring sick bodies to health (Evans 1994)—a belief that is still extant in many parts of the world today. Ancient Greek settlements were located in karst areas, and springs' issues were often monumentalized in order to protect their purity (Crouch 1993).

There seems to have been little doubt in ancient minds as to the definition of a spring. However, contemporary scientists have had greater difficulty, and this has led to a plethora of terms such as *springbrook, spring-fed stream, spring source, headwater spring, spring seep, spring run,* and so on. This has produced several attempts to standardize definitions, some based on characteristics of the communities (biocoenoses) living there, others based on the physical/chemical environment.

From a strict hydrological perspective, springs may be regarded as concentrated points of natural groundwater discharge, at a rate high enough to maintain flow on the surface (van Everdingen 1991). Some springs, however, are fed by very shallow or small aquifers which, when exhausted, may result in intermittent surface flow. From an aquatic organism's perspective, therefore, springs present a range of habitats, from the "classical" highly stable (especially temperature and discharge) type to those that have a highly variable hydroperiod such as those found in arid regions or, more commonly, those associated with wetlands. In wetlands, springs at peak flow are associated with zones of saturated soil and mosses that provide important habitats for many aquatic and semiaquatic invertebrate species (Danks and Rosenberg 1987).

Factors influencing habitat conditions in springs include those associated with the terrain (e.g., local geology, topography, and groundwater reserve) and geographic position (e.g., latitude and glacial history), together with smaller-scale influences such as vegetation (Table 20.1). In an individual

TABLE 20.1. Summary of Some of the Major Factors Influencing Habitat Conditions in Springs[a]

Factor	Longevity of Spring	Stability of Water Temperature	Microhabitat Diversity for the Fauna	Food in the Form of		Availability of Adjacent Spring Habitats[b]
				Algae	Detritus	
Geology/topography						
Extensive groundwater reservoir	+	+				+
Long groundwater residence	+	+		+		−
Small, shallow aquifers	−	−				
High elevation, steep slopes			−	−	−	
Bedrock			−	+	−	
Heterogeneous sediments			+		+	
Glacial activity						
Spring characteristics						
Large area of spring zone		+	+	+	+	
High discharge			−	−	−	
Stable discharge		+	+	+	+	
Vegetation in spring			+	+	+	
Riparian vegetation						
More vegetation		+	+	−	+	
Higher vegetation (trees)		+		−	+	
Extreme type of spring						
High minerals				+		
Thermal spring		+		−		
Saline spring				−		
Intermittent spring		−	−	+	+	
Human activity						
Agriculture	−	−	−	+		−
Silviculture	−	−	−	+/−	+/−	−
Spas			−	−	−	−

Source: Adapted from H. V. Danks and D. D. Williams, Arthropods of springs, with particular reference to Canada: and needs for research, Memoirs of the Entomological Society of Canada 155:203–217, 1991.

[a] + = positive influence, − = negative influence.

[b] Important in the colonization of new habitats and in maintaining gene flow among metapopulations.

449

spring all or some of these factors come together to characterize the size, rate and pattern of flow, temperature regime, and chemical composition of the habitat. The community of organisms that subsequently develops reflects tolerance of/adaptation to these factors, together with colonization abilities, the latter being of particular importance in regions affected by Pleistocene glaciation. Because of the extensive suite of factors affecting spring types and the relatively early stage of springs research, Danks and Williams (1991) favor the use of a minimum of five *key descriptors* to characterize each habitat (Table 20.2), rather than a rigid classification to which there are always exceptions.

CANADA AND ITS SPRINGS

Canada is the second-largest country in the world, extending approximately 4800 km from east to west (longitude 52°W to 141°W) and about 4500 km from south to north (latitude 42°N to 83°N). Whereas about one third of its total land area is arctic in nature, the majority of its lands support a great diversity of environments. Some 20 percent of the world's fresh waters are contained within its borders (EMR 1974), and interaction with regional bedrock, soils, physiography, vegetation, and climate have produced a large number of subsurface water types. Present diversity and distribution of the Canadian biota have been strongly influenced by climatic changes that took place during the Quaternary Period (~ the last 1.6 million years), a period believed to have been the most influential in the development of modern floras and faunas (Ross 1965). Even though many species may have been derived from Tertiary times, their present distribution is primarily the outcome of Quaternary events (Matthews 1979). The physiographic regions of Canada, together with the locations of major springs, are shown in Fig. 20.1. The ranges of environmental conditions measured in Canadian springs are summarized in Table 20.3.

In 1982 the Scientific Committee of the Biological Survey of Canada identified springs as habitats worthy of special study. The Survey, set up in 1977 under the auspices of the National Museum of Natural Sciences and the Entomological Society of Canada, has as its general goals the development and coordination of national initiatives in systematic and faunistic entomology in Canada. A project on springs was initiated that has as its specific goal the furtherance of knowledge of the systematics, distribution, and general ecology of invertebrates from these widespread but little-known aquatic habitats (Williams 1983).

INVERTEBRATE COMMUNITIES OF CANADIAN COLDWATER SPRINGS

Coldwater springs in Canada support basically the same major groups of invertebrates that are seen in springs in the United States and Europe, regions

TABLE 20.2. Recommended Minimal Key Descriptors of Springs Sampled for Biological Data

Key Descriptor	Status Observed		
	Low	Intermediate	High
Nature of source	Helocrene (discharge or seep into marshy substrate)	Limnocrene (discharge to basin)	Rheocrene (rapid flow on discharge)
Discharge at source	Low volume (<0.01 m$^3 \cdot$ s^{-1})	Medium volume (0.01–0.5 m$^3 \cdot$ s^{-1})	High volume (>0.5 m$^3 \cdot$ s^{-1})
Water temperature	Cold ($<10°$C)	Warm (10–$40°$C)	Hot ($>40°$C)
Chemistry; as total dissolved solids	Fresh water ($<1,000$ ppm)	Mineral ($1,000$–$35,000$ ppm)	Saline ($>35,000$ ppm)
Persistence[a] (approximate interval between major disturbances)	Intermittent (typically annual, but up to five years)	Apparently permanent ($>$ five years; i.e., some signs of inconsistency)	Permanent (>50 years; no disturbance)

Source: Based on H. V. Danks and D. D. Williams, Arthropods of springs, with particular reference to Canada: Synthesis and needs for research, Memoirs of the Entomological Society of Canada 155:203–217, 1991.

[a]Unless historical records are available, interpretation here is likely to be largely subjective based on observations of physical and biological conditions made at the site.

Fig. 20.1. The physiographic regions of Canada, together with the locations of major springs. (After R. O. van Everdingen, Physical, chemical, and distributional aspects of Canadian springs, Memoirs of the Entomological Society of Canada 155:7–28 1991.)

of the world where crenobiological research has been underway for some time. Among insects, the Diptera, Coleoptera, and Trichoptera predominate, although the Ephemeroptera, Odonata, Plecoptera, Hemiptera, and Collembola are also represented (Fig. 20.2), often depending on locality. Among the noninsect arthropods, water mites are especially abundant and diverse, as are ostracods. Amphipods and copepods may occur at high densities, but generally only one or two species are present in any particular spring. Nonarthropod metazoans include turbellarian flatworms, mollusks, and rotifers.

Table 20.4 details the three best-studied insect groups, together with the mites, found in Canadian springs. The most salient points of crenobiological interest of these groups are as follows.

Trichoptera

It appears that the number of caddisfly species specifically adapted to cold-water springs in Canada is rather small. The total number of trichopterans

TABLE 20.3. Ranges in Environmental Conditions Observed in Canadian Springs

Parameter	Range
Discharge	0.002 to 11,200 $l \cdot s^{-1}$
Water temperature	−2.9 to 82.2°C
pH	2.8 to >10
Conductivity	26 to 33,700 $\mu S \cdot cm^{-1}$
Total dissolved solids	32 to >75,000 ppm
Redox potential (Eh)	−252 to +683 mV
Bicarbonate	2.0 to 5,961 ppm
Sulphate	1.0 to 17,520 ppm
Chloride	0.1 to 44,300 ppm
Nitrate	0.003 to 55.0 ppm
Phosphate	0.003 to 8.2 ppm
Dissolved oxygen	0 to 8.8 ppm
Dissolved hydrogen sulphide	0 to 108.6 ppm
Calcium	1.2 to 1,823 ppm
Magnesium	0.05 to 1,190 ppm
Sodium	0.1 to 27,100 ppm
Potassium	0.1 to 1,568 ppm
Iron	0.001 to 2,600 ppm

Source: Adapted from R. O. Everdingen, Physical, chemical, and distributional aspects of Canadian springs, Memoirs of the Entomological Society of Canada 155:7–28, 1991.

known from North America exceeds 1200 species (in 20 families) (Wiggins 1984). Only 87 species have been recorded from springs in Canada, and few appear to be restricted to springs per se, or even to the first 50 m below the spring source. Quite a number of species are predominantly *associated* with springs, but not exclusively so.

The following synopsis is based on larval caddisfly collections from 25 small springs across Canada (Williams 1991), together with data from several regional studies and from records held at the Royal Ontario Museum (Williams and Williams 1987). The mean number of caddisfly species found per spring was 3.9 (range 0 to 9 species), with springs in western Canada supporting an average of 4.4 species compared with 3.4 species in the east. Species richness was positively correlated with the diversity of microhabitats present in individual springs. In particular, limnocrenes and rheocrenes with low current (<15 cm/s) and small-grain bed materials (sand and mud, <2 mm in diameter) supported few caddisfly species, although the species that were present often had large populations. Compared with the east, more species, for example *Anagapetus debilis* Ross, *Neothremma alicia* Banks, and *Parapsyche elsis* Milne, were specifically adapted to spring and/or high-altitude habitats in the west. Although eastern and western Canada had many genera in common, few species were distributed from coast to coast. Exceptions were *Hesperophylax designatus* (Walker), *Psychoglypha subborealis*

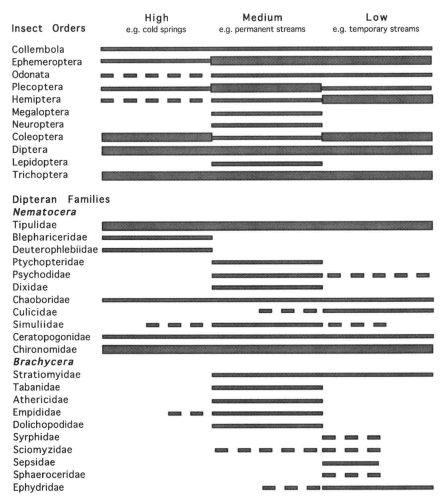

Fig. 20.2. Comparison of the relative importance of different aquatic insect orders and dipteran families in cold, stable freshwater springs with other types of lotic habitat (thick line indicates "well represented," thin line indicates "moderately well represented," broken line indicates "present but not significantly so"; information obtained mainly from Merritt and Cummins 1988, Williams and Feltmate 1992).

(Banks), *Onocosmoecus unicolor* Banks, and *Chyranda centralis* (Banks), all of which are large limnephilids. The latter four species are all strong fliers with less specialized requirements, especially in terms of temperature range tolerated and/or food.

Ordination techniques applied to the 25-spring data set confirmed an east/

TABLE 20.4. Details of the Families or Subfamilies of Insects and Mites Found Commonly in Canadian Springs

Trichoptera Families	Number of Species	Coleoptera Families	Number of Species	Chironomidae Subfamilies	Number of Genera[a]	Acari Families	Number of Species[b]
Philopotamidae	2	Haliplidae	1	Podonominae	3	Hydrovolziidae	3
Hydropsychidae	5	Dytiscidae	38	Tanypodinae	6	Piersigiidae	1
Rhyacophilidae	9	Hydraenidae	8	Diamesinae	3	Hydryphantidae	20
Glossosomatidae	4	Hydrophilidae	9	Prodiamesinae	1	Sperchontidae	6
Hydroptilidae	4	Dryopidae	1	Orthocladiinae	20	Teutoniidae	1
Phryganeidae	4	Chrysomelidae	6	Chironominae	8	Rutripalpidae	1
Brachycentridae	3					Anisitsiellidae	10
Limnephilidae	37					Lebertiidae	4
Uenoidae	9					Torrenticolidae	2
Goeridae	1					Limnesiidae	2
Lepidostomatidae	7					Hygrobatidae	7
Beraeidae	1					Feltriidae	11
Molannidae	1					Pionidae	10
						Aturidae	12
						Momoniidae	2
						Nudomideopsidae	2
						Mideopsidae	4
						Uchidastygacaridae	1
						Chappuisididae	1
						Neoacaridae	1
						Acalyptonotidae	2
						Athienemanniidae	5
						Laversiidae	1
						Arrenuridae	6

Source: Data from Williams and Williams 1987, Williams 1991, Larson 1991, Colbo 1991, and Smith 1991.

[a]This analysis was based on chironomid data from a series of 38 samples from natural and man-made springs and seeps in Labrador.

[b]The mites include several species that have a terrestrial or riparian phase in their life cycle.

west geographic difference in caddisfly assemblages and identified elevation, extent of groundwater source, and summer temperature as the environmental factors most strongly correlated with the overall species distributions observed (Williams 1991). At the regional level (east or west) factors that were most influential upon spring assemblage composition were riparian vegetation, current, substrate particle size, microhabitat diversity, and pH. Detritus-rich springs in the east were dominated by *Frenesia* and *Lepidostoma,* and those in the west by *Homophylax.* Scraper and predator species were abundant only in springs that had relatively high microhabitat diversity, current speed, and pH. Filter-feeding species were comparatively rare.

Glacial history has undoubtedly contributed to the paucity of Canada's spring fauna, as a relatively short period has elapsed since the retreat of the Laurentide Ice Sheet (approximately 8,000–12,000 years B.P.) for Canadian fresh waters to become repopulated from unglaciated sources (Matthews 1979). Colonization from major western routes (e.g., Beringia and the Columbia River system) and major eastern routes (e.g., the Mississippi and Ohio river systems), together with the limited migration abilities of many individual caddisfly species, have undoubtedly resulted in the east-west dichotomy seen today (Patrick and Williams 1990).

Coleoptera

Of the 663 species of aquatic beetle recorded from Canada and Alaska, 63 (8 percent) are known to occur in springs. These species are distributed among six families (Table 20.4), with the majority (38 species) belonging to the Dytiscidae (Roughley and Larson 1991). Representation of these families in springs very roughly approximates that seen in the total fauna. Families that are notably absent from springs include the Gyrinidae, Psephenidae, Elmidae, Amphizoidae, Curculionidae, and Scirtidae, despite their diverse occurrence in other aquatic habitats. As in the Trichoptera, there is a definite east/west split in distribution. For example, about 50 percent of the spring species are western (Manitoba to British Columbia), whereas about 25 percent are eastern (Manitoba to Newfoundland). Seventeen species are transcontinental. Western Canada has more species restricted to springs (47 species), in contrast to the east, which has a greater proportion of more broadly distributed species, including fewer true crenophiles (31 species). Roughley and Larson (1991), however, caution that these findings are preliminary due to the inadequacy of the Canadian database.

Latitudinally, the majority of spring species in Canada are to be found in the south. Further, the number of spring species is greater still in the continental United States, from which the Canadian fauna is believed to be slowly developing (Roughley and Larson 1991). By way of an example, Young (1955) recorded some 55 species (from 29 genera), representing five families of beetles, from a series of spring habitats in Florida. This single location thus supported almost the same diversity of spring-associated species as is known from the whole of Canada.

Sixty percent of Canadian spring beetles belong to one family, the Dytiscidae. Of these, 9 (24 percent) are exclusively crenophilic, 11 (29 percent) also occur in other types of running water, 6 (16 percent) occur in standing waters, and 12 (31 percent) are species known to live in a broad range of aquatic habitats. The dytiscid beetle assemblage living in Canadian springs thus seems derived from a mix of source habitats and phylogenetic elements (Roughley and Larson 1991).

Chironomidae

From a search of the literature, Colbo (1991) reports that of the 235 chironomid genera known to occur in the Holarctic region, 73 (31 percent) inhabit springs or seeps (the latter being defined as water oozing from soil to form a film of water spreading over a broad area). Of these, 65 were thought likely to occur in springs in northeastern North America. In a survey at 18 sites in Labrador, in eastern Canada, Colbo (1991) found 41 of these genera in either natural or man-made springs and seeps (Table 20.4). In general, the chironomid faunas of the seeps were less diverse than those of the springs.

The man-made habitats were serendipitously created as a result of cracks forming in the bases of earth-filled dikes and contained 38 genera, of which 35 were deemed to be normal spring inhabitants. A greater diversity of genera was found in these man-made habitats, a finding thought to reflect the greater heterogeneity of the latter's substrate (ranging from fine, high-pH, clay-like grout to iron-rich flocculent-coated silt and gravel, to bare silt, gravel, and bedrock). Even though the overall diversity of chironomids was greater in the man-made springs and seeps, the number of genera present in individual habitats was about the same as in the natural habitats. For example, 21 genera were found in a natural spring-seep at Churchill Falls, compared with 22 in a nearby dike spring-seep complex (Colbo 1991). The qualitative composition of the chironomid faunas was very similar in the natural and man-made habitats, as were the trichopteran faunas with which they cooccurred.

Acari

Over 115 species of hydrachnid water mite (representing 57 genera and 24 families; Table 20.4) have been recorded from Canadian springs. This represents about 23 percent of the total freshwater mite fauna presently known from Canada. Individual springs may contain upwards of 25 species, representing 20 genera, and densities that exceed 1000 animals/m^2 (Smith 1991). Springs that support a rich moss carpet at their edges may host up to 40 species from 25 or more genera. Although not as well studied as the spring-dwelling mites of Europe (e.g., Schwoerbel 1959, Viets 1978), in comparison with other Canadian crenophilic taxa they are moderately well known. They have their origins in four zoogeographic groups (Table 20.5).

Smith (1991) has proposed that the Canadian spring mite fauna is divisible into three reasonably distinct ecological groups based on the habitat prefer-

TABLE 20.5. The Zoogeographical Origins of the Canadian Spring-Dwelling Mite (Hydrachnida) Fauna

Zoogeographical Origin	Characteristic Features, Present Distribution and Habitat
Southern, relict taxa	A large group of remnant, cool-adapted forms probably widespread in North America during the Tertiary. Became restricted to southern and coastal refugia in the Pleistocene. Currently have highly disjunct distributions in southern Canada, especially in montane regions of the east and west. Cool helocrenes or rheocrenes; many adapted for hyporheic existence (e.g., Nudomideopsidae, Chappuisididae, Athienemanniidae, Anisitsiellidae, Momonidae).
Northern, recent taxa	Relatively cold-adapted species that dispersed through northern regions during the Pleistocene from late Tertiary or Quaternary origins. Many belong to large genera that underwent recent and extensive adaptive radiation throughout the Holarctic. Presently widespread in Canada. Typically found in cold rheocrenes or limnocrenes (e.g., Sperchontidae, Lebertiidae, Feltriidae, most Pionidae).
Arctic-alpine, recent taxa	A small group of exclusively cold-adapted stenophiles that apparently arose during the Pleistocene. They belong to holophyletic species groups with circumpolar distributions. In southern Canada they are restricted to high alpine regions. They are typically found in cold limnocrenes (e.g., *Pionacercus* [Pionidae], *Neobrachypoda* [Aturidae], Teutoniidae, Acalyptonotidae).
Neotropical taxa	A relatively small group of species belonging to genera of South American origin. They are warm-adapted forms that have rapidly extended northwards into temperate regions (since the appearance of the Panamanian Isthmus in the Pliocene). Most are specialized for a hyporheic life, but some live in eurythermic helocrenes (e.g., *Tyrrellia* [Limnesiidae]).

Source: After I. M. Smith, Water mites (Acari: Parasitengona: Hydrachnida) of spring habitats in Canada, Memoirs of the Entomological Society of Canada 155:141–167, 1991.

ences of the adults and deuteronymphs. These groups correspond to the three basic categories of spring types:

1. Mites characteristic of *helocrenes* tend towards being specialized for life just below the surface water film. They are adapted for crawling on substrates subject to the effects of surface tension. Frequently they be-

long to genera typically associated with moss mats and leaf litter, especially in seepage areas or splash zones.

2. Mites characteristic of *rheocrenes* tend to be adapted for living in running waters, including the interstitial habitat of the hyporheic zone. They exhibit clinging, crawling, and walking behaviors.

3. Mites characteristic of *limnocrenes* are generally adapted to living in standing waters. They belong to genera or species groups that also inhabit arctic and alpine shallow pools, and exhibit walking and swimming behaviors.

The relatively high number of hydrachnid species that are adapted to living in spring habitats is thought to point to the origin of this group in seepage water and shallow pools (Mitchell 1957, Smith and Cook 1991). Some elements in the fauna have failed to move out of these habitats (e.g., many of the Hydryphantidae), whereas others represent more recently evolved forms that have secondarily reinvaded springs from other fresh waters (e.g., several genera of Thyadinae). Both groups, however, exhibit adaptations suited to spring dwelling, such as short and broad leg sections, thick and spinose limb setae, sclerotised plates on the dorsal and ventral surface, and rotated leg axes (for enhanced crawling on a substrate through a thin film of water) (Mitchell 1960, Smith 1991).

Many mites may have survived the Pleistocene glaciation in Canada in marginal refugia, such as coldwater springs, at the edge of the glaciers. As the ice receded, they were able to radiate out to repopulate adjacent newly established habitats. In species with limited powers of dispersal, such a reinvasion scenario may have resulted in the highly disjunct distribution patterns seen today in or near Pleistocene refugial areas (Smith 1991).

The fact that many crenophilic mite species exhibit very narrow tolerance ranges for environmental factors such as dissolved chemicals and pH, and not just temperature, has led several authors (e.g., Schwoerbel 1959, Young 1969, Smith 1991) to advocate their potential as biomonitors of environmental change, both man-made and natural.

Faunal Composition and Spring Permanence

Mention was made earlier of habitat changes arising from different degrees of spring permanence. These should translate into faunal differences, although the paucity of studies of temporary springs makes comparison difficult and generalities tentative. Some data are available, however, at the regional level. In southern Ontario permanent coldwater springs are dominated by nemourid stoneflies, chironomids, caddisflies, mites, copepods, ostracods, and amphipods (Fig. 20.3). With the exception of the mites and chironomids (e.g., 20 species and 10 genera, respectively, present in Valley Spring, Ontario; Williams and Hogg 1988), diversity within these taxa is not high. For example,

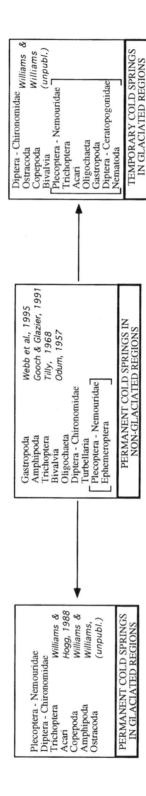

Fig. 20.3. Comparison of the dominant invertebrate taxa known to occur in permanent and temporary, coldwater springs in southern Ontario together with those found in permanent springs in nonglaciated regions. In each box the taxa are ranked from top to bottom, in the approximate order of numerical dominance seen in that spring type. Taxa enclosed within the square brackets are less common but usually present. It is possible that due to nonstandard collecting mesh size the microcrustaceans may have been underrepresented in some of the studies from which the data were gleaned.

Content of the three boxes:

Box 1 (left):

Plecoptera - Nemouridae
Diptera - Chironomidae
Trichoptera — *Williams &*
Acari — *Hogg, 1988*
Copepoda — *Williams &*
Amphipoda — *Williams,*
Ostracoda — *(unpubl.)*

PERMANENT COLD SPRINGS
IN GLACIATED REGIONS

Box 2 (middle):

Gastropoda — *Webb et al., 1995*
Amphipoda — *Gooch & Glazier, 1991*
Trichoptera — *Tilly, 1968*
Bivalvia — *Odum, 1957*
Oligochaeta
Diptera - Chironomidae
Turbellaria
[Plecoptera - Nemouridae]
[Ephemeroptera]

PERMANENT COLD SPRINGS IN
NON-GLACIATED REGIONS

Box 3 (right):

Diptera - Chironomidae — *Williams &*
Ostracoda — *Williams*
Copepoda — *(unpubl.)*
Bivalvia
[Plecoptera - Nemouridae]
Trichoptera
Acari
Oligochaeta
Gastropoda
Diptera - Ceratopogonidae
[Nematoda]

TEMPORARY COLD SPRINGS
IN GLACIATED REGIONS

there are usually only one or two species of stonefly and amphipod per spring, although population densities can be quite high (e.g., up to 2325 nymphs of the stonefly *Nemoura trispinosa* m^{-2} of bed in Valley Spring). Further, in this same spring *N. trispinosa* contributes greater than 17 percent to the total secondary production in the system, with the three most abundant caddisflies, *Diplectrona modesta, Parapsyche apicalis,* and *Lepidostoma vernale,* contributing more than 30 percent (Williams and Hogg 1988).

Intermittently flowing coldwater springs in southern Ontario have a fauna taxonomically similar to that found at permanent sites; however, there is a predominance of chironomids, ostracods, copepods, and sphaeriid clams (Fig. 20.3). Although moderate numbers of caddisfly larvae are sometimes present, they do not match the dense populations recorded from a temporary spring in California in which trichopterans dominated the fauna (Resh 1982).

Figure 20.3 also compares the faunas of these southern Ontario springs with those from adjacent regions of eastern North America that were largely unaffected by recent glacial activity (see Matthews 1979). The latter, geologically more stable habitats are dominated numerically by gastropod and bivalve mollusks, triclads, amphipods, oligochaetes, and trichopteran and chironomid larvae. Nymphs of nemourid stoneflies and mayflies are also often present. The diversity of species within some of these taxa may be very high—for example, 17 species of Oligochaeta in Old Driver Spring, southern Illinois (Webb et al. 1995). We suggest that the insect/noninsect dominance seen between permanent coldwater springs in Ontario and those to the south are related to respective glacial histories, with the more vagile insects predominating in regions that have been subject to recent glacial activity.

Thermal and Saline Springs in Canada

Brief mention should be made of two extreme types of spring found in Canada, thermal springs and salt springs. Thermal springs are found only in the west, in the southern Rocky Mountains (Waring et al. 1965). In this region thermal springs have been defined as never having a water temperature of less than 15°C at the source, the downstream limit of a spring's influence being set at the position where the annual mean temperature of the water is 5°C above the annual mean air temperature (i.e., 9°C; Pritchard 1991). The upper end of the range in water temperature measured in this region is 54°C, with the modal temperature being 32°C (van Everdingen 1972). Comparatively little is known of the communities of invertebrates that live in these springs, although as far as the insects are concerned they are believed to be dominated, as in thermal springs elsewhere, by four orders: the Diptera (especially the Chironomidae, Ephydridae, Simuliidae, and Stratiomyidae), the Coleoptera (Dytiscidae and Hydrophilidae), the Hemiptera (Saldidae and Corixidae), and the Odonata (Coenagrionidae and Libellulidae). Mites are also likely to be common (see list of possible species given in Smith 1991). Again, as is the norm for thermal springs, the communities of individual springs are likely to be very simple (e.g., Collins et al. 1976). Odonates recorded from thermal springs in Canada include *Amphiagrion abbreviatum* and *Argia vivida* (Zygoptera) and *Cordulegaster dorsalis* and *Libellula quadrimaculata* (Anisoptera) (Pritchard 1991).

The fauna of only one group of saline springs in Canada has been studied, those on Saltspring Island in southwest British Columbia. Other parts of Canada support these habitats, but they have not been studied from a biological perspective; for example, those located on top of salt beds near the boundary of the Canadian Shield and the Interior, those in the St. Lawrence Lowlands, and those in the Appalachian Region (see Fig. 20.1) (van Everdingen 1991). The Saltspring Island springs bring water to the surface, from a depth of >1,000 m, that is about 2.2 times the salinity of seawater. The annual water temperature range is from 7 to 21°C. The flora and fauna contains known halophilic species. There are seven species of bacteria, a cyanophyte, a filamentous green alga, together with salt-loving grasses and higher plants. The insects are represented by the hemipteran *Saldula comatula* (Saldidae), the chironomid *Thalassosmittia marina,* two unidentified species of Ephydridae, and two types of cyclorrhaphan Diptera. There are also five species of Hymenoptera, a collembolan (*Anurida* sp.), and three species of beetle (*Bembidion indistinctum,* Carabidae; *Ochthebius lecontei,* Hydraenidae; and *Thicanus mimus*, Anthicidae). Two species of spider are active in the riparian zone (Ring 1991).

CONCLUDING REMARKS

Canada is a large and diverse country with many different spring types supporting animal communities that are in general poorly known. This is especially true for many of the invertebrate taxa, although three insect groups and the hydrachnid mites have received some attention. Table 20.6 presents a summary of the information given in this paper, in an attempt to characterize the arthropod portion of the Canadian fauna. Overall, Canadian springs commonly support limnephilid caddisflies, dytiscid beetles, orthocladine chironomids, and five families of mites (especially hydryphantoid and hygrobatoid forms). Species richness within individual springs varies according to taxon, as does specificity to spring type, and microhabitat diversity within a spring has a strong effect on its fauna. Water temperature, substrate composition, and several other factors also are important. With the exception of the mites, the incidence of true crenophiles is rather low. Community composition has quite a strong east/west dichotomy and has been markedly influenced by the recent glacial history of the country in combination with differential dispersal abilities among species.

The present-day Canadian spring fauna is essentially an interglacial one that, along with the faunas of other freshwater habitats, can be expected to change in the near geological future as the climate of Canada continues to

TABLE 20.6. Summary of the Known Characteristics of Four Invertebrate Taxa Commonly Found in Canadian Springs Together with Factors that Influence Their Occurrence in These Habitats

Characteristic/Factor	Trichoptera	Coleoptera	Chironomidae	Acari
Percent occurrence of species or genera[a] in springs	~7 percent	~8 percent	~17 percent[a]	~23 percent
Species richness in springs	relatively low	relatively low	relatively high	moderately high
Number of species per spring	low (3.9)	low	high (up to 22)	high (up to 40)
Dominant taxon	Limnephilidae	Dytiscidae	Orthocladiinae	Five families[b]
Taxa restricted to springs	few	several	several	many
Specificity to spring type	moderate	moderate	low	high
East-west difference in fauna	strong	strong	? moderate	strong
Influence of glacial history	strong	strong	moderate	strong
Microhabitat/species richness link	yes	probably	yes	yes
Influential spring features:				
Water temperature	yes	yes	yes	yes
Substrate (size/texture)	yes	probably	yes	yes
Aquatic/emergent vegetation	yes	probably	probably	yes
Riparian vegetation	yes	?	?	yes
Current speed	yes	probably	probably	yes
Water depth	probably	probably	probably	yes
pH	yes	?	?	yes
Water chemistry	probably	?	?	yes

Source: Conclusions are based on information found in Colbo 1991, Danks and Williams 1991, Merritt and Cummins 1988, Roughley and Larsen 1991, Smith 1991, Williams and Feltmate 1992, Williams 1991.

[a]Data for the Chironomidae apply to the genus level only.

[b]See Table 20.4.

change. Such changes, however, are likely to be modified by human-induced global warming. In the latter case, since coldwater springs have temperature profiles that center around the regional mean annual air temperature, the prospect of even a few degrees rise may have dire consequences for cold stenotherms, including decrease of total spring invertebrate densities, promotion of precocial breeding (amphipods) and emergence (insects), and alteration of sex ratios, growth rates, and size at maturity (Hogg and Williams 1996). The seriousness of such shifts lies not in the absolute changes per se, but in the rapidity with which human-induced changes occur (measured in decades) relative to those resulting from natural oscillations in climate (measured in millennia), and whether or not organisms can adapt to them.

Canada has no legislation to protect spring habitats per se, despite the importance of their water in many national and regional developments (for example, as sources of drinking water, nature appreciation, tourism, etc.), and a brief on their scientific worth published by the Biological Survey of Canada (Williams et al. 1990). When surveyed in the mid-nineteen-eighties, federal and provincial agencies admitted to having very little, or no, information on spring locations or their properties. At that time Cherry (1987) warned that the existing government policy in Canada afforded little protection to groundwater resources in general, even though 25–30 percent of the nation's population rely on them for drinking. Traditionally the federal government has avoided a major role in managing or solving groundwater problems for reasons of: (1) ownership—the groundwater is regarded as being beneath *provincial* lands; (2) an "abundance mentality"; and (3) groundwater not falling within the mandate of a single government department, thus posing an administrative problem (Morgan 1993). Recent protection through provincial and regional legislation has been highly variable in its extent and effectiveness.

It is becoming clear that Canadian aquifers, together with the springs that they support, are being increasingly contaminated by chemicals from industry, agriculture, and the urban landscape (Pearse et al. 1985, Taylor 1992). Uncertainty in the disposal, and subsequent subsurface transport, of industrial, municipal, and household wastes is creating major and long-term problems for future generations (Cherry 1987). For example, in many urban areas salt is used each winter to clear highways of snow and ice. In southern Ontario, it has been shown that while 45 percent of the salt applied annually is removed by overland flow, the remaining 55 percent enters temporary storage in shallow groundwaters (Howard and Haynes 1993). In this region the chloride levels measured in springs can be as much as 1148 mg l^{-1}, and this level is sustained throughout the year (Williams et al. 1997). The effects that such abnormally high levels (>100 times the background level in pristine springs) have on spring invertebrates is currently unknown, but there are indications that several invertebrate taxa are sensitive to even moderate levels of salt and are lost from the communities of contaminated springs. Contaminant transport modeling suggests that these levels will be maintained for decades or, in the case of springs supplied by deepwater aquifers, for centuries.

Currently several factors have placed groundwater issues on the federal agenda. These include new national legislation on the environment, international forces in trade, concern for global water resources, and agreed international principles on environmental protection. Arising from this federal involvement are, for example, a proposed new Canada Drinking Water Act, a goal for comprehensive environmental management and sustainability (Canada's Green Plan—recently canceled for budgetary reasons), and ratification of the Rio Conference on the World Environment. Rapid growth of the Canadian groundwater industry (supply services and product manufacturing) also has contributed to a higher profile. However, the effectiveness of government and industry in protecting and managing Canada's groundwater resources is hampered by lack of basic groundwater data (e.g., inventories) and governmental policy (Morgan 1993).

ACKNOWLEDGMENTS

We thank those coworkers on Canadian springs on whose data and ideas we have drawn heavily, especially Drs. Hugh Danks, Ian Smith, Gordon Pritchard, Murray Colbo, Dave Larson, Rob Roughley, Robert van Everdingen, and Richard Ring. We are grateful also to the Natural Sciences and Engineering Research Council of Canada for funding, and to Marilyn Smith for preparation of the figures.

LITERATURE CITED

Cherry, J. A. 1987. Groundwater occurrence and contamination in Canada. Pages 387–426 *in* M. C. Healey and R. R. Wallace (eds.), Canadian Aquatic Resources. Canadian Bulletin of Fisheries and Aquatic Sciences 215.

Colbo, M. H. 1991. A comparison of the spring-inhabiting genera of Chironomidae from the Holarctic with those from natural and man-made springs in Labrador, Canada. Memoirs of the Entomological Society of Canada 155:169–179.

Crouch, D. P. 1993. Water Management in Ancient Greek Cities. Oxford University Press, New York.

Danks, H. V., and D. M. Rosenberg. 1987. Aquatic insects of peatlands and marshes in Canada: Synthesis of information and identification of needs for research. Memoirs of the Entomological Society of Canada 140:163–174.

Danks, H. V., and D. D. Williams. 1991. Arthropods of springs, with particular reference to Canada: Synthesis and needs for research. Memoirs of the Entomological Society of Canada 155:203–217.

EMR (Energy, Mines and Resources). 1974. The National Atlas of Canada (4th ed.). Energy, Mines and Resources, Ottawa, and Macmillan, Toronto, ON.

Evans, H. B. 1994. Water Distribution in Ancient Rome. University of Michigan Press, Ann Arbor, MI.

Gooch, J. L., and D. S. Glazier. 1991. Temporal and spatial patterns in mid-Appalachian springs. Memoirs of the Entomological Society of Canada 155:29–49.

Hogg, I. D., and D. D. Williams. 1996. Response of stream invertebrates to a global-warming thermal regime: An ecosystem-level manipulation. Ecology 77:395–407.

Howard, K. W. F., and J. Hayes. 1993. Groundwater contamination due to road de-icing chemicals—salt balance implications. Geoscience Canada 20:1–8.

Matthews, J. V. 1979. Tertiary and Quaternary environments: Historical backgound for an analysis of the Canadian insect fauna. Memoirs of the Entomological Society of Canada 108:31–86.

Merritt, R. W., and K. W. Cummins (eds.). 1988. An Introduction to the Aquatic Insects of North America (2nd ed.). Kendall/Hunt, Dubuque, IA.

Mitchell, R. D. 1957. Major evolutionary lines in water mites. Systematic Zoology 6: 137–148.

———. 1960. The evolution of thermophilous water-mites. Evolution 14:361–377.

Morgan, A. V. (ed.). 1993. Groundwater Issues and Research in Canada. Canadian Geosciences Council, Waterloo, ON.

Odum, H. T. 1957. Trophic structure and productivity of Silver Springs, Florida. Ecological Monographs 27:55–112.

Patrick, R., and D. D. Williams. 1990. Aquatic biota in North America. Pages 233–254 in M. G. Wolman and H. C. Riggs, (eds.), Surface Water Hydrology. Geological Society of America, Boulder, CO.

Pearse, P. H., F. Bertrand, and J. W. MacLaren. 1985. Currents of Change. Final Report, Inquiry on Federal Water Policy, Ottawa, ON.

Pritchard, G. 1991. Insects in thermal springs. Memoirs of the Entomological Society of Canada 155:89–106.

Resh, V. H. 1982. Age structure alteration in a caddisfly population after habitat loss and recovery. Oikos 38:280–284.

Ring, R. A. 1991. The insect fauna and some other characteristics of natural salt springs on Saltspring Island, British Columbia. Memoirs of the Entomological Society of Canada 155:51–61.

Ross, H. H. 1965. Pleistocene events and insects. Pages 583–596 in H. E. Wright and D. G. Frey (eds.), The Quaternary of the United States. Princeton University Press, Princeton, NJ.

Roughley, R. E., and D. J. Larson. 1991. Aquatic Coleoptera of springs in Canada. Memoirs of the Entomological Society of Canada 155:125–140.

Schwoerbel, J. 1959. Ökologische und tiergeographische Untersuchung über die Milben (Acari, Hydrachnellae) der Quellen und Bäche des südlichen Schwarzwaldes und seiner Randgebiete. Archiv für Hydrobiologie Supplement 24:385–546.

Smith, I. M. 1991. Water mites (Acari: Parasitengona: Hydrachnida) of spring habitats in Canada. Memoirs of the Entomological Society of Canada 155:141–167.

Smith, I. M., and D. R. Cook. 1991. Water mites. Pages 523–592 in J. Thorp and A. Covich (eds.), Ecology and Classification of North American Freshwater Invertebrates. Academic Press, New York.

Taylor, L. C. 1992. The response of spring-dwelling ostracods to intra-regional differences in groundwater chemistry associated with road-salting practices in southern

Ontario: A test using an urban-rural transect. Master's Thesis, University of Toronto, Toronto, ON.

Tilly, L. J. 1968. The structure and dynamics of Cone Spring. Ecological Monographs 38:169–197.

van Everdingen, R. O. 1972. Thermal and Mineral Springs in the Southern Rocky Mountains of Canada. Environment Canada Report EN36-415/1972.

———. 1991. Physical, chemical, and distributional aspects of Canadian springs. Memoirs of the Entomological Society of Canada 155:7–28.

Viets, K. O. 1978. Hydracarina. Pages 154–181 *in* J. Illies (ed.), Limnofauna Europea. G. Fischer, Stuttgart, Germany.

Waring, G. A., R. R. Blankenship, and R. Bentall. 1965. Thermal Springs of the United States and Other Countries of the World—a Summary. Geological Survey Professional Paper 492. U.S. Government Printing Office, Washington, DC.

Webb, D. W., M. J. Wetzel, P. C. Reed, L. R. Phillippe, and M. A. Harris. 1995. Aquatic biodiversity in Illinois springs. Journal of the Kansas Entomological Society 68:93–107.

Wiggins, G. B. 1984. Trichoptera. Pages 271–311 *in* R. W. Merritt and K. W. Cummins (eds.), An Introduction to the Aquatic Insects of North America. Kendall/Hunt, Dubuque, IA.

Williams, D. D. 1983. National survey of freshwater springs. Bulletin of the Entomological Society of Canada 15:30–34.

Williams, D. D., H. V. Danks, I. M. Smith, R. A. Ring, and R. A. Cannings. 1990. Freshwater springs: A national heritage. Bulletin of the Entomological Society of Canada Supplement 22:1–9.

Williams, D. D., and B. W. Feltmate. 1992. Aquatic Insects. C.A.B. International, Wallingford, Oxford, UK.

Williams, D. D., and I. D. Hogg. 1988. Ecology and production of invertebrates in a Canadian coldwater spring-springbrook system. Holarctic Ecology 11:41–54.

Williams, D. D., and N. E. Williams. 1987. Trichoptera from cold freshwater springs in Canada: Records and comments. Proceedings of the Entomological Society of Ontario 118:13–23.

Williams, D. D., N. E. Williams, and Y. Cao. 1997. Spatial differences in macroinvertebrate community structure in springs of southeastern Ontario in relation to their chemical and physical environments. Canadian Journal of Zoology 75:1404–1414.

Williams, N. E. 1991. Geographic and environmental patterns in caddisfly (Trichoptera) assemblages from coldwater springs in Canada. Memoirs of the Entomological Society of Canada 155:107–124.

Young, W. C. 1969. Ecological distribution of Hydracarina in North Central Colorado. American Midland Naturalist 82:367–401.

PART 3
Wetlands of the Central Prairies
and Mississippi River Basin

21 Wetlands of the Prairie Pothole Region

Invertebrate Species Composition, Ecology, and Management

NED H. EULISS, Jr., DALE A. WRUBLESKI, and
DAVID M. MUSHET

*T*he Prairie Pothole Region (PPR) of the United States and Canada
is a unique area where shallow depressions created by the scouring action of Pleistocene glaciation interact with midcontinental
climate variations to create and maintain a variety of wetland
classes. These wetlands possess unique environmental and biotic characteristics that add to the overall regional diversity and production of aquatic
invertebrates and the vertebrate wildlife that depend upon them as food.
Climatic extremes in the PPR have a profound and dynamic influence on
wetland hydrology, hydroperiod, chemistry, and ultimately the biota. Available knowledge on aquatic invertebrates in the PPR suggests that the diversity of invertebrates within each wetland class is low. Harsh environmental
conditions range from frigid winter temperatures that freeze wetlands and
their sediments to hot summer temperatures and drought conditions that
create steep salinity gradients and seasonally dry habitats. Consequently,
the invertebrate community is composed mostly of ecological generalists
that possess the necessary adaptations to tolerate environmental extremes.
In this review we describe the highly dynamic nature of prairie pothole
wetlands and suggest that invertebrate studies be evaluated within a conceptual framework that considers important hydrologic, chemical, and climatic events.

Invertebrates in Freshwater Wetlands of North America: Ecology and Management, Edited by
Darold P. Batzer, Russell B. Rader, and Scott A. Wissinger
ISBN 0-471-29258-3   © 1999 John Wiley & Sons, Inc.

PRAIRIE POTHOLE WETLANDS

The prairie pothole region (PPR) of North America covers approximately 715,000 km², extending from north-central Iowa to central Alberta (Fig. 21.1). The landscape of the PPR is largely the result of glaciation events during the Pleistocene Epoch. The last glaciers retreated from the PPR approximately 12,000 years ago, leaving behind a landscape dotted with many small depressional wetlands called *potholes* or *sloughs*. The present climate of the midcontinent PPR is dynamic, characterized by interannual variation between wet and dry periods in which abundant rainfall can be followed by drought (i.e., the wet/dry cycle). The association between prairie wetlands and groundwater tables of the region is complex and dynamic. Hydrologically, prairie wetlands can function as groundwater recharge sites, flow-through systems, or groundwater discharge sites. The hydrologic function a particular wetland performs is determined by variations in climate, its position in the landscape, the configuration of the associated water table, and the type of underlying geological substrate. The unique hydrology and climate of this region have a profound influence on the water chemistry, hydroperiod, and ultimately the biotic communities that inhabit prairie wetlands.

Fig. 21.1. The prairie pothole region (PPR) of North America.

Aquatic invertebrates inhabiting prairie pothole wetlands are well suited to cope with the highly dynamic and harsh environmental conditions of the PPR. Because of midcontinent temperature and precipitation extremes, wetlands of the PPR periodically go dry, freeze solidly in winter, and exhibit steep salinity gradients. These salinity gradients are due to the interrelation among precipitation, evapotranspiration, interaction with groundwater, and variation in the composition of soils. Due to the harsh environmental conditions of the PPR, the overall diversity of aquatic invertebrates within each wetland class is low because taxa are mostly restricted to a few ecological generalists. Those that do occur possess the necessary adaptations that allow them to exploit the naturally high productivity of prairie wetlands.

Despite the harsh climate, the PPR is an extremely productive area for both agricultural products and wildlife. The landscape has been substantially altered since settlement of the PPR in the late 1800s. Economic incentives to convert natural landscapes to agriculture are great and have resulted in the loss of over half of the original 8 million ha of wetlands (Tiner 1984, Dahl 1990, Dahl and Johnson 1991). Land-use impacts on wetland biota include enhanced siltation rates, contamination from agricultural chemicals, altered hydrology, the spread of exotic plants, and habitat fragmentation due to wetland drainage and conversion of native prairie grasslands into agricultural fields.

The highly dynamic PPR is a unique area that is of critical importance to migratory birds in North America and to the aquatic invertebrates that supply them with dietary nutrients. Despite the value of the knowledge generated thus far, a critical need still exists to expand our knowledge of wetland invertebrates to better our understanding of the PPR ecosystem and its susceptibility to anthropogenic influence. Given the highly dynamic nature of PPR wetlands and the extreme variation in chemical characteristics and hydroperiod, it is essential that invertebrate studies be placed in the proper conceptual framework to maximize the application of study results. Herein we describe the highly dynamic nature of prairie wetlands and suggest that invertebrate studies be evaluated within the context of hydrologic, chemical, and climatic events that characterize the region.

Geology

Glaciation events during the Pleistocene Epoch were the dominant forces that shaped the landscape of the PPR (Winter 1989). When the glaciers retreated, a landscape dotted with numerous small, saucer-like depressions was exposed. These depressions, caused by the uneven deposition of glacial till, the scouring action of glaciers, and the melting of large, buried ice blocks, are known today as prairie potholes or sloughs.

The deposition of glacial till was unevenly distributed throughout the PPR. Large moraines accumulated along the terminal ends of glaciers and formed ridges of low, rolling hills in a northwest to southeast orientation, such as the

Missouri Coteau. Where glaciers retreated quickly, large, gently rolling areas of glaciated plains were formed, and extremely flat lake beds developed where glaciers dammed meltwater. Due to the geologically young nature of the landscape and moderate rainfall, there are few natural surface drainage systems. Consequently, most wetlands in the PPR are not connected by overland flow.

Climate

The PPR is in the midcontinent of North America and is subject to the climatic extremes of this region (Winter 1989). Temperatures can exceed 40°C in summer and −40°C in winter. Isolated summer thunderstorms may bring several inches of rain in small localized areas while leaving adjacent habitats entirely dry. Also, winds of 50 to 60 km hr^{-1} can quickly dry wetlands during the summer or create windchills below −60°C during winter.

Besides the normal seasonal climatic extremes, the semiarid PPR also undergoes long periods of drought followed by long periods of abundant rainfall. These wet/dry cycles can persist for 10 to 20 years (Duvick and Blasing 1981, Karl and Koscielny 1982, Karl and Riebsame 1984, Diaz 1983, 1986). During periods of severe drought, most wetlands go dry during summer, and many remain completely dry throughout the drought years. Exposure of mud flats upon dewatering is necessary for the germination of many emergent macrophytes, and it facilitates the oxidation of organic sediments and nutrient releases that maintain high productivity. When abundant precipitation returns, wetlands fill with water and much of the emergent vegetation is drowned. Changes in water permanence and hydroperiod by normal seasonal drawdown and long interannual wet/dry cycles has a profound influence on all PPR biota, but is most easily observed in the hydrophytic community (van der Valk and Davis 1978a).

The PPR has a north-to-south and a west-to-east precipitation gradient, with areas to the north and west receiving less precipitation than those to the south and east. However, even in the wetter southeastern portion of the region, wetlands have a negative water balance. Evaporation exceeds precipitation by 60 cm in southwestern Saskatchewan and eastern Montana and by 10 cm in Iowa (Winter 1989). Despite this negative water balance, many wetlands contain water throughout the year and go dry only during periods of extended drought.

Hydrology

Although PPR wetlands receive the majority of their water from snowmelt runoff in the spring and rarely as summer precipitation, the association between prairie wetlands and groundwater tables of the region is complex and dynamic (Winter and Rosenberry 1995). Hydrologically, prairie wetlands can function as groundwater recharge sites, flow-through systems, or groundwater discharge sites. Groundwater recharge wetlands receive their water primarily from the atmosphere and there is little or no groundwater inflow. As a result, the mineral content of water in recharge wetlands is extremely low. Wetlands that function as flow-through systems both receive and discharge water and solutes from and into the ground. Water in flow-through wetlands generally reflects the chemical composition of groundwater. Wetlands that function as groundwater discharge sites receive the bulk of their solutes from groundwater, and their principal water loss is from evapotranspiration. As a result, the salinity of water in discharge wetlands can be highly variable and in some cases can exceed the salinity of seawater. The hydrologic function (recharge, discharge, flow-through) of a particular wetland is determined by variations in climate and by their position in the landscape, the configuration of the associated water table, and the type of underlying geologic substrate. The hydrologic function of individual wetlands defines their unique hydroperiod and chemical characteristics and ultimately the plant community they support. Hence, wetland classes based on vegetation (Stewart and Kantrud 1971) reflect the range of hydrologic function within any given wetland class. Temporary wetlands tend to recharge groundwater, seasonal wetlands can have either a recharge or flow-through function, semipermanent wetlands tend to have either a flow-through or discharge function, and saline tend to function mostly as discharge sites.

Chemistry

The chemical characteristics of prairie wetlands also vary in relation to fluctuations in climate and hydrology. Prairie wetlands have dissolved solid concentrations that span the gradient from fresh to extremely saline (LaBaugh 1989). Hydrologic processes, especially those that define how individual wetlands receive and lose water, largely determine the salt concentration of individual wetlands at any given point in time. Wetlands range in specific conductance from 42 (Petri and Larson 1973) to 472,000 μS cm^{-1} (Swanson et al. 1988, LaBaugh 1989), while fluctuations of individual wetlands can range from 1,160 to 43,600 μS cm^{-1} (LaBaugh et al. 1996 and unpublished data). Most wetlands in the PPR are alkaline (pH > 7.4) (LaBaugh 1989), with values as high as 10.4 in North Dakota marshes (Swanson et al. 1988).

Vegetation

Plant communities in prairie wetlands are dynamic and continually changing as a result of short- and long-term fluctuations in water levels, salinity, and anthropogenic disturbance. van der Valk and Davis (1978a) propose a general model describing how wetland plant communities respond to water level fluctuations due to the wet/dry cycle. Four wetland stages are identified: dry marsh, regenerating marsh, degenerating marsh, and lake marsh.

During drought periods marsh sediments and seed banks are exposed. During this dry marsh phase seeds of many mudflat annual and emergent plant species germinate on exposed mudflats, with annual species usually forming the dominant component (van der Valk and Davis 1976, 1978a, Davis and van der Valk 1978a, Galinato and van der Valk 1986, Welling et al. 1988a, b). When water returns, the annuals are lost but emergent macrophytes survive and expand by vegetative propagation (i.e., regenerating marsh). Depth and duration of the flooding period, combined with the tolerances of individual species of macrophytes, will determine how these wetland communities develop. If the wetland experiences only shallow flooding, the emergent macrophytes will eventually dominate the entire wetland. However, prolonged deep-water flooding results in the elimination of emergent macrophytes (i.e., degenerating marsh) due to the direct effects of extended inundation and the expansion of muskrats and their consumption of macrophytes. If water levels remain high, the lake marsh stage is eventually reached. Submersed macrophytes become established and dominate in the open water areas. A drawdown of the wetland will be necessary for reestablishment of emergent macrophytes.

Salinity modifies vegetation responses to water level fluctuations. Increasing salinity results in a loss of diversity, with the most saline wetlands having the fewest plant species (Kantrud et al. 1989). Soil salinity is also very important during the dry marsh phase, regulating the germination of emergent macrophytes on exposed mudflats (Galinato and van der Valk 1986). Kantrud et al. (1989) present information describing the salinity tolerances of many prairie wetland plant species, as well as the predicted changes that may occur as salinity changes over the course of the wet/dry cycle.

Until recently, little was known about the algal communities of prairie wetlands or their responses to changes in wetland hydrology (Crumpton 1989, Goldsborough and Robinson 1996, Robinson et al. 1997a, b). Algal biomass may be lower than macrophyte biomass, but their overall productivity may be similar due to high turnover rates (Murkin 1989). Four algal communities are recognized within wetland habitats: epipelon (motile algae inhabiting soft sediments), epiphyton (algae growing on submersed surfaces such as macrophytes), metaphyton (floating or subsurface mats of filamentous algae), and phytoplankton (algae of the water column) (Goldsborough and Robinson 1996). A conceptual model describing wetland stages where each of these four communities is dominant has been developed by Goldsborough and Robinson (1996).

Wetland Classes

A number of wetland classification systems are available (Stewart and Kantrud 1971, Millar 1976, Cowardin et al. 1979, Brinson 1993), but Stewart and Kantrud (1971) is specific to the glaciated prairies. Using Stewart and Kantrud (1971), there are seven wetland classes, based on the vegetational zone occupying the central, deepest part of the wetland basin and occupying at least 5 percent of the total wetland area. The seven vegetational zones identified by Stewart and Kantrud (1971) are the wetland-low-prairie, wet-meadow, shallow-marsh, deep-marsh, permanent-open-water, intermittent-alkali, and fen zones.

Stewart and Kantrud's (1971) wetland classification, while based on vegetational characteristics, reflects differences in water permanence and can be related to the water regime modifiers used by Cowardin et al. (1979). Wet-meadow vegetation (e.g., *Poa palustris, Hordeum jubatum*) dominates areas that typically contain water for several weeks after spring snowmelt. Shallow-marsh vegetation (e.g., *Eleocharis palustris, Carex antherodes*) dominates areas where water typically persists for a few months each spring, and deep-marsh vegetation (e.g., *Typha latifolia, Scirpus acutus*) occupies areas where water persists throughout the year. The permanent-open-water zone, characterized by the lack of vascular plants, dominates the central part of wetlands that rarely dry, even during periods of extended drought.

In terms of total area, wetlands of the temporary (Class II), seasonal (Class III), and semipermanent (Class IV) classes comprise the majority of the wetlands in the PPR (Fig. 21.2). Ephemeral (Class I) wetlands, while numerous, are small and are not considered wetlands by Cowardin et al. (1979). Permanent (Class V) and alkali (Class VI) wetlands, although usually large in size, are few in number (Stewart and Kantrud 1971). Fens (Class VII) are not common in the PPR, but their unique biota has raised some concern to preserve existing sites as areas of special ecological significance.

Within wetlands, vegetational zones frequently alternate between two or more distinct phases. These phases are identified by changes in the plant communities brought about by fluctuations in water levels or by changes in land-use practices. The six phases identified in Stewart and Kantrud's (1971) classification are the normal emergent, open-water, drawdown bare-soil, natural drawdown emergent, cropland drawdown, and cropland tillage phases. Although phase changes do not effect wetland classification, they do alter the vegetative structure available to invertebrate communities.

In addition to normal shifts between phases, vegetational zones also shift from one type to another in response to extended drought or above-normal precipitation. If this change occurs in the central, deepest part of a wetland, it can change its classification. During extended drought, wet-meadow vegetation often expands and dominates the central, shallow-marsh zone of seasonal wetlands. Conversely, during extended periods of above normal precipitation, shallow-marsh vegetation frequently expands into the wet-meadow zone. Thus, a seasonal (class III) wetland may shift to a temporary (Class II) wetland; the converse may occur during extended periods of above normal precipitation. As wetlands shift among phases and classes, the characteristic shift in vegetation effects a complementary shift in the invertebrate community; in general, enhanced vegetative diversity results in an increase in invertebrate richness (Driver 1977).

Land Use Influences

The agricultural value of the PPR has tremendously impacted prairie pothole wetlands. Wetland drainage (both surface and tile) to enhance agricultural

(*a*)

(*b*)

Fig. 21.2. Common wetland classes in the prairie pothole region: (*a*) temporary (Class II); (*b*) seasonal (Class III); (*c*) semipermanent (Class IV); alkali (Class VI).

(*c*)

(*d*)

Fig. 21.2. Continued.

production has been the primary factor resulting in the loss of wetlands in this region (Tiner 1984, Canada-United States Steering Committee 1985, Millar 1989, Dahl 1990, Dahl and Johnson 1991). Remaining wetlands are impacted by a number of agricultural practices that result in elevated sedimentation rates (Martin and Hartman 1986, Gleason and Euliss 1996a), drift of agricultural chemicals into wetlands (Grue et al. 1989), large inputs of nutrients (Neely and Baker 1989), unnatural variance in water-level fluctuation (Euliss and Mushet 1996), and altered vegetative communities (Kantrud and Newton 1996). Major nonagricultural impacts include alteration of hydrologic and chemical regimes due to road construction (Swanson et al. 1988) and urban development. The extent to which land use has altered the ecology of aquatic invertebrates is poorly studied but must be understood to facilitate effective management of prairie wetlands.

PAST INVERTEBRATE RESEARCH

As pointed out by Rosenberg and Danks (1987), invertebrates of freshwater wetlands are poorly studied and existing information is limited and scattered. As with many other regions in North America, our knowledge of PPR wetland invertebrates is incomplete, but significant work has been conducted, especially over the past several decades. Interestingly, the bulk of our knowledge of wetland invertebrates in the PPR has resulted from research conducted in scientific disciplines other than those concerned solely with invertebrate biology.

Food habit studies stressing the dietary value of invertebrates to waterfowl during the breeding season (e.g., Bartonek 1968, Bartonek 1972, Bartonek and Hickey 1969, Dirschl 1969, Swanson and Bartonek 1970, Swanson et al. 1977) provided the impetus for much of the past research on aquatic invertebrates in prairie pothole wetlands. Migratory waterfowl are of considerable economic value and are the subjects of international treaties. Additionally, the PPR is a critical breeding area for many North American species. As a consequence, much of our information on wetland invertebrates in the PPR has been directed towards developing a better understanding of waterfowl ecology and management. Invertebrate species lists and distribution studies also have significantly contributed to our knowledge of prairie wetland invertebrates, but the work has been patchy and is largely incomplete. Overall, basic ecology has provided the smallest contribution to our current knowledge of prairie wetland invertebrates, but development of more holistic perspectives of the critical role that invertebrates play in wetland ecology and function (Murkin and Wrubleski 1988) will likely stimulate such work. Aside from their obvious role in the feeding ecology of waterfowl and other birds, invertebrates provide critical food chain support for a wide variety of other organisms and play significant roles in nutrient cycling and overall wetland productivity

(Murkin and Batt 1987). Further, invertebrates are sensitive to agricultural chemicals that accumulate in wetlands (Grue et al. 1989) and there is a growing interest in using them as indicators of wetland and landscape condition in the PPR (Adamus 1996) and elsewhere in the United States.

Invertebrate and Habitat Diversity

There have been relatively few studies providing species lists and habitat distributions of invertebrate fauna in prairie wetlands (Table 21.1). However, those few suggest that diversity within specific wetland classes is low in comparison to many other areas of North America. Low diversity is likely attributable to the adverse conditions of the region, including extremely harsh winters and fluctuating hydrology and chemistry. Unlike migratory wildlife, invertebrates inhabiting this area must have adaptations that allow them to withstand these environmental extremes. On a regional scale, however, the overall diversity of invertebrates may be comparable to other areas in North America due to the diversity of wetland classes within the PPR.

Invertebrate Responses to Hydroperiod

The dynamic hydroperiods of prairie wetlands, in relation to both seasonal and the longer-term wet/dry cycle, also exert a positive influence on wetland productivity, including production of aquatic invertebrates. Prolonged inundation results in the decreased availability of plant nutrients, whereas oxidative processes that accompany drawdowns facilitate nutrient release and ultimately foster the development of plant communities that make significant contributions to the nutrient and detritus pool upon reflooding. The sharp increase in wetland productivity when wetlands reflood following a dry phase is the reason for artificially flooding and draining wetlands to enhance waterfowl populations (Cook and Powers 1958, Kadlec and Smith 1992), and it is the basis for the modern day practice of moist soil management (Fredrickson and Taylor 1982). The basic underpinnings of this phenomenon relate to nutrient releases from the aerobic decomposition of accumulated macrophyte litter by terrestrial hyphomycetes (Bärlocher et al. 1978).

Seasonal drawdown, as well as the longer-term wet/dry cycle, impacts the hydroperiod of prairie wetlands and has special significance to the ecology of aquatic invertebrates. During extreme drought, nearly all wetlands are dry, including many permanent habitats. Under such conditions invertebrate faunas undergo severe spatial reduction, and they are composed mostly of specialists that can tolerate hypersaline waters or those that have such short life cycles that reproduction can be completed within an extremely short time. Wetlands can shift hydrologic function during severe drought (Winter and Rosenberry 1995), and viable, but normally dormant, eggs in semipermanent wetlands can hatch producing an invertebrate community that more closely resembles

TABLE 21.1. Partial Species List of Invertebrates Present in Wetlands of the Prairie Pothole Region[a]

Taxon	Reference[b]
Gastropoda	
Hydrobiidae	
Cincinnatia cincinnatiensis (Anthony)	1
Amnicola limosa (Say)	1
Lymnaeidae	
Fossaria obrussa (Say)	1
Lymnaea stagnalis (L.)	1, 21
Stagnicola caperata (Say)	1, 8
Stagnicola elodes (Say)	1, 8
Physidae	
Aplexa hypnorum (L.)	1
Physa heterostropha (Say)	6
Physa gyrina (Say)	1
Physa integra Haldeman	1
Physa jennessi Dall	8
Planorbidae	
Armiger crista (L.)	1, 6, 8
Gyraulus circumstriatus (Tryon)	1, 9
Gyraulus parvus (Say)	1, 6, 9
Helisoma anseps (Mente)	1
Helisoma trivolvis (Say)	1
Planorbula armigera (Say)	1
Planorbula campestris (Dawson)	1
Promenetus exacuous (Say)	1, 6, 9
Promenetus umbilicatellus (Cock.)	9
Valvatidae	
Valvata tricarinata (Say)	1
Annelida	
Hirudinea	
Glossiphoniidae	
Alboglossiphonia heteroclita (L.)	6
Helobdella fusca (Castle)	6
Marvinmeyeria lucida (Moore)	6
Erpobdellidae	
Erpobdella punctata (Leidy)	17
Nephelopsis obscura (Verrill)	17
Oligochaeta	
Lumbriculidae	
Lumbriculus variegatus (Muller)	6
Naididae	
Chaetogaster diaphanus (Gruit.)	6
Tubificidae	
Tubifex tubifex (Muller)	17
Rotifera	
Brachionidae	
Keratella cochlearis (Gosse)	6

TABLE 21.1. (Continued)

Taxon	Reference[b]
Keratella quadrata (Muller)	6
Filiniidae	
Filinia longiseta	6
Lecanidae	
Monostyla lunaris	6
Crustacea	
Amphipoda	
Gammaridae	
Gammarus lacustris Sars	20
Talitridae	
Hyalella azteca (Saus.)	9, 20
Anostraca	
Artemiidae	
Artemia salina Leach	21
Branchinectidae	
Branchinecta lindahli Pack	8
Chirocephalidae	
Eubranchipus bundyi Forbes	6, 9
Eubranchipus ornatus Holm.	9
Cladocera	
Bosminidae	
Bosmina longirostris (Muller)	6
Daphnidae	
Daphnia pulex Leydig	6
Daphnia rosea Sars	6
Conchostraca	
Lynceidae	
Lynceus brachyurus Muller	6
Lynceus mucronatus (Pack.)	6
Copepoda	
Diaptomidae	
Aglaodiaptomus leptopus (Forb.)	6
Hesperodiaptomus franciscanus (Lillj.)	6
Ostracoda	
Cyclocyprididae	
Cyclocypris ampla Furtos	6
Insecta	
Ephemeroptera	
Baetidae	
Callibaetis pallidus Banks	16
Caenidae	
Caenis youngi Roem.	16
Odonata	
Aeshnidae	
Aeshna constricta Say	1, 3
Aeshna eremita Scudder	3

TABLE 21.1. (Continued)

Taxon	Reference[b]
Aeshna interrupta Walker	1, 3, 16
Anax junius Drury	1, 3, 16
Aeshna umbrosa	1, 3
Agrionidae	
Agrion aequabile (Say)	1
Hetaerina americana (Fabricius)	1, 3
Coenagrionidae	
Amphiagrion abbreviatum (Selys)	1
Argia apicalis (Say)	1, 3
Coenagrion angulatum Walker	1, 3, 6, 16, 18
Coenagrion resolutum (Hagen)	1, 3, 6, 16, 18
Enallagma antennatum (Say)	1,3
Enallagma boreale Selys	1, 3, 6, 16
Enallagma carunculatum Morse	1, 3
Enallagma civile (Hagan)	1, 3
Enallagma clausum Morse	1, 3
Enallagma cyathigerum (Charp.)	1, 3, 16
Enallagma ebrium (Hagen)	1, 3, 16
Enallagma hageni (Walsh)	1, 3
Ischnura damula Calvert	1, 3
Ischnura verticalis (Say)	1, 3
Nehalennia irene (Hagan)	1, 3
Corduliidae	
Somatochlora ensigera Martin	1, 3
Gomphidae	
Gomphus cornutus Tough	1,3
Gomphus externus Hagen	3
Gomphus graslinellus Walsh	1, 3
Lestidae	
Lestes congener Hagen	1, 3, 16, 19
Lestes disjunctus Selys	1, 3, 6, 16, 18
Lestes dryas Kirby	1, 3, 6, 18
Lestes forcipatus Rambur	3, 6
Lestes rectangularis Say	1, 3
Lestes unguiculatus Hagen	1, 3, 16, 18
Libellulidae	
Erythemis simplicicollis (Say)	1, 3
Leucorrhinia borealis Hagen	1, 3, 16
Leucorrhinia intacta (Hagan)	1, 3
Leucorrhinia proxima Calvert	3
Libellula luctuosa Burmeister	1, 3
Libellula pulchella Drury	1, 3
Libellula quadrimaculata L.	1, 3, 16
Plathemis lydia (Drury)	1, 3
Sympetrum corruptum (Hagen)	3, 16
Sympetrum costiferum (Hagen)	1, 3, 16
Sympetrum danae (Sulzer)	1, 3, 16

TABLE 21.1. (Continued)

Taxon	Reference[b]
Sympetrum internum Mont.	1, 3, 16
Sympetrum madidum (Hagen)	1, 3
Sympetrum obtrusum (Hagen)	1, 3
Sympetrum occidentale fasciatum Walker	1, 3
Sympetrum rubicundulum (Say)	1, 3
Sympetrum vicinum (Hagan)	1, 3
Coleoptera	
Dytiscidae	
Agabus ajax Fall	10
Agabus antennatus Leech	10
Agabus bifarius (Kirby)	10
Agabus canadensis Fall	10
Agabus falli Zimm.	10
Agabus punctulatus Aube	10
Colymbetes sculptilis Harris	10
Coptotomus longulus LeConte	10
Dytiscus alaskanus Balf.-Brow.	10
Dytiscus circumcinctus Ahrens	10
Dytiscus cordieri Aube	10
Dytiscus hybridus Aube	10
Graphoderus occidentalis Horn	10
Graphoderus perplexus Sharp	10
Hydaticus modestus Sharp	10
Hydaticus piceous LeConte	10
Hydroporous fuscipennis Schaum	10
Hydroporous notabilis LeConte	10
Hydroporous pervicinus Fall	10
Hydroporous tenebrosus LeConte	10
Hygrotus acaroides (LeConte)	10
Hygrotus canadensis (Fall)	10
Hygrotus compar (Fall)	10
Hygrotus impressopunctatus (Schall.)	10
Hygrotus patruelis (LeConte)	10
Hygrotus picatus (Kirby)	10
Hygrotus sayi Balf.-Brow.	10
Hygrotus sellatus (LeConte)	10
Hygrotus turbidus (LeConte)	10
Ilybius fraterculus LeConte	10
Ilybius biguttulus Germar	10
Laccophilus biguttatus Kirby	10
Laccophilus maculosus Say	10
Laccornis conoideus (LeConte)	10
Liodessus affinis (Say)	10
Rhantus consimilis Motsch.	10
Rhantus frontalis (Marsh.)	10
Rhantus suturellus (Harris)	10

TABLE 21.1. (Continued)

Taxon	Reference[b]
Gyrinidae	
Gyrinus maculiventris LeConte	10
Gyrinus minutus Fab.	10
Haliplidae	
Haliplus hoppingi Wallis	10
Haliplus immaculicollis Harris	10
Haliplus strigatus Roberts	10
Haliplus subguttatus Roberts	10
Peltodytes edentulus (LeConte)	10
Peltodytes tortulosus Roberts	10
Hydrophilidae	
Berosus fraternus LeConte	10
Berosus hatchi (Miller)	10
Berosus striatus (Say)	10
Helophorus oblongus LeConte	10
Helophorus linearis LeConte	10
Helophorus lineatus Say	10
Hydrobius fuscipes (L.)	10
Hydrochara obtusata (Say)	10
Hydrochus squamifer LeConte	10
Hydropnilus triangularis Say	10
Tropisternis lateralis (Say)	10
Hemiptera	
Belostomatidae	
Belostoma fluminea Say	4, 11
Benacus griseus (Say)	4
Lethocerus americanus (Leidy)	4, 11
Corixidae	
Callicorixa audeni Hungerford	2, 4, 11
Cenocorixa bifida (Hungerford)	4, 11
Cenocorixa dakotensis (Hungerford)	2, 4, 11
Cenocorixa expleta (Uhler)	4, 11
Cenocorixa utahensis (Hungerford)	2, 4, 11
Corisella tarsalis (Fieber)	2, 4, 11
Cymatia americana Hussey	2, 4, 11
Dasycorixa hybrida (Hungerford)	4
Dasycorixa johanseni (Walley)	4
Dasycorixa rawsonii Hungerford	4
Hesperocorixa atopodonta Hungerford	4, 11
Hesperocorixa laevigata Hungerford	4, 11
Hesperocorixa michiganensis (Hungerford)	4, 11
Hesperocorixa minorella (Hungerford)	4, 11
Hesperocorixa scabricula (Walley)	4
Hesperocorixa vulgaris (Hungerford)	2, 4, 11
Palmacorixa buenoi Abbott	4, 11
Palmacorixa gillettei Abbott	4, 11

TABLE 21.1. (Continued)

Taxon	Reference[b]
Palmacorixa janeae Brooks	4
Sigara alternata (Say)	2, 4, 11
Sigara bicoloripennis (Walley)	2, 4, 11
Sigara conocephala (Hungerford)	2, 4, 11
Sigara decorata (Abbott)	4, 11
Sigara decoratella (Hungerford)	2, 4, 11
Sigara fallenoidea (Hungerford)	4
Sigara grossolineata (Hungerford)	4, 11
Sigara lineata (Forster)	4, 11
Sigara mathesoni Hungerford	4
Sigara mullettensis (Hungerford)	4, 11
Sigara penniensis (Hungerford)	4
Sigara solensis (Hungerford)	2, 4, 11
Trichocorixa borealis Sailer	2, 4, 11
Trichocorixa naias (Kirkaldy)	2, 4, 11
Trichocorixa verticalis (Fieber)	2, 4, 11
Gerridae	
Gerris buenoi Kirkaldy	4, 11
Gerris comatus Drake & Hottes	4, 11
Gerris dissortis Drake & Harris	4, 11
Gerris notabilis Drake & Hottes	4
Gerris pingreensis Drake & Hottes	4, 11
Gerris remigis Say	4, 11
Rheumatobates rileyi Bergroth	4, 11
Hydropetridae	
Hydrometra martini Kirkaldy	4, 11
Mesovelidae	
Mesovelia mulsanti White	4, 11
Nepidae	
Nepa apiculata Uhler	4
Ranatra fusca Palisot deBeauvois	4, 11
Notonectidae	
Buenoa confusa Truxal	4, 11
Buenoa macrotibialis Hungerford	11
Buenoa margaritacia Torre-Bueno	11
Notonecta borealis Bueno-Hussey	4, 11
Notonecta kirbyi Hungerford	4, 11
Notonecta undulata Say	4, 11
Pleidae	
Plea striola Fieber	4, 11
Saldidae	
Lampracanthia crassicornis Uhler	4, 11
Saldula bassingeri Drake	4, 11
Saldula comatula (Parshley)	4, 11
Saldula nigrita Parshley	4, 11
Saldula opacula Zetterstedt	4, 11
Saldula opiparia Drake & Hottes	4, 11

TABLE 21.1. (Continued)

Taxon	Reference[b]
Saldula orbiculata (Uhler)	4
Saldula palustris (Douglas)	4, 11
Saldula separata (Uhler)	4
Veliidae	
Microvelia buenoi Drake	4, 11
Microvelia pulchella Westwood	4, 11
Neuroptera	
Sisyra vicaria (Walker)	16
Trichoptera	
Hydroptilidae	
Agraylea multipunctata Curtis	6, 16
Leptoceridae	
Triaenodes nox Ross	16
Triaenodes grisea (Banks)	6, 16
Limnephilidae	
Anabolia bimaculata (Walker)	16
Limnephilus externus Hagen	16
Limnephilus hyalinus Hagen	16
Limnephilus janus Ross	16
Limnephilus labus Ross	16
Philarctus quaeris Milne	16
Molannidae	
Molanna flavicornis Banks	16
Phryganeidae	
Agrypnia pagetana Curtis	6, 16
Agrypnia straminae Hagen	16
Polycentropodidae	
Polycentropus picicornis Steph.	16
Psychomyiidae	
Psychomyia flavida Hagen	16
Diptera	
Certapogonidae	
Alluaudomyia needhami Thom.	16
Bezzia cockerelli Mall.	16
Bezzia bicolor Meigen	16
Bezzia glabra Coq.	16
Forcipomyia monilicornis	16
Palpomyia lineata (Meig.)	16
Chaoboridae	
Chaoborus americanus (Joh.)	16
Chironomidae	
Ablablesmyia annulata (Say)	7
Ablabesmyia illinoensis (Mall.)	13, 16, 23
Ablabesmyia monilis (L.)	16
Ablabesmyia pulchripennis (Lund.)	7, 13, 15, 16, 23
Acricotopus lucens (Zetter.)	16
Acricotopus nitidellus (Mall.)	7, 13, 15, 22

TABLE 21.1. (Continued)

Taxon	Reference[b]
Apsectrotanypus johnsoni (Coq.)	7
Chironomus atrella (Town.)	13, 15, 16
Chironomus atroviridis (Town.)	13
Chironomus attenuatus Walker	7
Chironomus riparius Meig.	7, 13, 15, 16, 17, 22
Chironomus staegeri Lund.	7
Chironomus tentans Fab	6, 7, 13, 14, 15, 16, 22
Chironomus pallidivittatus Mall.	7, 22
Cladopelma viridula (L.)	7, 16, 23
Corynoneura scutellata Winn.	13
Cricotopus pilitarsis (Zetter.)	16
Cricotopus cylindraceus (Kieff.)	16
Cricotopus flavocinctus (Kieff.)	16
Cricotopus ornatus (Meig.)	15, 22
Cryptochironomus digitatus (Mall.)	15
Cryptochironomus fulvus (Joh.)	7
Cryptochironomus psittacinus (Meig.)	23
Demeijeria brachialis (Coq.)	23
Derotanypus alaskensis (Mall.)	15, 16, 23
Dicrotendipes lobiger (Kieff.)	23
Dicrotendipes modestus (Say)	7
Dicrotendipes nervosus (Staeg.)	7, 13, 16, 23
Einfedlia pagana (Meig.)	6, 7, 16
Endochironomus nigricans (Joh.)	7, 13, 15, 16, 22
Endochironomus subtendens (Town.)	23
Glyptotendipes amplus Town.	7
Glyptotendipes barbipes (Staeg.)	7, 13, 15, 16, 22
Glyptotendipes dreisbachi Town.	7
Glyptotendipes lobiferus (Say)	7, 16, 22
Glyptotendipes meridionalis Den.&Sub.	23
Glyptotendipes paripes (Edwards)	16, 17
Glyptotendipes senilis (Joh.)	23
Glyptotendipes unacus Town.	7
Guttipelopia rosenbergi Bilyj	16
Kiefferulus dux (Joh.)	7, 23
Labrundinia pilosella (Loew)	16, 23
Nanocladius distinctus (Mall.)	7
Orthocladius smolandicus Brund.	16
Parachironomus danicus Lehmann	23
Parachironomus forceps (Town.)	16, 23
Parachironomus monochromus (Wulp)	13, 16, 23
Parachironomus parilis Walk.	23
Parachironomus potamogeti (Town.)	13
Parachironomus tenuicaudatus (Mall.)	13, 23

TABLE 21.1. (Continued)

Taxon	Reference[b]
Parachironomus varus (Goet.)	16
Paratanytarsus laccophilus (Edw.)	7, 16
Phaenopsectra punctipes (Wied.)	7, 16
Polypedium halterale (Coq.)	7
Procladius bellus (Loew)	7, 16, 22
Procladius culiciformis (L.)	7, 16
Procladius deltaensis Roback	23
Procladius denticulatus Sub.	23
Procladius freemani Sub.	23
Procladius nietus Roback	15, 23
Procladius riparius (Mall.)	13
Psectrotanypus dyari (Coq.)	7, 13, 15, 16, 22
Psectrocladius barbimanus (Edw.)	7
Psectrocladius flavus (Joh.)	15, 16, 23
Psectrocladius simulans (Joh.)	16
Pseudochironomus middlekauffi Town.	16
Tanypus punctipennis Meig.	15, 22
Tanypus nubifer Coq.	23
Culicidae	
Aedes campestris D. & K.	12
Aedes dorsalis (Meig.)	12
Aedes flavescens (Mull.)	12
Aedes vexans (Meig.)	12
Anopheles earlei Vargas	16
Culex tarsalis Coq.	12
Culex territans Walker	16
Culiseta inornata (Will.)	12
Dixidae	
Dixella serrata (Garrett)	16
Stratiomyidae	
Hedriodiscus vertebratus Say	16
Tipulidae	
Tipula illustris Doane	16
Hymenoptera	
Pteromalidae	
Cyrtogaster trypherus (Walker)	16

[a]Large prairie marshes such as the Delta Marsh are not included in this list; see Rosenberg and Danks (1987) for species lists from larger marshes in Canada.

[b]References: 1, Alby (1968); 2, Applegate (1973); 3, Bick et al. (1977); 4, Brooks and Kelton (1967); 5, Cvancara (1983); 6, Daborn (1974); 7, Driver (1977); 8, Euliss and Mushet (unpublished data); 9, Gleason (1996); 10, Hanson and Swanson (1989); 11, Jacobson (1969); 12, Meyer and Swanson (1982); 13, Morrill (1987); 14, Nelson (1989); 15, Parker (1985); 16, Parker (1992); 17, Rasmussen (1983); 18, Sawchyn and Gillott (1974a); 19, Sawchyn and Gillott (1974b); 20, Swanson (1984); 21, Swanson et al. (1988); 22, Wrubleski (1996); 23, Wrubleski (unpublished data).

those that characterize temporary or seasonal marshes (Euliss and Mushet, unpublished data), especially when salt concentrations decrease due to deflation (LaBaugh et al. 1996) or dilution from overland flow.

As wetlands refill following drought, wetland invertebrate communities develop along predictable temporal guidelines. As pointed out by Wiggins et al. (1980), the early colonizing invertebrate community is structurally simple and is composed mostly of r-selected detritivorous invertebrates. Later the invertebrate community becomes more complex as predators and other major functional groups of invertebrates become established as water permanence persists and the habitat becomes more complex in vegetative structure. Reintroduction of invertebrates that cannot tolerate hydroperiods of less than a year (e.g., the amphipods *Hyalella* and *Gammarus*) likely occurs from dispersal on mammals and birds (Segeratröle 1954, Peck 1975, Rosine 1956, Daborn 1976, Swanson 1984). Accompanying this successional pattern are changes in the spatial distribution and abundance of invertebrates as water permanence persists. For example, fairy shrimp are very abundant during the initial reflooding of semipermanent wetlands, but their numbers drop considerably as the hydroperiod lengthens (Euliss and Mushet, unpublished data). Like mosquitoes, fairy shrimp are vulnerable to predation by carnivorous insects in structurally complex communities (Pennak 1989). Increased invertebrate diversity associated with increased water permanence was noted by Driver (1977), who used chironomid diversity to separate wetland classes of prairie potholes in Canada, and by Euliss et al. (unpublished data), who used recalcitrant remains of invertebrates to identify, classify, and delineate wetlands in the PPR.

Invertebrate Responses to Salinity

Some aquatic insects inhabiting prairie wetlands are adapted to a wide range of salt concentrations, while others have narrow tolerances. Insect taxa that tolerate wide variations in salt concentrations are the most common and widely distributed in prairie wetlands. Most adult Coleoptera and Hemiptera can withstand a wide range in salinity and are well adapted to exploit the spatial and temporal dynamics of prairie wetlands. Exposure to habitats with unsuitable features such as unfavorable salinities is avoided by flight. Larval insects and noninsect invertebrates that cannot fly must rely on other adaptations to facilitate survival under adverse conditions.

Nonflying invertebrates possess a different suite of adaptations that have allowed them to exploit highly productive saline prairie wetlands. Adaptations may include eggs and cysts, waterproof secretions, burrowing into substrates, and physiological adaptations (Scudder 1987). Swanson et al. (1988) reports that *Lymnaea stagnalis* was the dominant gastropod in permanent or semipermanent wetlands having specific conductances of $<5,000$ μS cm^{-1}. However, as salt concentrations exceeded 5,000 μS cm^{-1}, *Lymnaea elodes* replaced *Lymnaea stagnalis* but was unable to persist at higher concentrations

($>$10,000 μS cm^{-1}). Similarly, anostracans like *Branchinecta lindahli* are common invertebrates in seasonal and semipermanent wetlands in the initial stages of the wet/dry cycle when salt concentrations are low (Euliss and Mushet, unpublished data), but they cannot tolerate extremely high salt concentrations that develop in the drought stage of the wet/dry cycle. At salt concentrations $>$35,000 μS cm^{-1} the anostracan fauna may shift entirely to the brine shrimp, *Artemia salina* (Swanson et al. 1988). Amphipods are common in semipermanent and permanent wetlands of intermediate salinity; most species are found in waters of low or medium carbonate content (Pennak 1989). Insects also show this trend, with most taxa giving way to brine flies (Ephydridae) and certain water boatmen at higher salt concentrations. Increasing salt typically results in lower diversity, although the productivity of the few specialists capable of tolerating the osmotic stress may be high (Euliss et al. 1991).

Invertebrate Responses to Vegetation

Habitat structure provided by hydrophytes changes between years and within seasons as plant communities respond to hydrology, climate, and human alterations. Different invertebrate communities are often associated with different plant species or plant communities (e.g., Voigts 1976, McCrady et al. 1986, Wrubleski 1987, Olson et al. 1995). Macrophytes increase habitat structural complexity, providing additional food and living space within the water column for species that would otherwise not be present (e.g., Berg 1949, 1950, Krull 1970, Gilinsky 1984, Bergey et al. 1992). These plants also function as sites for oviposition (Sawchyn and Gillott 1974a, b, 1975), emergence (Sawchyn and Gillott 1974a, b), respiration (Batzer and Sjogren 1986), attachment (Campbell et al. 1982), and pupation (Butcher 1930). By increasing structural habitat complexity, these plants also modify predator-prey interactions (e.g., Rabe and Gibson 1984, Gilinsky 1984, Batzer and Resh 1991).

Areas with aquatic macrophytes have been reported to support higher numbers of invertebrates than bare areas (Gerking 1957, 1962, Krull 1970). However, Olson et al. (1995) found that nektonic invertebrates were more numerous in open-water areas with dense filamentous algae, but biomass was greater in *Typha* stands. Wrubleski (1991) observed no difference in insect emergence between areas with and without submersed macrophytes. Voigts (1976) reports that invertebrate groups respond differently to changes in macrophyte communities, but in general maximum numbers of aquatic invertebrates occurred where beds of submersed vegetation were interspersed with emergent vegetation. As the aquatic macrophyte communities change as a result of natural or anthropogenic alterations, so do their associated aquatic invertebrate communities (Driver 1977, Wrubleski 1991, Hanson and Butler 1994).

Aside from increasing the amount of habitat available to aquatic invertebrates, aquatic macrophytes also contribute to marked changes in the physical and chemical environment. These changes, in turn, may modify invertebrate responses to vegetation and the habitat they provide. Macrophytes restrict water circulation and contribute to gradients in light, temperature, and dissolved oxygen in very shallow standing waters (e.g., Kollman and Wali 1976, Carpenter and Lodge 1986, Rose and Crumpton 1996). Anoxic conditions can prevail within stands of emergent vegetation (Suthers and Gee 1986, Rose and Crumpton 1996) or beneath beds of submersed macrophytes (Kollman and Wali 1976), and this can impact invertebrate abundance, movement, and behavior (e.g., Murkin and Kadlec 1986a, Murkin et al. 1992).

Invertebrate Responses to Weather

Temperature. Winters in the PPR are very cold, and many shallow-water bodies freeze completely. These habitats have been referred to as "aestival ponds" (Welch 1952, Daborn and Clifford 1974). They are effectively winter-dry habitats, differing from other temporary aquatic habitats because the dry phase is a function of temperature rather than water supply; all biological activity is restricted to the summer period. The severity of freezing varies greatly between years and is dependent upon water depth, snow cover and the extent of low temperatures (Danks 1971a, Daborn and Clifford 1974), and this in turn can impact invertebrate community structure. The amphipods *Hyalella* and *Gammarus,* for example, are not found in wetlands that freeze completely (Daborn 1969). However, the relative importance of freezing and overwintering on invertebrate communities in these habitats has received little attention.

Many wetland invertebrates avoid the risk of freezing by migrating to habitats that do not freeze completely. Most Hemiptera and Coleoptera overwinter as adults, and many migrate from shallow habitats to deeper ponds and lakes (Danks 1978). However, Danell (1981) did recover a live corixid and an adult beetle from the frozen ice and the upper part of the frozen bottom sediments of a shallow lake in northern Sweden. Water striders (*Gerris* spp.) overwinter as adults in terrestrial vegetation adjacent to ponds, frequently in aggregations (Nummelin and Vespalainen 1982). Some mosquitoes also overwinter as adults (*Culiseta, Culex,* and *Anopheles* spp.) in rodent burrows, hollow trees, and unheated buildings (Wood et al. 1979).

In wetlands that do not freeze solid, some invertebrates will move to deeper water to avoid freezing. Several studies have reported invertebrate movements to deeper water habitats in the fall (Wodsedalek 1912, Eggleton 1931, Moon 1935, 1940, Gibbs 1979, Davies and Everett 1977, Boag 1981). However, those invertebrates that migrate to areas that do not freeze may experience potentially harmful anoxic conditions and increased levels of hydrogen sulfide and other toxic dissolved substances (Daborn and Clifford 1974, Danks 1971a). Invertebrate adaptations to these conditions include anaerobic metabolism (Reddy and Davies 1993), reduced activity and feeding (Davies and

Gates 1991), and movement to microhabitats offering better conditions (Brittain and Nagell 1981).

Those invertebrates that do not migrate must possess a means of tolerating freezing conditions. This is accomplished physiologically through freezing resistance (avoiding freezing) or freezing tolerance (Block 1991, Duman et al. 1991). Many benthic invertebrates are able, while encased in ice, to resist freezing by means of supercooling and the production of various antifreeze agents. Daborn (1971) and Sawchyn and Gillott (1975) both describe how coenagrionid damselflies were collected encased in ice, but were not frozen. Freeze-tolerant invertebrates are those that can survive extracellular ice formation within their bodies. Chironomids are a well-known example of this group (Danks 1971b).

Snow insulates and protects wetland invertebrates from the severe temperatures experienced above the ice (Danks 1971b). Emergent vegetation is important in holding this snow. Therefore removal of emergent vegetation would result in lower temperatures and possibly greater invertebrate mortality (Dineen 1953). Sawchyn and Gillott (1974a) report that adequate snow cover was necessary to prevent egg mortality in three species of *Lestes* damselflies. Sawchyn and Gillott (1975) suggest that overwintering mortality of coenagrionid damselfly nymphs observed by Daborn (1969, 1971) may have been due to lethal ice temperatures caused by an absence of snow cover. Further research is needed to determine how important overwintering conditions are in structuring PPR wetland invertebrate communities and the role that vegetation plays in mitigating temperature extremes.

Wind and Rain. Weather can be an important factor modifying wetland invertebrate activity and behavior. Inclement weather can reduce emergence of adult chironomids and other insects (Swanson and Sargeant 1972, Wrubleski and Ross 1989) and can force flying insects to seek shelter in stands of emergent vegetation (King and Wrubleski 1998). Rasmussen (1983) reports that windy, rainy weather conditions during the emergence period resulted in a reduction in mating and the subsequent production of chironomid larvae in a prairie pond. Reductions in invertebrate abundances or activity will impact the foraging behavior and survival of waterfowl, particularly the youngest ducklings, which are dependent upon flying insects as an important food resource (Chura 1961, Sugden 1973, Roy 1995, Cox et al. 1998).

Invertebrate Responses to Anthropogenic Disturbances

The PPR is far different today than in presettlement times, primarily because of modern agriculture. Agricultural activities on the uplands that surround prairie wetlands have impacted and altered aquatic invertebrate communities. Agrichemicals are the most obvious anthropogenic influence, and they have been shown to cause significant mortality in aquatic invertebrates (Borthwick 1988, Grue et al. 1989). Less obvious, however, are the physical effects related to cultivation, erosion, and sedimentation. Euliss and Mushet (unpublished data) found that Cladocera ephippia were less abundant in the tilled basins of temporary wetlands in agricultural fields compared to wetlands in grassland landscapes with no prior tillage history. Suspended silt and clay are known to be toxic to zooplankton and to reduce the foraging and assimilation rate of food items consumed by invertebrates (Robinson 1957, McCabe and O'Brien 1983, Newcombe and McDonald 1991). Other sediment effects include the clogging of filtering apparatuses, impacts on aquatic food chains through shading and covering of primary producers (Gleason and Euliss 1996b), and the burial of associated seed banks (Jurik et al. 1994, Wang et al. 1994). Although poorly studied, the burial of invertebrate eggs by sediments washing into wetlands may exert a significant influence on wetland invertebrate communities. Even seemingly innocuous influences like haying and burning of wetlands may negatively impact invertebrates if the vegetation is removed late in the growing season. As noted above, emergent vegetation holds snow, which provides thermal cover to protect invertebrates from severe freezing conditions. Wetland drainage, another land use associated with agriculture, has focused mostly on shallow temporary and seasonal wetlands within agricultural fields. The result has been a shift in the proportion of available wetland classes and alteration of hydrologic regimes of many non-drained wetlands. The nonintegrated PPR watersheds facilitated the drainage of shallow wetlands into larger semipermanent wetlands, resulting in semipermanent wetlands that are much deeper and more expansive than in pristine times. Lastly, road construction has severely altered the chemical and hydrologic characteristics of prairie potholes, the most important to invertebrates being the creation of hypersaline wetlands when construction projects isolate areas of wetlands from groundwater inflow (Swanson et al. 1988).

Recolonization and Dispersal Mechanisms

Recolonization and dispersal mechanisms for prairie pothole invertebrates are poorly studied, but clearly the dynamic and harsh environmental conditions of the area have influenced this region's naturally low invertebrate diversity. Flight is one of the most important dispersal mechanisms of insects. Flying insects rapidly disperse into temporary and seasonal wetlands following normal seasonal flooding, but the recolonization of wetlands following extreme drought may be slower because fewer flooded wetlands during the drought are available to provide refugia for recolonizing stocks (Swanson 1984).

Flightless life stages of insects and noninsect invertebrates face even greater challenges to recolonize previously dry wetlands. Common recolonization mechanisms include eggs and cysts resistant to drying and freezing, diapause, aestivation, waterproof secretions, epiphragms (snails), burrowing, and even the use of invertebrate and vertebrate wildlife. Wiggins et al. (1980) outline a temporal sequence, strongly influenced by climatic conditions, in which various taxa of invertebrates invade newly flooded habitats, using a variety of recolonization mechanisms. In general, the diversity of invertebrates is low because relatively few taxa possess the necessary physiological or behavioral adaptations that allow them to exploit the rich food resources available in prairie wetlands. Such systems favor ecological generalists that are early colonizers and exploit resources unavailable to other taxa lacking the necessary adaptations. The temporal sequence involves early detritivorous invertebrates with mostly r-selected characteristics and later support predatory invertebrates that recolonize habitats that persist for a sufficient length of time.

Passive dispersal mechanisms are important. They include dispersal by wind (Pennak 1989), by being carried in the digestive tracts of birds (Proctor 1964, Proctor et al. 1967, Swanson 1984), and by clinging to more mobile fauna. Ostracods and clams have been observed clinging to migrating Hemiptera and Coleoptera (Fryer 1974), and amphipods can be carried in the feathers of waterfowl (Segerströle 1954, Rosine 1956, Swanson 1984). Peck (1975) observed *Hyalella azteca* and *Gammarus lacustris* on the fur of muskrat (*Ondatra zibethicus*) and beaver (*Castor canadensis*). Although not documented, epizoochory is a common means of seed dispersal on feathers of waterfowl, and it is conceivable that invertebrate propagules are transported in that fashion as well. Cladocera ephippia float on the waters surface and sometimes form extensive mats. Some ephippia have elaborate appendages that may facilitate adhesion to feathers as has been described for epizoochory on the feathers of waterfowl (Vivian-Smith and Stiles 1994). However, wetland drainage increases the distance between wetlands and may disrupt transporting mechanisms or delay introductions.

BIOTIC INTERACTIONS

Aquatic Macrophytes and Algae

Herbivory by aquatic invertebrates on macrophytes has generally been reported to be unimportant (Crow and Macdonald 1978, Polunin 1984, Mann 1988), although this view has been questioned by Lodge (1991) and Newman (1991). Isolated accounts of terrestrial insect damage to wetland macrophytes have been reported (e.g., van der Valk and Davis 1978b, Klopatek and Stearns 1978, Penko and Pratt 1986, 1987, Sheldon 1987, Foote et al. 1988). At the Woodworth Study Area (Higgins et al. 1992) in south-central North Dakota, *Polygonum amphibium* was observed to be heavily grazed by the chrysomelid beetle *Galerucella nymphaeae* (Wrubleski and Detenbeck, unpublished data). Two European *Galerucella* species, *calmariensis* and *pusilla,* are currently being released as biocontrol agents for the introduced *Lythrium salicaria* (Malecki et al. 1993; see Blossey, this volume).

Although the relative extent of macrophyte herbivory is not well documented in prairie pothole wetlands, it is generally recognized that most macrophyte production eventually ends up as detritus (Davis and van der Valk

1978b). The abundant production of detritus may be the most important source of nutrients and energy for invertebrates in wetland habitats (Mann 1988, Murkin 1989). Either through direct consumption of decaying macrophyte tissue or the consumption of associated microbial fauna, this litter is generally thought to be extremely important to wetland invertebrates (Mann 1988). However, recent stable isotope studies have indicated that algae may be a more important food resource for invertebrates than macrophyte litter. Neill and Cornwell (1992) present evidence, based on stable isotope signatures, that aquatic macrophytes were not important sources of carbon for aquatic invertebrates in the Delta Marsh. This has been confirmed by a recent study in wetlands at the Woodworth Study Area (Wrubleski and Detenbeck, in prep.). $\delta^{13}C$ signatures for most aquatic invertebrates closely matched algae, but not macrophytes (Fig. 21.3). This has been confirmed by a recent study by Wrubleski and Detenbeck (in prep.). Except for epipelon, $\delta^{13}C$ signatures of algae were more deplete (more negative) than the aquatic macrophytes (Fig. 21.3). Most aquatic invertebrates also had $\delta^{13}C$ signatures that were more

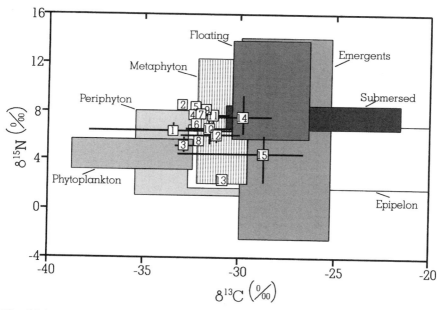

Fig. 21.3. Stable carbon and nitrogen isotope ratios for the algae, macrophytes, and aquatic invertebrates sampled in 10 wetlands at the Woodworth Study Area, North Dakota, 1994–95. Shaded boxes for the algae (phytoplankton, periphyton, epipelon, and metaphyton) and macrophytes (floating, emergents, and submersed) represent the complete range of values obtained. Mean and ranges are presented for the following aquatic invertebrates: 1, Cladocera; 2, copepods; 3, *Glyptotendipes;* 4, Tanypodinae; 5, Hydracarina; 6, Ephemeroptera; 7, Chaoboridae; 8, *Chironomus;* 9, Zygoptera; 10, Corixidae; 11, Notonectidae; 12, Dytiscidae; 13, Hydrophilidae; 14, Anisoptera; and 15, Gastropoda.

deplete than the macrophytes and closely matched the algal $\delta^{13}C$ signatures, indicating their reliance upon algae as their principal source of carbon.

Algal communities in prairie wetlands have generally been ignored (Crumpton 1989, Murkin 1989, Goldsborough and Robinson 1996), and consequently the effects of herbivory on algae have not been studied. In other wetland habitats invertebrate grazing has been found to be important in structuring algal communities and overall productivity (Cattaneo 1983, Hann 1991, Botts 1993). The relative importance of the different algal communities is unknown. Metaphyton has been reported to be an important habitat for many aquatic invertebrates (Ross and Murkin 1993, Olson et al. 1995), but does not appear to be a food resource (Goldsborough and Robinson 1996). Evidence from stable isotopes suggests that phytoplankton and periphyton are important food resources to aquatic invertebrates during the summer period (Wrubleski and Detenbeck, in prep.). More effort is needed to determine the relative importance of each algal community and how these values change over seasonal and longer-term wetland cycles.

Trophic Functions of Invertebrates in Prairie Wetlands

Invertebrates are widely recognized as an important food resource for waterfowl and other aquatic birds (see reviews by Murkin and Batt 1987, Swanson and Duebbert 1989, Krapu and Reinecke 1992, Sedinger 1992). They supply the necessary proteins and lipids for successful production of eggs by hens and for growth by ducklings (Swanson et al. 1974, 1977, Driver et al. 1974, Krapu 1979). Invertebrates are rich in protein and several essential amino acids that cannot be obtained from plants or seeds (Sugden 1973, Driver et al. 1974). They are also excellent sources of lipids and energy (Driver et al. 1974, Driver 1981, Afton and Ankney 1991). Due to their importance to waterfowl, invertebrate abundances can influence wetland use and feeding behavior (Kaminski and Prince 1981, Talent et al. 1982, Murkin and Kadlec 1986b, Sedinger 1992).

Aside from waterfowl, aquatic invertebrates are important food resources for passerines (Willson 1966, Mott et al. 1972, Voigts 1973, Twedt et al. 1991), shorebirds (Eldridge 1987), grebes, and other wetland birds. Adult aquatic insects (e.g., chironomids, dragonflies, mayflies) originating from wetland habitats provide an important food resource for many nonwetland birds as well (Busby and Sealy 1979, Sealy 1980, Guinan and Sealy 1987).

Tiger salamanders are a common amphibian found in prairie wetlands (Larson 1968, Buchli 1969, Deutschman 1984, Peterka 1987). Olenick and Gee (1981) report that tiger salamanders were benthic and fed primarily on *Gammarus*. Deutschman (1984) reports that tiger salamanders mostly consumed Cladocera, chironomids, amphipods, ephemeropterans, and hemipterans and that larger prey (i.e., large amphipods and chironomids) were preferred over smaller prey such as cladocerans and copepods. While the consumption of large prey maximizes growth (Deutschman 1984), it is likely that most large invertebrates are consumed when they are available in prairie wetlands. At night, tiger salamanders float up in the water column to feed on invertebrates (Anderson and Graham 1967, Branch and Altig 1981). They use deep portions of wetlands as refugia from avian predators during the day. Much of the nocturnal movement of tiger salamanders is apparently from the center and deeper portion of wetlands towards the shoreline. Lannoo (unpublished data) found that funnel traps with openings oriented towards the deeper wetland center caught twice the number of salamanders as traps with openings oriented parallel to the shoreline. Interestingly, Corkum (1984) notes that over 60 percent of the movements of aquatic invertebrates also occurred at night and towards the deeper water of the wetland center.

Conditions in most prairie wetlands are not favorable for fish (Peterka 1989). Frequent drying, nonintegrated watersheds, and harsh winter conditions generally prevent fish from establishing permanent populations. However, they can become established through deliberate introductions. For example, fathead minnows (*Pimphales promealas*) are released in wetlands for rearing by the baitfish industry (Hanson and Riggs 1995), and rainbow trout (*Salmo gairdneri*) are released for sport and commercial harvest (Peterka 1989). As in other aquatic habitats, fish can be very important predators of aquatic invertebrates and potentially compete with waterfowl and other marsh birds for food (Swanson and Nelson 1970). Hanson and Riggs (1995) report marked reductions in invertebrate abundance, biomass, and taxa richness in wetlands stocked with fathead minnows. Recently, Bouffard and Hanson (1997) have concluded that fish in wetlands were incompatible with objectives established for waterfowl management, primarily due to the negative impact of fish on invertebrate communities.

Leeches, dragonflies, beetles, and other predaceous invertebrates are abundant in prairie wetlands, but there have been few studies of predator-prey relationships and competition among aquatic invertebrates within these habitats. The importance of these interactions in other aquatic habitats is widely recognized (e.g., Bay 1974, Kerfoot and Sih 1987). In prairie wetlands vertebrate predators such as fish are often absent, making predaceous invertebrates the top aquatic predators. Anderson and Raasveldt (1974) report that *Gammarus* and *Chaoborus* were important predators of zooplankton in prairie lakes and ponds. Rasmussen and Downing (1988) observe that the spatial distributions of benthic dwelling chironomids were determined by the presence of the predatory leech, *Nephelopsis obscura*. Clearly, invertebrate predators may play an important role in structuring wetland invertebrate communities, but further research is needed to determine their relative importance.

Ecological Functions of Invertebrates in Prairie Wetlands

Aquatic invertebrates in prairie wetlands probably perform ecological functions similar to those documented in other shallow water habitats (Murkin and Wrubleski 1988). The trophic importance of aquatic invertebrates has been elaborated above, but other ecological functions are not as well known or studied. Benthic invertebrates influence sediment chemistry, structure, and nutrient dynamics (Gallepp 1979, Granéli 1979, Gardner et al. 1981, 1983). Their burrowing and mixing activities within the sediments can potentially impact seeds and seedling establishment of aquatic macrophytes (Grace 1984).

Aquatic invertebrates are often assumed to play a major role in decompositional pathways in wetlands (Polunin 1984, Mann 1988). However, there is relatively little evidence to support this assumption other than some generalized knowledge of feeding habits for some groups. In a study of macrophyte litter decomposition in the Delta Marsh, Bicknese (1987) reports that aquatic invertebrates had little influence on litter decomposition dynamics. This was further corroborated with stable isotope evidence which indicates that most invertebrates are not feeding on wetland macrophyte detritus (Neill and Cornwell 1992, Wrubleski and Detenbeck, in prep.). Mining and burrowing activities by invertebrates within litter undoubtedly contribute to increased litter decomposition, but direct consumption does not appear to be important.

Wetlands are often regarded as sources of insect pests such as mosquitoes, horseflies, and deerflies. Most mosquito production in the PPR actually originates from temporary snowmelt and rain pools. For example, *Aedes* spp. lay their eggs in shallow dry depressions, and the eggs hatch only when flooded by snowmelt or rainwater (Wood et al. 1979). One mosquito that does occur in more permanent waters and is of concern is *Culex tarsalis*. This species is the principal vector of western equine encephalitis, a serious viral disease for horses and humans (Wood et al. 1979).

Leeches are common ectoparasites on waterfowl (Trauger and Bartonek 1977). Amphipods, snails, leeches, and other wetland invertebrates act as intermediate hosts for a variety of bird intestinal parasites (e.g., LaBerge and McLaughlin 1989). Biting flies are also vectors of a variety of blood parasites for marsh birds (e.g., Meyer et al. 1974, Bennett et al. 1982). The relative impact of these parasites on wetland bird mortality is not well known (Sargeant and Raveling 1992).

CONCEPTUAL FRAMEWORK FOR RESEARCH AND MANAGEMENT

Considering the dynamic climate, hydrology, and chemistry of the prairies, it is significant that past research has shown no evidence that wetland invertebrates have developed adaptations specific to the PPR. Rather, the available evidence suggests that systems such as the PPR have a naturally low diversity of invertebrate fauna, comprised mostly of ecological generalists. Those taxa present already possessed the necessary physiological and behavioral adaptations needed to exploit the rich food resources of prairie wetlands (Wiggins et al. 1980). In support of this concept, Hamrum et al. (1971) reports low diversity of dragonflies in the PPR of western Minnesota relative to wetlands in eastern Minnesota outside of the PPR. Low diversity and no endemic species of odonates are also reported by Bick et al. (1977) for the Dakotas, and they conclude that the fauna was primarily (49 percent) transcontinental with many northern species. Apparently the Dakotas represent a greater obstacle to the northward movement of species than the converse.

The general characteristics of prairie pothole invertebrates and the dynamic nature of prairie wetlands warrant special consideration to facilitate effective research and conservation efforts. The critical roles that climate and parent geology have for wetland hydrology, hydroperiods, chemistry, and ultimately the biota need to be considered by both managers and scientists. Taxa that utilize prairie wetlands face seasonal dewatering, increasing osmotic tensions as salts concentrate, and winter temperatures that solidly freeze many prairie wetlands. On a broader temporal scale, invertebrates must be able to maintain populations in the face of long-term drought that often results in many wetlands being entirely dry. Recolonization often occurs from limited refugia, and many different adaptations have enabled invertebrates to withstand the rigors of the PPR.

It is important that researchers and managers work within a framework that considers the impact of the dynamic climate, hydrology, and chemistry on prairie wetland biota. Studies or conservation efforts that do not consider the spatial variability of wetlands (i.e., different chemical and biological characteristics of wetlands due to different hydrologic functions) will have a low probability of yielding satisfactory results, as will efforts that do not consider temporal variability (i.e., annual freezing, seasonal drawdowns, and widespread drawdowns in relation to long-term drought). Because the normal wet/dry cycle of the PPR is on a 10–20 year schedule (Duvick and Blasing 1981, Karl and Koscielny 1982, Karl and Riebsame 1984, Diaz 1983, 1986), it is clear that long-term research projects are needed to develop a complete understanding of how these systems function and how wetland invertebrates respond. Alternatively, researchers conducting shorter-term studies need to carefully define the class of wetland(s) studied, their basic chemical characteristics (a measure of salinity at a minimum), their vegetative characteristics, and, if possible, their hydrologic function. Finally they need to define the time frame of their studies within the context of the wet/dry cycle.

Likewise, managers of prairie wetlands need to manage their wetlands within the context of the long-term drought cycle. Periodic drawdown and oxidation of sediments is a natural process that is necessary to maximize the overall productivity of prairie wetlands. Managed wetlands are frequently drawn down to enhance production for waterfowl populations (Cook and Powers 1958, Fredrickson and Taylor 1982, Kadlec and Smith 1992). Such management also benefits invertebrates when wetlands are reflooded, due to nutrient release and its effect on primary production. Additionally, vegetation growing in dry wetland basins provides an additional source of nutrients, substrate, and fresh carbon for the detritus pool when reflooded. Despite this knowledge, however, most restored and constructed wetlands in the PPR do not have water control structures and hence are totally dependent upon natural wet/dry cycles to rejuvenate wetland productivity. In addition, many managed wetlands that have water control structures are not drawn down frequently enough to mimic the natural drought cycle and enhance productivity of invertebrates and other biota (Leigh Fredrickson, personal communication). Others, such as the Delta Marsh in south-central Manitoba, have stabilized water levels that have resulted in reduced wetland productivity and value as wildlife habitat (Batt, in press). Future research and management of PPR wetlands needs to focus more intensively on maximizing wetland functions, including the production of aquatic invertebrates, to offset prior wetland losses (Bellrose and Low 1978).

LITERATURE CITED

Adamus, P. R. 1996. Bioindicators for Assessing Ecological Integrity of Prairie Wetlands. United States Environmental Protection Agency, National Health and Environmental Effects Research Laboratory, Western Ecology Division, Corvallis, OR.

Afton, A. D., and C. D. Ankney. 1991. Nutrient-reserve dynamics of breeding lesser scaup: A test of competing hypotheses. Condor 93:89–97.

Alby, M. E. 1968. The Odonata (dragonflies and damselflies) of North Dakota. Master's Thesis, North Dakota State University, Fargo, ND.

Anderson, J. D., and R. E. Graham. 1967. Vertical migration and stratification of larval *Ambystoma*. Copeia 1967:371–374.

Anderson, R. S., and L. G. Raasveldt. 1974. *Gammarus* predation and the possible effects of *Gammarus* and *Chaoborus* feeding on the zooplankton composition in some small lakes and ponds in western Canada. Canadian Wildlife Service, Occasional Paper Number 18, Ottawa, ON.

Applegate, R. L. 1973. Corixidae (Water Boatmen) of the South Dakota glacial lake district. Entomological News 84:163–170.

Bärlocher, F., R. J. Mackay, and G. B. Wiggins. 1978. Detritus processing in a temporary vernal pool in southern Ontario. Archiv für Hydrobiologie 81:269–295.

Bartonek, J. C. 1968. Summer foods and feeding habits of diving ducks in Manitoba. Ph.D. Thesis, University of Wisconsin, Madison, WI.

———. 1972. Summer foods of American widgeon, mallards, and a green-winged teal near Great Slave Lake, N.W.T. Canadian Field Naturalist 86:373–376.

Bartonek, J. C., and J. J. Hickey. 1969. Food habits of canvasbacks, redheads, and lesser scaup in Manitoba. Condor 71:280–290.

Batt, B. D. J. in press. The Delta Marsh. *In* H. R. Murkin, A. G. van der Valk, and W. R. Clark (eds.), Prairie Wetland Ecology: The State of our Understanding and the Contribution of the Marsh Ecology Research Program. Iowa State University Press, Ames, IA.

Batzer, D. P., and V. H. Resh. 1991. Trophic interactions among a beetle predator, a chironomid grazer, and periphyton in a seasonal wetland. Oikos 60:251–257.

Batzer, D. P., and R. D. Sjogren. 1986. Larval habitat characteristics of *Coquillettidia perturbans* (Diptera: Culicidae) in Minnesota. Canadian Entomologist 118:1193–1198.

Bay, E. C. 1974. Predator-prey relationships among aquatic insects. Annual Review of Entomology 19:441–453.

Bellrose, F. C., and J. B. Low. 1978. Advances in waterfowl management research. Wildlife Society Bulletin 6:63–72.

Bennett, G. F., D. J. Mieman, B. Turner, E. Kuyt, M. Whiteway, and E. C. Greiner. 1982. Blood parasites of prairie anatids and their implications in waterfowl management in Alberta and Saskatchewan. Journal of Wildlife Diseases 18:287–296.

Berg, C. O. 1949. Limnological relations of insects to plants of the genus *Potamogeton*. Transactions of the American Microscopical Society 68:279–291.

———. 1950. Biology of certain Chironomidae reared from *Potamogeton*. Ecological Monographs 20:83–101.

Bergey, E. A., S. F. Balling, J. N. Collins, G. A. Lamberti, and V. H. Resh. 1992. Bionomics of invertebrates within an extensive *Potamogeton pectinatus* bed of a California marsh. Hydrobiologia 234:15–24.

Bick, G. H., J. C. Bick, and L. E. Hornuff. 1977. An annotated list of the Odonata of the Dakotas. Florida Entomologist 60:149–165.

Bicknese, N. A. 1987. The role of invertebrates in the decomposition of fallen macrophyte litter. Master's Thesis, Iowa State University, Ames, IA.

Block, W. 1991. To freeze or not to freeze? Invertebrate survival of sub-zero temperatures. Functional Ecology 5:284–290.

Boag, D. A. 1981. Differential depth distribution among freshwater pulmonate snails subjected to cold temperatures. Canadian Journal of Zoology 59:733–737.

Borthwick, S. M. 1988. Impacts of agricultural pesticides on aquatic invertebrates inhabiting prairie wetlands. Master's Thesis, Colorado State University, Fort Collins, CO.

Botts, P. S. 1993. The impact of small chironomid grazers on epiphytic algal abundance and dispersion. Freshwater Biology 30:25–33.

Bouffard, S. H., and M. A. Hanson. 1997. Fish in waterfowl marshes: Waterfowl managers' perspective. Wildlife Society Bulletin 25:146–157.

Bousfield, E. L. 1958. Fresh-water amphipod crustaceans of glaciated North America. Canadian Field-Naturalist 72:55–113.

Branch, L., and R. A. Altig. 1981. Nocturnal stratification of three species of *Ambystoma* larvae. Copeia 1981:870–873.

Brinson, M. M. 1993. A hydrogeomorphic classification for wetlands. Wetlands Research Program TR-WRP-DE-4. U.S. Army Corps of Engineers, Waterways Experiment Station, Vicksburg, MS.

Brittain, J. E., and B. Nagell. 1981. Overwintering at low oxygen concentrations in the mayfly *Leptophlebia vespertina*. Oikos 36:45–50.

Brooks, A. R., and L. A. Kelton. 1967. Aquatic and semiaquatic Heteroptera of Alberta, Saskatchewan and Manitoba (Hemiptera). Memoirs of the Entomological Society of Canada 51.

Buchli, G. L. 1969. Distribution, food and life history of tiger salamanders in Devils Lake, North Dakota. Master's Thesis, University of North Dakota, Grand Forks, ND.

Busby, D. G., and S. G. Sealy. 1979. Feeding ecology of a population of nesting yellow warblers. Canadian Journal of Zoology 57:1670–1781.

Butcher, F. G. 1930. Notes on the cocooning habits of *Gyrinus*. Journal of the Kansas Entomological Society 3:64–66.

Campbell, J. M., W. J. Clark, and R. Kosinski. 1982. A technique for examining microscopic distribution of Cladocera associated with shallow water macrophytes. Hydrobiologia 97:225–232.

Canada-United States Steering Committee. 1985. North American Waterfowl Management Plan—Draft, November 1985. Canadian Wildlife Service, Ottawa, ON, and United States Fish and Wildlife Service, Washington, DC.

Carpenter, S. R., and D. M. Lodge. 1986. Effects of submersed macrophytes on ecosystem processes. Aquatic Botany 26:341–370.

Cattaneo, A. 1983. Grazing on epiphytes. Limnology and Oceanography 28:124–132.

Chura, N. J. 1961. Food availability and preferences of juvenile mallards. Transactions of the North American Wildlife Conference 26:121–134.

Cook, H. H., and C. F. Powers. 1958. Early biochemical changes in the soils and waters of artificially created marshes in New York. New York Fish and Game Journal 5:9–65.

Corkum, L. D. 1984. Movements of marsh-dwelling invertebrates. Freshwater Biology 14:89–94.

Cox, R. R., Jr., M. A. Hanson, C. C. Roy, N. H. Euliss, Jr., D. H. Johnson, and M. G. Butler. 1998. Mallard duckling growth and survival in relation to aquatic invertebrates. Journal of Wildlife Management 62:124–133.

Cowardin, L. M., V. Carter, F. C. Golet, and E. T. LaRoe. 1979. Classification of Wetlands and Deepwater Habitats of the United States. U.S. Fish and Wildlife Service, FWS/OBS-79/31, Washington, DC.

Crow, J. H., and K. B. Macdonald. 1978. Wetland values: Secondary production. Pages 146–161 *in* P. E. Greeson, J. E. Clark, and J. R. Clark (eds,), Wetland Functions and Values: The State of Our Understanding. American Water Resources Association, Minneapolis, MN.

Crumpton, W. G. 1989. Algae in northern prairie wetlands. Pages 188–203 *in* A. van der Valk (ed.), Northern Prairie Wetlands. Iowa State University Press, Ames, IA.

Cvancara, A. M. 1983. Aquatic Mollusks of North Dakota. North Dakota Geological Survey, Report of Investigation No. 78.

Daborn, G. R. 1969. Transient ecological succession in a shallow pond. Master's Thesis, University of Alberta, Edmonton, AB.

———. 1971. Survival and mortality of coenagrionid nymphs (Odonata: Zygoptera) from the ice of an aestival pond. Canadian Journal of Zoology 49:569–571.

———. 1974. Biological features of an aestival pond in western Canada. Hydrobiologia 44:287–299.

———. 1976. Colonization of isolated aquatic habitats. Canadian Field-Naturalist 90:56–57.

Daborn, G. R., and H. F. Clifford. 1974. Physical and chemical features of an aestival pond in western Canada. Hydrobiologia 44:43–59.

Dahl, T. E. 1990. Wetland losses in the United States 1780's to 1980's. U.S. Department of the Interior, Fish and Wildlife Service, Washington, DC.

Dahl, T. E., and C. E. Johnson. 1991. Status and trends of wetlands in the coterminous United States, mid-1970's to mid-1980's. U.S. Department of the Interior, Fish and Wildlife Service, Washington, DC.

Danell, K. 1981. Overwintering of invertebrates in a shallow northern Swedish lake. Internationale Revue der Gesamten Hydrobiologie 66:837–845.

Danks, H. V. 1971a. Overwintering of some north temperate and Arctic Chironomidae. I. The winter environment. Canadian Entomologist 103:589–604.

————. 1971b. Overwintering of some north temperate and Arctic Chironomidae. II. Chironomid biology. Canadian Entomologist 103:1875–1910.

————. 1978. Modes of seasonal adaptation in the insects. I. Winter survival. Canadian Entomologist 110:1167–1205.

Davis, C. B., and A. G. van der Valk. 1978a. Mineral release from the litter of *Bidens cernua* L., a mudflat annual at Eagle Lake, Iowa. Internationale Vereinigung für Theoretische und Angewandte Limnologie, Verhandlungen 20:452–457.

————. 1978b. Litter decomposition in prairie glacial marshes. Pages 99–113 *in* R. E. Good, D. F. Whigham, and R. L. Simpson (eds.), Freshwater Wetlands. Ecological Processes and Management Potential. Academic Press, New York.

Davies, R. W., and R. P. Everett. 1977. The life history, growth, and age structure of *Nephelopsis obscura* Verrill, 1872 (Hirudinoidea) in Alberta. Canadian Journal of Zoology 55:620–627.

Davies, R. W., and T. E. Gates. 1991. The effects of different oxygen regimes on the feeding and vertical distribution of *Nephelopsis obscura* (Hirudinoidea). Hydrobiologia 211:51–56.

Deutschman, M. R. 1984. Secondary production, habitat use and prey preference of the tiger salamander (*Ambystoma tigrinum*) in south-central North Dakota. Master's Thesis, North Dakota State University, Fargo, ND.

Diaz, H. F. 1983. Some aspects of major dry and wet periods in the contiguous United States, 1895–1981. Journal of Climate and Applied Meteorology 22:3–16.

Diaz, H. F. 1986. An analysis of twentieth century climate fluctuations in northern North America. Journal of Climate and Applied Meteorology 25:1625–1657.

Dineen, C. P. 1953. An ecological study of a Minnesota pond. American Midland Naturalist 50:349–376.

Dirschl, H. J. 1969. Foods of lesser scaup and blue-winged teal in the Saskatchewan River Delta. Journal of Wildlife Management 33:77–87.

Driver, E. A. 1977. Chironomid communities in small prairie ponds: Some characteristics and controls. Freshwater Biology 7:121–133.

————. 1981. Calorific values of pond invertebrates eaten by ducks. Freshwater Biology 11:579–581.

Driver, E. A., L. G. Sugden, and R. J. Kovach. 1974. Calorific, chemical and physical values of potential duck foods. Freshwater Biology 4:281–292.

Duman, J. G., D. W. Wu, L. Xu, D. Tursman, and T. M. Olsen. 1991. Adaptations of insects to subzero temperatures. Quarterly Review of Biology 66:387–410.

Duvick, D. N., and T. J. Blasing. 1981. A dendroclimatic reconstruction of annual precipitation amounts in Iowa since 1680. Water Resource Research 17:1183–1189.

Eggleton, F. E. 1931. A limnological study of the profundal bottom fauna of certain fresh-water lakes. Ecological Monographs 1:231–331.

Eldridge, J. L. 1987. Ecology of migrant sandpipers in mixed-species foraging flocks. Ph.D. Thesis, University of Minnesota, Saint Paul, MN.

Euliss, N. H., Jr., R. L. Jarvis, and D. S. Gilmer. 1991. Standing crops and ecology of aquatic invertebrates in agricultural drainwater ponds in California. Wetlands 11: 179–190.

Euliss, N. H., Jr. and D. M. Mushet. 1996. Water-level fluctuation in wetlands as a function of landscape condition in the prairie pothole region. Wetlands 16:587–593.

Ferkinhoff, W. D., and R. W. Gundersen. 1983. A Key to the Whirligig Beetles of Minnesota and Adjacent States and Canadian Provinces (Coleoptera: Gyrinidae). Science Publications of the Science Museum of Minnesota, New Series, Vol. 5, No. 3.

Flannagan, J. F., and S. R. Macdonald. 1987. Ephemeroptera and Trichoptera of peatlands and marshes in Canada. Memoirs of the Entomological Society of Canada 140:47–56.

Foote, A. L., J. A. Kadlec, and B. K. Campbell. 1988. Insect herbivory on an inland brackish wetland. Wetlands 8:67–74.

Fredrickson, L. H., and T. S. Taylor. 1982. Management of Seasonally Flooded Impoundments for Wildlife. U.S. Fish and Wildlife Service, Resource Publication 148.

Fryer, G. 1974. Attachment of bivalve molluscs to corixid bugs. Naturalist 928:18.

Galinato, M. I., and A. G. van der Valk. 1986. Seed germination traits of annuals and emergents during drawdown in the Delta Marsh, Manitoba, Canada. Aquatic Botany 26:89–102.

Gallepp, G. W. 1979. Chironomid influence on phosphorus release in sediment-water microcosms. Ecology 60:547–556.

Gardner, W. S., T. F. Nalepa, M. A. Quigley, and J. M. Malczyk. 1981. Release of phosphorus by certain benthic invertebrates. Canadian Journal of Fisheries and Aquatic Sciences. 38:978–981.

Gardner, W. S., T. F. Nalepa, D. R. Slavens, and G. A. Laird. 1983. Patterns and rates of nitrogen release by benthic Chironomidae and Oligochaeta. Canadian Journal of Fisheries and Aquatic Sciences 40:259–266.

Gerking, S. D. 1957. A method of sampling the littoral macrofauna and its application. Ecology 38:219-226.

———. 1962. Production and food utilization in a population of bluegill sunfish. Ecological Monographs 32:31–78.

Gibbs, K. E. 1979. Ovovivparity and nymphal seasonal movements of *Callibaetis* spp. (Ephemeroptera: Baetidae) in a pond in southwestern Quebec. Canadian Entomologist 111:927–931.

Gilinsky, E. 1984. The role of fish predation and spatial heterogeneity in determining benthic community structure. Ecology 65:455–468.

Gleason, R. A. 1996. Influence of agricultural practices on sedimentation rates, aquatic invertebrates, and bird-use in prairie wetlands. Master's Thesis, Humboldt State University, Arcata, CA.

Gleason, R. A., and N. H. Euliss, Jr. 1996a. Impact of agricultural land-use on prairie wetland ecosystems: Experimental design and overview. Proceedings of the North Dakota Academy of Science 50:103–107.

Gleason, R. A., and N. H. Euliss, Jr. 1996b. Sedimentation of prairie pothole wetlands: The need for integrated research by agricultural and wildlife interests. Proceedings of the 1996 wetlands seminar, Water for agriculture and wildlife and the environment—win-win opportunities. U.S. Committee on Irrigation and Drainage, Denver, CO.

Goldsborough, L. G., and G. G. C. Robinson. 1996. Pattern in wetlands. Pages 77–117 *in* R. J. Stevenson, M. L. Bothwell, and R. L. Lowe (eds.), Algal Ecology. Freshwater Benthic Ecosystems. Academic Press, San Diego, CA.

Grace, J. B. 1984. Effects of tubificid worms on the germination and establishment of *Typha*. Ecology 65:1689–1693.

Granéli, W. 1979. The influence of *Chironomus plumosus* larvae on the exchange of dissolved substances between sediment and water. Hydrobiologia 66:149–159.

Grue, C. E., M. W. Tome, T. A. Messmer, D. B. Henry, G. A. Swanson, and L. R. DeWeese. 1989. Agricultural chemicals and prairie pothole wetlands: meeting the needs of the resource and the farmer—U.S. perspective. Transactions of the North American Wildlife and Natural Resources Conference 54:43-58.

Guinan, D. M., and S. G. Sealy. 1987. Diet of house wrens (*Troglodytes aedon*) and the abundance of the invertebrate prey in the dune-ridge forest, Delta Marsh, Manitoba. Canadian Journal of Zoology 65:1587–1596.

Hamrum, C. L., M. A. Anderson, and M. Boole. 1971. Distribution and habitat preference of Minnesota dragonfly species (Odonata, Anisoptera) II. Journal of the Minnesota Academy of Science 37:93–96.

Hann, B. J. 1991. Invertebrate grazer—periphyton interactions in a eutrophic marsh pond. Freshwater Biology 26:87–96.

Hanson, B. A., and G. A. Swanson. 1989. Coleoptera species inhabiting prairie wetlands of the Cottonwood Lake area, Stutsman county, North Dakota. Prairie Naturalist 21:49–57.

Hanson, M. A., and M. G. Butler. 1994. Responses of food web manipulation in a shallow waterfowl lake. Hydrobiologia 279/280:457–466.

Hanson, M. A., and M. R. Riggs. 1995. Potential effects of fish predation on wetland invertebrates: A comparison of wetlands with and without fathead minnows. Wetlands 15:167–175.

Hilton, D. F. J. 1987. Odonata of peatlands and marshes in Canada. Memoirs of the Entomological Society of Canada 140:57–63.

Higgins, K. F., L. M. Kirsch, A. T. Klett, and H. W. Miller. 1992. Waterfowl production on the Woodworth Station in south-central North Dakota, 1965–1981. U.S. Fish and Wildlife Service, Resource Publication 180.

Jacobson, L. A. 1969. The aquatic and semiaquatic Hemiptera of North Dakota. Master's Thesis, North Dakota State University, Fargo, ND.

Jurik, T. W., S. C. Wang, and A. G. van der Valk. 1994. Effects of sediment load on seedling emergence from wetland seed banks. Wetlands 14:159–165.

Kadlec, J. A., and L. M. Smith. 1992. Habitat management for breeding areas. Pages 590–610 *in* B. D. J. Batt, A. D. Afton, M. G. Anderson, C. D. Ankney, D. H. Johnson, J. A. Kadlec, and G. L. Krapu (eds), Ecology and Management of Breeding Waterfowl. University of Minnesota Press, Minneapolis, MN.

Kaminski, R. M., and H. H. Prince. 1981. Dabbling duck activity and foraging responses to aquatic macroinvertebrates. Auk 98:115–126.

Kantrud, H. A., J. B. Millar, and A. G. van der Valk. 1989. Vegetation of wetlands of the Prairie Pothole Region. Pages 132–187 *in* A. van der Valk (ed.), Northern Prairie Wetlands. Iowa State University Press, Ames, IA.

Kantrud, H. A., and W. E. Newton. 1996. A test of vegetation-related indicators of wetland quality in the prairie pothole region. Journal of Aquatic Ecosystem Health 5:177–191.

Karl, T. R., and A. J. Koscielny. 1982. Drought in the United States: 1895–1981. Journal of Climatology 2:313–329.

Karl, T. R., and W. E. Riebsame. 1984. The identification of 10 to 20 year temperature and precipitation fluctuations in the contiguous United States. Journal of Climate and Applied Meteorology 23:950–966.

Kerfoot, W. C., and A. Sih. 1987. Predation: Direct and Indirect Impacts on Aquatic Communities. University Press of New England, Hanover, NH.

King, R. S., and D. A. Wrubleski. 1998. Spatial and diel availability of flying insects as potential duckling food in prairie wetlands. Wetlands 18:100–114.

Klopatek, J. M., and F. W. Sterns. 1978. Primary productivity of emergent macrophytes in a Wisconsin freshwater marsh ecosystem. American Midland Naturalist 100:320–332.

Kollman, A. L., and M. K. Wali. 1976. Intraseasonal variations in environmental and productivity relations of *Potamogeton pectinatus* communities. Archiv für Hydrobiologie Supplement 50:439–472.

Krapu, G. L. 1979. Nutrition of female dabbling ducks during reproduction. Pages 59–70 *in* T. A. Bookhout (ed.), Waterfowl and Wetlands—an Integrated Review. Proceeding of the 1977 Symposium at the North Central Section, The Wildlife Society, Madison, WI.

Krapu, G. L., and K. J. Reinecke. 1992. Foraging ecology and nutrition. Pages 1–29 *in* B. D. J. Batt, A. D. Afton, M. G. Anderson, C. D. Ankney, D. H. Johnson, J. A. Kadlec, and G. L. Krapu (eds.), Ecology and Management of Breeding Waterfowl. University of Minnesota Press, Minneapolis, MN.

Krull, J. N. 1970. Aquatic plant-macroinvertebrate associations and waterfowl. Journal of Wildlife Management 34:707–718.

LaBaugh, J. W. 1989. Chemical characteristics of water in northern prairie wetlands. Pages 56–90 *in* A. van der Valk (ed.), Northern Prairie Wetlands. Iowa State University Press, Ames, IA.

LaBaugh, J. W., T. C. Winter, G. A. Swanson, D. O. Rosenberry, R. D. Nelson, and N. H. Euliss, Jr. 1996. Changes in atmospheric circulation patterns affect midcontinental wetlands sensitive to climate. Limnology and Oceanography 41:864–870.

LaBerge, R. J. A., and J. D. McLaughlin. 1989. *Hyalella azteca* (Amphipoda) as an intermediate host of the nematode *Streptocara crassicauda*. Canadian Journal of Zoology 67:2335–2340.

Larson, D. W. 1968. The occurrence of neotenic salamanders, *Ambystoma tigrinum diaboli*, in Devils Lake, North Dakota. Copeia 3:620–621.

Lodge, D. M. 1991. Herbivory on freshwater macrophytes. Aquatic Botany 41:195–224.

Malecki, R. A., B. Blossey, S. D. Hight, D. Schroeder, L. T. Kok, and J. R. Coulson. 1993. Biological control of purple loosestrife. BioScience 43:680–686.

Mann, K. H. 1988. Production and use of detritus in various freshwater, estuarine, and coastal marine ecosystems. Limnology and Oceanography 33:910–930.

Martin, D. B., and W. A. Hartman. 1986. The effect of cultivation on sediment and deposition in prairie pothole wetlands. Water, Air, and Soil Pollution 34:45–53.

McCabe, G. D., and W. J. O'Brien. 1983. The effects of suspended silt on feeding and reproduction of *Daphnia pulex*. American Midland Naturalist 110:324–337.

McCrady, J. W., W. A. Wentz, and R. L. Linder. 1986. Plants and invertebrates in a prairie wetland during duck brood-rearing. Prairie Naturalist 18:23–32.

Meyer, C. L., G. F. Bennett, and C. M. Herman. 1974. Mosquito transmission of *Plasmodium* (Giovannolaia) *circumflexum* Kikuth, 1931, to waterfowl in the Tantramar Marshes, New Brunswick. Journal of Parasitology 60:905–906.

Meyer, M. I., and G. A. Swanson. 1982. Mosquitoes (Diptera: Culicidae) consumed by breeding Anatinae in south central North Dakota. Prairie Naturalist 14:27–31.

Millar, J. B. 1976. Wetland classification in western Canada: A guide to marshes and shallow open water wetlands in the grasslands and parklands of the prairie provinces. Canadian Wildlife Service Report Series Number 37.

———. 1989. Perspective on the status of Canadian prairie wetlands. Pages 829–852 *in* R. R. Sharitz and J. W. Gibbons (eds.), Freshwater Wetlands and Wildlife. Conf-8603101, DOE Symposium Series No. 61. USDOE Office of Scientific and Technical Information, Oak Ridge, TN.

Moon, H. P. 1935. Flood movements of the littoral fauna of Windmere. Journal of Animal Ecology 4:216–228.

———. 1940. An investigation of the movements of freshwater invertebrate faunas. Journal of Animal Ecology 9:76–83.

Morrill, P. K. 1987. Disturbance of pond Chironomidae communities by deltamethrin insecticide. Master's Thesis, University of Saskatchewan, Saskatoon, SK.

Mott, D. F., R. R. West, J. W. De Grazio, and J. L. Guarino. 1972. Foods of the redwinged blackbird in Brown County, South Dakota. Journal of Wildlife Management 36:983–987.

Murkin, E. J., H. R. Murkin, and R. D. Titman. 1992. Nektonic invertebrate abundance and distribution at the emergent vegetation-open water interface in the Delta Marsh, Manitoba, Canada. Wetlands 12:45–52.

Murkin, H. R. 1989. The basis for food chains in prairie wetlands. Pages 316–338 *in* A. van der Valk (ed.), Northern Prairie Wetlands. Iowa State University Press, Ames, IA.

Murkin, H. R., and B. D. J. Batt. 1987. The interactions of vertebrates and invertebrates in peatlands and marshes. Memoirs of the Entomological Society of Canada 140: 15–30.

Murkin, H. R., and J. A. Kadlec. 1986a. Responses by benthic macroinvertebrates to prolonged flooding of marsh habitat. Canadian Journal of Zoology 64:65–72.

———. 1986b. Relationships between waterfowl and macroinvertebrate densities in a northern prairie marsh. Journal of Wildlife Management 50:212–217.

Murkin, H. R., and D. A. Wrubleski. 1988. Aquatic invertebrates of freshwater wetlands: function and ecology. Pages 239–249 *in* D. D. Hook, and others (eds.), The Ecology and Management of Wetlands. Vol. 1. Croom Helm, London, UK.

Neely, R. K., and J. L. Baker. 1989. Nitrogen and phosphorus dynamics and the fate of agricultural runoff. Pages 92–131 *in* A. van der Valk (ed.), Northern Prairie Wetlands. Iowa State University Press, Ames, IA.

Neill, C., and J. C. Cornwell. 1992. Stable carbon, nitrogen, and sulfur isotopes in a prairie marsh food web. Wetlands 12:217–224.

Nelson, R. D. 1989. Seasonal abundance and life cycles of chironomids (Diptera: Chironomidae) in four prairie wetlands. Ph.D. Thesis, North Dakota State University, Fargo, ND.

Newcombe, C. P., and D. D. MacDonald. 1991. Effects of suspended sediments on aquatic ecosystems. North American Journal of Fisheries Management 11:72–82.

Newman, R. M. 1991. Herbivory and detritivory on freshwater macrophytes by invertebrates: A review. Journal of the North American Benthological Society 10:89–114.

Nummelin, M., and K. Vespalainen. 1982. Densities of wintering water striders *Gerris odontogaster* (Heteroptera) around a breeding pond. Annales Entomologici Fennici 48:60–62.

Olenick, R. J., and J. H. Gee. 1981. Tiger salamanders (*Ambystoma tigrinum*) and stocked rainbow trout (*Salmo gairdneri*): Potential competitors for food in Manitoba prairie pothole lakes. Canadian Field-Naturalist 95:129–132.

Olson, E. J., E. S. Engstrom, M. R. Doeringsfeld, and R. Bellig. 1995. Abundance and distribution of macroinvertebrates in relation to macrophyte communities in a prairie marsh, Swan Lake, Minnesota. Journal of Freshwater Ecology 10:325–335.

Parker, D. W. 1985. Biosystematics of Chironomidae (Diptera) inhabiting selected prairie ponds in Saskatchewan. Master's Thesis, University of Saskatchewan, Saskatoon, SK.

Parker, D. W. 1992. Emergence phenologies and patterns of aquatic insects inhabiting a prairie pond. Ph.D. Thesis, University of Saskatchewan, Saskatoon, SK.

Peck, S. B. 1975. Amphipod dispersal in the fur of aquatic mammals. Canadian Field-Naturalist 89:181–182.

Penko, J. M., and D. C. Pratt. 1986. Effects of *Bellura oblique* on *Typha latifolia* productivity. Journal of Aquatic Plant Management 24:24–27.

———. 1987. Insect herbivory in Minnesota *Typha* stands. Journal of Freshwater Ecology 4:235–244.

Pennak, R. W. 1989. Fresh-water Invertebrates of the United States (3rd ed.). John Wiley & Sons, New York.

Peterka, J. J. 1987. Tiger salamanders (*Ambystoma tigrinum*) in North Dakota prairie lakes. North Dakota Academy of Science 41:32:

———. 1989. Fishes in northern prairie wetlands. Pages 302–315 *in* A. van der Valk (ed.), Northern Prairie Wetlands. Iowa State University Press, Ames, IA.

Petri, L. R., and L. R. Larson. 1973. Quality of water in selected lakes of eastern South Dakota. South Dakota Water Resources Commission Report Number 1.

Polunin, N. V. C. 1984. The decomposition of emergent macrophytes in fresh water. Advances in Ecological Research 14:115–166.

Proctor, V. W. 1964. Viability of crustacean eggs recovered from ducks. Ecology 45:656–658.

Proctor, V. W., C. R. Malone, and V. L. DeVlaming. 1967. Dispersal of aquatic organisms: Viability of disseminules recovered from the intestinal tract of captive killdeer. Ecology 48:672–676.

Rabe, F. W., and F. Gibson. 1984. The effect of macrophyte removal on the distribution of selected invertebrates in a littoral environment. Journal of Freshwater Ecology 2:359–370.

Rasmussen, J. B. 1983. An experimental study of competition and predation and their effects on growth and coexistence of chironomid larvae in a small pond. Ph.D. Thesis, University of Calgary, Calgary, AB.

Rasmussen, J. B., and J. A. Downing. 1988. The spatial response of chironomid larvae to the predatory leech *Nephelopsis obscura*. American Naturalist 131:14–21.

Reddy, D. C., and R. W. Davies. 1993. Metabolic adaptations by the leech Nephelopsis obscura during long-term anoxia and recovery. Journal of Experimental Zoology 265:224–230.

Robinson, M. 1957. The effects of suspended materials on the reproductive rate of *Daphnia magna*. Publications of the Institute of Marine Science, University of Texas, Port Aransas, TX, 4:265–277.

Robinson, G. G. C., S. E. Gurney, and L. G. Goldsborough. 1997a. Response of benthic and plankton algal biomass to experimental water-level manipulation in a prairie lakeshore wetland. Wetlands 17:167–181.

———. 1997b. The primary production of benthic and planktonic algae in a prairie wetland under controlled water-level regimes. Wetlands 17:182–194.

Rose, C., and W. G. Crumpton. 1996. Effects of emergent macrophytes on dissolved oxygen dynamics in a prairie pothole wetland. Wetlands 16:495–502.

Rosenberg, D. M., and H. V. Danks. 1987. Aquatic insects of peatlands and marshes in Canada: Introduction. Memoirs of the Entomological Society of Canada 140: 1–4.

Rosine, W. 1956. On the transport of the common amphipod, *Hyalella azteca*, in South Dakota by the mallard duck. Proceedings of the South Dakota Academy of Science 35:203.

Ross, L. C. M., and H. R. Murkin. 1993. The effect of above-normal flooding of a northern prairie marsh on *Agraylea multipunctata* Curtis (Trichoptera: Hydroptilidae). Journal of Freshwater Ecology 8:27–35.

Roy, C. C. 1995. Impact of fathead minnows on invertebrate communities and mallard ducklings in experimental wetlands. Master's Thesis, North Dakota State University, Fargo, ND.

Sargeant, A. B., and D. G. Raveling. 1992. Mortality during the breeding season. Pages 396–422 in B. D. J. Batt, A. D. Afton, M. G. Anderson, C. D. Ankney, D. H. Johnson, J. A. Kadlec, and G. L. Krapu (eds.), Ecology and Management of Breeding Waterfowl. University of Minnesota Press, Minneapolis, MN.

Sawchyn, W. W., and C. Gillott. 1974a. The life histories of three species of *Lestes* (Odonata: Zygoptera) in Saskatchewan. Canadian Entomologist 106:1283–1293.

———. 1974b. The life history of *Lestes congener* (Odonata: Zygoptera) on the Canadian prairies. Canadian Entomologist 106:367–376.

———. 1975. The biology of two related species of coenagrionid dragonflies (Odonata: Zygoptera) in western Canada. Canadian Entomologist 107:119–128.

Scudder, G. G. E. 1987. Aquatic and semiaquatic Hemiptera of peatlands and marshes of Canada. Memoirs of the Entomological Society of Canada 140:65–98.

Sealy, S. G. 1980. Reproductive responses of northern orioles to a changing food supply. Canadian Journal of Zoology 58:221–227.

Sedinger, J. S. 1992. Ecology of prefledging waterfowl. Pages 109–127 *in* B. D. J. Batt, A. D. Afton, M. G. Anderson, C. D. Ankney, D. H. Johnson, J. A. Kadlec, and G. L. Krapu (eds.), Ecology and Management of Breeding Waterfowl. University of Minnesota Press, Minneapolis, MN.

Segerstråle, S. G. 1954. The freshwater amphipods, *Gammarus pulex* (L.) and *Gammarus lacustris* G. O. Sars, Denmark and Fennoscandia—a contribution to the late- and post-glacial immigration history of the aquatic fauna of northern Europe. Commentationes Biologicae Societas Scientiarum Fennica 15:1–91.

Sheldon, S. P. 1987. The effects of herbivorous snails on submerged macrophyte communities in Minnesota lakes. Ecology 68:1920–1931.

Stewart, R. E., and H. A. Kantrud. 1971. Classification of natural ponds and lakes in the glaciated prairie region. U.S. Fish and Wildlife Service, Resource Publication 92.

Sugden, L. G. 1973. Feeding ecology of pintail, gadwall, American widgeon and lesser scaup in southern Alberta. Canadian Wildlife Service Report Series 24.

Suthers, I. M., and J. H. Gee. 1986. Role of hypoxia in limiting diel spring and summer distribution of juvenile yellow perch (*Perca flavescens*) in a prairie marsh. Canadian Journal of Fisheries and Aquatic Sciences 43:1562–1570.

Swanson, G. A. 1984. Dissemination of amphipods by waterfowl. Journal of Wildlife Management 48:987–991.

Swanson, G. A., and J. C. Bartonek. 1970. Bias associated with food analysis in gizzards of blue-winged teal. Journal of Wildlife Management 34:739–746.

Swanson, G. A., and H. F. Duebbert. 1989. Wetland habitats of waterfowl in the Prairie Pothole Region. Pages 228–267 *in* A. van der Valk (ed.), Northern Prairie Wetlands. Iowa State University Press, Ames, IA.

Swanson, G. A., G. L. Krapu, and J. R. Serie. 1977. Foods of laying female dabbling ducks on the breeding grounds. Pages 47–57 *in* T. A. Bookout (ed.), Waterfowl and Wetlands—an Integrated Review. Proceedings of the 1977 Symposium at the North Central Section, The Wildlife Society, Madison, WI.

Swanson, G. A., M. I. Meyer, and J. R. Serie. 1974. Feeding ecology of breeding blue-winged teal. Journal of Wildlife Management 38:396–407.

Swanson, G. A. and H. K. Nelson. 1970. Potential influence of fish-rearing programs on waterfowl breeding habitat. Pages 65–71 *in* E. Schnebergen (ed.), Symposium on the Management of Midwestern Winterkill Lakes. North Central Division, American Fish Society, Bethesda, MD.

Swanson, G. A., and A. B. Sargeant. 1972. Observation of nighttime feeding behavior of ducks. Journal of Wildlife Management 36:959–961.

Swanson, G. A., T. C. Winter, V. A. Adomaitis, and J. W. LaBaugh. 1988. Chemical Characteristics of Prairie Lakes in South-central North Dakota—Their Potential for Influencing Use by Fish and Wildlife. United States Fish and Wildlife Service, Fish and Wildlife Technical Report 18, Washington, DC.

Talent, L. G., G. L. Krapu, and R. L. Jarvis. 1982. Habitat use by mallard broods in south-central North Dakota. Journal of Wildlife Management 46:629–635.

Tiner, R. W. 1984. Wetlands of the United States: Current Status and Recent Trends. U.S. Fish and Wildlife Service, Washington, DC.

Trauger, D. L., and J. C. Bartonek. 1977. Leech parasitism of waterfowl in North America. Wildfowl 28:143–152.

Twedt, D. J., W. J. Bleier, and G. M. Linz. 1991. Geographic and temporal variation in the diet of yellow-headed blackbirds. Condor 93:975–986.

van der Valk, A. G. and C. B. Davis. 1976. The seed banks of prairie glacial marshes. Canadian Journal of Botany 54:1832–1838.

———. 1978a. The role of seed banks in the vegetation dynamics of prairie glacial marshes. Ecology 59:322–335.

———. 1978b. Primary production of prairie glacial marshes. Pages 21–37 *in* R. E. Good, D. F. Whigham, and R. L. Simpson (eds.), Freshwater Wetlands. Ecological Processes and Management Potential. Academic Press, New York.

Vivian-Smith, G., and E. W. Stiles. 1994. Dispersal of salt marsh seeds on the feet and feathers of waterfowl. Wetlands 14:316–319.

Voigts, D. K. 1973. Food niche overlap of two Iowa marsh icterids. Condor 75:392–399.

———. 1976. Aquatic invertebrate abundance in relation to changing marsh vegetation. American Midland Naturalist 95:313–322.

Wang, S. C., T. W. Jurik, and A. G. van der Valk. 1994. Effects of sediment load on various stages in the life and death of cattail (*Typha* x *glauca*). Wetlands 14:166–173.

Welch, P. S. 1952. Limnology (2nd ed.). McGraw-Hill, New York, NY.

Welling, C. H., R. L. Pederson, and A. G. van der Valk. 1988a. Recruitment from the seed bank and the development of zonation of emergent vegetation during a drawdown in a prairie wetland. Journal of Ecology 76:483–496.

———. 1988b. Temporal patterns in recruitment from the seed bank during drawdowns in a prairie wetland. Journal of Applied Ecology 25:999–1007.

Wiggins, G. B., R. J. Mackay, and I. M. Smith. 1980. Evolutionary and ecological strategies of animals in annual temporary pools. Archiv für Hydrobiologie Supplement 58:97–206.

Willson, M. F. 1966. Breeding ecology of the yellow-headed blackbird. Ecological Monographs 36:51–77.

Winter, T. C. 1989. Hydrologic studies of wetlands in the Northern Prairie. Pages 16–54 *in* A. van der Valk (ed.), Northern Prairie Wetlands. Iowa State University Press, Ames, IA.

Winter, T. C., and D. O. Rosenberry. 1995. The interaction of ground water with prairie pothole wetlands in the Cottonwood Lake Area, east-central North Dakota, 1979–1990. Wetlands 15:193–211.

Wodsedalek, J. E. 1912. Natural history and general behavior of the ephemeridae nymphs, Heptagenia interpunctata (Say). Annals of the Entomological Society of America 5:31–40.

Wood, D. M., P. T. Dang, and R. A. Ellis. 1979. The Mosquitoes of Canada (Diptera: Culicidae). Publication Number 1686. Biosystematics Research Institute, Ottawa, ON.

Wrubleski, D. A. 1987. Chironomidae (Diptera) of peatlands and marshes in Canada. Memoirs of the Entomological Society of Canada 140:141–161.

———. 1991. Chironomidae (Diptera) community development following experimental manipulation of water levels and aquatic vegetation. Ph.D. thesis, University of Alberta, Edmonton, AB.

———. 1996. Chironomidae (Diptera) of the Woodworth Study Area, North Dakota. Proceedings of the North Dakota Academy of Sciences 50:115–118.

Wrubleski, D. A., and L. C. M. Ross. 1989. Diel periodicities of adult emergence of Chironomidae and Trichoptera from the Delta Marsh, Manitoba, Canada. Journal of Freshwater Ecology 5:163–169.

22 Prairie Wetlands of South-Central Minnesota

Effects of Drought on Invertebrate Communities

ANNE E. HERSHEY, LYLE SHANNON,
GERALD J. NIEMI, ANN R. LIMA, and
RONALD R. REGAL

*B*enthic macroinvertebrate and zooplankton diversity is high in the prairie wetlands of south-central Minnesota, but both abundance and diversity of these groups are strongly affected by drought. Mollusks were more abundant in dry years, whereas insects, especially the nematoceran dipterans Chironomidae and Ceratopogonidae, were less abundant during the drought. Abundances of many insect taxa were significantly correlated with the Palmer Hydrologic Drought Index, although the correlations were often stronger when lag times were incorporated. Insect richness was also significantly lower in drought than in nondrought years, primarily due to richness within the Chironomidae. Among zooplankton, rotifers and cladocerans, but not copepods and ostracods, were more abundant during drought. Richness showed the same pattern as abundance. The abundances of both rotifers and cladocerans but not copepods and ostracods were significantly correlated with the Palmer Hydrologic Drought Index for the previous month. The extremes of wet and dry phases of the hydrologic cycle in this region support invertebrate communities which are taxonomically distinct, have different life history characteristics, and perform different functional roles within the ecosystem. A climate change scenario that involves more prolonged drought cycles is likely to result in continually shrinking wetland habitat, and possibly loss of

Invertebrates in Freshwater Wetlands of North America: Ecology and Management, Edited by
Darold P. Batzer, Russell B. Rader, and Scott A. Wissinger
ISBN 0-471-29258-3 © 1999 John Wiley & Sons, Inc.

515

these habitats from the region. If this occurs, it also will have dramatic effects on resident and migratory vertebrate communities as well as other wetland functions.

INTRODUCTION

The prairie wetlands of south-central Minnesota are nestled in an agricultural setting where they are periodically encroached upon by row crops and pastureland and dissected by roads and drainage ditches. These habitats are recognized as important waterfowl habitat, accounting for a high proportion of duck production each year (Minnesota Wetland Conservation Plan 1997), much of which is based on consumption of invertebrates (e.g., Swanson and Duebbert 1989). However, approximately 9 million acres of these wetlands have been drained from this region of Minnesota (Minnesota Wetlands Conservation Plan 1997). Nearly all of the transition from wetland land cover can be attributed to agricultural land use. The wetlands of Wright Co., Minnesota, are similar to the prairie potholes located just to the west (Mitsch and Gosselink 1986; see Euliss et al., this volume), which can be described as seasonally flooded wetlands with the hydroperiod extending from snowmelt through late June or early July, depending on snowpack and rainfall.

Wetland invertebrates form an important link among water quality, hydrology, primary productivity, and migratory and resident vertebrate populations. However, factors that control their biodiversity and abundance are poorly known and undoubtedly vary across wetland types and regions. Wetland invertebrates are critical food supplies for a wide variety of bird species, including herons, egrets, bitterns, rails, waterfowl, grebes, shorebirds, and many songbirds (Martin et al. 1951, Tester 1995). Yet surprisingly little is known about the value of wetlands to the overall survival and conservation of birds (Erwin 1996). The tremendous losses in acreage of these high-productivity habitats over the past 100 years have had a considerable impact on wetland-associated bird species (Richardson 1979). As with island bird communities, bird diversity declines with area; thus shrinkage or loss of prairie wetlands has important consequences for bird diversity in the upper midwest (Weller 1988).

HYDROLOGIC VARIATION

Although hydrologic variability is very important to biotic communities in all aquatic ecosystems, such variability is perhaps most keenly felt in wetlands because they can vary from virtual terrestrial environments to essentially lentic systems over the course of natural hydrologic variation. The objective of this chapter is to illustrate the impact of this hydrologic variability on the invertebrate fauna though a case study of prairie wetlands in south-central Minnesota studied during the drought of 1987–1990 and the subsequent wet years of 1991–1993. These wetlands range from 2–53 ha and are primarily dominated by sedges, grasses, cattails, and open water with relatively sparse vegetation (Olson 1992, Niemi et al. 1995; Table 22.1). The emergent vegetation provides important habitat for songbirds (Niemi et al. 1995), whereas the mix of open water and emergent vegetation is critical for nesting and foraging of a variety of duck species (Swanson and Duebbert 1989).

The Upper Midwest experienced a severe drought from 1987–1990. This is exemplified in the Wright Co. wetlands, where water depth was lowest in 1988, then increased until 1991 (Table 22.1). After 1991 there were slight decreases in water depth through 1993, but 1991–1993 were all considered wet years (Fig. 22.1).

The Palmer Hydrologic Drought Index (PHDI), which ranges between +6 and −6, is a widely used index of drought, based on precipitation, temperature data, and soil water content data, including outflow and storage (Karl and Knight 1985, Heddinghaus and Sabol 1991). For the Wright Co., Minnesota, wetlands this index showed a distinct bimodal pattern in the years 1987–1990 and 1991–1993 (Fig. 22.1). This was reflected in lower water depth in the wetlands, a reduced wetted perimeter, and shorter period of inundation.

MACROINVERTEBRATE COMMUNITY

Nine randomly selected sites from Victor and Corrina townships were sampled for benthic macroinvertebrates in 1989–1993 (Hershey et al., 1998). Eight samples were taken from each site on each date, except when a site was dry. For each site, four randomly drawn samples were collected from each of two randomly drawn transects perpendicular to a 50-m transect along the site perimeter. Samples were taken with a hand-held plastic core tube (19.6-cm^2 area). A sample consisted of two pooled cores (total area 35.4 cm^2) taken from the same location. The eight samples were mixed together and one subsample equal to $\frac{3}{8}$ of the total volume for the site was removed for insect analyses (total area = 106 cm^2). Each subsample was preserved in 95% ethanol until processed. Thus, the site was the unit of replication, and material from the site was collected from eight randomly selected locations.

At least 179 genera of aquatic insects representing seven orders were collected from the prairie wetlands of Wright Co. (Table 22.2). Of these, 101 were Diptera, and 51 of the Diptera were Chironomidae (Table 22.2). Biting midges (Ceratopogonidae), crane flies (Tipulidae), and soldier flies (Stratiomyidae) were the other commonly sampled Diptera. Prairie wetlands are also important mosquito breeding habitat (e.g., Moon 1984/1985). Generic diversity of ceratopogonids was higher than that reported for combined bogs, fens,

TABLE 22.1. Summary of Environmental Variables Gathered for Wright Co. Wetlands

Site	Wetland Area, ha	Percent Grass	Percent Sedges	Percent Live Cattails	Percent Dead Cattails	Percent Trees/Shrubs	Percent Water/thin Vegetation	Median Water Depth, cm					
								1988	1989	1990[a]	1991	1992	1993
5	9.5	5	14	14	24	3	40	4.0	24.0	17.0	31.0	47.0	44.5
12	38.0	11	24	18	20	5	22	4.0	7.0	8.0	28.0	8.0	20.0
13	1.8	0	0	59	10	0	30	6.0	11.0	10.0	34.0	25.0	43.0
14	3.5	25	7	18	10	4	36	0.0	0.0	19.5	34.0	23.0	24.0
17	8.0	21	12	14	11	4	37	0.0	9.0	3.0	25.0	2.0	9.0
18	9.5	8	19	16	18	1	39	15.0	16.0	29.0	30.0	51.5	26.0
19	2.9	34	3	14	8	2	39	1.0	0.0	1.0	13.0	31.0	17.0
26	4.6	1	13	35	25	2	26	10.5	30.0	47.0	48.0	31.5	25.0
51	6.4	—	—	—	—	—	—	—	22.0	21.0	35.0	33.0	25.0

[a] Data gathered in June 1990.

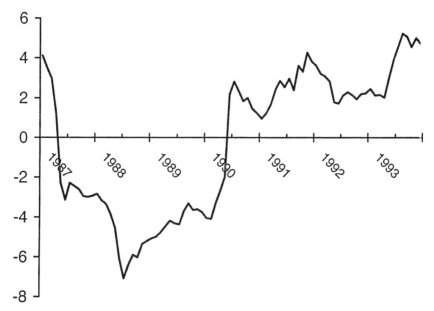

Fig. 22.1. Variation in Palmer Hydrological Drought Index in Wright Co., MN, from 1987–1993, illustrating the prolonged drought from 1987–1990 followed by the above normal hydric conditions of 1991–1993.

and marshes of Canada (Lewis 1987), and generic diversity for chironomids was similar or slightly higher than found in the Canadian wetland literature (Wrubleski 1987).

Coleoptera are also diverse inhabitants of prairie wetlands, and in Wright Co. were represented by at least 49 genera, but the Dytiscidae and Hydrophilidae showed highest diversity and the greatest number of individuals (Table 22.2). This compares to 70 genera collected from Canadian bogs, fens, and marshes, although Larson (1987) reports little difference among wetland types in the distribution of wetland beetles. Similarly, Hershey et al. (1995) found taxonomic diversity of beetles in temporary pond habitats of south-central Minnesota comparable to that in prairie wetlands (Table 22.2).

Collembola, Bivalvia, Isopoda, Annelida, and five genera of Gastropoda were also represented (Table 22.2). Noninsect groups generally were not studied to the same taxonomic resolution as insects in the Wright Co. study.

Effects of Drought

Drought had very different effects on insect and noninsect macroinvertebrates. Mollusks were more abundant in dry years, whereas insects were less abundant during the drought (Table 22.3). Gastropods were more than an order of magnitude more abundant in 1989 than in wetter subsequent years, and were

TABLE 22.2. Total Number of Each Taxon Collected and Number of Years in Which Taxon was Present on at Least One Site

Taxon	Number	Number of Years
Collembola	394	6
Odonata		
Lestidae		
Lestes	16	3
Coduliidae		
Somatochlora	9	3
Coenagrionidae		
Argia	1	1
Ephemeroptera		
Heptageniidae		
Heptagenia	1	1
Hemiptera		
Corixidae		
Sigara	3	2
Pleidae		
Plea striola	2	2
Notonectidae		
Notonecta	1	1
Gerridae		
Gerris	2	2
Hebridae		
Merragata	1	1
Belastomatidae		
Belastoma	4	2
Coleoptera		
Carabidae		
Carabidae	50	6
Dytiscidae		
Agabus	4	2
Bidessini	1	1
Celina	33	4
Colymbetes	2	2
Copelatus	2	2
Cybister	1	1
Desmopachria	5	3
Dytiscus	18	3
Hydaticus	23	4
Hydroporus	47	5
Hydrovatus	1	1
Hygrotus	120	6
Laccophilus	10	3
Laccornis	1	1
Liodessus	6	4

TABLE 22.2. (Continued)

Taxon	Number	Number of Years
Oreodytes	1	1
Rhantus	1	1
Unknown	9	1
Scirtidae		
Cyphon	24	4
Elodes	4	2
Prionocyphon	4	2
Unknown	6	1
Noteridae		
Pronoterus	1	1
Hydraenidae		
Hydraena	1	1
Hydraenidae unknown	1	1
Ochthebius	8	1
Hydrophilidae		
Anacaena	5	1
Berosus	43	4
Crentis	6	1
Cymbiodyta	10	2
Enochrus	234	5
Helocombus	23	5
Helophorus	34	4
Hydrobius	16	4
Hydrochus	34	5
Hydrophilus	33	4
Laccobius	15	5
Paracymus	22	5
Sperchopsis	20	3
Tropisternus	29	3
Unknown (larvae)	15	2
Lampyridae	9	3
Staphylinidae		
Stenus	29	3
Thinobius	19	2
Unknown	148	4
Haliplidae		
Haliplus	64	3
Melydridae	2	2
Curculionidae	59	6
Diptera (suborder Nematocera)		
Ceratopogonidae		
Ceratopogonidae unknown	4	2
Alluaudomyia	2	1

TABLE 22.2. (Continued)

Taxon	Number	Number of Years
Atrichopogon	166	5
Bezzia	340	6
Culicoides	334	6
Dasyhelea	3	1
Forcipomyia	4	3
Monohelea	2	1
Stillobezia	2	1
Chaoboridae		
Chaoborus	49	3
Chironomidae		
Acrictopus	1	1
Beckidia	1	1
Brillia	17	2
Cardiocladius	9	4
Chaetocladius	192	5
Chironomus	1071	4
Cladotanytarsus	4	3
Corynoneura	370	5
Crictopus	7	2
Dicrotendipes	7	3
Einfeldia	331	3
Endochironomus	130	4
Eukiefferiella	419	6
Heleniella	5	1
Heterotrissocladius	4	1
Hydrobaenus	16	3
Larsia	1152	5
nr *Larsia*	11	1
Lauterborniella	5	2
nr *Limnophyes*	14	4
Macropelopia	1	1
Mesocrictopus	3	1
Metriocnemus	4	3
Microtendipes	1	1
Orthocladius	307	5
Omisus	2	1
Parachironomus	36	4
Parakiefferiella	3	1
Paralimnophyes	4	1
Paraphaenocladius	413	5
nr *Paraphaenocladius*	2	1
nr *Parasmittia*	13	1
Paratanytarsus	16	2
Paratendipes	515	5
Parorthocladius	2	2

TABLE 22.2. (Continued)

Taxon	Number	Number of Years
Pentaneura	3	1
Phaenopsectra	11	4
Polypedilium	1364	6
Psectrocladius	921	4
Psectrotanypus	10	2
Pseudorthocladius	32	2
Pseudosmittia	19	2
Rheocricotopus	1	1
Smittia	21	5
Stictochironomus	2	1
Tanytarsus	685	4
Thienemannimyia	7	1
Thienemannia	7	2
Zavrelia	1	1
Unknown	44	5
Culicidae		
Aedes	72	6
Anopheles	2	2
Culex	65	5
Culiseta	45	5
Deinocerites	6	2
Coquillettidia	7	3
Wyeomyia	3	1
Unknown	1	1
Simuliidae		
Prosimulium	3	2
Tipulidae		
Gonomyia	1	1
Limonia	5	3
Limnophila	14	6
Molophilus	38	5
Ormosia	55	6
Pedicia	16	3
Pilaria	98	4
Prionocera	109	5
Pseudolimnophila	17	6
Tipula	142	6
Tipulidae Unknown1	24	2
Diptera (suborder Brachycera)		
Dixidae		
Dixella	2	1
Dolichopodidae	8	2
Empididae	48	3
Ephydridae	40	4

∠E 22.2. (Continued)

Taxon	Number	Number of Years
Muscidae		
Limnophora	5	3
Psychodidae		
Pericoma	49	4
Psychoda	9	1
Scyiomyzidae		
Sepedon	32	4
Stratiomyidae		
Allognosta	3	1
Caloparyphus	5	1
Nemotelus	1	1
Odontomyia	39	3
Oxycera	2	1
Stratiomys	408	6
Syrphidae		
Eristalis	6	4
Unknown	6	3
Tabanidae		
Hybomitra	3	2
Tabanus	50	6
Unknown	2	2
Trichoptera		
Hydroptilidae		
Unknown	5	1
Neotrichia	7	1
Hydropsychidae		
Arctopsyche	1	1
Lepidoptera		
Tortricidae	1	1
Pyralidae		
Acentria	6	2
Crambus	57	5
Noctuidae		
Archanara	1	1
Gastropoda		
Lymnaea	2610	6
Physa	1070	6
Gyraulus	4750	6
Helisoma	11	1
Aplexa	1150	6
Bivalvia	849	6
Isopoda	10	3
Annelida	705	6

TABLE 22.3. Back-Transformed Least-Square Estimates of Mean (no./m²) and 95 Percent Confidence Intervals for Macroinvertebrates in Wright Co. Wetlands[a]

Density	Year	Mean	95% CI	Contrast	F	P-value
Annelida	1989	277	(213–376)	Dry vs. 91–93	6.27	0.019
	1990	301	(213–455)	Dry vs. 92–93	1.14	0.295
	1991	192	(184–201)			
	1992	282	(216–383)			
	1993	247	(185–350)			
Gastropoda	1989	3556	(749–18,407)	Dry vs. 91–93	10.91	0.003
	1990	232	(193–288)	Dry vs. 92–93	10.69	0.003
	1991	299	(220–430)			
	1992	331	(222–534)			
	1993	295	(233–384)			
Sphaeriidae	1989	662	(282–1816)	Dry vs. 91–93	8.44	0.007
	1990	295	(210–444)	Dry vs. 92–93	12.20	0.002
	1991	229	(186–293)			
	1992	219	(185–266)			
	1993	202	(184–224)			
Total Insecta	1989	1007	(770–1329)	Dry vs. 91–93	21.97	0.000
	1990	1080	(722–1644)	Dry vs. 92–93	36.40	0.000
	1991	1490	(973–2309)			
	1992	3667	(2414–5597)			
	1993	3502	(2269–5433)			
Coleoptera	1989	389	(343–444)	Dry vs. 91–93	3.33	0.079
	1990	364	(250–562)	Dry vs. 92–93	2.83	0.105
	1991	285	(233–356)			
	1992	355	(266–490)			
	1993	249	(198–326)			
Chironomidae	1989	228	(191–280)	Dry vs. 91–93	312.3	0.000
	1990	365	(265–523)	Dry vs. 92–93	375.9	0.000
	1991	1234	(789–1963)			
	1992	2292	(1297–4109)			
	1993	2505	(1550–4089)			
Ceratopogonidae	1989	194	(184–203)	Dry vs. 91–93	35.17	0.000
	1990	236	(195–295)	Dry vs. 92–93	29.64	0.000
	1991	218	(191–251)			
	1992	315	(244–420)			
	1993	405	(260–676)			
Tipulidae	1989	216	(186–255)	Dry vs. 91–93	0.78	0.387
	1990	248	(199–320)	Dry vs. 92–93	2.85	0.103
	1991	201	(183–223)			
	1992	270	(201–386)			
	1993	332	(264–428)			

.tinued)

	Year	Mean	95% CI	Contrast	F	P-value
.dae	1989	285	(195–457)	Dry vs. 91–93	0.85	0.366
	1990	271	(226–330)	Dry vs. 92–93	0.74	0.397
	1991	192	(184–201)			
	1992	273	(230–329)			
	1993	338	(268–436)			
Total remaining insects	1989	309	(200–529)	Dry vs. 91–93	0.65	0.427
	1990	190	(133–328)	Dry vs. 92–93	0.84	0.368
	1991	214	(197–234)			
	1992	204	(187–223)			
	1993	214	(192–241)			

[a]Based on two sampling dates in each year. Data were analyzed using a repeated measures ANOVA with df = 1, 26. For 1989 and 1990, n = 7; for 1991–1993, n = 8.

the most abundant benthic invertebrates sampled in 1989. Fingernail clams were about three times more abundant in 1989 than subsequent years. Annelids were significantly more abundant in 1989–1990 than in 1991–1993, but the difference was very small (Table 22.3).

Insect abundance was low during the drought years of 1989 and 1990, as well as 1991, the first year postdrought (Table 22.3). However, in 1992 insect abundance attained densities 2 to 10 times higher than in 1989–1991 (Table 22.3). Average annual mean insect abundance was 3 to 5 times higher in postdrought years than in drought years (Table 22.3). In 1993 insect density was approximately 2 to 3 times higher than 1991 levels (Table 22.3).

However, within the insect community drought had differential effects on different components. The greatest effect was on the nematoceran dipterans Chironomidae and Ceratopogonidae. Both were much more abundant in wet years than in dry years (Table 22.3). Tipulidae (Diptera: Nematocera) were not affected (Table 22.3). Among chironomids and ceratopogonids there was a delayed response; these taxa increased each year postdrought (Table 22.3). Stratiomyidae, the dominant brachyceran dipteran, and Coleoptera were not significantly affected by the drought (Table 22.3). Remaining insects, which comprised about 7 percent of the postdrought insect fauna, also were not significantly affected by drought.

Noninsect macroinvertebrates were highly negatively correlated (p < 0.0001) with Palmer's drought index for the month that the samples were taken, but also with the previous month and the previous year (Table 22.4). However, they were not significantly correlated with Palmer's index from two years previous (Table 22.4). In the early season chironomids were most highly correlated with Palmer's index from the current year, although the correlation was significant for all years (Table 22.4). Late-season sampling showed higher correlation coefficients using one- and two-month lag times, and highest with a one-year lag (r = 0.84; Table 22.4).

TABLE 22.4. Correlations of Macroinvertebrates with Palmer Drought Index[a]

Early Season	Current Month	One Month Previous	Two Months Previous	Mean of 12 Months Previous	Mean of 13–24 Months Previous
Annelida	−0.17020	−0.20489	−0.19649	−0.19365	−0.10010
	0.3003	0.2109	0.2306	0.2375	0.5443
Gastropoda	−0.62020	−0.59131	−0.55390	−0.57392	−0.09829
	0.0001	0.0001	0.0003	0.0001	0.5517
Sphaeriidae	−0.57671	−0.54776	−0.53724	−0.54620	−0.29648
	0.0001	0.0003	0.0004	0.0003	0.0668
Total Insecta	0.51552	0.47018	0.47206	0.47157	0.41639
	0.0008	0.0025	0.0024	0.0024	0.0084
Coleoptera	−0.27732	−0.20630	−0.20526	−0.20925	−0.19786
	0.0874	0.2076	0.2100	0.2011	0.2273
Chironomidae	0.73307	0.68589	0.67625	0.68022	0.48036
	0.0001	0.0001	0.0001	0.0001	0.0020
Ceratopogonidae	0.57354	0.51405	0.54776	0.54205	0.65196
	0.0001	0.0008	0.0003	0.0004	0.0001
Tipulidae	−0.13382	−0.18002	−0.15895	−0.16020	0.02913
	0.4167	0.2728	0.3337	0.3299	0.8603
Stratiomyidae	−0.09314	−0.11993	−0.07124	−0.08252	0.26791
	0.5728	0.4671	0.6665	0.6175	0.0992
Remaining insects	−0.46017	−0.47652	−0.46690	−0.47369	−0.21077
	0.0032	0.0022	0.0027	0.0023	0.1978

Late Season	Current Month	One Month Previous	Two Months Previous	Mean of 12 Months Previous	Mean of 13–24 Months Previous
Annelida	−0.23448	−0.25871	−0.18373	−0.15622	0.08169
	0.1508	0.1118	0.2629	0.3423	0.6210
Gastropoda	−0.74334	−0.51239	−0.47194	−0.46627	−0.14532
	0.0001	0.0009	0.0024	0.0028	0.3774
Sphaeriidae	−0.58343	−0.55833	−0.55057	−0.55124	−0.32099
	0.0001	0.0002	0.0003	0.0003	0.0463
Total insects	0.40405	0.50222	0.59135	0.63993	0.77950
	0.0107	0.0011	0.0001	0.0001	0.0001
Coleoptera	−0.59972	−0.50242	−0.39804	−0.38238	−0.12394
	0.0001	0.0011	0.0121	0.0163	0.4522
Chironomidae	0.65321	0.76624	0.81363	0.84028	0.74098
	0.0001	0.0001	0.0001	0.0001	0.0001
Ceratopogonidae	0.38413	0.28535	0.27861	0.31614	0.48033
	0.0158	0.0783	0.0859	0.0499	0.0020
Tipulidae	0.24546	0.24153	0.24506	0.28637	0.53296
	0.1320	0.1385	0.1327	0.0772	0.0005
Stratiomyidae	0.20320	0.05408	0.01726	0.04873	0.27223
	0.2147	0.7437	0.9169	0.7683	0.0936
Remaining insects	−0.36364	−0.39817	−0.38393	−0.38468	−0.25862
	0.0229	0.0121	0.0158	0.0156	0.1119

[a] First line = Pearson correlation coefficient, second line = p-value; n = 39.

Insect densities in temporary ponds in the same region were very similar to those in the prairie wetlands during 1989 (Hershey et al. 1995). In the prairie wetlands of Wright Co., development of a dense fauna required a two-year lag time following the prolonged drought of 1987–1989. It is reasonable to expect lag times in recovery of aquatic insect populations to drought conditions, similar to other disturbances (Mackay 1984/1985, Cairns 1990). However, there is limited theoretical basis to predict what the lag time should be, nor are there comparable data in the literature.

Richness

Wetland insect communities are known to respond to drought with low density and diversity, and drought has been cited as the most important factor controlling wetland communities (Wiggins et al. 1980, Mackay 1984/85). Per-sample richness of the insect genera in Wright Co. prairie wetlands was significantly lower under drought conditions (Table 22.5). Nearly all of the difference in richness was due to chironomids; richness of other taxonomic groups did not appear to be affected by drought (Table 22.5). Temporary ponds in the same region showed richness under drought conditions (Hershey et al. 1995) very similar to that observed in the prairie wetlands.

Richness provides information on structure of the wetland communities. Because richness is correlated with density, the probability of encountering

TABLE 22.5. Back-Transformed Least-Square Estimates of Insect Richness and 95% Confidence Intervals in Wright Co. Wetlands[a]

Number of Taxa	Year	Mean	95% CI	Contrast	F	P-value
Total insects	1989	6.5	(5.5–7.8)	Dry vs. 91–93	14.77	0.001
	1990	8.3	(6.6–10.5)	Dry vs. 92–93	29.75	0.000
	1991	7.8	(6.2–9.9)			
	1992	11.1	(9.2–13.4)			
	1993	13.0	(10.8–15.8)			
Chironomidae	1989	2.4	(2.1–2.8)	Dry vs. 91–93	168.5	0.000
	1990	3.3	(2.6–4.2)	Dry vs. 92–93	152.2	0.000
	1991	5.7	(4.6–7.0)			
	1992	6.4	(5.0–8.2)			
	1993	7.9	(6.3–10.0)			
Tupulidae	1989	2.3	(2.1–2.5)	Dry vs. 91–93	0.52	0.476
	1990	2.1	(1.6–3.2)	Dry vs. 92–93	1.46	0.237
	1991	2.1	(1.9–2.3)			
	1992	2.5	(2.1–2.9)			
	1993	2.8	(2.5–3.0)			

[a] Based on two sampling dates in each year. Sample area was 106 cm^2, as described in Hershey et al. (1998). Data were analyzed using a repeated measures ANOVA with numerator df = 1, denominator df = 26. For 1989 and 1990, n = 7; for 1991–1993, n = 8.

new species in a site increases with the number of individuals examined (Pielou 1969). Thus, as insect density increased in 1992 and 1993, there was a corresponding increase in richness of insect taxa (Table 22.5). High richness is generally considered characteristic of healthy ecosystems (e.g., Solbrig 1991, Chapin and Korner 1995, Tilman 1996, Culotta 1996). Even when species perform similar trophic roles in an aquatic foodweb, redundancy may result in prolonged seasonal distribution of trophic function due to staggered developmental periods.

Noninsect macroinvertebrates thrived during the drought years, when insects were not abundant. Because insects, especially chironomids, became abundant during the postdrought cycle, while noninsect macroinvertebrates declined, the macroinvertebrate fauna was very different between these two phases of the wetland cycle (see Fig. 22.3).

ZOOPLANKTON COMMUNITY

Sampling zooplankton in a wetland requires different methods than those typically employed in open waters. Wetland sites are often characterized by thick patches of emergent vegetation and water depths of less than 1 m—conditions not conducive to the use of the various tow nets commonly used in zooplankton studies. As an alternative, we used funnel traps (Whiteside and Lindegaard 1980) to sample the zooplankton community. This technique relies on the diurnal vertical migration behavior of substrate-dwelling microfauna (Whiteside and Williams 1975). Organisms migrate up the stem of an inverted funnel and become trapped in a water-filled collecting jar. We placed 10 of these at each site and allowed them to remain in place overnight. Samples were preserved in alcohol and later identified to the lowest practical taxonomic level. Note that with this method the density of organisms is reported in terms of area (derived from the area of sediment enclosed by the trap) rather than volume.

The Wright Co. wetlands supported a diverse assemblage of zooplankton species in all sites. Most of these were species typically associated with small ponds (Hartland-Rowe 1966, Wiggins et al. 1980), although a few are more commonly found in larger lakes. A total of 102 zooplankton and eubranchiopod taxa were identified from these areas. The taxonomic level to which organisms were identified varied with the difficulty of classification. Cladocerans and copepods were generally identified to species, while most rotifers were recorded at the genus level. Ostracods were recorded only at the order level.

These wetland plankton communities included 43 cladoceran taxa, 22 copepod taxa, and 35 rotifer taxa. In addition, a conchostrachan (clam shrimp) species and an anostracan (fairy shrimp) species were present in a few of the sites. The relative abundance of these crustaceans varied from abundant, cosmopolitan species (e.g., *Cyclops vernalis*) which were found regularly in every site, to rare species which were found sporadically or at a single site (Table 22.6).

Cyclopoid copepods comprised a major proportion of the zooplankton community on nearly all sampling dates (Fig. 22.2). Densities were typically between 40,000 and 60,000 per m². Four taxa, *Cyclops navus, C. varicans, C. vernalis,* and *Canthocamptus* sp., were common to all sites. The harpacticoid *Canthocamptus* sp. was among the first species to appear in spring samples, while the cyclopoid *Cyclops* species appeared later. Immature stages (nauplii and copepodites or juvenile copepods) were abundant at every sampling period.

Cladocerans typically did not appear in the community until early May and were most abundant from late May through June. Maximum densities were generally between 15,000 and 20,000 per m². There was a significant seasonal effect on the abundance of most taxa. The most common cladocerans in the sites were *Chydorus sphaericus, Simocephalus vetulus, Pleuroxus denticulatus,* and *Ceriodaphnia* sp.

Rotifers were also common in all sites. Numbers were typically low in early-season samples and increased as the season progressed. Maximum seasonal densities were usually between 10,000 and 20,000 individuals per m², although higher numbers were found in the late-July samples that were collected only in 1990. Five taxa, *Testudinella patina, Lepadella* sp., *Platyias* sp., *Lecane* sp., and *Monostyla* sp., were common to all sites.

Ostracods were a consistent component of the zooplankton community during most sampling periods, although they were typically less abundant than other groups (Fig. 22.2). Maximum seasonal densities were usually between 4000 and 8000 per m². Although there were several different species present in the wetlands, they were difficult to separate routinely and were instead counted as a single taxon.

The relative abundance of these groups varied widely, although the predominant organisms were nearly always copepods. The average taxonomic composition for all samples was 13 percent rotifers, 12 percent cladocerans, 62 percent copepods, and 13 percent ostracods. In addition to the traditional zooplankton taxa, we also encountered occasional Eubranchiopod populations of *Eubranchipus bundyi* (fairy shrimp) and *Lynceus brachyurus* (clam shrimp). They were not collected frequently enough or in large enough numbers, however, for meaningful statistical analysis.

Factors Affecting the Abundance of Zooplankton Species

Many factors, including photoperiod, water temperature, water chemistry, food supply, oxygen, competition, and predation, have been shown to influence the distribution and abundance of zooplankton communities in permanent bodies of water (Beaver and Havens 1996, Shuter et al. 1997, Sternberger et al. 1996). In wetlands, which may be completely dry at certain times of the year, water availability is the ultimate determinant of whether active zo-

TABLE 22.6. The Zooplankton Community of Wright Co., Minnesota, Wetlands[a]

Taxon	Percent of Sites	Taxon	Percent of Sites	Taxon	Percent of Sites
Rotifers		Eubranchiopods		*Camptocercus rectirostris*	11.1
Lepadella sp.	100.0	*Lynceus brachyurus*	44.4	*Ceriodaphnia laticaudata*	11.1
Monostyla sp.	100.0	*Eubranchipus* sp.	22.2	*Ceriodaphnia veticulata*	11.1
Mytilina sp.	100.0			*Chydorus ovalis*	11.1
Platyias quadricornis	100.0	Cladocerans		*Daphnia dubia*	11.1
Testudinella patina	100.0	*Ceriodaphnia* sp.	100.0	*Daphnia galeata mendotae*	11.1
Trichocera sp.	100.0	*Chydorus sphaericus*	100.0	*Daphnia* sp.	11.1
Asplanchna sp.	88.8	*Simocephalus vetulus*	100.0	*Dunhevedia crassa*	11.1
Euchlanis sp.	88.9	*Alonella exisa*	88.9	*Eurycercus lamellatus*	11.1
Lecane sp.	88.9	*Daphnia pulex*	88.9	*Macrothrix laticornis*	11.1
Cephalodella sp.	77.8	*Pleuroxus denticulatus*	88.9	*Moina* sp.	11.1
Dipleuchanis sp.	77.8	*Pleuoxus procurvus*	88.9		
Keratella sp.	77.8	*Scapholeberis kingi*	88.9	Copepods	
Lopocharis sp.	66.7	*Simocephalus serrulatus*	88.9	*Canthocamptus* sp.	100.0
Asplanchopus sp.	55.6	*Alona setulosa*	66.7	*Cyclops bicuspidatus*	100.0
Gastropus sp.	55.6	*Kurzia lattissima*	55.6	*Cyclops navus*	100.0
Polyarthra sp.	55.6	*Alona circumfibriata*	44.4	*Cyclops varicans*	100.0
Trichotria tetractis	55.6	*Alona guttata*	44.4	*Cyclops vernalis*	100.0
Enteroplea sp.	44.4	*Alonella exigua*	44.4	*Eucyclops agilis*	100.0
Platyias patulus	44.4	*Simocephalus* sp.	44.4	*Macrocyclops albidis*	100.0
Platyias polycanthus	44.4	*Bosmina* sp.	33.3	*Paracyclops fimbriatus poppei*	88.9
Macrochaetus sp.	33.3	*Oxyurella tenuicaudis*	33.3	*Cyclops haueri*	33.3
Brachionus plicatilis	22.2	*Simocephalus expinosus*	33.3	*Bryocamptus* sp.	22.2

TABLE 22.6. (Continued)

Taxon	Percent of Sites	Taxon	Percent of Sites	Taxon	Percent of Sites
Brachionus quadridentata	22.2	Alona affinis	22.2	Diaptomus stagnalis	22.2
Filinia sp.	22.2	Alona rectangula	22.2	Ectocyclops phaleratus	22.2
Brachionus sp.	11.1	Pleuroxus hamulatus	22.2	Kellicottia sp.	22.2
Colurella sp.	11.1	Pleuroxus sp.	22.2	Mesocyclops edax	22.2
Enteroplea lacustris	11.1	Sida crystallina	22.2	Mesocyclops leuckarti	22.2
Epiphanes sp.	11.1	Simocephalus exspinosus	22.2	Orthocyclops modestus	22.2
Harringia sp.	11.1	Acroperus sp.	11.1	Cyclops bicolor	11.1
Manfredium eudactylotum	11.1	Alona costata	11.1	Cyclops crassicaudis brachycercus	11.1
Monommata sp.	11.1	Alona monocantha	11.1	Eucyclops speratus	11.1
Notommata sp.	11.1	Alona nana	11.1	Mesocyclops dybowski	11.1
Paracolurella aemula	11.1	Alona sp.	11.1	Paracyclops affinis	11.1
Squatinella sp.	11.1	Alonella sp.	11.1	Tropocyclops prasinus	11.1
Tripleuchlanis sp.	11.1	Bunops serricauda	11.1		
		Camptocercus macrurus	11.1	Ostracods	100.0

[a] Percentages indicate the percent of sites (out of nine) where each taxon was found at least once.

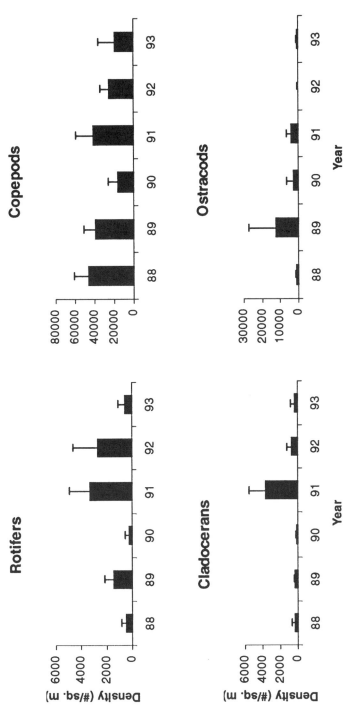

Fig. 22.2. Density of zooplankton taxa in drought (1988–1990) and postdrought (1991–1993) based on funnel trap samplers (Whiteside and Lindegaard 1980) set overnight (n = 10). Samplers rely on vertical migration behavior of substrate-dwelling microinvertebrates (Whiteside and Williams 1975). Sampling via a zooplankton net was not feasible due to dense emergent vegetation. Bars are geometric means, error bars represent 1 SE. Note that with this method, density of organisms is reported as number per unit area, representing the area of the substrate enclosed by the funnel, rather than per unit volume.

oplankton populations are present. Even in the absence of water, however, most zooplankton species are present in various dormant stages. Indeed, much work has been directed toward investigating the ecological and evolutionary significance of this zooplankton "egg bank" (Hairston 1996). Nearly all groups have been found to produce some form of long-lived dormant stage. Hairston et al. (1995) found viable cyclopoid copepod eggs over 100 years old. Moritz (1987) found viable *Ceriodaphnia* ephippia as old as 14 years. Nipkow (1961) hatched diapausing eggs of seven species of rotifers up to 35 years old.

When water is available, other factors act to determine the structure of wetland zooplankton communities. These were not thoroughly investigated in the Wright Co. sites, but the relationships among water depth, temperature, and the abundance of major groups were studied (Table 22.7). Both cladoceran and rotifer density were positively correlated with the related measures of water temperature and day of the year. Conversely, copepod density was negatively correlated with day of the year and ostracod density was negatively correlated with water temperature. We also found some strong negative correlations between groups. Copepods, for example, were negatively correlated with cladocerans, rotifers, and ostracods. Ostracods were negatively correlated with both rotifers and cladocerans. In no case, however, did we find any correlation between any zooplankton group and any macroinvertebrate group (Niemi et al. 1995).

Effects of Drought

Community Structure. There were distinctive differences in the species composition during and after the drought. Both the rotifer and cladoceran communities contained significantly more species in the wet years than in the dry years (Table 22.8). There was no difference in the number of copepod species between the wet and dry periods. These results are supported by the occurrence of 14 species of rotifers, 14 species of cladocerans, and 2 species of copepods in the wet years that were not sampled in the dry years. In addition, 3 cladoceran species and 2 copepod species were found only during the dry years.

Density. The density of several taxa was also significantly affected by the drought. During the dry years most sites were dry by early June, so one obvious effect of the drought was to shorten the active period for the entire zooplankton community. Both rotifer and cladoceran abundances were significantly correlated with the Palmer Drought Severity Index for the previous month (Table 22.9). Copepods and ostracods were not correlated with the drought index. Similarly, a comparison of the geometric mean densities for the dry and wet periods indicated that densities of cladocerans and rotifers were significantly higher in the wet period. There was no difference in the densities of copepods and ostracods (Fig. 22.2). For both rotifers and clado-

TABLE 22.7. Correlations of Zooplankton Abundance with Date, Water Depth, Temperature, and the Presence of Other Taxa

	Day of Year	Depth	Temperature	Rotifers	Cladocerans	Copepods	Ostracods
Day of Year	—						
Depth	0.09	—					
Temp	0.58**	0.15	—				
Rotifers	0.33**	0.11	0.16*	—			
Cladocerans	0.57**	0.04	0.35**	0.9	—		
Copepods	−0.46**	−0.07	−0.13	−0.57**	−0.53**	—	
Ostracods	−0.12	−0.03	−0.25**	−0.18*	−0.17*	−0.48**	—

TABLE 22.8. Back-Calculated Mean Number and 95 Percent Confidence Interval of Species for Major Taxa in Dry (1988–1990) and Wet (1991–1993) Years[a]

Taxon	Year	Mean Species	95% CI
Rotifers	88	2.1	(1.8–2.4)
	89	4.2	(3.8–4.7)
	90	3.0	(2.5–3.6)
	91	6.0	(5.0–7.3)
	92	4.9	(4.2–5.8)
	93	4.3	(3.8–4.9)
Cladocerans	88	2.9	(2.5–3.5)
	89	3.0	(2.6–3.4)
	90	2.5	(2.1–3.1)
	91	5.6	(4.9–6.5)
	92	4.1	(3.3–5.0)
	93	3.7	(3.1–4.4)
Copepods	88	4.7	(4.2–5.2)
	89	4.1	(3.7–4.7)
	90	4.3	(4.0–4.6)
	91	4.4	(4.3–4.6)
	92	5.2	(4.9–5.4)
	93	4.5	(4.3–4.8)

[a] Note that statistical comparisions (see text) were conducted using repeated measures analyses of variance.

cerans there was an immediate surge in population density in 1991, the first year following the drought. Densities in 1992 and 1993 declined somewhat, suggesting that the initial pulse may have resulted from the hatching of large numbers of resting stages accumulated in the sediments during the dry years.

The diverse assemblage of zooplankton is representative of groups that occupy several key positions in aquatic foodwebs. The largest percentage of these organisms are filter-feeders, obtaining food by filtering algae, protozoa, organic detritus, and bacteria from the water column. Others, including ostracods and harpacticoid copepods, are more substrate oriented and may feed

TABLE 22.9. Correlations between Major Zooplankton Taxa and Palmer Hydrological Drought Severity Index

Taxon	r	p
Rotifers	0.42042	0.004
Cladocerans	0.33007	0.027
Copepods	0.06258	0.683
Ostracods	−0.12823	0.401

on fine bits of detritus, algae, bacteria, and fungi swept or scraped from the surface of the sediments or aquatic vegetation. Some calanoid and cyclopoid copepods and several species of rotifers are predaceous. They may ingest algae, but also feed on other zooplankton taxa, utilizing both small species and the immature stages of larger species. In turn these zooplankton are preyed upon by insect predators. Some of the larger species, especially the eubranchiopods but also cladocerans, provide food for waterfowl (Swanson 1984/1985).

GENERAL IMPLICATIONS

There are an estimated 10.6 million acres of wetlands in Minnesota, compared to a presettlement estimate of over 20 million; most of the loss has been in the prairie wetland region (Minnesota Wetlands Conservation Plan 1997). Both macroinvertebrate and zooplankton communities of these wetlands are strongly affected by the interannual hydrologic variation characteristic of these habitats. The extremes of wet and dry phases of the hydrologic cycle support invertebrate communities which are taxonomically distinct (Fig. 22.3), have different life history characteristics, and perform different functional roles within the ecosystem. During the wet phase the benthic macroinverte-brate community is dominated by insects, with chironomids the most abun-

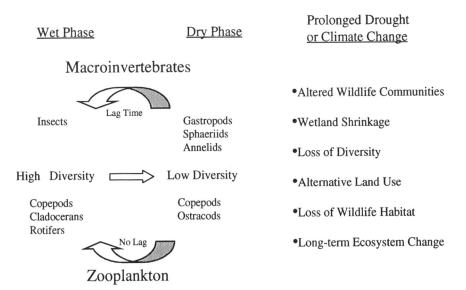

Fig. 22.3. Conceptual model of dominance of invertebrate groups during wet and dry phases in prairie wetlands, and implications for prolonged drought or climate change in the region.

dant group, whereas the zooplankton community is dominated by copepods, cladocerans, and rotifers. During the drought phase insects become depauperate in richness and abundance and mollusks and annelids become abundant within the benthos. In the zooplankton community rotifers and cladocerans decline and copepods and ostracods become more important (Fig. 22.3). The shift between phases does not occur at the same pace for the two invertebrate communities. Chironomid response to the return of the wet phase occurs gradually (Fig. 22.3); in the Wright Co. study maximum densities were seen two years postdrought. This lag time reflects the fact that most insects recruit from more permanent water bodies in the adult stage of the life cycle, although some have drought-resistant eggs or cocoons (see Wiggins et al. 1980, Mackay 1984/1985). In contrast, zooplankton species, which recruit from an egg bank in the soil, do not show a delay in recruitment when the wet phase returns (Fig. 22.3).

CONSERVATION AND CLIMATE CHANGE IMPLICATIONS

The sensitivity of prairie wetland macroinvertebrates to drought, combined with the importance of this habitat and food source to both migratory waterfall and resident vertebrate populations, suggest that if climate change in the region results in a trend toward lower precipitation, this will alter dramatically the biotic community of these wetlands. Wetland insects, including chironomids, are generally recognized to be of fundamental importance in the diets of waterfowl, especially ducks (see Brown and Hunter 1985, Swanson 1984/ 1985, Nummi 1993, Batzer et al. 1993, and see Batzer and Wissinger 1996). Many migratory bird species also utilize mollusks as a food source (e.g., Swanson 1984/1985, Swanson and Duebbert 1989), and these were considerably more abundant than insects during the recent drought. Because particular vertebrate species use mollusk and insect prey in different proportions, food availability would lead to different vertebrate community types under the two hydrologic extremes (e.g., Swanson 1984/1985).

The Wright Co. wetlands, like many prairie wetlands, occur in an agricultural setting, and during drought years their perimeters may be subjected to plowing and planting, resulting in a loss of habitat for wildlife and waterfowl. Although the wetlands are afforded some degree of protection under the Wetlands Conservation Act, this protection is conditional for wetlands on private land (Minnesota Wetlands Conservation Plan 1997). The Wright Co. study provides insight into variability of macroinvertebrate and zooplankton communities due to natural wet/dry cycles. However, it is not clear that the same model can be applied to recovery following tillage (Fig. 22.3); this area is in need of research. Furthermore, a climate change scenario that involves more prolonged drought cycles is likely to result in continually shrinking wetland habitat, and possibly loss of these habitats from the region (Fig. 22.3). If this occurs, it also will have dramatic effects on resident and migratory vertebrate communities as well as other wetland functions.

ACKNOWLEDGMENTS

We would like to thank the following individuals for their efforts in helping to gather field data, for compiling data, and aiding with analysis: Paul Mickelson, Jeff Schuldt, JoAnn Hanowski, Gloria Bly, Chris Larsen, Mike King, Nancy Kirsh, Phillip Monson, Frank Kaszuba, Amy Wold, and Chris Ernst. Special thanks also to also Anda Bellamy, Brenda Maas, Eulie Markham, and Margy Bell for their assistance in preparations of the manuscript. Financial support for this study was provided by the Metropolitan Mosquito Control District (MMCD) through its independent peer review panel. We thank the members of that panel, including Richard Anderson, Judy Helgen, Stuart Hurlbert, Robert Naiman, Roger Moon, William Schmid, Ken Simmons, Keith Solomon, Harrison Tordoff, and Mike Zicus, for their suggestions on the planning of this work and their reviews of the completed report to the MMCD.

LITERATURE CITED

Beaver, J. R., and K. E. Havens. 1996. Seasonal and spatial variation in zooplankton community structure and their relation to possible controlling variables in Lake Okeechobee. Freshwater Biology 36:45–56.

Batzer, D. P., M. McGee, V. H. Resh, and R. R. Smith. 1993. Characteristics of invertebrates consumed by mallards and prey response to wetland flooding schedules. Wetlands 13:41–49.

Batzer, D. P., and S. A. Wissinger. 1996. Ecology of insect communities in nontidal wetlands. Annual Review of Entomology 41:75–100.

Brown, P. W., and M. L. Hunter. 1985. Potential effects of insecticides on dabbling duck broods. Journal of the Minnesota Academy of Science 50:41–45.

Cairns, J. 1990. Lack of theoretical basis for predicting rate and pathways of recovery. Environmental Management 14:517–526.

Chapin, F. S., and C. Korner (eds.). 1995. Arctic and Alpine Biodiversity: Patterns, Causes, and Ecosystem Consequences. Springer-Verlag, Berlin.

Culotta, E. 1996. Exploring biodiversity's benefits. Science 273:1045–1046.

Erwin, R. M. 1996. The status of forested wetlands and waterbird conservation in North and Central America. Pages 61–109 in R. M. DeGraaf and R. I. Miller (eds.), Conservation of Faunal Diversity in Forested Landscapes. Chapman & Hall, London, UK.

Hairston, N. G., Jr., R. A. Van Brunt, and C. M. Kearns. 1995. Age and Survivorship of Diapausing Eggs in a Sediment Egg Bank. Ecology 76:6 1706–1714.

Hairston, N. G., Jr. 1996. Zooplankton egg banks as biotic reservoirs in changing environments. Limnology and Oceanography 41:1087–1092.

Hartland-Rowe, R. 1966. The fauna and ecology of temporary pools in western Canada. Internationale Vereinigung für Theoretische und Angewandte Limnologie, Verhandlungen, 16:577–584.

Heddinghaus, T. R., and P. Sabol. 1991. A review of the Palmer Drought Severity Index and where do we go from here? Pages 242–246 *in* Proceedings of Seventh Conference on Applied Climatology, September 10–13, 1991. American Meteorological Society, Boston.

Hershey, A. E., A. R. Lima, G. J. Niemi, and R. R. Regal. 1998. Effects of *Bacillus thuringiensis israelensis* (Bti) and methoprene on nontarget macroinvertebrates in Minnesota wetlands. Ecological Applications 8:41–60.

Hershey, A. E., L. Shannon, R. Axler, C. Ernst, and P. Mickelson. 1995. Effects of methoprene and Bti (*Bacillus thuringiensis* var. *israelensis*) on non-target insects: A divided pond study. Hydrobiologia 308:219–227.

Karl, T. R., and R. W. Knight. 1985. Atlas of Monthly Palmer Hydrological Drought Indices (1931–1983) for the Contiguous United States. Historical Climatology Series 3–7, National Climatic Data Center, Asheville, NC.

Larson, D. J. 1987. Aquatic Coleoptera of peatlands and marshes in Canada. Memoirs of the Entomological Society of Canada 140:99–132.

Lewis, D. J. 1987. Biting flies (Diptera) of peatlands and marshes in Canada. Memoirs of the Entomological Society of Canada 140:133–140.

Mackay, R. J. 1984/1985. Survival strategies of invertebrates in disturbed aquatic habitats. Journal of the Minnesota Academy of Science 50:28–30.

Martin, A. C., H. S. Zim, and A. L. Nelson. 1951. American Wildlife and Plants: A Guide to Wildlife Food Habits. Dover, New York.

Minnesota Wetlands Conservation Plan, Version 1. 1997. Minnesota Department of Natural Resources, St. Paul, MN.

Mitsch, W. J., and J. G. Gosselink. 1986. Wetlands. Van Nostrand Reinhold, New York.

Moon, R. D. 1984/1985. A brief overview of the life history, physiology, and ecology of Minnesota mosquitoes. Journal of the Minnesota Academy of Science 50:6–9.

Moritz, C. 1987. A note on the hatching and viability of *Ceriodaphnia* ephippia collected from lake sediments. Hydrobiologia 145:309–314.

Niemi, G., R. Axler, J. Hanowski, A. Hershey, and L. Shannon. 1995. Evaluation of the effects of methoprene and BTI on wetland birds and invertebrates in Wright County, MN 1988 to 1993. Metropolitan Mosquito Control District Final Report.

Nipkow, F. 1961. Die Radertiere im Plankton des Zurichsees und ihre Entwicklungsphasen. Schweizerische Zeitschrift für Hydrologie 23:398–461.

Nummi, P. 1993. Food-niche relationships of sympatric mallards and green-winged teal. Canadian Journal of Zoology 71:49–55.

Olson, S. 1992. Habitat heterogeneity and bird species composition in central Minnesota wetlands. Master's Thesis, University of Minnesota–Duluth.

Palmer, W. C. 1965. Meteorological Drought. Research Paper No. 45, U.S. Department of Commerce Weather Bureau, Washington, DC.

Pielou, E. C. 1969. An Introduction to Mathematical Ecology. Wiley-Interscience, New York.

Richardson, C. J. 1979. Primary productivity values in freshwater wetlands. Pages 131–145 *in* P. E. Greeson, J. R. Clark, and J. E. Clark (eds.), Wetland Function and Values: The State of Our Understanding. American Water Resources Association, Minneapolis, MN.

Shuter, B. J., and K. K. Ing. 1997. Factors affecting the production of zooplankton in lakes. Canadian Journal of Fisheries and Aquatic Sciences 54:358–377.

Solbrig, O. T. 1991. From Genes to Ecosystems: A Research Agenda for Biodiversity. IUBS-SCOPE-UNESCO Workshop Report. Harvard Forest, Petersham, MA.

Sternberger, R. S., A. T. Herlihy, and D. L. Kugler. 1996. Climatic forcing on zooplankton richness in lakes of the northeastern United States. Limnology and Oceanography 41:1093–1101.

Swanson, G. A. 1984/1985. Invertebrates consumed by dabbling ducks (Anatinae) on the breeding grounds. Journal of the Minnesota Academy of Science 50:37–40.

Tester, J. 1995. Minnesota's Natural Heritage: An Ecological Perspective. University of Minnesota Press, Minneapolis, MN.

Tilman, D. 1996. Biodiversity: Population versus ecosytem stability. Ecology 77:350–363.

Weller, M. W. 1988. Issues and approaches in assessing cumulative impacts on waterbird habitats in wetlands. Environmental Management 12:695–701.

Whiteside, M. C., and J. B. Williams. 1975. A new sampling technique for aquatic ecologists. Internationale Vereinigung für Theoretische und Angewandte Limnologie, Verhandlungen 19:1534–1539.

Whiteside, M. C., and C. Lindegaard. 1980. Complementary procedures for sampling small benthic invertebrates. Oikos 35:317–320.

Wiggins, G. B., R. J. Mackay, and I. M. Smith. 1980. Evolutionary and ecological strategies of animals in temporary pools. Archiv für Hydrobiologie Supplement 58: 97–206.

Willeke, G., J. R. M. Hosking, J. R. Wallis, and N. B. Guttman. 1994. The National Drought Atlas. Institute for Water Resources Report 94-NDS-4, U.S. Army Corps of Engineers.

Wrubleski, D. A. 1987. Chironomidae (Diptera) of peatlands and marshes in Canada. Memoirs of the Entomological Society of Canada 140:141–161.

23 Northern Prairie Marshes (Delta Marsh, Manitoba)

I. Macroinvertebrate Responses to a Simulated Wet/Dry Cycle

HENRY R. MURKIN and LISETTE C. M. ROSS

*T*he Marsh Ecology Research Program (MERP) was a long-term experiment to track the response of the marsh ecosystem to a simulated wet/dry cycle in the Delta Marsh. The Delta Marsh is a large lacustrine marsh on the south shore of Lake Manitoba in south-central Manitoba. The objective of this chapter is to review the changes in aquatic invertebrate distribution and abundance during the various stages of this experimental wet/dry cycle. During the dry stage aquatic invertebrates within the MERP experimental cells were dormant life stages and flying adult insects from the adjacent main marsh. Available habitat structure and food resources changed significantly following reflooding as the cells moved from the regenerating to the degenerating stages of the cycle. Invertebrate densities increased quickly following reflooding and remained fairly constant over the five years of flooding; however, there appeared to be a succession of taxa, with those groups responding to the conditions associated with early flooding giving way to those associated with the more open conditions characteristic of prolonged flooding. Habitat complexity and available food resources are important factors in determining the benthic and epiphytic invertebrate abundance and distribution during the wet/dry cycle in prairie wetlands. Habitat structure provided by marsh vegetation and associated litter combined with algal food resources contributed to the high invertebrate standing crops in these systems.

Invertebrates in Freshwater Wetlands of North America: Ecology and Management, Edited by Darold P. Batzer, Russell B. Rader, and Scott A. Wissinger
ISBN 0-471-29258-3 © 1999 John Wiley & Sons, Inc.

NORTHERN PRAIRIE WETLANDS

Northern prairie wetlands are dynamic systems that fluctuate from drought conditions with no standing water to periods of extended flooding during years of high precipitation (van der Valk and Davis 1978; also see Euliss et al., this volume) (Fig. 23.1). Because hydrologic regime (i.e., the depth and duration of flooding) is the principal factor determining the distribution and abundance of wetland plants (Kantrud et al. 1989), significant changes in the overall productivity and species composition of the vegetation communities occur during this cycle (Spence 1982, van der Valk 1981). During the dry marsh stage that occurs during years of low precipitation, a diverse array of both perennial and annual plant species germinates on the exposed mud flats. Species composition of the developing plant community is determined in part by the seed bank present in the substrate (van der Valk and Davis 1978) and the environmental conditions (e.g., soil temperature and moisture) on the substrate surface (Welling et al. 1988). Upon reflooding in subsequent years, the annual plants are killed by the standing water and the emergent vegetation beds expand through vegetative growth. During this period, referred to as the regenerating stage (van der Valk and Davis 1978), zonation patterns and distinct stands of emergent vegetation characteristic of these wetlands develop. With continued flooding, the wetland eventually enters the degenerating stage as the emergent species begin to die back due to a variety of factors, including intolerance to long-term flooding, disease, and muskrat feeding. At this point the marsh enters the lake marsh stage, where the only remaining vegetation is a narrow emergent fringe around the basin edge and possibly some beds of submersed vegetation in shallow sheltered areas of the basin.

These changes in wetland vegetation communities and associated litter production have direct implications for secondary consumers within these systems. Aquatic invertebrates are important secondary consumers in prairie wetlands (Murkin 1989). High macrophyte production and relatively low rates of direct herbivory in these systems (Murkin 1989, Smith and Kadlec 1985) ensure that considerable plant material enters the water column as detritus. This detritus serves as both habitat and food resources to a wide range of wetland invertebrates (Voigts 1976, Addicott et al. 1987, Murkin 1989). As a result, water level fluctuations associated with the wet/dry cycle will affect the availability and complexity of habitats and food resources provided for aquatic invertebrates by both living and dead plants (Schramm and Jirka 1989).

Algal communities also change considerably over the wet/dry cycle (Robinson et al. 1997a, b). Algae serve as food resources for a wide range of invertebrate species (Murkin 1989). Epipelon (algae associated with the sediment surface) predominate during the dry marsh stage, but only in those areas of the basin where shallow water remains or where surface sediments are saturated (Robinson et al. 1997a). With the return of water to the basin, epiphyton (algae growing on submersed plant surfaces) increases in abun-

REGENERATING MARSH

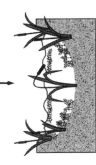

REGENERATING MARSH

Continued Flooding

- Reduction in emergent vegetation densities
- Decrease in epiphytic algae as submersed surfaces decrease
- Increase in metaphytic algae
- High densities of epiphytic invertebrates. These decline as submersed surfaces decline
- Establishment of benthic invertebrates as vegetation disappears

DEGENERATING MARSH

LAKE MARSH

Continued Flooding

- Vegetation from deeper parts of the basin disappears
- Decrease in metaphytic algae, increase in phytoplankton
- Decrease in epiphytic invertebrates, continued increases in benthic invertebrates

Return of Water Following Drawdown

- Die-off of terrestrial vegetation, establishment of aquatic vegetation, increase in epiphytic algae
- Initial response by drought resistant invertebrates and those invertebrates that have active dispersal
- Increase in epiphytic invertebrates in response to submersed habitats and epiphytic algae

DRY MARSH

Drawdown

- Death of remaining aquatic vegetation
- Establishment of epipelon algae on moist mudflat surfaces
- Germination of terrestrial and emergent vegetation
- Invertebrates with active dispersal move out of the wetland
- Drought-resistant invertebrates move to dormant stages
- Elimination of all other aquatic invertebrates
- Colonization by terrestrial invertebrates

Fig. 23.1. Schematic diagram of the wet/dry cycle in prairie wetlands (adapted from classifications devised by van der Valk and Davis 1978).

dance during the regenerating marsh stage in response to the available nutrients and submersed surfaces provided by newly flooded vegetation and plant litter. With continued flooding and the development of more open conditions during the degenerating marsh stage, metaphyton (mats of floating filamentous algae) dominate the algal community. Finally, with the elimination of vegetation from the center of the basin, phytoplankton (those algae free-floating in the water column) predominate. These changes in the algal community represent important changes in the food and cover resources available to aquatic invertebrates in these systems during the wet/dry cycle. An important consideration in assessing the importance of algae in supporting secondary production is that although algal standing crop at any point in time may not be large, production over the growing season can rival that of emergent and submersed vegetation (Murkin 1989, Robinson et al. 1997b). The magnitude of this production, the consistent supply of biomass over time, and the acceptability of that biomass to secondary consumers (Batzer and Wissinger 1996) indicate that algae likely play an important role in prairie wetland trophic dynamics (Robinson et al. 1997b).

In general, our understanding of the impacts of these changes in habitat availability and food resources on wetland invertebrates is limited (Murkin and Kadlec 1986). The objective of this chapter is to review changes in the distribution and abundance of nektonic (free-swimming) and benthic (associated with submersed surfaces) invertebrates in response to an experimental wet/dry cycle in a series of marsh impoundments within the Delta Marsh, a northern prairie marsh in south-central Manitoba. These changes will be related to changes in food, cover, and other resources available during the various stages of the cycle. Elsewhere in this volume, Dale Wrubleski reports on emerging insects from the Delta Marsh, including their response to experimental water level manipulations, and Brenda Hann reports on the distribution and abundance of invertebrates within unmanaged areas of the Delta Marsh.

DELTA MARSH

The Delta Marsh is a 15,000-ha lacustrine marsh on the south shore of Lake Manitoba (50°11′N, 98°19′W) in south-central Manitoba, Canada (Fig. 23.2). The main marsh is separated from Lake Manitoba by a forested barrier-beach ridge. The marsh consists of a matrix of emergent vegetation and open water with large bays (>1000 ha) up to 3 m deep and smaller shallow bays (<1000 ha) that are generally <1 m deep. Smaller isolated openings (1–5 ha, <0.5 m deep) are found in otherwise extensive stands of vegetation. Natural channels are present throughout the marsh, marking historical changes in drainage patterns between the lake and the marsh and the Assiniboine River, 20 km to the south.

The geological history and physiography of the Delta Marsh region is described by Teller and Last (1981), Rannie et al. (1989), and Sproule (1972).

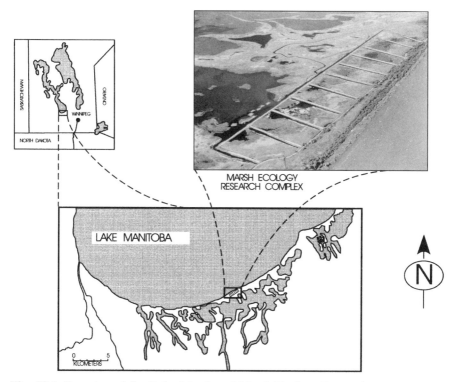

Fig. 23.2. Location of the Delta Marsh and Marsh Ecology Research Program study area, Manitoba, Canada.

The soils are gleysols or regosols (Walker 1965) consisting primarily of un-consolidated thick muck or peat deposits (Erlich et al. 1957, Walker 1959, Shay and Shay 1986) that result from incomplete decomposition of organic material produced in the marsh. The waters are moderately brackish, with conductivities averaging 2600 μS/cm at 25°C (Anderson 1978). pH ranges from 8.2 to 9.0 and total alkalinity averages 337.6 mg/l, largely as bicarbonate ions (Anderson and Jones 1976). Mean monthly temperatures range from −19.8°C in January to 19.1 in July. Mean annual precipitation is 498.6 mm, with 374.7 mm as rain from March to December and 135.0 mm water equiv-alent as snow from October to May.

The vegetation of the Delta Marsh has been described by Love and Love (1954), Walker (1959, 1965), and Anderson and Jones (1976). Changes in emergent vegetation over the past 40 years have been described by deGeus (1987). The dominant emergent vegetation in the deep marsh zone consists of hardstem bulrush (*Scirpus lacustris,* nomenclature follows Scoggan 1978) and cattail (*Typha* spp., primarily *T. latifolia, T. angustifolia,* and their hybrid, *T.* x *glauca*) (Fig. 23.3). Some shorelines exposed to heavy wave action are fringed with giant cane (*Phragmites australis*), but cane is usually found at higher elevations than bulrush or cattail. The shallow marsh zone is dominated

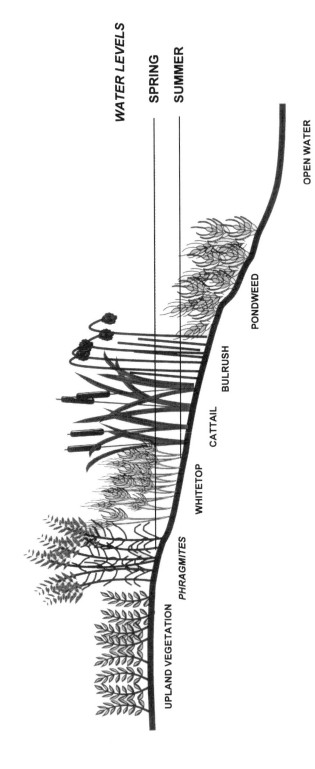

Fig. 23.3. Shoreline transect showing major vegetation zones, Delta Marsh.

by whitetop grass (*Scholochloa festucacea*) and, to a lesser degree, sedge species (*Carex* spp). Further up the elevation gradient, a zone of cane occurs which gives way to a mixture of perennial and annual forbs (Fig. 23.3).

The submersed vegetation of the marsh is surveyed by Anderson and Jones (1976). Three species or species-associations accounted for over 60 percent of the submersed vegetation present at that time. These included sago pondweed (*Potamogeton pectinatus*), sago pondweed/water milfoil (*Myriophyllum exalbescens*), and sheathed pondweed (*Potamogeton vaginatus*). The Delta Marsh also supports significant algal production, with all four algal communities (phytoplankton, epiphyton, epipelon, and metaphyton) represented in different areas of the marsh and during various seasons of the year (Hooper and Robinson 1976, Hooper-Reid and Robinson 1978a, b, Hosseini 1986, Gurney and Robinson 1988, Robinson et al. 1997a, b).

MARSH ECOLOGY RESEARCH PROGRAM

Located on the northern edge of the Delta Marsh, the Marsh Ecology Research Program (MERP) was initiated in 1979 as a joint project of Ducks Unlimited Canada and the Delta Waterfowl and Wetlands Research Station (Murkin et al. 1984). MERP was designed as a long-term replicated experiment to track the response of the marsh ecosystem through a series of water level treatments designed to simulate the wet/dry cycle in northern temperate marshes. Comprehensive long-term monitoring (see Murkin and Murkin 1989) was conducted on water budgets; vertebrates including waterfowl (Anatidae), coots (*Fulica americana*), blackbirds (Icteridae) and muskrats (*Ondatra zibethicus*); invertebrates; N, P, and C budgets; macrophyte and algae production; and macrophyte decomposition. Short-term studies were conducted (usually by graduate students) on other key questions that emerged as the work advanced.

MERP Study Area

The MERP experimental complex consisted of 10 contiguous marsh units or cells (approximately 5 ha each) created by building a series of dikes along the northern edge of the main marsh (Fig. 23.2). A 10-m buffer strip along all dikes and the borrow ditch on the western edge of each cell was observed throughout the study. No sampling took place within these strips to avoid the possible impacts of the dikes or ditch system. It is important to note, especially when considering the invertebrate communities, the presence of the borrow ditch (approximately 2 m deep) within each cell. Although they were not sampled during the study, these ditches held water throughout the study, including the drawdown period (see below). Besides the diked experimental cells, two undiked areas of similar size within the Delta Marsh were designated as controls (i.e., reference areas to the unmanipulated Delta Marsh).

TABLE 23.1. Schedule of Water Levels for MERP Experimental Cells, Delta Marsh, 1980–89

Year	Water Levels
1980	10 cells at normal levels of Delta Marsh (247.50 m AMSL)
1981	8 cells flooded ~1 m above normal (248.41 m AMSL)
	2 cells at normal levels of Delta Marsh (247.50 m AMSL)
1982	10 cells flooded ~1 m above normal (248.41 m AMSL)
1983	8 cells drawn down (247.00 m AMSL)
	2 cells flooded ~1 m above normal (248.41 m AMSL)
1984	10 cells drawn down (247.00 m AMSL)
1985–89	4 cells flooded at shallow level (247.50 m AMSL)
	3 cells flooded at medium level (247.80 m AMSL)
	3 cells flooded at high level (248.10 m AMSL)

Each diked experimental cell was equipped with a stop-log control structure and electric pump with automatic electrode controls to adjust and maintain (± 2 cm) water levels. The vegetation within the cells prior to the experimental treatments was comparable to the main Delta Marsh (Pederson 1981). The vegetation response to water level changes in the cells has been described by van der Valk (1992, 1994), and van der Valk et al. (1994). Effects of water levels on water budgets and water chemistry in the cells have been described by Kadlec (1986a, b, 1989, 1993).

Experimental Design

The basic design was to monitor the response of various components of the wetland ecosystem to a series of water level manipulations intended to simulate the wet/dry cycle of prairie marshes. The schedule for water levels and the number of replicates for each treatment are shown in Tables 23.1 and 23.2. Treatments were randomly applied to the experimental cells. The four phases of the experimental design were:

> *Baseline year:* During the first year of the study (1980), the control structures were left open and water levels within the cells were allowed to

TABLE 23.2. Number of Experimental Cells Within Each of the Water Levels Following Drawdown (1985–89)

	Water Level after Drawdown (AMSL)		
Drawdown Duration (Years)	Shallow (247.50 m)	Medium (247.80 m)	High (248.10 m)
1	2	–	–
2	2	3	3

track the levels of the Delta Marsh. This provided a year of baseline data before the actual water level manipulations began. Two of the cells (3 and 7) were kept at baseline levels for a second year (1981) to allow for comparison of different durations of drawdown later in the study (see Table 23.2).

Deep-flooding: Following the baseline year(s), the cells were flooded to ~1 m above normal for a period of two years (Table 23.1). These levels were designed to flood the existing cattail (*Typha* spp.) stands to a depth of 1 m and were similar to actual natural high-water levels that occurred in the Delta Marsh in the 1950s. This deep flooding was intended to eliminate the emergent vegetation from the central portions of the cells and thereby force them to the "lake marsh" stage by the end of the second year of flooding.

Drawdown: After the second year of deep flooding the cells were drawn down (completely dewatered) by adjusting the water levels to 50 cm below the long-term level of the Delta Marsh (Table 23.1). This was intended to simulate the dry marsh stage of the wet/dry cycle. Most cells were exposed to a two-year drawdown. Two cells (3 and 7) were drawn down for one year to allow comparisons between drawdowns of one- and two-year durations.

Reflooding: Following drawdown the cells were reflooded (Table 23.1) to three different levels (Table 23.2). It has been assumed that the rate at which prairie marshes proceed through the various stages of the wet/dry cycle is determined by water depth, with deeper marshes moving to the lake marsh stage more quickly than shallow-flooded marshes (van der Valk and Davis 1978).

MERP Invertebrate Sampling

Aquatic invertebrates were sampled throughout the MERP experiment. A variety of sampling techniques and sampling designs were used over the course of the MERP studies (see Ross and Murkin 1989). Details on specific techniques and sampling protocols are provided in the papers cited below. As mentioned earlier, no sampling took place within 10 m of a dike or the borrow ditch within each cell. In general, invertebrate sampling within the experimental cells and undiked reference areas were stratified by vegetation type based on the dominant emergent plants; for example, cattail (*Typha glauca*), hardstem bulrush (*Scirpus lacustris glaucus*), open water (standing water with no emergent vegetation). Because the dominant vegetation type varied over the course of the study, the number of sampling stations and vegetation types sampled varied as well. During the reflooding years (1985–89) sampling was also stratified by water depth (30 cm and 60 cm) within a vegetation type.

A variety of samplers were used to collect invertebrates within the nektonic, benthic, epiphytic, and emerging-insect groups. Early in the study, benthic samples were taken with a core sampler (see Murkin and Kadlec 1986). Activity traps were used at sampling sites with standing water to provide an index to the abundance of free-swimming invertebrates (nekton) (Murkin et al. 1983). Artificial substrates were placed at each sampling station with standing water to provide an index to benthic invertebrates (primarily Gastropoda and Chironomidae) colonizing submersed surfaces (Ross and Murkin 1989). Floating emergence traps were used to sample emerging insects and are described by Wrubleski (this volume).

All macroinvertebrates in the samples were sorted by taxa and counted. In most cases identification was to family. Notable exceptions occur in the Chironomidae, which were identified to subfamily. For many of the lower invertebrate groups (e.g., Oligochaeta, Nematoda) identification was to order. Biomass was determined by drying subsamples of each taxon and extrapolating to the entire sample (see Ross and Murkin 1989).

MACROINVERTEBRATE RESPONSES TO THE MERP WET/ DRY CYCLE

The Dry Marsh Stage

With the lack of water during the MERP dry marsh stage, aquatic invertebrates within the experimental cells were dormant life stages and flying adults (e.g., mosquitoes and chironomids) from the adjacent main marsh. Under normal conditions prairie marshes dry slowly during periods of low precipitation combined with losses by evapotranspiration (van der Valk and Davis 1978). As a result, those invertebrates dependent on standing water are concentrated in the remaining pools. In the MERP experimental marshes increased densities of aquatic invertebrates (primarily chironomid larva) were observed in the remaining pools as the marshes were pumped down. These concentrations of invertebrates attracted a variety of birds. An exclosure experiment during this period indicated that avian feeding activities reduced invertebrate densities in the remaining pools to the same levels as in the adjacent habitats (Peterson et al. 1989).

With the removal of water from the MERP experimental cells during the dry stage, those species (e.g., amphipods) that require standing water were eliminated. Many other species of aquatic invertebrates have the ability to survive dry periods through a variety of mechanisms (Wiggins et al. 1980, Neckles et al. 1990). For example, cladocerans and other crustaceans respond to falling water levels and/or crowding in the remaining pools by producing resting eggs (Wiggins et al. 1980). These drought-resistant eggs will remain dormant until the environmental stimuli associated with reflooding cause them to hatch. In another study in the Delta Marsh, species like *Daphnia pulex* were reduced when dry periods did not occur in experimental impoundments, thereby eliminating appropriate conditions (drying followed by reflooding) to allow the resting eggs to hatch (Neckles et al. 1990). Adult females of some mosquito (Culicidae) species lay their eggs on the litter and substrate in dry basins (Merritt and Cummins 1984), and the eggs then hatch upon reflooding. Neckles et al. (1990) found that *Aedes* sp. mosquitoes were eliminated from Delta Marsh impoundments by continuous flooding because dry egg-laying sites were not available. Some species survive the dry period as adults. Snails (Gastropoda) bury in the available litter and substrate and form mucilaginous sheaths over the shell opening to prevent drying.

The longevity of the dormant forms of the various invertebrates present in the dry basin is unknown. It is likely that there is some difference among the various species (e.g., the resting eggs of cladocerans versus dormant snails); however, there is little data available at the present time. Some snail species have been known to survive dry conditions for up to three years (Pennak 1978). During the MERP study there was no difference in the invertebrate response to the one- or two-year drawdowns (Murkin and Ross, unpublished data), indicating that most common species in the Delta Marsh can survive dry conditions for up to two years. Alternatively, it may indicate that the rate of recolonization was so rapid that the survival of existing invertebrates has little impact on subsequent population levels upon reflooding.

Terrestrial and semiterrestrial invertebrates will colonize the basin during the dry marsh stage (Murkin 1989). During the early stages of drawdown, rapid germination and growth of plants on the exposed substrates will provide both food and habitat for herbivorous terrestrial invertebrates. The algal community will be restricted to epipelon in low-lying moist areas of the basin (Robinson et al. 1997a) and would provide some food resources for terrestrial and semiterrestrial invertebrates within the basin. With continued dry conditions, plant litter will also become available to terrestrial detritivores.

The Regenerating/Degenerating Stages

Extensive stands of vegetation, both annual and perennial species, developed on the exposed substrates of the MERP experimental cells during the drawdown period (Welling et al. 1988, Merendino et al. 1990). With the return of standing water during the reflooding stage (1985–89) (Table 23.1), submersed leaves and stems of these plants provided diverse habitats for aquatic invertebrates. Five vegetation types (rayless aster, *Aster laurentianus;* red goosefoot, *Chenopodium rubrum;* whitetop, *Scolochloa festucacea;* softstem bulrush, *Scirpus lacustris validus;* cattail, *Typha* spp.) dominated the MERP cells during the early stages of reflooding. Based on their morphology, these vegetation types provided very different submersed habitats when initially flooded. With continued flooding, these habitats changed over time as the litter decomposed and the existing vegetation was replaced by plant species more tolerant of flooded conditions.

The depth of reflooding influenced the rate of these changes by affecting the rate of litter decomposition (Wrubleski, unpublished data), the species of vegetation surviving at a particular water depth, and the plant species able to colonize the sites following flooding (Squires and van der Valk 1992). Changes in the species composition and extent of the emergent vegetation stands during this period of the study are reported by Clark and Kroeker (1993) (includes vegetation cover maps) and Squires and van der Valk (1992). Basically, the five dominant vegetation types fell into three distinct groups: species which were unable to survive flooding (*Chenopodium rubrum* and *Aster laurentianus*), species whose peak growth is at shallow flooding depths (approximately 20 cm) (*Scolochloa festucacea* and *Scirpus lacustris validus*), and species whose peak growth is at slightly greater depths (45 cm deep) (*Typha glauca*) (van der Valk 1994). Emergent vegetation replaced the annual vegetation killed by flooding, especially in shallow areas of the cells. Initially, the emergent vegetation stands expanded through vegetative growth. However, with continuous flooding over the five years, plant densities in all areas of the cells began to decline. At the deep invertebrate sampling stations (60 cm) most of the emergent vegetation was eliminated after two to three years of flooding, while at the shallow sites (30 cm) there was some growth of emergent vegetation throughout the five years, albeit at much reduced densities in years 4 and 5.

Litter decomposed at a faster rate in areas flooded at 60 cm than in areas flooded at 30 cm (Wrubleski, unpublished data). After two years less than 20 percent of whitetop remained as fallen litter in the deeper areas, compared to 50 percent remaining in the shallower areas. Cattail, which decomposes more slowly than the other emergent species present, had 70 percent of its litter remaining after two years in shallow areas and 50 percent remaining in the deeper-flooded areas. Slower elimination and decomposition of plants at shallow-flooded sites meant that these areas remained in the regenerating stage longer than deep-flooded sites, which began to open up (i.e., enter the degenerating stage) two to three years after initial reflooding.

As emergent vegetation died back, first in 60-cm sites and later in 30-cm sites, increases in available light resulted in the development of both submersed vegetation and algal communities within the experimental cells. Extensive beds of submersed vegetation (primarily *Potamogeton pectinatus* and *Utricularia vulgaris)* developed among the surviving emergent vegetation at the shallow sites, while no or very sparse beds developed at the deep sites. Epiphytic algal biomass was initially higher in the deeper sites (Robinson et al. 1997a). However, by the end of the study epiphytic biomass was higher in the shallow sites, where both emergent and submersed vegetation were still available as a substrate. Metaphytic algae was slightly more productive in the shallower sites (Robinson et al. 1997b), where it was protected from wind action by emergent vegetation. Phytoplankton dominated deeper habitats once the emergent and submersed vegetation was eliminated from these sites.

In summary, habitat structure and available food resources changed significantly following reflooding as the cells moved through the regenerating and degenerating stages of the wet/dry cycle. The rate at which the cells moved through the cycle was influenced by water depth, with the shallow sites not opening up (i.e., entering the degenerating stage) until near the end of the five years of flooding. Deep sites moved through the regenerating stage quickly, entering the degenerating stage after two or three years of flooding.

An important step in the change from the regenerating to the degenerating stage is the hemi-marsh condition, where emergent vegetation and open water are available in approximately equal amounts in an interspersed pattern. The hemi-marsh has been shown to be the period of the marsh cycle supporting the most abundant and diverse bird communities (Weller and Spatcher 1965, Weller and Fredrickson 1974). High avian use of the hemi-marsh has been linked to the cover (Murkin et al. 1982) and food resources (primarily aquatic invertebrates) (Kaminski and Prince 1981) provided during this point in the wet/dry cycle. During the hemi-marsh stage, cover and food resources available to aquatic invertebrates will include those from both the regenerating stage (emergent vegetation/litter and epiphytic algae) and the degenerating stage (open water, unvegetated sediments, and phytoplankton), resulting in a diverse array of habitat conditions.

Nekton. Following the reflooding of the MERP cells in 1985, five taxa accounted for 85 percent of the nektonic invertebrates identified and 65 percent of the total nekton biomass. These dominant taxa were Cladocera (Order Cladocera), Copepoda (Order Eucopepoda), Corixidae (Order Hemiptera), Dytiscidae (Order Coleoptera), and *Hyalella azteca* (Order Amphipoda) (Murkin and Ross, unpublished data). Invertebrates that are drought-resistant with resting stages remaining within the dry basin will be among the first taxa groups to respond to reflooding and the early conditions in regenerating wetlands. This would account for the presence of cladocerans and copepods in the MERP cells soon after reflooding. Invertebrates that can actively disperse as flying adults, such as the hemipterans and coleopterans found early in the MERP reflooding experiment, will also be among the taxa to colonize newly flooded habitats (Fernando and Galbraith 1973). Corixids are common inhabitants of seasonally flooded ponds on the prairies (Neckles et al. 1990). Those taxa that require permanent water for development and overwintering, such as amphipods, will normally enter reflooded systems through passive dispersal (e.g., attachment to birds and mammals moving among wetlands (Swanson 1984)) or through connecting waterways and overland flow from permanent wetlands in the area. Surrounding deep wetlands and other permanent water areas provide important refugia for aquatic invertebrates during drought periods (Bataille and Baldassarre 1993). The rapid appearance of amphipod populations in the MERP experimental cells was likely the result of refugia provided by the borrow ditches within the cells. As mentioned earlier, these ditches had standing water throughout the drawdown period. The role of ditches in the survival and recolonization of other taxa is unknown.

Although significant changes in habitat occurred over time and between shallow and deep areas, total nektonic invertebrate levels remained fairly constant over the five years of flooding (Fig. 23.4). By years 4 and 5 of flooding, the spring peaks in abundance were somewhat reduced in the deeply flooded areas compared to the more shallowly flooded sampling sites. Although invertebrate densities remained relatively stable, a succession of taxa appears to have taken place as those groups responding to conditions of early flooding

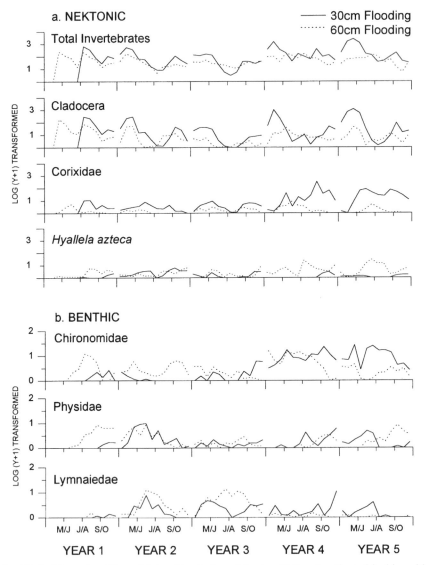

Fig. 23.4. Mean densities of (*a*) nektonic invertebrates (#/trap) collected in biweekly activity trap samples; (*b*) benthic invertebrates (#/sample) colonizing submersed artificial substrates in the MERP experimental cells, 1985–89.

gave way to those groups associated with more open conditions characteristic of prolonged flooding.

Cladocerans were the most abundant invertebrate group in the MERP cells and among the first to show significant increases in abundance during reflooding (Fig. 23.4). Cladocerans responded earlier in the deeply flooded sites in all vegetation types, indicating that the early spring conditions associated with deep flooding were more conducive to hatching of resting eggs. These conditions are likely related to water temperature and oxygen concentrations. In general, cladoceran levels were higher in the shallow sites throughout much of the five years of flooding, with the differences becoming more pronounced later in years 4 and 5. Shallow sites with the surviving emergent vegetation and developing beds of submersed vegetation appear to provide more suitable habitat than the deeper sites with little emergent and submersed vegetation present. Murkin et al. (1991) and Rabe and Gibson (1984) also found that cladocerans were more abundant within emergent vegetation stands than in open-water areas with little or no submersed vegetation. They suggest that the habitat structure provided by the submersed vegetation provides refuge from predators. Food is not likely an important factor determining cladoceran distribution because they are primarily planktivores and plankton populations would be higher in the more open unvegetated sites (Robinson et al. 1997a). The importance of avoiding predators is shown by physical adaptations of cladocerans that allow them to survive under the anoxic conditions which can often occur in heavily vegetated areas (Murkin et al. 1992). Certain species are able to increase the production of haemoglobin (Engle 1985), while others respond by decreasing locomotor activity, feeding rates, and energy output for reproduction. Reduced predation by fish and invertebrate predators which cannot tolerate low oxygen levels can result from the use of such habitats (Bennett and Streams 1986, Suthers and Gee 1986).

Corixidae levels in the MERP experimental cells remained fairly low until the submersed vegetation developed in the shallow sites during years 4 and 5 of flooding (Fig. 23.4). Murkin et al. (1991) also found higher densities of corixids in stands of submersed vegetation than in either open-water sites or stands of emergent vegetation. Primarily predators (Reynolds and Scudder 1987), corixids are likely responding to the habitat structure provided by the vegetation (an important feature to wait-and-pounce predators) and invertebrate prey (e.g., cladocerans) associated with the submersed vegetation. We observed corixids feeding on cladocerans in the Delta Marsh. It appears that corixids require the shelter and food provided by a mix of submersed and emergent vegetation to reach maximum population levels (Murkin 1983).

Like corixids, *Hyalella azteca* did not show significant increases in abundance until later in the flooding experiments (Fig. 23.4). Their densities, however, were higher in the deeper flooded areas during years 4 and 5, when much of the emergent and submersed vegetation had been eliminated and the sites were in the degeneration stage of the marsh cycle. *Hyalella azteca* are deposit feeders, feeding primarily on epiphytic and epibenthic algae associated with particulate litter (Hargrave 1970a, de March 1981). They ingest fine detrital particles and digest the algae and bacteria within the litter (Hargrave 1970b). As marshes moved into the early stages of degeneration cycle, the particle size produced from the decomposing vegetation would be fairly coarse. By years 4 and 5 of the flooding cycle the particle size would become much finer through decomposition and invertebrate processing and, in turn, more acceptable to *H. azteca* as a food source.

Benthos. Chironomidae and the three gastropod families, Lymnaeidae, Planorbidae, and Physidae, were the dominant benthic taxa sampled by the artificial substrates in the MERP experimental cells during the reflooding period 1985–89 (Murkin and Ross, unpublished data). Chironomids were by far the most abundant benthic taxa (Fig. 23.4), and therefore the total invertebrate response was similar to that of chironomids. Chironomids comprise the highest abundance and biomass in the benthos of the Delta Marsh in general (Wrubleski and Rosenberg 1990) and respond quickly to changes in water levels and available habitat (Murkin and Kadlec 1986). The chironomid response was earlier in the deeper (60 cm) sampling sites, probably due to greater amounts of submersed litter providing food and habitat and increasing availability of epiphytic algae in the newly flooded cells. The first chironomid taxa to respond in the deep sites were small epiphytic species colonizing the submersed litter (Wrubleski 1991). Later in the study, as the emergent litter was lost at the deep sites and with no submersed vegetation development, the dominant chironomids were larger benthic forms found in the bottom substrates. The chironomid response at the shallow sites was delayed compared to the deep sites and remained fairly low until later in the study (years 4 and 5). The dense vegetation and subsequent shading and therefore low algal abundance at these sites may have limited the chironomid response. Later, as the vegetation began to open up and beds of submersed vegetation developed, the chironomid densities increased significantly. The dominant chironomid species during these later years in the shallow sites were the small epiphytic species (Wrubleski 1991).

It is interesting to note that following the reflood in 1985, snail abundance was highest initially in those vegetation types that were flooded prior to the drawdown (i.e., the aquatic emergents, such as cattail, whitetop, and bulrush) (Murkin and Ross, unpublished data). As mentioned earlier, snails survive periods of drought by burrowing into the substrate and secreting a mucilaginous plug over their opening. When water returns, the plug is shed and the snail resumes activity. It appears that residual populations of snails present in habitats with aquatic vegetation before the drawdown resulted in a quicker response in those habitats following reflooding.

The response of the three gastropod families found in the MERP cells varied (Fig. 23.4). Physid snails responded very quickly to the reflooded deep sites, while both lymnaeids and planorbids did not respond to either flooding depth until the second year of flooding. Food resources likely played a key role in the gastropod response to the MERP reflooding. Functionally, physids, lymnaeids, and planorbids are classified as scrapers (Nelson 1982). Physids feed on epiphytic algae and bacteria attached to submersed vegetation (DeWitt 1955, Swamikannu and Hoagland 1989). In the MERP cells they were likely responding to submersed surfaces in the deeply flooded sites and the epiphytic algae associated with those surfaces. The delayed response of lymnaeids in the MERP cells may be a result of their preference for conditioned plant litter as a food resource (Bovjerg 1968). By the second year of flooding abundant conditioned litter was available in the MERP cells. This supply of litter would decrease over time as emergent plant production decreased with continued flooding and through decomposition of existing litter supplies, especially in the deeply flooded sites. As a result, the lymnaeid densities declined, especially in the deep sites.

Summary—Regenerating/Degenerating Stages

Habitat complexity and available food resources are important factors in determining benthic and epiphytic invertebrate abundance and distribution during the wet/dry cycle in prairie wetlands (Murkin 1989). The habitat structure provided by marsh vegetation and associated litter, combined with algal food resources, contribute to the high invertebrate standing crops in these systems. Changes in habitat structure and litter/algal food resources during the regenerating/degenerating stages had a marked effect on benthic and nektonic invertebrate populations within the MERP cells. Some taxa were adapted to the early stages of flooding (regenerating stage) (e.g., cladocerans and physid snails), while others increased in abundance later in flooding (degenerating stage) (e.g., amphipods, benthic chironomids). The hemi-marsh stage (i.e., the transition from regenerating to degenerating conditions) had diverse habitat conditions supporting the widest range of invertebrate taxa during the wet/ dry cycle. The high abundance and diversity of invertebrates typically found in interspersed marsh habitats is often attributed to the amount of emergent vegetation-open water interface (i.e., edge) in these habitats. However, Murkin et al. (1992) found little relationship between invertebrate densities and diversity and the amount of edge per se in various areas of the Delta Marsh. Rather, they suggest that the high invertebrate abundance and diversity in these interspersed habitats is the result of the mixture of habitat types present rather than the amount of edge. Each habitat type contributes its unique set of environmental conditions and invertebrate taxa, resulting in a diverse community within the wetland. The hemi-marsh stage, with its diverse habitat conditions as the wetland moves from the regenerating to the degenerating stages, would therefore support an abundant and diverse invertebrate community.

The invertebrate community during the hemi-marsh stage plays an important role in the high avian use and abundance observed during this stage of the marsh cycle. Most avian species in prairie marshes feed on invertebrates at various times of the year (Swanson and Meyer 1973, Murkin and Batt 1989). Different species feed on different invertebrates (Weller 1981), and within individual species a wide range of invertebrates are included in the diet (Krapu and Reinecke 1992). As a result, wetlands with a diversity of habitat types and associated invertebrate communities would support a greater abundance and diversity of marsh birds and other animals that feed on aquatic invertebrates (Murkin et al. 1982). It is important to note that as the MERP cells moved from the regenerating stage to the degenerating stage, the spring peaks in invertebrate numbers declined. This reduction in spring peaks has important implications for waterfowl that feed on invertebrates in the spring to meet the protein requirements of gonadal development and egg-laying (Murkin and Batt 1989).

The Lake Marsh Stage

With continued flooding, the marsh moves eventually to the lake marsh stage (Fig. 23.1). The submersed habitat available for aquatic invertebrates is reduced to remnant stands of submersed vegetation in sheltered areas of the marsh, the thin strip of emergent vegetation at the pond margin, and the unvegetated substrate in the deeper portions of the basin (see Murkin et al. 1992). The decline in emergent vegetation and the associated coarse litter production will affect those invertebrate groups that use emergent litter as habitat (e.g., burrowing chironomid larvae) and food (i.e., shredders in the detritivore functional group). With extended flooding, the available litter size will continue to be reduced through decomposition and processing by detritivores. The resulting fine organic particulate matter will serve as a food source for collector and filterer invertebrate groups, but its availability will also decline over time as processing and decomposition continue. Eventually the more recalcitrant material will be incorporated into the bottom sediments.

Besides the changing detrital resources, the available algal standing crops will change during the lake marsh stage. Loss of submersed surfaces provided by emergent vegetation and its associated litter would reduce the available habitat for epiphytic algal production. The dominant algae during the lake marsh stage is phytoplankton (Robinson et al. 1997a) and possibly epipelon in shallow unvegetated areas of the wetland. Robinson et al. (1997b) report that phytoplankton production in prairie wetlands is generally quite low compared to other algal groups and emergent vegetation. Epipelon reaches its maximum production in open pond situations with soft sediments and few macrophytes to shade the substrate. Although the epipelon does not have a large standing crop at any point in time, its rapid turnover rate results in high annual production, which in turn can support benthic invertebrates like chironomids (Murkin 1989).

The deep-flooding part of the MERP experiments was intended to set the experimental cells to the lake marsh stage; however, the first year of deep flooding was uncharacteristic of lake marsh conditions because of the large amount of litter present in the water column when the cells were initially flooded. During the second year of deep flooding, once a large portion of the litter had decomposed and much of the emergent vegetation had been killed, conditions would be more representative of the lake marsh stage of the wet/dry cycle.

Changes in available habitat and food resources provided by the plant litter and emergent vegetation will have a major impact on invertebrate populations during the deep flooding and subsequent development of the lake marsh stage. Nekton respond directly to litter resources and algae within the water column. Cladoceran levels increased during the first year of deep flooding and did not show the midsummer decline observed during the years with normal flooding (Fig. 23.5). The most common species of cladocerans in the Delta Marsh are within the planktivore group (Smith 1968) and therefore are likely responding to algae during this period. The phytoplankton levels were actually lower during this first year of flooding compared to the baseline year (Hosseini 1986). This may have been the result of grazing pressures from the high cladoceran populations during this period. Dense populations of cladocerans can potentially graze over 100 percent of the daily phytoplankton production (Porter 1977). Phytoplankton levels decreased further during the second year of deep flooding (Hosseini 1986), as did cladoceran levels (Fig. 23.5). Cladoceran levels during the lake marsh stage would appear to track phytoplankton levels, which are likely to be low in large, windswept (i.e., turbid) wetlands like the Delta Marsh.

Hyalella azteca levels declined in response to deep flooding and development of lake marsh conditions (Fig. 23.5). With the elimination of emergent and submersed vegetation and the subsequent loss of available litter through decomposition, *H. azteca* levels became very low during the second year of deep flooding and were much lower than found in the regenerating/degenerating stages of the MERP experiments. The lake marsh stage, with its open-water areas with little emergent or submersed plant litter production, appears to support low populations of this amphipod species. It is important to note that *H. azteca* levels were different between the degenerating stage and the lake marsh stage in the MERP experiments. The litter production during the degenerating stage as the emergent vegetation died back appears to provide conditions suitable for *H. azteca* production and survival. Later in the lake marsh stage, as emergent litter production was eliminated and the remaining litter was lost through decomposition, *H. azteca* numbers declined.

The loss of submersed surfaces will affect benthic invertebrates developing during the lake marsh stage (Wrubleski 1987). With increases in unvegetated substrates, the dominant invertebrate taxa during this stage are large benthic chironomid species. Both Driver (1977) and Wrubleski (1991) found that benthic chironomids increased with the increase in unvegetated substrate in prairie wetlands. Murkin and Kadlec (1986) found that large benthic chironomids became established when two years of above-normal flooding (1981–82) eliminated the emergent vegetation in the MERP cells, and were

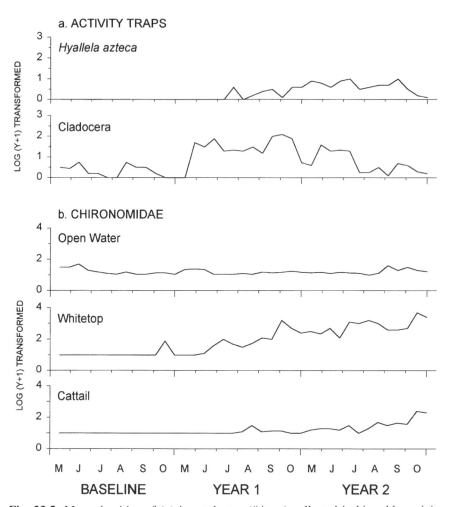

Fig. 23.5. Mean densities of (*a*) invertebrates (#/trap) collected in biweekly activity trap samples (summary of data from Murkin et al. 1991); (*b*) chironomids (#/sample) in core samples (summary of data from Murkin and Kadlec 1986) from the MERP experimental cells, 1980–82.

still increasing after two years when the cells had effectively entered the lake marsh stage (Fig. 23.5).

This chironomid response to deep flooding was not consistent across the various vegetation types. Murkin and Kadlec (1986) argue that differences were likely due to the amount and availability of dead roots and rhizomes within the substrates of the experimental cells. Deep flooding of the MERP cells eliminated much of the above-ground macrophyte production and eventually resulted in the death of below-ground components as well (Murkin and Kadlec 1986). Belowground biomass of aquatic macrophytes is quite high

(Hackney and de la Cruz 1980), and death of the plants would result in considerable below-ground material becoming available as both food and habitat. During deep flooding of the MERP cells chironomid levels did not increase in the previously open-water areas of the cells (Fig. 23.5). These areas did not support emergent vegetation prior to flooding and therefore did not have a large below-ground biomass available following flooding. In areas that supported emergent vegetation prior to flooding, chironomid levels did increase, but the timing of the response varied among vegetation types (Fig. 23.5). In former whitetop sites where the vegetation died very quickly after flooding (Murkin 1983), the chironomid response occurred soon after inundation. At cattail sites there was some initial survival and growth of the plants the first spring after flooding, but most of the plants died later that summer. At these sites the chironomid response was delayed until well into the second year of flooding. It appears that resources and habitat provided by the death of the below-ground roots and rhizomes played a role in the distribution and abundance of benthic chironomids during this period of deep flooding. With prolonged flooding and the subsequent decomposition of the below-ground components during development of lake marsh conditions, chironomid populations would likely decline in the former emergent vegetation areas until they were similar to the population levels found in open-water areas.

The Delta Marsh in recent times has been in the lake marsh stage. The marsh is connected to Lake Manitoba by several channels, and therefore water levels in the marsh are influenced directly by the lake. Prior to 1961, Lake Manitoba fluctuated within the prevailing climatic conditions, resulting in a 2.1-m range in water levels. These fluctuations resulted in the Delta Marsh experiencing the classic wet/dry cycle. In 1961 a dam was constructed on the Fairford River at the outlet of Lake Manitoba, and it has effectively stabilized water levels in the lake since that time. As a result, the Delta Marsh has proceeded to the lake marsh stage and stopped. The invertebrate communities in the marsh are characteristic of the lake marsh stage described above, with large benthic-dwelling chironomids dominating the invertebrate fauna of the large open bays. Hann (this volume) reviews invertebrate interactions in the current-day Delta Marsh.

SUMMARY AND RESEARCH NEEDS

Figure 23.1 presents a summary of the changing habitat conditions during the wet/dry cycle in prairie marshes. The changing habitat structure and available food resources during the various stages of the cycle have a major impact on the invertebrate community development. The high diversity and abundance of invertebrates found in these systems are, in part, a direct result of the cycle. Stable water levels will result in lake marsh conditions and relatively low invertebrate diversity and abundance associated with this stage of the marsh cycle. Only periodic drying to reestablish emergent vegetation in the central parts of the basin and subsequent reflooding will provide the habitat conditions for maximum invertebrate diversity and abundance.

Many factors will influence the invertebrate response to the wet/dry cycle. Our results are based on the conditions created by diking off areas of the Delta Marsh. There will be many variations in the invertebrate response to this cycle in other areas of the prairies. The water of the Delta Marsh is moderately brackish. Elevated salinity can play a major role in determining invertebrate community development in a particular wetland (see Euliss, this volume). The availability of adjacent permanent water areas will also affect invertebrate recolonization rates during the regenerating stage. Seed banks in basins will vary (van der Valk and Davis 1978), affecting the vegetation response to drawdown and the subsequent habitat structure available to aquatic invertebrates upon reflooding. Fish populations in the MERP experimental cells were low or nonexistent in some years. The presence of fish will certainly have an impact on the development of invertebrate populations (Hanson and Riggs 1995). In summary, the dynamic nature of the prairie environment will result in a great deal of variability in the invertebrate response to the wet/dry cycle.

Future research needs to address the many limitations affecting invertebrate research in prairie wetlands. Basic life history and taxonomic information is lacking for a wide range of invertebrate groups in these systems. The lack of basic ecological information, such as the trophic status of individual species, limits work on nutrient and energy budgets in these habitats. Investigation into the importance of algal resources as food for invertebrates and the role of aquatic invertebrates in litter processing will lead to a better understanding of the structure and function of these dynamic systems. An important consideration in all studies in prairie wetlands is the duration of the study. As shown in this chapter, conditions and invertebrate communities change dramatically over the wet/dry cycle, and therefore studies on invertebrates in prairie wetlands must include information on the stage of the wet/dry cycle. Important advances will come through long-term experiments that take into account the dynamic nature of the prairie environment and the spatial heterogeneity of prairie wetlands. Finally, the large scale of the MERP experiments precluded, in large part, isolation of mechanisms determining invertebrate distribution and abundance within the experimental cells. Experiments to elucidate these mechanisms in prairie wetlands are also required.

ACKNOWLEDGMENTS

This is Paper No. 90 of the Marsh Ecology Research Program, a joint project of Ducks Unlimited Canada and the Delta Waterfowl and Wetlands Research Station. Preparation of this contribution was made possible by Ducks Unlimited Canada through the Institute for Wetland and Waterfowl Research.

LITERATURE CITED

Addicott, J. F., J. M. Aho, M. F. Antolin, D. K. Padilla, J. S. Richardson, and D. A. Soluk. 1987. Ecological neighborhoods: Scaling environmental patterns. Oikos 49: 340–346.

Anderson, M. G. 1978. Distribution and production of sago pondweed *Potamogeton pectinatus* L. on a northern prairie marsh. Ecology 59:154–160.

Anderson, M. G., and R. E. Jones. 1976. Submerged Aquatic Vascular Plants of East Delta Marsh. Manitoba Department of Renewable Resources and Transportation Services Wildlife Report. Winnipeg, MB.

Bataille, K. J., and G. A. Baldassarre. 1993. Distribution and abundance of aquatic macroinvertebrates following drought in three prairie pothole wetlands. Wetlands 13:260–269.

Batzer, D. P., and S. A. Wissinger. 1996. Ecology of insect communities in nontidal wetlands. Annual Review of Entomology 41:75–100.

Bennett, D. V., and F. A. Streams. 1986. Effects of vegetation on Notonecta (Hemiptera) distribution in ponds with and without fish. Oikos 46:62–69.

Bovjerg, R. V. 1968. Responses to foods in lymnaeid snails. Physiological Zoology 41:412–423.

Clark, W. R., and D. W. Kroeker. 1993. Population dynamics of muskrats in experimental marshes at Delta, Manitoba. Canadian Journal of Zoology 71:1620–1628.

de Geus, P. M. 1987. Vegetation changes in the Delta Marsh, Manitoba between 1948–1980. Master's Thesis, University of Manitoba, Winnipeg, MB.

de March, B. G. E. 1981. *Hyalella azteca* (Saussure). Pages 61–77 in S. G. Lawrence (ed.), Manual for the Culture of Selected Freshwater Invertebrates. Canadian Special Publications of Fisheries and Aquatic Sciences 54.

DeWitt, R. M. 1955. The ecology and life history of the pond snail *Physa gyrina*. Ecology 36:40–44.

Driver, E. A. 1977. Chironomid communities in small prairie ponds: Some characteristics and controls. Freshwater Biology 7:121–133.

Engle, D. E. 1985. The production of haemoglobin by small pond *Daphnia pulex:* Intraspecific variation and its relation to habitat. Freshwater Biology 15:631–638.

Erlich, W. A., E. A. Poyser, and L. E. Pratt. 1957. Report on the reconnaissance soil survey of Carberry map sheet area. Manitoba Soils Survey, Soils Report 7:1–93.

Fernando, C. H., and D. Galbraith. 1973. Seasonality and dynamics of aquatic insects colonizing small habitats. Internationale Vereinigung für Theoretische und Angewandte Limnologie, Verhandlungen 18:1564–1575.

Gurney, S. E., and G. G. C. Robinson. 1988. VII. The influence of water level manipulation on metaphyton production in a temperate freshwater marsh. Internationale Vereinigung für Theoretische und Angewandte Limnologie, Verhandlungen 23: 1032–1040.

Hackney, C. T., and A. A. de la Cruz. 1980. In situ decomposition of roots and rhizomes of two tidal marsh plants. Ecology 61:226–231.

Hanson, M. A., and M. R. Riggs. 1995. Potential effects of fish predation on wetland invertebrates: A comparison of wetlands with and without fathead minnows. Wetlands 15:167–175.

Hargrave, B. T. 1970a. The distribution, growth, and seasonal abundance of *Hyalella azteca* (Amphipoda) in relation to sediment microflora. Journal of the Fisheries Research Board of Canada 27:685–699.

———. 1970b. The utilization of benthic microflora by *Hyallela azteca* (Amphipoda). Journal of Animal Ecology 39:427–437.

Hooper, N., and G. G. C. Robinson. 1976. Primary production of epiphytic algae in a marsh pond. Canadian Journal of Botany 54:2810–2815.

Hooper-Reid, N., and G. G. C. Robinson. 1978a. Seasonal dynamics of epiphytic algal growth in a marsh pond: Composition, metabolism and nutrient availability. Canadian Journal of Botany 56:2441–2448.

———. 1978b. Seasonal dynamics of epiphytic algal growth in a marsh pond: Productivity, standing crop and community composition. Canadian Journal of Botany 56:2434–2440.

Hosseini, S. M. 1986. The effects of water level fluctuations on algal communities of freshwater marshes. Ph.D. Dissertation, Iowa State University, Ames, IA.

Kadlec, J. A. 1986a. Effects of flooding on dissolved and suspended nutrients in small diked marshes. Canadian Journal of Fisheries and Aquatic Sciences 43:1999–2008.

———. 1986b. Input-output nutrient budgets for small diked marshes. Canadian Journal of Fisheries and Aquatic Sciences 43:2009–2016.

———. 1989. Effects of deep flooding and drawdown on freshwater marsh sediments. Pages 127–143 *in* R. R. Sharitz and J. W. Gibbons (eds.), Freshwater Wetlands and Wildlife. DOE Symposium Series No. 61. USDOE Office of Scientific and Technical Information, Oak Ridge, TN.

———. 1993. Effect of depth of flooding on summer water budgets for small diked marshes. Wetlands 13:1–9.

Kaminski, R. M., and H. H. Prince. 1981. Dabbling duck and aquatic macroinvertebrate responses to manipulated wetland habitat. Journal of Wildlife Management 45:1–15.

Kantrud, H. A., J. B. Miller, and A. G. van der Valk. 1989. Vegetation of wetlands of the prairie pothole region. Pages 132–187 *in* A. G. van der Valk (ed.), Northern Prairie Wetlands. Iowa State University Press, Ames, IA.

Krapu, G. L., and K. J. Reinecke. 1992. Foraging ecology and nutrition. Pages 1–29 *in* B. D. J. Batt, A. D. Afton, M. G. Anderson, C. D. Ankney, D. H. Johnson, J. A. Kadlec, and G. L. Krapu (eds.), Ecology and Management of Breeding Waterfowl. University of Minnesota Press, Minneapolis, MN.

Love, A., and D. Love. 1954. Vegetation of a prairie marsh. Bulletin of the Torrey Botanical Club 81:16–34.

Merendino, M. T., L. M. Smith, H. R. Murkin, and R. L. Pederson. 1990. The response of prairie wetland vegetation to seasonality of drawdown. Wildlife Society Bulletin 18:245–251.

Merritt, R. W., and K. W. Cummins. 1984. An Introduction to the Aquatic Insects of North America. Kendall/Hunt, Dubuque, IA.

Murkin, E. J., and H. R. Murkin. 1989. Marsh Ecology Research Program long-term monitoring procedures manual. Delta Waterfowl and Wetlands Research Station Technical Bulletin 2. Portage la Prairie, MB.

Murkin, E. J., H. R. Murkin, and R. D. Titman. 1992. Nektonic invertebrate abundance and distribution at the emergent vegetation-open water interface in the Delta Marsh, Manitoba, Canada. Wetlands 12:45–52.

Murkin, H. R. 1983. Responses by aquatic macroinvertebrates to prolonged flooding of marsh habitat. Ph.D. Dissertation, Utah State University, Logan, UT.

———. 1989. The basis for food chains in prairie wetlands. Pages 316–338 in A. G. van der Valk (ed.), Northern Prairie Wetlands. Iowa State University Press, Ames, IA.

Murkin, H. R., P. G. Abbott, and J. A. Kadlec. 1983. A comparison of activity traps and sweep nets for sampling nektonic invertebrates in wetlands. Freshwater Invertebrate Biology 2:99–106.

Murkin, H. R., and B. D. J. Batt. 1987. Interactions of vertebrates and invertebrates in peatlands and marshes. Pages 15–30 in D. M. Rosenberg and H. V. Danks (eds.), Aquatic Insects of Peatlands and Marshes. Memoirs of the Entomological Society of Canada 140.

Murkin, H. R., B. D. J. Batt, P. J. Caldwell, C. B. Davis, J. A. Kadlec, and A. G. van der Valk. 1984. Perspectives on the Delta Waterfowl Research Station—Ducks Unlimited Canada Marsh Ecology Research Program. Transactions of the North American Wildlife and Natural Resources Conference 49:253–261.

Murkin, H. R., and J. A. Kadlec. 1986. Responses by benthic macroinvertebrates to prolonged flooding of marsh habitat. Canadian Journal of Zoology 64:65–72.

Murkin, H. R., J. A. Kadlec, and E. J. Murkin. 1991. Effects of prolonged flooding on nektonic invertebrates in small diked marshes. Canadian Journal of Fisheries and Aquatic Sciences 48:2355–2364.

Murkin, H. R., R. M. Kaminski, and R. D. Titman. 1982. Responses by dabbling ducks and aquatic invertebrates to an experimentally manipulated cattail marsh. Canadian Journal of Zoology 60:2324–2332.

Neckles, H. A., H. R. Murkin, and J. A. Cooper. 1990. Influences of seasonal flooding on macroinvertebrate abundance in wetland habitats. Freshwater Biology 23:311–322.

Nelson, J. W. 1982. Effects of varying detrital nutrient concentration on macroinvertebrate abundance and biomass. Master's Thesis, Utah State University, Logan, UT.

Pederson, R. L. 1981. Abundance, distribution, and diversity of buried seed populations in the Delta Marsh, Manitoba, Canada. Ph.D. Dissertation, Iowa State University, Ames, IA.

Pennak, R. W. 1978. Freshwater Invertebrates of the United States. John Wiley & Sons, New York, NY.

Peterson, L. P., H. R. Murkin, and D. A. Wrubleski. 1989. Waterfowl predation on benthic macroinvertebrates during fall drawdown of a northern prairie marsh. Pages 681–689 in R. R. Sharitz and J. W. Gibbons (eds.), Freshwater Wetlands and Wildlife. DOE Symposium Series No. 61. USDOE Office of Scientific and Technical Information, Oak Ridge, TN.

Porter, K. G. 1977. The plant-animal interface in freshwater ecosystems. American Scientist 65:159–170.

Rabe, F. W., and F. Gibson. 1984. The effect of macrophyte removal on the distribution of selected invertebrates in a littoral environment. Journal of Freshwater Ecology 2:359–371.

Rannie, W. F., L. H. Thorliefson, and J. T. Teller. 1989. Holocene evolution of the Assiniboine River paleochannels and Portage la Prairie alluvial fan. Canadian Journal of Earth Science 26:1834–1841

Reynolds, J. C., and G. G. E. Scudder. 1987. Experimental evidence of the fundamental feeding niche in *Cenocorixa* (Hemiptera: Corixidae). Canadian Journal of Zoology 65:967–973.

Robinson, G. G. C., S. E. Gurney, and L. G. Goldsborough. 1997a. Responses of benthic and planktonic algal biomass to experimental water level manipulation in a prairie lakeshore wetland. Wetlands 17:176–181.

———. 1997b. The primary productivity of benthic and planktonic algae in a prairie wetland under controlled water level regimes. Wetlands 17:182–194

Ross, L. C. M., and H. R. Murkin. 1989. Invertebrates. Pages 35–38 *in* E. J. Murkin and H. R. Murkin (eds.), Marsh Ecology Research Program Long-Term Monitoring Procedures Manual. Delta Waterfowl and Wetlands Research Station Technical Bulletin 2, Portage la Prairie, MB.

Schramm, H. L., Jr. and K. J. Jirka. 1989. Effects of aquatic macrophytes on benthic macroinvertebrates in two Florida lakes. Journal of Freshwater Ecology 5:1–12.

Scoggan, H. J. 1978. The Flora of Canada. Publication in Botany 7, National Museum of Natural Sciences, Winnipeg, MB.

Shay, J. M., and C. T. Shay. 1986. Prairie marshes in western Canada, with specific reference to the ecology of five emergent macrophytes. Canadian Journal of Botany 64:443–454.

Smith, L. M., and J. A. Kadlec. 1985. Fire and herbivory in a Great Salt Lake marsh. Ecology 66:259–265.

Smith, T. G. 1968. Crustacea of the Delta Marsh Region, Manitoba. Canadian Field-Naturalist 82:120–139.

Spence, D. H. N. 1982. The zonation of plants in freshwater lakes. Advances in Ecological Research 12:37–125.

Sproule, T. A. 1972. A paleoecological investigation into the postglacial history of the Delta Marsh. Master's Thesis, University of Manitoba, Winnipeg, MB.

Squires, L., and A. G. van der Valk. 1992. Water-depth tolerances of the dominant emergent macrophytes of the Delta Marsh, Manitoba. Canadian Journal of Botany 70:1860–1867.

Suthers, I. M., and J. H. Gee. 1986. Role of hypoxia in limiting diel spring and summer distribution of juvenile yellow perch (*Perca flavescens*) in a prairie marsh. Canadian Journal of Fisheries and Aquatic Sciences 43:1562–1570.

Swamikannu, X., and K. D. Hoagland. 1989. Effects of snail grazing on the diversity and structure of a periphyton community in a eutrophic pond. Canadian Journal of Fisheries and Aquatic Sciences 46:1698–1704.

Swanson, G. A. 1984. Dissemination of amphipods by waterfowl. Journal of Wildlife Management 48:988–991.

Swanson, G. A., and M. I. Meyer. 1973. The role of invertebrates in the feeding ecology of Anatinae during the breeding season. Pages 143–185 *in* Waterfowl Habitat Management Symposium, Moncton, NB.

Teller, J. T., and W. M. Last. 1981. Late Quarternary history of Lake Manitoba. Quarternary Research V 16:97–116.

van der Valk, A. G. 1981. Succession in wetlands: A Gleasonian approach. Ecology 62:688–696.

———. 1992. Response by wetland vegetation to a change in water level. Pages 7–16 *in* C. M. Finlayson and T. Larsson (eds.), Wetland Management and Restoration. Swedish Environmental Protection Agency Report 3492. Swedish Environmental Protection Agency, Solna, Sweden.

———. 1994. Effects of prolonged flooding on the distribution and biomass of emergent species along a freshwater wetland coenocline. Vegetatio 110:185–196.

van der Valk, A. G., and C. B. Davis. 1978. The role of seed banks in the vegetation dynamics of prairie glacial marshes. Ecology 59:322–335.

van der Valk, A. G., L. Squires, and C. H. Welling. 1994. Assessing the impacts of an increase in water level on wetland vegetation. Ecological Applications 4:525–534.

van der Valk, A. G., and C. H. Welling. 1988. The development of zonation in freshwater wetlands. Pages 145–158 *in* H. J. During, M. J. A. Werger, and J. H. Willems (eds.), Diversity and Pattern in Plant Communities. SPB Academic, The Hague, The Netherlands.

Voigts, D. K. 1976. Aquatic invertebrate abundance in relation to changing marsh vegetation. American Midland Naturalist 95:313–322.

Walker, J. M. 1959. Vegetation studies on the Delta Marsh, Delta, Manitoba. Master's Thesis, University of Manitoba, Winnipeg, MB.

———. 1965. Vegetation changes with falling water levels in the Delta Marsh, Manitoba. Ph.D. Dissertation, University of Manitoba, Winnipeg, MB.

Weller, M. W. 1981. Freshwater Marshes: Ecology and Wildlife Management. University of Minnesota Press, Minneapolis, MN.

Weller, M. W., and L. H. Fredrickson. 1974. Avian ecology of a managed glacial marsh. Living Bird 12:269–291.

Weller, M. W., and C. E. Spatcher. 1965. Role of Habitat in the Distribution and Abundance of Marsh Birds. Department of Zoology and Entomology Special Report 43. Agricultural and Home Economics Experiment Station, Iowa State University, Ames, IA.

Welling, C. H., R. L. Pederson, and A. G. van der Valk. 1988. Recruitment from the seed bank and the development of zonation of emergent vegetation during a drawdown in a prairie wetland. Journal of Ecology 76:483–496.

Wiggins, G. B., R. J. Mackay, and I. M. Smith. 1980. Evolutionary and ecological strategies of animals in annual temporary pools. Archiv für Hydrobiologie Supplement 58:97–207.

Wrubleski, D. A. 1987. Chironomidae (Diptera) of peatlands and marshes in Canada. Pages 141–161 *in* D. M. Rosenberg and H. V. Danks (eds.), Aquatic Insects of Peatlands and Marshes. Memoirs of the Entomological Society of Canada 140.

———. 1991. Chironomid Recolonization of Marsh Drawdown Surfaces Following Reflooding. Ph.D Dissertation, University of Alberta, Edmonton, AB.

Wrubleski, D. A., and D. M. Rosenberg. 1990. The Chironomidae (Diptera) of Bone Pile Pond, Delta Marsh, Manitoba, Canada. Wetlands 10:243–275.

Wylie, G. D., and J. R. Jones. 1986. Limnology of a wetland complex in the Mississippi alluvial valley of southeast Missouri. Archiv für Hydrobiologie Supplement 74:288–314.

24 Northern Prairie Marshes (Delta Marsh, Manitoba)

II. Chironomidae (Diptera) Responses to Changing Plant Communities in Newly Flooded Habitats

DALE A. WRUBLESKI

*T*his chapter describes the development of a prairie marsh chironomid community following changes in the plant community brought about by the reflooding of a series of experimental marshes within the Delta Marsh, Manitoba. Emergence traps were used to compare chironomid communities among three "preflood" vegetation types, the terrestrial annual* Aster laurentianus, *and two emergent macrophytes,* Scolochloa festucacea *and* Scirpus lacustris validus, *and two water depths, over a four-year period. Chironomids colonized the marshes rapidly, with high numbers of individuals and species found in the first year of flooding. Patterns of community development differed between the* Aster *habitat and the two emergent macrophytes. In* Aster *a diverse group of chironomids, including epiphytic and bottom-dwelling species, was present from the first application of water. In the two habitats with emergent macrophytes, epiphytic species, particularly* Corynoneura cf. scutellata, *dominated chironomid emergence during the first two years. Prolonged flooding resulted in the death of the macrophytes and a decline in epiphytic species. Benthic species, particularly* Chironomus tentans *and* Glyptotendipes barbipes, *subsequently increased in numbers. Comparisons are made in this chapter with developmental sequences reported for chironomid communities in reservoirs and an unmanaged area of the Delta Marsh.*

Invertebrates in Freshwater Wetlands of North America: Ecology and Management, Edited by Darold P. Batzer, Russell B. Rader, and Scott A. Wissinger
ISBN 0-471-29258-3 © 1999 John Wiley & Sons, Inc.

INTRODUCTION

The chironomids (midges) are one of the first groups to colonize newly created water bodies. Chironomid colonization and succession have been studied in a variety of new reservoirs (e.g., Armitage 1977, 1983, Sephton et al. 1983, Brown and Oldham 1984, Rosenberg et al. 1984) and shallow water habitats (e.g., Clement et al. 1977, Street and Titmus 1979, Titmus 1979, Danell and Sjöberg 1982, Barnes 1983, Matena 1990), but the dynamics of chironomid assemblages have not been examined in newly created wetland habitats.

Chironomids are particularly abundant in shallow freshwater wetlands and often dominate the insect community. For example, in three pothole wetlands of the Minnedosa area, Manitoba, Bataille and Baldassarre (1993) report that chironomids represented 71–78 percent of the total number of emerging insects. In a prairie pond in central Saskatchewan, Parker (1992) reports that chironomids (not including females and the Tanytarsini) represented 66–71 percent of emerging insect numbers. Chironomids accounted for 81 percent of the total numbers of insects and 21% of the insect species emerging from the Dundas Marsh, Ontario (Judd 1953), and they represented 75–78 percent of total numbers of insects and 41–43 percent of the insect species emerging from a bog pool, Redmond's Pond, in southern Ontario (Judd 1958, 1961). In two coastal wetlands of Lake Michigan, chironomids represented 76 percent of the emerging insects (McLaughlin and Harris 1990). In prairie pothole habitats of North Dakota chironomids represented 60 percent of the insects flying over these habitats and accounted for 32.9 percent of the total biomass (King and Wrubleski 1998).

Chironomids are important components in the trophic structure of wetland habitats. They are very important food resources for waterfowl (e.g., Siegfried 1973, Danell and Sjöberg 1977, Sjöberg and Danell 1982, Swanson et al. 1985, Euliss and Grodhaus 1987, Jacobsen 1991, Batzer et al. 1993) and other wetland birds (Voigts 1973). In addition, adult chironomids originating from wetlands provide an important food resource for many nonwetland birds (e.g., Busby and Sealy 1979, Sealy 1980, Guinan and Sealy 1987, St. Louis et al. 1990, Blancher and McNicol 1991). However, other ecological functions of chironomids within wetland habitats are not well known.

The flooding of a series of experimental marshes, as part of the Marsh Ecology Research Program (MERP) (see Murkin and Ross, this volume), in the Delta Marsh (50°11′N, 98°19′W), Manitoba, enabled me to study chironomid colonization and succession within these structurally complex habitats. During an artificial drawdown the dry marsh bottoms were colonized by a variety of mudflat annuals and emergent macrophytes (van der Valk et al. 1989, van der Valk and Welling 1988, Welling et al. 1988a, b). Flooding of this "preflood" vegetation provided a diverse array of habitats for chironomid colonization. Marsh vegetation does influence chironomid community composition (Wrubleski 1987, 1989, Wrubleski and Rosenberg 1990), but how these effects are manifested is not well understood. In the experimental marshes terrestrial plants were drowned and aquatic macrophytes survived for varying periods of time, depending upon depth of flooding (Clark and Kroeker 1993, van der Valk et al. 1994). Death of these plants contributed to an abundance of plant litter on the bottom and additional habitat for chironomids. Thus, within these experimental marshes, habitats available for colonization by the chironomids varied in form, abundance, and duration.

In this chapter, I describe changes in the abundance, diversity, and species composition of the chironomid community within these experimentally flooded marshes. Observations were made over a period of four years to evaluate changes in the chironomid community and their relationship to changes in the plant communities. Information about habitat requirements of the dominant species is used to explain underlying patterns of chironomid community development. Comparisons are made with sequences of community development reported for reservoir habitats (e.g., McLachlan 1974). Comparisons are also made with an unmanaged area of the Delta Marsh (Wrubleski and Rosenberg 1990) to determine the effects of this disturbance on chironomid community parameters such as diversity and dominance.

STUDY DESIGN

Field Sampling

The original design of the reflooding experiment of the Marsh Ecology Research Program (MERP) called for four experimental marshes to be flooded at the long-term average of the Delta Marsh ("low" treatment), three flooded to a depth 60 cm above average ("high" treatment), and three to be flooded to a depth 30 cm above average ("medium" treatment) (Murkin et al. 1984; see also Murkin and Ross, this volume). However, water depths following reflooding increased from north to south within the marshes, and this variation rendered the marshes unsuitable as "replicates" for a study of chironomid community responses. Prior to flooding, the experimental marshes varied widely in the development of plant communities on the drawdown surfaces (van der Valk et al. 1989, van der Valk and Welling 1988, Welling et al. 1988a, b). Therefore, rather than using marsh flooding depths as treatments, I used preflood vegetative cover as my treatments. *Aster, Scolochloa,* and *Scirpus* had formed extensive stands on the drawdown surfaces of the experimental marshes (Welling et al. 1988a, b). Each of these three habitats was sampled at two water depths, shallow (20–40 cm) and deep (50–70 cm). I anticipated that the deep-water areas would lose their emergent macrophyte cover faster than the shallow sites.

Aster laurentianus Fern. was one of the dominant mudflat annuals to grow on the dry marsh bottoms (van der Valk 1986, Welling et al. 1988b), but did not survive flooding. *Scolochloa festucacea* (Willd.) Link (whitetop) and *Scirpus lacustris* L. ssp. *validus* (Wahl.) Koyama (softstem bulrush) were among five emergent macrophytes that dominated the drawdown surfaces of the marshes (van der Valk and Welling 1988, Welling et al. 1988a, b). They were selected for study because they have different flooding tolerances. *Scolochloa festucacea* is tolerant to seasonal flooding but not to prolonged flooding (Millar 1973, Neckles et al. 1985). *Scirpus lacustris validus* is more tolerant of flooding than *S. festucacea,* but intolerant to long periods of flooding (Harris and Marshall 1963, Shay and Shay 1986).

Each of the six habitat-depth combinations was sampled with six emergence traps (n = 36 traps). I used a modified LeSage and Harrison (1979) model "week" trap (basal area = 0.5 m^2) as described in Wrubleski (1984). The same trap sites were used over the four years of this study (1985–88). Traps were set out when there was sufficient water at a site to float them. At each trap site three stakes spaced 2 m apart and in a line were driven into the substrate. The emergence trap was anchored to the center stake and one of the outside stakes. At monthly intervals traps were rotated to the opposite stake to reduce trap and disturbance effects below the trap.

Deep-water sites, which were at lower elevations, were flooded earlier than shallow-water sites. Initiation of sampling varied between May 16 and July 20, 1985. Within each habitat-depth combination the six emergence trap sites were not flooded at the same time, but most sites in a treatment were flooded within two weeks of each other. In 1987 and 1988 traps were set out the last week of April and emptied weekly through the third week of October, providing data for 24 weekly sample periods. In 1986 traps were set out the first week of May and data were collected over 23 weeks.

Lost Traps, Missing Data, and Data Summation

Some samples were lost in the fall of each year, particularly in 1985, because traps were damaged by muskrats. Throughout the year other traps were occasionally upset by strong winds. When possible, the traps were replaced immediately, and the number of individuals present in the sample was adjusted proportionately to estimate the catch as if the trap had been in place for the full seven days (e.g., a sample from a trap in place for three days was multiplied by 2.3 to equal a seven-day sampling period). Most traps, however, were not replaced until their next regularly scheduled visit. I used the relationship between the missing trap and the five other traps within that habitat for the three previous sample periods to estimate the numbers and species of chironomids that would have been anticipated in the lost sample. This mathematical procedure is described in detail in Wrubleski (1991). Highest number of complete sample losses was 23 of 138 samples (16.7 percent) for *Scolochloa*-shallow sites in 1986. Total sample losses over the four-year period were 0.7 percent in the shallow sites and 5.8 percent in the deep sites.

All adult chironomids were identified to species (see Wrubleski 1991 for the complete species list). Nomenclature follows Oliver et al. (1990). Numbers collected from each trap during every week were summed for the entire sampling season, multiplied by two to give no. m^{-2} and then used to determine mean annual emergence (no. m^{-2} yr^{-1} ± SE, n = 6) for each habitat-water depth combination. Chironomid biomass and size classes were also determined, and these data are reported in Wrubleski (1991).

Community Parameters

The chironomid assemblages collected within each habitat-depth combination were described several ways. Number of species (species richness) collected within each was determined. Species diversity was estimated using the Shannon-Wiener (H') function (Krebs 1989, p. 361). The relative dominance index (RDI) of McNaughton (1967, after Driver 1977), which is the proportion of emergence represented by the two most abundant species, was also determined for each habitat-depth combination.

CHIRONOMID RESPONSES

Total Chironomidae

Chironomids colonized the experimental marshes rapidly, and large populations were found during the first year of flooding (Fig. 24.1). The highest number of total chironomids collected in any one year, 9168 m^{-2} yr^{-1}, was during the initial year of flooding in the *Aster*-deep sites (Fig. 24.1). *Scolochloa* and *Scirpus* sites had their highest numbers emerging during the fourth year, although these numbers were not different from numbers observed in the first year. However, total numbers emerging over the four years differed between depths. In the shallow sites little change was found among years, except for significant increases in numbers emerging from the *Scirpus* and *Scolochloa* sites between the third and fourth years of flooding. In the deep sites there was a trend to high numbers emerging in the first year of flooding, followed by a decline in the second and/or third years and a subsequent increase in the fourth year of flooding.

Comparisons of total chironomid numbers emerging among the three plant communities or between depths showed few differences (Fig. 24.1). Among shallow sites, *Aster* tended to have higher numbers emerging than the two emergent vegetation habitats. Among the deep sites, chironomid emergence from *Aster* tended to be higher during the first two years, but by the fourth year the pattern reversed, with higher numbers of chironomids emerging from *Scolochloa* and *Scirpus*. In 1985 all three deep habitats had higher emergence than their corresponding shallow sites, but this was probably a function of the longer flooding times in the deeper sites (Wrubleski 1991). In 1987 and 1988 *Scirpus*-deep sites had higher numbers emerging than the shallow sites.

Seasonal trends in emergence for total chironomids varied among years (Fig. 24.2). In 1985 patterns of emergence differed among the three plant

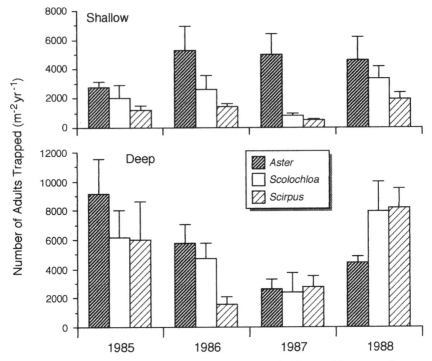

Fig. 24.1. Mean ($\pm$ SE, n = 6) annual emergence of chironomids over a four-year period from three habitats and two water depths in the experimental marshes, Delta Marsh, Manitoba.

communities. Emergence peaked in early August for both depths in the *Aster* habitat (Fig. 24.2*a*). In the *Scolochloa* habitat peak emergence was delayed until late August–early September in the deep sites, while emergence remained constantly low in the shallow sites. In the *Scirpus* habitat no readily obvious peaks were observed at either depth. Emergence in all three habitats shifted to a primarily early spring peak in subsequent years (Fig. 24.2), and the large July–August emergence peaks from the deep sites in 1985 were no longer observed. *Aster* had low continuous emergence throughout the summers of 1986, 1987, and 1988, particularly from the shallow sites. In 1986 and 1987 *Scolochloa* and *Scirpus* had very low emergence after the end of June, but in 1988 emergence from the deep sites showed an increase late in the summer.

Species Composition

Underlying the changes in total numbers of Chironomidae are changes in the abundances of the subfamilies, tribes, and species. Although patterns in total numbers were fairly similar among habitats over the four years (Fig. 24.1),

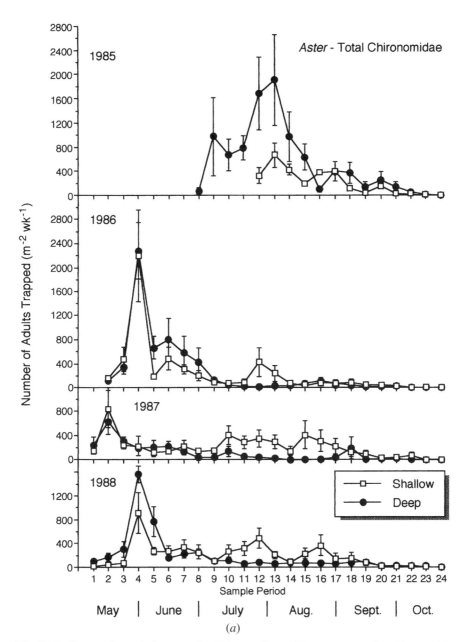

Fig. 24.2. Seasonal trends in mean (± SE, n = 6) weekly emergence for chironomids from the experimental marshes in (*a*) Aster, (*b*) *Scolochloa,* and (*c*) *Scirpus* habitats.

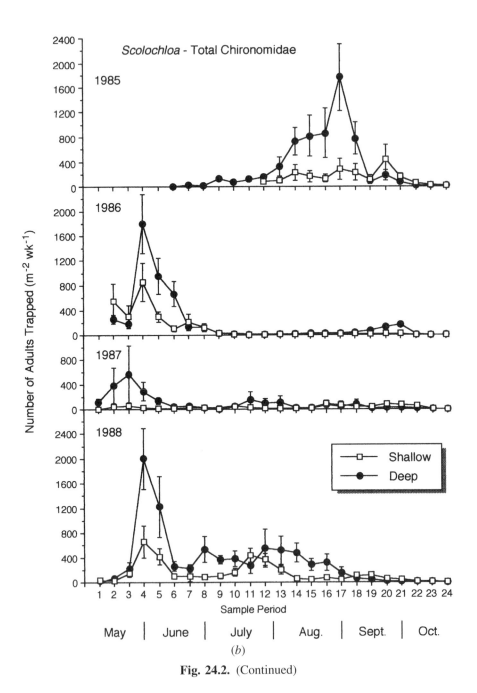

Fig. 24.2. (Continued)

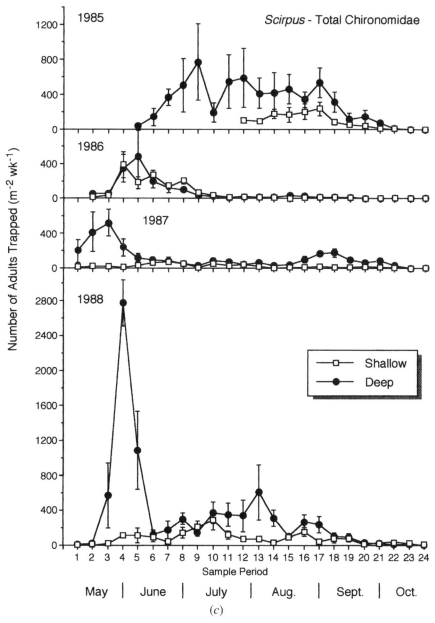

Fig. 24.2. (Continued)

species composition differed among habitats and varied over the duration of this experiment.

Chironomini. *Aster* sites were dominated by Chironomini during all four years (Fig. 24.3). In the first year of flooding, the *Aster*-deep sites were dominated by *Chironomus atrella* (Townes), which represented 39.7 percent of the chironomids collected (Table 24.1). In contrast, this species represented only 6.7 percent of the emergence from *Aster*-shallow sites. Here another member of this tribe, *Chironomus tentans* Fab., was the most abundant species. *Chironomus tentans* was the most abundant species from this habitat at both depths over the next three years, whereas *Chironomus atrella* emergence declined in subsequent years (Table 24.1, Wrubleski 1991). *Glyptotendipes barbipes* (Staeger) was also very abundant in *Aster* over the four years (Table 24.1).

In the two emergent macrophyte habitats the Chironomini were a much smaller proportion of the emergence in the first two years (Fig. 24.3, Tables 24.2, 24.3). *Dicrotendipes nervosus* (Staeger) was abundant during those years (Tables 24.2, 24.3). By the fourth year in the shallow sites and the third year in the deep sites, Chironomini dominated emergence from these two habitats (Fig. 24.3). This increase in abundance was due primarily to *Chironomus tentans* and *Glyptotendipes barbipes* (Tables 24.2, 24.3).

Orthocladiinae. During the initial few years in the *Scolochloa* and *Scirpus* habitats, Orthocladiinae, and specifically *Corynoneura* cf. *scutellata* Winn., dominated emergence numbers (Fig. 24.3, Tables 24.2, 24.3). However, by the third year this species declined considerably in number and proportion of emergence in these two habitats (Tables 24.2, 24.3). Orthocladiinae represented a small but constant proportion of chironomid emergence from the *Aster* habitat (Fig. 24.3). *Corynoneura* cf. *scutellata* was present in this habitat (Table 24.1, Wrubleski 1991) but did not dominate emergence as it had in the two emergent vegetation habitats.

Other orthoclads were also important components of emergence from the experimental marshes. *Cricotopus sylvestris* (Fab.) and/or *C. ornatus* (Meigen) were consistently among the most abundant species in *Aster* and *Scolochloa* sites but were not as abundant in the *Scirpus* habitat (Tables 24.1, 24.2, 24.3). *Limnophyes prolongatus* (Kieffer), a semiterrestrial species, had very low abundances in the *Aster* sites (Wrubleski 1991) but was abundant in *Scolochloa* and *Scirpus* sites in the first two years of flooding and then declined in abundance (Tables 24.2, 24.3, Wrubleski 1991).

Tanytarsini. In the shallow sites the Tanytarsini were less abundant in *Scirpus* than in *Aster* and *Scolochloa* habitats during the first two years (Fig. 24.3). The most abundant species was *Paratanytarsus* sp. 1, which was particularly numerous in the first few years of flooding (Tables 24.1–3). *Tanytarsus* sp. 1 was the most abundant species in *Scolochloa*-shallow sites in 1987 (Table

TABLE 24.1. The Five Most Abundant Species (mean m^{-2} yr^{-1} ± SE, n = 6) Collected from the *Aster* Habitat for Each Year and Depth, Listed in Decreasing Order of Abundance

Aster	1985	1986	1987	1988
Shallow	*Chironomus tentans* 725.0 ± 144.1	*Chironomus tentans* 2106.0 ± 684.0	*Chironomus tentans* 1954.3 ± 827.7	*Chironomus tentans* 1292.3 ± 414.5
	Paratanytarsus sp. 1 357.7 ± 96.6	*Glyptotendipes barbipes* 1322.7 ± 533.8	*Glyptotendipes barbipes* 999.3 ± 276.4	*Tanypus punctipennis* 847.0 ± 408.6
	Corynoneura cf. *scutellata* 232.7 ± 127.7	*Tanytarsus* sp. 1 344.0 ± 186.3	*Tanypus punctipennis* 338.7 ± 200.5	*Glyptotendipes barbipes* 782.3 ± 444.8
	Cricotopus ornatus 231.3 ± 64.8	*Corynoneura* cf. *scutellata* 273.3 ± 134.7	*Cricotopus sylvestris* 290.7 ± 92.2	*Cladotanytarsus* sp. 405.3 ± 270.1
	Glyptotendipes barbipes 206.0 ± 56.7	*Crictopus ornatus* 248.3 ± 157.7	*Crictopus ornatus* 278.0 ± 93.6	*Cricotopus ornatus* 267.3 ± 173.2
Deep	*Chironomus atrella* 3637.3 ± 972.6	*Chironomus tentans* 1391.7 ± 493.6	*Chironomus tentans* 850.7 ± 286.0	*Chironomus tentans* 1568.3 ± 303.2
	Chironomus tentans 1292.3 ± 809.6	*Glyptotitendipes barbipes* 1077.3 ± 544.7	*Tanytarsus* sp. 1 324.0 ± 185.5	*Glyptotendipes barbipes* 387.3 ± 183.0
	Tanytarsus sp. 1 1152.0 ± 655.9	*Tanytarsus* sp. 1 948.0 ± 750.6	*Paratanytarsus* sp. 3 297.7 ± 211.8	*Dicrotendipes nervosus* 357.3 ± 191.4
	Tanytarsus sp. 4 701.0 ± 441.9	*Corynoneura* cf. *scutellata* 883.0 ± 197.4	*Cladopelma viridula* 271.7 ± 165.7	*Cricotopus sylvestris* 238.7 ± 60.5
	Glyptotendipes barbipes 547.3 ± 235.7	*Chironomus atrella* 313.3 ± 146.6	*Glyptotendipes barbipes* 190.0 ± 104.1	*Paratanytarsus* sp. 3 226.3 ± 119.0

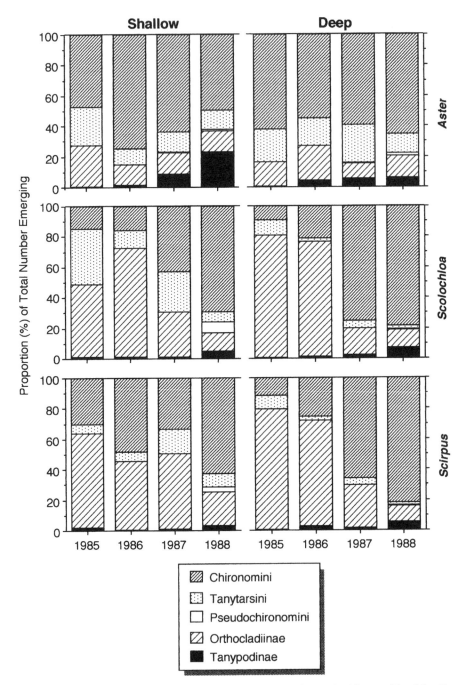

Fig. 24.3. Proportion of annual emergence contributed by each chironomid subfamily or tribe.

TABLE 24.2. The Five Most Abundant Species (mean $m^{-2} yr^{-1}$ ± SE, n = 6) Collected from the *Scolochloa* Habitat for Each Year and Depth, Listed in Decreasing Order of Abundance

Scolochloa	1985	1986	1987	1988
Shallow	*Corynoneura* cf. *scutellata* 782.3 ± 344.1	*Corynoneura* cf. *scutellata* 1175.3 ± 433.5	*Tanytarsus* sp. 1 163.7 ± 92.6	*Chironomus tentans* 1266.0 ± 317.9
	Paratanytarsus sp. 1 686.3 ± 446.1	*Limnophyes prolongatus* 275.3 ± 150.4	*Chironomus tentans* 162.3 ± 89.4	*Glyptotendipes barbipes* 849.7 ± 274.1
	Dicrotendipes nervosus 125.3 ± 74.9	*Paratanytarsus* sp. 1 257.0 ± 171.6	*Acricotopus nitidellus* 119.3 ± 85.9	*Pseudochiron. middlekauffi* 241.3 ± 105.9
	Limnophyes prolongatus 93.3 ± 37.2	*Chironomus riparius* 206.0 ± 111.7	*Corynoneura* cf. *scutellata* 99.7 ± 34.8	*Cricotopus sylvestris* 171.0 ± 65.0
	Parachironomus tenuicaudatus 78.7 ± 47.9	*Cricotopus ornatus* 186.0 ± 104.9	*Chironomus atrella* 68.0 ± 26.3	*Dicrotendipes nervosus* 114.0 ± 43.1
Deep	*Coiynoneura* cf. *scutellata* 3128.3 ± 1778.1	*Corynoneura* cf. *scutellata* 2739.3 ± 945.4	*Chironomus tentans* 978.3 ± 762.5	*Chironomus tentans* 4525.0 ± 1252.8
	Cricotopus sylvestris 1074.3 ± 452.5	*Dicrotendipes nervosus* 366.0 ± 79.7	*Glyptotendipes barbipes* 324.0 ± 266.2	*Glyptotendipes barbipes* 820.3 ± 367.5
	Paratanytarsus sp. 1 586.3 ± 198.6	*Glyptotendipes barbipes* 340.3 ± 208.9	*Corynoneura* cf. *scutellata* 187.7 ± 117.0	*Dictendipes nervosus* 549.3 ± 328.6
	Cricotopus ornatus 584.0 ± 312.0	*Cricotopus sylvestris* 282.3 ± 82.9	*Dicrotendipes nervosus* 166.3 ± 66.1	*Cricotopus sylvestris* 445.7 ± 178.2
	Dicrotendipes nervosus 218.3 ± 72.3	*Psectrocladius edwardsi* 272.3 ± 32.4	*Cricotopus sylvestris* 148.3 ± 35.9	*Tanypus punctipennis* 279.3 ± 174.3

TABLE 24.3. The Five Most Abundant Species (mean $m^{-2} yr^{-1} \pm$ SE, n = 6) Collected from the *Scirpus* Habitat for Each Year and Depth, Listed in Decreasing Order of Abundance

Scirpus	1985	1986	1987	1988
Shallow	*Corynoneura* cf. *scutellata* 596.0 ± 322.6	*Corynoneura* cf. *scutellata* 565.3 ± 123.9	*Corynoneura* cf. *scutellata* 120.7 ± 78.7	*Chironomus tentans* 582.0 ± 180.8
	Dicrotendipes nervosus 187.0 ± 88.1	*Dicrotendipes nervosus* 310.0 ± 90.5	*Acricotopus nitidellus* 101.0 ± 47.2	*Glyptotendipes barbipes* 361.7 ± 186.5
	Limnophyes prolongatus 94.0 ± 34.1	*Chironomus tentans* 143.3 ± 108.8	*Paratanytarsus* sp. 1 71.0 ± 25.0	*Acricotopus nitidellus* 307.3 ± 88.6
	Chironomus staegeri 56.7 ± 38.6	*Chironomus riparius* 117.3 ± 66.0	*Dicrotendipes nervosus* 44.7 ± 11.9	*Cladopelma viridula* 100.3 ± 74.9
	Paratanytarsus sp. 1 54.0 ± 22.5	*Paratanytarsus* sp. 1 83.3 ± 38.2	*Chironomus atrella* 28.7 ± 13.1	*Chironomus atrella* 85.7 ± 24.2
Deep	*Corynoneura* cf. *scutellata* 4600.0 ± 2415.6	*Corynoneura* cf. *scutellata* 895.0 ± 424.3	*Chironomus tentans* 678.0 ± 324.9	*Chironomus tentans* 4522.0 ± 575.2
	Paratanytarsus sp. 1 504.3 ± 162.0	*Dicrotendipes nervosus* 155.3 ± 67.8	*Chironomus atrella* 491.0 ± 289.2	*Glyptotendipes barbipes* 1523.3 ± 636.7
	Dicrotendipes nervosus 417.3 ± 90.0	*Chironomus atrella* 64.0 ± 27.4	*Corynoneura* cf. *scutellata* 341.3 ± 54.2	*Cricotopus sylvestris* 352.3 ± 175.3
	Parachironomus sp. 3 102.3 ± 25.8	*Limnophyes prolongatus* 48.7 ± 24.8	*Cricotopus sylvestris* 336.3 ± 66.8	*Dicrotendipes nervosus* 223.7 ± 55.9
	Parachironomus tenuicaudatus 101.7 ± 28.5	*Paratanytarsus* sp. 1 44.7 ± 25.0	*Dicrotendipes nervosus* 220.3 ± 31.7	*Cricotopus ornatus* 212.0 ± 144.0

24.2) and the second most abundant species in *Aster*-deep sites that same year (Table 24.1).

Tanypodinae. This group was rare in the first years of flooding, but numbers increased significantly in all habitats by the fourth year (Fig. 24.3). Much of the increase over the four years was due to *Tanypus punctipennis* Meigen (Tables 24.1, 24.2), although all species showed some increase in numbers over the four-year period (Wrubleski 1991).

Pseudochironomini. This tribe was represented by a single species, *Pseudochironomus middlekauffi* Townes (Wrubleski 1991). It showed consistent yearly increases in abundance in all three habitats and at both depths during the four years. By the fourth year it was the third most abundant species in *Scolochloa*-shallow sites (Table 24.2).

Chironomid Diversity

As with chironomid numbers, large numbers of species were present in the first year of flooding (Fig. 24.4*a*). During the first two years *Aster* had higher species richness than the two emergent macrophytes, and maintained this higher number of species in the shallow sites over the entire four-year period. In the *Scolochloa* and *Scirpus* sites richness was initially lower than for *Aster* but reached comparable numbers in the deep sites by the third or fourth years of flooding (Fig. 24.4*a*). In all three habitats numbers of species collected increased steadily in the deep sites over the four years, and by the fourth year were higher than in the shallow sites.

Aster had the highest Shannon-Wiener diversity in the first year of flooding (Fig. 24.4*b*) and the lowest relative dominance (Fig. 24.4*c*). The *Scirpus*-deep sites had particularly low diversity in the first year of flooding (Fig. 24.4*b*) due to the dominance of emergence by *Corynoneura* cf. *scutellata* (Table 24.3, Fig. 24.4*c*). In the *Aster*-shallow sites diversity dropped in the second year and then increased steadily in the third and fourth years (Fig. 24.4*b*), with relative dominance showing the opposite pattern (Fig. 24.4*c*). In *Aster*-deep sites diversity increased yearly. Both *Scolochloa* and *Scirpus* had increasing diversities up to the third year of flooding but then exhibited a decrease in diversity with a corresponding increase in dominance due to a large increase in the emergence of *Chironomus tentans* and *Glyptotendipes barbipes* at both depths (Tables 24.2, 24.3). Highest diversity and lowest dominance occurred when some emergent vegetation still remained and large amounts of plant litter were being deposited on the bottom. The most micro-habitat, both in terms of emergent vegetation and plant litter, was available at that time. With the subsequent elimination of the emergent macrophytes, diversity dropped and dominance by benthic species increased.

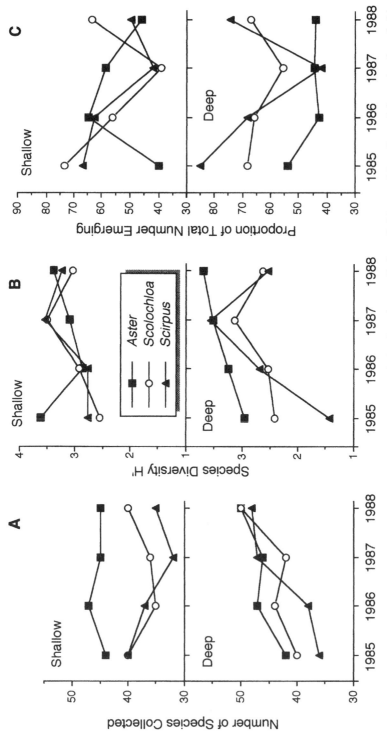

Fig. 24.4. (a) Numbers of species, (b) species diversity, and (c) the Relative Dominance Index of chironomids collected from the three habitats and two water depths in the experimental marshes.

MECHANISMS OF CHIRONOMID RESPONSES

Chironomid Habitat and Species Biology

The chironomid assemblage within the experimental marshes was dramatically influenced by changes in the aquatic macrophyte community following habitat reflooding. Very different patterns of chironomid community development were observed in the *Aster* versus the emergent macrophyte habitats.

Aster was killed following flooding, and the large input of plant litter supplied an abundance of structure. Additionally, abundant metaphyton developed in this habitat and emergent macrophytes invaded, resulting in a wide range of microhabitats becoming available for chironomid colonization. *Aster* had a diverse community, with both small epiphytic chironomids (e.g., *Corynoneura, Cricotopus, Paratanytarsus*) and larger benthic or mining species (*Chironomus, Glyptotendipes*) being abundant over the four years. *Aster* litter is persistent (Wrubleski et al. 1997), thereby providing a relatively stable habitat for many years. This may explain the lack of major changes within this community over the four-year period.

Scolochloa and *Scirpus* habitats showed more dynamic changes in chironomid community composition over time. During the first two years these habitats were dominated by *Corynoneura* cf. *scutellata*. Larvae of *Corynoneura* are very small free-living grazers dependent upon submersed surfaces (Table 24.4). The live stems and leaves of *Scolochloa* and *Scirpus* provided ideal habitat for this species, as well as other epiphytic species such as *Paratanytarsus* sp. 1 and several *Cricotopus* species (Table 24.4). The survival of these plants during the initial flooding delayed deposition of plant litter at that time (Wrubleski, personal observation, Murkin and Ross, this volume), and this may have prevented large populations of mining species such as *Chironomus tentans* and *Glyptotendipes barbipes* from becoming established in the first few years. Although both of these species have been reported to mine plant litter or other soft materials, they are more often reported to live in soft, highly organic bottom sediments (Table 24.4), which would have been present in both habitats. Their absence in emergent vegetation habitats suggests that factors other than availability of plant litter were responsible for the absence of bottom-dwelling species. Emergent macrophytes can produce unfavorable environmental conditions on the bottom. Through shading, they restrict algal growth, an important food resource for these species (Table 24.4) (Straskraba and Pieczynska 1970, Gurney and Robinson 1988). They also prevent mixing of the water column, leading to low oxygen conditions on the bottom (Dvorak 1969, Suthers and Gee 1986, Rose and Crumpton 1996).

Prolonged flooding, particularly at deeper depths, eliminated emergent macrophytes by the fourth year (van der Valk et al. 1994, Murkin and Ross, this volume). The death of these plants increased benthic plant litter and reduced shading of the bottom. Epiphytic species declined in numbers and benthic species dominated the chironomid assemblage. Driver (1977) and

TABLE 24.4. Literature References for Habitat and Food Habits of Dominant Chironomid Species Collected from the Experimental Marshes, Delta Marsh, Manitoba

Reference	Habitat	Food Habits
Tanypus punctipennis Fellton (1940)	• lake bottom	• predaceous; feeding on newly hatched and older chironomid larvae
Oláh (1976) Titmus and Badcock (1981)	• lake; open-water sediments • gravel pit lake bottom	• fed mainly on diatoms • fed mainly on unicellular algae
Corynoneura scutellata Kesler (1981)	• free-living on submerged surfaces	• grazers of periphyton
Cricotopus sylvestris Darby (1962)	• tubes on the surface of the mud, bottom debris, or submersed vegetation	• fed on diatoms, algal debris, and green algae
Menzie (1981)	• on submersed plants, when present, or on the bottom	• not reported
Cricotopus ornatus Swanson and Hammer (1983)	• tubes on the sediments, algal mats, and submersed vegetation	• not reported
Chironomus tentans Sadler (1935)	• tubes in the sediments or algae	• will eat whatever is offered, but where algae are present in sufficient quantities they comprise the main bulk of the diet
Palmen and Aho (1966)	• shallow water with large amounts of detritus as substratum	• not reported

Reference	Habitat	Feeding
Hall et al. (1970)	tubes in the sediment	filters plankton during the day and searches the sediment surface at night for larger food items
Topping (1971)	soft ooze and detritus	not reported
Mason and Bryant (1975)	within dead *Typha* stems	brown detrital material or rotting *Typha* and no living algae
Dicrotendipes nervosus		
Fellton (1940)	restricted to green algal mats	not reported
Moore (1980)	organically rich lake sediments	detritus, very little algae
Glyptotendipes barbipes		
Kimerle and Anderson (1971)	tubes in the sediments	irrigation of the tube supplies sestonic food particles, primarily algal cells
Driver (1977)	axiles of *Scolochloa*	not reported
Wrubleski and Rosenberg (1984)	miner of polystyrene foam	not reported
Paratanytarsus sp. 1		
Wrubleski (unpublished observations)	tubes on flooded vegetation	grazing algae

Murkin and Kadlec (1986) have reported similar increases in *Chironomus tentans* abundance when water levels increase on prairie wetlands. Driver (1977) suggests that the increase in this species can be attributed to its ability to use not only dead stems and roots of emergent macrophytes but also open areas created by the loss of vegetation. My results support that premise, but without information on larval distributions the basis for this response is unclear.

The death of *Aster, Scolochloa,* and *Scirpus* resulted in the deposition of abundant plant litter to the bottom. In wetland habitats plant litter has been thought to be an important food resource for invertebrate production (Mann 1988), but recent developments have indicated that algae may also be an important food resource for invertebrates (Murkin 1989, Campeau et al. 1994, Wrubleski and Detenbeck, in prep., see also Euliss et al., this volume). A review of food habits of the dominant chironomid species present in the experimental marshes indicates that algae are indeed important (Table 24.4). Street and Titmus (1982) were able to separate the effects of food and habitat provided by straw additions to a gravel-pit lake. Their findings indicated that habitat structure was a more important determinant of animal distributions than the availability of plant litter as food. Aquatic macrophytes and the associated plant litter provide an important habitat for chironomids, but it may be the highly productive algal communities of prairie wetlands (Crumpton 1989, Robinson et al. 1997a, b) that are the more important food resource for chironomids (Wrubleski and Detenbeck, in prep.).

While changes in the chironomid community in this study can be related to observed changes in habitat structure, interactions among species might also be important. Cantrell and McLachlan (1977), for example, found that competition between two benthic chironomids determined their distribution in a newly created reservoir. In the *Scolochloa* and *Scirpus* habitats of the experimental marshes, a striking shift from an assemblage dominated by *Corynoneura* cf. *scutellata* to one dominated by *Chironomus tentans* was observed (Fig. 24.5). However, published information indicates that these species use different microhabitats (Table 24.4). The negative relationship between them is more likely related to temporal changes in habitat structure than to competitive interaction. However, direct experiments with these and other species must be conducted before firm conclusions about interactions can be drawn.

Comparisons with Reservoir Studies

Several studies of aquatic invertebrate colonization and succession have been done in man-made reservoirs (e.g., Nursall 1952, Paterson and Fernando 1970, Sephton et al. 1983, Brown and Oldham 1984, Voshell and Simmons 1984). Invertebrate community development in reservoirs has been characterized into four phases (McLachlan 1974, after Murduchai-Boltovskoi 1961). The first phase is short and occurs during the initial filling of impoundments

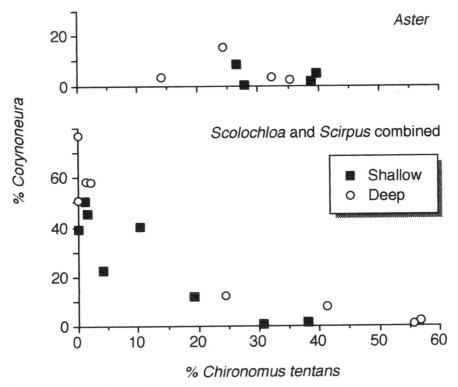

Fig. 24.5. Proportion of chironomid emergence represented by *Chironomus tentans* and *Corynoneura* cf. *scutellata* from the three habitats sampled in the experimental marshes.

when river fauna and terrestrial invertebrates are present within the reservoir. A second productive phase coincides with the final filling of the reservoir. High productivity has been attributed to the release of nutrients from the flooded vegetation and surface soils. The length of time needed to reach peak abundances (from one to four years) is determined by such factors as basin morphology, climate, water chemistry, and invertebrate population structure (Armitage 1977, Sephton et al. 1983). Chironomids, particularly *Chironomus plumosus*-type larvae (McLachlan 1974), often dominate this initial peak. A subsequent decline in invertebrate abundances typifies the third phase and has been attributed to the loss of the flooded terrestrial vegetation through decomposition, consumption by invertebrates, and sedimentation. Increasing populations of fish and predatory invertebrates (e.g., leeches, odonates) may also contribute to the decline in invertebrate densities (Andersson and Danell 1982). In the fourth phase, invertebrate abundances eventually reach an equilibrium. This general pattern of invertebrate community development is now considered part of the "reservoir paradigm" (Hecky et al. 1984). However,

Wiens and Rosenberg (1984) found little evidence of four phases when examining the responses of benthic invertebrates to reservoir formation in Southern Indian Lake, Manitoba.

The generalized pattern observed in reservoirs can be compared with chironomid community development in the diked marshes of this study. A first phase of faunal development in the experimental marshes was the presence of semiterrestrial chironomids, such as *Limnophyes prolongatus,* in *Scolochloa* and *Scirpus* during the first few years following flooding. With continued flooding this species declined in abundance.

A second or "productive" phase in the marshes resulted from high production by several dominant species. In *Scolochloa* and *Scirpus* habitats epiphytic *Corynoneura* cf. *scutellata* dominated during the first two years and benthic *Chironomus tentans* dominated in the fourth year. The high abundances of *Chironomus tentans,* following the death of the emergent macrophytes and the addition of litter to the bottom, are consistent with reservoir studies where *Chironomus* species often dominate following flooding of the terrestrial vegetation (McLachlan 1974). In the *Aster* habitat, however, the composition of the chironomid community did not show great differences over time, and the community was not dominated by one or two species, as was the case in the two emergent vegetation habitats (Fig. 24.4c).

Whether the plant litter serves as food or habitat, its eventual loss results in a decline in invertebrate numbers in reservoirs. In the experimental marshes chironomid abundances within the *Scolochloa* and *Scirpus* habitat, and probably *Aster* as well, can be expected to decline over the next few years. Decomposition of *Scirpus* and *Scolochloa* litter is rapid (Murkin et al. 1989), and little will remain as habitat for invertebrates. As more flood-tolerant emergent macrophytes or submersed macrophytes develop within the marshes and provide new habitat, numbers of chironomids and other invertebrates can be expected to recover.

A notable difference between reservoirs and the experimental marshes was the high numbers of Orthocladiinae found in this study. This may be due to a sampling artifact. Orthoclads prefer vegetated habitats, and most reservoir studies tend to focus on the deeper open-water or profundal habitats where few aquatic macrophytes are found. However, Paterson and Fernando (1970), Armitage (1977, 1983), and Ertlová (1980) sampled littoral habitats during reservoir formation and report high numbers of orthoclads. *Corynoneura* larvae, in particular, are extremely small and easily overlooked in benthic sampling programs. Street and Titmus (1979) report that very different results were obtained from larval sampling and emergence trap sampling programs. They suggest that the importance of larger benthic species as pioneers may have been overestimated in previous colonization studies where larval sampling alone had been done. If an emergence sampling program had not been used in the experimental marshes, this phase in the development of the chironomid community might have been missed entirely.

As in reservoir studies, my results suggest that the contribution of plant litter following flooding was the principal factor regulating the development of the chironomid community of the experimental marshes. In the experimental marshes, however, chironomid responses to the additions of litter took place in two phases. First was the death of the mudflat annuals, which provided an immediate pulse of coarse detritus to the bottom. Prolonged flooding caused the death of emergent macrophytes, and a second pulse of plant litter was added to the marsh bottom, resulting in a second pulse of chironomid production.

Comparisons with the Unmanaged Delta Marsh

Although many newly created water bodies are reported to produce higher numbers of chironomids than older, mature areas (e.g., McKnight and Low 1969, Whitman 1974, Street and Titmus 1979), this was not so for the experimental marshes. Numbers emerging from the marshes were comparable to those found in emergent vegetation habitats of Bone Pile Pond (BPP), an unmanipulated area of the Delta Marsh, but did not match the high numbers found in the *Potamogeton* habitat (Table 24.5, Wrubleski and Rosenberg 1990). Numbers of species were higher in the *Potamogeton* habitat of BPP than in the experimental marshes, but by year four species richness from the experimental marshes was similar to that found in the *Scirpus* and *Typha* habitats of BPP (Table 24.5, Fig. 24.4). Species diversities in the experimental marshes in the third year of flooding were comparable to values for BPP. Diversity was lower in the *Potamogeton* habitat of BPP in 1980 due to a large number of *Tanypus punctipennis* (Wrubleski and Rosenberg 1990). Relative dominance values tended to be lower in BPP, somewhat comparable to those in the *Aster*-deep sites in the second to fourth years of flooding.

The *Scirpus* and *Typha* habitats in BPP had an abundance of semiterrestrial chironomid species, including *Limnophyes immucronatus* and *L. prolongatus* (Table 24.5, Wrubleski and Rosenberg 1990). This was not so in the two emergent vegetation habitats of the experimental marshes. *Limnophyes prolongatus* was abundant initially in *Scolochloa* and *Scirpus* habitats but then declined in abundance with prolonged flooding (Tables 24.2, 24.3). Water levels within BPP fluctuated greatly, and occasionally there was no surface water present in the *Scirpus* and *Typha* habitats. This would explain the abundances of semiterrestrial species in these habitats. Flooding to depths of 20–40 cm greatly reduced the abundances of these species in the experimental marshes.

Tanypus punctipennis was a dominant species in the *Potamogeton* habitat of BPP (Table 24.5). *Tanypus* species prefer soft mud substrates through which they can move easily (e.g., Parkin and Stahl 1981, Titmus and Badcock 1981). Their gradual increase in numbers in the experimental marshes (Tables 24.1, 24.2, Wrubleski 1991) may be an indication of changing substrate conditions within the marshes; however, the exact reason remains unknown.

TABLE 24.5. **Mean Numbers of Chironomids Emerging (m^{-2} yr^{-1}), Species Richness, Diversity, Relative Dominance, and the Most Abundant Species of Chironomidae Collected from the Three Habitats Sampled in Bone Pile Pond, Delta Marsh, in 1980 and 1981**

		Potamogeton	Scirpus	Typha
Mean number per year		15,294.0–15,601.2	4,024.0–6,193.2	2,181.6–7,826.8
Number of species		53–62	42–58	49–49
Diversity (H')		2.76–3.65	3.36–3.45	3.52–3.69
Relative Dominance Index		61.2–45.2	44.9–46.6	46.4–40.2
Dominant Species[a]				
1980		*Tanypus punctipennis*	*Limnophyes immucronatus*	*Limnophyes prolongatus*
		Derotanypus alaskensis[b]	*Limnophyes prolongatus*	*Limnophyes immucronatus*
		Chironomus tentans	*Dicrotendipes nervosus*	*Chironomus tentans*
		Glyptotendipes barbipes	*Derotanypus alaskensis*	*Derotanypus alaskensis*
		Corynoneura cf. *scutellata*	*Paratanytarsus* sp. 1	*Tanypus punctipennis*
1981		*Cricotopus sylvestris*	*Paratanytarsus* sp. 1	*Cricotopus sylvestris*
		Tanypus punctipennis	*Limnophyes prolongatus*	*Paratanytarsus* sp. 1
		Glyptotendipes barbipes	*Glyptotendipes lobiferus*	*Limnophyes prolongatus*
		Cladotanytarsus sp.	*Corynoneura* cf. *scutellata*	*Corynoneura* cf. *scuellata*
		Corynoneura cf. *scutellata*	*Cricotopus sylvestris*	*Tanypus punctipennis*

Source: Calculated from data in Wrubleski and Rosenberg 1990.

[a] Abundant species are listed in decreasing order of abundance.

[b] As *Psectrotanypus alaskensis*.

CONCLUSIONS

Prairie marshes are dynamic habitats. The aquatic macrophytes, which typify these marshes, provide structurally complex habitats for aquatic invertebrates such as chironomids. Species composition and abundances of the chironomid assemblages within these habitats can change considerably among years (e.g., Morrill and Neal 1990, Wrubleski and Rosenberg 1990), but the factors responsible are not known. The results of my study indicate that changes in vegetation are important factors regulating chironomid communities in northern prairie wetlands.

The exact mechanisms of this regulation remain unknown. We do know that aquatic macrophytes provide habitat for many species. For example, in this study the two emergent macrophytes provided habitat for *Corynoneura* cf. *scutellata* and other epiphytic species. Death of these plants due to excessive flooding benefitted larger benthic species. This benefit may be in one or all of the following forms:

1. Plant litter provides soft material in which these species can burrow.
2. Algae and bacteria might colonize the abundant surface area provided by the litter and in turn provide an excellent food resource for chironomids.
3. The absence of emergent macrophytes might contribute to increased mixing of the water column and greater algal production on the bottom.

Further studies are needed to determine how dense stands of emergent macrophytes inhibit benthic chironomids. Field experiments, such as those described by Straskraba and Pieczynska (1970), are needed to tease apart the interactions between emergent macrophytes and chironomids. Much of this effort must use information about larvae. Emergence trap samples provide detailed descriptions of patterns, but are not capable of providing the information necessary to evaluate larval responses, the life cycle stage at which habitat manipulations have their true effects.

ACKNOWLEDGMENTS

This research was supported by the Marsh Ecology Research Program of Ducks Unlimited Canada and the Delta Waterfowl and Wetlands Research Station, the Department of Entomology at the University of Alberta, and the Natural Sciences and Engineering Research Council of Canada Grant A-2361 to John Spence. Lisette Ross and Karen Tome supervised field crews during the collection of the emergence trap samples. Field assistants were provided through the H. R. MacMillan and R. Howard Webster Fellowship Programs. Darold Batzer, Henry Murkin, Lisette Ross, and John Spence reviewed earlier drafts of the manuscript. Preparation of this contribution was made possible by Ducks Unlimited Canada through the Institute for Wetland and Waterfowl Research. This is Paper 91 of the Marsh Ecology Research Program, a joint project of Ducks Unlimited Canada and the Delta Waterfowl and Wetlands Research Station.

LITERATURE CITED

Andersson, A., and K. Danell. 1982. Response of freshwater macroinvertebrates to addition of terrestrial plant litter. Journal of Applied Ecology 19:319–325.

Armitage, P. D. 1977. Development of the macro-invertebrate fauna of Cow Green reservoir (Upper Teesdale) in the first five years of its existence. Freshwater Biology 7:441–454.

———. 1983. Chironomidae from Cow Green reservoir and its environs in the first years of its existence (1970–1977). Aquatic Insects 5:115–130.

Barnes, L. E. 1983. The colonization of ball-clay ponds by macroinvertebrates and macrophytes. Freshwater Biology 13:561–578.

Bataille, K. J., and G. A. Baldassarre. 1993. Distribution and abundance of aquatic macroinvertebrates following drought in three prairie pothole wetlands. Wetlands 13:260–269.

Batzer, D. P., M. McGee, V. H. Resh, and R. R. Smith. 1993. Characteristics of invertebrates consumed by mallards and prey response to wetland flooding schedules. Wetlands 13:41–49.

Blancher, P. J., and D. K. McNicol. 1991. Tree swallow diet in relation to wetland acidity. Canadian Journal of Zoology 69:2629–2637.

Brown, A. E., and R. S. Oldham. 1984. Chironomidae (Diptera) of Rutland Water. Archiv für Hydrobiologie Supplement 69:199–227.

Busby, D. G., and S. G. Sealy. 1979. Feeding ecology of a population of nesting yellow warblers. Canadian Journal of Zoology 57:1670–1681.

Campeau, S., H. R. Murkin, and R. D. Titman. 1994. Relative importance of algae and emergent plant litter to freshwater marsh invertebrates. Canadian Journal of Fisheries and Aquatic Sciences 51:681–692.

Cantrell, M. A., and A. J. McLachlan. 1977. Competition and chironomid distribution patterns in a newly flooded lake. Oikos 29:429–433.

Clark, W. R., and D. W. Kroeker. 1993. Population dynamics of muskrats in experimental marshes at Delta, Manitoba. Canadian Journal of Zoology 71:1620–1628.

Clement, S. L., A. A. Grigarick, and M. O. Way. 1977. The colonization of California rice paddies by chironomid midges. Journal of Applied Ecology 14:379–389.

Crumpton, W. G. 1989. Algae in northern prairie wetlands. Pages 188–203 in A. G. van der Valk (ed.), Northern Prairie Wetlands. Iowa State University Press, Ames, IA.

Danell, K., and K. Sjöberg. 1977. Seasonal emergence of chironomids in relation to egglaying and hatching of ducks in a restored lake (northern Sweden). Wildfowl 28:129–135.

Danell, K., and K. Sjöberg. 1982. Successional patterns of plants, invertebrates and ducks in a man-made lake. Journal of Applied Ecology 19:395–409.

Darby, R. E. 1962. Midges associated with California rice fields, with special reference to their ecology (Diptera: Chironomidae). Hilgardia 32:1–206.

Driver, E. A. 1977. Chironomid communities in small prairie ponds: Some characteristics and controls. Freshwater Biology 7:121–133.

Dvorak, J. 1969. Horizontal zonation of macrovegetation, water properties and macrofauna in a littoral stand of *Glyceria aquatica* (L.) Wahlb. in a pond in south Bohemia. Hydrobiologia 35:17–30.

Ertlová, E. 1980. Colonization of the littoral of the Liptovská Mara Reservoir by Chironomidae (Diptera) in the first two years after impoundment. Biológia (Bratislava) 35:311–319.

Euliss, N. H., Jr., and G. Grodhaus. 1987. Management of midges and other invertebrates for waterfowl wintering in California. California Fish and Game 73:238–243.

Fellton, H. L. 1940. Control of aquatic midges with notes on the biology of certain species. Journal of Economic Entomology 33:252–264.

Guinan, D. M., and S. G. Sealy. 1987. Diet of house wrens (*Troglodytes aedon*) and the abundance of the invertebrate prey in the dune-ridge forest, Delta Marsh, Manitoba. Canadian Journal of Zoology 65:1587–1596.

Gurney, S. E., and G. G. C. Robinson. 1988. The influence of water level manipulation on metaphyton production in a temperate freshwater marsh. Internationale Vereinigung für Theoretische und Angewandte Limnologie, Verhandlungen 23:1032–1040.

Hall, D. J., W. E. Cooper, and E. E. Werner. 1970. An experimental approach to the production dynamics and structure of freshwater animal communities. Limnology and Oceanography 15:839–928.

Harris, S. W., and W. H. Marshall. 1963. Ecology of water-level manipulations on a northern marsh. Ecology 44:331–343.

Hecky, R. E., R. W. Newbury, R. A. Bodaly, K. Patalas, and D. M. Rosenberg. 1984. Environmental impact prediction and assessment: The Southern Indian Lake experience. Canadian Journal of Fisheries and Aquatic Sciences 41:720–732.

Jacobsen, O. W. 1991. Feeding behaviour of breeding wigeon *Anas penelope* in relation to seasonal emergence and swarming behaviour of chironomids. Ardea 79:409–418.

Judd, W. W. 1953. A study of the population of insects emerging as adults from the Dundas Marsh, Hamilton, Ontario, during 1948. American Midland Naturalist 49:801–824.

———. 1958. Studies of the Byron Bog in southwestern Ontario. IX. Insects trapped as adults emerging from Redmond's Pond. Canadian Entomologist 90:623–627.

———. 1961. Studies of the Byron Bog in southwestern Ontario. XII. A study of the population of insects emerging as adults from Redmond's Pond in 1957. American Midland Naturalist 65:89–100.

Kesler, D. H. 1981. Grazing rate determination of *Corynoneura scutellata* Winnertz (Chironomidae: Diptera). Hydrobiologia 80:63–66.

Kimerle, R. A., and N. H. Anderson. 1971. Production and bioenergetic role of the midge *Glyptotendipes barbipes* (Staeger) in a waste stabilization lagoon. Limnology and Oceanography 16:646–659.

King, R. S., and D. A. Wrubleski. 1998. Spatial and diel availability of flying insects as potential duckling food in prairie wetlands. Wetlands 18:100–114.

Krebs, C. J. 1989. Ecological Methodology. Harper & Row, New York.

LeSage, L., and A. D. Harrison. 1979. Improved traps and techniques for the study of emerging aquatic insects. Entomological News 90:65–78.

Mann, K. H. 1988. Production and use of detritus in various freshwater, estuarine, and coastal marine ecosystems. Limnology and Oceanography 33:910–930.

Mason, C. F., and R. J. Bryant. 1975. Periphyton production and grazing by chironomids in Alderfen Broad, Norfolk. Freshwater Biology 5:271–277.

Matena, J. 1990. Succession of *Chironomus* Meigen species (Diptera, Chironomidae) in newly filled ponds. Internationale Revue der Gesamten Hydrobiologie 75:45–57.

McKnight, D. E., and J. B. Low. 1969. Factors affecting waterfowl production on a spring-fed salt marsh in Utah. Transactions of the North American Wildlife and Natural Resources Conference 34:307–314.

McLachlan, A. J. 1974. The development of chironomid communities in a new temperate impoundment. Entomologisk Tidskrift 95 Supplement:162–171.

McLaughlin, D. B., and H. J. Harris. 1990. Aquatic insect emergence in two Great Lakes marshes. Wetlands Ecology and Management 1:111–121.

McNaughton, S. J. 1967. Relationships among functional properties of California grasslands. Nature 216:168–169.

Menzie, C. A. 1981. Production ecology of *Cricotopus sylvestris* (Fabricius) (Diptera: Chironomidae) in a shallow estuarine cove. Limnology and Oceanography 26:467–481.

Millar, J. B. 1973. Vegetation changes in shallow marsh wetlands under improving moisture regime. Canadian Journal of Botany 51:1443–1457.

Moore, J. W. 1980. Factors influencing the composition, structure and density of a population of benthic invertebrates. Archiv für Hydrobiologie 88:202–218.

Morrill, P. K., and B. R. Neal. 1990. Impact of deltamethrin insecticide on Chironomidae (Diptera) of prairie ponds. Canadian Journal of Zoology 68:289–296.

Murduchai-Boltovskoi, F. B. 1961. Die Entwicklung der Bodenfauna in den Stauseen der Wolga. Internationale Vereinigung für Theoretische und Angewandte Limnologie, Verhandlungen 14:647–651.

Murkin, H. R. 1989. The basis for food chains in prairie wetlands. Pages 316–338 *in* A. G. van der Valk (ed.), Northern Prairie Wetlands. Iowa State University Press, Ames, IA.

Murkin, H. R., B. D. J. Batt, P. J. Caldwell, C. B. Davis, J. A. Kadlec, and A. G. van der Valk. 1984. Perspectives on the Delta Waterfowl Research Station—Ducks Unlimited Canada Marsh Ecology Research Program. Transactions of the North American Wildlife and Natural Resources Conference 49:253–261.

Murkin, H. R., and J. A. Kadlec. 1986. Responses by benthic macroinvertebrates to prolonged flooding of marsh habitat. Canadian Journal of Zoology 64:65–72.

Murkin, H. R., A. G. van der Valk, and C. B. Davis. 1989. Decomposition of four dominant macrophytes in the Delta Marsh, Manitoba. Wildlife Society Bulletin 17:215–221.

Neckles, H. A., J. W. Nelson, and R. L. Pederson. 1985. Management of Whitetop (*Scolochloa festucacea*) Marshes for Livestock Forage and Wildlife. Technical Bulletin 1. Delta Waterfowl and Wetlands Research Station, Portage la Prairie, MB.

Nursall, J. R. 1952. The early development of a bottom fauna in a new power reservoir in the Rocky Mountains of Alberta. Canadian Journal of Zoology 30:387–409.

Oláh, J. 1976. Energy transformation by *Tanypus punctipennis* (Meig.) (Chironomidae) in Lake Balaton. Annales Instituti Biologici (Tihany) Hungaricae Academiae Scientiarum 43:83–92.

Oliver, D. R., M. E. Dillon, and P. S. Cranston. 1990. A Catalog of Nearctic Chironomidae. Publication 1856/B. Research Branch, Agriculture Canada, Ottawa, ON.

Palmén, E., and L. Aho. 1966. Studies on the ecology and phenology of the Chironomidae (Dipt.) of the northern Baltic. 2 *Camptochironomus* Kieff. and *Chironomus* Meig. Annales Zoologici Fennici 3:217–244.

Parker, D. W. 1992. Emergence phenologies and patterns of aquatic insects inhabiting a prairie pond. Ph.D. Thesis, University of Saskatchewan, Saskatoon, SK.

Parkin, R. B., and J. B. Stahl. 1981. Chironomidae (Diptera) of Baldwin Lake, Illinois, a cooling reservoir. Hydrobiologia 76:119–128.

Paterson, C. G., and C. H. Fernando. 1970. Benthic fauna colonization of a new reservoir with particular reference to the Chironomidae. Journal of the Fisheries Research Board of Canada 17:213–232.

Robinson, G. G. C., S. E. Gurney, and L. G. Goldsborough. 1997a. Response of benthic and planktonic algal biomass to experimental water level manipulation in a prairie wetland. Wetlands 17:167–181.

——. 1997b. The primary productivity of benthic and planktonic algae in a prairie wetland under controlled water regimes. Wetlands 17:182–194.

Rose, C., and W. G. Crumpton. 1996. Effects of emergent macrophytes on dissolved oxygen dynamics in a prairie pothole wetland. Wetlands 16:495–502.

Rosenberg, D. M., B. Bilyj, and A. P. Wiens. 1984. Chironomidae (Diptera) emerging from the littoral zone of reservoirs, with special reference to Southern Indian Lake, Manitoba. Canadian Journal of Fisheries and Aquatic Sciences 41:672–681.

Sadler, W. O. 1935. Biology of the midge *Chironomus tentans* Fabricius, and methods for its propagation. Cornell University Agricultural Experimental Station Memoir 173:1–25.

Sealy, S. G. 1980. Reproductive responses of northern orioles to a changing food supply. Canadian Journal of Zoology 58:221–227.

Sephton, T. W., B. A. Hicks, C. H. Fernando, and C. G. Paterson. 1983. Changes in the chironomid (Diptera: Chironomidae) fauna of Laurel Creek Reservoir, Waterloo, Ontario. Journal of Freshwater Ecology 2:89–102.

Shay, J. M., and C. T. Shay. 1986. Prairie marshes in western Canada, with specific reference to the ecology of five emergent macrophytes. Canadian Journal of Botany 64:443–454.

Siegfried, W. R. 1973. Summer food and feeding of the ruddy duck in Manitoba. Canadian Journal of Zoology 51:1293–1297.

Sjöberg, K., and K. Danell. 1982. Feeding activity of ducks in relation to diel emergence of chironomids. Canadian Journal of Zoology 60:1383–1387.

St. Louis, V. L., L. Breebaart, and J. C. Barlow. 1990. Foraging behaviour of tree swallows over acidified lakes. Canadian Journal of Zoology 68:2385–2392.

Straskraba, M., and E. Pieczynska. 1970. Field experiments on shading effect by emergents on littoral phytoplankton and periphyton production. Rozpravy Ceskoslovenské Akademie Ved Rada Matematickych a Prirodnich Ved 80:7–32.

Street, M., and G. Titmus. 1979. The colonization of experimental ponds by Chironomidae (Diptera). Aquatic Insects 4:233–244.

Street, M., and G. Titmus. 1982. A field experiment on the value of allochthonous straw as food and substratum for lake macro-invertebrates. Freshwater Biology 12:403–410.

Suthers, I. M., and J. H. Gee. 1986. Role of hypoxia in limiting diel spring and summer distribution of juvenile yellow perch (*Perca flavescens*) in a prairie marsh. Canadian Journal of Fisheries and Aquatic Sciences 43:1562–1570.

Swanson, G. A., M. I. Meyer, and V. A. Adomaitis. 1985. Foods consumed by breeding mallards on wetlands of south-central North Dakota. Journal of Wildlife Management 49:197–203.

Swanson, S. M., and U. T. Hammer. 1983. Production of *Cricotopus ornatus* (Meigen) (Diptera: Chironomidae) in Waldsea Lake, Saskatchewan. Hydrobiologia 105:155–164.

Titmus, G. 1979. The emergence of midges (Diptera: Chironomidae) from a wet gravel-pit. Freshwater Biology 9:165–179.

Titmus, G., and R. M. Badcock. 1981. Distribution and feeding of larval Chironomidae in a gravel-pit lake. Freshwater Biology 11:263–271.

Topping, M. S. 1971. Ecology of larvae of *Chironomus tentans* (Diptera: Chironomidae) in saline lakes in central British Columbia. Canadian Entomologist 103:328–338.

van der Valk, A. G. 1986. The impact of litter and annual plants on recruitment from the seed bank of a lacustrine wetland. Aquatic Botany 24:13–26.

van der Valk, A. G., L. Squires, and C. H. Welling. 1994. Assessing the impacts of an increase in water level on wetland vegetation. Ecological Applications 4:525–534.

van der Valk, A. G., and C. H. Welling. 1988. The development of zonation in freshwater wetlands. An experimental approach. Pages 145–158 *in* H. J. During, M. J. A. Werger, and J. H. Willems (eds.), Diversity and Pattern in Plant Communities. SPB Academic, The Hague, The Netherlands.

van der Valk, A. G., C. H. Welling, and R. L. Pederson. 1989. Vegetation change in a freshwater wetland: a test of *a priori* predictions. Pages 207–217 *in* R. R. Sharitz and J. W. Gibbons (eds.), Freshwater Wetlands and Wildlife. Conf-8603101, DOE Symposium Series No. 61. USDOE Office of Scientific and Technical Information, Oak Ridge, TN.

Voigts, D. K. 1973. Food niche overlap of two Iowa marsh icterids. Condor 75:392–399.

Voshell, J. R., Jr., and G. M. Simmons, Jr. 1984. Colonization and succession of benthic macroinvertebrates in new reservoir. Hydrobiologia 112:27–39.

Welling, C. H., R. L. Pederson, and A. G. van der Valk. 1988a. Recruitment from the seed bank and the development of zonation of emergent vegetation during a drawdown in a prairie wetland. Journal of Ecology 76:483–496.

———. 1988b. Temporal patterns in recruitment from the seed bank during drawdowns in a prairie wetland. Journal of Applied Ecology 25:999–1007.

Whitman, W. R. 1974. The response of macro-invertebrates to experimental marsh management. Ph.D. Thesis, University of Maine, Orono, ME.

Wiens, A. P., and D. M. Rosenberg. 1984. Effect of impoundment and river diversion on profundal macrobenthos of Southern Indian Lake, Manitoba. Canadian Journal of Fisheries and Aquatic Sciences 43:638–648.

Wrubleski, D. A. 1984. Species composition, emergence phenologies, and relative abundances of Chironomidae (Diptera) from the Delta Marsh, Manitoba, Canada. Master's Thesis, University of Manitoba, Winnipeg, MB.

———. 1987. Chironomidae (Diptera) of peatlands and marshes in Canada. Pages 141–161 *in* D. M. Rosenberg and H. V. Danks (eds.), Aquatic Insects of Peatlands and Marshes in Canada. Memoirs of the Entomological Society of Canada 140.

———. 1989. The effect of waterfowl feeding on a chironomid (Diptera: Chironomidae) community. Pages 691–696 *in* R. R. Sharitz and J. W. Gibbons (eds.), Freshwater Wetlands and Wildlife. CONF-8603101, DOE Symposium Series No. 61. USDOE Office of Scientific and Technical Information, Oak Ridge, TN.

———. 1991. Chironomidae (Diptera) community development following experimental manipulation of water levels and aquatic vegetation. Ph.D. Thesis, University of Alberta, Edmonton, AB.

Wrubleski, D. A., and D. M. Rosenberg. 1984. Overestimates of Chironomidae (Diptera) abundances from emergence traps with polystyrene floats. American Midland Naturalist 111:195–197.

———. The Chironomidae (Diptera) of Bone Pile Pond, Delta Marsh, Manitoba, Canada. Wetlands 10:243–275.

Wrubleski, D. A., H. R. Murkin, A. G. van der Valk, and C. B. Davis. 1997. Decomposition of litter of three mudflat annual species in a northern prairie marsh during drawdown. Plant Ecology 129:141–148.

25 High Plains Wetlands of Southeast Wyoming

Salinity, Vegetation, and Invertebrate Communities

JAMES R. LOVVORN, WILFRED M. WOLLHEIM, and
E. ANDREW HART

Salinization of wetlands is an important issue in arid western North America. Despite the extreme value of saline wetlands to regional wildlife, there has been little work on mechanisms by which salinity alters foodweb structure and productivity. In lacustrine wetlands of the Laramie Basin, Wyoming, oligosaline wetlands (0.8–8 mS/cm) are vegetated mainly by dense beds of low macroalgae (Chara spp.), which are generally replaced in mesosaline wetlands (8–30 mS/cm) by less dense cover of erect angiosperms (Potamogeton pectinatus). These two plant growth forms support similar biomass of macroinvertebrates per unit area of vegetative cover. Taxonomic composition of invertebrates is strongly affected by salinity. Epiphytic snails and amphipods dominate invertebrate biomass in oligosaline wetlands, but are absent (snails) or greatly reduced (amphipods) in mesosaline wetlands. At the higher salinities they are replaced in predominant biomass by chironomid larvae and insect predators (odonates, hemipterans), along with greater densities of crustacean zooplankton (cladocerans, calanoid copepods). Based on preliminary analyses, we propose that these differences in community structure result mainly from direct salt toxicity to three competitive dominants—Chara, snails, and amphipods—with consequent indirect effects on other community components. More unstable water levels associated with higher salinities might also affect dominance by Chara.

Invertebrates in Freshwater Wetlands of North America: Ecology and Management, Edited by
Darold P. Batzer, Russell B. Rader, and Scott A. Wissinger
ISBN 0-471-29258-3 © 1999 John Wiley & Sons, Inc.

INTRODUCTION

Salinization of shallow lakes and wetlands is a serious problem in arid regions worldwide (Williams 1993, Lemly et al. 1993). In the United States salinity affects 23 percent of irrigated lands, and 77 to 91% in western regions such as the Columbia and Colorado Basins, Great Basin, and California's Central Valley (Ghassemi et al. 1995). In associated wetlands salinity affects community structure, diversity, and productivity (Hammer 1986, Hammer and Heseltine 1988, Hammer et al. 1990, Wollheim and Lovvorn 1995, 1996).

Taxonomic composition of invertebrates in saline wetlands has been described in detail (see Hammer 1986), but little has been published on their foodwebs, mechanisms by which salinity alters those foodwebs, and effects on upper trophic levels (Vareschi 1987, Wurtsbaugh 1992, Hart 1994, Cox and Kadlec 1995). Most previous studies have focused on either benthic or water-column habitats, alone or in pooled samples (Cannings and Scudder 1978, Lancaster and Scudder 1987, Hammer et al. 1990, Hammer and Forro 1992, Cox and Kadlec 1995). However, in many saline wetlands there is often much higher biomass of epiphytic invertebrates (Wollheim and Lovvorn 1995, 1996). In this study we determined how the biomass and taxonomic composition of benthic, planktonic, and epiphytic invertebrates differed between oligosaline and mesosaline wetlands in southeast Wyoming. We also evaluated mechanisms by which salinity might affect taxonomic composition and resulting trophic structure.

In North America invertebrates of inland saline waters have been studied in the western Canadian prairies (Thompson 1941, Rawson and Moore 1944, Tones 1976, Hammer et al. 1990), central British Columbia (Cannings and Scudder 1978, Lancaster and Scudder 1987, Hammer and Forro 1992), and the Central Valley of California (Severson 1987, Euliss et al. 1991, Parker and Knight 1992). Only a few published studies in southeast Wyoming (Wollheim and Lovvorn 1995, 1996) and the Great Salt Lake area (Huener and Kadlec 1992, Cox and Kadlec 1995), and ongoing studies in Colorado's San Luis Valley (Severn 1992, J. H. Gammonley, personal communication), have addressed invertebrate communities in saline wetlands of the Rocky Mountain region.

This relative lack of research belies the importance of saline wetlands in this region, which are in many cases the only wetlands over large areas. For example, Wyoming ranks sixth in the United States in waterfowl production; about 14 percent (51,800 in 1994) of these birds nest mostly in saline wetlands of the Laramie Basin, and an additional large fraction nest in saline wetlands throughout the state (Prenzlow 1994). About 18,700 ducks, mostly resident mallards *(Anas platyrhynchos)*, winter in Colorado's San Luis Valley, which is also a high, arid intermountain basin with many saline wetlands (Jeske 1993). The Monte Vista Wildlife Refuge in the southern San Luis Valley has the highest nesting densities and production of ducks reported for North America (Gilbert et al. 1996). Saline wetlands are important breeding, molting, migration, and wintering areas for a wide variety of birds throughout the Intermountain West (Kadlec and Smith 1989, Ratti and Kadlec 1992).

This chapter reviews results to date from a continuing study of community structure and foodweb function in shallow saline lakes (lacustrine wetlands) of the Laramie Basin in southeast Wyoming (Wollheim and Lovvorn 1995, 1996). Concepts emerging from this work are evolving, but for now will serve to stimulate debate and further research. In particular, we evaluate salinity effects on invertebrate communities through changes in plant habitat structure and direct salt toxicity to key species. We also propose a suite of mechanisms by which salinity alters foodweb structure and function in these wetlands.

HYDROLOGY, WATER CHEMISTRY, AND PLANT COMMUNITIES

The Laramie Basin is about 95 km long and 40 km wide and lies at an elevation of about 2200 m between the Laramie Range to the east and Medicine Bow Range to the west. The short, cool growing season restricts crop production, the main land use being cattle grazing and hay production through flood irrigation. Potential evapotranspiration of 52 cm/year is double the total precipitation of 26 cm/year (Martner 1986), so essentially all wetlands are saline (>0.5 g/L TDS or >0.8 mS/cm; Cowardin et al. 1979). The many scattered wetlands within the area are mostly wind-eroded playas, some of which have been enhanced by dikes, ditches, and water-control structures. As in many intermountain basins of western North America (e.g., Szymczak 1986), local surface runoff is negligible and snowmelt from adjacent mountains reaches wetlands via channelized irrigation flow and associated groundwater. Occurrence and amount of inflow to wetlands from hyporheic groundwater (about 0.75 g/L TDS) depend on the wind-scoured depth of a particular wetland's basin and its location relative to alluvial deposits recharged by leaking channels and irrigation return flows (Burritt 1962, Stanford and Ward 1993). The salinity of most area wetlands varies appreciably with frequency and amount of inflow from irrigation ditches and associated near-surface groundwater, mainly in May and June. Annual mountain snowpack and management of channelized flows thereby alter the arrangement of wetlands of different salinities in the landscape, which affects mobile consumers, such as birds, that move among different wetlands.

Salinity tolerance of both vertebrates and invertebrates can be altered by the ion composition of water, especially by predominance of Mg^{2+}, SO_4^{2-}, or both relative to Na^+ and Cl^- (Scudder 1969, Burnham and Peterka 1975, Mitcham and Wobeser 1988, Swanson et al. 1988, Goetsch and Palmer 1997). In the Laramie Basin in May 1996 we measured major ions in seven shallow lakes with a range of salinities (salinity classifications according to Cowardin et al. 1979). Oligosaline lakes George and Nelson (1.0–3.3 g/L TDS) were dominated by Ca^{2+} and SO_4^{2-}. Mesosaline lakes Soda, Hutton, and Creighton (7.5–15.5 g/L), polysaline Gibbs (27.9 g/L), and hypersaline Little Carroll

(131.6 g/L) were all dominated by Na^+ and SO_4^{2-}, with Creighton and Little Carroll also having appreciable Mg^{2+}. Concentration of Cl^- was far less than that of SO_4^{2-} in all lakes. Similar patterns have been reported for North Dakota (Swanson et al. 1988) and Saskatchewan (Rawson and Moore 1944).

Of wetlands we studied in summer 1992 (Wollheim and Lovvorn 1995, 1996), the oligosaline wetlands have had much more stable salinities over the last five years than have the mesosaline wetlands. The oligosaline wetlands generally receive flow from ditches, or shallow groundwater recharged by irrigation, every year. The mesosaline wetlands do not receive such flow every year and can vary between oligosaline (0.8–8 mS/cm, 0.5–5 g/L TDS) and mesosaline (8–30 mS/cm, 5–18 g/L), depending on annual water supply. One of our mesosaline lakes (Hoge) dried up in August 1992. Another mesosaline lake (Hutton) has received so much surface flow over the last few years that the formerly abundant submersed macrophytes were absent in 1996, probably owing to light attenuation by deeper water (Blindow 1992, Scheffer et al. 1993). Thus, wetlands classified as mesosaline have much more variable hydroperiods and salinities (cf. Brock and Lane 1983), and this greater variability probably affects biota in addition to absolute salinity values (Montague and Ley 1993).

In the Laramie Basin oligosaline lakes are dominated by the low-growing macroalgae *Chara* spp. (e.g., *C. globularis* in Lake George and *C. aspera* in Lake Nelson). During our initial invertebrate sampling in summer 1992 (Wollheim and Lovvorn 1995, 1996), the two oligosaline lakes studied (George and Nelson) also had appreciable areas of *Potamogeton pectinatus* and *Myriophyllum exalbescens*. However, after consistently high water levels in subsequent years, these plants were completely replaced by a monoculture of *Chara* by summer 1997. Oligosaline lakes have a fringe of the bulrush *Scirpus acutus*, which does not occur in mesosaline lakes due to salt intolerance (Hammer 1986). In contrast, mesosaline lakes are dominated by the erect submersed angiosperm *Potamogeton pectinatus,* with *Ruppia maritima* occurring in the upper mesosaline range. *Chara globularis* and *C. aspera* found in oligosaline lakes apparently cannot tolerate mesosaline waters (Hammer 1986). However, if mesosaline lakes (which do not receive inflow every year) are flooded to high levels and maintained in the lower mesosaline-upper oligosaline range over a series of years, *Chara longifolia* can become abundant offshore in deeper areas. At consistently higher salinities, marked shifts in dominance from *Chara* spp. to *P. pectinatus* suggest that salinity might affect invertebrates indirectly through changes in habitat structure of salt-sensitive plants, as well as by direct salt toxicity to the invertebrates themselves.

SAMPLING

To assess effects of salinity on invertebrates through altered habitat structure, Wollheim and Lovvorn (1995, 1996) studied benthic and epiphytic macroinvertebrates in stands of different wetland plants in the Laramie Basin in summer 1992. Study sites (and conductivity ranges in mS/cm) included oligosaline lakes Caldwell (0.8–1.1), George (2.2–2.4), and Nelson (2.9–4.2) and mesosaline lakes Hoge (9.3–19.0), Creighton (12.9–16.3), and Hutton (18.6–23.5). In oligosaline lakes macrophyte growth forms and species included thin-stemmed emergents *(Scirpus acutus)*, low macroalgae (*Chara* or benthic mats of filamentous algae), and erect aquatics *(Myriophyllum exalbescens* and *Potamogeton pectinatus)*. *Chara* spp. and benthic mats of filamentous algae were classified together because of similar height, density, and tangled growth form as opposed to erect angiosperms; low macroalgae were mainly *Chara globularis* in Lake George, *C. aspera* in Lake Nelson, and filamentous algae in Lake Caldwell. *Myriophyllum exalbescens* differs from *P. pectinatus* in having greater leaf dissection. Aside from sparse, late-season patches of *Chara* in Creighton Lake, the only macrophytes in mesosaline wetlands were *P. pectinatus* and *Ruppia maritima*. The latter two species were difficult to distinguish except by seed heads and fine structure of leaf sheaths, and represented a single growth form.

Epiphytic invertebrates were sampled by lowering a plexiglass tube 29.5 cm in diameter over the macrophytes (Wollheim and Lovvorn 1995, 1996). The tube was 50 cm tall, with a cylindrical net (100-μm mesh) attached to the top to increase height. Three legs supported the tube 10–15 cm above the sediments so that macrophytes could be cut at the sediment surface. A 100-μm net sewn to a flexible aluminum ring was then sealed around the tube bottom to retain severed macrophytes and invertebrates as the sample was brought to the water surface. Epiphytic samples included the water column up to the top of the plant canopy. Invertebrates were later rinsed from the plants, and those retained by a 500-μm sieve were analyzed. Data on epiphytic invertebrates were expressed as both biomass per unit area of lake bottom and biomass per unit volume of the water column (including the plants) up to the top of the plant canopy. Benthic invertebrates in the upper 5 cm of sediment were sampled directly beneath epiphytic samples with a 5-cm diameter plastic corer and washed through a 500-μm sieve, and the data were expressed as biomass per unit area of lake bottom.

To evaluate effects of macrophyte cover in general, we compared epiphytic and benthic invertebrates in macrophyte stands (including both erect aquatics and low macroalgae) with benthic and water-column invertebrates in unvegetated areas (Wollheim and Lovvorn 1995). Macroinvertebrates over unvegetated sediments were sampled by pulling a 19-cm diameter plankton net with 100-μm mesh from the sediment surface to water surface, sieving these samples through a 500-μm screen, and expressing the data as biomass per unit volume of water column.

For samples collected in summer 1992, recognized epiphytic taxa were classified as predators or grazers based on the literature, with no distinction between herbivores and detritivores (Wollheim and Lovvorn 1996). Because not all organisms were identified to species, we assumed that most members of given taxa belonged to the same consumer group. For example, Chironomidae include herbivores, detritivores, and predators, but most are herbivores and detritivores (Merritt and Cummins 1996, see below). Chironomids in 1992 samples were not identified below family. In summer 1997 we used the same plexiglass tube and core sampler to collect chironomid larvae for identification to subfamily and genus. Epiphytic (*Chara* or *P. pectinatus*) and core samples were collected from June 17–23 and July 28–29 in oligosaline lakes Nelson, George, and Hoge and from mesosaline Twin Buttes, Creighton, and Hutton (Hutton lacked macrophytes for epiphytic samples in 1992). Note that Hoge was mesosaline in 1992 but oligosaline in 1997. For 1997 samples, chironomid genera were classified as mainly primary consumers (collector-gatherers, filterers, scrapers, shredders, miners) or secondary consumers (engulfers and piercers) according to Merritt and Cummins (1996).

VEGETATION AND SALINITY EFFECTS ON INVERTEBRATE BIOMASS, TAXA, AND TROPHIC STRUCTURE

Our results indicated the following patterns:

1. In oligosaline lakes there were differences among plant growth forms in invertebrate biomass, but not in species composition.

Diversity of macrophyte species and growth forms declines from oligosaline to mesosaline lakes, so we expected greatest differentiation of invertebrate taxa among plant growth forms in oligosaline wetlands. However, within oligosaline lakes taxa did not differ appreciably among growth forms for either benthic or epiphytic invertebrates, except that epiphytic gastropods were more common in erect aquatics (Fig. 25.1). Thin-stemmed emergents (*Scirpus acutus*) supported lower biomass of epiphytic invertebrates, but about the same biomass of benthic invertebrates, as submersed forms (*M. exalbescens, P. pectinatus, Chara*/filamentous algae). Among submersed plants, low macroal-

---

25.1. Taxonomic composition (percent of total dry mass) of (*a*) epiphytic and (*b*) benthic macroinvertebrates (>500 μm) in low macroalgae (benthic mats of filamentous algae in Caldwell, *Chara globularis* in George, *C. aspera* in Nelson), *Myriophyllum exalbescens, Potamogeton pectinatus,* and *Scirpus acutus* in oligosaline lakes Caldwell, George, and Nelson of the Laramie Basin, Wyoming, June–August 1992. Predatory taxa are shown by exploded pie slices. In the legend for epiphytic invertebrates, other predators include mites, leeches, *Hydra*, and Ceratopogonidae; other grazers include Dixidae, Ephemeroptera, and Trichoptera. (Reproduced, with kind permission from Kluwer Academic Publishers, from W. M. Wollheim and J. R. Lovvorn, Salinity effects on macroinvertebrate assemblages and waterbird food webs in shallow lakes of the Wyoming High Plains, Hydrobiologia 310:207–223, 1995. ©1995 Kluwer Academic Publishers.)

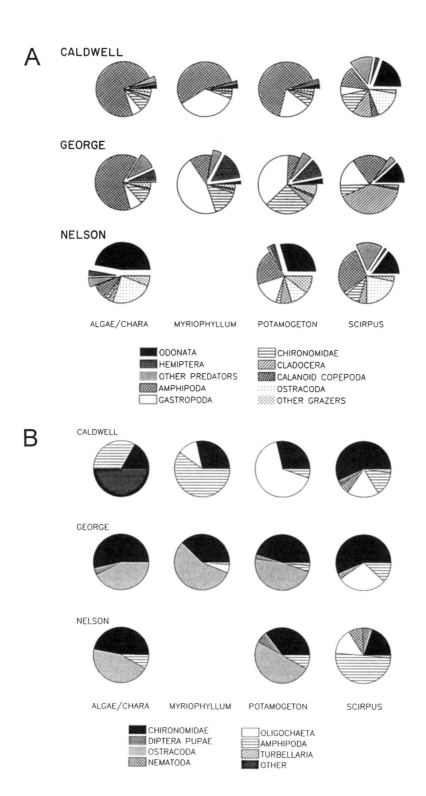

A

CALDWELL

GEORGE

NELSON

ALGAE/CHARA   MYRIOPHYLLUM   POTAMOGETON   SCIRPUS

■ ODONATA   ▤ CHIRONOMIDAE
▨ HEMIPTERA   ▧ CLADOCERA
▨ OTHER PREDATORS   ▨ CALANOID COPEPODA
▨ AMPHIPODA   ░ OSTRACODA
□ GASTROPODA   ▨ OTHER GRAZERS

B

CALDWELL

GEORGE

NELSON

ALGAE/CHARA   MYRIOPHYLLUM   POTAMOGETON   SCIRPUS

■ CHIRONOMIDAE   □ OLIGOCHAETA
▨ DIPTERA PUPAE   ▤ AMPHIPODA
▨ OSTRACODA   ▧ TURBELLARIA
▨ NEMATODA   ■ OTHER

gae had much higher biomass of epiphytic invertebrates per unit volume than did erect aquatics. However, the lower height of macroalgae in the water column resulted in biomass per unit area comparable to that of erect aquatics. In contrast, low macroalgae had less biomass of benthic invertebrates than erect angiosperms. Neither biomass or taxa of invertebrates differed significantly between *Myriophyllum exalbescens* and *Potamogeton pectinatus*.

2. Within the same growth form (*P. pectinatus/Ruppia maritima*) there were no differences between oligosaline and mesosaline lakes in total biomass of either epiphytic or benthic invertebrates, but taxonomic composition changed substantially.

Among epiphytic invertebrates (Fig. 25.2*a*), gastropods (snails) and amphipods were abundant in oligosaline lakes but were rare or absent in mesosaline lakes. Chironomid larvae and crustacean zooplankton (cladocerans and calanoid copepods), as well as their insect predators (odonates and hemipterans), were much more abundant in mesosaline lakes. Taxa of benthic invertebrates also changed at higher salinities (Fig. 25.2*b*). Ostracods were common in lakes of intermediate salinity (George, Nelson, and Hoge) but were unimportant in other lakes. Oligochaetes were abundant only in Caldwell, which had much higher nutrient levels than the other oligosaline lakes and was dominated by filamentous algae rather than *Chara*.

3. When lakes were pooled within salinity category, there were no differences between salinities in total biomass of epiphytic invertebrates or of benthic invertebrates in either vegetated or unvegetated areas. For zooplankton (macroinvertebrates in the water column in unvegetated areas), total biomass was greater in mesosaline than in oligosaline lakes (Wollheim and Lovvorn 1995, 1996).

4. Although total invertebrate biomass did not differ between oligosaline and mesosaline lakes, there were major differences in taxonomic and trophic composition that appeared not to result directly from changes in habitat structure.

Fig. 25.2. Taxonomic composition (percent of total dry mass) of (*a*) epiphytic and (*b*) benthic macroinvertebrates (>500 μm) beneath *Potamogeton pectinatus* in three oligosaline lakes (Caldwell, George, Nelson) and *P. pectinatus/Ruppia maritima* in three mesosaline lakes (Hoge, Creighton, Hutton) of the Laramie Basin, Wyoming, June–August 1992. Predatory taxa are shown by exploded pie slices. In the legend for epiphytic invertebrates, other predators include mites, leeches, *Hydra,* and Ceratopogonidae; other grazers include Dixidae, Ephemeroptera, and Trichoptera. (Reproduced, with kind permission from Kluwer Academic Publishers, from W. M. Wollheim and J. R. Lovvorn, Effects of macrophyte growth forms on invertebrate communities in saline lakes of the Wyoming High Plains, Hydrobiologia 323:83–96. ©1996 Kluwer Academic Publishers.)

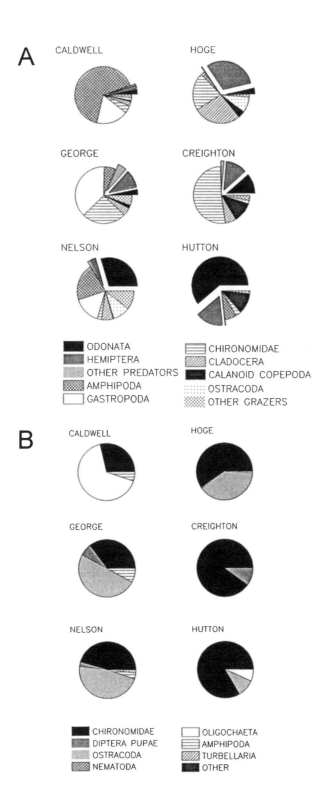

A

CALDWELL HOGE

GEORGE CREIGHTON

NELSON HUTTON

■ ODONATA ▤ CHIRONOMIDAE
▨ HEMIPTERA ▨ CLADOCERA
▦ OTHER PREDATORS ■ CALANOID COPEPODA
▨ AMPHIPODA ⦂ OSTRACODA
□ GASTROPODA ▨ OTHER GRAZERS

B

CALDWELL HOGE

GEORGE CREIGHTON

NELSON HUTTON

■ CHIRONOMIDAE □ OLIGOCHAETA
▨ DIPTERA PUPAE ▤ AMPHIPODA
▨ OSTRACODA ▨ TURBELLARIA
▨ NEMATODA ■ OTHER

Pooling lakes within salinity revealed major effects of salinity on invertebrate taxonomic composition (Fig. 25.3). Epiphytic snails and amphipods dominated total biomass in oligosaline lakes, but snails were absent and amphipods rare in mesosaline lakes. Chironomid larvae, crustacean zooplankton (cladocerans and calanoid copepods), and their insect predators (odonates and hemipterans) increased dramatically at higher salinities. In benthic habitats amphipods disappeared and chironomid larvae increased. In general, taxa richness in mesosaline lakes was much lower than in oligosaline lakes (Fig. 25.3, Table 25.1).

5. Chironomid larvae were mostly primary consumers (herbivores and detritivores), especially in mesosaline lakes.

For chironomid larvae in oligosaline lakes in 1997 (Table 25.1) the percentage of primary consumers as opposed to predators in cores was 37 percent (n = 89 larvae) for Nelson, 65 percent (n = 20) for George, and 93 percent (n = 14) for Hoge, for an average of 65 percent of larval numbers. In oligosaline epiphytic samples primary consumers made up 60 percent (n = 619) of larval numbers in Nelson and 100 percent (n = 245) in George, for a mean of 80 percent (no epiphytic samples were taken in Hoge in 1997). In core samples from mesosaline lakes, primary consumers made up 100 percent (n = 6) of larvae in Twin Buttes, 98 percent (n = 272) in Creighton, and 81 percent (n = 26) in Hutton, for an average of 93 percent. In mesosaline epiphytic samples primary consumers were 95 percent (n = 22) in Twin Buttes and 92 percent (n = 874) in Creighton, for a mean of 94 percent (there were no macrophytes in Hutton in 1997). Thus, except for core samples in Lake Nelson, all samples of chironomid larvae (including epiphytic samples from Nelson) were dominated by primary consumers. Primary consumers (collector-gatherers, scraper-herbivores) had overwhelming proportions in mesosaline lakes.

CONCEPTUAL MODEL OF COMMUNITY STRUCTURE IN SALINE WETLANDS

Our results showed major differences in community structure between oligosaline and mesosaline wetlands of the Laramie Basin in 1992. *Chara,* snails, and amphipods, which dominated macrophyte and invertebrate communities in oligosaline lakes, declined or disappeared at higher salinities. In shallow areas of mesosaline lakes these taxa were replaced as dominants by erect angiosperms, chironomid larvae, and insect predators, along with greater densities of crustacean zooplankton. To guide future research, we propose the following suite of mechanisms controlling these differences.

1. *Chara* spp. create dense benthic cover over the sediments in oligosaline wetlands. Due to salt toxicity or unstable water levels, *Chara* spp. do

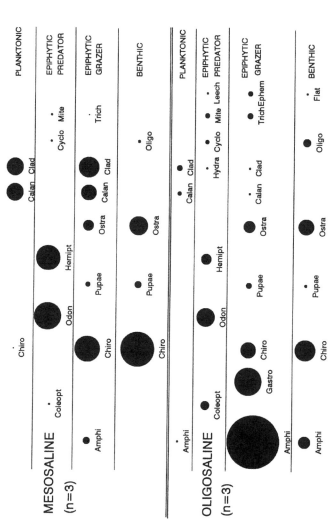

Fig. 25.3. Relative dry mass of major macroinvertebrate taxa (>500 μm) in oligosaline and mesosaline lakes of the Laramie Basin, Wyoming, June–August 1992. Planktonic, epiphytic predator, epiphytic grazer (all g/m^3), and benthic (g/m^2 for vegetated and unvegetated sediments combined) communities are shown. Relative biomass within each salinity category is combined over lake and month. Amphi = Amphipoda, Coleopt = Coleoptera, Gastro = Gastropoda, Chiro = Chironomidae, Odon = Odonata, Pupae = insect pupae (mainly Chironomidae), Hemipt = Hemiptera, Ostra = Ostracoda, Calan = Calanoid Copepoda, Clad = Cladocera, Cyclo = Cyclopoid Copepoda, Oligo = Oligochaeta, Trich = Trichoptera, Ephem = Ephemeroptera, Flat = flatworms. (Reproduced, with kind permission from Kluwer Academic Publishers, from W. M. Wollheim and J. R. Lovvorn, Salinity effects on macroinvertebrate assemblages and waterbird food webs in shallow lakes of the Wyoming High Plains, Hydrobiologia 310:207–223, 1995. © 1995 Kluwer Academic Publishers.)

613

TABLE 25.1. Genera and Species of Predominant Macroinvertebrate Orders in Three Oligosaline and Three Mesosaline Lakes of the Laramie Basin, Wyoming, June–August 1992[a]

Order	Family	Species	Oligosaline	Mesosaline
Gastropoda	Planorbidae	*Gyraulus parvus*	1	
	Physidae	*Physa gyrina*	3	
	Lymnaeidae	*Lymnaea caperata*	4	
Amphipoda	Hyalellidae	*Hyalella azteca*	1	1
	Gammaridae	*Gammarus lacustris*	3	
Hemiptera	Corixidae	*Cenocorixa* spp.	2	2
		Cenocorixa bifida	3	3
		Hesperocorixa laevigata	4	
		Graptocorixa		4
		Other	3	3
	Gerridae	*Gerris*	4	
		Trepobates	4	
	Hebridae	*Hebrus*	4	
	Macroveliidae	*Macrovelia*	4	
	Notonectidae	*Notonecta*	4	
Odonata	Coenagrionidae	*Enallagma*	2	1
	Lestidae	*Lestes*	3	
	Aeshnidae	*Aeshna*	4	
		Other	4	
Coleoptera	Dytiscidae	*Hygrotus*	2	2
		Hydaticus	3	
		Deronectes	4	3
		Liodessus	4	3
		Rhantus	4	
		Stictotarsus griseostriatus	4	

Order	Family	Genus/species		
	Haliplidae	*Haliplus*	2	3
	Curculionidae	*Hyperodes*		4
		Phytobius		4
	Gyrinidae	*Gyrinus*		4
	Hydrophilidae	*Helophorus*		4
		Paracymus terrestriae		4
	Staphylinidae	*Carpelimus*		4
Trichoptera	Leptoceridae	*Nectopsyche*	1	2
	Phryganeidae	*Phryganea*		2
Ephemeroptera	Baetidae	*Callibaetis*		2
	Caenidae	*Caenis*		2
Diptera	Chironomidae			
	(Tanypodinae)	*Larsia*		4
		Thienemannimyia		3
		Zavrelimyia		4
	(Orthocladiinae)	*Eukiefferiella*		4
		Parametriocnemus		4
		Orthocladius		3
	(Chironominae)	*Chironomus*		3
		Cryptochironomus		4
		Dicrotendipes		4
		Polypedilum		3
		Paratanytarsus		2

[a]See Wollheim and Lovvorn (1995, 1996). Branchiopoda, Copepoda, and Ostracoda were not identified to family level or below, and are not included. Data for Chironomidae are from core and epiphytic samples collected June–July 1997 (some lakes differ; see Sampling section). Within salinity category, the percentage of total numbers within an order comprised by different genera or species are expressed as the following ranks: 1 (> 70%), 2 (30–70%), 3 (5–30%), 4 (< 5%).

not occur in shallow areas of mesosaline lakes. Such areas are vegetated mainly by the erect submersed angiosperm *Potamogeton pectinatus,* which generally does not provide dense benthic cover.

2. In oligosaline lakes epiphyton growing on *Chara* spp. yielded 20–35 percent of total primary production, and epipelon (benthic algae) only 1–2 percent (Hart and Lovvorn 1999). However, in shallow areas of mesosaline lakes relative production by epipelon increases. This difference appears to result from replacement of *Chara* by *Potamogeton pectinatus* as dominant macrophyte, which greatly decreases macrophyte surface area for epiphyton and allows much greater light penetration to sediments (Hart and Lovvorn 1999).

3. The main epiphyton grazers in oligosaline wetlands are snails and amphipods (Fig. 25.3). These taxa disappear at higher salinities, probably due to salt toxicity.

4. In mesosaline wetlands epiphytic and benthic chironomid larvae increase (Fig. 25.3). We suggest that epiphytic chironomids (epiphyton grazers) increase due to loss or reduction of competing snails and amphipods. We propose that benthic chironomids (detritus and epipelon grazers) increase because of loss of *Chara,* increased light to sediments, and resulting increase of epipelon production.

5. In mesosaline wetlands insect predators (hemipterans, odonates) also increase (Fig. 25.3). It appears that these predators are food-limited at both salinities and increase at higher salinities due to greater availability of their main prey: chironomids and crustacean zooplankton.

6. In mesosaline wetlands there are increased crustacean zooplankton (cladocerans and copepods; Fig. 25.3). These zooplankton are probably food-limited in oligosaline wetlands. Zooplankton may increase in mesosaline wetlands because of:

   (a) greater phytoplankton production due to increased suspension of sediment-bound nutrients into the water column (due to loss of *Chara*);

   (b) greater production and entrainment of epipelon into the water column (due to loss of *Chara*);

   (c) More available P in the water column (due to of lack of sequestration of P in *Chara*).

By this construct, the pronounced differences in foodweb structure between oligosaline and mesosaline lakes result mainly from direct salt toxicity to *Chara,* snails, and amphipods, with resulting indirect effects on other community components. *Chara* controls zooplankton and benthos through its influence on algal production (by lowering nutrient availability and light to sediments). Snails and amphipods suppress chironomid biomass and production, and thus insect predators which eat mainly chironomids. Together, these three "foundation taxa" (Thorp and Bergey 1981) control the total amount and diversity of foods available to top vertebrate predators. If research confirms this construct, these communities are structured by suppression of three competitive dominants by salinity, which is in turn a function of annual hydrologic inputs. This scenario also suggests that *Chara*, snails, and amphipods are good indicators of the state of entire foodwebs as affected by freshwater inflow and salinization. What evidence currently exists to support the proposed mechanisms?

Salt Tolerance of Macrophytes, Epiphyte Grazers, and Insect Predators

Different species of *Chara* vary widely in salt tolerance (Hammer 1986). *Chara* is often not identified to species in wetland studies, because identification requires experience examining fine structures under magnification (Wood 1967). The dominant species of *Chara* in oligosaline lakes of the Laramie Basin, *C. globularis* and *C. aspera,* apparently cannot tolerate mesosaline waters, whereas *Potamogeton pectinatus* and *Ruppia maritima* are quite tolerant of higher salinities (Brock and Lane 1983, Hammer 1986, Hammer and Heseltine 1988). However, it is possible that *Chara* might replace *P. pectinatus* without salinity changes, depending on the frequency of disturbance by unstable water levels.

Salinity-driven dominance shifts between *Chara* and submersed angiosperms are complicated by different responses of *Chara* spp. and *P. pectinatus* to drawdown and flooding. Creighton Lake was mesosaline in 1992 (13–16 mS/cm), and *Chara* was uncommon there (Wollheim and Lovvorn 1995). However, in subsequent years this lake has been flooded to near-record levels, and in 1997 its salinity dropped from 11 mS/cm in May to 7 mS/cm in mid-June after continuous water inflow (mesosaline is defined as 8–30 mS/cm). By 1997 *Chara longifolia* had become quite abundant in offshore areas of the lake about 2 m deep. Meanwhile, flooded areas of former salt flat along shorelines had been colonized by pure stands of *P. pectinatus*. These patterns suggest that *P. pectinatus* is better able to colonize newly flooded or disturbed areas (often with higher soil salinities), whereas *C. longifolia* is able to dominate areas with less disturbance such as ice scour and dewatering of sediments. However, areas in Creighton Lake where *C. longifolia* became abundant retain at least a meter of water even during droughts, so its relative scarcity in 1992 probably resulted from salt intolerance rather than lowered water level per se.

Other studies have indicated successional shifts from *P. pectinatus* toward *Chara* during stable periods after water-level disturbances. Following very high or low water levels that greatly reduced macrophyte cover in Swedish lakes, *P. pectinatus* expanded first but was later replaced by *Chara tomentosa* (Blindow 1992). Similar replacement of *P. pectinatus* by *Chara* spp. has occurred in the Netherlands after nutrient reductions and resulting decreased turbidity (Coops and Doef 1996, van den Berg et al. 1997). Once established, *Chara* gradually improved its own light conditions and relative dominance by reducing sediment resuspension, tying up nutrients, and perhaps providing refugia for invertebrate grazers of phytoplankton and epiphyton (Coops and Doef 1996, van den Berg et al. 1997).

In the Laramie Basin, with consistently high water levels since 1992, *Chara* has expanded its coverage in several oligosaline lakes. In Lakes George and Nelson *C. globularis* and *C. aspera* had achieved total dominance by 1997, with complete disappearance of *P. pectinatus,* which was common in 1992. These lakes remain oligosaline in all years, so it appears that the disturbance of low water levels in the years preceding 1992 had provided *P. pectinatus* opportunities for establishment. Thus, unstable water levels more common in mesosaline lakes might favor *P. pectinatus* over *Chara* spp. regardless of salinity. Nevertheless, it appears that salinity alone is sufficient to prevent abundant growth of *C. globularis, C. aspera,* and *C. longifolia* in mesosaline lakes of the Laramie Basin, although more unstable water levels at higher salinities might play a role.

For snails, intolerance of most inland species to salinities above the oligosaline range is well known. In the littoral zones of Saskatchewan lakes, snails (taxa not given) decreased from several hundred per m^2 at salinities below 2.5 g/L TDS to 0–4 individuals per m^2 at ≥ 7.8 g/L, to none whatsoever at ≥ 13.4 g/L (Thompson 1941). Of the three snail species found in 18 saline lakes in Alberta and Saskatchewan, *Gyraulus deflectus* and *Physa gyrina* did not occur at salinities above 3.2 g/L, and *Lymnaea stagnalis* not above 5.9 g/L (Hammer et al. 1990). In North Dakota most snails such as *Lymnaea stagnalis* were restricted to salinities <5 mS/cm, and the unusually salt-tolerant *Stagnicola* spp. could not exist at sustained salinities >10 mS/cm (Swanson et al. 1988). Restriction to oligosaline waters was also reported for snails in Australia (Williams et al. 1990). In the Laramie Basin we found no snails above the oligosaline range (Fig. 25.3, Table 25.1) despite abundant periphyton foods (Hart and Lovvorn 1999).

Reduction of the amphipods *Gammarus lacustris* and *Hyalella azteca* in mesosaline wetlands was not as dramatic as for snails (Fig. 25.3), but still appeared to result from physiological limitation rather than reduction in food or habitat structure. In Saskatchewan and Alberta *G. lacustris* occurred mainly at salinities <9 g/L, whereas *H. azteca* was abundant at salinities up to 37 g/L (Hammer et al. 1990); *H. azteca* was generally much more common at all salinities (see also Thompson 1941, Rawson and Moore 1944, Timms et al. 1986). However, *H. azteca* densities in mesocosms were significantly lower at salinities of 8.0 and 11.0 g/L than at 5.6 g/L, and mortality increased dramatically at salinities >12 g/L (Galat et al. 1988). In other experiments (Nebeker and Miller 1988) production of young *H. azteca* declined appreciably between 7.5 and 10 g/L (11.8–15.2 mS/cm), with some decrease in survival of adults above 10 g/L (15.2 mS/cm). Severe reduction of young did not occur until 15 g/L (20.7 mS/cm), and of adult survival at about 20 g/L (27.7 mS/cm). The ion composition of experimental waters was not stated in either of these studies. Scudder (1969) suggests that variations in occurrence of *H. azteca* are strongly influenced by ion composition: in central British Columbia *H. azteca* occurred only up to 12 mS/cm in Na^+-dominated lakes, but up to 22 mS/cm in Mg^{2+}-dominated waters. As mesosaline wetlands in the Laramie Basin are all Na^+-dominated, it is likely that the much lower biomass of amphipods in mesosaline lakes (8–30 mS/cm; Fig. 25.3) results from direct salt toxicity to adults, young, or reproduction.

Amphipods are often especially abundant in beds of *Chara* (Thompson 1941, van den Berg et al. 1997), which might be attributed to its dense escape cover from predatory fish or tiger salamanders (*Ambystoma tigrinum*). In the Laramie Basin fish and tiger salamanders appear uncommon or absent in mesosaline lakes, and common in oligosaline lakes only in years of consistent water inflow—immigration of fish from irrigation ditches each year after winterkill appears important. The fish are mainly fathead minnows (*Pimephales promelas*) and Iowa darters (*Etheostoma exile*), which both eat amphipods in this area (Copes 1970). Tiger salamanders are known to be important predators of *H. azteca* and *Gammarus* (Dodson and Dodson 1971, Olenick and Gee 1981), but in 1992 appeared uncommon in our study wetlands. Given that fish and salamanders appeared rare or absent in mesosaline lakes in 1992, reduction of amphipods at the higher salinities probably did not result from loss of *Chara* as escape cover from these predators.

In contrast to salt toxicity to snails and amphipods at the higher salinities, dramatic increase of insect predators in mesosaline lakes (Fig. 25.3) probably did not result from physiological intolerance of lower salinities. In Saskatchewan, Alberta, and British Columbia Odonata (mainly *Enallagma*) and Hemiptera occurred commonly in both oligosaline and mesosaline lakes, and few species that occurred at higher salinities failed to occur or were less abundant at lower salinities (Thompson 1941, Rawson and Moore 1944, Lauer 1969, Scudder 1969, Tones 1976, Cannings et al. 1980, Timms et al. 1986, Lancaster and Scudder 1987, Hammer et al. 1990). Detailed analyses for one species of Corixidae (*Cenocorixa expleta*) revealed that its absence at lower salinities did not result from physiological intolerance of low salinities (Scudder 1983). Based on these and other studies (Knowles and Williams 1973, Williams et al. 1990), it is unlikely that the lower abundance of insect predators in oligosaline lakes of the Laramie Basin resulted from osmoregulatory factors.

These arguments also apply to chironomid larvae. A study of 86 lakes in British Columbia revealed that none of 36 chironomid genera that occurred in mesosaline waters failed to occur in oligosaline lakes, and only three genera (*Psectrocladius, Cricotopus, Orthocladius*) were notably more abundant in oligosaline than in mesosaline lakes (Walker et al. 1995). Also in British Columbia, Cannings and Scudder (1978) found that of 34 chironomid species occurring in the mesosaline range (8–12 mS/cm), none did not also occur in oligosaline waters, although the reverse was often true (see also Lauer 1969).

Further research on common species in the Laramie Basin is needed, but it appears that the markedly greater biomass of both chironomid larvae and insect predators in mesosaline wetlands (Fig. 25.3) did not result from physiological response to salinity.

Competition among Epiphyte Grazers: Snails and Amphipods versus Chironomids

We propose that epiphytic chironomid larvae increased in mesosaline wetlands because of elimination or strong reduction of competing snails and amphipods. In oligosaline wetlands chironomids might be suppressed by either competition for food ("consumptive competition," Schoener 1983) or direct interference competition by physical displacement or predation. Cuker (1983) suggests that in arctic lakes grazing by the snail *Lymnaea elodes* reduced densities of tube-dwelling chironomids through both competition for periphyton foods and physical displacement of larvae and adults. Hawkins and Furnish (1987) found that in Oregon streams removal of the competitive dominant snail *Juga silicula* resulted in increased numbers of chironomid larvae, especially tube-dwelling species. It appeared that *Juga* both competed with the chironomids for food and "bulldozed" over substrates, thereby displacing sessile larvae. Although such physical displacement has not been directly documented, food limitation and resulting competition among algivorous snails and larvae of chironomids and other insects have been widely reported (Eisenberg 1970, Mason and Bryant 1975, Lamberti et al. 1987, Batzer and Resh 1991, Hill 1992, Steinman 1996). As a variety of studies have shown snails to be competitive dominants over chironomid larvae, elimination of snails by salt toxicity is a likely contributor to increased chironomid biomass in mesosaline wetlands of the Laramie Basin.

Amphipods are probably also competitive dominants over chironomid larvae, through both competition for food and, in the case of *Gammarus*, direct predation. Although the major food of *H. azteca* in a British Columbia saline lake was detritus in bottom sediments, *H. azteca* assimilated epiphytes on *Chara* with far greater efficiency than leaf detritus or sediment organic matter (Hargrave 1970a). *H. azteca* was not carnivorous in experimental trials (Hargrave 1970b). Anderson and Raasveldt (1974) found that *Gammarus lacustris* rejected plant matter in favor of live animal prey (cladocerans *Daphnia* spp. and copepods *Diaptomus* spp.) and readily consumed midge larvae (*Clinotanypus* sp., *Chaoborus americanus*), whereas *H. azteca* ignored animal foods when offered. Thus, *G. lacustris* can readily exist on a plant or detritus diet, but unlike *H. azteca* it can also be predatory.

The relative importance of consumptive versus interference competition between *Gammarus lacustris* and insect larvae is unclear. Anderson and Raasveldt (1974) suggest, based on laboratory experiments, that predation by *G. lacustris* was an important reason that it rarely cooccurred with the midge

Chaoborus in 52 Alberta lakes studied. Menon (1966) also observed *G. lacustris* feeding on chironomid larvae in winter in Manitoba. Savage (1981) concludes that in British saline lakes the formerly abundant corixid *Sigara lateralis* was eliminated by *Gammarus tigrinus* because of the amphipod's competitive dominance for plant foods and its direct predation on early corixid instars; the relative importance of these mechanisms was not explored. Hart (1994) concludes, from mesocosm experiments on foodwebs from California's Salton Sea, that the scarcity of other invertebrates (*Trichocorixa, Artemia, Apocyclops, Balanus,* and nematodes) at lower salinities resulted from predation by *Gammarus mucronatus;* at higher salinities *G. mucronatus* was much reduced by salt toxicity and the other taxa became abundant. Similar release of chironomids and other insects (Hemiptera, Odonata) as salt toxicity reduces amphipod densities appears to occur in the Laramie Basin. If amphipods indeed suppress chironomids, the fact that *Hyalella azteca* (a mostly nonpredatory algal and detrital grazer) is much more abundant than *Gammarus lacustris* in these lakes (Table 25.1) suggests that competition for food is more important than predation.

The idea that snails and amphipods suppress chironomid larvae through competition for food depends on the chironomid species present being mostly periphyton grazers. In 1997 most chironomid larvae in both benthic and epiphytic samples were classified as collector-gatherers or scrapers, overwhelmingly so in mesosaline lakes. Work is underway in our study area to evaluate chironomid diets further with multiple stable isotopes (cf. Neill and Cornwell 1992).

Food Limitation of Insect Predators

As noted above, insect predators (Odonata, Hemiptera) might be limited in oligosaline wetlands by predation on early-instars by *Gammarus lacustris.* However, the amphipods present are mostly nonpredatory *Hyalella azteca,* and results of our foodweb simulation models indicate that insect predators in the Laramie Basin are strongly food-limited (see also Benke 1976, Lawton et al. 1980).

Odonates and hemipterans (mainly corixids in the Laramie Basin; Table 25.1) sometimes eat snails or amphipods when they are much more abundant than other foods, but in most cases chironomid larvae are overwhelmingly more important food items (often 70–80 percent by mass) (for Corixidae and Notonectidae see Reynolds 1975, Zalom 1978, Giller 1986, Reynolds and Scudder 1987; for Odonata see Pritchard 1964, Fischer 1966, Lawton 1970, Pearlstone 1973, Benke 1976, Thompson 1978). Both Pritchard (1964) and Pearlstone (1973) report experiments in which odonates had difficulty capturing gammarid amphipods because of their high speed and hard, smooth cuticle. Pearlstone states that *Enallagma* (the common genus in the Laramie Basin) never struck at mollusks in the laboratory and seldom ate amphipods even in lakes where *Hyalella azteca* was quite numerous. Thus, an increase in food-limited insect predators is expected in mesosaline wetlands where their chironomid foods are much more abundant (Fig. 25.3, Table 25.1).

Also, the productivity of chironomids is generally much higher than that of either snails or amphipods. Production:mean biomass (P:B) ratios of 30 or more have been commonly reported for chironomids (Benke 1984), whereas values for *Hyalella azteca* are 3.8–5.7 (Wen 1992, France 1993) and for pulmonate snails about 5.8 (Brown 1991). Benke (1976) notes that odonates often have standing stocks two to three times those of their prey (mainly chironomid larvae), so the prey must turn over at enormous rates to support their predators. Prey P:B ratios required to support odonates (around 30 in some studies) might be overestimated because odonates eat more in experiments than in the wild (Lawton 1971). However, it appears that production of snails or amphipods, even if they were eaten, would be inadequate to support large populations of odonates. Thus, the release of chironomid larvae from competition with snails and amphipods at higher salinities might result not only in more suitable prey for insect predators, but also in increased productivity of prey. Both these factors favor higher populations of insect predators. In this way salt toxicity to snails and amphipods might have important indirect effects on both community structure and total secondary production.

Alternatively, lack of fish in mesosaline lakes might have eliminated control of insect predators by fish (cf. Prejs et al. 1997). However, in 1992 water supply in the Laramie Basin was only 63 percent of the 80-year average and followed five similar years, including a 35-year low in 1989 (45 percent of average). Even oligosaline lakes received little inflow during this period. Fish, mainly fathead minnows and Iowa darters that apparently immigrate into the lakes via irrigation ditches each spring after winterkill, seemed rare in the oligosaline lakes we studied—they are usually quite conspicuous in these lakes when present. We do not feel that lack of fish in mesosaline lakes explained the higher abundance of insect predators there in 1992.

Increased Water-Column Production and Zooplankton

In mesosaline wetlands there was increased biomass of crustacean zooplankton (cladocerans and copepods; Fig. 25.3). Although species richness often declines (Hammer and Forro 1992, Green 1993), similar increases in zooplankton biomass with increasing salinity at intermediate ranges have been noted in other saline lakes (Whittaker and Fairbanks 1958, Edmondson et al. 1962, Reynolds 1979). This pattern might result from (a) salt toxicity to predators of zooplankton or (b) increased food for zooplankton at higher salinities.

Several authors have suggested that increased copepod abundance results at least partly from salt intolerance of important predators (Galat and Robinson 1983, Campbell 1995). Epibenthic and mainly grazing predators such

as amphipods, which are much reduced at higher salinities (Fig. 25.3), have
been shown to affect densities of pelagic prey (Anderson and Raasveldt 1974,
Schwartz 1992, Hart 1994). Few studies have explored the relative roles of
zooplankton limitation by food availability versus predators, but one detailed
study indicates that predation functioned mainly to influence the timing of
food limitation (Neill 1981). Given that insect predators of crustacean zooplankton increase at higher salinities in the Laramie Basin (Fig. 25.3), it seems
unlikely that reduction of amphipods results in lowering of overall predation.
Thus, increased zooplankton at higher salinities does not seem to result from
salt toxicity to predators.

Alternatively, food available to these zooplankton might increase. Our computer simulations suggest that zooplankton are food-limited in the oligosaline
wetlands, and other studies have shown that phytoplankton standing crops
often increase with salinity at intermediate ranges (Bierhuizen and Prepas
1985). We propose that zooplankton increased in mesosaline wetlands because of (a) greater phytoplankton production due to increased suspension of
sediment-bound nutrients into the water column (due to loss of *Chara*) (Crawford 1977); (b) greater production and entrainment of epipelon into the water
column (due to loss of *Chara*) (Goldsborough and Robinson 1996); and/or
(c) more available (assimilable) P in the water column (due to lack of sequestration of P in *Chara*) (Kufel and Ozimek 1994).

Chara spp. carpet the sediments in oligosaline wetlands of the Laramie
Basin, and our measurements show that *Chara* restricts light penetration to
sediments much more than does *P. pectinatus* in mesosaline lakes (Hart and
Lovvorn 1999). In summer 1997 total phosphorus in the water column averaged 46 ppb in mesosaline Creighton and Twin Buttes, versus only 30 ppb
in oligosaline George and Nelson. The percentage of total phosphorus that
was particulate (versus dissolved) was 55–71 percent in the mesosaline lakes
versus only 34–36 percent in the oligosaline lakes. Chlorophyll-*a* in the water
column was 67 percent higher in mesosaline lakes, and DOC concentrations
were 59 percent higher (Hart and Lovvorn 1999). These data all indicate
greater food available for zooplankton in mesosaline lakes, and suggest that
Chara might suppress water-column production in oligosaline lakes. Suppression of microalgae by *Chara* via allelopathy has been proposed (Hootsman and Blindow 1994).

SALINITY AND FOUNDATION SPECIES

Are there species that characterize salinity effects on wetlands in the Rocky
Mountain region, especially regarding trophic pathways to top vertebrate consumers? Identifying such species is important for predicting functional
changes due to species loss or other perturbations, assessing wetland functions
and values, finding indicators of water quality or ecosystem integrity, and
prioritizing research and management efforts (Dayton 1972, Walker 1992).

Although the desirability of identifying such species is recognized, definitions and appropriate applications have been strongly debated (Mills et al. 1993). The original concept of "keystone" species refers to those that prey on competitive dominants that would otherwise overwhelm and exclude less effective competitors (Paine 1966). Power et al. (1996) expanded this usage by defining keystone species as those exerting strong effects on community structure through large per capita or per biomass impacts, versus "dominant" species that exert strong effects because of their large biomass per se or large fraction of the system's energy passing through them (see also Raffaelli and Hall 1996). Both types are subsumed under Dayton's (1972) "foundation" species, defined as those disproportionately important to maintaining existing community structure, including species that provide most of the habitat structure (e.g., macrophytes), competitive dominants, and disturbers preventing their domination.

In the Laramie Basin there appear to be three major competitive dominants in oligosaline lakes: snails, amphipods, and *Chara*. According to our initial model, snails and amphipods outcompete and suppress chironomids in oligosaline wetlands, and *Chara* spp. gradually outcompete and replace *Potamogeton pectinatus* in the absence of disturbance at low salinities. There seem to be no consumers that regulate relative dominance among these taxa. Instead, salinity exerts abiotic constraints on mesosaline communities by (1) suppressing snails and amphipods with resulting release of chironomids and their insect predators and (2) raising food limits on crustacean zooplankton by eliminating *Chara* with its negative effects on their particulate foods in the water column. Jeppesen et al. (1994) note similar salinity effects on competitive dominance among grazing zooplankton, resulting in reduced grazing capacity and reduced diversity of fish predators. As Thorp and Bergey (1981) suggest generally for soft-bottom lentic habitats, there appear to be no major keystone consumers in wetlands of the Laramie Basin in the classic sense (Mills et al. 1993). Snails, amphipods, and chironomid larvae are foundation species, but not keystone species.

Fish and salamanders appear rare in mesosaline wetlands of the Laramie Basin, and can be uncommon in oligosaline wetlands depending on water inflows in a given year (E. A. Hart and J. R. Lovvorn, unpublished data). Thus, birds are often the main top consumers (Wollheim and Lovvorn 1995). Impacts of highly mobile consumers on community structure are generally quite patchy in space and time, depending on movements in response to foraging profitability (Lovvorn and Gillingham 1996). Such effects are difficult to measure and are probably underestimated (Edwards et al. 1982, Paine 1988). Computer simulations of species removals can suggest possible impacts (Pimm 1980), but results of simulated perturbations depend on self-fulfilling assumptions about species response to altered communities. Raffaelli and Hall (1996) obtained important insights into the role of birds in a soft-bottom estuarine community with field exclosures. In that community birds did not appear to have keystone effects, although eider ducks ate almost the entire production of their main prey (blue mussels) every year.

COMMUNITY CHANGE AND TOP CONSUMERS

In oligosaline lakes of the Laramie Basin 20–35 percent of primary production is by epiphytic algae (Hart and Lovvorn 1999). Snails and amphipods are the main grazers of this algae in oligosaline lakes, but appear to be replaced by chironomid larvae in mesosaline lakes due to salt intolerance of snails and amphipods. Does this major change in invertebrate grazers affect transfer of the system's main carbon source to higher trophic levels?

Most wetland birds are capable of opportunistic diet shifts among invertebrate taxa (Wollheim and Lovvorn 1995, Skagen and Oman 1996). Both *Chara* and *P. pectinatus* yield edible foliage and oogonia or seeds consumed by ducks. However, differences in vegetation structure (erect canopy of *P. pectinatus* versus low growth form of *Chara;* Wollheim and Lovvorn 1996) might make both plant and animal foods more available to surface-feeding birds in mesosaline wetlands. Crome (1986) presents evidence that both larval and emerging chironomids are more available and easily captured by surface-feeding birds, especially chicks, than are snails, amphipods, or insect predators. Total standing stock of invertebrates can be higher in either oligosaline or mesosaline wetlands, depending on nutrient levels (Wollheim and Lovvorn 1995). However, chironomid larvae generally have much higher *production* than do snails and amphipods (see above discussion). Whatever the mechanisms, Kingsford and Porter (1994) found that waterbird numbers were much higher on a saline wetland than on an adjacent freshwater wetland. Ongoing studies in the Laramie Basin will examine such relations for different bird species across wetlands of different salinity.

FUTURE WORK

Severson (1987) suggests that in any management program for invertebrates a decision must be made about which groups of invertebrates to favor. Although community structure and perhaps secondary production vary appreciably between oligosaline and mesosaline wetlands in the Laramie Basin, there are insufficient data on bird diets to judge which community is better for foraging birds (Wollheim and Lovvorn 1995). An important research priority is to obtain diet information for different bird species in wetlands of different salinity.

Foodwebs are complex systems, and it is hard to study interactions throughout intact foodwebs with experiments at reasonable cost. Thus, foodweb function is often evaluated with computer models such as network analyses (e.g., Baird and Ulanowicz 1989). Integration of empirical results with modeling, from study design to analysis and interpretation, provides a valuable tool for identifying critical information gaps and future research priorities. The rather abrupt transition in community structure between oligosaline and mesosaline wetlands (Fig. 25.3; Hammer 1986, Hammer et al. 1990) allows us to treat these systems in a discontinuous fashion with static models, as done successfully in other environments (Stone et al. 1993, Deegan et al. 1994, Gaedke and Straile 1994). In the Laramie Basin we are using multiple stable isotopes to verify and refine the trophic models. Field mesocosms from which snails and amphipods are removed (Raffaelli and Hall 1996) would also be helpful in testing our proposed model of foodweb control.

Salinity is an easily measured and regulated parameter that can be used to predict wetland function at individual to landscape scales and can be directly linked to water distribution systems in arid areas (Powell 1958). Computer models have been developed to predict effects of salinity changes on fish and entire foodwebs in estuaries (Powell and Matsumoto 1994, Baird and Heymans 1996), and such approaches have much potential for guiding future work in inland wetlands. Integration of empirical and modeling studies seems likely to produce the best linkages between research and management of invertebrates in saline wetlands of the Rocky Mountain region.

ACKNOWLEDGMENTS

We thank M. Z. Derby, R. Dillon, and B. C. Kondratiev for identifying taxa in Table 25.1. S. L. Hill, A. L. Hamel, and L. A. Fretz-Stotts did important field and laboratory work on primary producers and chironomid collections. V. W. Proctor identified species of *Chara*, and L. D. DeBrey identified chironomid genera. S. H. Anderson arranged funding for the initial study. This work was supported by the Wyoming Game and Fish Department and the Wyoming Water Resources Center.

LITERATURE CITED

Anderson, R. S., and L. G. Raasveldt. 1974. *Gammarus* Predation and the Possible Effects of *Gammarus* and *Chaoborus* Feeding on the Zooplankton Composition in Some Small Lakes and Ponds in Western Canada. Canadian Wildlife Service Occasional Paper 18.

Baird, D., and J. J. Heymans. 1996. Assessment of ecosystem changes in response to freshwater inflow of the Kromme River estuary, St. Francis Bay, South Africa: A network analysis approach. Water SA 22:307–318.

Baird, D., and R. E. Ulanowicz. 1989. The seasonal dynamics of the Chesapeake Bay ecosystem. Ecological Monographs 59:329–364.

Batzer, D. P., and V. H. Resh. 1991. Trophic interactions among a beetle predator, a chironomid grazer, and periphyton in a seasonal wetland. Oikos 60:251–257.

Benke, A. C. 1976. Dragonfly production and prey turnover. Ecology 57:915–927.

———. 1984. Secondary production of aquatic insects. Pages 289–322 *in* V. H. Resh and D. M. Rosenberg (eds.), The Ecology of Aquatic Insects. Praeger New York.

Blindow, I. 1992. Long- and short-term dynamics of submerged macrophytes in two shallow eutrophic lakes. Freshwater Biology 28:15–27.

Bierhuizen, J. F. H., and E. E. Prepas. 1985. Relationship between nutrients, dominant ions, and phytoplankton standing crop in prairie saline lakes. Canadian Journal of Fisheries and Aquatic Sciences 42:1588–1594.

Brock, M. A., and J. A. K. Lane. 1983. The aquatic macrophyte flora of saline wetlands in Western Australia in relation to salinity and permanence. Hydrobiologia 105:63–76.

Brown, K. M. 1991. Mollusca: Gastropoda. Pages 285–314 *in* J. H. Thorp and A. P. Covich (eds.), Ecology and Classification of North American Freshwater Invertebrates. Academic Press, New York.

Burnham, B. L., and J. J. Peterka. 1975. Effects of saline water from North Dakota lakes on survival of fathead minnow *(Pimephales promelas)* embryos and sac fry. Journal of the Fisheries Research Board of Canada 32:809–812.

Burritt, E. C. 1962. A ground water study of part of the southern Laramie Basin, Albany County, Wyoming. M.A. Thesis, University of Wyoming, Laramie, WY.

Campbell, C. E. 1995. The influence of a predatory ostracod, *Australocypris insularis*, on zooplankton abundance and species composition in a saline lake. Hydrobiologia 302:229–239.

Cannings, R. A., S. G. Cannings, and R. J. Cannings. 1980. The distribution of the genus *Lestes* in a saline lake series in central British Columbia, Canada (Zygoptera: Lestidae). Odonatologica 9:19–28.

Cannings, R. A., and G. G. E. Scudder. 1978. The littoral Chironomidae (Diptera) of saline lakes in central British Columbia. Canadian Journal of Zoology 56:1144–1155.

Coops, H., and R. W. Doef. 1996. Submerged vegetation development in two shallow, eutrophic lakes. Hydrobiologia 340:115–120.

Copes, F. A. 1970. A study of the ecology of the native fishes of Sand Creek, Albany County, Wyoming. Ph.D. Dissertation, University of Wyoming, Laramie, WY.

Cox, R. R., and J. A. Kadlec. 1995. Dynamics of potential waterfowl foods in Great Salt Lake marshes during summer. Wetlands 15:1–8.

Cowardin, L. M., V. Carter, F. C. Golet, and E. T. LaRoe. 1979. Classification of Wetlands and Deepwater Habitats of the United States. United States Fish and Wildlife Service, FWS/OBS-79/31.

Crawford, S. A. 1977. Chemical, physical and biological changes associated with *Chara* succession in farm ponds. Hydrobiologia 55:209–217.

Crome, F. H. J. 1986. Australian waterfowl do not necessarily breed on a rising water level. Australian Wildlife Research 13:461–480.

Cuker, B. E. 1983. Competition and coexistence among the grazing snail *Lymnaea*, chironomidae, and microcrustacea in an arctic epilithic lacustrine community. Ecology 64:10–15.

Dayton, P. K. 1972. Toward an understanding of community resilience and the potential effects of enrichments to the benthos at McMurdo Sound, Antarctica. Pages 81–95 *in* B. C. Parker (ed.), Conservation Problems in Antarctica. Allen Press, Lawrence, KS.

Deegan, L. A., J. T. Finn, C. S. Hopkinson, A. E. Giblin, B. J. Peterson, B. Fry, and J. E. Hobbie. 1994. Flow model analysis of the effects of organic matter-nutrient interactions on estuarine trophic dynamics. Pages 273–281 *in* K. R. Dyer and R. J. Orth (eds.), Changes in Fluxes in Estuaries: Implications from Science to Management. Olsen & Olsen, Fredensborg, Denmark.

Dodson, S. I., and V. E. Dodson. 1971. The diet of *Ambystoma tigrinum* from western Colorado. Copeia 1971:614–624.

Edmondson, W. T., G. W. Comita, and G. C. Anderson. 1962. Reproductive rates of copepods in nature and its relation to phytoplankton dynamics. Ecology 43:625–634.

Edwards, D. C., D. O. Conover, and F. Sutter. 1982. Mobile predators and the structure of marine intertidal communities. Ecology 63:1175–1180.

Eisenberg, R. M. 1970. The role of food in the regulation of the pond snail, *Lymnaea elodes*. Ecology 51:680–684.

Euliss, N. H., R. L. Jarvis, and D. S. Gilmer. 1991. Standing crops and ecology of aquatic invertebrates in agricultural drainwater ponds in California. Wetlands 11:179–190.

Fischer, Z. 1966. Food selection and energy transformation in larvae of *Lestes sponsa* (Odonata) in astatic waters. Internationale Vereinigung für Limnologie, Verhandlungen 16:600–603.

France, R. L. 1993. Production and turnover of *Hyalella azteca* in central Ontario, Canada compared with other regions. Freshwater Biology 30:343–349.

Gaedke, U., and D. Straile. 1994. Seasonal changes of trophic transfer efficiencies in a plankton food web derived from biomass size distributions and network analysis. Ecological Modelling 75/76:435–445.

Galat, D. L., M. Coleman, and R. Robinson. 1988. Experimental effects of elevated salinity on three benthic invertebrates in Pyramid Lake, Nevada. Hydrobiologia 158:133–144.

Galat, D. L., and R. Robinson. 1983. Predicted effects of increasing salinity on the crustacean zooplankton community of Pyramid Lake, Nevada. Hydrobiologia 105:115–131.

Ghassemi, F., A. J. Jakeman, and H. A. Nix (eds.). 1995. Salinisation of Land and Water Resources. University of New South Wales Press, Sydney, Australia.

Gilbert, D. W., D. R. Anderson, J. K. Ringelman, and M. R. Szymczak. 1996. Response of nesting ducks to habitat and management on the Monte Vista National Wildlife Refuge, Colorado. Wildlife Monographs 131:1–44.

Giller, P. S. 1986. The natural diet of the Notonectidae: field trials using electrophoresis. Ecological Entomology 11:163–172.

Goetsch, P.-A., and C. G. Palmer. 1997. Salinity tolerances of selected macroinvertebrates of the Sabie River, Kruger National Park, South Africa. Archives of Environmental Contamination and Toxicology 32:32–41.

Goldsborough, L. G., and G. G. C. Robinson. 1996. Pattern in wetlands. Pages 77–117 *in* R. J. Stevenson, M. L. Bothwell, and R. L. Lowe (eds.), Algal Ecology. Academic Press, New York.

Green, J. 1993. Zooplankton associations in East African lakes spanning a wide salinity range. Hydrobiologia 267:249–256.

Hammer, U. T. 1986. Saline Lake Ecosystems of the World. Dr. W. Junk, Dordrecht, The Netherlands.

Hammer, U. T., and L. Forro. 1992. Zooplankton distribution and abundance in saline lakes of British Columbia, Canada. International Journal of Salt Lake Research 1:65–80.

Hammer, U. T., and J. M. Heseltine. 1988. Aquatic macrophytes in saline lakes of the Canadian prairies. Hydrobiologia 158:101–116.

Hammer, U. T., J. S. Sheard, and J. Kranabetter. 1990. Distribution and abundance of littoral benthic fauna in Canadian prairie saline lakes. Hydrobiologia 197:173–192.

Hargrave, B. T. 1970a. Utilization of benthic microflora by *Hyalella azteca* (Amphipoda). Journal of Animal Ecology 39:427–437.

———. 1970b. Distribution, growth, and seasonal abundance of *Hyalella azteca* (Amphipoda) in relation to sediment microflora. Journal of the Fisheries Research Board of Canada 27:685–699.

Hart, C. M. 1994. Salinity and fish effects on Salton Sea invertebrates: A microcosm experiment. Master's Thesis, San Diego State University, San Diego, CA.

Hart, E. A., and J. R. Lovvorn. 1999. Vegetation dynamics and primary production in saline, lacustrine wetlands of a Rocky Mountain basin. Aquatic Botany, in press.

Hawkins, C. P., and J. K. Furnish. 1987. Are snails important competitors in stream ecosystems? Oikos 49:209–220.

Hill, W. R. 1992. Food limitation and interspecific competition in snail-dominated streams. Canadian Journal of Fisheries and Aquatic Sciences 49:1257–1267.

Hootsman, M. J. M., and I. Blindow. 1994. Allelopathic limitation of algal growth by macrophytes. Pages 175–192 *in* W. van Vierssen, M. Hootsman, and J. Vermaat (eds.), Lake Veluwe, a Macrophyte-Dominated System under Eutrophication Stress. Kluwer Academic, Dordrecht, The Netherlands.

Huener, J. D., and J. A. Kadlec. 1992. Macroinvertebrate response to marsh management strategies in Utah. Wetlands 12:72–78.

Jeppesen, E., M. Sondergaard, E. Kanstrup, B. Petersen, R. B. Eriksen, M. Hammershoj, E. Mortensen, J. P. Jensen, and A. Have. 1994. Does the impact of nutrients on the biological structure and function of brackish and freshwater lakes differ? Hydrobiologia 275/276:15–30.

Jeske, C. W. 1993. Factors influencing size of the wintering mallard population in the San Luis Valley, Colorado. Southwestern Naturalist 38:155–181.

Kadlec, J. A., and L. M. Smith. 1989. The Great Basin marshes. Pages 451–474 *in* L. M. Smith, R. L. Pederson, and R. M. Kaminski (eds.), Habitat Management for Migrating and Wintering Waterfowl in North America. Texas Tech University Press, Lubbock, TX.

Kingsford, R. T., and J. L. Porter. 1994. Waterbirds on an adjacent freshwater lake and salt lake in arid Australia. Biological Conservation 69:219–228.

Knowles, J. N., and W. D. Williams. 1973. Salinity range and osmoregulatory ability of corixids (Hemiptera: Heteroptera) in south-east Australian inland waters. Australian Journal of Marine and Freshwater Research 24:297–302.

Kufel, L., and T. Ozimek. 1994. Can *Chara* control phosphorus cycling in Lake Luknajno (Poland)? Hydrobiologia 275/276:277–283.

Lamberti, G. A., J. W. Feminella, and V. H. Resh. 1987. Herbivory and intraspecific competition in a stream caddisfly population. Oecologia 73:75–81.

Lancaster, J., and G. G. E. Scudder. 1987. Aquatic Coleoptera and Hemiptera in some Canadian saline lakes: Patterns in community structure. Canadian Journal of Zoology 65:1383–1390.

Lauer, G. J. 1969. Osmotic regulation of *Tanypus nubifer, Chironomus plumosus*, and *Enallagma clausum* in various concentrations of saline lake water. Physiological Zoology 42:381–387.

Lawton, J. H. 1970. Feeding and food energy assimilation in larvae of the damselfly *Pyrrhosoma nymphula* (Sulz.) (Odonata: Zygoptera). Journal of Animal Ecology 39:669–689.

———. 1971. Maximum and actual field feeding-rates in larvae of the damselfly *Pyrrhosoma nymphula* (Sulzer) (Odonata: Zygoptera). Freshwater Biology 1:99–111.

Lawton, J. H., B. A. Thompson, and D. J. Thompson. 1980. The effects of prey density on survival and growth of damselfly larvae. Ecological Entomology 5:39–51.

Lemly, A. D., S. E. Finger, and M. K. Nelson. 1993. Sources and impacts of irrigation drainwater contaminants in arid wetlands. Environmental Toxicology and Chemistry 12:2265–2279.

Lovvorn, J. R., and M. P. Gillingham. 1996. Food dispersion and foraging energetics: a mechanistic synthesis for field studies of avian benthivores. Ecology 77:435–451.

Martner, B. E. 1986. Wyoming Climate Atlas. University of Nebraska Press, Lincoln, NE.

Mason, C. F., and R. J. Bryant. 1975. Periphyton production and grazing by chironomids in Alderfen Broad, Norfolk. Freshwater Biology 5:271–277.

Menon, P. S. 1966. Population ecology of *Gammarus lacustris lacustris* Sars in Big Island Lake. Ph.D. Dissertation, University of Alberta, Edmonton, AB.

Merritt, R. W., and K. W. Cummins (eds.). 1996. An Introduction to the Aquatic Insects of North America (3rd ed.). Kendall/Hunt, Dubuque, IA.

Mills, L. S., M. E. Soule, and D. F. Doak. 1993. The keystone-species concept in ecology and conservation. BioScience 43:219–224.

Mitcham, S. A., and G. Wobeser. 1988. Effects of sodium and magnesium sulfate in drinking water on mallard ducklings. Journal of Wildlife Diseases 24:30–44.

Montague, C. L., and J. A. Ley. 1993. A possible effect of salinity fluctuation on abundance of benthic vegetation and associated fauna in northeastern Florida Bay. Estuaries 16:703–717.

Nebeker, A. V., and C. E. Miller. 1988. Use of the amphipod crustacean *Hyalella azteca* in freshwater and estuarine sediment toxicity tests. Environmental Toxicology and Chemistry 7:1027–1033.

Neill, C., and J. C. Cornwell. 1992. Stable carbon, nitrogen, and sulfur isotopes in a prairie marsh food web. Wetlands 12:217–224.

Neill, W. E. 1981. Impact of *Chaoborus* predation upon the structure and dynamics of a crustacean zooplankton community. Oecologia 48:164–177.

Olenick, R. J., and J. H. Gee. 1981. Tiger salamanders (*Ambystoma tigrinum*) and stocked rainbow trout (*Salmo gairdneri*): Potential competitors for food in Manitoba prairie pothole lakes. Canadian Field-Naturalist 95:129–132.

Paine, R. T. 1966. Food web complexity and species diversity. American Naturalist 100:65–75.

Paine, R. T. 1988. Food webs: Road maps of interactions or grist for theoretical development? Ecology 69:1648–1654.

Parker, M. S., and A. W. Knight. 1992. Aquatic invertebrates inhabiting saline evaporation ponds in the southern San Joaquin Valley, California. Bulletin of the Southern California Academy of Sciences 91:39–43.

Pearlstone, P. S. M. 1973. The food of damselfly larvae in Marion Lake, British Columbia. Syesis 6:33–39.

Pimm, S. L. 1980. Food web design and the effect of species deletions. Oikos 35: 139–149.

Powell, G. L., and J. Matsumoto. 1994. Texas estuarine mathematical programming model: A tool for freshwater inflow management. Pages 401–406 *in* K. R. Dyer and R. J. Orth (eds.), Changes in Fluxes in Estuaries: Implications from Science to Management. Olsen & Olsen, Fredensborg, Denmark.

Powell, W. J. 1958. Ground-water Resources of the San Luis Valley, Colorado. United States Geological Survey Water-Supply Paper 1379.

Power, M. E., D. Tilman, J. A. Estes, B. A. Menge, W. J. Bond, L. S. Mills, G. Daily, J. C. Castilla, J. Lubchenco, and R. T. Paine. 1996. Challenges in the quest for keystones. BioScience 46:609–620.

Prejs, A., P. Koperski, and K. Prejs. 1997. Food-web manipulation in a small, eutrophic Lake Wirbel, Poland: The effect of replacement of key predators on epiphytic fauna. Hydrobiologia 342/343:377–381.

Prenzlow, D. M. 1994. Design and evaluation of a waterfowl breeding population survey for Wyoming. Master's Thesis, University of Wyoming, Laramie, WY.

Pritchard, G. 1964. The prey of dragonfly larvae (Odonata: Anisoptera) in ponds in northern Alberta. Canadian Journal of Zoology 42:785–800.

Raffaelli, D. G., and S. J. Hall. 1996. Assessing the relative importance of trophic links in food webs. Pages 185–191 *in* G. A. Polis and K. O. Winemiller (eds.), Food Webs: Integration of Patterns and Dynamics. Chapman & Hall, New York.

Ratti, J. T., and J. A. Kadlec. 1992. Concept plan for the preservation of wetland habitat of the Intermountain West. United States Fish and Wildlife Service, Portland, OR.

Rawson, D. S., and J. E. Moore. 1944. The saline lakes of Saskatchewan. Canadian Journal of Research D 22:141–201.

Reynolds, J. D. 1975. Feeding in corixids (Heteroptera) of small alkaline lakes in central B.C. Internationale Vereinigung für Limnologie, Verhandlungen 19:3073–3078.

———. 1979. Crustacean zooplankton of some saline lakes of central British Columbia. Syesis 12:169–173.

Reynolds, J. D., and G. G. E. Scudder. 1987. Serological evidence of realized feeding niche in *Cenocorixa* species (Hemiptera: Corixidae) in sympatry and allopatry. Canadian Journal of Zoology 65:974–980.

Savage, A. A. 1981. The Gammaridae and Corixidae of an inland saline lake from 1975 to 1978. Hydrobiologia 76:33–44.

Scheffer, M., S. H. Hosper, M.-L. Meijer, B. Moss, and E. Jeppesen. 1993. Alternative equilibria in shallow lakes. Trends in Ecology and Evolution 8:275–279.

Schoener, T. W. 1983. Field experiments on interspecific competition. American Naturalist 122:240–285.

Schwartz, S. S. 1992. Benthic predators and zooplanktonic prey: predation by *Crangonyx shoemakeri* (Crustacea; Amphipoda) on *Daphnia obtusa* (Crustacea; Cladocera). Hydrobiologia 237:25–30.

Scudder, G. G. E. 1969. The fauna of saline lakes on the Fraser Plateau in British Columbia. Internationale Vereinigung für Limnologie, Verhandlungen 17:430–439.

———. 1983. A review of factors governing the distribution of two closely related corixids in the saline lakes of British Columbia. Hydrobiologia 105:143–154.

Severn, C. 1992. Distribution of aquatic insects along a hydrologic gradient in the Russell Lakes area, San Luis Valley, Colorado. Master's Thesis, Colorado School of Mines, Golden, CO.

Severson, D. J. 1987. Macroinvertebrate populations in seasonally flooded marshes in the northern San Joaquin Valley of California. Master's Thesis, Humboldt State University, Arcata, CA.

Skagen, S. K., and H. D. Oman. 1996. Dietary flexibility of shorebirds in the Western Hemisphere. Canadian Field-Naturalist 110:419–444.

Stanford, J. A., and J. V. Ward. 1993. An ecosystem perspective of alluvial rivers: connectivity and the hyporheic corridor. Journal of the North American Benthological Society 12:48–60.

Steinman, A. D. 1996. Effects of grazers on freshwater benthic algae. Pages 341–373 *in* R. J. Stevenson, M. L. Bothwell, and R. L. Lowe (eds.), Algal Ecology. Academic Press, New York.

Stone, L., T. Berman, R. Bonner, S. Barry, and S. W. Weeks. 1993. Lake Kinneret: A seasonal model for carbon flux through the planktonic biota. Limnology and Oceanography 38:1680–1695.

Swanson, G. A., T. C. Winter, V. A. Adomaitis, and J. W. LaBaugh. 1988. Chemical Characteristics of Prairie Lakes in South-central North Dakota—Their Potential for Influencing Use by Fish and Wildlife. United States Fish and Wildlife Service, Fish and Wildlife Technical Report 18.

Szymczak, M. R. 1986. Characteristics of Duck Populations in the Intermountain Parks of Colorado. Colorado Division of Wildlife Technical Publication 35.

Thompson, D. J. 1978. Prey size selection by larvae of the damselfly, *Ischnura elegans* (Odonata). Journal of Animal Ecology 47:769–785.

Thompson, J. S. 1941. Study of the macroscopic bottom fauna of eight Saskatchewan lakes of varying salinities. M.A. Thesis, University of Saskatchewan, Saskatoon, SK.

Thorp, J. H., and E. A. Bergey. 1981. Field experiments on responses of a freshwater, benthic macroinvertebrate community to vertebrate predators. Ecology 62:365–375.

Timms, B. V., U. T. Hammer, and J. W. Sheard. 1986. A study of benthic communities in some saline lakes in Saskatchewan and Alberta, Canada. Internationale Revue der Gesamten Hydrobiologie 71:759–777.

Tones, P. I. 1976. Factors influencing selected littoral fauna in saline lakes of Saskatchewan. Ph.D. Dissertation. University of Saskatchewan, Saskatoon, SK.

van den Berg, M. S., H. Coops, R. Noordhuis, J. van Schie, and J. Simons. 1997. Macroinvertebrate communities in relation to submerged vegetation in two *Chara*-dominated lakes. Hydrobiologia 342/343:143–150.

Vareschi, E. 1987. Saline lake ecosystems. Pages 347–364 *in* E.-D. Schulze and H. Zwolfer (eds.), Potentials and Limitations of Ecosystem Analysis. Springer-Verlag, New York.

Walker, B. H. 1992. Biodiversity and ecological redundancy. Conservation Biology 6: 18–23.

Walker, I. R., S. E. Wilson, and J. P. Smol. 1995. Chironomidae (Diptera): Quantitative paleosalinity indicators for lakes of western Canada. Canadian Journal of Fisheries and Aquatic Sciences 52:950–960.

Wen, Y. H. 1992. Life history and production of *Hyalella azteca* (Crustacea: Amphipoda) in a hypereutrophic prairie pond in southern Alberta. Canadian Journal of Zoology 70:1417–1424.

Whittaker, R. H., and C. W. Fairbanks. 1958. A study of plankton copepod communities in the Columbia Basin, southeastern Washington. Ecology 39:46–65.

Williams, W. D. 1993. Conservation of salt lakes. Hydrobiologia 267:291–306.

Williams, W. D., A. J. Boulton, and R. G. Taaffe. 1990. Salinity as a determinant of salt lake fauna: A question of scale. Hydrobiologia 197:257–266.

Wollheim, W. M., and J. R. Lovvorn. 1995. Salinity effects on macroinvertebrate assemblages and waterbird food webs in shallow lakes of the Wyoming High Plains. Hydrobiologia 310:207–223.

Wollheim, W. M., and J. R. Lovvorn. 1996. Effects of macrophyte growth forms on invertebrate communities in saline lakes of the Wyoming High Plains. Hydrobiologia 323:83–96.

Wood, R. D. 1967. Charophytes of North America. Stella's Printing, West Kingston, RI.

Wurtsbaugh, W. A. 1992. Food-web modification by an invertebrate predator in the Great Salt Lake (USA). Oecologia 89:168–175.

Zalom, F. G. 1978. Backswimmer prey selection with observations on cannibalism (Hemiptera: Notonectidae). Southwestern Naturalist 23:617–622.

26 Playas of the Southern High Plains

The Macroinvertebrate Fauna

DIANNE L. HALL, ROBERT W. SITES,
ERNEST B. FISH, TONY R. MOLLHAGEN,
DARYL L. MOORHEAD, and MICHAEL R. WILLIG

*P*laya lakes of the semiarid southern Great Plains are shallow, ephemeral wetlands that usually contain water from late spring to fall. They provide critical wildlife habitat in an area with a paucity of permanent water and are important foci for recharge to the Ogallala Aquifer. These lakes support a diversity of macroinvertebrates, including more than 124 taxa. Resident taxa (those with drought-resistant life stages) are dominated by phyllopod and ostracod crustaceans, as well as snails, whereas insects are the major constituents of the transient taxa (those that must immigrate after inundation). Residents are most abundant during early stages of playa succession, and consist mainly of detritivores. Transients dominate later successional stages and are mostly herbivores and predators. To date, most conservation studies of playa lakes have focused on management strategies that increase habitat quality for migrating and breeding waterfowl. Much of the research has focused on moist-soil management of vegetation as food resources for waterfowl. Little has been done to explore the impact of these conservation measures on the aquatic macroinvertebrate fauna. Similarly, little research has focused on the effects of surrounding land use practices on the fauna of playa lakes. Macroinvertebrates occupy many of the intermediate trophic levels in foodwebs of playas, and consequently may directly or indirectly affect taxa at higher and lower trophic levels. Future research should focus on the effects of landscape characteristics on aquatic macroinvertebrates from the joint perspectives of conservation biology and agroecology.

Invertebrates in Freshwater Wetlands of North America: Ecology and Management, Edited by
Darold P. Batzer, Russell B. Rader, and Scott A. Wissinger
ISBN 0-471-29258-3   © 1999 John Wiley & Sons, Inc.

SOUTHERN HIGH PLAINS

The Southern High Plains, or Llano Estacado, is an extensive, semiarid tableland (82,000 km²) lying south of the Canadian River in Texas and New Mexico (Fig. 26.1). It is devoid of permanently flowing surface water, but contains numerous ephemeral lakes or playas (Reddell 1965, Wood and Osterkamp 1984). These playas are the most important elements of surface hydrology and ecological diversity in a landscape that otherwise is dominated by agricultural activities.

The Texas High Plains is a large component of the Southern High Plains, extending from the northern Panhandle to the Trans-Pecos and Edwards Plateau. Like the rest of the Llano Estacado, it was dominated by shortgrass prairies (Stoddart and Smith 1955) until agricultural development began in the early 1900s. Irrigation from groundwater began in the area following World War II, and by 1977 more than 70,000 wells had tapped the Ogallala Aquifer in Texas (New 1979).

Agricultural development is pervasive on the Southern High Plains, producing one third of the nation's cotton. Irrigated crops are grown on about

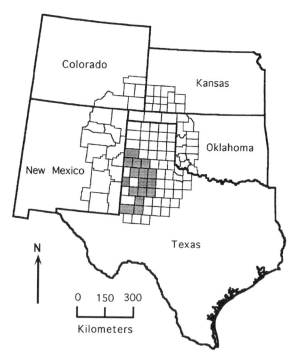

Fig. 26.1. Map of the 111 counties on the southern Great Plains that comprise the Playa Lakes Region (Playa Lake Joint Venture 1993). Playas considered in this chapter occur in the 16 shaded counties on the Southern High Plains of Texas.

20,000 km² (Palacios 1981), and the bulk of remaining land is used primarily for nonirrigated crop production (dry-land farming) or rangeland for grazing. Playas often are integrated into irrigation programs (e.g., pits are excavated to collect tailwater for reuse) and 65–70 percent of playas larger than 4 ha are used for grazing. The number of playas on the Southern High Plains that have been modified for agricultural use increased from 150 to 10,800 between 1965 and 1980 (Guthery et al. 1981). These practices directly alter standing biomass and species composition of floral and faunal communities, increase runoff and soil erosion, and provide fertilizer, pesticide, and nutrient inputs (Bolen et al. 1989b). Because playas are material sinks in these landscapes, accumulating both water and materials carried by water, they are affected greatly by surrounding land use practices.

The term *playa* has been used throughout the world, and even in this region, to refer to many different types of waterbodies. The playa lakes considered in this chapter conform to the description of freshwater, Type 2 playas of Reeves (1990), which are the most common lakes on the Southern High Plains. This chapter does not include those Type 2 playas that have been modified within urban stormwater management plans (see Wolf et al., this volume), municipal waste treatment facilities, or cattle feedlot operations.

PLAYA GEOMORPHOLOGY AND HYDROLOGY

Playa lakes are ephemeral wetlands (Cowardin et al. 1979) that provide critical wildlife habitat throughout the Southern Great Plains of Colorado, Kansas, Oklahoma, New Mexico, and Texas (Bolen et al. 1989b). Mechanisms for the origin of playas have been studied and debated for several decades. Reeves (1990) concludes that most playas probably formed through a combination of eolian activity at the surface and dissolution and subsidence of underlying strata. In addition, Osterkamp and Wood (1987) present evidence that playas can originate wherever water collects in a surficial depression (see also Wood and Osterkamp 1987). Moreover, Finley and Gustavson (1981) note lineament orientation of playas and attribute lake location to control exerted by underlying geologic structures. In a recent, comprehensive review, playa formation was attributed to integration of complex geomorphic, pedogenic, hydrochemical, and biological processes (Gustavson et al. 1995).

Playas have flat bottoms, resulting in a relatively uniform water depth throughout most of the basin. Generally they are less than 1 m in depth, and over 80 percent are smaller than 12 ha in surface area (Haukos and Smith 1992). The lakes occur on montmorillionitic clay lenses, which expand and seal the wetland when inundated. Typical of these soils is Randall clay, a fine, semectitic, thermic Udic Pellustert occupying the lowest elevational position in the internally drained watersheds of the Texas and New Mexico portion of the Southern High Plains (in the thermic temperature zone; Zartman and Fish 1992). The drainage basin for individual lakes ranges from less than

1 ha to more than 300 ha (Osterkamp and Wood 1987). Basin size is a primary determinant of water volume because groundwater throughout the Southern High Plains is generally quite deep and normally does not affect water levels in playa lakes. Rare exceptions may occur in urban areas and other locations where excessive recharge has created artificially elevated groundwater surfaces.

Estimates of the number of playa lakes throughout the Southern Great Plains range from 30,000 (Wood and Osterkamp 1984) to 37,000 (Reddell 1965). Recently, the Playa Lakes Joint Venture (1993), under the North American Waterfowl Management Plan, identified slightly more than 25,500 lakes in five states (Texas, 19,340; Kansas, 2,800; New Mexico, 2,640; Oklahoma, 580; Colorado, 200; see also Guthery and Bryant 1982). One hundred and eleven counties in the five-state area contain playas (Fig. 26.1), with 11,234 playas occurring in the 16 counties of Texas considered in this chapter.

In addition to their importance as wetlands, playas are primary recharge areas for the Ogallala Aquifer (Wood and Osterkamp 1984, Zartman 1987). Traditionally, most water in playas was assumed to evaporate because of low infiltration rates through the clay floor of the basin. More recent research has shown that extensive recharge occurs on the margins of clay lenses when water levels are at their highest, after storms (Zartman 1987, Zartman et al. 1994).

Playas usually contain water from late spring through fall, depending on the pattern and amount of rainfall, evaporation rates, and the extent to which the clay lining retards water loss. In Lubbock County, Texas, peak water volume usually occurs in June and playas typically are dry from December through March (Ward and Huddleston 1972). However, Curtis and Beierman (1980) found that only 33 percent of the playas in this area retained water during spring, summer, and fall, and most of these were modified for irrigation.

Several studies have examined playa water quality on the Southern High Plains (Lotspeich et al. 1969, Reeves 1970, Wells et al. 1970, U.S. Department of the Interior 1982, Wood and Osterkamp 1984). In this study we concentrate on those in which water quality was examined in conjunction with determinations of macroinvertebrate species composition (Sublette and Sublette 1967, Parks 1975, Hall 1997).

Chemical and Physical Properties of Water in Playa Lakes

Water quality differs greatly among playas. It varies temporally within playas as infiltration and evaporation occur (Table 26.1; Hall et al. 1995, Willig et al. 1995). Water quality also varies between years within the same playa, depending on the frequency and intensity of rainfall within each year (Ward 1964). Such temporal variation may compromise our ability to understand the effect of land use on spatial variability of water quality in playas.

Although extremely variable, playa water is quite turbid as estimated by either secchi depth or turbidity (Table 26.1). The organic and inorganic matter,

TABLE 26.1. Water Quality Parameters of Freshwater Playas that were Sampled for Invertebrates on the Southern High Plains of Texas

Parameter		Number of Samples	Minimum	Maximum
Physical properties	Turbidity (NTU)	41	20	2860
	Turbidity (ppm SiO_2)	17	45.0	1300.0
	Secchi depth (cm)	76	2.0	70.0
	Temperature (°C)	161	10.5	35.1
Aggregate properties	Total solids (mg/L)	41	195	2340
	Total suspended solids (mg/L)	41	10	1130
	Total dissolved solids (mg/L)	41	120	1570
	Total volatile suspended solids (mg/L)	41	<1	185
	Specific conductance (μmhos)	80	60.7	1176.7
	Hardness (mg/L)	41	6.2	81.5
	Hardness, total (mg/L $CaCO_3$)	77	40	218
	Hardness, calcium (mg/L $CaCO_3$)	17	20	182
	Alkalinity, methyl-orange (mg/L $CaCO_3$)	97	58	228
Nutrients	Total organic carbon (mg/L)	41	6.5	66.8
	Total inorganic carbon (mg/L)	41	<1.0	53.2
	Total carbon (mg/L)	41	12.3	77.5
	Total Kjeldahl nitrogen (mg/L)	41	0.36	3.36
	Ammonia-nitrogen (mg/L)	132	ND[a]	4.85
	NO_2/NO_3-nitrogen (mg/L)	41	<0.02	1.11
	Nitrate nitrogen (mg/L)	91	ND	0.21
	Total phosphorus (mg/L)	41	0.13	2.49
	ortho-phosphate phosphorus (mg/L)	115	ND	1.67
	Silica (mg/L)	41	6.8	28.2
Cations, anions and metals	pH	111	5.0	9.2
	Calcium (mg/L)	41	1.8	41.5
	Magnesium (mg/L)	41	1.5	25.4
	Sodium (mg/L)	41	0.8	22.5

TABLE 26.1. (Continued)

	Parameter	Number of Samples	Minimum	Maximum
	Potassium (mg/L)	41	0.1	76.8
	Chloride (mg/L)	58	1.25	162.00
	Sulfate (mg/L)	58	0.3	140.0
	Arsenic (µg/L)	41	<5	118
	Copper (µg/L)	41	9	123
DO and oxygen demands	Dissolved oxygen, surface (% Sat.)	76	2.3	186.6
	Dissolved oxygen, surface (mg/L)	137	0.20	15.34
	Dissolved oxygen, bottom (% Sat.)	76	ND	186.6
	Dissolved oxygen, bottom (mg/L)	136	0.02	15.34
	Biochemical oxygen demand (mg/L)	41	<3	84
	Chemical oxygen demand (mg/L)	41	5	130
	Chlorophyll-a (µg/L)	41	2	104
Pesticides	Aldicarb, total (µg/L)	41	<1.00	1.68
	Carbofuran, total (µg/L)	41	<0.06	0.12
	Cyanazine, total (µg/L)	41	<0.04	5.20
	Triazines, total (µg/L)	41	<0.1	15.9
	Alachlor, total (µg/L)	41	<0.10	0.35
	2,4-D, total (µg/L)	41	<0.7	15.1
	Captan, total (µg/L)	41	<0.005	0.100
	Carbaryl, total (µg/L)	41	<0.25	0.62
	Metolachlor, total (µg/L)	41	<0.1	16.6
	Chlorpyrifos, total (µg/L)	41	<0.10	0.31
	Pentachloraphenol, total (µg/L)	41	<0.05	0.20
	BenomylCarbendazim, total (µg/L)	41	<0.10	<0.10

Source: Sublette and Sublette 1967, Parks 1975, Hall et al. 1995, Willig et al. 1995, Hall 1997. Pesticide levels were assessed via Ohmicron immunoassays (Hall et al. 1995, Willig et al. 1995, Hall 1997).

[a]ND = none detected

which contribute to turbidity, are maintained in suspension because of basin shape (i.e., shallow with a smooth, flat bottom) and high regional winds.

Because water volume is low relative to playa surface area, water temperatures reflect daily and seasonal changes in air temperature. In addition, because of high water circulation in these shallow basins, many playas are isothermal, with mean temperature differences of less than 1°C from basin sediment to water surface (Hall et al. 1995, Willig et al. 1995). Although standing vegetation within playas coupled with high turbidity can confine the effects of solar heating to the top few centimeters of water, vertical temperature differences commonly do not exceed 5°C (Hall et al. 1995, Willig et al. 1995).

Measurements of hardness and solids (total, suspended, dissolved, and volatile) yield information on erosional processes, as well as autochthonous and allochthonous inputs. The high levels of solids found in playas are to be expected, given the extensive agricultural use of playa watersheds and their attendant clay basins (Table 26.1). However, the maximum concentrations of total and suspended solids in playas are greater than those normally found in untreated sewage (Thompson 1974), although the constituents of playa and sewage solids differ.

Most playas are nutrient-rich and fulfill criteria for eutrophic status (Table 26.1, Zafar 1959, Uttormark and Wall 1975). This reflects the high concentration of nutrients available for macroinvertebrates immediately after flooding. Inorganic and organic nutrient fractions probably arise from autochthonous and allochthonous sources. Inorganic enrichment (e.g., NO_2/NO_3-N, orthophosphate P) probably originates from allochthonous sources (e.g., agricultural runoff). However, organic enrichment (total organic C, total Kjeldahl N, ammonia-N) chiefly may be autochthonous, given that most playa basins support rich floral communities when dry.

The profile of dissolved oxygen (DO) is probably the most important expression of the trophic state of a waterbody. In natural waterbodies DO is determined by competition between agents supplying DO (diffusion and photosynthesis) and agents requiring DO (simple chemical oxidation and biological consumption). As is true for temperature, the shallow depth of playas and near-continuous wind probably account for the similarity between DO at the water surface and bottom for both minimum and maximum values (Table 26.1).

Most playas (Hall et al. 1995, Willig et al. 1995) maintained oxygen deficits. Twenty-seven playas (66 percent) had biochemical oxygen demand (BOD) values >8.0 mg/L DO, the saturation level at this elevation. Similarly, 36 playas (78 percent) had chemical oxygen demand (COD) values >8.0 mg/L DO. This deficit suggests that anaerobic conditions may occur. However, because of the high regional winds and presence of aquatic vegetation, DO levels, at least at the surface, have never been measured at <0.20 mg/L (Table 26.1). Concomitantly, vertical mixing of the water column precludes long periods of anaerobic conditions at the playa bottom.

Variation among playas in ratios of BOD to COD ranged from 13.5 to 100 percent, but most were <50 percent. High ratios suggest the presence of highly labile resources (e.g., fertilizer) rather than more recalcitrant compounds (e.g., cellulose). Indeed, playas surrounded by agriculture have higher ratios of BOD to COD, on average, than do playas surrounded by rangeland or Conservation Reserve Program grasslands (Hall et al. 1995, Willig et al. 1995).

Although pH ranges from 5.0 to 9.2, playa water is slightly alkaline on average (Table 26.1). The dominant cations typically are calcium and potassium, followed by magnesium and sodium. These four elements usually account for 95 percent of the cations of regional playa waters. Chlorides, sulfates, bicarbonate, and carbonate account for more than 95 percent of the anions. Only chloride and sulfate concentrations are reported herein (Table 26.1). Given the above-neutral pH of most playa water, most of the inorganic carbon is in the form of bicarbonate.

Some metals commonly are found in playa lakes. Arsenic and copper occur in herbicides, defoliants, and insecticides that have been used in this region. Arsenic was recovered from more than half of 41 basins, whereas measurable copper was found in all of them (Hall et al. 1995, Willig et al. 1995). However, it is unclear if these metals naturally occur at such concentrations in this region (Irwin and Dodson 1991); their levels are probably at least partially attributable to anthropogenic sources.

In the highly agrarian environment of the Southern High Plains, pesticide use is extremely common. The pesticides most commonly found in playas were cyanazine, other triazines, alachlor, and metolachlor (Table 26.1; Hall et al. 1995, Willig et al. 1995). This is to be expected given the high regional use of products bearing these active ingredients, the results of previous water tests (Mollhagen et al. 1993), and the cross-reactivity of chemicals in these particular assays. Cross-reactivity can occur, especially for triazine and acetanilide herbicides (e.g., cyanazine, alachlor, metolachlor), resulting in uncertainty in the specific identity of the parent compounds. Because of cross-reactivity, a result above the detection limit should be interpreted to mean that one or more similar compounds (including the nominate compound, congeners, or their metabolites) were encountered.

VEGETATION

Playa lakes are dominated by annual plants with life cycles that reflect the unpredictable moisture regime during the growing season. Guthery et al. (1982) identified 14 physiognomic types of playa vegetation, based on consideration of moisture regime and physical disturbance (grazing, cultivation, and irrigation). Because quality of wintering habitat is so critical for populations of waterfowl, recent research has focused on moist-soil management of playa lakes (Haukos and Smith 1993a). Managed lakes with water supplementation had higher seed biomass and greater numbers of ducks than did unmanaged lakes.

Occurrence and dominance of plant species within a particular playa is a function of moisture regimes in prior years, which developed particular seed banks, and the moisture pattern of the current year, which controls germination and subsequent growth (Haukos and Smith 1993b). In biomass studies on four playa lakes, Comer (1994) encountered two monotypic communities, one composed entirely of curly dock (*Rumex crispus*) and the other of pink smartweed (*Persicaria pensylvanica*). Mixed communities are more common (Haukos and Smith 1993b) and characterized the other two playas, one dominated by pink smartweed, barnyard grass (*Echinochloa crusgalli*), and ragweed (*Ambrosia grayi*) and the other by pink smartweed and ragweed. Other frequently occurring species (Haukos and Smith 1997) include spike rush (*Eleocharis macrostachya*), cheeseweed (*Malvella leprosa*), blue sunflower (*Helianthus ciliaris*), and evening primrose (*Oenothera canescens*).

MACROINVERTEBRATES

Playa wetlands provide essential habitat for more than 124 taxa of macroinvertebrates (Table 26.2) that spend all or part of their life cycles in water. Although intimately and inextricably associated with the aquatic system, little information is available concerning the macroinvertebrate fauna of the playas. In addition to their inherent importance as components of biodiversity in aquatic ecosystems (Pennak 1989, Merritt and Cummins 1996), macroinvertebrates represent critical links in foodwebs (Fig. 26.2) supporting many vertebrate populations (Merickel and Wangberg 1981). Macroinvertebrates are particularly crucial food sources for breeding ducks and their broods (Krapu and Swanson 1977, Bolen et al. 1989a), as well as for migrating shorebirds (Baldassarre and Fischer 1984, Davis 1996).

Diversity

Playa lakes contain a diverse macroinvertebrate fauna that can be divided into two groups, based on life history strategies for living in an ephemeral environment. Resident species have drought-resistant life stages and usually no autonomous means of movement among playas. Transient species do not have drought-resistant life stages and are capable of movement among playas some time during their life cycles, usually as adults. The transient species are mostly insects, whereas the resident fauna is composed mostly of noninsect species.

The insect community of playa lakes is actually quite rich. Of the approximately 124 taxa of macroinvertebrates in playa lakes, 84 are insects (Table 26.2). In total, 6 orders and 31 families of insects have been recorded. Families contributing the greatest number of species to the fauna are Libellulidae (ca 5 species), Corixidae (6 species), Dytiscidae (ca 12 species), Hydrophil-

TABLE 26.2. Invertebrate Taxa Collected from Freshwater Playas on the Southern High Plains of Texas[a]

Taxa	Sublette and Sublette (1967)	Parks (1975)	Merickel and Wangberg (1981)	Neck and Schramm (1992)	Horne (1996)	Hall (1997)
Ectoprocta (bryozoans)						x
Oligochaeta						
Naididae	x					
Lumbricidae						x
Tubificidae						
Limnodrilus hoffmeisteri (Claparède)						x
Limnodrilus sp.						x
Hirudinea	x	x				
Erpobdellidae						
Erpobdella punctata (Leidy)						x
Glossiphonidae						
Helobdella triserialis (Blanchard)						x
Gastropoda						
Lymnaeidae						
Fossaria cockerelli Pilsbry & Ferriss				x		x
Fossaria bulimoides (Lea)	x	x				
Physidae						
Physella bottimeri (Clench)				x		
Physella virgata Lea						x
Planorbidae						
Gyraulus parvus Say				x		
Planorbella tenuis (Dunker)						x
Planorbella trivolvis (Say)	x	x	x	x		

Taxon	1	2	3
Pelecypoda			
Sphaeriidae			
Sphaerium striatinum (Lamarck)		x	
Anostraca			
Branchinectidae			
Branchinecta lindahli Packard			x
Branchinecta packardi Pearse	x		x
Streptocephalidae			
Streptocephalus dorothae Mackin	x		x
Streptocephalus texanus Packard	x		x
Thamnocephalidae			
Thamnocephalus platyurus Packard	x	x	x
Notostraca			
Triopsidae			
Triops longicaudatus (LeConte)	x	x	x
Conchostraca			
Caenestheriidae			
Caenestheriella setosa (Pearse)	x		x
Eocyzicus concavus (Mackin)	x		
Leptestheriidae			
Leptestheria compleximanus (Packard)	x		x
Lynceidae			
Lynceus brevifrons (Packard)	x		x
Ostracoda			

TABLE 26.2. (Continued)

Taxa	Sublette and Sublette (1967)	Parks (1975)	Merickel and Wangberg (1981)	Neck and Schramm (1992)	Horne (1996)	Hall (1997)
Candoniidae						
Candona patzucaro Tressler					x	
Cyprididae						
Cyprinotus antillensis (Broodbakker)					x	x
Megalocypris gnathostomata (Ferguson)	x	x			x	x
Megalocypris pseudoingens Delorme					x	
Physocypria globula (Furtos)					x	
Cypridopsidae						
Cypridopsis vidua (Müller)					x	
Potamocypris unicaudata Schaefer					x	
Ilyocyprididae						
Pelocypris tuberculatum (Ferguson)						x
Limnocytheridae						
Limnocythere sanctipatricii (Brady and Robertson)					x	
Copepoda			x			
Cladocera			x			
Odonata						
Anisoptera						
Gomphidae	x	x	x			
Aeshnidae						
Anax junius (Drury)			x			x

Libellulidae			
Orthemis ferruginea (Fabricius)	x	x	
Pantala flavescens (Fabricius)		x	
Pantala sp.			
Plathemis lydia Drury	x	x	
Sympetrum corruptum (Hagen)		x	x
Sympetrum sp.	x		
Tramea sp.	x		
Zygoptera			
Coenagrionidae			
Enallagma civile (Hagen)	x	x	
Lestidae			
Lestes alcer Hagen		x	x
Lestes disjunctus Selys	x		
Ephemeroptera			
Baetidae			
Callibaetis sp.			x
Cloeon sp.	x	x	
Caenidae		x	
Heteroptera			
Belostomatidae			
Belostoma flumineum Say	x	x	x
Corixidae			
Corisella edulis (Champion)	x	x[b]	x
Corisella tarsalis (Fieber)	x	x	x

TABLE 26.2. (Continued)

Taxa	Sublette and Sublette (1967)	Parks (1975)	Merickel and Wangberg (1981)	Neck and Schramm (1992)	Horne (1996)	Hall (1997)
Rhamphocorixa acuminata (Uhler)	x		x			x
Sigara alternata (Say)	x	x				x
Sigara sp.	x					
Trichocorixa reticulata (Guérin-Méneville)			x			x
Trichocorixa verticalis (Fieber)						x
Gerridae						
Gerris marginatus Say						x
Gerris sp.			x			
Mesoveliidae						
Mesovelia mulsanti White						x
Notonectidae						
Buenoa margaritacea Torre-Bueno			x[c]			x
Notonecta undulata Say		x	x			x
Notonecta unifasciata Guérin-Méneville			x			
Notonecta sp.	x	x				
Saldidae						
Saldula interstitialis (Say)			x			
Saldula pallipes (Fabricius)						x
Veliidae						
Microvelia beameri McKinstry			x			
Microvelia sp.						x

Coleoptera

Curculionidae

Taxon	1	2	3
Bagous sp.	x		
Lissorhoptrus simplex (Say)	x		
Listronotus filiformis (LeConte)	x		
Listronotus grypidioides (Dietz)	x		
Listronotus scapularis (Casey)	x		
Notiodes aeratus (LeConte)	x		

Dytiscidae

Taxon	1	2	3
Brachyvatus sp.	x		
Copelatus chevrolati Aubé		x	
Copelatus sp.	x		
Cybister fimbriolatus (Say)	x	x	
Eretes sticticus (Linnaeus)	x	x	x
Hygrotus nubilus (LeConte)	x	x	
Laccophilus fasciatus terminalis Aubé	x	x	x
Laccophilus q. quadrilineatis Horn	x	x	
Laccophilus sp.			x
Liodessus affinis (Say)		x	
Neobidessus sp.	x		
Thermonectus nigrofasciatus Aubé		x	
Thermonectus nigrofasciatus ornaticollis (Aubé)	x		
Uvarus texanus (Sharp)		x	
Uvarus lacustris Say	x		

Gyrinidae

Taxon	1	2	3
Dineutus assimilis (Kirby)	x	x	
Dineutus sp.	x		x

TABLE 26.2. (Continued)

Taxa	Sublette and Sublette (1967)	Parks (1975)	Merickel and Wangberg (1981)	Neck and Schramm (1992)	Horne (1996)	Hall (1997)
Gyrinus parcus Say						
Gyrinus sp.			x			x
Haliplidae						
Haliplus triopsis Say			x			x
Haliplus tumidus LeConte						x
Helophoridae						
Helophorus linearis LeConte			x			x
Helophorus sp.	x					
Hydrophilidae						
Berosus exiguus (Say)			x			x
Berosus infuscatus LeConte						x
Berosus miles LeConte	x		x			x
Berosus rugulosus Horn	x					
Berosus stramineus Knisch	x					
Berosus styliferus Horn	x	x	x			x
Enochrus hamiltoni (Horn)			x			x
Enochrus sp.			x			
Hydrophilus triangularis Say		x	x			x
Octhebius sp.	x					
Paracymus confusus Wooldridge						x
Tropisternus lateralis (Fabricius)	x		x			x
Trichoptera	x					
Leptoceridae						
Oecetis sp.			x			
Diptera						
Ceratopogonidae	x					x
Culicoides variipennis (Coquillett)			x			
Forcipomyia sp.						x

Taxon					
Chironomidae					
Ablabesmyia sp.					x
Apedilum sp.	x	x			
Chironomus stigmaterus Say				x	
Chironomus sp.	x	x		x	
Clinotanypus sp.	x	x	x		
Cricotopus sp.	x	x			
Cryptochironomus sp.	x	x			
Dicrotendipes californicus Johannsen				x	
Dicrotendipes sp.	x	x			
Endochironomus nigricans (Johannsen)	x	x			
Labrundinia sp.	x	x			
Parachironomus sp.	x	x			
Polypedilum sp.	x	x			
Procladius bellus (Loew)	x	x			
Procladius sp.	x	x			
Tanypus sp.	x	x			
Tanytarsus sp.	x	x			
Tanytarsini			x		
Tanypodinae			x		
Culicidae					
Aedes nigromaculis (Ludlow)	x	x			
Culex tarsalis Coquillett	x	x	x		

TABLE 26.2. (Continued)

Taxa	Sublette and Sublette (1967)	Parks (1975)	Merickel and Wangberg (1981)	Neck and Schramm (1992)	Horne (1996)	Hall (1997)
Dolichopodidae						x
Ephydridae						x
Notophila sp.						x
Psychodidae						x
Stratiomyidae						
Odontomyia sp.			x			x
Syrphidae						
Eristalis sp.						x
Tabanidae						
Tabanus sp.						x
Tipulidae						
Tipula sp.		x				
Acarina	x					
Arrenuridae						
Arrenurus dentipetiolatus Marshall			x			x
Arrenurus new sp.						x
Eylaidae						
Eylais sp.			x			x
Hydrachnidae						
Hydrachna sp.						x
Pionidae						
Piona floridana Cook						x

[a] See Fig. 26.1.
[b] Species identified by Merickel and Wangberg (1981) as *Corisella inscripta* (Uhler) and later determined to be *C. edulis* (Champion) by RWS.
[c] Individual identified by Merickel and Wangberg (1981) as *Buenoa scimitra* Bare and later determined to be *B. margaritacea* Torre-Bueno by RWS.

Macroinvertebrate Food Web of Playa Lakes

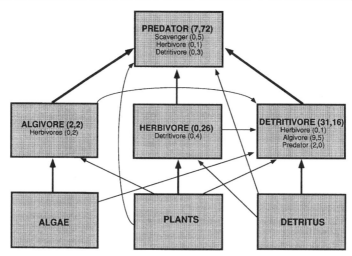

Fig. 26.2. Idealized, composite model of the foodweb of playa lakes on the Southern High Plains of Texas, illustrating the contribution of insect and noninsect macroinvertebrates. Dominant trophic classifications are indicated in capital letters (e.g., PREDATOR). In some cases a taxon may consume food from secondary sources as well, resulting in secondary trophic classifications which are indicated by upper and lower case letters (e.g., Herbivore). The number of noninsect and insect taxa, respectively, comprising a particular trophic category is represented by the numbers in parentheses adjacent to the trophic category (after Merritt and Cummins 1996, Pennak 1989, Thorp and Covich 1996). For example, 47 taxa were primarily detritivores, 9 of the 31 noninsect taxa secondarily consumed algae, and 5 of the 16 insect taxa secondarily consumed algae.

idae (11 species), and Chironomidae (ca 15 species). Because of the poor state of knowledge concerning taxonomy of chironomid larvae, assigning specific identities is especially problematic. Studies generally report chironomid presence in playas at familial, subfamilial, tribal, or generic levels (Table 26.2); thus, measures of species richness are underestimates of the true values.

In contrast to the insect fauna, the noninsect community is not especially rich (Table 26.2), being composed of only 28 families in 4 phyla (Ectoprocta [bryozoans], Annelida [oligochaetes and leeches], Mollusca [snails and clams], and Arthropoda [crustaceans and water mites]). However, the paucity of reported noninsect species may be a reflection of the historically weak taxonomic resolution of these taxa. The actual number of annelid and ectoproct species is unknown because of limited collections and improper preservation, which makes identification problematic (Sublette and Sublette 1967, Hall 1997). Similarly, the molluscan fauna has been documented poorly. Four families represented by eight species have been identified (Sublette and Sublette 1967, Parks 1975, Merickel and Wangberg 1981, Neck and Schramm 1992, Hall 1997).

The most abundant noninsect taxa of playa lakes occur in the Arthropoda, numbering at least 25 species in approximately 18 families (Sublette and Sublette 1967, Parks 1975, Merickel and Wangberg 1981, Hall 1997). The most abundant arthropod taxa are crustaceans, specifically branchiopods, ostracods, and copepods. Although each subphylum is abundant during initial stages of playa succession, copepods and branchiopod cladocerans (water fleas) have been studied inadequately, with occurrences noted only at the class or ordinal level (Merickel and Wangberg 1981). Conversely, the other branchiopods (fairy, clam, and tadpole shrimp) and ostracods (seed shrimp) have been relatively well studied (Moore and Young 1964, Ferguson 1967, Belk 1975, Sissom 1976, Horne 1993, 1996) and represent 19 species in 12 families.

The final arthropod taxon represented in playa lakes is the Acari. Four families of water mites represented by five species have been collected (Sublette and Sublette 1967, Merickel and Wangberg 1981, Hall 1997). However, water mites are relatively scarce and often are recorded only at the ordinal level (Sublette and Sublette 1967).

Modes of Persistence and Immigration/Emigration

Because of the ephemeral nature of playas, their aquatic inhabitants must have life stages that are resistant to desiccation or capable of immigration to and emigration from these habitats. As mentioned previously, macroinvertebrate species of playa lakes have disparate means of coping with the dry period of their ephemeral environment. Relatively few of the aquatic insect species are year-round residents of playas. The only insects that are known to be permanent occupants are members of the dipteran families Culicidae and Chironomidae. Many species of the culicine genus *Aedes* lay eggs in the soil of low-lying areas where they may remain dormant for long periods (Merritt and Cummins 1996). *Aedes nigromaculis* has been reared successfully in the laboratory from dry playa soil (Harrell unpublished data) and occurs in the Western and Central Plains, where it is a dominant mosquito of irrigated pastures. It is well adapted for life in playas because adults can appear as early as four days after the eggs have been flooded in the spring (Harwood and James 1979).

Overall, 14 genera of chironomids have been found to possess drought-resistant larval stages (Wiggins et al. 1980). One playa-inhabiting species, *Endochironomus nigricans*, is drought-resistant (Johannsen 1905). Congeners of other playa chironomids (*Chironomus* spp., *Cricotopus* sp., *Parachironomus* sp. [Johannsen 1905], and *Polypedilum* [Hinton 1960, 1968]) have been reported to survive periods of drought, but without supporting data we can only infer that these particular species have drought-resistant stages.

Congeners of a variety of other insect species inhabiting playas have been shown to withstand periods without surface water. The haliplid beetles *Hal-*

iplus immaculicollis and *H. strigatus* were found aestivating in an inch of dried soil of prairie ponds (Wallis 1933), and Young (1960) suggests that *H. ohioensis* must be resistant to summer drought. However, it is unknown whether haliplids can survive the extended dry periods of most playas.

The damselflies *Lestes dryas* and *L. unguicaulatus,* as well as several species of the dragonfly genus *Sympetrum,* oviposit endophytically above the waterline. Eggs hatch in spring and larval development accelerates with increasing ambient temperatures, ensuring that maturation occurs before the pool is dry. *Lestes disjunctus,* which occurs in playas, does not possess physiological adaptations to increase developmental rates with temperature (Corbet 1962), and probably is a spring immigrant from permanent waterbodies.

Odontomyia (Stratiomyidae) may be able to survive eight months in dry temporary pools in southern Ontario (Wiggins et al. 1980). Because most stratiomyids are terrestrial, this is not surprising. Nonetheless, such behavior has not been reported in semiarid environments. Similarly, the horse fly, *Tabanus similis,* has been reported to overwinter as larvae in dry pool basins (Wiggins et al. 1980). However, *Tabanus subsimilis,* the most common horse fly of the Southern High Plains, has not been reported to share this behavior.

Several species of leptophlebiid and siphlonurid mayflies can survive as drought-resistant eggs in vernal pools (Burks 1953, Landa 1968, Clifford 1970, Edmunds et al. 1976). However, *Callibaetis* (Baetidae), a common mayfly of playas, migrates to vernal pools from permanent bodies of water for oviposition (Ward 1992). A generation is completed in the pools, and the next generation of adults oviposits in permanent bodies of water. Generations alternate between permanent and temporary bodies of water.

Most heteropteran and coleopteran inhabitants of playas also alternate between permanent and temporary bodies of water, although a second generation in the permanent body of water is not required, because adults can overwinter and recolonize temporary bodies of water. They are considered to be true migrants because the same individual makes the return trip to the temporary body of water in the spring.

Literature regarding overwintering of Curculionidae in temporary pools is virtually nonexistent. In playas weevils apparently emigrate as soon as pools dry and the vegetation dies. In other ephemeral pools aquatic weevils move to high ground for the winter or fly to shelter in hedgerows and small woodlots or in the soil around the bases of bunch grasses (C. W. O'Brien, personal communication).

In contrast to most insect species of playa lakes, most noninsect taxa have life stages which are resistant to desiccation. Freshwater bryozoans can persist for long periods as dormant statoblasts that are resistant to freezing and drying (Brusca and Brusca 1990). Similarly, annelids can survive periods of desiccation as cocooned eggs or as adults in cysts or tests (Wiggins et al. 1980, Pennak 1989, Thorp and Covich 1991).

Several snail species burrow, as adults, into the substrate, where they endure periods of desiccation (Pennak 1989, Thorp and Covich 1991). However, it is unclear how the single clam species (Table 26.2) persists during long periods of desiccation. Although adult members of the family Sphaeriidae burrow into mud or under damp leaves during times of drought (Wiggins et al. 1980, Pennak 1989), neither of these substrates persists for long in the semiarid environment of the Southern High Plains. Similarly, water mites of the families Arrenuridae and Pionidae persist as deutonymphs in the soil of temporary ponds (Wiggins et al. 1980). However, drought-resistant life stages of elyaid and hydrachnid water mites are not known. Consequently, sphaeriid clams, as well as eylaid and hydrachnid mites, most likely are transported from permanent waterbodies to playas after they flood.

Crustacea is the most highly drought-adapted taxon of playa lakes. All constituent species have resting eggs, cysts, or juvenile stages that are capable of withstanding extended periods of drying and freezing (Hartland-Rowe 1972, Horne 1993). The absence of fairy, clam, and tadpole shrimp from permanent water bodies suggests that they lack predator defenses or require drought or freezing to initiate development of resting eggs (Prophet 1963). However, there is some argument as to whether drying or freezing is necessary for hatching, and it appears in some cases that temperature and abrupt changes in oxygen levels may be the key signals (Belk and Cole 1975, Wiggins et al. 1980).

Colonization Patterns

The pattern of macroinvertebrate colonization of playas is reflective of the aforementioned life history strategies. Transient species immigrate in following inundation, whereas resident species emerge from the basin soil. Consequently, resident species usually are the most abundant taxa during initial stages of lake succession (first four to six weeks), with transient species dominating later.

Although insects colonize temporary waterbodies readily (Fernando 1960), few details are known regarding the mechanisms whereby they detect suitable bodies of water (Ward 1992). Most of the aquatic insect inhabitants of the playas likely colonize from nearby sources of permanent water. Deep-water pits associated with irrigation practices undoubtedly serve as a primary source of biotic recruitment in many cases.

Many aquatic insects immigrate to bodies of water via aerial dispersal (Fernando 1958, 1959). Some flying aquatic insects select colonization sites based on size and reflected color (Popham 1943, Fernando 1959). Records of flight directed into reflective surfaces that were interpreted as water have been reported for a variety of insects. For example, a dozen specimens of the corixid *Hesperocorixa laevigata* mistook a turquoise blue Datsun for water in Arizona (Shaefer and Shaefer 1979). Similar episodes of substrate confusion have been recorded for other corixids (Fattig 1932), hydrophilids (Benham 1976, Last 1976), and other taxa. Indiscriminately directed flight of dispersing aquatic insects would be ill-adaptive on the Southern High Plains because little surface water is available. These anecdotal records demonstrate that some aquatic insects employ navigational criteria in their choice of colonization sites. Under normal circumstances these criteria facilitate an efficient and rapid colonization of appropriate habitats (i.e., playas).

Indeed, colonization of playas by insects is rapid, with many taxa appearing within the first weeks following inundation, albeit not in high abundance. Nonetheless, several studies have shown that a slight increase in diversity might occur in playas in the three to four months after vernal filling, but diversity during the first week typically is not the lowest (Parks 1975, Merickel and Wangberg 1981).

Colonization patterns of insects in playas are diverse. Species richness for some coleopterans (curculionids, gyrinids, and dytiscids) is highest in the first few weeks after filling (Hall 1997). Larvae of most dytiscid genera are present within one two two weeks of inundation. Densities of adult dytiscids can be quite high (Hall 1997), peaking approximately one month after initial arrival. Within two months most dytiscids have progressed to the adult stage, with the exception of *Cybister fimbriolatus,* which is large and requires more time to complete development.

The generation time of dytiscids in playas is much faster than that of other coleopterans, such as haliplids or hydrophilids. Both colonize playas as adults. Because haliplids feed chiefly on vegetation, which is sparse during initial stages of inundation, this taxon is not particularly common in playas immediately after flooding. Haliplid oviposition occurs soon after adult colonists arrive; after several weeks larval density increases and adult density decreases (Hall 1997).

Most heteropteran life cycles are similar to those of the haliplids. Adult density decreases through attrition, whereas nymphal density increases until late summer, when recently emergent adults emigrate. The majority of hemipteran taxa are most abundant approximately six to eight weeks following playa inundation (Hall 1997).

Dipteran taxa are abundant throughout the wet period of the playa (Hall 1997). Syrphids and some culicids, as mentioned earlier, are most abundant immediately after playa inundation. Other culicids and ephydrids are more common approximately one month after playas fill. Abundances of other dipteran taxa (chironomids, tabanids, and psychodids) increase throughout the wet period.

Unlike most insect species of playa lakes, the noninsect taxa rapidly attain high abundances once playas flood. Some species, especially phyllopods, hatch within 48 hours and complete their life cycles in less than a month (Prophet 1963, Loring et al. 1988, Thorp and Covich 1991). They are the numerically dominant taxa during early colonization. They consist mostly of detritivores, and their abundances usually peak within the first six weeks of playa succession (Sublette and Sublette 1967, Hall 1997) when detrital resources are abundant.

Not all noninsect taxa are present within the first month of playa succession. Some species of mites only appear approximately six weeks after playa inundation (Sublette and Sublette 1967, Hall 1997). Delayed colonization might reflect the timing of immigration of their insect hosts. Colonization by other species (e.g., physid snails and glossiphonid leeches) may reflect dependence for transport by waterfowl and wading birds (Maguire 1959, Maguire 1963, Proctor 1964, Proctor and Malone 1965, Proctor et al. 1967).

Although most noninsect taxa can withstand periods of desiccation, they also use passive means to disperse between playas. With the exception of water mites, most noninsect taxa of playas are dispersed by birds, either adhering to feathers and other body parts (Maguire 1959, 1963; Pennak 1989) or passing unscathed as resting eggs through the digestive tract (Proctor 1964, Proctor and Malone 1965, Proctor et al. 1967). Insects are used as agents of transport by some species (Pennak 1989), especially eylaid and hydrachnid water mites which have no drought-resistant life stages (Wiggins et al. 1980, Thorp and Covich 1991). Moreover, aerial dispersal of noninsect taxa is probably quite common among playa lakes (average density $>1/\text{mile}^2$) as a result of high winds (Wiggins et al. 1980).

SUCCESSIONAL CHANGES

Time has a pervasive effect on macroinvertebrate communities in playa lakes. Species richness and diversity increase with time (although both usually decrease immediately before drying), but the abundances of particular taxa either increase or decrease, depending on life history strategy and the rate at which the playa dries (Parks 1975, Merickel and Wangberg 1981, Neck and Schramm 1992, Moorhead et al. in press). Detritivores and omnivores dominate early stages but become less abundant in later stages, when herbivores and predators become dominant (Table 26.3). These data suggest a general

TABLE 26.3. Proportional Contributions of Particular Trophic Groups to Invertebrate Communities in Playas for each Sampling Period Based on Abundances[a]

	Sampling Period					
	1		2		3	
Trophic Group	Mean	SD	Mean	SD	Mean	SD
Detritivores	41.66	25.67	45.76	15.84	25.42	14.51
Filter Feeders	36.14	27.28	7.82	13.60	0.48	0.63
Herbivores	2.30	4.83	2.64	2.94	3.25	2.42
Omnivores	3.96	6.21	0.07	0.21	0.00	0.00
Predators	14.30	22.05	42.31	12.87	69.99	14.28

Source: D. L. Moorhead, D. L. Hall, and M. R. Willig, Succession of macroinvertebrates in playas of the Southern High Plains, USA. Journal of the North American Benthological Society, in press.
[a]SD = standard deviation; n = 10.

pattern of trophic succession beginning with early dominance of rapidly developing species (especially Crustacea) that utilize detritus and dead organic matter carried into the playa by runoff from the watershed or from the inundation of autochthonous sources. Numbers of herbivores increase as algae and aquatic vegetation grow, and predator numbers increase in later stages of succession (Table 26.3), apparently in response to an increase in available prey. This pattern is consistent with that found by Schneider and Frost (1996) for temporary ponds in Wisconsin. They suggest that the effect of predation on community structure increases with the duration of the wet period. However, the short duration of the wet phase of playas, in comparison to that of most of the Wisconsin ponds studied by Schneider and Frost (1996), may preclude any pervasive effect of predation on community structure.

Moorhead et al. (in press) also discovered temporal trends in the composition of macroinvertebrate communities in playas: (1) composition becomes more similar among playas over time and (2) differences in composition between communities within particular playas decrease over time. These results suggest that random events play a larger role in determining community structure at early stages of colonization and that more deterministic processes, such as biotic interactions, may exert greater influence on community composition during later stages of development.

CONSERVATION

As the primary wetlands in a semiarid landscape dominated by agriculture, playas represent an important source of invertebrate diversity that would otherwise be absent from the region. In addition, invertebrate production from playas represents a critical food source for vertebrate wildlife such as waterfowl (Bolen 1989a), shorebirds (Baldassarre and Fischer 1984), amphibians (Anderson 1997), and mammals (Schmidly 1991). Because playa wetlands are an important portion of the Central Flyway of the United States, they indirectly contribute to the health of wildlife populations and the diversity of avian communities in North and South America (U.S. Fish and Wildlife Service 1988).

Modifications of playa basins for conservation, recreation, and agricultural purposes have significant implications for the biodiversity of playas. For example, lands entered into the Conservation Reserve Program often are associated with playa basins. Although the seeded grasses assist in reducing erosion, they also reduce the rate at which playas fill during spring and summer convection storms, and shorten the duration of inundation. This likely has a major impact on invertebrate populations and communities and on the wildlife that rely on them for food. Similarly, modification of playas so that they are permanently filled for use in recreation or directly interconnected as part of a flood control program, as is done in a number of urban areas on the Southern High Plains (Wolf et al., this volume), changes the nature of playa ecosystems and alters the abundance and species composition of invertebrates. Finally, agricultural practices (e.g., rowcrop cultivation of cotton and sorghum) can exacerbate sedimentary filling of playas and reduce their contribution to regional biodiversity (Luo et al. 1997).

The flora and fauna of playas are not sufficiently well studied to understand the underlying mechanisms that enhance their diversity and that of communities occupying adjacent terrestrial environments. Even fewer comparative studies have addressed the consequences of human activities on playa invertebrates or documented the consequences of altered invertebrate communities on other wildlife species. Despite the importance of playa wetlands from a conservation perspective, they are not sufficiently well represented in areas protected as part of the nation's system of nature reserves. Almost all playas are situated on privately owned land, diminishing the likelihood that they will be well studied from a long-term research perspective or sufficiently protected from development. We reiterate the concerns of Vandermeer and Perfecto (1997) as they relate to the conservation status of agroecosystems. Playas represent critical components of a biodiversity-rich agroecosystem. Understanding the abiotic and biotic factors responsible for the origin and maintenance of their biodiversity, as well as their interconnections with the surrounding agroecosystem's flora and fauna, remains an important task that should assume high priority as the foci of agroecology and conservation research merge in the coming decades.

LITERATURE CITED

Anderson, A. M. 1997. Habitat use and diet of amphibians breeding in playa wetlands on the Southern High Plains of Texas. Master's Thesis, Texas Tech University, Lubbock, TX.

Baldassarre, G. A., and D. H. Fischer. 1984. Food habits of fall migrant shorebirds on the Texas High Plains. Journal of Field Ornithology 55:220–229.

Belk, D. 1975. Key to the Anostraca (fairy shrimps) of North America. The Southwestern Naturalist 20:91–103.

Belk, D., and G. A. Cole. 1975. Adaptational biology of desert temporary-pond inhabitants. Pages 207–226 in N. F. Hadley (ed.), Environmental Physiology of Desert Organisms. Dowden, Hutchinson & Ross, Stroudsburg, PA.

Benham, B. R. 1976. Swarming of *Helophorus brevipalpis* Bedel (Col., Hydrophilidae) in North Devon. Entomologists' Monthly Magazine 111:127–128.

Bolen, E. G., G. A. Baldassarre, and F. S. Guthery. 1989a. Playa lakes. Pages 341–365 in L. M. Smith, R. L. Pederson, and R. M. Kaminski (eds.), Habitat Management for Migrating and Wintering Waterfowl in North America. Texas Tech University Press, Lubbock, TX.

Bolen, E. G., L. M. Smith, and H. L. Schramm, Jr. 1989b. Playa lakes: Prairie wetlands of the Southern High Plains. BioScience 39:615–623.

Brusca, R. C., and G. J. Brusca. 1990. Invertebrates. Sinauer Associates, Sunderland, MA.

Burks, B. D. 1953. The mayflies, or Ephemeroptera, of Illinois. Illinois Natural History Survey Bulletin 26:1–216.

Clifford, H. F. 1970. Analysis of a northern mayfly (Ephemeroptera) population with special reference to allometry of size. Canadian Journal of Zoology 48:305–316.

Comer, G. L., Jr. 1994. Remote sensing of vegetation biomass in playa wetlands. Master's Thesis. Texas Tech University, Lubbock, TX.

Corbet, P. S. 1962. A biology of dragonflies. Quadrangle Books, Chicago, IL.

Cowardin, L. M., V. Carter, F. C. Golet, and E. T. LaRoe. 1979. Classification of Wetlands and Deepwater Habitats of the United States. Office of Biological Services, U.S. Fish and Wildlife Service, Washington, D.C.

Curtis, D., and H. Beierman. 1980. Playa Lakes Characterization Study. U.S. Fish and Wildlife Service, Division of Ecological Service, Fort Worth, TX.

Davis, C. A. 1996. Ecology of spring and fall migrant shorebirds in the playa lakes region of Texas. Ph.D. Dissertation, Texas Tech University, Lubbock, TX.

Edmunds, G. F., Jr., S. L. Jensen, and L. Berner. 1976. The Mayflies of North and Central America. University of Minnesota Press, Minneapolis, MN.

Fattig, P. W. 1932. Water-boatmen try to land on auto top (Hemipt.: Corixidae). Entomological News 44:152.

Ferguson, E., Jr. 1967. New ostracods from the playa lakes of eastern New Mexico and western Texas. Transactions of the American Microscopical Society 86:244–250.

Fernando, C. H. 1958. The colonization of small freshwater habitats by aquatic insects. 1. General discussion, methods, and colonization in the aquatic Coleoptera. Ceylon Journal of Science (Bio. Sci.) 1:117–154.

———. 1959. The colonization of small freshwater habitats by aquatic insects. 2. Hemiptera (The water bugs). Ceylon Journal of Science (Bio. Sci.) 2:5–32.

———. 1960. Colonization of freshwater habitats with special reference to aquatic insects. Pages 182–186 *in* Centenary and Bicentenary Congress, Singapore 1958.

Finley, R. J., and T. C. Gustavson. 1981. Lineament Analysis Based on Landsat Imagery, Texas Panhandle. Circular 85-5, Bureau of Economic Geology, Austin, TX.

Gustavson, T. C., V. T. Holliday, and S. D. Hovorka. 1995. Origin and Development of Playa Basins, Sources of Recharge to the Ogallala Aquifer, Southern High Plains, Texas and New Mexico. Report of investigations No. 229, Bureau of Economic Geology, Austin, TX.

Guthery, F. S., and F. C. Bryant. 1982. Status of playas in the Southern Great Plains (USA). Wildlife Society Bulletin 10:309–317.

Guthery, F. S., F. C. Bryant, B. Kramer, A. Stoecker, and M. Dvoracek. 1981. Playa lake assessment study for U.S. Department of the Interior. Water and Power Resources Service, Southwest Region, Amarillo, TX.

Guthery, F. S., J. M. Pates, and F. A. Stormer. 1982. Characterization of playas of the north-central Llano Estacado in Texas. Transactions of the North American Wildlife and Natural Resources Conference 47:516–527.

Hall, D. L. 1997. Species diversity of aquatic macroinvertebrates in playa lakes: Island biogeographic and landscape influences. Ph.D. Dissertation, Texas Tech University, Lubbock, TX.

Hall, D. L., M. R. Willig, D. L. Moorhead, and T. R. Mollhagen. 1995. Variation in playa lakes: Islands of diversity in a sea of agriculture and aridity. Bulletin of the North American Benthological Society 12:221.

Hartland-Rowe, R. 1972. The limnology of temporary waters and the ecology of the Euphyllopoda. Pages 15–31 *in* R. B. Clark and R. J. Wootton (eds.), Essays in Hydrobiology. University of Exeter, England.

Harwood, R. F., and M. T. James. 1979. Entomology in Human and Animal Health. (7th ed.). Macmillan, New York.

Haukos, D. A., and L. M. Smith. 1992. Ecology of Playa Lakes. Fish and Wildlife Leaflet 13.3.7. Waterfowl Management Handbook. U.S. Fish and Wildlife Service, Washington, DC.

———. 1993a. Moist-soil management of playa lakes for migrating and wintering ducks. Wildlife Society Bulletin 21:288–298.

———. 1993b. Seed bank composition and predictive ability of field vegetation in playa lakes. Wetlands 13:32–40.

———. 1997. Common flora of the playa lakes. Texas Tech University Press, Lubbock, TX.

Hinton, H. E. 1960. Cryptobiosis in the larva of *Polypedilum vanderplanki* Hint. (Chironomidae). Journal of Insect Physiology 5:286–300.

———. 1968. Reversible suspension of metabolism and the origin of life. Proceedings of the Royal Society of London (Series B) 171:43–57.

Horne, F. R. 1993. Survival strategy to escape desiccation in a freshwater ostracod. Crustaceana 65:53–61.

———. 1996. Ostracods of Texas playas. The Southwestern Naturalist 41:450–455.

Irwin, R. J., and S. Dodson. 1991. Contaminants in Buffalo Lake National Wildlife Refuge, Texas. Ecological Services, Arlington Field Office, U.S. Fish and Wildlife Service, Washington, DC.

Johannsen, O. A. 1905. Aquatic Nematocerous Diptera. II. Chironomidae. Mayflies and midges of New York. New York State Museum Bulletin 86:76–315.

Krapu, G. L., and G. A. Swanson. 1977. Foods of juvenile, brood hen, and post-breeding pintails in North Dakota. Condor 79:504–507.

Landa, V. 1968. Developmental cycles of central European Ephemeroptera and their relation to the phylogeny and systematics of their order. Sveriges International Congress Zoological Papers 54:1–2.

Last, H. 1976. Swarming of *Helophorus brevipalpis* Bedel (Col., Hydrophilidae). Entomologists' Monthly Magazine 111:128.

Loring, S. J., W. P. MacKay, and W. G. Whitford. 1988. Ecology of small desert playas. Pages 89–113 *in* J. L. Thames and C. D. Ziebell (eds.), Small Water Impoundments in Semi-arid Regions. University of New Mexico Press, Albuquerque, NM.

Lotspeich, F. B., V. L. Hauser, and O. R. Lehman. 1969. Quality of waters from playas on the Southern High Plains. Water Resources Research 5:48–58.

Luo, Hong-Ren, L. M. Smith, B. L. Allen, and D. A. Haukos. 1997. Effects of sedimentation on playa wetland volume. Ecological Applications 7:247–252.

Maguire, B., Jr. 1959. Passive overland transport of small aquatic organisms. Ecology 40:312.

————. 1963. The passive dispersal of small aquatic organisms and their colonization of isolated bodies of water. Ecological Monographs 33:161–185.

Merickel, F. W., and J. K. Wangberg. 1981. Species composition and diversity of macroinvertebrates in two playa lakes on the Southern High Plains, Texas. The Southwestern Naturalist 26:153–158.

Merritt, R. W., and K. W. Cummins. 1996. An Introduction to the Aquatic Insects of North America (3rd ed.). Kendall/Hunt, Dubuque, IA.

Mollhagen, T. R., L. V. Urban, R. H. Ramsey, A. W. Wyatt, C. D. McReynolds, and J. T. Ray. 1993. Assessment of Non-point Source Contamination of Playa Basins in the High Plains of Texas (Brazos Basin Watershed, Phase 1). Water Resources Center, Texas Tech University, Lubbock, TX.

Moore, W. G., and J. B. Young. 1964. Fairy shrimps of the genus *Thamnocephalus* (Branchiopoda, Anostraca) in the United States and Mexico. The Southwestern Naturalist 9:68–77.

Moorhead, D. L., D. L. Hall, and M. R. Willig. Succession of macroinvertebrates in playas of the Southern High Plains, USA. Journal of the North American Benthological Society, in press.

Neck, R. W., and H. L. Schramm, Jr. 1992. Freshwater molluscs of selected playa lakes of the Southern High Plains of Texas. The Southwestern Naturalist 37:205–209.

New, L. 1979. 1977 High Plains Irrigation Survey. Texas Agricultural Extension Service, College Station, TX.

Osterkamp, W. R., and W. W. Wood. 1987. Playa-lake basins on the Southern High Plains of Texas and New Mexico: I. Hydrologic, geomorphic, and geologic evidence for their development. Geological Society of America Bulletin 99:215–223.

Palacios, N. 1981. Llano Estacado playa lake water resources study. Pages 15–20 *in* J. S. Barclay and W. V. White (eds.), Playa Lakes Symposium. Biology Service Program, U.S. Fish and Wildlife Service, Fort Worth, TX.

Parks, L. H. 1975. Some trends of ecological succession in temporary aquatic ecosystems (playa lakes). Ph.D. Dissertation, Texas Tech University, Lubbock, TX.

Pennak, R. W. 1989. Fresh-water Invertebrates of the United States (3rd ed.). John Wiley & Sons, New York.

Playa Lakes Joint Venture. 1993. Accomplishment Report 1989-1992. North American Waterfowl Management Plan, U.S. Fish and Wildlife Service, Washington, DC.

Popham, E. J. 1943. Ecological studies of the commoner species of British Corixidae. Journal of Animal Ecology 12:124–136.

Proctor, V. W. 1964. Viability of crustacean eggs recovered from ducks. Ecology 45:656–658.

Proctor, V. W., and C. R. Malone. 1965. Further evidence of the passive dispersal of small aquatic organisms via the intestinal tract of birds. Ecology 46:728–729.

Proctor, V. W., C. R. Malone, and V. L. DeVlaming. 1967. Dispersal of aquatic organisms: Viability of disseminules recovered from the intestinal tract of captive killdeer. Ecology 48:672–676.

Prophet, C. W. 1963. Some factors influencing the hatching of anostracan eggs. Transactions of the Kansas Academy of Science 66:150–159.

Reddell, D. L. 1965. Water resources of playa lakes. Cross Section 13:1.

Reeves, C. C., Jr. 1970. Location, Flow and Water Quality of Some West Texas Playa Lake Springs. Water Resources Center, Texas Tech University, Lubbock, TX.

————. 1990. A proposed sequential development of lake basins, Southern High Plains, Texas and New Mexico. Pages 209–232 *in* T. C. Gustavson (ed.), Geologic Framework and Regional Hydrology, Upper Cenozoic Blackwater Draw and Ogallala Formations, Great Plains. University of Texas, Austin, TX.

Schmidly, D. J. 1991. The bats of Texas. Texas A & M University Press, College Station, TX.

Schneider, D. W., and T. M. Frost. 1996. Habitat duration and community structure in temporary ponds. Journal of the North American Benthological Society 15:64–86.

Shaefer, C. W., and M. I. Shaefer. 1979. Corixids (Hemiptera: Heteroptera) attracted to automobile roof. Entomological News 90:230.

Sissom, S. L. 1976. Studies on a new fairy shrimp from the playa lakes of West Texas (Branchiopoda Anostraca, Thamnocephalidae). Crustaceana 30:39–42.

Stoddart, L. A., and A. D. Smith. 1955. Range Management (2nd ed.). McGraw-Hill, New York.

Sublette, J. E., and M. S. Sublette. 1967. The limnology of playa lakes on the Llano Estacado, New Mexico and Texas. The Southwestern Naturalist 12:369–406.

Thompson, G. B. 1974. Variation of urban runoff quality and quantity with duration and intensity of storms. Master's Thesis, Texas Tech University, Lubbock, TX.

Thorp, J. H., and A. P. Covich. 1991. Ecology and Classification of North American Freshwater Invertebrates. Academic Press, San Diego, CA.

U.S. Department of the Interior. 1982. Llano Estacado Playa Lake Water Resources Study. Bureau of Reclamation, Southwest Regional Office, Amarillo, TX.

U.S. Fish and Wildlife Service. 1988. Playa Lakes Region Waterfowl Habitat Concept Plan, Category 24 of the North American Waterfowl Management Plan. U.S. Fish and Wildlife Service, Albuquerque, NM.

Uttormark, P. D., and J. P. Wall. 1975. Lake Classification—a Trophic Characterization of Wisconsin Lakes. National Environmental Research Center, Office of Research and Development, U.S. Environmental Protection Agency, Corvallis, OR.

Vandermeer, J., and I. Perfecto. 1997. The agroecosystem: A need for the conservation biologist's lens. Conservation Biology 11:591–592.

Wallis, J. B. 1933. Revision of the North American species (north of Mexico) of the genus *Haliplus*, Latreille. Transactions Royal Canadian Institute 19:1–76.

Ward, C. R. 1964. Ecological changes in modified playa lakes with special emphasis on mosquito production. Master's Thesis, Texas Technological College, Lubbock, TX.

Ward, C. R., and E. W. Huddleston. 1972. Multipurpose Modification of Playa Lakes. Pages 203–286 *in* C. C. Reeves, Jr., (ed.), Playa Lakes Symposium, 29-30 October 1970. International Center for Arid and Semi-Arid Land Studies and Department of Geosciences, Publication No. 4, Texas Tech University, Lubbock, TX.

Ward, J. V. 1992. Aquatic Insect Ecology: Biology and Habitat. John Wiley & Sons, New York.

Wells, D. M., E. W. Huddleston, and R. G. Rekers. 1970. Potential Pollution of the Ogallala Aquifer by Recharging Playa Lake Water; Pesticides. Water Pollution Control Research Series, Project No. 16060 DCO, U.S. Environmental Protection Agency, Washington, DC.

Wiggins, G. B., R. J. MacKay, and I. M. Smith. 1980. Evolutionary and ecological strategies of animals in annual temporary pools. Archiv für Hydrobiologie Supplement 58:97–206.

Willig, M. R., D. L. Hall, D. L. Moorhead, T. R. Mollhagen, and E. B. Fish. 1995. Variation in water quality among playa lakes: A multivariate approach and landscape perspective. Bulletin of the Ecological Society of America Supplement 76: 285.

Wood, W. W., and W. R. Osterkamp. 1984. Recharge to the Ogallala Aquifer from playa lake basins on the Llano Estacado (an outrageous proposal?). Pages 337–349 *in* G. Whetstone (ed.), Proceedings of the Ogallala Aquifer Symposium II. Water Resources Center, Texas Tech University, Lubbock, TX.

———. 1987. Playa lake basins on the Southern High Plains of Texas and New Mexico: II. A hydrologic model and mass-balance arguments for their development. Geological Society of America Bulletin 99:224–230.

Young, F. N. 1960. The water beetles of a temporary pond in southern Indiana. Proceedings of the Indiana Academy of Science 69:154–164.

Zafar, A. R. 1959. Taxonomy of lakes. Hydrobiologia 13:287–299.

Zartman, R. E. 1987. Playa lakes recharge aquifers. Crops Soils 39:20.

Zartman, R. E, and E. B. Fish. 1992. Spatial characteristics of playa lakes in Castro County, Texas. Soil Science 153:62–68.

Zartman, R. E, P. W. Evans, and R. H. Ramsey. 1994. Playa lakes on the Southern High Plains in Texas: Reevaluating infiltration. Journal of Soil and Water Conservation 49:299–301.

27 Urban Playas of the Southern High Plains

The Influence of Water Quality on Macroinvertebrate Diversity and Community Structure

CRAIG F. WOLF, DARYL L. MOORHEAD, and
MICHAEL R. WILLIG

*T*he urban playas of Lubbock, Texas, are hydrologically isolated and yet exist within a small geographical area, thus minimizing differences among impoundments due to differences in soils, climate, species pools, and evolutionary history that normally occur over larger distances. We hypothesized that differences in macroinvertebrate abundances, diversity, and community composition could be attributed to differences in surrounding landscape and water quality characteristics, but found little evidence to support this notion. In contrast, some differences in macroinvertebrate diversity among playas may be related to general habitat features. For example, macroinvertebrate abundances and diversity were greater in playas with extensive macrophyte communities. Conversely, a basin with no littoral vegetation had the highest numbers of zooplankton but the lowest number of insect taxa and lowest overall diversity of macroinvertebrates. Few relationships were found to exist between water quality characteristics and composition of aquatic macroinvertebrate communities, but ammonia concentrations were highest in the playa that was dominated by zooplankton, perhaps indicating a more eutrophic status than for the other playas. Analyses of community structure revealed few consistent relationships among playas, and similarities that did emerge appeared to be determined by abundances of relatively few taxa (e.g., palaemonids, chironomids,*

Invertebrates in Freshwater Wetlands of North America: Ecology and Management, Edited by
Darold P. Batzer, Russell B. Rader, and Scott A. Wissinger
ISBN 0-471-29258-3 © 1999 John Wiley & Sons, Inc.

notonectids, baetids, and gerrids). Thus, macroinvertebrate community com-position and species dominance differed among urban playas even though species diversity was comparable.

URBAN PLAYAS

The Southern High Plains of West Texas contain between 25,000 and 30,000 playas that dominate the wetland and wildlife habitats of the region (Haukos and Smith 1992). Playas are shallow, ephemeral pools in largely isolated watersheds that capture approximately 88 percent of the local precipitation, which subsequently is lost through groundwater seepage and evaporation (Bolen et al. 1989). Playas provide critical habitat for more than 115 species of birds (including 20 species of waterfowl), 10 species of mammals, 14 species of amphibians, and 60 taxa of macroinvertebrates (Sublette and Sub-lette 1967, Haukos and Smith 1992, Neck and Schramm 1992, Smith 1993). The importance of playas as centers of biodiversity extends beyond the local level because they provide nesting and wintering habitats for more than 2 million waterfowl that use the central migratory flyway (U.S. Fish and Wild-life Service 1988).

Urban areas of the Southern High Plains incorporate modified playas in management plans for stormwater and urban runoff. Modifications usually consist of deepening playas to retain greater volumes of water, often resulting in permanent inundation. Thus, the hydrological regimes of rural and urban playas differ in duration; rural playas fill with spring rains and dry during the summer, while urban playas usually remain inundated throughout the year. Urban and rural playas also differ with respect to surrounding land use pat-terns and characteristics of entering runoff water. Areas immediately sur-rounding playas within the City of Lubbock, Texas, often are dedicated to city parks and open areas that host a wide range of recreational activities, embedded within a matrix of commercial, residential, and industrial lands. In contrast, most rural playas are surrounded by agricultural lands dedicated to row-crop agriculture, livestock grazing, and conservation grasslands. These landscape features undoubtedly affect physical attributes of the playas, such as water chemistry (see below), as well as biological communities, although comparative studies have not yet been performed.

Some differences between the biological communities of rural and urban playas are obvious. For example, most urban playas are stocked with fish, which can greatly alter the structure of invertebrate communities. In contrast, few rural playas have fish because they usually retain water for only a few months of each year. Large differences in vegetation also exist among urban and rural playas. Urban playas tend to have little vegetation, although scattered stands of submerged and emergent macrophytes sometimes occur. Conversely, rural playas usually have stands of terrestrial plants that grow when basins are dry and are inundated when basins fill. For these reasons, the invertebrate communities of urban and rural playas might be expected to differ.

Although playas in Lubbock serve as centers of biodiversity in an urban setting, the diversity of macroinvertebrates in these playas has not been examined until now. Moreover, relationships among macroinvertebrates, land use patterns, and water quality have not been evaluated, despite ongoing investigations of the quality of stormwater runoff entering playas (Ennis 1994). The primary objective of this chapter is to evaluate the diversity and composition of aquatic macroinvertebrate communities of urban playas as they relate to water quality characteristics and surrounding land use patterns.

INVERTEBRATE BACKGROUND

Although research has focused on species diversity and community composition of aquatic macroinvertebrates in rural playas on the Southern High Plains of the United States (Sublette and Sublette 1967, Merickel and Wangberg 1981, Smith 1993), little attention has been given to macroinvertebrates in urban playas. Because of physical and morphological differences between rural and urban playas (see Hall et al., this volume), invertebrate community composition may differ between these habitats, especially with regard to species that require alternating wet and dry periods or cannot survive predation by fish that are present in most urban playas. Other factors that may differentially affect macroinvertebrate communities in rural and urban playas include habitat complexity (i.e., presence and absence of littoral vegetation), trophic status (i.e., oligotrophy versus eutrophy), and surface water quality. Water quality can be a particularly important factor affecting aquatic organisms in urban impoundments.

Urban Water Quality

Urban runoff typically contains high levels of suspended solids and organic matter that enhance oxygen demand of recipient waters, often increasing mortality rates of sensitive aquatic organisms (Mason 1991). Moreover, stormwater runoff accounts for the bulk of many inputs. For example, stormwater runoff provided 85 percent of metals, 90 percent of oxygen demand material, over 50 percent of nutrients, and 99 percent of suspended solids entering waterbodies in Tallahassee, Florida (Livingston and Cox 1989). Similar findings have been reported for Lubbock, Texas, where chemical oxygen demand, as well as total and suspended solids in surface water runoff entering playas, have exceeded standards for raw sewage (Thompson et al. 1974, Wells et al. 1975).

Chemical inputs to surface water impoundments differ as functions of surrounding land use (i.e., residential, commercial, or agricultural) and the area of adjacent, impervious surfaces, such as parking lots and streets (Porcella and Sorensen 1980, Jones and Clark 1987). Rimer and Nissen (1978) found that chemical oxygen demand, suspended solids, and total phosphorus concentrations in urban impoundments were correlated positively with the extent of impervious surfaces. Other factors that influence the type and quantity of materials entering surface water impoundments include the time of year in which the runoff occurs, elapsed time between inputs, and conditions prior to runoff, including human activities within the watershed (Wells et al. 1975).

Macroinvertebrate Communities and Water Quality

The influence of physicochemical conditions on the structure of invertebrate communities is a central focus of aquatic ecology (Macan 1963, Hynes 1970, Rosenberg and Resh 1993). Although much research has focused on the influences of productivity and trophic status (Macan 1949, Kullberg 1992, France 1995, Hanson and Butler 1994), other studies have demonstrated significant impacts of water quality on aquatic invertebrates (Hershey 1985, Hammer et al. 1990, Rasmussen 1993, Wollheim and Lovvorn 1995). Physicochemical conditions and habitat attributes may play even more dominant roles in structuring macroinvertebrate communities when human activities alter habitat quality (Ward 1992). Indeed, diversity and community composition of macroinvertebrates have been used to evaluate environmental stress resulting from urbanization and industrialization (Cairns and Pratt 1993). Playas located within urban areas are subject to many human activities, including direct modification of habitat, hydrological regime, and characteristics of drainage basins. Because macroinvertebrate communities in lentic ecosystems often are influenced by physicochemical characteristics of water, and because physicochemical characteristics of runoff entering urban impoundments are related to land use patterns and human activities, macroinvertebrate communities of urban playas should show some response to land use characteristics.

EXPERIMENTAL APPROACH

Details of experimental approaches are described elsewhere (Wolf 1996). Consequently, only a brief outline of methods and materials is included herein. Macroinvertebrate communities, water quality characteristics, and surrounding landscape features were examined for 8 of the 32 playas occurring within the City of Lubbock, Texas, USA (101°52′N, 33°35′W). Physical attributes and surrounding landscape characteristics for playas are given in Table 27.1. Macroinvertebrates were collected via net and dredge sampling in March, April, June, July, August, and November–December 1993 from the benthic, pleustonic, nektonic, and emergent vegetation regions of the littoral

TABLE 27.1. Total Watershed Area, Allocation among Land Use Categories, and Aquatic Vegetation of Selected Urban Playas in Lubbock, Texas

Playa	Watershed (ha)[a]	Single Family Residence (%)	Multiple Family Residence (%)	Commercial (%)	Parks (%)	Vegetation
Rushland	105	76.3	3.6	2.3	17.8	Seasonal[b]
Higinbotham	128	87.9	0.0	0.0	12.1	Seasonal
Wendover	139	68.9	3.8	9.3	18.1	Seasonal
Maxey	262	46.3	5.5	35.1	13.0	None
Leroy Elmore	196	76.3	6.2	2.2	15.4	None
Buster Long	162	14.0	23.4	53.3	9.3	None
Jack Stevens	148	78.7	3.8	6.2	11.3	Perennial[c]
Quaker & Brownfield	114	62.3	14.4	23.3	0.0	Perennial

[a]Total watershed area that provided direct surface water runoff to playas.
[b]Mostly summer annuals, e.g., smartweed (*Persicaria pennsylvanica*) and dockweed (*Rumex cripsus*).
[c]Extensive stands of cattails (*Typha* sp.) and persistent pondweed (*Potamogeton* sp.).

zone in each playa. Not all individuals could be identified to the specific level, although distinct morphotypes could be identified and were used in subsequent taxonomic analyses. To avoid confusion, herein we use the term *taxon-level* analysis to refer to analyses based on distinct morphotypes.

The suite of water quality characteristics determined for each playa was alkalinity, biological oxygen demand (BOD), chemical oxygen demand (COD), conductivity, dissolved oxygen, hardness, pH, temperature, total Kjeldahl nitrogen (TKN), ammonia, nitrate-nitrite nitrogen, total phosphorus, ortho-phosphorus, total carbon, total organic carbon (TOC), and total inorganic carbon (Table 27.2). Analyses were conducted at monthly intervals from February 1993 through March 1994. Temperature and pH were measured *in situ*, but for other assays grab-samples were taken from each playa, placed on ice, and transported to the laboratory for analyses. Chemical analyses were performed according to U.S. Environmental Protection Agency (1974) and American Public Health Association (1992) standards.

Urban land use surrounding each playa was classified as single-family housing, multiple-family housing, commercial, or parks and vacant lots. Total surface area of each land use category that drained directly into each lake was estimated from a topographical map of the U.S. Geological Survey. Only

TABLE 27.2. Water Quality Attributes of Urban Playas in Lubbock, Texas

Water Quality Parameter[a, b]	Units	Mean	Standard Deviation
Temperature[*, ns]	°C	15.39	7.51
pH[*, *]		8.80	0.57
Dissolved oxygen[*, ns]	mg O_2/L	8.58	2.76
Conductivity[*, ns]	mmhos	2.48	20.49
Alkalinity[*, *]	mg $CaCO_3$/L	103.27	30.42
Hardness[*, *]	mg $CaCO_3$/L	131.87	58.58
Chemical oxygen Demand[*, *]	mg O_2/L	67.05	24.32
Biological oxygen Demand[*, *]	mg O_2/L	8.30	5.50
Total Kjeldahl nitrogen[*, *]	mg N/L	2.92	2.47
Total phosphorus[*, ns]	mg P/L	0.73	2.55
Nitrate-nitrogen[*, *]	mg N/L	0.22	0.16
Ortho-phosphorus[*, ns]	mg P/L	0.29	0.16
Ammonia-nitrogen[*, *]	mg N/L	0.70	1.11
Total carbon[*, *]	mg C/L	27.94	7.81
Inorganic carbon[*, ns]	mg C/L	9.36	6.47
Total organic carbon[*, *]	mg C/L	18.58	5.63

Source: Wolf 1996.

[a] significant differences among dates indicated by an asterisk (*).

[b] significant differences among playas indicated by an asterisk (*).

areas producing direct runoff of surface water into each lake were estimated. We did not include possible, but infrequent and unpredictable, inputs from storm sewers, or surface overflow drainage between playas.

PHYSICOCHEMICAL ATTRIBUTES OF URBAN PLAYAS

Ten of the 16 characteristics of water quality showed significant differences among playas (Table 27.2). Pairwise tests detected differences between playas in eight of the parameters, with some playas showing consistent differences from others (Wolf 1996). For example, BOD ranged from ca 3.7 mg O_2/L in Leroy Elmore and Jack Stevens playas to over 11 mg O_2/L in Rushland and Higinbotham playas; total organic carbon also was higher in Higinbotham (23.3 ± 5.3 mg C/L) than in Leroy Elmore and Jack Stevens playas (14.2 ± 1.9 and 15.6 ± 4.1 mg C/L, respectively). Ammonia concentrations were highest in Buster Long playa (1.9 ± 2.3 mg N/L), and TKN was significantly higher in Higinbotham than in Quaker & Brownfield playas (4.6 ± 2.7 versus 2.1 ± 2.1 mg N/L, respectively). Higinbotham was one of the most eutrophic playas, whereas Jack Stevens and Leroy Elmore were among the least eutrophic. Alkalinity, hardness, pH, and COD also showed significant differences between playas, although temporal patterns in alkalinity and hardness were similar among playas. Concentrations of nitrate-nitrogen and phosphorus showed few differences between playas and remained near detection limits throughout most of the study.

In general, urban playas in Lubbock are turbid, well mixed, and somewhat alkaline. Flat topography, continual winds, and alkaline soils probably contribute to these characteristics. Temperatures change in a roughly seasonal pattern, and infrequent stormwater runoff cause short-term fluctuations in concentrations of organic matter and nutrients. Although significant differences in a number of water quality characteristics existed among playas, temporal patterns were similar for all playas in this study. Detailed comparisons of water chemistry between urban and rural playas have not been made, but rural playas tend to be lower in hardness, chemical oxygen demand, total Kjeldahl nitrogen, and ammonia and higher with respect to ortho-phosphate concentrations (Hall et al., this volume). Urban and rural playas have similar levels of biological oxygen demand, nitrate-nitrogen, total carbon, total inorganic carbon, and total phosphorus content.

The areal extent of watersheds providing runoff to playas ranged from 105 ha (Rushland playa) to 262 ha (Leroy Elmore playa), averaging about 157 ha (Table 27.1). Patterns of land use also differed among watersheds, although single-family housing was the dominant land use category surrounding most playas (averaging 64 percent). Only the watershed of Buster Long playa was dominated by another land use, with 53 percent of the land surface associated with commercial operations. Open areas and parks represented about 12 percent, on average, of the landscapes around these playas, whereas multiple family dwellings (apartment complexes) accounted for about 8 percent of the land surface.

MACROINVERTEBRATE COMMUNITIES IN URBAN PLAYAS

Macroinvertebrate Richness and Abundance

Over 10,500 individuals representing at least 94 taxa and 45 families of aquatic invertebrates were collected from playas during this study (Table 27.3). Taxa and individuals were not distributed evenly among playas, with 29–43 percent of the total taxa present in Rushland (33 percent), Higinbotham (32 percent), Wendover (38 percent), Quaker & Brownfield (43 percent), and Maxey (29 percent) playas. Jack Stevens playa contained 63 percent of the taxa, whereas Leroy Elmore and Buster Long playas were species-poor, containing 16 percent and 19 percent of the taxa, respectively. Although taxonomic richness was similar in five of the eight playas, abundances of invertebrates were highest in three playas, with over 70 percent of the total number of individuals collected from Rushland (19 percent), Wendover (21 percent), and Jack Stevens (30 percent) playas. Far fewer individuals were obtained from Higinbotham (6 percent), Quaker & Brownfield (9 percent), Maxey (7 percent), Buster Long (7 percent) and Leroy Elmore (<1 percent) playas.

Rank abundance plots illustrated similar patterns in the distributions of taxa in Rushland, Wendover, Quaker & Brownfield, Jack Stevens, and Higinbotham playas (Fig. 27.1), each of which contained > 30 taxa. However, numerically dominant taxa differed among playas. Crustaceans were numerically most abundant in three playas, with *Palaemonetes kadiakensis* (a freshwater shrimp) representing 33 percent and 61 percent of the invertebrates from Quaker & Brownfield and Maxey playas, respectively, and cladocerans accounting for 62 percent of individuals from Buster Long playa. In contrast, the Notonectidae (back-swimmers) and Corixidae (water boatmen) accounted for 38 percent and 44 percent of the individuals from Rushland and Wendover playas, respectively. The Coenagrionidae (damselfly) accounted for 26 percent of the individuals from Jack Stevens playa, whereas the Chironomidae (midges) accounted for 71 percent and 28 percent of the individuals from Leroy Elmore and Higinbotham playas, respectively. Thus, playas can be separated into two groups according to the numerical dominance of crustaceans or insects.

Macroinvertebrate Diversity

Fisher's log series α was used to estimate macroinvertebrate diversity in each playa (Taylor et al. 1976). This metric is robust with respect to deviations

TABLE 27.3. Systematic List of Aquatic Invertebrate Taxa Collected from Selected Urban Playas in Lubbock, Texas

Annelida	ODONATA
HIRUDINEA	Aeshnidae
Gnathobdellida	*Anax* sp.
Hirudinidae[a]	*Gynacantha* sp.
sp. 1	Libellulidae
sp. 2	*Belonia* sp.
Pharyngobdellida	*Erythrodiplax* sp.
Erpobdellidae[a]	*Orthemis* sp.
sp. 1	*Pachydiplax* sp.
Mollusca	*Perithemis* sp.
GASTROPODA	*Plathemis* sp.
Basommatophora	*Tramea* sp.
Lymnaeidae[c]	Coenagrionidae
sp. 1	*Enallagma* sp.
Physidae[c]	HEMIPTERA
sp. 1	Hydrometridae[f]
sp. 2	*Hydrometra martini*
sp. 3	Macroveliidae[f]
sp. 4	*Macrovelia* sp.
Planorbidae[c]	Gerridae
sp. 1	*Gerris marginatus*
sp. 2	*Rhematobates* sp.
Arthropoda	Belostomatidae
CRUSTACEA	*Belostoma flumineum*
Cladocera[b]	Nepidae[f]
sp. 1	*Ranatra nigra*
Eucopepoda[b]	Corixidae
Calanoid sp.	*Corisella edulis*
Cyclopoid sp.	*Corisella tarsalis*
Ostracoda[b]	*Ramphocorixa* sp.
sp. 1	*Sigara alternata*
sp. 2	Notonectidae
Decopoda	*Buenoa* sp. 1
Palaemonidae	*Buenoa* sp. 2
Palaemonetes kadiakensis	*Notonecta undulata*
INSECTA	Mesoveliidae
Collembola	*Mesovelia mulsanti*
Isotomidae	Hebridae[f]
sp. 1	*Hebrus* sp.
Ephemeroptera	Saldidae
Baetidae	sp. 1
Callibaetis sp.	*Saldula* pallipes
Caenidae	*Saldula* sp. 2
Caenis punctatus	

(*continued*)

TABLE 27.3. (Continued)

Coleoptera	*Stenus* sp. 1
Gyrinidae[d]	*Stenus* sp. 2
Dineutus sp.	Salpingidae[d]
Haliplidae	*Limnebius* sp.
Haliplus sp. 1	Georyssidae[d]§
Haliplus sp. 2	*Georyssus* sp.
Peltodytes sp.	Chrysomelidae
Dytiscidae	*Disonycha* sp.
Brachyvatus sp.	*Donacia* sp.
Copelatus sp.	sp. 1
Laccophilus fasciatus	Diptera
Laccophilus proximus	Tipulidae[e]
Liodessus sp.	sp. 1
Neobidessus sp.	Culicidae[e]
Thermonectus sp.	*Culex* sp.
Uvarus sp.	Chaoboridae[e]
Sphaeridae[d]	*Chaoborus* sp.
Sphaerius sp.	Psychodidae[e]
Hydrophilidae	*Pericoma* sp.
Berosus sp. 1	Ceratopogonidae[e]
Berosus sp. 2	sp. 1
Berosus sp. 3	Chironomidae
Berosus sp. 4	adult sp. 1
Berosus sp. 5	adult sp. 2
Enochrus sp.	adult sp. 3
Helophorus sp.	larvae spp.
Hydrochous sp.	Stratiomyidae
Hydrophilus triangularis	*Ondontomyia* sp.
Laccobius sp.	Tabanidae[e]
Paracymus sp.	*Tabanus* sp.
Tropisternus lateralis	Ephydridae[e]
Staphylinidae[d]	sp. 1
Micaralymma sp.	Muscidae[e]
	sp. 1

[a] Families were combined to form the Annelida group for familial data analyses.
[b] Taxa in these orders were combined to form the Crustacea group for familial data analyses.
[c] Families were combined to form the Gastropoda group for familial data analyses.
[d] Families were combined to form the Coleoptera (other) group for familial data analyses.
[e] Families were combined to form the Diptera (other) group for familial data analyses.
[f] Families were combined to form the Hemiptera (other) group for familial data analyses.

from log series distributions, as well as to variation in the total number of individuals sampled (Magurran 1988). In particular, the index (α) is given by:

$$\alpha = N(1 - x)/x$$

where N is the total number of individuals and x is the ratio of taxon richness

(S) to total number of individuals (N). More specifically, x can be estimated by iteration (Taylor et al. 1976):

$$S/N = [(1 - x)^{-1}][-\ln(1 - x)]$$

Chi-square goodness-of-fit tests were used to compare observed versus expected species distributions for each playa and to evaluate whether the observed distribution was significantly different from the expected log series model (Fig. 27.1; Sokal and Rohlf 1995). For all playas except Buster Long ($P = 0.029$), distributions of macroinvertebrate taxa collected from all dates adhered to the log series model, suggesting that one or few factors dominate the composition of the community (Magurran 1988). Alternatively, Magurran (1988) suggests that a log-series distribution can result from sequential arrival of species that occupy a random proportion of the unoccupied niche space.

To compare diversity between playas, Fisher's log series α for particular playas was considered to be significantly different if their 95 percent confidence limits did not overlap (Table 27.4). Although this method inflates experiment-wise error rate, results can be useful in a heuristic context. Based on all invertebrates collected from March through December, α was significantly greater for Jack Stevens than Buster Long and Rushland playas; α for Buster Long playa was significantly lower than that of Jack Stevens and Quaker & Brownfield playas (final column, Table 27.4).

Fisher's log series α also was calculated for each playa in each sampling season, based on macroinvertebrate collections made in spring (March and April), summer (June, July, and August), and fall (November–December). Jack Stevens playa consistently had the highest α throughout all seasons (Table 27.4). During the spring diversity of Jack Stevens playa was significantly higher than of all other playas except Quaker & Brownfield playa. Fisher's log series α was not calculated for Higinbotham and Leroy Elmore playas during the spring, because species abundances were evenly distributed among the few collected species. For the summer, diversity of Buster Long playa was significantly lower than Quaker & Brownfield and Jack Stevens playas. For the fall, diversity of Leroy Elmore and Buster Long playas was each significantly lower than Rushland, Higinbotham, Quaker & Brownfield, or Jack Stevens playas.

Macroinvertebrate Community Composition

Two communities could exhibit the same richness and diversity, but their composition could be completely different. In contrast, community composition reflects species richness and relative abundances, but retains a taxonomic component. Invertebrate community composition was evaluated at two levels of taxonomic resolution: the taxon level (defined previously) and the familial level (Table 27.3). Invertebrates were grouped according to familial or higher taxonomic level (i.e., order, class) so that each group contained at

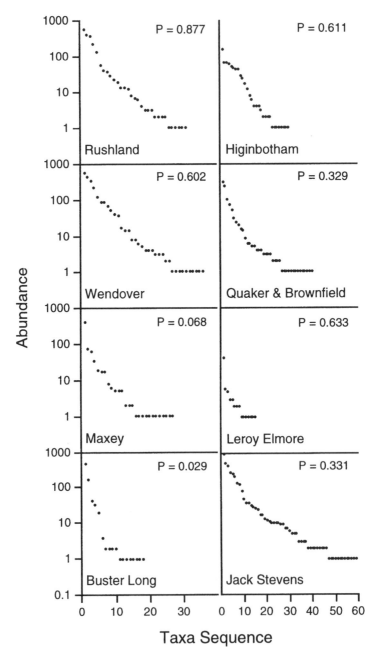

Fig. 27.1. Rank abundance of aquatic invertebrates in urban playas of Lubbock, Texas. Probability values for chi-square goodness-of-fit tests compare observed to expected log series distributions.

TABLE 27.4. Fisher's Log Series α (Mean ± Standard Deviation) for Each Playa[a]

Play	Spring	Summer	Autumn	Combined
Rushland	1.5 ± 0.8a	4.7 ± 0.9ab	3.7 ± 1.0a	5.2 ± 0.9ab
Higinbotham		5.5 ± 1.1ab	3.2 ± 1.1a	6.4 ± 1.2abc
Wendover	1.5 ± 0.8a	5.1 ± 0.9ab	1.6 ± 0.5ab	6.1 ± 1.0abc
Quaker & Brownfield	2.4 ± 0.8ab	7.3 ± 1.3ab	3.3 ± 0.9a	8.4 ± 1.3ac
Maxey	1.5 ± 0.6a	5.3 ± 1.1ab	1.4 ± 0.7ab	5.6 ± 1.1abc
Leroy Elmore		4.9 ± 1.4ab	0.3 ± 0.3b	5.6 ± 1.4abc
Buster Long	1.1 ± 0.5a	3.0 ± 0.8a	0.5 ± 0.3b	3.3 ± 0.8b
Jack Stevens	6.2 ± 1.3b	8.7 ± 1.2b	4.0 ± 1.0a	10.3 ± 1.3c

[a]Same letters following means denote playas that are indistinguishable (within season) based on species diversity.

least 1 percent of the total number of collected organisms. This approach minimizes the influence exerted by a few individuals representing a single taxon on measures of community similarity, and permits an assessment of whether the taxonomic level of identification affects the conclusions of community analyses. The term familial is used throughout the remainder of this chapter in reference to composite groupings at which quantitative analyses were undertaken.

Four similarity measures (Euclidean distance, Cosine, Jaccard, Ochai) were used to evaluate the relationships among urban playas based on macroinvertebrate community composition (Ludwig and Reynolds 1988). Organisms collected at all dates were pooled for these analyses (separate analyses performed by date provided no additional insights and thus are not included herein).

Euclidean distance and Cosine similarity are based on species abundance data, but differ in the degree to which they are sensitive to abundance. Euclidean distance (E_{jk}) is defined as:

$$E_{jk} = \sqrt{\sum_{i=1}^{s} (x_{ij} - x_{ik})^2}$$

where x_{ij} and x_{ik} are the abundances of the ith species (s) for playa j and k, respectively. This measure emphasizes differences in the abundances of taxa between playas. Conversely, the Cosine index (C_{jk}) is defined by:

$$(C_{jk}) = \frac{\sum_{i=1}^{s} (x_{ij} x_{ik})}{\sqrt{\left(\sum_{i}^{s} x_{ij}^2\right)\left(\sum_{i}^{s} x_{ik}^2\right)}}$$

and places greater importance on the relative abundances of taxa.

Jaccard and Ochai measures are based on presence-absence data and differ in the manner in which shared taxa are considered. The Jaccard index (J_{jk}) is given by:

$$J_{jk} = \frac{a}{a + b + c}$$

where a is the number of taxa in common to playas j and k, b is the number of taxa that occur in j but not k, and c is the number of species that occur in k but not j. The Ochai index (C_{jk}) is the binary form of the Cosine index, and is given by:

$$O_{jk} = \frac{a}{\sqrt{a + b} \, \sqrt{a + c}}$$

where values are as previously defined.

The Jaccard, Ochai, and Cosine similarity values were transformed into dissimilarity values for subsequent analysis. An 8-by-8 playa matrix was created for each measure and subjected to cluster analysis with a unweighted pair-group method average algorithm (UPGMA) using taxon-level and familial-level data separately (SPSS 1990).

At the taxon level the Jaccard and Ochai indexes, based on data from all sampling dates, showed a number of similarities (Fig. 27.2a, b). Both indexes identified Wendover and Maxey playas as being most similar to each other and placed them in a larger cluster with Rushland, Higinbotham, Quaker & Brownfield, and Jack Stevens playas. Although these methods differed with respect to the degree of similarity estimated between particular playas within the cluster, both approaches also identified Buster Long and Leroy Elmore playas as being least similar to the other six playas. Even fewer differences in interplaya relationships existed between analyses based on Jaccard and Ochai measures using familial-level data (Fig. 27.2c, d). Interplaya patterns were identical with differences limited to relative strengths of associations. In general, taxon-level and familial-level analyses were similar in the way that they clustered playas. For example, Wendover and Maxey playas were consistently identified as being most similar to each other, while Buster Long and Leroy Elmore playas were placed outside the cluster comprising the other six playas (Fig. 27.2). While modest differences existed among analyses at the taxon and familial levels, the greatest was that the Jaccard index identified Jack Stevens playa as being most similar to Higinbotham playa at the familial level, but the two playas were much less similar at the taxon level (Fig. 27.2a, c).

Euclidean distance and Cosine indexes, based on taxon-level data, similarly clustered Wendover, Leroy Elmore, Quaker & Brownfield, and Maxey playas, but showed little consistency in relationships among the other playas (Fig.

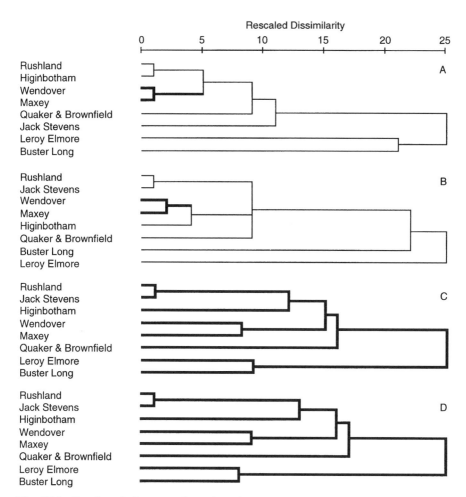

Fig. 27.2. Results of cluster analyses based on Jaccard and Ochai similarity indexes for macroinvertebrate communities in urban playas of Lubbock, Texas. Consensus between dendrograms represented by bold lines. (*a*) Jaccard's index based on taxon-level data. (*b*) Ochai's index based on taxon-level data. (*c*) Jaccard's index based on familial data. (*d*) Ochai's index based on familial data.

27.3*a*, *b*). Analyses based on familial-level data showed greater correspondence between indexes (Fig. 27.3*c*, *d*), consistently pairing Quaker & Brownfield and Maxey playas, Higinbotham and Leroy Elmore playas, and Jack Stevens and Rushland playas. Analyses at both taxon and familial levels clustered Leroy Elmore, Quaker & Brownfield and Maxey playas. However, taxon-level data indicated that Leroy Elmore playa was similar to Wendover playa and distant from Higinbotham playa, whereas familial-level data suggested that Leroy Elmore and Higinbotham playas were more similar.

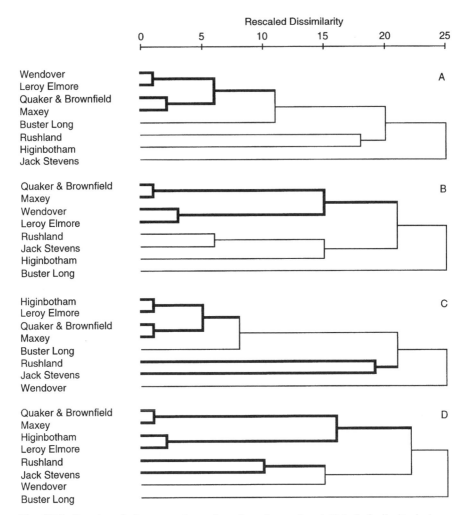

Fig. 27.3. Results of cluster analyses based on Jaccard and Ochai similarity indexes for macroinvertebrate communities in urban playas of Lubbock, Texas. Consensus between dendrograms represented by bold lines. (*a*) Euclidian distance based on taxon-level data. (*b*) Cosine index based on taxon-level data. (*c*) Euclidian distance index based on familial data, (*d*) Cosine index based on familial data.

Comparisons between approaches based on presence-absence of taxa (Jaccard and Ochai indexes) and those based on abundances of taxa (Euclidean distance and Cosine indexes) revealed that Wendover and Maxey playas consistently were placed within a cluster of two to four playas, based on taxon-level analyses. Buster Long playa was consistently outside this cluster of two to four playas, based on taxon-level data. At the familial level all indexes paired Rushland and Jack Stevens playas, whereas Quaker & Brownfield, Higinbotham, and Maxey playas were grouped within clusters that variously consisted of five to six playas. The only consistent relationship between playas identified by all indexes at both levels of taxonomic resolution was that Quaker & Brownfield and Maxey playas always occurred within a cluster composed of the six most similar playas.

A principal components analysis (PCA) was conducted to identify macroinvertebrate families that contributed the greatest variation among urban playas (SPSS 1990). The first three PC scores accounted for 75 percent of the variation in abundances of familial groups. The first PC accounted for 44 percent of the variation among urban playas in macroinvertebrate familial abundances and was loaded positively with densities of gastropods, baetid mayflies, annelids, and aeshnid dragonflies. This PC separated Jack Stevens playa from the other playas. The second PC accounted for 16 percent of the variation, was loaded positively with densities of chrysomelid beetles and gerrids (water striders), and distinguished Quaker & Brownfield playa. The third PC accounted for 15 percent of the variation, was loaded positively with notonectids (back-swimmers) and corixids (water boatmen), and separated Wendover playa from all the others. These results were most consistent with cluster analysis based on the Cosine index and familial-level data (Fig. 27.3*d*). A cluster consisted of Rushland, Jack Stevens, and Wendover playas, which contained relatively greater numbers of notonectids, baetid mayflies, caenid mayflies, coenagrionids (damselflies), and chironomids (midges). A second cluster consisted of Quaker & Brownfield, Maxey, Leroy Elmore, and Higinbotham playas, which contained greater numbers of palaemonids (freshwater shrimp), corixids, and gerrids.

Macroinvertebrate-Environment Relationships

Mantel's nonparametric test (Manly 1985, 1986) evaluated whether differences in water chemistry among playas were correlated with differences in community composition among playas (Smouse et al. 1986). Because macroinvertebrates were collected on six dates (March, April, June, July, August, and November–December), only physicochemical data for these dates were used in the analysis. Thus, two similarity matrices were constructed for each sampling date, one characterizing pair-wise differences in macroinvertebrate composition (based on abundances of individual taxa) and the other representing pair-wise differences in water chemistry. Because water chemistry attributes have different units (i.e., mg/L, mmho, °C), data were transformed to Z scores for analysis. A significant positive relationship existed between interplaya differences in physicochemical attributes and invertebrate community composition during November ($P = 0.048$), but no significant relationships occurred at other dates ($P > 0.2$ in all cases). Thus, little evidence supports the hypothesis that differences in macroinvertebrate communities are related to differences in water quality characteristics among urban playas.

Spearman rank correlations between particular physicochemical parameters and land use categories revealed few significant relationships, despite interplaya variation in water quality parameters and surrounding land use patterns

(Wolf 1996). Conductivity and ammonia concentration were correlated positively to area of multiple-family housing and commercial land use; total organic carbon also was correlated positively to multiple-family housing. Nevertheless, the general lack of correlation between land use patterns and water quality characteristics does not preclude the possible importance of water quality parameters as a factor influencing macroinvertebrate communities.

IMPLICATIONS

The urban playas of Lubbock provide an uncommon context in which to study aquatic macroinvertebrate communities. Playas are isolated hydrologically but remain comparable in many respects. A number of communities can be examined within a small geographical area, minimizing differences among impoundments due to differences in soils, climate, surrounding landscape, drainage relationships, species pools, and evolutionary history that normally occur over larger distances. Although some significant differences in physicochemical and biological characteristics exist among playas, little indicates that these characteristics are correlated.

Few consistent differences among playas characterize macroinvertebrate diversity, but Jack Stevens playa usually showed significantly higher diversity than did Buster Long and Leroy Elmore playas (Table 27.4). This was evident in overall diversity (combined data from all seasons) as well as diversity for each season. Fisher's log series α was highest for Jack Stevens playa at all times, whereas α for Buster Long playa was usually the lowest. Some of these differences may be related to general habitat characteristics; Jack Stevens playa is shallow and supports an extensive emergent and submerged macrophyte community, whereas Buster Long playa is a steep-sloping basin with no littoral vegetation and high phytoplankton populations (Brownlow 1994). Indeed, Wollheim and Lovvorn (1995) report a direct relationship between biomass of aquatic invertebrates and extent of macrophyte habitat in shallow lakes of the Wyoming High Plains. Moreover, an inverse relationship existed between biomass of zooplankton and that of other aquatic invertebrates. Characteristics of urban playas of Lubbock are consistent with these observations; the highest numbers of zooplankton and lowest number of insect taxa occurred in Buster Long playa, which lacks macrophytes and has high phytoplankton populations (Brownlow 1994). Wollheim and Lovvorn (1995) also found that littoral macrophyte habitats supported a greater diversity and abundance of aquatic invertebrates. This greater abundance of invertebrates probably occurs because macrophytes increase structural and resource dimensions of niche space (Rosine 1955, Rooke 1984) by enhancing food quality (Carpenter and Lodge 1986) and cover from predators (Crowder and Cooper 1982). The higher diversity and abundance of invertebrates in Jack Stevens and Quaker & Brownfield playas may result from the greater extent of macrophyte habitat in these playas (Table 27.1).

Analyses of community structure also revealed few consistent relationships among playas. Only Quaker & Brownfield and Maxey playas were always among the six most similar playas—possibly because only these two playas had relatively high abundances of palaemonids and chironomids. However, the PCA and cluster analysis based on the Cosine similarity measure (familial-level data) shared some common features. The Cosine index produced two main clusters of playas with one (including Jack Stevens and Wendover playas) having relatively greater numbers of notonectids and baetid mayflies and a second group (including Quaker & Brownfield playa) having relatively greater numbers of gerrids. In comparison, the PC-1 separated Jack Stevens playa from the others, partly based on densities of baetid mayflies; the PC-2 separated Quaker & Brownfield playa from the rest, in part due to densities of gerrids; the PC-3 separated Wendover playa from the others, partly based on abundances of notonectids.

Species-Environment Relationships

Urban playas were variable with respect to some water quality measures, but Mantel analyses demonstrated little overall relationship between water quality and composition of aquatic macroinvertebrate communities, although many water quality attributes that differed among playas have been shown to influence community composition of aquatic invertebrates elsewhere (Effler et al. 1990, Hammer et al. 1990, Gower et al. 1994, Malmqvist and Eriksson 1995, Wollheim and Lovvorn 1995). For example, many invertebrates are sensitive to levels of dissolved oxygen (Wetzel 1983). Dissolved oxygen ranged between 14.35 and 1.87 mg/L in Lubbock playas, with the latter value approaching a stressful level for some organisms (e.g., mayflies, dragonflies, and damselflies; Merritt and Cummins 1984). Also, invertebrate communities in prairie lakes can be affected by salinity (Wollheim and Lovvorn 1995, Hammer et al. 1990). Salinity was not measured directly for urban playas of Lubbock, but relatively high conductivity and alkalinity illustrate high ionic concentrations.

The level of eutrophication or trophic status in lakes long has been used as a predictor of species distributions and abundances (Macan 1963, Saether 1979). In fact, Scheffer (1990) proposed two types of shallow freshwater lakes: those with clear water (oligotrophic) dominated by macrophytes and those with turbid waters (eutrophic) dominated by phytoplankton. Nutrient enrichment can cause a shift from a macrophyte-dominated community to a phytoplankton-dominated community, with attendant changes in invertebrate communities (Saether 1979, Growns et al. 1992). In the present study total organic carbon, ammonia, TKN and nitrate concentrations differed significantly between playas (Wolf 1996) and may be responsible for some differences in macroinvertebrate communities. For example, ammonia can have significant impacts on aquatic invertebrates (Effler et al. 1990, Richards et al. 1993). In the present study ammonia concentrations were highest in Buster Long playa (1.9 ± 2.3 mg N/L; Wolf 1996), which also differed from other playas in that its macroinvertebrate community was dominated by cladocerans and copepods. This is consistent with large phytoplankton populations reported for this playa (Brownlow 1994). A relationship between ammonia and abundances of cladocerans may be mediated by phytoplankton in this playa.

In conclusion, urban playas exhibited similar macroinvertebrate species diversity, although invertebrate community composition and dominance differed greatly. Differences in community composition were not related differences in water quality characteristics, although some taxa are affected by particular physicochemical characteristics. Macroinvertebrate communities of urban playas in the Southern High Plains region of West Texas may be affected most strongly by vegetational characteristics, and secondarily by particular water quality parameters.

ACKNOWLEDGMENTS

Voucher specimens were identified by Dr. Robert Sites (University of Missouri). Financial support was provided by the Office of Research Services and Department of Biological Sciences, Texas Tech University.

LITERATURE CITED

American Public Health Association (APHA). 1992. Standard Methods for the Examination of Water and Waste Water. (18th ed.). American Public Health Association, Washington, DC.

Bolen, E. G., L. M. Smith, and H. L. Schramm, Jr. 1989. Playa lakes: Prairie wetlands of the Southern High Plains. Bioscience 39:615–623.

Brownlow, C. J. 1994. Trophic state assessment of an urban lake, Lubbock, Texas. Master's Thesis, Texas Tech University, Lubbock, TX.

Cairns, J., Jr., and J. R. Pratt. 1993. A history of biological monitoring using benthic macroinvertebrates. Pages 10–27 *in* D. M. Rosenberg and V. H. Resh (eds.), Freshwater Biomonitoring and Benthic Macroinvertebrates. Chapman & Hall, New York.

Carpenter, S. D., and D. M. Lodge. 1986. Effects of submersed macrophytes on ecosystem processes. Aquatic Botany 26:341–370.

Crowder, L. B., and W. E. Cooper. 1982. Habitat structural complexity and the interaction between bluegills and their prey. Ecology 63:1802–1813.

Effler, S. W., C. M. Brooks, M. T. Auer, and S. M. Doerr. 1990. Free ammonia and toxicity criteria in polluted urban lake. Research Journal of Water Pollution Control Federation 62:771–779.

Ennis, T. E. 1994. City of Lubbock stormwater NPDES permit data analysis. Master's Thesis, Texas Tech University, Lubbock, TX.

Environmental Protection Agency (EPA). 1974. Methods for Chemical Analysis Of Water and Wastes. Environmental Protection Agency, Water Quality Office, Analytical Quality Control Laboratory, Cincinnati, OH.

France, R. L. 1995. Macroinvertebrate standing crop in littoral regions of allochthonous detritus accumulation: Implications for forest management. Biological Conservation 75:35–39.

Gower, A. M., G. Myers, M. Kent, and M. E. Foulkes. 1994. Relationships between macroinvertebrate communities and environmental variables in metal-contaminated streams in south-west England. Freshwater Biology 32:199–221.

Growns, J. E., J. A. Davis, F. Cheal, L. G. Schmidt, R. S. Rosich, and S. J. Bradley. 1992. Multivariate pattern analysis of wetland invertebrate communities and environmental variables in Western Australia. Australian Journal of Ecology 17:275–288.

Hall, D. L., R. W. Sites, E. B. Fish, T. R. Mollhagen, D. L. Moorhead, and M. R. Willig. Playas of the Southern High Plains: The macroinvertebrate fauna. This volume.

Hammer, U. T., J. S. Sheard, and J. Kranabetter. 1990. Distribution and abundance of littoral benthic fauna in Canadian prairie saline lakes. Hydrobiologia 197:173–192.

Hanson, M. A., and M. G. Butler. 1994. Responses of food web manipulation in a shallow waterfowl lake. Hydrobiologia 279/280:457–466.

Haukos, D. A., and L. M. Smith. 1992. Ecology of Playa Lakes. Waterfowl management handbook leaflet 13.3.7, Office of Information Transfer, U.S. Fish and Wildlife Service, Fort Collins, CO.

Hershey, A. E. 1985. Littoral chironomid communities in an arctic Alaskan lake. Holarctic Ecology 8:39–48.

Hynes, H. B. N. 1970. The ecology of Running Waters. University of Toronto, Toronto, ON.

Jones, R., and C. C. Clark. 1987. Impact of watershed urbanization on stream insect communities. Water Resources Bulletin 23:1047–1055.

Kullberg, A. 1992. Benthic macroinvertebrate community structure in 20 streams of varying pH and humic content. Environmental Pollution 78:103–106.

Livingston, E., and J. Cox. 1989. Florida Development Manual. Vol. 1., Florida Department of Environmental Regulation, Tallahassee, FL.

Ludwig, J. A., and J. F. Reynolds. 1988. Statistical ecology: A Primer on Methods and Computing. John Wiley & Sons, New York.

Macan, T. T. 1949. Corixidae (Hemiptera) of an evolved lake in the English Lake District. Hydrobiologia 2:1–23.

Macan, T. T. 1963. Freshwater ecology. John Wiley & Sons, New York.

Magurran, A. E. 1988. Ecological Diversity and Its Measurement. Princeton University Press, Princeton, NJ.

Malmqvist, B., and A. Eriksson. 1995. Benthic insects in Swedish lake-outlet streams: Patterns in species richness and assemblage structure. Freshwater Biology 34:285–296.

Manly, B. F. 1985. The Statistics of Natural Selection on Animal Populations. Chapman & Hall, New York.

Manly, B. F. 1986. Multivariate Statistical Methods: A Primer. Chapman & Hall, New York.

Mason, C. F. 1991. Biology of freshwater pollution (2nd ed.). Longman Scientific & Technical, Harlow, Essex, UK.

Merickel, F. W., and J. K. Wangberg. 1981. Species composition and diversity of macroinvertebrates in two playa lakes on the Southern High Plains, Texas. Southwestern Naturalist 26:153–158.

Merritt, R. W., and K. W. Cummins. 1984. An Introduction to the Aquatic Insects of North America (2nd ed.). Kendall/Hunt, Dubuque, IA.

Neck, R. W., and H. L. Schramm, Jr. 1992. Freshwater molluscs of selected playa lakes of the Southern High Plains of Texas. Southwestern Naturalist 37:205–209.

Porcella, D. B., and D. L. Sorensen. 1980. Characteristics of nonpoint source urban runoff and its effects on stream ecosystems. Corvallis Environmental Research Laboratory, Office of Research and Development (U.S. EPA), Corvallis, OR.

Rasmussen, J. B. 1993. Patterns in the size structure of littoral zone macroinvertebrate communities. Canadian Journal of Aquatic Sciences 50:2192–2207.

Richards, C., G. E. Host, and J. W. Arthur. 1993. Identification of predominant environmental factors structuring stream macroinvertebrate communities within a large agricultural catchment. Freshwater Biology 29:285–294.

Rimer, A. E., and J. A. Nissen. 1978. Characterization and impact of stormwater runoff from various land cover types. Water Pollution Control Federation 50:252–264.

Rooke, B. J. 1984. The invertebrate fauna of four macrophytes in a lotic system. Freshwater Biology 14:507–513.

Rosenberg, D. M., and V. H. Resh. 1993. Freshwater Biomonitoring and Benthic Macroinvertebrates. Chapman & Hall, New York.

Rosine, W. N. 1955. The distribution of invertebrates on submerged aquatic plant surfaces in Muskee Lake, Colorado. Ecology 36:308–314.

Saether, O. A. 1979. Chironomid communities as water quality indicators. Holarctic Ecology 2:65–74.

Scheffer, M. 1990. Multiplicity of stable states in freshwater systems. Hydrobiologia 200/201:475–486.

Smith, C. L. 1993. Water boatmen (Hemiptera: Corixidae) faunas in the playa lakes of the Southern High Plains. Master's Thesis, Texas Tech University, Lubbock, TX.

Smouse, P. E., J. C. Long, and R. R. Sokal. 1986. Multiple regression and correlation extensions of the Mantel test of matrix correspondence. Systematic Zoology 35:627–632.

Sokal, R. R., and F. J. Rohlf. 1995. Biometry: The Principles and Practice of Statistics in Biological Research (3rd ed.). W. H. Freeman & Co., New York.

SPSS. 1990. Statistical Package for the Social Sciences. SPSS, Chicago, IL.

Sublette, J. E., and M. S. Sublette. 1967. The limnology of playa lakes on the Llano Estacado, New Mexico and Texas. Southwestern Naturalist 12:369–406.

Taylor, L. R., R. A. Kempton, and I. P. Woiwod. 1976. Diversity and the log series model. Journal of Animal Ecology 45:255–271.

Thompson, G. B., D. M. Wells, R. M. Sweazy, and B. J. Claborn. 1974. Variation of Urban Runoff Quality and Quantity with Duration and Intensity of Storms. Interim Report, Water Resources Center, Texas Tech University, Lubbock, TX.

U.S. Fish and Wildlife Service. 1988. Playa Lakes Region Waterfowl Habitat Concept Plan, Category 24 of the North American Waterfowl Management Plan. U.S. Fish and Wildlife Service, Albuquerque, NM.

Ward, J. V. 1992. Aquatic Insect Ecology: Biology and Habitat. John Wiley & Sons, New York.

Wells, D. M., R. M. Sweazy, B. J. Claborn, and R. H. Ramsey. 1975. Variation of Urban Runoff Quality and Quantity with Duration and Intensity of Storms: Phase III. Vol. 4, Project Summary. Final report of the office of Water Research and Technology, Water Resources Center, Texas Tech University, Lubbock, TX.

Wetzel, R. G. 1983. Limnology. CBS College, New York.

Wolf, C. F. 1996. Aquatic macroinvertebrate diversity and water quality of urban lakes. Master's Thesis, Texas Tech University, Lubbock, TX.

Wollheim, W. M., and J. R. Lovvorn. 1995. Salinity effects on macroinvertebrate assemblages and waterbird food webs in shallow lakes of the Wyoming High Plains. Hydrobiologia 310:207–223.

28 Temporarily Flooded Wetlands of Missouri

Invertebrate Ecology and Management

PATRICK A. MAGEE, FREDERIC A. REID, and
LEIGH H. FREDRICKSON

O *n a geographic scale, Missouri wetlands can be characterized by a climatic gradient of cooler, drier conditions in the northwest to warmer, wetter conditions in the southeast. Missouri overlaps the deciduous forest and tallgrass prairie biomes and has a diversity of wetland types. Historically, the most extensive wetlands were the bottomland hardwoods of the southeastern bootheel region. Invertebrate faunas are diverse, with forested wetlands being dominated by crustaceans, mollusks, or oligochaetes. On relatively high-elevation forested wetlands chironomid midges are better adapted to the more ephemeral wet periods. Herbaceous wetlands have a greater diversity of taxa and include more insects. Few inventories of Missouri wetland invertebrate faunas have been completed, and more detailed taxonomic treatments are necessary. Hydrology tends to have the greatest effect on local abundance and distribution of invertebrates. Multivoltine taxa are more likely to be found on ephemeral sites, whereas univoltine taxa occur where hydroperiods are longer. Many wetlands in Missouri are detrital based, and invertebrate productivity is high during the dormant season for plants (November to April). In forested wetlands peaks in the abundance of invertebrate detritivores match the life cycle events of numerous avian predators, as well as other wildlife. These predators have high protein and calcium requirements during molt, egg laying, and growth periods, and wetland invertebrates are key components of waterbird diets. Most wetlands in the state have been altered, but numerous sites are restored and intensively managed by federal, state, and private entities. In intensively managed wetlands it is possible to manipulate hydrology to mimic historical water regimes to promote invertebrate production.*

Invertebrates in Freshwater Wetlands of North America: Ecology and Management, Edited by
Darold P. Batzer, Russell B. Rader, and Scott A. Wissinger
ISBN 0-471-29258-3   © 1999 John Wiley & Sons, Inc.

Also, problem plant control can be tied to invertebrate management by disking or mowing treatments followed by shallow flooding. Where wetland complexes exist, production of invertebrates for multiple vertebrate predators can be achieved within separate management units by understanding the life cycles of invertebrates and matching them with the life cycles of vertebrate predators.

PHYSIOGRAPHY AND HUMAN IMPACTS

Missouri is diverse in physiography and geological origins. Four regional watersheds are represented (Fig. 28.1). Most of western and central Missouri are part of the Missouri River watershed, which comprises the largest portion of the state. Southwestern and south-central Missouri feed the Arkansas-White-Red River watershed, whereas northeastern and southeastern Missouri drain to the Upper and Lower Mississippi River watersheds, respectively (U.S. Geologic Survey 1974).

Grasslands in northern Missouri began to establish 26 million years ago and were probably well established in the Pliocene about 2.5 to 13 million years ago, as evidenced by numerous herbivore fossil finds (Collins and Glenn 1995). This region of Missouri (north of the Missouri River) was glaciated 400,000 years ago during the Kansan glaciation of the Pleistocene (Johnson 1987). The region is characterized by rolling hills and a mix of grassland and deciduous forest (Collins and Glen 1995). Few native grasses and forests remain. This region is a major agricultural area as the soils are deep (loess deposits), erosional, and rich. Wetlands in this region are primarily riparian woodlands or marshes associated with streams.

Western Missouri (south of the Missouri River) is characterized as tallgrass prairie and was indirectly influenced by glaciation, as it was the outwash plain during glacial recession. The soils in this region are shallow and rocky, making cropping unproductive. Therefore, agriculture is dominated by livestock production. Wetlands are an important component of the tallgrass prairie habitat mosaic and occur at the extreme end of a soil moisture gradient associated with streams, impoundments, and low-lying runoff sites (Ryan 1990).

Central and southern Missouri comprise the Ozark Plateau, characterized by ancient mountains, karst topography, and deciduous/mixed forest types. Most of the forests have been harvested, and few old-growth stands are present. The most notable aquatic habitats in this region are the spring-fed rivers and their adjacent riparian corridors. Further, subsurface waters are common in caves and have unique attributes, including the presence of some rare invertebrates such as Salem and Bristly cave crayfish (Missouri Natural Heritage Program rare and endangered species database, ⟨www.tnc.heritage.org⟩).

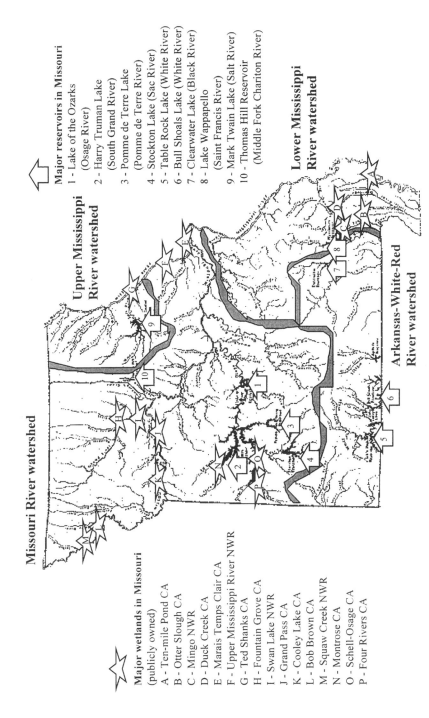

Missouri River watershed

Upper Mississippi River watershed

Lower Mississippi River watershed

Arkansas-White-Red River watershed

Major reservoirs in Missouri

1 - Lake of the Ozarks
   (Osage River)
2 - Harry Truman Lake
   (South Grand River)
3 - Pomme de Terre Lake
   (Pomme de Terre River)
4 - Stockton Lake (Sac River)
5 - Table Rock Lake (White River)
6 - Bull Shoals Lake (White River)
7 - Clearwater Lake (Black River)
8 - Lake Wappapello
   (Saint Francis River)
9 - Mark Twain Lake (Salt River)
10 - Thomas Hill Reservoir
   (Middle Fork Chariton River)

Major wetlands in Missouri
(publicly owned)

A - Ten-mile Pond CA
B - Otter Slough CA
C - Mingo NWR
D - Duck Creek CA
E - Marais Temps Clair CA
F - Upper Mississippi River NWR
G - Ted Shanks CA
H - Fountain Grove CA
I - Swan Lake NWR
J - Grand Pass CA
K - Cooley Lake CA
L - Bob Brown CA
M - Squaw Creek NWR
N - Montrose CA
O - Schell-Osage CA
P - Four Rivers CA

Fig. 28.1. Watersheds and wetlands of Missouri. 3-D lines represent watershed boundaries.

Southeastern Missouri is part of the Mississippi River lowlands, with flat topography, and is adjacent to the Ozark Uplift. Approximately 45,000 years ago the area served as the main river channel of both the Mississippi and Ohio rivers (Korte and Fredrickson 1977). Overflow flooding by the Mississippi, Ohio, Castor, St. Francis, Whitewater, and Little rivers and surface puddling as a result of precipitation are the key hydrological sources for wetlands in this vast, ancient floodplain region. Although this region was historically the richest for wetlands in Missouri, most of the 940,000 ha of bottomland forests have been drained and harvested and replaced by agricultural crops such as soybeans, corn, rice, and cotton. Currently, fewer than 24,000 ha are still bottomland hardwoods, representing a 98 percent loss of this ecosystem (Vaught and Bowmaster 1983). Channelization on the St. Francis River reduced the length of the active channel in some stretches by 58 percent and reduced the area of riparian forests by 78 percent (Fredrickson 1980). The largest remaining tract of bottomland forest, the 11,336-ha Mingo Swamp, is protected by state and federal agencies and includes a 4,800-ha wilderness area (Korte and Fredrickson 1977).

The Missouri River and Mississippi River (north of the Ohio River confluence) corridors represent a fifth major physiographic region in Missouri, characterized by large rivers with extensive floodplains. These river systems were historically dynamic wetland complexes with a river channel meandering across the floodplain. The Missouri River, for example, migrated across one third of its floodplain in a 50-year period from 1879–1930 (Schmudde 1963, in Galat et al. 1996). These river channels have been extensively modified by dredging, channelization, levee construction, drainage, removal of timber, and construction of wing dikes (Galat et al. 1996, Humburg et al. 1996). The once wide and variable channels have been narrowed and straightened for navigation and flood control. As an example of the impacts on these rivers and their floodplain wetlands, the Missouri River was shortened from 876 km in 1879 to 802 km by 1972 (Humburg et al. 1996). The Mississippi River corridor was dominated by forested wetlands, particularly in the southern parts of the state, whereas the Missouri River had extensive marshes and willow/cottonwood riparian forests.

Hydrology has been altered in most of Missouri's wetlands. Of the original 1.9 million ha of wetlands in Missouri, less than 13 percent remain (Dahl 1990). Historically, 11 percent of Missouri was wetland; currently, less than 2 percent of Missouri is wetland (Epperson 1992). Many wetlands in Missouri are restored and intensively managed (Fig. 28.1). Many rivers throughout the state have been dammed, creating large reservoirs and degrading stream habitat (Fig. 28.1). Despite riparian losses, these reservoirs have extensive shorelines and large surface areas and have created wetlands associated with reservoir margins.

Missouri is in the major agricultural region of the United States. Many native communities were altered dramatically by conversion to row cropping, hay production, or livestock grazing. Consequently, as a result of normal agricultural practices, many cropped fields are flooded during either the growing season (April to September) or the dormant season (October to April). Upon flooding, these fields are considered farmed wetlands.

Characteristics of Missouri Wetlands

Missouri has a temperate climate with hot, humid summers and cold, wet winters. Despite this generalization, a climatic gradient occurs in Missouri from the northwest to the southeast (Fig. 28.2). The highest precipitation and longest growing season occur in the southeast, which is dominated by forested wetlands. Mean annual precipitation in the bootheel (Caruthersville) is 31 cm greater than in Maryville in the northwest corner. Precipitation is generally highest from November to May in the south, but long-term averages do not differ dramatically among months. Further, large rainfall events can occur in any month of the year in southeastern Missouri (Fredrickson and Reid 1990). Large annual and monthly variations in precipitation are common throughout the state (Table 28.1). However, a pronounced wet/dry season is characteristic of higher latitudes, with the wettest period occurring from May through September (Table 28.1). In extreme northwest Missouri the growing season av-

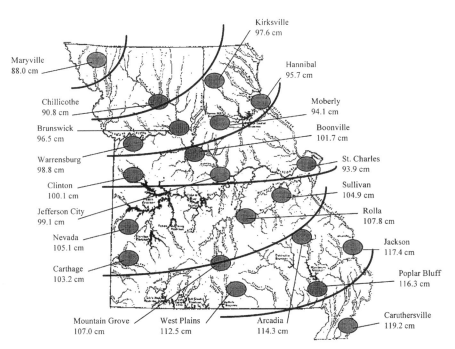

Fig. 28.2. Lines represent rough approximations of climatic contours. Mean annual precipitation gradient (northwest to southeast) from 21 cities in Missouri (⟨www.ncdc.noaa.gov/coop-precip.html⟩).

TABLE 28.1. Monthly Precipitation (cm) Data for Northwestern, Central, and Southeastern Missouri, Illustrating Northwest to Southeast Climatic Gradient

Location	Variable	January	March	May	July	September	November	Annual
Maryville	Mean	2.4	5.7	12.2	11.1	10.3	4.9	88.1
40.3° N latitude 94.5° W longitude	Range	0–10.8	0.4–17.3	3.0–27.0	0.9–65.3	0.7–32.7	0–18.2	
	Number of years	61	62	65	65	64	61	
Boonville	Mean	4.1	7.3	13.2	10.4	10.8	7.8	101.7
39.0° N latitude 92.8° W longitude	Range	0–12.8	1.1–21.6	3.0–37.9	1.2–34.3	0.7–33.6	0.1–42.9	
	Number of years	59	59	59	59	59	58	
Caruthersville	Mean	10.0	12.3	11.7	9.1	8.4	10.8	119.2
36.2° N latitude 89.7° W longitude	Range	1.4–40.0	2.5–36.0	0.9–35.4	1.0–24.0	0.5–21.5	1.5–31.3	
	Number of years	63	65	65	65	66	66	

Source: data from ⟨www.ncdc.noaa.gov/coop-precip.html⟩.

erages a month shorter than in southern Missouri (Fredrickson and Reid 1990). The combination of shortened growing seasons and lower precipitation favors herbaceous marshes in northwestern Missouri.

In addition to precipitation patterns and duration of growing season, Missouri wetlands also show a northwest to southeast gradient in temperature, with cooler temperatures in the northwest warming toward the southeastern portion of the state. Throughout the state seasonal changes in temperature occur. Although ice commonly forms from December to mid-February, it varies among years within a wetland and among wetlands geographically.

In summary, Missouri can be characterized by a climatic gradient from northwest to southeast. The amount and timing of precipitation interacting with other climatic variables (temperature) are key abiotic factors influencing length of the growing season. Marsh ecosystems of the northwest receive most precipitation during the summer growing season, whereas swamps and bottomland hardwood wetlands in the southeast receive most precipitation from November to May, corresponding with the winter dormant season for plants. Most of the invertebrates are active and growing during winter.

CLASSIFICATION

According to Epperson (1992), Missouri wetlands can be classified into eight types:

1. Swamps (forests with long hydroperiod including flooding during growing season, dominated by bald cypress and water tupelo)
2. Shrub swamps (similar hydrology to swamps but with shrubby vegetation such as buttonbush)
3. Forested wetlands (floodplain or riparian wetlands mainly flooded during the dormant season and dominated by oaks, willows, cottonwoods, maples, elms, sycamore, and ash)
4. Marshes (standing water or saturated soils, dominant plants include grasses, sedges, and robust emergents)
5. Wet meadows (shorter hydroperiod than marshes, may only have saturated soils, dominant plants include sedges, rushes, grasses)
6. Fens and seeps (groundwater-influenced, may have saturated soils, localized sites dominated by grasses and sedges)
7. Lacustrine wetlands associated with pond, lake, and reservoir borders (water depth is less than 2 m, dominant plants are emergents) and
8. Riverine wetlands associated with stream beds (wetlands within the stream channel, sand and gravel bars included).

In addition to this classification, constructed wastewater treatment wetlands also occur in Missouri.

Many wetlands in Missouri have been restored and are diverse complexes dominated by moist-soil habitats (managed as seasonal marshes), greentree reservoirs (managed as forested wetlands), and farmed wetlands (flooded crops or crop residue). Many are intensively managed sites, whereas others are relatively unmanaged, remnant natural wetlands (Fig. 28.1). The focus of this chapter is on the invertebrates of managed, temporarily flooded wetlands occurring throughout the state along this environmental gradient.

MAJOR INVERTEBRATE TAXA IN MISSOURI WETLANDS

Marshes

Herbaceous wetlands (marshes) are characterized by a diverse macroinvertebrate fauna. These taxa have numerous strategies for negotiating dry periods (Wiggins et al. 1980) and represent a complete range of functional groups. Herbaceous wetlands are typically dominated by insects, crustaceans, annelids, and gastropods. A diverse representation of insects includes Diptera (chironomids, culicids, tabanids, tipulids), Odonata, Hemiptera (corixids, notonectids, belostomatids, gerrids), Coleoptera (hydrophilids, dytiscids), and Ephemeroptera. The crustaceans consist primarily of Amphipoda (gammarids, talitrids) and Decapoda, as well as numerous microcrustaceans (Cladocera, Eucopepoda). The gastropods consist of pulmonate snails (planorbids, physids, and lymnaeids), whereas annelids are dominated by Oligochaeta and Hirundinea (leeches).

Forested Wetlands

In ephemeral, seasonal, or semipermanent forested wetlands the major invertebrate taxa are Bivalvia, Gastropoda, Oligochaeta, Crustacea, and dipteran Insecta. Community composition tends to be more consistent among forested wetlands and restricted to a smaller number of species than in herbaceous wetlands. The majority of these taxa are classified as overwintering permanent residents with low mobility (Wiggins et al. 1980). These permanent residents, particularly the Bivalvia, Amphipoda, and Isopoda, have relatively few adaptations for drought resistance and therefore are less commonly found in ephemeral wetlands. They are more likely found in wetlands with longer hydroperiods, where moist refugia are available during dry periods.

In the Mingo Swamp 72 taxa occurred (White 1985), although many of these were semiaquatic insects and arachnids. Five taxa were common to three distinct study sites differing primarily in hydrology: Cladocera, Isopoda, Amphipoda, Bivalvia, and Oligochaeta. Between 84–97 percent of all invertebrates collected in bottomland hardwood swamps in southeastern Missouri consisted of Chironomidae, *Caecidotea* isopods, *Crangonyx* amphipods, *Pisidium* bivalves, and Oligochaeta (Batema et al. 1985). In a black willow (*Salix nigra*) wetland in northeastern Missouri the dominant taxa were similar. Of the 40 invertebrate families collected from willow wetlands, seven were dominant in frequency of occurrence, density, and biomass: Sphaeriidae, Planorbidae, Physidae, Gammaridae, Asellidae, Chironomidae, and one family of Oligochaeta (Magee 1989). Of the above seven taxa, all but chironomids are considered overwintering, permanent residents (Group 1 by Wiggins et al. 1980).

Agricultural Wetlands

Although diversity tends to be low, the density and biomass of invertebrates in farmed wetlands may be high. Few data are available, but invertebrate collections in farmed wetlands on a Mississippi River floodplain were dominated by Cladocera (Daphnia) and Oligochaeta in corn and Eucopepoda, Cladocera (Daphnia), and Oligochaeta in soybeans (Bundy and Magee, unpublished data). Fairy shrimp (Anostraca) occur in flooded corn fields adjacent to willow wetlands (Magee, unpublished data). Adult fairy shrimp were present in an autumnally flooded corn field, frozen in ice, and yet survived.

FACTORS INFLUENCING THE DISTRIBUTION AND ABUNDANCE OF INVERTEBRATES

Abiotic Factors

Hydrology and Hydroperiod. Hydrology consists of the timing, rate, duration, and depth of flooding and drawdown, and has a key influence on the presence, distribution, and abundance of aquatic invertebrates (Reid 1985, Fredrickson and Reid 1990). Precipitation is the main hydrologic source for Missouri wetlands and the amount and timing of precipitation is key to wetland function. Precipitation patterns (long-term hydrology) influence life history adaptations of invertebrates and dictate the geographical distribution of invertebrates. Short-term hydrology (annual precipitation) influences the microdistribution and abundance of invertebrates locally (Reid 1985). It is important to emphasize that annual rainfall in Missouri varies substantially among years and that complex interactions occur in time and space. For example, during the flood of 1993, rainfall from northern Missouri (Maryville recorded its highest monthly precipitation on record during July 1993, 65 cm) and adjacent states was responsible for flooding in southern Missouri. During this time Mississippi River floodplain crops in southeastern Missouri were inundated by overbank flooding, while upland crops were withering in a drought.

Missouri has ephemeral, seasonal, semipermanent, and permanently flooded wetlands. The hydroperiod affects the taxonomic composition and abundance of aquatic invertebrates in herbaceous wetlands. In semipermanent wetlands in northeastern Missouri, physid snails (*Physa gyrina*) occurred in lower density than in a moist-soil wetland with a shorter hydroperiod (Reid et al. 1996). In wetlands with prolonged flooding, submergent vegetation may be limited because of factors such as turbidity (caused by fish). These wetlands tend to have lower invertebrate abundance and diversity when compared to more seasonally flooded sites (Reid et al. 1996). The absence of living plants as a substrate and the presence of fish in deeply flooded marsh systems tend to reduce invertebrate abundance and diversity.

Littoral habitats are confined to depths less than 2 m, the normal extent of light penetration and the zone where most emergent and submergent plants grow. The direct influence of water depth on plant community composition plays a major role in controlling invertebrate distribution. In relatively shallow waters, where the water column is filled with a complex plant structure, invertebrates occur in greater diversity and abundance (Reid 1983).

In forested wetlands of southern Missouri, lower sites that remain flooded during the growing season are semipermanent or permanent wetlands. Many of these sites harbor dense stands of submergents. However, flooding of seasonal habitats overlaps with a precipitation pattern characterized by a wet dormant season. Dormant season flooding coincides with a low level of photosynthetic activity. Therefore, the base of invertebrate food chains is not composed chiefly of primary producers, but the fuel for secondary production is largely detritus. Many of Missouri's wetlands have high invertebrate productivity during winter, which emphasizes the importance of detritus as a base for energy flow.

Forest macroinvertebrates are also influenced by depth, timing, and duration of flooding. Experiments in the Mingo Swamp focused on invertebrate distribution and abundance in forest types, which differed in relation to topography and hydrology. Pin oak (*Quercus palustris*)/sweetgum (*Liquidamber styraciflua*) forests dominated at slightly higher elevations than overcup oak (*Quercus lyrata*)/red maple (*Acer rubrum*) forests. Therefore, the pin oak/sweetgum sites were more shallowly flooded and became flooded later in the winter than overcup oak/red maple sites (Batema 1987). Further, managed greentree reservoirs (consisting of overcup oak/red maple in deep and pin oak/sweetgum in more shallow sites) were artificially flooded in October, whereas naturally flooded bottomland hardwoods (also consisting of both forest types) were flooded by precipitation starting in December. Timing of drawdown also differed, with greentree sites being dewatered earlier than sites with natural drawdowns. Invertebrate comparisons were between artificial and natural hydrologies with respect to deep versus shallow forest types.

Although invertebrate responses were complex, some trends were evident. Invertebrates responded rapidly to hydrologic inputs and were present in all plots on both forest types within two weeks after flooding, despite the timing of flooding (White 1982, Batema 1987). In greentree reservoirs invertebrates (largely Chironomidae) reached peak densities (13,800/m^2) in the pin oak/sweetgum sites within four weeks after initiation of flooding in mid-October

(Batema 1987). The invertebrate peak coincided with water depths of 20–40 cm and nutrient inputs associated with decomposing litter (Wylie 1985, Batema 1987). No fall peak of invertebrates occurred in the more deeply flooded (40–70 cm) overcup oak/red maple sites in greentree reservoirs.

In the spring peak invertebrate populations in greentree reservoirs (consisting of Isopoda, Amphipoda, and Bivalvia in densities ranging from 2,000–11,000/m^2) occurred in late April, when water depths were 1–10 cm (Batema 1987). This trend was similar in both forest types, although the overcup oak/red maple sites had greater populations of benthic organisms (Batema 1987). Greentree reservoirs were typically drawndown in spring prior to the natural drawdown of wetlands in the Mingo Swamp, and therefore, peak invertebrate abundance occurred earlier than on naturally flooded sites and was associated with the concentration of invertebrates in shrinking pools. Fast-growing multivoltine dipterans (Chironomidae) showed a dual peak in fall and spring, whereas the univoltine, group 1 permanent residents (Wiggins et al. 1980) showed a single peak in abundance following a relatively long overwintering growth period. Isopods demonstrate the typical pattern for permanent residents in forested wetlands in Missouri (Fig. 28.3). Forests on higher sites were flooded shallowly or were dry and did not support many invertebrates in the spring. Low-lying sites, however, have potentially large concentrations of invertebrates following drawdown and are important foraging habitats for certain waterbirds.

On naturally flooded sites in the Mingo Swamp, flooding rarely occurred in the fall. The peak number of invertebrates (5,250–7,300/m^2) composed largely of Isopoda and Amphipoda, occurred in May for both forest communities following spring flooding, when water depth was 4–35 cm (Batema 1987). Naturally flooded forests remained flooded longer into the spring than greentree reservoir sites and supported more invertebrates.

Invertebrate taxa respond in specific ways to water depth, and each taxa may have an optimal range of depths under which it thrives. In black willow wetlands in northeastern Missouri, the sites with the greatest water depths were associated with the lowest invertebrate abundance. Peak abundance coincided with depths in the range 11–20 cm in May and June, when water temperature was 27°C (Magee 1989). Water depths were generally greater than 60 cm in willows in the fall, and invertebrate numbers were low. Six of the seven dominant taxa occurred most frequently in depths ranging from 11–20 cm, whereas, the seventh taxa (Bivalvia) occurred most frequently in depths from 26–50 cm. The greatest mean density of invertebrates (range from 1151–1293/m^2) was associated with depths from 1–30 cm, intermediate densities (range from 913–921/m^2) coincided with depths from 31–50 cm, and low densities (135–588/m^2) occurred at depths greater than 50 cm (Magee 1989).

In northern Missouri forested wetlands consist primarily of low-lying willow sites or willow/cottonwood riparian habitats. Although these forest communities differ substantially from the extensive bottomland hardwoods of

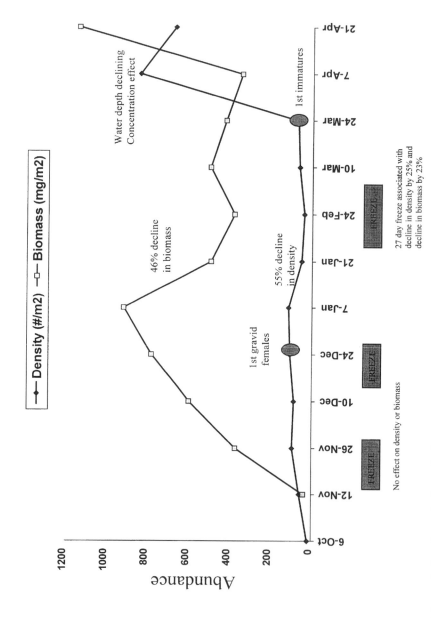

Fig. 28.3. Pattern of abundance in Isopoda from a greentree reservoir in Mingo Swamp, 1979–80 (data from White 1985).

southern Missouri, invertebrate responses to hydrology follow similar trends. Univoltine taxa characterize these wetlands, especially where flood duration is relatively long. The most ephemeral sites tend to lack group 1 resident invertebrates except during wet years or under managed flooding regimes. Although insects are less common in forested wetlands, chironomids tend to dominate on sites with short hydroperiods. Further studies are needed to determine relationships between hydroregimes and specific taxa; however, it is clear that the greater the variety of hydrological regimes within a wetland complex, the greater the diversity of invertebrates.

Other Abiotic Factors. In addition to hydrology, chemical properties of water, climate, soil types, and fire also influence invertebrate distribution and abundance. Sedimentation may also render certain invertebrate communities depauperate. In willow wetlands in northeastern Missouri, sites with high levels of sedimentation (5–10 cm deposited over detritus) had lower taxa diversity, density, and biomass than nonsilted sites (Magee 1989). Oligochaeta and Chironomidae comprised a large proportion of the total invertebrate community in sedimented sites. Bivalvia (Sphaeriidae, *Musculium partumeium* Say) dominated the invertebrate community in a willow wetland that did not receive heavy sedimentation; they accounted for 18 percent of the frequency of occurrence, 28 percent of the density, and 39 percent of the biomass. In contrast, Bivalvia collected in the willow habitat with heavy sedimentation accounted for only 4 percent of the frequency of occurrence, 1 percent of density, and 3 percent of biomass (Magee 1989). Like bivalves, Amphipoda were nearly absent from the sedimented site, but were one of the dominant invertebrates in the nonsilted willow site. These relationships point to the dramatic effects of sedimentation on invertebrate taxa composition.

During relatively cold winters with prolonged ice cover in southeast Missouri, invertebrate density peaks were lower and shorter in duration than in milder years (White 1985, Batema 1987). Certain invertebrates may be more sensitive to cold than others. For example, adult rat-tailed maggots were not present in flooded bottomland hardwoods in southeastern Missouri by December; numerous adults were found frozen in the ice (White 1985). Further, Amphipoda in the Mingo Swamp experienced a 50 percent decline in density and a 31 percent decline in biomass during the first cold period (ice cover) in November and experienced a 66 percent decline in density and a 52 percent decline in biomass after a 27-day freeze from late January to late February (White 1985). In contrast, numbers of Chironomidae seemed unaffected by ice cover and biomass of Isopoda was relatively stable during cold periods (White 1985).

Biotic Factors

Vegetation Type and Structure. A restored Mississippi River floodplain marsh in northeastern Missouri was characterized by a variety of plant communities, but invertebrates preferred water smartweed (*Polygonum coccinium*) to other stands of vegetation (Reid 1983). Invertebrates were also collected from rice cutgrass (*Leersia oryzoides*), barnyard grass (*Echinochloa* spp.), nutgrass (*Cyperus* spp.), and American lotus (*Nelumbo lutea*) in two impoundments. One impoundment was a moist-soil wetland with a late June drawdown, and the other was semipermanently flooded. The effects of hydrology and plant communities on invertebrates are interrelated and difficult to separate. The moist-soil habitat had stands of all species accept American lotus, whereas the semipermanent impoundment was dominated by American lotus but also contained water smartweed and rice cutgrass. The greatest density and diversity of nektonic invertebrates was associated with water smartweed in both impoundments, with a peak abundance exceeding $1000/m^3$ (Reid 1983). Rice cutgrass in both impoundments had intermediate densities $(300–400/m^3)$ and diversity. In the more ephemeral moist-soil wetland barnyard grass and nutgrass had the lowest density and diversity of aquatic macroinvertebrates. In the semipermanently flooded wetland, invertebrate abundance in American lotus was also low (Reid 1983).

The high diversity and abundance of invertebrates associated with water smartweed can be attributed to the complex structure and high surface area of this plant. In water smartweed stands much of the water column is filled, thus providing an extensive substrate for invertebrate colonization. In contrast, a relatively monotypic stand of American lotus dominated the semipermanent impoundment. American lotus had simple structure, and consequently this wetland had a depauperate invertebrate assemblage (also, fish predators may have contributed to low invertebrate numbers in the semipermanent wetland).

Leaf Drop. In herbaceous wetlands the interactions of abiotic factors such as climate (e.g., drought) and biotic factors such as structure and vigor of vegetation can also affect invertebrate diversity and abundance. For example, drought-stressed water smartweed had early and reduced flowering in a northeastern Missouri moist-soil wetland (Reid 1983). These conditions led to early leaf senescence in June. Leaf drop occurred as the site was flooded in July. This physiological adaptation to a summer drought simplified water smartweed structure and, by the following spring, reduced invertebrate diversity and density (Reid 1983). Despite the reduced invertebrate production associated with water-column invertebrates, drought-induced leaf drop did not negatively impact benthic invertebrates (Reid 1983). This interaction between plant structure and nektonic invertebrates is key to understanding how the dynamic nature of wetlands can influence the response of invertebrates to constantly changing conditions.

Quantity and Quality of Detritus. The quantity and quality of detritus present in wetlands may play a key role in determining aquatic invertebrate abundance and diversity. Whereas herbaceous wetlands with diverse plant communities usually have complex substrates taking up much of the water column (including detritus during the dormant season), forested wetlands are characterized by relatively simple habitats comprised of leaf litter layers. Much of the water column may be open water above the benthic litter accumulations.

Bottomland hardwoods in southeastern Missouri consist of a diverse mix of deciduous trees, including major associations of pin oak and sweetgum, and red maple and overcup oak. These forested wetlands have relatively high leaffall rates (2900–4300 Kg/ha) and relatively slow decay rates, thus leading to the development of large, diverse accumulations of detritus (Batema 1987). The half lives range from 0.71–1.53 years on naturally flooded sites and 0.86–3.25 years in greentree reservoirs, and require 2–7 years for complete breakdown. Upon flooding in the fall, "old" litter layers provide preconditioned detritus immediately available for invertebrate colonization. For example, in the Mingo Swamp, invertebrates were present within two weeks of flooding and reached peak numbers within four weeks of flooding (Batema 1987). In another study bottomland hardwood invertebrates reached peak numbers within two and a half weeks of flooding (White 1985). These rapid invertebrate responses partly reflect the availability of palatable detritus.

In contrast, leaffall production is lower and litter decay rates are more rapid in willow wetlands of northeast Missouri than in the bottomland hardwood oak forests. Annual leaffall was 2500–2900 Kg/ha and decay rate (half life) averaged 1.4 to 2.3 years (Magee 1989). Much of the annual leaffall decomposed within a year, and more leaffall was covered with sediments or was lost by lateral transport. Therefore, litter layers did not develop and preconditioned litter was not available in willow wetlands. Invertebrates in these wetlands did not reach peak numbers and biomass until spring, six months following flooding (Magee 1989). The absence of a fall invertebrate peak may have reflected the unavailability of conditioned litter. In addition, large willow trees on drier sites with greater leaffall had a greater diversity and abundance of aquatic invertebrates than smaller willows in habitats with a longer hydroperiod (Magee 1989). The relationships between quantity and quality of litter and invertebrate abundance need further investigation.

Invertebrate Community Succession and Influence of Fish. In Missouri wetlands invertebrate functional groups colonize litter in a successional pattern. For example, in the willow detritus of northeastern Missouri amphipod and isopod shredders reached peak abundance prior to collector groups, Chironomidae and Oligochaeta (Magee 1989). Studies emphasizing detailed taxonomic identification are needed to understand such community dynamics.

Historically, most semipermanently flooded habitats in low-lying areas of Missouri were lacustrine or riverine oxbows. More frequently, low sites were seasonally or ephemerally flooded and therefore were used by fish infrequently or for specific life cycle events. The pin oak/sweetgum wetlands of Mingo Swamp were sporadically available during spring and, for example, were never flooded for longer than one month in 1981 and 1982 (Stewart 1983). These wetlands were used by numerous larval fish (e.g., banded pygmy sunfish, mosquitofish, starhead topminnow, and flier) during spring because newly flooded habitats were highly oxygenated and, upon flooding, detrital decomposition and nutrient leaching were prompt (Stewart 1983, Wylie 1985). Larval fish grew rapidly in shallow, cool, nutrient-rich water that lacked large predators. These ephemeral wetlands did not support large predaceous fish, but predaceous diving beetles were likely predators of larval banded pygmy fish (Stewart 1983). Despite the short hydroperiods, adult banded pygmy sunfish spawned in pin oak sites during a one-week flood event in March 1982 (Stewart 1983). Adult slough darters, starhead topminnows, and mosquitofish also exploited ephemeral pin oak forests in preparation for breeding in more permanent wetlands (Stewart 1983).

In contrast to pin oak/sweetgum forests, the lower elevation overcup oak/ red maple sites were generally available for fish during spring, and spawning was common for banded pygmy sunfish, mosquitofish, and slough darters. The lower-lying dead tree habitats were flooded throughout spring and summer, and many fish progressively moved from the well-oxygenated ephemeral pin oak sites to deeper overcup oak forests to permanent dead tree habitats (Stewart 1983). The dynamic changes in habitat conditions in bottomland hardwood systems constantly influence fish distribution and abundance.

Interactions among fish, waterbird, and invertebrate distribution and abundance have been observed. In some seasonally flooded forested wetlands invertebrate production is sufficiently large to support both avian and fish predators. However, some waterbirds, most notably wood ducks, exploit drying basins where invertebrate prey are concentrated, but where conditions are less favorable for fish.

Artificial flooding has created more permanent fish habitat suitable for large insectivorous fish. Originally fish were included along with waterbirds in the management of the older "man-made" wetlands in Missouri (Swan Lake NWR and Fountain Grove CA). However, in many of these habitats, as well as in remnant natural, semipermanent wetlands, rough fish (e.g., carp) were introduced. Carp and other large fish have a direct negative impact on invertebrate productivity in wetlands as a result of foraging. Indirectly, they remove vegetation and increase turbidity, which causes invertebrate density and diversity to decline. This led to drainage of wetland basins in favor of moist-soil management. Draining would eliminate fish and provide plant and invertebrate resources for numerous waterbirds, but diving ducks and other waterbirds that favor more permanently flooded sites and exploit submerged aquatic beds would be negatively impacted. No information is available from Missouri regarding the role of submerged aquatic vegetation as refugia for aquatic invertebrates from fish predators.

INTEGRATING INVERTEBRATES INTO WETLAND MANAGEMENT STRATEGIES

Protection and management of wetland complexes will continue to be the best way to support diverse invertebrate faunas. This protection may include acquisition or easement of lands, but alteration of watershed hydrology must also be considered. Efforts should attempt to replicate regional hydrologic patterns. For example, in Missouri, where winter rains have historically inundated shallow wetlands, management strategies should favor similar flooding patterns. A mosaic of wetland basins can offer alternatives that no single basin can provide. In addition, restoration of degraded wetlands has its greatest potential in landscapes of marginal agricultural use.

More waterbird species forage on invertebrates during some period of their annual cycle, than on any other trophic level (Fredrickson and Reid 1986, Reid 1993). These avian predators generally do not have a consistent dietary requirement for invertebrates, but invertebrate prey helps meet high protein or calcium requirements associated with critical growth, molt, or egg-laying periods when gross energy is less important (Driver 1981). For example, female mallards (*Anas platyrhynchos*) in the Mingo Swamp consumed invertebrates (29–41 percent of diet) during the prebasic molt in midwinter (Heitmeyer 1985). Most of the diet (79 percent) of laying wood ducks (*Aix sponsa*) in forested wetlands in southeastern Missouri consisted of invertebrates (Drobney and Fredrickson 1979). Further, invertebrates accounted for one third of the diet of female wood ducks in the fall and one third of the diet of male wood ducks in all seasons (Drobney 1980). Wood ducks fed primarily on amphipods, isopods, and snails in the Mingo Swamp (Drobney 1977). Because these overwintering, resident invertebrates have low mobility and few defense tactics, they are easily captured.

Integrating invertebrates into wetland management strategies requires knowledge of the life histories of both avian predators and invertebrate prey. Hydrologic manipulation is the key to influencing invertebrate numbers and availability to coincide with waterbird life cycle events. Further, this philosophy can be extended to more comprehensive management objectives which include herptiles, fishes, and mammals.

Timing of flooding can directly influence invertebrate availability. Fall peaks in invertebrate abundance overlap with periods of rapid litter decomposition and peak use of forested wetlands by ducks (Heitmeyer 1985, Wylie 1985). Each of these events occurs within two to three months of initial flooding. Also, both the timing and depth of flooding will influence the temperature of the water, which in turn will strongly influence growth and metabolism. Further, rate and depth of flooding are important since rapid or deep flooding tends to "dilute" invertebrate numbers. Duration of flooding will influence adult survival and reproduction and the structure of predator populations. Maintaining high water levels will result in reduced invertebrate biomass over time and the potential to establish predatory fish populations. In many Missouri wetlands flood duration was historically a function of the interaction between topography and annual precipitation. A key to managing wetlands is to mimic the dynamics of these interactions and thereby produce variable hydroperiods similar to natural patterns.

In shallow waters temperatures and dissolved oxygen may often reach lethal levels during hot summer months in Missouri. Holding water in drainage ditches for a "reserve" of invertebrates is not a viable strategy under such conditions. This strategy has been suggested in debates regarding the placement of borrow ditches inside or outside of levees.

Availability of prey initially increases with declining water levels, provided the invertebrates do not migrate. Some of the most productive foraging sites are in drying pools where invertebrates become concentrated. For example, common grackles (*Quiscalus quiscula*) and killdeer (*Charadrius vociferus*) fed on physid snails, causing 94 percent mortality in eight days (Reid 1983). This feeding frenzy occurred in a moist-soil wetland in northeastern Missouri and coincided with a drawdown. Differential mortality occurred based on snail size. *Physa gyrina*, the largest snail, experienced the highest mortality (96 percent), *Aplexa hyphorum,* an intermediate-sized genus, experienced 85 percent mortality, and the smallest snail, *P. integra,* experienced the lowest mortality (68 percent; Reid et al. 1996). In general, effective management involves slow drawdowns that concentrate invertebrate prey and are timed to match certain life cycle events of predators (e.g., shorebird migration).

Another key to making invertebrates available to waterbirds is to flood at depths that are within the predator's foraging range. The majority of waterbirds feed in water less than 25 cm (Fredrickson and Reid 1986). Mallards foraged in depths of 20–40 cm at Mingo Swamp (Heitmeyer 1985), and foraging efficiency decreased for mallards at depths greater than 30.5 cm (Strong 1986). Sora (*Porzana carolina*) prefer foraging depths of 5–15 cm, although they can extend this range to nearly 50 cm by walking on top of robust emergent vegetation (Rundle 1980, Sayre and Rundle 1984).

Wetland managers often spend much time and money trying to control invasive plant species. Control of problem plants can be effectively linked to invertebrate management activities. For example, American lotus, which tends to form monocultures on semipermanent wetlands, has been successfully controlled in northeastern Missouri using a combination of summer drawdown and early fall shallow disking or mowing. This physical disturbance creates a litter with high surface area, and upon shallow flooding early in the fall it is rapidly colonized by microorganisms which initiate detrital processing. The artificial fragmentation of organic matter allows invertebrates to respond more rapidly and at a higher abundance.

Hydrologic manipulation of wetland basins can have direct and indirect influence on invertebrates through the duration of aquatic habitat and through the physiological response of hydrophytes. Wetland management should strive to replicate regional hydrologic patterns because invertebrates have adapted to such dynamic conditions. In midlatitude wetlands such as Missouri, fall flooding will stimulate the hatching of many insect and crustacean species, and larval forms may continue to develop over the winter. Our present knowledge of water manipulations suggests that management for specific hydrophyte communities (e.g., water smartweed) may be the most practical means of increasing invertebrate production.

LITERATURE CITED

Batema, D. L., G. S. Henderson, and L. H. Fredrickson. 1985. Wetland invertebrate distribution in bottomland hardwoods as influenced by forest type and flooding regime. Pages 196–202 *in* J. O. Dawson and K. A. Majeros (eds.), Proceedings of Fifth Central Hardwoods Forest Conference Department of Forestry, University of Illinois, Champaign-Urbana, IL.

Batema, D. L. 1987. Relations among wetland invertebrate abundance, litter decomposition, and nutrient dynamics in a bottomland hardwood ecosystem. Ph.D. Dissertation, University of Missouri, Columbia, MO.

Collins, S. L., and S. M. Glenn. 1995. Grassland ecosystems and landscape dynamics. Pages 128–156 *in* A. Joern and K. H. Keeler (eds.), The Changing Prairie, North American Grasslands. Oxford University Press, New York.

Dahl, T. E. 1990. Wetland losses in the United States 1780s to 1980s. U.S. Department of the Interior, Fish and Wildlife Service, Washington, DC.

Driver, E. A. 1981. Calorific values of pond invertebrates eaten by ducks. Freshwater Biology 11:579–582.

Drobney, R. D. 1977. The feeding ecology, nutrition and reproductive bioenergetics of wood ducks. Ph.D. Dissertation, University of Missouri, Columbia, MO.

———. 1980. Reproductive bioenergetics of wood ducks. Auk 97:480–490.

Drobney, R. D., and L. H. Fredrickson. 1979. Food selection by wood ducks in relation to breeding status. Journal of Wildlife Management 43:109–120.

Epperson, J. E. 1992. Missouri wetlands: A Vanishing Resource. Missouri Department of Natural Resources: Division of Geology and Land Survey. Water Resources Report No. 39.

Fredrickson, L. H. 1980. Impact of water management on the resources of lowland hardwood forests. Pages 51–64 *in* R. H. Chabreck and R. H. Mills (eds.), Integrating Timber and Wildlife Management in Southern Forests. Twenty-ninth Forest Symposium, Louisiana State University, Baton Rouge, LA.

Fredrickson, L. H., and F. A. Reid. 1986. Wetland and riparian habitats: A nongame management overview. Pages 59–96 *in* J. B. Hale, L. B. Best, and R. L. Clawson (eds.), Management of Nongame Wildlife in the Midwest: A Developing Art. North Central Section, The Wildlife Society, Chelsea, MI.

———. 1990. Impacts of hydrologic alteration on management of freshwater wetlands. Pages 71–90 *in* J. M. Sweeney (ed.), Management of Dynamic Ecosystems. North Central Section, The Wildlife Society, West Lafayette, IN.

Galat, D. L., J. W. Robinson, and L. W. Hesse. 1996. Restoring aquatic resources to the Lower Missouri River: Issues and initiatives. Pages 49–71 *in* D. L. Galat and A. G. Frazier (eds.), Overview of River-Floodplain Ecology in the Upper Mississippi River Basin, vol. 3 of J. A. Kelmelis (ed.), Science for Floodplain Management into the 21st century, Government Printing Office, Washington, DC.

Heitmeyer, M. E. 1985. Wintering female mallards in the Mississippi Delta. Ph.D. Dissertation, University of Missouri, Columbia, MO.

Humburg, D. D., D. A. Graber, S. P. Havera, L. H. Fredrickson, and D. Helmers. 1996. What did we learn from the Great Flood of 1993? Pages 139–148 *in* J. T. Ratti (ed.), Seventh International Waterfowl Symposium, Memphis, TN.

Johnson, T. R. 1987. The Amphibians and Reptiles of Missouri. Missouri Department of Conservation, Jefferson City, MO.

Korte, P. A., and L. H. Fredrickson. 1977. Loss of Missouri's lowland hardwood ecosystem. Transactions of the Forty-second North American Wildlife and Natural Resource Conference 42:31–46.

Magee, P. A. 1989. Aquatic macroinvertebrate association with willow wetlands in northeastern Missouri. Master's Thesis, University of Missouri, Columbia, MO.

Magee, P. A., D. D. Humburg, and L. H. Fredrickson. 1993. Aquatic macroinvertebrate association with willow wetlands in northeastern Missouri. Wetlands 13:304–310.

Reid, F. A. 1983. Aquatic macroinvertebrate response to management of seasonally-flooded wetlands. Master's Thesis, University of Missouri, Columbia, MO.

———. 1985. Wetland invertebrates in relation to hydrology and water chemistry. Pages 72–79 *in* M. D. Knighton (ed.), Water Impoundments for Wildlife: A Habitat Management Workshop. USDA-Forest Service, St. Paul, MN.

———. 1993. Managing wetlands for waterbirds. Transactions of the North American Wildlife and Natural Resource Conference 58:345–350.

Reid, F. A., L. H. Fredrickson, and D. D. Humburg. 1996. Density and growth of the pond snail, *Physa gyrina*, in a managed Mississippi River floodplain wetland. Western Wetlands 2:1–11.

Rundle, W. D. 1980. Management habitat selection and feeding ecology of migrant rails and shorebirds. Master's Thesis, University of Missouri, Columbia, MO.

Ryan, M. R. 1990. A dynamic approach to the conservation of the prairie ecosystem in the Midwest. Pages 91–106 *in* J. M. Sweeney (ed.), Management of Dynamic Ecosystems. North Central Section, The Wildlife Society, West Lafayette, IN.

Sayre, M. W., and W. D. Rundle. 1984. Comparison of habitat use by migrant soras and Virginia rails. Journal of Wildlife Management 48:599–605.

Schmudde, T. H. 1963. Some aspects of the Lower Missouri River floodplain. Annals of the Association of American Geographers 53:60–73.

Stewart, E. M. 1983. Distribution and abundance of fishes in natural and modified bottomland hardwood wetlands. Master's Thesis, University of Missouri, Columbia, MO.

Strong, A. M. 1986. Foraging decisions of captive mallards. Master's Thesis, University of Missouri, Columbia, MO.

U.S. Geologic Survey. 1974. Hydrologic Unit map—1974, State of Missouri.

Vaught, R., and J. Bowmaster. 1983. Missouri Wetlands and Their Management. Missouri Department of Conservation.

White, D. C. 1982. Leaf decomposition, macroinvertebrate production and wintering ecology of mallards in Missouri lowland hardwood wetlands. Master's Thesis, University of Missouri, Columbia, MO.

———. 1985. Lowland hardwood wetland invertebrate community and production in Missouri. Archiv für Hydrobiologie 103:509–533.

Wiggins, G. B., R. J. MacKay, and I. M. Smith. 1980. Evolutionary and ecological strategies of animals in annual temporary pools. Archiv für Hydrobiologie Supplement 58:97–206.

Wylie, G. D. 1985. Limnology of lowland hardwood wetlands in Missouri. Ph.D. Dissertation, University of Missouri, Columbia, MO.

29 Prairie Floodplain Ponds

Mechanisms Affecting Invertebrate Community Structure

STEVEN L. KOHLER, DEBORAH CORTI,
MICHAEL C. SLAMECKA, and
DANIEL W. SCHNEIDER

We present a simple conceptual model of the roles of hydroperiod (pond duration) and fish predation in determining the structure of invertebrate communities in floodplain ponds. Fish have access to all pond types (temporary, permanent) during flooding and remain in floodplain ponds after floodwaters recede. Therefore, we argue that in contrast to other models of the mechanisms structuring lentic freshwater communities, fish predation should strongly affect the organization of floodplain pond communities throughout the gradient in pond permanence. Large, active invertebrate predators should be excluded from floodplain ponds by fish predation and consequently should not influence invertebrate community structure. We examine predictions of the model using field experiments conducted in ponds in two temperate prairie riverfloodplain systems (the Mississippi and Sangamon Rivers). In both systems predictions of the model were well supported. Effects of fish predation were strong in both temporary and permanent ponds, and large, active invertebrate predators were rare in all pond types. In the Mississippi River ponds, where the experiment was maintained for a full year, hydroperiod had pronounced effects on the distribution and abundance of many taxa. Although hydroperiod and fish predation should have strong and largely predictable effects on community structure, a significant stochastic component, resulting from effects of the flood regime on colonization of ponds, should contribute appreciably to observed variation among ponds in invertebrate communities.

Invertebrates in Freshwater Wetlands of North America: Ecology and Management, Edited by
Darold P. Batzer, Russell B. Rader, and Scott A. Wissinger
ISBN 0-471-29258-3   © 1999 John Wiley & Sons, Inc.

INTRODUCTION

From a hydrologic perspective a river and its floodplain function as a single unit because their water, sediment, and organic budgets are inseparable. Numerous workers have argued similarly that, from an ecological perspective, rivers cannot be divorced from their floodplains, and refer to this hydrological and ecological unit as the river-floodplain system (Welcomme 1979, Junk et al. 1989). For example, a growing body of evidence, largely from tropical systems, suggests that aquatic productivity in river floodplain systems is strongly dependent upon their flooding regimes (Junk et al. 1989, Bayley 1991). As floodwaters spill over a river's banks onto the floodplain, the entire area inundated can be considered a wetland. As floodwaters recede, river water may remain in depressions on the floodplain, forming floodplain pools. These environments are the subjects of this chapter. Our focus is on invertebrate communities in floodplain ponds, largely from temperate systems in agricultural (prairie) landscapes. More general reviews of floodplain wetlands and their functional significance in river-floodplain systems can be found in Hillman (1986), McArthur (1988), and Ward (1989).

Floodplain ponds vary greatly in size and hydroperiod, ranging from small, ephemeral pools that may hold water for only a few days to large, permanent lakes (e.g., oxbow lakes). Despite being referred to as pools or ponds, floodplain ponds are not exclusively lentic environments. As floodwaters spread over the floodplain, ponds lose their identity and become interconnected, with the hydrologic regime in the "ponds" being more lotic than lentic (Drago 1989). As floodwaters recede, ponds assume their identity, but they may be connected to the river for days to weeks, even after the river is confined to its channel (Halyk and Balon 1983, Drago 1989). The position of a pond in the floodplain and the magnitude of the flood determine whether a floodplain pond receives floodwaters and becomes connected to other ponds during a flood, and the duration of its connection to the river. Consequently, there can be appreciable variation among years in the hydrologic regime of individual ponds.

Both temporary and permanent floodplain ponds may change drastically in size because of evaporation after floodwaters have left the floodplain (see Shiel 1976). Because of this, physical and chemical parameters tend to have much broader ranges in ponds than in nearby main channel areas. Hillman (1986) compared physical/chemical parameter values from five floodplain ponds (billabongs) on the River Murray, Australia, and from several nearby mainstream sites. All parameters, which included temperature, turbidity, soluble reactive and total phosphorus, nitrate, and conductivity, had a wider range of values in ponds than in the river. Extremely high values for most parameters were observed during later stages of dehydration. Floodplain ponds were richer, in terms of plant nutrient concentrations, than mainstream sites.

Larimore et al. (1973) did not observe similar patterns in the Kaskaskia River, Illinois, and several of its permanent floodplain ponds that were intensively sampled for one year. In general, plant nutrient concentrations did not differ significantly between the mainstream and the ponds; concentrations of P and N were generally high in both environments, as is typical of many prairie streams in Illinois (Wiley et al. 1990). A major difference between the river and the permanent floodplain ponds involved dissolved oxygen concentrations. Pools with well-developed mats of bottom dwelling (*Ceratophyllum*) and/or surface-dwelling plants (*Lemna, Wolffia*) had much lower dissolved oxygen levels than ponds lacking such vegetation and the river. In summer, with vegetation mats present, dissolved oxygen was often observed to decline from supersaturation ($\sim$20 mg/L) at the surface to <0.5 mg/L at 10 cm below the surface. Therefore, even permanent floodplain ponds can present harsh environmental conditions.

This brief review suggests that floodplain ponds are unique freshwater environments that share characteristics with lentic and lotic habitats and present difficult challenges to their inhabitants. In the following we discuss invertebrate communities of floodplain ponds, especially mechanisms affecting their structure. We first give a general description of the invertebrate fauna of floodplain ponds and consider the extent to which the fauna resembles that of the adjacent river. We then present a general conceptual model of how prominent abiotic and biotic features of floodplain pond environments should determine the structure of invertebrate communities. Finally, we examine two case studies that address predictions of the model. Throughout, we illustrate areas in need of additional research.

PATTERNS IN FAUNAL COMPOSITION

In their classic review of the fauna of temporary pools, Wiggins et al. (1980) consider temporary floodplain pools to be distinct from "true" temporary (annual) pools because they may receive water, and hence animals, from the adjacent river. They acknowledge that floodplain pools may have a fauna that is similar to true temporary pools, but they should also include species from the river. This raises several questions. To what extent does the fauna of floodplain ponds resemble that of, first, the river and second, other, isolated temporary pools in the area, but not on the floodplain? Third, how does the fauna of floodplain pools change with hydroperiod (i.e., pool duration)? Unfortunately, we are unaware of any studies that address the second of these three questions, although such comparisons should prove interesting in light of the conceptual models presented below.

It is clear, from the few available studies addressing the first question (Boulton and Lloyd 1991, Hillman and Shiel 1991), that macroinvertebrate communities of floodplain ponds are quite distinct from riverine communities. Hillman and Shiel (1991) surveyed macroinvertebrate communities along the River Murray in southeast Australia, downstream of a reservoir. They sampled four riverine sites and five floodplain ponds in a primarily agricultural catchment. None of the ponds dried completely during the eight-year study. They found a total of 390 macroinvertebrate taxa, of which 102 were found only in the river sites and 156 only in the floodplain ponds; only about one third of the taxa (132) were found in both habitats. A cluster analysis of the data, based on a measure of community dissimilarity, indicated that assemblages in the ponds were similar to each other but quite dissimilar from the main channel communities. They also found substantial variation in pond assemblages among years, even though the ponds were permanent during the study.

Boulton and Lloyd (1991) also worked on the River Murray system, but much farther downstream, below its confluence with the Darling. They sampled a temporary pond, a permanent pond, and 11 additional sites representing three main habitats: backwaters that had little or no flow, anabranch channels that were connected to the main river and had either relatively slow or fast flow, and the main channel. Each site was sampled once, shortly after floodwaters had receded. Multivariate ordination (detrended correspondence analysis) indicated that the temporary pond's macroinvertebrate assemblage was quite distinct from that at all of the other sites. The permanent pond's fauna was similar to that of some backwater sites, but quite different from the main channel sites. The floodplain ponds supported considerably more taxa and, in general, larger populations than the other habitats. The temporary pond was distinctive in having a high proportion of its fauna composed of predators (especially dytiscid beetles) and detritivores, whereas lotic sites (main channel, slow- and fast-flowing anabranches) were dominated by detritivores.

These studies indicate that the invertebrate fauna of floodplain ponds is more similar to assemblages found in lentic than in lotic habitats. In a few cases the composition of the micro- and macroinvertebrate fauna has been thoroughly surveyed and described, especially for permanent ponds (Paloumpis and Starrett 1960, Monakov 1969, Larimore et al. 1973, Shiel 1976, Parsons and Sharton 1978, Shiel and Koste 1983, Maher 1984, Boulton and Lloyd 1991, Hillman and Shiel 1991, Corti 1994). However, there have been too few studies of how faunal composition and community structure change with variation in hydroperiod (cf. Schneider and Frost 1996 for woocdland vernal pools) to determine if general patterns exist. We discuss a particular example below.

In summary, our attempt to address the questions posed at the beginning of this section emphasizes that, with available information, few generalizations can be made regarding the structure of floodplain pond invertebrate communities. In the following section we outline a simple model that makes general predictions for ways that invertebrate community structure in floodplain ponds should differ from other temporary ponds in a region, and for changes in community structure along a gradient in hydroperiod.

Community Structure: Conceptual Models

A General Model. Wellborn et al. (1996) propose that hydroperiod and the distribution of invertebrate and vertebrate predators across the gradient in hydroperiod are the dominant forces affecting the structure of lentic freshwater communities. In their conceptual model these two factors interact to produce three main classes of freshwater habitat, each having relatively distinct animal communities. One class is temporary habitats that lack fish or large invertebrate predators. In general, periodic drying and its associated effects on the physical environment of ponds (e.g., water temperature, dissolved oxygen concentration) strongly restrict the pool of species able to successfully use such habitats (Wiggins et al. 1980). Such temporary habitats hold water for too short a duration to support animals with relatively long life cycles, unless the animals can survive in dry conditions for prolonged periods. A second class of habitats identified by Wellborn et al. (1996) is permanent waterbodies that lack fish but support populations of large, active invertebrate predators. Although permanent, these habitats are generally shallow and consequently exhibit severe fluctuations in dissolved oxygen (e.g., anoxic conditions during winter) and temperature, which prevent fish from maintaining populations. The absence of fish and relatively long-duration hydroperiods allow for the establishment of large, long-lived invertebrate predators (Eyre et al. 1992, Jeffries 1994). The final class of habitats is permanent waterbodies that contain predaceous fish (Wellborn et al. 1996). Large, active invertebrate predators are effectively excluded from such habitats by fish predation.

Wellborn et al. argue that these three classes of habitats should support relatively distinct animal communities because traits of individuals that allow a species to be successful in one habitat type will effectively prevent it from occupying other habitat types. For example, for taxa inhabiting temporary habitats the necessity to complete development rapidly before the habitat dries will often dictate that individuals exhibit high activity rates to maximize feeding and growth rates. However, high activity levels make individuals extremely susceptible to visual predators such as fish (McPeek 1990), leading to their exclusion from habitats occupied by fish. Wellborn et al. (1996) should be consulted for a thorough and lucid discussion of the mechanisms leading to changes in community structure across the permanence gradient in freshwater lentic habitats.

Floodplain Ponds. Although the conceptual model of Wellborn et al. (1996) likely applies to a broad range of freshwater environments, we suggest that a simpler model may be more applicable to floodplain wetlands. The main reason to depart from their model is that all pond types (temporary, permanent) should be accessible to fish during flooding. Fish should be present in temporary floodplain ponds for at least part, if not all, of the wet phase. It is well known that many fish species readily move onto the floodplain during floods (Guillory 1979, Ross and Baker 1983, Kwak 1988). It is perhaps less well appreciated that, as floodwaters recede, many fish remain on the floodplain in floodplain pools.

Halyk and Balon (1983) provide an excellent example of this. They studied the fish fauna in 29 floodplain pools of a 4th order warmwater stream in Southern Ontario. Pools were sampled in August, over two months after floodwaters had receded. At that time individual pools on the floodplain were isolated from each other and from the river. Only 19 of the 29 pools still contained water in mid-August. All of these 19 pools contained fish, even though several pools were quite small (area <5 m²; volume <0.5 m³). Even the smallest pool sampled (area = 1 m²; volume = 0.1 m³) contained 19 fish representing 7 species with a total biomass of 11.4 g. The 10 pools that dried had also contained fish earlier in the summer.

The composition of the invertebrate fauna inhabiting temporary and permanent floodplain pools further supports the hypothesis that fish are regularly present in these habitats. Characteristic components of vernal ponds are the large, noncladoceran branchiopod crustaceans (i.e., fairy shrimps [Anostraca], tadpole shrimps [Notostraca], and clam shrimps [Laevicaudata and Spinicaudata]) (Pennak 1978). These groups are highly vulnerable to predation by fish and large invertebrates and are generally excluded from waterbodies containing such predators (Kerfoot and Lynch 1987). Therefore, if fish were present throughout the permanence gradient in floodplain ponds, we would expect fairy, tadpole, and clam shrimps to be absent even from temporary floodplain ponds. Unfortunately, little data is available to address this hypothesis. Noncladoceran branchiopods were not present in the surveys of temporary and permanent floodplain ponds reported for the Kaskaskia River, Illinois (Larimore et al. 1973), Alcovy River, Georgia (Parsons and Wharton 1978), River Murray, Australia (Crome and Carpenter 1988, Boulton and Lloyd 1991), and the Mississippi River (Corti et al. 1997), or in our samples of four pools on the Sangamon River, Illinois (see below). Boulton and Lloyd (1992) observed several fairy, shield, and clam shrimps emerging from experimentally inundated dry floodplain sediments on the River Murray, suggesting that at least some ponds had not been colonized by fish. Therefore, although the available data are not extensive, fish and invertebrate faunas of floodplain pools generally support the hypothesis that fish predation should influence the structure of floodplain pond invertebrate communities throughout the gradient in pond permanence.

Patterns observed by Winemiller (1996) in oxbow lakes on the Brazos River, Texas, provide further observational support for the importance of fish predation in floodplain ponds. He found strong evidence of trophic cascades in each of three oxbows that differed markedly in hydroperiod (i.e., zooplankton abundance was relatively low in a temporary oxbow having relatively high abundance of planktivorous fish, but relatively high in more permanent oxbows having low planktivore density).

According to the logic outlined by Wellborn et al. (1996), because fish should be present in floodplain ponds throughout the permanence gradient, there should be two main classes of floodplain wetland habitat: temporary and permanent. The presence of fish in both habitat types suggests that large, active invertebrate predators, such as the larvae of the dragonfly Anax, should be rare in floodplain ponds and should not influence the structure of their animal communities.

Thus, predictions of our model differ in two main ways from those of Wellborn et al. (1996). First, large, active invertebrate predators should not be present in floodplain ponds. Second, because taxa must coexist with fish throughout the permanence gradient, many species present in temporary ponds should also be present in permanent ponds. This contrasts sharply with the pattern observed frequently in lentic freshwater habitats, where species inhabiting temporary ponds are generally absent from permanent ponds (Wiggins et al. 1980, Williams 1987). It should be noted though that factors other than predation could prevent species from occupying a broad range of habitats. Two such factors that are known to be important are life history characteristics and biotic interactions other than predation. A number of taxa appear to require a dry period during their life cycle (e.g., the resting eggs of many noncladoceran branchiopods (Dodson and Frey 1991)), which would prevent them from inhabiting more permanent ponds. Second, species could be excluded from permanent habitats by interspecific competition (e.g., McLachlan 1985).

Finally, although we expect fish to have strong effects on invertebrate community structure in floodplain ponds throughout the permanence gradient, determination of the composition of the community in any given pond is likely to have a strong stochastic component. The reason for this expectation largely stems from the diverse ways that ponds can be colonized and the way the flood regime can interact with these mechanisms (see also Winemiller 1996). The major ways that floodplain habitats are colonized are:

1. By riverine animals transported in floodwaters
2. By aerial insects from nearby permanent aquatic habitats
3. From desiccation-resistant resting stages in floodplain sediments
4. By transport on other organisms, such as waterfowl (e.g., passive transport of pond snails by waterfowl [Boag 1986]; active transport of ostracods by toads [Seidel 1995]) (Maher 1984, Crome and Carpenter 1988, Boulton and Lloyd 1991).

Further complicating matters is that, during flooding, ponds may become interconnected, directly on the floodplain and/or indirectly via the main channel. This would permit considerable movement, especially by highly mobile animals such as fish, among "ponds" during flooding and while floodwaters recede but ponds are still connected to the main channel (Kwak 1988). Thus, depending upon the magnitude and frequency of flooding, there is potential for marked, frequent shifts in community composition, especially at higher trophic levels. We suspect that these factors will result in considerably greater long-term temporal variation in community structure in temporary floodplain ponds than in other temporary lentic habitats. Unfortunately, we are unaware of any studies in temporary freshwater ponds that would allow analysis of concordance in community composition among years (but see Jeffries 1994, Schneider, this volume).

Community Structure: Field Experiments

In contrast to other freshwater habitats (Wellborn et al. 1996), in floodplain ponds the roles of hydroperiod and biotic interactions, especially predation, in structuring communities have received very little attention. Below we consider two recent studies conducted on temperate prairie floodplain rivers by members of our research group at the Illinois Natural History Survey/University of Illinois.

Mississippi River. Corti et al. (1997) studied the invertebrate community of mudflat habitat in floodplain ponds on the Mississippi River near St. Charles, Missouri. The study site was just upstream of a lock and dam on the river in an area that had been restored to wetland prairie from farmland in 1989. Four ponds were studied that varied in hydroperiod due to variation in size, topography, and proximity to the river. During the 12-month study pond 1 dried out five times, pond 2 dried out four times, pond 3 dried out once, and pond 4 always had measurable water. Ponds 1 and 2 had measurable standing water for slightly >200 days (standing water was absent during most of July, August, and September), while ponds 3 and 4 had standing water for >325 days. For some of our statistical analyses we refer to ponds 1 and 2 as the short-duration ponds and ponds 3 and 4 as the long-duration ponds.

Invertebrates were sampled monthly with a core sampler from seven 1 × 1-m plots located in the deepest part of each pond. Three treatments were applied at random to two replicate plots in each pond to assess effects of fish and waterfowl predation on benthic invertebrates. One treatment, referred to as "all-access," was open to all predators. The other two treatments employed exclusion cages to restrict access by predators. Cages were constructed of 0.4-cm plastic mesh attached to a wooden frame. The top and three sides of each cage were covered with this mesh. The fourth side of a cage depended upon treatment. It was covered with 0.4-cm mesh in the treatment intended to exclude all vertebrate predators ("no access"). In the "small-fish access" treatment it was covered with 5.08 × 7.62-cm mesh that was intended to exclude large fish and waterfowl. The seventh plot in each pond received a cage in which the fourth side was left uncovered to serve as a control for cage effects. For most of the study the top of the cages extended above the water surface.

In their analyses of the experimental results Corti et al. (1997) focus on the responses of individual taxa to the treatments and several summary community-level responses (e.g., taxa richness, total invertebrate abundance, total biomass). In this chapter we consider in greater detail effects of hydroperiod and predation on invertebrate community structure and highlight responses of select taxa to the treatments.

We first explore variation in community structure among ponds and among predator treatments using an analysis of community similarity. Comparisons of community similarity were made for all possible pond-predator treatment combinations using a measure of similarity (S) (Schoener 1968) that is based on the presence or absence of taxa in a community and their relative abundances:

$$S = 1 - 1/2 \sum |P_{ij} - P_{ik}|$$

where P_{ij} is the proportion of species i in community j and the summation is taken over the n species in the community. Similarity values can range from 0 (complete dissimilarity) to 1 (complete similarity). Although nearly 100 invertebrate taxa were observed during the study, most of these were quite rare (i.e., mean abundance per core sample <1 individual in all treatments on most dates). Therefore, we have restricted our analyses to the 16 most commonly encountered taxa (listed in Table 29.1). Data used in the analysis were

TABLE 29.1. Loadings of the Mean Abundance of Each Taxon on the First Four Principal Components[a]

Taxon	PC1	PC2	PC3	PC4
Coleoptera				
Berosus	0.752	0.459	0.020	−0.037
Chironomidae				
Chironomus	−0.224	0.734	0.395	0.249
Dicrotendipes	0.654	0.249	0.570	−0.343
Endochironomus	0.528	−0.323	0.684	0.055
Polypedilum	0.569	0.453	0.424	0.161
Tanytarsini	0.756	0.503	−0.122	0.262
Orthocladiinae	0.805	0.432	−0.207	0.195
Tanypodinae	−0.526	0.524	0.126	0.548
Ceratopogonidae				
Palpomyia	−0.833	0.262	0.242	−0.008
Crustacea				
Ostracoda	−0.654	0.187	0.442	−0.400
Cladocera				
Diaphanosoma	−0.305	−0.429	0.107	0.473
Copepoda	0.955	−0.056	−0.026	0.230
Nematoda	−0.271	−0.402	0.757	0.211
Oligochaeta				
Branchiura sowerbyi	−0.804	0.123	−0.155	0.465
Other oligochaetes	0.457	−0.649	0.098	0.390
Gastropoda				
Physella	0.373	−0.861	0.050	0.076

[a] All abundance values were transformed by ln (x + 1). The four principal components accounted for 40.1 percent, 21.7 percent, 13.0 percent, and 9.2 percent of the variance, respectively.

the abundance of each taxon averaged over the duration of the study, including only sampling dates where the mean density, over all ponds, was ≥ 1 individual/sample in at least one predator treatment.

Community similarity did not differ appreciably among the predator treatments within each pond, but exhibited more striking differences when comparisons were made among ponds. These results are well illustrated by cluster and principal components analysis of the data. The cluster analysis was performed on a dissimilarity matrix, using $1 - S$ as a measure of dissimilarity. The results (Fig. 29.1) show clear separation of the four ponds, with pond 1, which dried five times during the study, being strongly separated from the other three ponds. The results suggest that invertebrate communities in the ponds that dried at least once (ponds 1–3) were more similar to each other than they were to communities in pond 4, which did not dry out. Finally, although community structure was clearly more similar within ponds than among ponds, there are indications, especially in the longer-duration ponds, of differences in community similarity between the treatment that allowed predator access to plots (all access) and the treatments that restricted predator access (i.e., the no-access and small-fish-access treatments).

The principal components analysis provides some explanations for the patterns observed in Figure 29.1. The analysis was performed on the mean abundance, over the duration of the experiment, in each pond-predator treatment combination for each of the 16 most commonly encountered taxa. All values were transformed by $\ln (x + 1)$. The first four principal components were

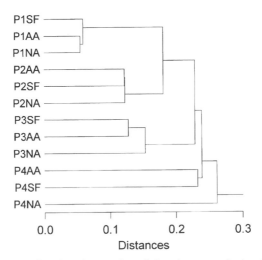

Fig. 29.1. Dendogram showing the results of the cluster analysis of the dissimilarity between invertebrate communities in the predator treatments in each pond. The distance metric is Euclidean distance. Clusters were formed using the single-linkage method. Ponds 1 through 4 are labeled P1–P4, respectively. Predation treatments are labeled: AA = all access, SF = small fish access, NA = no access.

extracted, each of which had an eigenvalue >1. The four ponds are well separated by PC1 and PC2 (Fig. 29.2). Taxa with strong positive loadings on PC1 were most abundant in ponds 1 and/or 2 and relatively rare in the longer-duration ponds (3 and 4) (Table 29.1). PC2 further separated the most temporary pond (pond 1) from the other ponds (Fig. 29.2). Taxa with strong negative loadings on PC2 (i.e., *Physella,* other oligochaetes) were abundant in pond 1 and much less abundant elsewhere (Table 29.1). *Chironomus,* which had a strong positive loading on PC2, was rare in pond 1 relative to the other ponds.

With one exception, the cladoceran *Diaphanosoma,* the taxa considered in Table 29.1 exhibited highly significant variation in abundance among ponds (Corti et al. 1997). For most taxa, except the nematodes, this significant variation allowed taxa to be readily classified as being more abundant in shorter-duration (1 and 2) or longer-duration (3 and 4) ponds (Table 29.2).

Patterns in the distribution of taxa over the gradient in hydroperiod are consistent with available information on life cycles, abilities to withstand desiccation, and mechanisms for recolonizing habitats that have dried. This information has been extensively reviewed elsewhere (Wiggins et al. 1980, Williams 1987, 1996, Boulton 1989, Boulton et al. 1992), so we will focus on observations most relevant to the pattern presented in Table 29.2.

The tubificid oligochaete *Branchiura sowerbyi* was found only in the pond that did not dry out during the study. *Branchiura* has a one- or two-year life cycle (Carroll and Dorris 1972, Casellato 1984), and no mechanisms of desiccation resistance have been reported for it or other tubificids (Wiggins et

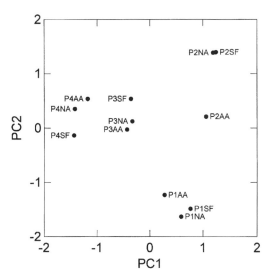

Fig. 29.2. Scores for each of the pond-predation treatment combinations on the first two principal components axis. Ponds and predation treatments are labeled as in Fig. 29.1.

TABLE 29.2. Taxa Occurring in Mississippi River Floodplain Ponds That Were Significantly More Abundant in Short-Duration or Long-Duration Ponds[a]

Pond Type with Greatest Abundance	
Short Duration	Long Duration
Berosus	*Chironomus*
Dicrotendipes	Tanypodinae
Endochironomus	*Palpomyia*
Polypedilum	Ostracoda
Tanytarsini	*Branchiura sowerbyi*
Orthocladiinae	
Copepoda	
Other oligochaetes	
Physella	

[a] Only two common taxa, *Diaphanosoma* and Nematoda, did not exhibit significant patterns in abundance associated with pond permanence. See Corti et al. (1997) for details of the statistical analyses.

al. 1980), so it would be rapidly eliminated from temporary habitats. By contrast, oligochaetes other than *Branchiura* (largely Naididae, especially *Dero digitata, D. nivea,* and *Nais elinguis,* and the tubificid *Limnodrilus* spp.) were present in all ponds, but were considerably more abundant in pond 1 than elsewhere. Naidids and lumbriculids may be able to survive extended periods in cocoons or as cysts (Wiggins et al. 1980, Brinkhurst and Gelder 1991). Boulton (1989) observed appreciable numbers of oligochaetes to emerge from "dry" sediments (about 10 percent water by weight) that were collected from riffles and pools in dry temporary streams and rewetted in the laboratory. Stanley et al. (1994) found live oligochaetes in the sediments of a desert stream for over two weeks after surface water had disappeared. Finally, it is likely that oligochaetes rapidly recolonize temporary floodplain habitats from floodwaters (see Stanley et al. 1994).

Several Chironomidae (e.g., *Dicrotendipes, Endochironomus, Polypedilum,* Tanytarsini [especially *Tanytarsus* and *Paratanytarsus*], Orthocladiinae [especially *Cricotopus*]) were abundant in the shorter-duration ponds. Representatives of some of these genera (*Dicrotendipes, Endochironomus, Polypedilum*), among others, are known to produce larval cocoons during the winter after surface ice formation (Danks and Jones 1978). Larvae in cocoons are dormant. Larvae have been observed in cocoons in dry sediments (Grodhaus 1980), and cocoons may allow larvae to survive periods of desiccation by maintaining a humid environment around the larvae (Pinder 1995). We do not know if chironomid larvae we studied utilized this strategy for desiccation resistance. Wiggins et al. (1980) observed all of the genera listed above in temporary woodland pools and found live larvae of several genera in soil samples taken from dry basins and flooded with water in the laboratory. This suggests that these taxa are well equipped to survive in habitats that may be dry for several months each year. Of course, it is also likely that oviposition by adults produced in nearby permanent habitats contributed to the rapid recolonization of dried ponds by midges (Batzer and Wissinger 1996).

By contrast, adaptations for desiccation resistance are unknown for the subfamily Tanypodinae, whose larvae are predaceous and do not burrow or build tubes (Wiggins et al. 1980). Therefore, their low abundance in short duration ponds is not unexpected.

The ostracods are well equipped to survive in dry conditions because of their highly desiccation-resistant eggs (Wiggins et al. 1980). However, most species require a month or more to complete development (DeLorme 1991). Consequently, drying of a pond before eggs are laid could decimate a population. This may account for the very low ostracod populations in the short duration ponds.

Effects of the predator treatments on the abundance and size distributions of taxa were considered in detail in Corti et al. (1997). We will review some of the major findings here. First, over half of the taxa considered in Table 29.1 were strongly affected by the predator treatments. In nearly all cases the abundance and/or mean individual biomass of individuals was lower in the all-access treatment than in the restricted-access treatments. Corti et al. (1997) argued that because the small-fish access and no-access treatments generally did not differ and because waterfowl were rarely observed foraging at the ponds, significant predator effects were primarily due to the activities of large fish. For some taxa these effects were evident in all ponds (e.g., Fig. 29.3: *Dicrotendipes*). Other taxa were strongly affected by vertebrate predators in only a subset of the ponds (i.e., there were significant pond-by-predator interaction effects) (e.g., Fig. 29.3: *Polypedilum*). This was most frequently observed when predator effects were weak in ponds where a taxon occurred in low abundance. Finally, as predicted by our conceptual model, strong effects of vertebrate predation were observed for several taxa even in ponds that dried frequently (i.e., *Dicrotendipes, Chironomus, Endochironomus, Polypedilum, Physella*).

Sangamon River. In 1994 we conducted a similar study in four floodplain ponds on the Sangamon River in the Sangamon State Conservation Area, Cass County, Illinois. Two ponds were relatively close to the river, while the others were separated from the river by a man-made levee. Floodwaters reached all ponds during typical spring floods. Two of the ponds dried out by late summer, while the others retained water through the fall.

In mid-July, after floodwaters had receded, we installed six 1 × 1-m fish exclusion cages in each pond. Cages were constructed of 0.4-cm plastic mesh on all sides, with mesh extending 10 cm into the sediments and above the water surface. In early August we sampled invertebrate populations inside

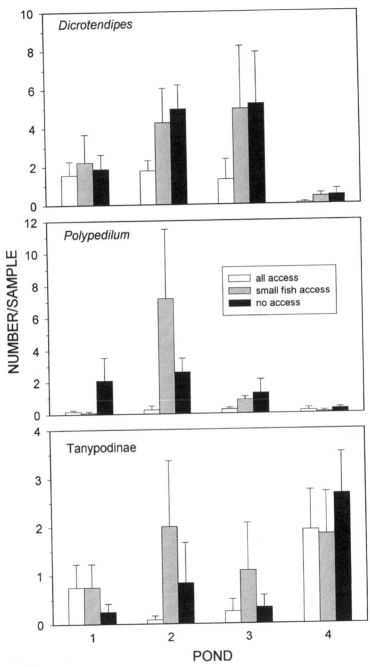

Fig. 29.3. Abundance (mean ± SE over the duration of the study) of *Dicrotendipes, Polypedilum,* and Tanypodinae larvae in the predation treatments in each of four floodplain ponds on the Mississippi River. Results of statistical analyses are presented in Corti et al. (1997).

and outside the cages in two habitats: sediments and the water column. Each water-column sample consisted of a 1-m pass using a D-frame net (1.0-mm mesh) held just above the bottom. Benthic invertebrates in the sediments were sampled with a core sampler (area = 0.0035 m²) penetrated to a depth of 5 cm (sediments were typically anaerobic at greater depths). Two water-column and two sediment samples were taken from each cage and from outside the cages at randomly selected locations.

Seining indicated that fish were present in all of the ponds; a total of 14 species were collected. These included several species that are largely benthic feeders (e.g., yellow bullhead, channel catfish, tadpole madtom, mud darter, logperch) and species that likely fed more from the water column (e.g., bluegill, warmouth, black striped topminnow).

We observed strong effects of fish predation on invertebrates in the water-column habitat, but not in the sediment habitat (Fig. 29.4). Aquatic insects, especially Corixidae and Hydrophilidae, and water mites dominated water column samples. Most taxa were more abundant inside the fish exclusion cages than outside the cages (total invertebrates: $F_{1,11}$ = 16.86, P = 0.002; *Trichocorixa* [Corixidae]: $F_{1,11}$ = 6.03, P = 0.032; Chironomidae: $F_{1,11}$ = 4.62, P = 0.05; Ceratopogonidae: $F_{1,11}$ = 6.63, P = 0.026).

In contrast to the observations of Corti et al. (1997), insect larvae were rare in sediment samples from Sangamon River floodplain ponds. Harpacticoid copepods, ostracods, water mites, and nematodes dominated these communities. Fish predation did not affect the abundance of any of these taxa (all P values > 0.15). Corti et al. also did not observe significant effects of fish predation on ostracods or copepods, but nematodes exhibited strong responses in the longer-duration ponds.

CONCLUSIONS

Although there have been few experimental studies, there appear to be striking differences between floodplain ponds and other freshwater lentic environments in the mechanisms determining the structure of invertebrate communities. A major factor contributing to these differences is the presence of fish in floodplain ponds throughout the pond permanence gradient. As a consequence, taxa highly vulnerable to fish predation (e.g., noncladoceran branchiopods, large and active invertebrate predators) should be rare in floodplain ponds. The limited available data appear to support this hypothesis. Because fish predation should be an important environmental factor in most floodplain ponds, regardless of hydroperiod, shifts in invertebrate community structure over the gradient in habitat permanence should be more subtle than in other lentic systems (cf. Schneider and Frost 1996, Wellborn et al. 1996). In floodplain ponds changes in community structure should result largely from shifts in the relative abundance of species rather than striking changes in the composition of communities. In part, this pattern should be evident because, dur-

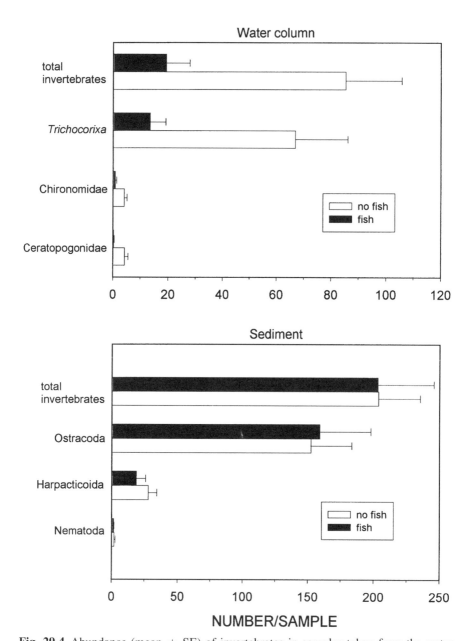

Fig. 29.4 Abundance (mean ± SE) of invertebrates in samples taken from the water column (top panel) and sediments (bottom panel) inside ("no fish") and outside ("fish") fish exclusion cages in floodplain pools on the Sangamon River.

ing flooding, taxa can colonize even highly ephemeral ponds from the river or from longer-duration ponds on the floodplain, if ponds are connected. Because of this, we expect that, in contrast to other freshwater lentic systems, in temporary floodplain ponds it will be common to encounter taxa that cannot complete development before the pond dries and have limited ability to survive dry conditions.

Considerably more experimental and observational work is needed to determine if these general expectations are upheld. In general, floodplain ponds have received considerably less attention than other freshwater habitats. Given that these systems are often highly productive (e.g., Halyk and Balon 1983) and appear to be critical to the overall function of river-floodplain systems (e.g., Bayley 1991, 1995), this is especially unfortunate. In the temperate zone most river-floodplain systems have been highly modified in terms of their flow regimes and the extent of connection between the main channel and its floodplain (e.g., through construction of levees) (Sparks 1995). Nonetheless, nearly "natural" conditions still exist in segments of large rivers like the Illinois and Mississippi (Sparks 1995), presenting important research opportunities. Such opportunities need to be exploited to better understand the ecology of floodplain wetlands and their role in the ecology of temperate river-floodplain systems.

LITERATURE CITED

Batzer, D. P., and S. A. Wissinger. 1996. Ecology of insect communities in nontidal wetlands. Annual Review of Entomology 41:75–100.

Bayley, P. B. 1991. The flood pulse advantage and the restoration of river-floodplain systems. Regulated Rivers: Research and Management 6:75–86.

———. 1995. Understanding large river-floodplain ecosystems. BioScience 45:153–158.

Boag, D. A. 1986. Dispersal in pond snails: Potential role of waterfowl. Canadian Journal of Zoology 64:904–909.

Boulton, A. J. 1989. Over-summering refuges of aquatic macroinvertebrates in two intermittent streams in central Victoria. Transactions of the Royal Society of Southern Australia 113:23–34.

Boulton, A. J., and L. N. Lloyd. 1991. Macroinvertebrate assemblages in floodplain habitats of the lower River Murray, South Australia. Regulated Rivers: Research and Management 6:183–201.

———. 1992. Flooding frequency and invertebrate emergence from dry floodplain sediments of the River Murray, Australia. Regulated Rivers: Research and Management 7:137–151.

Boulton, A. J., E. H. Stanley, S. G. Fisher, and P. S. Lake. 1992. Over-summering strategies of macroinvertebrates in intermittent streams in Australia and Arizona. Pages 227–237 in R. D. Robarts and M. L. Bothwell (eds.), Aquatic Ecosystems in Semi-arid regions: Implications for Resource Management. N. H. R. I. Symposium Series 7, Environment Canada, Saskatoon, Saskatchewan, SK.

Brinkhurst, R. O., and S. R. Gelder. 1991. Annelida: Oligochaeta and Branchiobdellida. Pages 401–435 *in* J. H. Thorp and A. P. Covich (eds.), Ecology and Classification of North American Freshwater Invertebrates. Academic Press, San Diego, CA.

Carroll, J. H., Jr., and T. C. Dorris. 1972. The life history of *Branchiura sowerbyi*. American Midland Naturalist 87:413–422.

Casellato, S. 1984. Life-cycle and karyology of *Branchiura sowerbyi* Beddard (Oligochaeta, Tubificidae). Hydrobiologia 115:65–69.

Corti, D. 1994. Assessing the effects of hydroperiod and predation on benthic invertebrate community structure in Mississippi River floodplain wetlands. Master's thesis, University of Illinois at Urbana-Champaign, Urbana, IL.

Corti, D., S. L. Kohler, and R. E. Sparks. 1997. Effects of hydroperiod and predation on a Mississippi River floodplain invertebrate community. Oecologia 109:154–165.

Crome, F. H. J., and S. M. Carpenter. 1988. Plankton community cycling and recovery after drought dynamics in a basin on a floodplain. Hydrobiologia 164:193–211.

Danks, H. V., and J. W. Jones. 1978. Further observations on winter cocoons in Chironomidae (Diptera). Canadian Entomologist 110:667–669.

DeLorme, L. D. 1991. Ostracoda. Pages 691–722 in J. H. Thorp and A. P. Covich (eds.), Ecology and Classification of North American Freshwater Invertebrates. Academic Press, San Diego, CA.

Dodson, S. I., and D. G. Frey. 1991. Cladocera and other Branchiopoda. Pages 723–786 *in* J. H. Thorp and A. P. Covich (eds.), Ecology and Classification of North American Freshwater Invertebrates. Academic Press, San Diego, CA.

Drago, E. C. 1989. Morphological and hydrological characteristics of the floodplain ponds of the Middle Parana River (Argentina). Review of Tropical Hydrobiology 22:183–190.

Eyre, M. D., R. Carr, R. P. McBlane, and G. N. Foster. 1992. The effects of varying site-water duration on the distribution of water beetle assemblages, adults and larvae (Coleoptera: Haliplidae, Dytiscidae, Hydrophilidae). Archive Für Hydrobiologie 124:281–291.

Grodhaus, G. 1980. Aestivating chironomid larvae associated with vernal pools. Pages 315–322 *in* D. A. Murray (ed.), Chironomidae: Ecology, Systematics, Cytology and Physiology. Pergammon Press, Oxford, UK.

Guillory, V. 1979. Utilization of inundated floodplain by Mississippi River fishes. Florida Scientist 42:222–228.

Halyk, L. C., and E. K. Balon. 1983. Structure and ecological production of the fish taxocene of a small floodplain system. Canadian Journal of Zoology 61:2446–2464.

Hillman, T. J. 1986. Billabongs. Pages 457–470 *in* P. De Deckker and W. D. Williams (eds.), Limnology in Australia. CSIRO, Melbourne, Australia.

Hillman, T. J., and R. J. Shiel. 1991. Macro- and microinvertebrates in Australian billibongs. Internationale Vereinigung für Theoretische und Angewandte Limnologie 24:1581–1587.

Jeffries, M. 1994. Invertebrate communities and turnover in wetland ponds affected by drought. Freshwater Biology 32:603–612.

Junk, W. J., P. B. Bayley, and R. E. Sparks. 1989. The flood-pulse concept in river-floodplain systems. Special Publication of the Canadian Journal of Fisheries and Aquatic Sciences 106:110–127.

Kerfoot, W. C., and M. Lynch. 1987. Branchiopod communities: associations with planktivorous fish in space and time. Pages 367–378 *in* W. C. Kerfoot and A. Sih (eds.), Predation: Direct and Indirect Impacts on Aquatic Communities. University Press of New England, Hanover, NH.

Kwak, T. J. 1988. Lateral movement and use of floodplain habitat by fishes of the Kankakee River, Illinois. American Midland Naturalist 120:241–249.

Larimore, R. W., E. C. Doyle, and A. R. Brigham. 1973. Ecology of Floodplain Pools in the Kaskaskia River Basin of Illinois. University of Illinois Water Resources Center Report Number 75.

Maher, M. 1984. Benthic studies of waterfowl breeding habitat in southwestern New South Wales. I. The fauna. Australian Journal of Marine and Freshwater Research 35:85–96.

McArthur, J. V. 1988. Aquatic and terrestrial linkages: Floodplain functions. Pages 107–116 *in* D. Hook and R. Lea (eds.) Proceedings of the Symposium on Forested Wetlands of the Southern U.S. U.S.D.A. Forest Service.

McLachlan, A. J. 1985. What determines the species present in a rain-pool? Oikos 45: 1–7.

McPeek, M. A. 1990. Behavioral differences between *Enallagma* species (Odonata) influencing differential vulnerability to predators. Ecology 71:1714–1726.

Monakov, A. V. 1969. The zooplankton and the zoobenthos of the White Nile and adjoining waters in the Republic of the Sudan. Hydrobiologia 33:161–185.

Paloumpis, A. A., and W. C. Starrett. 1960. An ecological study of benthic organisms in three Illinois River flood plain lakes. American Midland Naturalist 64:406–435.

Parsons, K., and C. H. Wharton. 1978. Macroinvertebrates of pools on a piedmont river floodplain. Georgia Journal of Science 36:25–33.

Pennak, R. W. 1978. Freshwater invertebrates of the United States. John Wiley & Sons, New York.

Pinder, L. C. V. 1995. The habitats of chironomid larvae. Pages 107–135 *in* P. Armitage, P. S. Cranston, and L. C. V. Pinder (eds.), The Chironomidae: Biology and Ecology of Non-biting Midges. Chapman & Hall, London, UK.

Ross, S. T., and J. A. Baker. 1983. The response of fishes to periodic spring floods in a southeastern stream. American Midland Naturalist 109:1–14.

Schneider, D. W., and T. M. Frost. 1996. Habitat duration and community structure in temporary ponds. Journal of the North American Benthological Society 15:64–86.

Schoener, T. W. 1968. The *Anolis* lizards of Bimini: Resource partitioning in a complex fauna. Ecology 49:704–726.

Seidel, B. 1995. Behavioral and ecological aspects of the association between *Cyclocypris globosa* (Sars, 1863) (Cypridoidea, Cyclocypridinae) and *Bombina variegata* (L., 1758) (Anura, Bombinatoridae) in temporary pools in Austria. Crustaceana 68: 813–823.

Shiel, R. J. 1976. Associations of Entomostraca with weedbed habitats in a billabong of the Goulburn River, Victoria. Australian Journal of Marine and Freshwater Research 27:533–549.

Shiel, R. J., and W. Koste. 1983. Rotifer communities of billabongs in northern and south-eastern Australia. Hydrobiologia 104:41–47.

Sparks, R. E. 1995. Need for ecosystem management of large rivers and their floodplains. BioScience 45:168–182.

Stanley, E. H., D. L. Buschman, A. J. Boulton, N. B. Grimm, and S. G. Fisher. 1994. Invertebrate resistance and resilience to intermittency in a desert stream. American Midland Naturalist 131:288–300.

Ward, J. V. 1989. Riverine-wetland interactions. Pages 385–400 *in* R. R. Sharitz and J. W. Gibbons (eds.), Freshwater Wetlands and Wildlife. USDOE Office of Scientific and Technical Information, DOE Symposium Series No. 61, Oak Ridge, TN.

Welcomme, R. L. 1979. Fisheries Ecology of Floodplain Rivers. Longman, London, UK.

Wellborn, G. A., D. K. Skelly, and E. E. Werner. 1996. Mechanisms creating community structure across a freshwater habitat gradient. Annual Review of Ecology and Systematics 27:337–363.

Wiggins, G. B., R. J. Mackay, and I. M. Smith. 1980. Evolutionary and ecological strategies of animals in annual temporary ponds. Archiv für Hydrobiologie Supplement 58:97–206.

Wiley, M. J., L. L. Osborne, and R. W. Larimore. 1990. Longitudinal structure of an agricultural prairie river system and its relationship to current stream ecosystem theory. Canadian Journal of Fisheries and Aquatic Sciences 47:373–384.

Williams, D. D. 1987. The Ecology of Temporary Waters. Timber Press, Portland, OR.

———. 1996. Environmental constraints in temporary fresh waters and their consequences for the insect fauna. Journal of the North American Benthological Society 15:634–650.

———Winemiller, K. O. 1996. Factors driving temporal and spatial variation in aquatic floodplain food webs. Pages 298–312 *in* G. A. Polis and K. O. winemiller (eds.), Food Webs: Integration of Patterns and Dynamics. Chapman & Hall, New York.

PART 4
Wetlands of the Western Mountains, Deserts, and Valleys

30 Wetlands of Grand Teton and Yellowstone National Parks

Aquatic Invertebrate Diversity and Community Structure

WALTER G. DUFFY

*A*quatic invertebrates were sampled monthly from May through September in six wetlands in Yellowstone and Grand Teton National Parks during 1995. Wetlands sampled exhibited semipermanent, seasonal, and temporary hydrologic regimes. Physical and chemical characteristics of wetland water were similar, except for specific conductance, which was either low (<60 μS cm^{-1}), intermediate (270–550 μS cm^{-1}), or high (>900 μS cm^{-1}). A total of 187 taxa of aquatic invertebrates were identified from all samples. Almost 70 percent of the taxa collected had not been reported previously in either park. Taxa richness was greatest in semipermanent wetlands and least in temporary wetlands. Among all wetlands, monthly mean abundance ranged from 18,041 m^{-2}–335,975 m^{-2}. Foodwebs of semipermanent wetlands were most complex, while those in temporary wetlands were most simple.

GREATER YELLOWSTONE ECOSYSTEM

Yellowstone and Grand Teton National Parks form the core of the Greater Yellowstone Ecosystem (GYE), a 77,000-km^2 area in the north-central Rocky Mountains. Yellowstone National Park was established in 1872 and was the first national park in the world. The park covers 898,349 ha of the northern Rocky Mountains, encompassing a series of high volcanic plateaus and spec-

Invertebrates in Freshwater Wetlands of North America: Ecology and Management, Edited by Darold P. Batzer, Russell B. Rader, and Scott A. Wissinger
ISBN 0-471-29258-3 © 1999 John Wiley & Sons, Inc.

tacular river valleys straddling the continental divide. Altitude in the GYE varies considerably and influences local climate since mean annual temperature in the region varies inversely with elevation. Elevation in the central Yellowstone Plateau averages 2377 m and ranges from 1615 m near Gardiner, Montana, at the park's north entrance to 3462 m at Eagle Peak along the southeast boundary (Craighead et al. 1995). Mean annual temperature at Mammoth (2070 m), near the north entrance of the park, is 4°C but is cooler throughout much of the region (Craighead et al. 1995). Precipitation varies from 40–50 cm in the northern range of Yellowstone National Park to 150–175 cm in the northern part of Grand Teton National Park (Farnes 1997). Much of this precipitation is received during late winter snowstorms. Defining features of Grand Teton National Park are the Teton Mountains and Jackson Hole, both the products of a block fault which runs along the base of the Teton Range (Love and Reed 1995). Elevation in Grand Teton National Park ranges from 1981 m at Jackson Hole to 4186 m at Grand Teton Mountain a short distance away. Seven mountain peaks in the Teton Range are above 3658 m (Knight 1994). As in Yellowstone, this spectacular landscape is the product of volcanism, glaciation, and continued geological activity (Love and Reed 1995).

The many palustrine wetlands with temperate thermal regimes in Yellowstone and Grand Teton National Parks have received little attention. Plant communities of wetlands in northern Yellowstone (Chadde et al. 1988), central Yellowstone (Mattson 1984), and parts of Grand Teton (Reed 1952) have been described. The distribution of amphibian species in both parks is relatively well known (Koch and Peterson 1995), and the biology of trumpeter swans (*Cygnus buccinator*) using wetlands in the GYE has been the subject of several investigations (Shea 1981, Tuggle 1986, Squires and Anderson 1995). However, with the exception of species distributions for mosquitoes in Yellowstone National Park (Nielsen and Blackmore 1996) and mollusks in both parks (Beetle 1989), the invertebrate taxa of these wetlands remain largely unknown.

Geothermal wetlands have attracted greater attention (Brock 1967, Brock et al. 1969, Collins 1975, Fraleigh and Wiegert 1975) than those with temperate thermal regimes and, as a consequence, knowledge of their aquatic invertebrate communities is more complete. Source water in these geothermal wetlands ranges from 31–75°C, exhibit extreme variation in pH (2.15 to 8.9) and have high concentrations of Ca (430 ppm), Si (215–400 ppm), and S (120 ppm) (Brock 1967, Collins 1975, Fraleigh and Wiegert 1975). Dense algal mats consisting of one of a few species of blue-green algae develop in these wetlands, reaching maximum biomass (687 mg l^{-1}) at water temperatures of 50–60°C (Brock 1967). Algal biomass is low below 40°C and near the thermal maxima 73–75°C. Two species of brine flies (Diptera: Ephydridae) are common in the thermal wetlands, *Paracoenia turbida* and *Ephydra bruesi* (Brock et al. 1969). Both species colonize and feed on the blue-green algal mat, but neither species colonizes mats where water temperatures are

>41.7°C (Brock et al. 1969, Collins 1977). The brine flies are prey for two water mites, *Partnuniella thermalis* (Acarina: Protsiidae) and *Thermacarus nevadensis* (Acarina: Hydrodromidae), as well as a nonthermophilic fly, *Tachytrechus angustipennis* (Diptera: Dolichopodidae), whose adults feed on the eggs and larvae of Ephydridae near the surface (Collins et al. 1976). Other species colonize these wetlands at water temperatures <30°C (e.g., Diptera: Chironomidae).

The algal-bacterial mat in these thermal wetlands is nutrient-limited and the high biomass of algae persists because most of the mat is in water too hot for Ephydridae to colonize. At water temperatures of 30–40°C feeding by larvae of Ephydridae reduces algal biomass to roughly 50 percent of maximum, algal production is actually greatest in this temperature zone. Through this series of experiments Collins et al. (1976) suggest that feeding by Ephydridae larvae in these systems increases net primary production, increases the efficiency of energy transfer between first and second trophic levels, and affects ecosystem processes (energy transfer) but not prey densities.

In this chapter I present data for nonthermal wetland aquatic invertebrate communities from Yellowstone and Grand Teton National Parks. Data presented are drawn from a longer ongoing study of energy flow in wetlands of the GYE. My objectives are (1) to document the contribution of the wetland aquatic invertebrate community to overall biological diversity in the GYE and (2) to evaluate community composition, diversity, and abundance of wetlands with different hydrologic regimes.

STUDY WETLANDS

Study areas were located in both Yellowstone and Grand Teton National Parks. The study area in Yellowstone is part of the northern range and consists of the lower Lamar River Valley from its confluence with the Yellowstone River, upstream about 6 km to the single bridge over the river. Elevation in the area ranges from about 2400 to 2500 m. Vegetation at lower elevations of the Lamar Valley is sagebrush (*Artemisia* spp.)-grassland steppe, which grades to lodgepole pine (*Pinus contorta* var. *latifolia*) interspersed with grasses (*Poa* spp.) at higher elevations. The area receives heavy use by elk (*Cervus elephus*) and bison (*Bison bison*). Bison graze sedges (*Carex* spp.) in the wet meadow zone of these wetlands during May and September. Elk presumably also use these wetlands, since feces of both are common in their basins.

The primary study area in Grand Teton National Park is located on the Snake River Plain east of Jackson Lake. Surficial geology of this study area is glacial outwash or moraine deposits from the Pinedale, Bull Lake, and older glaciers (Love and Reed 1995) at elevations of 2200–2500 m. Vegetation on the Snake River riparian area consists of an overstory of narrow-leaf cottonwood (*Populus angustifolia*), aspen (*P. tremuloides*), and clumps of blue-spruce (*Picea pungens*) with an understory of willows (*Salix* spp.) and other species. Lodgepole pine predominates on the southern exposure of larger moraines such as Signal Mountain, but is replaced by Engelman spruce (*Picea engelmannii*) and subalpine fir (*Abies lasiocarpa*) on northern exposures. The vegetation colonizing the glacial till outwash area consists of a sagebrush-grassland community. An additional study area was established at the 3200-m elevation in Bridger-Teton National Forest adjacent to Grand Teton. This area is an alpine meadow surrounded by whitebark pine (*Pinus albicaulis*) and Engelman spruce.

Wetlands in the GYE may be classified as riverine, lacustrine, or palustrine (Cowardin et al. 1979). The focus of this chapter is primarily on wetlands that are typically classified as palustrine (i.e., area < 8 ha, water depth < 2 m, lacking wave-formed shoreline features). Larger wetlands (area > 8 ha) may also be classified as palustrine if they are dominated by emergent vegetation, trees, or shrubs. The hydrology of palustrine wetlands is particularly useful for grouping them into similarly functioning units. Cowardin et al. (1979) recognize eight distinct hydrologic regimes. In this chapter aquatic invertebrate data are presented for wetlands having three hydrologic regimes: semipermanently, seasonally, and temporarily flooded. In semipermanently flooded palustrine wetlands, surface water is present during the growing season of most years. Surface water is present for extended periods of the growing season in seasonally flooded palustrine wetlands, particularly early in the growing season. However, surface water is absent by the end of the growing season in most years. In temporarily flooded wetlands surface water is present for only brief periods during the growing season, usually early in the season.

SAMPLING

Aquatic invertebrate samples were collected from six wetlands monthly from June through September 1995, when water was present in basins. However, sampling at wetland SP-sage was not initiated until July, and the temporary wetlands were sampled only during July since they did not contain water on other dates. Aquatic invertebrates were sampled with a 7.6-cm diameter core that was 1.5 m long so that each sample contained surficial sediments and the entire water column. A sliding door with a window of 149-μm nitex was installed 5 cm above the bottom of the core. For collecting a sample, the core was pushed through the water column or an aquatic macrophyte bed until it penetrated the sediment 10–15 cm, the door was closed, and the sample was lifted from the water. The sample was then concentrated either on the nitex door panel or in a 149-μm sieve before being preserved with 90 percent ethanol. Four randomly placed samples were collected from each wetland on each sampling date. Aquatic invertebrates were removed from samples under a stereo microscope at 10× magnification and preserved in 90 percent ethanol until they could be identified. Identification of specimens was to the lowest feasible taxonomic level. Voucher specimens have been prepared and are available by contacting the author.

Foodwebs were constructed using both direct observations of gut contents and information from published literature. The stomachs of ≤10 individuals of common invertebrate predator taxa from each wetland were examined for prey each month. Common invertebrate predators included Odonata, Chaoboridae, predaceous Chironomidae such as *Procladius* spp., and the oligochaete *Cheatogaster* spp. Other invertebrate taxa were assigned to a trophic status from diet information in Merritt and Cummins (1996) or Pennak (1994). Diet of *Ambystoma tigrinum* was inferred from data presented by Sprules (1972) and Holomuzki and Collins (1987).

Physical and chemical data were recorded each time samples were collected. Water temperature was measured with a hand-held thermometer, dissolved oxygen concentration was measured using the Azide modification of the Winkler method, and pH and specific conductance were measured electrometrically. A water sample was collected and frozen for later nutrient analyses. Total phosphorus was measured using the Stannous chloride method and total nitrogen as the sum of ammonia, nitrate, and nitrite nitrogen (APHA 1992).

Density of wetlands in the study areas was determined from 1:24,000 scale draft wetland maps of the area (U.S. Fish and Wildlife Service, Denver, Colorado). Data presented here on wetland density should be considered tentative since only draft maps were available for use.

Cluster analysis was used to group wetlands whose aquatic invertebrate communities were similar. Only the 20 most abundant taxa from each wetland were used in analyses since rare taxa were often associated with a single wetland and confounded results. Statistical analyses were performed with SYSTAT v. 5 software (Wilkinson et al. 1992).

RESULTS

Wetland Distribution

Wetlands exhibiting five and four different hydrologic regimes were identified in the Yellowstone and Grand Teton study areas, respectively (Table 30.1). Seasonally flooded wetlands were the most common type of wetland in both study areas, particularly at higher elevations. Semipermanent wetlands were found only at low and middle elevations. Saturated wetlands were distributed throughout the Lamar Valley but were not found in the Grand Teton study area. Few temporary and permanent wetlands were found in either study area.

Water Chemistry and Basin Morphometry

Water in the wetlands sampled was similar physically and chemically, except for specific conductance and maximum depth. Three ranges of specific con-

TABLE 30.1. Density (no. km^{-2}) of Palustrine Wetlands within Three Elevation Zones in the Study Areas of Yellowstone and Grand Teton National Parks

Wetland Class	Yellowstone			Grand Teton		
	<2500 m	2500–2700 m	>2700 m	<2500 m	2500–2700 m	>2700[a] m
Saturated	0.8	0.4	0.4	0	0	0
Temporary	0.1	0	0	0.1	0	0
Seasonal	2.3	1.8	4.4	2.8	1.8	5.0
Semipermanent	0.3	0.9	0	1.7	0.5	0
Permanent	0.1	0	0	0	0	6.0
All classes	3.6	3.1	4.8	4.6	2.3	11.0
Area sampled	49.0	29.0	2.5	13.5	8.0	2.0

[a]Represents density of wetlands above 2700 m elevation in a meadow of the Teton Range, not the entire region.

ductance were observed. Low-conductance water (≤ 60 μS cm^{-1}) was found in the SP-spruce wetland and the T-meadow1 and T-meadow2 wetlands (Table 30.2). Intermediate-conductance (270–550 μS cm^{-1}) water was found in the SP-sage wetland and the S-fir wetland. Relatively high-conductance (930 μS cm^{-1}) water was found in the T-aspen wetland. Maximum depth of the wetlands sampled was consistent with their hydrologic classification. Semipermanent wetlands retained water throughout the period, the seasonal wetland became dry between August and September, and the temporary wetlands contained water only from late June through early July. pH of water in most wetlands was neutral or slightly basic. However, pH of SP-fir ranged from 6.3 to 6.9 S.U. Other chemical and physical characteristics varied little among wetlands (Table 30.2). Total nitrogen concentration exceeded 1.0 mg l^{-1} in all wetlands sampled, and total phosphorus concentration ranged from 0.01– 0.05 mg l^{-1}.

Invertebrate Community Composition and Diversity

A total of 187 taxa representing 24 orders of aquatic invertebrates were collected from the six wetlands sampled during 1995 (Appendix 30.A). This included 43 taxa (genera or species) of flies, making Diptera the most diverse order collected. In addition to Diptera, 25 species of water fleas (Cladocera), 21 species or genera of beetles (Coleoptera), 14 species of worms (Tubificida), and 11 species of snails (Gastropoda) were found. Almost 70 percent (129) of the 187 taxa collected had not been previously recorded from either park.

Number of taxa recorded in wetlands ranged from 2 in the T-aspen wetland to 98 in the SP-spruce wetland (Table 30.3). Most taxa were collected with quantitative methods. However, as many as 18 taxa not recorded in quantitative samples were collected with qualitative methods. Taxa richness, as mean per sample, mean per month, and total, was greater in semipermanent and seasonal wetlands than in temporary wetlands.

Cluster analysis using all 187 taxa suggested patterns of aquatic invertebrate community composition among types of wetlands (Fig. 30.1). Communities in the 2 T-meadow wetlands sorted distinctly from those in other wetlands. The community in the SP-sage wetland located in the Lamar Valley also differed from the other wetlands. K-means clustering using the 20 most abundant taxa in each wetland suggested the two semipermanent wetlands (SP-sage, F = 160.10, and SP-spruce, F = 197.01) were similar and differed from the S-fir (F = 57.16) and both T-meadow wetlands (F = 0.18). Two or three taxa were associated with three of the clusters. High densities of two midge larvae, *Cladopelma* spp. and *Paratanytarsus* spp., were uniquely associated with the SP-sage wetland. Similarly, high densities of two seed shrimp (Ostracoda), *Cypridopsis vidua* and *Cypris palustra*, distinguished the invertebrate community of the SP-spruce wetland from other wetlands. A third cluster consisted of two water fleas (*Simocephalus serrulatus* and *Chydorus sphaericus*) and immature cyclopoid copepods that were common in SP-sage,

TABLE 30.2. Water Quality Parameters Measured at Six Wetlands in Yellowstone and Grand Teton National Parks during 1995[a]

Parameter	Abbreviated Wetland Name				
	SP-sage	SP-spruce	S-fir	T-meadow	T-aspen
Hydrologic regime	Semipermanent	Semipermanent	Seasonal	Temporary	Temporary
Vegetation zone	Big sagebrush	Engelmann spruce	Subalpine fir	Subalpine meadow	Aspen
n	1	1	1	2	1
Water depth (cm)	50–104	100–166	0–65	0–30	0–20
Water temperature (°C)	13.0–22.0	8.5–19.0	10.8–15.2	19–21.5	27.5
Dissolved oxygen (mg $\times$ l^{-1})	2.6–5.5	1.2–3.4	1.6–4.5	—	—
pH (S.U.)	8.2–10.0	6.3–6.9	7.2–8.5	7.5–8.5	9.2
Conductance (μS $\times$ cm^{-1})	450–550	10–20	270–390	50–60	930
Maximum depth (cm)	50–104	100–166	0–65	0–30	0–20
Total nitrogen	1.21–1.44	0.94–1.34	1.31	—	—
Total phosphorus	0.01–0.05	0.01–0.05	0.01	—	—

[a]Data are ranges recorded during the sampling.

TABLE 30.3. Mean and Total Taxa Richness at Six Wetlands in Yellowstone and Grand Teton National Parks during 1995[a]

	Abbreviated wetland name				
Parameter	SP-sage	SP-spruce	S-fir	T-meadow (1 and 2)	T-aspen
Mean richness · sample^{-1}	22.7 (4.6)	18.8 (4.3)	19.2 (4.9)	10.8 (2.5)	2 (0.0)
Mean richness · month^{-1}	35.7 (0.6)	34.0 (2.2)	39.3 (5.0)	21 (—)	2 (—)
Total richness · wetland^{-1}	63	98	92	21	2

[a] Numbers in parentheses are standard deviations.

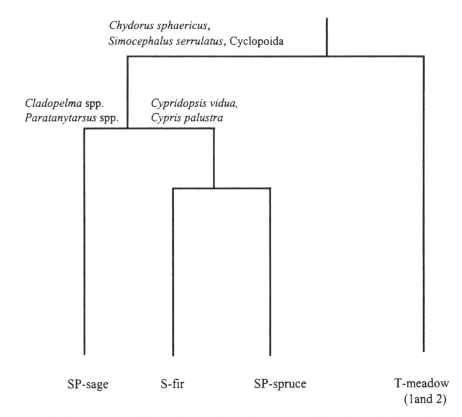

Fig. 30.1. Dendogram illustrating results of cluster analysis of aquatic invertebrate community composition and mean abundance in four types of wetlands in Yellowstone and Grand Teton National Parks.

SP-spruce, and S-fir, but absent from or uncommon in the T-meadow wetlands (Tables 30.4–7).

Invertebrate Abundance

Abundance of aquatic invertebrates was greatest in the SP-sage wetland in the Lamar River Valley and least in the T-meadow wetlands. Abundance in the SP-sage wetland ranged from about 68,000 organisms $\cdot$ m^{-2} in July to almost 336,000 organisms $\cdot$ m^{-2} in August (Table 30.4). Benthic organisms comprised more than 65 percent of the invertebrates collected from this wetland, with midge larvae being particularly abundant. In contrast, 90 percent

TABLE 30.4. Seasonal Abundance (no. m^{-2}) of Aquatic Invertebrates in Wetland SP-sage in the Lamar River Valley, Yellowstone National Park during 1995

Taxon	July	August	September
Annelida			
Cheatogaster diaphanus	2,806	1,751	907
Limnodrilus udekemianus	1,751	4,009	696
Other Annelida	3,017	528	1,688
Cladocera			
Ceriodaphnia reticulata	63	17,787	3,165
Daphnia pulex	7,069	1,372	317
Simocephalus serrulatus	1,118	13,399	20,108
Chydorus sphaericus	422	11,500	17,260
Other Cladocera	22,282	38,824	5,064
Copepoda			
Macrocyclops albidis	0	1,118	3,904
Eucyclops agilis	10,023	2,743	2,743
Aglaodiaptomus forbesi	0	1,329	63
A. leptopus	802	3,904	1,857
Other Copepoda	211	7,026	7,619
Ostracoda			
Cypridopsis vidue	2,638	1,372	63
Cypris palustra	3,545	5,549	1,751
Other Ostracoda	0	0	63
Insecta			
Cladopelma sp.	1,013	4,431	146,814
Paratanytarsus sp.	5,127	163,483	10,339
Procladius sp.	1,435	28,865	30,068
Psectrocladius sp.	0	10,656	1,646
Other Insecta	2,110	13,989	6,204
Mollusca			
Sphaerium nitidum	1,435	4,811	3,376
Other Mollusca	1,582	63	0
Total	67,794	335,975	267,337

TABLE 30.5. Seasonal Abundance (no. m^{-2}) of Aquatic Invertebrates in Wetland SP-spruce on Signal Mountain, Grand Teton National Park during 1995

Taxon	June	July	August	September
Annelida				
Cheatogaster diaphanus	274	3,650	0	0
Ophidonais serpentina	633	0	2,638	2,532
Other Annelida	169	63	654	506
Cladocera				
Ceriodaphnia reticulata	0	1,013	0	0
Simocephalus serrulatus	4,748	1,857	7,258	6,035
Chydorus sphaericus	696	18,948	10,086	11,056
Other Cladocera	2,827	8,335	6,056	4,368
Copepoda				
Acanthocyclops vernalis	211	0	106	127
Macrocyclops albidis	0	106	2,743	2,891
Aglaodiaptomus forbesi	169	63	928	886
Harpactaoida	2,743	9,136	971	612
Other Copepoda	7,069	4,959	5,043	4,368
Ostracoda				
Cypridopsis vidue	5,170	10,930	55,788	43,951
Cypris palustra	0	118,223	35,575	40,660
Other Ostracoda	2,595	1,076	8,524	9,263
Insecta				
Microspectra sp.	0	380	823	928
Procladius sp.	0	591	169	190
Psectrocladius sp.	0	422	5,402	6056
Chaoborus americanus	169	105	528	506
Other Insecta	485	1,097	3,165	2,638
Total	31,439	184,625	150,654	141,581

of the organisms collected from the SP-spruce wetland were water-column invertebrates (Cladocera, Copepoda, and Ostracoda), and midge larvae were uncommon (Table 30.5). Predators, primarily dragonfly and phantom midge larvae (Odonata and *Chaoborus americanus*), were regularly collected in low numbers from the SP-spruce wetland.

Abundance in the S-fir wetland reached more than 143,000 organisms · m^{-2} in August, before the wetland dried out. Water fleas were the most common organisms collected from this seasonal wetland, particularly *Simocephalus serrulatus* and *Chydorus sphaericus* (Table 30.6). Two species of aquatic worms (Oligochaeta) were abundant in June, while another predatory worm (*Chaetogaster diaphanus*) was abundant in August. Midge larvae were most abundant in this seasonal wetland during August.

Aquatic invertebrate abundance was lower in the temporary wetlands than in other wetlands sampled. Mean abundance in the T-meadow1 and T-meadow2 wetlands was about 18,000 organisms · m^{-2} (Table 30.7). Midge

TABLE 30.6. Seasonal Abundance (no. m^{-2}) of Aquatic Invertebrates in a Seasonal Wetland S-fir in Grand Teton National Park during 1995

Taxon	June	July	August
Annelida			
Cheatogaster diaphanus	0	63	11,352
Ophidonais serpentina	528	0	0
Paranais littoralis	7,912	0	0
Limnodrilus udekemianus	6,921	0	0
Other Annelida	5,824	950	63
Cladocera			
Ceriodaphnia reticulata	1,118	1,794	9,875
Simocephalus serrulatus	1,161	7,343	16,247
Chydorus sphaericus	106	211	54,818
Other Cladocera	5,612	1,794	10,445
Copepoda			
Acanthocyclops vernalis	0	169	1,224
Harpactaoida	7,913	63	0
Other Copepoda	22,113	7,448	14,728
Ostracoda			
Cypridopsis vidue	950	739	2,954
Cypris palustra	422	2,701	0
Other Ostracoda	29,561	5,781	63
Crustacea			
Lynceus brachyurus	633	63	0
Insecta			
Heterotrissocladius spp.	0	0	6,710
Microspectra spp.	1,899	0	4,853
Procladius spp.	0	274	844
Psectrocladius spp.	1,561	5,191	5,528
Other Insecta	2,743	7,385	6,773
Mollusca			
Sphaerium securis	380	591	317
Other Mollusca	105	421	211
Total	104,234	40,934	143,480

larvae (*Psectrocladius* spp.) and a calanoid copepod (*Hesperodiaptomus shoshone*) were the most abundant taxa in these temporary wetlands. Abundance was greater in the T-aspen wetland, but only two taxa occurred in this wetland (Table 30.7).

Aquatic Foodwebs

Foodweb complexity was greatest in the two semipermanent wetlands and most simple in the temporary alpine meadow wetlands (Fig. 30.2). Although the T-aspen wetland contained only two taxa, it was considered somewhat of

TABLE 30.7. Mean Abundance (no. m⁻²) of Aquatic Invertebrates in Subalpine Temporary Wetlands T-meadow in Shoshone National Forest and T-aspen in Grand Teton National Park during July 1996

Taxon	T-meadow1 and T-meadow2	T-aspen
Cladocera		
Daphnia pulex	106	47,475
D. schodleri	106	0
Other Cladocera	0	0
Copepoda		
Acanthocyclops vernalis	106	0
Hesperodiaptomus shoshone	3,376	0
Leptodiaptomus coloradensis	950	0
Other Copepoda	633	0
Ostracoda		
Candona spp.	1,372	0
Cypridopsis vidue	1,266	12,660
Other Ostracoda	0	0
Crustacea		
Brachinecta coloradensis	422	0
Insecta		
Aedes cataphylla	528	0
A. fitchii	317	0
Psectrocladius spp.	6,753	0
Other Insecta	106	0
Total	16,041	60,135

an anomaly. Top predators in the SP-sage wetland were tiger salamander (*Ambystoma tigrinum*) larvae. Tiger salamander larvae were not found in the SP-spruce wetland, where an intermediate predator guild included damselfly, dragonfly, and predaceous diving beetle larvae as well as true aquatic bugs (Hemiptera). Prey consumed by the odonate predators included midge larvae, other insects, seed shrimp, Cladocera, and copepods of varying sizes. Although I did not determine the diets of predaceous diving beetles and true aquatic bugs (Notonectidae, Gerridae), their diets are reported to be similar to those of most odonates (Holomuzki 1985, Merritt and Cummins 1996). A third predator guild was represented by predaceous genera of midge larvae and aquatic worms. Stomachs of these taxa often included small species of water fleas, immature copepods, or early-instars of midge larvae. Phantom midge larvae represented a fourth predator guild in the semipermanent wetlands, but occurred only in the SP-spruce wetland. Prey of phantom midges included nekton ranging in size from larger seed shrimp to small water fleas such as *Chydorus sphaericus*. Two taxa, the fingernail clam *Sphaerium nitidum* and the aquatic worm *Limnodrilus udekemianus*, coexisted with these aquatic predators. The foodweb of the seasonal wetland S-fir was similar to

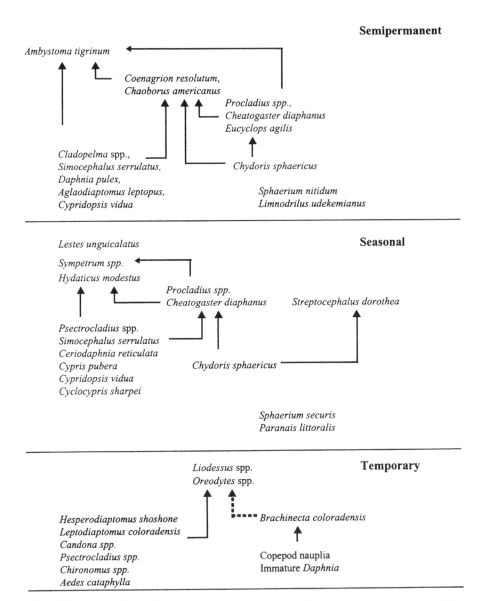

Fig. 30.2. Generalized foodweb for semipermanent (top panel), seasonal (middle panel), and temporary subalpine wetlands (bottom panel) in the GYE.

the foodweb of semipermanent wetlands, but lacked a vertebrate predator. Filter-feeding fairy shrimp (*Streptocephalus dorothea*) capable of ingesting smaller zooplankton also appeared in seasonal wetlands. Top predators in the foodweb of the temporary (T-meadow) wetlands were two genera of predaceous diving beetles (Dytiscidae) (Fig. 30.2). A fairy shrimp capable of ingesting smaller zooplankton was also found in the T-meadow wetlands, with the remainder of the community consisting of filter-feeding copepods, water fleas, mosquito larvae, and collector-gatherer midge larvae.

DISCUSSION

Taxonomic Composition and Diversity

The aquatic invertebrate communities of wetlands in Yellowstone and Grand Teton National Parks were found to be quite diverse, with a total of 187 taxa recorded from only one year of sampling. The diversity of aquatic invertebrates in these wetlands should not be surprising, since they provide habitats that are structurally complex, productive and of varying hydroperiod and have water that varies in concentrations of dissolved constituents. Life history strategies employed by 148 aquatic invertebrate genera to colonize wetlands in eastern North America have been categorized by Wiggins et al. (1980). However, the difficulty of sampling aquatic invertebrates in wetlands and extracting animals from organic detritus has hampered developments in this field. Consequently, the contribution of wetland aquatic invertebrate communities to ecosystem biodiversity is often overlooked, even by specialists. For example, Reice (1994) characterizes aquatic macroinvertebrate communities in "ponds" worldwide as being impoverished of species. While fewer taxa are typically found in wetlands than in clean running waters, wetlands are capable of supporting diverse aquatic invertebrate communities.

Where entire wetland aquatic invertebrate communities have been studied, they have typically been diverse. In Australia 137 taxa were recorded from 33 wetlands during one year (Growns et al. 1992), while sampling a similar number of wetlands (29) during early summer in Scotland produced 125 taxa (Jeffries 1994). In North America Olson et al. (1995) record 93 aquatic macroinvertebrate taxa from a northern prairie wetland. Studies of wetlands bordering a Great Lakes connecting river also documented a diverse aquatic invertebrate community having 171 taxa (Duffy et al. 1987). Forty-eight taxa were found in a southern bottomland forest that retained water only during the winter and early spring (Duffy and Labar 1992). Together, these data suggest that wetlands in regions around the globe may support aquatic invertebrate communities composed of from 100–200 taxa during seasons favorable for growth and development.

Taxonomic composition of wetland aquatic invertebrate communities in Yellowstone and Grand Teton National Parks differed among wetlands of varying hydroperiod. Diversity was generally greater in wetlands having longer hydroperiods. Predominant taxa in the SP-sage and SP-spruce, two semipermanent wetlands, also differed. The SP-sage wetland in the Lamar River Valley was characterized by high densities of two genera of midge larvae, *Cladopelma* and *Paratanytarsus*. In contrast, midge larvae were not abundant in the SP-spruce wetland, which was characterized taxonomically by two species of seed shrimp, *Cypris palustra* and *Cypridopsis vidua*. Three of these four taxa are recognized as associated with more permanent water (Wiggins et al. 1980, Bazzanti 1997). The presence of tiger salamander larvae in the SP-sage wetland may be responsible for observed differences in community composition between these two wetlands. The prey of tiger salamanders varies with ontogeny (Leff and Bachmann 1986), with small larvae consuming primarily small crustacean zooplankton, then incorporating larger invertebrates or prey such as small fish into their diet as they grow (Leff and Bachmann 1986, Holomuzki and Collins 1987). Sprules (1972) demonstrated that tiger salamanders are capable of restructuring the foodwebs in subalpine wetlands. Primary predators in the SP-spruce wetland were invertebrates (larvae of phantom midges, dragonflies, predaceous diving beetles) that are considered incapable of regulating prey populations (Johnson and Crowley 1980). Other factors possibly contributing to differences in invertebrate community composition between these semipermanent wetlands include surrounding vegetation type and water chemistry. The SP-spruce wetland has a small drainage basin within a Englemann spruce/lodgepole pine forest. These factors likely contribute to the low concentrations of dissolved salts and low pH in the water from this wetland, compared with other wetlands.

The seasonal wetland in the subalpine fir zone (S-fir) was characterized as taxonomically similar to the SP-spruce wetland, with high densities of seed shrimp. Similarities in the communities of these two wetlands may have also been influenced by surrounding upland vegetation and small drainage basin. This seasonal wetland was also located within 200 m of a semipermanent wetland, and migration of some taxa could be expected. In addition to the seed shrimp, predominant taxa in this seasonal wetland during spring included immature cyclopoid copepods, aquatic worms (*Paranais littoralis*), and a fairy shrimp (*Streptocephalus dorothea*). The aquatic worms were found almost exclusively mining the stems of dead emergent vegetation. Several genera of midges associated either with seasonal hydroperiods (*Psectrocladius*) or clean water (*Hetertrissocladius*) became common in late summer (Beck 1977, Bazzanti et al. 1997).

The invertebrate community found in the temporary alpine meadow wetlands (T-meadow1 and T-meadow2) was characterized by the lack of *Simocephalus serrulatus, Chydorus sphearicus,* and cyclopoid copepodids. All three of these taxa were considered overwintering residents by Wiggins et al. (1980). Lack of copepods has also been reported as characteristic of some aquatic invertebrate communities in temporary Australian wetlands (Growns et al. 1992). Predominant species in these temporary wetlands were large fairy shrimp (*Brachinecta coloradensis*) and the calanoid copepod *Hesperodiaptomus shoshone,* a species that is restricted to elevations of >3000 m (Pennak 1994). Larvae of two genera (*Laccodytes* and *Oreodytes*) of predaceous diving beetles also inhabited these wetlands, and individuals collected with a dip net were 80–100 mm long. The presence of these rather large predators was surprising since temporary wetlands often lack predators or have very few (Schneider and Frost 1996).

Only two species were found in the T-aspen temporary wetland. However, these wetlands were used as mineral licks by moose (*Alces alces*). Moose visited these wetlands daily, and their activities had denuded vegetation from the wetland and surrounding area, leaving the water extremely turbid.

Invertebrate Abundance

Abundance of aquatic invertebrates in wetlands from Yellowstone and Grand Teton National Parks was considered to be generally high. However, comparisons with published information from other regions are made difficult by the variety of approaches taken and sampling methods employed. For example, in prairie wetlands Duffy and Birkelo (1997) report aquatic macroinvertebrate densities from core samples ranging from $4700-20,000 \cdot m^{-2}$, while Neckles et al. (1990) report densities of total aquatic invertebrates from activity traps of up to $210,000 \cdot m^{-2}$. Densities reported from Great Lakes coastal wetlands also vary. Density of all aquatic macroinvertebrates in Gerking samples from wetlands along the St. Mary's River flowing from Lake Superior ranged from $7000-20,000 \cdot m^{-2}$ (Duffy et al. 1987). Using a core sampler, Botts (1997) found that density of midge larvae in ponds along the Lake Erie shore ranged from about 5000 to $> 200,000 \cdot m^{-2}$. The upper range in abundance from these and the present study suggests that density of aquatic invertebrates in wetlands may frequently approach or exceed $100,000-200,000 \cdot m^{-2}$. Sampling both water-column and benthic habitats and using fine mesh sieves, as was done in this study and by Botts (1997), presents difficulties. However, more quantitative methods should be encouraged, since they will reveal the true diversity and abundance of aquatic invertebrate communities in wetlands.

The abundance, distribution, and diversity of aquatic invertebrate communities reported here seems consistent with the growing body of literature for wetlands worldwide. However, the interpretations were based on data from the initial year of a long-term study, and limited replication precluded the application of more rigorous statistical procedures.

Invertebrate Diversity and Ecosystem Function

The role of wetlands and the aquatic invertebrates inhabiting them in the processes or functioning of the Greater Yellowstone Ecosystem remains largely unknown. Wetland aquatic invertebrate communities are generally recognized for their role in food chain support. They are the primary link between primary production/detrital resources and vertebrates. Birds such as waterfowl require a diet high in protein during specific life stages and may feed exclusively on aquatic invertebrates to meet these demands (Swanson and Duebbert 1989). Other wetland-associated birds, such as red- and yellow-

headed blackbirds, marsh wrens, and shorebirds, also consume aquatic invertebrates (Weller 1994). Fish may occur naturally in some wetlands, but are
usually absent from those having irregular hydrologic regimes. However, im-
prudent introductions of fish into wetlands can disrupt aquatic invertebrate
communities and the link between primary consumers and consumers (Duffy
1998). Other values that wetland aquatic invertebrates provide to the GYE
should be documented.

As one of the last relatively intact ecosystems in the continental United
States, the GYE may be a reservoir for biological diversity (Pace 1997).
Conserving biological diversity is one of the missions of the National Park
System (Clark and Minta 1994). However, research in Yellowstone and Grand
Teton National Parks has primarily addressed a few charismatic, economically
important, or endangered species and their habitats (Yellowstone National
Park 1997). The astonishing proportion (~70 percent) of taxa identified in
this study that were new records for the parks supports the contention by
Clark and Minta (1994) that species representing most of the biological diversity in the parks are either assumed to be doing well or ignored. Biologically diverse wetland aquatic invertebrate communities may contribute
substantially to ecosystem functions other than foodweb support. The role of
water fleas (Cladocera) and other filter-feeding aquatic invertebrates in maintaining water quality of lakes by removing algae is well documented (Sommer
et al. 1986). These functional groups should function similarly in wetlands.
Aquatic invertebrates also may play a more important role in wetland nutrient
cycling than has been recognized. Aquatic invertebrates can regenerate as
much as 35–60 percent of their own nutrient content each day (Lehman 1980),
helping fuel algal growth and influencing algal competition by altering nutrient ratios (Sterner 1990).

LITERATURE CITED

American Public Health Association. 1989. Standard Methods for the Examination of
   Water and Wastewater (17th ed.). American Public Health Association, Washington,
   DC.

Batzer, D. P., and S. A. Wissinger. 1996. Ecology of insect communities in nontidal
   wetlands. Annual Review of Entomology 41:75–100.

Bazzanti, M., M. Seminara, and S. Baldoni. 1997. Chironomids (Diptera: Chironomidae) from three temporary ponds of different wet phase duration in central Italy.
   Journal of Freshwater Ecology 12:89–99.

Beck, M. W., Jr. 1977. Environmental Requirements and Pollution Tolerance of Common Freshwater Chironomidae. EPA-600/4-77-024, U.S. Environmental Protection
   Agency, Cincinnati, OH.

Beetle, D. E. 1989. Checklist of recent Mollusca of Wyoming, USA. Great Basin
   Naturalist 49:637–645.

Botts, P. S. 1997. Spatial pattern, patch dynamics and successional change: chironomid
   assemblages in a Lake Erie coastal wetland. Freshwater Biology 37:277–286.

Brock, M. L. 1967. Relationship between standing crop and primary productivity along a hot spring thermal gradient. Ecology 48:566–571.

Brock, M. L., R. G. Wiegert and T. D. Brock. 1969. Feeding by *Paracoenia* and *Ephydra* (Diptera: Ephydridae) on the microorganisms of hot springs. Ecology 50: 192–200.

Chadde, S. W., P. L. Hansen, and R. D. Pfister. 1988. Wetland Plant Communities of the Northern Range, Yellowstone National Park. Final Report to National Park Service, Yellowstone National Park, WY.

Clark, T. W., and S. C. Minta. 1994. Greater Yellowstone's Future: Prospects for Ecosystem Science, Management and Policy. Homestead, Moose, WY.

Collins, N. C. 1975. Population biology of a brine fly (Diptera: Ephydridae) in the presence of abundant algal food. Ecology 56:1139–1148.

———. 1977. Mechanisms determining the relative abundance of brine flies (Diptera: Ephydridae) in Yellowstone thermal spring effluents. Canadian Entomologist 109: 415–422.

Collins, N. C., R. Mitchell, and R. G. Wiegert. 1976. Functional analysis of a thermal spring ecosystem, with an evaluation of the role of consumers. Ecology 57:1221–1232.

Cowardin, L. M., V. Carter, F. C. Golet, and E. T. LaRoe. 1976. Classification of Wetlands and Deepwater Habitats of the United States. FWS/OBS/-79/31, U.S. Fish and Wildlife Service, Washington, DC.

Craighead, J. J., J. S. Sumner, and J. A. Mitchell. 1995. The Grizzly Bears of Yellowstone: Their Ecology in the Yellowstone Ecosystem, 1959–1992. Island Press, Washington, DC.

Despain, D. G. 1973. Major Vegetation Zones of Yellowstone National Park. Information Paper No. 19, U.S. Department of the Interior, Yellowstone National Park, WY.

Duffy, W. G. 1998. Population dynamics, production and prey consumption of fathead minnows in prairie wetlands: A bioenergetics approach. Canadian Journal of Fisheries and Aquatic Sciences 54:1–13.

Duffy, W. G., T. R. Batterson, and C. D. McNabb. 1987. The St. Mary's River, Michigan: An Ecological Profile. Biological Report 85(7.10), U.S. Fish and Wildlife Service, Washington, DC.

Duffy, W. G., and C. S. Birkelo. 1997. Aquatic macroinvertebrate production in prairie wetlands. In prep.

Duffy, W. G., and D. J. LaBar. 1994. Aquatic invertebrate production in southeastern USA wetlands during winter and spring. Wetlands 14:88–97.

Farnes, P. 1997. The snows of Yellowstone. Yellowstone Science 5:8–11.

Fraleigh, P. C., and R. G. Wiegert. 1975. A model explaining successional change in standing crop of thermal blue-green algae. Ecology 56:656–664.

Frank, D. A., and S. J. McNaughton. 1993. Evidence for the promotion of above-ground grassland production by native large herbivores in Yellowstone National Park. Oecologia 96:157–161.

———. 1993. The ecology of plants, large mammal herbivores and drought in Yellowstone National Park. Ecology 73:2043–2058.

Growns, J. E., J. A. Davis, F. Cheal, L. G. Schmidt, R. S. Rosich, and S. J. Bradley. 1992. Multivariate pattern analysis of wetland invertebrate communities and environmental variables in western Australia. Australian Journal of Ecology 17:275–288.

Holomuzki, J. R., and J. P. Collins. 1987. Trophic dynamics of a top predator, *Ambystoma tigrinum nebulsoum* (Caudata: Ambystomatidae), in a lentic community. Copeia 1987:949–957.

Jeffries, M. 1994. Invertebrate communities and turnover in wetland ponds affected by drought. Freshwater Biology 32:603–612.

Johnson, D. M., and P. H. Crowley. 1980. Odonate "hide and seek": Habitat specific rules? Pages 569–579 *in* W. C. Kerfoot (ed.), Evolution and Ecology of Zooplankton Communities. University Press of New England, Hanover, NH.

Knight, D. H. 1994. Mountains and Plains: the Ecology of Wyoming Landscapes. Yale University Press, London, CT.

Koch, E. D. and C. R. Peterson. 1995. Amphibians and Reptiles of Yellowstone and Grand Teton National Parks. University of Utah Press, Salt Lake City, UT.

Leff, L. G., and M. D. Bachmann. 1986. Ontogenetic changes in predatory behavior of larval tiger salamanders (*Ambystoma tigrinum*). Canadian Journal of Zoology 64:1337–1344.

Lehman, J. T. 1980. Release and cycling of nutrients between planktonic algae and herbivores. Limnology and Oceanography 25:620–632.

Love, J. D., and J. C. Reed, Jr. 1995. Creation of the Teton Landscape: The Geologic Story of Grand Teton National Park. Grand Teton Natural History Association, Moose, WY.

Mattson, D. J. 1984. Classification and environmental relationships of wetland vegetation in central Yellowstone National Park, Wyoming. Master's Thesis, University of Idaho, Moscow, ID.

Merritt, R. W., and K. W. Cummins. 1996. An introduction to the Aquatic Insects of North America (3rd ed.). Kendall/Hunt, Dubuque, IA.

Neckles, H. A., H. R. Murkin, and J. A. Cooper. 1990. Influences of seasonal flooding on macroinvertebrate abundance in wetland habitats. Freshwater Biology 23:311–322.

Nielsen, L. T., and M. S. Blackmore. 1996. The mosquitoes of Yellowstone National Park (Diptera, Culicidae). Journal of the American Mosquito Control Association 12:695–700.

Olson, E. J., E. S. Engstrom, M. R. Doeingsfeld, and R. Bellig. 1995. Abundance and distribution of macroinvertebrates in relation to macrophyte communities in a prairie marsh, Swan Lake, Minnesota. Journal of Freshwater Ecology 10:325–335.

Pace, N. 1997. A molecular view of microbial diversity and the biosphere. Science 276:734–740.

Pennak, R. W. 1994. Fresh-water Invertebrates of the United States: Protozoa to Mollusca (3rd ed.). John Wiley & Sons, New York.

Reed, J. F. 1952. The vegetation of the Jackson Hole Wildlife Park, Wyoming. American Midland Naturalist 48:700–729.

Reice, S. R. 1994. Nonequilibrium determinants of biological community structure. American Scientist 82:424–435.

Schullery, P. (ed.). 1995. Thermophile conference focuses on science and policy. Yellowstone Science 3:19.

Schneider, D. W., and T. M. Frost. 1996. Habitat duration and community structure in temporary ponds. Journal of the North American Benthological Society 15:64–86.

Shea, R. E. 1981. Nesting ecology of the trumpeter Swan in Yellowstone National Park. Proceedings of the Trumpeter Swan Society Conference 6:33–34.

Smith, R. B., and R. L. Christiansen 1980. Yellowstone Park as a window on the earth's interior. Scientific American 242:104–117.

Sommer, U., Z. M. Gliwics, W. Lampert, and A. Duncan. 1986. The PEG-model of seasonal succession of planktonic events in fresh waters. Archiv für Hydrobiologie 106:433–471.

Sprules, W. G. 1972. Effects of size-selective predation and food competition on high altitude zooplankton communities. Ecology 53:375–386.

Squires, J. R., and S. H. Anderson. 1995. Trumpeter swan (*Cygnus buccinator*) food habits in the Greater Yellowstone Ecosystem. American Midland Naturalist 133: 274–282.

Sterner, R. W. 1990. The ratio of nitrogen to phosphorus resupplied by herbivores: zooplankton and the algal competitive arena. American Naturalist 136:209–229.

Swanson, G. A., and H. F. Duebbert. 1989. Wetland habitats of waterfowl in the prairie pothole region. Pages 228–267 *in* A. Van der Valk (ed.), Northern Prairie Wetlands. Iowa State University Press, Ames, IA.

Tuggle, B. N. 1986. The occurrence of *Theromyzon rude* (Annelida: Hirudinea) in association with mortality of trumpeter swan cygnets (*Cygnus buccinator*). Journal of Wildlife Disease 22:279–280.

Weller, M. W. 1994. Freshwater Marshes: Ecology and Wildlife Management (3rd ed.). University of Minnesota Press, Mineapolis, MN.

Wiggins, G. B., R. J. MacKay, and I. M. Smith. 1980. Evolutionary and ecological strategies of animals in annual temporary pools. Archiv für Hydrobiologie Supplement 58:97–206.

Wilkinson, L., M. Hill, J. P. Welna and G. K. Birkenbeuel. 1992. SYSTAT for Windows: Statistics, Version 5 Edition. Evanston, IL.

Yellowstone National Park. 1997. Yellowstone's Northern Range: Complexity and Change in a Wildland Ecosystem. National Park Service, Mammoth Hot Springs, WY.

APPENDIX 30.A. AQUATIC INVERTEBRATE TAXA COLLECTED FROM WETLANDS IN THE GREATER YELLOWSTONE ECOSYSTEM

Order
Family
Genus species

Turbellaria
Phagocata sp.

Rotifera
*Asplanchna brightwelli**
*Euchlanis dilatata**
*Trichotria pocillium**

Nematoda
Mesodorylaimus sp.*

Hydroida
*Hydra americana**

Rhynchobdellidae
Glossiphoniidae
*Glossiphonia complanata**
*Helobdella fusca**
*Helobdella stagnalis**
*Placobdella ornata**

Pharyngobdellida
Erpobdellidae
*Mooreobdella fervida**

Tubificida
Naididae
*Chaetogaster diaphanus**
*Chaetogaster limnaei**
Nais communis *
*Nais simplex**
*Ophidonias serpentina**
*Pristina breviseta**
*Pristina idrensis**
*Pristina longisoma**
*Pristina osborni**
*Paranias litoralis**
*Specaria josinae**
*Vejdovskella comata**
Tubificidae
*Limnodrilus udekemianus**
*Limnodrilus hoffmeisteri**

Cladocera
Daphnidae
*Ceriodaphnia reticulata**
Daphnia pulex
Daphnia rosea
Daphnia schodleri
*Scapholebaris mucronata**
*Simocephalus serrulatus**
*Simocephalus vetulus**
Chydoridae
*Acroperus harpae**
*Alona guttata**
*Alona quadrangula**
*Biaptura affinis**
*Chydorus gibbus**
*Chydorus sphaericus**
*Graptoleberis testudinaria**
*Leydigia leydigia**
*Pleuroxus aduncus**
*Pleuroxus procurvatus**
Holopedidae
*Holopedium gibberum**
Macrothricidae
*Eurycercus lamellatus**
*Ilyocryptus sordidus**
*Lathonura rectirostris**
*Macrothrix montana**
*Streblocerus serricaudatus**
Sididae
*Diaphanosoma brachyurum**
*Sida crystallina**

Copepoda
Cyclopoida
*Acanthocyclops vernalis**
*Eucyclops agilis**
*Eucyclops speratus**
*Macrocyclops albidus**
*Tropocyclops prasinus**
Harpactacoida
Canthocamptus sp.*
Calanoida
*Aglaodiaptomus leptopus**
*Aglaodiaptomus forbesi**
*Hesperodiaptomus shoshone**
*Leptodiaptomus coloradensis**

Anostraca
Brachinecta coloradensis*
Streptocephalus dorothea*

Conchostraca
Lynceus brachyurus*

Amphipoda
Hyalella azteca

Ostracoda
Candona stagnalis*
Candona decora*
Cyclocypris ampula*
Cyclocypris sharpei*
Cypridopsis aculeata*
Cypridopsis vidua*
Cyprinotus incongruens*
Cypris palustera*
Cypris pubera*
Eucypris rava*
Eucypris hystrix*
Limnocythere reticulata*

Hyradracarina
Arrenurus sp.*
Hydrachna sp*

Diptera
Ceratopogonidae
Alluaudomyia sp.*
Bezzia/Probezzia sp.
Palpomyia tibialis*
Stilobezzia sp.*
Chironomidae
Ablabesmyia sp.*
Procladius sp.*
Chironomus sp.*
Cladopelma sp.*
Clinotanypus sp.*
Corynoneura sp.*
Cryptochironomus sp.*
Dicrotendipes sp.*
Derotanypus sp.*
Heterotrissocladius sp.*
Labrundinea sp.*

Macropelopia sp.*
Microspectra sp.*
Nanocladius sp.*
Orthocladius annectens*
Paratanytarsus sp.*
Pentaneura sp.*
Procladius sp.*
Psectrocladius sp.*
Psectrotanypus cf. varius*
Pseudokiefferiella sp.*
Rheocricotopus sp.*
Rheotanytarsus sp.*
Thienemanniella sp.*
Tribelos sp.*
Trissopelopia sp.*
Chaoboridae
Chaoborus americanus*
Culicidae
Aedes cataphylla
Aedes fitchii
Anopheles earlei
Anopheles punctipennis
Culiseta alaskaensis
Culiseta inornata*
Dixidae
Dixella sp.
Ephidridae
Hydrellia sp.
Parydra sp.
Setacera sp.
Tipulidae
Prionocera cf. primoveris
Tipula sp.

Collembola
Podura aquatica*
Entomobryinae
Corynothrix sp.*
Smithuinus aureus*

Ephemeroptera
Caenidae
Caenis youngi
Baetidae
Callibaetis ferrugineus

Coleoptera
Carabidae
Chrysomelidae
Donacia subtilus *
Donacia tubercalifrons *
Dytiscidae
Cybister sp. *
Coptotomus cf. longulus *
Deronectes griseostriatus *
Eretes sp.
Graphoderus sp.
Hydaticus modestus *
Hygrotus patruelis *
Laccodytes sp.
Oreodytes sp. *
Rhantus zimmermanni *
Gyrinidae
Gyrinus affinis *
Gyrinus latilimbus *
Haliplidae
Haliplus immaculicollis *
Haliplus longulus *
Haliplus subguttatus *
Helophoridae
Helophorus lineatus *
Hydrophilidae
Tropisternus sp.
Staphylinidae
Stenus sp.

Hemiptera
Corixidae
Callicorixa audeni
Cenocorixa sp.
Hesperocorixa atopodonta *
Sigara alternata
Gerridae
Gerris buenoi
Gerris notabilis
Gerris remigis
Limnoporus notabilus
Notonectidae
Notonecta irrorata *
Notonecta undulata

Odonata
Coenagrion resolutum
Enallagma carunculatum
Enallagma civile
Enallagma vernale
Lestes congener
Lestes unguicalatus
Aeshna sp.
Sympetrum sp.

Trichoptera
Eocosmoecus spp. *
Grammotaulis sp. *
Limnephilus sp.
Philarctus quaeris *
Trianodes sp.

Mecoptera
Bittacidae
Bittacus sp. *

Gastropoda
Armiger crista *
Gyraulus circumstriatus
Stagnicola (Lymnaea) caperata
Stagnicola (Lymnaea) elodes
Lymnaea stagnalis jugularis
Helisoma cf. anceps anceps
Helisoma trivolvis subcrenatum
Planorbula campestris
Promenetus exacuous exacuous
Pseudosuccinea columella *
Valvata sincera

Bivalvia
Sphaeriidae
Pisidium milium
Pisidium subtruncatum
Sphaerium lacustre
Sphaerium nitidum
Sphaerium occidentale

---

* = taxon not previously recorded in either Grand Teton or Yellowstone National Parks (129 of 187 taxa not previously recorded).

31 Subalpine Wetlands in Colorado

Habitat Permanence, Salamander Predation, and Invertebrate Communities

SCOTT A. WISSINGER, ANDY J. BOHONAK,
HOWARD H. WHITEMAN, and WENDY S. BROWN

*W*etlands are abundant in wet subalpine and montane valleys on the western slopes of the Rocky Mountains. Although geological and vegetational characteristics are well studied for a diversity of wetland types and incorporated into a regional classification, there is a paucity of information about the invertebrates in these habitats and the factors that affect their distribution and abundance. In this chapter we report on work at a subalpine wetland complex in the Elk Mountains of central Colorado composed of numerous adjacent basins that vary considerably in size, depth, water chemistry, and hydroperiod. We identified over 100 planktonic and benthic invertebrates at the site and found that different basins have quite different species assemblages. We used comparative data on the abiotic environment (area, depth, hydroperiod, chemistry) in 40 basins and information about the density of tiger salamanders, the only vertebrate predator, to construct a path analytic model to explore hypotheses about the relative importance of these factors on invertebrate community composition. Basin area, hydroperiod, and salamander densities have significant and collinear effects on species diversity and community composition. Large, permanent basins with multiple year-classes of salamander larvae are most diverse and are dominated by small-bodied zooplankton, chironomid midges, small dytiscid beetles, and cased caddisflies. Semipermanent, autumnal basins are less diverse than permanent basins and dominated by species that are absent or rare in permanent basins (large-bodied cladocerans and copepods, fairy shrimp, large dytiscid bee-

Invertebrates in Freshwater Wetlands of North America: Ecology and Management, Edited by
Darold P. Batzer, Russell B. Rader, and Scott A. Wissinger
ISBN 0-471-29258-3   © 1999 John Wiley & Sons, Inc.

tles). In semipermanent basins beetles replace salamander larvae as the top predators on benthic invertebrates, although salamander adults and hatchlings remain the top predators on large-bodied zooplankton. Small, vernal basins lack salamanders and are dominated by a subset of autumnal-basin species (several zooplankton, mosquitoes, a beetle, a caddisfly, a corixid water boatman) that are able to complete development before they dry in early summer. Although a few species are restricted to permanent habitats because they cannot tolerate drying, most of the invertebrates in these wetland are physiologically capable of, and have life cycles amenable for, exploiting both permanent and nonpermanent basins. Thus, biotic interactions, especially the direct and indirect effects of salamander predation, are viable hypotheses for explaining patterns of invertebrate distribution and abundance along the permanence gradient from vernal to semipermanent to permanent wetland basins. Human activities that threaten salamander populations should have cascading effects on the invertebrate communities in permanent subalpine wetlands.

INTRODUCTION

Mountain valleys on the wet western slopes of the Rocky Mountains contain numerous wetlands that occur in a diversity of geomorphologic and hydrologic settings. For example, within just a few kilometers of our study site in central Colorado are, from high to low elevations, lacustrine marshes bordering alpine lakes (>3500 m), subalpine (2800–3500 m) ponds and fens in bedrock and soil depressions, subalpine riparian meadows, montane (2100–2800 m) beaver ponds and riparian willow thickets, and montane kettle ponds and fens in broad glacial valleys. The geological origins, hydrology, and vegetational characteristics of these and other types of montane and subalpine habitats have been used to construct a detailed classification system for the wetlands of the Rocky Mountain Region (see review by Windell et al. 1986). However, as is often the case in general treatments of wetlands ecology (see Batzer and Wissinger 1996), there is a paucity of information about invertebrate communities in all of the wetland types discussed in Windell et al. (1986). This is not an oversight—with the exception of several early studies by Pennak and his students (e.g., Neldner and Pennak 1955, Schmitz 1956; also see Blake 1945), there is little information about benthic invertebrates in Rocky Mountain wetlands. Similarly, with the exception of the zooplankton studies at our site (Dodson 1970, 1974, Sprules 1972, Maly 1973, Maly and Maly 1974, Maly et al. 1980, Willey and Threlkeld 1993), most research on high-elevation zooplankton assemblages has focused on deep alpine lakes (e.g., Larson et al. 1996 and references therein).

In this chapter we focus on the invertebrate communities in a complex of subalpine ponds located near the Rocky Mountain Biological Lab (RMBL) in Gunnison Co., Colorado. We first describe the habitat characteristics (geomorphology, hydrology, chemistry, plant structure, salamander predators) of the many small basins at our study site. We then discuss the invertebrate fauna and patterns of distribution and abundance among basins. We use path analysis to help disentangle the collinear relationship among the various habitat variables and to compare hypotheses about how those variables directly or indirectly affect invertebrate community composition. Finally, we discuss the results of previous experimental work conducted at this study site (Dodson 1970, 1974, Sprules 1972, Wissinger et al. 1996, Wissinger et al., 1999) that examined how different habitat factors affect the distributional patterns for several of the dominant taxa in these communities.

HABITAT DESCRIPTION

Our study site is located in the Mexican Cut Nature Preserve (MCNP), a 960-acre inholding within the White River National Forest that is owned by The Nature Conservancy and managed by the RMBL for ecological research. The wetlands occur at or just below tree line (3400–3500 m elevation) on two flat shelves on the northeast slopes of Galena Mountain in Gunnison Co., Colorado (Fig. 31.1). Although many of the basins at Mexican Cut are only a few meters or tens of meters apart, they often contain quite distinct assemblages of invertebrates. Because all of the basins occur at about the same elevation and have the same macroclimate and the same pool of potential colonists, the site provides an excellent ecological setting for isolating the effects of basin characteristics (area, size, hydroperiod, water chemistry) on community composition. These wetlands typify many of the other subalpine habitats that we have been studying in this region in terms of their abiotic setting, the surrounding terrestrial habitat, and the dominant aquatic flora and fauna.

Bedrock and Water Chemistry

There are over 50 small (10 m^2–4682 m^2 surface area), shallow (0.3–2.5 m depth) basins in the wetland complex at MCNP (Fig. 31.1, Table 31.1). The site is underlain by upper Paleozoic rocks, and basins in the main cluster on the lower shelf lie on contact-metamorphosed sandstone (quartzite) that provides no source of alkalinity (99 percent silica) (Prather 1982). Thus, the water is extremely soft and poorly buffered in most basins (Table 31.2; Dodson 1982, Harte et al. 1985), as is generally true in the region (e.g., Neldner and Pennak 1955, Schmitz 1956, Landers et al. 1987). All basins are extremely oligotrophic, with total nitrogen and phosphorus concentrations typically below detection limits (<1 μg/l; conducted at USDA Forest Service

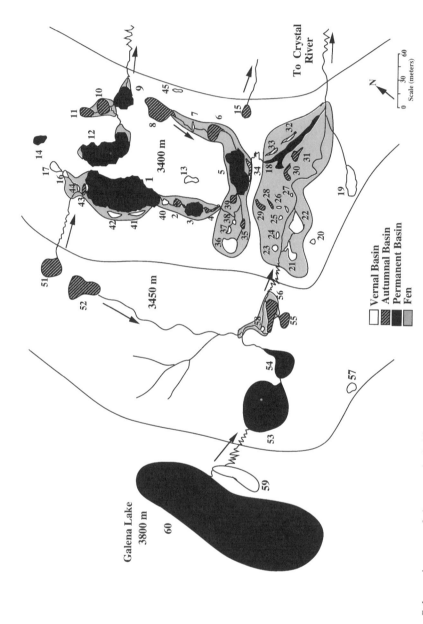

Fig. 31.1. Schematic map of the aquatic habitats at the Mexican Cut Nature Preserve. The three vertical contour lines separate from right to left, the lower cut at 3400 m, the upper cut at 3450 m, and Galena Lake at 3800 m elevation.

TABLE 31.1. Basin Characteristics (Area, Depth, Hydroperiod) and Invertebrate Community Composition for 41 Wetland Basins at the Mexican Cut Nature Preserve[a]

	Vernal																Autumnal																					Permanent				
Basin	20	22	42	44	23	41	13	16	36	19	25	21	24	7	26	37	15	38	29	27	39	43	35	30	31	28	4	51	52	55	56	6	8	10	11	2	3	9	5	12	1
Area (m²)	16	90	25	20	18	25	89	30	400	230	15	160	44	37	12	18	64	10	64	20	28	60	20	20	12	10	30	600	750	400	750	206	928	283	213	15	175	585	1267	1688	4632
Depth (m)	0.3	0.3	0.3	0.5	0.3	0.3	0.6	0.6	0.5	0.6	0.5	0.5	0.3	0.3	0.5	0.8	0.6	0.4	0.6	0.5	0.5	0.8	0.5	0.5	0.7	0.7	0.8	0.7	0.7	1.0	0.7	1.2	0.7	1	1	15	1.5	1.5	1.5	2.0	2.5
Wet (open water days)	40	41	42	51	56	58	58	59	59	59	59	61	63	64	64	64	74	74	75	76	81	84	85	88	88	89	99	100	102	104	105	110	112	114	115	116	130	140	138	135	135
Keratella	C	C	C	C	C	C	C	C	C	C	C	C	C	C	C	C	C	C	C	C	C	C	C	C	C	C	C	C	C	C	C	C	C	C	C	C	C	C	C	C	C
Stictotarsus	C	C	C	C	C	C	C	C	C	C	C	C	C	C	C	C	C	C	C	C	C	C	C	C	C	C	C	C	C	C	C	C	C	C	C	C	C	C	C	C	C
Tanytarsus	C	C	C	C	R	C	C	C	C	C	C	C	C	C	C	C	C	C	C	C	C	C	C	C	C	C	C	C	C	C	C	C	C	C	C	C	C	C	C	R	C
Asynarchus	R	R	R	C	R	C	R	C	C	C	C	C	C	C	C	C	C	C	C	R	C	C	C	C	C	C	C	C	C	R	C	C	C	C	C	C	C	C	C	C	C
L.coloradensis	C	C	C	C	C	C	R	C	C	C	C	C	C	C	C	C	C	C	C	C	C	C	C	C	C	C	C	C	C	C	C	C	C	C	C	C	C	C	C	C	C
Aedes spp.		C	C	C	C	C	C	C	C	C	C	C	C	C	C	R	C	C	C	R	C	C	C	C	C																
Arctocorixa			C	C		C	C	C	C	C	C	C	C	C			C	C	C	C	C	C	C															R	R	R	R
C. riparius				C	C	C	C	C	C	C	C	C	C	C			C	C	R	C	C	C																R	C	C	C
Acanthocyclops							C	R	C	C	C	C	C				C	C	C	C	C	C	C	C	C	C															
D. middendorfiana									R	C	C	R	C	R			C	C	R	C	C	R	C																		
Branchinecta										C	C	C	C	R			R	C	C	R	C	R	R			C	C														
Paratanytarsus										C	C	C	C	R			C	C	C	C	C																		C		
Psectrocladius										C	C	C	C	C	C		R	C	C	C	C						C														C
Conochilus										C	C	C	C	C			R	C	C	C																	C	C	C	C	C
Euchlanis											R	R	C	C			C	R	R	C	C	C	C																		
Limnophila											R	R	C	C	C		R	R	C	R	R	R	R		R																
Lumbricolus											R	R	C	C	C	C	C	C	C	C	C	C	C	R	C		R	R	R	R	C	C	C	C	R	C	C	C	C	C	C
Sanfilippodytes																	R	R	C	R	R	R		R																	
Hydropous																	R	R	C	C	R							R	R	R	R	R	R	R	R	R		R	R	R	R
Chydorus																	C	C	C	C	C						C	C													
Limnephilus picturatus																	R	R	R	R	C			C				R							R	C		R	R	R	R
Simocephalus																	R	R	C	R	R	C	R	R						R				R	C	R		R	R	R	R
Chironomus salinarus																							C																		
Agabus kootenai																	C		C	C		C																			
Limnodrilus																			R	R		R	R														C			C	R
Callicorixa																				R		R															C			C	C
Gerris																			R	R	R	R								R							C	C	C	C	C
Pisidium																				C		R							C								C	C	C	C	C

761

TABLE 31.1. (Continued)

Basin	20	22	42	44	23	41	13	16	36	19	25	21	24	7	26	37	15	38	29	39	27	43	35	30	31	28	4	51	52	55	56	6	8	10	11	2	3	9	5	12	1
Area (m²)	16	90	25	20	18	25	89	30	400	230	15	160	44	37	12	18	64	10	64	20	28	60	20	22	10	12	30	600	750	400	750	206	928	283	213	15	175	585	1267	1688	4632
Depth (m)	0.3	0.3	0.3	0.5	0.3	0.3	0.6	0.5	0.5	0.6	0.5	0.5	0.3	0.3	0.5	0.8	0.6	0.4	0.6	0.5	0.5	0.8	0.5	0.5	0.7	0.7	0.8	0.7	0.7	1.0	0.7	1.2	0.7	1	1	1	1.5	1.5	1.5	2.0	2.5
Wet (open water days)	40	41	42	51	56	58	58	59	59	59	59	61	63	64	64	64	74	74	75	76	81	84	85	88	88	89	99	100	102	104	105	110	112	114	115	116	130	140	138	135	135
	⟵ Vernal ⟶																⟵ Autumnal ⟶																				⟵ Permanent ⟶				
Hesperophylax																																							C		C
Agabus tristus																								R	R		R				R						R	R	C	R	R
Procladius																												C	C	C	R	C	C	C	C	C	C	C	C	R	R
Mesostoma																												C	C	C	C	C	C	C	C		C	C	R	R	R
Helobdella																												C	C	C	C	C	C	C	C	C	C	C	C	C	C
Chydorus																											C	C	C	C	C	C	R	C	R	C	C	C	C	C	C
Cricotopus																								R			R	R	R	R	R	R	C	C	C	R	R	C	R	C	C
Limnephilus externus																															R	R	R	C	R	R	C	C	C	C	C	C
Agabus strigulosus																												R	R	R	C	R	R	C	C	C	C	C	R	R	R	R
Acilius																														C			C	C	R	C	C					
Dytiscus																													R	R	R	R	R	R	R	R		C	C	C	C	C
Somatochlora																												R	R	R	R	R	R	R	R	R	R	R	R	C	R	R
Limnesia																																		R	R				C	R	R		R
Piona																																					R	R	C	C	C	C	C
Bezzia																																					R	C	C	C	R	C	C
Nephelopsus																																						C	C	R	R	C
Agrypnia																																						R	R	C	C	C
Daphnia rosea																																						C	C	C	C	C
Scapholoberis																																						C	C	C	C	C
Lestes																																						C	C	C	C	R
Coenagrion																																						C	C	C	C	C
Plateumaris																																						C	C	C	C	C
Ilybius																																						R	R	R	R	R
Rhantus																																						R	R	R	R	R
Haliplus																																						R	R	C	R	R
Ablabesmyia																																						C	C	C	C	C
Culicoides																																						C	C	C	C	C
Cladopelma																																						C	C	C	C	C
Enallagma																																						C	C	C	C	C
Cladotanytarsus																																						C	C	C	C	C
Dicrotendipes																																						C	C	C		C
Endochironomus																																						C	C	C		C

762

Microtendipes																																					C	C	C	C	C
Pagastiella																																					C	C	C	C	C
Coryneura																																					C	C	C	C	C
Eukiefferella																																					C	C	C	C	C
Arrenurus																																					C	C	C	C	C
Lebertia																																					C	R	C	R	C
Hydrozetes																																					C	R	C	C	C
Glossiphonia																																					R	R	R	R	R
Daphnia pulex																																						R		C	R
Aeshna																																						R	R	R	R
Coenocorixa																																						R		R	R
Haliplus longullus																																						R	R	R	R
Hydrobious																																						R	R	R	R
Gyrinus																																								R	R
Callibaetis																																								R	C
Stenus																																								C	C
Pseudodiamesa																																									R
Notonecta																																									R
Stratiomyia																																									R
Helophorus																																									R
Basin #	20	22	42	44	23	41	13	16	36	19	25	21	24	7	26	37	15	38	29	39	27	43	35	30	31	28	4	51	52	55	56	6	8	10	11	2	3	9	5	12	1

[a] Area is maximum coverage (m^2) and depth is maximum depth (m) when the basins are filled. Wet represents the average number of ice-free days with water above the substrate. Vernal basins have dried every year between 1989 and 1996. Autumnal basins dry in most years but have remained wet in one to three of the years during this time period. Permanent basins have not dried in at least the last 50 years. C = commonly encountered and R = rare species. See Appendix 31.A for taxonomic affiliations.

TABLE 31.2. Water Chemistry Data for Select Basins at the Mexican Cut Nature Preserve for July 1993[a]

Basin	PH	Conduct. (μS/cm)	Al [μg/L]	Ca	Mg	Na	K	NH4	Cl	NO3	SO4	PO4	Sum Bases	Sum Acids	Difference = alkalinity	ANC
								[μeq/L]								
1	7.27	35.19	23.47	303.59	90.52	9.48	2.53	0.00	1.41	0.00	53.93	0.00	406.12	55.34	350.79	286.70
2	6.96	29.89	19.41	254.34	77.93	5.87	1.13	0.00	1.38	0.00	45.22	0.00	339.27	46.60	292.66	248.50
3	7.17	35.71	22.57	303.59	91.26	9.79	2.53	0.00	1.69	0.00	52.78	0.00	407.17	54.47	352.70	294.30
5	7.06	38.52	22.65	336.68	89.04	8.61	2.53	0.00	1.38	0.00	59.03	0.00	436.86	60.41	376.45	333.70
6	6.01	8.04	20.34	53.59	14.73	9.92	3.43	0.00	4.23	0.52	19.51	0.00	81.67	24.26	57.41	44.20
7	6.26	10.33	23.51	98.70	26.99	12.14	6.68	2.44	4.26	0.00	13.03	0.00	144.50	17.29	127.21	87.90
8	6.52	7.29	20.24	68.21	19.83	6.83	5.01	1.28	2.71	0.00	12.14	0.00	99.89	14.85	85.04	86.30
9	6.06	7.05	22.72	46.36	12.84	9.92	2.28	0.00	2.71	1.66	13.70	0.00	71.39	18.07	53.32	44.60
10	6.46	8.56	28.19	77.40	25.59	6.83	6.68	0.00	3.86	0.00	9.97	0.00	116.49	13.84	102.65	101.70
11	6.37	6.90	26.97	59.28	19.83	5.05	6.27	0.55	2.43	0.00	9.08	0.00	90.42	11.50	78.92	77.60
12	6.77	4.93	11.80	42.17	14.07	3.87	2.53	0.00	2.91	0.00	7.91	0.00	62.64	10.82	51.82	105.60
13	5.66	6.23	48.59	41.17	12.10	3.26	2.97	0.00	1.41	0.00	2.04	0.00	59.49	3.45	56.04	32.70
15	5.94	8.04	17.72	48.40	12.84	10.18	1.53	0.00	2.62	0.65	20.68	0.00	72.95	23.94	49.01	45.30
16	7.15	37.68	21.41	271.26	80.07	9.18	6.19	0.00	5.22	0.00	21.22	0.00	366.69	26.43	340.26	377.20
18	7.03	29.34	24.58	221.61	54.31	9.35	2.28	1.50	2.23	1.40	41.16	0.00	287.54	44.79	242.75	226.90
20	6.05	9.06	29.66	65.32	14.07	8.31	3.79	0.00	3.19	3.45	15.82	0.00	91.48	22.46	69.02	52.90
21	6.97	34.48	17.65	309.03	55.87	6.74	3.53	0.00	0.82	0.00	119.57	0.00	375.18	120.39	254.78	205.90
22	6.17	16.28	32.35	144.41	24.69	9.35	3.79	1.50	2.91	2.55	33.77	0.00	182.23	39.22	143.01	123.30
23	6.74	34.04	17.85	230.49	44.76	4.13	1.36	0.00	1.13	0.23	98.65	0.00	280.74	100.00	180.74	230.90
24	6.55	27.64	30.05	249.65	51.10	6.74	2.17	0.00	0.93	0.00	24.21	0.00	309.67	25.15	284.52	242.00
26	6.93	39.07	27.91	357.93	94.22	6.09	3.89	0.00	1.47	0.00	47.41	0.00	462.13	48.88	413.25	369.20
51	7.29	42.89	25.72	332.98	66.24	6.35	4.58	0.00	2.12	0.00	57.76	0.00	410.15	59.87	350.28	415.10
52	6.74	9.27	12.12	55.49	14.73	0.74	2.15	0.00	1.18	0.00	14.30	0.00	73.11	15.49	57.62	79.80
55	7.49	49.03	34.58	412.97	42.71	4.13	1.36	0.00	1.41	0.00	31.86	0.00	461.17	33.27	427.90	545.00
56	7.67	60.90	28.31	508.03	83.69	3.83	2.56	0.00	1.41	0.00	36.85	0.00	598.11	38.26	559.84	689.80

[a] All analyses were conducted at the USDA Forest Service Rocky Mountain Forest and Range Experiment Station, Fort Collins, Colorado. See Wissinger and Whiteman 1992 for additional data for other months and other years.

Rocky Mountain Forest and Range Experiment Station, Fort Collins, Colorado; also see Wissinger and Whiteman 1992).

Two linear sequences of ponds (51–1–5–18; 52–56–21; see Fig. 31.1) receive runoff that flows over a local outcrop of metamorphosed limestone, and these basins therefore have up to 100-fold greater total dissolved solids and acid neutralizing capacity (ANC) than those that do not receive this runoff (Table 31.2). This local variation in ANC provided the opportunity to isolate the impact of water chemistry on community composition and monitor the potential effects of acidification (Dodson 1982, Harte et al. 1985, Wissinger and Whiteman 1992).

Geomorphology and Hydrology

Many of the basins are embedded in a series of saturated peat fens that have formed in bedrock depressions on the lower shelf of the MCNP. Remote basins are in bedrock (e.g., basins 13, 14, 57) or soil depressions (e.g., basins 55, 15, 19, 20; see Fig. 31.1) formed from slumping and/or snowpack compression (see Windell et al. 1986). Two basins (53, 54) are avalanche impactponds with physical and biological characteristics more similar to that of Galena Lake than to the wetlands on the lower shelves (Fig. 31.1).

Climate monitoring stations near our site document that over the past 10 years (1986–96) the area has received an annual average of 130 cm of rain-equivalent precipitation, about 100 cm of which falls as snow (USDA National Resource Conservation Service Snotel Station at Elko Park and NOAA reporting station at RMBL). Total snowfall accumulations range from 1000–1500 cm, with an average maximum snowpack of about 3 m depth. These data probably slightly underestimate the snowfall and snowpack at MCNP, which is 300 m and 600 m higher than Elko Park and RMBL, respectively (A. Meers, B. Barr, personal communication).

Snowmelt is the primary source of hydrological input to all basins and enters mainly as overland flow under the melting snowpack. During peak runoff basins in the main drainages are connected by stream flow and peripheral basins are connected to the largest ponds by overland flow (e.g., 42–1; Fig. 31.1). After snowmelt, the basins within the three main drainages (1–5; 8–5; 12–9) are separated at the surface by saturated fen but retain subsurface connections via the bedrock-confined groundwater in each drainage.

Although there is a gradation in hydroperiods, the basins can be grouped into three hydroperiod categories (Table 31.1). Shallow, hydrologically isolated basins are vernal (sensu Wiggins et al. 1980); they dry by midsummer and usually do not refill until the following spring (Fig. 31.1). Semipermanent, autumnal habitats (after Wiggins et al. 1980) often dry for a short time in late summer but then refill in autumn before ice cover. In two of the past nine years many autumnal basins (51, 52, 55, 6, 4, 6, 8, 10, 11, 15) remained filled throughout the growing season as a result of unusually heavy summer rains (also see Wissinger and Whiteman 1992). Finally, the largest and deepest basins are permanent ponds that have not dried completely in the past 50 years (1, 3, 5, 9, 12, 18; R. L. Willey, personal communication). All basins are covered with ice and snow from late October to early summer. Thus, even permanent ponds are filled and free of ice for only about four months, and most temporary basins have a considerably shorter openwater growing season (Table 31.1). During winter the deepest (1–2 m) ponds (1, 5, 9, 12) probably do not freeze solid, although early spring cores suggest that a slushy mixture of water and "frazil ice" (see Oswood et al. 1991) can extend to within 5 cm of the substrate (Whiteman and Wissinger, unpublished data). Freezing in autumnal basins may vary among years. Salamander larvae survive the winter in autumnal ponds when they do not dry (Wissinger and Whiteman 1992), suggesting that either they can survive freezing or there is a thin layer of water similar to that observed in the permanent ponds. However, in most years the nonpermanent basins freeze from the top down to and into the substrate, and therefore are probably most accurately characterized as "aestival" (i.e., liquid water is not available during winter (see Welch 1952, Danks 1971a, Daborn 1974, Daborn and Clifford 1974, Lee 1989, Ward 1992). Freezing appears to be an important source of mortality for aquatic invertebrates in aestival habitats at high latitudes and altitudes (Andrews and Rigler 1985, Daborn 1971, Danks 1971b, Duffy and Liston 1985), and physiological studies suggest that there is considerable variation among species in their tolerance to subfreezing temperatures and physical contact with external ice (Oswood et al. 1991, Frisbee and Lee 1997).

During summer, water temperatures fluctuate considerably on a daily basis, especially in small basins where they can drop at night to 5–10°C and reach daytime highs of 20–30°C (also see Willey 1974). The basins receive relatively high levels of UV radiation as a result of the thin atmosphere, low humidity, and shallow, extremely clear water, and several of the most common zooplankton species in the temporary habitats contain photoprotective carotenoid pigments (see Hairston 1976, 1979).

The various physical habitat characteristics described above can be summarized and compared to other wetlands by evaluating them in the context of Southwood's (1977, 1988) habitat templet, which considers temporal stability (degree of permanence), variability (predictability in drying), and harshness (also see Williams 1987). As in many wetland complexes, the different basins at our study site span the range of temporal stability from vernal to autumnal to permanent. Permanent and vernal basins are the most predictable habitats in that they never or always dry on a seasonal basis, respectively. As in all wetlands, the dry phase of the hydroperiod is a relatively harsh environment for aquatic organisms. However, in high-elevation (and high-latitude [tundra]) wetlands the wet phase of the hydroperiod can also be relatively harsh because of high levels of UV radiation and fluctuating temperatures during the growing season and the frozen, aestival conditions during winter. Because of the year-round harsh conditions, several authors have suggested that the invertebrate fauna of high-elevation wetlands should be impoverished relative to lower-elevation habitats (Neldner and Pennak 1955, Schmitz 1956, Daborn and Clifford 1974).

Primary Production, Vegetation, and Detrital Inputs

The quillwort, *Isoetes bolanderi,* is the most common submergent plant, and by midsummer this species forms a short (3–10 cm) lawn of vegetation on the soft substrates of many basins. Several ponds (e.g., 10, 11, 56) also contain local patches of starwort (*Calitriche palustris*) and pondweed (*Potamogeton gramineus).* Emergent vegetation is patchily abundant along the edges of the large ponds and throughout the surrounding fens (Fig. 31.1). The dominant shoreline emergents are sedges, rushes, and grasses (*Carex aquatilis, Carex nebraskensis, Juncus confusus,* and *Deschampsia caespitosa),* which are often rooted in a saturated carpet of *Sphagnum* and other mosses (*Polytrichium* sp.) that form floating mats on some pond margins. Thickets of willows (*Salix planifolia*) and other woody vegetation (mainly *Vaccinium caespitosum*) are scattered around the major basins and throughout the peat fens (Fig. 31.1). Emergent species grade landward at the edges of these fens into a diverse transitional plant community dominated by *Carex nova, Caltha leptosepala, Trollius laxus, Pedicularis groenlandica, Sedum rhodanthum, Erigeron peregrinus, Gentian thermalis, Bistorta bistortoides, Swertia perennis, Phleum commutatum, Erythronium grandiflora, Veronica wormskjodlii, Valeriana capitata,* and *Castilleja rhexifolia* (also see Buck 1960). This particular plant assemblage is typical for subalpine fens in this region and has been termed the "*Deschampsia caespitosa/Caltha leptosepala* fen community" (Windell et al. 1996). The surrounding terrestrial habitat on the lower shelf is an open spruce-fir (*Picea engelmanni-Abies lasiocarpa*) woodland that opens into subalpine meadow on the upper shelf. The surrounding terrestrial habitat at MCNP is typical for many of the subalpine wetlands that we have studied at similar elevations throughout the region (see Langenheim 1962).

Benthic substrates are the main site of algal primary production, and a community (30 species identified) dominated by diatoms and desmids (mainly *Frustulia rhomboides, Cymbella lunata, Pinnularia biceps, Cosmarium* spp., and *Staurastrum* spp) is attached to both hard and soft substrates. Qualitative surveys do not suggest major differences in the abundance and taxonomic composition of the periphyton community across basins (Blake, Brown, and Wissinger, unpublished data). Water-column primary production by phytoplankton is extremely low (<50 mg C/m^3 based on chlorophyll-a analysis), and the water is therefore extremely clear in all but a few basins with organic staining from peat (e.g., 27–29).

The bedrock and inorganic substrate in most basins is covered with organic sediments that vary from a thin veneer in small isolated basins (e.g., 19, 20, 57) to a thick organic ooze in the centers of the largest ponds (e.g., 1, 5, 12). Peat substrates are common in basins embedded in the fens. Autochthonous inputs are mainly from *Isoetes,* and along pond margins, from patches of sedges and other emergent vegetation. The latter are of primary importance as a food source for detritivorous aquatic insects that congregate on emergent detritus (Wissinger et al. 1996). Terrestrial organic matter (needles, twigs, branches, leaves) dominates the detritus in basins that are lined with spruce trees (e.g., 8, 14, 15, 19) or willows (e.g., 29, 35, 41, 43).

Salamander Predators

The only aquatic vertebrate in the wetlands at MCNP is the tiger salamander, *Ambystoma tigrinum nebulosum*. Metamorphic (terrestrial) adults of this salamander migrate from terrestrial overwintering sites in late spring and congregate in the largest permanent and semipermanent ponds (1, 3, 5, 6, 8, 9–12), where they mate and deposit eggs. After breeding, metamorphs enter autumnal and vernal basins, where they feed for six to eight weeks before returning to the surrounding forest to overwinter (Whiteman et al. 1994). Eggs typically hatch in mid-July, but, unlike larvae in lower elevation populations, the larvae cannot complete development before the onset of winter and are therefore only likely to survive at this elevation in permanent waters (Wissinger and Whiteman 1992, Whiteman et al. 1994). Metamorphosis occurs after two to five growing seasons, but many individuals forgo it completely and become sexually mature as larvae—i.e., become paedomorphic adults (Whiteman 1994, Whiteman et al. 1996). Paedomorphs and several year-classes of larvae are the top predators in permanent basins at Mexican Cut and other subalpine habitats in the region. In contrast, hatchling larvae and metamorphs are the only salamander stages found in autumnal basins (Table 31.3).

Dietary analyses from stomach pumping indicate that nearly every aquatic invertebrate in these wetlands is consumed by one or more stages of these salamanders (see methods and data in Whiteman et al. 1994; Wissinger and Whiteman, unpublished manuscript). Metamorphs are specialists on fairy shrimp (*Branchinecta coloradensis*), which comprise up to 99 percent of their diet. Larvae exhibit an ontogenetic dietary niche shift in size and in composition, from plankton to benthos. Hatchlings feed mainly on copepods and cladocerans, one-to-two-year-old larvae on larger zooplankton (including fairy shrimp) and small benthic invertebrates (mainly dipteran larvae), and over two-year-old larvae and paedomorphs (together referred to hereafter as "branchiates") on a wide size range of prey, including the largest invertebrate species and smaller salamander larvae. The importance of salamanders as keystone predators on zooplankton has been experimentally demonstrated and undoubtedly underlies patterns of zooplankton distribution and abundance in these wetlands (Dodson 1970, 1974, Sprules 1972). The role of salamander predation in determining benthic community composition is less well understood and has been the focus of much of our experimental work on species interactions.

TABLE 31.3. Distribution and Abundance of Tiger Salamanders at the Mexican Cut Nature Reserve in 1990, Based on Multiple Mark-Recapture Censuses[a]

Pond	Hatchling Larvae	Branchiates	Metamorphs
1	350	1143	2
3	10	61	0
5	0	733	1
6	40	0	4
7	0	0	0
8	450	0	29
9	80	81	4
10	38	0	6
11	40	0	18
12	120	107	8
13	0	0	2
15	0	0	0
18	0	2	0
52	0	0	2

[a]Branchiates are >1-year-old larvae + paedomorphs (see text).

INVERTEBRATE FAUNA

Total Abundance and Biomass

Benthic invertebrate data presented here are based on (1) qualitative surveys from D-net sweeps taken over the past eight years in all basins and (2) replicate quantitative samples using a 0.25-m² drop box (see Wissinger 1989) taken on a biweekly basis during the open-water months in 1989–91 in five permanent (1, 3, 5, 9, 12) and six autumnal (6, 8, 10, 11, 55, 56) basins. We avoided taking replicate quantitative samples in the small vernal basins because of the potential for decimating populations in those habitats. Zooplankton samples were taken with an 80 μm mesh net and a 2.2-L Van Dorn sampling bottle mounted on a pole (see Bohonak and Whiteman, manuscript). Biomass estimates from the quantitative data were calculated as the product of number of individuals and taxon-specific individual dry weights (see methods in Whiteman et al. 1996).

Permanent basins have, on average, significantly higher peak densities of invertebrates (3408 ± 1680 S.D. individuals/m²) than autumnal basins (1263 ± 305 S.D. individuals per m²). However, because the numerically dominant taxa in permanent ponds (cladocerans, copepods, and chironomids) are smaller than those in autumnal ponds (fairy shrimp, caddisfly larvae), the peak standing biomass of autumnal ponds (1864 ± 630 S.D. mg/m²) is actually greater than that of permanent ponds (1002 ± 236 S.D. mg/m²). Peak abundances in all basins occur in late June to late July, depending on the timing of ice melt in spring. Densities decline in late summer as adults emerge and remain low throughout the fall because many taxa overwinter as eggs that hatch the following spring.

Overall, the invertebrate abundances in these high-elevation wetlands are one to four orders of magnitude lower than is typically reported for low-elevation habitats (e.g., Voigts 1976, Mittlebach 1981, Duffy and Labar 1994, Brinkman and Duffy 1996). Many low-elevation wetlands are eutrophic, with high nitrogen and phosphorus inputs and high autotrophic and/or high detrital production, depending on wetland type (see summary data in Mitsch and Gosselink 1993). In contrast, the subalpine wetlands described in this chapter are, by classical limnological standards, oligotrophic to ultraoligotrophic (see Wetzel 1995). Nitrogen and phosphorus inputs are negligible, autotrophic production is low, and allochthonous detritus are relatively nonnutritious (spruce needles; see Anderson and Cargill 1987). However, based on the few alpine lake studies from western North America that report benthic data (e.g., Taylor and Erman 1980, Donald and Anderson 1980, Hoffman et al. 1996), and on our own surveys of alpine lakes at or near our study site (e.g., Galena Lake, Yule Lakes, Copper Lake, Emerald Lake), the invertebrate communities in the MCNP wetlands appear to be more productive than those in deep lentic habitats at comparable elevations. Most biological limnology of alpine lakes has focused on plankton (e.g., Larson et al. 1996 and references therein), and there is clearly a need for comparative benthic data on production levels and species composition in high-elevation wetlands and lakes.

Community Diversity and Composition

Despite low overall densities, there is a surprising diversity of invertebrates, with approximately 100 taxa identified to date at this single wetland complex (Appendix 31.A). This diversity belies the notion that high-elevation wetlands necessarily have impoverished invertebrate faunas (Schmitz 1956, Daborn and Clifford 1974). Our survey work in other high-elevation wetlands in the vicinity of the RMBL indicates that the diversity at this site is representative for wetlands in the region. Emerging work from other locations in the central Rocky Mountains corroborates our findings that subalpine and montane wetlands are more diverse than previously appreciated (Duffy, this volume, Rader, unpublished data).

Invertebrate diversity at our study site is strongly dependent on hydroperiod (Table 31.2; Fig. 31.2a; $F = 1458.5$, $P < 0.001$). This dependence has been observed in other comparative wetland studies (see reviews by Williams 1987, Batzer and Wissinger 1996, and recent papers by Schneider and Frost 1996, Schneider 1997). However, as Schneider and Frost (1996) have noted, interpreting the causal relationships between hydroperiod and diversity can be confounded by other variables, including habitat size. At our study site, basin area and degree of permanence were positively related ($F = 26.5$, $p < 0.001$;

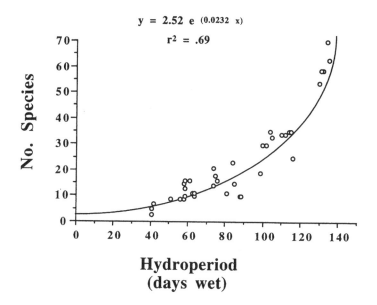

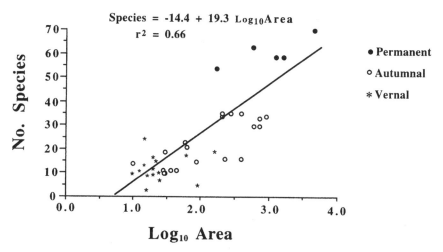

Fig. 31.2. The effects of (*a*) hydroperiod and (*b*) basin area on invertebrate species richness in wetlands at the Mexican Cut Nature Preserve. Data from Table 31.1. See text for statistical analyses.

$r^2 = 0.41$); hence, invertebrate diversity was also strongly associated with basin area (F = 76.7, p < 0.001; Fig. 31.2b). This species-area effect, while not universally observed (e.g., Ebert and Balko 1987, Schneider and Frost 1996), is well documented for a variety of taxa in freshwater habitats (protozoa, Cairns et al. 1969; zooplankton, Barbour and Brown 1974, Hebert 1986, Dodson 1992; snails, Lassen 1975, Bronmark 1985; beetles, Nilsson 1984; fish, Tonn and Magnuson 1982, Angemeier and Schlosser 1989; and mixed-taxa assemblages, Browne 1981, Roth and Jackson 1987). A second confounding factor is that vertebrate predation is also, as in other wetlands (see Batzer and Wissinger 1996) and lentic habitats in general (see Wellborn et al. 1996), a covariate of permanence. Vertebrate predators (fish or amphibians) are often more abundant in permanent than in temporary basins. Thus, direct effects of habitat drying on diversity are confounded by the effects of predator-mediated coexistence (see reviews by Schneider and Frost 1996, Wellborn et al. 1996).

The species composition of the invertebrate assemblage at MCNP is typical for the region (Wissinger and Brown, unpublished data) and, although more diverse, is similar to that reported in small aestival ponds at similar elevations in Colorado and Wyoming (Blake 1945, Neldner and Pennak 1955, Schmitz 1959, Duffy, this volume). Across all habitats, dipterans, particularly chironomid larvae, are the most abundant and species-rich (26 species) group of benthic taxa, followed by beetles (17 species), caddisflies (5 species), water mites (5 species), odonates (5 species), water bugs (5 species), and leeches (3 species) (Appendix 31.A). Cladocerans (7 species), copepods (3 species), and, in temporary habitats, the fairy shrimp *Branchinecta coloradensis* dominate the zooplankton. The dominance of chironomids, and high relative abundances of taxa such as limnephilid caddisflies and dytiscid beetles, appear to be characteristic of high-elevation (see references above) and high-latitude (tundra) wetlands in North America (e.g., Daborn 1974, Butler et al. 1980, Butler 1982). The relatively high proportional abundance of chironomids and other common taxa in alpine and arctic wetlands may reflect their tolerance to freezing conditions during winter (Oswood 1989, Oswood et al. 1991). Alpine and tundra lakes are also dominated by chironomids, but differ from shallow lentic habitats in the relatively high diversity and abundance of mayflies and stoneflies (Taylor and Erman 1980, Donald and Anderson 1980, Andrews and Rigler 1985, Hershey 1985), two groups of taxa that are rare or absent in subalpine wetlands. One exception is the mayfly *Callibaetis ferrugineus*, which, although rare at MCNP, is locally abundant in montane beaver ponds.

Patterns of distribution and abundance among basins are taxon-specific and at least superficially correlated with hydroperiod. In the most diverse groups patterns of distribution and abundance are species-specific, with different taxa restricted or abundant in only one or two of the hydroperiod categories. Among the beetles, only *Strictotarsus* (= *Deronectes*) *grieostriatus* is ubiquitous, with other species restricted mainly to permanent (*Haliplus* spp., *Rhantus gutticollis, Ilybius fraterculus, Agabus strigilosus*) or autumnal habitats (*Acilius semisulcatus, Dytiscus dauricus, Hydroporus* sp., *Hygrotus* sp., *Agabus strigulosus*) (Table 31.1). Among the chironomids several taxa are ubiquitous (e.g., *Paratanytarsus, Psectrocladius, Tanytarsus*), several occur mainly in permanent (e.g., *Chironomus salinaris, Dicrotendipes, Microtendipes, Ablabesmyia*), and several mainly in temporary (vernal + autumnal; e.g., *Chironomus riparius, Procladius*) basins. Zooplankton distribution are similar to those reported at this study site in the 1970s (Dodson 1970, 1974, Sprules 1972). Fairy shrimp (*Branchinecta coloradensis*), the large cladoceran *Daphnia middendorfiana,* and the large copepod *Hesperodiaptomus shoshone* are most abundant in autumnal habitats, whereas smaller species (*Daphnia pulex* and *Daphnia rosea, Chydorus sphaericus*) dominate in permanent habitats. These particular zooplankton assemblages are characteristic of many subalpine ponds near RMBL (R. L. Willey, personal communication) and perhaps across much of the central Rockies (e.g., compare to Duffy, this volume).

In summary, large, permanent basins are most diverse and contain the greatest number of species with restricted distributions (Table 31.1, Fig. 31.3). The primary consumers in permanent-basin foodwebs are dominated by planktonic and epibenthic crustaceans, numerous chironomid midges, and several species of detritus-shredding caddisflies. Dragonflies, beetles, and water bugs are intermediate predators on all of the smaller taxa (zooplankton and chironomids). Several year-classes of larval salamanders (including paedomorphs) are top predators in these basins and are known to prey on nearly all of the dominant invertebrates in the underlying foodweb (Dodson 1970, Dodson and Dodson 1971, Collins and Holomuzki 1984, Holomuzki and Collins 1987, Holomuzki 1989a, b, Zerba and Collins 1992, Whiteman et al. 1994, Whiteman et al. 1996) (Fig. 31.3*a*). In autumnal basins beetle larvae (especially large taxa such as *Dytiscus* and *Acilius*), hemipterans, and caddisflies replace salamander larvae as the top predators on other benthic invertebrates, whereas metamorphic adults and hatchling salamander larvae are the top predators on the large-bodied zooplankton, and occasionally on large benthic taxa such as caddisflies (Fig. 31.3*b*). Finally, vernal foodwebs are the least complex and are dominated by a subset of the fauna found in autumnal basins that can, through a variety of adaptations, complete their life cycles in these extremely ephemeral habitats (Fig. 31.3*c*). These adaptations include:

1. Desiccation-resistant resting stages such as the diapausing eggs of crustaceans (*Leptodiaptomus coloradensis, Daphnia middendorfiana, Acanthocyclops*), rotifers (*Keratella*), mosquitoes (*Aedes communis*), and the adults of the beetles *Strictotarsus grieostriatus* (see reviews of desiccation resistance by Wiggins et al. 1980, Williams 1987, Batzer and Wissinger 1996)

(a) Permanent Wetlands

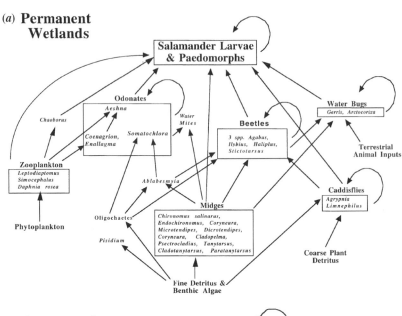

(b) Autumnal Wetlands

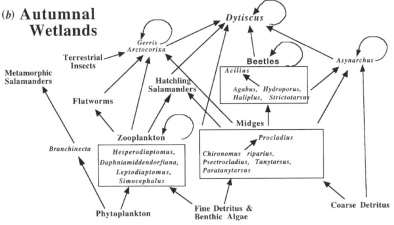

(c) Vernal Wetlands

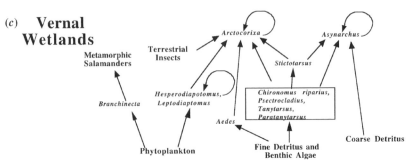

2. Rapid colonization and development, as in midges (*Chironomus ripar-ius*)

3. Rapid larval development and a diapausing adult stage timed to the dry phase of the hydroperiod, as in caddisflies (e.g., *Asynarchus nigriculus,* Wissinger and Brown, ms.)

4. Seasonal cyclic migrations between adjacent permanent and temporary habitats, as we have observed for *Arctocorixa* water boatmen (Dodson and Dodson 1971)

A Path Analytic Approach to the Problem of Multiple Correlations

As with patterns of diversity, the apparent correlation of particular species and clusters of species with hydroperiod is potentially confounded by other abiotic (water chemistry, area, depth) and biotic (salamander predation) factors that covary with hydroperiod. We used path analysis to further explore the collinear relationships among these habitat variables and to disentangle their direct and indirect effects on community composition. A path analysis is essentially a sequence of multiple regressions structured in a manner that reflects a priori hypotheses about how the variables are related (see general treatments by Tukey 1954, Li 1975, Pedhazur 1982 and ecological applications by Schemske and Horvitz 1988, Edwards and Armbruster 1989, Kingsolver and Schemske 1991, Mitchell 1992, Wooten 1994). We chose this technique over a direct or indirect ordination (e.g., Gower et al. 1994, Larson et al. 1995, Romo et al. 1996, Botts 1997) because we wanted to incorporate biologically relevant information about the habitats into alternative hypotheses

---

Fig. 31.3. Summary of the dominant taxa and energy pathways in permanent, autumnal, and vernal wetlands at the Mexican Cut Nature Reserve. With a few exceptions, the links are inferred from dietary data that we or others have collected at this particular site. In particular, data for all stages of salamanders are from published and unpublished analyses of stomach contents collected from a nonlethal stomach pumping technique (Wissinger and Whiteman 1992, Whiteman et al. 1994, Whiteman et al. 1996, Wissinger and Whiteman, unpublished ms.). Connections to, from, and within the zooplankton guilds (including *Chaoborus* and flatworms) are based on published data from this site (Dodson 1970, 1974, Sprules 1972, Maly 1973, Maly and Maly 1974, 1980). Caddisfly diets are based on gut dissections (Sparks 1993), midge diets on microscopic examination of gut contents (Wissinger and Brown, unpublished data), odonate diets on fecal pellet analyses (Wissinger, unpublished data), and adult and hemipteran diets on observations of prey capture in natural and experimental (lab and field) settings (also see Reynolds 1975). Inferences about oligochaete and clam diets are based on general literature for these groups and are the least well established for the study site. Loops represent intraguild predation and/or cannibalism within groups and are based on field and laboratory observations (caddisflies, beetles, Hemiptera, salamanders), corroborated in some cases by the dietary data cited above.

(Edwards and Armbruster 1989). For example, the directionality of cause-and-effect relationships is known and not reciprocal for some pairs of variables (e.g., hydroperiod might affect water chemistry, but not vice versa). A path analysis also can extricate and quantify the relative importance of direct and indirect effects in a nexus of collinear variables (for an example, see Schemske and Horvitz 1988).

The first step was to order the basins according to each of the major factors of interest (most to least permanent, largest to smallest, most to fewest salamander predators, gradients in community composition). For some of the variables this ordering could be accomplished with univariate data. Thus, for hydroperiod we used the average (1989–1996) number of days of ice-free standing water (Table 31.1), and for salamanders we used the separate densities of each life stage (hatchling larvae, metamorphs, large branchiates $\geq$ one-year-old larvae + paedomorphs; Table 31.3).

For the other major variables (water chemistry, basin morphometry, and invertebrate community composition) we combined collinear (redundant) measurements into one or a few multivariate dimensions using appropriate ordination techniques. We used principal components analysis (PCA) to combine data on water chemistry (12 variables, see Table 31.2) and basin morphometry (area, depth; Table 31.1). Exploratory analyses (e.g., scatterplots, covariance tests, correlation matrices) indicated that linearity assumptions for PCA were met in both cases. The first PCA axis explained 52 percent and 83 percent of the variance in water chemistry and basin morphometry, respectively. Secondary axes explained little more of the variation for both categories of data and were thus eliminated from the analyses, a broken-stick model being used as a guide (Jackson 1993). The first multivariate axis in the chemistry PCA expressed a gradient from high-alkalinity (relatively high pH, TDS, Ca, Mg, ANC) to low-alkalinity water (relatively low pH, TDS, Ca, Mg, ANC). The primary axis for basin morphometry represented a gradient from large, deep basins to small, shallow basins.

We summarized the invertebrate data in two ways. We first conducted a PCA for the 11 basins for which we had replicate quantitative data and found strong primary and secondary axes. However, our confidence in these axes was compromised by the low number of replicates (11 basins) relative to the large number of species. In order to independently verify the biological relevance of these axes, we also analyzed the presence-absence data from surveys of 41 of the basins (Table 31.1) using multidimensional scaling (MDS), a technique with fewer assumptions than PCA and more appropriate for binary data (see Digby and Kempton 1987). The major axes from the MDS on survey data correlated well with the major axes obtained from the PCA analysis on quantitative data and generally reflected the gradation in taxa observed along the size-hydroperiod-salamander gradient discussed above (Table 31.1, Fig. 31.3).

Having ordered the ponds for all of the variables, we then constructed a series of path analyses in which the overall dependent variable was the first axis from the MDS summary of invertebrate community composition (Fig.

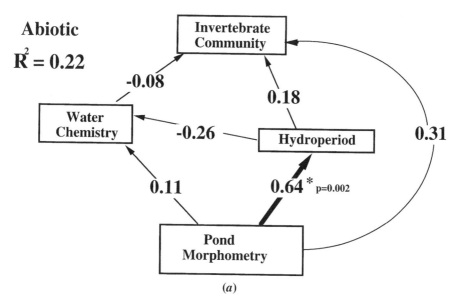

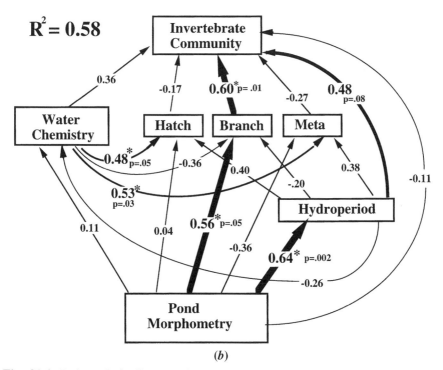

Fig. 31.4. Path analysis diagrams for (*a*) abiotic only effects on invertebrate community composition and (*b*) abiotic + salamander effects on invertebrate community composition. Path coefficients are the partial regression coefficients from multiple regression (see text).

31.4). Path coefficients between each pair of variables were the standardized partial regression coefficients and represent the direct effect of the independent (causal) factor on the dependent (affected) variable for that pair (Pedhazur 1982). We calculated effect coefficients as the sum of direct and indirect path effects (Table 31.4). An explanation for the logic of calculating effect coefficients is given in Schemske and Horvitz (1988). Of the several possible alternative configurations for the abiotic variables, we found that the "best" abiotic model explained only 22 percent of the overall variation in invertebrate community composition (Fig. 31.4a). In all of the models,

1. Hydroperiod was directly and significantly ($p = 0.002$) affected by pond morphometry—relatively large, deep basins were more permanent than small, shallow basins.
2. Water chemistry was not related to either basin morphometry or to hydroperiod.
3. Invertebrate community composition was not significantly affected by any of the abiotic variables, including hydroperiod.

We then added salamander predation to the "best" abiotic model and found that this new model explained 58 percent of the variation in invertebrate community composition, more than twice that of analysis with only abiotic effects (Fig. 31.4b). In this model the density of branchiate salamanders had the strongest direct effect on invertebrate community composition ($p = 0.01$). Based on the overall effect coefficients (direct + indirect path coefficients), basin morphometry and branchiate salamanders are the two factors that have the greatest overall effect on invertebrate community composition (Table 31.5). The effect of branchiate salamanders is direct, whereas that of basin morphometry is indirect and is translated through its influence on hydroperiod and salamander distributions.

TABLE 31.4. Summary of the Effect Coefficients Calculated from Direct and Indirect Path Effects in Path Analysis with Salamanders[a]

	Direct Effect	Indirect Effect(s)	Effect Coefficient
Pond morphometry	−0.11	0.54	**0.43**
Water chemistry	0.36	−0.53	−0.17
Hydroperiod	**0.48**	−0.29	0.19
Salamanders	−0.17	—	−0.17
Hatchlings			
Metamorphs	−0.27	—	−0.27
Branchiates (larvae paedomorphs)	**0.60**	—	**0.60**

[a] See text and Fig. 31.4.

Mechanisms and the Roles of Life History Studies and Manipulative Experiments

The analysis above suggests that salamander predation is the single most important variable that is directly correlated with patterns of invertebrate community structure. However, we know from life history studies that the distributions of some species are directly due to their inability to complete development in temporary habitats. For example, the caddisfly *Agrypnia deflata*, which occurs only in permanent ponds, emerges in midsummer, mates immediately, and deposits aquatic, desiccation-vulnerable egg masses at a time of the year when only permanent habitats are typically filled (Wissinger and Brown, ms). Similarly, the dragonflies *Aeshna palmata* and *Somatochlora semicircularis* require several years to complete larval development at this elevation, as evidenced by multiple year-classes of larvae (also see Willey 1973). *Somatochlora* larvae can resist short periods of desiccation and are therefore occasionally present in autumnal basins (Willey 1974, Johannsson and Nilsson 1991), but *Aeshna* larvae cannot and are therefore restricted to permanent habitats. Clearly, no additional mechanisms that involve biotic interactions need to be invoked for explaining the distributional patterns of species like *Agrypnia* and *Aeshna*.

However, many of the other taxa with restricted distributions at our study site can complete their life cycles across a range of hydroperiods, and explanations related to species-area relationships and salamander predation are therefore viable alternative hypotheses. Experimental studies implicate salamander predation as the direct or indirect causal factor that underlies patterns of distribution and abundance for two groups of taxa. Field experiments with zooplankton and salamanders at MCNP have demonstrated that the largest species (e.g., *Daphnia middendorfiana, Hesperodiaptomus shoshone,* and *Branchinecta coloradensis*) are dominant competitors over small species and should exclude them in all habitats in the absence of salamander predation (Dodson 1970, 1974, Sprules 1972). However, large species are preferentially preyed on by branchiate salamanders (larvae + paedomorphs) and are therefore eliminated from most permanent ponds, hence allowing small species to flourish (Dodson 1970, 1974; Sprules 1972; Bohonak and Whiteman, ms.).

Field experiments suggest a similar scenario for explaining the distributional patterns of caddisflies. In the absence of salamander predation, a dominant competitor (*Asynarchus nigriculus*) should exclude less aggressive species via intraguild predation (e.g., *Limnephilus externus*). However, preferential predation by salamanders on this dominant competitor indirectly benefits other species which predominate in permanent habitats (Wissinger et al. 1996; Wissinger et al., in press). It is not the large size (as in the zooplankton example) of *Asynarchus* that makes it more vulnerable to salamander predation, but rather its frenetic foraging activities, which appear to be necessary for the timely completion of development in temporary habitats (Wissinger et al. 1999). The tradeoff between high activity levels, which are necessary for rapid development, and low activity levels, which are necessary for co-existence with vertebrate predators, is an important component of a general conceptual model for shifts in species composition along permanence gradients in lentic habitats (Wellborn et al. 1996). A second tradeoff along this gradient is between prey vulnerability to vertebrate versus large invertebrate predators that replace vertebrates at the top of temporary wetland foodwebs (as in Fig. 31.3). Although numerous examples from the amphibian literature document this tradeoff for amphibian larvae, only McPeek's (1990a, b, 1996) work with *Enallagma* damselflies has demonstrated such a tradeoff for macro-invertebrates.

Given the collinear nature of the effects of basin size, hydroperiod, and shifts in predator effects, attributing patterns of distribution and abundance to any particular mechanism, in the absence of detailed, site-specific life history studies or experimental data can only be speculative. For example, at our study site it is surprising that damselflies occur only in permanent habitats. Closely related species are known to be desiccation and/or freeze tolerant at other locations (Daborn 1971, Sawchyn and Gillot 1974, 1975, Baker and Clifford 1981, Norling 1984), and all three damselflies are abundant in autumnal habitats at lower elevations near our study site (personal observation). Their restriction to permanent habitats at MCNP could be due to (1) the extremely short growing season; (2) the severity of winter conditions in shallow autumnal basins; (3) a species-area effect in which small populations near the limits of their physiological range are likely, at any point in time, to be found only in the largest habitats (also see Jeffries 1994); and/or (4) vulnerability to one or more of the invertebrate predators (e.g., *Dytiscus*) that occur mainly in autumnal habitats at this elevation. Comparative data that might help support or refute these alternatives could be obtained from habitats along a hydroperiod gradient that is comparable in size, or from a series of basins in which hydroperiod and vertebrate predation are completely cross-classified (e.g., permanent but salamander-free habitats). However, few comparative studies are able to cleanly separate alternative hypotheses, and a definitive understanding of the mechanisms that underlie community patterns will be gained most efficiently through replicated field experiments (see review by Wilbur 1997). The well-defined habitat boundaries of subalpine wetlands, their relative simplicity in habitat structure, and their ease of survey and manipulation are all advantages for using them as model systems for understanding the relative importance of biotic and abiotic influences on invertebrate community composition.

ANTHROPOGENIC THREATS AND CONSERVATION

Despite the remote location of many high-elevation wetlands in western North America, these relatively pristine habitats are potentially vulnerable to a variety of human impacts, including mining, deforestation, grazing, water diversions, and other activities that occur on both private and public lands, including wilderness areas (see summary by Windell et al. 1986). One regional threat to the western slopes of the Rockies has been the decline in precipitation pH from emissions of coal-fired power plants and cities in the intermountain deserts to the west. Although acid rain in the west is relatively benign compared to that in northeastern North America (Baker et al. 1991, Shannon and Sisterton 1992), many montane habitats are especially vulnerable because of low buffering capacities associated with crystalline bedrock and thin soils (Melack et al. 1982, Kling and Grant 1984, Tonnessen 1984, Harte et al. 1985), and there is some evidence that surface waters in this region are episodically acidified during peak snowmelt in spring (Blanchard et al. 1987, Vertucci and Corn 1996). Wetlands such as those described in this study should be especially vulnerable to pulses of acidity because of their lack of buffering capacity and their small size. In addition to the potential for being directly affected physiologically, invertebrate communities should be affected by any change in the abundance of salamanders, which can be quite vulnerable to acidification (see review by Freda et al. 1991). Harte and Hoffman (1989) hypothesized that a population decline in the salamanders at our study site during the 1980s was related to egg mortality during pulses of acidification in spring. However, since 1988 we have found no evidence that chronic or episodic acidification has had any impact on the salamander population at MCNP (Wissinger and Whiteman 1992; Whiteman and Wissinger, unpublished data). Summer precipitation events during this time period have been circumneutral, and although pH in some years has fallen briefly to below 5.0 during peak runoff in early spring, it has always returned to circumneutral levels by the time salamanders become active and begin to breed (Wissinger and Whitman 1992). There is no evidence to date at this or other study sites in the region to suggest that recent declines in amphibian populations are related to acidification (Vertucci and Corn 1996). Nonetheless, many high-elevation wetlands in the central Rockies remain, by virtue of low buffering capacity, vulnerable to acid inputs and should continue to be closely monitored.

A second threat to the biotic integrity of these habitats is the introduction (via helicopters and fixed-wing aircraft) of trout into high alpine lakes for recreational fishing (e.g., Horton and Ronayne 1996). Many of these lakes and permanent wetlands associated with outflows below these lakes were historically fishless and, like the habitats at MCNP, supported populations of tiger salamanders. It is well known that ambystomatid salamander larvae and paedomorphs are vulnerable to fish predation (Semlitsch 1987, 1988, Jackson and Semlitsch 1993, Figiel and Semlitsch 1990, Sexton et al. 1994 and references therein), and the introduction of trout to these lakes might not only eliminate salamanders but should, based on the results presented in this chapter, lead to changes in invertebrate community composition. In particular, we have hypothesized that permanent habitats with salamanders provide a refuge for invertebrate species that, in the absence of salamanders, would be out-competed by or preyed on by invertebrates that cannot coexist with salamander predators. Fish introductions have been implicated in community-level changes in high-elevation communities in the northern Cascade Mountains (Liss et al. 1995, Hoffman et al. 1996), but we know of no comparable studies in the central Rockies. Future studies that explore the effects of fish introductions in this region should consider both the direct effects of trout predation on invertebrates and the indirect effects of reducing or eliminating salamanders from the previously fishless habitats.

ACKNOWLEDGMENTS

We thank J. Jannot, K. Buhn, G. Marucca, P. Ode, G. Sparks, D. Weigle, and G. Rouse for field assistance and the Rocky Mountain Biological Laboratory and Alan Carpenter of the Colorado Field Office of The Nature Conservancy for facilitating access to the Mexican Cut Nature Preserve. We are grateful to D. Ruiter and B. Kondratieff for verifying caddisfly and beetle taxonomy, respectively. We also thank Darold Batzer for his comments which greatly improved an earlier version of the manuscript. This work was supported by grants from the Colorado Field Office of The Nature Conservancy and the National Science Foundation (BSR 8958253 and DEB 9407856).

LITERATURE CITED

Anderson, N. H., and A. S. Cargill. 1987. Nutritional ecology of detritivorous aquatic insects. Pages 903–925 *in* F. Slanksy and J. G. Rodriquez (eds.), Nutritional Ecology of Insects, Mites, Spiders, and Related Invertebrates. John Wiley & Sons, New York.

Andrews, D., and F. H. Rigler. 1985. The effects of an Arctic winter on benthic invertebrates in the littoral zone of Charl Lake, Northwest Territories. Canadian Journal of Zoology 63:2825–2834.

Angemeier, O. L., and I. J. Schlosser. 1989. Species area relationships for stream fishes. Ecology 70:1450–1462.

Baker, L. A., A. T. Herlihy, P. R. Kaufmann, and J. M. Eilers. 1991. Acidic lakes and streams in the United States: The role of acidic deposition. Science 252:1151–1154.

Baker, R. L., and H. F. Clifford. 1981. Life histories and food of *Coenagrion resolutum* (Coenagrionidae: Odonata) and *Lestes disjunctus disjunctus* (Lestidae: Odonata) populations from the boreal forest in Alberta, Canada. Aquatic Insects 3:179–191.

Barbour, C. C., and J. H. Brown. 1974. Fish species diversity in lakes. American Naturalist 108:473–489.

Batzer, D. P., and S. A. Wissinger. 1996. Ecology of insect communities in nontidal wetlands. Annual Review of Entomology 41:75–100.

Blake, I. H. 1945. An ecological reconnaissance in the Medicine Bow Mountains. Ecological Monographs 15:208–242.

Blanchard, C., H. Michaels, A. Brodman, and J. Harte. 1987. Episodic Acidification of a Low Alkalinity Pond in Colorado. Publication 88-1, Energy Resource Group, University of California, Berkeley, CA.

Bohonak, A. J., and H. H. Whiteman. Dispersal of the fairy shrimp *Branchinecta coloradensis* (Anostraca): Effects of hydroperiod and salamanders. Ms.

Botts, P. S. 1997. Spatial pattern, patch dynamics, and successional change: Chironomid assemblages in a Lake Erie coastal wetland. Freshwater Biology 37:277–286.

Brinkman, M. A., and W. G. Duffy. 1996. Evaluation of four wetland aquatic invertebrate samplers and four sample sorting methods. Journal of Freshwater Ecology 11:193–200.

Bronmark, C. 1985. Freshwater snail diversity: Effects of pond area, habitat heterogeneity, and isolation. Oecologia 67:127–131.

Browne, R. A. 1981. Lakes as islands: The biogeographic distribution, turnover rates, and species composition in the lakes of central New York. Journal of Biogeography 8:75–83.

Buck, P. 1960. Vegetational succession in subalpine ponds in the Rockies. Proceedings of the Oklahoma Academy of Science 49:2–6.

Butler, M. G. 1982. A 7-year life cycle for two *Chironomus* species in arctic Alaskan tundra ponds (Diptera: Chironomidae). Canadian Journal of Zoology 60:58–70.

Butler, M. G., M. C. Miller, and S. Mozley. 1980. Macrobenthos. Pages 297–449 *in* J. E. Hobbie (ed.), Limnology of Tundra Ponds. Dowden, Hutchinson, & Ross, Stroudsburg, PA.

Cairns, J. C., M. L. Dahlberg, K. L. Dickson, N. Smith, and W. T. Waller. 1969. The relationship of freshwater protozoan communities to the MacArthur-Wilson equilibrium model. American Naturalist 103:439–454.

Collins, J. P., and J. R. Holomuzki. 1984. Intraspecific variation in diet within and between trophic morphs in larval tiger salamanders (*Ambystoma tigrinum nebulosum*). Canadian Journal of Zoology 62:168–174.

Daborn, G. R. 1971. Survival and mortality of coenagrionid nymphs (Odonata: Zygoptera) from the ice of an aestival pond. Canadian Journal of Zoology 57:2143–2152.

———. 1974. Biological features of an aestival pond in western Canada. Hydrobiologia 44:287–299.

Daborn, G. R., and H. F. Clifford. 1974. Physical and chemical features of an aestival pond in western Canada. Hydrobiologia 44:43–59.

Danks, H. V. 1971a. Overwintering of some north temperate and arctic Chironomidae. I. The winter environments. Canadian Entomologist 103:589–604.

———. 1971b. Overwintering of some north temperate and arctic Chironomidae. II. Chironomid biology. Canadian Entomologist. 103:1875–1910.

Digby, P. G. N., and R. A. Kempton 1987. Multivariate Analysis of Ecological Communities. Chapman & Hall, London.

Dodson, S. I. 1970. Complementary feeding niches sustained by size-selective predation. Limnology and Oceanography 15:131–137.

———. 1974. Zooplankton competition and predation: An experimental test of the size-efficiency hypothesis. Ecology 55:605–613.

———. 1982. Chemical and biological limnology of six west-central Colorado mountain ponds and their susceptibility to acid rain. American Midland Naturalist 107:1733–1739.

————. 1992. Predicting crustacean zooplankton species richness. Limnology and Oceanography 37:848–856.

Dodson, S. I., and V. E. Dodson. 1971. The diet of *Ambystoma tigrinum* larvae from western Colorado. Copeia 1971:614–624.

Donald, D. B., and R. S. Anderson. 1980. The lentic stoneflies (Plecoptera) from the Continental Divide region of southwestern Canada. Canadian Entomologist 112: 753–758.

Duffy, W. G., and D. J. LaBar. 1994. Aquatic macroinvertebrate production in southeastern USA wetlands during winter and spring. Wetlands 14:88–97.

Duffy, W. G., and C. R. Liston. 1985. Survival following exposure to subzero temperatures and respiration in cold acclimatized larvae of *Enallagma boreale*. Freshwater Invertebrate Biology 4:1–7.

Ebert, T. A., and M. L. Balko. 1987. Temporary pools as islands in space and time: the biota of vernal pools in San Diego, Southern California, USA. Archiv für Hydrobiologie 110:101–123.

Edwards, M. E., and W. S. Armbruster. 1989. A tundra–steppe transition of Kathul Mountain, Alaska, USA. Arctic and Alpine Research 21:296–304.

Fiegel, C. R., and R. D. Semlitsch. 1990. Population variation in survival and metamorphosis of larval salamanders (*Ambystoma maculatum*) in the presences and absence of fish populations. Copeia 1990:818–826.

Freda, J., W. J. Sadinski, and W. A. Dunson. 1991. Long term monitoring of amphibian populations with respect to the effects of acidic deposition. Water, Air, and Soil Pollution 30:439–450.

Frisbee, M. P., and R. E. Lee. 1997. Inoculative freezing and the problem of winter survival for freshwater macroinvertebrates. Journal of the North American Benthological Society 16:636–650.

Gower, A. M., G. Mers, M. Kent, and M. E. Foulkes. 1994. Relationships between macroinvertebrate communities and environmental variables in metal-contaminated streams in south-west England. Freshwater Biology 32:199–221.

Hairston, N. G. 1976. Photoprotection by carotenoid pigments in the copepod *Diaptomus nevadensis*. Proceedings of National Academy Science. USA 73:971–974.

————. 1979. The relationship between pigmentation and reproduction in the two species of *Diaptomus* (Copepoda). Limnology and Oceanography 24:38–44.

Harte, J., and E. Hoffman. 1989. Possible effects of acidic deposition on a Rocky Mountain population of the tiger salamander, *Ambystoma tigrinum*. Conservation Biology 3:149–158.

Harte, J., G. P. Lockette, R. A. Schneider, H. Michaels, and C. Blanchard. 1985. Acid precipitation and surface water vulnerability on the western slope of the high Colorado Rockies. Water, Air, and Soil Pollution 25:313–320.

Hebert, P. D. N. 1986. Patterns in the composition of arctic tundra pond microcrustacean communities. Canadian Journal of Fisheries and Aquatic Sciences 43:1416–1425.

Hershey, A. E. 1985. Littoral chironomid communities in an arctic Alaskan lake. Holarctic Ecology 8:39–48.

Hoffman, R. L., W. J. Liss, G. L. Larson, E. K. Deimling, and G. Lomnicky. 1996. Distribution of nearshore macroinvertebrates in lakes of northern Cascade mountains, Washington, U.S.A. Archiv für Hydrobiologie 136:363–389.

Holomuzki, J. R. 1989a. Salamander predation and vertical distributions of zooplankton. Freshwater Biology 21:461–472.

———. 1989b. Predatory behavior of larval *Ambystoma tigrinum nebulosum* on *Limnephilus* (Trichoptera) larvae. Great Basin Naturalist 43:475–476.

Holomuzki, J. R., and J. P. Collins. 1987. Trophic dynamics of a top predator, *Ambystoma tigrinum nebulosum* (Caudata: Ambystomatidae), in a lentic community. Copeia 1987:949–957.

Horton, B., and D. Ronayne. 1996. Taking stock of stocking: The alpine lakes fisheries program comes under scrutiny. Idaho Wildlife 15:22.

Jackson, D. A. 1993. Stopping rules in principal components analysis: A comparison of heuristical and statistical approaches. Ecology 74:2204–2214.

Jackson, M. E., and R. D. Semlitsch. 1993. Paedomorphosis in the salamander *Ambystoma talpoideum*: Effects of a fish predator. Ecology 74:342–350.

Jeffries, M. 1994. Invertebrate communities and turnover in wetland ponds affected by drought. Freshwater Biology 32:603–612.

Johannsson, F., and A. N. Nilsson 1991. Freezing tolerance and drought resistance of *Somatochlora alpestris* larvae in boreal temporary pools (Anisoptera: Corduliidae). Odonatologica 20:245–252.

Kingsolver J. G., and D. W. Schemske. 1991. Path analysis of selection. Trends in Ecology and Evolution 6:276–280.

Kling, C. W., and M. C. Grant. 1984. Acid precipitation in the Colorado Front Range: An overview with time predictions for signficiant effects. Arctic and Alpine Research 16:321–329.

Landers, D. H., J. M. Eilers, D. F. Brakke, W. S. Kellar, P. E. Silverstein, M. E. Schonbrod, R. E. Crowe, R. A. Linthurst, J. M. Omernik, S. A. Teatue, and E. P. Meier. 1987. Characteristics of Lakes in the Western United States: Vol. 1. Populations, Descriptions, and Physiochemical Relationships. EPA-600/3-86/054a. U.S. Environmental Protection Agency, Washington, DC.

Langenheim, J. H. 1962. Vegetation and environmental patterns in the Crested Butte area, Gunnison County, Colorado. Ecological Monographs 32:249–285.

Larson, G. L., C. D. McIntire, E. Karnaugh-Thomas, and C. Hawkins-Hoffman. 1996. Limnology of isolated and connected high-mountain lakes in Olympic National Park, Washington State, USA. Archiv für Hydrobiologie 134:75–92.

Lassen, H. H. 1975. The diversity of freshwater snails in view of the equilibrium theory of island biogeography. Oecologia 19:1–18.

Lee, R. E. 1989. Insect cold hardiness: To freeze or not to freeze? Bioscience 39:308–313.

Li, C. C. 1975. Path Analysis—a Primer. Boxwood, Pacific Grove, CA.

Liss, W. J., G. L. Larson, E. K. Deimling, R. Gresswell, R. Hoffman, M. Kiss, G. Limnicky, C. D. McIntire, R. Truitt, and T. Tyler. 1995. Ecological Effects of Stocked Trout in Naturally Fishless High Mountain Lakes: North Cascades National Park Service Complex, WA, USA. Technical Report NPS/PNROSU/NRTR-95-03. National Park Service, Pacific Northwest Region, Seattle, WA.

Maly, E. J. 1973. Density, size, and clutch of two high altitude diaptomid copepods. Limnology and Oceanography 18:840–847.

Maly, E. J., and M. P. Maly. 1974. Dietary differences between two co-occurring calanoid copepod species. Oecologia 17:325–333.

Maly, E. J., S. Schoenholtz, and M. T. Arts. 1980. The influence of flatworm predation on zooplankton inhabiting small ponds. Hydrobiologia 76:233–240.

McPeek, M. A. 1990a. Determination of species composition in the *Enallagma* damselfly assemblages of permanent lakes. Ecology 71:83–98.

———. 1990b. Behavioral differences between *Enallagma* species (Odonata) influencing differential vulnerability to predators. Ecology 71:1714–1726.

———. 1996. Tradeoffs, food web structure, and the coexistence of habitat specialists and generalists. American Naturalist 148:S124–S138.

Melack, J. M., J. L. Stoddard, and D. R. Dawson. 1982. Acid precipitation and buffering capacity of lakes in the Sierra Nevada. Pages 465–470. Proceedings of American Watershed Research Association on Hydrometrics.

Mitsch, W. J., and J. G. Gosselink. 1993. Wetlands (2nd ed.). Van Nostrand Reinhold, New York.

Mitchell, R. J. 1991. Testing evolutionary and ecological hypotheses using path analysis and structured equation modeling. Functional Ecology 6:123–129.

Mittlebach, G. G. 1981. Patterns of invertebrate size and abundance in aquatic habitats. Canadian Journal of Fisheries and Aquatic Sciences 38:896–904.

Neldner, K. H., and R. W. Pennak. 1955. Seasonal faunal variations in a Colorado alpine pond. American Midland Naturalist 53:419–430.

Nilsson, A. N. 1984. Species richness and succession of aquatic beetles in some kettle-hole ponds in northern Sweden. Holarctic Ecology 7:149–156.

Norling, U. 1984. Life history patterns in the northern expansion of dragonflies. Advances in Odonatology 2:127–156.

Oswood, M. W. 1989. Community structure of benthic invertebrates in interior Alaskan streams and rivers. Hydrobiologia 172:97–110.

Oswood, M. W., L. K. Miller, and J. G. Irons. 1991. Overwintering of freshwater macro-invertebrates. Pages 360–375 *in* R. E. Lee and D. L. Denlinger (eds.), Insects at Low Temperature. Chapman & Hall, New York.

Pedzhauer, E. J. 1982. Multiple Regression in Behavioral Research. Holt, Rinehart, & Winston, New York.

Pierce, B. A. 1985. Acid tolerance in amphibians. Bioscience 35:239–243.

Prather, T. 1982. Geology of the Gunnison Country. B & B Press, Gunnison, CO.

Reynolds, J. 1975. Feeding in corixids of small alkaline lakes in central British Columbia. Internationale Vereinigung für theocetiche und angewarte Limnologie, Verhandlungen 19:3073–3078.

Romo, S., E. Van Donk, R. Gylstra, and R. Gulatis. 1998. A multivariate analysis of phytoplankton and food web changes in a shallow biomanipulated lake. Freshwater Biology 36:683–696.

Roth, A. H., and J. F. Jackson. 1987. The effect of pools size on recruitment of predatory insects and on mortality in a larval anuran. Herpetologica 43:224–232.

Sawchyn, W. W., and N. S. Church. 1973. The effects of temperature and photoperiod on diapause development in the eggs of four species of *Lestes* (Odonata:Zygoptera). Canadian Journal of Zoology 51:1257–1266.

Sawchyn, W. W., and C. Gillott. 1974. The life histories of three species of *Lestes* (Odonata: Zygoptera) in Saskatchewan. Canadian Entomologist 106:1283–1293.

Schemske, D. W., and C. C. Horvitz. 1988. Plant-animal interactions and fruit production in a neotropical herb: A path analysis. Ecology 69:1128–1137.

Schmitz, E. H. 1956. Seasonal biotic events in two Colorado alpine tundra ponds. American Midland Naturalist 61:424–446.

Schneider, D. W. 1997. Predation and food web structure along a habitat duration gradient. Oecologia 110:567–575.

Schneider, D. W., and T. M. Frost. 1996. Habitat duration and community structure in temporary ponds. Journal of the North American Benthological Society 15:64–86.

Semlitsch, R. D. 1987. Interactions between fish and salamander larvae. Oecologia 72: 482–486.

———. 1988. Allopatric distribution of two salamanders: Effects of fish predation and competitive interactions. Copeia 1988:290–298.

Sexton, O. J., C. A. Phillips, and E. Routman. 1994. The response of naive breeding adults of the spotted salamander to fish. Behaviour 130:113–121.

Shannon, J. D., and D. L. Sisterson. 1992. Estimation of S and NO_x deposition budgets for the United States and Canada. Water, Air, and Soil Pollution 63:211–215.

Southwood, T. R. E. 1977. Habitat, the templet for ecological strategies. Journal of Animal Ecology 46:337–365.

———. 1988. Tactics, strategies, and templets. Oikos 52:94–100.

Sparks, G. A. 1993. Competition and intraguild predation between two species of caddisfly (Trichoptera) larvae in permanent and semipermanent high elevation ponds. B.S. Thesis, Allegheny College, Meadville, PA.

Sprules, W. G. 1972. Effects of size-selective predation and food competition on high altitude zooplankton communities. Ecology 53:375–386.

Taylor, T. P., and D. C. Erman. 1980. The littoral bottom fauna of high elevation lakes in Kings Canyon National Park, California. California Fish and Game 66:112–119.

Tonn, W., and J. J. Magnuson. 1982. Patterns in the species composition and richness of fish assemblages in northern Wisconsin lakes. Ecology 63:1149–1166.

Tonnessen, K. A. 1984. Potential for aquatic ecosystem acidification in the Sierra Nevada, California. Pages 147–169 *in* G. Henry (ed.), Early Biotic Response to Advancing Lake Acidification. Butterworth, Boston.

Tukey, J. W. 1954. Causation, regression, and path analysis. Pages 35–66 *in* O. Kepthorne, T. A. Bancroft, J. W. Gowen, H. L. Lush (eds.), Statistics and Mathematics in Biology. Iowa State College Press, Ames, IA.

Vertucci, F. A., and P. S. Corn 1996. Evaluation of episodic acidification and amphibian declines in the Rocky Mountains. Ecological Applications 6:449–457.

Voigts, D. K. 1976. Aquatic invertebrate abundance in relation to changing marsh vegetation. American Midland Naturalist 95:313–322.

Ward, J. V. 1992. Aquatic Insect Ecology. Vol. 1. John Wiley & Sons, New York.

Welch, P. S. 1952. Limnology. McGraw-Hill, New York.

Wellborn, G. A., D. K. Skelly, and E. E. Werner. 1996. Mechanisms creating community structure across a freshwater habitat gradient. Annual Review of Ecology and Systematics 27:337–363.

Wetzel, R. G. 1983. Limnology (2nd ed.). W. B. Saunders, Philadelphia, PA.

Whiteman, H. H. 1994. Evolution of facultative paedomorphosis in salamanders. Quarterly Review of Biology 69:205–221.

Whiteman, H. H., S. A. Wissinger, and A. Bohonak. 1994. Seasonal movement patterns and diet in a subalpine population of the tiger salamander *Ambystoma tigrinum nebulosum*. Canadian Journal of Zoology 72:1780–1787.

Whiteman, H. H., S. A. Wissinger, and W. S. Brown. 1996. Growth and foraging consequences of facultative paedomorphosis in the tiger salamander, *Ambystoma tigrinum nebulosum*. Evolutionary Ecology 10:429–422.

Wiggins, G. B., R. J. Mackay, and I. M. Smith. 1980. Evolutionary and ecological strategies of animals in annual temporary pools. Archiv für Hydrobiologie Supplement 58:97–206.

Wilbur, H. M. 1997. Experimental ecology of food webs: complex systems in temporary ponds. Ecology 78:2279–2302.

Willey, R. L. 1973. Emergence patterns of the subalpine dragonfly *Somatochlora semicircularis* (Odonata: Corduliidae). Psyche 81:121–133.

———. 1974. Drought resistance in subalpine nymphs of *Somatochlora semicircularis* (Odonata: Corduliidae). American Midland Naturalist 87:215–221.

Willey, R. L., and S. T. Threlkeld. 1993. Organization of crustacean epizoan communities in a chain of subalpine ponds. Limnology and Oceanography 38:623–627.

Williams, D. D. 1987. The Ecology of Temporary Waters. Timber Press, Portland, OR.

Windell, J. T., B. E. Willard, D. J. Cooper, S. Q. Foster, C. F. Knud-Hansen, L. P. Rink, and G. N. Kiladis. 1986. An Ecological Characterization of Rocky Mountain Montane and Subalpine Wetlands. U.S. Fish and Wildlife Service Biology Report 86(11).

Wissinger, S. A. 1989. Life history and size variation in larval dragonfly populations. Journal of the North American Benthological Society 7:13–28.

Wissinger, S. A., and W. S. Brown. Habitat permanence, life histories and the distribution of five caddisfly species in subalpine wetlands in central Colorado. Unpublished manuscript.

Wissinger, S. A., G. B. Sparks, G. L. Rouse, W. S. Brown, and H. Steltzer. 1996. Intraguild predation and cannibalism among larvae of detritivorous caddisflies in subalpine wetlands. Ecology 77:2421–2430.

Wissinger, S. A., and H. H. Whiteman. 1992. Fluctuation in a Rocky Mountain population of salamanders: Anthropogenic acidification or natural variation? Journal of Herpetology 26:377–391.

Wissinger, S. A., H. H. Whiteman, G. B. Sparks, G. L. Rouse, and W. S. Brown. 1998. Tradeoffs between competitive superiority and vulnerability to predation in caddisflies along a permanence gradient in subalpine wetlands. Ecology. In press.

Wooten, J. T. 1994. Predicting direct and indirect effects: an integrated approach using experiments and path analysis. Ecology 75:151–165.

Zerba, K. E., and J. P. Collins. 1992. Spatial heterogeneity and individual variation in the diet of an aquatic top predator. Ecology 73:268–279.

APPENDIX 31.A. TAXONOMIC LIST OF MACROINVERTEBRATES AT MEXICAN CUT

Turbellaria
  Mesostoma ehrenbergii
Rotifera
  Keratella cochlearis
  Conochilus hippocrepis
  Euchlanis sp.
Oligochaeta
  Lumbriculus variegatus
  Limnodrilus
Hirudinea
  Nephelopsis obscura
  Glossiphonia complanata
  Helobdella stagnalis
Mollusca
  Pisidium
Crustacea
  Anostraca
    Branchinecta coloradensis
  Cladocera
    Chydorus sphaericus
    Ceriodaphnia quadrangula
    Daphnia middendorffiana
    Daphnia pulex
    Daphnia rosea
    Scaphaloberis mucronata
    Simocephalus vetulus
  Copepoda
    Leptodiaptomus coloradensis
    Hesperodiaptomus shoshone
    Acanthocyclops vernalis
  Ostracoda
    Cypris palustera
Hydracarina
  Arrenurus
  Limnesia
  Piona
  Lebertia
  Hydrozetes
Ephemeroptera
  Callibaetis ferrugineus hageni

Odonata
  Anisoptera
    Aeshna palmata
    Somatochlora semicircularis
  Zygoptera
    Enallagma cyathigerum
    Coenagrion resolutum
    Lestes disjunctus
Hemiptera
  Corixidae
    Arctocorixa lawsonii
    Callicorixa audeni
    Coenocorixa bifida
  Gerridae
    Gerris gillettei
  Notonectidae
    Notonecta
Trichoptera
  Phyrganeidae
    Agrypnia deflata
  Limnephilidae
    Limnephilus picturatus
    Limnephilus externus
    Asynarchus nigriculus
    Hesperophylax occidentalis
Coleoptera
  Chyrsomelidae
    Plateumaris pusilla
  Dytiscidae
    Acilius semisulcatus
    Agabus tristus
    Agabus strigulosus
    Agabus kootenai
    Dytiscus duricus
    Dytiscus alaskanus
    Hydroporus
    Sanfilippodytes
    Ilybius fraterculus
    Rhantus gutticolis
    Stictotarsus grieostriatus

Haliplidae
  Haliplus leechi
Hydrophilidae
  Helophorus parasplendidus sperryi
  Helophorus eclectus
  Hydrobius fuscipes
Gyrinidae
  Gyrinus affinis
Staphylinidae
  Stenus
Diptera
Ceratopogonidae
  Bezzia
  Culicoides
Chaoboridae
  Chaoborus americanus
  Eucorethra underwoodi

Chironomidae
  Chironomus salinarius
  Chironomus riparius
  Cladopelma
  Dicrotendipes
  Endochironomous
  Microtendipes
  Pagastiella
  Pseudodiamesia
  Corynoneura
  Cricotopus
  Eukiefferella
  Paraphaenocladius
  Psectrocladius
  Ablabesmyia
  Procladius
  Cladotanytarsus
  Paratanytarsus
  Tanytarsus
Culicidae
  Aedes communis
  Aedes pullatus
Tipulidae
  Limnophila sp.
Stratiomyidae
  Stratiomyia sp.

32 Tinajas of Southeastern Utah

Invertebrate Reproductive Strategies and the Habitat Templet

CHESTER R. ANDERSON, BARBARA L. PECKARSKY, and SCOTT A. WISSINGER

*T*inajas are small (<30 m diameter) rain-filled wetlands that have been eroded out of sandstone bedrock. In this chapter we first describe the physical, chemical, and biological attributes of 460 tinajas. We then show how hydroperiod, predictability of hydroperiod, productivity, and species composition are correlated with pool depth and consider the various invertebrate reproductive strategies that are associated with these habitat parameters. Dissolved oxygen was used to estimate primary productivity, and the growth rate of the fairy shrimp Branchinecta lindahli *and the biomass of chironomids were used to estimate secondary productivity. Experiments were also conducted in artificial pools with cladocerans (Daphnia) and fairy shrimp to determine the response of secondary productivity to water depth. Pool depth appears to be the single most important abiotic factor affecting tinaja structure and function, governing hydroperiod (duration and predictability), primary and secondary productivity, and ultimately invertebrate abundance and diversity. Shallow pools had greater productivity (which resulted in faster larval maturation rates) and lower rates of predation and interspecific competition, but had shorter, less predictable hydroperiods. Deeper pools had longer, more predictable hydroperiods but lower productivity and greater rates of predation and interspecific competition. Macroinvertebrate communities in the shallow pools were dominated by crustaceans and snails, whereas the deep pools were dominated by a variety of wetland insects. The efficacy of desiccation tolerance or cyclic colonization, the strategies typically used by wetland invertebrates for colonizing temporary wetlands, should be compromised by the unpredictability of filling and drying schedules in shallow tinajas and*

Invertebrates in Freshwater Wetlands of North America: Ecology and Management, Edited by Darold P. Batzer, Russell B. Rader, and Scott A. Wissinger
ISBN 0-471-29258-3 © 1999 John Wiley & Sons, Inc.

greater predation and competition intensity in deeper pools. We argue that these habitat parameters should instead favor spatial and temporal bet-hedging strategies. Adult flying insects should partition their reproductive effort in multiple bouts of egg laying across different pools and at different times (annual iteroparity). Flightless crustaceans should partition their reproductive effort through passive dispersal of eggs, prolonged diapause, and variable egg hatching rates to ensure that at least one batch of propagules can complete development before pool drying in shallow pools or before being consumed by predators in deeper pools.

INTRODUCTION

Tinajas are small, rain-filled wetlands that develop in depressions of sandstone bedrock created by the action of wind and water (Spence and Henderson 1993). They occur in small drainages that flow only during rainstorms, and have particularly strong diurnal fluctuations of dissolved oxygen and temperature and hydroperiods that vary from a few days to permanent (Scholnik 1994). Tinajas are completely surrounded by bedrock, and most tinajas have no vegetation associated with them. Some tinajas contain low meadow (flat sandy areas downstream from the pools), with riparian obligate species, and a few tinajas have riparian trees and/or emergent vegetation (Spence and Henderson 1993, Haefner and Lindahl 1991).

In this chapter we first describe how basin morphometry (pool depth and pool exposure) affects hydroperiod, temperature, and productivity and thus the potential for predation and interspecific competition. We then consider the various life history strategies that may have evolved in response to this habitat templet (sensu Wellborn et al. 1996, Southwood 1977). Although a few tinajas occur as isolated pools, most are part of a system existing in close proximity along and between drainages (Fig. 32.1), and we argue that effective strategies for exploiting these habitats rely on the staggered and multiple hydroperiods of different pools within this system.

GEOGRAPHY, GEOLOGY, AND CLIMATE

Tinajas are found mostly on the Colorado Plateau, where large expanses of sandstone bedrock are exposed to arid weathering and wind erosion (see Dodson 1987, Grahm 1994, Scholnik 1994, LaFrancois 1996). They are most common on the eastern slope of the Waterpocket Fold of southeastern Utah, the location of our studies. The Fold is a monocline deformation that extends 170 km from Glen Canyon north to the base of Thousand Lake Mountain. The southern third of the monocline lies in Glen Canyon National Recreation

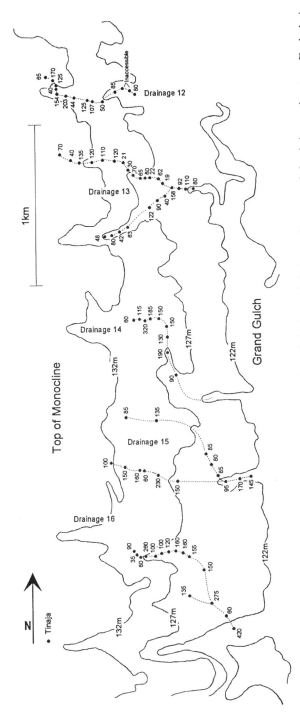

Fig. 32.1. A contour map showing drainages 12–16 and the distribution of tinajas and their maximum pool depth in centimeters. Each drainage empties into Grand Gulch, which contains an intermittent stream with sandy substrate that flows only when there is substantial rainfall.

Area, and the northern two thirds is in Capitol Reef National Park (CRNP). The monocline extends from an elevation of 1070 m to 2000 m from south to north. The majority of the tinajas lie on the Navajo sandstone formation of the monocline (Smith et al. 1963).

Data were collected in fall 1993 and spring 1994 as part of a larger survey conducted by the Division of Resource Management at CRNP (Borthwick 1993, Berghoff 1994, 1995). This survey focused in the area south of the Burr Trail, where a preliminary survey of aerial photographs and previous fieldwork indicated that tinajas were the most concentrated. We found 80 drainages in the 29-km section of the monocline that was surveyed, containing a total of 460 tinajas, with a range of 1–35 tinajas per drainage and a mean of 7 (Fig. 32.1). The distance between drainages averaged 0.36 km, and the distance between tinajas within a drainage ranged from <1 to >200 m.

The average annual precipitation at the headquarters of CRNP (elevation = 1676 m) is 20.07 cm. The greatest rainfalls are derived from monsoonal thundershowers that occur in July (2 cm), August (5 cm), September (1.6 cm), and October (1.7 cm). Average high air temperatures range from 33°C in July to 5°C in January, while average minimum temperatures range from 17°C in July to −7.5°C in January.

BASIN MORPHOMETRY, WATER TEMPERATURE, AND CHEMISTRY

Maximum pool volume was calculated as 1/2 the volume of an ellipsoid $[(2/3)*\pi*(L/2)*(W/2)*MPD)]$ (Dodson 1987, Haefner and Lindahl 1991) where MPD = maximum pool depth, L = maximum pool length, and W = pool width. We also estimated the height of the inlet of each tinaja, the percentage of the substratum consisting of sand (small (<10 percent), medium (10–50 percent), or large (>50 percent)), and the color of the water, and noted the presence of rust on the outlet or an oily film on the water surface. Inlet and outlet pour-offs of tinajas ranged in height from a few centimeters to >70 m. The lengths and widths of each tinaja (delineated by the surrounding sandstone) were also measured. Pool exposures were categorized, following Brown and Johnson (1983) and Spence and Henderson (1993), as exposed (receiving almost no shade during the day), protected (receiving some shade in the morning or afternoon), or very protected (rarely receiving sunlight).

The average surface area of the pools was 9.56 ± 0.86 m² (range 0.004–306 m²), the mean maximum pool depth was 1.07 ± 0.03 m (range 0.19–4.2 m), and the mean maximum pool volume was 13.43 ± 2.19 m³ (range 0.003–858 m³). All tinajas were surrounded by bare sandstone, and the total length (open water plus low meadow) of the tinajas averaged 12.4 ± 0.6 m (range 1.1–94 m), with a mean width of 6.5 ± 0.3 m (range 0.5–45 m). Fifteen percent of the tinajas had less than 10 percent sand substratum, 30 percent had 10–50 percent sand, and 55 percent had greater than 50 percent sand. Twenty percent of the pool's surface appeared oily, and 11 percent of tinajas had rusty stains on their outlets. Twenty-six of the tinajas surveyed were exposed, 59 percent were protected, and 12 percent were very protected.

The maximum (late afternoon) and minimum (at first light of the day) temperatures were measured monthly (March–June) in several pools at mid-depth. Water temperatures ranged from a low of 0°C in February to a high of 30.5°C in June. Diurnally, water temperatures increased faster and maximum temperatures were greater in the shallower pools than in the deeper pools. Shallow pools also cooled faster at night and minimum temperatures were lower than in the deeper pools. In winter exposed and protected tinajas had surface ice, and very protected pools were frozen solid.

Conductivity of the water in pools ranged from 3–134 μS cm^{-1}. The pH ranged from 6.6 to 9.5, with a mean of 7.5. Mean mg L^{-1} of PO$_3^-$ NH$_4^+$, NO$_3^-$, Cl, Fe, MgCO$_3$, and CaCO$_3$ were calculated from data previously collected by Haefner and Lindahl 1991 (Table 32.1).

RIPARIAN HABITATS

Twenty-five percent of tinajas surveyed had meadows containing low-growing herbs, grasses, or grass-like species (e.g., *Solidago* spp., *Elymus* spp., and *Juncus* spp.). Seventeen percent of tinajas had emergent vegetation found adjacent to or in the pools. This vegetation was either *Typha latifolia* (14.3 percent of the tinajas) or *Phragmites australis* (5.6 percent), and interestingly, those two species never cooccurred in the same tinaja. Beyond the emergent vegetation and mixed with the low meadow were riparian trees such as *Populus fremonti* (16.2 percent), *Salix nigra* (14 percent), *Tamarix ramisissima* (7.3 percent), *Salix exigua* (5 percent), and *Acer negundo* (0.7 percent). A few tinajas had pinyon pine and Utah juniper surrounding the riparian vegetation.

TABLE 32.1. Summary of the Mean, Maximum and Minimum Levels of Various Chemical Parameters in Tinajas

	N	Mean	Max	Min	Standard Error
Conductivity	408	43.34	134.00	3.00	1.17
pH	135	6.59	9.60	5.40	0.04
NH$_4^+$	88	0.32	2.80	0.05	0.04
PO$_3^-$	94	0.18	3.00	0.00	0.04
NO$_3^-$	70	0.13	3.00	0.01	0.04
Fe	174	0.43	8.50	0.00	0.07
MgCO$_3$	169	5.73	30.00	0.23	0.33
CaCO$_3$	168	19.35	103.00	3.09	1.16

Source: Data from J. W. Haefner and A. M. Lindahl, The Ecology of Small Pools in Capitol Reef National Park, UT, NPS Contract No. PX-1350-8-0187, 1991.

HYDROPERIOD

Unless aquatic insect larvae can diapause through dry periods, the time required for larvae to mature must be less than the hydroperiod (McLachlan 1983). Hydroperiods (days duration) of each tinaja were estimated as the product of evaporation rate and maximum pool depth, using the rate of change in pool depth (zm) as a linear estimate of evaporation rate. Evaporation rates were calculated three times in 1994 (February 21–March 9 [46 pools], March 20–April 3 [7 pools], April 30–May 9 [10 pools]), and once in 1995 (April 10–15 [29 pools]). We also conducted multiple regression analyses to predict evaporation rate and hydroperiod from maximum pool depth and degrees of pool exposure. Degrees of pool exposure were estimated as a continuous variable (different from the discrete exposure categories above) by measuring the degrees of open sky between the eastern and western horizons.

During periods when the sun was high and air temperatures were greatest (April and May), a significant proportion of the variation in evaporation rates could be explained by maximum pool depth and exposure ($r = 0.735$, $p = 0.066$; $r = 0.583$, $p = 0.005$, for April 30–May 9, 1994, and April 10–15, 1995, respectively). In those cases shallower, more exposed pools had faster evaporation rates. However, in February and March, when the sun was lower and temperatures were lower, evaporation rates could not be predicted by maximum pool depth. In contrast, maximum pool depth and pool exposure explained a large proportion of the variance in pool hydroperiod (duration in days) even when air temperatures and sun angles were lower ($r = 0.569$, $p < 0.001$; $r = 0.770$, $p = 0.105$; $r = 0.672$, $p = 0.066$; and $r = 0.735$, $p = 0.027$ for February 21–March 9, March 10–April 3–May 9, 1994, and April 10–15, 1995, respectively). Thus, over the time of the study, shallow exposed pools consistently had shorter hydroperiods than deeper pools.

PREDICTABILITY OF HYDROPERIOD

To test hypotheses concerning the evolution of reproductive strategies in response to the predictability of the tinaja habitat, we needed measures of habitat predictability that adequately reflected differences among habitats (Stearns 1981). We were particularly interested in the predictability of (1) the timing of pool filling and (2) the hydroperiod. We hypothesized that the timing of pool filling would select for invertebrates with efficient diapause stages or mobile stages that could select and remain in permanent refugia (Batzer and Wissinger 1996, Wissinger 1997). Further, unpredictable hydroperiods that result in high juvenile mortality should favor invertebrates with a diapausing larval stage or bet-hedging strategies (iteroparity and variable egg-hatching rates).

A model based on the Stanford Watershed Model (Crawford and Linsley 1966) was developed for tinajas for observation of hydroperiod over long periods of time. The depth of a pool at time t is dependent on the frequency and intensity of precipitation, the area of the watershed, the proportion of runoff, and evaporation rates. The change in pool depth from time t to t − 1 = $d_t - d_{t-1} = -r (t - 1) + 12ARP_{t-1}/\pi sa$, where d_t = depth of pool (cm) at time t (0 < d_t < Max Pool Depth), d_{t-1} = depth of pool (cm) at time t − 1, r = evaporation rate (cm/d) for time t − 1, A = area of the watershed above the pool, R = proportion of precipitation that reaches the pool (0 < R < 1), P = total precipitation in time t + 1, and sa = the surface area of the pool; 12/sa is a conversion of volume to depth based on a volume measurement that is 1/2 the volume of an ellipsoid (see Fig. 32.2 for a graphical explanation of the model).

The evaporation rate was determined as above for a continuum of maximum pool depths, and the amount of runoff was measured directly (Dunne and Leopold 1978). A rain gauge was placed next to one deep cylindrical pool with sa = 139,141 cm² in a drainage where A = 18.35 km². After each rain the amount of precipitation was recorded as well as z_m of the pool. We fit the equation $d_t - d_{t-1} = -0.37 + 1^{-2.14+19.01P}$ to these data points and $d_t - d^{t-1} = 12ARP/\pi sa$ was then solved for R for various amounts of precipitation.

Model predictions of pool hydroperiod were most sensitive to the area of the watershed and the proportion of runoff reaching tinajas. We determined that it takes only between 3 and 4 mm of rain to fill all the tinajas in the system. A 3-mm rainfall event between April and November occurred on average every 17 days, but the monthly variation for precipitation was very high.

Hydroperiods during the winter were much greater because of very low evaporation rates, and therefore pools were much less dependent on precipitation to maintain their water. This usually resulted in a large percentage of

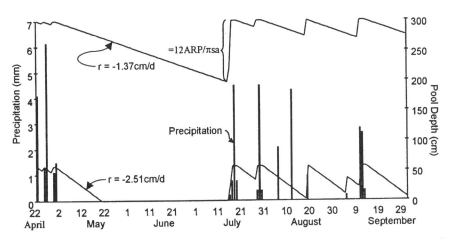

Fig. 32.2. Precipitation and modeled pool depths for the driest summer on record (1942). Maximum pool depths are 300 and 50 cm. r = evaporation rate = 0.064 ln(Maximum Pool Depth).

the pools having water at the beginning of the breeding season (Fig. 32.2). But pools became much more dependent on precipitation to maintain their water as evaporation rates increased throughout the summer, and since shallow pools had higher rates of evaporation as well as less depth, they were particularly dependent on precipitation events (Fig. 32.2). Therefore, we quantified predictability of hydroperiod during the breeding season (May–September) using long-term precipitation data (39 years) and Colwell's (1974) contingency (M) and constancy (C) estimations.

Contingency measures strong predictable fluctuations, and constancy will be 1 for unchanging measures. Predictability was the sum of C and M, and for tinajas constancy equaled 0.21 and contingency equaled 0.06, and therefore predictability equaled 0.27, which is considered very low (Colwell 1974). From this analysis we concluded that the predictability of pool filling entering the breeding season (May) was high, but throughout the rest of the season the predictability of hydroperiod and the predictability of pool filling were extremely low.

PRIMARY PRODUCTIVITY

Rapid growth and development are important for the exploitation of temporary habitats, and levels of primary production should therefore have an important effect on whether invertebrate consumers can or cannot successfully colonize tinajas. Productivity influences the maturation rate of aquatic larvae, and for reproduction to be successful aquatic larvae must be able to mature to the adult stage before the pool dries, unless they have the ability to diapause through the dry period (McLachlan 1983).

As an estimate of primary productivity among tinajas we used the difference between the diel maximum (late afternoon) and minimum (first light of day) dissolved oxygen (DO). These differences were regressed against pool depths (z_m) and exposures to determine whether they correlated with primary productivity. The production of DO depended on the balance among photosynthesis, total respiration, and exchanges with the atmosphere (Ginot and Herve 1994) and was a function of light intensity (which decays exponentially in water) and water temperature, which were in turn influenced by water depth and exposure (Bierman et al. 1994, Ginot and Herve 1994, Haeder and Figueroa 1997, Clarke 1939). To compare production among tinajas of different depths and exposures we assumed that:

1. Water was mixed daily (Ginot and Herve 1994).
2. The night-time transfer of oxygen to and from the atmosphere was constant among pools.
3. Community metabolism could be accounted for by subtracting the minimum dissolved oxygen measured at first light of day.

4. The increase in DO from first light to midafternoon was due to primary productivity from photosynthesis
5. The rate of oxygen loss to the atmosphere increased during the day as oxygen exceeded the saturation point in the shallow pools, conservatively skewing the results towards failure to reject the null hypothesis of no difference in productivity between shallow and deep pools (Ginot and Herve 1994).

We found that DO at first light was the diel minimum and was below the 100 percent saturation point in all of the pools. By approximately 1000 h DO reached saturation and continued to supersaturation in the shallow and medium depth pools but not in the deep pools. It peaked between 1300 and 1700 h (e.g., Horne 1972, Morton and Bayly 1977, Hama and Handa 1992). At water depths of 15, 65, and 110 cm the DO was the same at first light (supporting our assumption of daily mixing) but diverged as the day advanced (Fig. 32.3). The DO in the deeper layers did not increase as much as it did in the shallower layers, indicating that benthic algae were not contributing as much to primary productivity in the deeper pools as in the shallow pools (e.g., Markager 1994). Furthermore, the differences between the diel maximum and minimum DO measured at first light showed a logarithmic decline with pool depth. This decline, however, was only significant for the dates later in the season when sunlight was more direct and temperatures were higher ($r = 0.343$, $p = 0.392$; $r = 0.603$, $p = 0.105$; $r = 0.610$, $p = 0.006$; $r = 0.797$, $p < 0.001$, for March, April, May, and June 1994, respectively). These data indicate that primary productivity per unit volume in the tinajas was greater in the shallow pools than in the deeper pools, especially during the reproductive periods of the invertebrates (the summer months).

We tested this hypothesis experimentally by estimating primary productivity in artificial shallow and deep pools. Ten plastic tubs (surface area = 1568 cm^2, 35 cm deep) were set 25 cm into the ground so that temperatures were comparable to natural tinajas. Five tubs were filled on April 1, 1994, to 30 cm depth (deep) with filtered water from a second-order stream. The other five tubs were filled to 15 cm depth (shallow). These depths were maintained throughout the experiment by adding filtered, boiled, and cooled water. The tubs were inoculated on April 4 with algae and *Daphnia* (see below for estimates of secondary productivity). *Daphnia* were collected from a nearby tinaja and added to the experimental pools at densities of seven *Daphnia* L^{-1}. On April 17 the maximum DO was measured in each tub at 1500 h, and on April 18 the minimum DO was measured at 0615 h and the mean difference between the max and min DO was used again as an estimate of primary productivity. Maximum and minimum temperatures were not significantly different between the treatments. As in the field data, primary productivity was higher in shallow tubs than deep tubs, supporting the field observations (deep tubs = 24.15 ± 0.37 cm, DO = 1.54 ± 0.11 mg/L; shallow tubs = 9.65 ± 0.29 cm, DO = 2.54 ± 0.31 mg/L; $t = 2.78$, $p = 0.05$).

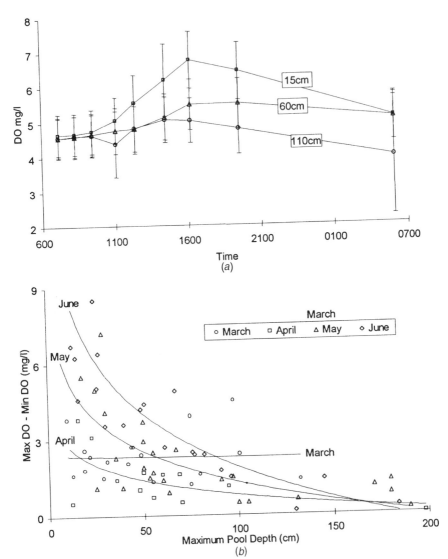

Fig. 32.3. (*a*) Mean (± SE) diel dissolved oxygen at 15, 65, and 110 cm depth in three deep pools (pool depths = 130, 145, and 180 cm). Measurements taken on June 2–3, 1994. (*b*) The difference between the diel maximum and minimum dissolved oxygen taken at middepth regressed against pool depth for four dates. r = 0.343, p = 0.392; r = 0.603, p = 0.105; r = 0.610, p = 0.006; r = 0.797, p < 0.001, for March, April, May, and June 1994, respectively.

SECONDARY PRODUCTIVITY

We used several different methods to estimate secondary productivity in experimental and natural tinajas to determine whether the differences in primary productivity were reflected in higher trophic levels. In general, results of secondary productivity experiments conducted in artificial habitats (tubs) agreed closely with primary productivity data from tubs and tinajas; i.e., secondary production was greater in shallow than in deep pools. These higher productivity rates should have facilitated the rapid development of species, making it more likely that they could emerge before these extremely ephemeral habitats dried. In natural pools this difference in secondary production, as measured by invertebrate biomass, should have been exaggerated by the lower densities (hence lower prey removal) of predators in temporary versus permanent wetlands. That the comparative data were less conclusive might have been the result of variations in predator densities among pools with comparable hydroperiods.

Zooplankton Biomass

The biomass of zooplankton in tinajas was estimated in the field on April 13, 1994, in 11 tinajas that did not have fairy shrimp and where *Daphnia* made up a mean of 98.6 ± 0.9% of the zooplankton biomass L^{-1}. The remainder of the samples consisted of cyclopoid copepods. Biomass of *Daphnia* was calculated from vertical tows and regressed against pool depth and exposure. These data revealed no relationship between biomass L^{-1} and pool depth. In contrast, *Daphnia* collected from the shallow and deep tubs on April 28 had greater densities and biomass per liter in the shallow than in the deep experimental tubs. The mean density (no. L^{-1}) in deep tubs was 5.59 ± 0.45 and in shallow tubs was 24.47 ± 3.48 (ANOVA F = 28.94, p < 0.01; 1, 8 df); and the mean biomass (mg L^{-1}) in deep tubs was 0.14 ± 0.02 and in shallow tubs was 0.60 ± 0.11 (ANOVA F = 17.45, p < 0.01; 1, 8 df). However, individual sizes of *Daphnia* did not differ between deep and shallow treatments (ANOVA F = 0.00, p = 0.96; 1, 8 df).

Fairy Shrimp Growth Rates

We measured the growth rate per liter for two cohorts of fairy shrimp, one in March and one in April 1994, that were sampled with vertical tows from 6 and 7 pools, where fairy shrimp made up 97.0 ± 0.2 percent of the biomass. Total dry mass per liter of water was calculated and adjusted for temporal decreases in pool volume. Dry mass gained per day was regressed against mean pool depth and pool exposure to examine the relationship between secondary productivity and pool depth. As in the pools dominated by *Daphnia,* there was no significant relationship between growth rate of fairy shrimp and either pool depth or exposure. As an independent estimate of secondary productivity in pools, we estimated the effects of pool depth on fairy shrimp growth rates experimentally. Experimental tubs were set up as in the *Daphnia* experiment and incubated for one week with phytoplankton and periphyton obtained from a nearby tinaja. On April 27, 1995, fairy shrimp that averaged 3 mm in length and were hatched from a single cohort were placed into tubs at densities of 1 L^{-1} (48 per deep tub, 24 per shallow tub, mean dry mass per fairy shrimp = 0.019 mg). After two weeks the mean densities and changes in dry mass per liter were calculated and compared between treatments with a t-test. As in the tub experiments with *Daphnia,* fairy shrimp biomass was greater in the shallow (0.67 ± 0.07 mg L^{-1}) than in the deep (0.47 ± 0.04 mg L^{-1}) tubs (t = 2.36, p = 0.09).

Total Benthic and Limnetic Biomass in Tinajas

Finally, on May 15, 1995, we estimated total secondary productivity for both the benthos and the limnetic zone in each of 22 pools. This sampling was conducted before the reproductive periods of most of the predators (to reduce the effects of predation on biomass; Pajunen and Salmi 1991) but late enough in the year that solar radiation and temperature could stimulate primary productivity. The limnetic zone was sampled as above, and biomass of invertebrates (mg dry mass per liter) was calculated. Benthic organisms were scraped from the substratum, separated from the mud, rinsed, dried at 60°C for 36 hours, then weighed. As in most of the other field data, we found no relationship between total dry mass of limnetic organisms and pool depth. However, the regression between mg dry mass of chironomids, which made up the entire benthic biomass, and pool depth was significant (r = 0.701, p < 0.001), with higher benthic midge biomass in the shallow pools.

INVERTEBRATE COMMUNITY SAMPLING

To determine the composition of the entire invertebrate community, we qualitatively surveyed all 460 tinajas by sampling with a D-net in each habitat type (the water column, substrate, and emergent vegetation) of each pool. Organisms captured were identified in the field to at least the generic level, and unknown specimens were preserved and later identified in the laboratory. We collected 44 invertebrate taxa during this survey and four amphibian taxa (Table 32.2), which was similar to past sampling results by Haefner and Lindahl (1991). The invertebrates ranged from typical permanent water species to those found only in temporary pools (Wiggins et al. 1980, Williams 1985). Cumulative relative frequencies of maximum pool depth were calculated for predators versus nonpredators. Those frequency distributions were statistically compared using a Kolmogorov-Smirnov Two Sample Test, indicating that predators used the deeper pools more than the shallow pools (D = 0.044; $D_{0.05}$ = 0.042). A regression between maximum pool depth and

TABLE 32.2. Aquatic Taxa of Tinajas

Order/Family	Genus/Species
Copepoda	*Eucyclops agilis*
Anamopoda	*Daphnia*
	Ceriodaphnia quadrangula
Anostraca	*Branchinecta lindahli*
	Streptocephalus texanus
Conchostraca	*Eulimnida texana*
Notostraca	*Triops longicaudus*
Baetidae	*Callibaetis*
Aeshnidae	*Aeshna walkeri*
	Anax epiashna
Libellulidac	*Pantala hymenea*
	Sympetrum
	Pachydiplex
	Belonia
	Tramea
Corduliidae	*Somatochlora semicircularis*
Lestidae	*Archilestes grandis*
Coenagrionidae	*Enallagma praevanun*
Notonectidae	*Notenecta kirbyi*
Corixidae	*Graptocorixa abdominalis*
	Corisella edulis
Gerridae	*Gerris remigis*
Limnephilidae	*Limnephilus atercus*
Dytiscidae	*Deronectes striatellus*
	Rhantus gutticollis
	Laccophilus maculosus
	Agabus ilybiiformis
	Agabus disintigratus
	Derovatellus
	Oreodytes
Hydrophyllidae	*Tropisternus ellipticus*
	Berosus
Gyrinidae	*Gyrinus*
Chironomini	*Stictochironomus*
	Phaenospectra
Tanypodinae	*Ablabesmyia*
	Tanytarsini
Orthocladiinae	*Tanytarsus*
	Psectrocladius
Culicidae	*Culiseta maccrackenae*
	Culex tarsalis
Tabanidae	*Tabanus*
Ceratopogonidae	*Bezzia*
Gastropoda	*Physella*
Anura	*Scaphiopus intermontanus*
	Bufo punctatus
	Bufo woodhousei
	Hyla arenicolor

number of taxa per tinaja indicated there were more taxa in deeper than in shallower pools ($r = 0.285$, $p < 0.001$), suggesting that interspecific competition might be greater in deeper than in shallow pools.

THE HABITAT TEMPLET AND REPRODUCTIVE STRATEGIES

Southwood (1977, 1988) distinguishes among three main components of habitat templet: durational stability (ephemerality), temporal variability (predictability), and magnitude of fluctuation (harshness). Tinajas exist on a continuum from shallow to deep pools that determines their ephemerality, predictability, and harshness. Shallower pools have shorter, less predictable hydroperiods but greater productivity and lower rates of predation and interspecific competition. Deeper pools have longer, more predictable hydroperiods but lower production rates and greater rates of predation and interspecific competition.

Invertebrates exhibit two general strategies for exploiting temporary wetlands (Batzer and Wissinger 1996). The first strategy is desiccation tolerance, which is well documented in flightless organisms (mollusks, crustaceans, worms) and in a variety of aquatic insects in which eggs, immatures, and/or adults are known to be able to survive in dried substrates (Wiggins et al. 1980, Williams 1985, Batzer and Wissinger 1996). However, in tinajas the dry phase of the hydroperiod is so harsh that this strategy might often not be effective. Much of the substrate is bedrock, and where sandy alluvium has collected the sediments often dry completely. The dried sand is also susceptible to wind erosion, and our observations suggest that immature stages such as larvae or pupae would be vulnerable to predation, exposure, and even wind transport (Maguire 1963). Based on rehydration experiments and sampling of dry sediments during the dry phase of tinaja hydroperiods, desiccation-tolerant eggs of crustaceans were the only common invertebrate life stage that we found. Thus, it does not appear that desiccation tolerance is as important a strategy for survival in tinajas as it is in other types of temporary wetlands (Wiggins et al. 1980, Williams 1985, Batzer and Wissinger 1996).

A second strategy for exploiting temporary wetlands that is employed by aquatic insects is seasonal migration between permanent and temporary habitats (Batzer and Wissinger 1996, Wissinger 1997). This strategy, termed *cyclic migration,* should be especially effective in situations in which wetlands with different hydroperiods are clustered in wetland complexes, as in the linear sequences of tinajas with various hydroperiods that occur at our study site. Cyclic colonizers such as predaceous diving beetles, a variety of hemipteran water bugs (water boatmen, back-swimmers, water striders), and dragonflies typically exploit predictable ephemeral waters by colonizing them at the beginning of the wet phase of the hydroperiod and then completing one or more generations before emerging and emigrating back to permanent refugia before the temporary habitats dry (Wissinger 1997). However, the unpredictability of tinajas could compromise this strategy, especially if the duration of the wet phase of hydroperiods is shorter than the development time of larval insects. Our observations of dead or dying larvae in drying pools suggest that this unpredictability is an important component of the habitat templet of tinajas.

Because tinajas have unpredictable hydroperiods, it will be difficult for adults to predict whether a particular basin is likely to remain wet long enough for the completion of offspring development. Thus, rather than risk their entire reproductive effort on one particular habitat through one bout of reproduction (semelparity), adults should distribute propagules among pools as a spatial bet-hedging strategy, or distribute propagules over time through repeated bouts of reproduction as a temporal bet-hedging strategy, such that at least one batch of propagules will be deposited in a habitat that remains wet long enough for the completion of larval development (Hairston and Bohonak 1998, Hairston et al. 1996). Life history theory predicts that repeated bouts of reproduction, or iteroparity, will be favored over semelparity when juvenile mortality is high relative to adult mortality, as should be the case in unpredictably drying wetlands (Murphy 1968, Shaffer 1974, Stearns 1976). Because of the close proximity of tinajas, the high variability in precipitation within a season, the high runoff, and the consequential filling of pools in the tinaja system, this type of annual iteroparity (see terminology in Begon et al. 1985) in combination with spatial bet-hedging should be effective for both fugitive species that opportunistically colonize tinajas and cyclic colonizers that rely on source populations in permanent tinajas. We have observed the adults of beetles, back-swimmers, water boatmen, and dragonflies sequentially leaving and entering different tinajas within a drainage, suggesting that all of these taxa employ this bet-hedging strategy. For taxa with short-lived adults iteroparity and spatial bet-hedging should be less important. However, even short-lived semelparous adults (e.g., microcrustaceans and chironomid dipterans) can spatially bet-hedge by passive dispersal of diapausing eggs and temporally bet-hedge by producing diapausing eggs that hatch at different times in response to different sequences of wetting and drying (Hairston et al. 1985, Batzer and Wissinger 1996). Regardless of the colonization strategies employed by the invertebrates that inhabit tinajas, rapid development will be a coadaptation that is necessary for the successful completion of development both for taxa that deposit desiccation-resistant eggs and for those that emigrate as adults (Wiggins et al. 1980, Williams 1985, Wellborn et al. 1996). Such rapid development may be facilitated by the greater temperatures and productivity and lower rates of predation and interspecific competition in the more ephemeral, shallow pools, allowing for greater exploitation of these environments (Anderson and Cummins 1979, Richardson and Mackay 1984, Peckarsky and McIntosh 1998, Twombly and Burns 1998).

As in other small aquatic ecosystems, crustaceans used the shallower, more ephemeral tinajas, presumably to avoid greater predation rates and interspecific competition found in the deeper, more permanent pools (Table 32.3,

TABLE 32.3. Summary of Results

The Habitat Templet	
Shallow Pools	Deep Pools
Ephemeral	Permanent
Unpredictable	Predictable
Greater temperature fluctuations	Smaller temperature fluctuations
More productive	Less productive
Less predation	More predation
Less interspecific competition	More interspecific competition
Life History Strategies	
Shallow Pools	Deep Pools
Faster larval growth rate	Slower larval growth rate
Nondiapausing larvae	Nondiapausing larvae
Diapausing eggs	Mobile adults
Variable egg hatching	Longer-lived adults
Annual semelparity	Annual iteroparity
Passive bet-hedging	Active bet-hedging

Woodward 1983, Wellborn et al. 1996). Crustaceans in tinajas are semelparous, without a diapausing larval stage. However, with a diapausing egg stage, crustaceans can exploit highly ephemeral tinajas by having phenomenal maturation rates made possible by the high primary productivity in these environments.

Aquatic insect larvae apparently do not have the ability to take advantage of the greater productivity in shallow tinajas because they are unable to mature as quickly as crustaceans (Table 32.4). Therefore, their life history strategies must include ways to cope with the greater predation rates and interspecific competition that is expected in deeper, more permanent pools. Predation in deeper, more permanent pools also increases juvenile mortality in aquatic habitats and, like unpredictable hydroperiods, selects for bet-hedging strategies. Unlike crustaceans, once insects mature to adults, they have the ability to actively escape from deteriorating environments and migrate to other habitats, both terrestrial and aquatic, and wait for another favorable breeding period. In contrast, crustaceans escape deteriorating environments and wait for a favorable breeding period using diapausing eggs (Hairston and Munns 1984, Danks 1987).

CONCLUSIONS

In tinajas hydroperiod, predictability of hydroperiod, and primary productivity are all functions of pool depth. Shallow pools have greater productivity and

TABLE 32.4. The Relationship between Life History Characteristics and Maximum Pool Depth[a]

Taxon		Bet-hedge over Time Active = A	Bet-hedge over Space Passive = P	Diapause?	Adult Mobility	Mean MPD	N	S.E.
Snails (*Physella*)	Semelparous	P	P	Eggs	No	1.10	9	0.1
Fairy shrimp (*Branchinecta*)	Semelparous	P	P	Eggs	No	1.15	115	0.1
Cladocerans	Semelparous	P	P	Eggs	No	1.19	15	0.1
Damselflies	Iteroparous	A	A	Eggs?	Yes	1.30	15	0.1
Mosquitoes	?	?	?	No	Yes	1.35	46	0.1
Water boatmen (*Graptocorixa*)	Iteroparous	A	A	No	Yes	1.63	292	0.1
Back-swimmers (*Notonecta*)	Iteroparous	A	A	No	Yes	1.65	224	0.1
Caddisflies (*Limnephilus*)	?	A	A	Eggs	Yes	1.66	88	0.2
Mean Max Pool Depth	—	—	—	—	—	1.70	463	0.1
Water striders (*Gerris*)	Iteroparous	A	A	No	Yes	1.73	181	0.1
Midges	?	?	?	Eggs?	Yes	1.76	239	0.1
Mayflies (*Callibaetis*)	?	?	?	No	Yes	1.78	125	0.1
Tadpoles	Iteroparous	A	A	No	Yes	1.82	44	0.3
Beetles	Iteroparous	A	A	Eggs?	Yes	1.85	103	0.2
Dragonflies	Iteroparous	A	A	Eggs?	Yes	1.93	28	0.4

[a]The table is sorted by ascending mean maximum pool depth (MPD).

temperatures (which resulted in faster larval maturation rates), shorter, less predictable hydroperiods, and lower rates of predation and interspecific competition. Deeper pools have longer, more predictable hydroperiods, lower productivity, and greater rates of predation and interspecific competition. We also found that immatures of semelparous species (crustaceans) predominated in the shallower, more unpredictable habitats and larvae of annually iteroparous species (insects) were most abundant in the deeper, more predictable habitats. We predict that the persistence of many of the invertebrate populations in the tinaja system depends on a complex interplay between within-pool population dynamics and on the connectivity of subpopulations across clusters of pools with different hydroperiods. We argue that the evolution of reproductive strategies in invertebrates living in tinajas is partly a response to high juvenile mortality caused by short and unpredictable hydroperiods in shallow pools, as well as greater productivity and faster maturation rates in these habitats and high predation intensity in deeper pools. We propose that these habitat parameters, along with the spatial mosaic of the pools, should lead to the evolution of bet-hedging strategies that include the partitioning of reproductive effort among pools and over time (spatial and temporal bet-hedging). Spatial bet-hedging strategies include passive dispersal of diapausing eggs and active dispersal of eggs by mobile adults. Temporal bet-hedging strategies include variable egg-hatching rates and annual iteroparity. Clearly, an understanding of population and community dynamics will require a metapopulation and even a metacommunity approach, and we suggest that these habitats provide an excellent opportunity for the development of spatially explicit models for metapopulations that depend on unidirectional and cyclic movements between patchily distributed subpopulations.

LITERATURE CITED

Anderson, N. H., and K. W. Cummins. 1979. Influences of diet on the life histories of aquatic insects. Journal of the Fisheries Research Board of Canada 36:335–342.

Batzer, D. P., and S. A. Wissinger. 1996. Ecology of insect communities in nontidal wetlands. Annual Review of Entomology 41:75–100.

Begon, M., J. L. Harper, and C. R. Townsend. 1990. Ecology: Individuals, Populations and Communities (2nd ed.). Blackwell Scientific Publications, Boston.

Berghoff, K. 1994. Capitol Reef National Park Tinaja Wetland Survey, Summary Report. Unpublished.

———. 1995. Capitol Reef National Park Tinaja Wetland Survey, Summary Report. Unpublished.

Bierman, V. J., Jr, N. N. Rabalais, S. C. Hinz, D. W. Zhu, R. F. Turner, and W. J. Wiseman, Jr. 1994. A preliminary mass balance model of primary productivity and dissolved oxygen in the Mississippi River plume/inner Gulf Shelf region. Estuaries 17:886–899.

Borthwick, S. 1993. Capitol Reef National Park Tinaja Wetland Survey Proposal. Unpublished.

Brown, B. T., and R. R. Johnson. 1983. The distribution of bedrock depressions (tinajas) as sources of surface water in Organ Pipe Cactus National Monument, Arizona. Journal of the Arizona-Nevada Academy of Science 18:61–68.

Charnov, E. L., and W. M. Schaffer. 1973. Life history consequences of natural selection: Cole's result revisited. American Naturalist 107:791–793.

Clarke, G. L. 1939. The utilization of solar energy by aquatic organisms. *In* Problems of Lake Biology. Publications of the American Association for the Advancement of Science 10:27–38.

Colwell, R. K. 1974. Predictability, constancy, and contingency of periodic phenomena. Ecology 55:1148–1153.

Crawford, N. H., and R. K. Lindsley. 1966. Digital Simulation in Hydrology: Stanford Watershed Model IV. Stanford University Department of Civil Engineering, Palo Alto, CA.

Danks, H. V. 1987. Insect dormancy: An Ecological Perspective. Biological Survey of Canada Monograph Series. Number 1. Ottawa, ON.

Dodson, S. I. 1987. Animal assemblages in temporary desert rock pools: Aspects of ecology of *Dasyhelea sublettei* (Diptera: Ceratopogonidae). Journal of the North American Benthological Society 6:65–71.

Dunne, T., and L. B. Leopold. 1978. Water in Environmental Planning. W. H. Freeman & Co., San Francisco.

Ginot, V., and J. Herve. 1994. Estimating the parameters of dissolved-oxygen dynamics in shallow ponds. Ecological Modelling 73:169–187.

Grahm, T. B. 1994. Predation by dipteran larvae on fairy shrimp (Crustacea: Anostraca) in Utah rock-pools. Southwestern Naturalist 39:206–207.

Haeder, Donat-P., and F. L. Figueroa. 1997. Photoecophysiology of marine macroalgae. Photochemistry and Photobiology 66:1–14.

Haefner, J. W., and A. M. Lindahl. 1991. The Ecology of Small Pools in Capitol Reef National Park, UT. NPS Contract No. PX-1350-8-0187.

Hairston, N. G., Jr., and A. J. Bohonak 1997. Copepod reproductive strategies: Life-history theory, phylogenetic pattern and invasion on inland waters. Journal of Marine Systems 15:23–34.

Hairston, N. G., Jr., C. M. Kearns, and S. P. Ellner. 1996. Phenotypic variation in a zooplankton egg bank. Ecology 77:2382–2392.

Hairston, N. G., Jr., and W. R. Munns, Jr. 1984. The timing of copepod diapause as an evolutionarily stable strategy. American Naturalist 123:733–751.

Hama, J., and N. Handa. 1992. Diel photosynthetic production of cellular organic matter in natural phytoplankton populations, measured with ^{13}C and gas chromatography/mass spectrometry. Marine Biology 112:183–190.

Horne, F. R. 1971. Some effects of temperature and oxygen concentration on phyllopod ecology. Ecology 52:343–347.

LaFrancois, T. 1995. Biology and ecology of rock pools in Capitol Reef National Park, Utah. Master's Thesis, Colorado State University, Fort Collins, CO.

Maguier, B., Jr. 1963. The passive dispersal of small aquatic organisms and their colonization of isolated bodies of water. Ecological Monographs 33:161–185.

Markager, S. 1994. Open-water measurement of areal photosynthesis in a dense phytoplankton community. Archiv für Hydrobiologie 129:405–424.

McLachlan, A. 1983. Life-history tactics of rain-pool dwellers. Journal of Animal Ecology 52:545–561.

Morton, D. W., and I. A. E. Bayly. 1977. Studies on the ecology of some temporary freshwater pools in Victoria with special reference to microcrustaceans. Australian Journal of Marine and Freshwater Research 28:439–454.

Murphy, G. 1. 1968. Pattern in life history and the environment. American Naturalist 106:581–588.

Peckarsky, B. L., and A. R. McIntosh. 1998. Fitness and community consequences of avoiding multiple predators. Oecologia 113:565–576.

Pajunen, V. I., and J. Salmi. The influence of corixids on the bottom fauna of rock-pools. Hydrobiologia 222:77–84.

Richardson, J. S., and R. J. Mackay. 1984. A comparison of the life history and growth of *Limnephilus indivisus* (Trichoptera: Limnephilidae) in three temporary pools. Archiv für Hydrobiologie 99:515–528.

Schaffer, W. M. 1974. Selection for optimal life histories: the effects of age structure. Ecology 55:291–303.

Scholnick, D. A. 1994. Seasonal variation and diurnal fluctuation in ephemeral desert pools. Hydrobiologia 294:111–116.

Smith, F., Jr., L. D. Huff, E. N. Hinrichs, and R. G. Luedke. 1963. Geology of the Capitol Reef Area, Wayne, Garfield Counties, Utah. Geological Survey Professional Paper No.363.

Southwood, T. R. E. 1977. Habitat, the templet for ecological strategies? Journal of Animal Ecology 46:337–365.

Spence, J. R., and N. R. Henderson. 1993. Tinaja and hanging garden vegetation of Capitol Reef National Park, southern Utah, U.S.A. Journal of Arid Environments 24:21–36.

Stearns, S. C. 1976. Life-history tactics: A review of the ideas. Quarterly Review of Biology 51:3–47.

Stearns, S. C. 1981. On measuring fluctuating environments: Predictability, constancy and contingency. Ecology 62:185–199.

Twombly, S., and C. W. Burns. 1998. Effects of food quality on individual growth and development in the fresh-water copepod *Boeckella triarticulata.* Journal of Plankton Research 18:2179–2196.

Wellborn, G. A., D. K. Skelly, and E. E. Werner. 1996. Mechanisms creating community structure across a freshwater habitat gradient. Annual Review of Ecology and Systematics 27:337–363.

Wiggins, G. B., R. J. Mackay, and I. M. Smith. 1980. Evolutionary and ecological strategies of animals in annual temporary tanks. Archiv für Hydrobiologie Supplement 58:97–206.

Williams, D. D. 1985. Biotic adaptations in temporary lentic waters with special reference to those in semi-arid regions. Hydrobiologia 125:85–110.

Wissinger, S. A. 1997. Cyclic colonization in predictably ephemeral habitats: A template for biological control in annual crop systems. Biological Control 10:4–15.

Woodward, B. D. 1983. Predator-prey interactions and breeding-pond use of temporary pond species in a desert anuran community. Ecology 64:1549–1555.

33 Spring-Formed Wetlands of the Arid West

Islands of Aquatic Invertebrate Biodiversity

MARILYN J. MYERS and VINCENT H. RESH

*T*he Great Basin is the largest desert in the United States, but numerous wetlands associated with springs are scattered across its arid landscape. Studies of the Great Basin aquatic invertebrate fauna indicate that while individual spring habitats may or may not be species-rich, the diverse range of spring types (e.g., lentic, lotic, hot water, cold water) provide an array of conditions that can be used by many species of aquatic invertebrates. Even with limited surveys, several endemic species have been discovered, especially in less motile groups such as spring snails (Gastropoda: Hydrobiidae). Because of the isolation and harshness of the intervening landscape, aquatic invertebrates in the Great Basin face unusual challenges for dispersal and colonization compared with those of temperate systems. Although the Great Basin is sparsely populated by humans, few spring wetlands have not been impacted by human-directed activities, especially cattle grazing, water development or diversion, and recreational activities.*

INTRODUCTION

Throughout much of the arid West, springs are the primary source of permanent water that provides critical habitat for a variety of endemic fish (e.g., Sigler and Sigler 1994), amphibians (e.g., Hovingh 1993, Sada et al. 1995),

Invertebrates in Freshwater Wetlands of North America: Ecology and Management, Edited by Darold P. Batzer, Russell B. Rader, and Scott A. Wissinger
ISBN 0-471-29258-3 © 1999 John Wiley & Sons, Inc.

and aquatic invertebrates (e.g., Hovingh 1993, Hershler 1998). The springs and surrounding riparian areas provide water for introduced and native mammals (e.g., bighorn sheep, antelope, burros) and habitat for birds (Carothers et al. 1974, Szaro and Jakle 1985); they also serve as important links between aquatic and terrestrial ecosystems (Jackson and Fisher 1986, Murkin and Batt 1987, Gray 1993). Because of the species richness of the plants, vertebrates, and invertebrates that are concentrated in these habitats, spring-formed wetlands are "hotspots" of biodiversity (Shepard 1993, Sada et al. 1995). Although the vertebrates that occur in the Great Basin have received attention (e.g., Brown 1978, Sigler and Sigler 1987), invertebrates (and especially aquatic invertebrates) have been largely neglected.

The geographical area considered in this chapter is the Great Basin (a subset of the Basin and Range Province in which all drainages end in closed basins), which includes nearly all of Nevada, portions of eastern California, southeastern Oregon, southern Idaho, and the western half of Utah (Fig. 33.1). Although there are lakes (e.g., Great Salt Lake, Lake Topaz, Pyramid Lake) and rivers (e.g., Sevier, Humboldt, Truckee, and Carson Rivers) that have associated wetlands in this region, the focus of this chapter is on wetlands formed by springs.

For the purposes of this chapter, a "spring" refers to the spring source and the downstream habitat it creates; together they create the spring wetland.

Fig. 33.1. Great Basin (as determined by Fenneman and Johnson 1946) and surrounding states.

This definition of a spring is less restrictive than definitions that make a distinction between the spring source and the downstream areas, which is usually based on fluctuations in water temperature (e.g., Erman and Erman 1995). We choose this approach because even though there is a change in habitat, species composition, and temperature between the source and downstream areas, the springs and their surrounding wetlands in the arid West represent isolated "islands" of water (e.g., a springbrook only 200 m in total length or a wetland habitat of less than 1 ha) that we will consider as one unit.

In this chapter we give a brief description of the Great Basin, discuss spring classification, and contrast dispersal and colonization of spring invertebrates in desert versus temperate systems. The limited information on biodiversity of aquatic invertebrates in spring habitats of the Great Basin is presented, and factors important in determining species richness in these sites are discussed. Finally, conservation issues and future needs for research are suggested.

GEOLOGY

The Basin and Range Province is a series of basins (grabens) and mountain ranges (horsts). The mountain ranges extend 64–128 km (40–80 miles) in a north-south orientation, are 10–24 km (6–15 miles) wide, and reach elevations of 4000 m, although elevations of 2100–3000 m (7,000–10,000 feet) are most common. The basins occur at an average elevation of about 1500 m and are filled with depositional material from the mountains. During the Pleistocene over 100 of the basins were filled with lakes (Benson et al. 1990) and large expanses of land were covered with huge lakes, such as the ancient Lakes Bonneville, Lahontan, Russell, and Manly. The distribution of springs in the Great Basin is not random. Clusters of springs occur along fault lines at the base of ranges (e.g., Deep Springs, California; Ash Meadows, Nevada) and in the canyons of the mountains; however, individual, isolated springs located in either the basins or ranges are also common.

CLIMATE

Located in the rain shadow of the Sierra Nevada mountain range, the Great Basin receives limited precipitation and is the largest desert region in the United States (Rumney 1987). Estimated annual precipitation over the entire region averages about 27.9 cm (Eakin et al. 1976), with about 10–15 cm of rain in the basins and 40–150 cm of snow in the mountain ranges. Most precipitation falls from November through March and is associated with cyclonic fronts. However, thunderstorms in the summer can produce intense downpours in localized areas. Evaporation greatly exceeds precipitation in the Great Basin: in Las Vegas annual precipitation averages 10.2 cm but the evaporation rate averages 106.7 cm, creating a deficit of 96.5 cm; in Reno annual precipitation averages 17.8 cm but the evaporation rate equals 61 cm, creating a 44-cm deficit (Planert and Williams 1995). This deficit accounts for the high levels of salinity found in many of the surface waters.

CLASSIFICATION OF SPRINGS

Although several classification schemes have been suggested for springs, there is no unified or generally accepted system. The majority of proposed classifications (Table 33.1) are based on physical characteristics of the springs and tend to be organized over disciplines (e.g., geology, chemistry) or distinctive characteristics (e.g., discharge, temperature).

All spring types in Table 33.1 are found in the Great Basin and, regardless of the water temperature, chemistry, or geological origin, nearly all permanent springs in the Great Basin have associated wetlands. The emergent water merely represents one point in the hydrologic continuum of groundwater–surface water–groundwater recharge. Wetland classification usually depends on vegetation type, plant or animal species, or hydrologic conditions (Mitsch and Gosselink 1986) rather than the source of water. For this reason, there is no distinct class for spring-formed wetlands; rather, springs form wetlands of all types.

TABLE 33.1. Examples of Spring Classification Systems

Category	Examples of Types	References
Typology	Limnocrene	Steinmann 1915
	Rheocrene	Theinemann 1922
	Helocrene	
Discharge	Magnitude 1 (>10 m^3/s)	Meinzer 1923
	Magnitude 8 (<10 ml/s)	
Geology	Contact	Fetter 1994
	Fault	Bryan 1919
	Depression	
	Artesian	
	Tube	
Chemistry	Hard/soft	Van Everdingen 1991
	Fresh/mineral	
	Salt	
	Sulfur	
	Soda	
Temperature	Cold	Waring 1965
	Warm	Pritchard 1991
	Hot	

DISPERSAL OF SPRING INVERTEBRATES

Aquatic invertebrates in isolated desert springs face a variety of conditions that differ significantly from habitats in more mesic regions. First, the landscape between adjacent springs is not a moist, forested landscape; it is arid and inhospitable. Thus, dispersal between desert springs poses a greater physiological challenge than dispersal between springs in mesic regions. Unfortunately, little is known about the flight capabilities of most aquatic insects or the means of dispersal for nonaerial invertebrates. Because the number of adult aquatic insects decreases as one moves away from a stream (e.g., Jackson and Resh 1989) or lake (e.g., Kovats et al. 1996) even in more hospitable habitats, it seems likely that few adult aquatic insects make long-distance flights across the arid, inhospitable landscape.

A second, related issue is that there is little to no connectivity among spring-fed wetland habitats of the Great Basin. In a forested landscape several first-order streams may join and form, in succession, larger (i.e., higher-order) streams. Besides allowing for drift and upstream movements by immature stages of insects (and all stages of nonflying invertebrates), this connectedness facilitates dispersal within and between watersheds and provides a source of colonists after disturbance.

In an arid landscape such as the Great Basin a first-order spring brook may only flow for 200 m before infiltrating back into the ground. Two or more springs may occur at different elevations in one canyon, but the channel between springs is composed of dry, alluvial material with no riparian vegetation. Although insects may be directed from one spring to another because of confinement within the canyon, there is no continuous riparian corridor for them to follow and no connection to neighboring canyons.

Finally, the next-closest source of water may or may not be suitable for colonization by a particular species, since neighboring springs can vary drastically in their temperature, conductivity, and chemical characteristics. Thus, an adult aquatic insect leaving its natal spring source in the mountains and flying even a short distance downstream to the basin below will most likely find a spring with a higher temperature, higher conductivity, and a water chemistry very unlike that in the habitat it left, and in many instances its eggs or larvae would not survive in the new habitat.

Aquatic invertebrates vary greatly in their ability to move from one spring to another. At one extreme of the spectrum are strong fliers like Odonata, Coleoptera, Hemiptera, and some limnephilid caddisflies (Sheldon 1984, Pennak 1989); at the other are nonaerial invertebrates such as flatworms, hydrobiid snails, and clams. During the Pleistocene, when water was plentiful and riparian areas were lush, dispersal of aquatic invertebrates among habitats was probably facilitated through connections between streams and basins. As the climate changed and the lakes shrank or dried completely, aquatic habitats became increasingly isolated. All of these factors lead to the isolation of invertebrates in springs, and we expect that this isolation resulted in high numbers of endemic species and relict species.

Many spring insects have reduced flight abilities, suggesting that there are advantages in remaining at a spring rather than venturing to new habitats. For example, some members of elmid beetles (Shepard 1992), Plecoptera (personal obs.), and Trichoptera (Erman 1984) that occur in springs are brachypterous or apterous. It has been noted before that insects living in permanent habitats are more likely to have lost their ability to fly than those in temporary habitats (e.g., Johnson 1969, Roff 1990). Using mtDNA, we are investigating the pattern of dispersal of Trichoptera among springs. Preliminary results show that strong fliers like *Hesperophylax designatus* are part of one metapopulation (a population of populations), whereas poorer fliers like *Lepidostoma* spp. form isolated populations.

COLONIZATION OF SPRING INVERTEBRATES

The lack of connectivity probably has its greatest influence on the recolonization of aquatic habitats after disturbance. Colonists are thought to come from four sources: downstream drift, upstream migration, oviposition from aerial colonists, and upward movement from the hyporheos (Williams and Hynes 1976). For spring ecosystems that have been scoured by a flash flood, there is no upstream source of colonists because there is no water above the spring source; likewise, there is no downstream source of colonists because insects that survived the flood itself are left to dry on an alluvial fan. The hyporheos of these systems has not been studied, and its contribution to recolonization is unknown. Thus, aerial oviposition is probably the primary means of recolonization. In addition, refuges (i.e., springs on hillsides as opposed to in-channel locations) are probably important sources of colonists for disturbed sites, but this topic has not been examined in detail.

During 1994–1997 we saw the extirpation of invertebrate populations as a result of three types of disturbance: grazing, dewatering, and flash floods. For example, a *Limnephilus* caddisfly population was eliminated when the fence protecting Antelope Spring (Mono Co., California) was illegally cut, allowing the area to be overgrazed by cattle. Two years after the fence was repaired the spring was recolonized by *Limnephilus* adults (because there are no fluvial connections to this spring). Likewise, a pond supplied by spring flow (Barrel Spring, Inyo Co., California) was dewatered when the water source to the pond was turned off. An isolated population of a semiaquatic tettrigid grasshopper (*Paratettix mexicanus*) was eliminated. Although the pond has since been rewatered, it has not been recolonized by the tettrigid.

As a final example, in July 1997 intense rainstorms created flash floods in selected canyons in the White Mountains, California. In Marble Canyon water depth was as deep as 5 m in confined sections of the canyon, and it appeared that the bottom substrate was thoroughly scoured and rearranged. As is typical in flash floods, the water receded quickly and within two days the only surface flow was from water supplied by spring sources that had existed before the flood. No aquatic invertebrates could be found in samples taken in two spring-brooks within the channel one week after the flood; however, several crushed terrestrial insects were found in the samples, and damselflies (*Argia*) were observed ovipositing in the water at the time the benthic samples were taken. Benthic samples taken four months after the flood contained baetids, simu-liids, empidids, stratiomyids, muscids, and several species of chironomids. Because of the extent of bottom disturbance during the flood, it is most likely that aerial adults were responsible for the recolonization. Six species of cad-disflies occurred in these springs before the flood (Myers and Resh, unpub-lished data), but it remains to be seen if the same or different species will recolonize the springs, and in what time frame.

Great Basin systems are similar to the cold desert spring systems in Wash-ington, with one notable difference: the spates in the cold desert systems occur in winter as a consequence of warm chinook winds rapidly melting snow over frozen ground (Cushing and Gaines 1989). There the disturbance occurs when few aerial colonists are present, in contrast to the summertime flash floods of the Great Basin. However, what these systems have in common is the absence of drift or upstream migration and an unknown but probably limited contri-bution from the hyporheos to recolonization (Cushing and Gaines 1989). Ovi-position by aerial adults is the most likely source of colonists in both systems, although the magnitude of this contribution is constrained by the distance adults can fly, the suitability of the potential recolonized habitat, and the pool of species available in the region.

BIODIVERSITY OF SPRING INVERTEBRATES

The fish that occur in springs of the Basin and Range Province have been studied and the fauna documented since the end of the last century (Gilbert 1893) through today (Miller 1948, Soltz and Naiman 1978, Echelle and Echelle 1993, Sigler and Sigler 1994, Sada et al. 1995). By comparison, the aquatic invertebrates are poorly known. Brues (1928, 1932) contributed early inventories on aquatic invertebrates in hot springs, and La Rivers (1948, 1950, 1953) did extensive work on naucorids in Great Basin springs. Recently, in-terest in the biogeography of aquatic invertebrates in springs has grown, and our knowledge of specific groups, including hydrobiid snails (i.e., Hershler and Sada 1987, Hershler 1989, Herschler and Pratt 1990, Hershler 1998), elmid beetles (Shepard 1992), stoneflies (Sheldon 1979, Nelson and Baumann 1989), and Diptera (Anderson and Anderson 1995) has increased. However, the Great Basin is large (373,000 km^2), and the access to water sources dif-ficult. Consequently, only a small portion of the total area has been surveyed.

An exception to this is the hydrobiid snails that have been both intensively and extensively studied (Hershler and Sada 1987, Hershler 1989, Herschler and Pratt 1990, Hershler 1998).

Even with the renewed interest in biodiversity, factors that determine species richness at a site are generally unknown. A definitive discussion of the determinants of biodiversity in springs is beyond the scope of this chapter, and several recent texts discuss broad aspects of the topic (e.g., Huston 1994, Gaston 1996). Here we discuss the factors that we believe are most important in determining species richness of aquatic invertebrates in spring-fed wetlands: habitat size; distance between springs; habitat harshness; intensity and frequency of disturbance; and the number of species in the regional pool.

Habitat Size

Although springs are the primary source of water for a few large wetlands in the Great Basin (e.g., Fish Springs, Utah, 4860 ha; Ruby Lake, Nevada, 4050 ha), most spring-formed wetlands are small, supporting a wetland of 1 ha or less (e.g., Sand Spring, Death Valley; Montenegro Spring, White Mountains, California). Small springs likely have fewer species than larger ones, as predicted by the species-area relationship of the theory of island biogeography (MacArthur and Wilson 1967).

Distance between Springs

Springs in the Great Basin are often separated by expanses of arid scrubland. More isolated springs will likely have fewer species than those that are (1) part of a closely associated cluster of springs or (2) part of a connected drainage network of springbrooks. Thus, a drainage network is likely to have more species because of the increased area and habitat complexity. We would also expect the species composition of a spring within a network to be more similar to that of a neighboring spring than to that of a more distant spring.

Adverse Conditions

Harsh conditions can limit the number of species found in a spring. As water temperature increases, the number of aquatic invertebrate species decreases (Lamberti and Resh 1983, Pritchard 1991). Salt content (e.g., Colburn 1988), low dissolved oxygen (e.g., Sloan 1956), or other limiting factors (e.g., pH, sulfur) may also exclude particular species, reducing the total number of species occurring in a spring. Many of the springs in the Great Basin, especially those situated in basins where the temperatures and salinity tend to be high and the dissolved oxygen levels low, present physiological challenges to aquatic invertebrates and limit their species richness (Colburn 1988).

Intensity and Frequency of Disturbance

Two types of disturbances that have major consequences for aquatic invertebrates in the Great Basin are dewatering (e.g., from drought or agricultural diversion) and flash flooding. Dewatering (especially to the point that intermittancy occurs) would be expected to be a major determinant of species richness and composition. Erman and Erman (1995) found that spring permanence was the most important factor in determining species richness of Trichoptera in springs of the Sierra Nevada.

Flash floods confined in canyons can have a devastating effect on the aquatic invertebrate fauna. Although the effects of flash flooding have been documented on streams in Arizona (Meffe and Minckley 1987, Gray and Fisher 1981, Grimm and Fisher 1989) and Washington (Cushing and Gaines 1989), no research has been published on the impact of flash floods on aquatic invertebrates in the Great Basin. In the White Mountains we have observed that springs within a confined channel and subject to flash floods have fewer species of Trichoptera than adjacent springs on hillsides, and that the species composition of each is different in spite of similar water chemistry (Myers and Resh, unpublished data). For example, permanent springs confined in a channel often lack hydrobiid snails and flatworms, two groups that would be slow to recolonize a disturbed habitat because of their lack of mobility and the lack of connectivity to nonaffected channels, as described above.

Regional Pool of Species

The number of aquatic species in the regional pool of a desert area such as the Great Basin is likely less than in temperate regions because of the overall scarcity of water and the harshness of the environment. Because the number of species in the regional pool sets the upper limit on species richness at a site (Caley and Schluter 1997), we would expect that local species richness may be less than in more moderate environments; however, insufficient data is available to verify this hypothesis.

Current research has demonstrated that endemicity and species richness of aquatic invertebrates can be high in isolated springs (Table 33.2) (Herschler and Sada 1987, Erman and Erman 1990, Hershler 1998). The same factors that can lead to low species richness (small size, isolation, adverse conditions) can also lead to the evolution of endemic species or to the value of springs as refugia for relict species (Erman and Erman 1990). Even though one individual spring may have relatively few species, some are often unique to the region: thus, the value of each spring is also in its contribution to the regional pool. Collectively, the range of habitat diversity created in spring-formed wetlands of the Great Basin can be exploited by many species of aquatic invertebrates.

Because complete information is unavailable for any group of aquatic invertebrates in the Great Basin, we choose here to discuss the biodiversity of

TABLE 33.2. Number of Species and Number of Endemics Found in Selected Studies of Springs in the Great Basin

Group	Common Name	Number of Species in Study Area	Number of Endemics	References	Location
Gastropoda					
Hydrobiidae	Spring snails	22	10	Herschler 1989	Death Valley
Trichoptera	Caddisflies	58	2	Myers and Resh (unpub.)	Inyo, Mono Counties
Coleoptera					
Elmidae	Riffle beetles	4	2	Shepard 1992	Death Valley
Hemiptera					
Naucoridae	Creeping water bugs	8	4	Polhemus and Polhemus 1988	Great Basin

a few specific groups. Anderson and Anderson (1995), using emergence traps at five spring wetlands in central Oregon, found that Diptera was the most diverse group of aquatic insects (77 species in 11 families) and that chironomids had the most species (28). Trichoptera were the next most diverse, with 11 species in 10 families. In our work in springs of eastern Inyo and Mono Counties, California, we have found 58 species of caddisflies in 15 families (Myers and Resh, unpublished data); two of these are newly described species of *Lepidostoma* (Weaver and Myers 1998). A new species of amphipod (*Stygobromus*) has also been found (Holsinger, personal communication) at three sites. As our work continues, we expect that the species list for caddisflies will increase and more undescribed species will be discovered.

Hershler (1989) reports that 20 of 22 species of spring snails in the Death Valley System are known only from that area and 10 are restricted to single springs or spring complexes. In more recent work, Hershler (1998) describes 58 new species of *Pyrgulopsis* found in spring habitats throughout Nevada and Utah. Spring snails are generally found in sites below 2100 m (D. Sada, personal communication), but they are not restricted to basins; they are present in mountain springs as well. In contrast, Plecoptera are most often found at higher elevations and in cooler water. Sheldon (1979) lists 40 species in 9 families that he found in streams on 15 mountain ranges. However, it is not known which of these species were found in spring habitats. Finally, Hovingh (1993) reports 23 species of Mollusca and 7 species of leeches within the spring systems of Snake Valley, Tule Valley, and Fish Springs Flat in the southwest Bonneville Basin of Utah.

CONSERVATION

Although the Basin and Range Province is a remote region with very low human population density, few springs remain unimpacted by such anthropogenic activities as: groundwater pumping, diversion, or spring manipulation (e.g., excavation of spring source to increase flow); grazing by cows, horses, and burros; wildlife improvements; recreational activities; and introduction of exotic fish and amphibians (Williams et al. 1985, Hovingh 1993, Shepard 1993).

Groundwater pumping and diversion have clear and dramatic effects on springs and their associated wetlands because the wetted area shrinks or disappears altogether, as has occurred in the Owens Valley to support the water demand in Los Angeles (DeDecker 1992, Scheidlinger 1992). The most thoroughly documented example of the impact of groundwater pumping on aquatic fauna may be the case of Devil's Hole, Nevada. Groundwater pumping near Ash Meadows, to support agricultural development, was reducing the level of water in Devil's Hole to the point that an endemic species of pupfish (*Cyprinodon diabolis*) was threatened with extinction. An elmid beetle (*Stenelmis calida*) that also occurs in this spring and is endemic to a few springs in Ash Meadows (Shepard 1992) would likely have also been threatened had the water levels dropped. A decision from the U.S. Supreme Court eventually stopped the pumping that threatened this ecosystem (Deacon and Williams 1991).

The Great Basin supports three introduced species of ungulates: cows, burros, and horses. No studies have been published on the impacts of feral burros or wild horses on aquatic invertebrates. Although the impact of cattle on a variety of watershed characteristics has been well documented (Meehan and Platts 1978, Kaufman and Kruger 1984, Fleischner 1994), only a few studies have looked at the impact of livestock on aquatic invertebrates (e.g., Rinne 1988, Quinn et al. 1991, Scott et al. 1994), and only two have investigated spring ecosystems specifically (Taylor et al. 1989, Hall and Amy 1990). Cattle congregate in riparian areas, where the best forage and the only source of water are located. Because the riparian areas are small and flows are typically quite low, the wastes from the animals are not flushed away as they are in larger systems. Quinn et al. (1992) found that aquatic invertebrates in streams less than 6 m in width were most impacted by grazing activities. Most of the springs in the Great Basin have widths of less than 1 m; when they are subjected to heavy grazing pressure, the number and richness of aquatic invertebrates decline (Myers and Resh, unpublished data).

Both recreation and wildlife improvements are generally seen as constructive uses of public land within the Great Basin; however, both can have negative impacts on the wetland habitat. For instance, at Waucoba Spring in Saline Valley, California, the spring has been tapped and the entire flow goes into a water trough to provide water for game animals (e.g., deer). Although a water trough may benefit some wildlife, it greatly reduces the riparian habitat that once existed for aquatic invertebrates. Bleich (1992) discusses a variety of water developments that have been installed for wildlife in Inyo and Mono Counties, California, and notes that the impacts of these developments on nontarget organisms were not considered at the time of implementation.

Recreationalists often modify hot springs by creating soaking pools; these can have a marked effect on spring ecosystems. For instance, in Palm Springs, Saline Valley, California, a series of cement tubs has been constructed, the native riparian vegetation drastically altered, and exotic aquatic species (goldfish) introduced. Management of such habitats has been lax, and presumably recreationalists do not realize the impacts of their activities. Active management and/or education of the recreationalists is needed to protect these sensitive and unique habitats.

Introduced aquatic species can also have a negative impact on the native fauna. Species that have been introduced into the warm waters of the Great Basin include *Tilapia* (Cichlidae), largemouth bass (*Macropterous salmoides*), sunfish (*Lepomis* spp.), mosquitofish (*Gambusia affinis*), guppies and mollies (*Poecilia* spp.), and bullfrogs (*Rana catesbiana*) (Shepard 1993). While the impact of exotic fish on native fish has been documented (e.g., Williams et al. 1985), their impact on aquatic invertebrates, especially in this region, is not known.

One species of aquatic insect is protected under the Endangered Species Act—the Ash Meadows naucorid (*Ambrysus amargosus*). This species is currently listed as threatened, and its case history is discussed in detail by Polhemus (1993). Its habitat (Point of Rock Springs) received protection through the creation of the Ash Meadows National Wildlife Refuge. However, early efforts to construct a refugium at this site for another species, the endangered Devil's Hole pupfish (*Cyprinodon diabolis*), degraded the habitat for the Ash Meadows naucorid. Fortunately, recent restoration designed specifically for the naucorid has recreated the riffle habitat they require and their population is being monitored.

Endemic species that occur at only one site are at greatest risk of extirpation. Unfortunately, many springs have been dewatered or heavily impacted by cattle before surveys could be conducted, and many species may now have more restricted distributions or have been extirpated before their presence could be documented. Restoration of the spring-formed wetland habitat does not guarantee restoration of the endemic fauna that once occurred there, especially for species that have limited dispersal capabilities. In this region of vast expanses of drought-tolerant shrubs and grasses, springs are a small yet significant resource that supports many vertebrates and invertebrates. The challenge is to document the biodiversity that these riparian islands contain and to educate land managers and the public about the value of these unique ecosystems.

NEEDS FOR FUTURE RESEARCH

The aquatic invertebrates of the Great Basin are still largely unknown. The trend in research conducted thus far has been to study one taxonomic group (e.g., hydrobiid snails; Hershler 1998), a limited geographical area (e.g., Bridge Creek drainage, Oregon; Anderson and Anderson 1995), or one group within a limited geographical area (e.g., elmid beetles in Death Valley; Shepard 1993). The size of the region, remoteness of the water sources, extremes in temperature, and lack of specialists working in this area indicate that the aquatic invertebrates of the Great Basin will remain a fruitful area of research for years to come. Because of the isolation of the spring wetlands, they offer an ideal setting in which to test hypotheses related to the theory of island biogeography (MacArthur and Wilson 1967). Furthermore, "connectedness" (or lack of it) among spring habitats and its relationship to species richness and to rates, patterns, and modes of recolonization is also a valuable, unexplored area of research. Finally, one could view the water resources of this region as having been naturally fragmented from the once-extensive water systems that occurred during the Pleistocene. Consequently, studies of the dynamics of the metapopulations that have been since created would be a valuable area of research.

ACKNOWLEDGMENTS

This research has been supported in part by grants from the White Mountain Research Station and the Margaret C. Walker Fund for teaching and research in systematic entomology. We thank John Holsinger and John Weaver for their assistance in the identifications of *Stygobromus* and *Lepidostoma,* respectively.

LITERATURE CITED

Anderson, T. M., and N. H. Anderson. 1995. The insect fauna of spring habitats in semiarid rangelands in central Oregon. Journal of the Kansas Entomological Society Supplement 68(2):65–76.

Benson, L. V., D. R. Currey, R. I. Dorn, K. R. Lajoie, C. G. Oviatt, S. W. Robinson, G. I. Smith, and S. Stine. 1990. Chronology of expansion and contraction of four Great Basin lake systems during the past 35,000 years. Palaeogeography, Palaeoclimatology, Palaeoecology 78:241–286.

Bleich, V. C. 1992. History of wildlife water development, Inyo County, California. Pages 100–106 *in* C. A. Hall Jr., V. Doyle-Jones, and B. Widawski (eds.), The History of Water: Eastern Sierra Nevada, Owens Valley, White-Inyo Mountains. White Mountain Research Station, San Diego, CA.

Brown, J. H. 1978. The theory of insular biogeography and the distribution of boreal birds and mammals. Great Basin Naturalist 2:209–227.

Brues, C. T. 1928. Studies on the fauna of hot springs in the Western United States and the biology of thermophilous animals. Proceedings of the American Academy of Arts and Sciences 63:139–227.

―――. 1932. Further studies on the fauna of North American hot springs. Proceedings of the American Academy of Arts and Sciences 67:185–303.

Bryan, K. 1919. Classification of springs. Journal of Geology 27:522–561.

Caley, M. J., and D. Schluter. 1997. The relationship between local and regional diversity. Ecology 78:70–80.

Carothers, S. W., R. R. Johnson, and S. W. Aitchison. 1974. Population structure and social organization of southwestern riparian birds. American Zoologist 14:97–108.

Colburn, E. A. 1988. Factors influencing species diversity in saline waters of Death Valley, USA. Hydrobiologia 158:215–226.

Cushing, C. E., and W. L. Gaines. 1989. Thoughts on recolonization of endorheic cold desert spring-streams. Journal of the North American Benthological Society 8:277–287.

Deacon, J. E., and C. Deacon Williams. 1991. Ash Meadows and the legacy of the Devil's Hole pupfish. Pages 69–87 *in* W.L. Minckley and J. E. Deacon (eds.), Battle against Extinction. University of Arizona Press, Tucson, AZ.

DeDecker, M. 1992. The death of a spring. Pages 223–226 *in* C. A. Hall Jr., V. Doyle-Jones, and B. Widawski (eds.), The History of Water: Eastern Sierra Nevada, Owens Valley, White-Inyo Mountains. White Mountain Research Station, San Diego, CA.

Eakin, T. E., D. Price, and J. R. Harrill. 1976. Summary Appraisals of the Nation's Ground-water Resources—Great Basin Region. Geological Survey Professional Paper 813-G.

Echelle, A. A., and A. F. Echelle. 1993. Allozyme perspective on mitochondrial DNA variation and evolution of the Death Valley pupfishes (Cyprinodontidae: *Cyprinodon*). Copeia 1993:275–287.

Erman, N. A. 1984. The mating behavior of *Parthina linea* (Trichoptera: Odontoceridae), a caddisfly of springs and seeps. Pages 131–136 *in* J. C. Morse (eds.), Proceedings of the Fourth International Symposium on Trichoptera. Dr. W. Junk, The Hague, The Netherlands.

Erman, N. A., and D. C. Erman. 1990. Biogeography of Caddisfly (Trichoptera) Assemblages in Cold Springs of the Sierra Nevada (California, USA). California Water Resources Center Contribution 200.

———. 1995. Spring permanence, Trichoptera species richness, and the role of drought. Journal of the Kansas Entomological Society 68(2) Supplement 50–64.

Fenneman, N. M., and D. W. Johnson. 1946. Physical Divisions of the United States. Map 1:700,000.

Fetter, C. W. 1994. Applied Hydrogeology. Prentice-Hall, Englewood Cliffs, NJ.

Fleischner, T. L. 1994. Ecological costs of livestock grazing in western North America. Conservation Biology 8:629–644.

Gaston, K. J. 1996. Biodiversity: A biology of numbers and difference. Blackwell Science, Cambridge, MA.

Gilbert, C. H. 1893. Report on the fishes of the Death Valley expedition collected in southern California and Nevada in 1891, with descriptions of new species. North American Fauna 7:229–234.

Gray, L. J. 1993. Response of insectivorous birds to emerging aquatic insects in riparian habitats of a tallgrass prairie stream. American Midland Naturalist 129:288–300.

Gray, L. J., and S. G. Fisher. 1981. Postflood recolonization pathways of macroinvertebrates in a lowland Sonoran desert stream. American Midland Naturalist 106:249–257.

Grimm, N. B., and S. G. Fisher. 1989. Stability of periphyton and macroinvertebrates to disturbance by flash floods in a desert stream. Journal of the North American Benthological Society 8:293–307.

Hall, D. A., and P. S. Amy. 1990. Microbiology and water chemistry of two natural springs impacted by grazing in south central Nevada. Great Basin Naturalist 50:289–294.

Hershler, R. 1998. A systematic review of the hydrobiid snails (Gastropoda: Rissooidea) of the Great Basin, western United States. Part I. Genus *Pyrgulopsis*. Veliger 41:1–132.

———. 1989. Springsnails (Gastropoda: Hydrobiidae) of Owens and Amargosa River (Exclusive of Ash Meadows) drainages, Death Valley System, California-Nevada. Proceedings of the Biological Society of Washington 102:176–248.

Hershler, R., and W. L. Pratt. 1990. A new *Pyrgulopsis* (Gastropoda: Hydrobiidae) from southeastern California, with a model for historical development of the Death Valley hydrographic system. Proceedings of the Biological Society of Washington 103:279–299.

Hershler, R., and D. W. Sada. 1987. Springsnails (Gastropoda: Hydrobiidae) of Ash Meadows, Amargosa Basin, California-Nevada. Proceedings of the Biological Society of Washington 100:776–843.

Hovingh, P. 1993. Zoogeography and paleozoology of leeches, molluscs, and amphibians in Western Bonneville Basin, Utah, USA. Journal of Paleolimnology 9:41–54.

Huston, M. A. 1994. Biological Diversity. Cambridge University Press, Cambridge, UK.

Jackson, J. K., and S. G. Fisher. 1986. Secondary production, emergence, and export of aquatic insects of a Sonoran desert stream. Ecology 67:629–638.

Jackson, J. K., and V. H. Resh. 1989. Distribution and abundance of adult aquatic insects in the forest adjacent to a northern California stream. Environmental Entomology 18:278–283.

Johnson, C. G. 1969. Migration and dispersal of insects by flight. Methuen, London, UK.

Kaufman, J. B., and W. C. Kruger. 1984. Livestock impacts on riparian ecosystems and streamside management implications . . . a review. Journal of Range Management 37:430–438.

Kovats, Z. E., J. J. H. Ciborowski, and L. D. Corkum. 1996. Inland dispersal of adult aquatic insects. Freshwater Biology 36:265–276.

La Rivers, I. 1948. A new species of *Plecoris* from Nevada, with notes on the genus in the United States (Hemiptera: Naucoridae). Annals of the Entomological Society of America 41:371–376.

———. 1950. A new naucorid genus and species from Nevada. Annals of the Entomological Society of America 43:368–373.

———. 1953. New gelastocorid and naucorid records and miscellaneous notes, with a description of the new species, *Ambrysus amargosus*. Wasmann Journal of Biology 11:83–96.

Lamberti, G. A., and V. H. Resh. 1983. Geothermal effects on stream benthos: Separate influences of thermal and chemical components of epiphyton and macroinvertebrates. Canadian Journal of Fisheries and Aquatic Sciences 40:1995–2009.

MacArthur, R. H., and E. O. Wilson. 1967. The Theory of Island Biogeography. Priceton University Press, Princeton, NJ.

Meehan, W. R., and W. S. Platts. 1978. Livestock grazing and the aquatic environment. Journal of Soil and Water Conservation 33:274–278.

Meffe, G. K., and W. L. Minckley. 1987. Persistence and stability of fish and invertebrate assemblages in a repeatedly disturbed Sonoran desert stream. American Midland Naturalist 117:177–191.

Meinzer, O. E. 1923. Outline of Ground-water Hydrology. USGS Water Supply Paper 494. U.S. Government Printing Office, Washington, DC.

Miller, R. R. 1948. The Cyprinodont fishes of Death Valley system of eastern California and Southwestern Nevada. Miscellaneous Publications, Museum of Zoology, University of Michigan, Ann Arbor, MI.

Mitsch, W. J., and J. G. Gosselink. 1986. Wetlands. Van Nostrand Reinhold, New York.

Murkin, H. R., and B. D. J. Batt. 1987. The interactions of vertebrates and invertebrates in peatlands and marshes. Pages 15–33 *in* D. M. Rosenberg and D. V. Danks (eds.), Aquatic Insects of Peatlands and Marshes of Canada. Memoirs of the Entomological Society of Canada No. 140.

Nelson, C. R., and R. W. Baumann. 1989. Systematics and distribution of the winter stonefly genus *Capnia* (Plecoptera: Capniidae) in North America. Great Basin Naturalist 49:289–363.

Pennak, R. W. 1989. Fresh-water Invertebrates of the United States. Third Edition. John Wiley & Sons, New York.

Planert, M., and J. S. Williams. 1995. The Ground Water Atlas of the United States. Segment 1 California, Nevada. Hydrologic Investigations Atlas HA-730 B.

Polhemus, D. A. 1993. Conservation of aquatic insects: Worldwide crisis or localized threats? American Zoologist 33:588–598.

Polhemus, D. A., and J. T. Polhemus. 1988. Family Naucoridae Leach, 1815. The creeping water bugs. Pages 521–527 *in* T. J. Henry and R. C. Froeschner (eds.), Catalog of the Heteroptera, or True Bugs, of Canada and the Continental United States. E. J. Brill, New York.

Pritchard, G. 1991. Insects in thermal springs. Pages 89–106 *in* D. D. Williams and H. V. Danks (eds.), Arthropods of Springs, with Particular Reference to Canada. Memoirs of the Entomological Society of Canada No. 155.

Quinn, J. M., R. B. Williamson, R. K. Smith, and M. L. Vickers. 1992. Effects of riparian grazing and channelisation on streams in Southland, New Zealand. 2. Benthic invertebrates. New Zealand Journal of Marine and Freshwater Resources 26: 259–273.

Rinne, J. N. 1988. Effects of livestock grazing exclosure on aquatic macroinvertebrates in a montane stream, New Mexico. Great Basin Naturalist 48:146–153.

Roff, D. A. 1990. The evolution of flightlessness in insects. Ecological Monographs 60:389–421.

Rumney, G. R. 1987. Climate of North America. Pages 613–623 *in* J. E. Oliver and R. W. Fairbridge (eds), The Encyclopedia of Climatology. Encyclopedia of Earth Sciences, vol. 11. Van Nostrand Reinhold, New York.

Sada, D. W., H. B. Britten, and P. F. Brussard. 1995. Desert aquatic ecosystems and the genetic and morphological diversity of Death Valley system speckled dace. American Fisheries Society Symposium 17:350–359

Scheidlinger, C. R. 1992. Owens Valley wetlands: Inventory and description. Page 445 *in* C. A. Hall, Jr., V. Doyle-Jones, and B. Widawski (eds.), The History of Water: Eastern Sierra Nevada, Owens Valley, White-Inyo Mountains. White Mountain Research Station, San Diego, CA.

Scott, D., J. W. White, D. S. Rhodes, and A. Koomer. 1994. Invertebrate fauna of three streams in relation to land use in Southland, New Zealand. New Zealand Journal of Marine and Freshwater Research 28:277–290.

Sheldon, A. L. 1979. Stonefly (Plecoptera) records from the basin ranges of Nevada and Utah. Great Basin Naturalist 39:289–292.

———. 1984. Colonization dynamics of aquatic insects. Pages 401–429 *in* V. H. Resh and D. M. Rosenberg (eds.), The Ecology of Aquatic Insects. Praeger, New York.

Shepard, W. D. 1992. Riffle beetles (Coleoptera: Elmidae) of Death Valley National Monument, California. Great Basin Naturalist. 52:378–381.

———. Desert springs—both rare and endangered. Aquatic Conservation: Marine and Freshwater Ecosystems 3:351–359.

Sigler, J. W., and W. F. Sigler. 1994. Fishes of the Great Basin and the Colorado Plateau. Past and Present Forms. Pages 163–237 *in* K. T. Harper, L. L. St. Clair, K. H. Thorne, and W. M. Hess (eds.), Natural History of the Colorado Plateau and Great Basin. University of Colorado Press, Niwot. CO.

Sigler, W. F., and J. W. Sigler. 1987. Fishes of the Great Basin: A natural history. University of Nevada, Reno, NV.

Sloan, W. C. 1956. The distribution of aquatic insects in two Florida springs. Ecology 37:81–98.

Soltz, D. L., and R. J. Naiman. 1978. The natural history of native fishes in the Death Valley system. Natural History Museum of Los Angeles County Science Series 30: 1–76.

Steinmann, P. 1915. Praktikum der Süsswasserbiologie. I Teil: Die Organismen des fliessenden Wassers. Gebrüder Brontraeger, Berlin.

Szaro, R. C., and M. D. Jakle. 1985. Avian use of a desert riparian island and its adjacent scrubhabitat. Condor 87:511–519.

Theinemann, A. 1922. Hydrobiologische Untersuchungen an Quellen. Archiv für Hydrobiologie 14:151–190.

Taylor, F. R., L. A. Gillman, and J. W. Pedretti. 1989. Impact of cattle on two isolated fish populations in Pahranagat Valley, Nevada. Great Basin Naturalist 49:491–495.

Van Everdingen, R. O. 1991. Physical, chemical and distributional aspects of Canadian springs. Pages 7–28 *in* D. D. Williams and H. V. Danks (eds.), Arthropods of Springs, with Particular Reference to Canada. Memoirs of the Entomological Society of Canada No. 155.

Waring, G. A. 1965. Thermal springs of the United States and Other Countries of the World—a summary. Geological Survey Professional Paper 492, U.S. Government Printing Office, Washington, DC.

Weaver, J. S., III, and M. J. Myers. 1998. Two new species of caddisflies of the genus *Lepidostoma* Rambur (Trichoptera: Lepidostomatidae) from the Great Basin. Aquatic Insects 20:189–195.

Williams, D. D., and H. B. N. Hynes. 1976. The recolonization mechanisms of stream benthos. Oikos 27:265–272.

Williams, J. E., D. B. Bowman, J. E. Brooks, A. A. Echelle, R. J. Edwards, D. A. Hendrickson, and J. J. Landye. 1985. Endangered aquatic ecosystems in North American deserts with a list of vanishing fishes of the region. Journal of the Arizona-Nevada Academy of Science 20:1–48.

34 Seasonal and Semipermanent Wetlands of California

Invertebrate Community Ecology and Responses to Management Methods

FERENC A. DE SZALAY, NED H. EULISS, Jr., and
DAROLD P. BATZER

*O*ver 90 percent of California's original wetlands have been
drained or converted to agricultural uses (e.g., rice fields), and
most of the remaining wetlands are found in the Central Valley.
Semipermanent and seasonal marshes are vital overwintering
habitats for many species of waterfowl and shorebirds. Aquatic invertebrates are an important component of the fauna of these habitats because
populations of pestiferous mosquitoes are often very high, and other taxa
are important food resources for migratory birds. Although mosquito ecology has been extensively studied, factors influencing most wetland invertebrates are still poorly understood. In this chapter we begin by examining
how colonization, trophic relationships, and abiotic interactions influence
invertebrate distributions. Species assemblages change after wetlands are
flooded because food resources, habitat structure, and predation vary temporally. We also discuss aspects of mosquito ecology. Predation by fish and
invertebrates and abiotic factors are the most important causes of larval
mosquito mortality. We also review recent research conducted at managed
wildlife habitats in the Central Valley. These studies demonstrate that invertebrate populations are affected by vegetational and water management
practices and by water quality and plant physical structure. Because these
habitats are artificially filled from canals and hydroperiods are different
from naturally flooded wetlands, many of the observed patterns may be
unique to these intensively managed ecosystems. We also discuss studies examining interactions between aquatic invertebrates and avian botulism be-

Invertebrates in Freshwater Wetlands of North America: Ecology and Management, Edited by
Darold P. Batzer, Russell B. Rader, and Scott A. Wissinger
ISBN 0-471-29258-3   © 1999 John Wiley & Sons, Inc.

cause invertebrates may be an important mechanism to transfer botulism toxin from infected bird carcasses to healthy waterfowl. We discuss how invertebrate responses may affect management goals in wildlife habitats and conclude by listing some research topics that need further study.

INTRODUCTION

Invertebrate communities in stream and lake ecosystems have been extensively studied, leading to overarching concepts of invertebrate ecology in these habitats (e.g., trophic-level interactions, Carpenter and Kitchell 1988; the River Continuum Concept, Vannote et al. 1980; invertebrates as bioindicators, Rosenberg and Resh 1993). In contrast, most research in wetland habitats has focused on efforts to control pest mosquito populations, and much less is known about other wetland invertebrates. As a result, few ecological paradigms have been developed in wetland ecosystems. However, recent research indicates that environmental constraints and colonization dynamics are important factors regulating invertebrate communities in temporarily flooded habitats (Wiggins et al. 1980, Williams 1996).

In California most wetlands are managed as overwintering habitat for waterfowl and shorebirds (Gilmer et al. 1982, Helmers 1992). Migratory waterbirds begin returning from their breeding grounds in the northern United States and Canada in late August, and numbers increase until they peak in late December. Approximately 60 percent of all ducks and geese migrating along the Pacific flyway use wetlands in the Central Valley (6–7 million birds), and this comprises about 18 percent of all North American waterfowl (Gilmer et al. 1982). Both ducks and shorebirds feed heavily on wetland invertebrates during the winter months (Heitmeyer et al. 1989, Helmers 1992). Moreover, some species of shorebirds (Helmers 1992) and waterfowl (McLandress et al. 1996) nest in California, and juvenile waterbirds feed primarily on aquatic insects (Sugden 1974, Helmers 1992).

The length and timing of flooding affect the distribution and composition of wetland plant communities and can be used to classify these ecosystems. For example, permanent wetlands are flooded year-round, but seasonal and semipermanent wetlands are dry during portions of each year. By definition, seasonal wetlands are usually dry before the end of the plant's growing season, but semipermanent wetlands remain filled throughout most of the growing season (Cowardin et al. 1979). In California most habitats are seasonally or semipermanently flooded, and these range from inland freshwater wetlands to brackish estuarine marshes in the Sacramento-San Joaquin River Delta near the San Francisco Bay (Heitmeyer et al. 1989).

Most existing marshes are impounded and flooded via canals starting in late summer, with drawdowns occurring in late winter. These flooding schedules differ dramatically from those of naturally flooded marshes. Managers of wildlife habitats control flooding schedules to promote species of plants that produce seeds eaten by waterfowl, and use burning, mowing, or disking to reduce dense stands of emergent vegetation to increase accessibility to feeding waterbirds. Although these methods are used to enhance waterfowl and shorebird habitats, the effects of these practices on invertebrates and plants important in waterbird diets are relatively unknown (e.g., Kantrud 1986).

In this chapter we discuss research examining invertebrate ecology in nontidal wetlands in California. We begin by reviewing characteristics of invertebrate communities in seasonal and semipermanent wetlands and discuss factors influencing invertebrate distribution and abundance, including colonization, trophic relationships, and abiotic interactions. We also discuss the ecology of invertebrates consumed by waterfowl and control of pestiferous mosquito populations. Finally, we review research conducted at Grizzly Island Wildlife Area (Solano Co.) and Sacramento National Wildlife Refuge (Glenn Co. and Colusa Co.) that are managed as overwintering habitat for migratory birds. We discuss research examining invertebrate responses to various management practices and suggest topics for future study.

SEASONAL AND SEMIPERMANENT WETLANDS OF CALIFORNIA

Climate

California has a Mediterranean climate with cool, wet winters and warm, dry summers; over 80 percent of the annual precipitation occurs from November to March (Kahrl 1979). Historically, winter rains and spring snowmelt from the Sierra Nevada Mountains flooded low-lying wetland habitats along river beds and in the Central Valley, and many of these areas would dry out during the nonrainy summer months. Consequently, wetland hydroperiods are affected by interactions of precipitation and river discharge. Rain-filled wetlands typically flood from November through March. River discharge is regulated by seasonal changes in precipitation, and the lowest stream flows are recorded during the months of July through October (Fig. 34.1). Wetlands flooded by rivers can have various hydroperiods because rivers that receive runoff primarily from rainfall (e.g., Eel River in the northern Coast Ranges) have maximum discharge from December through March, but rivers that also receive runoff from snowmelt (e.g., Sacramento and San Joaquin Rivers) can have high flows through early summer.

Annual precipitation is highest along the coast and at higher elevations, and precipitation generally decreases from north to south (Kahrl 1979). Highest precipitation occurs in the northern Coast Range and Sierra Nevada mountains (mean precipitation = 97–240 cm year^{-1}), and lowest precipitation

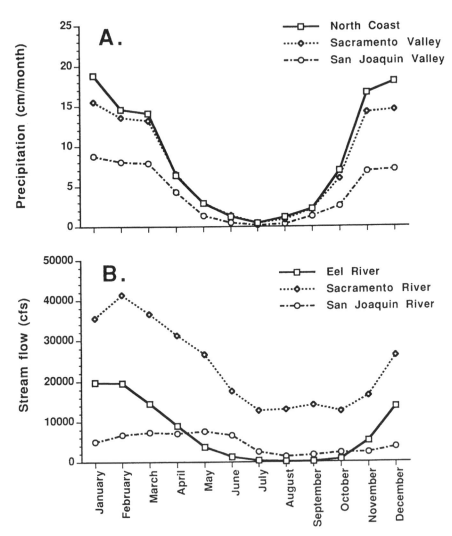

Fig. 34.1. Mean precipitation (cm per month) and stream flow (cubic feet per second) in California. (*a*) Precipitation in northern Coast Range mountains and the Sacramento and San Joaquin Valleys. Values are average precipitation from 1961–1990 (Anonymous 1996). (*b*) The Eel River is located within the northern Coast Ranges, and stream flow was measured near Scotia, California; stream flow in the Sacramento River was measured near Sacramento, California; and stream flow in the San Joaquin River was measured near Vernalis, California. Stream flows are averages of 66–83 years of USGS data.

occurs in the Imperial Valley and the Mojave Desert (mean precipitation = 6–11 cm year^{-1}). Rainfall in the Central Valley ranges from 56 cm year^{-1} in the north to 16 cm year^{-1} in the south.

Seasonal temperatures vary widely within California because the state extends >1,200 km from north to south and ranges in elevation from below sea level in the Imperial Valley to >3,000 m in the Sierra Nevada mountains. At lower elevations winter temperatures are usually above freezing throughout most of the state (Heitmeyer et al. 1989). Local temperature regimes affect evaporation rates, and annual evaporation is much greater than precipitation in the southern portions of the state and in the Central Valley. For example, mean annual evaporation is 152–178 cm year^{-1} in the Central Valley and is greater than 254 cm year^{-1} in the Imperial Valley (Kahrl 1979)! Wetland hydroperiods are affected by regional differences in precipitation and evaporation rates. For example, wetlands with short hydroperiods (e.g., seasonal wetlands such as playas and vernal pools) are common in regions with low precipitation and high evaporation rates (e.g., San Joaquin Valley and the southern California deserts), whereas wetlands with longer hydroperiods (e.g., semipermanent and permanent marshes) are more common in regions with high precipitation and low evaporation rates (e.g., northern Sacramento Valley and the Coast Ranges).

The intensity and duration of the rainy season is highly variable, and precipitation in a given year is often 100 percent greater or less than the average annual amount. For this reason, Kahrl (1979; p. 6) says, "variations in precipitation are so great that the state rarely enjoys a 'normal' year . . ." As a result, wetlands historically varied in size from year to year. However, more than 1400 dams and thousands of miles of levees have been built throughout California to control runoff during winter rains and spring snowmelt (Mount 1995), and flooding in most wetlands is now controlled.

Physiography and Vegetation

It is estimated that >90 percent of California's original wetlands have been drained or converted to agricultural uses such as rice farming (Bertoldi and Swain 1996). Only 182,000 ha of tidal and nontidal wetlands remain in California, and the majority (115,000 ha) are in the Central Valley (Heitmeyer et al. 1989; Fig. 34.2). An additional 3,700 ha of nontidal wetlands were formed in the Imperial Valley in the early 1900s when an irrigation diversion of the Colorado River collapsed during winter flooding and created the present Salton Sea (Kahrl 1979). Nontidal wetlands also occur in the Basin and Range region in northeastern California, the southern California deserts, and the Sierra Nevada and Coast Range mountains. However, little research has been conducted in these wetlands, and the total area comprised by these habitats has not been estimated (Bertoldi and Swain 1996).

The Central Valley is approximately 640 km long from north to south and 64 km wide from east to west. It is bordered on the east by the Sierra Nevada

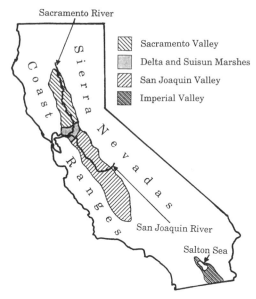

Fig. 34.2. Major inland wetland regions in California. Wetland regions were determined from the following sources: Gilmer et al. (1982), Heitmeyer et al. (1989), and Bertoldi and Swain (1996).

mountains and on the west by the Coast Range mountains. Because of differences in climate, the valley can be divided into three ecoregions: the Sacramento Valley is the northern portion and is drained by Sacramento River flowing south; the San Joaquin Valley is drained by the San Joaquin River flowing north; and the Delta and Suisun Marshes are located where the Sacramento and San Joaquin Rivers join and flow into the San Francisco Bay (Gilmer et al. 1982).

In the Sacramento Valley only 32,000 ha of mostly semipermanent marshes remain from the original >600,000 ha of semipermanent and riverine wetlands (Heitmeyer et al. 1989). Some common plants in semipermanent marshes include tule (*Scirpus acutus*), cattail (*Typha* spp.), and submergent plants such as widgeon grass (*Ruppia maritima*), horned pondweed (*Zannichellia palustris*), and sago pondweed (*Potamogeton pectinatus*). Moist-soil vegetation in seasonal marshes includes alkali bulrush (*Scirpus robustus*), smartweed (*Polygonum* spp.), common barnyard grass (*Echinochloa crusgalli*), swamp timothy (*Heleochloa schoenoides*), and sedges (*Carex* spp.). Common trees in riverine habitats include alder (*Alnus* spp.), willow (*Salix* spp.), and cottonwood (*Populus* spp.).

In the San Joaquin Valley about 51,000 ha of wetlands remain of the original 650,000 ha (Heitmeyer et al. 1989). Most are seasonally flooded freshwater or alkaline marshes, and there are relatively few riverine wetlands. Plants in these habitats include species common in the Sacramento Valley and also species tolerant of alkaline soils such as iodine bush (*Allenrolfea occidentalis*) and saltgrass (*Distichlis spicata*).

In the Delta and Suisun Marshes 29,000 ha of wetlands remain of the original 308,000 ha, and most existing wetlands are in Suisun Marsh (Heitmeyer et al. 1989). In the past these were tidally flooded brackish and freshwater marshes, but today most are protected from tidal flooding by levees and are seasonally flooded. Vegetation in brackish marshes is dominated by salttolerant plants such as pickleweed (*Salicornia virginica*), alkali bulrush, saltgrass, fathen (*Atriplex patula*), baltic rush (*Juncus balticus*), and brass buttons (*Cotula coronopifolia*). Freshwater marshes support plant communities that are similar to those in Sacramento Valley marshes.

INVERTEBRATES

Population and Community Ecology

Because seasonal and semipermanent wetlands are dry during portions of most years, aquatic invertebrates must recolonize these habitats after they are flooded. Invertebrates use various colonization mechanisms including (1) aerial colonization from nearby flooded habitats; (2) passive colonization in flood waters; and (3) a drought-tolerant life cycle stage (Wiggins et al. 1980). All of these mechanisms are used by invertebrate colonists in wetlands in California. For example, we used sticky traps to collect flying aquatic insects that colonized and emerged from seasonal wetlands. We sampled from September through March and collected damselflies, dragonflies, corixid water boatmen, dytiscid and hydrophilid beetles, mosquitoes, chironomid midges, syrphid flies, and brine flies (Batzer et al. 1997, F. de Szalay, unpublished data). Numbers of most taxa declined in early winter (i.e., mid-December) and then increased again in spring, but some species (e.g., *Culiseta inornata* mosquitoes and *Cricotopus sylvestris* midges) were abundant in winter. Flightless invertebrates (e.g., amphipods) inhabiting permanently flooded slough channels can colonize wetlands during flooding (Batzer and Resh 1992a). Garcia and Hagen (1987) report that a dytiscid beetle, *Agabus disintegratus*, aestivated in plant litter and among the roots of plants in dry wetland ponds. *Agabus* beetles collected in the summer months were in a state of reproductive diapause and would not fly; they could live up to five months without food or water, relying on fat reserves (Garcia et al. 1990). Reproductive diapause was broken with short-daylength light regimes, indicating that these beetles are active in late summer and can emerge and mate soon after the wetlands are flooded.

Invertebrate communities change over time in both semipermanent and seasonal wetlands. In a semipermanent marsh densities of most insects and crustaceans were highest during summer, but copepod numbers fluctuated widely throughout the flooding period (Feminella and Resh 1989, Bergey et al. 1992). In seasonal marshes, populations of some taxa increase rapidly after flooding and decrease after a relatively short time (e.g., syrphid, mosquito, and brine fly larvae, and tadpole shrimp), but other taxa persist as long as water is available (e.g., water boatmen, beetles, midges, and amphipods) (Walton et al. 1991, Batzer and Resh 1992a). Temporal changes in populations of detritivores, predators, and herbivores are different after flooding and can be correlated with the expected availability of their food resources. Detritivores rapidly colonize newly flooded wetlands (within one week) because high amounts of decaying plant matter are available immediately upon flooding, and predator and herbivore numbers increase later as their food resources become available (de Szalay and Resh 1996).

Predatory invertebrates can affect distributions of their prey. For example, numbers of mayflies sampled within 10 days of initial flooding were lower in ponds with the predaceous tadpole shrimp, *Triops longicaudatus,* than in ponds without shrimp (Walton et al. 1991). Although mayflies were negatively correlated with shrimp, they were positively correlated with other predatory invertebrates (e.g., dragonfly, damselfly, and dytiscid beetle larvae), indicating that shrimp were the most important predator. Furthermore, smaller mayflies were more susceptible to predation by shrimp because they moved relatively short distances when disturbed and could be easily overtaken. Predators can also indirectly influence algal biomass when they decrease herbivore numbers (Batzer and Resh 1991).

Although relatively few aquatic invertebrate taxa feed on living macrophytes (Batzer and Wissinger 1996), herbivores feeding on algae can be abundant in wetland habitats (Batzer and Resh 1991, de Szalay and Resh 1996). Furthermore, when herbivores that feed on macrophytes are present, they can have a pronounced effect on plant dynamics. For example, the introduced red swamp crayfish, *Procambarus clarkii,* was shown to affect the distribution of sago pondweed in a semipermanent marsh (Feminella and Resh 1989). In this marsh submergent stands of sago pondweed typically formed thick floating mats from May through October that covered up to 80 percent of the open-water areas, but in years when crayfish were abundant they reduced sago pondweed stands to 5 percent of the surface area. When crayfish were excluded from bare areas, sago pondweed regenerated in the exclosures, and when crayfish were added to enclosures with dense pondweed, the amount of vegetation remaining after 45 days was negatively correlated with crayfish densities.

Emergent and submergent plants are often very abundant in wetlands, and these provide both food (Severson 1987, Feminella and Resh 1989, Batzer and Resh 1992a, de Szalay and Resh 1996) and attachment substrates (Bergey et al. 1992, de Szalay and Resh 1996) for invertebrates. Abundance and composition of invertebrate communities can be different between different plant species. For example, higher invertebrate abundances were found in plant stands that decomposed rapidly after flooding than in plant stands that decomposed slowly, and these differences may have been a result of availability of detrital food resources (Severson 1987). Furthermore, proportions of detritivores, herbivores, and predators were different in benthic versus epiphytic habitats within stands of pickleweed (de Szalay and Resh 1996). Most invertebrates collected in benthic habitats were detritivores, but herbivores were sometimes the most abundant trophic group collected in epiphytic habitats. These patterns may also reflect differences in food availability because most detritus will eventually sink to the bottom and algae biomass is higher on plant stems than in benthic sediments.

Plant density can also affect invertebrate numbers. In stands of sago pondweed, densities of many invertebrate taxa were positively associated with plant biomass, except during times of maximum plant growth, when few taxa were correlated with plant biomass (Bergey et al. 1992). Plant cover may also affect invertebrate colonization because densities of some taxa were higher in vegetation adjacent to open water than in densely vegetated areas (Batzer and Resh 1992a), and numbers of some taxa that aerially colonized seasonal wetlands were correlated with amount of emergent plant cover (de Szalay 1995).

Evaporation during the warm, dry summer months can be high in California wetlands, affecting physicochemical conditions that impact invertebrates. For example, dissolved ionic concentrations in vernal pool wetlands are correlated with the number of times the wetlands dry up during the year (Ebert and Balko 1987). Invertebrate species richness is correlated with pool size and the number of times the wetlands are dry; small pools dry up more frequently and generally have higher ionic concentration and lower species richness than larger pools (Ebert and Balko 1987, Gallagher 1996). Physicochemical conditions can also be affected by geology. Solute levels (alkalinity, total dissolved solids, and conductivity), pH, water surface area, and elevation were sampled in 58 vernal pools in the Sacramento Valley, and these factors were strongly correlated with regional geological and floristic characteristics (King et al. 1996). Crustacean species assemblages in these vernal pools clustered into four well defined groups that were correlated with geologic/floristic characteristics but not with geographical proximity of the wetlands. This indicated that the geology of wetland basins had a greater effect on the composition of invertebrate communities than locality.

Abiotic factors can affect invertebrate distribution and survival. For example, effects of water quality were estimated to kill 9–90 percent of larval and pupal mosquitoes in six sampled habitats (Reisen et al. 1989). An example of extremely harsh abiotic conditions inhabited by aquatic invertebrates are shallow drainwater basins used to evaporate agricultural wastewater. Water enters these basins at about 20 percent the ionic concentration of seawater, then flows through a series of condensing ponds, and is about 600 percent the concentration of seawater before evaporating completely (Euliss et al. 1991a). Invertebrate communities in these basins are very simple; a midge, *Tanypus grodhausi,* and a water boatman, *Trichocorixa reticulata,* comprised >96 percent of the invertebrate biomass. Although invertebrate abundance was high (mean density was 4600 invertebrates m^{-1} and mean biomass was

790 mg m^{-1}), it varied with ionic concentration. Midge abundance was negatively correlated with ionic concentration, whereas water boatmen had a unimodal distribution; water boatman biomass increased until ionic concentrations were approximately equal to seawater and then decreased at higher concentrations.

Mosquito Ecology and Control

In California, as in other regions of the United States, more research has examined the ecology of mosquitoes than of any other wetland invertebrate, largely because wetlands afford excellent breeding habitats for pestiferous mosquitoes. For example, Garcia and Des Rochers (1984) estimate that a single 24-ha wetland produced over 144 million *Aedes* mosquitoes! In California mosquitoes are nuisance pests, and some species vector viral diseases, including western equine encephalomyelitis and St. Louis encephalitis (Reeves and Hammon 1962). Some comprehensive sources on biology and control of California mosquitoes include Bohart and Washino (1978), Collins and Resh (1989), Reeves (1990), and Proceedings and Papers of the Annual Conferences of the California Mosquito and Vector Control Association (1932 to present, Sacramento, California). Here we will discuss additional information examining mosquito ecology in wetland habitats.

Although predaceous fish, especially the aptly named mosquitofish, *Gambusia affinis,* are widely used as biological control agents of immature mosquitoes in permanently flooded habitats (Coykendall 1980, Chapman 1985), they are not effective in seasonal and semipermanent wetlands because they must be reintroduced each time the habitats are flooded. In seasonal wetlands predaceous invertebrates have received attention as potential biological control agents because predatory species naturally colonize newly flooded wetlands. However, invertebrate predators are not always effective, since some mosquito species (e.g., *Aedes* spp. and *Psorophora columbiae*) colonize and emerge as adults before most predatory invertebrates are abundant (Walton et al. 1990, Batzer and Resh 1992b, Fry et al. 1994).

However, once predaceous invertebrate populations are established they can reduce mosquito production. In one study predators killed 4–85 percent of larval and pupal mosquitoes and were the most important source of mortality in five of six sites sampled (Reisen et al. 1989). Walton et al. (1990) reports that species assemblages of potential mosquito predators changed after flooding; tadpole shrimp colonized experimental mesocosms within days, and numbers of predaceous beetles, hemipterans, and odonates increased later. Dytiscid beetle larvae appeared to be the most effective mosquito predators, and mosquito numbers generally decreased and were low 20–30 days post-flooding. Although tadpole shrimp are sometimes pests in agricultural rice fields and can impact populations of beneficial insect species (e.g., midges and mayflies; Walton et al. 1991, Fry et al. 1994), they may be useful as biological control agents in some temporarily flooded habitats because they

hatch from desiccation-resistant eggs after 48–72 hours, reproduce parthen-
ogenically after only six days, and can eliminate populations of some mosquito species (Tietze and Mulla 1991, Fry et al. 1994).

The physical structure of wetland plants will affect survival and recruitment
of mosquitoes. *Gambusia* fish predation of *Anopheles* mosquitoes is lower in
dense stands of submerged vegetation than in open water (Orr and Resh
1989). *Anopheles* survival is also higher in dense vegetation in the absence
of fish predators, suggesting that plant stands provide a better microhabitat
than open-water areas (Orr and Resh 1989, 1992). For example, emergent
vegetation will reduce the effects of wind action and water currents that can
carry mosquito larvae into less favorable habitat (Garcia and Des Rochers
1984). *Anopheles* recruitment was highest when plant stem density was 500–
1000 stems m^{-2}, but it decreased at higher stem densities (Orr and Resh
1992).

Some research has examined whether vegetational management practices
that are used to remove plant stands (e.g., prescribed burning and using
tractor-pulled mowers or diskers) can also be used to control mosquitoes. Results show that mosquito numbers are generally lower in areas where the vegetation is burned, mowed, or disked than in unmanipulated control areas
(de Szalay et al. 1995, Schlossberg and Resh 1997), and mosquito populations
may be easier to control in mowed areas because they are concentrated along
wetland edges (Batzer and Resh 1992b). Moreover, deeply flooded wetlands
have more open water because standing macrophytes are submerged, and
these areas have fewer mosquitoes and more mosquito predators (e.g., dytiscid
beetle larvae) than shallowly flooded areas (Batzer and Resh 1992b).

Invertebrates Consumed by Waterfowl

Although dabbling ducks (Anatinae) feed heavily on plant foods during the
entire year, invertebrate foods are a vital source of nutrients for growth, molting, and reproduction (Sugden 1974, Krapu 1981). Studies throughout California have shown that immature chironomid midges are often the most
important invertebrate taxa in waterfowl diets. Other important taxa include
crustaceans (e.g., amphipods, copepods), insects (e.g., odonates, water boatmen, hydrophilid and dytiscid beetles, and brine flies), and snails (Connelly
and Chesemore 1980, Euliss and Grodhaus 1987, Euliss and Harris 1987,
Miller 1987, Euliss et al. 1991b, Batzer et al. 1993). Although we found no
studies examining diets of shorebirds in inland wetlands of California, shorebirds and waterfowl often feed on similar invertebrate taxa (Wetmore 1925,
Baldassarre and Fischer 1984).

The proportion of invertebrates in waterfowl diets generally increases from
fall through spring, and this may be related to temporal changes in invertebrate populations (Connelly and Chesemore 1980, Euliss and Harris 1987,
Miller 1987, Batzer et al. 1993). For example, invertebrates made up only 1–
2 percent of food eaten by northern pintails (*Anas acuta*) in August and

September, but invertebrates were more abundant in February and March, when they comprised 29–66 percent of pintail diets (Miller 1987).

Waterfowl morphology and feeding methods will affect the types of invertebrates eaten. Duck species with wider bill lamellae consume larger invertebrates than species with finely spaced bill lamellae (Euliss et al. 1991b, Batzer et al. 1993). For example, northern shovelers have very finely spaced bill lamellae and can prey on rotifers and copepods as small as 100 microns in length (Euliss et al. 1991b). Most managed waterfowl habitats are shallowly flooded because dabbling ducks typically feed in <20 cm of water (Euliss and Harris 1987, Euliss et al. 1991b). Although benthic invertebrates in deep water may not be immediately available to foraging dabbling ducks, many invertebrates (e.g., mayflies, caddisflies, midges) will migrate to the water surface to emerge. These newly emerged adults are vulnerable to foraging ducks. Additionally, when seasonal marshes are drawn down, benthic invertebrates are exposed and other mobile invertebrates are concentrated into small, shallow pools (Euliss and Grodhaus 1987, Euliss et al. 1991b).

Feeding waterfowl have been shown to affect invertebrate abundances. In one study densities of some invertebrates eaten by ducks were lower in open plots than in waterfowl exclosures, but other taxa showed the reverse pattern (Severson 1987), indicating that waterfowl can affect invertebrate populations directly (by predation) and indirectly (by habitat disruption). For example, mosquitoes are rarely eaten by waterfowl, but larval densities are lower where feeding ducks eliminate stands of floating vegetation (Collins and Resh 1989).

INVERTEBRATE RESPONSES TO WILDLIFE MANAGEMENT PRACTICES

Research Conducted at Grizzly Island Wildlife Area

Study Site Description. Portions of Suisun Marsh are managed by state and federal agencies, but the majority (~85 percent) is privately owned (Rollins 1981). Most of the private land is managed by duck hunting clubs, the earliest of which were established in the 1880s (Mall 1969). Grizzly Island Wildlife Area (WA) comprises 3356 ha in Suisun Marsh (Heitmeyer et al. 1989) and is managed by the California Department of Fish and Game. Grizzly Island WA includes many seasonal marshes and uplands and some semipermanently flooded wetlands and sloughs. These habitats are used by many species of wildlife, and most seasonal marshes are managed for migratory waterfowl and shorebirds.

Levees have been built around most wetlands to regulate flooding, and wetlands are filled via canals from the Sacramento–San Joaquin River delta. Most seasonal marshes are flooded from October through February, and water depths are kept shallow (<30 cm) to facilitate feeding by dabbling ducks.

Many wetlands are periodically irrigated during the summer to promote the growth of hydrophytes that are food for waterfowl (Rollins 1981).

The climate at Suisun Marsh is relatively mild; mean air temperature ranges from 8°C in January to 22°C in July (Mall 1969). Mean annual precipitation is 53 cm (McLandress et al. 1996). Water salinity varies seasonally; the lowest water salinities are in February (~ 3 ppt) during the rainy season, and salinities increase during the dry season, reaching 15 ppt in August (Rollins 1981). Salinities have increased throughout Suisun Marsh because upstream diversions for agriculture have decreased mean annual outflow in the river delta from 33.6 million acre feet to 15.9 million acre feet from 1900–1960, allowing salt water intrusions to extend farther upstream into the delta (Mall 1969).

Responses to Mowing, Disking, and Prescribed Burning. Wetlands that include a 50:50 mixture of open water:emergent vegetation are known as "hemi-marshes" (Weller and Spatcher 1965), and waterfowl use of these areas is greater than in areas with unbroken stands of dense emergent plant cover (Kaminski and Prince 1981, Murkin et al. 1982). In California wetland managers use various cultural practices, including mowing and disking with tractors, prescribed burning, and cattle grazing in dense plant stands, to provide hemi-marsh habitats for overwintering waterfowl (Rollins 1981, Heitmeyer et al. 1989). Because little is known about how these practices affect invertebrate and plant food resources of waterfowl, we conducted a series of experiments examining the effects of mowing, disking, and prescribed burning on wetland invertebrates and plants in seasonal marshes at Grizzly Island WA.

Initial experiments examined the effects of mowing in stands of pickleweed. This salt-tolerant plant forms dense monotypic stands in these brackish wetlands, and wetland managers commonly use mowing or disking to reduce stands of pickleweed (Rollins 1981). Before the wetlands were flooded, we removed 50 percent of the vegetation by mowing strips within pickleweed stands. After flooding, these mowed portions were mostly open water, but nonmowed portions of these treatment areas and adjacent control areas were mostly dense emergent vegetation. Although temporal changes of aquatic invertebrate communities in the mowed treatment areas and control areas were similar, abundances of many taxa were different in several ways (Batzer and Resh 1992a, de Szalay et al. 1996). First, densities of epiphytic invertebrates were lower in the open-water portions of the 50 percent mowed treatment areas than in control areas. However, densities of water boatmen, midges, dytiscid and hydrophilid beetles, and amphipods were generally higher in the unmowed emergent vegetation within the 50 percent mowed areas than in control areas. Second, total invertebrate numbers in 50 percent mowed treatment areas (combining densities in open water and vegetated portions) compared to control areas either increased or decreased depending on the taxa. For example, mosquitoes, epiphytic midges, and brine flies had lower total numbers in mowed versus control habitats, whereas water boatmen, hydrophilid and dytiscid beetles, benthic midges, and amphipods had higher total numbers in mowed versus control habitats.

Batzer and Resh (1991) show that epiphytic algal biomass was indirectly affected by mowing because high numbers of the predatory beetle, *Berosus ingeminatus,* colonized 50 percent mowed treatment areas after the wetlands were flooded in autumn, and these decreased numbers of their herbivorous midge prey, *Cricotopus sylvestris.* In contrast, nonmowed control areas were colonized by fewer *Berosus* beetles, and *Cricotopus* midges in these areas were abundant and positively correlated with algal resources. In late winter beetle populations declined, and midge numbers increased in mowed treatment areas as they exploited the abundant algal food resources in these areas. Midge numbers in control areas did not increase, since algal biomass had already been heavily grazed. Although predators reduced midge populations in mowed areas during the fall, the total number of midges collected in mowed and control areas during the entire sampling period was similar because midge populations increased markedly in mowed areas after predator numbers decreased.

The effects of cultural practices were also examined in stands of saltgrass, which form dense monotypic stands along shallow wetland margins (Mall 1969). Saltgrass does not produce seeds eaten by waterfowl, and wetland managers at Grizzly Island WA suppress it with burning, mowing, or disking (Rollins 1981). The effects of burning and mowing were tested in small-scale (100 m^2) experimental plots (de Szalay and Resh 1997), and the effects of burning or tractor-pulled disking were also tested in large-scale (600 m^2) experimental plots (Schlossberg and Resh 1997). Densities of water boatmen, dytiscid and hydrophilid beetles, midges, and oligochaetes were generally higher in saltgrass stands that had been burned or disked than in unmanipulated control areas (de Szalay and Resh 1997, Schlossberg and Resh 1997). However, densities of mosquitoes, other midges, and copepods were sometimes lower in burned or disked treatment areas than in control areas. Invertebrate densities in mowed treatment areas never differed from those in control areas because saltgrass grew back quickly after mowing.

Plants colonized the treatment areas after the wetlands were drawn down, and plant communities were sampled in burned, disked, and mowed areas in the following year. Areas that were mowed were largely recolonized by saltgrass, but burned and disked areas had more diverse plant communities, including goosefoot (*Chenopodium rubrum*), brass buttons, alkali bulrush, and western sea-purslane (*Sesuvium verrucosum*). Unlike saltgrass, these plants did not grow in dense stands, and more bare ground was visible in burned and disked areas than in mowed areas or control areas (de Szalay and Resh 1997, Schlossberg and Resh 1997). Changes to the plant communities in burned and disked areas lasted at least two years posttreatment, and invertebrate assemblages were different during the second flooding season in treatment and control areas (Schlossberg and Resh 1997).

Invertebrate communities may be different between created hemi-marsh habitats (e.g., mowed, burned, and disked areas) and dense plant stands (e.g., unmanipulated control areas) for several reasons. For example, populations of epiphytic invertebrates (e.g., mosquitoes, some midges, brine flies) may be lower in hemi-marshes because there are fewer plant stems. Burning and mowing may also impact the amount of plant litter and algal biomass and thereby affect food resources of detritivorous and herbivorous species. Furthermore, densities of some predatory invertebrates (e.g., water boatmen and hydrophilid and dytiscid beetles) are higher in areas where the vegetation is removed, and they can reduce numbers of their prey. Finally, these practices may affect cues used by some invertebrates to select oviposition or colonization sites. For example, densities of some taxa are correlated with plant cover (de Szalay 1995), which is lower after burning, disking, or mowing. However, most of these mechanisms have not been examined in these habitats, and their relative importance is not known.

Responses to Flooding Depth. Although most seasonal marshes are kept shallow to enable dabbling ducks to forage on benthic invertebrates and seeds, ample evidence exists that areas with deeper water support the most invertebrates. When we examined invertebrate responses to flooding, densities of midge larvae, water boatmen, and hydrophilid beetle larvae were usually greater at deep (>30 cm) compared to shallow depths (Batzer and Resh 1992b, Batzer et al. 1993, Batzer et al. 1997). Deep water covers more plant surfaces and may increase the availability of substrate for epiphytic invertebrates (Batzer and Resh 1992b). Organic matter can also accumulate in deep sections of marshes, and densities of detritivorous taxa (e.g., *Chironomus stigmaterus* midge larvae) may be higher in these areas because of the increased availability of food. Also, predatory beetle larvae are more abundant in deeper water (Batzer et al. 1997), probably because they are tracking an increase in their midge prey.

Responses to Flooding Schedules. Most California wetlands are dry in the summer, and managers first flood seasonal marshes as early as August or as late as October. It is important to note that flooding schedules in managed marshes deviate strongly from hydroperiods in naturally flooded wetlands in California; even late flooding (late September–October) precedes natural flooding cycles. In a previous study Batzer et al. (1993) found that the date when wetlands are initially flooded can influence invertebrate populations. Therefore, we conducted an additional experiment in 1991–92 to examine invertebrate responses to various flooding schedules. Twelve wetlands were selected and two were flooded on one of the following six dates: August 5, 19, September 2, 16, 30, and October 14. All marshes remained flooded until they were drawn down in March. For purposes of analysis, we divided the 12 wetlands into two groups: (1) early-flooded sites (those first filled August

5, 19, or September 2); and (2) late-flooded sites (those first filled September 16, 30, or October 14).

We sampled in February–March, when waterfowl forage extensively on invertebrates, and compared invertebrate communities between early and late flooded sites (Table 34.1). Odonates and midges were much more abundant in early- versus late-flooded sites. Batzer et al. (1997) provide a detailed

TABLE 34.1. Numbers of Epiphytic and Benthic Invertebrates Collected During February and March in 12 Seasonally Flooded Marshes of Suisun Marsh[a]

Taxa	Early-Flooded Sites Number/sample (SE)	Late-Flooded Sites Number/sample (SE)	P[b]
Epiphytic sweeps			
Odonates[c]	26.3 (12.0)	0.2 (0.2)	<0.0001
Water boatmen[d]	30.8 (22.5)	46.3 (17.2)	0.5960
Water beetles			
Dytiscidae[e]	6.9 (2.3)	12.6 (3.9)	0.2413
Hydrophilidae[f]	12.7 (4.9)	43.2 (31.0)	0.4862
Fly larvae			
Midges[g]	143.9 (28.0)	37.2 (11.9)	0.0057
Brine flies[h]	6.1 (4.5)	16.7 (4.5)	0.1256
Amphipods[i]	20.5 (16.2)	4.5 (3.6)	0.3568
Ostracods[j]	4.1 (2.8)	43.6 (39.6)	0.6304
	Number/m^2 (SE)	Number/m^2 (SE)	P
Benthic cores			
Hydrophilid beetles[f]	91.2 (24.2)	123.0 (60.4)	0.6358
Midge larvae[k]	2569.2 (369.7)	740.8 (309.3)	0.0854
Oligochaetes[l]	1132.8 (369.1)	642.4 (368.8)	0.3691

[a]Habitats designated as early-flooded sites were first flooded on August 5, 19, or September 2, and those designated as late-flooded sites were first flooded on September 16 or 30 or October 14. Epiphytic samples were collected with a 30-cm diameter D-frame aquatic net, and numbers represent pooled collections from three 1-m horizontal sweeps through the plants and water column. Benthic samples were collected using a 10-cm diameter corer, and data were converted to numbers per m^2 from 18 pooled samples per site.

[b]P values are from t-tests. When variances were not equal, data were $\log_{10}(x + 1)$ transformed prior to tests.

[c]Aeshnidae, Libellulidae, and Coenagrionidae.

[d]*Trichocorixa verticalis.*

[e]*Agabus* and *Rhantus.*

[f]*Berosus ingeminatus.*

[g]*Cricotopus* and *Microtendipes.*

[h]*Scatella* and *Brachydeutera.*

[i]*Eogammarus confervicolus.*

[j]Unknown species.

[k]*Chironomus stigmaterus.*

[l]Tubificidae and Naididae.

analysis of the response of the benthic midge, *Chironomus stigmaterus,* to these flooding manipulations. Since adult *Chironomus* were not active during the late fall and winter, they concluded that larval densities were higher in early- versus late-flooded sites because most females had completed oviposition before the late-flooded sites were available for colonization. *Chironomus* larvae were abundant throughout the winter, but these populations remained after autumn oviposition. A summer peak in adult odonate oviposition and colonization might also explain the low numbers of odonate larvae we found in late-flooded marshes.

Other taxa unaffected by flooding date were aerial colonizers such as water boatmen and brine flies, which probably breed later into the year than odonates and midges. Densities of taxa known to aestivate in dry wetlands (e.g., dytiscid beetles and possibly ostracods; Garcia et al. 1990, F. de Szalay, unpublished data) were also similar in early- and late-flooded sites. Thus, flooding date may have minimal effect on their populations. Although a previous study found that *Berosus ingeminatus* hydrophilid beetles were more abundant in early- versus late-flooded wetlands (Batzer et al. 1993), late-winter densities of these beetles did not differ between sites in this study (Table 34.1). However, a detailed analysis of their populations during the entire season showed that densities in early-flooded sites peaked earlier in the year, suggesting that these beetles, like midges and odonates, are reproductively less active in late autumn and winter (Batzer et al. 1997). Amphipods and oligochaetes reproduce after colonizing newly flooded wetlands (Batzer and Resh 1992a, Batzer et al. 1993). Their somewhat lower numbers in late-flooded sites may be explained by their populations having had less time to increase.

Historically, most seasonal marshes would remain dry until the onset of heavy winter rains in November, and only permanently flooded wetlands would be available in late summer. Consequently, many invertebrates common in managed seasonal marshes are probably typical permanent water taxa (e.g., *Chironomus* and *Cricotopus* midges and water boatmen) that opportunistically colonize newly flooded sites. Because the nonnatural flooding schedules used in managed seasonal marshes can affect invertebrate communities, it may be necessary to test whether the ecological and evolutionary paradigms developed for invertebrates in natural temporary water wetlands (e.g., Wiggins et al. 1980, Williams 1996) can be applied to California's intensively managed seasonal marshes.

Research Conducted at Sacramento National Wildlife Refuge

Study Site Description. The 4360-ha Sacramento National Wildlife Refuge (NWR) is located 10 km south of Willows in Glenn Co. The Sacramento Valley has been extensively developed for agriculture, and rice is the major crop grown in the area. Sacramento NWR was initially established in 1937 to provide overwintering habitat for migratory birds and reduce crop depredation of nearby rice fields by waterfowl. Most wetlands are seasonally flooded during the winter and periodically irrigated during the summer to culture moist-soil plants. Other wetlands are permanently flooded to provide habitat for nesting waterfowl in summer and migrant waterbirds in the winter. These habitats are drawn down every three to five years to recycle nutrients (Cook and Powers 1958, Kadlec and Smith 1992) and control undesirable vegetation.

The climate at the Sacramento NWR is similar to that of most areas within the Sacramento Valley. Winter temperatures average 3–15°C and summer temperatures average 14–32°C. Annual precipitation averages 48 cm (National Weather Service Files, Willows, California).

Responses to Abiotic Factors and Patterns of Spatial Distribution. As in much of the arid West, competition with urban and agricultural interests for limited water resources has resulted in insufficient quantities of water being available for wetland management. These shortfalls are especially acute during drought years. To compensate for water shortages, wetland managers sometimes use agricultural drainwater to augment their water allotments. Compared to the high-quality water from the Sacramento River, drainwater from agricultural fields is warmer, has lower dissolved oxygen concentrations, is higher in dissolved salts, trace elements and minerals, and sometimes contains agrichemicals (California Regional Water Quality Control Board, Central Valley Region 1989). Although environmental contaminants in drainwater have caused embryonic mortality and abnormalities in waterbirds in the San Joaquin Valley (e.g., Ohlendorf et al. 1986a, 1986b, 1987), the impact of water quality on wetland invertebrates is not well understood. Recent legislative action has increased the availability of high-quality water for wetland management, but much of the water used in wetlands is still commingled with agricultural drainwater because both are conveyed through the same canal network. Recent emphasis on improving and separating water delivery systems may result in less use of agricultural drainwater in the future.

On the Sacramento NWR Sefchick (1992) compared invertebrate faunas in wetlands filled with agricultural drainwater and wetlands filled with Sacramento River water and found that water quality did not significantly alter invertebrate communities. However, the agricultural drainwater received at the Sacramento NWR is of fairly high quality, and studies in other habitats in the Central Valley showed that invertebrate diversity is low in saline drainwater basins (Euliss et al. 1991a).

Other studies conducted at Sacramento NWR have examined factors influencing invertebrate communities in different microhabitats. Mackay (1993) conducted an extensive study of the spatial distribution of wetland invertebrates and compared them with the chemical and physical conditions. Mackay et al. (in review) developed logistic regression models that explained 52–74 percent of the observed variability of seven invertebrate taxa (calanoid copepods, cyclopoid copepods, Cladocera, chironomid midges, water boatmen, damselflies, and oligochaetes). Mean water depth, turbidity, temperature, ionic concentration, pH, sediment organic matter, and redox potential were the most important variables identified in the analyses. Taxa occupied different micro-habitats within these wetlands. For example, calanoid copepods were more abundant in areas with high turbidity, but other taxa (e.g., cyclopoid copepods, cladocerans, midges, water boatmen, and damselflies) were more abundant in areas with low turbidity. Likewise, odonates were more abundant in areas with low sediment organic matter, but other taxa (e.g., calanoid copepods, cladocerans, water boatmen, and oligochaetes) were more abundant in areas with high sediment organic matter. In general, most taxa responded to phys-ical and chemical variables affecting primary production within the micro-habitats they used.

A recent study (Gleason et al., in review) examined distributions of aquatic invertebrates in stands of swamp timothy, prickle grass (*Crypsis niliaca*), Ber-muda grass (*Cynodon dactylon*), spike rush (*Eleocharis macrostachya*), spran-gletop (*Leptochloa fascicularis*), joint grass (*Paspalum distichum*), common barnyard grass, cocklebur (*Xanthium strumarium*), nodding smartweed (*Po-lygonum lapathifolium*), tule, and common cattail (*Typha latifolia*). These plant species are all important to wetland managers, either because they are cultured to produce seeds eaten by migratory waterfowl or because they are considered to be weeds. Although density and biomass of invertebrates were similar among plant species, taxon richness was greater in smartweed, tule, cattail, Bermuda grass, and sprangletop than in prickle grass and swamp tim-othy. Plant structure also affected the abundance of invertebrate feeding groups; density and biomass of collector-gatherer and scraper functional groups was greater in forbs (i.e., smartweed and cocklebur) and robust emer-gents (i.e., tule and cattail) than in short (i.e., swamp timothy and prickle grass) and tall grasses (i.e., barnyard grass and sprangletop). In contrast, fil-terers, the numerically dominant functional group, and predators did not show any preference for specific plant species. The underlying causes of these re-lationships (e.g., using plants as a direct food source and habitat, effects of plant density on algal biomass and dissolved oxygen levels) were not deter-mined. However, recent studies using stable isotopes have found algae to be a more important food resource than macrophyte litter for invertebrates in other North American wetland habitats (Neill and Cornwell 1992), and further research is needed to determine which factors are important in California wetlands.

Association with Botulism. Avian botulism is a type of food poisoning caused by the ingestion of a neuroparalytic toxin produced by the anaerobic bacterium *Clostridium botulinum* (Wobeser 1981). Botulism was first reported in California about 1890 (Hobmaier 1932), and it is currently regarded as one of the most persistent waterfowl diseases in the state. In 1969 >141,000 waterfowl were killed in a single epizootic (Hunter 1970) and losses in the tens to hundreds of thousands are common each year. The bacteria are widely distributed within wetlands. In one study conducted at Sacramento NWR, *C.*

botulinum bacteria were found in the benthic sediments of all 10 sampled wetlands, and bacteria were present throughout the year (Sandler et al. 1993). Invertebrates have long been associated with avian botulism, although defining their role has presented a major challenge to researchers. The importance of fly maggots in concentrating botulism toxin in decomposing waterfowl carcasses has been shown (Rosen 1971), but the relationship between free-living wetland invertebrates and botulism is not well understood. The microenvironment concept of avian botulism (Bell et al. 1955) is based on the premise that toxin produced within carcasses of dead aquatic invertebrates may somehow precipitate outbreaks. Botulism spores are found in the tissues of most wetland fauna, and toxic cultures of the bacteria have been developed from carcasses of invertebrates incubated in the laboratory (Jensen and Allen 1960). However, the exact mechanism for the transfer of the toxin to birds has not been determined, and whether or not the transfer actually occurs is not clear.

There is some evidence to suggest that aquatic invertebrates may be attracted to decomposing duck carcasses, and botulism toxin has been isolated from free-living aquatic invertebrates in wetlands collected near infected, decomposing waterfowl carcasses (Duncan and Jensen 1976). Hence, it is possible that free-living aquatic invertebrates may become exposed through an association with infected waterfowl carcasses; such an association may transfer the toxin to healthy birds and cause outbreaks. However, a recent study at Sacramento NWR failed to detect botulism toxin in aquatic invertebrates collected 40 cm from infected duck carcasses (Hicks 1992). Furthermore, duck carcasses placed in wetlands did not affect taxonomic composition, biomass, or numbers of aquatic invertebrates (Hicks et al. 1997). However, it is conceivable that the 40-cm sampling distance used was not close enough to carcasses for significant effects in toxin accumulation in invertebrate tissues or changes in the invertebrate community to be detected (Hicks 1992, Hicks et al. 1977). The fact that Rocke et al. (1999) found a significant association between the probability of botulism outbreaks and invertebrate abundance demonstrates the complexity of the relationship of invertebrates and botulism and highlights the need for additional research.

Implications of Invertebrate Responses to Wildlife Management Practices

The responses of invertebrates to management practices have important implications for wetland managers. The results of these experiments clearly indicate that some methods can be used to enhance wildlife habitats by increasing invertebrate food resources used by migratory waterbirds. For example, many taxa (e.g., water boatmen, midges, and beetles) that are abundant in burned, disked, or mowed areas are important in waterfowl and shorebird diets. We also found that the biomass of large midges was highest in burned areas (de Szalay and Resh 1997), indicating that this practice increased midges in the size class readily captured by feeding dabbling ducks (see Batzer et al. 1993). Plant food resources were also increased by these cultural practices because plant species that were more abundant in burned and disked areas produce many seeds eaten by ducks (de Szalay and Resh 1997, Schlossberg and Resh 1997). However, it is important to note that some invertebrate taxa that are important in shorebird and waterfowl diets are negatively affected by these cultural practices (e.g., brine flies and some midges), and care must be taken to assess the applicability of these practices in other wetland habitats. Because some negative effects are possible, these practices should only be used on a subset of the available wetland habitats, and portions of plant stands should be left to provide habitat for beneficial epiphytic species. Furthermore, plant structure and physicochemical factors also affect invertebrate distributions, and some taxa important in waterfowl diets have opposite responses to these factors (e.g., abundances of water boatmen and odonates showed opposite associations with sediment organic matter). Therefore, management for a range of habitat types (e.g., permanent, semipermanent, and seasonal wetlands) is desirable because monospecific plant stands may lack microhabitats used by important invertebrate taxa.

Our results also indicate that invertebrates are affected by water management practices. Although many wetland managers assume that deep-water habitats are unimportant to foraging dabbling ducks because benthic foods are beyond their reach, midge numbers were higher in deep water than in shallow water, and these become available as pupae migrate to the water's surface to emerge. Furthermore, these areas are heavily used by foraging waterfowl after late-season drawdowns (Batzer et al. 1993). Thus, no real change in current management is required to provide deep water areas for invertebrates, although managers should realize the virtues of those habitats. Moreover, early flooding of seasonal marshes can be beneficial in terms of increasing invertebrate food resources for waterfowl. *Chironomus stigmaterus* is probably the single most important invertebrate food for California waterfowl (Euliss and Grodhaus 1987, Batzer et al. 1993), and it thrives in early-flooded sites (Batzer et al. 1997). Many of the other invertebrates that prosper in early-flooded marshes are also valuable waterfowl foods (e.g., epiphytic midges, odonates, amphipods, and water beetles).

Batzer et al. (1997) found that adult *Chironomus* midges colonized summer flooded wetlands from nearby previously filled habitats. However, midge populations in isolated early-flooded sites will build slowly because of a lack of aerial colonists. Invertebrate food resources for foraging waterfowl may be low in isolated habitats until invertebrate populations have time to reproduce. Euliss and Grodhaus (1987) suggest filling subunits (termed *brood-stock ponds*) of isolated seasonal marshes one to two months before the remainder of wetlands are flooded. Adult midges and other invertebrates produced in brood-stock ponds can inoculate the surrounding wetlands when they are flooded and reduce the time needed to repopulate these habitats.

FUTURE RESEARCH NEEDS

Although recent research has expanded our knowledge of the ecology of invertebrates inhabiting California's wetlands, many aspects still need further study. For example:

1. Most research has examined factors influencing invertebrate communities in seasonal marshes in the Central Valley. Little is known about other habitats such as vernal pools and riverine wetlands and other areas such as the Basin and Range region and the Imperial Valley.
2. Many of the processes that regulate invertebrate communities in wetland habitats are still poorly understood. For example, different invertebrate assemblages are associated with stands of different plant species, but the underlying causes for these patterns have not been determined.
3. Wetlands in California provide vital overwintering and nesting habitat for waterfowl and shorebirds. Therefore, additional research should examine how management practices affect invertebrate food resources. The importance of aquatic invertebrates in the epidemiology of avian botulism also needs further study. Moreover, California's wetlands support many other wildlife species, and it is important that management methods that enhance the habitat for these species are also developed.

LITERATURE CITED

Anonymous. 1996. Climatological Data, Annual Summary, California. Vol. 100, No. 13. National Oceanic and Atmospheric Administration, National Climatic Data Center, Asheville, NC.

Baldassarre, G. A., and D. H. Fischer. 1984. Food habits of fall migrant shorebirds on the Texas high plains. Journal of Field Ornithology 55:220–229.

Batzer, D. P., F. de Szalay, and V. H. Resh. 1997. Opportunistic response of a benthic midge (Diptera: Chironomidae) to management of California seasonal wetlands. Environmental Entomology 26:215–222.

Batzer, D. P., M. McGee, and V. H. Resh. 1993. Characteristics of invertebrates consumed by mallards and prey response to wetland flooding schedules. Wetlands 13: 41–49.

Batzer, D. P., and V. H. Resh. 1991. Trophic interactions among a beetle predator, a chironomid grazer, and periphyton in a seasonal wetland. Oikos 60:251–257.

———. 1992a. Macroinvertebrates of a California seasonal wetland and responses to experimental habitat manipulation. Wetlands 12:1–7.

———. 1992b. Wetland management strategies that enhance waterfowl habitats can also control mosquitoes. Journal of the American Mosquito Control Association 8: 117–125.

Batzer, D. P., and S. A. Wissinger. 1996. Ecology of insect communities in nontidal wetlands. Annual Review of Entomology 41:75–100.

Bell, J. F., G. W. Sciple, and A. A. Hubert. 1955. A microenvironment concept of the epizootiology of avian botulism. Journal of Wildlife Management 19:352–357.

Bergey, E. A., S. F. Balling, J. N Collins, G. A. Lamberti, and V. H. Resh. 1992. Bionomics of invertebrates within an extensive *Potamogeton pectinatus* bed of a California marsh. Hydrobiologia 234:15–24.

Bertoldi, G. L., and W. C. Swain. 1996. California wetland resources. Pages 127–134 *in* J. D. Fretwell, J. S. Williams, and P. J. Redman (eds.), National Water Summary on Wetland Resources. USGS Water-Supply Paper No. 2425, U.S. Geological Survey, Reston, VA.

Bohart, R. M., and R. K. Washino. 1978. Mosquitoes of California. University of California, Division of Agricultural Sciences, Berkeley, CA.

California Regional Water Quality Control Board, Central Valley Region. 1989. Nonpoint Source Water Quality Impacts at Sacramento National Wildife Refuge, Willows, California. September.

Carpenter, S. R., and J. F. Kitchell. 1988. Consumer control of lake productivity. Bioscience 38:764–769.

Chapman, H. C. 1985. Biological Control of Mosquitoes. Bulletin No. 6, American Mosquito Control Association, Fresno, CA.

Collins, J. N., and V. H. Resh. 1989. Guidelines for the ecological control of mosquitoes in non-tidal wetlands of the San Francisco Bay Area. Special Publication, California Mosquito and Vector Control Association, Inc. and The University of California Mosquito Research Program, Davis, CA.

Connelly, D. P., and D. L. Chesemore. 1980. Food habits of pintails, *Anas acuta*, wintering on seasonally flooded wetlands in the northern San Joaquin Valley, California. California Fish and Game 66:233–237.

Cook, H. H., and C. F. Powers. 1958. Early biochemical changes in the soils and waters of artificially created marshes in New York. New York Fish and Game Journal 5:9–65.

Cowardin, L. M., V. Carter, F. C. Golet, E. T. LaRoe. 1979. Classification of Wetlands and Deepwater Habitats of the United States. FWS/OBS-79/31, U.S. Fish and Wildlife Service, Washington DC.

Coykendall, R. L. (ed.). 1980. Fishes in California Mosquito Control. Biological Control Commission, California Mosquito Control Association, Sacramento, CA.

de Szalay, F. A. 1995. Ecological interactions of wetland management practices and aquatic invertebrates in Suisun Marsh, California. Ph.D. Dissertation, University of California, Berkeley, CA.

de Szalay, F. A., D. P. Batzer, and V. H. Resh. 1996. Mesocosm and macrocosm experiments to examine effects of mowing emergent vegetation on wetland invertebrates. Environmental Entomology 25:303–309.

de Szalay, F. A., D. P. Batzer, E. B. Schlossberg, and V. H. Resh. 1995. A comparison of small and large scale experiments examining the effects of wetland management practices on mosquito densities. Proceedings of the California Mosquito and Vector Control Association 63:86–90.

de Szalay, F. A., and V. H. Resh. 1996. Spatial and temporal variability of trophic relationships among aquatic macroinvertebrates in a seasonal marsh. Wetlands 16:458–466.

————. 1997. Responses of wetland invertebrates and plants important in waterfowl diets to burning and mowing of emergent vegetation. Wetlands 17:149–156.

Duncan, R. M., and W. I. Jensen. 1976. A relationship between avian carcasses and living invertebrates in the epizootiogy of avian botulism. Journal of Wildlife Diseases 12:116–126.

Ebert, T. A., and M. L. Balko. 1987. Temporary pools as islands in space and time: The biota of vernal pools in San Diego, southern California, USA. Archiv für Hydrobiologie 110:101–123.

Euliss, N. H., and G. Grodhaus. 1987. Management of midges and other invertebrates for waterfowl wintering in California. California Fish and Game 73:238–243.

Euliss, N. H., and S. W. Harris. 1987. Feeding ecology of northern pintails and green-winged teal wintering in California. Journal of Wildlife Management 51:724–732.

Euliss, N. H., R. L. Jarvis, and D. S. Gilmer. 1991a. Standing crops and ecology of aquatic invertebrates in agricultural drainwater ponds in California. Wetlands 11: 179–190.

————. 1991b. Feeding ecology of waterfowl wintering on evaporation ponds in California. Condor 93:582–590.

Feminella, J. W., and V. H. Resh. 1989. Submersed macrophytes and grazing crayfish: an experimental study of herbivory in a California freshwater marsh. Holarctic Ecology 12:1–8.

Fry, L. L., M. S. Mulla, and C. W. Adams. 1994. Field introductions and establishment of the tadpole shrimp, *Triops longicaudatus* (Notostraca: Triopsidae), a biological control agent of mosquitoes. Biological Control 4:113–124.

Gallagher, S. P. 1996. Seasonal occurrence and habitat characteristics of some vernal pool Branchiopoda in northern California, U.S.A. Journal of Crustacean Biology 16:323–329.

Garcia, R., and B. Des Rochers. 1984. Studies of the biology and ecology of mosquitoes in relation to the development of integrated control measures at Gray Lodge Wildlife Refuge, Butte County, California. Miscellaneous Publication, University of California Mosquito Control Research Program, Davis, CA.

Garcia, R., and K. S. Hagen. 1987. Summer dormancy in adult *Agabus disintegratus* (Crotch) (Coleoptera: Dytiscidae) in dried ponds in California. Annals of the Entomological Society of America 80:267–271.

Garcia, R., K. S. Hagen, and W. G. Voigt. 1990. Life history, termination of summer diapause, and other seasonal adaptations of *Agabus disintegratus* (Crotch) (Coleoptera: Dytiscidae) in the Central Valley of California. Quaestiones Entomologicae 26:139–149.

Gilmer, D. S., M. R. Miller, R. D. Bauer, and J. R. LeDonne. 1982. California's Central Valley wintering waterfowl: Concerns and challenges. Transactions of the North American Wildlife and Natural Resources Conference 47:441–452.

Gleason, R. A., S. W. Cordes, N. H. Euliss, Jr., and W. E. Newton. Aquatic invertebrates associated with hydrophytes in seasonally impounded wetlands in California. In review.

Heitmeyer, M. E., D. P. Connelly, and R. L. Pederson. 1989. The Central, Imperial, and Coachella Valleys of California. Pages 475–505 *in* L. M. Smith, R. L. Pederson, and R. M. Kaminski (eds.), Habitat Management for Migrating and Wintering Waterfowl in North America. Texas Tech University Press, Lubbock, TX.

Helmers, D. L. 1992. Shorebird management manual. Western Hemisphere Shorebird Reserve Network, Manomet, MA.

Hicks, J. M. 1992. Aquatic invertebrate community structure near waterfowl carcasses in a Sacramento Valley wetland. Master's Thesis, Humboldt State University, Arcata, CA.

Hicks, J. M., N. H. Euliss, Jr., and S. W. Harris. 1997. Aquatic invertebrate ecology during a simulated botulism epizootic in a Sacramento Valley wetland. Wetlands 17:157–162.

Hobmaier, M. 1932. Conditions and control of botulism (duck disease) in waterfowl. California Fish and Game 18:5–21.

Hunter, B. F. 1970. Ecology of waterfowl botulism toxin production. Transactions of the North American Wildlife Conference 35:1–9.

Jensen, W. I., and J. P. Allen. 1960. A possible relationship between aquatic invertebrates and avian botulism. Transactions of the North American Wildlife Conference 25:171–180.

Kadlec, J. A., and L. M. Smith. 1992. Habitat management for breeding areas. Pages 590–610 in B. D. J. Batt, A. D. Afton, M. G. Anderson, C. D. Ankney, D. H. Johnson, J. A. Kadlec, and G. L. Krapu (eds.), Ecology and Management of Breeding Waterfowl. University of Minnesota Press, Minneapolis, MN.

Kahrl, W. L. (ed.). 1979. The California Water Atlas. California Governer's Office of Planning and Research, Sacramento, CA.

Kaminski, R. M., and H. H. Prince. 1981. Dabbling duck and aquatic macroinvertebrate responses to manipulated wetland habitat. Journal of Wildlife Management 45:1–15.

Kantrud, H. A. 1986. Effects of Vegetation Manipulation on Breeding Waterfowl in Prairie Wetlands—a Literature Review. Technical Report No. 3, U.S. Fish and Wildlife Service, Washington, DC.

King, J. L., M. A. Simovich, and R. C. Brusca. 1996. Species richness, endemism and ecology of crustacean assemblages in northern California vernal pools. Hydrobiologia 328:85–116.

Krapu, G. L. 1981. The role of nutrient reserves in mallard reproduction. Auk 98:29–38.

Mackay, J. 1993. The influence of biotic and abiotic factors on microhabitat selection by aquatic invertebrates in managed wetlands. Master's Thesis, Humboldt State University, Arcata, CA.

Mackay, J., N. H. Euliss, Jr., and S. W. Harris. Abiotic influences on aquatic invertebrates in California wetlands. In review.

Mall, R. E. 1969. Soil-Water-Salt Relationships of Waterfowl Food Plants in the Suisun Marsh of California. Wildlife Bulletin No. 1, California Department of Fish and Game, Sacramento, CA.

McLandress, M. R., G. S. Yarris, A. E. H. Perkins, D. P. Connelly, and D. G. Raveling. 1996. Nesting biology of mallards in California. Journal of Wildlife Management 60:94–107.

Miller, M. R. 1987. Fall and winter foods of northern pintails in the Sacramento Valley, California. Journal of Wildlife Management 51:405–414.

Mount, J. F. 1995. California Rivers and Streams. University of California Press, Berkeley, CA.

Murkin, H. R., J. M. Kaminski, and R. D. Titman. 1982. Responses by dabbling ducks and aquatic invertebrates to an experimentally manipulated cattail marsh. Canadian Journal of Zoology 60:2324–2332.

Neill, C., and J. C. Cornwell. 1992. Stable carbon, nitrogen, and sulfur isotopes in a prairie marsh food web. Wetlands 12:217–224.

Ohlendorf, H. M., D. J. Hoffman, M. K. Saiki, and T. W. Aldrich. 1986a. Embryonic mortality and abnormalities of aquatic birds: Apparent impacts by selenium from irrigation drainwater. Science of the Total Environment 52:49–63.

Ohlendorf, H. M., R. L. Hothem, T. W. Aldrich, and A. J. Krynitsky. 1987. Selenium contamination of the grasslands, a major California waterfowl area. Science of the Total Environment 66:169–183.

Ohlendorf, H. M., R. L. Hothem, C. M. Bunck, T. W. Aldrich, and J. F. Moore. 1986b. Relationships between selenium concentrations and avian reproduction. Transactions of the North American Wildlife and Natural Resources Conference 51:330–342.

Orr, B. K., and V. H. Resh. 1989. Experimental test of the influence of aquatic macrophyte cover on the survival of *Anopheles* larvae. Journal of the American Mosquito Control Association 5:579–585.

———. 1992. Influence of *Myriophyllum aquaticum* cover on *Anopheles* mosquito abundance, oviposition, and larval microhabitat. Oecologia 90:474–482.

Reeves, W. C. (ed.). 1990. Epidemiology and Control of Mosquito-Borne Arboviruses in California, 1943–1987. California Mosquito and Vector Control Association, Sacramento, CA.

Reeves, W. C., and W. M. Hammon. 1962. Epidemiology of the Arthropod-Borne Viral Encephalitidies in Kern County, California, 1943–1952. University of California, Berkeley, Publications in Public Health 4:1–257.

Reisen, W. K., R. P. Meyer, J. Shields, and C. Arbolante. 1989. Population Ecology of Preimaginal *Culex tarsalis* (Diptera: Culicidae) in Kern County, California. Journal of Medical Entomology 26:10–22.

Rocke, T. E., N. H. Euliss, Jr., and M. D. Samuel. 1999. Environmental characteristics associated with avian botulism epizootics in California wetlands. (Wildlife Society Bulletin). 63:in press.

Rollins, G. L. 1981. A Guide to Waterfowl Habitat Management in Suisun Marsh. California Department of Fish and Game Publication, Sacramento, CA.

Rosen, M. N. 1971. Botulism. Pages 100–117 *in* J. W. Davis, R. C. Anderson, L. Karstad, and D. O. Trainer (eds.), Infectious and Parasitic Diseases of Wild Birds. Iowa State University Press, Ames, IA.

Rosenberg, D. M., and V. H. Resh (eds.). 1993. Freshwater Biomonitoring and Benthic Macroinvertebrates. Chapman & Hall, New York.

Sandler, R. J., T. E. Rocke, M. D. Samuel, and T. M. Yuill. 1993. Seasonal prevalence of *Clostridium botulinum* Type C in sediments of a northern California wetland. Journal of Wildlife Diseases 29:533–539.

Schlossberg, E. B., and V. H. Resh. 1997. Mosquito control and waterfowl habitat enhancement by vegetation manipulation and water management: A two year study. Proceedings of the California Mosquito and Vector Control Association. 65:11–15.

Sefchick, J. A. 1992. Composition and stability of aquatic invertebrate populations at the Sacramento National Wildlife Refuge, California. Master's Thesis, Humboldt State University, Arcata, CA.

Severson, D. J. 1987. Macroinvertebrate populations in seasonally flooded marshes in the northern San Joaquin Valley of California. Master's Thesis, Humboldt State University, Arcata, CA.

Sugden, L. G. 1974. Feeding Ecology of Pintail, Gadwall, American Widgeon and Lesser Scaup Ducklings. Canadian Wildlife Service Report Series No. 24.

Tietze, N. S., and M. S. Mulla. 1991. Biological control of *Culex* mosquitoes (Diptera: Culicidae) by the tadpole shrimp, *Triops longicaudatus* (Notostraca: Triopsidae). Journal of Medical Entomology 28:24–31.

Vannote, R. L., G. W. Minshall, K. W. Cummins, J. R. Sedell, and C. E. Cushing. 1980. The river continuum concept. Canadian Journal of Fisheries and Aquatic Science 37:130–137.

Walton, W. E., N. S. Tietze, and M. S. Mulla. 1990. Ecology of *Culex tarsalis* (Diptera: Culicidae): Factors influencing larval abundance in mesocosms in Southern California. Journal of Medical Entomology 27:57–67.

Walton, W. E., N. S. Tietze, and M. S. Mulla. 1991. Consequences of tadpole shrimp predation on mayflies in some Californian ponds. Freshwater Biology 25:143–154.

Weller, M. W., and C. E. Spatcher. 1965. Role of habitat in the distribution and abundance of marsh birds. Special Report No. 43, Iowa Agricultural and Home Economics Experimental Station, Iowa State University, Ames, IA.

Wetmore, A. 1925. Food of American Phalaropes, Avocets, and Stilts. Department Bulletin No. 1359, United States Department of Agriculture, Washington, DC.

Wiggins, G. B., R. J. Mackay, and I. M. Smith. 1980. Evolutionary and ecological strategies of animals in annual temporary pools. Archiv für Hydrobiologie Supplement 58:97–206.

Williams, D. D. 1996. Environmental constraints in temporary fresh waters and their consequences for the insect fauna. Journal of the North American Benthological Society 15:634–650.

Wobeser, G. A. 1981. Diseases of Wild Waterfowl. Plenum Press, New York.

35 Agricultural Wetland Management for Conservation Goals

Invertebrates in California Ricelands

RACHEL EMERSON O'MALLEY

---

*T**his chapter reviews some of the effects of agricultural conversion on wetland invertebrates, the ways that invertebrates are commonly used to assess conservation goals, and the link between these two perspectives. It then uses the example of the invertebrate fauna of a large-scale agricultural wetland, the rice ecosystem of northern California, to illustrate several options for enhancing the scope of conservation assessment for invertebrates in agricultural wetlands.*

CONSERVATION OF WETLAND COMMUNITIES

In the past few decades wetlands have achieved dubious distinction as primary targets of many conservation efforts, primarily due to their high productivity, disproportionate share of threatened and endangered species, and rapid rate of destruction. Wetland invertebrates, like invertebrates in general, have only recently been perceived as important targets for conservation, but they are increasingly known to be important indicators of conservation values such as biodiversity and productivity (Kremen et al. 1993, Kellert 1993, Clark and Samways 1996, Samways et al. 1996). Furthermore, some wetland invertebrates are recognized as threatened species in their own right (Pedersen and Holmen 1994, Palmer and Palmer 1995, Williams et al. 1995, Polhemus 1996,

Invertebrates in Freshwater Wetlands of North America: Ecology and Management, Edited by Darold P. Batzer, Russell B. Rader, and Scott A. Wissinger
ISBN 0-471-29258-3   © 1999 John Wiley & Sons, Inc.

Mierzwa and Smyth 1996). Agricultural development can degrade wetland ecosystems. Therefore, the development of effective conservation strategies for wetland invertebrates—and wetlands in general—may hinge upon the modification of agriculture to accommodate biodiversity protection.

In this chapter I first review some of the effects agricultural conversion has had on wetlands generally and the ways that invertebrates are commonly used to assess conservation goals. I then link the two, reviewing how conservation and agricultural programs have interacted both in theory and in practice as well as strategies to improve agricultural wetlands for conservation of invertebrate communities. Finally, I present my own data assessing arthropod communities in a large-scale wetland conservation project in the rice fields of California's Sacramento Valley.

AGRICULTURE AND WETLANDS

Both globally and historically, the greatest loss of wetlands has been to agricultural uses. In Asia and Europe most historical wetlands were put into crop production centuries or millennia ago. In other parts of the world, including the United States, this trend has continued throughout the current century. From the mid-1950s through the mid-1970s, agricultural conversion accounted for 87 percent of wetlands lost in the continental United States (Tiner 1984, Frayer et al. 1989); from the mid-1970s to the mid-1980s, 54 percent of wetland conversion was due to agriculture (Dahl and Johnson 1991). Only recently has urban development outpaced agricultural uses as the prime cause of wetland loss. Indeed, perhaps owing to changes in agricultural legislation designed to reduce wetland loss, from 1982 to 1987 37 percent of converted wetlands were put into agricultural use, while 48 percent of wetland loss was due to urban development (Brady and Flather 1994). Nowhere has loss of natural wetlands to agriculture been more acute than in California. Originally supporting over a million hectares of marshes and other wetlands, the state has lost 95 percent of this habitat over the past 200 years, with the vast majority being converted to agriculture (Frayer et al. 1989). However, as is the case in other areas, much of this land is still maintained as modified wetlands for agricultural production (Fig. 35.1). However, the ultimate questions for this and other agricultural wetlands are, how useful is this modified habitat for wetland conservation, and how should we monitor for and improve its conservation value, especially for invertebrate communities?

CONSERVATION GOALS FOR INVERTEBRATES

Invertebrates are considered targets of conservation efforts in two ways. First, because they are locally adapted or dependent on a limited habitat, growing numbers of rare species of invertebrates have been identified as conservation

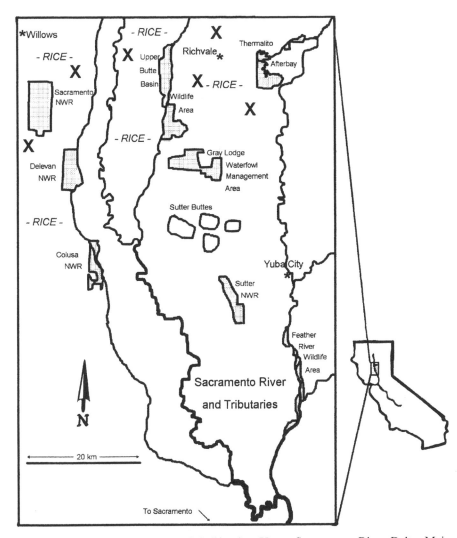

Fig. 35.1. Rice-growing region of California—Upper Sacramento River Delta. Main waterways and wildlife habitat are indicated. NWR = National Wildlife Refuge. Study sites are marked with (X). Essentially all the matrix within which the indicated NWRs are embedded was once natural wetland and is currently mostly rice production.

targets in and of themselves (Wilcove et al. 1993, Deyrup 1994). Although the majority of these are charismatic lepidopterans (Britten et al. 1994, Launer and Murphy 1994, Gaskin 1995, Aubert et al. 1996, Herms et al. 1996, Wahlberg 1997, Varshney 1997, among many others), a variety of threatened odonates, ephemeropterans, carabids, arachnids, shrimp, crayfish, and snails have been identified in wetlands (France and Collins 1993, McKillop et al. 1993,

Voigt 1994, Light et al. 1995, Wassermann and Schmidt-Kloiber 1996, Dillon and Ahlstedt 1997). In recent years these relatively popular invertebrate orders have even been joined on the endangered roles by such prosaic creatures as dipterans (Palmer and Palmer 1995, Kingsley 1996, Kato and Miura 1996, Malmqvist 1996).

Invertebrates also play a second set of roles in the conservation literature. As easily assayed organisms, invertebrates are often used as important indicators of ecosystem and community processes. These processes fall into four rough categories:

1. *Biodiversity.* Invertebrates are among the most speciose groups of organisms on the planet, and they are well represented across many trophic levels. Thus, impacts on the biodiversity of invertebrates in an ecosystem of either in terms functional groups or species richness, are often considered indicative of likely effects on other types of organisms (Oliver and Beattie 1993, 1996a, b, Burkett and Thompson 1994, Rykken et al. 1997). Because of their range of dispersal habits and wide variety of life histories, invertebrates are equally well suited for measurement of alpha (within site and within habitat), beta (within site but between habitat), and gamma (between site but within habitat) diversity (Myers 1996, Oliver and Beattie 1996b, Niemela 1997).

2. *Abundance and biomass.* Particularly in wetland systems, the abundances of invertebrates are perhaps most often treated in sum as important proteinaceous food sources supporting community foodwebs, and thus as important indicators of ecosystem productivity (Niemela et al. 1993, Colwell and Dodd 1995, Persson 1996, Chavez-Ramirez et al. 1996, France 1997). After bacteria, invertebrates often represent the largest biomass in wetland ecosystems; disruption of invertebrate productivity thus has the potential to undermine consumers much further up the trophic web. As a result, research interested in vertebrate conservation often uses invertebrate abundance and biomass to gauge habitat quality (Batzer and Resh 1992a,b, McCracken and Foster 1993a, Batzer et al. 1993, 1997).

3. *Relationships among species.* Only in the last decade has the importance of relationships among species, particularly involving invertebrates, been recognized as an important conservation value. Understanding and conserving the mutualistic relationships between arthropods and plants they pollinate is now widely acknowledged to be at least as important as the maintenance of rare populations of either mutualist in isolation (Suzan et al. 1994, Sipes and Tepedino 1995, Buchmann 1996). Advances in foodweb techniques have also led to greater recognition of the conservation value of trophic interactions among organisms (Cohen et al. 1990, Batzer and Resh 1991, Wootton et al. 1996a, b, Power 1996, Power et al. 1996a, b).

4. *Temporal and spatial dynamics.* Finally, measuring the changes through time of biodiversity, invertebrate abundance, and the relationships among invertebrates is rarely done, but it is certainly critical to understanding outcomes of and mechanisms behind conservation efforts (Resh and Rosenberg 1989). For example, it is now widely acknowledged that foodwebs are not static through time and that at any moment only a small subset of a larger web of organisms will be contributing to the dynamics of a community (Cohen et al. 1993, Power 1996). Depending on the circumstances, either greater or lesser variability of the foodweb may in itself be considered desirable from a conservation standpoint. Similarly, the spatial scale of available habitat, or the size of the unit of analysis, has been shown to influence assessments of value of even high-quality habitat for target organisms (Mladenoff and Stearns 1993, Donovan et al. 1995). Increasingly, studies of invertebrates are taking this kind of variability into account with observations taken at several spatial scales (de Szalay et al. 1996).

Lamentably few studies actually assay invertebrate communities for all of these functional and indicator roles. A survey of studies that measure different combinations of them, however, strengthens the objective basis for conservation assessments, in particular for species-rich communities such as wetlands, including wetlands converted to agricultural use.

CONSERVATION AND WETLAND AGRICULTURE

Many ecologists have called for conservation biology and agricultural management to be linked (Orr 1991, Gilpin et al. 1992, Gall and Orians 1992). However, a debate rages in the conservation literature about the intrinsic compatibility—or incompatibility—of agriculture and conservation goals (Curtin 1993, Knight et al. 1994, Wuerthner 1994, Siegel 1996). Significantly, many of the arguments made in this debate rest more upon general perceptions of agricultural effects and ecological processes than upon clear data. Where data do exist on the compatibility of agricultural land uses with the advancements of conservation goals, the conclusions range from highly optimistic (Salafsky 1993, Safina 1993, Thiollay 1995, Ndubisi et al. 1995, Bignal and McCracken 1996, Perfecto et al. 1996, de Jong 1997) to catastrophic (Blankespoor 1991, Canaday 1996, Herkert 1997), depending on the context of the study and the conservation goals pursued. Particularly for wetland systems, it is important to distinguish the type of agricultural use being considered. Without doubt, many agricultural production methods in and around natural wetlands decimate associated invertebrate populations (Schuler 1987). This outcome is due in part to the direct effects of insecticides on target and nontarget invertebrates (Robinson 1993, Solbert and Higgins 1993, McCracken and Foster 1993b, Frampton and Cilgi 1994, Klenner 1994) and in part to indirect effects of habitat loss (including drainage), host plant destruction, and desiccation (Tamisier and Grillas 1994). Yet, where intact wetland preservation is not economically or politically achievable, some kinds of agricultural uses may prove preferable to other kinds of development. Careful agricultural uses can support or augment invertebrate biodiversity and abundance, even for populations of certain scarce invertebrate species (Samways et al. 1996). Agricultural modifications comprise a broad spectrum of land uses, ranging from chemical-intensive upland crops requiring drained and filled wetlands, to inundated rice production in partially modified wetland ecosystems, to harvest of naturally occurring food or peat crops from minimally disturbed wetlands. Therefore, the choice of agricultural crop, as well as management decisions for that crop, can be expected to significantly affect the impacts that cultivation will have on the natural wetland invertebrate community. Thus, whether agriculture writ large has positive or negative effects on conservation goals is not the issue. Rather, the appropriate questions are how to enhance specific agricultural practices for specific conservation purposes and whether such changes can substantially improve conservation value while still meeting agricultural needs. The first major distinction in the extent to which different agricultural uses will affect wetland invertebrates is between draining wetlands for upland agriculture and modifying wetlands to greater or lesser extents for inundated agricultural activities.

Upland Crops in Wetlands

For many wetlands agricultural conversion has meant that habitats are tiled or ditched, fields drained, and the "reclaimed" lands used for the production of upland crops. In these cases, for wetland-associated invertebrates agricultural production usually represents near complete destruction of most communities and populations (Zaidelman 1996). Less catastrophically, upland crops are sometimes produced during the dry season or at the margins of otherwise intact wetlands. Wetland invertebrate communities may be significantly degraded by such proximity to upland crops if they receive destructive agricultural runoff (Schuler 1987). But in some cases peripheral cropping is believed to have relatively few or even positive effects on wildlife, including invertebrates (Samways 1989, Granval et al. 1993, Rief 1996). Because of the alternately catastrophic (in the case of conversion) or highly unpredictable (in the case of marginal production) effects of using wetlands to produce upland crops, I will focus more attention on agricultural uses that offer concrete promise of fruitful collaboration with conservation—namely, wetland crops.

Wetland Crops

Wetland agriculture can logically be divided into two categories: relatively modest activities associated with harvest of natural species, and intensive wetland crop management, best exemplified by short-statured rice production.

Harvest of Natural Species

Even seemingly moderate agricultural activity (harvesting naturally occurring crops such as crayfish, peat moss, *Spartina* grasses, and *Zizania* wild rice from lightly managed wetlands) can alter invertebrate community composition substantially. Crayfish cultivation and harvest is quite common through much of the world (Holdich 1993). Studies of ecological impacts of crayfish cultivation in areas where they are native are generally limited to direct effects on the harvested crayfish populations themselves. Evidence for multispecies effects of crayfish cultivation comes largely from studies of the impacts on endemic fauna of cultivation of exotic crayfish species. In at least one case this common practice has been shown to threaten endemic crayfish species to the point that their populations may not be able to recover (Skurdal 1995). Effects of crayfish introduction and cultivation on other wetland invertebrates, if any exist, are not well documented.

For some invertebrate species agricultural expansion of host or habitat-forming plants may be beneficial. For example, the creation of peat pools in Denmark at the turn of the century probably promoted the expansion of certain rare dragonfly species (Pedersen and Holmen 1994), and in Sweden growing peat moss has been shown to increase acid-tolerant invertebrates in acidified lakes (Henrikson 1993). However, harvesting and agricultural use of such peat bogs has been demonstrated to threaten wildlife (see Giberson and Hardwick, this volume), despite some evidence that the impact of agricultural peat use on wildlife is minimal (Smith et al. 1995, Rubec 1995).

Harvest of other natural species, such as *Spartina* for animal bedding in New England and *Zizania* wild rice in Minnesota, has undoubtedly contributed to the maintenance of the associated wetland ecosystems in the landscape. However, remarkably little is known about the direct effects of management and harvest of these crops on associated wetland invertebrates.

Wetland Rice Production. Cultivated ricelands represent 15 percent of the total wetland area around the globe, as well as 10 percent of arable land; rice is a staple grain for 40 percent of the world's people (Hook 1993). In recent years rice has overtaken corn as the primary staple for Latin America. Because of its importance to humans, the invertebrate foodweb for rice may be better documented than that of any other ecosystem. The International Rice Research Institute database of arthropods in rice has been used by Schoenly et al. (1996) to carefully examine constancy of wetland invertebrate food webs in the Philippines. Settle et al. (1996) also use rice invertebrate foodweb analysis to assess impacts of insecticide use on wetland invertebrates in Indonesia. Its global distribution makes rice agriculture ideal for comparative studies of the effects of agricultural practices on wetland invertebrates.

The crop, really a complex of hundreds of cultivated varieties of several species of *Oryza* grasses, experiences a wide range of management regimes. These vary from dryfarmed to deep-water, from traditional low-input to chemical-intensive, and from short (four-month) to long (6-month) growing seasons. Traditional low-intensity methods of cropping *Oryza* rice species in locations where they are native—or even where they are introduced but mimic ecosystem functions of native species—in many cases may incidentally preserve wetland habitats for sensitive invertebrate use. However, intensive rice cultivation practices are far more common in North America, Europe, and much of Asia. Like low-input cultivation, intensive rice cultivation impedes wetland filling and drainage, thus providing some benefit to invertebrates and other wildlife (Hanson and Brode 1980, Strong et al. 1990, Loughman and Batzer 1992, Andrews and Rose 1992, Simpson et al. 1994, Mesleard 1994, Fasola and Ruiz 1996). However, this benefit may be offset by high levels of pesticide use and dramatic habitat alteration common to conventional intensive management practices (Fasola and Alieri 1992, Mathias and Moyle 1992, Robinson 1993, Tamisier and Grillas 1994).

Common Strategies for Improving Agricultural Lands for Invertebrates

Given the wide range of wetland cropping methods and the preponderance of agricultural wetlands in the landscape, the implementation and testing of agricultural practices to make cropped wetlands more hospitable for wetland invertebrates should be a central goal of conservation. Most often, however, the effects of different kinds of agricultural management on wetland invertebrates are assumed but not documented. While many different agricultural techniques have been suggested and put into practice for conservation goals, few data exist to document their efficacy. Broadly speaking, studies of the effects of particular agricultural management strategies on invertebrate conservation fall into two overlapping categories: direct insecticide reduction and invertebrate habitat management enhancement. The latter includes a wide variety of practices, including plant species or architectural diversification, water, nutrient, or climate management, and off-season land husbandry strategies.

Direct Effects: Insecticide Reduction. As urban areas encroach upon farmland and the human and wildlife costs of insecticides are exposed, many farmers are striving to reduce insecticide use in their fields. Abundant evidence exists demonstrating direct effects of insecticide applications on non-target invertebrates in wetland farming systems (Grigarick et al. 1990, Roger et al. 1994). In addition, recent innovative foodweb studies have documented nonintuitive changes in temporal dynamics as well as in species composition of rice invertebrate foodwebs as a direct result of eliminating pesticides (Cohen et al. 1994, Schoenly 1996, Settle et al. 1996). These studies reveal that only through considering changes in the invertebrate community through time, in particular the build-up or dampening of insecticide effects as a result of interspecific interactions, is it possible to detect patterns of wetland invertebrate response to insecticide reduction.

Indirect Effects: Habitat Management for Invertebrates. Several agricultural practices are commonly advocated to improve habitat for invertebrates in farm fields. Historically, these methods have largely been promoted to increase natural biological control by drawing in or conserving natural enemies of arthropod pest species. More recently, many have also been advocated as wildlife conservation measures to specifically provide habitat for invertebrates and other wildlife. Many of these habitat manipulations fall under the broad category of diversification of the agricultural landscape.

Diversification usually takes the form of either taxonomic diversification of crops and border flora or architectural diversification through multiple canopy development or habitat mosaics. Study of these practices has generated a rich literature demonstrating the benefits of biodiversity in agricultural settings, both for natural arthropod pest control (Root 1973, Risch et al. 1983, Andow and Risch 1985, Kemp and Barrett 1989, Russell 1989, Letourneau 1990a, b, Andow 1991) and as habitat for invertebrate and vertebrate wildlife (Blooij 1994, Altieri 1994, Marino and Landis 1996, Marc and Canard 1997). This topic has generated two special issues of the journal *Agriculture, Ecosystems and the Environment* within the past five years (1992, 1997), an indication of the widespread interest in diversification of agricultural ecosystems. In sum, these studies document how agricultural diversification affects invertebrate behavior, populations, and communities through time. Many of these studies have been derived from interest in agricultural management styles practiced by traditional small farmers, which tend to incorporate both taxonomic and architectural diversity in the cropping landscape (Altieri et al. 1983, Letourneau 1986, 1987, Perfecto et al. 1996, Salafsky 1993, Thiollay 1995). Others result from the long-term interest in European literature in the effects on fauna of hedgerows, conservation headlands, strip cropping, and other practices that specifically increase variability of the agricultural landscape (Duelli et al. 1990, Hassall et al. 1992, Lys 1994, Saville et al. 1997, Charrier et al. 1997, Dover 1997).

In wetland agriculture diversification for the benefit of arthropod fauna has primarily consisted of creation of wetland reserves or set-asides mixed with rice fields. While the effects of such practices on wetland invertebrates are not well documented, they are assumed to be comparably beneficial (Payne and Wentz 1992). At a smaller scale, the genetic diversity of traditional wetland rice systems has been shown to benefit arthropod pest control (Andow and Hidaka 1989), as has spatial variability of traditional rice planting methods (Ishii-Eiteman and Power 1997). In this country recent reduction of topographic diversity of most rice fields by laser-assisted leveling technologies has been implicated in the loss of habitat diversity needed to support invertebrates and other wildlife (although little documentation of this effect is available).

Other habitat manipulations to influence invertebrates generally consist of management of abiotic factors such as light, nutrients and water level, and off-season management practices. In at least one case the addition of nitrogen to rice fields has been shown to enhance invertebrate populations (Simpson et al. 1994, but see Letourneau and Fox 1988, Fox et al. 1990), but systematic nutrient management for conservation values is rare. For wetland agriculture and invertebrate conservation, water is likely the key abiotic factor that influences population and community dynamics (Hesler and Grigarick 1992). Agricultural wetlands differ from natural wetlands both in the quantities of water available to invertebrates and in timing of flow. For example, in Mediterranean climates such as California the hydrologic cycle of agricultural wetlands is exactly the reverse of natural wetland cycles (Table 35.1). However, because of its phenological versatility, widespread cultivation, and unique adaptation for wetland production, rice is ideally suited for water manipulations designed to enhance the value of wetland agriculture for conservation of wetland invertebrates.

Combined Effects: Organic Management. One of the most common strategies farmers adopt to conserve wildlife in their farmland is to follow a set of guidelines, including elimination of synthetic chemicals, that allows them to market their product as "organic" and obtain a market premium. This management regime confers direct benefits on the invertebrate community through insecticide reduction or elimination, and it provokes indirect invertebrate habitat alterations due to herbicide elimination and changes in nutrient and organic matter availability. Because of their widespread use, the effects of organic production methods on invertebrate communities are better documented than are other conservation management strategies (Moreby et al. 1994, Pearsall and Walde 1995, Drinkwater et al. 1995, Phelan et al. 1995, Letourneau et al. 1996). In wetland agricultural systems, however, only one study has contrasted invertebrates in organic and in conventionally managed farms. Hesler et al. (1993) detected an increase in abundance of three predators, *Belostoma flumineum, Notonecta* spp., and *Thermonectus basillaris,* in organically managed fields, but no differences in biodiversity or overall abundance of arthropods, despite the absence of synthetic insecticides. However, the data used in that study were collected only once during the season, providing a snapshot, but no indication of differences in temporal dynamics or in interspecific relationships between the two management regimes.

TABLE 35.1. Hydrologic Cycle of Natural versus Agricultural Wetlands in Mediterranean Climates, as Exemplified by the Pattern in the Sacramento Valley of California

Habitat	Winter (~ September–April)	Summer (~ May–August)
Native tule marshes	Floody and marshy	Marshy to dry
California rice fields	Well drained between rains	Flooded and marshy

CASE STUDY: MANAGING CALIFORNIA RICELANDS FOR INVERTEBRATE CONSERVATION

Over the past four years, I have studied the effects of two habitat manipulations of rice in the Sacramento Valley, California, on wetland arthropod communities. Rice fields currently predominate in rural areas surrounding Sacramento (see Fig. 35.1), and to some extent they have staved off urbanization, providing much of the remaining habitat for declining and threatened wetland species in the region. Ricelands currently host 21 sensitive and endangered vertebrate species, including the giant garter snake, *Thamnophis gigas,* and 20 bird species (WESCO 1991), although their habitat value for these organisms is subject to debate. The invertebrate populations of these fields are important in the diets of many of these endangered vertebrates, as indicators of the overall health of the biotic community, and as potential arthropod pests and natural enemies in the agricultural system. Invertebrate diversity is also high, with over 500 morphospecies of arthropods associated with rice fields (personal observation).

As a result of concern over the loss of wetland habitat for migratory waterfowl, as well as human health considerations, rice farmers in the Sacramento area have implemented two major conservation management strategies on a wide spatial scale. In 1988 the Central Valley Habitat Joint Venture (CVHJV), a coalition of conservation groups such as Ducks Unlimited, the Nature Conservancy, and the California Waterfowl Association, joined with the Rice Industry Association, the U.S. Fish and Wildlife Service, and the University of California to advocate experimental winter flooding of rice fields (Paullin 1992). Winter flooding of rice fields was anticipated to enhance the agricultural environment by improving winter habitat for migratory waterfowl while reducing the need for outside inputs and burning residual straw in the rice ecosystem (CRHP 1992). Similarly, organic rice farms are increasingly common in the Sacramento Valley; they now encompass more than 1000 ha of land.

Although these innovations demonstrate that there is great interest in increasing the value of rice lands for conservation purposes, few systematic studies have been conducted that address either the conservation or the agronomic implications of such efforts. Over the course of three rice growing seasons (June–September 1993–1995) I repeatedly sampled the plant and arthropod communities in 7 to 15 separate rice fields, each of 60 to 100 ha, in the upper Sacramento Valley of California (Fig. 35.1). Arthropod sampling included pitfall traps on the levees and sticky traps both floating within fields and on levees. Each year I included both organic and conventionally managed sites that had either been flooded during the preceding winter or were unflooded controls.

Invertebrate Foodweb Analysis

As emphasized in the review above, most efforts to assess wetland invertebrate communities have used a limited subset of analytic techniques. Such simplistic measures and analyses do not provide the complex and mechanistic information needed to evaluate conservation efforts effectively. To more fully examine the impacts of habitat manipulations on wetland invertebrate communities, I have used a variety of techniques to look for both static and dynamic changes in community function (see O'Malley 1997, 1999a, b). To illustrate the differing value of these methods, here I describe three efforts I have used to explore the effects of habitat manipulations on invertebrate communities. The first and most commonly used method is analysis of variance (ANOVA) to detect differences in arthropod diversity in treatment versus control fields (see O'Malley 1997, 1999a). The second is more exploratory: I show how structural equation models (SEM) can reflect the most likely hypotheses for how arthropod functional groups are interacting and how habitat manipulations influence each group and the interactions among functional groups (see O'Malley 1997, 1999b). Finally, I illustrate how these apparently static treatment effects can vary over time, as reflected by repeated-measures polynomial tests and contrasting of foodweb models at different points in the season. I finally summarize the message these diverse results hold for managing agricultural land for wildlife conservation.

Direct Responses to Habitat Manipulations. To assess the direct effects of habitat manipulations on wetland invertebrates, researchers commonly use standard ANOVA. This method revealed a number of effects of conservation management methods in my study system (reported elsewhere, O'Malley 1997, 1999a), but here I describe only the most frequent measure of conservation effects on the wetland arthropod community, species diversity. To broadly assess changes in this parameter, I used ANOVA techniques to contrast arthropod species richness through the season in winter-flooded and organic rice fields with that in conventionally managed controls. Using either the number of morphospecies or broader functional and taxonomic categories in each sample as the dependent variable, I performed repeated measures MANOVAs over four sampling dates, with year, treatments (organic/conventional and winter-flooded/control), sample location (interior/levee), and interactions among these parameters as explanatory variables in the statistical model. This strategy is useful to detect overall differences in the level of biodiversity in the field over the series of dates for each treatment, as well as to assess interactions between the treatments. I found that both winter flooding and organic management significantly increased arthropod richness (Fig. 35.2a, b) within rice fields, but that the magnitude of the flooding effect is greater in organic fields than in conventional fields (Fig. 35.2c). While these results are valuable in that they suggest that both habitat manipulations do in fact have positive effects on wetland invertebrate communities, this information served more to intrigue than to explain.

Indirect Interactions Among Invertebrates. Structural equation modeling is an extension of path analysis that is useful in testing suites of linked interactions among different entities (Wright 1921, Browne et al. 1994). Here I

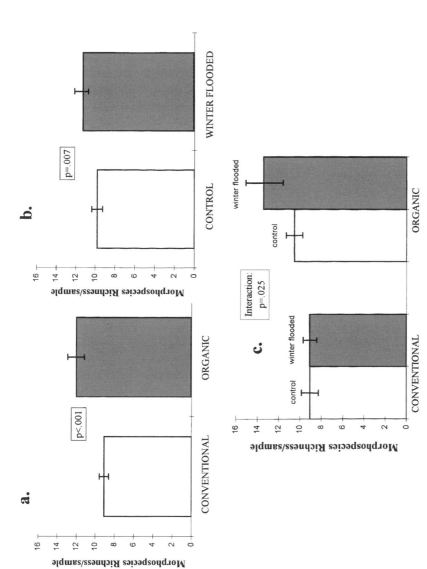

Fig. 35.2. Direct effects of winter flooding, organic management, and the combination of both on wetland arthropod diversity. P-values reflect between-subject results of repeated measures MANOVAs for main treatment effects or interaction between winter flooding and organic management, as indicated. Means and error bars represent standard error around predicted means for each treatment effect and for the interaction between the two habitat manipulations obtained from repeated measures MANOVA.

describe how this method can be used to further elucidate the relationship between the habitat manipulations and arthropod diversity (described fully in O'Malley 1997, 1999b). I first developed a basic hypothetical model to explore how wetland invertebrate trophic interactions were affected by the two manipulations. To do this, I incorporated fundamental natural history information into contrasting hypothetical structural equation models. I then used goodness-of-fit measures—here the estimated cross validation index (ECVI)—to evaluate which of several trophic models better fit my data (Hillborn and Mangel 1997) (Fig. 35.3).

The models I present here aggregate the 558 morphospecies I collected in the fields into six trophic groupings: three sizes of predators and three sizes of prey (see O'Malley 1997 for details). I used SEM to assess the impact of winter flooding and organic management methods, as well as the interaction between the two, on early-season (mid- to late June) counts of each of these trophic groups (Fig. 35.3). I also overlaid this test of habitat effects on two different assumptions about foodweb structure—either that all the predators can eat all the prey, but predators do not eat each other (Fig. 35.3a), or that larger predators can eat smaller predators as well as prey (Fig. 35.3b). Both models indicated that winter flooding had mildly positive effects on populations of the largest predators (mostly carabids, spiders, and predaceous flies), as well as on the smallest prey species (primarily small dipterans and cicadellids). Organic management appeared to greatly enhance these populations as well, and it also directly augmented small predators and medium-sized and large prey. From these models it should be evident that the main advantage of structural equation models is their ability to elucidate indirect, as well as direct, effects. In this case both habitat manipulations show interesting indirect effects on prey and predator groups. The direct effects of these two treatments on small prey ripple up to indirectly increase small predator populations further through the interaction between small prey and small predators. For the foodweb models with intratrophic level predation, the direct and indirect effects of these treatments on small predators further enhance populations of large predators both directly and indirectly (through medium-sized predators). Thus, both winter flooding and organic management appear to contribute directly and indirectly to the enhancement of large predator populations.

Why test two different foodweb structures? Testing multiple models, in combination with conducting experiments, is a way to tease apart the important and unimportant causal links in a community (Wootton 1994). In this case, using ECVI, we see that a model with intratrophic level predation better fits the observed data (i.e., has a lower ECVI) than does one without it. Examining multiple models can also illustrate the robustness of our conclusions. For the models shown here, direct habitat manipulation effects seem essentially unchanged by changes in foodweb structure, increasing confidence in the estimated effects. A caution is in order: SEM methods are not intended to be conclusive; rather, they allow a more exploratory approach to data analysis. Structural equation modeling is better suited to refining of hypotheses

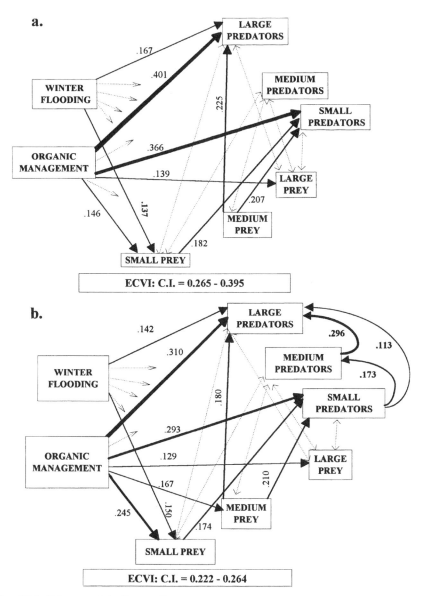

Fig. 35.3. Direct and indirect effects of habitat manipulations on suites of wetland arthropods; using structural equation models to detect the effects of intratrophic-level predation on a wetland invertebrate foodweb. ⟷ indicates that pathway was included in model but path coefficient did not differ significantly from 0. → indicates significant positive interaction between model components; arrow emits from explanatory variable and points toward dependent variable. Greater line width indicates greater magnitude of path coefficient (indicated numerically as well). ECVI = estimated cross validation index. Smaller values reflect better fit of model to data. Error reflects 90 percent confidence intervals around these values.

and teasing out likely causation for effects on complex foodweb interactions than is the ANOVA.

Temporal Dynamics and Invertebrate Conservation. Given the potential for complex interactions among members of large wetland invertebrate communities such as this one, it is conceivable that effects of habitat manipulations will build up or dampen over time. In this case the effects of winter habitat manipulations may be seen in different trajectories of population change over the growing season, rather than in a difference in season-long mean abundances or in population levels on any given date. One simple way to detect changes in the temporal trajectories of wetland invertebrates due to habitat manipulations is to use repeated-measures polynomial contrasts between communities in manipulated fields and those in control fields—results that come from the same MANOVA tests mentioned above. Here I show one example, for the effects of winter-flooding (Fig. 35.4a) and organic management methods (Fig. 35.4b) on the ratio of predators to prey. This measure can be of interest to farmers who want to maintain control of arthropod pests by their natural enemies, or to conservationists attempting to make agricultural fields better mimic predator:prey ratios of natural wetlands. Over the course of the entire rice growing season, there were no significant differences in mean predator:prey ratios between either winter-flooded or organic fields and their conventional management references. However, as the first-order polynomial contrasts show, both treatments strongly influence the trajectories of these measurements through time (Fig. 35.4). While nonflooded and conventional fields experience decreasing wetland invertebrate predator:prey ratios as the season progresses, winter-flooded and organic fields both experience increasing ratios through time.

If it is assumed that patterns and strengths of interaction among species are relatively static, as many foodweb ecologists have done (Cohen et al. 1990), it might be tempting to construct likely foodweb structures (e.g., Fig. 35.3) and then use them to project manipulation-induced changes in population dynamics through time (e.g., Fig. 35.4). However, recently many scientists have acknowledged that foodwebs are themselves dynamic. A final example from my study helps illustrate this observation. If the same structural equation model is used to assess wetland arthropod data sets from the same rice fields in early summer (Figs. 35.4 and 35.5a), midsummer (Fig. 35.5b), and late summer (Fig. 35.5c), the changes in interactions among community members through the growing season can be explored in more detail. The conspicuous observation from this exercise was that habitat manipulations concluded in the winter (in the case of winter flooding) or at the beginning of the summer (organic practices primarily differ from conventional management in field preparation and off-season treatment) continue to influence the wetland arthropod community long after the manipulation is complete. Winter flooding initially has positive effects on large predators and small prey (described above), but these direct effects soon dampen or disappear (Fig. 35.5b),

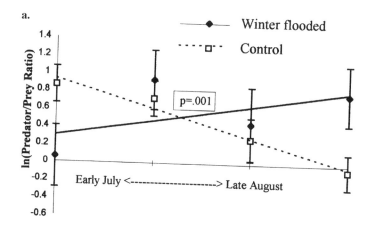

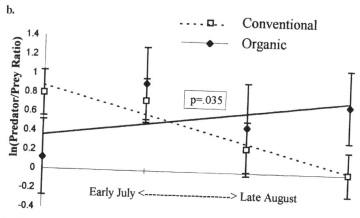

Fig. 35.4. Change through the summer in arthropod predator:prey ratios. P-values represent first-order-polynomial contrasts of predator:prey ratios for main treatment effects from repeated-measures MANOVA. Regression lines are plotted to illustrate significant differences in linear trends through time.

later shifting to large prey (Fig. 35.5c). Organic management has consistently positive direct effects on large and small predators in rice fields, but by mid-season its effect on prey dissipates. By the end of the season, however, direct effects of organic management on medium-sized and large prey cause large predators to escalate further through indirect effects mediated through medium-sized and large predators. Thus, the indirect effects of organic management on large arthropod predators augment its direct influence on their numbers, possibly explaining the change in predator:prey ratio through the season (detected through other methods above) (Fig. 35.4). In sum, these models show that the nature and strength of interactions among members of

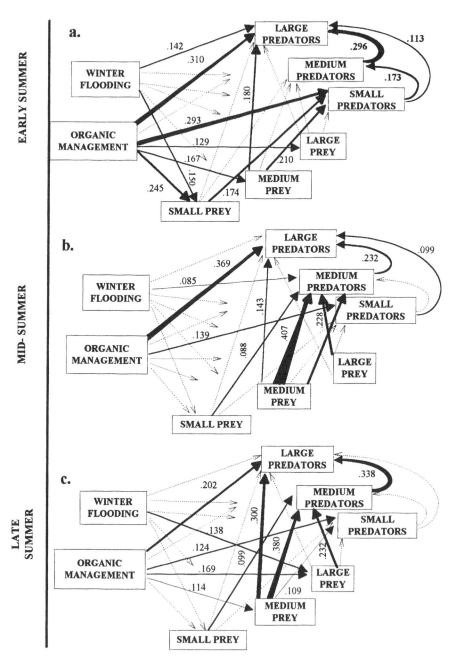

Fig. 35.5. Change in direct effects of winterflooding and organic management and in interactions among wetland arthropods in rice fields through the summer. Notation as in Fig. 35.3.

the wetland arthropod community change throughout the growing season, and in fact interactions between prey and predators appear to strengthen visibly as the season progresses. Thus, it may not be appropriate to project effects of one foodweb structure through time to explain temporal dynamics of wetland invertebrate communities. Furthermore, these analyses show that the principal cause of shifting predator:prey ratios in this species-rich community appears to be altered indirect effects mediated through multiple functional groups.

Messages for Conservation Management of Agricultural Lands. I have presented a small subset of the effects of winter flooding and organic management on wetland invertebrates in rice fields to illustrate a key point: static measures and simple questions may be misleading. Using ANOVA to assess season-long effects of these strategies on wetland arthropod biodiversity gives an optimistic picture: diversity appears to be enhanced by both methods. Other data demonstrate that effects of these treatments on invertebrate productivity may be similarly beneficial (unpublished data). However, these are both admittedly simplistic measures of conservation value. Wildlife advocates are increasingly interested in the effects of habitat management on processes and functions of the community and ecosystem in general. To understand *why* the wetland invertebrate community responds in a particular way to a change in management, it is necessary to look at interactions among species by testing models of community interactions. If this methodology is used, it appears that both winter flooding and organic management have directly positive effects on some groups of predatory and prey arthropods, but that interactions among the arthropods through time may be the driving force behind a predator:prey ratio that increases through time. Understanding this kind of detail about the interactions among community members is especially important in highly productive, species-rich ecosystems such as wetlands, for which the assessment of interrelated effects of treatments on a suite of organisms over time will allow better analysis and prediction of how to improve the management to achieve specific conservation goals. Generation of such predictions is fundamentally assisted by the ability to depict community interactions in a graphically intuitive way, such as with a path diagram, although these depictions must be directly tested and confirmed before their conclusions can be trusted enough to guide large-scale management changes.

CONCLUSIONS

Given the importance of agriculture and the state of wetland conservation globally, it is critical that better links be forged between wetland agriculture and wetland conservation and that better analysis be done to evaluate current wetland management and suggest even better management innovations. As my own work shows, even intensive wetland agriculture can support dynamic, species-rich invertebrate communities. While recent innovations to merge agriculture and conservation are promising, outcomes of such efforts are sometimes counterintuitive and perhaps always complex. Careful analysis of a variety of effects on different spatial and temporal scales is needed to evaluate their potential for success before conservation management measures are implemented on a wide scale. Part of the value of wetland invertebrate communities is their high richness and productivity. But these very features ensure that agricultural wetland invertebrate communities will be dynamic and complicated; understanding their responses to manipulations will require careful probing using multiple techniques. However, and in spite of these complexities, there is every reason to hope for a better merging of agriculture uses and the conservation of wetland communities in the future.

LITERATURE CITED

Altieri, M. A. 1994. The influence of adjacent habitats on insect populations in crop fields. Biodiversity and Pest Management in Agroecosystems. Food Products Press, New York.

Altieri, M. A., D. K. Letourneau, and J. R. Davis. 1983. Developing sustainable agroecosystems. Bioscience 33:45–49.

Andow D. A, and S. Risch. 1985. Predation in diversified agroecosystems: Relations between a coccinelid predator *Coleomagilla maculata* and its food. Journal of Applied Ecology 22:357–372.

Andow, D. A., and K. Hidaka. 1989 Experimental Natural History of Sustainable Agriculture—Syndromes of Production. Agriculture Ecosystems and Environment. 27:447–462.

Andow, D. A. 1991. Vegetational diversity and arthropod population response. Annual Review of Entomology 36:561–586.

Andrews, E., and T. Rose. 1992. An Assessment of the Feasibility of Integrating Wetland Management, Water Storage and Rice Farming in the Sacramento Valley. Report 817, Philip Williams & Associates, San Francisco, CA.

Aubert, J. H. Descimon, and F. Michel. 1996. Population biology and conservation of the Corsica swallowtail butterfly *Papilio hospiton* gene. Biological Conservation 78:247–255.

Batzer, D. P., F. de Szalay, and V. H. Resh. 1997. Opportunistic response of a benthic midge (Diptera: Chironomidae) to management of California seasonal wetlands. Environmental Entomology 26:216–222.

Batzer, D. P., R. M. McGee, V. H. Resh, and R. R. Smith. 1993. Characteristics of invertebrates consumed by mallards and prey response to wetland flooding schedules. Wetlands 13:41–49.

Batzer, D. P., and V. H. Resh. 1991. Trophic interactions among a beetle predator, a chironomid grazer and periphyton in a seasonal wetland. Oikos 60:251–257.

———. 1992a. Wetland management strategies that enhance waterfowl habitats can also control mosquitoes. Journal of the American Mosquito Control Association 8: 117–125.

————. 1992b. Macroinvertebrates of a California seasonal wetland and responses to experimental habitat manipulation. Wetlands 12:1–7.

Bignal, E. M., and D. I. McCracken. 1996. Low-intensity farming systems in the conservation of the countryside. Journal of Applied Ecology 33:413–424.

Blankespoor, G. W. 1991. Slash-and-burn shifting agriculture and bird communities in Liberia, West Africa. Biological Conservation 57:41–71.

Blooij, K. 1994. Diversity patterns in carabid assemblages in relations to crops and farming systems. Pages 425–431 *in* K. Desender et al. (eds.), Carabid Beetles: Ecology and Evolution. Kluwer Academic, Dordrecht, The Netherlands.

Brady, S. J., and C. H. Flather. 1994. Changes in wetlands on nonfederal rural land of the conterminous United States from 1982 to 1987. Environmental Management 18:693–705.

Britten, H. B., P. F. Brussard, and D. D. Murphy. 1994. The pending extinction of the uncompahgre fritillary butterfly. Conservation Biology 8:86–94.

Browne, M. W., G. Mels, and M. Coward. 1994. Path analysis: RAMONA. Pages 163–224 *in* L. Wilkinson and M. Hill (eds.), SYSTAT for DOS: Advanced Applications, Version 6 Edition. SYSTAT, Evanston, IL.

Buchmann, S. L., and G. P. Nabhan. 1996. The Forgotten Pollinators. Island Press/Shearwater Books, Washington, DC.

Burkett, D. W., and B. C. Thompson. 1994. Wildlife association with human-altered water sources in semiarid vegetation communities. Conservation Biology 8:682–690.

California Ricelands Habitat Partnership (C.R.H.P.). 1992. A Cooperative Conservation Effort for Waterfowl, Fisheries and Sustainable Agriculture. Pamphlet.

Canaday, C. 1996. Loss of insectivorous birds along a gradient of human impact in Amazonia. Biological Conservation 77:63–77.

Charrier, S., S. Petit, and F. Burel. 1997. Movements of *Abax parallelepipidus* (Coleoptera, Carabidae) in woody habitats of a hedgerow network landscape: A radiotracing study. Agricultural Ecosystems and Environment 61:133–144.

Chavez-Ramirez, F., H. E. Hunt, R. D. Slack, and T. V. Stehn. 1996. Ecological correlates of whooping crane use of fire-treated upland habitats. Conservation Biology 10:217–223.

Clark, T. E., and M. J. Samways. 1996. Dragonflies (Odonata) as indicators of biotope quality in the Kruger National Park, South Africa. Journal of Applied Ecology 33:1001–1012.

Cohen, J. E., F. Briand, C. M. Newman, and Z. J. Palka. 1990. Community Foodwebs: Data and Theory. Springer-Verlag, New York, Berlin.

Cohen, J. E., R. A. Beaver, S. H. Cousins, D. L. DeAngelis, L. Goldwasser, K. L. Heong, R. D. Holt, A. J. Kohn, J. H. Lawton, N. Martinez, R. O'Malley, L. M. Page, B. C. Patten, S. L. Pimm, G. A. Polis, M. Rejmanek, T. W. Schoener, K. Schoenly, W. G. Sprules, J. M. Teal, R. E. Ulanowicz, P. H. Warren, H. M. Wilbur, and P. Yodzis. 1993. Improving food webs. Ecology 74:252–258.

Cohen, J. E, K. Schoenly, K. L. Heong, H. Justo, G. Arida, A. T. Barrion, and J. A. Litsinger. 1994. A food web approach to evaluating the effect of insecticide spraying on insect pest population dynamics in a Philippine irrigated rice ecosystem. Journal of Applied Ecology 31:747–763.

Colwell, M. A., and S. L. Dodd. 1995 Waterbird communities and habitat relationships in coastal pastures of northern California. Conservation Biology 9:827–834.

Curtin, C. G. 1993. The evolution of the U.S. National Wildlife Refuge System and the doctrine of compatibility. Conservation Biology 7:29–38.

Dahl, T. E., and C. E. Johnson. 1991. Status and Trends of Wetlands in the Conterminous United States, Mid-1970s to Mid-1980s. U.S. Department of the Interior, Fish and Wildlife Service, Washington, DC.

de Jong, W. 1997. Developing swidden agriculture and the threat of biodiversity loss. Agriculture Ecosystems and Environment 62:187–197.

de Szalay, F. A., D. P. Batzer, and V. H. Resh. 1996. Mesocosm and macrocosm experiments to examine effects of mowing emergent vegetation on wetland invertebrates. Environmental Entomology 25:303–309.

Deyrup, M. 1994. Invertebrates. Vol. 4 of R. E. Ashton (ed.), Rare and Endangered Biota of Florida. University of Florida Press, Gainesville, FL.

Dillon, R. T., Jr., and S. A. Ahlstedt. 1997. Verification of the specific status of the endangered Anthony's river snail, *Athearnia anthonyi,* using allozyme electrophoresis. Nautilus 110:97–101.

Donovan, T. M., F. R. Thompson III, J. Faaborg, and J. R. Probst. 1995. Reproductive success of migratory birds in habitat sources and sinks. Conservation Biology 9: 1380–1395.

Dover, J. W. 1997. Conservation headlands: Effects on butterfly distribution and behaviour. Agriculture, Ecosystems and Environment 63:31–49.

Drinkwater, L. E., D. K. Letourneau, F. Workneh, H. C. van Bruggen, and C. Shennan. 1995. Fundamental differences between conventional and organic tomato agroecosystems in California. Ecological Applications 5:1098–1112.

Duelli, P., M. Studer, I. Marchand, and S. Jakob. 1990. Population movements of arthropods between natural and cultivated areas. Biological Conservation 54:193–207.

Fasola, M., and R. Alieri. 1992. Conservation of heronry Ardeidae sites in north Italian agricultural landscapes. Biological Conservation 62(3):219–228.

Fasola, M., and X. Ruiz. 1996. The value of rice fields as substitutes for natural wetlands for waterbirds in the Mediterranean region. Colonial Waterbirds 19:122–128.

Frampton, G. K., and T. Cilgi. 1994. Long-term effects of pesticides on Carabidae in U. K. farmland: Some initial results from the SCARAB project. Pages 439–444 *in* K. Desender et al. (eds.), Carabid Beetles: Ecology and Evolution. Kluwer Academic, Dordrecht, The Netherlands.

France, R. L., and N. C. Collins. 1993. Extirpation of crayfish in a lake affected by long-range anthropogenic acidification. Conservation Biology 7:184–188.

France, R. L. 1997. Macroinvertebrate colonization of woody debris in Canadian shield lakes following riparian clearcutting. Conservation Biology 11:513–521.

Frayer, W. E., D. D. Peters, and H. R. Pywell. 1989. Wetlands of the California Central Valley: Status and Trends 1939 to mid-1980's. U.S. Fish and Wildlife Service Region 1, Portland, OR.

Gall, G. A. E., and G. H. Orians. 1992. Agriculture and biological conservation. Agriculture, Ecosystems and Environment 42:1–8.

Gaskin, D. E. 1995. Butterfly conservation programs must be based on appropriate ecological information. Proceedings of the Entomological Society of Ontario 126: 15–27.

Gilpin, M., G. A. E. Gall, and D. S. Woodruff. 1992. Ecological dynamics and agricultural landscapes. Agriculture, Ecosystems and Environment 42:27–52.

Granval, P., R. Aliaga, and P. Soto. 1993. The impact of agricultural management on earthworms Lumbricidae, common snipe *Gallinago-Gallinago* and the environmental value of grasslands in the Dives Marshes Calvados. Gibier faune sauvage 10: 59–73.

Grigarick, A. A., R. K. Webster, R. P. Meyer, F. G. Zalom, and K. A. Smith. 1990. Effect of pesticide treatments on nontarget organisms in California rice paddies. Hilgardia 58(1):1–35.

Hanson, G. E., and J. M. Brode. 1980. Status of the Giant Garter Snake *Thamnophis couchi Gigas* (Fitch). Inland Fisheries Endangered Species Program Special Publication 80–5. California Department of Fish and Game.

Hassall, M., A. Hawthorne, M. Maudsley, P. White, and C. Cardwell. 1992. Effects of headland management on invertebrate communities in cereal fields. Agriculture Ecosystems and Environment 40:155–178.

Henrikson, B. I. 1993. *Sphagnum* mosses as a microhabitat for invertebrates in acidified lakes and the color adaptation and substrate preference in *Leucorrhinia-Dubia* Odonata Anisoptera. Ecography 16:143–153.

Herkert, J. R. 1997. Bobolink *Dolichonyx oryzivorus* population decline in agricultural landscapes in the Midwestern USA. Biological Conservation 80:107–112.

Herms, C. P., D. G. McCullough, D. L. Miller, L. S. Bauer, and R. A. Haack. 1996. Laboratory rearing of *Lycaeides melissa* samuelis (Lepidoptera: Lycaenidae), an endangered butterfly in Michigan. Great Lakes Entomologist 29:63–75.

Hesler, L. S., A. A. Grigarick, M. J. Oraze, and A. T. Palrang. 1993. Arthropod fauna of conventional and organic rice fields in California. Journal of Economic Entomology 86:149–158.

Hook, D. D. 1993. Wetlands: History, current status, and future. Environmental Toxicology and Chemistry 12:2157–2166.

Ishii-Eiteman, M. J, and A. G. Power. 1997. Response of green rice leafhoppers to rice-planting practices in northern Thailand. Ecological Applications 7:194–208.

Kato, M., and R. Miura. 1996. Flowering phenology and antophilous insect community at a threatened natural lowland marsh at Nakaikemi in Tsuruga, Japan. Contributions from the Biological Laboratory Kyoto University 29:1–48.

Kellert, S. R. 1993. Values and perceptions of invertebrates. Conservation Biology 7: 845–855.

Kemp, J. C., and G. W. Barrett. 1989. Spatial patterning: Impact on uncultivated corridors on arthropod populations within soybean agroecosystems. Ecology 70:114–128.

Kingsley, K. J. 1996. Behavior of the Delhi sands flower-loving fly (Diptera: Mydidae), a little-known endangered species. Annals of the Entomological Society of America 89:883–891.

Klenner, M. F. 1994. The carabid fauna of diflubenzuron-sprayed and unsprayed plots in Westphalian oak forests—a post-treatment comparison. Pages 439–444 *in* K.

Desender et al. (eds.), Carabid Beetles: Ecology and Evolution. Kluwer Academic, Dordrecht, The Netherlands.

Knight, R. L., G. N. Wallace, and W. E. Riebsame. 1995. Ranching the view: Subdivisions versus agriculture. Conservation Biology 9:459–461.

Kremen, C., R. K. Colwell, T. L. Erwin, D. D. Murphy, R. F. Noss, and M. A. Sanjayan. 1993. Terrestrial arthropod assemblages: their use in conservation planning. Conservation Biology 7:796–808

Launer, A. E, and D. D. Murphy. 1994. Umbrella species and the conservation of habitat fragments: A case of a threatened butterfly and a vanishing grassland ecosystem. Biological Conservation, 69:145–153.

Letourneau, D. K. 1986. Associational resistance in squash monocultures and polycultures in tropical Mexico. Environmental Entomology 15:285–292.

———. 1987. The enemies hypothesis: Tritrophic interactions and vegetational diversity in tropical agroecosystems. Ecology 68:1616–1622.

———. 1990a. Abundance patterns of leafhopper enemies in pure and mixed stands. Environmental Entomology 19:505–509.

———. 1990b. Mechanisms of predator accumulation in a mixed crop system. Ecological Entomology 15:63–69.

Letourneau, D. K., L. E. Drinkwater, and C. Shennan. 1996. Effects of soil management on crop nitrogen and insect damage in organic vs. conventional tomato fields. Agriculture Ecosystems and Environment 57:179–187.

Light, T., D. Erman, C. Myrick, and J. Clarke. 1995. Decline of the Shasta crayfish (*Pacifastacus fortis* Faxon) of northeastern California. Conservation Biology 9: 1567–1577.

Loughman, D. L., and D. P. Batzer. 1992. Assessment of Rice Fields as Habitats for Ducks Wintering in California. Final Report. California Department of Fish and Game.

Lys, J. A. 1994. The positive influence of strip-management on ground beetles in a cereal field: increase, migration and overwintering. Pages 439–444 *in* K. Desender et al. (eds.), Carabid Beetles: Ecology and Evolution. Kluwer Academic, Dordrecht, The Netherlands.

Malmqvist, B. 1996. The ibis fly, *Atherix ibis* (Diptera: Athericidae), in Sweden: The distribution and status of a red-listed fly species. Entomologisk Tidskrift 117:23–28.

Marc, P., and A. Canard. 1997. Maintaining spider biodiversity in agroecosystems as a tool in pest control. Agriculture, Ecosystems and Environment 62:229–235.

Marino, P. C., and D. A. Landis. 1996. Effects of landscape structure on parasitoid diversity and parasitism in agroecosystems. Ecological Applications 6:276–284.

Mathias, M. E., and P. Moyle. 1992. Wetland and aquatic habitats. Agriculture, Ecosystems and Environment 42:165–176.

McCracken, D. I., and G. N. Foster. 1993a. Surface-active invertebrate communities and the availability of potential food for the chough *Pyrrhocorax-Pyrrhocorax* L. on pastures in Northwest Islay. Pedobiologia 37:141–158.

———. 1993b. The effect of ivermectin on the invertebrate fauna associated with cow dung. Environmental Toxicology and Chemistry 12:73–84.

McKillop, W. B., W. M. McKillop, and A. C. Conroy. Observations on the life history and distribution of the showy pond snail, *Bulimnea megasoma* (Say) (Gastropoda: Pulmonata) in southeastern Manitoba. Canadian Field-Naturalist 107:192–195.

Mesleard, F. 1994. Agricultural abandonment in a wetland area: Abandoned ricefields in the Camargue, France: Can they be of value for conservation? Environmental Conservation 21:354–357.

Mierzwa, K., and A. P. Smyth. 1996. A preliminary assessment of metapopulation structure in an endangered dragonfly (Somatochlora, hineana). Bulletin of the Ecological Society of America 77:306.

Mladenoff, D. J., and F. Stearns. 1993. Eastern hemlock regeneration and deer browsing in the northern Great Lakes region: A re-examination and model simulation. Conservation Biology 7:889–900.

Moreby, S. J., N. J. Aebischer, S. E. Southway, and N. W. Sotherton. 1994. A comparison of the flora and arthropod fauna of organically and conventionally grown winter wheat in southern England. Annals of Applied Biology 125:13–27

Myers, A. A. 1996. Species and generic gamma-scale diversity in shallow-water marine amphipoda with particular reference to the Mediterranean. Journal of the Marine Biological Association of the United Kingdom 76:195–202.

Ndubisi, F., T. Demeo, and N. D. Ditto. 1995. Environmentally sensitive areas: A template for developing greenway corridors. Landscape and Urban Planning 33:159–177.

Niemela, J. 1997. Invertebrates and boreal forest management. Conservation Biology 11:601–610.

Niemela, J., D. Langor, and J. R. Spence. 1993. Effects of clear-cut harvesting on boreal ground-beetle assemblages (Coleoptera: Carabidae) in western Canada. Conservation Biology 7:551–561.

Oliver, I., and A. J. Beattie. 1993. A possible method for the rapid assessment of biodiversity. Conservation Biology 7:562–568.

———. 1996a. Invertebrate morphospecies as surrogates for species: A case study. Conservation Biology 10:99–109.

———. 1996b. Designing a cost-effective invertebrate survey: A test of methods for rapid assessment of biodiversity. Ecological Applications 6:594–607.

O'Malley, R. E. 1997. Evaluating Wildlife Conservation Strategies for an Agricultural Wetland: Dynamics of Top-Down Versus Bottom-up Influences, Omnivory and Spatial Scale. Ph.D. Dissertation, University of California, Santa Cruz.

———. 1999a. Farm management for conservation goals: Efficacy vs. adoption in California rice production. In preparation.

———. 1999b. Using path analysis to disentangle the web: Top-down vs bottom-up forces and omnivory in a semi-aquatic arthropod community. In preparation.

Orr, D. 1991. Biological diversity, agriculture, and the liberal arts. Conservation Biology 5:268–270.

Palmer, R. W., and A. R. Palmer. 1995. Impacts of repeated applications of *Bacillus thuringiensis* var *israelensis* de Barjac and Temephos, used in blackfly (Diptera: Simuliidae) control, on macroinvertebrates in the middle Orange River, South Africa. Southern African Journal of Aquatic Sciences 21:35–55.

Paullin, D. G. 1992. Central Valley Habitat Joint Venture. Proceedings of the California Mosquito Vector Control Association 60:10–11.

Payne, J. M., and W. A. Wentz. 1992. Private property and wetland conservation. Transactions of the Fifty-seventh North American Wildlife and Nature Resources Conference 57:225–233.

Pearsall, I. A., and S. Walde. 1995. A comparison of epigaeic Coleoptera assemblages in organic, conventional and abandoned orchards in Nova Scotia, Canada. Canadian Entomologist 127:641–648.

Pedersen, H., and M. Holmen. 1994. Protected insects in Denmark: Part 4. Dragonflies. Entomologiske Meddelelser 62:33–58.

Perfecto, I., R. A. Rice, R. Greenberg, and M. E. van Doort. 1996. Shade coffee: A disappearing refuge for biodiversity. Bioscience 46:598–608.

Persson, L. 1996. Productivity and consumer regulation: Concepts, patterns, and mechanisms. Pages 396–434 *in* G. A. Polis and K. O. Winemiller (eds.), Food Webs: Integration of Patterns and Dynamics. Chapman & Hall, New York.

Phelan, P. L., J. F. Mason, and B. R. Stinner. 1995. Soil-fertility management and host preference by European corn borer, *Ostrinia nubilalis* (Huebner), on *Zea mays* L.: A comparison of organic and conventional chemical farming. Agriculture Ecosystems and Environment 56:1–8.

Polhemus, D. A. 1996. The Orangeblack Hawaiian Damselfly, *Megalagrion xanthomelas* (Odonata: Coenagrionidae.) Bishop Museum Occasional Papers 45: 30–53.

Power, M. E. 1996. Disturbance and food chain length in rivers. Pages 286–297 *in* G. A. Polis and K. O. Winemiller (eds.), Food Webs: Integration of Patterns and Dynamics. Chapman & Hall, New York.

Power, M. E., D. Tilman, J. A. Estes, B. A. Menge, W. J. Bond, L. S. Mills, G. Daily, J. C. Castilla, J. Lubchenco, and R. T. Paine. 1996a. Challenges in the quest for keystones: Identifying keystone species is difficult-but essential to understanding how loss of species with affect ecosystems. Bioscience 46:609–620.

Power, M. E., W. E. Dietrich, and J. C. Finlay. 1996b. Dams and downstream aquatic biodiversity: Potential food web consequences of hydrologic and geomorphic change. Environmental Management 20:887–895.

Resh, V. H., and D. M. Rosenberg. 1989. Spatial temporal variability and the study of aquatic insects. Canadian Entomologist 121:941–963.

Rief, S. 1996. The influence of cultivation on selected Diptera (Nemotocera: Limoniidae, Tipulidae, Trichoredse, Brachycera: Empididae, Hybotidae, Dolichopodidae) of different ecosystems on low peat bogs. Faunistisch-Oekologische Mitteilungen Supplement 20:47–76.

Risch, S. J., D. Andow, and M. A. Altieri. 1983. Agroecosystem diversity and pest control: Data, tentative conclusions, and new research directions. Environmental Entomology 12:625–629.

Robinson, T. 1993. Fate and Transport of Agricultural Contaminants from Rice Paddies, Impact Sampling Strategies and the Potential Environmental Degradation to Dry Tropical Coastal Wetlands—Guanacaste, Costa Rica. Ph. D. Dissertation. University of California—Santa Barbara.

Roger, P. A., I. Simpson, R. Oficial, S. Ardales, and R. Jimenez. 1994. Effects of pesticides on soil and water microflora and mesofauna in wetland ricefields: A summary of current knowledge and extrapolation to temperate environments. Australian Journal of Experimental Agriculture 34:1057–1068.

Root, R. B. 1973. Organization of plant-arthropod association in simple and diverse habitats: The fauna of collards (*Brassica oleracea*). Ecological Monographs 43:94–124.

Rubec, C. 1995. Canada: Peatland sustainability and resources use. Gunneria 70:251–263.

Russell, E. P. 1989. Enemies hypothesis: A review of the effect of vegetational diversity on predatory insects and parasitoids. Environmental Entomology 18:590–599.

Rykken, J. J., D. E. Capen, and S. P. Mahabir. 1997. Ground beetles as indicators of land type diversity in the Green Mountains of Vermont. Conservation Biology 11:522–530.

Safina, C. 1993. Population trends, habitat utilization, and outlook for the future of the sandhill crane in North America: A review and synthesis. Bird Population 1:1–27.

Salafsky, N. 1993. Mammalia use of a buffer zone agroforestry system bordering Gunung Palung National Park, West Kalimantan, Indonesia. Conservation Biology 7:928–933.

Samways, M. J. 1989. Insect conservation and the disturbance landscape. Agriculture Ecosystems and Environment 27:183–194.

Samways, M. J., P. M. Caldwell, and R. Osborn. 1996. Spatial patterns of dragonflies (Odonata) as indicators for design of a conservation pond. Odonatologica 25:157–166.

Saville, N. M., W. E. Dramstad, G. L. A. Fry, and S. A. Corbet. 1997. Bumblebee movement in a fragmented agricultural landscape. Agriculture Ecosystems and Environment 61:145–154.

Schoenly, K. G. 1996. Quantifying the impact of insecticides on food web structure of rice-arthropod populations in a Philippine farmer's irrigated field: A case study. Pages 343–351 *in* G. A. Polis and K. O. Winemiller (ed.), Food Webs: Integration of Patterns and Dynamics. Chapman & Hall, New York.

Schoenly, K., J. E. Cohen, K. L. Heong, J. A. Litsinger, G. B. Aquino, A. T. Barrion, and G. Arida. 1996. Food web dynamics of irrigated rice fields at five elevations in Luzon, Philippines. Bulletin of Entomological Research 86:451–466.

Schuler, C. A. 1987. Impacts of Agricultural Drainwater and Contaminants on Wetlands at Kesterson Reservoir, California. Ph.D. Dissertation, Oregon State University, Corvallis, OR.

Settle, W. H., H. Ariawan, E. T. Astuti, W. Cahyana, A. L. Hakim, D. Hindayana, A. S. Lestari, and S. Pajarningsih. 1996. Managing tropical rice pests through conservation of generalist natural enemies and alternative prey. Ecology 77:1975–1988.

Siegel, J. S. 1996. "Subdivisions versus Agriculture": From false assumptions come false alternatives. Conservation Biology 10:1473–1474.

Simpson, I. C., P. A. Roger, R. Oficial, and I. F. Grant. 1994. Effects of nitrogen fertilizer and pesticide management on floodwater ecology in a wetland ricefield:

II. Dynamics of microcrustaceans and dipteran larvae. Biology and Fertility of Soils 17:138–146.

Sipes, S. D., and V. J. Tepedino. 1995. Reproductive biology of the rare orchid, *Spiranthes diluvialis*: Breeding system, pollination, and implications for conservation. Conservation Biology 9:929–938.

Skurdal, J. 1995. Human impact on natural populations of European crayfish. Fauna (Oslo) 48:134–143.

Smith, R. S., A. G. Lunn, and M. D. Newson. 1995. The Border Mires in Kielder Forest: A review of their ecology and conservation management. Forest Ecology and Management 79:47–61.

Solberg, K. L., and K. F. Higgins. 1993. Effects of glyphosate herbicide on cattails, invertebrates, and waterfowl in South Dakota wetlands. Wildlife Society Bulletin 21:239–307.

Strong, M. A., J. G. Mensik, and D. S. Walsworth. 1990. Converting rice fields to natural wetlands in the Sacramento Valley of California. Transactions of the Western Section of the Wildlife Society 26:29–35.

Suzan, H., G. P. Nabhan, and D. T. Patten. 1994. Nurse plant and floral biology of a rare night-blooming cereus *Peniocereus striatus* (Brandegee) F. buxbaum. Conservation Biology 8:461–470.

Tamisier, A., and P. Grillas. 1994. A review of habitat changes in the Camargue: An assessment of the effects of the loss of biological diversity on the wintering waterfowl community. Biological Conservation 70:39–47.

Thiollay, J. M. 1995. The role of traditional agroforests in the conservation of rain forest bird diversity. Conservation Biology 9:335–353.

Tiner, R. W. 1984. Wetlands of the Conterminous United States: Current Status and Recent Trends. U.S. Department of the Interior, Fish and Wildlife Service, Washington D.C.

Varshney, R. K. 1997. Index Rhopalocera Indica part III. Genera of butterflies from India and neighbouring countries (Lepidoptera: (C) Lycaenidae). Oriental Insects 31:83–138.

Voigt, N. 1994. Importance of heathland ecosystems for the invertebrate fauna. Faunistisch-Oekologische Mitteilungen Supplement 16:1–11.

Wahlberg, N. 1997. The life history and ecology of *Melitaea diamina* (Nymphalidae) in Finland. Nota Lepidopterologica 20(1–2):70–81.

Wassermann, G., and A. Schmidt-Kloiber. 1996. Ephemeroptera and Odonata of an artificial Danube backwater irrigation system. Archiv für Hydrobiologie Supplement 113:493–496.

Western Ecological Services Company, Inc. (WESCO). 1991. Environmental Attributes of Rice Cultivation in California. California Rice Promotion Board, Yuba City, CA.

Wilcove, D. S., M. McMillan, and K. C. Winston. 1993. What exactly is an endangered species? An analysis of the U.S. Endangered Species List 1985–1991. Conservation Biology 7:87–93.

Williams, R. N., M. S. Ellis, and D. S. Fickle. 1995. Insects in the Killbuck Marsh Wildlife Area: 1993 survey. Ohio Journal of Science 95:226–232.

Wootton, J. T. 1994. Predicting direct and indirect effects: An integrated approach using experiments and path analysis. Ecology 75:151–165.

Wootton, J. T., M. S. Parker, and M. E. Power. 1996a. Effects of disturbance on river food webs. Science 273:1558–1561.

Wootton, J. T., M. E. Power, R. T. Paine, and C. A. Pfister. 1996b. Effects of productivity, consumers, competitors, and El Niño events on food chain patterns in a rocky intertidal community. Proceedings of the National Academy of Sciences of the United States of America 93:13855–13858.

Wuerthner, G. 1994. Subdivisions versus agriculture. Conservation Biology 8:905–908.

Zaidelman, F. R. 1996. Drainage and agromelioration effects on soil biota. Pochvovedenie 6:728–734.

PART 5
Coastal Freshwater Wetlands

36 Tidal Freshwater Wetlands

Invertebrate Diversity, Ecology, and Functional Significance

DAVID J. YOZZO and ROBERT J. DIAZ

*T*idal freshwater wetlands are vegetated intertidal habitats characterized by measurable tidal fluctuation and, on occasion, measurable salinity (usually <0.5 practical salinity units). They are unique endpoint habitats created by a combination of terrestrial-riverine and oceanic-estuarine influences. Vascular floral composition is species-rich, among the highest of any wetland type. In contrast, the invertebrate faunal composition of tidal freshwater wetlands is species-poor relative to nontidal rivers, lakes, or estuaries. Tidal freshwater wetlands are characterized by high primary and secondary production and provide critical nursery habitat for many freshwater and estuarine fishery species. Major habitat types found in tidal freshwater wetlands include submerged and floating aquatic macrophyte beds, intertidal emergent marshes, unvegetated intertidal mudflats, and tidal creeks. Macroinvertebrate communities of tidal freshwater wetlands are dominated by annelids (Tubificidae, Naididae, Enchytraidae) and insect larvae (Chironomidae). Meiofaunal communities are dominated by nematodes, microcrustaceans (Ostracoda, Copepoda), naidid oligochaetes, and tardigrades. Despite the potential ecological importance of tidal freshwater wetland invertebrate communities, we know relatively little about how they function in comparison to those of nontidal freshwater and/or estuarine wetlands and how they may respond to both natural and man-induced disturbance.

Invertebrates in Freshwater Wetlands of North America: Ecology and Management, Edited by
Darold P. Batzer, Russell B. Rader, and Scott A. Wissinger
ISBN 0-471-29258-3 © 1999 John Wiley & Sons, Inc.

TIDAL FRESHWATER WETLANDS

Tidal freshwater wetlands are vegetated intertidal habitats characterized by measurable tidal fluctuation and at times measurable salinity (usually <0.5 practical salinity units (psu); during droughts salinity can be >5 psu). This physically places tidal freshwater wetlands between the fall line, the transition point from nontidal river to tidal river-estuary, and the upstream limit of salt intrusion (Fig. 36.1). Although they perform many of the same functions as tidal salt or brackish marshes, their floral and faunal composition is similar to that of nontidal freshwater wetlands. In contrast to salt marshes, in which plant distribution and zonation are usually dominated by one or two species (e.g., *Spartina*), tidal freshwater wetlands frequently include mixed communities of many species, including grasses, shrubs, rushes, and herbaceous plants. Floristic intertidal zonation occurs along an elevation gradient, but may

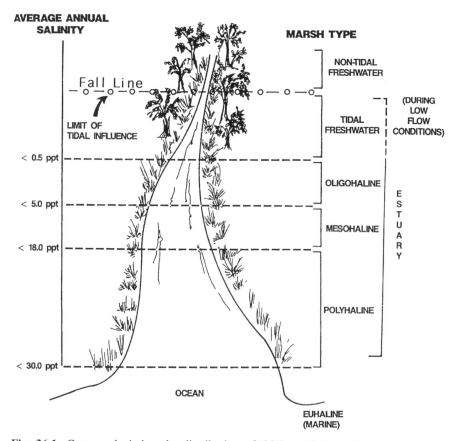

Fig. 36.1. Cartoon depicting the distribution of tidal marsh type along an estuarine salinity gradient. (Modified from Odum et al. 1984.)

not be as sharply defined as in saline coastal marshes. Submerged or floating aquatic plants may be present in tidal creeks or shallow embayments. Seasonal variation in plant community composition is pronounced, with annual and perennial plant species alternating in both aerial extent and biomass. Primary production and above-ground macrophyte standing crop in tidal freshwater wetlands is among the highest reported for wetlands (Odum et al. 1984). Soils of tidal freshwater wetlands are predominantly fine silts and clays. Organic matter content ranges from 10–45 percent, depending on location along an elevation gradient from creekbank to upland edge. In the high marsh, or upper intertidal zone, sediment deposition rates are low and organic matter accumulates. In the lower intertidal zone, sediments contain less organic matter and are more dynamic, with considerable erosion and deposition taking place (Simpson et al. 1983). Nutrient cycling in tidal freshwater wetlands is similar to that of saline coastal marshes. They are relatively open systems and may act either as sources, sinks, or transformers of nutrients, depending on time of year, geomorphology, and other factors.

The ecology of tidal freshwater wetlands has been reviewed in considerable detail by Simpson et al. (1983), Odum et al. (1984), and Odum (1988). However, tidal freshwater wetlands invertebrate communities are still poorly studied in comparison to salt marshes, which have received intensive study for decades. Odum (1988) says, regarding why tidal freshwater wetlands have not received the attention of freshwater or estuarine scientists, "Historically, tidal freshwater environments have been ignored by limnologists because of the presence of oceanic tidal influence, and neglected by marine ecologists because they are bathed by freshwater and inhabited primarily by freshwater organisms."

GEOGRAPHIC DISTRIBUTION

Tidal freshwater wetlands are found in all coastal areas of North America where rivers exhibit moderate to strong tidal influence. They are most extensive along the Atlantic coast between Georgia and New England, especially in southern New Jersey and the Chesapeake Bay region and along the coastal plain rivers of South Carolina and Georgia. Odum et al. (1979) estimate that there were 500,000–1,000,000 ha of tidal freshwater wetlands along the Atlantic and Gulf coasts. Many of the major population centers in the mid-Atlantic region, including Philadelphia, Washington, D.C., and Richmond, started as settlements located near the fall line delineating the upper extent of tidal influence and in close proximity to tidal freshwater wetlands (Simpson et al. 1983).

Tidal freshwater wetlands appear to develop best in areas which have a major influx of fresh water, a tidal amplitude in excess of 0.5 m, and a basin geomorphology which constricts and magnifies the tidal wave in the upstream portion of the estuary (Odum et al. 1984). For example, the highest average vertical tidal range reported for the entire Chesapeake Bay and its tidal tributaries, which do have extensive tidal freshwater wetlands, occurs in the tidal freshwater Mattaponi River, near Walkerton, Virginia, at 1.2 m (Williams 1992). Tidal freshwater wetlands are noticeably absent from the microtidal Albemarle-Pamlico Sound estuary, North Carolina. Restricted tidal range as a result of the Outer Banks barrier island chain prevents the development of extensive tidal freshwater reaches in this region. The Cape Fear Estuary, located south of the Outer Banks in southeastern North Carolina, has extensive areas of tidal freshwater marsh in its upper reaches. Extensive tidal freshwater wetlands are found in the northeast Gulf of Mexico (see Gaston, this volume), particularly in the upper reaches of Mobile Bay and the Mobile River, Alabama. Extensive freshwater wetlands can be found in coastal Louisiana, but these systems are characterized by irregular, low-amplitude, wind-driven tides. On the West Coast, tidal freshwater wetlands are most extensive in the Pacific Northwest, where significant freshwater flow and a high tidal amplitude provide for their development. The Columbia River has the most extensive tidal freshwater wetland system on the West coast.

PREVIOUS RESEARCH

Invertebrate communities of tidal freshwater wetlands have received little attention from estuarine ecologists. The existing literature on invertebrate communities in tidal freshwater ecosystems focuses primarily on river bottom benthos. Diaz (1978, 1989, 1994) reports on collections of tidal freshwater macrobenthos from the James River Estuary, Virginia, and concludes that no single species is specifically adapted for existence in tidal freshwater conditions; rather, the tidal freshwater macrobenthos is comprised of species which are broadly adaptable to changing environmental conditions and remarkably resilient to environmental perturbations and pollutants. Studies of tidal freshwater benthic communities from the Schuylkill River, Pennsylvania (Ettinger 1982), and Delaware River, New Jersey (Crumb 1977), support the general contention of Diaz (1978) and attribute an observed lack of faunal diversity in tidal freshwaters to a lack of habitat diversity. A comprehensive survey of benthic and epibenthic invertebrates in the Hudson River estuary by Ristich et al. (1977) found that the tidal freshwater portion of the Hudson River estuary supported the lowest number of species. The tidal freshwater Columbia River appears to have a higher number of species than any of the East Coast systems (Clairain et al. 1978). Some of the earlier surveys of tidal freshwater macrobenthos include information on shallow vegetated shorelines or emergent marsh communities. However, detailed descriptions of intertidal freshwater wetland invertebrate communities are few and are generally restricted to Hudson River and James River systems on the East Coast and Columbia River systems on the West Coast.

SIGNIFICANCE OF TIDAL FRESHWATER WETLANDS

A Transition Zone between Nontidal Freshwater and Marine Habitats

Tidal freshwater wetlands are transitional between the brackish and salt marshes of the lower estuary and nontidal freshwater conditions above the fall line. Organisms which are unable to adapt to varying tidal conditions are excluded, but many freshwater invertebrate taxa commonly found in low-gradient rivers and streams and the littoral zone of ponds and lakes are likely to be encountered in tidal freshwater wetlands. A surprising number of eury-haline estuarine invertebrate taxa make their way up into tidal freshwater. Many estuarine crustaceans (isopods, amphipods, caridean shrimp, portunid crabs) are tolerant of low-salinity to freshwater conditions for a considerable portion of their life history. The transition zone function of tidal freshwater wetlands is the primary determinant of the interesting and seasonally varying mix of invertebrate and fish species.

A Nursery and Predation Refuge for Migratory and Resident Fish Species

Fish community composition in tidal freshwater wetlands can be best de-scribed as a mixture of warm freshwater species, estuarine species that extend their range into freshwater habitats, and anadromous marine species that use tidal freshwater as spawning and nursery habitat.

On the East Coast, resident freshwater species are represented primarily by three families: cyprinids (minnows and shiners), centrarchids (black bass, sunfishes and crappies), and ictalurids (catfishes). Estuarine residents include white perch (*Morone americana*) and hogchokers (*Trinectes maculatus*). Mi-gratory species occurring in tidal freshwater wetlands along the Atlantic coast include juvenile striped bass (*Morone saxatilis*), menhaden (*Brevoortia tyr-annus*), and summer flounder (*Paralichthys dentatus*) (Odum et al. 1979). The tidal freshwater marsh surface is used as a spawning habitat and nursery for several small marsh-resident species, including mummichogs and mosquito-fish (*Gambusia affinis*). A diverse assemblage of freshwater and estuarine nekton species access the lower intertidal marsh surface to forage during flood tides (McIvor and Odum 1988). Lush beds of submerged vegetation in tidal creeks provide feeding habitat and a predation refuge for many species (Rozas and Odum 1987a, 1988) including bluespotted sunfish (*Enneacanthus glorio-sus*) and banded killifish (*Fundulus diaphanus*).

On the West Coast, anadromous fish dominate and in the Columbia River include salmon (*Onchorhynchus* spp.), American shad (*Alosa sapidissima*), smelt (Osmeridae), and Pacific lamprey (*Entosphenus tridentatus*). Freshwater species include sturgeon (*Acipenser* spp.) and trout (*Salmo* spp.). Juvenile Chinook salmon (*Onchorhynchus tshawytscha*) and starry flounder (*Pla-

tichthys stellatus) use tidal freshwater marsh habitats extensively as nursery grounds. Peamouth (*Mylocherlus caurinus*) and threespine stickleback (*Gasterosteus aculeatus*) are resident tidal freshwater species in the Columbia River (Clairain et al. 1978).

Benthic invertebrate prey resources on the intertidal marsh surface and the shallow subtidal benthos of marsh creeks are clearly of importance in maintaining the nursery function of tidal freshwater wetlands for many fish species. Common benthic prey taxa available to foraging nektonic organisms include insect larvae (Diptera, Ephemeroptera), oligochaetes (Tubificidae, Naidae, Enchytraeidae), and crustaceans (Amphipoda, Isopoda). Epiphytic invertebrate prey taxa associated with submerged or floating vegetation in marsh creeks include naidid oligochaetes, dipteran larvae, damselfly naiads (Zygoptera), mayflies (Ephemeroptera), and amphipods. Microcrustaceans (ostracods, copepods, cladocerans) are critical prey resources for post-larval and early juvenile fish and are abundant in shallow intertidal marsh pools and rivulets and among submerged aquatic vegetation in tidal creeks and embayments (Dias et al. 1978, Diaz et al. 1978, Munson 1980).

TIDAL FRESHWATER WETLAND HABITATS

Submerged and Floating Aquatic Macrophyte Beds

Submerged and floating aquatic macrophyte communities are a common feature in tidal freshwater wetlands. Common submerged macrophyte species in tidal freshwater include Eurasian water milfoil (*Myriophyllum spicatum*), various pondweeds (*Potamogeton* spp.), naiads (*Najas* spp.), waterweeds (*Elodea* spp.), tapegrass (*Vallisneria americana*), water stargrass (*Zosterella dubia*), and coontail (*Ceratophyllum demersum*). Floating macrophytes in tidal freshwater include water chestnut (*Trapa natans*) and duckweeds (*Lemna* spp.). Aquatic macrophyte beds may be extensive in shallow riverine bays, such as in the Upper Chesapeake Bay and the tidal freshwater portion of the Hudson River Estuary and Columbia River.

Dense beds of submerged aquatic macrophytes are often present in tidal freshwater marsh creeks. Rozas and Odum (1987b) document a relationship among stream order, submerged aquatic vegetation (SAV) density, and juvenile fish utilization in tidal freshwater marshes of the Chickahominy River, Virginia. Small first- and second-order tidal creeks supported denser SAV due to increased light penetration relative to larger, higher-order creeks; juvenile fish use was highest in creeks which supported denser assemblages of SAV. The surfaces of submerged and floating aquatic plants are colonized by a variety of epiphytic invertebrates, especially aquatic insect larvae, and are of particular importance as a food resource for juvenile fishes and macrocrustaceans in tidal freshwater wetlands. Rozas and Odum (1988) conducted foraging experiments using enclosures with and without SAV in order to determine the relative profitability of foraging among SAV compared to foraging on unvegetated substrates. Banded killifish (*Fundulus diaphanus*) were found to consume more prey in vegetated enclosures, but mummichogs (*F. heteroclitus*) and blue-spotted sunfish (*Enneacanthus gloriosus*) consumed larger prey items among SAV. Invertebrate prey taxa recovered from fish stomachs included insect larvae (Ephemeroptera, Odonata, and Chironomidae), cladocerans, amphipods, grass shrimp (*Palaemonetes pugio*), copepods, isopods, and gastropods.

McIvor and Odum (1988) investigated the composition, distribution, and abundance of benthic invertebrate prey communities on erosional versus depositional subtidal creekbanks in the Chickahominy River marshes. Litter bags containing terrestrial plant detritus were placed directly upon the substrate to be colonized by benthic invertebrates. The taxonomic composition of prey taxa differed between the two subtidal habitats. Erosional substrates were characterized by high densities of chironomids, ceratopogonids, oligochaetes, mud crabs (*Rithropanopeus* sp.), and harpacticoid copepods. In contrast, depositional substrates were colonized mostly by amphipods, ephemeropterans, trichopterans, and turbellarians. Erosional substrates supported a greater abundance of benthic invertebrates; however, the depositional substrate was characterized by greater invertebrate biomass and is apparently a more energetically favorable foraging habitat for resident and transient marsh fishes.

Findlay et al. (1989) studied the epiphytic and benthic invertebrate communities associated with water chestnut (*Trapa natans*) at Tivoli South Bay, Hudson River, New York. Sediment and plant-associated invertebrates were sampled using corers, emergence traps, artificial substrates, and macrophyte collections, with emphasis on the Chironomidae (the numerically dominant taxa and an important component of the diet of juvenile fish in the study area). In sediment cores the subfamily Chironominae was most abundant, with two genera, *Tanytarsus* and *Polypedilum,* dominating the collections. *Chironomus* sp., *Harnischia,* the tanypodine midge *Procladius,* and the naidid oligochaetes *Stylaria, Chaetogaster,* and *Dero* were also common in sediment cores. Additional taxa present included gastropods, cladocerans, ostracods, and collembolids. Densities of benthic invertebrates at Tivoli South Bay were comparable to infaunal densities from nontidal freshwater habitats. Artificial substrates and the surfaces of water chestnut leaves in Tivoli South Bay were colonized mostly by Chironominae (*Endochironomus, Dicrotendipes, Tanytarsus, Ablabesmyia,* and *Polypedilum*). Tanypodine larvae (*Procladius*) and Orthocladinae (*Cricotopus*) were common on artificial substrates. A marked midsummer decline in invertebrate abundance was reported in Tivoli South Bay during both years of the study. Possible explanations include predation by juvenile fishes and other invertebrates (e.g., damselfly larvae, predaceous chironomids), change in habitat structure/suitability resulting in a reduction in available surface area for colonization, and emergence of adult insects.

An earlier study of the benthic and epiphytic invertebrate fauna of a submerged vegetation bed in Hudson River was conducted in 1975–1976 (Menzie

1980). This study was conducted at a location in the lower estuary (Haverstraw Bay) which is considered oligohaline (0.5-5.0 psu). However, Eurasian water milfoil (*Myriophyllum spicatum*) was the dominant vegetation. Like Findlay et al. (1989), this earlier effort focused on the Chironomidae and other insect larvae. Menzie (1980) reasons that the shallow vegetated coves along the shoreline of Haverstraw Bay supported greater densities of benthic and epiphytic invertebrates due to increased surface area for colonization relative to deeper, unvegetated habitats. These areas provide good foraging habitat for juvenile fishes, which were often observed to congregate in the vegetated shallows. Plant collections and sediment cores were used to compare benthic and epiphytic macroinvertebrate faunas. Emergence of adult insects was estimated using sweep nets and emergence traps. Twenty-three species of chironomids were identified in benthic and plant samples. Sediment-dwelling taxa included *Chironomus, Tanytarsus, Procladius,* and *Polypedilum.* Plant-dwelling genera included *Cricotopus, Dicrotendipes, Polypedilum,* and *Rheotanytarsus.* Copepods (mostly harpacticoids) and oligochaetes (mostly Tubificidae) were abundant in sediments. Copepods (mostly cyclopoids) and cladocerans were common on *Myriophyllum* stems and leaves. Additional invertebrates associated with *Myriophyllum* included bivalves, naidid oligochaetes, ostracods, water mites, turbellarians, and hydrozoans.

Typically estuarine taxa present in this transition-zone habitat included the polychaetes *Hobsonia* (*Hypaniola*) *florida* and *Marenzelleria* (*Scolecolepides*) *viridis*; the cumacean *Almyracuma proximoculi;* the isopods *Chiridotea almyra* and *Cyathura polita;* and the decapod *Rithropanopeus harrisi.* Documented invertebrate predators of chironomids in the study area included the hydrozoans *Hydra* sp. and *Cordylophora lacustris,* the amphipod *Gammarus daiberi,* and the damselfly naiad *Enallagma durum.* Juvenile fish collected in the study area included striped bass (*Morone saxatilis*) and bluegill sunfish (*Lepomis macrochirus*), the juveniles of which feed primarily on small benthic invertebrates, including chironomid larvae.

Yozzo and Odum (1993) studied the epiphytic invertebrate community associated with water chestnut (*Trapa natans*) in Tivoli South Bay, with emphasis on the microcrustacea. Characteristic taxa included several common littoral cladocerans (*Sida crystallina, Holopedium gibberum, Bosmina longirostris, Ceriodaphnia* sp., *Alona* sp.), the cyclopoid copepod *Eucyclops agilis,* and several species of ostracods, including *Cypridopsis vidua, Darwinula stevensoni, Candona* sp., and *Physocypria* sp. Epiphytic macroinvertebrates observed during this study include many that were reported by Findlay et al. (1989) and some additional taxa, including gastrotrichs, bryozoans (Cristatellidae, Plumatellidae), leeches (Hirudinea), and amphipods (Talitridae, Gammaridae). Aquatic insect larvae and adults included chironomids, ceratopogonids, aquatic bugs (Hemiptera), aquatic beetles (Coleoptera), and gastropods (Lymnaidae, Ancylidae, Physidae, Planorbidae). Predator exclusion experiments failed to document the effect of juvenile fish predation on epiphytic microcrustacean communities in Tivoli South Bay; however, ostracods, cyclopoid copepods, cladocerans and early instar chironomid larvae were numerically dominant prey items in the guts of juvenile and adult banded killifish (*Fundulus diaphanus*) and several other resident fish species in the study area.

Emergent Intertidal Marshes

Emergent intertidal marshes are the most common tidal freshwater wetland type. Extensive tidal freshwater marshes occur along many areas of the Atlantic coast from New England to Northern Florida. Extensive tidal freshwater marshes can be found in the Hudson River Estuary; the Delaware River and its tidal tributaries in southern New Jersey; the Chesapeake Bay and its tidal tributaries in Maryland and Virginia, and many of the coastal plain rivers in South Carolina and Georgia. On the Pacific Coast the Columbia River has the most extensive tidal freshwater marshes.

Vegetation zones in tidal freshwater marshes are not as clearly defined as in salt and brackish marshes. In some cases local geomorphology and microtopographic conditions result in no discernible pattern along an elevation gradient from upland to marsh edge. In other cases a clear zonation pattern may be evident, with plants occurring in distinct bands or zones (Table 36.1, Odum et al. 1984). Typically, along the Atlantic coast the lower intertidal zone is dominated by fleshy, broad-leaved emergent perennials such as spatterdock (*Nuphar luteum*) and pickerelweed (*Pontedaria cordata*). Arrow arum (*Peltandra virginica*) occurs in the low-to-mid intertidal. In the tidal tributaries of lower Chesapeake Bay the arrow arum community type is extensive and dominates the broad marsh landscape. In the mid- to upper reaches of the marsh, additional species are encountered, both annuals and perennials, including wild rice (*Zizania aquatica*) and giant cutgrass (*Zizaniopsis miliacea*), cattails (*Typha* spp.), arrowhead (*Sagittaria latifolia*), sweet flag (*Acorus calamus*), smartweeds (*Polygonum* spp.), and bur marigold (*Bidens* spp.). Over 50 species have been reported from a single marsh (Odum et al. 1984). Seed availability and germination potential may determine, in part, the distribution of plants across the marsh landscape. Various species are restricted to upper intertidal locations as a result of intolerance to extended periods of inundation (Leck and Simpson 1987). Competition and shading may also contribute to the determination of plant zonation and distribution patterns.

Along the Pacific Northwest coast, intertidal areas are dominated by common spike-rush (*Eleocharis palustris*), Lyngbye's sedge (*Carex lyngbyei*), and tufted hairgrass (*Deschampsia caespitosa*). In high intertidal areas willows (*Salix* spp.) and reed canary grass (*Phalaris arundinacea*) are common (Clairain et al. 1978).

Seasonal variation in the plant community is evident. In late spring–early summer the broad-leaved emergent plant species dominate. By late summer grasses and herbaceous species dominate. Latitudinal variation in climate,

TABLE 36.1. Tidal Freshwater Marsh Vegetation Community Types

Spatterdock	Generally occurs in pure stands below mean low water. Often forms dense clonal colonies on submerged point bars.
Arrow arum/ pickerelweed	Occurs throughout intertidal zone; purest stands are most abundant in lower intertidal marsh in spring/ early summer.
Wild rice	May dominate intertidal marsh during mid–late summer, overtopping discontinuous subcanopy of broad leafed emergents.
Cattail	Principal component of many tidal freshwater marshes; may form monospecific stands or co-occur with other species. Found mostly in upper intertidal zone.
Giant cutgrass	Primarily occurs south of Maryland/Virginia. Found in pure stands or in association with other emergent species.
Mixed aquatic	Extremely variable species composition; primarily found in upper intertidal zone. Includes arrow arum, smartweeds, bur marigolds, sweetflag, cattails, arrowhead, jewelweed, and many others. Species dominance varies seasonally.
Big cordgrass	Occurs mostly in pure stands along tidal creeks and sloughs, tidal freshwater to oligohaline marshes. May intermix with cattails, wild rice, and common reed in upper intertidal zone.
Bald cypress/black gum	Ecotonal community between marsh and swamp/ upland forest. Mixture of herbs, shrubs and trees. Understory includes some emergent marsh species; however, species diversity and density is reduced due to shading.

Source: Modified from Odum et al. 1984.

physiography, and geomorphology result in considerable heterogeneity in zonation and plant community composition, especially in comparison to the more homogeneous plant communities of salt and brackish marshes. Microalgae are a conspicuous component of the plant community in tidal freshwater wetlands, particularly in winter, when the effects of shading by emergent vegetation are diminished. Benthic diatoms and filamentous green algae are common in intertidal pools and rivulets on the marsh surface and along exposed creekbanks in the lower intertidal. Very little is known about the ecological significance of microalgae in tidal freshwater wetlands (Diaz 1982).

The diverse vegetation of tidal freshwater marshes provides multiple niches for exploitation by fauna, especially benthic and epiphytic aquatic invertebrates. Larval and adult insects, amphipods, isopods, oligochaete worms, and mollusks are the dominant benthic macroinvertebrate taxa present (Fig. 36.2).

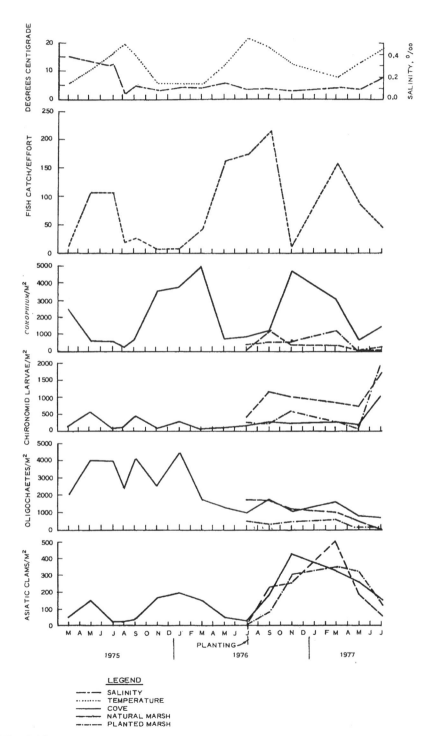

Fig. 36.2. Seasonal trends of aquatic biota and water quality at Miller Sands tidal freshwater marshes, Columbia River, Oregon. (Modified from Clairain et al. 1978.)

A diverse meiofaunal community exists, and the distribution of taxa is largely determined by microtopographic features (Munson 1980, Yozzo and Smith 1995). Intertidal pools and rivulets are dominated by microcrustaceans such as ostracods and copepods. These shallow microhabitats are frequented by larval and juvenile fishes (e.g., *Fundulus* spp.) which forage intensively on the resident meiofauna. Distinct vegetative hummocks formed by marsh macrophytes such as *Peltandra* support dense populations of nematodes, tardigrades, and naidid oligochaetes.

Decomposition of plant material occurs rapidly in tidal freshwater marshes. Odum and Heywood (1978) report that *Nuphar luteum* leaves decomposed completely in as little as 25 days. Other tidal freshwater marsh species break down within 50–150 days, in contrast to the refractory plant material of brackish and salt marshes (e.g., *Spartina*), which may take over a year to decompose completely. Broad-leaved emergent perennials such as *Peltandra, Pontedaria, Sagittaria,* and *Nuphar* are characterized by high nitrogen concentrations. The resultant detritus is of high nutritional quality to detritivores such as amphipod and isopod crustaceans (Odum et al. 1984). In contrast to brackish and salt marshes, very little plant-derived litter accumulates on the freshwater intertidal marsh surface at the beginning of the growing season (Simpson et al. 1983).

One of the earliest and most comprehensive studies of tidal freshwater mud flat and marsh benthic communities was conducted in the mid-1970s at Windmill Point, a dredged material habitat development project located in the James River, Virginia (Diaz and Boesch 1977, Diaz et al. 1978). This site consisted of 4.9 ha of intertidal freshwater marsh and low upland habitat partially enclosed by a sand dike, and an intertidal mudflat habitat located on the outside of the dike. Benthic invertebrates, nekton utilization, wildlife utilization, and plant and soil characteristics were monitored from summer 1976 to fall 1977 and compared to nearby reference locations in order to assess the feasibility of using fine-grained dredged material to create functionally equivalent wetland habitats.

Overall, tubificid oligochaetes (primarily *Limnodrilus* spp.) and chironomid larvae dominated the benthos at both the Windmill Point created marsh and the Herring Creek natural marsh (Table 36.2). There were few differences between the marshes in terms of total taxa and H′ diversity (Table 36.3). The two biggest faunal differences between the marshes were:

1. The oligochaete *Quistadrilus* (*Peloscolex*) *multisetosus* was abundant in Herring Creek marsh, but only rarely encountered at the Windmill Point marsh.
2. The Asiatic clam (*Corbicula fluminea*) was abundant in the lower unvegetated intertidal zone of the Windmill Point marsh, but occurred only rarely at Herring Creek marsh.

A few species of chironomids (*Dicrotendipes nervosus* and *Tanypus* spp.) exhibited strong seasonal abundance patterns, but most organisms exhibited

TABLE 36.2. General Faunal Composition of Macrobenthos from the Windmill Point and Herring Creek, James River, Virginia, Tidal Freshwater Marshes

	Percent of Taxa					
Taxonomic Group	July 1976	November 1976	January 1977	April 1977	July 1977	Total
Platyhelminthes	2.1	3.8	2.8	0.0	0.0	2.6
Nemertea	0.0	1.9	0.0	2.9	0.0	1.3
Mollusca	12.8	13.5	13.8	5.8	13.7	11.7
Bivalvia	6.4	5.8	8.3	2.9	7.8	5.2
Gastropoda	6.4	7.7	5.5	2.9	5.8	6.5
Annelida	25.5	36.5	33.3	41.1	33.3	28.6
Oligochaeta	23.4	26.9	27.8	38.2	27.4	22.1
Polychaeta	0.0	1.9	0.0	0.0	0.0	1.3
Hirudinea	2.1	7.7	5.5	2.9	5.8	5.2
Arthropoda	59.6	44.2	50.0	50.0	51.0	55.8
Insecta	46.8	34.6	38.8	42.8	43.1	50.6
Chironomidae	29.7	21.2	30.5	28.6	29.7	27.3

	Percent of Individuals					
Taxonomic Group	July 1976	November 1976	January 1977	April 1977	July 1977	Total
Platyhelminthes	0.0[a]	0.8	0.2	0.0	0.0	0.1
Nemertea	0.0	0.1	0.0	0.4	0.0	0.1
Mollusca	3.6	13.4	1.3	1.3	5.5	4.9
Bivalvia	3.4	12.0	1.2	1.3	5.0	4.5
Gastropoda	0.2	1.4	0.1	0.0[a]	0.5	0.4
Annelida	80.9	74.1	70.5	86.7	73.3	77.6
Oligochaeta	80.8	73.2	69.0	86.7	72.7	77.1
Polychaeta	0.0	0.0[a]	0.0	0.0	0.0	0.0[a]
Hirudinea	0.1	0.9	1.5	0.0	0.6	0.5
Arthropoda	15.5	11.6	28.1	11.6	21.2	17.2
Insecta	15.4	10.2	25.6	11.4	21.0	16.6
Chironomidae	15.0	7.6	24.0	9.9	19.6	15.3

Source: Modified from Diaz et al. 1978.
[a] Less than 0.03 percent of total.

a constant frequency of occurrence throughout the study period. Highest macrofaunal densities were reported from the lower intertidal marsh and subtidal channels at Windmill Point (Table 36.3). Highest densities occurred in summer and lowest in winter. A comparison of the Windmill Point marsh fauna to a nearby open-water subtidal bottom in the James River revealed similar community composition between marsh and open water habitats, numerical dominants in the open water being *Limnodrilus* spp., *Ilyodrulus tempeltoni, Corbicula fluminea,* and *Coelotanypus scapularis.* In general, taxa which were

TABLE 36.3. Descriptive Statistics for Macrobenthic Community Structure Parameters for the Windmill Point and Herring Creek, James River, Virginia, Tidal Freshwater Marshes[a]

		Number of Individuals		Number of Species		Diversity (H')	
Stratum	Date	Mean	SE	Mean	SE	Mean	SE
High marsh—cattail (*Typha*) zone							
WP	July	63	26	4.4	0.7	1.21	0.22
	November	40	14	4.6	0.6	1.51	0.11
	January	9	4	2.2	0.6	0.80	0.32
	April	51	14	2.7	0.5	0.54	0.15
	July	94	77	4.1	1.2	1.15	0.30
HC	July	54	16	6.6	1.0	1.85	0.25
	November	41	14	7.2	0.8	2.02	0.16
	January	45	10	9.0	1.0	2.45	0.13
	April	56	20	5.8	0.7	2.00	0.18
	July	93	20	10.3	1.6	2.30	0.19
Low marsh—pickerelweed, arrow arum, and arrowhead zone							
WP	July	267	88	7.6	0.6	1.68	0.12
	November	90	11	7.2	0.4	1.59	0.08
	January	27	5	3.1	0.1	1.35	0.06
	April	54	13	3.6	0.5	1.33	0.07
	July	221	27	7.4	0.8	1.83	0.11
HC	July	56	18	7.1	1.0	1.72	0.12
	November	40	7	7.1	0.8	2.13	0.22
	January	122	25	7.5	0.7	2.12	0.06
	April	49	11	7.1	0.8	2.05	0.13
	July	60	12	8.8	0.7	2.22	0.08
Low intertidal mudflat—unvegetated							
WP	July	74	9	7.9	0.7	2.12	0.15
	November	46	5	5.1	0.5	1.57	0.11
	January	7	3	3.0	1.5	0.74	0.24
	April	7	4	3.9	0.7	1.11	0.35
	July	32	4	6.3	0.6	2.09	0.16
HC	July	23	4	5.1	0.8	2.22	0.80
	November	20	2	6.5	0.4	2.31	0.12
	January	42	5	8.0	0.5	2.52	0.14
	April	38	14	4.9	0.6	1.82	0.17
	July	65	11	9.6	0.7	2.48	0.17

commonly encountered in the marsh but rare in open water were chironomids (except *Coelotanypus scapularis*), tipulids, tabanids, and ceratopogonids (Table 36.4).

The meiobenthos at Windmill Point was composed mostly of nematodes, tardigrades, cladocerans, ostracods, and copepods (Table 36.5). Highest meio-

TABLE 36.3. (Continued)

Stratum	Date	Number of Individuals		Number of Species		Diversity (H')	
		Mean	SE	Mean	SE	Mean	SE
Subtidal marsh channel—unvegetated							
WP	July	125	18	5.9	0.5	1.61	0.19
	November	68	11	6.6	0.6	1.62	0.18
	January	36	16	4.0	0.8	1.37	0.18
	April	165	37	6.5	0.5	2.01	0.08
	July	160	16	7.6	0.6	1.90	0.11
HC	July	19	5	4.3	0.8	1.48	0.32
	November	21	5	6.5	0.6	2.21	0.09
	January	64	19	9.8	1.1	2.41	0.17
	April	19	4	5.8	0.8	2.02	0.22
	July	27	4	8.0	0.7	2.52	0.13

Source: Modified from Diaz et al. 1978.

[a] WP = Windmill Point, created marsh. HC = Herring Creek, natural marsh. Seasonal sampling for 1976–1977. All values are mean and standard error, based on eight replicate 160-cm^2 cores.

faunal densities were reported from the low intertidal marsh at Windmill Point, dominated by pickerelweed, arrow arum, and arrowhead, and from the high marsh at Herring Creek, dominated by *Typha* (Table 36.6). Deposit-feeding nematodes (Monohysteridae) occurred at all locations and were the most abundant taxon. Predatory and omnivorous nematodes (Dorylaimidae and *Anatonchus* sp.) were present at all locations, but were less abundant than the deposit-feeders. Tardigrades were most abundant in the upper intertidal marsh. Cladocerans were common in lower intertidal habitats but were absent from high marsh collections. *Ilyocryptus* was the most commonly encountered genus. Ostracods occurring at Windmill Point included *Candona* sp., *Cypridopsis* sp., and *Physocypria* sp. *Darwinula stevensoni* was common at the reference sites only. Cyclopoid copepods were more abundant than harpacticoids; both free-swimming and benthic forms were encountered, including *Paracyclops affinis* and *P. fimburiatus*. Harpacticoids were mostly of the genus *Canthocamptus* (Munson 1980).

Analysis of fish dietary habits in conjunction with the Windmill Point study demonstrated the numerical importance of small food items, principally meiofaunal crustaceans, in the diets of resident marsh fishes (Dias et al. 1978). Spottail shiners (*Notropis hudsonius*), creek chubsuckers (*Erimyzon oblongus*), juvenile channel catfish (*Ictalurus punctatus*), mummichogs (*Fundulus heteroclitus*), and juvenile white perch (*Morone americana*) all preyed heavily upon copepods, ostracods, cladocerans, and early-instar chironomid larvae (Table 36.7). Oligochaetes in the fish stomachs were in conspicuously low numbers, despite their abundance in the community and potential availability. However, since they lack an exoskeleton or hard integument, they were likely being consumed and digested rapidly. Oligochaete setae were often present

TABLE 36.4. Macrobenthic Taxa Found in Tidal Freshwater Marshes of the James River, Virginia

Phylum	Class	Order	Family	Genus Species
Platyhelminthes	Turbellaria		Plagiostomidae	*Hydrolimax grisea* Haldeman
			Planaridae	*Cura foremanii* (Girard)
Nemertea				*Prostoma rubrum* (Leidy)
Mollusca	Bivalvia		Corbiculidae	*Corbicula fluminea* (Phillippi)
			Sphaeriidae	*Sphaerium transversum* (Say)
				Pisidium sp.
	Gastropoda		Unionidae	*Elliptio complanata* Lightfoot
			Physidae	*Physa* sp.
			Lymnaeidae	*Lymnaea stagnalis* (Linnaeus)
			Planorbidae	*Gyraulus* sp.
			Ancylidae	*Ferrissia* sp.
			Pomatiopsidae	*Pomatiopsis* sp.
Annelida	Polychaeta		Sabellidae	*Manyunkia speciosa* Leidy
	Oligochaeta		Tubificidae	*Tubifex* sp.
				Aulodrilus limnobius Bretscher
				Aulodrilus pigueti Kowalewski
				Branchiura sowerbyi Beddard
				Ilyodrilus templetoni (Southern)
				Limnodrilus spp.
				Limnodrilus cervix Brinkhurst
				Limnodrilus hoffmeisteri Claparede
				Limnodrilus udekemianus Verrill
				Quistadrilus (*Peloscolex*) *multisetosus* Brinkhurst
				Isochaetides (*Peloscolex*) *freyi* Brinkhurst
				Peloscolex ferox
				Potamothrix nr. *hammonensis* Michaelsen

Phylum	Class	Order	Family	Species
			Naididae	*Chaetogaster* sp.
				Nais spp.
				Nais variabilis Piguet
				Dero digitata (Muller)
				Dero furcatus (Muller)
				Slavenia appendiculata d'Udekem
				Stylaria lacustris (Linnaeus)
				Enchytraeus sp.
	Hirudinea		Enchytraeidae	*Helobdella elongate* (Castle)
			Lumberliculidae	*Helobdella stagnalis* (Linnaeus)
			Piscicolidae	*Helobdella puntatalineata* Moore
				Batracobdella phalera Graf
Arthropoda	Arachnida	Spiders		
	Crustacea	Isopoda	Asellidae	*Caecidotea* (*Asellus*) sp.
		Amphipoda	Gammaridae	*Gammarus fasciatus* Say
			Hyalellidae	*Hyalella azteca* (Saussure)
	Insecta	Collembola	Isotomidae	
			Sminthuridae	
		Ephemeroptera	Ephemeridae	*Hexagenia mingo* Walsh
			Caenidae	*Caenis* sp.
			Ephemerellidae	*Ephemerella* sp.
		Odonata	Gomphidae	*Gomphus* sp.
			Protoneuridae	*Anomalagrion* sp.
			Libellulidae	*Perithemis* sp.
		Trichoptera	Leptoceridae	*Oecetis* sp.
		Hemiptera	Corixidae	*Trichocorixa* sp.
			Veliidae	*Velia* sp.
			Mesoveliidae	*Mesovelia* sp.
			Tipulidae	*Helius* sp.
				Tipula sp.
		Diptera	Chaoboridae	*Chaoborus punctipennis* (Say)
			Tabanidae	*Chrysops* sp.
				Anacimas sp.

TABLE 36.4. (Continued)

Phylum	Class	Order	Family	Genus Species
			Chironomidae	*Ablabesmyia* sp. E
				Ablabesmyia nr. *monilis*
				Chironomus spp.
				Coelotanypus concinnus
				Coelotanypus scapularis (Loew)
				Cryptochironomus spp.
				Cryptochironomus fulvus
				Dicrotendipes spp.
				Dicrotendipes nervosus (Staeg.)
				Einfeldia sp.
				Endochironomus sp.
				Glyptotendipes sp.
				Harnischia sp.
				Polypedilum spp.
				Procladius bellus (Loew)
				Pseudochironomus sp.
				Rheotanytarsus sp.
				Stictochironomus devinctus (Say)
				Cryptocladopelma sp.
				Tanypus spp.
				Tanypus neopunctipennis
				Tanytarsus sp.
				Trichocladius sp.
				Lauterborniella sp.
				Cricotopus sp.
			Ceratopogonidae	*Palpomyia* sp.
			Dolichopodidae	*Argyra* sp.
				Hydrophorus sp.
		Coleoptera	Chrysomelidae	*Donacia* sp.

Source: From Diaz et al. 1978, unpublished data.

TABLE 36.5. Meiofaunal Taxa Found in Tidal Freshwater Marshes of James River, Virginia

Phylum	Class	Order	Family	Genus Species
Aschelminthes	Nematoda	Monhysteridae	Monhysteridae	*Monhystera* sp.
				Monhystrella spp.
				Prismatolaimus sp.
		Dorylaimida	Alaimoidae	*Alaimus* sp.
			Dorylaimidae	*Dorylaimus* sp.
				Mesodorylaimus sp.
				Amphidorylaimus sp.
				Thornenema sp.
			Mononchidae	*Anatonchus* sp.
			Bathyodontidae	*Alaimus* sp.
		Araeolaimida	Plectidae	*Paraplectonema* sp.
			Camaculaimidae	*Aphalolaimus* sp.
		Tylenchida	Tylenchidae	*Tylenchus* sp.
			Aphelenchididae	*Aphelenchoides* sp.
		Enoplida	Dorylaimidae	*Dorylaimus* sp.
				Mesodorylaimus sp.
			Alaimoidae	*Alaimus* sp.
Annelida	Polychaeta		Sabellidae	*Manyunkia* sp.
Tardigrada	Eutardigrada		Macrobiotidae	*Isohypsibius saltursus* Schuster et al.
				Macrobiotus richtersii J. Murray
				Macrobiotus dispar J. Murray
				Macrobiotus furcatus Ehrenberg
				Macrobiotus hufelandii S. Schultze
	Heterotardigrada		Scutechiniscidae	*Hypsibius* sp.
				Echiniscus sp.

907

TABLE 36.5. (Continued)

Phylum	Class	Order	Family	Genus Species
Arthropoda	Crustacea	Cladocera	Sididae	*Sida crystallina* O. F. Muller
				Latona setifera O. F. Muller
				Diaphanosoma sp.
			Daphnidae	*Moina micrura* Kurz
				Ceriodaphnia sp.
			Bosminidae	*Bosmina longirostris* O. F. Muller
			Macrothricidae	*Ilyocryptus* spp.
				Macrothrix sp.
			Chydoridae	*Diaphanosoma agilis* Fischer
				Kurzia latissima Kurz
				Leydigia leydigi Leydia
				Leydigia acanthocercoides Fischer
				Alona costata Sars
				Alona affinis Leydig
				Alona quadrangularis O. F. Muller
				Pleuroxus denticulatus Birge
				Chyodorus sphaericus O. F. Muller
	Copepoda	Cyclopoida	Cyclopidae	*Eucyclops agilis* Koch
				Eucyclops spearatus Lilljeborg
				Paracyclops affinis Sars
				Paracyclops fimbriatus Fischer
				Macrocyclops fuscus Jurine
				Halicyclops magniceps Lilljeborg
				Mesocyclops edax S.A. Forbes
				Cyclops vernalis Fischer

Harpacticoida	Canthocamptidae	*Canthocamptus staphlinoides* Pearse
		Canthocamptus robertcokeri M. S. Wilson
		Canthocamptus sp.
		Bryocamptus sp.
		Moraria sp.
Ostracoda	Cyclocypridae	*Physocypria* sp.
	Cypridae	*Physocypria* sp.
		Cypridopsis sp.
		Candona sp.
	Darwinulidae	*Darwinula stevensoni* Brady and Robertson
	Cytherideidae	*Ilyocypris gibba* (Ramdohr)
	Cypridopsidae	*Cypridopsis vidua* O. F. Muller

Source: From Diaz et al. 1978, Munson 1980, and Yozzo and Smith 1995.

TABLE 36.6. Meiofauna Species and Abundance from the Windmill Point and Herring Creek, James River, Virginia, Tidal Freshwater Marshes[a]

Marsh Zone	Marsh	Taxa	Abundance
High marsh—Cattail	WP	15	294
(*Typha*)	HC	20	332
Low marsh—pickerelweed,	WP	11	462
arrow arum, arrowhead	HC	19	269
Low intertidal mudflat—	WP	19	325
unvegetated	HC	19	244
Subtidal marsh channel—	WP	16	301
unvegetated	HC	14	232

Source: Modified from Diaz et al. 1978 and Munson 1980.
[a] WP = Windmill Point, created marsh. HC = Herring Creek, natural marsh. Sampled July 1977. All values are total taxa and individuals for 30 cm², 8 replicate 3.75 cm² × 5 cm long cores.

in fish stomachs, supporting this contention, but there was no way to quantify their trophic importance.

Assessment of the dietary habits of shorebirds was conducted as part of the Windmill Point study. Over 20 species of birds were observed to feed principally on aquatic invertebrates in the lower intertidal zone, including semipalmated sandpipers (*Calidris pusillus*) and western sandpipers (*Calidris mauri*). Primary food items included oligochaetes and insect larvae. Common snipe (*Capella gallnago*) preyed upon the snail *Physa*, insect larvae, and oligochaetes (Wass and Wilkens 1978).

Yozzo and Steineck (1994) report on the ostracod fauna of tidal freshwater marshes at Stockport, Hudson River, New York. A variety of habitat types were sampled, including emergent low marsh, unvegetated shallows, and floating aquatic macrophyte beds. Both sediment-dwelling and epiphytic ostracods were collected. The fauna was dominated by one species (*Physocypria* sp.), and several other widely distributed taxa were common (*Darwinula stevensoni, Candona* sp. and *Cypridopsis vidua*). Several additional taxa were represented by the occurrence of a few specimens, including *Candona caudata, Limnocythere* sp., and *Ilyocypris gibba*. Although comprehensive surveys of freshwater ostracods are rare, the results of this study suggest that ostracod faunas of tidal freshwater wetlands may be species-poor relative to those of nontidal freshwater habitats. Intensive, repeated collections from other tidal freshwater wetlands might reveal additional species, but it seems likely that tidal freshwater ostracod communities are primarily comprised of a few broadly distributed taxa.

Yozzo and Smith (1995) studied seasonal abundance patterns and microhabitat distribution patterns of meiofauna at a tidal freshwater marsh on the Chickahominy River, a subestuary of Chesapeake Bay, Virginia. Meiofauna were sampled from intertidal pools and vegetative (*Peltandra virginica*) hummocks at both upper and lower intertidal locations. Nematodes, ostracods, tardigrades, oligochaetes (Naididae), copepods (Harpacticoida and Cyclopoida), and the sabellid polychaete *Manyunkia* were numerically dominant in monthly collections (Fig. 36.3). Less commonly occurring taxa included rotifers, turbellarians, water mites (Hydracarina), early-instar chironomid larvae (temporary meiofauna), and cladocerans (mostly *Bosmina longirostris*). Bryozoan statoblasts were commonly observed in samples collected from intertidal pools.

Highest densities of meiofauna were reported during August–September, coincident with recruitment of oligochaetes and *Manyunkia* sp. to the surfaces of low marsh hummocks. Diaz et al. (1978) report this species as occurring infrequently at Windmill Point, Virginia. However, they may have missed the seasonal recruitment event, since meiofauna were sampled at only one time during the Windmill Point study. Nematodes were significantly more abundant on the surfaces of *Peltandra* hummocks, perhaps in association with microoxygenated zones occurring within the dense, fibrous root mat of *Peltandra*. Previous workers in salt marshes (Osenga and Coull 1983) have reported a similar association between nematodes and *Spartina* clumps in southeastern

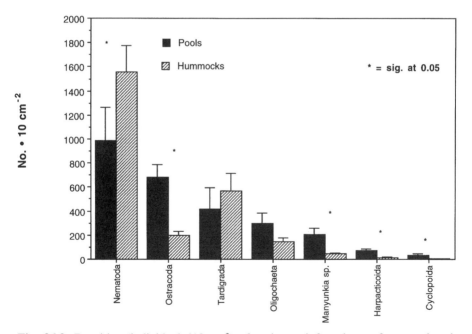

Fig. 36.3. Densities (individuals/10 cm²) of major meiofaunal taxa from pool and plant hummock microhabitats, Eagle Bottom Marsh, Chickahominy River, Virginia. (Modified, with kind permission from Kluwer Academic Publishers, from D. J. Yozzo and D. E. Smith, Seasonality, abundance, and microhabitat distribution of meiofauna from a Chickahominy River, Virginia freshwater marsh, Hydrobiologia 310:197–206, Fig. 2. © 1995.)

U.S. salt marshes. Ostracods were significantly more abundant in intertidal pools. Greatest densities of ostracods occurred in May. The taxonomic composition of ostracods at Eagle Bottom Marsh was quite similar to that reported from Hudson River tidal freshwater marshes (Yozzo and Steineck 1994) and included *Physocypria* sp., *Candona* sp., *Darwinula stevensoni*, *Cypridopsis vidua*, and *Ilyocypris gibba*. Tardigrades were significantly more abundant on *Peltandra* hummocks and were generally more common in winter collections. Copepods were common, although never particularly abundant, in both pool and hummock samples. This is in sharp contrast to salt marshes, where harpacticoid copepods are a numerically dominant component of meiofaunal communities (Coull 1985, Fleeger 1985) and are a primary food item for juvenile estuarine fishes. Both cyclopoids (*Eucyclops agilis*) and harpacticoids (*Canthocamptus* sp., *Bryocamptus* sp.) were collected at Eagle Bottom Marsh. Female harpacticoids were gravid during winter months; female cyclopoids bore eggs throughout the year. Harpacticoid densities decreased from May to October at both low and high marsh sites. Cyclopoid densities were more variable throughout the sampling period, but lowest densities were recorded in June–July, coincident with the period of greatest abundance of larval and juvenile fishes on the intertidal marsh surface.

Another potential predator on tidal freshwater marsh meiofauna is the grass shrimp *Palaemonetes pugio*. Grass shrimp are seasonally abundant on the surface of Virginia tidal freshwater marshes in late summer and fall. As submerged vegetation beds in adjacent tidal creeks die off, grass shrimp emigrate from the subtidal creeks up onto the intertidal marsh surface, seeking refuge from predation themselves. At this time grass shrimp may potentially affect the densities of intertidal freshwater meiofauna. In a southeastern U.S. salt marsh Bell and Coull (1978) found that predation effects by *P. pugio* were significant in influencing the densities of oligochaetes, polychaetes, and harpacticoid copepods in intertidal habitats.

Yozzo and Smith (1995) also report briefly on the dynamics and species composition of benthic macrofauna on tidal freshwater marsh surfaces. The amphipod *Gammarus fasciatus* is numerically dominant in the intertidal and subtidal. The isopods *Caecidotea* sp. and *Cyathura polita* (a widely distributed estuarine form) were commonly observed in intertidal microhabitats, and along with *G. fasciatus* were often collected in small fish traps placed on the marsh surface. Juvenile blue crabs (*Callinectes sapidus*) and mud crabs (*Rithropanopeus harrisi*) were frequently observed in intertidal pools and rivulets on the marsh surface. Rozas and Odum (1987a) report that *R. harrisi* was abundant in submerged vegetation beds contiguous with marshes in the Chickahominy River. Additional macrofaunal taxa from Virginia tidal freshwater marshes include the amphipod *Corophium* sp., oligochaetes (Tubificidae, Enchytraidae), leeches (*Placobdella ornata, Mooreobdella* sp.), dipteran larvae (Chironomidae, Ceratopogonidae), gastropods (*Physa integra, Gyraulus parvus*), bivalves (*Pisidium* sp.), megalopterans (*Sialis* sp.), and odonates (*Somatochlora* sp., *Enallagma* spp.). A number of these taxa are known predators on meiofauna and smaller macrofauna, especially *Sialis* and the Odonata.

Although most of our information on invertebrate faunas of tidal freshwater marshes is from either Hudson River or James River systems, some information is available from other East Coast systems. McCormick (1970) reports that aquatic beetles, isopods, and snails were commonly observed at Tinicum Marsh, near Philadelphia. Springtails were the most commonly observed insect in and around debris at the low-water mark. Fingernail clams were common in tidal creeks, and large masses of oligochaetes (Tubificidae) were present in shallow subtidal areas. Earthworms (*Ophisthopora* sp.) were common in the intertidal marsh, and the leeches *Glossiphonia coplanata* and *Erpobdella punctata* were collected, along with the snail *Lioplax* sp.

In a companion study at Tinicum Marsh, Grant and Patrick (1970) report high invertebrate diversity, listing such taxa as damselflies and dragonflies, various beetles, water striders, water scorpions, and dipteran larvae. Freshwater shrimp, colonial bryozoans, and flatworms were abundant. Snails (*Physa*), amphipods, isopods, fingernail clams, freshwater mussels, a sponge, leeches, crayfish, and various worms were moderately abundant in the streams draining the intertidal marsh.

Simpson et al. (1979) investigated the terrestrial insect community associated with emergent vegetation in a Delaware River, New Jersey, tidal freshwater marsh. Insects were sampled among three species of marsh vegetation (bur marigold, arrow arum, and jewelweed). In comparison to the case in some terrestrial plant-insect associations, species diversity and insect density was generally low. Ladybird beetles (Coccinellidae) were abundant among all three plant types. Other commonly occurring taxa included snout beetles (Curculionidae), fireflies (Lampyridae), lizard beetles (Lagnuridae), long-legged flies (Dolichopodidae), picture-winged flies (Otitidae), flower flies (Syrphidae), deer flies (Tabanidae), minute pirate bugs (Anthocoridae), aphids (Aphidae), and leafhoppers (Cicadellidae). These investigators found little evidence of insect herbivory and concluded that tidal freshwater marshes are probably only lightly grazed. The trophic significance of terrestrial insects in tidal freshwater marshes remains unstudied.

Macrobenthos of the Miller Sands system, in the Pacific Northwest, are overwhelmingly dominated by the amphipod *Corophium salmonis,* tubificid and enchytraeid oligochaetes, chironomid larvae, and the Asiatic clam (Fig. 36.2). Highest faunal densities occurred in the fall and winter. Heavy predation from fish in the spring and summer may be responsible for the decline in benthic fauna. Stomach content analysis of the four numerically dominant fish (starry flounder, peamouth, three-spine stickleback, Chinook salmon) indicated that the principal food items were cladocerans (*Daphnia longispina*), copepods (*Eurytemora hirundoides*), amphipods (*Corophium salmonis*), and chironomid pupae and larvae (Clairain et al. 1978).

SUMMARY AND SUGGESTIONS FOR FUTURE RESEARCH

Tidal freshwater wetlands are highly productive ecosystems which combine functional, structural, and faunal characteristics of estuarine intertidal marshes and nontidal, freshwater wetlands. Although they occur on all three coasts of the United States, the vast majority of our information on these wetlands is derived from sites in the Northeast and mid-Atlantic regions. The invertebrate fauna of tidal freshwater wetlands is composed mostly of freshwater species, although a few euryhaline estuarine species are able to colonize these transitional wetland environments. Species richness is moderate to low, depending on the specific habitat type and perhaps geographic location. For example, the species richness for oligochaetes at Miller Sands, Oregon, appears to be at least twice that of Atlantic coast tidal freshwater wetlands. Unfortunately, broad generalizations are difficult to make, given the very limited geographic coverage of studies conducted thus far.

Perhaps the most significant function of tidal freshwater wetlands, and the one which has received the most attention, is provision of nursery habitat for juvenile fishes and macrocrustaceans. Both resident and seasonally migrating fishery species use tidal freshwater wetlands extensively, as a predation refuge and feeding ground, and benefit from the abundance of epiphytic and benthic invertebrate prey resources.

A variety of habitat types occur in tidal freshwater wetlands, including submerged and floating aquatic macrophyte beds, emergent intertidal marshes, unvegetated mudflats, and tidal creeks. Each of these may provide critical nursery or refuge habitat for resident and migratory juvenile fishes and macrocrustaceans. For example, larval and juvenile killifishes, juvenile blue crabs, and grass shrimp are abundant in shallow pools and intertidal rivulets on the marsh surface; larger, predatory species (e.g., black bass, catfish, chain pickerel) are common in deeper, subtidal creeks and open waters adjacent to the marsh.

In general, the epiphytic and benthic invertebrate faunas associated with submerged or floating macrophytes in tidal freshwaters are dominated by aquatic insect larvae, especially chironomids and other Diptera, and oligochaetes (mostly Naididae). Gastropods and microcrustaceans (ostracods, copepods, cladocerans) are also commonly associated with submerged or floating macrophytes in tidal freshwaters.

Benthic macroinvertebrate communities on the intertidal freshwater marsh surface are dominated by tubificid and enchytraid oligochaetes along with chironomids. Gammaridean amphipods and isopods (both freshwater and brackish species) are abundant in shallow marsh pools and intertidal rivulets. Other taxa commonly found on the intertidal marsh surface include odonate larvae, gastropods, and leeches. Adult insects, especially beetles and leafhoppers, may be seasonally abundant among intertidal emergent vegetation.

Meiofaunal communities of the intertidal freshwater marsh surface are dominated by nematodes, tardigrades, harpacticoid and cyclopoid copepods, ostracods, naidid oligochaetes, and the polychaete *Manyunkia*. Distinct seasonal peaks in abundance of the dominant meiofaunal taxa are evident, and microtopographic features are strong determinants of local abundance and distribution patterns.

Suggestions for Future Research

Future research in this area should emphasize faunal comparisons between tidal freshwater wetlands in the Northeast and mid-Atlantic (from which most of our information is derived) and other areas such as the southeast Atlantic coast, the northeast Gulf of Mexico, and the Pacific Northwest. Information currently available suggests that geographically disparate tidal freshwater wetlands function similarly, but intensive studies are needed to further elucidate the role of invertebrate populations in supporting fishery species.

The invertebrate fauna of forested tidal freshwater wetlands (tidal swamps) should be documented and comparison should be made to that of intertidal marshes and adjacent terrestrial habitats. These adjacent habitats are linked via mobile organisms and tidal exchange; such linkages are thought to be of considerable importance in the structure and function of coastal ecosystems (e.g., seagrass-mangrove-coral reef). Population dynamics, secondary production and trophic interactions of invertebrates in submerged aquatic macrophyte beds, intertidal freshwater marshes, and tidal swamps should be investigated in order to improve our understanding of the structure, functional significance, and interconnectedness of wetland habitats in tidal freshwaters.

Despite the potential importance of tidal freshwater habitats, we still know very little regarding the structure and function of tidal freshwater wetland invertebrate communities, especially in comparison to other systems such as salt marshes and nontidal freshwater wetlands, which have received intensive study over several decades. Understanding the role of invertebrate communities in maintaining trophic links and the apparent ability of these communities to withstand both natural and man-induced environmental perturbations will benefit efforts to manage and preserve tidal freshwater wetlands. Recent initiatives to restore and/or create tidal freshwater wetlands in large Atlantic Coast estuaries (e.g., Hudson River, Delaware Bay, upper Chesapeake Bay) can benefit from a detailed understanding of the role of invertebrate communities in support of higher consumers (fish, herpetofauna, avifauna). This information can be critically important in the formulation of ecologically sound and technically feasible goals and objectives for tidal freshwater wetland restoration projects. This is certainly an exciting area for future research.

ACKNOWLEDGMENTS

This chapter is Contribution 2119 of the Virginia Institute of Marine Science.

LITERATURE CITED

Bell, S. S., and B. C. Coull. 1978. Field evidence that shrimp predation regulates meiofauna. Oecologia 35:141–148.

Clairain, E. J., R. A. Cole, R. J. Diaz, A. W. Ford, R. T. Huffman, L. J. Hunt, and B. R. Wells. 1978. Summary Report. Habitat Development Field Investigations Miller Sands Marsh and Upland Habitat Development Site, Columbia River, Oregon. Technical Report TR D-77-38. U. S. Army Engineer Waterways Experiment Station, Vicksburg, MS.

Coull, B. C. 1985. Long-term variability of estuarine meiobenthos: An 11 year study. Marine Ecology Progress Series 24:205–218.

Crumb, S. E. 1977. Macrobenthos of the tidal Delaware River between Trenton and Burlington, New Jersey. Chesapeake Science 18:253–265.

Dias, R. K., J. V. Merriner, and M. Hedgepeth. 1978. Aquatic biology—nekton. Pages 55–78 in Habitat Development Field Investigations Windmill Point Marsh Development Site James River, Virginia. Appendix D: Environmental Impacts of Marsh Development with Dredged Material: Botany, Soils, Aquatic Biology and Wildlife. Technical Report TR D-77-23. U.S. Army Engineer Waterways Experiment Station, Vicksburg, MS.

Diaz, R. J. 1978. Ecology of tidal freshwater and estuarine Tubificidae (Oligochaeta). Pages 319–330 in R. O. Brinkhurst and D. G. Cook (eds.), Aquatic Oligochaete Biology. Plenum Press, New York.

Diaz, R. J. 1982. Examination of Tidal Flats. Vol. 2. A Review of Identified Values. FHWA/RD-80/182, Federal Highway Administration, Washington, DC.

———. 1989. Pollution and tidal benthic communities of the James River Estuary, Virginia. Hydrobiologia 180:195–211.

———. 1994. Response of tidal freshwater macrobenthos to sediment disturbance. Hydrobiologia 278:201–212.

Diaz, R. J., and D. F. Boesch. 1977. Habitat Development Field Investigations Windmill Point Marsh Development Site James River, Virginia. Appendix C: Environmental Impacts of Marsh Development with Dredged Material: Acute Impacts on the Macro-benthic Community. Technical Report TR D-77-23. U.S. Army Engineer Waterways Experiment Station, Vicksburg, MS.

Diaz, R. J., D. F. Boesch, J. L. Hauer, C. A. Stone, and K. Munson. 1978. Part II, Aquatic biology—benthos. Pages 18–54 in Habitat Development Field Investigations Windmill Point Marsh Development Site James River, Virginia. Appendix D: Environmental Impacts of Marsh Development with Dredged Material: Botany, Soils, Aquatic Biology and Wildlife. Technical Report TR D-77-23. U.S. Army Engineer Waterways Experiment Station, Vicksburg, MS.

Ettinger, W. S. 1982. Macrobenthos of the freshwater tidal Schuylkill River at Philadelphia, Pennsylvania. Journal of Freshwater Ecology 1:599–606.

Findlay, S., K. Schoerberl, and B. Wagner. 1989. Abundance, composition, and dynamics of the invertebrate fauna of a tidal freshwater wetland. Journal of the North American Benthological Society 8:140–148.

Fleeger, J. W. 1985. Meiofaunal densities and copepod species composition in a Louisiana, USA, estuary. Transactions of the American Microscopical Society 104:321–332.

Grant, R. R., and R. Patrick. 1970. Tinicum Marsh as a water purifier. Pages 105–123 *in* J. McCormick, R. R. Grant, and R. Patrick (eds.), Two Studies of Tinicum Marsh, Delaware and Philadelphia Counties, Pennsylvania. Conservation Foundation, Washington, DC.

Leck, M. A., and R. L. Simpson. 1987. Seed bank of a freshwater tidal wetland: turnover and relationship to vegetation change. American Journal of Botany 74: 360–370.

McCormick, J. 1970. The natural features of Tinicum Marsh, with particular emphasis on the vegetation. Pages 1–104 *in* J. McCormick, R. R. Grant, and R. Patrick (eds.), Two Studies of Tinicum Marsh, Delaware and Philadelphia Counties, Pennsylvania. Conservation Foundation, Washington, DC.

McIvor, C. C., and W. E. Odum. 1988. Food, predation risk, and microhabitat selection in a marsh fish assemblage. Ecology 69:1341–1351.

Menzie, C. A. 1980. The chironomid (Insecta: Diptera) and other fauna of a *Myriophyllum spicatum* L. plant bed in the lower Hudson River. Estuaries 3:38–54.

Munson, K. E. 1980. A study of the distribution and habitat preference of meiobenthos of a freshwater tidal marsh on the James River, Virginia. Master's Thesis, George Washington University, Washington, DC.

Odum, W. E. 1988. Comparative ecology of tidal freshwater and salt marshes. Annual Review of Ecology and Systematics 19:147–176.

Odum, W. E., and M. A. Heywood. 1978. Decomposition of intertidal freshwater marsh plants. Pages 89–97 *in* R. E. Good, D. F. Whigham, and R. L. Simpson (eds.), Freshwater Wetlands: Ecological Processes and Management Potential. Academic Press, New York.

Odum, W. E., M. L. Dunn, and T. J. Smith, III. 1979. Habitat value of tidal freshwater wetlands. Pages 248–255 *in* P. E. Greeson, J. R. Clark, and J. E. Clark (eds.), Wetland Functions and Values: The State of our Understanding. American Water Resources Association, Minneapolis, MN.

Odum, W. E., T. J. Smith, III, J. K. Hoover, and C. C. McIvor. 1984. The Ecology of Tidal Freshwater Marshes of the United States East Coast: A Community Profile. FWS/OBS-87/17. Fish and Wildlife Service, U.S. Department of the Interior, Washington, DC.

Osenga, G. A., and B. C. Coull. 1983. *Spartina alterniflora* Loisel root structure and meiofaunal abundance. Journal of Experimental Marine Biology and Ecology 67: 221–224.

Ristich, S. S., M. Crandall, and J. Fortier. 1977. Benthic and epibenthic macroinvertebrates of the Hudson River, I. Distribution, natural history and community structure. Estuarine, Coastal, and Shelf Science 5:255–266.

Rozas, L. P., and W. E. Odum. 1987a. Fish and macrocrustacean use of submerged plant beds in tidal freshwater marsh creeks. Marine Ecology Progress Series 38: 101–108.

———. 1987b. Use of tidal freshwater marshes by fishes and macrofaunal crustaceans along a marsh stream-order gradient. Estuaries 10:36–43.

———. 1988. Occupation of submerged aquatic vegetation by fishes: Testing the roles of food and refuge. Oecologia 77:101–106.

Simpson, R. L., D. F. Whigham, and K. Brannigan. 1979. The mid-summer insect communities of freshwater tidal wetland macrophytes, Delaware River Estuary, NJ. Bulletin of the New Jersey Academy of Science 24:22–28.

Simpson, R. L., R. E. Good, M. A. Leck, and D. F. Whigham. 1983. The ecology of freshwater tidal wetlands. BioScience 33:255–259.

Wass, M., and E. Wilkens. 1978. Wildlife resources. Pages 89–102 *in* Habitat Development Field Investigations Windmill Point Marsh Development Site James River, Virginia. Appendix D: Environmental Impacts of Marsh Development with Dredged Material: Botany, Soils, Aquatic Biology and Wildlife. Technical Report TR D-77-23. U.S. Army Engineer Waterways Experiment Station, Vicksburg, MS.

Williams, J. P. 1992. Exploring the Chesapeake in Small Boats. Tidewater, Publishers, Centreville, MD.

Yozzo, D. J., and W. E. Odum. 1993. Fish predation on epiphytic microcrustacea in Tivoli South Bay, a Hudson River tidal freshwater wetland. Hydrobiologia 257:37–46.

Yozzo, D. J., and D. E. Smith. 1995. Seasonality, abundance, and microhabitat distribution of meiofauna from a Chickahominy River, Virginia tidal freshwater marsh. Hydrobiologia 310:197–206.

Yozzo, D. J., and P. L. Steineck. 1994. Ostracoda from tidal freshwater wetlands at Stockport, Hudson River Estuary: Abundance, distribution, and composition. Estuaries 17:680–684.

37 Bayous of the Northern Gulf of Mexico

Distribution and Trophic Ecology of Invertebrates

GARY R. GASTON

*L*ow-salinity streams that flow through northern Gulf of Mexico swamps are referred to as bayous. *They differ from tidal-freshwater streams of the Atlantic coast by their physical attributes, but are distinguished most by their aquatic macrophytes and associated fauna. The invertebrates of vegetated bayous sometimes occur in numbers that exceed density and diversity of adjacent freshwater and estuarine habitats. This faunal diversity of bayous contradicts a well-established paradigm that brackish habitats have the lowest species diversity of all aquatic ecosystems. Sediments below floating macrophytes and unvegetated bayous are dominated by oligochaetes and chironomids, which occur in densities similar to those in tidal-freshwater streams of the Atlantic coast.*

Bayou trophic ecology reflects local variations in habitat heterogeneity. Greatest trophic diversity occurs in vegetated regions, where motile detritivores (especially amphipods, isopods, and snails) dominate. Subsurface-deposit feeders (burrowers) are most abundant in unvegetated bayous. Few suspension feeders inhabit bayous, perhaps because tidal currents necessary to deliver nutrients are weak and sediments are too loosely consolidated for habitation by sessile taxa. It appears that food is seldom limiting to bayou detritivores, which means we should focus on other factors in order to characterize their ecosystems.

Many bayous are contaminated or otherwise impacted by activities of humans. The rapid population growth along coastal regions and lack of fo-

Invertebrates in Freshwater Wetlands of North America: Ecology and Management, Edited by Darold P. Batzer, Russell B. Rader, and Scott A. Wissinger
ISBN 0-471-29258-3 © 1999 John Wiley & Sons, Inc.

cus on management of bayous have led to degraded conditions and an un-certain future for bayou ecosystems.

BAYOUS

Low-salinity streams that wind through swamps of the northern Gulf of Mexico are called *bayous,* pronounced "by yous" in Louisiana and "by ohs" elsewhere. They are broadly defined as small, sluggish tidal streams of river deltas. Bayous are often downstream components of Gulf Coast blackwater streams or drainages of swamps and shallow freshwater lakes. They represent an ill-defined portion of the continuum of habitats between low-gradient freshwater streams and open-water estuaries. The term *bayou* is most commonly used in Louisiana and Texas, but is applied widely to coastal streams along much of the Gulf Coast.

Bayous are typified by near-stagnant water and anoxic sediments. Sometimes the water is milky-white with sediments, but even without sediments the water is darkened with dissolved organic matter. Few invertebrates are endemic to bayous. Rather, they typically include a combination of estuarine fauna too far inland and freshwater fauna too far seaward. The plants that surround bayous are typical of freshwater swamp habitats, especially the swollen trunks of tupelo gum and bald cypress, stands of palmetto, and floating mats of duckweed and alligator weed. Thus, bayous elude our familiar classifications of either freshwater or estuarine. They are both. Blue crabs and largemouth bass, water hyacinths and duck potato, owls, egrets, and wood storks all inhabit bayous.

Louisiana has 78 percent of the freshwater wetlands of the continental United States (Sklar 1985, Dardeau et al. 1992) and most of the bayous of the Gulf Coast. The Gulf of Mexico receives 50 percent of the total discharge draining the contiguous United States, and much of it enters the Gulf via Louisiana estuaries (Isphording and Fitzpatrick 1992), so it is not surprising that much of the data concerning bayous were collected in Louisiana. Many Florida streams rapidly grade from blackwater streams or spring-fed streams to tidal creeks (Livingston 1992, Patrick 1994). The Mobile Bay delta region of Alabama includes extensive wetlands, swamps, and bayous, but the term is seldom used there (Bault 1972, Dardeau et al. 1990, Stout 1990). Many of the habitats referred to as bayous in Mississippi (McBee and Brehm 1979, Sullivan 1981, Peterson 1997) and Texas (Darville and Harrel 1980, Petrick 1981, Twidwell 1981) are actually tidal creeks.

Until recently, we knew little of the functional ecology of bayous, about the interactions of the invertebrates, or even about the densities of the most-abundant species. Our perceptions of bayous were based on research of Atlantic tidal-freshwater ecosystems, which many people assume to be similar to bayous. They are not. Bayous are unique. Low-salinity streams of the Gulf Coast have not been well studied, but it is apparent from the limited data available that they differ substantially from their Atlantic counterparts.

The purpose of this chapter is to review what we know about bayous and discuss the functional ecology of bayou invertebrates. We will explore the paradigms established for low-salinity habitats to see if they adequately characterize bayous, and propose new paradigms for several aspects of bayou ecosystems.

BACKGROUND

It is widely held that diversity of invertebrates decreases up-estuary to a minimum in low-salinity habitats, then increases again in freshwater streams (Ax and Ax 1970, Remane and Schlieper 1971, Boesch et al. 1976, McLusky 1981, Wetzel 1983, Day et al. 1989, Diaz 1989, 1994, Pennak 1989, Moore 1992). This pattern has been attributed to many interactive factors of the physical, chemical, and biological environment, such as a general lack of habitat diversity (Diaz and Boesch 1977, Diaz 1989, Mitsch and Gosselink 1993), salinity stress (Remane and Schlieper 1971, Odum 1988, Odum et al. 1984, 1988), and release of toxic metabolites (Moshiri et al. 1978). The instability of a complex of variables in bayous (and brackish waters in general) makes the environment extremely selective (Cognetti 1994). Many of the marine and freshwater invertebrates that find their way into tidal freshwater (termed the "tidal river zone" by Day et al. 1989) cannot tolerate the effects of a harsh environment, resulting in low species diversity and low densities of fauna.

Bayous are special types of low-salinity habitats and pose challenges to the invertebrates that colonize them. To generalize, darkly shaded bayous with overhanging trees support few species and low densities of invertebrates. But the pattern does not characterize all bayous. Many bayous of the Gulf Coast support dense and diverse communities of invertebrates. They are wide bayous with ample sunlight for dense growth of aquatic plants (macrophytes), phytoplankton, and benthic algae (reviewed by Sullivan 1981, Felley 1992, Moore 1992). The floating plants are especially rich in invertebrate species and provide habitat for abundant populations. This influence of aquatic vegetation on invertebrates is widely known for freshwater habitats (O'Hara 1967, Watkins et al. 1983, Livingston 1991a), seagrass communities (Day et al. 1989, Livingston 1991b), and tidal-freshwater embayments (Findlay et al. 1989), but is not recognized as a characteristic of tidal-freshwater streams.

SIGNIFICANCE OF AQUATIC MACROPHYTES

The most salient feature of bayous along the northern Gulf of Mexico, and the feature that distinguishes them from low-salinity habitats of the U.S. Atlantic Coast, is the presence and abundance of floating plants in sunny regions. Few tidal-freshwater streams of the U.S. East Coast or Europe support dense populations of aquatic macrophytes, and hardly any have extensive growths of floating plants. There are exceptions. The tidal Potomac River supported an abundance of submersed macrophytes before 1930, virtually none from the late 1930s to 1982, then a resurgence of submersed species after 1982 (especially the exotic, *Hydrilla verticillata;* Carter and Rybicki 1990, Carter et al. 1994, Thorp et al. 1997).

Dense growths of macrophytes cover the surface of many Gulf Coast tidal-freshwater streams (see Platt 1988, Eleuterius 1990, Felley 1992, Moore 1992, Mitsch and Gosselink 1993), especially wide portions of bayous where sunlight is available most of the day. Many bayous are choked by water hyacinth (*Eichhornia crassipes*), duckweed (*Lemna minor, Spirodela polyrhiza, Wolffia* sp., *Wolffiella* sp.), and alligator weed (*Alternanthera philoxeroides*), and almost all bayous have some floating plants along their length. Water hyacinths and duckweed seem to be the most important habitats for invertebrates, likely due to their dense below-surface growth, but submersed species (such as widgeongrass, *Ruppia maritima,* tapegrass, *Vallisneria americana,* sago pondweed, *Potamogeton perfoliatus,* and water milfoil, *Myriophyllum spicatum*) also may provide habitat for dense invertebrate populations (McBee and Brehm 1979, Livingston 1992, Moore 1992). Dense populations of amphipods inhabit floating and submersed plants (O'Hara 1967, McBee and Brehm 1979, Sikora and Sklar 1987, Thorp et al. 1997; Tables 37.1, 37.2). Aquatic plants along the Gulf Coast often persist through winter months, although duckweed may die back after flowering during the fall. Water hyacinths are somewhat susceptible to frost, but winter cold is seldom severe enough to kill them. Floating macrophytes are less conspicuous or absent in tidal-freshwater habitats of the Atlantic, perhaps because of greater tidal ranges (Mitsch and Gosselink 1993) and more severe winters there (Menzel and Cooper 1992).

INVERTEBRATE DENSITY AND DISTRIBUTION

Yozzo and Diaz (this volume) show that tidal-freshwater habitats of Virginia are transitional areas between riverine and estuarine habitats, with unique features and important characteristics for energy flow to adjacent ecosystems. Similarly, bayous are transitional areas with a unique set of physical and biological attributes and an ecosystem that includes diverse and abundant populations of invertebrates (Table 37.1). Invertebrate species density and distribution differ by habitat, so it is important to distinguish between bayous with and without floating plants, to discuss the ecology of characteristic fauna, and to clarify the role of the physical environment in structuring bayou communities.

TABLE 37.1. Diversity (number of taxa) and Mean Density (number/m²) of Invertebrates in Bayous and Tidal Freshwater Habitats of the Northern Gulf of Mexico and United States East Coast[a]

Location	Richness	Density	Reference
Atchafalaya Basin, LA	–	2885.0[b]	Beck 1977
Bayou D'Inde, LA	27	1207.0[b]	Gaston 1987a
Bayou LaBranche, LA	27	30,744.1[b,d]	Gaston 1988
	47	7500.0[c]	Gaston 1988
	20	5977.7[b]	Gaston 1990
Bayou Trepagnier, LA	18	5672.0[b]	LaDEQ 1989
Bayou Verdine, LA	23	300.0[b]	Gaston 1987a
Blind River, LA	26	9258.7[b,c]	Watson et al. 1981
Choupique Bayou, LA	20	~2500[b]	Gaston 1987a, Gaston and Nasci 1988
	27	~7000[b,c]	Gaston 1987a
Contraband Bayou, LA	15	1502.0[b]	Gaston 1987a, Gaston and Nasci 1988
James River, VA	–	1800–4000[b]	Diaz and Boesch 1977
James River, VA	10–14	1847–3433[b]	Diaz 1994
Lac des Allemands, LA	66	16,198.0[b,c]	Sklar 1985
	–	6110.0[b]	Sikora and Sklar 1987
	–	16,903.0[c]	Sikora and Sklar 1987
Rockefeller Refuge, LA	54	13,760.0[b,c]	Gaston 1987b
Old Fort Bayou, MS	–	~10,000.0[b,c]	Peterson 1997
Pine Island Bayou, TX	54	4042.0[b]	Darville and Harrel 1980
Simmons Bayou, MS	45		McBee and Brehm 1979
summer	–	>17,653[b,c]	
fall	–	>9859[b,c]	
winter	–	>65,100[b,c]	
spring	–	>45,464[b,c]	

[a] Some values estimated from figures (~).
[b] Invertebrates in sediments.
[c] Invertebrates in macrophytes.
[d] Includes ostracods (17,688.7/m²).

Bayous with Floating Plants

Bayous with floating plants support higher densities of invertebrates (>10,000/m²; Table 37.1) than most open-estuarine bays and lagoons of the region (2846/m², Gaston et al. 1995; 2590/m², Dardeau et al. 1992), and similar densities to adjacent intermediate marshes (8000/m², Gaston 1987b). The Lac des Allemands region of southern Louisiana is a wetland ecosystem of tidal-freshwater lakes and swamps and has a community composition typical of many vegetated bayous. Sikora and Sklar (1987) provide an excellent analysis of the region, which supported extremely high densities of inverte-

TABLE 37.2. Numerically Dominant and Large-Biomass Taxa in Bayous and Tidal Freshwater Habitats of the Northern Gulf of Mexico[a]

Taxa	Density	Location	Reference
Cnidaria			
Hydra sp.	306.5	Lac des Allemands, LA	Sikora and Sklar 1987
Nematoda	3705.5	Bayou LaBranche, LA	Gaston 1988
Annelida			
Oligochaeta			
Naididae	3998.4	Lac des Allemands, LA	Sikora and Sklar 1987
	552.8	Bayou LaBranche, LA	Gaston 1988
	312.0	North Island Canal, LA	Gaston 1987b
Tubificidae[b]	3694.4	Bayou LaBranche, LA	Gaston 1988
	2707.4	Bayou LaBranche, LA	Gaston 1990
	1851.2	Pine Island Bayou, TX	Darville and Harrel 1980
	695.1	Lac des Allemands, LA	Sikora and Sklar 1987
	178.0	Contraband Bayou, LA	Gaston and Young 1992
	194.0	North Island Canal, LA	Gaston 1987b
	159.7	Mississippi Bayou, LA	LaDEQ 1989
	*****[c]	Escambia River, FL	Patrick 1994
Polychaeta			
Hobsonia florida	1016.7	Bayou LaBranche, LA	Gaston 1988
	1503.7	Bayou LaBranche, LA	Gaston 1990
	187.6	Contraband Bayou, LA	Gaston and Young 1992
	108.8	Bayou D'Inde, LA	Gaston and Young 1992
	*****	Dog River; Miflin Creek, AL	Bault 1972
	*****	Fowl River; Fish River, AL	Bault 1972
Streblospio benedicti	220.8	Bayou Verdine, LA	Gaston and Young 1992
	92.2	Contraband Bayou, LA	Gaston and Young 1992
Laeonereis culveri	93.2	Bayou Verdine, LA	Gaston and Young 1992
	*****	Fowl River; Fish River, AL	Bault 1972

Taxon		Location	Reference
Hirudinea	7.4	Bayou LaBranche, LA	Gaston 1990
	5.1	Lac des Allemands, LA	Sikora and Sklar 1987
	*****	Escambia River, FL	Patrick 1994
Mollusca			
Gastropoda			
Physidae	222.0	North Island Canal, LA	Gaston 1987b
	180.7	Pine Island Bayou, TX	Darville and Harrel 1980
	36.1	Bayou LaBranche, LA	Gaston 1988
	*****	Escambia River, FL	Patrick 1994
Hydrobiidae	792.0	North Island Canal, LA	Gaston 1987b
	16.7	Bayou LaBranche, LA	Gaston 1988
Bivalvia			
Rangia cuneata	10–15	Dog River, AL	Swingle and Bland 1974
	8.3	Bayou LaBranche, LA	Gaston 1988
	7.4	Bayou LaBranche, LA	Gaston 1990
	*****	Neches River, TX	Harrel and McConnell 1995
	*****	Bay Minette Creek, AL	Bault 1972
	*****	Tensaw River, AL	Bault 1972
Sphaerium sp.	39.3	Pine Island Bayou, TX	Darville and Harrel 1980
	*****	Escambia River, FL	Patrick 1994
Crustacea			
Conchostraca	1236.1	Bayou LaBranche, LA	Gaston 1988
Ostracoda	17,688.7	Bayou LaBranche, LA	Gaston 1988
Branchiopoda	208.0	North Island Canal, LA	Gaston 1987b
Amphipoda			
Grandidierella			
bonnieroides	> 10,000.0	Simmons Bayou, MS	McBee and Brehm 1979
	8022.0	Old Fort Bayou, MS	Peterson 1997
	1148.1	Bayou LaBranche, LA	Gaston 1990
	780.5	Bayou LaBranche, LA	Gaston 1988

TABLE 37.2. (Continued)

Taxa	Density	Location	Reference
Corophium louisianum	314.8	Bayou LaBranche, LA	Gaston 1990
Gammarus mucronatus[d]	630.0	Old Fort Bayou, MS	Peterson 1997
Gammarus spp.	25.9	Bayou LaBranche, LA	Gaston 1990
	1134.0	Old Fort Bayou, MS	Peterson 1997
	87.9	Pine Island Bayou, TX	Darville and Harrel 1980
	*****	Fowl River; Fish River, AL	Bault 1972
Hyalella azteca	5337.0	Lac des Allemands, LA	Sikora and Sklar 1987
	317.0	North Island Canal, LA	Gaston 1987b
	*****	Fowl River, AL	Bault 1972
Crangonyx gracilis	*****	Escambia River, FL	Patrick 1994
Isopoda			
Cyanthura polita	40.7	Bayou LaBranche, LA	Gaston 1990
Asellus[e] obtusus	553.8	Lac des Allemands, LA	Sikora and Sklar 1987
Asellus attenatus	*****	Escambia River, FL	Patrick 1994
Mysidacea			
Taphromysis louisianae	55.6	Bayou LaBranche, LA	Gaston 1990
Decapoda			
Procambarus clarkii	6.3	Lac des Allemands, LA	Sikora and Sklar 1987
Insecta			
Corixidae	213.0	North Island Canal, LA	Gaston 1987b
Ceratopogonidae	172.2	Bayou LaBranche, LA	Gaston 1988
	170.9	Lac des Allemands, LA	Sikora and Sklar 1987
Chironomidae	1767.9	Lac des Allemands, LA	Sikora and Sklar 1987
	1644.4	Bayou LaBranche, LA	Gaston 1988
	909.5	Pine Island Bayou, TX	Darville and Harrel 1980
	770.0	North Island Canal, LA	Gaston 1987b

[a] Values are mean densities (number/m^2) of region sampled.
[b] Several authors listed *Limnodrilus cervix* and/or *L. hoffmeisteri* as dominant tubificid oligochaetes.
[c] Noted as numerical dominant (no densities given).
[d] *Gammarus mucronatus* is now *Macrogammarus mucronatus*.
[e] *Asellus* is now *Caecidotea*.

brates in aquatic-plant communities (Table 37.3). Indian River Lagoon, Florida, is another well-studied region that differs from bayous in physical characteristics but supports similar invertebrate communities in its vegetated habitats (see Moore 1992).

The numbers of invertebrates in vegetated bayous repudiate the well-established premise that low-salinity habitats (5–6 ppt) are areas of "lowest number of species between marine and freshwater," as proposed by Remane and Schlieper (1971). Lowest diversity and density of invertebrates in Gulf estuaries occur in unvegetated oligohaline bayous or farther down-estuary in the fine-sediment, mesohaline, open-water embayments where habitat heterogeneity is minimal.

Floating-plant communities are dominated by a diverse invertebrate fauna, including more amphipods and a greater variety of insects, than occur in sediments. The fauna are similar to invertebrates of backswamp habitats near bayous (Sklar 1985). Amphipods composed 46 percent of the invertebrates in vegetation of Louisiana backswamps, but only 21 percent in sediments (Sklar 1985). Free-living amphipods are often the most abundant taxa among bayou plants (Table 37.2). They actively swim and feed on detritus among the roots and submerged leaves. *Hyalella azteca* and *Crangonyx gracilis* are most com-

TABLE 37.3. Mean Density (number/m²) of Taxa Collected in Sediment and Floating Vegetation in Lac des Allemands, Louisiana

Taxa	Sediment	Floating Vegetation
Amphipoda	1311	7768
Collembola	6	7
Coleoptera	21	91
Decapoda	3	2
Diptera	1000	2409
Ephemeroptera	0	45
Gastropoda	118	854
Bivalvia	500	8
Hemiptera	13	174
Hydroidia	2	278
Isopoda	484	333
Lepidoptera	15	212
Neuroptera	2	7
Odonata	11	148
Oligochaeta	2606	4062
Rhynchobdellida	7	8
Tricladida	11	427
Totals	6110	16903

Source: From Sikora and Sklar 1987.

mon in freshwater areas, while *Mucrogammarus mucronatus* and some un-described species of *Gammarus* (R. Heard, personal communication) numerically dominate areas with more salinity influence. Other amphipods, such as *Corophium louisianum, C. lacustre,* and *Grandidierella bonnieroides,* live in tubes associated with the vegetation or attached to submerged structure (Heard 1982). *Grandidierella bonnieroides* periodically leaves the tubes to selectively feed on detritus among the macrophytes (S. LeCroy, personal communication), which accounts for the high densities of this species (a tube dweller) in vegetation samples. *Gammarus* spp., *Corophium lacustre,* and other "benthic" crustaceans vertically migrate into surface waters of bayous at night (Dauer et al. 1982, Macquart-Moulin 1984, Stearns and Dardeau 1990) and may densely populate vegetation. Physid snails commonly graze the epiphytes of water hyacinth and alligator weed, and isopods (mainly *Asellus attenatus* and *A. obtusus*) feed on accumulating or sloughing detritus among the vegetation. Predatory *Hydra* and planaria are often present on the undersides of duckweed and water-lily leaves. Benthos in sediments below floating macrophytes occur at much lower densities than invertebrates associated with the plants (Table 37.3). Invertebrate densities in sediments below plants are similar to values reported for unvegetated tidal freshwater of the James River, Virginia (Diaz 1994, Diaz and Boesch 1977).

The most conspicuous bayou invertebrates are associated with floating plants, but significant numbers of species live in association with other water-column structure (submersed and emergent plants, snags, or trees) or belong to zooplankton communities (Vecchione 1989, Felley 1992, Moore 1992). Snags provide ample structure for many invertebrates (Hauer and Benke 1991, Neuswanger et al. 1982) and are significant habitats in bayous without macrophytes. Trunks, roots, and knees of cypress trees support many species of invertebrates. The bayous adjacent to Lake Pontchartrain (Louisiana) are inhabited by *Sphaeroma terebrans,* a wood-boring isopod that occurs in tropical and subtropical habitats throughout much of the world but appears to be limited by the distribution of cypress trees in Louisiana (C. Franze and M. Poirrier, personal communication).

Bayous without Floating Plants

Bayous without floating plants support relatively few invertebrate species and low population densities. These bayous fit the model proposed by Remane and Schlieper (1971). In fact, unvegetated bayous have the lowest species diversity and lowest densities of benthic invertebrates in any Gulf of Mexico estuarine habitats (Gaston and Nasci 1988, Gaston and Young 1992). Unvegetated bayous also have fewer invertebrates than most freshwater habitats of the Gulf Coast. Gaston (1987b) sampled a series of low-salinity habitats in an impounded research area of Rockefeller Refuge (Cameron Parish, Louisiana) and found densities of bayou invertebrates (without floating plants) were often below $400/m^2$. Invertebrates of adjacent estuarine habitats (intermediate, brackish, and salt marshes; mudflats; tidal creeks) averaged 2016.3/ m^2, and those of freshwater (impounded) marshes averaged 3740/m^2.

The fauna of unvegetated bayous are similar to those of Atlantic tidal freshwater. They are usually dominated by sessile or discretely motile species. Oligochaetes may be modestly abundant, especially naidids in more freshwater areas and tubificids in oligohaline areas. *Limnodrilus cervix* and *L. hoffmeisteri* are the most common tubificids reported. Oligochaete abundance and diversity is inversely related to salinity in bayous, and oligochaetes are replaced by polychaetes down-estuary. Taxonomy of bayou oligochaetes is still poorly known despite much recent work. Many investigators provide oligochaete identifications only to family level. Brackish regions of bayous may also support moderate densities of polychaetes (especially *Hobsonia florida* and *Streblospio benedicti;* Table 37.2). Bayou sediments with more freshwater influence are dominated by chironomids and ceratopogonids (Diptera; Table 37.2). Midge larvae (chironomids) may reach densities in excess of 1500/m^2 in the upper reaches of bayous (Table 37.2). The most complete listing of bayou dipterans is provided by Sikora and Sklar (1987). As with oligochaetes, dipteran taxonomy is poorly known for bayous. It is likely that many new species of bayou dipterans and oligochaetes remain to be described. Ostracod and nematode densities may exceed 15,000/m^2 in unvegetated bayous, and Conchostraca and Branchiopoda may be common (Table 37.2), but most authors do not report them.

Invertebrate Motility

Another salient characteristic of bayous is their motile fauna. Atlantic tidal-freshwater habitats are dominated by relatively sessile taxa (oligochaetes and chironomids, Diaz 1994, Yozzo and Diaz, this volume), but vegetated bayous support large populations of crustacea (amphipods, isopods, mysids) that actively move within the water column and among habitats (Stearns and Dardeau 1990). Bayous may also be inhabited by numerous species of adult insects, insect larvae, and snails (physids and hydrobiids). There are commercially important species in bayous, such as blue crabs (*Callinectes sapidus*) and crayfish (especially *Procambarus clarkii* and *P. acutus;* Penn 1960, Albaugh 1973), as well as juvenile penaeid shrimp (Hackney and de la Cruz 1981), and grass shrimp (*Palaemonetes* spp.; Dardeau et al. 1992, Moore 1992).

The invertebrates associated with plants are significant to our understanding of the role that bayous play in estuarine and freshwater ecosystems. We think of most low-salinity invertebrates as sessile, but that view may be overly simplistic. Bayous are relatively shallow systems, and there is an opportunity for invertebrates to move quickly between bottom sediments and the surface waters. Many motile bayou species (mainly crustacea) vertically migrate with flood tides, especially at night (Stearns and Dardeau 1990). While the reasons for these vertical migrations are not clear, they do not appear to be in response to physiological stress. The timing of nocturnal, flood-tide migrations may help retain these species in upper reaches of the estuary, enhance reproductive contact, help them avoid visual predation, or provide alternative feeding opportunities (Stearns and Dardeau 1990).

Unique and Characteristic Fauna

Most bayou fauna are well-studied euryhaline species or characteristic tidal-freshwater taxa, but many bayou invertebrates are less familiar. There is at least one species of sponge (*Corvospongilla becki*) whose type locality is a bayou of the lower Atchafalaya (Louisiana) basin (Poirrier 1978). Other sponges, *Eunapius fragilis, Trochospongilla horrida,* and *T. leidyi,* were found in the same habitat. *Trochospongilla horrida* also is common on hard substrates of the Escambia River, Florida (Patrick 1994), but remains to be studied in bayous elsewhere. Bayous near Lake Pontchartrain, Louisiana, supported a different sponge species (*Ephydatia fluviatilis*) that grew on alligator weed, wood snags, and other hard structure of bayous (Poirrier 1974). A species of entoproct (*Urnatella gracillis*) was first reported in Louisiana bayous (drainage canals) by Poirrier and Johnson (1970). It typically inhabits more freshwater habitats, especially eutrophic and highly impacted sites, but may be common in bayous. Poirrier and Denoux (1973) found the colonial hydroid, *Cordylophora caspia,* indicative of brackish-water conditions, including bayous of southern Louisiana.

The only abundant bivalves in bayous of the northern Gulf of Mexico are the fingernail clams (Sphaeriidae) and the common brackish-water clam, *Rangia cuneata* (Table 37.2). *Rangia* grows large, is one of the few long-lived species in bayous, and may be important in the bayou trophic dynamics (Tarver and Dugas 1973). Empty shells along the bank attest to predation of *Rangia* by raccoons. Other bivalves, such as clams (*Mulinia lateralis*) and oysters (*Crassostrea virginica*), occasionally inhabit oligohaline reaches of bayous, but seldom persist there. *Corbicula fluminea,* the exotic Asian clam, dominates tidal reaches of the James River, Virginia (Diaz 1989), and occurs in many rivers of the southeast (Odum et al. 1984, Livingston 1992), but has not been reported in bayous of the northern Gulf of Mexico. *Arcidens confragosus* is a unionid bivalve that occurs throughout the Mississippi River drainage, including Bayou Teche, Louisiana (Thorp and Covich 1991). Other freshwater mollusks, especially Unionidae, are abundant in spring-fed streams of Alabama, Mississippi, Florida, Louisiana, and Texas (Johnson 1970, Livingston 1992), but rarely occupy bayou habitats.

Many studies list dominant zooplankton of Gulf Coast estuaries (see Dardeau et al. 1992, Moore 1992), but few sampled in bayous. Vecchione (1987, 1989, 1991) found that bayous of southwestern Louisiana supported dense populations of zooplankton, especially the copepod *Acartia tonsa.* Adult oysters (*Crassostrea virginica*) seldom occurred in harvestable densities in low-salinity bayous, even though oyster larvae reached the bayous in ample numbers (Vecchione 1987).

Taxonomy and systematics of low-salinity habitats is in its infancy, but it appears that no invertebrate species occur exclusively in tidal freshwater (Mitsch and Gosselink 1993). Several interrelated physical, chemical, and biological factors could account for this. Physical conditions in bayous are harsh, and dramatic changes occur in the thermodynamic equilibrium of dissolved substances as salinity approaches 1 ppt (Moore 1992). Also, diversity of bayou communities may be limited by genetic homogeneity. Homogenous populations produce few new species. Homogeneity is especially common in populations that inhabit brackish regions with extreme variations in physical parameters, "above all where salinity is at its lowest" (Cognetti 1994). Perhaps few endemic species inhabit bayous because few originate there.

PHYSICAL ENVIRONMENT

Most authors attribute the low diversity and absence of specialized fauna in tidal freshwater to severe physical conditions, especially salinity and temperature stress (Ax and Ax 1970, Remane and Schlieper 1971, Diaz 1980, Wetzel 1983). Like all low-salinity habitats, bayous experience irregular changes in salinity and current flow, periodic episodes of high sedimentation, poor water circulation, and radical seasonal changes in nutrient input and temperature (reviewed by Moore 1992). These rigors are compounded by stresses of pollution (see below). Additionally, two physical factors limiting species diversity in bayous received little attention in past studies: light intensity and dissolved oxygen.

Light intensity is a limiting factor to distribution of floating macrophytes in bayous. Long periods of sunlight are required to support dense communities of floating vegetation. Where sunlight is adequate, floating plants cover bayou surfaces (Smock and Gilinsky 1992) and invertebrates flourish. Lower reaches of bayous in Louisiana, where tree-lined bayous grade into marshes, are densely populated by floating plants. The plants become increasingly less common in shaded regions, and often are absent in the upper bayou. Invertebrate populations of those upper regions decline with the plants (Table 37.1).

Dissolved oxygen appears to be a common factor in survival and persistence of invertebrates in estuarine habitats, but remains largely unstudied in bayous. Dissolved oxygen was only recently recognized as a widespread problem in Gulf of Mexico estuaries (Petrick 1981, Livingston 1984, Whitledge 1985, Sikora and Sklar 1987, Moore 1992, Summers et al. 1992). Most past measurements of water-column dissolved oxygen were taken during the daytime, so the critical period when dissolved oxygen values were lowest (overnight) was missed. The importance of overnight measurements of bayou dissolved oxygen was demonstrated by Petrick (1981), and it now appears that many bayous, perhaps all of them, undergo dissolved-oxygen stress during warm months. Furthermore, Moshiri et al. (1978) provided evidence that algal metabolites cause fish kills in low-salinity habitats, many of which were previously attributed to low dissolved oxygen. When one considers all of these environmental factors and pollution (see below), it is not surprising that benthic fauna of bayous are sometimes depauperate, nor is it surprising that most invertebrates live near the air-water interface (on macrophytes).

INVERTEBRATE ADAPTATIONS TO PHYSICAL CONDITIONS

How do the fauna that inhabit bayous cope with such harsh conditions? Many species have elaborate physiological, morphological, and behavioral adaptations. Bayou polychaetes, chironomids, and oligochaetes have evolved respiratory pigments, especially intracellular and extracellular hemoglobin, which aid their habitation of hypoxic waters and anoxic sediments (Mangum 1973, 1976, 1985, 1990). Species with these pigments remain in hypoxic conditions. Other species, blue crabs for instance, can osmoregulate in order to survive daily and seasonal changes in salinity (Tagatz 1971), though this adaptation may be limited to a narrow range of salinity. As conditions change, species may insulate themselves from salinity and temperature variations by occupying sediment burrows (annelids), closing their valves (clams and snails), or climbing snags (insects and snails). Some invertebrates may leave the area by swimming (penaeid shrimps and portunid crabs) or drifting downstream (insects). Bayou species migrate great distances (blue crabs) to reproduce in more optimal physical conditions, and their young must find a mechanism to return to bayous as they mature.

Seasonal patterns of invertebrate abundance in bayous result from many variations in habitat conditions, especially winterkill of macrophytes, seasonal invertebrate reproduction patterns, and seasonal changes in salinity and other physical conditions. Many species survive harsh conditions by reproducing before harsh conditions arrive, leading to temporal patterns in distribution. Most bayou invertebrates are short-lived species that occupy bayou habitats for less than a year, reproduce once, and die (Gaston and Nasci 1988, Gaston and Young 1992). McBee and Brehm (1979) attribute their observed seasonal population changes in a Mississippi bayou (Table 37.1) primarily to invertebrate spawning cycles. Amphipods were more abundant during summer, chironomids dominated during spring, and polychaetes and oligochaetes were most numerous during winter and spring. Amphipods collected by McBee and Brehm had their greatest densities during summer months as macrophytes (mainly widgeongrass, *Ruppia maritima*) grew. As plants died, invertebrate populations associated with them decreased proportionately (McBee and Brehm 1979, Sikora and Sklar 1987). Rainfall events may also affect invertebrate populations. Benthic invertebrates of Louisiana bayous reached their lowest diversities and lowest densities during winter (Gaston and Nasci 1988), perhaps because salinity values were lowest then. Bayou salinities change on several temporal scales, certainly with tidal cycles and sometimes with season. Peterson (1997) found seasonal upstream shifts in dominant crustacea of low-salinity habitats following hurricanes. He found peak abundance of amphipods occurred between July and September, matching patterns reported in bayous of southwestern Louisiana (Gaston and Nasci 1988), the Neches River of Texas (Harrel et al. 1976), and the Wolf River estuary of Mississippi (Milligan 1979).

Bayou plankton populations generally reflect spatial, rather than seasonal, conditions. Remane and Schlieper (1971) describe brackish river mouths as areas of a "wealth of plankton," due to accumulations of rich organic nutrients. Many bayous are eutrophic, so population patterns related to nutrients are less obvious. Populations of phytoplankton (Maples 1987) and zooplankton (Vecchione 1989, 1991) in southwestern Louisiana bayous did not correlate closely with seasonal factors. They were more influenced by water-column stratification (especially copepods), wind-driven tides (salinity effects), shading, spatial-temperature variation, and contaminants. This is not to imply that seasonality does not occur in planktonic communities. Weak seasonal patterns were reported for oyster larvae in bayous (Vecchione 1987), and most meroplankton had some seasonality in their population densities. But zooplankton patterns in bayous are complex and should not be simplified by generalizations about seasonality. Zooplankton respond to many variables and differ in density over short temporal scales, and their patterns are confounded by their motile planktonic nature and their distribution in three dimensions. For instance, Stearns and Dardeau (1990) demonstrate the importance of tidal stage in composition of zooplankton (due to vertical migration). In short, spatial variables explain more of the variance in bayou plankton communities than broad-scale temporal (seasonal) variables.

TROPHIC ECOLOGY

Diaz (1994) likens depauperate tidal-freshwater ecosystems to those of the Great Lakes, noting that the two have similar benthic species composition and diversity. He attributes the depauperate fauna of tidal freshwater to limited habitat diversity and a harsh physical environment, but stops short of describing the trophic structure of tidal freshwater, and little is found about trophic structure of tidal-freshwater ecosystems elsewhere in the literature. Bayou trophic structure has been described in only a few papers (Gaston and Nasci 1988, Gaston and Young 1992).

Lakes and tidal freshwater (including bayous) have another characteristic in common. Neither supports many species of suspension feeders, one of the most common trophic groups of invertebrates in marine, estuarine, and freshwater riverine habitats. Why? Suspension feeders generally need flowing water to provide their food. Bayous lack the currents that typify freshwater streams dominated by suspension-feeding species, and they lack the tidal energy characteristic of estuarine habitats dominated by suspension feeders. Bayous are slow-moving, almost stagnant streams. Bayous also lack the water-column production and benthic-algae production characteristic of estuaries and necessary to nourish suspension feeders (Gaston et al. 1997). Furthermore, bayous generally have loosely consolidated sediments that are not amenable to the sessile existence of most suspension feeders.

Analysis of invertebrate trophic structure in bayous poses a problem. There is not an adequate trophic-structure classification system to characterize low-salinity habitats. Classifications of invertebrate trophic structure in freshwater (Merritt and Cummins 1984), marine (Jumars and Fauchald 1977, Fauchald and Jumars 1979), and estuarine habitats (Gaston 1987c, Gaston and Nasci 1988, Gaston et al. 1985, 1988, 1995) are designed to describe the dominant feeding-biology characteristics of those specific systems. Bayous have characteristics of both estuarine and freshwater habitats. Their combination of fauna confounds classification systems designed for either one, so investigators must borrow from both systems. Trophic-structure characteristics used in freshwater habitats (shredders, scrapers, collectors, and predators) are based on river-continuum generalizations of dependence on organic matter derived from leaf litter, wood, and other plant matter that originated in the riparian zone. This classification works for upper reaches of bayous. The classification used to describe detritus-based marine and estuarine ecosystems (surface-deposit feeders, subsurface-deposit feeders, suspension feeders, herbivores, carnivores) are most applicable to brackish portions of bayous where detritivores dominate. In the most simple characterization, bayous are dominated by motile detritivores above the sediment surface and burrowing detritivores below it. The importance of motile detritivores increases with complexity (heterogeneity) of habitat structure, especially submersed and floating vegetation. Amphipods, isopods, snails, and mysids are active deposit feeders of bayous. They inhabit floating or submersed vegetation, live on snags, or crawl over leaf litter. Vegetation and other structure provide bayous with three-dimensional habitats in an ecosystem that otherwise provides only muddy sediments and anoxic leaf litter. Snags are especially important for invertebrates in tidal freshwater without plants because snags allow benthic species to separate from the harsh bottom conditions and occupy near-surface structure. Bayou plants, especially near-surface or floating macrophytes, provide better-oxidized habitats than bottom sediments or leaf litter, provide excellent habitat for foraging, can be occupied temporarily during low-oxygen events, and provide refuge from predation.

Some motile crustaceans use the entire water column of shallow bayous (Stearns and Dardeau 1990). This ability to diversify and expand feeding habitat may be a widespread characteristic of low-salinity crustacea (Dauer et al. 1982, Macquart-Moulin 1984), but it remains to be studied in many areas. It is easy to understand why such a behavior is successful. It allows species to seek food in a variety of habitats and vary their behavior with tidal stage, such as vertical migration on nocturnal flood tides. Feeding in diverse habitats confounds assessment of the energy flow through the system and brings to light the limitations of trophic-structure classification defined by habitat.

Invertebrates of bayous with muddy substrates are more easily character-ized because they are closely allied with the invertebrate fauna of Atlantic tidal-freshwater ecosystems. Both systems are typified by high densities of chironomids and oligochaetes, and sometimes polychaetes, snails, and tube-dwelling amphipods. All of these taxa can be characterized as burrowing or surface detritivores (deposit feeders), but that tells little of the complexity and diversity of their behavior. For instance, tubificid oligochaetes feed head-down in the sediments (subsurface-deposit feeders), usually below the redox-potential discontinuity (RPD), and defecate on the sediment surface. This feeding, termed *conveyer-belt feeding,* is important in its effects on the geo-chemical nature of surficial sediments. The most abundant chironomids are subsurface-deposit feeders (collector-gatherers), which are physiologically well adapted to reducing conditions below the sediment surface. They carry a reservoir of oxygen in their hemoglobin and actively oxidize or irrigate their temporary burrows as necessary. Such behavior allows them to feed below the sediment surface where the risk of predation is slight. A common poly-chaete of bayous, *Hobsonia florida,* also uses respiratory pigments (chloro-cruorin) to counter the reducing sediments, and uses tentacles to feed on detritus at the sediment-water interface from a mucous tube (surface-deposit feeding). Hydrobiid snails plow the sediments above the RPD to feed on detritus (subsurface-deposit feeding), and physid snails feed on periphyton or detritus in living and dead structure (herbivory and/or surface-deposit feeding; Thorp and Covich 1991).

Much has been written about the predatory role of blue crabs in estuaries, but few studies acknowledge them also as detritivores. Juvenile blue crabs often detritus feed, and they are more common in bayous than adults. Most adult blue crabs in low-salinity habitats are males (Moore 1992). Many other predatory species inhabit bayous, most notably other decapod crustaceans (crabs and shrimp), fish (Felley 1987, 1989, 1992, Hastings et al. 1987, Cash-ner et al. 1994), and other vertebrates (Moore 1992, Smock and Gilinsky 1992). All of these taxa provide vectors to transport nutrients to and from bayous.

Rangia cuneata is a suspension-feeding clam that warrants mention, not for its abundance, but for its biomass. *Rangia cuneata* is one of the few large-biomass species of bayous, is one of the few suspension feeders in bayous, and is unique in its apparent longevity. Although *R. cuneata* occurs in bayous, it is most common in open-water oligohaline regions of Gulf estuaries (Stout 1990). It inhabits sandy-mud regions of bayous, especially where bayous grade into tidal creeks and currents are adequate for suspension feeding. It apparently can osmoregulate at salinities below 10 ppt, but is tolerant of salinities between 0 and 25 ppt (Moore 1992). Much of the detritus ingested by *R. cuneata* of the Atlantic coast may originate as riverine organic matter (Tenore et al. 1968), but it feeds mostly on planktonic detritus and near-bottom seston in Gulf estuaries (Gaston et al. 1997). There is a temptation to assume that *R. cuneata,* as a large-biomass species, provides significant organic-matter export from bayous. However, its longevity and modest population densities (Table 37.2) may suggest that it plays a minor role in productivity of bayous compared to smaller invertebrates.

There are other functional-feeding groups, besides suspension feeders, that are poorly represented or lacking in bayous. Bayous lack the water-column oxygen and stream energy characteristic of freshwater streams and necessary to support dense populations of filter-feeding insects (filtering collectors), the water clarity and hard substrates (rocks) required for *in situ* benthic algae production to support diverse species of snails (scrapers), and debris dams that are amenable to a diversity of invertebrate fauna (especially shredders, predators, and filtering collectors; Smock and Gilinsky 1992). The absence of trophic diversity supports observations by Diaz and Boesch (1977) and Diaz (1989, 1994) that lack of habitat heterogeneity accounts in part for the low diversity of many tidal-freshwater habitats. Some of the trophic diversity lost by absence of suspension feeders is countered by specialized invertebrates associated with bayou macrophytes.

PRODUCTIVITY

Productivity of estuaries is summarized by Dardeau et al. (1992), but productivity of low-salinity habitats has not been studied for most regions of the Gulf Coast. The first extensive research was provided by Day et al. (1977), who studied a tidal-freshwater region in southern Louisiana. They found bayou productivity (net daytime photosynthesis) was 316 g DM m^{-2}. Nighttime respiration was 446 g DM m^{-2}. Annual export to the estuary was 8016 metric tons of organic C, 1047 metric tons of N, and 154 metric tons of P. Greatest export occurred during the spring. The areas studied were Bayou Chevreuil and Bayou Boeuf, part of the headwaters of the Barataria Basin. They were turbid (average Secchi depth 30 cm), sluggish, acidic, shallow streams. Day et al. (1977), in describing these habitats, were the first to demonstrate the discrepancy of invertebrate populations between vegetated and unvegetated tidal freshwater of the Gulf Coast.

Chironomids in the southeastern United States are highly productive; they may produce six or more life cycles per year (Cooper 1987). Amphipods and associated populations of invertebrates in vegetated habitats are similarly productive. Sikora and Sklar (1987) report invertebrate biomass values of 12.6 g AFDM m^{-2} (43.8 percent amphipods) in vegetated regions of swamps adjacent to Bayou des Allemands, Louisiana. These values closely matched invertebrate biomass of a crayfish farm (14.7 g AFDM m^{-2}), managed for high production of *Procambarus clarkii* (21 percent of biomass). Furthermore, amphipods may be especially important in nutrient cycling and foodwebs of bayous because they are major prey items of many fish and blue crabs (Thomas 1976, Overstreet and Heard 1978, 1982, Matlock and Garcia 1983, Hines et al. 1990).

The general productivity model for vegetated bayous characterizes invertebrates as dependent almost exclusively on the detrital food chain of decomposing plants, rather than the planktonic food chain of most open-estuarine habitats or the shredded leaf litter of freshwater streams. Some organic matter arrives in the bayou via stream flow as dissolved (DOM) and particulate (POM) forms, and the importance of these sources is greatest in bayous without plants. Bottom sediments of bayous are often covered by leaf litter and decaying organic matter reflective of the surrounding swamps, providing ample (albeit poor-quality) food for detritivores. Floating macrophytes support high densities of detritivores, which apparently feed on decaying macrophyte leaves and root matter. Some nutrients leave bayous through the aquatic food chain (see Felley 1992), but much is transported downstream via detrital export. Stern et al. (1991) demonstrate seasonal patterns of nutrient transport in Willow Bayou, Louisiana. They found that bayous trapped nutrients (nitrate) during summer months and exported them (ammonium) during winter. We know little else about the production, energy flow, or ecosystem function of bayous or other low-salinity Gulf Coast ecosystems (Smock and Gilinsky 1992, Mitsch and Gosselink 1993).

There is a temptation to focus on nutrient input as the key to understanding invertebrate ecology in low-salinity habitats such as bayous. Certainly that is the lesson taught from the comprehensive research in nutrient-poor freshwater and marine ecosystems. Invertebrates are food-limited in those systems. But such a premise may not be applicable to bayous. When discussing ecosystems, we often begin and end discussions with food input and nutrient output. Bayous are seldom nutrient-poor ecosystems, and many are eutrophic. The key to understanding distributions of bayou invertebrates lies in understanding their habitat quality and heterogeneity, as well as nutrient flow. Most low-salinity ecosystems likely provide adequate quantities and quality of organic matter to support the detritus-based foodweb, and undoubtedly the highest-quality detritus is provided by macrophytes and other *in situ* production. Unvegetated bayous likely receive much of their organic matter from upstream, from the riparian zone (especially leaf litter and wood), and from downstream via tidal inflow.

HABITAT DEGRADATION

Much of what we know about low-salinity habitats of the Gulf Coast results from studies of pollution effects. Summaries of such studies in low-salinity habitats of the Gulf Coast are provided by Dardeau et al. (1992), Felley (1992), and Moore (1992). Many bayous receive effluent from industry, municipal sewage discharge, and agricultural runoff, but they are seldom perceived as habitats that should receive priority for conservation. They have suffered from a history of drilling operations, especially in Texas and Louisiana, which led to brine and contaminant discharge, dramatic salinity increases, and contaminant loadings (Ziser 1978). Even today, a wide variety of chemical stresses reach bayous, including contaminant metals, organic contaminants, synthetics, highly acidic waters from swamps, nutrients, pathogens, and radionuclides (Dardeau et al. 1992).

There are only a few multidisciplinary or comprehensive studies of contaminant effects on bayous. One such study was conducted on the Calcasieu River ecosystem of southwestern Louisiana to determine effects of petrochemical industries, agricultural runoff, and municipal discharge on bayou communities. Surface sediments were contaminated by metals (Ramelow et al. 1989, Mueller et al. 1989) and organics (Murray and Beck 1990), and contaminants were present in tissues of commercial shrimp (Murray and Beck 1989) and crabs (Murray et al. 1992). Gaston and Nasci (1988), Gaston et al. (1988), and Gaston and Young (1992) use benthic trophic structure of bayous as a mechanism to assess contaminant impacts there. Benthic communities of contaminated bayous supported a poor mix of macrobenthic trophic groups and were limited to low numbers of invertebrates. Predatory invertebrates and suspension feeders were much less common at the contaminated sites than they were elsewhere. Vecchione (1989) concludes that zooplankton communities of the region suffered as well. Rather than a typical increase in abundance of zooplankton near treated-sewage discharge, Vecchione found a scarcity of copepods, perhaps indicative of the toxic nature of substances released into the bayous with treated sewage (chlorine, toxic metals, ammonia). Concentrations of contaminants and impacts on the benthic communities of the Calcasieu River region led to posting of a human health advisory, which limited recreational activities, shrimp trawling, and commercial fishing. Samples of commercial fish, shrimp, and crabs collected in contaminated bayous of the region contained contaminants in edible tissues, which confirmed the suspected linkage between benthos and commercial harvests (Pereira et al. 1988, Ramelow et al. 1989, Murray et al. 1992).

Sewage discharge worsens the low-oxygen conditions of bayous and leads to increased contaminant and nutrient loading. Sewage is much more than nutrients and decomposing fecal matter. It includes high levels of chlorine applied during disinfection, soaps and surfactants, and an unknown quantity of other household and industrial chemicals. All of these products potentially affect sediment and water quality and may limit habitation of the effluent region by invertebrates and fish. Indeed, fish kills are common in bayous contaminated by sewage (Felley 1987, Gaston and Nasci 1988).

Agricultural runoff may adversely affect bayou ecosystems. For instance, water-level manipulation is required for rice farming along coastal Louisiana and Texas, which leads to increased nutrient loading and potentially high contaminant concentrations. We know little about these and other specific agricultural effects on bayous, which are difficult to ascertain in a background of harsh physical conditions, periodic contamination, and widely varying nutrient loadings.

Presence or absence of contaminant-sensitive (indicator) species was commonly used to assess pollution impacts. Amphipods were absent from contaminated bayous in southwestern Louisiana (Gaston and Nasci 1988, Gaston and Young 1992). Gammarid amphipods were especially sensitive to contaminants, whereas corophiid amphipods, especially *Corophium louisianum,* were more tolerant of pollution. Similarly, an ectoproct (*Victorella pavida*) and a barnacle (*Balanus subalbidus*) inhabit bayous of Mississippi and Louisiana and may be good indicators of water quality within their ranges (Mulino 1975, Poirrier and Mulino 1977, Poirrier and Partridge 1979). Most researchers now recognize the many shortcomings in usage of indicator species for pollution assessment. For instance, species may evolve resistance to contaminants, as demonstrated for oligochaetes exposed to high concentrations of toxic metals (Klerks and Levinton 1992). Biotic indices that include ecosystem-wide response variables, not single indicator species, are now widely used (Weisberg et al. 1997).

Other impacts to the bayou ecosystems are more difficult to quantify. Habitat loss is the major cause for concern (Dardeau et al. 1992), especially in Louisiana, where land subsidence coupled with human impacts have led to rapid wetland loss (Deegan et al. 1984, Wiseman et al. 1990, Day et al. 1995, Turner 1997). Channelization is closely followed by salinity intrusion (Hackney and Adams 1992), which results in loss of freshwater marshes and aquatic vegetation and development of brackish open-water conditions where wetlands stood just a decade before. Costanza et al. (1990) provide a model characterizing changes in Louisiana wetlands associated with salinity intrusion (coastal ecological landscape spatial simulation).

CONCLUSIONS

It is hoped that this information about bayous will put to rest the notion that low-salinity habitats are necessarily the regions of lowest diversity. Vegetated bayous support higher diversity and a greater abundance of invertebrates than adjacent freshwater or estuarine habitats. With such harsh conditions, why do so many species occupy bayous? Bayous are appealing habitats to species with adequate adaptations. Predation pressure is generally low in bayous, allowing species that can tolerate the conditions to grow and flourish. Detrital food is plentiful and probably not limiting during any season for detritivores, which dominate all bayou habitats. Although conditions are harsh, many species have evolved necessary respiratory pigments and other physiological adaptations, allowing them to colonize bayous.

Bayous are productive habitats that warrant protection, but not because they harbor rare or endemic species. Bayous are transitory residences or habitats at the edge of distribution for most invertebrate taxa, the extreme border between conditions of tolerance and intolerance, and the habitat where freshwater species wandering too far downstream mix with saltwater species wandering too far upstream. Little is known of the interactions between these two groups of fauna, or of whether there is adaptation, cooperation, or dependence of species during encounters. What we know of bayous provides the basis for a unique opportunity to investigate these interactions. Bayou ecology is in its infancy, and future ecological investigations are needed to provide insight into the unique bayou habitats of the northern Gulf of Mexico.

ACKNOWLEDGMENTS

The author gratefully acknowledges many people for their contributions to this work. C. Cleveland (University of Mississippi) helped gather data and edited an earlier version of the manuscript. The following people provided data or suggestive comments: R. Hastings (Southeastern Louisiana University), R. Cashner and M. Poirrier (University of New Orleans), K. Konzen, B. Libman, R. Baca, S. Threlkeld (University of Mississippi), E. Turner, N. Rabalais, J. Fleeger, and J. Day (Louisiana State University), F. Sklar (South Florida Water Management District), W. Sikora (Athens, Georgia), R. Heard, S. LeCroy, and C. Rakocinski (University of Southern Mississippi), and M. Dardeau (Dauphin Island Sea Lab, Alabama).

LITERATURE CITED

Albaugh, D. W. 1973. Life histories of the crayfish *Procambarus acutus* and *Procambarus hinei* in Texas. Ph.D. Dissertation, Texas A&M University, College Station, TX.

Ax, P., and R. Ax. 1970. Das Verteilungsprinzip des subterranen Psammon am Übergang Meer-Süsswasser. Mikrofauna des Meeresbodens 1:1–51.

Bault, E. I. 1972. A survey of the benthic organisms in selected coastal streams and brackish waters of Alabama. Alabama Department of Conservation, Seafoods Division, Alabama Marine Resources Laboratory, Dauphin Island, AL.

Beck, L. T. 1977. Distribution and relative abundance of freshwater macroinvertebrates of the lower Atchafalaya River Basin, Louisiana. Master's Thesis, Louisiana State University, Baton Rouge, LA.

Boesch, D., M. L. Wass, and R. W. Virnstein. 1976. The dynamics of estuarine benthic communities. Pages 177–196 *in* M. Wiley (ed.), Estuarine Processes. Vol. 1. Academic Press, New York.

Carter, V., and N. B. Rybicki. 1990. Light attenuation and submersed macrophyte distribution in the tidal Potomac River and estuary. Estuaries 13:441–452.

Carter, V., N. B. Rybicki, J. M. Landwehr, and M. Turtora. 1994. Role of weather and water quality in population dynamics of submersed macrophytes in the tidal Potomac River. Estuaries 17:417–426.

Cashner, R. C., F. P. Gelwick, and W. J. Matthews. 1994. Spatial and temporal variation in the distribution of fishes of the LaBranche wetlands area of the Lake Pontchartrain estuary, Louisiana. Northeast Gulf Science 13:107–120.

Cognetti, G. 1994. Colonization of brackish waters. Marine Pollution Bulletin 28:583–586.

Cooper, C. M. 1987. Benthos of Bear Creek, Mississippi: Effects of habitat variation and agricultural sediments. Journal of Freshwater Ecology 4:101–113.

Costanza, R., F. H. Sklar, and M. L. White. 1990. Modeling coastal landscape dynamics. Bioscience 40:91–107.

Darville, R. G., and R. C. Harrel. 1980. Macrobenthos of Pine Island Bayou in the Big Thicket National Preserve, Texas. Hydrobiologia 69:213–223.

Dardeau, M. R., R. L. Shipp, and R. K. Wallace. 1990. Faunal components. Pages 89–114 *in* Mobile Bay: Issues, Resources, and Management. NOAA Estuary-of-the-Month Seminar No. 15. U.S. Department of Commerce, NOAA Estuarine Programs Office, Washington, DC.

Dardeau, M. R., R. F. Modlin, W. W. Schroeder, and J. P. Stout. 1992. Estuaries. Pages 615–744 *in* C. T. Hackney, S. M. Adams, and W. A. Martin (eds.), Biodiversity of the Southeastern United States: Aquatic communities. John Wiley & Sons, New York.

Dauer, D. M., R. M. Ewing, J. W. Sourbeer, W. T. Harlan, and T. L. Stokes, Jr. 1982. Nocturnal movements of the macrobenthos of the Lafayette River, Virginia. Internationale Revue der Gesamten Hydrobiologie 67:761–775.

Day, J. W., Jr., T. J. Butler, and W. H. Conner. 1977. Productivity and nutrient export studies in a cypress swamp and lake system in Louisiana. Pages 255–269 *in* M. Wiley (ed.), Estuarine Processes. Vol. 2. Academic Press, New York.

Day, J. W., Jr., C. A. S. Hall, W. M. Kemp, and A. Yanez-Arancibia. 1989. Estuarine Ecology. John Wiley & Sons, New York.

Day, J. W., Jr., D. Pont, P. F. Hensel, and C. Ibanez. 1995. Impacts of sea-level rise on deltas in the Gulf of Mexico and the Mediterranean: The importance of pulsing events to sustainability. Estuaries 18:636–647.

Deegan, L. A., H. Kennedy, and C. Neill. 1984. Natural factors and human modifications contributing to marsh loss in Louisiana's Mississippi River Deltaic Plain. Environmental Management 8:519–528.

Diaz, R. J. 1980. Ecology of tidal freshwater and estuarine Tubificidae (Oligochaeta). Pages 319–330 *in* R. O. Brinkhurst and D. G. Cook (eds.), Aquatic Oligochaete Biology. Plenum, New York.

———. 1989. Pollution and tidal benthic communities of the James River estuary, Virginia. Hydrobiologia 180:195–211.

———. 1994. Response of tidal freshwater macrobenthos to sediment disturbance. Hydrobiologia 278:201–212.

Diaz, R. J., and D. F. Boesch. 1977. Habitat Development Field Investigations Windmill Point Marsh Development Site, James River, Virginia. Waterways Experimental Station Technical Report D-77-23. U.S. Army Corps of Engineers, Vicksburg, MS.

Eleuterius, L. N. 1990. Tidal Marsh Plants. Pelican, Gretna, LA.

Fauchald, K., and P. A. Jumars. 1979. The diet of worms: A study of polychaete feeding guilds. Oceanography and Marine Biology Annual Review 17:193–284.

Felley, J. D. 1987. Nekton assemblages of three tributaries to the Calcasieu Estuary, Louisiana. Estuaries 10:321–329.

———. 1989. Nekton assemblages of the Calcasieu Estuary. Contributions in Marine Science 31:95–117.

———. 1992. Medium-low-gradient streams of the Gulf coastal plain. Pages 233–269 *in* C. T. Hackney, S. M. Adams, and W. A. Martin (eds.), Biodiversity of the Southeastern United States: Aquatic Communities. John Wiley & Sons, New York.

Findlay, S., K. Schoeberl, and B. Wagner. 1989. Abundance, composition, and dynamics of the invertebrate fauna of a tidal freshwater wetland. Journal of the North American Benthological Society 8:140–148.

Gaston, G. R. 1987a. Benthic ecology of Calcasieu Lake, Louisiana. Pages 5A1–5E16 *in* L. R. DeRouen and L. H. Stevenson (eds.), Ecosystem Analysis of the Calcasieu River/Lake Complex. Vol. 2. U.S. Department of Energy (DE-F601-83EP31111), Washington, DC.

————. 1987b. Relative Abundances of Invertebrate Species by Habitat Types on Rockefeller Refuge. Louisiana Board of Regents, Research and Development Program, Baton Rouge, LA.

————. 1987c. Benthic Polychaeta of the Middle Atlantic Bight: Feeding and distribution. Marine Ecology-Progress Series 36:251–262.

————. 1988. Benthic and epibenthic communities of Bayou Trepagnier and Bayou LaBranche (St. Charles Parish, Louisiana). Shell Oil Company, Houston, TX.

————. 1990. Benthic and epibenthic communities of Bayou Trepagnier and Bayou LaBranche (St. Charles Parish, Louisiana). Shell Oil Company, Houston, TX.

Gaston, G. R., S. S. Brown, C. F. Rakocinski, R. W. Heard, and J. K. Summers. 1995. Trophic structure of macrobenthic communities in northern Gulf of Mexico estuaries. Gulf Research Reports 9:111–116.

Gaston, G. R., C. M. Cleveland, S. S. Brown, and C. F. Rakocinski. 1997. Benthic-pelagic coupling in northern Gulf of Mexico estuaries: Do benthos feed directly on phytoplankton? Gulf Research Reports 9:231–237.

Gaston, G. R., D. L. Lee, and J. C. Nasci. 1988. Estuarine macrobenthos in Calcasieu Lake, Louisiana: Community and trophic structure. Estuaries 11:192–200.

Gaston, G. R. and J. C. Nasci. 1988. Trophic structure of macrobenthic communities in the Calcasieu estuary, Louisiana. Estuaries 11:201–211.

Gaston, G. R., P. A. Rutledge, and M. L. Walther. 1985. The effects of hypoxia and brine on recolonization by macrobenthos off Cameron, Louisiana (USA). Contributions in Marine Science 28:79–93.

Gaston, G. R., and J. C. Young. 1992. Effects of contaminants of macrobenthic communities in the upper Calcasieu Estuary, Louisiana. Bulletin of Environmental Contamination and Toxicology 49:922–928.

Hackney, C. T., and S. M. Adams. 1992. Aquatic communities of the southeastern United States: Past, present, and future. Pages 747–760 *in* C. T. Hackney, S. M. Adams, and W. A. Martin (eds.), Biodiversity of Southeastern United States: Aquatic Communities. John Wiley & Sons, New York.

Hackney, C. T., and A. A. de la Cruz. 1981. Some notes on the macrofauna of an oligohaline tidal creek in Mississippi. Bulletin of Marine Science 31:658–661.

Harrel, R. C., J. Ashcraft, R. Howard, and L. Patterson. 1976. Stress and community structure of macrobenthos in a Gulf Coast riverine estuary. Contributions in Marine Science 20:69–81.

Harrel, R. C., and M. A. McConnell. 1995. The estuarine clam *Rangia cuneata* as a biomonitor of dioxins and furans in the Neches River, Taylor Bayou, and Fence Lake, Texas. Estuaries 18:264–271.

Hastings, R. W., D. A. Turner, and R. G. Thomas. 1987. The fish fauna of Lake Maurepas, an oligohaline part of the Lake Pontchartrain estuary. Northeast Gulf Science 9:89–98.

Hauer, F. R., and A. C. Benke. 1991. Rapid growth of snag-dwelling chironomids in a blackwater river: The influence of temperature and discharge. Journal of the North American Benthological Society 10:154–164.

Hines, A. H., A. M. Haddon, and L. A. Wiechert. 1990. Guild structure and foraging impact of blue crabs and epibenthic fish in a subestuary of Chesapeake Bay. Marine Ecology-Progress Series 67:105–126.

Heard, R. W. 1982. Guide to common tidal marsh invertebrates of the northeastern Gulf of Mexico. Mississippi-Alabama Sea Grant Consortium, Ocean Springs, MS.

Isphording, W. C., and J. F. Fitzpatrick, Jr. 1992. Geologic and evolutionary history of drainage systems in the southeastern United States. Pages 19–56 in C. T. Hackney, S. M. Adams, and W. A. Martin (eds.), Biodiversity of Southeastern United States: Aquatic Communities. John Wiley & Sons, New York.

Johnson, R. I. 1970. The systematics and zoogeography of the Unionidae (Mollusca: Bivalvia) of the southern Atlantic slope region. Bulletin of the Museum of Comparative Zoology, Harvard University 148:239–320.

Jumars, P. A., and K. Fauchald. 1977. Between-community contrasts in successful polychaete feeding strategies. Pages 1–20 in B. C. Coull (ed.), Ecology of Marine Benthos. University of South Carolina Press, Columbia, SC.

Klerks, P., and J. S. Levinton. 1992. Evolution of resistance and changes in community composition in metal-polluted environments. Pages 223–241 in R. Dallinger and P. S. Rainbow (eds.), Ecotoxicology of Metals in Invertebrates. Lewis, Boca Raton, FL.

Livingston, R. J. 1984. The Ecology of the Apalachicola Bay system: An Estuarine profile. U.S. Fish and Wildlife Service Technical Report FWS/OBS/82-05:1–148.

Livingston, R. J. (ed.). 1991a. The Rivers of Florida. Springer-Verlag, New York.

———. 1991b. Historical relationships between research and resource management in the Apalachicola River estuary. Ecological Applications 1:361–382.

———. 1992. Medium-sized rivers of the Gulf coastal plain. Pages 351–385 in C. T. Hackney, S. M. Adams, and W. A. Martin (eds.), Biodiversity of Southeastern United States: Aquatic Communities. John Wiley & Sons, New York.

Louisiana Department of Environmental Quality (LaDEQ). 1989. Impact Assessment of Bayou Trepagnier. Final Report OWR/02/89/001. Louisiana Department of Environmental Quality, Baton Rouge, LA.

Macquart-Moulin, C. 1984. La phase pelagique et les comportements migratoires des amphipodes benthiques (Mediterranee nord-occidentale). Tethys 11:171–196.

Mangum, C. P. 1973. Evaluation of the functional properties of invertebrate hemoglobins. Netherlands Journal of Sea Research 7:303–315.

———. 1976. Primitive respiratory adaptations. Pages 191–278 in R.C. Newell (ed.), Adaptation to Environment: Essays on the Physiology of Marine Animals. Butterworth's, London, UK.

———. 1985. Oxygen transport in invertebrates. American Journal of Physiology 248: R505–R514.

———. 1990. The fourth Riser lecture: The role of physiology and biochemistry in understanding phylogeny. Proceedings of the Biological Society of Washington 103: 235–247.

Maples, R. 1987. Phytoplankton ecology of Calcasieu Lake, Louisiana. Pages 3A1–3E39 *in* L. R. DeRouen and L. H. Stevenson (eds.), Ecosystem Analysis of the Calcasieu River/Lake Complex. Vol. 2. U.S. Department of Energy (DE-F601-83EP31111), Washington, DC.

Matlock, G. C., and M. A. Garcia. 1983. Stomach contents of selected fishes from Texas bays. Contributions in Marine Science 26:95–110.

McBee, J. T., and W. T. Brehm. 1979. Macrobenthos of Simmons Bayou and an adjoining residential canal. Gulf Research Reports 6:211–216.

McLusky, D. S. 1981. The Estuarine Ecosystem. John Wiley & Sons, New York.

Menzel, R. G., and C. M. Cooper. 1992. Small impoundments and ponds. Pages 389–420 *in* C. T. Hackney, S. M. Adams, and W. A. Martin (eds.), Biodiversity of the Southeastern United States: Aquatic Communities. John Wiley & Sons, New York.

Merritt, R. W., and K. W. Cummins. 1984. An Introduction to the Aquatic Insects of North America (2nd ed.). Kendall/Hunt, Dubuque, IA.

Milligan, M. R. 1979. Species composition and distribution of benthic macroinvertebrates in a Mississippi estuary. Master's Thesis, University of Southern Mississippi, Hattiesburg, MS.

Mitsch, W. J., and J. G. Gosselink. 1993. Wetlands (2nd ed.). Van Nostrand Reinhold, New York.

Moore, R. H. 1992. Low-salinity backbays and lagoons. Pages 541–614 *in* C. T. Hackney, S. M. Adams, and W. A. Martin (eds.), Biodiversity of the Southeastern United States: Aquatic Communities. John Wiley & Sons, New York.

Moshiri, G. A., W. G. Crumpton, and D. A. Blaylock. 1978. Algal metabolites and fish kills in a bayou estuary: An alternative explanation to the low dissolved oxygen controversy. Journal of the Water Pollution Control Federation 50:2043–2046.

Mueller, C. S., G. J. Ramelow, and J. N. Beck. 1989. Mercury in the Calcasieu River/Lake complex. Bulletin of Environmental Contamination and Toxicology 42:71–80.

Mulino, M. M. 1975. The influence of brackish-water intrusion on macroinvertebrate associations of the lower Tchefuncte River, Louisiana. Master's Thesis, University of New Orleans, New Orleans, LA.

Murray, H. E., and J. N. Beck. 1989. Halogenated organic compounds found in shrimp from the Calcasieu Estuary. Chemosphere 19:1367–1374.

———. 1990. Concentrations of selected chlorinated pesticides in shrimp collected from the Calcasieu River/Lake complex, Louisiana. Bulletin of Environmental Contamination and Toxicology 44:798–804.

Murray, H. E., G. R. Gaston, and C. Murphy. 1992. Presence of hexachlorobenzene in blue crabs of the Calcasieu Estuary, Louisiana. Journal of Environmental Science and Health A27:1095–1101.

Neuswanger, D. J., W. W. Taylor, and J. B. Reynolds. 1982. Comparison of macroinvertebrate herpobenthos and haptobenthos in side channel and slough in the upper Mississippi River. Freshwater Invertebrate Biology 1:13–24.

Odum, W. E. 1988. Comparative ecology of tidal freshwater and salt marshes. Annual Review of Ecology and Systematics 19:147–176.

Odum, W. E., T. J. Smith, J. K. Hoover, and C. C. McIvor. 1984. The ecology of tidal freshwater marshes of the United States east coast: A community profile. U.S. Fish and Wildlife Service Technical Report FWS/OBS-83/17:1–177.

Odum, W. E., L. Rozas, and C. McIvor. 1988. A comparison of fish and invertebrate community composition in tidal freshwater and oligohaline marsh systems. Pages 561–568 *in* D. D. Hook et al. (eds.), The Ecology and Management of Wetlands. Timber Press, Portland, OR.

O'Hara, J. 1967. Invertebrates found in water hyacinth mats. Quarterly Journal of the Florida Academy of Science 30:73.

Overstreet, R. M., and R. W. Heard. 1978. Food of the red drum, *Sciaenops ocellata,* from Mississippi Sound. Gulf Research Reports 6:131–135.

———. 1982. Food contents of six commercial fishes from Mississippi Sound. Gulf Research Reports 7:137–149.

Patrick, R. 1994. Rivers of the United States. Vol. 1. John Wiley & Sons, New York.

Penn, G. H. 1960. An illustrated key to the crawfishes of Louisiana with a summary of their distribution within the state (Decapoda, Astacidae). Tulane Studies in Zoology 7:3–20.

Pennak, R. W. 1989. Freshwater Invertebrates of the United States. Protozoa to Mollusca (3rd ed.). John Wiley & Sons, New York.

Pereira, W. E., C. E. Rostad, C. T. Chiou, T. I. Brinton, L. B. Barber, and D. K. Demcheck. 1988. Contamination of estuarine water, biota, and sediment by halogenated organic compounds: A field study. Environmental Science and Technology 22:772–778.

Peterson, M. S. 1997. Spatial and temporal changes in subtidal benthic crustaceans along a coastal river-estuarine gradient in Mississippi. Gulf Research Reports 9: 321–326.

Petrick, D. 1981. Intensive Survey of Hunting Bayou. Texas Department of Water Resources, Austin, TX.

Platt, S. G. 1988. A checklist of the flora of the Manchac Wildlife Management Area, St. John the Baptist Parish, Louisiana. Proceedings of the Louisiana Academy of Sciences 51:15–20.

Poirrier, M. A. 1974. Ecomorphic variation in gemmoscleres of *Ephydatia fluviatilis* Linnaeus (Porifera: Spongillidae) with comments upon its systematics and ecology. Hydrobiologia 44:337–347.

———. 1978. *Corvospongilla becki* n. sp., a new fresh-water sponge from Louisiana. Transactions of the American Microscopical Society 97:240–243.

Poirrier, M. A., and G. J. Denoux. 1973. Notes on the distribution, ecology and morphology of the colonial hydroid *Cordylophora caspia* (Pallas) in southern Louisiana. Southwestern Naturalist 18:253–255.

Poirrier, M. A., and S. A. Johnson. 1970. Notes on the distribution and ecology of *Urnatella gracilis* Leidy, 1851 (Entoprocta) in Louisiana. Proceedings of the Louisiana Academy of Sciences 33:43–45.

Poirrier, M. A., and M. M. Mulino. 1977. Effects of environmental factors on the distribution and morphology of *Victorella pavida* (Ectoprocta) in Lake Pontchartrain, Louisiana, and vicinity. Chesapeake Science 18:347–352.

Poirrier, M. A., and M. R. Partridge. 1979. The barnacle, *Balanus subalbidus,* as a salinity bioindicator in the oligohaline estuarine zone. Estuaries 2:204–206.

Ramelow, G. J., C. L. Webre, C. S. Mueller, J. N. Beck, J. C. Young, and M. P. Langley. 1989. Variations of heavy metals and arsenic in fish and other organisms from the Calcasieu River and Lake, Louisiana. Archives of Environmental Contamination and Toxicology 18:804–818.

Remane, A., and C. Schlieper. 1971. Biology of Brackish Water. Wiley Interscience, New York.

Sikora, W. B., and F. H. Sklar. 1987. Benthos. Pages 58–79 *in* W. H. Conner and J. W. Day (eds.), The Ecology of Barataria Basin, Louisiana. U.S. Fish and Wildlife Service Biological Report 85(7.13):1–165.

Sklar, F. H. 1985. Seasonality and community structure of the backswamp invertebrates in a Louisiana cypress-tupelo wetland. Wetlands 5:69–86.

Smock, L. A., and E. Gilinsky. 1992. Coastal plain blackwater streams. Pages 271–311 *in* C. T. Hackney, S. M. Adams, and W. A. Martin (eds.), Biodiversity of the Southeastern United States: Aquatic Communities. John Wiley & Sons, New York.

Stearns, D. E., and M. R. Dardeau. 1990. Nocturnal and tidal vertical migrations of "benthic" crustaceans in an estuarine system with diurnal tides. Northeast Gulf Science 11:93–104.

Stern, M. K., J. W. Day, Jr., and K. G. Teague. 1991. Nutrient transport in a riverine-influenced, tidal freshwater bayou in Louisiana. Estuaries 14:382–394.

Stout, J. P. 1990. Estuarine habitats. Pages 63–88 *in* Mobile Bay: Issues, Resources, and Management. NOAA Estuary-of-the-Month Seminar No. 15. U.S. Department of Commerce, NOAA Estuarine Programs Office, Washington, DC.

Sullivan, M. J. 1981. Relationship between Water Quality and the Community Structure of Epiphytic Diatoms in a Mississippi bayou. Water Resources Research Institute Report OWRT-A-133-MISS(2):1-27. Mississippi State University, Starkville, MS.

Summers, J. K., J. M. Macauley, P. T. Heitmuller, V. D. Engle, A. M. Adams, and G. T. Brooks. 1992. Annual Statistical Summary: EMAP-estuaries Louisianian Province—1991. EPA/600/R-93/001. U.S. Environmental Protection Agency, Gulf Breeze, FL.

Swingle, H. A., and D. G. Bland. 1974. Distribution of the estuarine clam, *Rangia cuneata* Gray in coastal waters of Alabama. Alabama Marine Resources Bulletin 10:9–16.

Tagatz, M. E. 1971. Osmoregulatory ability in blue crabs in different temperature-salinity combinations. Chesapeake Science 12:14–17.

Tarver, J. W., and R. J. Dugas. 1973. A study of the clam, *Rangia cuneata* in Lake Pontchartrain and Lake Maurepas, Louisiana. Louisiana Wildlife and Fisheries Commission Technical Bulletin 5:1–99.

Tenore, K. R., D. B. Horton, and T. W. Duke. 1968. Effects of bottom substrate on the brackish water bivalve, *Rangia cuneata*. Cheasapeke Science 3:238–248.

Thomas, J. D. 1976. A survey of gammarid amphipods of the Barataria Bay, Louisiana region. Contributions in Marine Science 20:87–100.

Thorp, A. G., R. C. Jones, and D. P. Kelso. 1997. A comparison of water-column macroinvertebrate communities in beds of differing submersed aquatic vegetation in the tidal freshwater Potomac River. Estuaries 20:86–95.

Thorp, J. H., and A. P. Covich. 1991. Ecology and Classification of North American Freshwater Invertebrates. Academic Press, New York.

Turner, R. E. 1997. Wetland loss in the northern Gulf of Mexico: Multiple working hypotheses. Estuaries 20:1–13.

Twidwell, S. R. 1981. Intensive survey of Armand Bayou. Publication IS-20. Texas Department of Water Resources, Austin, TX.

Vecchione, M. 1987. Variability in the distribution of late-stage oyster larvae in the Calcasieu estuary. Contributions in Marine Science 30:77–90.

———. 1989. Zooplankton distribution in three estuarine bayous with different types of anthropogenic influence. Estuaries 12:169–179.

———. 1991. Long-term trends in the abundance of the copepod *Acartia tonsa* in the Calcasieu estuary. Contributions in Marine Science 32:89–101.

Watkins, C. E., II, J. V. Shireman, and W. T. Haller. 1983. The influence of aquatic vegetation upon zooplankton and benthic macroinvertebrates in Orange Lake, Florida. Journal of Aquatic Plant Management 21:78–83.

Watson, M. B., C. J. Killebrew, M. H. Schurtz, and J. L. Landry. 1981. A preliminary survey of Blind River, Louisiana. Warmwater Streams Symposium. American Fisheries Society 1981:303–319.

Weisberg, S. B., J. A. Ranasinghe, D. M. Dauer, L. C. Schaffner, R. J. Diaz, and J. B. Frithsen. 1997. An estuarine benthic index of biotic integrity (B-IBI) for Chesapeake Bay. Estuaries 20:149–158.

Wetzel, R. G. 1983. Limnology (2nd ed.). W.B. Saunders, Philadelphia, PA.

Whitledge, T. E. 1985. Nationwide review of oxygen depletion in estuarine and coastal waters. Brookhaven National Lab, Brookhaven, NY.

Wiseman, W. J., Jr., E. M. Swenson, and J. Power. 1990. Salinity trends in Louisiana estuaries. Estuaries 13:265–271.

Ziser, S. W. 1978. Seasonal variations in water chemistry and diversity of the phytophilic macroinvertebrates of three swamp communities in southeastern Louisiana. Southwestern Naturalist 23:545–562.

38 Coastal Wetlands of the Upper Great Lakes

Distribution of Invertebrate Communities in Response to Environmental Variation

JOSEPH P. GATHMAN, THOMAS M. BURTON, and
BRIAN J. ARMITAGE

W*etland invertebrate communities have been studied in only a few Great Lakes coastal regions: Lake St. Clair, the St. Mary's River, Green Bay, and, more recently, Saginaw Bay and the Les Cheneaux Islands region of northern Lake Huron. These communities are diverse compared to the nonwetland benthos and are characterized by certain abundant and diverse taxa, such as Gastropoda, Crustacea, and especially Insecta. Direct and indirect effects of water movements especially influence invertebrates. Wave energy is attenuated by emergent vegetation, creating mixing gradients between pelagic and littoral waters to which invertebrates appear to respond. Complex vegetation structure corresponds with complex invertebrate communities, so where wave and wind exposure reduce plant diversity, invertebrate diversity is relatively low. Sediment quality is strongly influenced by waves and also appears to affect invertebrate communities. Characteristic abiotic factors, such as interannual and annual water level changes, seiches, storm surges, and ice scour, are likely to have profound influences on invertebrate communities, but have not been studied. We conclude that:

1. The current state of knowledge of coastal wetland communities is rudimentary, with basic descriptive work still lacking.

Invertebrates in Freshwater Wetlands of North America: Ecology and Management, Edited by
Darold P. Batzer, Russell B. Rader, and Scott A. Wissinger
ISBN 0-471-29258-3 © 1999 John Wiley & Sons, Inc.

2. *Hydrology is an important factor distinguishing these wetlands and deserves much more study.*

3. *Interactions between hydrology and other factors are proximate causes of invertebrate distributions within wetlands.*

Descriptive work is still necessary, but comparability among sites should be emphasized. Experimental studies of mechanisms are sorely needed, but are hindered by the lack of basic descriptive data.

INTRODUCTION

Knowledge of invertebrate communities of coastal wetlands of the Laurentian Great Lakes is based on studies of only a few sites, most of which are located along Lakes Erie and Ontario and their connecting waterways (see reviews by Krieger 1992, Botts, this volume). Herdendorf et al. (1981a, b, c) review the literature on the coastal wetlands of the Great Lakes and find no wetland-only or site-specific reports of invertebrate studies from Lakes Superior, Huron, Michigan, or St. Clair, or from their connecting channels, including the Detroit River. Krieger's (1992) review includes data from two wetland areas of the upper Great Lakes shores: the St. Mary's River (Duffy et al. 1987) and Green Bay (McLaughlin and Harris 1990). Since that time, invertebrate communities of the *Scirpus americanus*-dominated zone of a Saginaw Bay wetland have received considerable attention (Brady 1992, 1996, Brady and Burton 1995, Brady et al. 1995, Cardinale 1996, Cardinale et al. 1997), but those of most other sites have remained unstudied. It remains an enigma that so little is known about these systems, even though many general discussions of their high value are included in literature and policy about the Great Lakes.

Since 1993 we have participated with several other researchers in studies designed to develop baseline invertebrate community data for the upper Great Lakes. Most data collected on these projects are yet to be published. The objective of this review is to draw preliminary conclusions about invertebrate ecology of coastal wetlands of the upper Great Lakes based on summaries of some of these unpublished data, published data from Saginaw Bay wetlands, and general field observations.

WETLANDS OF THE UPPER GREAT LAKES

While a distinction between the "upper" and "lower" Great Lakes may seem artificial, Cook and Johnson (1974) conclude that the deepwater benthos of the upper Great Lakes was significantly different from the deepwater benthos of the lower lakes. Further, Smith et al. (1991) report that Lake Huron wetlands in Canada had more swamp and fen habitats, and more complex plant communities, than did sites on Lakes Erie, Ontario, and St. Clair. Lake Erie wetlands tended to occur in sheltered bays behind barrier beaches, while Lake Huron was characterized by lacustrine wetlands that were exposed to some wave action and storm events, apparently precluding development of organic soils.

Distribution and General Description

About 50 percent of pre-settlement wetlands of the upper Great Lakes have been converted to other uses (Brazner 1997, Comer et al. 1995, Jaworski and Raphael 1978). The majority were lost to agricultural drainage in the late 1800s, especially in southern Michigan. Wet meadows, lake-plain prairie wetlands, and adjacent swamps have been especially vulnerable, with losses approaching 100 percent in certain areas (Comer et al. 1995, Jaworski and Raphael 1978). Many of these wetlands extended inland for several kilometers, and their loss, combined with increased nutrient, sediment, and toxic inputs to the remaining littoral wetlands, has probably had major impacts on the biota.

Prince et al. (1992) identify 65,547 ha of remaining coastal wetland area in eight major areas of the upper Great Lakes, mostly on Lakes Michigan and Huron. This estimate should be viewed as conservative because it includes only major wetland complexes of known importance to waterfowl. The largest of these complexes are located along Saginaw Bay (12,140 ha) and Georgian Bay (12,600 ha) on Lake Huron, Green Bay (9,980 ha) and Big Bay de Noc (7,720 ha) on Lake Michigan, and Chequamegon Bay (4,170 ha) on Lake Superior.

Almost 90 percent of the Great Lakes' shoreline consists of sandy and rocky environments that are exposed to high-energy wave activity (Prince et al. 1992). Coastal wetlands develop only where shoreline morphology offers at least partial protection from the direct force of waves and storm surges. Maynard and Wilcox (1997) classify Great Lakes coastal wetlands into eight types, based on shoreline form:

1. Open shoreline wetlands
2. Unrestricted bay wetlands
3. Shallow sloping beach wetlands
4. River delta wetlands
5. Restricted riverine wetlands
6. Lake-connected inland wetlands
7. Barrier beach wetlands
8. Diked wetlands (Fig. 38.1)

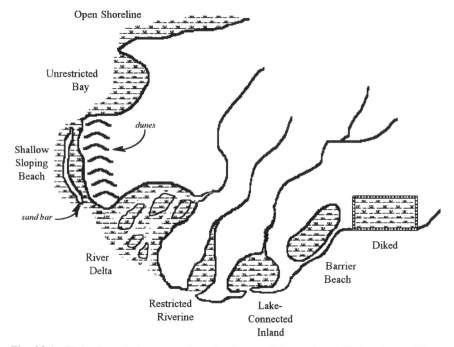

Fig. 38.1. Eight Great Lakes coastal wetland types (Maynard and Wilcox 1997; Prince et al. 1992).

Diked wetlands are among the most common types of wetlands along the Lake Erie shoreline, but are less common in the upper Great Lakes.

Minc (1996) reports that a transect extending from deep water to the shoreline, in most coastal wetlands, would encounter most or all of the following vegetation zones (Fig. 38.2):

1. A submergent marsh zone
2. An emergent marsh zone, with widths varying from less than 10 m to more than 500 m, depending on how protected the site is (personal observation)
3. A narrow but diverse shoreline zone, 5 to 30 m wide (personal observation)
4. A sedge and grass dominated herbaceous zone (wet meadow) of variable width
5. A shrub-dominated zone
6. A swamp forest

The submergent plant zone (Fig. 38.2) usually does not extend lakeward beyond the emergent zone except in marshes with the greatest protection from

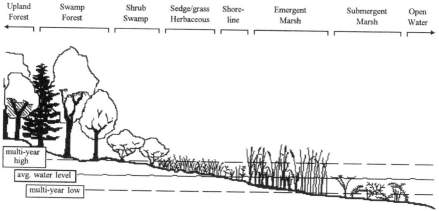

Fig. 38.2. Profile of typical coastal wetland vegetation zones in relation to water levels (Minc 1996).

waves (Minc 1996). The submergent zone may also be absent where suspended sediment loads are very high, as they are in southern Green Bay (H. J. Harris, personal communication). In most wetlands submergent plant zones were found primarily in erosional openings created by wind and wave action within the emergent marsh (Minc 1996).

Physical Environment

The areal extent of Great Lakes coastal wetlands depends on mean lake levels, which rise and fall by as much as 150 cm or more over periods of 7 to 15 years in response to variation in precipitation and evapotranspiration. For example, Saginaw Bay levels recorded in 1986 were the highest of the last several decades, but then decreased by about 1.3 m to the low levels of 1990 (Fig. 38.3). By 1997 Lake Huron had risen to only about 15 cm below the 1986 highs (personal observation). Historically, these hydrological "cycles" have caused plant communities to move inland as water levels rose and lakeward as water levels fell (Burton 1985, Mitsch 1992). The inland migration of plant communities is still possible in many northern areas of the upper three Great Lakes but is restricted in many wetlands by structures such as dikes, seawalls, and road embankments.

Superimposed on long-term lake level fluctuations are annual fluctuations with amplitudes of 20 to 40 cm (Fig. 38.3). The time required for snowmelt water from the entire upper Great Lakes basin to accumulate delays high lake levels until July. As a result of this lag, maximum flooding in coastal wet meadows occurs much later than in nearby inland wetlands, which generally reach their highest levels shortly after spring thaw. Spatial variation in flooding regime determines plant community distributions. For example, the narrow shoreline (strand) zone usually occurs near the mean annual water level,

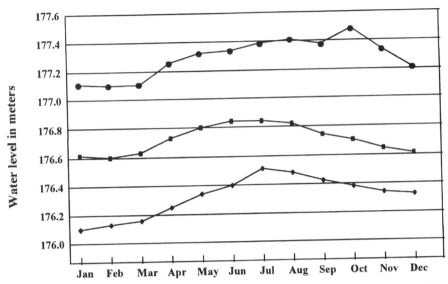

Fig. 38.3. Seasonal water level fluctuations in Saginaw Bay, 1983 through 1992 (Prince and Burton 1996). ◆—Low water year, 1990; ■—10-year average, 1983–1992; ●—High-water year, 1986.

while wet meadows are found above this level. When predicted patterns are not observed, it is often because changes in vegetation distribution lag behind long-term changes in mean annual water level.

Short-term water-level fluctuations occurring over scales of minutes to hours are superimposed on the annual and multiyear cycles in coastal wetlands (see Bedford 1992). Wind-driven seiches cause sinusoidal water-level fluctuations with amplitudes in the 10–20 cm range and periods that vary from less than an hour (personal observation) to as much as 14 hours (Bedford 1992). These fluctuations result in temporal variation in habitat conditions as shoreline areas alternate between inundation and dewatering and as submerged macrophyte stem area increases and decreases.

Seiche activity combined with wave energy results in mixing of lake and wetland water as much as 150 m or more into the emergent plant zone (Suzuki et al. 1995, Brady et al. 1995, Cardinale 1996, Cardinale et al. 1997). Because many wetlands on Saginaw Bay are more than 500 m wide, water in the inner 350+ m does not completely mix with bay water when emergent plant (*Scirpus*) density is high enough to substantially reduce wave energy. This results in measurable water quality differences between inner and outer parts of these marshes from mid-June through the rest of the growing season. How widely this phenomenon occurs has not yet been determined, but researchers conducting invertebrate studies of fringing wetlands of large lakes should be aware of these potential gradients.

Storm surges, unlike seiches, cause major short-term fluctuations in water level of as much as 90 cm or more (personal observation, Whitt 1996) over periods of 24 hours or less. Whitt (1996) documents three such storm surges in Saginaw Bay in 1994 and 1995. One such event destroyed almost all muskrat lodges in the emergent zone and killed many of the newborns (personal communication, V. Brady). The surge also destroyed bird nests in the wetland (Whitt 1996) and sheared the tops off of many of the emergent plants (personal observation). These storm surges would have led to a comparable dewatering of wetlands on the opposite shoreline of the bay. The effects of storm surges on invertebrate communities are unknown, but are likely to be important in determining community structure.

Fringing wetlands of the littoral zone freeze to depths of 50 cm or more during most winters, making these shallow areas particularly harsh environments for benthic organisms. As ice breaks up, it is moved by waves and water level fluctuations, scouring plants and animals from the substrate and contributing to low early-season invertebrate abundance and richness. Immigration and egg hatches allow communities to recover and reach peak density from July to September.

The combination of ice scour, storm surges, and other water movements alters emergent plant community structure. Ice scour and storm surges also transport much of the detritus out of the marsh into deeper water or deposit it as thick mats along exposed shorelines. In many *Scirpus*-dominated areas the bottom is composed of sand and silt over deposits of clay and is relatively low in organic matter, although "islands" of *Typha* can trap organic detritus within and shoreward of themselves. Cattail stands and patches of submergent and floating-leaved plants within the *Scirpus* zone combine with substrate variations to make the emergent marsh a very patchy environment.

INVERTEBRATE FAUNA OF GREAT LAKES COASTAL WETLANDS

Coastal wetlands are distinguished from other upper Great Lakes' coastal and deepwater habitats by high microhabitat and taxonomic diversity, but the many regional and local differences among these habitats and their resident fauna make it inappropriate to discuss "the" coastal wetland fauna. In fact, Cook and Johnson (1974) suggest that the harshness and variety of inshore environments, and their relative isolation from each other, could result in relatively rapid speciation, citing records of gastropod subspecies that are unique to Lake Superior and adjacent inland waters in the Hudson Bay drainage.

The biota of the Great Lakes connecting channels have received considerable attention (e.g., Kauss 1991, Leach 1991, Manny and Kenaga 1991), as have the nearshore nonwetland benthos (e.g., Barton and Griffiths 1984, Griffiths 1991, Mozley and Garcia 1972). Many of these studies focus on the loss and recovery of indicator taxa, such as mayfly and caddisfly larvae, resulting from pollution and subsequent abatement efforts in the Great Lakes in the recent past (e.g., Davis et al. 1991, Mozley and LaDronka 1988, Schloesser

1988). Such studies are of limited use in understanding wetland communities because they appear not to have included wetlands and because commonly used indicator taxa are not necessarily wetland residents. For example, burrows of pollution-sensitive mayflies, such as *Hexagenia limbata,* are quite common just offshore of the emergent zones of the Les Cheneaux Islands area of northern Lake Huron (P. Hudson, personal communication), but our observations in the same region suggest that they are rare in wetlands.

Microinvertebrates

Even though coastal wetlands are important habitats for many Great Lakes planktivorous fishes (Liston and Chubb 1986, Jude and Pappas 1992, Brazner 1997), the only published data available for zooplankton in upper Great lakes wetlands are from the St. Mary's River (Duffy et al. 1987) and a Saginaw Bay wetland (Brady 1992, Brady and Burton 1995). The distinction between plankton and benthos blurs in wetlands, with "zooplankton" often being associated with plant stems and benthic detritus and many "benthic" invertebrates found well above the sediments, either on stems and floating debris or swimming. "Zooplankton" are included in macroinvertebrate samples if fine-mesh sampling devices are used or if water above the sediment surface is included in sediment cores. Brady and Burton (1995) quantify microcrustacea that were collected by incidental capture during macroinvertebrate sampling using 250-μm mesh plant stem samplers and sediment corers that included up to 35 cm of overlying water (Table 38.1).

Comparisons between the Saginaw Bay and St. Mary's River wetland zones must be viewed with caution because different sampling procedures were used, and Duffy et al.'s (1987) data from the St. Mary's River appear not to include benthic microcrustaceans as Brady's (1992) Saginaw Bay samples do. Even so, there are interesting similarities between these two very different coastal wetland areas. Cladocera was the dominant group of microcrustaceans collected from both Saginaw Bay and St. Mary's River coastal wetlands (Table 38.1). Duffy et al. (1987) report 27 species of Cladocera from St. Mary's River wetlands, while Brady and Burton (1995) report 26 species from Saginaw Bay. The numerically dominant cladocerans were in the genus *Chydorus,* with *C. sphaericus* being the dominant species in the St. Mary's River (Duffy et al. 1987). The genus *Chydorus* was the most abundant taxon present in Saginaw Bay during spring and early summer, decreased in abundance to low levels in July and early August, and increased to annual highs (>100,000 individuals/m^3) in late October (Brady and Burton 1995). *Acroperus harpae* was the second most abundant taxon in both wetlands (Table 38.1) but tended to be present in significant numbers only from August through October. *Sida crystallina* and *Eurycercus lamellatus* were abundant large-bodied Cladocera at both sites and often comprised the largest proportion of biomass along with various species of *Alona* and *Simocephalus.* Additional information on changes in species composition through the year in Saginaw Bay is available in graphical form (Brady and Burton 1995).

TABLE 38.1. Crustacean Zooplankton Abundance (# m⁻³) in St. Mary's River and Saginaw Bay Wetlands

Taxon	St. Mary's		Saginaw Bay	
	Open Water[a]	Emergent[a]	Emergent[b]	Sediment[b]
Copepoda				
Cyclopoida				
Acanthocyclops vernalis			410–27,168	0–28,303
Cyclops bicuspidatus thomasi	6			
C. vernalis	143			
Eucyclops agilis			74–24,109	0–44,446
Macrocyclops albidis		137,000		
M. edax	1,272			
Calanoida				
Diaptomus ashlandi	41			
D. minutus	7			
D. sicilis	61			
Epischura lacustris	4			
Eurytemora affinis			0–11,012	
Limnocalanus mucrurus	16			
Skistodiaptomus oregonensis			0–447	
Harpacticoida			14–8,792	858–61,880
Cladocera				
Acroperus harpae		365,000	0–59,189	0–4,359
Alona guttata		92,000		
A. exigua		5,000		
A. intermedia		11,000		
A. quadrangularis		29,000	42–17,901	0–36,740
A. rectangularis		87,000		
Alonella acutirostris		46,000		

957

TABLE 38.1. (Continued)

Taxon	St. Mary's		Saginaw Bay	
	Open Water[a]	Emergent[a]	Emergent[b]	Sediment[b]
Bosmina longirostris	6	13,000	0–9,955	
Camptocerus rectirostris		64,000	0–25,916	
Ceriodaphnia (total)	<1		0–13,396	
Ceriodaphnia lacustris	<1			
C. megalops		<1,000		
C. quadrangula		41,000		
Chydorus (total)		1,342,000	0–107,156	0–83,592
Chydorus gibbosus		3,000		
C. sphaericus		1,042,000		
Daphnia galeata mendotae	5			
D. retrocurva	3			
Eurycercus lamellatus	13	106,000	0–14,360	0–4,902
Graptoleberis testudinella		29,000		
Holopedium gibberum	22			
Ilyocryptus spinifer		<1,000	0–11,090	0–20,110
Lathonura rectirostris		7,000		
Latona parviremus	<1			
L. setifera			0–1,721	
Leptadora kindtii	92			
Macrothrix laticornis		<1,000	0–5,785	0–14,582
M. rosea		32,000	0–8,134	
Pleuroxus denticulatus		38,000	0–2,483	
P. procurvus		<1,000	0–7,402	
P. truncatum		<1,000		
P. uncinatus		<1,000		

Polyphemus pediculus	1	3,000		
Ophryoxus gracilis		<1,000		
Rhynchotalona falcata		<1,000		
Scapholeberis kingi			0–1,593	
Sida crystallina	138	44,000	0–30,693	0–3,094
Simocephalus (total)		222,000	0–33,752	
Simocephalus serrulatus		222,000		
Ostracoda		201,000	1,392–36,868	25,193–137,283

[a] Duffy et al. (1987); "open water" refers to nonvegetated, soft-bottom areas.
[b] Brady (1992).

Overall, microcrustacean abundance was lower in Saginaw Bay than in St. Mary's River sites. Ostracods were fairly abundant in both sites but not identified to species (Table 38.1). Only one species of cyclopoid Copepoda was reported in the St. Mary's emergent wetlands, while two different species were reported from Saginaw Bay. These represent only 3 of the 13 genera of Cyclopoida known to occur in the upper Great Lakes (Table 38.2). Duffy et al. (1987) found no Calanoida in the emergent zone of the St. Mary's River wetlands, while Brady and Burton (1995) found two species from Saginaw Bay (Table 38.1). Brady and Burton (1995) also found significant numbers of Harpactocoida copepods in sediment samples from Saginaw Bay emergent wetlands, but Duffy et al. (1987) does not report any from the St. Mary's wetlands, even though 10 genera are known from the Upper Great lakes (Table 38.2). Zooplankton abundances in the emergent wetlands of the St. Mary's River were more than an order of magnitude greater, and Cladocera were much more dominant, than in open-water (soft-bottomed, nonvegetated) areas of the river (Duffy et al. 1987, Table 38.1).

Saginaw Bay wetland microcrustacea can be placed into categories according to seasonal patterns of abundance (Brady and Burton 1995). Two cladoceran species were abundant only in summer, six were abundant only in fall, and no species were abundant only in the spring. Several taxa were abundant in two seasons; the genus *Chydorus* and two other cladoceran species were abundant during spring and fall, and three cladoceran species and the ostracods were abundant in summer and fall. The authors suggest that the phenology observed may be related to seasonal changes in feeding strategies.

TABLE 38.2. Genera of Copepoda (Harpacticoida and Cyclopoida) from the Upper Great Lakes

Harpactacoida	Cyclopoida
Attheyella (4)	*Acanthocyclops* (5)
Bryocamptus (7)	*Cyclops* (2)
Canthocamptus (5)	*Diacyclops* (5)
Elaphoidella (1)	*Ectocyclops* (1)
Epactophanes (1)	*Eucyclops* (3)
Gulcamptus (1)	*Homocyclops* (1)
Moraria (5)	*Macrocyclops* (2)
Nitokra (3)	*Megacyclops* (1)
Parastenocaris (2)	*Mesocyclops* (2)
Phyllognathopus (1)	*Microcyclops* (2)
	Orthocyclops (1)
	Paracyclops (3)
	Tropocyclops (1)

Source: Hudson, personal communication. Numbers of species are in parentheses.

Copepods were abundant at all times, possibly because they are omnivorous. The spring-abundant cladoceran assemblage was a combination of filterers and sediment-associated species which may have been responding to early phytoplankton and periphyton production, whereas the summer-abundant cladocerans required macrophytes for attachment.

Rotifera and smaller species of microcrustaceans were not well represented in Saginaw Bay or St. Mary's River samples. In a follow-up study on the effects of zebra mussel invasion in the Saginaw Bay wetland, Brady (1996) found 1000–2000 rotifers/m^3 and 1000 or more individuals/m^3 of *Bosmina longirostris* and *Ceriodaphnia dubia*. Her data suggested that these taxa and many Ostracoda, smaller Copepoda, and their nauplii were significantly underrepresented in sampling reported by Brady (1992) and Brady and Burton (1995). These taxa were also rare in the Duffy et al. (1987) report on St. Mary's wetlands.

Macroinvertebrates

Comparisons of variables such as standing stocks, diversity, and production among the few wetland sites studied so far (Table 38.3) are of limited usefulness because of the variability in methodologies and variables measured. Taxa richness, for example, has a different meaning in each study, depending on the levels of taxonomic resolution achieved. Also, some studies sampled only one type of emergent vegetation (French 1988, Brady 1992) or were not specific about microhabitats sampled (Duffy et al. 1987). Brady's (1992) work is probably the most intensive study of a single site, but does not include annual productivity, while Duffy et al. (1987) report productivity estimates, but not invertebrate densities. Table 38.3 is probably best regarded as a demonstration of incompatibility among studies, rather than as a comparison of data from different sites.

Our unpublished data from the wetlands of Saginaw Bay (Tables 38.4 and 38.5) are from a study of several coastal-zone sites using similar quantitative and semiquantitative methods, so comparisons among sites should be reliable. Only two of the seven sites sampled were directly connected to Saginaw Bay (the littoral *Scirpus* zone and undisturbed littoral wet meadow sites). The other five sites were separated from the lake littoral emergent zone by a sandy ridge less than 200 m wide (inland swale forest, inland swale wet meadow, inland emergent zone, and inland wet meadow sites) or were several hundred meters inland from the shoreline (disturbed inland wet meadow), but their water levels appeared to be strongly influenced by lake levels through subsurface hydrologic connections. By far the most taxa (69) were found in the littoral *Scirpus* zone, but this may have resulted in part from this site being sampled more intensively than the others. Taxa richness was second highest in the inland swale wet meadow (27) and ranged from 12 to 23 in the other sites. Only Isopoda, Hydracarina, Planorbidae, Chironomidae, and Ceratopogonidae were common to all or most sites (Tables 38.4 and 38.5). Dragonflies were

TABLE 38.3. Summaries of Several Great Lakes Coastal Wetland Macroinvertebrate Communities

Author(s)	Study Location	Sampling Method	Microhabitat	Mesh Size (μm)	Invertebrate Density (# m^{-2})	Taxa Richness	Standing Stock (mg m^{-2})	Production (mg m^{-2} yr^{-1})
Brady (1992)	Saginaw Bay	sediment core	sediment	250	~20,000 2-year avg.[a]; 48,900 max.	34	~1,320 2-year avg.[a]; 2,300, max.	n.a.
Brady (1992)	Saginaw Bay	modified Gerking	plant stems/ water column	250	~13,437 2-year avg.[a]; 50,000 max.	91	~718 2-year avg.[a]; 1,500 max.	n.a.
Brown et al. (1988)	Anchor Bay, Lake St. Clair	modified KUG	sumbersed plants and water column	592	~2,379 avg.[a]; 4,947 max.	49	790 max.	n.a.
Cole and Weigman (1983)	Sebewaing Harbor, Saginaw Bay	Ponar dredge	sediment	600	~2,617 avg.[a]	28	18,915 max.	n.a.
Duffy (1985)	St. Mary's River	not reported	Sparganium	not reported	24,043	n.a.	n.a.	n.a.
Duffy et al. (1987)	St. Mary's River	not reported	"benthos"	not reported	>12,000 avg.; 19,267 max.	171	n.a.	24,682 max.
French (1988)	Belvidere Bay, Lake St. Clair	Ponar and KUG	sediment, plants, water column	visible to unaided eye	~4,499 avg.[a]; 8,181 max.	n.a.	n.a.	n.a.

Herdendorf et al. (1986)	Anchor Bay, Lake St. Clair	n.a.	"macrozoo-benthos"	n.a.	7,665	n.a.	n.a.	n.a.
McLaughlin and Harris (1990)	Green Bay, diked marsh	emergence	aerial	not reported	60.8/24 hrs	21 families; 46 genera	30.8/24 hrs	~11,242[a]
McLaughlin and Harris (1990)	Green Bay, undiked marsh	emergence	aerial	not reported	41.9/24 hrs	18 families; 36 genera	17.8/24 hrs	~6,497[a]

[a] Values not reported by author—estimated from published data.

TABLE 38.4. Insects Collected from Seven Wetland Habitats in the Coastal Zone of Saginaw Bay, Lake Huron[a]

Taxon	Littoral Scirpus Zone[b]	Inland Swale Forest	Inland Swale Wet Meadow	Littoral Wet Meadow	Inland Wet Meadow	Inland Emergent Zone	Disturbed Inland Meadow
Coleoptera							
Circulionidae							
Dytiscidae							
Colymbetes sp.							U
Copelatus gyphicus	U			C			
Deronectes sp.				C			C
Derovatelus sp.							C
Dytiscus sp.			U			U	
Hygrotus sp.	U						
Laccophilus sp.				C			
Uvarus sp.				C			
Elmidae							
Dubiraphia sp.	U						
Stenelmis sp.	U						
Haliplidae							
Haliplus sp.			U	U			
Hydrophilidae							
Berosus sp.				U			
Hydrobius sp.							U
Helophorus sp.							C
Unidentified				C			
Gyrinidae							
Gyrinus sp.	C						
Scirtidae							
Microcara sp.							U
Unidentified		U					
Staphylinidae							
Phloeonomus sp.	U	U		U			C
Collembola							
Isotomidae							
Isotomurus tricolor	U						
Sminthuridae							
Bourletiella sp.	U						
Pseudobourletiella sp.	U						

Taxon						
Diptera						
Ceratopogonidae	C		U	U		
Chironomidae	A	A	C	C	A	U
Culicidae						C
Aedes sp.					A	A
Unidentified		U			A	A
Dixidae						
Dixella sp.	U		U	U	U	
Ephydridae						
Ilythea sp.	U		C	C		
Notiphila sp.	U					
Muscidae						
Psychodidae					U	
Stratiomyidae						
Hedriodiscus sp.	U		C			
Tabanidae	A			C	U	
Tipulidae						
Ephemeroptera						
Baetidae						
Baetis levitans	U					
Caenidae						
Brachycercus sp.	U			U		
Caenis sp.	A		C			C
Hemiptera						
Corixidae						
Sigara decorata	C					
Hemiptera/ Corixidae						
Trichocorixa naias	U					U
Gerridae						
Trepobates sp.	C		C			
Gerris sp.	U					
Hebridae						
Merragata brunnea	U					
Merragata hebroides	U					
Macroveliidae						
Macrovelia sp.	U			U		
Mesoveliidae						
Mesovelia mulsanti	U					
Notonectidae						
Buenoa sp.	U					
Notonecta lunata	U					
Veliidae						
Microvelia sp.	U					

TABLE 38.4. (Continued)

Taxon	Littoral Scirpus Zone[b]	Inland Swale Forest	Inland Swale Wet Meadow	Littoral Wet Meadow	Inland Wet Meadow	Inland Emergent Zone	Disturbed Inland Meadow
Lepidoptera							
Noctuidae							
Bellura sp.	U						
Pyralidae							
Acentria niveus	U						
Parapoynx sp.	U						
Petrophila sp.	U						
Megaloptera							
Corydalidae							
Chauliodes sp.		C					
Odonata							
Coenagrionidae							
Enallagma sp.	C		U				
Ishnura verticalis	A		U		C		
Nehalennia sp.			A	A			
Lestidae							
Lestes sp.	U		U				
Gomphidae							
Gomphus sp.	U						
Libellulidae							
Erythemis sp.	U						
Unidentified			U				
Trichoptera							
Hydroptilidae							
Agraylea multipunctata	C						
Hydroptila sp.	C						
Orthotrichia sp.	U						
Oxyethira sp.	C						
Leptoceridae							
Nectopsyche sp.	U		U		U		
Oecetis sp.	C						
Limnephilidae							
Limnephilus sp.		A	A				U
Phryganeidae							
Agrypnia vestita	U				U		
Banksiola sp.			U				

[a]Each taxon has been qualitatively categorized as abundant (A), common (C), or uncommon (U).
[b]Some littoral emergent marsh data were taken from Brady (1992) and Cardinale (1996). This site was studied more intensively than the others.

TABLE 38.5. Noninsects Collected from Seven Wetland Habitats in the Coastal Zone of Saginaw Bay, Lake Huron[a]

Taxon	Littoral *Scirpus* Zone[b]	Inland Swale Forest	Inland Swale Wet Meadow	Littoral Wet Meadow	Inland Wet Meadow	Inland Emergent Zone	Disturbed Inland Meadow
Annelida							
Hirudinea	C						
Oligochaeta							
Lumbricidae	A		U			U	C
Naididae							
Dero sp.	A	A					
Naidum sp.	C						
Ophidonais serpentina	C						
Stylaria lacustris	A	U					
Other	A	A					
Tubificidae							
Branchiura sp.	C				U	C	
Other	C	U			U	U	
Arthropoda							
Amphipoda							
Gammaridae							
Gammarus sp.	A	C	A				
Hyalellidae							
Hyalella azteca	A	U	U		U		
Isopoda							
Asellidae							
Asellus sp.	C	A	A	C			
Hydracarina							
Hydrachna sp.	A	C	C	A	C	U	
Trombidium sp.				U	A	A	A
Cnidaria							
Hydrozoa							
Hydra sp.	C		U				

TABLE 38.5. (Continued)

Taxon	Littoral *Scirpus* Zone[b]	Inland Swale Forest	Inland Swale Wet Meadow	Littoral Wet Meadow	Inland Wet Meadow	Inland Emergent Zone	Disturbed Inland Meadow
Mollusca							
Bivalvia							
Dreissinidae							
Dreissina polymorpha	A						
Sphaeriidae							
Sphaerium sp.		A	A			U	
Unionidae	U						
Gastropoda							
Ancylidae							
Ferrissia parallela	C						
Bithyniidae							
Bithynia tentaculata				C			
Lymnaeidae							
Fossaria sp.	U	A	A				
Pseudosuccinea col.	U			C			A
Stagnicola elodes	U		C	A			C
Physidae							
Physa sp.	U			C			A
Aplexa elongata		U		A			
Planorbidae							
Helisoma sp.							
Gyraulus parvus	U	A			C	U	
Promenetus sp.		A	A	C	C	C	U
Nematoda	C						
Platyhelminthes							
Turbellaria							
Dugesia sp.	C						

[a] Each taxon has been qualitatively categorized as abundant (A), common (C), or uncommon (U).

[b] Some littoral emergent marsh data were taken from Brady (1992) and Cardinale (1996). This site was studied more intensively than the others.

surprisingly uncommon at all sites, while the damselfly *Ischnura verticalis* was abundant in the littoral *Scirpus* zone and common in the inland (non-swale) wet meadow, two very different habitats. The damselfly *Nehalennia* was abundant in the littoral and inland swale wet meadows. The limnephilid caddisflies were abundant in both the herbaceous and forested inland swale sites (different areas of the same swale), while the megalopteran *Chauliodes* was uniquely characteristic of the forested portion of the swale because it inhabited coarse woody debris found only at this site. Possible factors influencing these site-to-site differences include connectivity to the open bay, flooding phenology, vegetation structure, and predation pressure.

Dominance patterns in Saginaw Bay coastal zone wetlands varied substantially (Fig. 38.4). The seasonally flooded disturbed inland wet meadow (a former sheep pasture), without direct surface connection to the bay, was dominated by *Aedes* mosquitoes (57 percent of community) and other Diptera. The remainder of the community consisted of typical temporary-water fauna: snails, mites, and beetles. In contrast, the inland swale wet meadow, which dried down only in relatively dry years, was dominated by damselflies (36 percent), with snails, amphipods, caddisflies, and fingernail clams each comprising 10 percent or less of the community. The forest portion of the swale, with its leaf litter substrate, contained a community that was dominated by the large chironomid *Chironomus* (25 percent), and isopods (20 percent). The littoral wet meadow was seasonally flooded and largely dominated by snails (61 percent). These samples were sorted live in the field using RBP (Rapid Bioassessment Protocol) methods, so the data are likely biased toward larger animals, which may explain the low proportions of Chironomidae at most sites.

Trichoptera are good indicators of environmental diversity in wetlands and may be useful as indicator species for wetland conditions. The diversity and abundance of caddisflies are dependent upon many factors, but primary among these are vascular plant diversity and abundance, water quality, and predation. In recent studies of coastal wetlands in the Upper Great Lakes (Table 38.6), the lowest mean diversity of caddisflies occurred in relatively homogeneous Saginaw Bay wetlands dominated by *Scirpus*. These studies reveal many new state records for Trichoptera, including dramatic regional and continent-wide disjuncts, suggesting that much new species distribution information awaits discovery in coastal wetlands. While Trichoptera have received more attention than most taxa in these environments, a clear picture of their distributions and indicator status will require considerably more effort.

The low diversity of caddisflies in Saginaw Bay reflects low overall invertebrate diversity there as compared to some sites with different environmental characteristics. The St. Mary's River system includes many wetlands that are more protected from wind and waves than the southeastern Saginaw Bay shore. This is reflected in more diverse fauna and flora in the river sites, including many species characteristic of calm waters. For example, the soft-bottom burrowing Ephemeridae mayflies, several large Trichoptera, at least

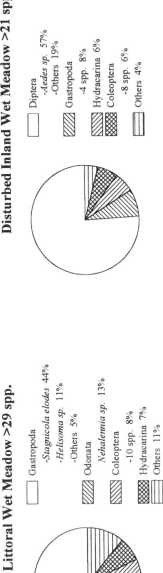

Inland Swale Wet Meadow >32 spp.

☐ Odonata
  -*Nehalennia sp.* 36%
  -Others 2%
▨ Gastropoda
  -*Promenetus sp.* 8%
  -Others 5%
▧ Amphipoda 10%
▧ Trichoptera
  -3 spp. 9%
▤ Bivalvia
  Sphaerium sp. 7%
▥ Others 23%

Disturbed Inland Wet Meadow >21 spp.

☐ Diptera
  -*Aedes sp.* 57%
  -Others 19%
▨ Gastropoda
  -4 spp. 8%
▧ Hydracarina 6%
▨ Coleoptera
  -8 spp. 6%
▥ Others 4%

Inland Swale Wet Forest >19 spp.

☐ Diptera
  -*Chironomus sp.* 25%
  -Others 6%
▨ Isopoda
  -*Asellus sp.* 20%
▧ Bivalvia
  -*Sphaerium sp.* 13%
▧ Trichoptera
  -*Limnephilus sp.* 11%
▥ Gastropoda
  -*Promenetus sp.* 4%
  -*Pseudosuccinea columella* 4%
  -Others 2%
▥ Others 15%

Littoral Wet Meadow >29 spp.

☐ Gastropoda
  -*Stagnicola elodes* 44%
  -*Helisoma sp.* 11%
  -Others 5%
▨ Odonata
  Nehalennia sp. 13%
▧ Coleoptera
  -10 spp. 8%
▧ Hydracarina 7%
▥ Others 11%

Fig. 38.4. Comparison of macroinvertebrate communities from four wetland habitats in the coastal zone of Saginaw Bay.

TABLE 38.6. Summary Data for Trichoptera Diversity in Upper Great Lakes Coastal Wetlands

Water Body	Mean Number of Species	Range	Number of Wetlands
Lake Huron—Saginaw Bay	17	14–22	4
Lake Superior	18	5–26	6
Lake Michigan	24	18–34	7

Source: B. Armitage, P. Hudson, and D. Wilcox, unpublished data.

three species of Anisoptera, and at least two species of Isopoda were found in St. Mary's River wetlands but were absent or uncommon in Saginaw Bay (Table 38.7). The polychaete *Manayunkia speciosa,* the crayfish *Orconectes propinquus,* and *Simulium,* a lotic blackfly, were only found at the St. Mary's site. The insect community of the wave-swept Saginaw Bay wetland (littoral *Scirpus* zone; Table 38.4) was numerically dominated by an assemblage of Diptera (Chironomidae, Ceratopogonidae, Tabanidae), Caenidae mayflies, the odonate *Ischnura verticalis,* and the microcaddisfly family, Hydroptilidae. Other characteristic taxa included Amphipoda, Isopoda, Nematoda, Hydracarina, Ancylidae, Tubificidae, and Naididae (Table 38.5). Large, mobile insects were notably uncommon, as were the Lymnaeidae, Planorbidae, and Physidae snails and Sphaeriidae clams. The modified Gerking vegetation sampler and core sampling devices used by Brady (1992) were probably not very efficient at capturing large, fast-swimming invertebrates. However, the low density of snails and fingernail clams may be due to a combination of wave action and low organic content of the sediments of the *Scirpus* zone where the samples were taken. Several species of snails and clams were present in more protected sites along Saginaw Bay (Table 38.5).

Although Diptera were the most abundant order of insects collected from most Great Lakes wetlands, substantial differences have been reported for numbers of families present. Brady (1992) found only six families of immature Diptera in Saginaw Bay *Scirpus*-dominated coastal wetlands, and the addition of several nearby habitats increases family richness only to 10 (Table 38.3). In contrast, McLaughlin and Harris (1990) record 19 families of Diptera emerging from 2 Green Bay wetlands, including several that are considered semiaquatic (Table 38.7). However, three Odonata families and only one family each of Hymenoptera, Lepidoptera, and Trichoptera were found, results similar to the low diversity reported for Saginaw Bay (Brady 1992). Trichoptera family diversity in most Great Lakes wetlands is usually higher than that reported by Brady (1992) and McLaughlin and Harris (1990) when blacklight sampling of adults is used (personal observation). Among insects the number of caddisfly species found in all habitats of the Great Lakes is probably exceeded only by the number of Chironomidae species (Table 38.8). Among the 11 caddisfly families found in Great Lakes habitats (Table 38.9), Lepto-

TABLE 38.7. Presence/Absence of Wetland Invertebrates of Saginaw Bay,[a] the St. Mary's River,[b] Lake St. Clair,[c] Sebawaing Harbor, Saginaw Bay[d] and Green Bay[e]

Taxon	Saginaw Bay Epiphyton	Saginaw Bay Sediment	St. Mary's River Vegetation Littoral	Lake St. Clair Epiphyton	Sebawaing Harbor Sediment	Green Bay Emerging Insects
Hydrozoa						
Hydridae						
Hydra sp.	X	X				
Nematoda	X	X				
Tardigrada	X	X				
Turbellaria	X	X				
Annelida						
Oligochaeta						
Naididae		X				
Stylaria fossularis			X			
Stylaria sp.	X					
others	X					
Tubificidae						
Branchiura sowerbyi					X	
Branchiura sp.		X				
Limnodrilus cervis					X	
Limnodrilus claparedeanus					X	
Limnodrilus hoffmeisteri					X	
Limnodrilus maumeensis					X	
Limnodrilus profundicola					X	
Limnodrilus ukedemianus					X	
Potamothrix vejdovskyi					X	
Ryacodrilus sp.					X	

Taxon	1	2	3	4	5
Others	X		X		
Hirudinea	X				
Erpobdellidae		X			
Dina sp.					
Erpobdella sp.					
Glossiphoniidae					X
Helobdella stagnalis					X
Polychaeta					
Manayunkia speciosa					X
Mollusca					
Gastropoda			X		
Ancylidae			X		
Ferrissia parallela					
Hydrobiidae					
Amnicola sp.	X				
Lymnacidae			X	X	
Fossaria humilis	X				
Pseudosuccinea columella	X				
Stagnicola catascopium	X				
Physidae					
Physa heterostropha	X	X			
Physa sp.				X	
Planorbidae					
Gyraulus parvus	X	X			
Gyraulus sp.				X	
Bivalvia			X		
Dreissenidae					
Dreissena polymorpha	X				
Sphaeriidae					
Pisidium sp.			X		

TABLE 38.7. (Continued)

Taxon	Saginaw Bay Epiphyton	Saginaw Bay Sediment	St. Mary's River Vegetation Littoral	Lake St. Clair Epiphyton	Sebawaing Harbor Sediment	Green Bay Emerging Insects
Arthropoda						
Acari	X	X	X			
Crustacea						
Amphipoda						
Gammaridae						
Gammarus fasciatus	X		X			
Gammarus pseudolimnaeus	X	X				
Gammarus sp.				X	X	
Hyalellidae						
Hyalella azteca	X		X	X	X	
Decapoda						
Astacidae						
Cambaridae						
Orconectes propinquus			X			
Isopoda						
Asellus forbesi	X					
Asellus intermedia			X			
Asellus sp.				X	X	
Lirceus sp.			X	X		

Taxon					
Insecta					
Coleoptera					
Chrysomelidae					
Donacia sp.					
Dytiscidae			X		
Copelatus gyphicus	X				
Hygrotus/Hydroporus sp.	X				
Elmidae					
Dubiraphia sp.	X				
Stenelmis sp.		X			
Gyrinidae					
Gyrinus sp.	X				
Staphylinidae					
Phloeonomus sp.		X			
Collembola					
Isotomidae				X	
Isotomurus tricolor	X	X			
Sminthuridae					
Bourletiella sp.	X				
Pseudobourletiella spinata	X				
Diptera					
Ceratopogonidae					
Atrichopogon sp.	X	X			X
Bezzia cockerelli					X
Bezzia glabra					X
Palpomyia lineata					X
Sphaeromias longipennis					X

TABLE 38.7. (Continued)

Taxon	Saginaw Bay Epiphyton	Saginaw Bay Sediment	St. Mary's River Vegetation Littoral	Lake St. Clair Epiphyton	Sebawaing Harbor Sediment	Green Bay Emerging Insects
Chironomidae						
Ablabesmyia sp.	X		X	X		X
Chironomus decorus						X
Chironomus tentans						X
Chironomus sp.	X				X	
Cladopelma laccophila	X					
Cladotanytarsus mancus	X					
Cladotanytarsus vandervulpi	X					
Coelotanypus sp.					X	
Corynoneura sp.	X					
Cricotopus bincinctus	X					
Cricotopus sylvestris	X					
Cricotopus sp.	X					
Cryptochironomus sp.			X		X	X
Cryptotendipes sp.	X					
Demijerea atrimana						
Demijerea brachialis						X
Dicrotendipes sp.	X					X
Einfeldia sp.	X					X
Endochironomus nigricans						X
Endochironomus sp.	X				X	
Glyptotendipes bellus						X
Glyptotendipes sp.	X				X	X
Guttipelopia sp.						X

Taxon			
Kiefferulus dux			X
Larsia sp.	X		
Microchironomus sp.	X		
Micropsectra sp.			X
Orthocladius sp.	X	X	
Parachironomus arcuatus			
Parachironomus hirtalatus			X
Parachironomus tenuicaudatus			X
Paracladopelma sp.	X		
Paratanytarsus sp.	X		
Polypedilum digitifer			X
Polypedilum sordens			X
Polypedilum trigonus			X
Polypedilum sp.	X		
Potthastia longimanus		X	
Procladius bellus	X	X	X
Procladius (*Holotanypus*) sp.			X
Psectrocladius sp			X
Pseudochironomus sp.		X	
Stictochironomus sp.	X		
Tanypus sp.	X		
Tanytarsus sp.	X		X
Thienemanniella sp.	X		
Virgatanytarsus sp.	X		
Zavreliella varipennis			X
Chloropidae			X
Culicidae			
Coquillettidia sp.			X
Culex sp.			X
Culiseta sp.			X

TABLE 38.7. (Continued)

Taxon	Saginaw Bay Epiphyton	Saginaw Bay Sediment	St. Mary's River Vegetation Littoral	Lake St. Clair Epiphyton	Sebawaing Harbor Sediment	Green Bay Emerging Insects
Dixidae						X
Dolichopodidae						X
Drosophilidae						X
Empididae						X
Ephydridae						
Ilythea sp.	X					
Notiphila sp.	X					
Lonchopterridae						X
Mycetophilidae						X
Muscidae	X					
Otitidae						X
Sciaridae						X
Sciomyzidae						X
Sepsidae						X
Simuliidae						X
Simulium sp.			X			
Sphaeroceridae						X
Stratiomyidae						X
Hedriodiscus sp.	X					
Syrphidae						X
Tipulidae						X

Taxon				
Ephemeroptera				
Ameletidae				
Ameletus sp.		X		
Baetidae				
Baetis levitans			X	X
Baetis sp.	X	X		
Caenidae				
Brachycercus sp.				
Caenis amica			X	X
Caenis latipennis			X	X
Caenis sp.		X		
Ephemerellidae				
Ephemerella sp		X		
Ephemeridae				
Ephemera simulans		X		
Hexagenia limbata		X		
Leptophlebiidae				
Leptophlebia sp.		X	X	
Hemiptera				
Corixidae				
Sigara cornuta		X		X
Sigara decorata				X
Trichocorixa naias		X	X	X

TABLE 38.7. (Continued)

Taxon	Saginaw Bay Epiphyton	Saginaw Bay Sediment	St. Mary's River Vegetation Littoral	Lake St. Clair Epiphyton	Sebawaing Harbor Sediment	Green Bay Emerging Insects
Gerridae						
Trepobates sp.	X					
Hebridae						
Merragata sp.	X					
Mesoveliidae						
Mesovelia mulsanti	X					
Notonectidae						
Buenoa sp.	X					
Notonecta lunata	X					
Veliidae						
Microvelia sp.	X					
Hymenoptera						
Braconidae						X
Lepidoptera				X		
Noctuidae						
Bellura sp.	X					
Pyralidae						
Acentria niveus	X	X				
Limnaecia phragmitella						X
Parapoynx sp.	X					
Petrophila sp.	X					
Synclita obliteralis						X

Taxon	1	2	3	4	5
Megaloptera					
Sialidae					
Sialis sp.		X			
Odonata					
Aeschnidae					
Aeschna sp.	X		X		
Anax junius	X				
Coenagrionidae					
Enallagma antennatum					X
Enallagma boreale	X		X		X
Enallagma hageni					X
Enallagma signatum					X
Enallagma sp.					X
Ishnura verticalis				X	
Nehallenia irene	X				
Gomphidae					
Arigomphus sp	X				
Gomphus sp.					X
Lestidae					
Lestes disjunctus			X		
Libellulidae					
Erythemis sp.					X
Libellula sp.			X		
Sympetrum vicinum	X				

TABLE 38.7. (Continued)

Taxon	Saginaw Bay Epiphyton	Saginaw Bay Sediment	St. Mary's River Vegetation Littoral	Lake St. Clair Epiphyton	Sebawaing Harbor Sediment	Green Bay Emerging Insects
Trichoptera						
Hydroptilidae						
Agraylea multipunctata	X					
Hydroptila sp.	X					
Orthotrichia sp.	X					
Oxyethira sp.	X	X				
Leptoceridae						
Ceraclea sp.			X		X	
Nectopsyche diarina	X					
Nectopsyche sp.	X	X		X		
Oecetis sp.				X		
Trianodes sp.			X			
Limnephilidae						
Grammotaulus sp.			X			
Phryganeidae						
Agrypnia vestita	X					
Phrygania sp.			X			
Polycentropodidae						
Phylocentropus sp.			X			
Polycentropus sp.			X			

[a] Brady 1992.
[b] Duffy et al. 1987.
[c] Brown et al. 1988.
[d] Cole and Weigman 1983.
[e] McLaughlin and Harris 1990.

TABLE 38.8. Summary Table of Aquatic Insects from the Upper Great Lakes

Order	Number of Families	Number of Genera	Number of Species
Plecoptera—stoneflies	6	13	14
Odonata—dragonflies and damselflies	8	21	38
Ephemeroptera—mayflies	9	27	77
Hemiptera—true bugs	7	14	21
Coleoptera—beetles	8	26	35
Trichoptera—caddisflies	11	43	132
Diptera (Chironomidae)—midges[a]	1	63	n.a.

Source: Hudson, personal communication.

[a] Wave shore zone of Lake Superior and Lake Huron only (Barton and Hynes, 1978).

ceridae, Limnephilidae, and Hydroptilidae (in descending order) are the most frequently encountered in coastal wetlands.

DISCUSSION

Role in Ecosystem Function

Studies in inland wetlands and other aquatic habitats suggest that invertebrates are likely to be very important in trophic interactions in coastal wetlands. D. Ewert (Michigan Nature Conservancy) and M. Hamas and students (Central Michigan University) have observed that several species of Chironomidae

TABLE 38.9. Trichoptera of the Upper Great Lakes: Number of Genera (Number of Species)

Family	Superior	Michigan	Huron	Totals
Brachycentridae	1(1)	1(1)	0	1(1)
Glossosomatidae	1(1)	0	0	1(1)
Helicopsychidae	1(1)	1(1)	1(1)	1(1)
Hydropsychidae	2(2)	3(10)	3(7)	3(10)
Hydroptilidae	5(7)	4(18)	4(6)	6(24)
Lepidostomatidae	1(2)	1(1)	1(1)	1(2)
Leptoceridae	6(22)	6(23)	8(18)	8(37)
Limnephilidae	10(18)	4(11)	7(14)	10(27)
Molannidae	1(4)	1(2)	1(3)	1(4)
Phryganeidae	4(8)	4(8)	6(8)	6(10)
Polycentropodidae	4(11)	4(7)	4(11)	5(15)
Totals	*36(77)*	*29(82)*	*35(69)*	*43(132)*

Source: Hudson and Armitage, unpublished data.

emerge from coastal wetlands and congregate in large swarms in adjacent forested habitats along Northern Lake Huron at the time of stopover for avian neotropical migrants (D. Ewert, personal communication). Their hypothesis, supported by several years of field observations, is that these Chironomidae swarms are key food sources for the neotropical migrants at a time when terrestrial insect populations are still quite low.

Coastal wetlands are important feeding and nursery habitats for waterfowl (Prince et al. 1992) and many species of fish (Brazner 1997, French 1988, Jude and Pappas 1992, Liston and Chubb 1986), most of which feed heavily on invertebrates during at least some part of their development. It seems safe to assume that fish and waterfowl rely on coastal wetland invertebrates, but we know of no specific studies on these trophic links. Similarly, the role of invertebrates as organic matter processors and in energy flows and nutrient cycling in these systems can only be inferred from studies of other systems.

Environmental Factors That Influence Invertebrate Distribution

Geomorphology, hydrology, and composition of bottom sediments of coastal wetlands are determined by climate and geology and appear to be key driving forces in determining location and plant community composition of coastal wetlands (Minc 1996). These same variables plus plant community composition are likely to be the key factors determining distributions of wetland invertebrates. Site-specific data on invertebrates of Great Lakes coastal wetlands are available only for riverine and unrestricted-bay types of wetlands. The riverine site may not be typical, because this site includes wetlands along a major shipping channel, the St. Mary's River, and thus any conclusions drawn here should be considered to be preliminary.

Hydrology. Long-term water-level fluctuations are known to alter wetland plant community structure and distribution, and should also influence invertebrates, but the effects of certain short-term water movements have been studied more. Ship traffic in the St. Mary's River creates currents and eddies which can carry away broken plant stems from coastal margins. An estimated 18.9 percent of a single species of coenagrionid damselfly was lost from these wetlands because of the loss of eggs in exported stems (Duffy et al. 1987). Brady (1996) concludes that waves were probably responsible for the immigration of zebra mussel veligers into Saginaw Bay marshes. Many taxa on wave-swept shorelines (Barton and Hynes 1978) and in the outer, more exposed zones of coastal wetlands (Brady 1992) are typically considered lotic-adapted, including heptageniid mayflies (personal observation), the chironomid *Rheotanytarsus* (Cardinale 1996), and elmid beetles (Brady 1992). Cardinale (1996) suggests that weaker-swimming invertebrates without adaptations for attachment to plant stems, such as the Ceratopogonidae and aquatic mites, were excluded from the outer littoral of Saginaw Bay, where wave action was strong. These observations suggest the possibility of a gradient from lotic taxa in hydrologically harsh wetland zones to lentic communities in more protected areas.

Short-term water movements can be expected to affect invertebrates indirectly by influencing many important habitat characteristics. Water movements sort sediment particles by size, determining spatial distribution of substrate patches. For example, waves and seiches in the Quanicassee wetland of Saginaw Bay prevented accumulations of organic sediment, except in the few protected areas near shore (personal observation), which were the only places in the marsh where Brady (1992) found isopods. Water quality factors, such as conductivity, turbidity, temperature, and nutrient concentrations, are affected by horizontal and vertical mixing caused by wave and seiche action, and correlate with invertebrate distributions (Brady 1996, Brady and Burton 1995, Brady et al. 1995, Cardinale 1996, Cardinale et al. 1997). Water fluctuations also alter colonizable surface area as stems of emergent plants are submerged and exposed (Brady 1992), and waves can alter plant density and community composition (Duffy et al. 1987).

Diking coastal wetlands isolates them, reducing natural hydrologic variation and preventing horizontal mixing. In a study of Green Bay coastal wetlands nearly 80 percent greater insect biomass (p < 0.01) and a richer assemblage emerged from a diked marsh than from a nearby nondiked marsh (McLaughlin and Harris 1990). These differences may have resulted from increased water clarity associated with the exclusion of carp and separation of the marsh from turbid bay water (Harris, personal communication). Lower Green Bay waters carry a heavy sediment load from riverine sources, preventing the growth of invertebrate-supporting submergent plants in the undiked marsh.

Water Quality. Coastal emergent vegetation can create water quality gradients by reducing the mixing and sediment-disturbing power of waves as they move toward shore. Cardinale et al. (1997) found such littoral gradients in six water quality variables: turbidity, dissolved oxygen, pH, alkalinity, conductivity, and chloride. Principal components analysis created a composite of these variables which formed a gradient from pelagic waters toward shore after emergent vegetation had developed. In June, before the gradients were established, mean invertebrate abundance was around 50 individuals per stem at nearshore and offshore transects. But as plant stem height and density increased, invertebrates near the pelagic zone increased in abundance to nearly 250 individuals per stem, while those near the shore did not increase. Though the composite water quality variable was highly correlated with epiphytic larval chironomid abundance and biomass, the authors conclude that chironomids probably did not respond directly to water quality differences. Instead, invertebrate abundance appeared to respond to epiphytic algal biomass, which was 140 μg Chlorophyll-*a* per stem higher offshore than nearshore at peak invertebrate biomass. The nearshore algae may have been less successful because of reduced boundary-layer diffusion rates caused by the reduction of water movement in areas farther from pelagic waters.

Turbidity appeared to influence the relative proportion of suspension feeders found in the littoral community. Cardinale et al. (1997) found many filter-feeding chironomids on plant stems in the well-mixed, turbid outer littoral waters, but few in the less turbulent inner littoral zone, where omnivorous invertebrates were numerically dominant. Although light attenuation from high turbidity undoubtedly affects primary producers, periphytic algal biomass was higher in the outer littoral than nearer shore. The authors conclude that the high nutrient availability in this well-mixed zone was responsible for the higher algal biomass, which was probably responsible for higher chironomid abundance. However, excessively high nutrient levels in Great Lakes waters may adversely affect wetland production. McLaughlin and Harris (1990) conclude that the separation of diked wetlands from the hypereutrophic waters of Green Bay was beneficial to invertebrate populations.

Sediment Composition. The relationships between sediment characteristics and Great Lakes littoral zoobenthos were studied by Cole and Weigman (1983) in Saginaw Bay, Lake Huron, and Brest Bay, Lake Erie. Biomass of tubificid worms and chironomids, which jointly comprised over 90 percent of the biomass collected, was significantly higher in fine, silty sediments than in coarse, sandy sediments ($p < 0.05$). However, invertebrate species diversity and total density were not significantly correlated with sediment size. Abundance of the amphipod *Gammarus pseudolimnaeus* was significantly correlated with percent organic matter. The authors conclude that highest benthic abundances were associated with small sediment particle size resulting from low to moderate water turbulence.

Similarly, in nonwetland nearshore zones Johnson et al. (1987) found that assemblages in the benthos of Georgian Bay, Lake Huron, were best explained by variations in temperature, sediment richness, pH, and Ca concentration, in that order, and Barton and Hynes (1978) determined that invertebrate abundance and richness were directly related to the propensity of the substrate to wave-induced instability. Barton and Griffiths (1984) concluded that the degree of wave exposure coupled with sediment heterogeneity determined which of nine community types occurred at a given site. In southeastern Lake Michigan, waters from 4 to 7 m deep were characterized by certain chironomid species, but below this zone the bottom sands were not subject to wave-induced shifting and an assemblage of *Pontoporeia hoyi* amphipods, sphaeriids, and oligochaetes was dominant, with total abundances rising exponentially to depths of 25 meters (Mozley and Garcia 1972).

The proximity of different types of sediment patches also appeared to influence the benthic community (Cole and Weigman 1983). When sandy patches were located near patches of finer sediment, invertebrate biomass, number of species, and percent organic matter in the sand approached the higher values common in fine sediment. Fine sediments appear to harbor more zoobenthic biomass than sands, and their apparent influence on nearby sand patches suggests that fauna may not vary continuously over sediment size gradients. Instead, it may be that when a threshold percent of fine sediments or sediment heterogeneity is reached, the benthos "switch" to a community type characteristic of fine, organic-rich sediments.

Vegetation. Aquatic plants provide habitat structure for invertebrates, but the relationships between the two are complex. Effects of vegetation density, community composition, leaf morphology, and surface area on Great Lakes invertebrate communities have been studied, but simple causal relationships have proven elusive. Krecker (1939) found that Lake Erie littoral invertebrates were most abundant on submersed macrophytes that had the most dissected leaves, hence the most complex three-dimensional structure. French (1988) reports similar results in Lake St. Clair, but Brown et al. (1988), working in the same region, concludes that invertebrate assemblages were not substantially affected by the degree of submersed macrophyte structural complexity. However, they found that invertebrate diversity was correlated to the diversity of physical structure provided by a species-rich plant community.

Emergent plant density appeared to influence insect abundance in Green Bay wetlands. McLaughlin and Harris (1990) found that 60 percent more insect individuals, and 65 percent greater insect biomass, emerged from a sparse emergent zone dominated by *Typha* and *Sparganium* than from the zone composed of a more dense assemblage of the same plant species ($p <$ 0.05). Insect abundance for both the sparse and dense zones was greater than that found in adjacent nonemergent open water areas. Cardinale (1996), however, found no correlation between *Scirpus americanus* submerged surface area and algal or invertebrate biomass in a Saginaw Bay marsh. Similarly, invertebrate density and biomass may not be correlated with submergent macrophyte surface area (French 1988) or biomass (Brown et al. 1988).

Brady and Burton (1995) report that plants in a Saginaw Bay coastal marsh supported greater microcrustacean biomass than did sediments. Abundances were actually higher in sediments, but were dominated by small species of Cladocera, while the high biomass associated with plants was the result of dominance by larger animals. In contrast, macroinvertebrate biomass was higher in the sediment samples than on plant stems, with peak values of 2,300 mg/m^2 and 1500 mg/m^2, respectively (Brady 1992). Plants supported 32 species of microcrustaceans, while sediments supported 23. The higher richness associated with the plant stems resulted from more Cladocera and Calanoida copepod species, while Ostracoda and Harpacticoida were more associated with sediments. Fifty-four insect species were collected in vegetation samples, while only 16 occurred in sediments.

Fish Predation. Many fish are known to inhabit Great Lakes coastal marshes for at least part of their life cycles (Liston and Chubb 1986), but their impacts on invertebrate communities in these wetlands have not been studied. Many fish are known to feed heavily on invertebrates in Great Lakes wetlands (Duffy et al. 1987, French 1988), and even shallow, seasonally flooded wet meadows can support many small fish species and juvenile stages of larger species (personal observation). Though some studies show that fish predation can influence invertebrate abundance (e.g., Bohanan and Johnson 1983, Morin 1984), this impact may be less important in large, dense emergent wetlands. High vegetation density can reduce fish foraging efficiency (Gilinsky 1984), and it increases resource partitioning between rock bass and pumpkinseed (*Lepomis gibbosus*) in Lake St. Clair (French 1988).

Spawning carp are found in large numbers in Saginaw Bay and Les Cheneaux Islands wetlands of Lake Huron (personal observation). Their feeding and energetic mating behavior disrupt sediments, disturbing resident invertebrates. Carp predation may have been partly responsible for lower emerging insect abundance and biomass in a coastal marsh compared to a diked marsh along the Green Bay shoreline (McLaughlin and Harris 1990). Possibly more influential, however, was the loss of insect habitat through the carp's uprooting and consumption of submergent vegetation.

Exotic Species. Eurasian water milfoil (*Myriophyllum spicatum*) and purple loosestrife (*Lythrum salicaria*) are exotic plants that have become established in several coastal wetland areas, but effects have not been studied in these sites. By contrast, the recent invasion of a Saginaw Bay wetland by the zebra mussel, *Dreissena polymorpha*, was documented as mussel densities rose from zero to 55,000 m^{-2} (Brady 1996). However, individuals attached to plant stems within 300 meters of the shore did not survive winter ice, so many mussels formed free-rolling benthic clusters, or "druses", of up to 1000 individuals which overwintered in the outer zone of the marsh, where waters did not freeze to the bottom (Brady et al. 1995).

The establishment of zebra mussel populations caused substantial changes in some Great Lakes aquatic environments (Nalepa and Fahnenstiel 1995). Probably the most noticeable change was a drastic increase in water clarity resulting from their filter feeding, which appeared to have caused increases in macroinvertebrate abundance in Lake Ontario (Stewart and Haynes 1994). The authors attribute this result to three factors: the transfer of energy from nekton to benthos through mussel feeding, the creation of habitat structure by mussel colonies, and the use of mussels as food by predators such as crayfish and leeches.

Brady (1996) investigated whether or not a recent invasion of zebra mussels appeared to have affected the resident invertebrate community in the Saginaw Bay littoral zone. Although the remains of several taxa of planktonic invertebrates were found in mussel feces, predation experiments indicated that ambient densities of mussels were not capable of reducing zooplankton populations. On the other hand, laboratory experiments indicated that zebra mussels were capable of reducing *Ceriodaphnia dubia* populations through competition for phytoplankton food, though ambient densities in the marsh appeared to be too low to produce this effect. Brady (1996) found no significant differences in microcrustacean communities before and after zebra mussel invasion (BACI design), but densities of epiphytic larval chironomids of the tribe Tanytarsini and sediment-associated tubificid oligochaetes were significantly higher after mussel invasion. She suggests that the filtering Chironomidae may benefit from feeding currents created by mussels, while tubificids may take advantage of the organic-rich mussel feces as a food source.

Nonwetland Nearshore Benthos

The invertebrate community of the nearshore soft-bottom macrobenthos is much less diverse than that of coastal wetlands and is dominated by common profundal groups, though many taxa overlap both habitats (Cook and Johnson 1974). The burrowing mayflies *Hexagenia* and *Ephemera* appear to characterize the soft-bottom environment just beyond emergent vegetation, but can overlap into wetlands. Similarly, many wetland animals also occur in the nearshore benthos, so these environments may be important to the maintenance of the wetland fauna, or may simply be zones of transition to deeper-water communities.

As in coastal wetlands, Trichoptera of the nearshore zone have been studied in some detail, and the patterns appeared to mirror those of invertebrate communities in general (personal observation). The diversity and abundance of caddisflies is dependent primarily on substrate (silt, sand, bedrock, etc.) and water quality. Leptoceridae, Molannidae, and Polycentropodidae are the most frequently encountered families, and some typically lotic caddisflies, such as the Hydropsychidae, are found only in the nearshore zone.

CONCLUSIONS

Investigations into the unique abiotic and biotic characteristics of coastal wetlands are needed in order to understand what distinguishes Great Lakes coastal wetlands from inland palustrine and small-lake littoral environments. The data collected to date suggest that water movements in the Great Lakes are largely responsible, through direct and indirect mechanisms, for the composition and distribution of biota in coastal wetlands. Short-term water movements create currents of varying velocities and directions and occasional catastrophic events. Interactions between waves and plants create physical and chemical gradients that appear to create spatial heterogeneity in invertebrate habitat quality through variations in food resources and environmental "harshness."

Coastal wetlands are ecotones between upland and lacustrine systems. At the lakeward end of the ecotone the nonvegetated soft-bottom environment appears to be a transition zone between wetland and profundal communities, though there appear to be some taxa fairly unique to these habitats. At the upper end of the ecotone distributions of terrestrial and aquatic taxa should overlap, depending on the spatial and temporal interspersion of standing water and "dry" substrate. Seasonal fluctuations of water levels in the Great Lakes create gradients of flood duration, especially where uninterrupted gentle slopes are found. This spatial variation in surface hydrology produces vegetation zones and is likely to affect invertebrates indirectly through habitat alteration and directly through life history and dispersal mechanisms. Animals that cannot survive the harshness of a seasonal environment or exploit such habitats by migrating in during favorable conditions would be excluded from these areas. Because inundation patterns differ between coastal and inland wetlands, it is also likely that local invertebrate subpopulations in coastal and inland wetlands differ in times of development and emergence.

Short-term and seasonal hydrologic variation also affect substrate quality. Currents and waves sort sediments in coastal zones just as they do in rivers and streams. Within-site and among-site variations in water velocity can cause spatial variations in sediment particle size, resulting in variations in invertebrate diversity and abundance. Differences in depth and flood duration should create gradients in oxygen availability and decomposition rates, which may affect the accumulation of detritus and the distribution and abundance of invertebrates.

While the influence of water movement is likely to be consistent throughout the upper Great Lakes, the actual causes and effects of these and other factors may vary. Many wetland types are near shorelines but lack surface water connections with littoral zones, so it is a great oversimplification to suggest that the same hydrological forces act on all coastal wetlands. The studies cited here simply reinforce this contention, because all the wetlands investigated are of different types, with substantial environmental differences between them. It follows, then, that an understanding of these wetlands will not come from focused studies on one or two "characteristic" sites, because no such place exists. Instead, much work will need to be done in a variety of sites in order to encompass the range of conditions found in coastal wetlands of the upper Great Lakes.

We recommend that future studies combine descriptive work with experimental investigations of forces that are characteristic of coastal wetlands. Emphasis should be placed on cooperative work so that descriptive data will be comparable among studies. Experimental work will be most fruitful where descriptive data have already been collected. Interactions between hydrology and other factors, especially vegetation, should be studied for their effects on water and sediment quality, emergence phenology, spatial heterogeneity, and vertebrate predators. Finally, researchers should cooperate with state and local agencies to encourage conservation and protection of these often-neglected systems.

ACKNOWLEDGMENTS

Funding for our research on Great Lakes coastal wetlands has been provided by the Land and Water Management Division, Michigan Department of Environmental Quality; Wildlife Division, Michigan Department of Natural Resources; The Michigan Chapter of the Nature Conservancy; Ohio Biological Survey; U.S. Environmental Protection Agency; and the Biological Resources Division, U.S. Geological Survey. We thank our cooperators and coinvestigators from these agencies who have provided field facilities, background information, unpublished data, or help with field sampling or taxonomic identification, including J. Brazner (U.S. EPA), D. Ewert and K. Gilges (Michigan Nature Conservancy), and P. Hudson and D. Wilcox (BRD-USGS, Great Lakes Science Center). We also thank coinvestigators on other aspects of these studies that are included indirectly in this chapter, including H. H. Prince (mammals and birds, Michigan State University) and Dennis Albert (plant community composition, Michigan Natural Features Inventory). We also thank the numerous undergraduate and graduate students who have contributed to these studies, including V. Brady, B. Cardinale, A. Conklin, R. Hollister, D. Kashian, B. Keas, K. Kintigh, T. Losee, J. Oles, S. Riffell, M. Scalabrino, R. Serbin, K. Stanley, C. Stricker, D. Swank, A. Vaara, and M. Whitt.

LITERATURE CITED

Barton, D. R., and M. Griffiths. 1984. Benthic invertebrates of the nearshore zone of eastern Lake Huron, Georgian Bay, and North Channel. Journal of Great Lakes Research 10:407–416.

Barton, D. R., and H. B. N. Hynes. 1978. Wave-zone macrobenthos of the exposed Canadian shores of the St. Lawrence Great Lakes. Journal of Great Lakes Research 4:27–45.

Bedford, K. W. 1992. The physical effects of the Great Lakes on tributaries and wetlands. Journal of Great Lakes Research 18:571–589.

Bohanan, R. E., and D. M. Johnson. 1983. Response of littoral invertebrate populations to a spring fish exclusion experiment. Freshwater Invertebrate Biology 2:28–40.

Brady, V. J. 1992. The invertebrates of a Great Lakes coastal marsh. Master's Thesis, Michigan State University, East Lansing, MI.

———. 1996. Zebra mussels (*Dreissena polymorpha*) in a Great Lakes coastal marsh. Ph.D. Dissertation, Michigan State University, East Lansing, MI.

Brady, V. J., and T. M. Burton. 1995. The microcrustacean community of a Saginaw Bay coastal emergent marsh. Pages 325–341 *in* M. Munawar, T. Edsall, and J. Leach (eds.), The Lake Huron Ecosystem: Ecology, Fisheries and Management. SPB Academic, Amsterdam, The Netherlands.

Brady, V. J., B. J. Cardinale, and T. M. Burton. 1995. Zebra mussels in a coastal marsh: The seasonal and spatial limits of colonization. Journal of Great Lakes Research 21:587–593.

Brazner, J. C. 1997. Regional, habitat, and human development influences on coastal wetland and beach fish assemblages in Green Bay, Lake Michigan. Journal of Great Lakes Research 23:36–51.

Brown, C. L., T. P. Poe, J. R. P. French III, and D. W. Schloesser. 1988. Relationships of phytomacrofauna to surface area in naturally occurring macrophyte stands. Journal of the North American Benthological Society 7:129–139.

Burton, T. M. 1985. The effects of water level fluctuations on Great Lakes coastal marshes. Pages 3–13 *in* H. H. Prince and F. M. D'Itri (eds.), Coastal Wetlands. Proceedings of the First Great Lakes Coastal Wetlands Colloquium. Lewis, Chelsea, MI.

Cardinale, B. J. 1996. The effects of a pelagic-littoral mixing gradient on an epiphytic invertebrate community. Master's Thesis, Michigan State University, East Lansing, MI.

Cardinale, B. J., T. M. Burton, and V. J. Brady. 1997. The community dynamics of epiphytic midge larvae across the pelagic-littoral interface: Do animals respond to changes in the abiotic environment? Canadian Journal of Fisheries and Aquatic Sciences 54:2314–2322.

Cole, R. A., and D. L. Weigman. 1983. Relationships among zoobenthos, sediments, and organic matter in littoral zones of western Lake Erie and Saginaw Bay. Journal of Great Lakes Research 9:568–581.

Comer, P. J., D. A. Albert, H. A. Wells, B. L. Hart, J. B. Raab, D. L. Price, D. M. Kashian, R. A. Corner, and D. W. Schuen. 1995. Michigan's Native Landscape, as Interpreted from General Land Office Surveys 1816–1856. Michigan Natural Features Inventory Report to Water Division, U.S. Environmental Protection Agency, and Wildlife Division, Michigan Department of Natural Resources, Lansing, MI.

Cook, D. G., and M. G. Johnson. 1974. Benthic macroinvertebrates of the St. Lawrence Great Lakes. Journal of the Fisheries Research Board of Canada 31:763–782.

Davis, B. M., P. L. Hudson, and B. J. Armitage. 1991. Distribution and abundance of caddisflies (Trichoptera) in the St. Clair-Detroit River system. Journal of Great Lakes Research 17:522–535.

Duffy, W. G. 1985. The population ecology of the damselfly *Lestes disjunctus disjunctus* (Zygoptera: Odonata) in the St. Mary's River, Michigan. Ph.D. Dissertation. Michigan State University, East Lansing, MI.

Duffy, W. G., and T. R. Batterson. 1987. The St. Mary's River, Michigan: An Ecological Profile. U.S. Fish and Wildlife Service Biological Report 85(7.10), Washington, DC.

French, J. R. P., III. 1988. Effect of submersed aquatic macrophytes on resource partitioning in yearling rock bass (*Ambloplites rupestris*) and pumpkinseeds (*Lepomis gibbosus*) in Lake St. Clair. Journal of Great Lakes Research 14:291–300.

Gilinsky, E. G. 1984. The role of fish predation and spatial heterogeneity in determining benthic community structure. Ecology 65:455–468.

Griffiths, R. W. 1991. Environmental quality assessment of the St. Clair River as reflected by the distribution of benthic macroinvertebrates in 1985. Hydrobiologia 219:143–164.

Herdendorf, C. E. 1986. The Ecology of Lake St. Clair Wetlands: A Community Profile. U.S. Fish and Wildlife Service, Biological Report 85(7.9). Washington DC.

Herdendorf, C. E., S. M. Hartley, and M. D. Barnes (eds.). 1981a. Fish and Wildlife Resources of the Great Lakes Coastal Wetlands within the United States. Vol. 4: Lake Huron. U.S. Fish and Wildlife Service FWS/OBS-81/02-v4, Washington, DC.

──────. 1981b. Fish and Wildlife Resources of the Great Lakes Coastal Wetlands within the United States. Vol. 5: Lake Michigan. U.S. Fish and Wildlife Service FWS/OBS-81/02-v5, Washington, DC.

──────. 1981c. Fish and Wildlife Resources of the Great Lakes Coastal Wetlands within the United States. Vol. 6: Lake Superior. U.S. Fish and Wildlife Service FWS/OBS-81/02-v6, Washington, DC.

Jaworski, E., and C. N. Raphael. 1978. Coastal wetlands value study in Michigan. Phase I and II. Fish, Wildlife, and Recreational Values of Michigan's Coastal Wetlands. Report to the Michigan Coastal Management Program, Michigan Department of Natural Resources, Lansing, MI.

Johnson, M. G., O. C. McNeill, and S. E. George. 1987. Benthic macroinvertebrate associations in relation to environmental factors in Georgian Bay. Journal of Great Lakes Research 13:310–327.

Jude, D. J., and J. Pappas. 1992. Fish utilization of Great Lakes coastal wetlands. Journal of Great Lakes Research 18:651–672.

Kauss, P. B. 1991. Biota of the St. Mary's River: Habitat Evaluation and Environmental Assessment. Hydrobiologia 219:1–35.

Krecker, F. H. 1939. A comparative study of the animal population of certain submerged aquatic plants. Ecology 20:553–562.

Krieger, K. A. 1992. The ecology of invertebrates in Great Lakes coastal wetlands: Current knowledge and research needs. Journal of Great Lakes Research 18:634–650.

Leach, J. H. 1991. Biota of Lake St. Clair: Habitat evaluation and environmental assessment. Hydrobiologia 219:187–202.

Liston, C. R., and S. Chubb. 1985. Relationships of water level fluctuations and fish. Pages 121–134 *in* H. H. Prince and F. M. D'Itri (eds.), Coastal Wetlands. Proceedings of the First Great Lakes Coastal Wetlands Colloquium. Lewis, Chelsea, MI.

Manny, B. A., and D. Kenaga. 1991. The Detroit River: Effects of contaminants and human activities on aquatic plants and animals and their habitats. Hydrobiologia 219:269–279.

Maynard, L., and D. Wilcox. 1997. Background Report on Great Lakes Wetlands Prepared for the State of the Great Lakes 1997 Report Issued Jointly by the U.S. Environmental Protection Agency Great Lakes National Program Office, Chicago, Illinois, and Environment Canada, Toronto, Ontario.

McLaughlin, D. B., and H. J. Harris. 1990. Aquatic insect emergence in two Great Lakes marshes. Wetlands Ecology and Management 1:111–121.

Minc, L. 1996. Michigan's Great Lakes Coastal Wetlands: Definition, Variability, and Classification. Report to Michigan Natural Features Inventory, East Lansing, MI.

Mitsch, W. J. 1992. Combining ecosystem and landscape approaches to Great Lakes wetlands. Journal of Great Lakes Research 18:552–570.

Morin, P. J. 1984. Odonate guild composition: Experiments with colonization history and fish predation. Ecology 65:1866–1873.

Mozley, S. C., and L. C. Garcia. 1972. Benthic macrofauna in the coastal zone of southeastern Lake Michigan. Pages 102–116 *in* Proceedings of the Fifteenth Conference on Great Lakes Research.

Mozley, S. C., and R. M. LaDronka. 1988. *Ephemera* and *Hexagenia* (Ephemeridae, Ephemeroptera) in the Straits of Mackinac, 1955–56. Journal of Great Lakes Research 14:171–177.

Nalepa, T. F., and G. L. Fahnenstiel. 1995. *Dreissena polymorpha* in the Saginaw Bay, Lake Huron ecosystem: Overview and perspective. Journal of Great Lakes Research 21:411–416.

Prince, H. H., and T. M. Burton. 1996. Wetland Restoration in the Coastal Zone of Saginaw Bay, 1995. Report to the Michigan Department of Natural Resources, Lansing, MI.

Prince, H. H., P. I. Padding, and R. W. Knapton. 1992. Waterfowl use of the Laurentian Great Lakes. Journal of Great Lakes Research 18:673–699.

Schloesser, D. W. 1988. Zonation of mayfly nymphs and caddisfly larvae in the St. Mary's River. Journal of Great Lakes Research 14:227–233.

Smith, P. G. R., V. Glooschenko, and D. A. Hagen. 1991. Coastal wetlands of three Canadian Great Lakes: Inventory, current conservation initiatives, and patterns of variation. Canadian Journal of Fisheries and Aquatic Sciences 48:1581–1594.

Stewart, T. W., and J. M. Haynes. 1994. Benthic macroinvertebrate communities of southwestern Lake Ontario following invasion of *Dreissena*. Journal of Great Lakes Research 20:479–493.

Suzuki, N., S. Endoh, M. Kawashima, Y. Itakura, C. D. McNabb, F. M. D'Itri, and T. R. Batterson. 1995. Discontinuity bar in a wetland of Lake Huron's Saginaw Bay. Journal of Freshwater Ecology 10:111–123.

Whitt, M. B. 1996. Avian breeding use of coastal wetlands on the Saginaw Bay of Lake Huron. Master's Thesis, Michigan State University, East Lansing, MI.

39 Lake Erie Coastal Wetlands

A Review and Case Study of Presque Isle Invertebrates

PAMELA SILVER BOTTS

*C**oastal wetlands in Lake Erie are physically heterogeneous and temporally dynamic habitat in which physical processes such as seiches, waves, sediment transport, and long-term water-level fluctuations create constantly changing habitat. The term* Lake Erie coastal wetland *encompasses habitat including such diverse wetland types as estuarine river mouths (e.g., Old Woman Creek, Ohio), open-water marshes (West Harbor, Ohio), lagoon wetlands (Presque Isle, Pennsylvania; Point Pelee, Ontario), as well as low, marshy backshore areas in parts of Michigan, Ohio, and Ontario. Published research on Lake Erie coastal wetlands has focused on hydrodynamics, nutrient dynamics, sediment transport, and vegetation. Relatively little is known, however, about the invertebrates in these wetlands. The first objective of this chapter is to briefly review the existing literature with reference to benthic invertebrates in Lake Erie coastal wetlands.*

Presque Isle is a recurved barrier sandspit located on the south shore of Lake Erie, opposite the city of Erie, Pennsylvania. The 13-km peninsula rests at the eastern end of a platform of glacial deposits, and it has been migrating toward the northeast along the coast of Lake Erie for at least 3500 years. Sand eroded from beaches southwest of the peninsula and from the neck of the peninsula is deposited at its distal tip, creating a recurved series of dune ridges and lagoons. Wetlands form on the peninsula when lagoons are isolated from the lake by barrier sand bars or when storms create sand berms on exposed beaches trapping water behind them. The second objective of this chapter is to highlight studies of invertebrates, par-

Invertebrates in Freshwater Wetlands of North America: Ecology and Management, Edited by Darold P. Batzer, Russell B. Rader, and Scott A. Wissinger
ISBN 0-471-29258-3 © 1999 John Wiley & Sons, Inc.

ticularly chironomids, and their responses to the temporal and spatial dynamics of Presque Isle wetlands.

LAKE ERIE WETLANDS

Coastal wetlands in Lake Erie form in locations protected by diverse geomorphic features, including barrier sand bars, deltas, beaches, and berms. Along the 1402-km coast of the lake, wetlands are as diverse and dynamic as the landforms that promote their development. Often the erosional and depositional processes that promote wetland formation lead to their displacement, alteration, or destruction, thereby creating a mosaic of wetland habitats that shift in space and change over time (Mitsch 1992, Botts and Donn 1996, Davidson-Arnott and Reid 1994).

Approximately 96 wetland systems covering 83 km^2 are found along the coast of Lake Erie (Herdendorf 1992). They constitute the remnants of more than 4000 km^2 of lake-associated wetlands that historically were present (Mitsch and Gosselink 1993). Lake Erie coastal wetlands include lagoons found on extensive sand spits (Long Point, Ontario), estuarine river mouths (Old Woman Creek, Ohio), open-water marshes (West Harbor, Ohio), delta wetlands (St. Clair River mouth), and solution lagoons (Bass Islands, western Lake Erie) (Herdendorf 1992).

Virtually all undiked Lake Erie coastal wetlands are exposed to high wave energy and to marked fluctuations in water level that result from changes in lake level as well as from seiches and storm surges (Bedford 1992). Water levels in the Great Lakes fluctuate as much as 1.75 m on a 7–10-year cycle that appears to be caused by long-term variation in precipitation, evaporation, and runoff (Rasid et al. 1992). Seasonal changes in water level typically amount to about 30 cm annually, with lows before snowmelt and highs during early summer. Superimposed on cyclic changes are more impressive short-term changes driven by storms that cause seiches and storm surges capable of raising or lowering water levels more than 1 m in a few hours (Burton 1985). The consequence of both long- and short-term fluctuations in water level seems to be a constant "resetting" of successional stages. Thus, the processes causing in-filling and the shift from an aquatic to a terrestrial environment constantly are being interrupted. Changes in water level strongly influence sediment chemistry, nutrient cycling, and vegetation dynamics (Burton 1985), and the result is a failure of the wetlands to senesce (Herdendorf 1987).

In this chapter I summarize the existing literature with reference to benthic invertebrates in Lake Erie coastal wetlands, and I highlight the spatial and temporal dynamics of wetlands on Presque Isle, located on the Pennsylvania coast of Lake Erie. Ongoing research in Presque Isle wetlands specifically addresses the relationship between wetland dynamics and chironomid species assemblages. Finally, I identify recurring themes and research needs in invertebrate ecology in Lake Erie coastal wetlands.

WHAT DO WE KNOW?

Lessons from Other Lake Erie Wetland Systems

Herdendorf (1987 and 1992) provides extensive descriptions of coastal marsh systems in Lake Erie. The hydrology and geomorphology (Bedford 1992, Herdendorf 1990, Lyon and Green 1992, Rasid et al. 1992, Davidson-Arnott and Reid 1994, Reeder and Eisner 1994, Burton 1985), nutrient dynamics (Mitsch 1992, Jackson et al. 1995), vegetation (Reeder and Mitsch 1989, Klarer and Millie 1992, Stuckey and Moore 1995), and vertebrates (Johnson 1989, Randall et al. 1996, Whillans 1996) of many Lake Erie or lower Great Lakes coastal systems are reasonably well known. What is most conspicuous in these descriptions is the dearth of data related to wetland invertebrates.

While the invertebrates of deeper waters of Lake Erie, particularly *Hexagenia,* have been the focus of numerous studies, coastal wetland invertebrates have received very little attention (Herdendorf 1987, Krieger 1992). Most of the invertebrate information for Lake Erie wetlands is found in U.S. Fish and Wildlife Service reports or remains unpublished. A few studies provide taxonomic lists and life history information (Hunt 1962, Hiltunen and Manny 1982, Herdendorf and Lindsay 1975, Lindsay 1976), but, except for oligochaetes (Brinkhurst and Jamieson 1972, Lindsay 1976, Schloesser et al. 1995), the ecology, population dynamics, and production of wetland invertebrates in Lake Erie remain unknown (but see Krieger and Klarer 1991).

The tantalizingly few studies published on the ecology of invertebrates in Lake Erie wetlands strongly suggest that habitat heterogeneity is a key to understanding the invertebrate assemblages. The high diversity of invertebrates in the coastal wetlands seems to be associated with the availability of vegetative substrata and distinctly different benthic microhabitats found in the more sheltered wetlands or coastal zones (Krecker and Lancaster 1933, Krecker 1939, Chilton et al. 1986, McLaughlin and Harris 1990, Krieger and Klarer 1991, Herdendorf 1992). In Old Woman Creek, Ohio, large numbers of usually benthic invertebrates are common in the water column (Krieger 1992) and, as in other Great Lakes wetlands (McLaughlin and Harris 1990, Krull 1970), the presence of aquatic vegetation has the effect of increasing invertebrate abundance and diversity (Krecker 1939). Riley and Bookhout (1990) show that sediment-dwelling invertebrates were more abundant in shallow water (20–30 cm) than deep water (>40 cm) sediments in marshes managed as waterfowl habitat, and early spring drawdown increased invertebrate densities.

Assemblages of invertebrates in sheltered Great Lakes coastal wetlands often are distinct from those found in the lakes themselves, suggesting either that permanent lake habitat is not a source for most wetland invertebrates or that environmental conditions are unsuitable for persistence of populations of lake species in the wetlands. On the other hand, in wetlands directly exposed to wave wash from the lake (e.g., Presque Isle beach ponds) species assemblages often include numerous invertebrates most often associated with permanent lake habitat (Campbell 1993). Whether these species are permanent or transient in the wetlands is unknown.

The advent of the zebra and quagga mussels, *Dreissena polymorpha* and *Dreissena bugensis,* has markedly affected benthic invertebrate assemblages in Great Lakes littoral zones and wetlands. Zebra mussels increase invertebrate abundances and alter species composition (Dermott et al. 1993, Griffith 1993, Stewart and Haynes 1994). The effect is thought to be due to a combination of increased habitat structure and the presence of abundant pseudofeces produced when mussels translocate phytoplankton from the water column into the benthic habitat. For example, zebra mussels in Presque Isle wetlands occur as unattached clusters called "druses" that support higher abundances of oligochaetes, chironomids, amphipods, and flatworms than bare sand substrate (Botts et al. 1996).

Zebra mussels also strongly influence aquatic macrophytes. In some Great Lakes coastal wetlands zebra mussels overwinter successfully in deeper portions of the marsh, providing a ready pool of colonists that attach to new macrophyte populations each spring (Brady et al. 1995). Mussel filtration of the water column and concomitant increases in water clarity in macrophyte beds appear to be responsible for increasing macrophyte abundance and the spread of beds into new, or historically occupied, areas of coastal littoral zones and wetlands (Stuckey and Moore 1995). The increase in macrophyte abundance also is likely to cause an increase in invertebrate abundance and diversity.

At the same time, however, the presence of zebra mussels has been associated with decimation of populations of native unionids in Great Lakes habitats (Mackie 1991, Haag et al. 1993, Schloesser and Nalepa 1994). Fouling of shells by *Dreissena* interferes with gamete release, host location, feeding, and burrowing. Similar declines of native unionids have been documented in Presque Isle coastal wetlands and littoral zones (Schloesser and Masteller, unpublished data).

Presque Isle Wetlands

The Habitat. Presque Isle is an important economic and recreational resource for the city of Erie, Pennsylvania. It encloses and protects Presque Isle Bay, a natural harbor that has played a prominent historical role in the development on Erie as a major port on the Great Lakes. Presque Isle is a state park, and its beaches and trails are used by more than 4 million visitors annually (PA

DER 1993). Habitats in the park currently are managed for both conservation and recreation (hunting, fishing, swimming, hiking, and biking).

Presque Isle is a recurved barrier sandspit located on the south shore of Lake Erie, opposite the city of Erie, Pennsylvania. The peninsula is 13 km long, and its width varies from less than 400 m at its "neck" to more than 2 km before curving into its final distal tip. The peninsula rests on a platform of glacial deposits and has been migrating along the coast of Lake Erie toward the northeast for at least 3500 years (PA DER 1993). Sand eroded from beaches southwest of the peninsula and from the neck is carried by littoral currents to the distal tip of Presque Isle, where it is deposited. Over time the deposition of sand creates a recurved series of dune ridges and lagoon wetlands (Jennings 1909, Kormondy 1969, Botts and Donn 1996). The neck of the peninsula has been breached as recently as the 1920s due to erosion during spring storms. In an effort to prevent breaching and extensive beach erosion, the Army Corps of Engineers has constructed a series of 58 "riprap" breakwalls off the lakeward shore of the peninsula. The breakwalls markedly reduce shoreline erosion and probably alter the delivery of sand to the tip of the peninsula.

The interior of the peninsula is characterized by dune ridges and palustrine sand plains. Dune ridges usually form during severe northeasterly storms (Jennings 1909). Eventually, the dunes are stabilized by vegetation, particularly cottonwoods and dune grasses, that form barriers for wind-borne particles, thereby entrapping more sand. During periods with no severe storms, sand accumulates more gradually in low-lying palustrine sand plains, which alternately support upland and wetland species, depending on lake water levels (e.g., Campbell 1992). Elevation ranges between 176 m above mean sea level at the lake shore to 182 m at the highest elevation, a maximum relief of only 6 m. Groundwater/wetland water levels are strongly influenced by changes in Lake Erie water level throughout the peninsula (PA DER 1993).

Wetlands form on Presque Isle in two ways (Jennings 1909, Kormondy 1969, Botts and Donn 1996). Lagoon wetlands form when barrier sand bars develop behind the recurving distal tip of the peninsula (Gull Point), isolating large, but shallow (<0.5 m) lagoons from the lake. Lagoons often shift and change conformation, and they may remain open to the lake for several years before assuming their final form. Once stabilized, lagoon wetlands generally persist for many years. During the ontogeny of lagoon wetlands, growth of emergent aquatic vegetation and aeolian deposition of sand in shallow areas of the pond may result in isolation of several "sibling" wetlands from a single lagoon. Beach ponds form parallel to the lake shore when northeasterly storms deposit large bars of sand on the exposed beach, trapping water behind them. Although there are notable exceptions, beach ponds are relatively unstable and often persist for less than a full year before destruction by new storms.

Wetland Biota. The sand dune/marsh/upland forest complexes on Presque Isle provide unique examples of the process of primary vegetation succession (Jennings 1909, Kormondy 1969), with seral stages from unvegetated to several-hundred-year-old climax forests found within 5 km of each other. Wetland vegetation on Presque Isle is characterized by the presence of red maple and blackgum in wooded swamps, buttonbush in shrub-scrub swamps, and cattails, bur reed, *Phragmites, Nuphar, Nymphaea,* as well as assorted sedges, in marshes and wet meadows. The peninsula supports 60 listed plant species of special concern in Pennsylvania.

The wetland vegetation was characterized in 1909 by Jennings as undergoing hydrarch succession requiring about 60 years to reach "climax," but Bissel (1993) shows that plant succession in many Presque Isle communities is not unidirectional. Instead, wetland vegetation shifts in response to changing lake levels. Thus, when lake levels rise, swamp forests and shrub swamps are replaced by emergent marshes, emergent marshes are replaced by the aquatic bed community, and dry sand plain communities are replaced by palustrine sand plain communities. Similarly, when lake levels fall, swamp forests and shrub swamps replace emergent marshes, palustrine sand plains replace aquatic beds, emergent marsh expands into palustrine sand plain as the palustrine sand plain moves into newly exposed aquatic beds, and dry sand plains replace palustrine sand plains.

A number of rare plant species are found in the wetlands themselves (Bissel 1993), but, as in most wetlands, these rare species and plant associations are threatened by invasive species. The most severe problems on Presque Isle are caused by *Phragmites australis,* which since the 1970s has spread to cover tens of hectares of emergent marshes. *Typha* × *glauca,* hybrid cattails, are causing similar problems. Both purple loosestrife and reed canary grass also pose a threat to the wetland communities on the peninsula, but they have yet to cover extensive areas of the habitat.

In addition to the extensive studies of the vegetation of Presque Isle wetlands, the invertebrate fauna has been sampled repeatedly. Seventeen species of mollusks were found in Kormondy's (1969) study, but only *Physa gyrinis, Lymnaea humilis, Lymnaea palustris, Sphaerium partumeium,* and *Sphaerium securis* were common and widely distributed. Kormondy (1969) reports the presence of the amphipods *Hyalella azteca,* abundant in all locations, and *Gammarus fasciatus,* found only in ponds that receive wave wash from Lake Erie. The peninsula supports exceptionally high species diversity (45 species, 10 rare) of Cladocera (Campbell 1993) as a consequence of high habitat heterogeneity in the wetlands. Nearly half of the species reported were chydorids characteristic of wetland benthic or phytal habitats, while the remainder were lake or open-water species.

The wetlands on Presque Isle have been sampled extensively for insects. Masteller (1993a) reports a high diversity of caddisflies (7 families, 34 species) from Presque Isle, including two rare species of Molannidae. He concludes that the potential diversity on the peninsula is even higher than reported, and he attributes the diversity of caddisflies to the diversity of habitats available in a confined area. Ninety-one species of Coleoptera were collected from Presque Isle ponds by Gamble (1931), but Kormondy (1969) reports only 26 species, a decline he attributes to less intensive sampling techniques and to a marked change in Ridge Pond, where, of the 32 species reported by Gamble, Kormondy found only 3. Odonates also have been collected extensively on Presque Isle (Emery 1933, Gower and Kormondy 1963, Kormondy 1969, Ahrens et al. 1968, Masteller 1993b). Odonate diversity was explained by the wide variety of aquatic habitats on the peninsula and by the fact that species present in older ponds are retained, with some additions and deletions, from species assemblages present in younger ponds. Odonate diversity potentially includes about 50 species, but turnover in individual wetlands is high. Masteller (1993b) raised concerns of extirpation of some species from the park when 16 species collected in earlier studies were not found in the 1993 survey. At least 42 species of Chironomidae were reported in Gull Point wetlands (Botts 1997). Species distribution and abundance were best explained by environmental heterogeneity, particularly the abundance of submerged macrophytes and surrounding stands of *Phragmites*.

The Gull Point Wetlands. A major focus of ecological research on Presque Isle has been plant primary succession. The spatial and temporal dynamics of the habitat make it an ideal site to address questions that pertain to patterns of species change over time and to the mechanisms that produce those patterns. The present state of understanding of succession on Presque Isle remains confined to patterns of change, however. Little has been done to address mechanisms of change. Here I describe the patterns that have been observed, relate those patterns to the idea, repeatedly expressed in the literature, that Great Lakes coastal wetlands do not senesce, and suggest future lines of research to clarify the mechanisms of change in Presque Isle wetlands.

If one examines a map of Presque Isle (Fig. 39.1) and superimposes dates, distances, and sizes associated with the wetlands on the peninsula, it becomes clear that Presque Isle wetlands are distributed along correlated temporal and spatial gradients (Table 39.1). Wetlands on Gull Point, the distal tip, are young (<25 years old), small (<0.25 ha), and close to the lake (<200 m). Farther back on the peninsula, within the broad region between the neck and Gull Point, wetlands are older (50–400 years old), larger (ca 0.1–2.7 ha), and more isolated from the lake (600–1300 m). The degree of isolation is more than a matter of distance in this part of the peninsula, because the inland wetlands are surrounded by mature or maturing forest rather than palustrine sand plain and stands of *Phragmites*. Finally, the oldest wetlands on the peninsula consist of a series of highly invaginated and interconnected lagoons that are open to the bay at Marina Lake, an artificial lake that was dredged in 1956, and Graveyard Pond, where Commodore Perry's dead were consigned to the deeps during the War of 1812. The lagoon system is highly disturbed and is subject to dredging and nonmotorized boat traffic. The Gull Point and inland wetlands together constitute an elegant system in which to address changes in species composition over time.

Jennings (1909) described the physical origins and estimated the ages of a large number of Presque Isle wetlands, using a combination of historical records and estimates based on plant species composition. He described

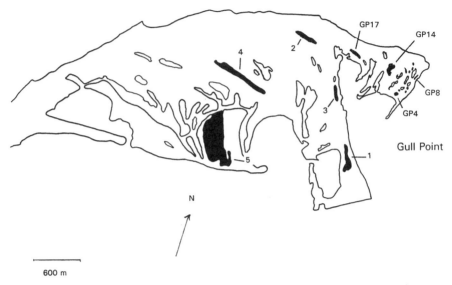

Fig. 39.1. Map of Presque Isle, Pennsylvania, showing locations of Gull Point wetlands (labelled GPxx) and five inland wetlands (1 = Near Horseshoe Pond, 2 = Dead Pond, 3 = Thompson Beach Pond, 4 = Ridge Pond, 5 = Big Pond).

changes in the wetland and upland vegetation that correspond to classical models of succession (Cowels 1899) on Great Lakes sand dunes. He based his assessment of vegetation dynamics on the concept that spatial patterns provide a snapshot picture of temporal changes in habitat patches that are similar but of different ages.

Kormondy (1969), working with wetland vegetation, used the fact that cottonwood seeds germinate at the extreme edges of new wetlands. The seed-

TABLE 39.1. Age, Approximate Surface Area, and Approximate Distance from the Lake of Selected Wetlands on Presque Isle, Pennsylvania[a]

Wetland	Age (years)	Area (ha)	Distance to Lake Erie (m)
Gull Point 8	4	0.03	91
Gull Point 4	11	0.06	61
Gull Point 14	24	0.18	152
Gull Point 17	49	0.25	183
Near Horseshoe Pond	95	0.11	670
Dead Pond	105	0.16	800
Thompson Beach Pond	125	0.13	731
Ridge Pond	295	0.81	411
Big Pond	325	2.69	1310

[a] Surface areas and distance from the lake vary with lake level.

lings grow quickly, stabilizing the sand and acting as traps for windborne sand, which is deposited at the edge of the wetland, building dune ridges that are defined by rows of cottonwoods. The ages of wetlands were estimated from incremental borings of the cottonwoods growing on the dune ridges surrounding the wetlands. On the basis of this chronology, Kormondy was able to confirm many of the successional patterns proposed by Jennings.

Ellenberger et al. (1973) examined one young beach pond and two older ponds that had been studied earlier by both Kormondy and Jennings. They analyzed diatoms from core samples in each pond and confirmed that diatom species composition differed among ponds, but they did not establish a clear successional sequence of diatom species composition that could be related to wetland age. On the basis of sediment chemistry and vegetation dry mass, however, they suggested that the ponds appeared to be on the general trajectory described by Jennings (1909) and Kormondy (1969).

Thus, temporal patterns in the biota of Presque Isle wetlands nearly always are described as exhibiting a classical pattern of primary succession from lagoon to marsh to thicket to forest, a unidirectional process requiring hundreds of years. Recently, these ideas have been challenged by Bissel (1993), who states that succession is not unidirectional in Presque Isle wetlands. He suggests that Lake Erie water level fluctuation and disturbance, particularly fire and trampling, strongly influence temporal patterns of change in wetland vegetation.

Chironomid Spatial and Temporal Patterns on Gull Point. Botts and Donn (1996) use a series of 32 aerial photographs taken between 1946 and 1993 to establish the absolute ages and origins of 21 wetlands on Gull Point. The area of Gull Point in 1946 was approximately 13.5 ha. By 1993 the study area on Gull Point encompassed 61.1 ha, of which 3.8 ha consisted of fully enclosed lagoon wetlands scattered through an upland matrix of palustrine and dry sand plain and dunes. An additional 3.9 ha on Gull Point consisted of shallow lagoons, formed in the 1940s, which never became completely isolated from the lake. The oldest of the 11 enclosed lagoon wetlands formed in 1946 (Botts and Donn 1996). Lagoon wetlands were formed and isolated from the lake in 1974, 1977, 1986, and 1987. Subsequent sibling wetland formation resulted in 10 fully isolated wetlands. In addition, one beach pond formed in 1992 appears stable, while one beach pond formed in 1992 and several formed in 1993, 1994, 1995, and 1996 have come and gone.

Vegetation patterns similar to that described by Jennings (1909) and Kormondy (1969) in young ponds were noted. The youngest ponds contain few macrophytes but abundant surface diatoms and some *Chara.* The rest of the ponds contain *Potamogeton, Myriophyllum, Najas, Elodea,* and *Ceratophyllum* and are surrounded by *Phragmites.* In wetlands that formed prior to the ongoing invasion by *Phragmites australis, Typha* spp. and *Sparganium* often are present as well. None of the ponds support *Vallisneria, Nuphar,* or *Nymphaea,* but *Utricularia* spp. is abundant in most ponds. Except for the very youngest ponds, the ages of the Gull Point wetlands cannot be deduced reliably from the vegetation present in or around the pond (Botts, personal observation).

In spite of the rapidity with which the spatial configuration of the habitat on Gull Point changes, the response of benthic invertebrates to the dynamics of the Gull Point habitat remained virtually unstudied until 1994, when Botts (1997) initiated a study of the dynamics of assemblages of chironomid midges in the wetlands. The goal of the research was to determine whether chironomid species composition was a function of wetland age, position in the wetland/upland habitat matrix, or environmental characteristics.

A hand-held 48-mm inner diameter core sampler was used to collect two benthic samples monthly from 22 locations on Gull Point. Sampling locations included every wetland present on the sampling date (13–17 wetlands) and four stations on Gull Point beaches (in Lake Erie). In the laboratory, samples were rinsed through a 61-μm sieve and larvae were separated from the samples using sugar flotation. All larvae were mounted in CMC-10 for identification to lowest practical taxon (usually genus or species group) using the keys in Wiederholm (1983). Annual mean abundances for each taxon were computed and used to generate a species by location matrix. An environmental factor by location matrix that included position, age, and physical, chemical, and vegetation parameters at each sampling location also was generated. Data were analyzed using CANOCO (ter Braak 1988).

A total of 42 chironomid taxa were collected from 22 sampling locations on Gull Point (Botts 1997). Three taxa, *Tanytarsus lugens* gp., *Paratanytarsus* sp., and *Cladotanytarsus mancus,* were both abundant and widespread. *Stictochironomus* sp., *Cladotanytarsus mancus, Cladotanytarsus* sp. gp. A, and *Cryptochironomus* sp. were more abundant in sandy, open habitats than in wetlands with abundant submerged macrophytes, while wetlands with abundant macrophytes and higher organic content in the substrate tended to support chironomid assemblages containing *Chironomus* spp., *Ablabesmyia* spp., and *Paratanytarsus* sp. Environmental characteristics of the wetlands explained 52 percent of the chironomid species composition in the wetlands, suggesting that differential larval survival or female choice of oviposition sites along an environmental gradient resulted in the pattern of larval distribution seen in the wetlands. The relationship, while statistically significant, was weak, and Botts (1997) suggests that because chironomid adults are good dispersers it was more likely that ovipositing females had access to any wetland in the mosaic and that wetlands did not represent spatially isolated habitats on Gull Point. A number of environmental characteristics, including macrophyte abundance, *Phragmites* cover around the wetland, and sediment organic matter, were strongly correlated with pond age, but pond age per se explained only 3 percent of the variability in the species/environment relationship.

The initial study of the spatial distribution of chironomid larvae across Gull Point wetlands raised the question of why the influence of wetland age could not be detected in the Gull Point wetlands. Several alternative hypotheses could explain the apparent absence of an age effect.

1. Pond age does influence chironomid species composition, but the effect requires more than a 50-year span of time to become apparent. That is, had sampling been conducted in very old wetlands, an age effect would have been noticed (see below).
2. Succession of chironomid species is a "colonization" process, and the age-influenced dynamics occur very early during pond formation. That is, only very young ponds support different species of chironomids from all other wetlands.
3. Chironomids, probably because of good aerial dispersal capabilities and possibly because of nondiscriminating oviposition behavior, use the wetlands on Presque Isle as a single, spatially discontinuous habitat. That is, habitat "preference" is expressed as differential larval survival in wetlands with different ecological conditions, and the chironomid assemblages in any given wetland are characterized by repeated colonizations and extinctions or by a large proportion of "transient" species.

At present, access to the Gull Point wetlands is restricted between April 1 and November 31 to protect migrating and nesting shore birds. Thus, experimental tests of hypotheses 2 and 3 are not presently possible. The spatial and temporal scales of the initial study have been extended to help distinguish whether hypothesis 1 may be correct. A long-term study designed to document chironomid species composition in Gull Point wetlands was instituted formally in 1996, although sampling was begun in 1994.

In March 1997 benthic core samples were collected in each of the Gull Point wetlands and in six older ponds in the interior of the peninsula. The ponds ranged in age from <1 month to 325 years and in distance from the lake from <20 m to 1300 m. Preliminary analysis of the 1997 data (Botts, unpublished) shows that even across such a wide range of ages, chironomid species composition does not reflect differences due to pond age (canonical correlation analysis, trace F statistic = 1.54, p = 0.14). Two of the ponds sampled by Botts in 1997, Ridge Pond and Dead Pond, were sampled in 1909 by Jennings and in 1969 by Kormondy. Kormondy (1969) reports much lower chironomid diversity than Botts (1997), but the species present were similar, and no consistent pattern of change in species composition with pond age was apparent.

Conclusions for Presque Isle Wetlands

What do the Presque Isle wetlands tell us about wetland senescence in Great Lakes coastal wetlands? Based on data collected from wetlands of precisely known age, neither vegetation nor chironomid species composition clearly reflects wetland "age." The Gull Point wetlands all are less than 50 years of age, and the lack of clear differences may reflect the fact that they are not yet old nor isolated enough for marked differences in vegetation and chironomid taxonomic composition to have developed. However, data from older wetlands on the peninsula also fail to show a pattern of species composition that is dependent upon wetland age.

A number of environmental characteristics, including macrophyte abundance, *Phragmites* cover, sediment organic content, wetland area, and distance of the wetland from the lake, were correlated with chironomid species composition. These environmental characteristics are correlated with wetland age over short time spans (Botts 1997), but only for as long as water levels in the lake remain stable. For example, Pond 16 was isolated from the lake in 1974 and was 53 m from the lake in 1995. It was completely surrounded by a dense stand of *Phragmites*, had high sediment organic content, and supported abundant submerged macrophytes. In 1996–1997 Lake Erie water levels were 50–60 cm above long-term means, resulting in shoreline erosion extensive enough that Pond 16 is now < 5 m from the lake and receives wave overwash from the lake during every storm event. The dense *Phragmites* stand that once surrounded the wetland has been destroyed. Half of the pond has filled with sand washed in from the lake, resulting in a dramatic reduction of pond area and of sediment organic content (Botts, unpublished data). Thus, although this pond is 25 years old, it now has the environmental characteristics of a newly formed beach pond. Botts (1997) suggests that environmental age (defined by habitat characteristics such as macrophyte abundance, *Phragmites* cover, sediment organic matter, etc.) is more important than the absolute age (time from the original formation of the wetland) in determining the biotic composition of the wetlands.

Mitsch and Gosselink (1993) raise serious issues regarding the generality of linear, directed changes in wetland biota. They suggest, instead, that environmental gradients are steep in wetlands, and the biota respond to the gradients, or to changes in the gradients. One of the environmental gradients that influences chironomid spatial pattern is wetland surface area (Botts 1997). On Gull Point, wetland area is strongly correlated with Lake Erie water level (Fig. 39.2) and the wetlands are hydrologically connected to the lake. Examination of the spatial distribution of macrophytes, organic matter, and sand patches in individual wetlands shows clear evidence of areas of upwelling of lake water into the wetlands (Botts, personal observation). In addition to the consequences already noted for Wetland 16, record high water levels in 1997 resulted in extremely high water levels in the wetlands and palustrine sand plains, reflooding many of the lagoon sibling ponds back to their original undivided levels. The 1997 spring sampling reflects a marked reduction in chironomid abundances and an apparent rearrangement of species composition that may be partially explained by the dramatic changes in the habitat itself. If so, this is further support for the idea that the wetlands fail to "age" unidirectionally.

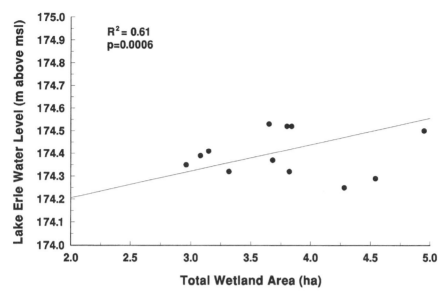

Fig. 39.2. Regression of total wetland surface area on Gull Point, Presque Isle, Pennsylvania versus Lake Erie water levels (m above mean sea level) between 1946 and 1995.

The initial (Botts 1997) study of Gull Point chironomid assemblages generated two important sets of questions that can be addressed in long-term or manipulative studies of spatial and temporal patterns of species composition in the Presque Isle wetlands. First, how are the spatial and temporal dynamics of the wetlands reflected in the (meta)population dynamics of the chironomid assemblages on Gull Point? Are the wetlands really used as a single, but spatially discontinuous, habitat? Can we identify source-sink or mainland-island dynamics in colonization/extinction dynamics of chironomids in the wetlands? Second, are there interesting "successional" dynamics of invertebrates in Presque Isle wetlands occurring at temporal scales much finer (i.e., short-term in brand new ponds) or much coarser (i.e., long-term over a span of several hundred years) than initially thought? What are the colonization dynamics of new ponds? What are the effects of long-term spatial isolation on inland pond assemblages?

WHAT'S NEXT?

To reiterate the conclusion reached by Krieger (1992), "Given the relative paucity of information on invertebrate ecology in Great Lakes coastal wetlands, research on almost any aspect of their ecology will significantly advance our state of knowledge about these communities." This statement is true for both basic and applied questions. In fact, given the present state of our knowledge, virtually everything we learn will be new to our understanding of invertebrates in Lake Erie wetlands. The recurrent theme in the literature is that Great Lakes wetlands are "different" than inland freshwater wetlands and coastal saltwater wetlands as measured by virtually any meter stick. Yet the literature is replete with statements that begin something like "We do not know this or that for the Great Lakes, but in inland wetlands we see . . ." Themes that repeatedly occur in the literature regarding Great Lakes wetlands in general are:

1. The hydrology and geomorphology of Great Lakes coastal wetlands create physical and chemical conditions unique to these wetlands, and processes that occur in inland wetlands do not necessarily apply to freshwater coastal wetlands. For example, Great Lakes coastal wetland vegetation may exhibit very different successional dynamics than found in inland wetlands because of constant disturbance from water level fluctuations on several spatial and temporal scales.

2. Exotic species are altering the physical, chemical, and biological characteristics of virtually every habitat in Lake Erie wetlands. Zebra mussels are altering trophic dynamics, providing extensive colonizable habitat rich in organic matter. *Phragmites, Typha × glauca,* and purple loosestrife are altering vegetative communities and seriously threatening rare plant assemblages.

3. The role of benthic and epiphytic wetland invertebrates in coastal marshes in trophic ecology, while assumed to be significant, has been addressed in very few studies of either fish or waterfowl ecology and virtually no studies of the ecology of other vertebrate and invertebrate groups. Much of what is reported for Lake Erie wetlands is based on data collected in other locations.

4. The landscape ecology of Lake Erie coastal wetland invertebrates is virtually unknown, in spite of the highly dynamic nature of many of the wetland habitats. This is particularly true for Presque Isle, Point Pelee, and Long Point, all highly mobile sand spits.

5. Lake Erie wetlands are strongly influenced by anthropogenic effects, including sediment and water contamination by point and nonpoint pollutants, recreational use, and wetland alteration and destruction. Except for a few studies in diked and managed wetlands, little is known of the effects of human activities on the wetlands or the invertebrates they support.

LITERATURE CITED

Ahrens, C., G. H. Beatty, and A. F. Beatty. 1968. A survey of the Odonata of western Pennsylvania. Pennsylvania Academy of Science 42:103–109.

Bedford, K. W. 1992. The physical effects of the Great Lakes on tributaries and wetlands. Journal of Great Lakes Research 18:571–589.

Bissel, J. K. 1993. Rare plants and rare plant communities of Presque Isle. Bartonia 57:2–8.

Botts, P. S. 1997. Spatial pattern, patch dynamics and successional change: Chironomid assemblages in a Lake Erie coastal wetland. Freshwater Biology 37:277–286.

Botts, P. S., and R. Donn. 1996. Temporal delineation of wetlands on Gull Point, Presque Isle. Pages 177–192 *in* G. Mulamoottil (ed.), Wetlands: Environmental Gradients, Boundaries, and Buffers. CRC Press, Boca Raton, FL.

Botts, P. S., B. A. Patterson, and D. W. Schloesser. 1996. Zebra mussel effects on benthic invertebrates: Physical or biotic. Journal of the North American Benthological Society 15:179–184.

Brady, V. J., B. J. Cardinale, and T. M. Burton. 1995. Zebra mussels in a coastal marsh: the seasonal and spatial limits of colonization. Journal of Great Lakes Research 21:587–593.

Brinkhurst, R. O., and B. G. Jamieson. 1972. Aquatic Oligochaetes of the World. University of Toronto Press, Toronto, ON.

Burton, T. M. 1985. The effects of water level fluctuations on Great Lakes coastal marshes. Pages 3–27 *in* H. H. Prince and F. M. D'Itri (eds.), Coastal Wetlands. Lewis, Chelsea, MI.

Campbell, J. M. 1992. Habitat and population ecology of *Lobelia kalmii* on Presque Isle, Pa. correlated with changes in Lake Erie water levels. Journal of the Pennsylvania Academy of Sciences 66:123–127.

Campbell, J. M. 1993. Species diversity of littoral cladoceran communities of Presque Isle. Journal of the Pennsylvania Academy of Sciences 67:115–119.

Chilton, E. W., R. B. Lowe, and K. M. Schurr. 1986. Invertebrate communities associated with *Bangia atropurpurea* and *Cladophora glomerata* in western Lake Erie. Journal of Great Lakes Research 12:149–153.

Cowels, H. C. 1899. The ecological relations of the vegetation on the sand dunes of Lake Michigan. Botanical Gazette 27:95–117.

Davidson-Arnott, R. G. D., and H. E. Conliffe Reid. 1994. Sedimentary processes and the evolution of the distal bayside of Long Point, Lake Erie. Canadian Journal of Earth Science 31:1461–1473.

Dermott, R., J. Mitchell, I. Murray, and E. Fear. 1993. Biomass and production of zebra mussels (*Dreissena polymorpha*) in shallow waters of northeastern Lake Erie. Pages 399–413 *in* T. F. Nalepa and D. W. Schloesser (eds.), Zebra Mussels: Biology, Impacts, and Control. Lewis, Boca Raton, FL.

Ellenberger, R. S., H. R. Laube, R. A. Walter, and J. R. Wholer. 1973. Pond succession on Presque Isle, Erie, Pennsylvania. Proceedings of the Pennsylvania Academy of Sciences 47:133–135.

Emery, R. J., Sr. 1933. A systematic survey of the dragonflies and damselflies of Presque Isle, with notes on their ecology and distribution. Master's Thesis, University of Pittsburgh, Pittsburgh, PA.

Gamble, J. T. 1931. Studies on the ecology and distribution of aquatic beetles of Presque Isle, Lake Erie, Pennsylvania. Proceedings of the Pennsylvania Academy of Science 5:97–100.

Gower, J. L., and E. J. Kormondy. 1963. Life history of *Lestes rectangularis,* with special reference to seasonal regulation. Ecology 44:398–402.

Griffiths, R. W. 1993. Effects of zebra mussels (*Dreissena polymorpha*) on benthic fauna of Lake St. Clair. Pages 415–438 *in* T. F. Nalepa and D. W. Schloesser (eds.), Zebra Mussels: Biology, Impacts, and Control. Lewis, Boca Raton, FL.

Haag, W. R., D. J. Berg, D. W. Garton, and J. L. Ferris. 1993. Reduced survival and fitness in native bivalves in response to fouling by the introduced zebra mussel (*Dreissena polymorpha*) in western Lake Erie. Canadian Journal of Fisheries and Aquatic Sciences 50:13–19.

Herdendorf, C. E. 1987. Ecology of the Coastal Marshes of Western Lake Erie: A Community Profile. U.S. Fish and Wildlife Service Biological Report 85(79), Washington, DC.

———. 1990. Great Lakes Estuaries. Estuaries 4:493–503.

———. 1992. Lake Erie coastal wetlands: An overview. Journal of Great Lakes Research 18:533–551.

Herdendorf, C. E., and W. K. Lindsay. 1975. Benthic macroinvertebrate populations of Sandusky Bay. Pages 359–377 *in* D. B. Baker, W. B. Jackson, and B. L. Prater (eds.), Proceedings Sandusky River Basin Symposium. PLUARG/International Joint Commission, Great Lakes Regional Office, Windsor, ON.

Hiltunen, J. K., and B. A. Manny. 1982. Distribution and Abundance of Macrozoobenthos in the Detroit River and Lake St. Clair, 1977. U.S. Fish and Wildlife Service, Great Lakes Fisheries Laboratory. Administrative Report 80-2. Ann Arbor, MI.

Hunt, G. S. 1962. Water pollution and the ecology of some aquatic invertebrates in the lower Detroit River. Pages 29–49 *in* Proceedings Fifth Conference of Great Lakes Research, Publication No. 9, Great Lakes Research Division, University of Michigan, Ann Arbor, MI.

Jackson, C. R., C. M. Foreman, and R. L. Sinsabaugh. 1995. Microbial enzyme activities as indicators of organic matter processing rates in a Lake Erie coastal wetland. Freshwater Biology 34:329–342.

Jennings, O. E. 1909. A botanical survey of Presque Isle, Erie County, Pennsylvania. Annals of the Carnegie Museum 5:145–159.

Johnson, D. L. 1989. Lake Erie wetlands: Fisheries considerations. Pages 257–273 *in* K. L. Krieger (ed.), Lake Erie Estuarine Systems: Issues, Resources, Status and Management. U.S. Department of Commerce, National Oceanic and Atmospheric Administration, Washington, DC.

Klarer, D. M., and D. F. Millie. 1992. Aquatic macrophytes and algae at Old Woman Creek estuary and other Great Lakes coastal wetlands. Journal of Great Lakes Research 18:622–633.

Kormondy, E. J. 1969. Comparative ecology of sandspit ponds. American Midland Naturalist 82:28–61.

Krecker, F. H. 1939. A comparative study of the animal population of certain submerged aquatic plants. Ecology 20:553–562.

Krecker, F. H., and L. Y. Lancaster. 1933. Bottom shore fauna of western Lake Erie: A population study to a depth of six feet. Ecology 14:79–93.

Krieger, K. A. 1992. The ecology of invertebrates in Great Lakes coastal wetlands: Current research knowledge and research needs. Journal of Great Lakes Research 18:634–650.

Krieger, K. A., and D. M. Klarer. 1991. Zooplankton dynamics in a Great Lakes coastal marsh. Journal of Great Lakes Research 17:255–269.

Krull, J. N. 1970. Aquatic plant-macroinvertebrate associations and waterfowl. Journal of Wildlife Management 34:707–718.

Lindsay, W. K. 1976. Sandusky Bay, Ohio as a distinct freshwater estuary. Ph.D. Dissertation, Ohio State University, Columbus, OH.

Lyon, J. G., and R. G. Greene. 1992. Use of aerial photographs to measure the historical areal extent of Lake Erie coastal wetlands. Photogrammetric Engineering and Remote Sensing 58:1355–1360.

Mackie, G. L. 1991. Biology of the exotic zebra mussel, *Dreissena polymorpha,* in relation to native bivalve and its potential impact in Lake St. Clair. Hydrobiologia 219:251–268.

Masteller, E. C. 1993a. The Trichoptera (caddisflies) of Presque Isle State Park and Lake Erie, Erie County, Pennsylvania. Journal of the Pennsylvania Academy of Sciences 67:132–136.

———. 1993b. Odonata (dragonflies and damselflies) of Presque Isle State Park, Erie County, Pennsylvania. Journal of the Pennsylvania Academy of Sciences 67:137–140.

McLaughlin, D. B., and H. J. Harris. 1990. Aquatic insect emergence in two Great Lakes marshes. Wetlands Ecology and Management 1:111–121.

Mitsch, W. J. 1992. Combining ecosystem and landscape approaches to Great Lakes wetlands. Journal of Great Lakes Research 18:552–570.

Mitsch, W. J., and J. G. Gosselink. 1993. Wetlands. Van Nostrand Reinhold, New York.

Pennsylvania Department of Environmental Resources (PA DER). 1993. Resource management plan Presque Isle State Park. Commonwealth of Pennsylvania, Harrisburg, PA.

Randall, R. G., C. K. Minns, V. W. Cairns, and J. E. Moore. 1996. The relationship between an index of fish production and submerged macrophytes and other habitat features at three littoral areas in the Great Lakes. Canadian Journal of Fisheries and Aquatic Sciences 53:35–44.

Rasid, H., D. Baker, and R. Kreutzwiser. 1992. Coping with Great Lakes flood and erosion hazards: Long Point, Lake Erie, vs. Minnesota Point, Lake Superior. Journal of Great Lakes Research 18:29–42.

Reeder, B. C., and W. R. Eisner. 1994. Holocene biogeochemical and pollen history of a Lake Erie, Ohio, coastal wetland. Ohio Journal of Science 94:87–93.

Reeder, B. C., and W. J. Mitsch. 1989. Seasonal patterns of planktonic and macrophyte productivity of a freshwater coastal wetland. Pages 49–136 in W. J. Mitsch (ed.), Wetlands of Ohio's Coastal Lake Erie: A Hierarchy of Systems. Ohio Sea Grant, Columbus, OH.

Riley, T. Z., and T. A. Brookhout. 1990. Response of aquatic macroinvertebrates to early-spring drawdown in nodding smartweed marshes. Wetlands 10:173–185.

Schloesser, D. W., and T. F. Nalepa. 1994. Dramatic decline of unionid bivalves in offshore waters of western Lake Erie after infestation by the zebra mussel, *Dreissena polymorpha*. Canadian Journal of Fisheries and Aquatic Sciences 51:2234–2242.

Schloesser, D. W., T. B. Reynoldson, and B. A. Manny. 1995. Oligochaete fauna of western Lake Erie 1961 and 1982: Signs of sediment quality recovery. Journal of Great Lakes Research 21:294–306.

Stewart, T. W., and J. M. Haynes. 1994. Benthic macroinvertebrate communities of southwestern Lake Ontario following invasion of *Dreissena*. Journal of Great Lakes Research 20:497–493.

Stuckey, R. L., and D. L. Moore. 1995. Return and increase in abundance of aquatic flowering plants in Put-in-Bay Harbor, Lake Erie, Ohio. Ohio Journal of Science 95:261–266.

ter Braak, C. J. F. 1988. CANOCO—a FORTRAN Program for Canonical Community Ordination. Microcomputer Power, Ithaca, NY.

Whillans, T. H. 1996. Historic and comparative perspectives on rehabilitation of marshes as habitat for fish in the lower Great Lakes basin. Canadian Journal of Fisheries and Aquatic Sciences 53:58–66.

Wiederholm, T. 1983. Chironomidae of the Holarctic Region, Part 1. Keys and diagnoses: Larvae. Entomologica Scandinavia Supplement 19.

40 A Prairie Coastal Wetland (Lake Manitoba's Delta Marsh)

Organization of the Invertebrate Community

BRENDA J. HANN

*T*he invertebrate community in Delta Marsh, a coastal wetland, is a vital component of the habitat for all marsh inhabitants, but especially waterfowl and young-of-the-year fish. Species composition, seasonal phenology, and spatial distributions of the microcrustacean and nektonic macroinvertebrate components of the community are presented for areas of the marsh with a diverse wetland fish assemblage, as well as for areas that are fishless.

The microcrustacean community (39 taxa) is similar to that in the littoral zones of north temperate lakes, with a few species present occasionally that are typically considered to be "lake" species, common in the adjacent Lake Manitoba. Cladocerans comprise the predominant component of the microcrustacean community in a fishless portion of the wetland. Cyclopoid copepods are the most abundant microcrustaceans in the presence of fish, both inshore and offshore, in the water column and among vegetation. Planktonic species are most abundant in spring, whereas phytophilous species are prevalent in summer and autumn. Seasonal phenology of nektonic macroinvertebrates is complex, with highest densities in open water areas. Preferences for open-water or Typha margins are shown. Invertebrate-macrophyte associations are numerous and intricate and are only partially explained in terms of relative available surface areas of plants.

Invertebrates in Freshwater Wetlands of North America: Ecology and Management, Edited by Darold P. Batzer, Russell B. Rader, and Scott A. Wissinger
ISBN 0-471-29258-3 © 1999 John Wiley & Sons, Inc.

INTRODUCTION

Wetlands are refuges for biodiversity, due in part to the seasonal and spatial habitat heterogeneity. Diversity and productivity of aquatic invertebrates in wetlands is remarkably high, yet little studied (Batzer and Wissinger 1996) compared with extensive studies of zooplankton in the pelagic zone of lakes. In wetlands interannual vagaries of water depth (flooding to drought conditions) and presence or absence of fish interact with seasonal subtleties of predator-prey relations and population dynamics (demographic stochasticity) to accentuate the complexity of invertebrate communities.

The invertebrate community has direct relevance to the planktivorous fish which undergo seasonal migrations into coastal wetlands like Delta Marsh on the shore of Lake Manitoba because the diets of both adult and larval fish consist largely of cladocerans (Whiteside 1988, Whiteside et al. 1985). Invertebrates are important to waterfowl in northern prairie marshes (Murkin and Kadlec 1986). The densities and biomass of spring aquatic macroinvertebrates, in particular, influence waterfowl densities in marsh habitats. Swanson and Meyer (1977) found that waterfowl feed extensively on invertebrates during the spring to meet the high protein requirements of breeding and egg laying. Waterfowl may use fish presence/absence as cues to estimate invertebrate densities in selecting breeding sites (Mallory et al. 1994). Differential responses exist among waterfowl species in strength of invertebrate-waterfowl associations for breeding and feeding (Murkin and Kadlec 1986) and invertebrate-macrophyte-waterfowl associations (Krull 1970, Lillie and Evrard 1994).

Invertebrates constitute vital elements in the marsh foodweb in coastal wetlands, as has been shown in recent studies of energy and biomass transfer using stable isotope techniques (Neill and Cornwell 1992, Keough et al. 1996). There is often considerable energy and biomass transfer between the aquatic and terrestrial habitats, since many macroinvertebrates spend portions of their life cycle in each niche (Merritt and Cummins 1984).

Microinvertebrates, in particular, are important players in the wetland foodweb, as their grazing activity may regulate biomass of phytoplankton and epiphyton on submersed macrophytes (Timms and Moss 1984). Of practical relevance, phytophilous grazers play a key role in the success of biomanipulation of shallow ecosystems (Reynolds 1994). They have been shown to be capable of ameliorating the effects of high nutrient loading by controlling algal populations and facilitating the persistence of submersed macrophytes, thereby maintaining a macrophyte-dominated clear-water stable state (Scheffer et al. 1993). The importance of the macrophyte-dominated state for waterfowl has been clearly demonstrated (Hargeby et al. 1994, Hanson and Butler 1994a, b).

The seasonal pattern of abundance of microinvertebrates, especially Cladocera and Copepoda, has been investigated in the littoral zone of lakes (Pennak 1966, Whiteside 1974, Whiteside et al. 1978, Paterson 1993), but seldom in wetlands (Smyly 1952, Straskraba 1965, Daggett and Davis 1974, Schoenberg 1988, Irvine et al. 1990, Hann and Zrum 1998). Similarly, seasonal and spatial variability in composition and structure of the macroinvertebrate community throughout the open-water season in wetlands has been investigated in few studies (Macan 1949, 1965, 1966, Dvorak 1970a, b, Korinkova 1971, Murkin et al. 1992). The paucity of such studies is attributable, at least in part, to difficulties associated with sampling in a heterogeneous habitat, ranging from open water to dense submersed macrophytes to flocculent sediments, and in part to difficulties with taxonomic identifications of an extremely diverse fauna as well as lack of information on life histories of many species. Thus, priorities in wetland ecology can be clearly delineated:

1. Taxonomy—the need to know species of invertebrates, diversity, life histories, generation times, etc.
2. Functional ecology—the need to know diet, nutrient and energy cycling; food web dynamics; use of fine particulate organic matter (FPOM) and coarse particulate organic matter (CPOM), both occurring as detritus derived from macrophytes; habitat complexity; and trophic linkages (Murkin and Kadlec 1986, Murkin and Batt 1987, Murkin and Wrubleski 1988, Krieger 1992)

Physical and biotic factors are important influences on invertebrate communities, but how they affect community structure and dynamics is poorly investigated in wetlands. Physical disturbances peculiar to coastal wetlands (e.g., seiches) such as Delta Marsh, adjacent to large, shallow Lake Manitoba (and similarly, Netley Marsh, located on the southern shore of Lake Winnipeg) can have substantial effects on wetland habitat, including wind-induced turbulence and turbidity, rapid and large changes in water level, dislodgement of extensive sections of macrophyte rhizomes, producing drifting "islands" that override and damage submersed vegetation, and ultimately, modification of *Typha* shorelines. Prairie wetlands, whether coastal or prairie pothole, are also subject to anthropogenic disturbances due to the proximity of agricultural development, with attendant use of fertilizers and pesticides.

Disturbance events in prairie coastal wetlands may have varying impacts on the invertebrate fauna at particular times of year. For example, a nutrient pulse in early spring, prior to migration of fish in the marsh from the lake, emergence of invertebrates from resting stages, and germination of macrophytes, may produce a phytoplankton bloom that may attenuate light to such an extent that macrophyte development is precluded and, in their absence, sediments are easily resuspended, maintaining high turbidity and further reducing the quality and quantity of habitat for invertebrates. In contrast, such a nutrient pulse a month later, after macrophytes are well established, may stimulate their growth, thereby providing additional habitat and refuge for invertebrates.

DELTA MARSH, MANITOBA, CANADA

Delta Marsh is a 22,000-ha lacustrine wetland 98°23'W, 50°11'N in south-central Manitoba, Canada, surrounded to the south by agricultural land and aspen parkland and bordered by Lake Manitoba to the north. Further details are presented in Murkin et al. (this volume). Crescent Pond is a small (8-ha) littoral pond within the Delta Marsh, separated from Lake Manitoba by a forested beach ridge and a remnant oxbow of the Assiniboine River which formerly flowed northward into Lake Manitoba from 4500 to 2200 years BP (Teller and Last 1981). Although the pond has probably existed for 2000–2500 years, Goldsborough (1987) has documented its progressive in-filling and reduction in open-water area over the last 40 years. Water level regulation of the adjacent Lake Manitoba, which began in 1960–1961, may have contributed to the process. Blind Channel is a long, meandering paleo-channel of the Assiniboine River, with indirect exchange of water with Lake Manitoba via Cram Creek.

The water chemistry in Delta Marsh has been characterized at several sites throughout the open-water season; parameters are summarized in Table 40.1. Two partially isolated ponds within the marsh, Crescent Pond and West Dike Pond, diverge from the main marsh in several chemical parameters. Crescent Pond, often fishless, and occasionally flooded in the spring with water from the marsh, has values of almost all chemical parameters that are lower than in the rest of the marsh. In contrast, West Dike Pond has generally substantially elevated chemical parameters. The high conductivity and concentrations of several ions (e.g., Mg^{++}, SO_4^{--}, Na^+, Cl^-) are similar to values documented in many prairie potholes (Barica 1978).

The most prevalent submersed macrophytes in Delta Marsh include *Potamogeton zosteriformis, P. pectinatus, Ceratophyllum demersum, Myriophyllum spicatum* var. *exalbescens,* and *Utricularia vulgaris,* and the macroalga *Chara* sp. *Lemna minor* and *Lemna trisulca* float at the surface. Emergent vegetation is predominantly *Typha latifolia, Typha glauca, Scirpus lacustris, Carex atherodes, Scolochloa festucacea,* and *Phragmites australis* (Shay and Shay 1986). Algal communities in Delta Marsh, including phytoplankton, epiphyton, epipelon, and metaphyton, have been recently reviewed by Goldsborough and Robinson (1996).

Littoral microcrustaceans occupy three distinct habitats: true zooplankton in the open water, phytophilous (or epiphytic) microinvertebrates living in association with aquatic vegetation, and adjacent sediment-associated (epi-benthic or meiofaunal) microinvertebrates. Littoral microinvertebrates in Delta Marsh are exposed to a diverse array of invertebrate and vertebrate predators, including insect larvae and nymphs, *Mesostoma* sp., *Chaetogaster* sp., *Hydra* sp., leeches, water mites (Hann 1996a), salamanders (*Ambystoma tigrinum*), and planktivorous, piscivorous, and detritivorous and omnivorous fish (Kiers and Hann 1996).

TABLE 40.1. Water Chemistry Parameters for Delta Marsh, 1994[a]

	Delta Marsh	Crescent Pond	West Dike Pond
Water temperature (°C)	18.9	20.7	19.2
Water depth (m)	0.8	0.7	0.7
Secchi depth (m)	0.4	>0.60	>0.66
pH (field)	8.6	8.8	9.1
pH (lab)	8.6	9.0	9.1
Conductivity (μS/cm)	1875.8	1087.7	4543.3
Calcium (mg/L)	44.6	26.1	51.1
Magnesium (mg/L)	71.8	41.7	279.7
Sodium (mg/L)	241.8	138.0	540.3
Potassium (mg/L)	21.5	20.1	22.2
Manganese (mg/L)	0.1	0.0	0.1
Iron (mg/L)	0.0	<0.01	0.0
Sulphate (mg/L)	153.7	58.2	636.0
Chloride (mg/L)	345.2	193.3	1083.3
Carbonate (mg/L)	24.3	30.3	62.8
Bicarbonate (mg/L)	328.1	201.3	201.3
Nitrate + nitrite (mg/L)	<0.05	<0.05	<0.05
Fluoride (mg/L)	0.3	0.2	0.2
Alkalinity (mg/L)	306.3	215.7	269.7
Hardness (mg/L)	407.3	237.0	1280.0
TDS (mg/L)	1061.9	606.7	2776.7
Ionic balance (%)	100.1	101.7	100.4
Ammonia-N (mg/L)	0.1	0.1	0.0
TKN (mg/L)	1.9	1.6	1.7
Total phosphorus (mg/L)	0.1	<0.05	<0.05
Total carbon (mg/L)	93.5	71.7	91.4
TOC (mg/L)	21.7	19.7	34.2
TIC (mg/L)	69.4	60.0	63.0
True color (Co units)	32.9	56.7	38.3
Turbidity (NTU)	14.1	0.9	1.5
Silicon (mg/L)	3.2	0.7	2.0
Total chlorophyll (μg/L)	15.9	2.3	2.9

[a]For details see Goldsborough 1995.

In wetlands and other shallow water ecosystems invertebrate distribution patterns are strongly mediated by interrelated spatial and seasonal components of environmental variation, governed largely by patterns in development of submersed macrophytes. In early spring, prior to the germination and growth of emergent and submersed macrophytes, the open-water region is extensive. Because most of the wetland habitat is less than 1–2 m in depth, with adequate light penetration, submersed macrophytes can grow over large expanses of substratum, extend upwards through the water column, and eventually form dense beds or even a canopy of macrophytes at the water surface that can strongly attenuate penetration of light. With senescence of submersed macrophytes in late summer, the open-water region expands again.

SEASONAL ABUNDANCE OF MICOINVERTEBRATES

A total of 28 species of cladocerans, 6 species of cyclopoid copepods, 4 species of calanoid copepods, and the harpacticoid copepod *Onychocamptus mohammed* comprise the microcrustacean fauna of Delta Marsh (Table 40.2). Detailed patterns of seasonal species abundance and phenology, and community structure, are described in Zrum and Hann (1995) and Hann and Zrum (1997).

Mean seasonal abundance and percentage composition of Cladocera, Copepoda, and Rotifera are provided in Table 40.3. The proportional microinvertebrate community composition, assessed using two sampling methods, an acrylic cylinder to sample the water column zooplankton, and funnel traps to sample macrophyte-associated microinvertebrates (see Zrum and Hann 1995, Hann and Zrum 1997 for details of sampling protocols), is similar within each site. However, cladocerans and calanoid copepods are substantially more abundant in Crescent Pond (fishless), whereas cyclopoid copepods are predominant in Blind Channel (Table 40.3). Estimates of microcrustacean abundance in the water column and associated with macrophytes are not directly comparable, since planktonic abundance is a volumetric estimate and phytophilous abundance is based on a behavioral response (i.e., nocturnal vertical migration) from a unit area over a period of time.

Crescent Pond

Cladocerans comprise the predominant component of the microcrustacean community throughout the open-water period, except in late May in the shallow littoral zone. Cyclopoid and calanoid copepods are more abundant in the spring than cladocerans, but copepod densities decline in June and remain at low densities until autumn.

Daphnia rosea is the dominant planktonic cladoceran from late May through the end of June. *Simocephalus vetulus* is abundant among submersed macrophytes throughout June and July. In July *Ceriodaphnia dubia* increases in abundance, becoming the most abundant species in the planktonic cladoceran community for the remainder of the summer. *Chydorus* spp., *C. dubia*, and *Pleuroxus denticulatus* are the dominant phytophilous species from early June to mid-July.

The dominant cyclopoid in early spring is *Diacyclops thomasi,* followed later in the season by *Acanthocyclops vernalis* and *Eucyclops agilis. Macrocyclops albidus* also appears among vegetation in August. Calanoids occur at low density throughout the summer.

TABLE 40.2. Microcrustacean Species Reported from Delta Marsh

	Taxon			
			Copepoda	
Cladocera	Cyclopoida	Calanoida	Harpacticoida
Leptodora kindti (Focke) 1844	*Microcyclops varicans rubellus* (Lilljeborg) 1901	*Epischura lacustris* Forbes 1882	*Onychocamptus mohammed* (Blanchard and Richard) 1891
Polyphemus pediculus (Linné) 1761	*Eucyclops agilis* (Koch) 1838	*Leptodiaptomus nudus* Marsh 1904	
Diaphanosoma birgei Korinek 1981	*Macrocyclops albidus* (Jurine) 1820	*Skistodiaptomus oregonensis* (Lilljeborg) 1889	
Latona parviremis Birge 1910	*Acanthocyclops vernalis* (Fischer) 1853	*Aglaodiaptomus leptopus* (Forbes) 1882	
Sida crystallina (O. F. Müller) 1785	*Diacyclops thomasi* (S. A. Forbes) 1882		
Bunops serricaudata Birge 1893	*Diacyclops navus* (Herrick) 1882		
Bosmina longirostris (O. F. Müller) 1785			
Ceriodaphnia dubia Richard 1894			
Daphnia magna Straus 1820			
Daphnia pulex Leydig 1860 emend. Richard 1896			
Daphnia rosea Sars 1862 emend. Richard 1896			
Daphnia retrocurva Forbes 1882			
Scapholeberis kingi Sars 1903			
Simocephalus vetulus Schödler 1858			
Simocephalus serrulatus (Koch) 1841			
Alona circumfimbriata Megard 1967			
Alona guttata Sars 1862			
Alonella cf. *excisa*			
Camptocercus sp.			
Chydorus sp. 1			
Chydorus sp. 2			
Eurycercus longirostris Hann 1982			
Graptoleberis testudinaria (Fischer) 1848			
Kurzia latissima (Kurz) 1874			
Leydigia leydigi (Leydig) 1860			
Pleuroxus denticulatus Birge 1878			
Pleuroxus procurvus (O. F. Müller) 1785			
Pleuroxus trigonellus (O. F. Müller) 1785			
Pseudochydorus globosus (Baird) 1850			

1019

TABLE 40.3. Seasonal Mean Densities of Planktonic and Phytophilous Microinvertebrates

	Crescent Pond (Fishless)		Blind Channel (Fish Present)	
(a) Water Column	Number/L	Percent	Number/L	Percent
Cladocera	89	60.5	36	34.7
Cyclopoida	36	24.5	64	43.8
Calanoida	21	14.3	9	6.2
Rotifera	1	0.7	37	25.3
(b) Funnel Traps	Number/m^2	Percent	Number/m^2	Percent
Cladocera	70395	67.2	19941	27.7
Cyclopoida	32905	31.4	46923	65.2
Calanoida	1503	1.4	173	0.2
Rotifera	28	>1	4968	6.9

Blind Channel

Planktonic cladocerans increase in density by mid-June, then, with the influx of planktivorous fish from Lake Manitoba, remain at low density for the rest of the summer. *Daphnia retrocurva* occasionally occurs in June in low numbers and is likely a temporary visitor from Lake Manitoba. Cyclopoid copepods are generally more abundant than cladocerans. The planktonic cladoceran peak in late June consists typically of *Bosmina longirostris,* found exclusively in the water column, a much smaller zooplankter than those species (*C. dubia, D. rosea,* and *Chydorus* spp.) which comprise the cladoceran peak in Crescent Pond at this time. *Ceriodaphnia dubia, S. vetulus,* and *Alona* sp. increase in abundance among macrophytes in July, and *P. denticulatus* reaches higher density at the end of August.

The dominant cyclopoid copepods throughout the season are *D. thomasi* and *A. vernalis,* carnivorous species that prey on other microcrustaceans, rotifers, dipteran larvae, and oligochaetes (Gliwicz and Umana 1994). The numbers of calanoid copepods in Blind Channel are consistently low. *Epischura lacustris* is an occasional early-season visitor from Lake Manitoba, where it occurs much more commonly.

SPATIAL DISTRIBUTION OF INVERTEBRATES

The distribution of invertebrates in open-water and littoral regions of lakes, ponds, and wetlands is governed by factors such as water depth and chemistry, macrophyte abundance and composition, and life history characteristics of the organisms.

The spatial distribution of invertebrates in wetlands has often been examined with respect to proximity to the *Typha*-open water interface (Dvorak

1970a, Mason and Bryant 1974, Murkin et al. 1992). The open-water habitat generally is more homogeneous, with gradual temperature and dissolved oxygen gradients and light-attenuation profiles. In contrast, the *Typha*-open water interface, and moreso shoreward from it, is characterized by abrupt temperature, dissolved oxygen, and light-attenuation gradients. Emergent macrophytes, standing litter, and fallen, decomposing leaves and detritus predominate, with few submersed macrophytes. *Lemna minor, L. trisulca,* and filamentous green algal mats (metaphyton) commonly occur in this habitat, where there is less wind-generated turbulence and, as a consequence, lower turbidity and often hypoxic conditions (Suthers and Gee 1986). Submersed macrophytes (*Potamogeton, Myriophyllum, Ceratophyllum*) and *Chara* develop as patches or as continuous cover of the soft substratum extending outward from the *Typha.* In early stages of development macrophytes form an open understory, with free water movement among plants and gradients of physical and chemical variables similar to those found in the open water. However, the expansion of submersed macrophyte beds over the season results in a blurring of the demarcation between open-water and littoral regions. Eventually the dense beds of submersed vegetation result in greatly reduced water movement and steep physical and chemical gradients, akin to those documented in the zone of emergent vegetation.

Generally, true zooplankton (filter-feeding cladocerans, copepods, and rotifers) are more abundant in the open water than in the adjacent macrophyte zone (Smyly 1952, Pennak 1966, Murkin et al. 1992). Higher densities of phytophilous invertebrates have been found associated with submersed macrophytes than in open water (Straskraba 1965, Dvorak 1970a, Soszka 1975, Gerking 1957, Pardue and Webb 1985); this is perhaps attributable to greater habitat heterogeneity or increased surface area available for colonization, habitat, food resources, or refuges (Korinkova 1971, Dvorak 1987).

In Delta Marsh the spatial distribution of invertebrates has been examined using intensive seasonal sampling of an array of sites in the open water and at the *Typha*-open water interface in Crescent Pond and Blind Channel (Hann 1996a, Zrum and Hann 1995, Hann and Zrum 1997). Microinvertebrates were sampled weekly, May–August, using an acrylic cylinder for collecting in the water column and funnel traps for collecting among macrophytes (Zrum and Hann 1995, Hann and Zrum 1997). Macroinvertebrates were sampled one to two times per week, May–September, in Crescent Pond only, using activity traps (Murkin et al. 1983).

Microinvertebrates

In the water column of Crescent Pond, inshore densities at the *Typha*-open water interface of both planktonic and phytophilous cladocerans are higher than offshore densities, 20 m from the *Typha*-open water interface (Table 40.4*a*). In comparison, in Blind Channel, offshore densities of both planktonic and phytophilous cladocerans (Table 40.4*a*) are higher than inshore. Inshore densities of Cladocera in Crescent Pond are six times higher than those in

TABLE 40.4. Mean Seasonal Microcrustacean Densities in Inshore and Offshore Regions in Crescent Pond and Blind Channel, Delta Marsh, Manitoba

(a) *In Water Column*	Crescent Pond		Blind Channel	
INSHORE	Number/L	Percent	Number/L	Percent
Cladocera	122.4	70.8	20.8	23.7
Planktonic	82.4		16.0	
Phytophilous	40.0		4.7	
Cyclopoida	36.1	20.9	63.4	72.3
Calanoida	14.4	8.3	3.5	4.0
OFFSHORE				
Cladocera	54.3	47.7	49.6	38.5
Planktonic	41.9		45.0	
Phytophilous	12.4		4.7	
Cyclopoida	34.1	30.0	64.3	49.9
Calanoida	25.4	22.3	15.0	11.6
(b) *On Macrophytes*				
INSHORE	Number/m^2	Percent	Number/m^2	Percent
Cladocera	57156.5	62.8	24590.4	29.2
Planktonic	16112.4		7212.7	
Phytophilous	41044.1		17377.7	
Cyclopoida	32763.8	36.0	59470.3	70.6
Calanoida	1076.7	11.8	117.6	0.2
OFFSHORE				
Cladocera	82037.5	70.3	15276.2	34.2
Planktonic	12279.1		7759.8	
Phytophilous	69765.5		7531.7	
Cyclopoida	32746.7	28.1	29171.5	65.3
Calanoida	1913.6	1.6	226.3	0.5

Blind Channel, but offshore densities are similar. Inshore and offshore densities of cyclopoid copepods are similar in Crescent Pond, as they are in Blind Channel. However, absolute densities of cyclopoids in Blind Channel are twice those in Crescent Pond. Calanoid copepods are much more dense in Crescent Pond than in Blind Channel and are more dense offshore at both sites.

In association with macrophytes (Table 40.4b), cladocerans (especially those designated as phytophilous species) are more abundant offshore than inshore in Crescent Pond, and the reverse is true in Blind Channel. Inshore densities in Crescent Pond are double those in Blind Channel, whereas offshore densities in Crescent Pond are five times those in Blind Channel. Cyclopoid copepod densities inshore in Blind Channel are double those offshore, while inshore and offshore densities in Crescent Pond are similar. Calanoid copepods are more abundant offshore than inshore in both Crescent Pond and Blind Channel, with offshore densities generally double the inshore densities.

In Crescent Pond Cladocera are generally predominant in terms of percentage composition (Table 39.4), both inshore and offshore, in the water column as well as among vegetation. Only in the water column offshore does the percentage of cladocerans drop below 60 percent of the total community, and calanoid copepods are present in their highest percentage (and absolute abundance) representation.

In Blind Channel, cyclopoid copepods predominate both inshore and offshore, in the water column and among vegetation. Again, in the water column offshore the cyclopoid percentage drops as the calanoid and cladoceran components of the assemblage increase, perhaps reflecting the effects of predation by planktivorous fish. The cladocerans that increase in abundance include small, transparent species, e.g., *Bosmina longirostris* and *Diaphanosoma birgei*. The littoral zone cladoceran assemblage in Delta Marsh is characteristic of other temperate, shallow-water lakes and ponds (Keen 1973, Lim and Fernando 1978, Whiteside et al. 1978, Kelso and Ney 1985, Paterson 1994), and overlaps considerably with the previously known regional species pool (Smith 1968, Salki 1981). The seasonal pattern of abundance of littoral microcrustaceans in Delta Marsh generally conforms to that identified by Whiteside (1988) as the most frequently observed pattern in northern lakes, in which there is a spring–early summer peak, followed by a midsummer decline. Only phytophilous microcrustaceans in fishless Crescent Pond exhibited a prolonged midsummer peak and a late summer decline, a pattern consistent with that observed in other northern lakes (Lehtovaara and Sarvala 1984, Whiteside et al. 1985). Planktivorous fish predation has been demonstrated to play a role in the midsummer decline in littoral microcrustaceans (Doolittle 1982, Bohanan and Johnson 1983, Whiteside 1988). Invertebrate predation may also be influential (Goulden 1971), but other studies have been unable to demonstrate that seasonal microcrustacean dynamics were affected by invertebrate predators (Paterson 1994, Johnson et al. 1996). Mean seasonal abundances of planktonic and phytophilous cladocerans and copepods were similar to those determined in other littoral ecosystems (Paterson 1993, Whiteside et al. 1978, Lim and Fernando 1978, Williams 1982).

The dominance of littoral planktonic species (e.g. *Daphnia* spp., *B. longirostris*) in spring and phytophilous species throughout the summer and autumn is a pattern observed previously in lakes (Pennak 1966, Kelso and Ney 1985). Correlated with this shift in dominance seasonally is a shift in diversity from low spring values to higher summer and autumn values (Kelso and Ney 1985). In general, there appears to be an early-season assemblage composed of a few large, planktonic species, with subsequent development of a more species-rich assemblage of smaller-bodied, phytophilous species.

Many species of Cladocera and Copepoda in wetlands occurred in both open water and associated with submersed macrophytes. Although some species demonstrated clear habitat preferences (e.g., *D. rosea* and *D. birgei* in the water column and *S. vetulus* in vegetation), many others exhibited considerable overlap in habitat. Such habitat flexibility has also been observed in other studies of littoral zones of lakes (Smyly 1952, Pennak 1966, Paterson 1993). Seasonal patterns of abundance of several chydorid cladocerans, with peaks in June and July, corresponded closely with those documented previously in the littoral zone of a north temperate lake (Whiteside et al. 1978), predominantly associated with submersed macrophytes and sediments but observed to occur in the water column as well.

Despite the markedly higher cladoceran abundances in fishless Crescent Pond, planktonic cladoceran species richness is similar for the two locations. Species diversity of phytophilous Cladocera is higher in Crescent Pond, perhaps due to increased habitat complexity (Whiteside and Harmsworth 1967) arising from the greater number of macrophyte species present throughout the majority of the sampling period compared to Blind Channel.

Macroinvertebrates

Spatial and temporal patterns of nektonic macroinvertebrate abundance and community composition, assessed both in open water and at the *Typha* interface in Crescent Pond (Table 40.5), are detailed in Hann (1996a). A comprehensive list of taxa collected is provided in Table 40.6. In 1986 corixids were overwhelmingly dominant throughout the season. Amphipods, conchostracans, and water mites were much more abundant in 1988. In Crescent Pond water mites and corixid nymphs are particularly abundant in open water in comparison with densities at the *Typha* margins, whereas zygopteran naiads and dytiscid adults are more abundant near the *Typha*.

Dytiscid larvae, actively swimming predators, reach high densities in May around the pond margins and throughout June in the open water. As dytiscid larvae typically seek moist soil in which to pupate, those larvae at the pond margins in May pupate, then reappear in July as adults. Dytiscid adults are present in highest numbers in early July around the pond margins. Several other Coleoptera adults are present throughout the season (e.g., *Acilius, Agabus, Hydaticus, Haliplus, Hydrobius*).

Ephemeropteran naiads (e.g., *Caenis* sp.) are most abundant from May to early July both at pond margins and in open water. Univoltine taxa typically have a peak in adult emergence in June and early July. Naiads remain present throughout the season, with a slight increase in abundance again in late August and September. Bivoltine taxa have one summer generation, then the naiads that are produced will overwinter. Emergence patterns vary with abundance and locality and from year to year within the same species (Brittain 1982).

Corixid nymphs (e.g., *Hesperocorixa, Trichocorixa*) peak in late June to early July, then decline, with a second, more substantial rise in abundance from mid-August throughout September, particularly in the open water. Corixid adults reach the highest abundance at pond margins from late August

TABLE 40.5. Mean Densities (Number per trap) of Macroinvertebrates in Crescent Pond in Open Water and at the *Typha*-Open Water Interface in 1986 and 1988

	May–June		July		August–September	
	Typha	Open	*Typha*	Open	*Typha*	Open
1986 (fish present)						
Acari	0.94	1.67	0.80	0.92	0.93	1.26
Anisoptera naiads	0.04	0	0.03	0	0.39	0
Coleoptera	0.17	0.15	5.53	3.67	5.24	2.51
Corixid adults	21.65	18.75	19.26	25.57	21.65	18.75
Corixid nymphs	14.0	39.06	5.17	22.0	10.74	29.30
Zygoptera naiads	0.47	0.11	0.68	0.08	0.47	0.11
Dytiscid adults	1.06	0.29	0.95	0.51	0.80	0.43
Dytiscid larvae	0.001	0.003	0.003	0.005	0.002	0.003
Ephemeroptera naiads	0.18	0.17	0.11	0.27	0.18	0.17
Hirudinea	0.02	0.003	0.009	0.005	0.02	0.003
Amphipoda	1.01	0.71	0.75	1.41	1.01	0.71
Notonectidae	0.16	0.09	0.13	0.12	0.16	0.09
Conchostraca	0	0	0	0	0	0
1988 (fishless)						
Acari	2.35	3.33	1.94	5.98	2.91	5.61
Anisoptera naiads	0.08	0	0	0.02	0.06	0.11
Coleoptera	0.08	0.19	0.06	0.02	0.09	0.07
Corixid adults	0.55	0.39	0.38	0.36	0.81	0.23
Corixid nymphs	1.53	0.84	0.97	1.66	2.25	0.77
Zygoptera naiads	0.18	0.13	0.03	0.04	0	0.04
Dytiscid adults	2.00	1.04	0.53	0.36	0.72	0.23
Dytiscid larvae	1.70	2.81	0.38	0.59	0.03	0
Ephemeroptera naiads	0.63	0.39	0.09	0.20	0.41	1.45
Hirudinea	0.23	0.14	0.78	0.50	0.53	0.11
Amphipoda	7.93	4.59	12.28	17.34	50.63	138.91
Notonectidae	0.13	0.27	0.28	0.30	0.22	0.18
Conchostraca	59.75	131.59	9.66	55.13	16.69	8.89

to October. Notonectids, which lurk at the water surface awaiting prey, peak in abundance in late August and September and are more frequently trapped at pond margins.

Odonates, voracious "sit-and-wait" predators, are present throughout the season among vegetation, although dragonfly (Anisoptera) naiads (e.g., *Aeschna*) are most abundant from May to early August. Adults typically emerge in August and September and lay eggs, then either eggs or naiads overwinter. Damselfly (Zygoptera) naiads (e.g., *Enallagma, Lestes*) peak in late August to October and probably overwinter as late-instar naiads, emerge in spring,

TABLE 40.6. Macroinvertebrates reported from Delta Marsh, Manitoba

INSECTA
EPHEMEROPTERA
  Caenidae
    Caenis sp.
  Baetidae
    Baetis sp.
  Heptageniidae
    Heptagenia sp.
  Isonychiidae
    Isonychia sp.
  Ephemeridae
    Hexagenia sp.
    Ephemera sp.
ODONATA
Anisoptera
  Aeschnidae
    Aeschna mutata
  Libellulidae
    Libellula sp.
    Sympetrum sp.
    Leucorrhinia sp.
Zygoptera
  Lestidae
    Lestes sp.
  Coenagrionidae
    Enallagma sp.
    Nehalennia sp.
HEMIPTERA
  Corixidae
    Hesperocorixa sp.
    Trichocorixa sp.
  Notonectidae
    Notonecta sp.
  Mesoveliidae
    Mesovelia sp.
  Nepidae
    Ranatra sp.
  Gerridae
    Gerris sp.
  Belostomatidae
    Lethocerus sp.
    Belostoma sp.

TRICHOPTERA
  Hydroptilidae
    Agraylea multipunctata
  Phryganeidae
    Agrypnia sp.
    Ptilostomis sp.
    Phryganea sp.
  Limnephilidae
    Hesperophylax sp.
  Leptoceridae
    Mystacides sp.
    Ceraclea sp.
    Oecetis sp.
  Lepidostomatidae
    Lepidostoma sp.
LEPIDOPTERA
  Pyralidae
    Acentria niveus
COLEOPTERA
  Gyrinidae
    Gyrinus sp.
  Dytiscidae
    Acilius sp.
    Dytiscus sp.
    Hydroporus sp.
    Colymbetes sp.
    Graphoderus occidentalis
    Rhantus binotatus
    Rhantus suturellus
    Neoporus sp.
    Laccophilus maculosus
    Hygrotus sp.
    Uvarus sp.
    Coptotomus sp.
    Hydaticus sp.
  Haliplidae
    Peltodytes sp.
    Haliplus sp.
  Hydrophilidae
    Hydrophilus triangularis
    Enochrus sp.
    Hydrochara sp.
    Hydrobius sp.
  Staphylinidae
    Stenus sp.

TABLE 40.6. (Continued)

DIPTERA
  Chaoboridae
    Chaoborus sp.
  Chironomidae
    Chironominae
      Chironomini
        Polypedium (Polypedilum) sp.
        Glyptotendipes sp.
      Pseudochironomini
        Pseudochironomus sp.
        Cryptochironomus sp.
        Dicrotendipes sp.
        Parachironomus sp.
    Tanypodinae
    Tanytarsini
    Orthocladiinae
      Abiskomyia sp.
  Culicidae
    Aedes flavescens
    Culex tarsalis
    Culex restuans
    Culex nigripalpus
    Mansonia perturbans
    Anopheles sp.
  Psychodidae
    Psychoda sp.
  Heleidae (= Ceratopogonidae)
  Tipulidae
    Tipula sp.
    Prionocera sp.
MEGALOPTERA
    Sialis sp.
PLECOPTERA
  Perlodidae
    Cascadoperla sp.
ANNELIDA
OLIGOCHAETA
    Stylaria lacustris
    Chaetogaster sp.
HIRUDINEA
    Glossiphonia complanata
    Erpobdella punctata
    Helobdella stagnalis

MOLLUSCA
GASTROPODA
    Lymnaea stagnalis
    Physa gyrina
    Physa jennessi
    Bulimnea megasoma
    Helisoma trivolvis
    Helisoma campanulatus
    Aplexa hypnorum
    Pseudosuccinea columnella
    Gyraulus circumstratus
    Gyraulus parvus
    Promenetus umbilicatellus
    Stagnicola elodes
CNIDARIA
HYDROZOA
    Hydra oligactis
PLATYHELMINTHES
TURBELLARIA
    Mesostoma ehrenbergii
PORIFERA
    Spongilla lacustris
BRYOZOA
    Plumatella sp.
    Pectinatella sp.
CRUSTACEA
CONCHOSTRACA
    Lynceus brachyurus
    Caenestheriella setosa
AMPHIPODA
    Hyalella azteca
HYDRACARINA
NEMOTODA
GASTROTRICHA
    Chaetonotus sp.

lay eggs, then mature over the summer and early autumn. Emergence periods differ between species and for the same species between years (Kormondy and Gower 1965).

Amphipods (*Hyalella azteca*), both prereproductive and reproductive adults, are present early in the spring, probably overwintering as adults. They release their young and die, so large individuals are absent from the population by May. Juveniles produced during spring reproduction are steadily recruited into the population, mature as young adults in July and August, and produce a second generation of young, which overwinter as juveniles. Similar life history patterns have been observed in West Blue Lake, Manitoba (Biette 1969) and in a northern Ontario lake (Lindeman and Momot 1983). Water mites (Acari) are consistently abundant at all sites throughout the season. Leeches (Hirudinea) (e.g., *Erpobdella punctata, Glossiphonia complanata*) are present throughout the season. Mollusca, not quantitatively sampled in this study, include *Lymnaea stagnalis, Physa gyrina, Stagnicola elodes, Gyraulus circumstriatus,* and *Helisoma* sp.

Several groups of macroinvertebrates are underrepresented as a consequence of the sampling methods employed, including funnel traps (see Murkin et al. 1983), which rely on active swimming behavior of the individual macroinvertebrate organism. Surface swimmers (e.g., Gyrinidae, Velidae) and those taxa in which individuals are relatively immobile or firmly attached to substrata (e.g. case-building trichopterans, flatworms, phytophilous chironomids, snails, bryozoans, *Hydra,* and leeches) cannot be accurately quantified.

In years when fish are present in Crescent Pond, *Pimephales promelas* (fathead minnows) and *Culaea inconstans* (five-spined stickleback) are found throughout the season in the open-water areas, with peak abundance in late June and early July, then may migrate back into Lake Manitoba (if passage from the pond is possible), returning again in September and October. Several taxa, such as amphipods, conchostracans, dytiscid larvae, and ephemeropteran larvae, occur in markedly lower abundance in years when fish are present. Of these, only dytiscid larvae are themselves predators, but all are typically consumed readily by planktivorous fish. Macan (1966) similarly found that coleopteran larvae were decimated after the introduction of fish into a pond. Further detailed examination of effects of fish predation on the macroinvertebrate fauna awaits their identification to species.

Murkin et al. (1991) found that all nektonic invertebrates and functional groups (predator-parasite, herbivore-detritivore) occurred at higher densities, biomass, and species richness in open-water sites with dense submersed vegetation at normal water depths, as well as when subjected to fluctuating water levels (Murkin et al. 1992), in comparison with densities at the emergent macrophyte-open water interface. The elevated abundance and biomass of invertebrates in open water despite continued fish predation has been attributed to several factors:

1. More favorable food supply (epiphyton, detritus)

2. Higher dissolved oxygen concentrations, particularly higher DO minima

3. Fine detritus derived from submersed macrophytes in the open water in comparison with the coarse emergent macrophyte detritus inshore

4. Selective fish predation, leading to overall increased secondary production

5. Exclusion of large benthivores, e.g., carp, maintaining stagnant conditions inshore (Dvorak 1978, 1987)

Macrophyte-Invertebrate Associations

Diverse invertebrate communities exist among submersed vegetation of ponds. Abundance of phytophilous invertebrates is related to a suite of factors, including plant morphology, surface texture, epiphytic algal growth and community composition, nutrient content of the plant tissues, and the presence of defensive chemicals (Cyr and Downing 1988, Downing and Cyr 1985, Wiüm-Anderson et al. 1982). Aquatic plants provide invertebrates with shelter from predators (Dvorak and Best 1982, Diehl 1992) and act as spawning sites and sites for attachment (Rooke 1984, 1986a, b). Invertebrates can consume part of the plant (Lodge 1991) or its associated periphyton (Rooke 1984, 1986a, b). In eutrophic waters grazing invertebrates may prevent algal blooms, thereby allowing submersed macrophytes to persist (Irvine et al. 1990).

Different submerged plants within a pond create various microhabitats, which result in different assemblages of invertebrates. Phytophilous invertebrates are not equally abundant on all plant species (Cyr and Downing 1988, Downing and Cyr 1985), and associations between invertebrates and specific aquatic plant species in lakes have frequently been examined (Quade 1969, Gerrish and Bristow 1979, Rooke 1986a, b). Difonzo and Campbell (1988) found that relative abundance and composition of littoral cladoceran communities varied depending on the type of microhabitat (e.g., plant species, rocks, or water column). Some species may be specialized for certain microhabitats, allowing resource partitioning in the community. Other studies have shown associations between macrophytes and macroinvertebrates, particularly insects (Gerrish and Bristow 1979, Dvorak and Best 1982, Rooke 1984, Chilton 1990). The densities of organisms also varied depending on the type of plant, yet invertebrate species composition was generally similar for all macrophytes (Gerrish and Bristow 1979).

The morphology of the plant may play an important role in determining invertebrate community composition and preferred associations. Abundance of invertebrates per unit macrophyte biomass may vary with plant species and degree of leaf dissection (Krecker 1939, Gerking 1957, Gerrish and Bristow 1979, Dvorak and Best 1982, Rooke 1986a, b). Plants with finely dissected leaves may have higher invertebrate abundance per unit biomass or surface area than broad-leaved plants (Pardue and Webb 1985, Chilton 1990). Macrophytes with dissected leaves could provide more substratum for periphyton

(Dvorak and Best 1982) or could trap FPOM and CPOM from the water column (Rooke 1984, 1986a, b), thereby enriching the food supply for plant-associated invertebrates. Despite strong support for this hypothesis, largely based on the premise that plant surface area increases with extent of leaf dissection, other studies have been equivocal. Complex-leaved macrophytes may act as better refuges from predation than finely dissected leaves on plants (Irvine et al. 1990). Although abundances of different taxa varied depending on plant species, macrophytes such as *Ceratophyllum demersum* and *Myriophyllum* spp. in general do not support more invertebrates per unit plant biomass than broad-leaved plants (Cyr and Downing 1988).

Most ecological research has focused on open-water habitats rather than on the littoral zone of lakes and ponds. Littoral studies have been hampered by difficulties with quantitative sampling of vegetational and sediment substrata (Lodge et al. 1988, Irvine et al. 1990). Sampling methods often give inaccurate, imprecise estimates of invertebrate abundance, and collection, processing, and counting can be laborious (Whiteside and Williams 1975, Downing and Cyr 1985). Various techniques have been used for sampling phytophilous invertebrates: vacuum pumps (Campbell et al. 1982), activity traps (Whiteside 1974, Whiteside and Lindegaard 1980, Murkin et al. 1983), and plastic bags (DiFonzo and Campbell 1988). Many box-like samplers which clip macrophytes and retain the sample have been devised (Macan 1949, Gerking 1957, McCauley 1975, Minto 1977). The Downing box sampler attempts to minimize habitat disturbance and loss of invertebrates during collection (Downing 1984, 1986).

Associations between plants (macrophytes *Ceratophyllum demersum* and *Potamogeton zosteriformis* and the macroalga *Chara*) and invertebrates have been examined in Crescent Pond (see Hann 1996b for details) to determine if the structure and relative abundances of invertebrate communities associated with these aquatic plants differed among plant taxa.

In general, invertebrate species composition is similar for *Ceratophyllum demersum, Chara vulgaris,* and *Potamogeton zosteriformis* (Table 40.7). The most abundant taxonomic groups are Cladocera, Copepoda, Rotifera, Chironomidae, and Ostracoda. These microinvertebrates occur in similar abundance (numbers per gram dry mass of plant tissue) in association with *Ceratophyllum* and *Potamogeton,* but abundance is generally lowest with *Chara.* Percentage composition of major groups is most similar for *Potamogeton* and *Chara.* Other numerically important taxa are *Chaetogaster, Hyalella azteca, Mesostoma,* Gastropoda, *Hydra,* and *Agraylea.* Numbers for macroinvertebrates such as Ephemeroptera, Odonata, Dytiscidae, Corixidae, and Notonectidae are too low to permit quantitative comparisons among plants. Cladocera are the most diverse group, with 17 species found associated with aquatic plants. Chydoridae predominate, with 10 species, of which *Chydorus* is most abundant. *Ceriodaphnia dubia* occurs in consistently high numbers on both *Potamogeton* and *Ceratophyllum. Diaphanosoma birgei* occurs predominantly on *Potamogeton,* and *Simocephalus vetulus* on *Ceratophyllum.*

TABLE 40.7. Invertebrates (numbers per gm macrophyte biomass) Associated with Macrophytes in Crescent Pond, Delta Marsh, Manitoba

	Potamogeton spp.	*Chara vulgaris*	*Ceratophyllum demersum*
CLADOCERA			
Alona spp.	21	3	15
Alonella excisa	0	0	0
Bosmina longirostris	0	1	0
Camptocercus sp.	21	8	14
Ceriodaphnia dubia	245	49	288
Chydorus sp.	353	226	823
Daphnia pulex	5	0	2
Diaphanosoma birgei	162	36	39
Eurycercus longirostris	16	2	2
Graptoleberis testudinaria	105	8	84
Pleuroxus aduncus	4	3	5
Pleuroxus denticulatus	0	3	0
Pleuroxus procurvus	31	9	74
Pseudochydorus globosus	1	0	3
Scapholeberis kingi	0	0	2
Simocephalus vetulus	1	2	15
Total	965	349	1366
COPEPODA			
Calanoida	186	28	10
Cyclopoida	784	226	330
Total	970	253	340
ROTIFERA	1031	338	1010
OSTRACODA	338	207	201
CONCHOSTRACA	1	0	1
AMPHIPODA	7	10	32
ANNELIDA (Oligochaeta)			
Stylaria lacustris	3	5	7
Chaetogaster sp.	39	10	37
CHIRONOMIDAE			
Chironomini	108	37	122
Tanypodinae	11	7	7
Tanytarsini	9	9	14
Orthocladiinae	0	0	6
Total	127	53	148
EPHEMEROPTERA	0	0	3
ODONATA	1	0	1
TRICHOPTERA	0	2	5
COLEOPTERA	0	0	1
HEMIPTERA	2	0	1
PLATYHELMINTHES	3	1	11
HYDRACARINA	5	1	4
GASTROPODA	30	26	11
CNIDARIA	18	6	22

TABLE 40.7. (Continued)

	Potamogeton spp.	*Chara vulgaris*	*Ceratophyllum demersum*
Grazers	553	1020	262
Filterers	413	346	86
Herbivores/detritivores	41	80	43
Predators	68	86	18
Percent Cladocera	28	29	44
Percent Copepoda	28	21	11
Percent Rotifera	30	28	33
Percent Ostracoda	10	17	7
Percent Chironomidae	4	5	5

Besides cladocerans, rotifers form an important and large part of invertebrate communities associated with submerged vegetation in Crescent Pond. Cyclopoid copepods are dominant on all macrophytes. Highest numbers of copepods occur on *Potamogeton,* with fewer on both *Ceratophyllum* and *Chara. Diaptomus nudus,* a calanoid copepod, occurs in moderate numbers, chiefly on *Potamogeton.* Ostracod abundance is similar for all macrophytes.

Among chironomids, Chironominae are most abundant, followed by Tanypodinae, Tanytarsini, and Orthocladiinae. The oligochaete *Chaetogaster* occurs in large numbers on *Ceratophyllum* and *Potamogeton. Stylaria lacustris* is found mainly on *Ceratophyllum. Hyalella azteca* (Amphipoda) and *Mesostoma* predominate on *Ceratophyllum.* In general, *Hydra* is found on *Potamogeton* and *Ceratophyllum* rather than on *Chara.* Finally, *Agraylea,* the principal trichopteran, occurs abundantly only on *Ceratophyllum,* but is often associated with filamentous green algal mats, of which filaments it constructs its case (Wiggins 1977).

Cladocera can be separated according to feeding mode into filter feeders, typically ingesting phytoplankton from the water column, and scraper-grazers, feeding on epiphyton on the surfaces of submersed plants. Filterers occur in highest abundance associated with *Ceratophyllum,* in slightly lower numbers with *Potamogeton,* and in substantially lower numbers with *Chara* (Table 40.7). Grazers occur most abundantly on *Ceratophyllum,* in lower numbers on *Potamogeton,* and in somewhat lower abundance on *Chara.* Overall, numbers of filter feeders are considerably lower than of grazers in association with submersed plants.

Macroinvertebrates can be similarly separated into trophic groups: predators and herbivore-detritivores. Predators utilize plants as surfaces only (e.g., "sit-and-wait predators"), whereas herbivore-detritivores feed on epiphyton on plant surfaces. Herbivore-detritivores and predators occur in highest abundance in association with *Ceratophyllum,* in lower numbers with *Potamogeton,* and in lowest numbers with *Chara* (Table 40.7).

Rooke (1984) showed that although each macrophyte does not appear to have a characteristic fauna associated with it, different submerged plants do provide a specific substratum or resource that can be utilized by different types of invertebrates. Although invertebrate species composition in Crescent Pond appears to be generally similar in association with *Ceratophyllum demersum, Chara vulgaris,* and *Potamogeton zosteriformis,* differences in invertebrate abundance exist among taxonomic groups.

Fewer planktonic organisms (e.g., daphniid cladocerans, copepods, rotifers) are associated with submersed vegetation, particularly dense beds of macrophytes which develop as the season progresses. Low concentrations of dissolved oxygen may occur in these weed beds, especially at night, when plants are not photosynthesizing, but physical conditions that are detrimental to filtering and feeding activities of zooplankton may be of greater importance in restricting their distribution (Irvine et al. 1990).

RELEVANCE TO WETLAND CONSERVATION, BIODIVERSITY, AND MANAGEMENT ISSUES

When much of a wetland is maintained in the macrophyte-dominated stable state, diversity and abundance of invertebrates are generally higher, thereby increasing its value to waterfowl. The likelihood of maintenance of this stable state can be increased by reducing the impact of bottom-feeding fish, such as carp, that have negative effects on habitat quality in many ways (e.g., uprooting submersed macrophytes and rhizomes of emergent macrophytes, increasing turbidity).

Timing and location of perturbation/disturbance is critical in terms of potential impact on the wetland ecosystem. Overland flow of nutrients or pesticides passes through the zone of dense emergent vegetation and litterfall, which acts as a filter, minimizing damage to the foodweb. In contrast, aerial input (via spraydrift) or water entry into the wetland will have much more substantial and direct effects on the foodweb, based in the open water and associated with the submersed macrophytes.

FUTURE RESEARCH

As yet, there are few testable hypotheses concerning the mechanisms by which wetland invertebrate assemblages are organized. Significant gaps remain in our fundamental knowledge of the seasonal and spatial distribution patterns of wetland invertebrate species which must be addressed prior to initiating serious attempts to understand the community dynamics and organization. The few observed patterns of invertebrate abundance and distribution require explanation in terms of possible biotic and abiotic factors. Complex foodweb interactions involving several trophic levels need to be elucidated, and are perhaps best approached by experiments conducted within wetlands. The eventual goal must be to construct a comprehensive model of invertebrate community organization in coastal prairie wetlands.

ACKNOWLEDGMENTS

Alvin Dyck, Brian Grantham, Gayle Stilkowski, and Leanne Zrum assisted in sample collection and data analysis. I thank the Portage Country Club for access to their property to conduct research. The staff of the University Field Station (Delta Marsh) have provided continuing support. Ken Sandilands and Steven Ellis assisted in identification of insects. This is University Field Station (Delta Marsh) Publication Number 282.

LITERATURE CITED

Barica, J. 1978. Variability in ionic composition and phytoplankton biomass of saline eutrophic prairie lakes within a small geographic area. Archiv für Hydrobiologie 81:304–326.

Batzer, D. P., and S. A. Wissinger. 1996. Ecology of insect communities in nontidal wetlands. Annual Review of Entomology 41:75–100.

Biette, R. M. 1969. Life history and habitat differences between *Gammarus* and *Hyalella azteca* in West Blue Lake, MB. Master's Thesis, University of Manitoba, Winnipeg, MB.

Bohanan, R. E., and D. M. Johnson. 1983. Response of littoral invertebrate populations to a spring fish exclusion experiment. Freshwater Invertebrate Biology 2:28–40.

Brittain, J. E. 1982. The biology of mayflies. Annual Review of Entomology 27:119–147.

Campbell, J. M., W. J. Clark, and R. Kosinski. 1982. A technique for examining microspatial distribution of Cladocera associated with shallow water macrophytes. Hydrobiologia 97:225–232.

Chilton, E. W., II. 1990. Macroinvertebrate communities associated with three aquatic macrophytes (*Ceratophyllum demersum, Myriophyllum spicatum,* and *Vallisneria americana*) in Lake Onalaska, Wisconsin. Journal of Freshwater Ecology 5:455–466.

Cyr, H., and J. A. Downing. 1988. The abundance of phytophilous invertebrates on different species of submerged macrophytes. Freshwater Biology 20:365–374.

Daggett, R. F., and C. C. Davis. 1974. A seasonal quantitative study of the littoral Cladocera and Copepoda in a bog pond and acid marsh in Newfoundland. Internationale Revue der gesamten Hydrobiologie 59:667–683.

Diehl, S. 1992. Fish predation and benthic community structure: The role of omnivory and habitat complexity. Ecology 73:1646–1661.

Difonzo, C. D., and J. M. Campbell. 1988. Spatial partitioning of microhabitats in littoral cladoceran communities. Journal of Freshwater Ecology 4:303–313.

Doolittle, W. L. 1982. The nature and cause of the midsummer decline of littoral zooplankton in Lake Itasca, Minnesota. Ph.D. Thesis, University of Tennessee, Knoxville, TN.

Downing, J. A. 1984. Sampling the benthos of standing waters. Pages 87–130 *in* J. A. Downing and F. H. Rigler (eds.), A Manual on Methods for the Assessment of Secondary Productivity in Fresh Waters (2nd ed.). IBP Handbook 17. Blackwell Scientific, Oxford, UK.

———. 1986. A regression technique for the estimation of epiphytic invertebrate populations. Freshwater Biology 16:161–173.

Downing, J. A., and H. Cyr. 1985. Quantitative estimation of epiphytic invertebrate populations. Canadian Journal of Fisheries and Aquatic Sciences 42:1570–1579.

Dvorak, J. 1970a. Horizontal zonation of macrovegetation, water properties and macrofauna in a littoral stand of *Glyceria aquatica* (L.) Wahlb, in a pond in South Bohemia. Hydrobiologia 35:17–30.

———. 1970b. A quantitative study on the macrofauna of stands of emergent vegetation in a carp pond of south-west Bohemia. Rozpravy Československé Akademie Věd Řada Matematických Přírodních Věd 80:63–114.

———. 1978. Macrofauna of invertebrates in helophyte communities. Pages 389–395 *in* D. Dykyjova and J. Kvet (eds.), Pond Littoral Ecosystems. Structure and Functioning. Methods and Results of Quantitative Ecosystem Research in the Czechoslovakian IBP Wetland Project. Springer-Verlag, Berlin.

———. 1987. Production-ecological relationships between aquatic vascular plants and invertebrates in shallow waters and wetlands—a review. Archiv für Hydrobiologie Beiheft Ergebnisse Limnologie 27:181–184.

Dvorak, J., and E. P. H. Best. 1982. Macro-invertebrate communities associated with the macrophytes of Lake Vechten: Structural and functional relationships. Hydrobiologia 95:115–126.

Gerking, S. D. 1957. A method of sampling the littoral macrofauna and its application. Ecology 38:219–226.

Gerrish, N., and J. M. Bristow. 1979. Macroinvertebrate associations with aquatic macrophytes and artifical substrates. Journal of Great Lakes Research 5:69–72.

Gliwicz, Z. M., and G. Umana. 1994. Cladoceran body size and vulnerability to copepod predation. Limnology and Oceanography 39:419–424.

Goldsborough, L. G. 1987. Ontogeny of a small marsh pond: Revisited. University Field Station (Delta Marsh) Annual Report 22:37–39.

———. 1995. Weather and water quality data summary (1994), University Field Station (Delta Marsh). University Field Station (Delta Marsh) Annual Report 29:11–19.

Goldsborough, L. G., and G. G. C. Robinson. 1996. Pattern in wetlands. Pages 77–117 *in* R. J. Stevenson, M. L. Bothwell and R. L. Lowe (eds.), Algal Ecology: Freshwater Benthic Ecosystems. Academic Press, San Diego, CA.

Goulden, C. E. 1971. Environmental control of the abundance and distribution of the chydorid Cladocera. Limnology and Oceanography 16:320–331.

Hann, B. J. 1996a. Nektonic macroinvertebrates in a wetland pond (Crescent Pond, Delta Marsh, MB). University Field Station (Delta Marsh) Annual Report 30:68–75.

———. 1996b. Invertebrate associations with submersed aquatic plants in a prairie wetland. University Field Station (Delta Marsh) Annual Report 30:76–82.

Hann, B. J., and L. Zrum. 1997. Littoral microcrustaceans (Cladocera, Copepoda) in a prairie coastal wetland: Seasonal abundance and community structure. Hydrobiologia 357:37–52.

Hanson, M. A., and M. G. Butler. 1994a. Responses to food web manipulation in a shallow waterfowl lake. Hydrobiologia 279/280:457–466.

————. 1994b. Responses of plankton, turbidity, and macrophytes to biomanipulation in a shallow prairie lake. Canadian Journal of Fisheries and Aquatic Sciences 51: 1180–1188.

Irvine, K., H. Balls, and B. Moss. 1990. The entomostracan and rotifer communities associated with submerged plants in the Norfolk Broadland—effects of plant biomass and species composition. Internationale Revue der gesamten Hydrobiologie 75:121–141.

Johnson, D. M., T. H. Martin, P. H. Crowley, and L. B. Crowder. 1996. Link strength in lake littoral food webs: Net effects of small sunfish and larval dragonflies. Journal of the North American Benthological Society 15:271–288.

Keen, R. 1973. A probabilistic approach to the dynamics of natural populations of Chydoridae (Cladocera, Crustacea). Ecology 54:524–534.

Kelso, W. E., and J. J. Ney. 1985. Seasonal dynamics and size structure of littoral Cladocera in Claytor Lake, VA. Journal of Freshwater Ecology 3:211–221.

Keough, J. R., M. E. Sierszen, and C. A. Hagley. 1996. Analysis of a Lake Superior coastal food web with stable isotope techniques. Limnology and Oceanography 41: 136–146.

Kiers, A., and B. J. Hann. 1996. Seasonal abundance of fish in Delta Marsh, MB. University Field Station (Delta Marsh) Annual Report 30:83–90.

Korinkova, J. 1971. Sampling and distribution of animals in submerged vegetation. Véstnick Československé Společnosti Zoologické 35:209–221.

Kormondy, E. J., and J. L. Gower. 1965. Life history variations in an association of Odonata. Ecology 46:882–886.

Krecker, F. H. 1939. A comparative study of the animal population of certain submerged aquatic plants. Ecology 20:553–562.

Krieger, K. A. 1992. The ecology of invertebrates in Great Lakes coastal wetlands: Current knowledge and research needs. Journal of Great Lakes Research 18:634–650.

Krull, J. N. 1970. Aquatic plant-macroinvertebrate associations and waterfowl. Journal of Wildlife Management 34:707–718.

Lehtovaara, A., and J. Sarvala. 1984. Seasonal dynamics of total biomass and species composition of zooplankton in the littoral of an oligotrophic lake. Internationale Vereinigung für Theoretische und angewandte Limnologie, Verhandlungen 22:805–810.

Lillie, R. A., and J. O. Evrard. 1994. Influence of macroinvertebrates and macrophytes on waterfowl utilization of wetlands in the Prairie Pothole Region of northwestern Wisconsin. Hydrobiologia 279/280:235–246.

Lim, R. P., and C. H. Fernando. 1978. Production of Cladocera inhabiting the vegetated littoral of Pinehurst Lake, Ontario, Canada. International Vereinigung für Theoretische und angewandte Limnologie, Verhandlungen 20:225–231.

Lindeman, D. M., and W. T. Momot. 1983. Production of the amphipod *Hyalella azteca* in a northern Ontario lake. Canadian Journal of Zoology 61:2051–2059.

Lodge, D. M., J. W. Barko, D. Strayer, J. M. Melack, G. G. Mittelbach, R. W. Howarth, B. Menge, and J. E. Titus. 1988. Spatial Hetergeneity and Habitat Interactions in Lake Communities. Pages 181–208 *in* S. R. Carpenter, Complex Interactions in Lake Communities. Springer-Verlag, New York.

Macan, T. T. 1949. A survey of a moorland fishpond. Journal of Animal Ecology 18: 160–187.

———. 1965. The fauna in the vegetation of a moorland fishpond. Archiv für Hydrobiologie 61:273–310.

———. 1966. The influence of predation on the fauna of a moorland fishpond. Archiv für Hydrobiologie 61:432–452.

Mallory, M. L., P. J. Blancher, P. J. Weatherhead, and D. K. McNicol. 1994. Presence or absence of fish as a cue to macroinvertebrate abundance in boreal wetlands. Hydrobiologia 279/280:345–351.

Mason, C. F., and R. J. Bryant. 1974. The structure and diversity of the animal communities in a broadland reedswamp. Journal of Zoology, London 172:289–302.

McCauley, V. J. E. 1975. Two new quantitative samplers for aquatic phytomacrofauna. Hydrobiologia 47:81–89.

Merritt, R. W., and K. W. Cummins. 1984. An Introduction to the Aquatic Insects of North America (2nd ed.). Kendall/Hunt, Dubuque, IA.

Minto, M. L. 1977. A sampling device for invertebrate fauna of aquatic vegetation. Freshwater Biology 7:425–430.

Murkin, H. R., P. G. Abbott, and J. A. Kadlec. 1983. A comparison of activity traps and sweep nets for sampling nektonic invertebrates in wetlands. Freshwater Invertebrate Biology 2:99–106.

Murkin, H. R., and B. D. J. Batt. 1987. The interactions of vertebrates and invertebrates in peatlands and marshes. In D. M. Rosenberg and H. V. Danks (eds.), Aquatic Insects of Peatlands and Marshes of Canada. Memoirs of the Entomological Society of Canada 140:15–30.

Murkin, H. R., and J. A. Kadlec. 1986. Responses of benthic macroinvertebrates to prolonged flooding of marsh habitat. Canadian Journal of Zoology 64:65–72.

Murkin, H. R., and D. A. Wrubleski. 1988. Aquatic invertebrates of freshwater wetlands: Function and ecology. Pages 239–249 in D. D. Hook et al. (eds.), The Ecology and Management of Wetlands. Vol. 1. Croon Helm, London, UK.

Murkin, H. R., J. A. Kadlec, and E. J. Murkin. 1991. Effects of prolonged flooding on nektonic invertebrates in small diked marshes. Canadian Journal of Fisheries and Aquatic Sciences 48:2355–2364.

Murkin, E. J., H. R. Murkin, and R. D. Titman. 1992. Nektonic invertebrate abundance and distribution at the emergent vegetation-open water interface in the Delta Marsh, Manitoba, Canada. Wetlands 12:45–52.

Neill, C., and J. C. Cornwell. 1992. Stable carbon, nitrogen, and sulfur isotopes in a prairie marsh food web. Wetlands 12:217–224.

Pardue, W. J., and D. J. Webb. 1985. A comparison of aquatic macroinvertebrates occurring in association with Eurasian watermilfoil (*Myriophyllum spicatum* L.) with those found in the open littoral zone. Journal of Freshwater Ecology 3:69–79.

Paterson, M. 1993. The distribution of microcrustacea in the littoral zone of a freshwater lake. Hydrobiologia 263:173–183.

Paterson, M. J. 1994. Invertebrate predation and the seasonal dynamics of microcrustacea in the littoral zone of a fishless lake. Archiv für Hydrobiologie Supplement 99:1–36.

Pennak, R. W. 1966. Structure of zooplankton populations in the littoral macrophyte zone of some Colorado lakes. Transactions of the American Microscopical Society 85:329–349.

Quade, H. W. 1969. Cladoceran faunas associated with aquatic macrophytes in some lakes in Northwestern Minnesota. Ecology 50:170–179.

Reynolds, C. S. 1994. The ecological basis for the successful biomanipulation of aquatic communities. Archiv für Hydrobiologie 130:1–33.

Rooke, J. B. 1984. The invertebrate fauna of four macrophytes in a lotic system. Freshwater Biology 14:507–513.

———. 1986a. Macroinvertebrates associated with macrophytes and plastic imitations in the Eramosa River, Ontario, Canada. Archiv für Hydrobiologie 106:307–325.

———. 1986b. Seasonal aspects of the invertebrate fauna of three species of plants and rock surfaces in a small stream. Hydrobiologia 134:81–87.

Salki, A. G. 1981. The crustacean zooplankton communities of three prairie lakes subject of varying degrees of anoxia. Master's Thesis, University of Manitoba, Winnipeg, MB.

Scheffer, M., S. H. Hosper, M.-L. Meijer, B. Moss, and E. Jeppesen. 1993. Alternative equilibria in shallow lakes. Trends in Ecology and Evolution 8:275–279.

Schoenberg, S. A. 1988. Microcrustacean community structure and biomass in marsh and lake habitats of the Okefenokee Swamp: Seasonal dynamics and responses to resource manipulations. Holarctic Ecology 11:8–18.

Shay, J. M., and C. T. Shay. 1986. Prairie marshes in western Canada, with special reference to the ecology of five emergent macrophytes. Canadian Journal of Botany 64:443–454.

Smith, T. G. 1968. Crustacea of the Delta Marsh region, Manitoba. Canadian Field-Naturalist 82:120–139.

Smyly, W. J. P. 1952. The Entomostraca of the weeds of a moorland pond. Journal of Animal Ecology 21:1–11.

Soszka, G. J. 1975. The invertebrates on submerged macrophytes in three Masurian lakes. Ekologia Polska 3:371–391.

Straskraba, M. 1965. Contributions to the productivity of the littoral region of pools and ponds. I. Quantitative study of the littoral zooplankton of the rich vegetation of the backwater Labicko. Hydrobiologia 26:421–443.

Suthers, I. M., and J. H. Gee. 1986. Role of hypoxia in limiting diel spring and summer distribution of juvenile yellow perch *Perca flavescens* in a prairie marsh. Canadian Journal of Fisheries and Aquatic Sciences 43:1562–1570.

Swanson, G. A., and M. I. Meyer. 1977. Impact of fluctuating water levels on feeding ecology of breeding blue-winged teal. Journal of Wildlife Management 41:426–433.

Teller, J. T., and W. M. Last. 1981. Late Quaternary history of Lake Manitoba, Canada. Quaternary Research 16:97–116.

Timms, R. M., and B. Moss. 1984. Prevention of growth of potentially dense phytoplankton populations by zooplankton grazing, in the presence of zooplanktivorous fish in a shallow wetland ecosystem. Limnology and Oceanography 29:472–486.

Whiteside, M. C. 1974. Chydorid (Cladocera) ecology: Seasonal patterns and abundance of populations in Elk Lake, Minnesota. Ecology 55:538–550.

————. 1988. 0+ fish as major factors affecting abundance patterns of littoral zooplankton. Internationale Vereinigung für Theoretische und angewandte Limnologie, Verhandlungen 23:1710–1714.

Whiteside, M. C., W. L Doolittle, and C. M. Swindoll. 1985. Factors affecting the early life-history of yellow perch, *Perca flavescens*. Environmental Biology of Fishes 12:47–56.

Whiteside, M. C., and R. V. Harmsworth. 1967. Species diversity in chydorid (Cladocera) communities. Ecology 48:664–667.

Whiteside, M. C., and J. B. Lindegaard. 1980. Complementary procedures for sampling small benthic invertebrates. Oikos 35:317–320.

Whiteside, M. C., and J. B. Williams. 1975. A new sampling technique for aquatic ecologists. International Vereinigung für Theoretische und angewandte Limnologie, Verhandlungen 19:1534–1539.

Whiteside, M. C., J. B. Williams, and C. P. White. 1978. Seasonal abundance and pattern of chydorid, Cladocera in mud and vegetative habitats. Ecology 59:1177–1188.

Wiggins, G. B. 1977. Larvae of the North American Caddisfly Genera (Trichoptera). University of Toronto Press, Toronto, ON.

Williams, J. B. 1982. Temporal and spatial patterns of abundance of the Chydoridae (Cladocera) in Lake Itasca, Minnesota. Ecology 63:345–353.

Wiüm-Anderson, S., U. Anthoni, C. Christophersen and G. Jouen. 1982. Allelopathic effects on phytoplankton by substances isolated from aquatic macrophytes (Charales). Oikos 39:187–190.

Zrum, L., and B. J. Hann. 1995. The impact of invertebrate and vertebrate predation on littoral zooplankton in a wetland ecosystem. University Field Station (Delta Marsh) Annual Report 29:132–144.

PART 6
Synthesis

41 Ecology of Wetland Invertebrates

Synthesis and Applications for Conservation and Management

SCOTT A. WISSINGER

*T*he purpose of this chapter is to summarize themes from the collected papers in this book and consider their importance to the overall ecology, management, and conservation of wetlands. Wetland invertebrate communities are dominated by a distinctive group of taxa, many of which do not occur in terrestrial or aquatic (>2 m depth) ecosystems. The species that dominate in wetlands can intermix with aquatic and terrestrial taxa, especially at margins, but in and of themselves most wetland communities are not ecotonal. The most often cited factor that affects species composition and diversity in wetland invertebrate communities is hydroperiod (duration, timing, and predictability of drying and filling, and permanence), and wetland invertebrates exhibit a variety of adaptations for desiccation resistance and recolonization that are effective for different types of hydroperiods. Recolonization of temporary wetlands depends on several types of dispersal between wetland basins, and the persistence of populations in one wetland depends on the presence and proximity of other wetlands and aquatic habitats. Because hydroperiod is correlated with other abiotic (dissolved oxygen, salinity, pH, nutrients, harshness of drying or freezing) and biotic (primary production, detritus breakdown, types of vertebrate predators) factors that affect invertebrates, separating the direct effects of hydroperiod from indirect effects translated through these other factors will require carefully designed manipulative experiments. Spatial heterogeneity also plays an important role in determining invertebrate community structure. Both diversity and abundance of inverte-

Invertebrates in Freshwater Wetlands of North America: Ecology and Management, Edited by Darold P. Batzer, Russell B. Rader, and Scott A. Wissinger
ISBN 0-471-29258-3 © 1999 John Wiley & Sons, Inc.

brates are higher in vegetated than in nonvegetated subhabitats, and because different taxa are associated with different plants, assemblages will be most diverse in wetlands with heterogeneous plant communities. Invertebrates play a major role in recycling nutrients and are the primary consumers through which plant production becomes accessible to higher trophic levels. Themes in this book related to the roles of invertebrates in ecosystem function include:

1. *Most below-water line macrophyte production is consumed as detritus, but herbivory on above-waterline parts of plants is probably more important than previously expected.*
2. *CPOM shredders are rare in most wetlands, and the nutrients and energy in detritus probably enter consumer webs via FPOM collectors and algal/biofilm grazers.*
3. *Benthic algae is probably more important for secondary production than previously thought in some wetlands.*
4. *Fluxes of nutrients, detritus, and organisms across wetland-aquatic and wetland-terrestrial ecotones should affect ecosystem function in wetlands and adjacent ecosystems.*

Effective conservation and management will rely on an understanding of the effects of hydroperiod and spatial heterogeneity on wetland invertebrates. Making temporary wetlands permanent will lead to changes in invertebrate species, and restored wetlands should be designed to mimic the hydroperiods in natural wetlands that perform functions comparable to those desired in restored habitats. Different species of waterfowl and other wildlife use different subhabitats that support different types of vegetation and hence different invertebrate communities; thus, basin heterogeneity (depth, substrate, hydroperiod) should be a priority for mitigation and restoration projects. Because of the importance of dispersal among wetlands, conservation strategies should account for the interdependence of different wetlands across the landscape. Finally, given their important roles in the cycling of nutrients and transformation of primary to secondary production, invertebrates should be included in functional assessment protocols for assessing the ecological integrity of natural and restored wetlands.

INTRODUCTION

Information about the ecology of wetlands has exploded during the past 25 years as ecologists, government regulators, and policy-makers have endeavored to define and delineate wetlands and to characterize their hydrologic and geomorphologic position in the landscape. The natural services that wetlands provide societies (water purification, groundwater recharge, waste processing, storm buffers, consumptive and nonconsumptive natural resources) are among the most frequently cited examples of how conservation and restoration of ecosystems can be economically beneficial (e.g., Daily 1996). Invertebrates are widely recognized as key links between wetland primary production and higher trophic levels, yet they have historically received minimal treatment in conceptual models of ecosystem function and classification. Compared to other topics such as hydrology, biogeochemistry, soils, and wetland plants, they receive minimal attention in general treatments of wetlands ecology (e.g., Good et al. 1978, Lugo et al. 1990, Mitsch and Gosselink 1993, Fretwell et al. 1996) and wetlands restoration (Kusler and Kentual 1990, Wheeler et al. 1995). Invertebrates are strikingly absent in standard methodologies for habitat integrity and functional assessment (Adamus et al. 1991, Stiehl 1994, Smith et al. 1994, Brinson and Rheinhart 1996, Wilson and Mitsch 1996) and for defining, characterizing, and delineating jurisdictional wetlands (Environmental Laboratory 1987, Sippel 1988, Federal Interagency Committee for Wetland Delineation 1989, Government of Canada 1991). Several authors in this book note that information about invertebrates is also strikingly absent in edited volumes and regional summaries of wetland ecology (e.g., Windell et al. 1986, Dammon and French 1987, Wharton et al. 1982, Myers and Ewel 1990, Epperson 1992, Playa Lake Joint Venture 1993, Davis and Ogden 1994). Much of the historical knowledge about the ecology of wetland invertebrates is based on the importance of invertebrates as food for ducks and other types of game and nongame wildlife and as biological control agents for mosquitoes (review by Batzer and Wissinger 1996, Chapters 21, 34, recent articles by Nordstrom and Ryan 1996, de Szalay et al. 1997, King and Wrubleski 1998, Turner and McCarty 1998, Safran et al. 1998). This book brings together for the first time a compilation of the current state of knowledge about the ecology of invertebrates from a wide array of wetland habitats across North America. My goal is to summarize and synthesize the major themes that emerge from this volume and extend the application of this information to include other issues of wetlands conservation and management.

WETLANDS—UNIQUE HABITATS WITH UNIQUE INVERTEBRATE SPECIES

Wetlands have often been viewed as spatial and/or temporal ecotones between water and land. However, most wetlands do not occur at the boundary between aquatic and terrestrial habitats, nor are they transitional stages in a succession from aquatic to terrestrial ecosystems (Tiner 1993; fuller explanation below). The recent shift towards viewing wetlands as unique and enduring habitats rather than as transition zones has in part been stimulated by the recognition that the dominant plants in wetlands are specialists with distributions and adaptations that are distinct from terrestrial and aquatic species. Most emergent plants are not tolerant of extended and/or deep inundation (>2 m = aquatic habitat; Cowardin et al. 1979), nor do they survive or compete well in the unsaturated, aerobic soils of terrestrial habitats. Thus, the presence of wetland plants has become one of the main criteria used for delineating wetlands for ecological and jurisdictional purposes (Tiner 1989, 1991). Below I argue that the fauna of wetlands is similarly dominated by invertebrates uniquely adapted to the shallow and often fluctuating water levels in wetlands. Rather than referring to these species as "semiaquatic"or "semiterrestrial," it would be simpler and more accurate to use the term *wetland invertebrates* to distinguish this group of taxa from those that inhabit aquatic and terrestrial habitats.

Wetland invertebrates can be distinguished from terrestrial and aquatic invertebrates at multiple taxonomic levels. Among insects and crustaceans there are entire families of wetland specialists such as marsh beetles (Helodidae), marsh-loving beetles (Limnichidae), minute bog beetles (Sphaeriidae), burrowing water beetles (Noteridae), marsh flies (Sciomyzidae), phantom crane flies (Ptychopteridae), velvet water bugs (Hebridae), water measurers (Hydrometridae), fairy shrimp (Anostraca), clam shrimp (Conchostraca), and tadpole shrimp (Nostrostraca), all of which are rarely or never encountered in aquatic or terrestrial habitats (Table 41.1). In nearly all "aquatic" families of invertebrates, specialization to wetlands has occurred at the level of subfamily, tribe, genus, and species. Some species are generalists that occur in both aquatic and wetland habitats, but in some families most species are wetland specialists (e.g., water boatmen, back-swimmers, predaceous diving beetles; Table 41.1). Several groups of "terrestrial" arthropods include families, tribes, genera, or species that occur mainly in wetlands, including a variety of plant bugs, moths, springtails, lacewings, crickets, grasshoppers, diving wasps, and spiders. Many of these taxa live on the above-water parts of wetland vegetation and are obligately associated with particular species of wetland plants as herbivores, stem borers, plant parasites, and pollinators, or are obligately associated with each other as parasitoids, predators, or commensals. Invertebrate communities in dense stands of emergent marsh vegetation (Chapters 2, 4, 15), in the trees of swamps (Chapters 5, 7, 28), in the *Sphagnum* of peatlands (Chapter 17, Rosenberg and Danks 1987), and in the phytotelm of pitcher plants (Chapters 11, 18) include many wetland specialists as well as generalist species that also occur in terrestrial habitats. Emphasizing the uniqueness of wetland taxa and their diverse taxonomic affiliations has important consequences for accurately assessing the biological diversity in wetlands (e.g., Richter et al. 1997), and for understanding the various roles that invertebrates play in wetland ecosystem function.

Wetlands often include several types of subhabitats each of which can have both above- and below-waterline invertebrate assemblages. Because many of the investigators who study wetland invertebrates have been trained as aquatic ecologists, the types of samplers, sampling strategies, and general knowledge

TABLE 41.1. Wetland Invertebrates in North America and Their Taxonomic Affinities[a]

Insects

Bugs (Hemiptera)

1. Families in which most or all species are wetland specialists; Velvet water bugs (Hebridae), water measurers (Hydrometridae), water treaders (Mesoveliidae), pygmy back-swimmers (Pleidae), shore bugs (Ochteridae), toad bugs (Gelastocoridae), and shorebugs (Saldidae)

2. Families of aquatic bugs with wetland specialists, aquatic specialists, and wetland-aquatic generalists; e.g., Nepidae (waterscorpions), waters striders (Gerridae), water boatmen (Corixidae), back-swimmers (Notonectidae), creeping water bugs (Naucoridae), giant water bugs (Belostomatidae)

Beetles (Coleoptera)

1. Families in which most or all species are wetland specialists; e.g., marsh bettles (Helodidae), moss beetles (Hydraenidae), minute mudbeetles (Georyssidae), burrowing water beetles (Noteridae)

2. Families with wetland and aquatic species; e.g., predaceous diving beetles (Dytiscidae), whirligig beetles (Gyrinidae), crawling water beetles (Haliplidae), water scavenger beetles (Hydrophilidae)

3. Families of mainly terrestrial beetles with wetland species; e.g., weevils (Curculionidae), leaf beetles (Chrysomelidae), rove beetles (Staphylinidae), hister beetles (Histeridae), ground beetles (Carabidae)

Flies (Diptera)

1. Families in which most or all species are wetland specialists; e.g., marsh flies (Sciomyzidae), mosquitoes (Culicidae), biting midges (Ceratopogonidae), horse and deer flies (Tabanidae), long-legged flies (Dolichopodidae), shore flies (Ephydridae)

2. Families with wetland and aquatic species; e.g., phantom midges (Chaoboridae), midges (Chironomidae), soldier flies (Stratiomyiidae), crane flies (Tipulidae), moth flies (Psychodidae), phantom crane flies (Ptychopterdae), dance flies, (Empididae), dixid midges (Dixidae)

3. Families of mainly terrestrial flies with some wetland specialists; e.g., flesh flies (Sarcophagidae), snipe flies (Rhagonidae), root maggot flies (Anthomyiidae), dung flies (Scathophagidae), small dung flies (Sphaeroceridae), rattail maggots (Syrphidae)

TABLE 41.1. (Continued)

Dragonflies (Odonata)	Families with aquatic and wetland species; e.g., common skimmers (Libellulidae), green-eyed skimmers (Corduliidae), belted skimmers (Macromiidae), darners (Aeshnidae), spread-winged damselfies (Lestidae), narrow-winged damselflies (Coenagrionidae)
Mayflies (Ephemeroptera)	Aquatic families with a few wetland species: e.g., small square-gills (Caenidae); small minnow mayflies (Baetidae), primitive minnow mayflies (Siphlonuridae), pronggills (Leptophlebiidae)
Caddisflies (Trichoptera)	Familes with aquatic and wetland species; northern case-makers (Limnephilidae), giant case-makers (Phryganeidae) long-horned case-makers (Leptoceridae)
Springtails (Collembola)	1. Families with mainly wetland species; e.g., elongate springtails (Poduridae) 2. Terrestrial families with wetland species; e.g., globular springtails (Sminthuridae, Entomobryidae)
Moths (Lepidoptera)	Terrestrial families with species having caterpillars that specialize on wetland plants; e.g., pyralid moths (Pyralidae), noctuid moths (Noctuidae), nepticulid moths (Nepticulidae), cosmopterigid moths (Cosmopterygidae), tortricid moths (Tortricidae)
Plant bugs (Homoptera)	Terrestrial families with specialists on wetland macrophytes; e.g., aphids (Aphididae), spittlebugs (Cercopidae), leafhoppers (Cicadellidae), plant hoppers (Delphacidae), sedge and grass scales (Aclerdidae)
Grasshopper and crickets (Orthoptera)	1. Families with mainly wetland species; e.g., pygmy locusts (Tetrigidae) 2. Terrestrial families with wetland species; e.g., meadow grasshoppers (Tettigoniidae), crickets (Gryllidae), grasshoppers (Acrididae)
Ants, bees, wasps (Hymenoptera)	1. Parasitic wasp families with wetland hosts; e.g., Diapriidae, Scellionidae, Ichneumonidae 2. Families with aquatic and wetland species; e.g., diving wasps (Trichogrammidae, Mymaridae)
Crustaceans	
Water fleas (Cladocera and other small branchiopods)	Many epibenthic and phytophilous species in the Daphniidae, Chydoridae, Macrothericidae, and Bosminidae occur only or mainly in wetlands.
Fairy shrimp (Anostraca)	Most species are wetland specialists.

Taxon	Description
Clam shrimp (Conconstraca)	Most species are wetland specialists.
Tadpole shrimp (Nostostroca)	Most species are wetland specialists.
Seed shrimp (Ostracoda)	Several genera and species are wetland specialists.
Copepods (Copepoda)	1. Many large calanoids are wetland specialists. 2. Many harpacticoids and cyclopoids are generalists that occur in both aquatic and wetland habitats.
Sow bugs (Isopoda)	Several species are wetland specialists and others are generalists that occur across a wide range of aquatic and wetland habitats.
Scuds (Amphipoda)	Several species are wetland specialists and others are generalists that occur across a wide range of aquatic and wetland habitats.
Crayfish, shrimp, Prawns (Decapoda)	1. Several shrimps and prawns (e.g., *Palaeomontes paludosus*) 2. Crayfish species in the family Cambaridae
Annelids	
Segmented worms	1. Many in the aquatic families Tubificidae, Naididae, Enchytraeidae 2. Many in the terrestrial family Lumbriculidae
Leeches (Hirudinea)	Temporary habitat specialists reported
Mollusca	
Snails (Gastropoda)	Species in the families Physidae, Lymnaeidae, and Planorbidae
Clams (Bivalvia)	Species in the Pisididae genera *Sphaerium, Musculium, Pisidium*
Miscellaneous Taxa	
Rotifera	Many colonial rotifers appear to be temporary wetland specialists.
Flatworms (Turbellaria)	Many restricted to temporary habitats (e.g., *Typhloplanida*).
Miscellaneous Meiofauna	Water bears (Tardigrada) and Nematoda. Common, but specialization unknown.

[a]Most taxa are listed in one or more chapters in this book. Descriptions of the particular types of habitats and microhabitats can be found in McCafferty (1983), Borror et al. (1989), Williams and Feltmate (1992), Thorp and Covich (1994), and Peckarsky et al. (1990), Merritt and Cummins (1996).

of the habits and habitats of invertebrates are biased towards collecting and studying the subset of wetland species that belong to aquatic taxa. Many authors recognize the inadequacy of coverage across all subhabitats and note that their taxonomic lists are incomplete (Chapters 7, 9, 10, 15, 21, 22, 23, 24, 28, 30, 37). Habitats that are most likely to be undersampled include dense stands of emergent vegetation, mats of floating plants, saturated soils of peatlands, fens, and wet meadows, understories and canopies of shrub thickets and wooded wetlands, and marginal habitats at wetland-terrestrial ecotones.

The spatial and temporal relationships among wetland, aquatic, and terrestrial invertebrates become most apparent when one realizes that most wetlands have, but are not themselves, ecotones (Fig. 41.1a). Spatially, many wetlands are surrounded completely by terrestrial habitats, and although there will be a wetland-terrestrial ecotone, the wetland itself is a unique habitat that is not adjacent to an aquatic habitat (Fig. 41.1a). Most wetlands described in this book are "non-ecotonal" (after Tiner 1993) (Chapters 5, 7, 8, 9, 10, 12, 13, 14, 15, 16, 19, 21, 22, 23, 24, 25, 26, 27, 28, 29, 30, 31, 32, 34, 35, 39). In large wetland complexes ecotones can occur at the boundaries of different subhabitats, but the different subhabitats are typically not aquatic-terrestrial boundaries (Chapters 2, 4, 6, 17). The taxonomic lists in this book provide abundant evidence that the invertebrate communities in these wetlands are dominated by wetland specialists (Table 41.1) and that aquatic generalists that occur in the deepest and most permanent subhabitats are not likely to overlap in time or space with terrestrial generalists that occur at the wetland-terrestrial ecotone (Fig 41.1a).

Some wetlands are classic examples of ecotones in the sense that they occur between land and water; e.g., those on low-lying lake shores, in riparian zones of rivers, and adjacent to estuaries, and those associated with springs (Chapters 3, 7, 17, 20, 28, 33, 36, 37, 38, 40, Carter et al. 1994). However, many "ecotonal wetlands" (after Tiner 1993) include unique plant communities with wetland specialists that are not found in adjacent aquatic and terrestrial habitats (Fig 41.1b; Chapters 3, 7, 6, 37, 40, Findlay et al. 1989). Wetland invertebrates can intermix with aquatic species at aquatic-wetland ecotones and with terrestrial species at a terrestrial-wetland ecotones, but terrestrial and aquatic species do not typically overlap. Some wetland complexes will include both ecotonal and non-ecotonal habitats (Chapters 2, 4, 7, 17, 29; Naiman and Decamps 1997).

In a temporal context, wetlands have historically been viewed as transitional stages in a "hydrarch" succession in which aquatic habitats eventually fill in and become terrestrial habitats (e.g., Gates 1926). However, for most wetlands there is little botanical, geomorphological, or hydrological evidence to support this myth. Ecotonal and nonecotonal wetlands can be considered temporal ecotones in the sense that seasonal fluctuations in the water table or interannual cycles of drought are accompanied by temporal shifts in the de-

Non-Ecotonal Wetlands

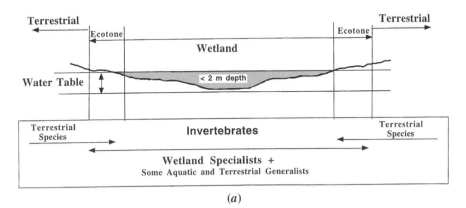

(a)

Ecotonal Wetlands

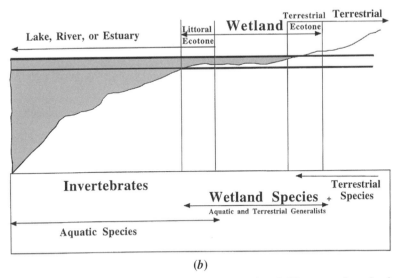

(b)

Fig. 41.1. The landscape position of *(a)* nonecotonal and *(b)* ecotonal wetlands and the distribution of wetland, aquatic, and terrestrial invertebrates (see text).

gree to which they are dominated by wetland or terrestrial plants and animals (Chapters 7, 12, 15, 17, 22, 26, van der Valk 1981, van der Valk 1978, Welling et al. 1988, Haukos and Smith 1993, Tiner 1993). However, with a few exceptions (Chapters 39, 40), wetlands exhibit cyclical patterns of change rather than linear progressions towards becoming terrestrial (Tiner 1993). In the absence of human perturbation, wetlands in North America are probably being created by natural geomorphological processes faster than they are being destroyed (Tiner 1993).

AN INVERTEBRATE-SENSITIVE PERSPECTIVE ON HYDROPERIODS

Surface water levels in most natural wetlands, even those that are permanent in most years, fluctuate seasonally, and many wetlands dry completely, either on a seasonal cycle or as a result of interannual variation in climate (i.e., drought). The schedule of filling and drying, or hydroperiod, is the most often cited form of disturbance in wetlands and is probably the single most important component of the "habitat template" (Southwood 1977, 1988). The concept of habitat template has been particularly useful to invertebrate ecologists for organizing information about tradeoffs associated with different life history and colonization strategies (reviews by Wiggins et al. 1980, Williams 1987, Batzer and Wissinger 1996, Wissinger 1997). A review of the chapters in this book suggests that a biologically relevant classification of wetland habitat templates should include the following components:

1. *Permanence* (permanent, semipermanent [dry in some years], temporary)
2. *Predictability* of drying and filling (seasonal, interannual climate cycles, relatively unpredictable)
3. *Phenology* (seasonal timing) of drying and filling
4. *Duration* of the dry and wet phases
5. *Harshness* during both phases

Several of these biologically relevant components of hydroperiod are not included in wetland classifications based on a hydrogeological perspective of wetlands (Cowardin et al. 1979, Mitsch and Gosselink 1993, Tiner 1996). Wiggins et al. (1980) distinguish temporary habitats that dry only for a short period in summer and refill in autumn ("autumnal") from those that are typically filled for only a short time in spring ("vernal"). Although some wetlands are clearly autumnal (Chapters 13, 14, 15, 19, 21, 22) or vernal (Chapter 28), many other types of temporary hydroperiods in North America reflect regional differences in climates (Fig. 41.2). For example, playas in Texas and marshes in California have a relatively short wet phase to their hydroperiod, like vernal pools in the Northeast, but the wet phases are in the summer and winter rather than the spring and therefore might be called "summer-wet" and "winter-wet" hydroperiods, respectively (Fig. 41.2). Temporary wetlands in the Southeast are wet for most of the year, but the dry phase is at a different time of year than in autumnal wetlands of the North (Figure 41.2). I am not necessarily in favor of a proliferation in jargon, but it is not clear that these other patterns of drying and filling are any less important than autumnal and vernal patterns, and distinguishing among them is certainly biologically relevant. The duration and seasonal timing of the hydroperiod dry phase will affect invertebrate *population and community ecology*—e.g., community composition and diversity will depend on which and how many species can tolerate particular dry phases; and the roles of invertebrates in *ecosystem processes,* e.g., rates of primary production, grazing, detritus breakdown, and nutrient cycling, will all depend on which season and for how long wetlands dry (see Invertebrates and Ecosystem Function, below).

The predictability of drying and filling is a component of hydroperiod that is superimposed on permanence, phenology, and duration. Among temporary wetlands, tidal wetlands are among the most predictable (Chapters 36, 37) and those that fill only from hit-or-miss convectional storms are among the

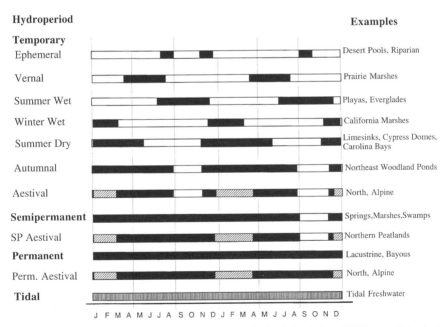

Fig. 41.2. A summary of the various hydroperiods (schedule of filling and drying) described in the chapters of this book for wetlands of North America. Differences in the degree of permanence and the duration, seasonal timing, and predictability of drying and filling can all affect invertebrate colonization strategies, community composition, and ecosystem function (see text).

least predictable (Chapter 32; Fig. 41.2). Interannual climatic variation can also create relatively unpredictable disturbances, either as drought or as extended inundation. The latter can be as important of a disturbance to invertebrate communities in temporary wetlands as drought can be to permanent habitats (Chapters 4, 5, 8; also Leslie et al. 1997, Golladay et al. 1997). Cycles of drought are perhaps best described in the Great Plains and Upper Midwest (including the Great Lakes), where wetland hydroperiods shift during droughts from permanent to autumnal and from autumnal to vernal (Chapters 21, 22, 23, 24, 38, 39). Most studies on wetland invertebrate communities are relatively short in duration, and patterns of species composition, even in "permanent" wetlands, might often reflect (unbeknownst to the investigator) differences in the time since the last extraordinary drought or inundation event associated with interannual climate variation (Chapters 15, 11, 26). Long-term studies are necessary for understanding postdrought patterns of community assembly and succession and how invertebrate communities change with cyclic changes in vegetation (Chapters 21, 22, 23, van der Valk and Davis 1978).

Temporary wetlands (Chapters 15, 22, 23, 24, 31, 32, 34), and microhabitats in them (Chapters 11, 18), are excellent model ecosystems for studying the dynamics of community assembly (e.g., Drake 1991, Lawlor and Morin 1993, Wilson and Whittaker 1995, Morin and Lawlor 1995), and for understanding the community attributes that lead to stability (resistance to and resilience after change) in communities (Tilman and Downing 1994, Naeem et al. 1994). Wetland permanence gradients provide an excellent opportunity to study the relative importance of stochastic and deterministic processes in ecological communities (Connell and Lawton 1992). In wetlands that dry frequently population dynamics and community composition should depend more on colonization dynamics than on the outcome of population interactions (competition, predation, mutualism).

Finally, the degree of harshness during both the wet and dry phase of a hydroperiod is a component of habitat templet that is often cited as having an important effect on invertebrates. Harshness during the dry phase will depend on the type of substrate, vegetation cover, and local climate. Certainly the parched, vegetation-less substrates of wetlands in relatively arid climates (Chapters 25, 26, 27, 30, 32) present different challenges for desiccation tolerance than the moist litter of woodland pools (Chapters 5, 9, 12, 14) or soils of wet meadows in humid climates (Chapters 8, 9, 13, 15). Such regional variation in harshness during the dry phase of temporary hydroperiods could underlie geographical patterns in the relative importance of different colonization strategies and the dominant species in wetland invertebrate communities (see Life History and Colonization Strategies, below). Wetlands can also vary in the degree of physical harshness during the wet phase of hydroperiods, and factors such as anoxia and freezing can affect community composition and invertebrate activities, which in turn will affect ecosystem function (Chapters 17, 18, 22, 23, 24, 25, 30, 31; Fig. 41.2).

In summary, hydroperiod classifications used for jurisdictional and definitional purposes (Tiner 1996) and functional assessments of created and restored wetlands (e.g., Brinson 1993a, b) do not provide enough information for an ecologically relevant description of wetland habitats. An invertebrate-sensitive classification scheme should include components such as predictability, timing, harshness, and duration, all of which can affect the species composition, diversity, and abundance of invertebrates and their roles in ecosystem function.

LIFE HISTORY AND COLONIZATION STRATEGIES

Given the range of disturbance regimes that can be found in wetlands (Fig. 41.1), it is not surprising that invertebrate ecologists have been interested in matching different life history and colonization strategies to particular hydroperiods (reviews by Wiggins et al. 1980, Williams 1987, Batzer and Wissinger 1996). The broad coverage of wetland types and hydroperiods in this book provides an opportunity to identify geographic and habitat-specific patterns for the importance of five basic invertebrate colonization strategies:

1. Desiccation tolerance
2. Timed emergence with adult survival in adjacent terrestrial habitat
3. Aerial dispersal from other wetland or aquatic habitats
4. Immigration and emigration in flood waters
5. Passive dispersal on other animals or by wind

There is little information on passive dispersal in this book, and our understanding of its importance continues to be constrained by the logistical difficulties of detecting it (reviews by Williams 1985, Williams and Feltmate 1992).

Desiccation Tolerance

Drying is the most important fundamental niche axis in temporary wetlands, and desiccation tolerance is perhaps the most important and defining life history characteristic of invertebrates that inhabit them (Fig. 41.2). Wetland ecologists often assume the taxa they encounter in temporary and semipermanent wetlands are able to tolerate desiccation in one or more life stages in or near dried basins (Chapters 2, 3, 4, 7, 8, 12, 13, 14, 15, 21, 22, 23, 24, 26, 28, 29, 30, 31, 32, 34, 35). Desiccation tolerance is well described for some taxa (see reviews by Wiggins et al. 1980, Williams 1987, Batzer and Wissinger 1996), but not for others, and is often only inferred from the rapid appearance of species after inundation. In many cases this inference is probably reasonable, but in others the rapid appearance of species can also be from eggs deposited by opportunistic dispersers (e.g., chironomids), cyclic colonizers (e.g., water striders), and passive dispersers on waterbirds, all of which can arrive within a few days of inundation (references in Batzer and Wissinger 1996).

Desiccation tolerance has been especially well studied in micro- and macrocrustaceans (e.g., water fleas, copepods, fairy shrimp, clam shrimp). In some species, desiccation-tolerant eggs are known to survive several wet-dry cycles, thus forming an "egg bank" that is analogous to plant seed banks (Hairston 1995, and references in Delorme 1991, Dodson and Frey 1991, Williamson 1991). Even in permanent habitats, copepods are know to remain quiescent in the sediments for a variable number of years and time emergence seasonally to avoid fish predators (Hairston 1987, Hairston et al. 1990, 1995). The presence of egg banks in other groups of wetland invertebrates (insects, annelids, mollusks) is poorly documented, and is one challenge for future research (see Batzer and Wissinger 1996 for a few examples). A second challenge is to understand the importance of different desiccation-resistant strategies (life stage, length of diapause, hatching cues) for recolonization under different seasonal and interannual drought cycles. Although patterns of plant colonization after drought are well understood in some wetland systems (e.g., Welling et al. 1988, van der Valk and Davis 1989, Haukos and Smith 1994), comparable data are wanting for invertebrates. A third challenge for future research is to understand the relative importance of different types of substrates and refuges as sites for resisting desiccation. In addition to burrowing in sediments during drought, organisms can also be found under mats of plant detritus (Chapters 15, 34), in the stems of live or dead macrophytes (Chapters 2, 4, 15, 21, 22, 23, 38, 39), among macrophyte root masses (e.g., Chapter 34), in/under woody debris (Chapters 5, 9, 15), and in burrows of large invertebrates (e.g., crayfish, shrimp) and vertebrates (e.g., alligators, muskrats, beavers; Chapters 2, 4). We do not yet understand the relative importance of these microhabitats or whether their presence underlies interbasin differences in the composition of wetland invertebrate communities.

A fourth challenge in northern and high-elevation wetlands is to understand the importance of freezing. Winter-freeze habitats have been termed "aestival" because liquid water is not available (Welch 1952). Aestival conditions are likely in most northern and high-elevation wetlands, including those that are otherwise permanent (e.g., Chapters 12, 13, 14, 15, 17, 18, 21, 22, 23, 24, 30, 31, 32, 38, 40). Although winter mortality associated with aestival conditions (low oxygen, desiccation, physical damage from ice) is documented for particular species (e.g., Danks 1971, Daborn 1971, 1974, Daborn and Clifford 1974, Andrews and Rigler 1985), the importance of invertebrate winterkill and strategies for avoiding it have received little attention from wetlands ecologists. Between-basin differences in invertebrate community composition in northern and alpine wetland could be related to variation in tolerance to aestival conditions that has been observed among different taxa (Lee 1989, Oswood et al. 1991, Frisbee and Lee 1997). Winterkill should also play an indirect role in invertebrate community composition because it can eliminate or reduce populations of fish and amphibian predators (see Predator-Permanence Gradients, below).

The relative importance of desiccation resistance and aerial dispersal appears to depend in part on the harshness of dried substrates and in part on conditions that favor other colonization modes. A common theme in chapters that focus on wetlands in arid regions of North America is that aerial dispersal (both opportunistic and cyclic) is as important as or more important than desiccation resistance for the recolonization of temporary wetlands (Chapters 25, 27, 30, 32, 33, 34). In several studies this observation can be directly linked to the extremely harsh substrate conditions between periods of inundation (Chapters 26, 32). In contrast, desiccation resistance appears to be most important in eastern North America, where high humidities, relatively cool temperatures, and moist soils favor the survival of desiccation-tolerant life stages (Chapters 8, 9, 10, 11, 13, 14, 15). This may be especially true in wooded wetlands, where the forest canopy reduces substrate drying to the point that special adaptations for desiccation tolerance are unnecessary (Chapter 14) and at least some invertebrates can remain active long after surface water has disappeared (Chapter 7).

Timed Emergence and Recolonization by Adults

Many insects that inhabit permanent wetlands recolonize them annually from adults that emerge and then reoviposit in the same wetland. Although this is usually assumed to be the dominant mode of insect colonization in permanent wetlands, we actually know relatively little about the site fidelity of adult insects and rates of immigration and emigration (but see McPeek 1989, Michiels and Dhondt 1991). Recolonization by adult insects is also important in temporary wetlands. Many insects in temporary habitats complete development rapidly and emerge to the adult stage of the life cycle just before habitats dry (reviews by Wiggins et al. 1980, Williams 1987, Wissinger and Batzer 1986; Chapters 12, 13, 15, 31, 32). Some species deposit desiccation-resistant eggs in or near the dried basins, but others remain in the adult stage during the dry phase of the hydroperiod and deposit eggs when basins refill (Wiggins et al. 1980). Insects that are strong flyers and feed as adults (e.g., dragonflies) can remain active until basins refill (Chapter 15, Corbet 1980, Williams and Feltmate 1992). In other taxa (e.g., caddisflies, craneflies, mosquitoes) adults enter an ovarian diapause in adjacent terrestrial vegetation or soil. Temperature and photoperiod cues trigger arousal from diapause and females then oviposit (examples in Gower 1967, Wiggins 1973, Otto 1981, Pritchard 1983, Williams and Feltmate 1992, Wissinger and Brown, in press). This mode of colonization should be most important in permanent wetlands and in temporary wetlands that dry predictably and for a relatively short time (Tauber et al. 1986). For most wetlands it is difficult to assess the importance of this strategy without detailed life history information because ovipositing adults can either be immigrants or residents. In extremely isolated, temporary habitats where immigration is unlikely, oviposition by resident adults should be the main mode of insect colonization (Chapter 33). Because this strategy relies on the presence of appropriate terrestrial habitat for feeding and/or ovarian diapause, these insects will be vulnerable to human impacts on adjacent lands. Understanding the terrestrial habitat requirements of wetland insects that spend the dry phase of hydroperiods as adults will be important for establishing criteria for terrestrial buffer zones around wetlands (e.g., Burke and Gibbons 1995).

Aerial Dispersal by Insects

Colonization by winged adult insects is an often-cited mode of wetland colonization (nearly all chapters that address colonization) that includes several different types of movements that have important consequences for understanding the genetic structure of wetland invertebrate populations and metapopulations (see Gilpin and Hanski 1991 and Hanski and Gilpin 1997). Three types of between-wetland dispersal are (1) density-dependent emigration from permanent wetlands to other habitats; (2) opportunistic colonization between temporary wetlands; and (3) cyclic colonization between permanent and temporary wetlands. Because permanent wetlands are relatively stable habitats, density-dependent emigration is a likely mechanism of population regulation, and many insects have developmental switches that trigger dispersal to other habitats in response to crowding (e.g., Denno and Grissell 1979, Denno and Roderick 1992). This type of dispersal can lead to "source-sink" population dynamics if species cannot complete development in the temporary wetlands to which they emigrate (see Pulliam 1988, Pulliam and Danielson 1988).

Opportunistic colonization is a second type of dispersal that is most likely to be important for species that colonize temporary wetlands. Opportunistic colonizers are "r-selected" species with life history characteristics that are likely to evolve when mortality is density-independent (review by Boyce 1984), as will be the case when wetlands dry rapidly and unpredictably (e.g., Chapters 12, 17, 16, 31, 32). Opportunistic colonizers can be an important component of communities in any wetland during the early stages of recolonization. Rapidly developing taxa such as mosquitoes and midges can complete development within a few weeks and escape the increasingly density-dependent conditions (competition and/or predation) created by later arriving species or the drying of extremely temporary pools (review by Batzer and Wissinger 1986). In wetlands that dry unpredictably, opportunistic colonizers can use a bet-hedging strategy that involves distributing eggs among different basins to insure that at least some propagules will survive (Chapter 32). Several authors suggest that opportunistic aerial colonizers should be most important in south temperate and subtropical wetlands in North America, where insect life histories are more likely to be multivoltine, with adult dispersers always available when habitats refill (Chapters 2, 10, 11, 28).

A third type of aerial dispersal is cyclic colonization, a strategy that involves seasonal migrations by winged adults between temporary and permanent habitats (Chapters 2, 3, 4, 8, 12, 14, 15, 17, 21, 22, 26, 31, 32, 34, 35; see review by Wissinger 1997). Although this phenomenon is well described for water bugs (water striders, water boatmen, back-swimmers; e.g., Fernando 1959, Young 1965, Landin and Vepsalainen 1977, Pajunen and Salmi 1988, Spence and Anderson 1994) and beetles (predaceous diving beetles, water scavenger beetles; see Fernando and Galbraith 1973, Landin 1976, Nilsson 1986), its importance has been largely overlooked by North American wetlands ecologists (but see Wiggins et al. 1980). Some water bugs and beetles that alternate seasonally between temporary and permanent habitats exhibit "flight polymorphisms" that involve an alternation of different types of adults that develop under different environmental conditions (reviews by Harrison 1980, Spence and Anderson 1994, Wissinger 1997). Long-winged adults in ovarian diapause typically leave permanent habitats in spring and give rise to one to several generations of highly fecund but sedentary adults (poorly developed wings or wing musculature) in temporary habitats. Long-winged adults return to permanent refuges to overwinter or when temporary habitats dry. Concurrent changes in wing morphology, reproductive state, and behavior emphasize that these developmental polymorphisms involve a suite of adaptations that are coordinated by a common set of neuro-endocrinological events. Cyclic colonizers do not necessarily exhibit wing polymorphisms, but most do exhibit seasonal or generational shifts in their allocation of resources to reproduction versus flight. Cyclic colonization should be especially important in wetland complexes with many temporary and permanent basins (Chapters 2, 4, 15, 17, 31, 32, 34, 39). Surprisingly, cyclic colonization is also documented between relatively isolated wetland basins (Chapters 8, 9, 10, 12, 13, 14, 21, 26, 28), suggesting that its importance is not limited to wetland complexes. One challenge for future research will be to quantify the degree to which the persistence of invertebrate populations in one type of wetland is dependent on the presence of other types of wetlands across the landscape (Jeffries 1994)

Colonization of Ecotonal Wetlands with Flood Waters

Invertebrates can move with the flow of water between ecotonal wetlands and adjacent aquatic habitats. In some cases riparian wetlands are population sinks for lotic species that are washed in during floods (Chapter 7), but in other cases flood-borne immigrants are important for the colonization of these habitats (Chapter 29). Some invertebrates in riparian wetlands exhibit life history strategies that involve ontogenetic habitat shifts between lotic to lentic habitats (Chapters 7, 16). This phenomenon is best described for mayflies that oviposit and complete early larval development in streams and then migrate with flood waters to wetlands where they complete development and emerge before the wetlands dry in summer (Chapters 7, 16). The movement of invertebrates with flood and ebb tides is reported to be important in some (Chapter 37) but not in other freshwater tidal wetlands (Chapter 36). Finally, flood waters can be an important source of early colonists in wetlands that are managed as waterfowl habitat or for agriculture (Chapters 34, 35 and references therein).

COMMUNITY ECOLOGY OF WETLAND INVERTEBRATES

Correlates of Abundance and Diversity

Physiochemical factors that are most often cited in this volume as having an important influence on community composition and invertebrate abundance in wetlands include *hydroperiod* (most chapters), *dissolved oxygen* (Chapters 2, 26, 32, 37), *salinity* (Chapters 20, 21, 25, 33, 36, 37), *pH* (Chapters 4, 11, 17, 18), *suspended sediment* (Chapters 21, 27, 37), and *nutrient levels* (hypereutrophy, Chapters 2, 5, 21; hyperoligotrophy, Chapters 31, 32; dystrophy, Chapter 17). These data complement a recent review of the importance of these factors as fundamental niche constraints on the distribution and abundance of wetland invertebrates (Batzer and Wissinger 1996). In addition, two biological factors are repeatedly cited as being important to invertebrate communities in many types of wetland communities. The first is the diversity and particular species of plants in wetland basins (Chapters 2, 8, 10, 11, 14, 15, 16, 17, 18, 21, 23, 36, 37, 40), and the second is the presence or absence of fish or salamanders as top predators (Chapters 2, 8, 13, 19, 21, 24, 28, 29, 30, 31, 40).

Because hydroperiod can affect water and substrate chemistry, nutrient availability, vegetation, primary production, detritus processing, and presence of vertebrate predators, it is difficult to isolate the direct effects (drying, inundation) from the indirect effects translated through these other factors. For some factors, reasonable hypotheses can be generated for the directionality of cause and effect (Fig. 41.3). For example, hydroperiod can have a major influence on water chemistry (Mitsch and Gosselink 1993) or on the presence of top vertebrate predators such as fish and amphibians (Wellborn et al. 1996), but not vice versa. However, relationships among other factors can be reciprocal. For example, invertebrates can be affected by and affect vegetation, primary production, detritus, and nutrient availability (Fig. 41.3). My purpose here is not to attempt a synthesis of all these relationships, but simply to emphasize that there are many direct and indirect effects that wetland environments can have on invertebrates and vice versa, and it will therefore be difficult to interpret causal relationships from correlational data.

Direct and Indirect Effects of Hydroperiod on Invertebrate Communities

In many types of wetlands invertebrate diversity increases along a permanence gradient (Chapters 2, 4, 8, 12, 13, 15, 19, 20, 21, 22, 26, 31, 32, 33, 34; but

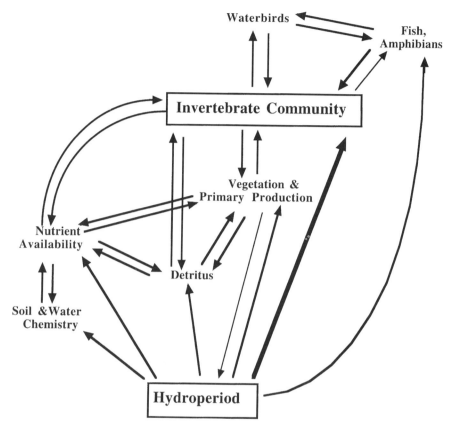

Fig. 41.3. A schematic model for the biologically relevant direct and indirect effects of hydrology on wetland invertebrates. Because many of the covariates of hydroperiod can also affect and be affected by invertebrates, it is difficult to infer the direct of effects of hydroperiod (drying) on invertebrates from comparative data.

see Chapters 5 and 9). There are several hypotheses for this observation. The most parsimonious is that many species lack adaptations for tolerating or escaping the dry phase of temporary hydroperiods. Consistent with this hypothesis is the clear permanence-diversity relationship observed in wetlands in relatively arid climates where conditions in dried basins are likely to be especially harsh (Chapters 13, 30, 31, 32). Two problems with this hypothesis are that (1) permanent-habitat species are often replaced by closely related species that are able to complete their life cycles in temporary habitats (Wellborn et al. 1996) and (2) more species are than are not desiccation tolerant in many taxa of wetland invertebrates. Two alternative hypotheses focus on the interaction between disturbance frequency and population interactions. There is experimental evidence from other ecosystems that predators become increasingly important agents of community structure along a declining gradient of disturbance. In relatively benign habitats predators are thought to be most likely to mediate the coexistence of a diverse prey community (Chapter 13, Menge et al. 1986, Schneider and Frost 1997; but see Chesson and Huntly 1997). However, the strong correlation between the presence of top vertebrate predators and habitat permanence makes it difficult to test this hypothesis with comparative data (Predator-Permanence Gradient, below). The "intermediate disturbance hypothesis" differs in that it predicts relatively low diversity in stable habitats because competitive dominants exclude other species (e.g., lakes—see Chapter 39), and relatively low diversity in highly disturbed (temporary) habitats that contain only rapid colonizers. Semipermanent and shallow permanent wetlands that are disturbed at intermediate frequencies (occasional drought, anoxia, winterkill) should, according to this hypothesis, have the highest diversity. The observation that extended inundation in semipermanent wetlands is accompanied by a decline in invertebrate diversity (Chapters 4, 9, 21, 38, 39, 40) is consistent with this hypothesis. A fourth potential explanation is that permanent basins are on average larger than temporary basins and should therefore have more species, regardless of permanence status (references in Chapter 31; Larson and House 1990). Several mechanisms can underlie species-area correlations, including (1) a dynamic equilibrium between colonization and local extinction predicted by island-biogeography theory (MacArthur and Wilson 1967) and (2) habitat heterogeneity—that is, larger habitats contain more types of subhabitats which support different types of species. Although studies that follow changes in diversity during community assembly after drought can provide data that support one or more of the alternative explanations for diversity-permanence correlations (Chapters 13, 15, 12, 22, 23, 24, 34), separating their relative importances will require experimental manipulation of predators, temporal variability, and spatial heterogeneity in habitats of the same size (e.g., see Wilbur 1997).

Predator-Permanence Gradients

The top predators in wetland foodwebs change along permanence gradients, and these predator shifts are correlated with diversity (Chapters 2, 13), abundance (Chapters 2, 19; Mallory et al. 1994), and invertebrate community composition (Chapters 2, 4, 8, 13, 15, 19, 21, 30, 31, 40). One exception is in some floodplain wetlands where fish that colonize during floods have the same impact on invertebrates in permanent and temporary basins (Chapter 28). However, in most nonecotonal wetlands it appears there are species replacements along this predator-permanence gradient consistent with those observed in lentic aquatic habitats (Wellborn et al. 1996). The predator pattern for most non-ecotonal wetlands in North America is that:

1. Fish (e.g., centrarchid sunfish, ictalurid catfish, cyprinid minnows) are the top aquatic predators in permanent habitats that do not freeze.

2. Predatory amphibians (e.g., larvae of ambystomatid salamanders) and freeze/drought-tolerant fishes (e.g., mudminnows, green sunfish) are the top predators in permanent aestival wetlands and in temporary wetlands with a long wet-phase hydroperiod.
3. Large invertebrates (e.g., dragonfly nymphs, dytiscid beetle larvae) are the top predators in temporary wetlands with a short wet-phase hydroperiod (Chapters 2, 7, 8, 13, 15, 21, 28, 30, 31; reviews by Batzer and Wissinger 1996).

Mechanisms that underlie the shifts in top aquatic predators in wetlands include:

1. A lack of drought and freeze tolerance in most fish
2. Drying time constraining amphibian larval development (e.g., Pechman et al. 1989, Rowe and Dunson 1995)
3. The exclusion of salamanders by fish (e.g., Semlitsch 1987, Sexton et al. 1994)
4. Selective predation by fish and amphibians on conspicuous (large and/ or active) invertebrate predators (reviews by Batzer and Wissinger 1996, Wellborn et al. 1996)

Most experimental data that link predator shifts with prey species-replacements along permanence gradients involve amphibians. Alternative morphologies or behaviors that are effective against different types of predators (Werner and McPeek 1994), or between traits that facilitate the timely escape from drying habitats (rapid growth) and those that facilitate coexistence with predators (e.g., Werner and Anholt 1993, Skelly 1997), are responsible for amphibian species replacements along predator-permanence gradients. The few experimental studies that causally link predator effects to invertebrate species replacements along this continuum include Dodson (1970, 1974) and Sprules (1972), who showed that size-selective predation by salamanders on zooplankton underlies differences in community composition in permanent versus temporary wetlands; McPeek (1990a, b, 1996), who demonstrated that the different species of damselflies in fish and fishless habitats exhibit antipredatory behaviors that are alternatively effective against fish and large invertebrates; and Wissinger et al. (in press), who found that caddisfly species-replacements along a predator-permanence gradient in alpine wetlands reflect a tradeoff between high activity rates for rapid development and low activity rates to facilitate coexistence with salamanders.

Although these studies explain patterns of distribution and abundance for some taxa, the generality of these results and the community-wide impacts of predation have not been separated from other hydroperiod correlates. The replacement of predator-intolerant with predator-tolerant invertebrates along permanence gradients should in and of itself not lead to a change in diversity, suggesting that the often-observed diversity-permanence gradient is not easily explained by predator effects alone (see Direct and Indirect Effects of Hydroperiod on Invertebrate Communities, above). Understanding the cause-effect relationships of predators on wetlands invertebrate community structure will require an experimental approach that untangles predator effects from covarying hydroperiod correlates (on chemistry, nutrients, vegetation; Fig. 41.3) as well as the direct effects of hydroperiod (as in Wilbur 1997). In addition, future studies of wetland invertebrate communities should include the influence of waterfowl, wading birds, and wetland-associated songbirds as top predators on wetland invertebrates. The impacts of these predators are potentially strong, especially in small nonecotonal wetlands, where they can include both the direct effects of predation and indirect effects of waterfowl-induced changes in vegetation (Chapter 34, Batzer and Wissinger 1996).

Habitat Heterogeneity and Invertebrate Diversity and Abundance

Invertebrate diversity and abundance are often higher in vegetated than in open-bottom habitats (Chapters 2, 3, 4, 5, 10, 15, 16, 23, 24, 28, 34, 36, 37, 38; review by Batzer and Wissinger 1996 and Olson et al. 1995). The manipulation of vegetation structure is a common practice in managed wetlands and provides experimental evidence that vegetation structure is an important causal factor that affects invertebrate composition, diversity, and abundance (Chapter 34, references in Batzer and Wissinger 1996, Kirkman and Sharitz 1994, Foster and Procter 1995, de Szalay and Resh 1997). A variety of mechanisms have been offered to explain the empirical relationships between vegetation and invertebrate diversity and abundance, including that vegetation provides (1) greater surface area per se = higher densities; (2) greater surface area, hence more epiphytic biofilm for grazers = higher densities; (3) refuge from predators = higher densities (e.g., Heck and Crowder 1991, Jordan et al. 1996); and (4) more types of spatial niches = greater diversity. These explanations are not mutually exclusive, and which subset is responsible for the empirical relationship between vegetation and invertebrate abundance and diversity in appears to vary among plant species and wetland types (Batzer and Wissinger 1996). Several authors note that increased diversity and abundance of invertebrates in dense vegetation is largely due to an increase in phytophilous microcrustaceans (Chapters 38, 40).

Invertebrate diversity within wetlands also appears to be strongly affected by the number of different types of subhabitats or vegetation zones. Different types and zones of vegetation contain different invertebrate species (Chapters 2, 3, 4, 7, 8, 9, 10, 17, 19, 21, 24, 25, 34, 35), and wetlands with the greatest diversity of plant species or types of vegetation have the most diverse invertebrate faunas (Chapters 2, 4, 7, 10, 15, 16, 17, 21, 22, 23, 24, 25, 34, 35, 37, 38; also Kirkman and Sharitz 1994). Studies in which multiple sampling strategies include all vegetation zones, above- and below-water parts of emergent and floating plants, and year-round sampling have found an incredible diversity of invertebrate species (e.g., Chapter 17, Williams et al. 1994). However, most studies do not include multiple sampling strategies, all spatial habitats, and/or all hydroperiod phases or seasons, and several authors note that for one or more of these reasons their species lists are incomplete (Chapters 2, 7, 9, 15, 22, 28, 36, 37, 38; also see McElligott and Lewis 1994, Brinkman and Duffy 1996, Turner and Trexler 1997, Soumille and Thiery 1997).

In summary, it is clear that habitat heterogeneity plays a major role in determining the overall diversity of invertebrates and patterns of invertebrate distribution and abundance in a wetland. Mechanisms that underlie invertebrate-vegetation relationships are understood for some plants and some types of vegetation assemblages (review by Batzer and Wissinger 1996; also de Szalay et al. 1996), and a challenge for the future will be to broaden our understanding of the relative importance of the structural (niches, predator refuge) and trophic (as food and as a substrate for food) roles of vegetation and other sources of habitat heterogeneity in wetland ecosystems (e.g., snags and other woody debris; France 1997).

INVERTEBRATES AND ECOSYSTEM FUNCTION

Wetlands are typically autotrophic ecosystems in which primary production exceeds total community respiration. The relative importance of different primary producers (phytoplankton, herbaceous macrophytes, trees and shrubs, and/or benthic algae) varies among wetland types, but autochthonous production is usually more important than allochthonous inputs (Mitsch and Gosselink 1993; see Chapter 31 for an exception). For example, in marshes, which are among the most productive ecosystems on earth, the historical paradigm is that macrophyte production is most important and that most of this energy enters consumer foodwebs as detritus (Brisson et al. 1981; Mitsch and Gosselink 1993). Although the general tenets of ecosystem function are reasonably well understood for many wetlands (Mitsch and Gosselink 1993 and references therein), there is a paucity of detailed information about the roles that particular species or functional groups of invertebrates play in the flow of energy and cycling of nutrients in wetlands. This is somewhat surprising given the general recognition that invertebrates are the major "primary consumers" and "critical links" between primary producers and the vertebrates that are often the focus of wetlands management (waterfowl, wading birds, shore birds, fish, amphibians, mammals).

Information about the roles of different species in a foodweb is often organized by combining species into functional groups such as trophic levels or guilds. One approach is to categorize species by functional feeding groups (FFG) based on the dominant type of food consumed and their feeding mechanism (Cummins and Klug 1979, Merritt and Cummins 1996a, b, Minshall

1996). The FFG concept was developed in the context of stream ecosystems and is therefore familiar to many wetlands investigators who were trained as aquatic ecologists (e.g., Chapters 2, 3, 7, 8, 9, 10, 12, 14, 15, 16, 28, 29, 31, 34, 39, 40; also see Fredrickson and Reid 1988). Spatial and temporal variations in the abundance of different trophic groups in streams (e.g., *scrapers* of attached algae and biofilms, *shredders* of coarse particular organic matter [CPOM]), various types of *collectors* [burrowers, gatherers, filterers] of fine particulate organic matter [FPOM], and *predators*), has been important for the development of many of the current paradigms for energy flow and nutrient cycling in lotic systems (e.g., Allan 1995). Below I summarize the information presented in the various chapters in this book and/or cited in those chapters to (1) depict a general FFG model for invertebrate pathways through which primary production enters wetland foodwebs (Fig. 41.4); (2) describe patterns of spatial and temporal variation in those pathways within

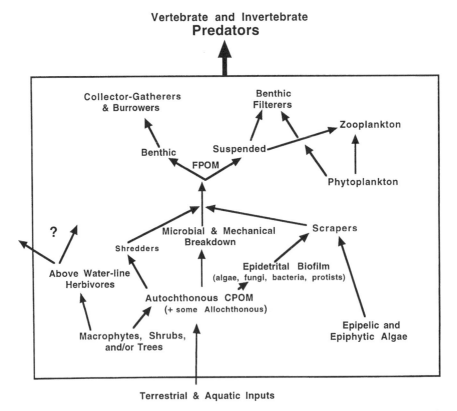

Fig. 41.4. A functional feeding group model for the various invertebrate pathways by which primary production is converted to secondary (animal) production in wetland foodwebs. The relative importance of different pathways will differ among wetlands (see Invertebrates and Ecosystem Function, above).

and among wetlands; and (3) consider areas for future research to address current contradictions and gaps in our understanding of the roles that invertebrates play in energy flow and nutrient recycling in wetlands. A caveat to this approach is that it ignores the importance of species variability and richness within these FFGs and the positive effects that redundancy can have on ecosystem stability and sustainability (Naeem 1998).

Above-Waterline Macrophyte Herbivory

In aquatic ecosystems it is generally assumed that most macrophyte production is consumed as detritus (e.g., Wetzel 1983, Jeffries and Mills 1990, Allan 1995). In wetland ecosystems this is probably true for the below-waterline parts of macrophytes (but see Feminella and Resh 1989). However, the abundance and diversity of herbivores (e.g., leaf and stem herbivores, stem borers, gall-forming insects) on the above-waterline parts of floating, emergent herbaceous, and woody vegetation suggest that live macrophyte tissue is an important invertebrate food source in many wetlands (Chapters 2, 4, 7, 10, 15, 17, 21, 22, 23, 24, 30, 38). Several studies have quantified the importance of the various types of herbivores that occur in the canopy of dense stand of emergent macrophytes (review by Batzer and Wissinger 1996), but there is little information about the importance of this herbivory in terms of energy flow and nutrient movement (Fig. 41.4). Similarly, few studies have quantified the importance of herbivory in the canopies of trees and its ecosystem consequences, including the potential for the fallout of frass and feces into the water and the effects of defoliation on light penetration (Goyer et al. 1990). Clearly one challenge for the future is to quantify above-ground herbivory and the degree to which nutrients and energy are returned to the below-waterline foodweb or exported to adjacent habitats (Fig. 41.4).

Relative Importance of Algal and Macrophyte Production

The relative importance of macrophyte detritus versus benthic algae for invertebrate production appears to vary considerably among wetland types. Leaves and wood are certainly important food sources in forested wetlands (Chapters 5, 9, 12, 14, 28) where invertebrate communities are dominated by CPOM detritivores, especially micro- and macrocrustaceans (isopods, amphipods). In some marshes invertebrate production is also thought to be driven mainly by the abundance of macrophyte detritus, but in marshes, it is insects that are typically the major detritivores (Chapters 15, 16, 17, 31). In some marshes, there is evidence that benthic algae are far more important than previously suspected. For example, the use of stable isotopes of carbon and nitrogen has revealed that much of the energy entering consumer foodwebs is from benthic algae rather than macrophytes (Goldsborough and Robinson 1996, Robinson et al. 1997a, b). Isotopic signatures suggest that in some wetlands epiphytic and epidetrital algae is more important to invertebrate grazers than mats of filamentous algae (Chapters 4, 21; but see Chapter 30). Algal blooms that occur after reinundation in temporary wetlands are thought to be especially important for the remobilization of nutrients mineralized during the dry phase of the hydroperiod (Chapters 2, 3, 4, 21, 23, 24, 34; references in Batzer and Wissinger 1996). Clearly there is a need for future research to determine the relative importance of algal and macrophyte production across different types of marsh habitats and at different times during hydroperiods (Chapters 2, 4, 10, 19, 25, 28, 34, 36, 38, 39).

Detritus Processing by Invertebrates in Wetlands

There is an obvious paucity of shredders in many wetland invertebrate communities, even in wetlands with dense stands of macrophytes (Chapters 10, 12, 21, 22, 23, 24, 37, 40; but see Chapter 2). This is curious because the general paradigm for the breakdown of vascular plant detritus is that (1) soluble constituents leach as dissolved organic matter; (2) microbes (first fungi and then bacteria) "condition" CPOM by breaking down refractory constituents (cellulose, hemicellulose, lignin, etc.) and incorporating dissolved nutrients (especially nitrogen and phosphorus); (3) shredders then consume this conditioned CPOM-microbial complex; and (4) this shredding produces both local and exported sources of food that is then available for a variety of FPOM collectors (gatherers, burrowers, filterers) (Anderson and Sedell 1979, Barlocher and Kendrick 1985, Cummins et al. 1989, Webster and Benfield 1986, Allan 1995).

There are several potential explanations for the dearth of shredders in vegetated wetlands. First, in northern wetlands some dominant shredders (e.g., limnephilid caddisflies) have life cycles timed such that larvae hatch in autumn when macrophyte CPOM is at a maximum. Larvae are present only from late autumn to early spring, when sampling is least likely. Although this might explain the underrepresentation of shredders in northern wetlands (12, 13, 14, 15, 17, 21, 22, 23, 24, 40), it does not provide a general explanation for low shredder diversity and abundance in other wetlands. A second potential explanation is that low oxygen levels in the detritus inhibit microbial conditioning and/or shredder colonization, which should lead to the accumulation of organic matter. While organic matter does accumulate in some ombrotrophic wetlands, it does not in other types of wetlands that also lack shredders. For example, many ecotonal wetlands are open to adjacent aquatic systems and presumably better oxygenated (Mitsch and Gosselink 1993). Export to adjacent aquatic habitats is one potential explanation for the absence of both shredders and accumulated organic matter (Chapters 3, 7, 16, 29, 33, 36, 37, 38, 40). Organic matter accumulation is less likely in temporary than in permanent wetlands because detritus can be mineralized during the dry phase of the hydroperiod. Algal blooms occurring after reinundation are thought to be stimulated by the abundance of nutrients released during dry-phase mineralization of macrophyte detritus (Chapters 2, 3, 4, 21, 23, 24, 34;

also Batzer and Wissinger 1996). Macrophytes are still the ultimate primary source of the energy and nutrients, but the pathway by which the nutrients enter the consumer web is through grazers on benthic algae rather than through the CPOM shredder-FPOM collector pathway (Fig. 41.4).

A third explanation for the paucity of shredders in many wetlands is that the rapid breakdown of herbaceous plant material by mechanical, microbial, and chemical processes precludes shredding. This would be especially true for submergent and floating plants with short half-life decay rates (review by Webster and Benfield 1986; also Chapters 21, 23, Findley et al. 1990). Annual plants that rapidly invade dried basins might also produce an abundance of fast-decaying detritus that, upon inundation, should support dense populations of grazers and FPOM collectors, especially chironomid midges (Chapter 24). The importance of this detritus as a basis of secondary production is no less important than if they had been eaten by shredders, but the energy and nutrients enter the consumer web via a different route than that suggested by the shredder paradigm (Fig. 41.4). One route could be epidetrital biofilms, which are important substrates for both scrapers and epibenthic microcrustaceans in wetlands. Relatively little is known about the movements of nutrients and energy from macrophyte detritus to the autotrophic (algae, protists), heterotrophic (bacteria, fungi) and nonliving (trapped FPOM, extracellular microbial matrices) components of attached biofilms. The importance of algae and other biofilm constituents as a trophic basis for secondary production in wetlands could be tightly linked to this alternative pathway for detritus mineralization.

Energy and Nutrient Fluxes across Wetland Ecotones

Spatial subsidies of energy across ecosystem boundaries can have important effects on foodweb dynamics (review by Polis et al. 1997). Invertebrates can play various roles in the flux of energy and nutrients across both terrestrial-wetland and aquatic-wetland ecotones (Fig. 41.1). Several authors have suggested that invertebrates in open, ecotonal wetlands (e.g., Chapters 3, 7, 16, 29, 33, 36, 37, 38, 39) are less likely to be food-limited than those in non-ecotonal wetlands. Historically, invertebrates in riparian wetlands (including freshwater tidal wetlands) have been considered to be of fundamental importance for the modification, storage, and transfer of dissolved and particulate organic matter en route from terrestrial and wetland ecosystems to stream channels (e.g., Odum 1988, Cummins et al. 1989, Allan 1995; Naiman and Decamps 1997). More recently, the movement of the invertebrates themselves (both aerially and in flood and ebb water) has been identified as having potentially important impacts on the energy budgets of both the wetland and adjacent aquatic ecosystem (e.g., Chapters 7, 16, 37). Although these links are generally assumed to be important, there is little quantitative evidence for how such spatial subsidies influence ecosystem function in ecotonal wetlands (Odum 1988).

It is generally assumed that both the energy and the nutrients in aquatic and aerial stages of invertebrates that are consumed by birds, mammals, and amphibians are transferred to adjacent terrestrial habitats, but the energetic importance for either ecosystem has not been documented (Batzer and Wissinger 1996). It is perhaps not surprising that freshwater wetlands have yet to play a major role in the development of general models for the importance of spatial subsidies to foodweb dynamics (Polis et al. 1996, 1997).

Integrating Community and Ecosystem Ecology— The Trophic Basis of Production

Clearly there is still much to learn about the roles that invertebrates play in the movements of energy and nutrients within wetlands and between different wetland subhabitats and adjacent ecosystems. The use of stable isotopes will play an important role in testing various hypotheses about the relative importance of different trophic pathways and the relative importance of different energy sources for consumer foodwebs in wetland ecosystems (general reviews by Peterson and Fry 1987, Fry 1991, Lajtha and Michener 1994). Constructing functional foodwebs requires population and community data about which species interact (connectivity webs) and quantitative information about the sources and amounts of energy transferred across links (Paine 1992). One approach is to quantify rates at which different sources of primary production are transformed to secondary production (as in Benke and Wallace 1997 and references therein). Understanding the trophic basis of invertebrate production in wetland foodwebs will be important for interpreting experiments designed to determine the degree to which primary production in wetlands is controlled by top-down (the cascading effects of predator removal/addition at the tops of foodwebs) versus bottom-up processes (nutrient limitation) (Batzer and Resh 1991, Batzer 1998).

APPLICATIONS FOR CONSERVATION AND MANAGEMENT

Wetlands are threatened by various human activities, and these threats and the effects on wetland invertebrates are reviewed in nearly every chapter in this book. Other applied issues that are addressed in individual chapters include:

1. The roles that invertebrates play in the conservation and management of waterfowl and other wildlife such as shorebirds, songbirds, and amphibians (Chapters 2, 10, 21, 22, 23, 24, 25, 26, 28, 30, 31, 34, 35, 36, 37, 38, 39, 40)
2. The role of invertebrates for pest control management, especially mosquitoes and invasive alien plants (Chapters 28, 34, 35)

3. The conservation of invertebrate biodiversity in wetlands (Chapters 8, 11, 13, 17, 18, 21, 26, 30, 33)

4. Invertebrates as bioindicators of ecosystem health (Chapters 2, 3)

5. Recommendations for encouraging the development of invertebrate communities in restored (Chapters 3, 4, 5, 15) and created wetlands (Chapter 4, 5, 19)

6. The effects of land use impacts on adjacent aquatic and terrestrial ecosystems (Chapters 3, 7, 12, 16, 28, 29, 36, 37, 38, 39)

7. The interdependency of wetlands linked by dispersal (Chapters 14, 15, 21, 26, 31)

The importance of wetland invertebrates has long been appreciated by those interested in wetland habitats for the conservation and management of waterfowl and for controlling pests (Eldridge 1990), and much of the rapidly growing literature on this subject is referenced in the individual chapters listed above and in Batzer and Wissinger (1996). Below I will focus on the application of the information in this synthesis to wetlands regulations, for rationales and strategies for conservation, and for the design of constructed (created, restored, enhanced) wetland habitats.

Protecting Wetlands as Enduring Habitats with Unique Species

Conceptualizing wetlands and wetland plant and animal communities as unique and enduring features of the landscape has important ramifications for conservation and management. The message that most wetlands are not transient stages in an aquatic to terrestrial succession (So why not just fill them in now!) and not just boundaries between terrestrial and aquatic habitats is a particularly important one for wetland ecologists to impart to regulators, policy-makers, and the public. The recognition of the uniqueness of wetland invertebrate communities (Table 41.1) has lagged behind a similar recognition for wetland plant communities. Just as hydrophytic plants have been classified (obligate wetland, facultative wetland, facultative, facultative upland) for recognizing and delineating wetlands (Tiner 1991, 1993a, b), so too could invertebrates and invertebrate communities be used for determining boundaries between wetland and upland habitats. Such corroborative evidence would be especially useful along relatively flat gradients where delineations based on the three primary indicators can be difficult (see Tiner 1993, Davis et al. 1996).

Also to be emphasized in distinguishing wetland from aquatic and terrestrial invertebrates is that the destruction of wetlands will lead to the local extinction of these species. Most wetland species cannot just move to terrestrial or aquatic habitats when wetlands are filled or excavated. Invertebrate biodiversity typically exceeds that of wetland plants and vertebrates combined, and pointing out that the loss of wetland habitats will lead to the loss of invertebrate species is an under-appreciated rationale for the preservation and conservation of wetland habitats (Foster and Procter 1995). The loss of wetlands will also affect terrestrial organisms that depend on them as local concentrations of food (Gibbs 1993, Turner and McCarty 1997).

Changing the Hydroperiods of Natural Wetlands

There is an abundance of evidence that wetlands with different hydroperiods contain distinctly different invertebrate communities that are either directly or indirectly dependent on particular temporal patterns of drying (or not) and refilling (see Direct and Indirect Effects of Hydroperiod, above). Temporary wetlands in particular have disappeared at an alarming rate because these shallow basins and depressions are (1) sensitive to changes in the water table associated with human activities; (2) easily drained, filled, and/or converted to agricultural land (Chapters 3, 8, 9, 17, 21, 25, 26, 35); and (3) often excavated and deepened into permanent aquatic habitats for recreation (Chapters 2, 3, 8, 21, 27, 37; also Tiner 1984, Abernathy and Turner 1987, Dahl 1990, Cashin et al. 1992, Fretwell et al. 1996). Converting wetlands to deep-water habitats (>2 m) will result in the replacement of temporary with permanent wetland species, which in turn can affect waterfowl and other wildlife dependent on particular species, types, and sizes of invertebrates (Chapters 4, 5, 8, 21, 34, Batzer and Wissinger 1996, Collinson et al. 1995). Making wetlands permanent will also reduce invertebrate abundances as a result of fish predation (e.g., Chapters 2, 19, Hanson and Butler 1994, Mallory et al. 1994). In terms of ecosystem function, the importance of seasonal drying for the remobilization of nutrients and subsequent flushes of invertebrates after inundation is a well-known phenomenon that is an important component of waterfowl conservation and management (Chapters 21, 22, 23, 24, 28, 34, Fredrickson 1991). Although some state laws regulate the conversion of wetlands to deep-water habitats, federal wetland laws do not. They also do not give regulators jurisdictional powers related to changes in adjacent land use, buffer zones, alterations to waterways, or withdrawals from local aquifers, which can all lead to changes in the hydroperiods of natural wetland basins (e.g., Richardson and McCarthy 1994, Findlay and Houlahan 1997). The information presented in this book makes it abundantly clear that wetland invertebrate communities are not all alike and that particular assemblages of wetland animals will be associated with particular wetland hydroperiods.

Constructing Wetlands—Temporal Variability and Spatial Heterogeneity

Although constructed wetlands have been increasingly used to mitigate for jurisdictional encroachments into natural habitats, few are monitored after construction to assess the degree to which they mimic the natural wetlands they were intended to replace (review by Bedford 1996; Mitsch and Wilson

1996). When follow-up studies have been conducted on mitigation projects, the most important problems have been related to hydroperiod, and in particular to depth and to making habitats permanent (e.g., Erwin 1991, Confer and Niering 1992). Many of the chapters in this book provide evidence that permanent constructed habitats will not contain the same invertebrate communities, support the same types of game and nongame wildlife, or function in the same ways as the natural temporary habitats that they were intended to replace. There is increasing interest in the development of standards for evaluating the degree to which mitigation wetlands act as functional replacements for the impacted sites they are designed to replace (Smith et al. 1994, Brinson and Rheinhardt 1996, Wilson and Mitsch 1996). Unfortunately, neither the reference standards that are being developed nor the assessment tools that have historically been used to evaluate the functional integrity of wetlands (e.g., wetland evaluation technique = WET, Adamus et al. 1991; habitat evaluation procedure = HEP, Stiehl 1994) rely on any information about wetland invertebrate communities or their role in ecosystem function. Some restoration efforts are sensitive to the importance of hydroperiod. For example, wetlands constructed by the U.S. Fish and Wildlife Service's Partners for Wildlife Program are designed to restore wetland habitats that previously drained for agriculture (Brown 1994). A primary objective of this program is to restore waterfowl habitat and the basins are designed to be shallow and to fluctuate seasonally with groundwater levels or via manipulation with control structures (as in Reid and Fredrickson, 1988, Fredrickson 1991, Chapter 34 and references therein).

A second theme in this book is the importance of spatial heterogeneity, and in particular vegetation as habitat and food or food substrate. The different types of invertebrate communities that develop in different wetland microhabitats attract different species of birds, which require different types and sizes of invertebrate prey and different habitat types and depths for efficiently exploiting those prey (divers, dabblers, waders, etc.). Thus, wetlands should be constructed with subhabitats that differ in hydroperiod, depth, and types of substrates to support a greater diversity of plants, hence invertebrates, and hence wildlife, than those with one homogeneous basin (Chapters 15, 21, 24, 26, 28, and 34 and recent articles by Grover and Baldassarre 1995, Galatowitsch and van der Valk 1996, VanRees-Siewert and Dinsmore 1996, Turner and McCarty 1997). Habitat heterogeneity is easily quantifiable and could easily be incorporated into functional assessment protocols as an indicator of invertebrate diversity (as in Brinson and Rheinhardt 1996).

Metapopulations of Wetland Invertebrates and Landscape Management

A second level of habitat heterogeneity that is important for the establishment and persistence of diverse invertebrate communities is related to the landscape position of wetlands with respect to each other and to adjacent terrestrial and/or aquatic habitats. Because some wetland insects that live in temporary habitats use adjacent terrestrial habitats as sites for adult diapause or feeding during the dry phase of the hydroperiod, buffer zones of native vegetation will be important for the maintenance of those wetland communities (see Timed Emergence and Recolonization by Adults, above). Current federal and state/provincial regulations that protect wetlands often do not prescribe buffer zones widths based on any knowledge of their importance to the animals that live in those wetlands (Findlay and Houlahan 1997, but see Brown et al. 1990). Invertebrate communities in ecotonal wetlands depend on movements of organisms across the wetland-aquatic ecotone, and those fluxes depend on natural hydrologic fluctuations (tides, seiches, floods) and on the geomorphologic integrity of the landscape (Chapters 3, 7, 16, 28, 29, 36, 37, 38, 40; also see O'Neill et al. 1997).

Dispersal by opportunists and cyclic colonizers is an important source of colonists in all wetland communities. Discrete populations and communities that are connected by dispersal are likely to function as metapopulations and even metacommunities (see introduction by Hanski and Simberloff 1997). Theoretical and empirical studies of metapopulations suggest that the removal of one or more habitats (subpopulations) will affect whether other subpopulations or the entire metapopulation will persist (see reviews by Gilpin and Hanski 1991, Hanski and Gilpin 1997). Even small satellite populations can be important for the long-term persistence of metapopulations (Doak and Mills 1994). Because wetlands are often discrete habitats embedded in terrestrial landscapes, effective conservation and restoration strategies are especially likely to depend on an understanding of how the persistence of viable populations in one habitat depend on the proximity to other wetlands and particular types of wetlands (Jeffries 1994, Harrison and Fahrig 1995, Moloney and Levin 1996, Bell et al. 1997).

Species Diversity and Ecosystem Functions and Services

Two themes that have emerged from this book are that (1) many interacting factors can affect community composition and diversity in wetlands (Fig. 41.1); and (2) there are many gaps in our knowledge about the relative importance of different functional groups and pathways by which invertebrates process primary production and how that varies among wetland types and across hydroperiod gradients (Fig. 41.3). Lumping taxa into functional groups is useful for structuring our thinking about ecosystem function and simplifying the taxonomic resolution that is necessary for assessing ecosystem health (e.g., Chapter 3; also Korner 1993, Gitay et al. 1996). However, biomonitoring approaches that lump species by functional groups ignore the importance of species richness within these groups. Historically the term *species redundancy* has had a negative connotation among conservation biologists because it has been used to imply that some species within a functional group might be superfluous. However, there has been a recent flurry of research that empirically links biodiversity to ecosystem stability (resistance to perturbation and resilience after perturbation) and to the sustainability and reliability of ecosystem function (material cycling and energy flow) (Tilman 1994, Naeem et al. 1995, Tilman et al. 1996, McGrady-Steed et al. 1997, Naeem and Li 1997, Naeem 1998). Because invertebrates are the primary consumers of plant production in wetlands, diversity within functional groups (i.e., species redundancy) could be necessary for the reliability of ecosystem function and hence the ecosystem services that wetlands provide societies.

ACKNOWLEDGMENTS

Comments by Darold Batzer and Russ Radar greatly improved the manuscript. Thanks to Sue and AJ for their patience. I gratefully acknowledge the support of the National Science Foundation (DEB-9407856).

LITERATURE CITED

Abernathy, Y., and R. E. Turner. 1987. U.S. Forested Wetlands: 1940–1980. Bioscience 37:721–727.

Adamus, P. R., L. T. Stockwell, E. J. Clairain, M. E. Morrow, L. P. Rozas, and R. D. Smith. 1991. Wetland Evaluation Technique (WET). Vol. I. Literature Review and Evaluation Rationale. Technical Report WRP-DE-2. U.S. Army Corps of Engineers Waterways Experiments Station, Vicksburg, MS.

Allan, J. D. 1995. Stream Ecology: Structure and Function of Running Water. Chapman & Hall, New York.

Anderson, H. H., and J. R. Sedell. 1979. Detritus processing by macroinvertebrates in stream ecosystems. Annual Review of Entomology 24:351–377.

Andrews, D. A., and F. H. Rigler. 1985. The effects of an Arctic winter on benthic invertebrates in the littoral zone of Char Lake, Northwest Territories. Canadian Journal of Zoology 63:2825–2834.

Barlocher, F., and B. Kendrick. 1985. Leaf-conditioning by micro-organisms. Oecologia 20:359–362.

Batzer, D. P. 1998. Trophic interactions among detritus, benthic midges, and predatory fish in a freshwater marsh. Ecology 79:1688–1698.

Batzer, D. P., and V. H. Resh. 1991. Trophic interactions among a beetle predator, a chironomid grazer, and periphyton in a seasonal wetland. Oikos 60:251–257.

Batzer, D. P., and S. A. Wissinger. 1996. Ecology of insect communities in nontidal wetlands. Annual Review of Entomology 41:75–100.

Bedford, B. L. 1996. The need to define hydrological equivalence at the landscape scale for freshwater wetland mitigation. Ecological Applications 6:57–68.

Bell, S. S., M. S. Fonseca, and L. B. Mooten. 1997. Linking restoration and landscape ecology. Restoration Ecology 5:318–323.

Benke, A. C., and J. B. Wallace. 1997. Trophic basis of production among riverine caddisflies: Implications for food web analysis. Ecology 78:1132–1145.

Boyce, M. S. 1984. Restitution of r- and K-selection as a model of density-dependent model natural selection. Annual Review Ecology Systematics 15:427–449.

Brinkman, M. A., and W. G. Duffy. 1996. Evaluation of four wetland aquatic invertebrate samplers and four sample sorting methods. Journal of Freshwater Ecology 11:193–200.

Brinson, M. M. 1993a. A Hydrogeomorphic Classification for Wetlands. Wetland Research Program Technical Report WRP-DE-4. U.S. Corps of Engineers Waterways Experiment Station, Vicksburg, MS.

———. 1993b. Changes in the functioning of wetlands along environmental gradients. Wetlands. 12:65–74.

Brinson, M. M., and R. Rheinhardt. 1996. The role of reference wetlands in functional assessment and mitigation. Ecological Applications 6:69–76.

Brisson, M. M., A. E. Logo, and S. Brown. 1981. Primary productivity, decomposition, and consumer activity in freshwater wetlands. Annual Review of Ecology and Systematics 12:123–161.

Brown, D. F. 1994. A partnership for wildlife. Pennsylvania Game News 65:28–32.

Brown, M. T., J. M. Schaefer, and K. H. Brandt. 1990. Buffer Zones for Water, Wetlands, and Wildlife in East Central Florida. Final Report, Florida Agricultural Experimental Station, Center for Wetlands, University of Florida, Gainesville, FL.

Burke, V. J., and J. W. Gibbons. 1995. Terrestrial buffer zones and wetland conservation: A case study of freshwater turtles in a Carolina Bay. Conservation Biology 9:1365–1369.

Carter, V., and P. T. Gammon, and M. K. Garrett. 1994. Ecotone dynamics and boundary determination in the Great Dismal Swamp. Ecological Applications 4:189–203.

Cashin, G. E., J. R. Dorney, and C. J. Richardson. 1992. Wetland alteration trends on the North Carolina coast. Wetlands 12:63–71.

Chesson, P., and N. Huntly. 1997. The roles of harsh and fluctuating conditions in the dynamics of ecological communities. American Naturalist 150:519–533.

Collinson, N. H., J. Bigg, A. Corfield, M. J. Hodson, D. Walker, M. Whitfield, and P. J. Williams. 1995. Temporary and permanent ponds: An assessment of the effects of drying out on the conservation value of aquatic macroinvertebrate communities. Biological Conservation 74:125–133.

Confer, S. R., and W. A. Niering. 1992. Comparison of created and natural freshwater emergent wetlands in Connecticut. Wetlands Ecology and Management 2:429–447.

Cornell, H. V., and J. H. Lawton. 1992. Species interactions, local and regional processes, and limits to the richness of ecological communities: A theoretical perspective. Journal of Animal Ecology 61:1–12.

Corbet, P. S. 1980. Biology of Odonata. Annual Review of Entomology 25:189–210.

Cowardin, L. M., V. Carter, G. C. Gloet, and E. T. LaRoe. 1979. Classification of Wetlands and Deepwater Habitats of the United States. U.S. Fish and Wildlife Service Report FWS/OBS-79-31.

Cummins, K. W., and M. J. Klug. 1979. Feeding ecology of stream invertebrates. Annual Review of Ecology and Systematics 10:147–172.

Cummins, K. W., M. A. Wilzbach, D. M. Gates, J. B. Perry, and W. B. Taliaferro. 1989. Shredders and riparian vegetation. Bioscience 39:24–30.

Daborn, G. R. 1971. Survival and mortality of coenagrionid nymphs (Odonata: Zygoptera) from the ice of an aestival pond. Canadian Journal of Zoology 57:2143–2152.

———. 1974. Biological features of an aestival pond in western Canada. Hydrobiologia 44:287–299

Daborn, G. R., and H. F. Clifford. 1974. Physical and chemical features of an aestival pond in western Canada. Hydrobiologia 44:43–59.

Dahl, T. W. 1990. Wetlands Losses in the United States, 1780s to 1980s. U.S. Department of the Interior, Fish and Wildlife Service, Washington, DC.

Dahl, T. W., C. E. Johnson, and W. E. Frayer. 1991. Wetland Status and Trend in the Conterminous United States, Mid-1970s to Mid-1980s. U.S. Department of the Interior, Fish and Wildlife Service, Washington, DC.

Daily, G. C. 1997. Nature's Services: Societal Dependence on Natural Ecosystems. Island Press, Washington, DC.

Damman, A. W. H., and T. W. French. 1987. The Ecology of Peat Bogs of the Glaciated Northeastern United States: A Community Profile. U.S. Fish and Wildlife Service Biological Report 85(7.16):100.

Danks, H. V. 1971. Overwintering of some north temperate and arctic Chironomidae. II Chironomid biology. Canadian Entomologist 103:1875–1910.

Davis, S. M., and J. C. Ogden (eds.). 1994. Everglades: The Ecosystem and Its Restoration. St. Lucie Press, Delray Beach, FL.

Davis, M. M., S. W. Sprecher, J. S. Wakely, and G. R. Best. 1996. Environmental gradients and identification of wetlands in north-central Florida. Wetlands 16:512–523.

de Szalay, F. A., D. P. Batzer, and F. H. Resh. 1996. Mesocosm and macrocosm experiments to examine effects of mowing emergent vegetation on wetland invertebrates. Environmental Entomology 25:203–209.

de Szalay, F. A., and V. H. Resh. 1997. Responses of wetland invertebrates and plants important in waterfowl diets to burning and mowing of emergent vegetation. Wetlands 17:149–156.

Delorme, L. D. 1991. Ostracoda. Pages 691–722 *in* J. H. Thorp and A. P. Covich (eds.), Ecology and Classification of North American Freshwater Invertebrates. Academic Press, New York.

Denno, R. F., and E. E. Grissell. 1979. The adaptiveness of wing-dimorphism in the salt-marsh inhabiting planthopper, *Prokelisia marginata* (Homoptera: Delphacidae). Ecology 6:221–236.

Denno, R. F., and G. K. Roderick. 1992. Density-related dispersal in planthoppers: Effects of interspecific crowding. Ecology 73:1323–1334.

Diehl, J. K., H. J. Holliday, C. J. Lindgren, and R. E. Roughly. 1997. Insects associated with purple loosestrife, *Lythrum salicaria*, in southern Manitoba. Canadian Entomologist 129:937–948.

Doak, D. F., and L. S. Mills. 1994. A useful role for theory in conservation. Ecology 75:15–626.

Dodson, S. I. 1970. Complementary feeding niches sustained by size-selective predation. Limnology and Oceanography 15:131–137.

————. 1974. Zooplankton competition and predation: An experimental test of the size-efficiency hypothesis. Ecology 55:605–613.

Dodson, S. I., and D. G. Frey 1991. Cladocera and other Branchiopoda. Pages 725–786 *in* J. H. Thorp and A. P. Covich (eds.), Ecology and Classification of North American Freshwater Invertebrates. Academic Press, New York.

Drake, J. A. 1991. Community-assembly mechanics and the structure of an experimental species ensemble. American Naturalist 137:1–26.

Eldridge, J. 1990. Aquatic invertebrates important for waterfowl production. Fish and Wildlife Leaflet 13.3.3. U.S. Fish and Wildlife Service, Washington, DC.

Environmental Laboratory. 1987. Corps of Engineers Wetlands Delineation Manual. Technical Report Y-87-1. U.S. Army Engineer Waterways Experiment Station, Vicksburg, MS.

Epperson, J. E. 1992. Missouri Wetlands: A Vanishing Resource. Missouri Department of Natural Resources: Division of Geology and Land Survey. Water Resources Report No. 39.

Erwin, K. 1991. An Evaluation of Wetland Mitigation in the South Florida Water Management District. Volume I. Report to the South Florida Water Management District, West Palm Beach, FL.

Federal Interagency Committee for Wetland Delineation. 1989. Federal Manual for Identifying and Delineating jurisdictional wetlands. Cooperative Technical Publication, U.S. Army Corps of Engineers, U.S. Environmental Protection Agency, U.S. Fish and Wildlife Service, and U.S.D.A. Soil Conservation Service, Washington, DC.

Feminella, J. W., and V. H. Resh. 1989. Submersed macrophytes and grazing crayfish: An experimental study of herbivory in a California freshwater marsh. Holarctic Ecology 12:1–18.

Fernando, C. H. 1959. The colonization of small freshwater habitats by aquatic insects. 2. Hemiptera. Ceylon Journal of Science 2:5–32.

Fernando, C. H., and D. Galbraith. 1983. Seasonality and dynamics of aquatic insects colonizing small habitats. Internationale Vereinigung für theoretische und angewandte Limnologie, Verhandlungen 18:1564–1575.

Findlay, C. S., and J. Houlahan. 1997. Anthropogenic correlates of species richness in south-eastern Ontario wetlands. Conservation Biology 11:1000–1009.

Findlay, S., K. Howe, and H. K. Austin. 1990. Comparison of detritus dynamics in two tidal freshwater wetlands. Ecology 71:288–295.

Foster, A. P., and D. A. Procter. 1995. The occurrence of some scarce East Anglian fen invertebrates in relation to vegetation management. Pages 223–240 *in* B. D. Wheeler, S. C. Shaw, W. J. Fojt, and R. A. Robertson (eds.), Restoration of Temperate Wetlands. John Wiley & Sons, New York.

France, R. L. 1997. The importance of beaver lodges in structuring littoral communities in boreal headwater lakes. Canadian Journal of Zoology 75:1009–1013.

Fredrickson, L. H. 1991. Strategies for Water Level Manipulations in Moist-soil Systems. U.S. Fish and Wildlife Leaflet 12.4.6, Washington, D.C.

Fredrickson, L. H., and F. A. Reid. 1988. Invertebrate Response to Wetland Management. U.S. Fish and Wildlife Leaflet 13.2.1, Washington, DC.

Fretwell, J. D., J. S. Williams, and P. J. Redman (eds.). 1996. National Summary on Wetland Resources. Water-Supply Paper 2425, U.S. Geological Survey, Washington, DC.

Frisbee, M. P., and R. E. Lee. 1997. Inoculative freezing and the problem of winter survival for freshwater macroinvertebrates. Journal of the North American Benthological Society 16:636–650.

Fry, B. 1991. Stable isotope diagrams of freshwater food webs. Ecology 72:2293–2297.

Galatowitsch, S. M., and A. G. van der Valk 1996. The vegetation of restored and natural prairie wetlands. Ecological Applications 6:102–112.

Gates, F. C. 1926. Plant succession about Lake Douglas, Cheboygan County, Michigan. Botanical Gazette 82:170–182.

Gibbs, J. P. 1993. Importance of small wetlands for the persistence of local populations of wetland-associated animals. Wetlands 13:25–31.

Gilpin, M. E., and I. A. Hanski 1991. Metapopulation Dynamics: Empirical and Theoretical Investigations. Academic Press, London.

Gitay, H., J. B. Wilson, and W. G. Lee. 1996. Species redundancy: A redundant concept. Journal of Ecology 84:121–124.

Goldsborough, L. G., and G. C. Robinson.1996. Algal ecology: Patterns in wetlands. Pages 77–117 *in* R. J. Stevenson, M. L. Bothwell, R. L. Lowe (eds.), Algal Ecology: Freshwater Benthic Systems. Academic Press, New York.

Golladay, S. W., B. W. Taylor, and B. J. Palik. 1997. Invertebrate communities of forested limesink wetland in southwest Georgia, USA: Habitat use and influence of extended inundation. Wetlands 17:383–393.

Good, R. E., D. F. Whigham, and R. L. Simpson. 1978. Freshwater Wetlands: Ecological Processes and Management Potential. Academic Press, New York.

Government of Canada. 1991. Federal Policy on Wetland Conservation. Environment Canada, Ottawa, ON.

Gower, A. M. 1967. A study of *Limnephilus lunatus* (Trichoptera: Limnephilidae) with reference to its life cycle in watercress beds. Transactions of Royal Entomological Society London 119:283–302.

Goyer, R. A., G. J. Lenhard, and J. D. Smith. 1990. Insect herbivores of a baldcypress/tupelo ecosystem. Forest Ecology and Management 33/34:517–521.

Grover, A. M., and G. A. Baldassarre. 1995. Bird species richness within beaver ponds in south-central New York. Wetlands 15:108–118.

Hairston, N. G., Jr. 1987. Diapause as a predator avoidance adaptation. Pages 281–290. *In* W. C. Kerfoot and A Sih (eds.), Predation: Direct and Indirect Impacts on Aquatic Communities. University Press of New England, Hanover, NH.

———. 1996. Phenotypic variation in a zooplankton egg bank. Ecology 77:2382–2393.

Hairston, N. G., Jr., T. A. Dillon, and B. T. DeStasio. 1990. A field test for the cues of diapause in a freshwater copepod. Ecology 71:2218–2223.

Hanski, I. A., and M. E. Gilpin 1997. Metapopulation Biology: Ecology, Genetics, and Evolution. Academic Press, New York.

Hanski, I., and D. Simberloff. 1997. The metapopulation approach, its history, conceptual domain, and application to conservation. Pages 5–26 *in* I. A. Hanski and

M. E. Gilpin (eds.), 1997. Metapopulation Biology: Ecology, Genetics, and Evolution. Academic Press, New York.

Hanson, M. A., and M. G. Butler. 1994. Responses to food web manipulation in a shallow waterfowl lake. Hydrobiologia 279/280:457–466.

Harrison, R. B. 1980. Dispersal polymorphisms in insects. Annual Review of Ecology Systematics 11:95–118.

Harrison, S., and L. Fahrig. 1995. Landscape pattern and population conservation. Pages 298–308 *in* L. Hansson, L. Fahrig, and G. Merriam (eds.), Mosaic Landscapes and Ecological Processes. Chapman & Hall, New York.

Haukos, D. A., and L. M. Smith. 1994. Composition of seed banks along an elevational gradient in playa wetlands. Wetlands 14:301–307.

Heck, K. L., and L. B. Crowder. 1991. Habitat structure and predator-prey interactions in vegetated systems. Pages 381–299 *in* S. S. Bell, E. D. McCoy, and H. R. Mushinsky (eds.), Habitat Structure: The Physical Arrangement of Objects in Space. Chapman & Hall, New York.

Jackson, M. E., and R. D. Semlitsch. 1993. Paedomorphosis in the salamander *Ambystoma talpoideum*: Effects of a fish predator. Ecology 74:342–350.

Jeffries, M. 1994. Invertebrate communities and turnover in wetland ponds affected by drought. Freshwater Biology 32:603–612.

Jeffries, M., and D. Mills. 1990. Freshwater Ecology: Principles and Applications. Belhaven Press, London, UK.

Jordan, F., C. J. DeLeon, and A. C. McCreary. 1996. Predation, habitat complexity, and distribution of the crayfish *Procambarus alleni* within a wetland habitat mosaic. Wetlands 16:452–457.

King, R. S., and D. A. Wrubeleski. 1998. Spatial and diel availability of flying insects as potential duckling food in prairie wetlands. Wetlands 18:100–114.

Kirkman, L. K., and R. R. Sharitz. 1994. Vegetation disturbance and maintenance of diversity in intermittently flooded Carolina bays in South Carolina. Ecological Applications 4:177–188.

Korner, C. 1993. Scaling from species to vegetation: The usefulness of functional groups. Pages 117–132 *in* E. D. Schulze and H. A. Mooney (eds.), Biodiveristy and Ecosystem Function. Springer-Verlag, New York.

Krapu, G. L., R. J. Greenwood, C. P. Dwyer, K. M. Draft, and L. M. Corwardin. 1997. Wetland use, settling in patterns, and recruitment in mallards. Journal of Wildlife Management 61:736–746.

Lamberti, G. A., and J. W. Moore. 1984. Aquatic insects as primary consumers. Pages 164–195 *in* V. H. Resh and D. M. Rosenberg (eds.), The Ecology of Aquatic Insects. Praeger, New York.

Kusler, J. A., and M. E. Kentual (eds.). 1990. Wetland Creation and Restoration: The Status of the Science. Island Press, Washington, DC.

Landin, J., and K. Vepsalainen. 1977. Spring dispersal flights of pond-skaters, *Gerris* sp. (Heteroptera). Oikos 29:156–160.

Lajtha, K., and R. H. Michener (eds.). 1996. Stable Isotopes in Ecology and Environmental Science (Methods in Ecology). Blackwell Scientific, Oxford, UK.

Lawler, S. P., and P. J. Morin. 1993. Food web architecture and population dynamics in laboratory microcosms of protists. American Naturalist 141:675–686.

Lee, R. E. 1989. Insect cold hardiness: To freeze or not to freeze? Bioscience 39:308–313.

Leslie, A. J., T. L. Crisman, J. P. Prenger, and K. C. Ewel. 1997. Benthic macroinvertebrates of small Florida pond cypress swamps and the influence of dry periods. Wetlands 17:447–455.

Lugo, A. E., M. M. Brinson, and S. L. Brown (eds.). 1990. Forested Wetlands, Ecosystems of the World 15. Elsevier, Amsterdam, The Netherlands.

MacArthur, R. H., and E. O. Wilson. 1964. The Theory of Island Biogeography. Princeton University Press, Princeton, NJ.

Magee, P. A. 1993. Detrital Accumulation and Processing in Wetlands. Fish and Wildlife Leaflet 12.3.14. U.S. Fish and Wildlife Service, Washington, DC.

Malecki, R. A., B. Blossey, S. D. Hight, D. Schroeder, and L. T. Kok. 1993. Biological control of purple loosestrife. Bioscience 43:680–686.

McCafferty, W. P. 1983. Aquatic Entomology: The Fisherman's and Ecologist's Illustrated Guide to Insects and Their Relatives. Jones & Bartlett, Boston.

McElligott, P. E. K., and D. J. Lewis. 1994. Relative efficiencies of wet and dry extraction techniques for sampling macroinvertebrates in a a subarctic peatland. Memoirs of the Entomological Society of Canada 169:285–289.

McGrady-Steed, J., P. M. Harris, and P. J. Morin. 1997. Biodiversity regulates ecosystem predictability. Nature 390:162–165.

McPeek, M. A. 1989. Differential dispersal tendencies among *Enallagma* damselflies (Odonata) inhabiting different habitats. Oikos 56:187–195.

―――. 1990a. Determination of species composition in the *Enallagma* damselfly assemblages of permanent lakes. Ecology 71:83–98.

―――. 1990b. Behavioral differences between *Enallagma* species (Odonata) influencing differential vulnerability to predators. Ecology 71:1714–1726.

―――. 1996. Tradeoffs, food web structure, and the coexistence of habitat specialists and generalists. American Naturalist 148:S124–S138.

Menge, B. A., J. Lubchenco, S. D. Gaines, and L. R. Ashkenas. 1986. A test of the Menge-Sutherland model of community organization in a tropical rocky intertidal food web. Oecologia 71:75–89.

Merritt, R. W., and K. W. Cummins (eds.). 1996. An Introduction to the Aquatic Insects of North America (3rd ed.). Kendall/Hunt, Dubuque, IA.

Michiels, N. K., and A. A. Dhondt. 1991. Characteristics of dispersal in sexually mature dragonflies. Ecological Entomology 16:449–459.

Minshall, G. W. 1996. Organic matter budgets. Pages 591–605 *in* F. R. Hauer and G. A. Lamberti (eds.), Methods in Stream Ecology. Academic Press, New York.

Mitsch, W. J., and J. G. Gosselink. 1993. Wetlands (2nd ed.). Van Nostrand Reinhold, New York.

Mitsch, W. J., and R. F. Wilson. 1996. Improving the success of wetland creation and restoration with know-how, time, and self-design. Ecological Applications 6:77–83.

Moloney, K. A., and S. A. Levin. 1996. The effects of disturbance architecture on landscape-level population dynamics. Ecology 77:375–394.

Morin, P. J., and S. P. Lawyer. 1995. Food web architecture and population dynamics: Theory and empirical evidence. Annual Review Ecology and Systematics 26:505–529.

Myers, R. L., and J. J. Ewel (eds.). 1990. Ecosystems of Florida. University of Central Florida, Orlando, FL.

Naeem, S. 1998. Species redundancy and ecosystem reliability. Conservation Biology 12:39–45.

Naeem, S., and S. Li. 1997. Biodiversity enhances ecosystem reliability. Nature 390: 507–509.

Naeem, S., L. J. Thompson, S. P. Lawyer, J. H. Lawton, and R. M. Goodwin. 1994. Declining biodiversity can alter the performance of ecosystems. Nature 369:734–737.

Naeem, S., L. J. Thompson, S. P. Lawyer, J. H. Lawton, and R. M. Goodwin. 1995. Biodiversity and ecosystem functioning: Empirical evidence from experimental microcosms. Philosophical Transactions of the Royal Society, London B 347:249–262.

Naiman, R. J., and H. Decamps 1997. The ecology of interfaces: Riparian zones. Annual Review of Ecology and Systematics 28:621–658.

Nilsson, A. N. 1986. Life cycles and habitats of the northern European Agabini (Coleoptera: Dytiscidae). Entomologica Basiliensia 11:391–347.

Nordstrom, L. H., and M. R. Ryan. 1996. Invertebrate abundance at occupied and potential piping plover nesting beaches: Great Plains alkali wetlands vs. the Great Lakes. Wetlands 16:429–435.

Odum, W. E. 1988. Comparative ecology of tidal freshwater and salt marshes. Annual Review of Ecology and Systematics 19:147–176.

Olson, E. J., E. S. Engstrom, M. R. Doeringsfeld, and R. Bellig. 1995. Abundance and distribution of macroinvertebrates in relation to macrophyte communities in a prairie marsh, Swan Lake, Minnesota. Journal of Freshwater Ecology 10:325–335.

O'Neill, M. P., J. C. Schmidt, J. P. Drobrowolski, C. P. Hawkins, and C. M. U. Neale. 1997. Identifying sites for riparian wetland restoration: Application of a model to the upper Arkansas River Basin. Restoration Ecology 5:85–102.

Osenburg, C. W., and G. G. Mittlebach. 1996. The relative importance of resource limitations and predator limitation in food chains. Pages 134–148 in G. A. Polis and K. O. Winemiller (eds.), Food Webs: Integration of Patterns and Dynamics. Chapman & Hall, New York.

Oswood, M. W., L. K. Miller, and J. G. Irons. 1991. Overwintering of freshwater macro-invertebrates. Pages 360–375 in R. E. Lee and D. L. Denlinger (eds.), Insects at Low Temperature. Chapman & Hall, New York.

Otto, C. 1981. Why does duration of flight periods differ in caddisflies? Oikos 27: 383–386.

Paine, R. T. 1992. Foodweb analysis through field measurement of per capita interaction strength. Nature 355:73–75.

Pajunen, V. I., and J. Salmi. 1988. The winter habitats and spring dispersal of rock-pool corixids (Heteroptera: Corixidae). Annales Entomologici Fennici 54:97–102.

Palmer, M. A., R. F. Ambrose, and N. L. Poff. 1997. Ecological theory and community restoration ecology. Restoration Ecology 5:291–300.

Pechman, J. H. K., D. E. Scott, J. W. Gibbons, and R. D. Semlitsch. 1989. Influence of wetland hydroperiod on diversity and abundance of metamorphosing juvenile amphibians. Wetlands Ecology and Management 1:3–11.

Peckarsky, B. L., P. R. Raissinet, M. A. Penton, and D. J. Conklin. 1990. Freshwater macroin-vertebrates of northeastern North America. Comstock Press, Ithaca, NY.

Peterson, B. J., and B. Fry. 1987. Stable isotopes in ecosystems studies. Annual Review of Ecology and Systematics. 18:293–320.

Playa Lakes Joint Venture. 1993. Accomplishment Report 1989–1992. North American Waterfowl Management Plan, U.S. Fish and Wildlife Service, Washington, DC.

Polis, G. A., W. B. Anderson, and R. D. Holt. 1997. Toward an integration of landscape and food web ecology: The Dynamics of Spatially Subsidized Food Webs. Annual Review of Ecology and Systematics 28:289–316.

Polis, G. A., and K. O. Winemiller (eds.). 1996. Food Webs: Integration of Patterns and Dynamics. Chapman & Hall, New York.

Pritchard, G. 1983. Biology of Tipulidae. Annual Review of Entomology 28:1–22.

Pulliam, H. R. 1988. Sources, sinks, and population regulation. American Naturalist 132:652–661.

Pulliam, H. R., and B. J. Danielson. 1988. Sources, sinks, and habitat selection: A landscape perspective on population dynamics. American Naturalist 137:S50–S66.

Reice, S. R. 1994. Nonequilibrium determinants of ecological community structure. American Scientist 82:424–435.

Richardson, C. J., and E. J. McCarthy. 1994. Effects of land development and forest management on hydrologic response in southeastern wetlands: A review. Wetlands 14:56–71.

Richter, B. D., D. P. Braun, M. A. Mendelson, and L. L. Master. 1997. Threats to imperiled freshwater fauna. Conservation Biology 11:1081–1093.

Robinson, G. G. C., S. E. Gurney, and L. G. Goldsborough. 1997a. The primary production of benthic and planktonic algae in a prairie wetland under controlled water-level regimes. Wetlands 17:182–194.

———. 1997b. Response of benthic and planktonic algal biomass to experimental water-level manipulation in a prairie lakeshore wetland. Wetlands 17:167–181.

Rosenberg, D. M., and H. V. Danks. 1987. Aquatic insects of peatlands and marshes in Canada. Memoirs of Entomological Society of Canada.

Rowe, C. L., and W. A. Dunson. 1995. Impacts of hydroperiod on growth and survival of larval amphibians in temporary ponds of central Pennsylvania, USA. Oecologia 102:397–403.

Safran, R. J., C. R. Isola, M. A. Colwell, and O. E. Williams. 1997. Benthic invertebrates at foraging locations of nine waterbird species in managed wetlands of the northern San Joaquin Valley, California. Wetlands 17:407–415.

Schneider, D. W. 1997. Predation and food web structure along a habitat duration gradient. Oecologia 110:567–575.

Schneider, D. W., and T. M. Frost. 1996. Habitat duration and community structure in temporary ponds. Journal of the North American Benthological Society 15:64–86.

Semlitsch, R. D. 1987. Interactions between fish and salamander larvae. Oecologia 72:482–486.

Sippel, W. S. 1988. Wetland Identification and Delineation Manual. Vol. I: Rationale, Wetland Parameters, and Overview of Jurisdictional Approach. April 1988 revised interim final. Office of Wetlands Protection, U.S. Environmental Protection Agency, Washington, DC.

Skelly, D. K. 1992. Field evidence for a behavioral antipredator response in a larval amphibian. Ecology 73:704–708.

———. 1995. A behavioral tradeoff and its consequences for the distribution of *Pseudacris* treefrog larvae. Ecology 76:150–164.

———. 1997. Tadpole communities. American Scientist 85:36–45.

Smith, R. D., A. Ammann, C. Bartoldus, and M. M. Brinson. 1994. An Approach for Assessing Wetland Functions Based on Hydrogeomorphic Classification, Reference Wetlands, and Functional Indices. Wetlands Research Program Technical Report WRP-DE-9. U.S. Army Engineers Waterways Experiment Station, Vicksburg, MS.

Soumille, H., and A. Thiery. 1997. A new quantitative sediment corer for sampling invertebrates across the mud water interface and soil of shallow rice fields. Annales de limnologie 33:197–203.

Southwood, T.R.E. 1977. Habitat, the templet for ecological strategies. Journal of Animal Ecology 46:337–365.

———. 1988. Tactics, strategies, and templets. Oikos 52:94–100.

Spence, J. R., and N. M. Anderson. 1994. Biology of water striders: Interactions between systematics and ecology. Annual Review of Entomology 39:101–128.

Sprules, W. G. 1972. Effects of size-selective predation and food competition on high altitude zooplankton communities. Ecology 53:375–386.

Stiehl, R. B. 1994. Habitat Evaluation Procedures (HEP) Workbook. National Biological Survey. National Ecological Resources Center, Fort Collins, CO.

Tauber, J. J., C. A. Tauber, and S. Masaki. 1986. Seasonal Adaptations of Insects. Oxford University Press, New York.

Thorp, J. H., and A. P. Covich (eds.). 1991. Ecology and Classification of North American Freshwater Invertebrates. Academic Press, New York.

Tilman, D., and J. A. Downing. 1994. Biodiversity and stability in grasslands. Nature 367:363–365.

Tilman, D., J. A. Downing, and D. A. Wedin. 1994. Does diversity beget stability? Nature 371:257–264.

Tilman, D., D. Wedin, and J. Knops. 1996. Biodiversity and sustainability influenced by biodiversity in grassland ecosystems. Nature 379:718–720.

Tiner, R. W. 1984. Wetlands of the United States: Current Status and Recent Trends. U.S. Fish and Wildlife Service, National Wetlands Inventory, Washington, DC.

———. 1989. Wetland boundary delineation. Pages 232–248 *in* R. P. Brooks, F. J. Brenner, and R. W. Tiners (eds.), Wetlands Ecology and Conservation: Emphasis in Pennsylvania. The Pennsylvania Academy of Science, Philadelphia, PA.

———. 1991. The concept of a hydrophyte for wetland identifications. BioScience 41: 236–247.

———. 1993a. Using plants as indicators of wetland. Proceedings of the Academy of Natural Sciences of Philadelphia 144:240–253.

———. 1993b. The primary indicators methods—a practical approach to wetland recognition and delineations in the United States. Wetlands 13:50–64.

———. 1996. Wetland definitions and classifications in the United States. Pages 27–34 *in* Water-Supply Paper 2425, J. D. Fretwell, J. S. Williams, and P. J. Redman (eds.), National Summary on Wetland Resources. U.S. Geological Survey, Washington, DC.

Turner, A. M., and J. P. McCarty. 1998. Resource availability, breeding site selection, and reproductive success of red-winged blackbirds. Oecologia 113:140–146.

Turner, A. M., and J. C. Trexler. 1997. Sampling aquatic invertebrates from marshes: Evaluating the options. Journal of the North American Benthological Society 16: 694–709.

Van der Valk, A. G. 1981. Succession in wetlands: A Gleasonian approach. Ecology 62:688–696.

Van der Valk, A. G., and C. B. Davis 1978. The role of seed banks in the vegetation dynamics of prairie glacial marshes. Ecology 59:322–325.

VanRees-Siewert, K. L., and J. L. Dinsmore 1996. Influence of wetland age on bird use of restored wetlands in Iowa. Wetlands 16:577–582.

Ward, J. V. 1992. Aquatic Insect Ecology. Vol. 1. Biology and Habitat. John Wiley & Sons, New York.

Webster, J. R., and E. F. Benfield. 1986. Vascular plant breakdown in freshwater ecosystems. Annual Review of Ecology and Systematics 17:567–594.

Welch, P. S. 1952. Limnology. McGraw-Hill, New York.

Wellborn, G. A. 1994. Size-biased predation and the evolution of prey life histories: A comparative study of freshwater amphipod populations. Ecology 75:2104–21017.

Wellborn, G. A., D. K. Skelly, and E. E. Werner. 1996. Mechanisms creating community structure across a freshwater habitat gradient. Annual Review of Ecology and Systematics 27:337–363.

Welling, C. H., R. L. Pederson, and A. G. Van der Valk. 1988. Recruitment from the seed bank and the development of zonation of emergent vegetation during a drawdown in a prairie wetland. Journal of Ecology 76:483–496.

Werner, E. E., and B. R. Anholt. 1993. Ecological consequences of the tradeoff between growth and mortality rates mediated by foraging activity. American Naturalist 142:242–272.

Werner, E. E., and M. A. McPeek 1994. Direct and indirect effects of predators on two anuran species along an environmental gradient. Ecology 75:1368–1382.

Wetzel. R. G. 1983. Limnology (2nd ed.), Saunders College, Philadelphia, PA.

Wharton, C. H., W. M. Kitchens, E. C. Pendleton, and T. W. Sipe. 1982. The Ecology of Bottom-land Hardwood Swamps of the Southeast: A Community Profile. U.S. Fish and Wildlife Service, Biological Services Program FWS/OBS-81/37.

Wheeler, B. D., S. C. Shaw, W. J. Fojt, and R. A. Robertson. 1995. Restoration of Temperate Wetlands. John Wiley & Sons, New York.

Wiggins, G. B. 1973. A contribution to the biology of caddisflies (Trichoptera) in temporary pools. Life Science Contributions of the Royal Ontario Museum 88:1–28.

———. 1996. Larvae of the North American caddisflies (Trichoptera) (2nd ed.). University of Toronto Press, Toronto, ON.

Wiggins, G. B., R. J. Mackay, and I. M. Smith. 1980. Evolutionary and ecological strategies of animals in annual temporary pools. Archive für Hydrobiologie Supplement 58:97–206.

Wilbur, H. M. 1997. Experimental ecology of food webs: Complex systems in temporary ponds. Ecology 78:2279–2302.

Williams, D. D. 1987. The Ecology of Temporary Waters. Timber Press, Portland, OR.

Williams, D. D., and B. W. Feltmate. 1992. Aquatic Insects. CAB International, Oxford, UK.

Williams, R. N., M. S. Ellis, and D. S. Fickle 1996. Insects in the Killbuck Marsh Wildlife Area, Ohio. 1994 Survey. Ohio Journal of Science 96:34–40.

Williamson, C. E. 1991. Copepoda. Pages 787–822 *in* J. H. Thorp and A. P. Covich (eds.), Ecology and Classification of North American Freshwater Invertebrates. Academic Press, New York.

Wilson, J. B., and R. J. Whittaker. 1995. Assembly rules demonstrated in a salt-marsh community. Journal of Ecology 83:801–807.

Wilson, R. F., and W. J. Mitsch. 1996. Functional assessment of five wetlands constructed to mitigate wetland loss in Ohio, USA. Wetlands 16:436–451.

Windell, J. T., B. E. Willard, D. J. Cooper, S. Q. Foster, C. F. Knud-Hansen, L. P. Rink, and G. N. Kiladis. 1986. An Ecological Characterization of Rocky Mountain Montane and Subalpine Wetlands. U.S. Fish and Wildlife Service Biology Report 86 (11).

Winemiller, K. O. 1996. Factors driving temporal and spatial variation in aquatic floodplain food webs. Pages 298–312 *in* G. A. Polis and K. O. Winemiller (eds.), Food Webs: Integration of Patterns and Dynamics. Chapman & Hall, New York.

Wissinger, S. A. 1997. Cyclic colonization and predictable disturbance: A template for biological control in ephemeral crop systems. Biological Control 10:4–15.

Wissinger, S. A., and H. H. Whiteman. 1992. Fluctuation in a Rocky Mountain population of salamanders: Anthropogenic acidification or natural variation? Journal of Herpetology 26:377–391.

Wissinger, S. A., and W. S. Brown. Habitat permanence, life histories and the distribution of five caddisfly species in subalpine wetlands in central Colorado (unpub. ms).

Wissinger, S. A., G. B. Sparks, G. L. Rouse, W. S. Brown, and H. Steltzer. 1996. Intraguild predation and cannibalism among larvae of detritivorous caddisflies in subalpine wetlands. Ecology 77:2421–2430.

Wissinger, S. A., H. H. Whiteman, G. B. Sparks, G. L. Rouse, and W. S. Brown. 1999. Tradeoffs between competitive superiority and vulnerability to predation in caddisflies along a permanence gradient in subalpine wetlands. Ecology. 80: In press.

Young, E. C. 1965. The incidence of flight polymorphism in British Corixidae and description of the morphs. Journal of Zoology 146:567–576.

SUBJECT INDEX

Because this book is organized around specific habitat types and geographic regions (see Contents), that kind of information is not listed in this index. The particular focus on invertebrates that is described in each chapter is indicated by each chapter title, and, thus, perusing titles will be useful for locating information of specific interest.

TAXONOMIC INDEX

Only animal and plant genera specifically listed in the text are listed here. Other invertebrate taxa that did not merit discussion by authors can be found in various tabular taxa lists (see "Community, invertebrate, structure" in the Subject Index for table locations).